STUDENT'S SOLUTIONS MA
TO ACCOMPANY
BARNETT / KEARNS

INTERMEDIATE ALGEBRA

Structure and Use

5TH EDITION

Fred Safier

City College of San Francisco

McGraw-Hill, Inc.

New York St. Louis San Francisco Auckland Bogotá Caracas
Lisbon London Madrid Mexico City Milan Montreal New Delhi
San Juan Singapore Sydney Tokyo Toronto

Student's Solutions Manual to Accompany Barnett/Kearns:
INTERMEDIATE ALGEBRA: Structure and Use

This book is printed on recycled paper containing 10%post consumer waste.

2 3 4 5 6 7 8 9 0 MAL MAL 9 0 9 8 7 6 5 4

ISBN 0-07-005106-2

The editor was Karen M. Minette;
the production supervisor was Leroy A. Young.
Malloy Lithographing, Inc., was printer and binder.

CONTENTS

CHAPTER 1 Preliminaries

Exercise 1-1 Basic Concepts

Key Ideas and Formulas

Collections of objects are called **sets.** Sets are often designated by capital letters: A, B, C.

Objects in a set are called **members** or **elements** of the set.

$a \in A$ means "a is an element of the set A"
$a \notin A$ means "a is not an element of set A"

Sets are usually described by **listing:** {-9, 9} or by a **rule:**
$\{x \mid x^2 = 81\}$.

Variables are symbols used to represent unspecified elements of a set.

Constants are symbols that name exactly one object.

Some sets are **finite,** some sets are **infinite,** some sets are **empty** or **null** sets. $\varnothing$ represents "the empty set".

If each element of set A is also an element of set B, we say that A is a subset of B. $A \subset B$.

If two sets have exactly the same elements (the order of listing does not matter), the sets are **equal.** A = B.

An **algebraic expression** is a meaningful symbolic form using constants, variables, mathematical operations, and grouping symbols.

Two or more algebraic expressions joined by plus or minus signs are called **terms.**

Two or more algebraic expressions joined by multiplication are called **factors.**

Important sets and their symbols:

SYMBOL	NUMBER SYSTEM	DESCRIPTION	EXAMPLES
N	Natural numbers	Counting numbers (also called positive integers)	1, 2, 3, …
W	Whole numbers	Natural numbers and 0	0, 1, 2, 3, …
Z	Integers	Set of natural numbers, their negatives, and 0	…, -2, -1, 0, 1, 2, …

Q	Rational numbers	Any number that can be represented as $\frac{a}{b}$, where a and b are integers and $b \neq 0$	$-4, \frac{-3}{5}, 0, 1, \frac{2}{3}, 3.67$
R	Real numbers	Set of all rational and irrational numbers (the irrational numbers are all the real numbers that are not rational)	$-4, \frac{-3}{5}, 0, 1, \frac{2}{3}, 3.67, \sqrt{2}, \pi, \sqrt[3]{5}$

Note: $N \subset W \subset Z \subset Q \subset R$

A line with a real number associated with each point, and vice versa, is called a **real number line.** Each number associated with a point is called the **coordinate** of the point. The point with coordinate 0 is called the **origin.**

---

1. 2 is an element of D, but not an element of B. F
3. 6 is an element of K, but not an element of B. F
5. Every element of G is an element of D. T
7. Every element of J is an element of E. T
9. 3 is an element of H, but not an element of E. F
11. 2 is an element of A, but not an element of B. F

13.

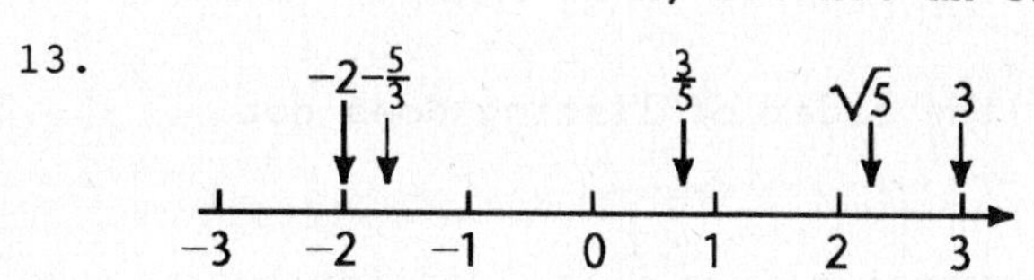

15.

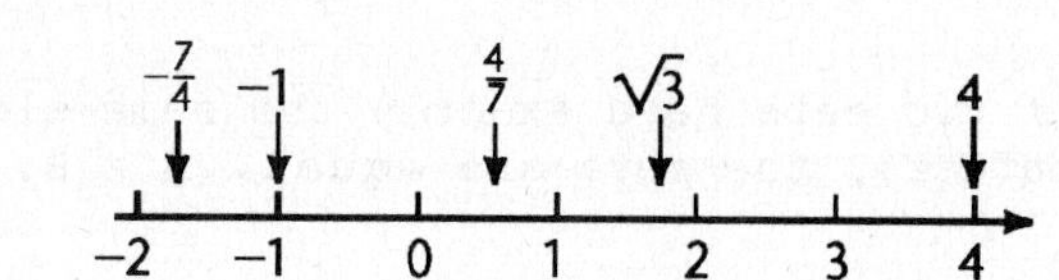

17.

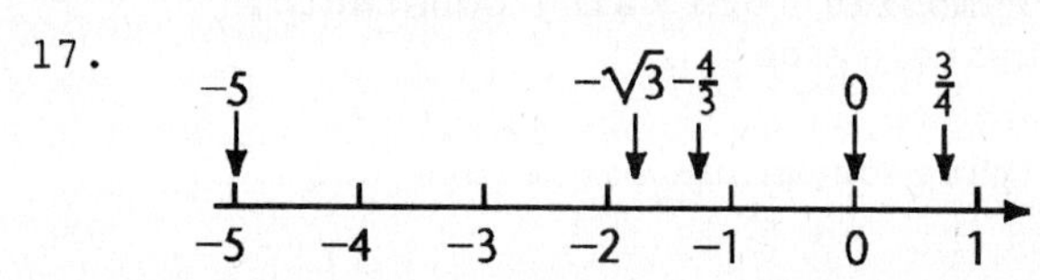

19. -3 is one of infinitely many examples of a negative integer. So are -7 and -90. 0 is the only possible example of an integer that is neither positive nor negative.

    5 is one of infinitely many examples of a positive integer. So are 12 and 150.

21. $\frac{2}{3}$, $-\frac{7}{8}$, 2.65 are three of infinitely many.

23. {6, 7, 8, 9}

25. {a, s, t, u}

27. Since there has been, so far, no woman president of the United States, the set is empty. ∅.

29. {5}

31. Since no numbers satisfy the condition $x + 9 = x + 1$, the set is empty. ∅.

33. {-2, 2}

35. (A) T (B) F (C) T

37. (A) Since $\frac{15}{4}$ has decimal representation 3.75, it lies between 3 and 4.

(B) Since $-\frac{4}{3}$ has decimal representation -1.33..., it lies between -2 and -1.

(C) Since $-\sqrt{7}$ has decimal representation -2.645..., it lies between -3 and -2.

39. (A) Since $\frac{12}{5}$ has decimal representation 2.4, it lies between 2 and 3.

(B) Since $-\frac{22}{5}$ has decimal representation -4.4, it lies between -5 and -4.

(C) Since $\sqrt{45}$ has decimal representation 6.708..., it lies between 6 and 7.

41. (A) Since $\frac{33}{2}$ has decimal representation 16.5, it lies between 16 and 17.

(B) Since $-\frac{41}{2}$ has decimal representation -20.5, it lies between -21 and -20.

(C) Since $\sqrt{32}$ has decimal representation 5.65..., it lies between 5 and 6.

43. (A) Since $\frac{48}{9}$ has decimal representation 5.33..., it lies between 5 and 6.

(B) Since $-\frac{32}{9}$ has decimal representation -3.55..., it lies between -4 and -3.

(C) Since $\sqrt{129}$ has decimal representation 11.35..., it lies between 11 and 12.

45. Three terms: $3x$, $-4y$, $5xy$

47. One term: $x(2y - 3z)$

49. Two terms: xyz, $w(x + y + z)$

51. Four factors: 2, a, b, c

53. Two factors: $x + 2$, $x - 3$

55. Three factors: $3x$, $x + 3y$, $3y + x$

57. $\sqrt{11}$ is an element of A, but not an element of Q. F

59. Every element of C is an element of Q. T

61. Every element of A is an element of R. T

63. -2 is an element of B, but not an element of W. F

65. -2 is an element of B, but not an element of W. F

67. $\frac{1}{8}$ is an element of C, but not an element of Z. F

69. Every element of Z is an element of Q. T

71. π is an element of R, but not an element of Q. F

73. (A) List the elements of M, followed by the elements of N that are not in M. {1, 2, 3, 4, 6}

(B) {2, 4}

75. (A) List the elements of M, followed by the elements of N that are not in M. {4, 6, 8, 10, 2} or {2, 4, 6, 8, 10}

(B) {4, 6}

77.
$$\begin{aligned} 100c &= 9.090909\ldots \\ c &= 0.090909\ldots \\ \hline 99c &= 9 \\ c &= \frac{1}{11} \end{aligned}$$

79.
$$\begin{aligned} 1000c &= 123.123123123\ldots \\ c &= 0.123123123\ldots \\ \hline 999c &= 123 \\ c &= \frac{123}{999} \text{ or } \frac{41}{333} \end{aligned}$$

81. $1000c = 567.6767\ldots$
$10c = 5.6767\ldots$
$990c = 562$
$c = \frac{562}{990}$ or $\frac{281}{495}$

83. $100{,}000c = 23{,}456.456456\ldots$
$100c = 23.456456\ldots$
$99{,}900c = 23{,}433$
$c = \frac{23{,}433}{99{,}900}$ or $\frac{7811}{33{,}300}$

85. (A) 0.88888888... (B) 0.27272727... (C) 2.23606797... (D) 1.37500000...

87. (A) 0.4000... (B) -0.075000... (C) 1.732050808... (D) 0.003 003 003...

89. {P, V}, {P, S}, {P, T}, {V, S}, {V, T}, {S, T}

91. {A, K, Q}, {A, K J}, {A, K, 10}, {A, Q, J}, {A, Q, 10}, {A, J, 10}, {K, Q, J}, {K, Q, 10}, {K, J, 10}, {Q, J, 10}

Exercise 1-2 Equality and Inequality

Key Ideas and Formulas

The use of an **equality sign** (=) between two expressions states that the two expressions are names or descriptions of exactly the same object.

The symbol ≠ means **is not equal to.**

If two algebraic expressions involving at least one variable are joined with an equal sign, the resulting form is called an **algebraic equation.** An algebraic equation is neither true nor false until the variables have been replaced by constants.

Basic Properties of Equality	
If a, b, and c are names of objects, then:	
1. $a = a$.	REFLEXIVE PROPERTY
2. If $a = b$, then $b = a$.	SYMMETRIC PROPERTY
3. If $a = b$ and $b = c$, then $a = c$.	TRANSITIVE PROPERTY
4. If $a = b$, then either may replace the other in any statement without changing the truth or falsity of the statement.	SUBSTITUTION PRINCIPLE

Definition of $a < b$ and $b > a$

For a and b real numbers, we say that **a is less than b** or **b is greater than a** and write
$a < b$ or $b > a$
if there exists a positive real number such that $a + p = b$ or equivalently $b - a = p$).

Inequality Symbols		On a number line:
$a < b$	a is less than b	a is to the left of b
$a > b$	a is greater than b	a is the right of b
$a \le b$	a is less than or equal to b	
$a \ge b$	a is greater than or equal to b	

Basic Inequality Properties	
For any real numbers a, b, and c:	
1. Either $a < b$, $a = b$, or $a > b$.	TRICHOTOMY PROPERTY
2. If $a < b$ and $b < c$, then $a < c$.	TRANSITIVE PROPERTY

A **double inequality** such as $a < x \le b$ means that $a < x$ and $x \le b$. To **graph an inequality statement** in one variable on a real line is to graph the **solution set** of the inequality statement, that is, the set of all real number replacements of the variable that make the statement true.

---

1. $11 - 5 = \frac{12}{2}$ 3. $4 > -18$ 5. $-12 < -3$ 7. $x \ge -8$

9. $-2 < x < 2$ 11. < 13. > 15. < 17. = 19. < 21. >

23. <, < 25. > 27. > 29. < 31. < 33. > 35. > 37. >

39. The solution set for $x \le 3$ is the set of all real numbers less than or equal to 3. This includes all the points to the left of 3. The bracket through 3 indicates that 3 is included. Graphically,

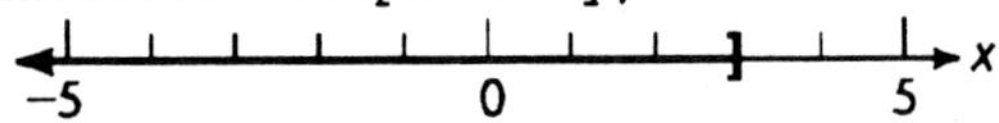

41. The solution set for $-5 < x \le -1$ is the set of all real numbers between -5 and -1, including -1 but not -5. Graphically,

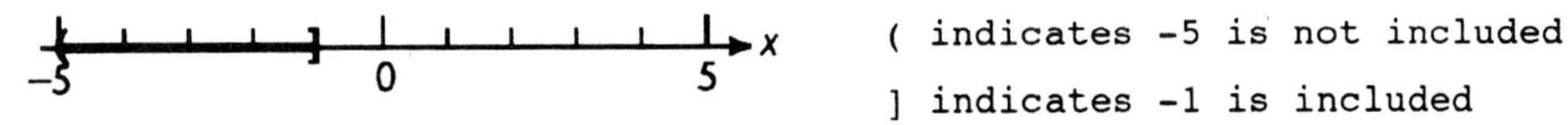

(indicates -5 is not included

] indicates -1 is included

43. The solution set for $x > -4$ is the set of all real numbers greater than -4. Graphically,

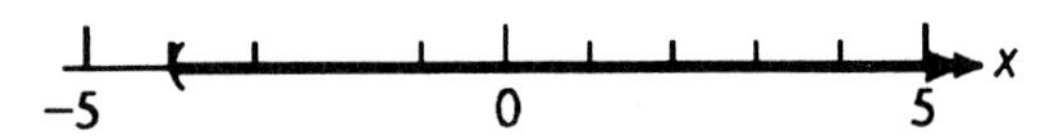

The (indicates -4 is not included.

45. The solution set for $-1 < x < 3$ is the set of all real numbers between -1 and 3. The parentheses indicate -1 and 3 are not included.

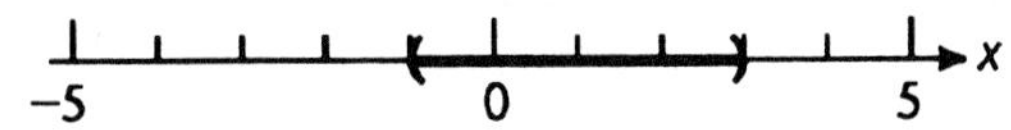

47. The solution set for $x \ge 0$ is the set of all positive or zero real numbers. This includes all points to the right of 0. The bracket through 0 indicates that 0 is included. Graphically,

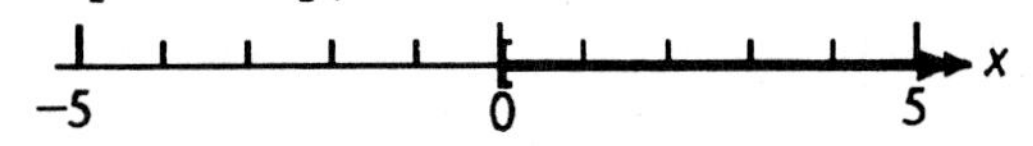

49. The solution set for $-4 \le x \le -1$ is the set of all real numbers between -4 and -1, including -4 and -1. Graphically,

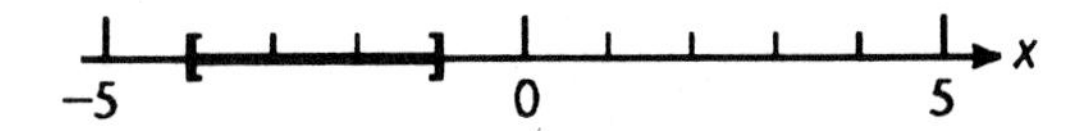

The brackets indicate that -4 and -1 are included.

51. The solution set for $-2 < x \le 5$ is the set of all real numbers between -2 and -5, including 5 but not -2. Graphically,

53. $x - 8 > 0$ 55. $x + 4 \ge 0$

57. Let x = the unknown number, then the statement translates as follows:

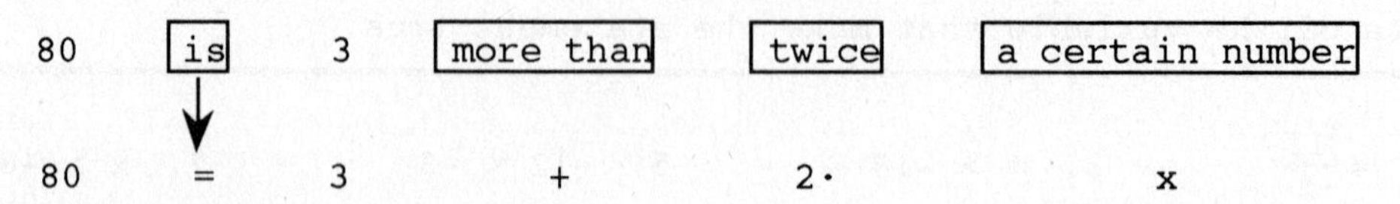

Or, more compactly:
$80 = 3 + 2x$

59. x is greater than or equal to -3 is understood to mean

-3 | is less than or equal to | x
-3 ≤ x

and less than 4 is understood to mean

x | is less than | 4
x < 4

These two statements are combined into the double inequality statement:
$-3 \le x < 4$.

61. Let x = a certain number, then the statement translates as follows:

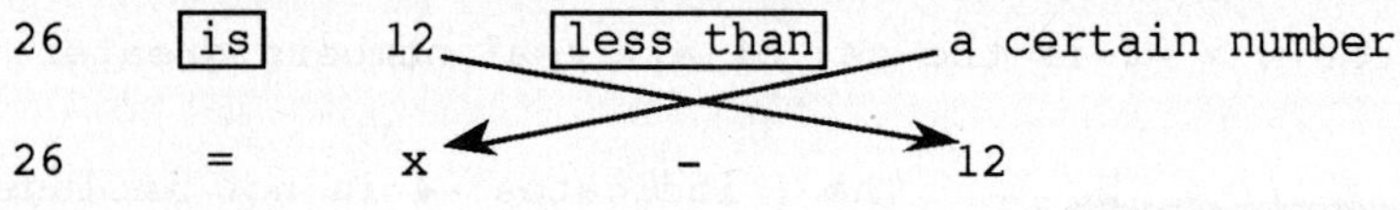

Or, more compactly:
$26 = x - 12$

63.

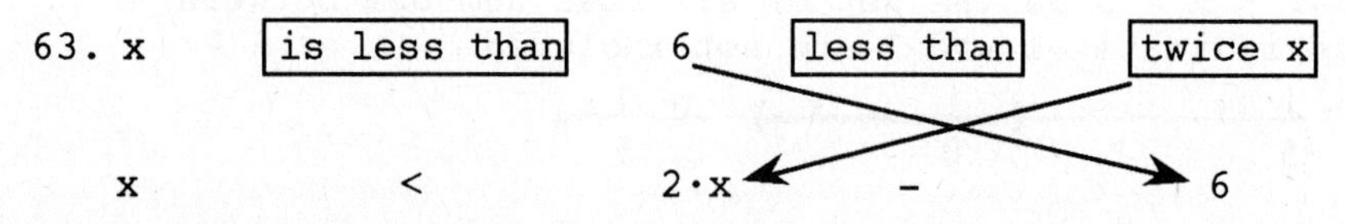

Or, more compactly:
$x < 2x - 6$

65. Let x = the unknown number, then the statement translates as follows:

6 | times | a number | is | 4 | more than | 3 | times | the number
6 · x = 4 + 3 · x

Or, more compactly:
$6x = 4 + 3x$

67. Let x = a certain number. Then
"the number which is 7 [more than] [the certain number]"
translates to 7 + x

Then the statement is understood to mean:

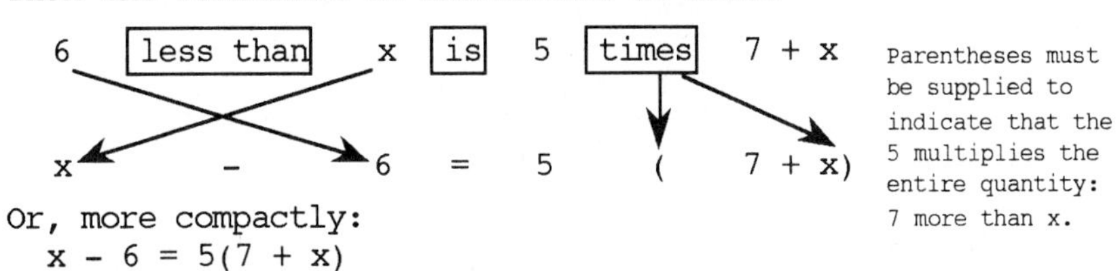

Parentheses must be supplied to indicate that the 5 multiplies the entire quantity: 7 more than x.

Or, more compactly:
$x - 6 = 5(7 + x)$

69. Inclusive means that both 63 and 72 are included, so the ≤ symbol is used.
$63 \le \frac{9}{5}C + 32 \le 72$

71. Let x = the first of the three consecutive natural numbers
Then x + 1 must be the next of the three
x + 2 must be the third of the three
Then the statement is understood to mean:

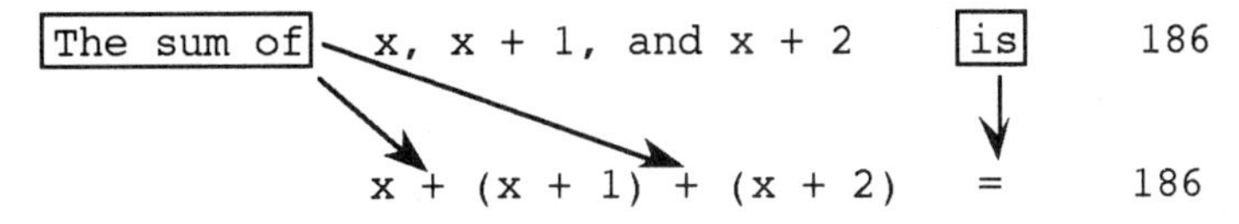

73. $t = -5$

75. $5x + 7x = 12x$

77. $3 - x$

79. Symmetric property

81. Transitive property for equality

83. Trichotomy property

85. Substitution property

87. Transitive property for equality or substitution property

89. Substitution property

91. Substitution property

93. "Is" does not translate into "equal" in this case. (8 is actually an element in the set of even numbers.) The properties of equality do not apply.

95. In this problem there are three quantities: area, length and width. They are related by the given statement: area = (length)(width). We use the variable x to represent the width. Then the length is represented by translating the phrase

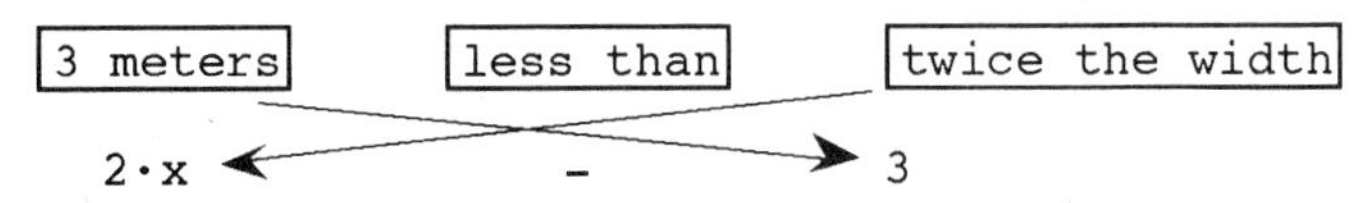

The area is represented by the constant 90. Substituting these symbols into the area relation, we have

$90 = (2x - 3)x$ or $90 = x(2x - 3)$

97. In this problem there are three quantities: perimeter, length and width. They are related by the formula $2w + 2\ell = P$, where P is perimeter, ℓ is length, and w is width. We use the variable x to represent the width. Then the length is represented by translating the phrase

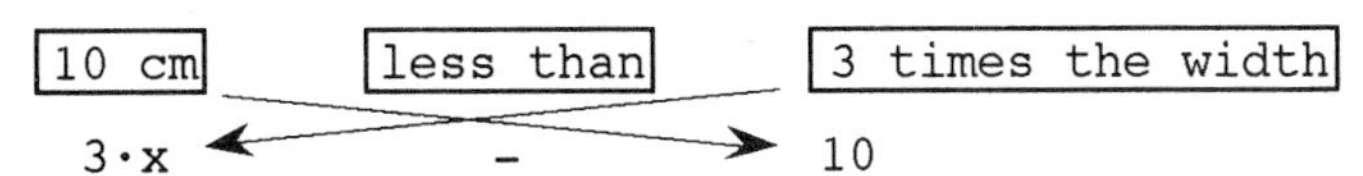

The perimeter is represented by the constant 210. Substituting these symbols into the perimeter formula, we have

$$2x + 2(3x - 10) = 210$$

99. (A)

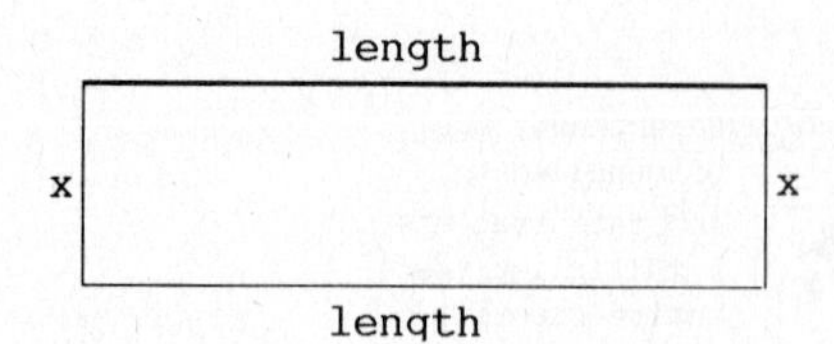

Since the length plus the width x is half the perimeter, we can write

$$\text{length} + x = 200$$
$$\text{or} \quad \text{length} = 200 - x$$

Therefore, using the area relation that area = (length)(width), we see that the area $= (200 - x)x$ or, using the variable A for area,

$$A = x(200 - x)$$

(B) Both the length and the width must be nonnegative; in other words: $0 \le x$, $0 \le 200 - x$. Therefore, adding x to both sides of the second inequality statement, $0 + x \le 200 - x + x$, that is,

$$x \le 200$$

Combining the two requirements for x, we can write

$$0 \le x \le 200$$

An intuitive argument: Neither the length $200 - x$, nor the width x can be more than half the perimeter, or 200. So x can range between a minimum of 0 and a maximum of 200. So $0 \le x \le 200$.

Exercise 1-3 Real Number Properties

Key Ideas and Formulas

Basic Properties of the set of Real Numbers

Let R be the set of real numbers and x, y, and z arbitrary elements of R.

ADDITION PROPERTIES		MULTIPLICATION PROPERTIES	
CLOSURE:	$x + y$ is a unique element in R, that is, the sum of two real numbers is also a real number.	CLOSURE:	xy is a unique element in R, that is, the product of two real numbers is also a real number.
ASSOCIATIVE:	$(x + y) + z = x + (y + z)$	ASSOCIATIVE:	$(xy)z = x(yz)$
COMMUTATIVE:	$x + y = y + x$	COMMUTATIVE:	$xy = yx$
IDENTITY:	0 is the additive identity; that is, $0 + x = x + 0 = x$ for all $x \in R$, and 0 is the only element in R with this property.	IDENTITY:	1 is the multiplicative identity; that is, for $x \in R$, $1x = x1 = x$, and 1 is the only element in R with this property.
INVERSE:	For each $x \in R$, $-x$ is its unique additive inverse; that is, $x + (-x) = (-x) + x = 0$, and $-x$ is the only element in R relative to x with this property.	INVERSE:	For each $x \in R$, $x \ne 0$, $\frac{1}{x}$ is its unique multiplicative inverse; that is, $x\left(\frac{1}{x}\right) = \left(\frac{1}{x}\right)x = 1$, and $\frac{1}{x}$ is the only element in R relative to x with this property.

COMBINED PROPERTY

DISTRIBUTIVE: $x(y + z) = xy + xz$

$(x + y)z = xz + yz$

The process of rewriting $x(y + z)$ as $xy + xz$ is usually called "multiplying (out)". The process of rewriting the sum $xy + xz$ as the product $x(y + z)$ is usually called "factoring out". The x is referred to as the **common factor.**

Order of Operations

Unless grouping symbols indicate otherwise, multiplication and division are to be done before addition and subtraction. Operations are done from left to right.

Relative to addition, commutativity and associativity permit us to change the order of addition at will and insert or remove parentheses as we please. The same is true for multiplication, but not for subtraction and division.

Additional Distributive Properties
For real numbers a, b, c, ...,
$(ab - ac) = a(b - c) = (b - c)a$
$a(b + c + d + \dots + f) = ab + ac + ad + \dots + af$
$ab + ac + ad + \dots + af = a(b + c + d + \dots + f)$
$= (b + c + d + \dots + f)a$

Simplifying algebraic expressions is the process of using the commutative, associative, distributive, and other properties to transform an algebraic expression into a simpler **equivalent** expression, that is, one which yields equal numbers for each replacement of the variables by numbers from the respective replacement sets.

---

1. $3 + 4\cdot5 = 3 + 20 = 23$
3. $5\cdot6 - 3\cdot4 = 30 - 12 = 18$
5. $4\cdot2 + 3(8 - 3\cdot2) = 8 + 3(8 - 6) = 8 + 3(2) = 8 + 6 = 14$
7. $3 + x$
9. $(5\cdot7)z$
11. mn
13. $(9 + 11) + M$
15. $3(x + 1) = 3\cdot x + 3\cdot1 = 3x + 3$
17. $x(2 + x) = x\cdot2 + x\cdot x = 2x + x^2$
19. $7x$
21. $x + y$
23. Commutative property for addition
25. Associative property for addition
27. Commutative property for multiplication
29. Distributive property
31. Associative property for multiplication
33. Identity property for addition
35. Identity property for multiplication
37. Distributive property
39. $(x + 7) + 2 = x + 7 + 2$
    $= x + 9$
41. $4(5y) = 4\cdot5y$
    $= 20y$
43. $12 + (u + 3) = 12 + u + 3$
    $= u + 12 + 3$
    $= u + 15$
45. $(3x)7 = 3x\cdot7$
    $= 7\cdot3x$
    $= 21x$
47. $0 + 1x = 1x$
    $= x$
49. Commutative property for addition
51. Commutative property for multiplication
53. Commutative property for addition
55. Associative property for multiplication

57. $(x + 7) + (y + 4) + (z + 1) = x + 7 + y + 4 + z + 1$
$= x + y + z + 7 + 4 + 1$
$= x + y + z + 12$

59. $(3x + 5) + (4y + 6) = 3x + 5 + 4y + 6$
$= 3x + 4y + 5 + 6$
$= 3x + 4y + 11$

61. $0 + (1x + 3) + (y + 2) = 0 + 1x + 3 + y + 2$
$= x + 3 + y + 2$
$= x + y + 3 + 2$
$= x + y + 5$

63. $(12m)(3n)(1p) = 12m \cdot 3n \cdot 1p$
$= 12 \cdot 3 \cdot m \cdot n \cdot p$
$= 36mnp$

65. $x(x + 5) = x \cdot x + x \cdot 5 = x^2 + 5x$

67. $2x + 18 = 2 \cdot x + 2 \cdot 9 = 2(x + 9)$

69. $(x + 2)8 = x \cdot 8 + 2 \cdot 8 = 8x + 16$

71. $6x + 12 = 6 \cdot x + 6 \cdot 2 = 6(x + 2)$

73. $ax + ay = a(x + y)$

75. $y + xy = 1 \cdot y + x \cdot y = y(1 + x)$

77. $a(a + b) = a \cdot a + a \cdot b = a^2 + ab$

79. $(x + 1)y = x \cdot y + 1 \cdot y = xy + y$

81. (A) is true—it states the commutative property for addition.
(B) is false—one of many examples showing its falseness is $12 - 4 \neq 4 - 12$
(C) is true—it states the commutative property for multiplication.
(D) is false—one of many examples showing its falseness is $12 \div 4 \neq 4 \div 12$

83. False—one of many examples showing its falseness is $8 \div (2 + 2) \neq 8 \div 2 + 8 \div 2$.

85. False—one of many examples showing its falseness is $12 \times (6 \div 2) \neq (12 \times 6) \div (12 \times 2)$

87. Since the difference of two integers is always an integer, Z is closed under −.

89. Since the sum of two natural numbers is always a natural number, N is closed under +.

91. Since the product of two positive rational numbers is always a positive rational number, the set is closed under ×.

93. Since the product of two negative integers is never negative, the set is not closed under ×.

95. 1. Commutative property for addition; 2. Associative property for addition; 3. Distributive property; 4. addition (substitution principle for equality)

97. 1. Commutative property for addition; 2. Associative property for addition; 3. Associative property for addition; 4. Substitution principle for equality; 5. Commutative property for addition; 6. Associative property for addition.

Exercise 1-4 Addition and Subtraction

Key Ideas and Formulas

For each real number x, we denote its additive inverse by −x, the **opposite of** x or the **negative of** x. The opposite of x is obtained from x by changing its sign. The opposite of 0 is 0.

Double Negative Property: $-(-a) = a$

Multiple Uses of the Minus Sign
1. As the operation "subtract": $9 \overset{\downarrow}{-} 3 = 6$
2. As the operation "the negative or opposite of": $\overset{\downarrow}{-}(-8) = 8$
3. As part of a number symbol: $\overset{\downarrow}{-}4$

$-x$ is not necessarily a negative number. It is negative if x is positive, but $-x$ is positive if x is negative.

The **absolute value** of a number x is an operation on x, denoted by $|x|$. The absolute value of x can be thought of geometrically as the distance of x from 0 on the number line. Symbolically,

$$|x| = \begin{cases} x \text{ if } x \text{ is positive} \\ 0 \text{ if } x \text{ is } 0 \\ -x \text{ if } x \text{ is negative} \end{cases}$$

The absolute value of a number is never negative.

Addition of Real Numbers: $a + b$

1. If a or b is zero: $0 + b = b$ $a + 0 = a$.
2. If a and b are both positive: add as you have learned in arithmetic.
3. If a and b are both negative: take the opposite of the sum of the absolute values: $(-4) + (-8) = -(4 + 8) = -12$.
4. If a and b have opposite signs: subtract the smaller absolute value from the larger, then attach the sign of the number with the larger absolute value: $(-7) + 3 = -(7 - 3) = -4$

Subtraction of Real Numbers: $a - b = a + (-b)$

Combined Operation: When three or more terms are combined by addition and subraction, and symbols of grouping are omitted, we convert (mentally) any subtraction to addition, then add.

To **evaluate** an algebraic expression for particular values of the variables means to replace each variable by a given value, then calculate the resulting arithmetic value.

---

1. -7 3. $+6$ or 6 5. $+2$ or 2 7. $+27$ or 27 9. 0

11. -10 13. $+4$ or 4 15. $(+3) - (+9) = (+3) + (-9) = -6$

17. $(+9) - (-3) = (+9) + (+3) = (+12)$ or 12

19. Since the negative of a positive number is negative and the negative of a negative number is positive, the correct choice is <u>sometimes</u> negative.

21. $-[-(-3)] = -(+3) = -3$ 23. $-|-(+2)| = -|-2| = -(+2) = -2$

25. $-(|-9| - |-3|) = -(9 - 3) = -(6) = -6$

27. $(-2) + (-6) + 3 = (-8) + 3 = -5$

29. $5 - 7 - 3 = 5 + (-7) + (-3) = 5 + (-10) = -5$

31. $-7 + 6 - 4 = (-7) + 6 + (-4) = (-7) + (-4) + 6 = (-11) + 6 = -5$

33. $-2 - 3 + 6 - 2 = (-2) + (-3) + 6 + (-2) = (-2) + (-3) + (-2) + 6 = (-7) + 6 = -1$

35. $6 - [3 - (-9)] = 6 - (3 + 9) = 6 - 12 = 6 + (-12) = -6$

37. $[6 - (-8)] - [(-8) - 6] = [6 + 8] - [(-8) + (-6)] = 14 - [-14] = 14 + 14 = 28$

39. $\{4 - [5 - (6 - 7)]\} = \{4 - [5 - (-1)]\} = \{4 - [5 + 1]\} = \{4 - 6\}$
$= \{4 + (-6)\} = -2$

41. $\left|2 - \left|(3 - |5 - 8|)\right|\right| = \left|2 - \left|(3 - |5 + (-8)|)\right|\right| = \left|2 - \left|(3 - |-3|)\right|\right|$
$= \left|2 - |(3 - 3)|\right| = \left|2 - |0|\right| = |2 - 0| = |2| = 2$

43. $|1 - [3 - (5 - 7)]| + |2 - [4 - (6 - 8)]|$
$= |1 - [3 - (5 + (-7))]| + |2 - [4 - (6 + (-8))]|$
$= |1 - [3 - (-2)]| + |2 - [4 - (-2)]| = |1 - [3 + 2]| + |2 - [4 + 2]|$
$= |1 - 5| + |2 - 6| = |1 + (-5)| + |2 + (-6)| = |-4| + |-4| = 4 + 4 = 8$

45. Since $-(-5) = 5$, -5 is correct.

47. Since $|+7| = 7$ and $|-7| = 7$, either 7 or -7 is correct.

49. Since $(-3) + (-5) = -8$, -5 is correct.

51. Since $(-3) - 5 = -8$, 5 is correct.

53. -29.191

55. 76.025

57. $(x - z) + y = [23.417 - (-13.012)] + (-52.608) = 36.429 - 52.608 = -16.179$

59. $|y - z| = |-52.608 - (-13.012)| = |-39.596| = 39.596$

61. $|z - x - y| = |-13.012 - 23.417 - (-52.608)| = |16.179| = 16.179$

63. $|z| - x = |-13.012| - 23.417 = 13.012 - 23.417 = -10.405$

65. $x + y = (3) + (-8) = -5$

67. $y - x = (-8) - (3) = -8 + (-3) = -11$

69. $(x - z) + y = [(3) - (-2)] + (-8) = [(3) + (2)] + (-8) = 5 + (-8) = -3$

71. $\left|(-z) - |y|\right| = \left|[-(-2)] - |-8|\right| = \left|(+2) - (+8)\right| = \left|(+2) + (-8)\right| = |-6| = 6$

73. $-\left||y| - |x|\right| = -\left||-8| - |3|\right| = -\left|(+8) - (+3)\right| = -|5| = -(5) = -5$

75. $x + y = \frac{1}{3} + \left(-\frac{1}{2}\right) = \frac{2}{6} + \left(-\frac{3}{6}\right) = \frac{-1}{6}$

77. $y - x = \left(-\frac{1}{2}\right) - \left(\frac{1}{3}\right) = -\frac{1}{2} - \frac{1}{3} = -\frac{3}{6} - \frac{2}{6} = -\frac{5}{6}$

79. $(x - z) + y = \left[\frac{1}{3} - \left(-\frac{1}{4}\right)\right] + \left(-\frac{1}{2}\right) = \left[\frac{1}{3} + \frac{1}{4}\right] - \frac{1}{2} = \frac{1}{3} + \frac{1}{4} - \frac{1}{2} = \frac{4}{12} + \frac{3}{12} - \frac{6}{12} = \frac{1}{12}$

81. $\left|(-z) - |y|\right| = \left|-\left(-\frac{1}{4}\right) - \left|-\frac{1}{2}\right|\right| = \left|\frac{1}{4} - \frac{1}{2}\right| = \left|\frac{1}{4} - \frac{2}{4}\right| = \left|-\frac{1}{4}\right| = \frac{1}{4}$

83. $-\left||y| - |x|\right| = -\left|\left|-\frac{1}{2}\right| - \left|\frac{1}{3}\right|\right| = -\left|\frac{1}{2} - \frac{1}{3}\right| = -\left|\frac{3}{6} - \frac{2}{6}\right| = -\left|\frac{1}{6}\right| = -\frac{1}{6}$

85. True—this states the commutative property for addition.

87. False—one of many examples showing its falseness is
$(+7) - (-3) = +10$ but $(-3) - (+7) = -10$, so
$(+7) - (-3) \neq (-3) - (+7)$

89. True—this states the associative property for addition.

91. False—one of many examples showing its falseness is
$|(+9) + (-3)| = +6$ but $|+9| - |-3| = +12$, so
$|(+9) + (-3)| \neq |+9| + |-3|$

93. True

95. False—one of many examples showing its falseness is
$5 - (3 + 2) = 5 - 5 = 0$ but $(5 - 3) + 2 = 2 + 2 = 4$, so
$5 - (3 + 2) \neq (5 - 3) + 2$

97. Commutative property for addition, associative property for addition, additive inverse property, identity property for addition

99. Definition of subtraction, Problem 98, associative property for addition, definition of subtraction

101. We are to represent a fall in price by addition of a negative number and a rise in price by addition of a positive number. Thus, we write:
$23.50 + (-3.25) + (-6.75) + (+2.50) + (+7.75) =$
$23.50 + (2.50) + (7.75) + (-3.25) + (-6.75) =$
$33.75 + (-10.00) = 23.75$ or \$23.75

103. We are to represent a bogie as +1, a birdie as -1, and par as 0. Hence, the score is
$0 + (+1) + (+1) + (-1) + 0 + 0 = +1$ or 1 over par

Exercise 1-5 Multiplication and Division of Real Numbers

Key Ideas and Formulas

Multiplication of Real Numbers: ab

1. If a or b is 0: $0 \cdot b = 0$ $a \cdot 0 = 0$
2. If a and b are both positive: multiply as you have learned in arithmetic.
3. If a and b are both negative: take the product of the absolute values:
$(-5) \cdot (-8) = 5 \cdot 8 = 40$
4. If a and b have opposite signs: take the opposite of the product of the absolute values: $5 \cdot (-8) = -(5 \cdot 8) = -40$

The product of two numbers with like signs is positive.

The product of two numbers with unlike signs is negative.

$(-1)a = -a \qquad (-a)b = -(ab) \qquad (-a)(-b) = ab$

$$-ab = \begin{cases} (-a)b \\ a(-b) \\ -(ab) \\ (-1)ab \end{cases}$$

Definition of Division
We write $\left.\begin{array}{r} a \div b = Q \\ b\,\overline{)\,a}^{\,Q} \end{array}\right\}$ if and only if $Qb = a$ and Q is unique
The quotient Q is the number that must be multiplied times b to produce a.

Zero cannot be used as a divisor. Division by zero is undefined.

Division of Real Numbers: $\frac{a}{b}$

1. If $b = 0$, $\frac{a}{b}$ is not defined.

   If $a = 0$, but $b \neq 0$, $\frac{a}{b} = 0$.

2. If a and b are both positive: divide as you have learned in arithmetic.
3. If a and b are both negative: take the quotient of the absolute values: $\frac{-20}{-5} = \frac{20}{5} = 4$.
4. If a and b have opposite signs: take the opposite of the quotient of the absolute values: $\frac{-20}{5} = -\left(\frac{20}{5}\right) = -4$

   $\frac{20}{-5} = -\left(\frac{20}{5}\right) = -4$.

   The quotient of two numbers with like signs is positive.
   The quotient of two numbers with unlike signs is negative.

   $\frac{-a}{-b} = \frac{a}{b} \qquad -\frac{-a}{b} = -\frac{a}{-b} \qquad \frac{-a}{b} = \frac{a}{-b} = -\frac{a}{b}$

---

1. $(-3)(-5) = 3 \cdot 5 = 15$

3. $(-18) \div (-6) = \frac{18}{6} = 3$

5. $(-2)(+9) = -(2 \cdot 9) = -18$

7. $\frac{-9}{+3} = -\frac{9}{3} = -3$

9. 0

11. 0

13. Not defined

15. Not defined

17. $\frac{-21}{3} = -\frac{21}{3} = -7$

19. $(-4)(-2) + (-9) = 8 + (-9) = -1$

21. $(+5) - (-2)(+3) = (+5) - (-6) = (+5) + (6) = 11$

23. $5 - \frac{-8}{2} = 5 - (-4) = 5 + 4 = 9$

25. Both are 8

27. $-12 + \frac{-14}{-7} = -12 + 2 = -10$

29. $\frac{6(-4)}{-8} = \frac{-24}{-8} = 3$

31. $\frac{22}{-11} - (-4)(-3) = -2 - (12) = -2 + (-12) = -14$

33. $\frac{-16}{2} - \frac{3}{-1} = (-8) - (-3) = (-8) + (3) = -5$

35. $(+5)(-7)(+2) = (-35)(+2) = -70$

37. Since $a0 = 0$, the answer is 0.

39. $[(+2) + (-7)][(+8) - (+10)] = [-5][-2] = 10$

41. $12 - 7[(-4)(5) - 2(-8)] = 12 - 7[(-20) - (-16)] = 12 - 7[-4]$
$= 12 + (-7)(-4) = 12 + 28 = 40$

43. $\frac{9}{-3} - \frac{3 + 9(-2)}{-2 - (-3)} = -3 - \frac{3 + (-18)}{-2 - (-3)} = -3 - \frac{-15}{+1} = (-3) - (-15) = 12$

45. $\left\{\left[\frac{8}{(-2)}\right] - \left[21 + 5(-3)\right]\right\} - (-2)(-4) = \{[-4] - [21 + (-15)]\} - (-2)(-4)$
$= \{[-4] - [6]\} - (+8)$
$= \{-10\} - (+8)$
$= -18$

47. $[8 - (9 - 7)]/6 - 3 = [8 - 2]/6 - 3 = 6/6 - 3 = 1 - 3 = -2$

49. $\frac{1 - 4 \div 2 \cdot 3}{6 \div 2 - 4 \div 2} = \frac{1 - 2 \cdot 3}{3 - 2} = \frac{1 - 6}{1} = \frac{-5}{1} = -5$

51. $12 \div \{4 - [3(2 \cdot 9) \div 27 - 1]\} = 12 \div \{4 - [3(18) \div 27 - 1]\}$
$= 12 \div \{4 - [54 \div 27 - 1]\} = 12 \div \{4 - [2 - 1]\} = 12 \div \{4 - 1\} = 12 \div 3 = 4$

53. $[5 \div (4 - 3) - 8] \div (6 - 3) = [5 \div 1 - 8] \div 3 = [5 - 8] \div 3 = (-3) \div 3 = -1$

55. $\frac{2 + 4 \div 2 - 4}{1 + 2 \cdot 3 - 4} = \frac{2 + 2 - 4}{1 + 6 - 4} = \frac{4 - 4}{7 - 4} = \frac{0}{3} = 0$

57. $[8 - (9 - 7)]/(6 - 4) + (4 + 2)/2 = [8 - 2]/(6 - 4) + (4 + 2)/2$
$= 6/2 + 6/2$
$= 3 + 3$
$= 6$

59. $8 - [(9 - 3)/2 - 2 + 4 - 4/2] = 8 - [6/2 - 2 + 4 - 4/2]$
$= 8 - [3 - 2 + 4 - 2]$
$= 8 - 3$
$= 5$

61. $\frac{(8 - 1) - 8/(6 - 2)}{4 + 2/2} = \frac{7 - 8/4}{4 + 2/2} = \frac{7 - 2}{4 + 1} = \frac{5}{5} = 1$

63. $\frac{1 - 2 \cdot 3 + 4}{6 - 4 \div 2 \cdot 3} = \frac{1 - 6 + 4}{6 - 2 \cdot 3} = \frac{-5 + 4}{6 - 6} = \frac{-1}{0}$ is not defined

65. $\frac{z}{w} = \frac{(-24)}{(2)} = -12$

67. $\frac{w}{y} = \frac{(2)}{(0)}$ is not defined

69. $\frac{z}{x} - wz = \frac{(-24)}{(-3)} - (2)(-24) = 8 - (-48) = 56$

71. $\frac{xy}{w} - xyz = \frac{(-3)0}{(2)} - (-3)(0)(-24) = \frac{0}{2} - 0 = 0 - 0 = 0$

73. $-|w||x| = -|2||-3| = -(2)(3) = -6$

75. $\dfrac{|z|}{|x|} = \dfrac{|-24|}{|-3|} = \dfrac{24}{3} = 8$

77. $$\begin{aligned}(wx - z)(z - 8x) &= [(2)(-3) - (-24)][(-24) - 8(-3)]\\ &= [(-6) - (-24)][(-24) + (+24)]\\ &= (18)(0)\\ &= 0\end{aligned}$$

79. $$\begin{aligned}wx + \frac{z}{wx} + wz &= (2)(-3) + \frac{(-24)}{(2)(-3)} + (2)(-24)\\ &= (-6) + \frac{(-24)}{(-6)} + (-48)\\ &= (-6) + (+4) + (-48)\\ &= (-54) + (+4)\\ &= -50\end{aligned}$$

81. $$\begin{aligned}\frac{8x}{z} - \frac{z - 6x}{wx} &= \frac{8(-3)}{(-24)} - \frac{(-24) - 6(-3)}{2(-3)}\\ &= \frac{(-24)}{(-24)} - \frac{(-24) + (+18)}{(-6)}\\ &= (+1) - \frac{(-6)}{(-6)}\\ &= (+1) - (+1)\\ &= 0\end{aligned}$$

83. $$\begin{aligned}\frac{24}{3w - 2x} - \frac{24}{3w + 2x} &= \frac{24}{3(2) - 2(-3)} - \frac{24}{3(2) + 2(-3)}\\ &= \frac{24}{(6) + (+6)} - \frac{24}{(6) + (-6)}\\ &= \frac{24}{12} - \frac{24}{0} \text{ is not defined}\end{aligned}$$

85. $$\begin{aligned}\frac{z}{wx} - 2[z + 3(2x - w)] &= \frac{(-24)}{(2)(-3)} - 2\{(-24) + 3[2(-3) - (2)]\}\\ &= \frac{(-24)}{(-6)} - 2\{(-24) + 3[(-6) - (2)]\}\\ &= (+4) - 2\{(-24) + 3[-8]\}\\ &= (+4) - 2\{(-24) + [-24]\}\\ &= (+4) - 2\{-48\}\\ &= (+4) + (+96)\\ &= 100\end{aligned}$$

87. $\dfrac{8y - w}{8x - z} = \dfrac{8 \cdot 0 - 2}{8(-3) - (-24)} = \dfrac{0 - 2}{-24 - (-24)} = \dfrac{-2}{0}$ is not defined

89. $$\begin{aligned}\frac{z}{x} - wz &= \frac{-\frac{5}{12}}{-\frac{1}{3}} - \left(\frac{1}{2}\right)\left(-\frac{5}{12}\right) = -\frac{5}{12} \div \left(-\frac{1}{3}\right) - \left(-\frac{5}{24}\right) = -\frac{5}{12} \cdot \frac{-3}{1} - \frac{-5}{24}\\ &= \frac{15}{12} + \frac{5}{24} = \frac{30}{24} + \frac{5}{24} = \frac{35}{24}\end{aligned}$$

91. $\dfrac{xy}{w} - xyz = \dfrac{\left(-\frac{1}{3}\right)\left(\frac{3}{4}\right)}{\frac{1}{2}} - \left(-\frac{1}{3}\right)\left(\frac{3}{4}\right)\left(-\frac{5}{12}\right) = \dfrac{-\frac{1}{4}}{\frac{1}{2}} - \left(-\frac{1}{4}\right)\left(-\frac{5}{12}\right)$

$= -\frac{1}{4} \div \frac{1}{2} - \frac{5}{48} = -\frac{1}{4}\cdot\frac{2}{1} - \frac{5}{48} = \frac{-2}{4} - \frac{5}{48} = \frac{-24}{48} - \frac{5}{48} = -\frac{29}{48}$

93. Since we cannot divide by 0, the correct choice is never.

95. Always

97. When x and y are of opposite signs.

99. 1. Identity property for addition; 2. Distributive property; 3. Addition property for equality; 4. Inverse and associative property for addition; 5. Inverse property for addition; 6. Identity property for addition; 7. Symmetric property for equality.

101. 1. Distributive property; 2. Inverse property for addition; 3. Definition of multiplication; 4. Inverse property for addition.

103. Recall: birdie—(-1), bogie—(+1), double bogie—(+2), par—(0). Hence the score would be 72 + 3(-1) + 7(+1) + 2(+2) + (18 - 3 - 7 - 2)(0) = 72 - 3 + 7 + 4 + 0 = 80, or 8 over par.

105. (A) Twice normal speed forward would appear as +2(+3) or +6 miles per hour forward.
(B) Twice normal speed backward would appear as -2(+3) or -6 miles per hour, that is 6 miles per hour backward.

Exercise 1-6 Exponents and Order of Operations

Key Ideas and Formulas

Natural Number Exponents: For n a natural number, and a real number:

$$a^n = \underbrace{a\cdot a\cdot \ldots \cdot a}_{n \text{ factors of } a}$$

$$a^m a^n = a^{m+n}$$

Extension of order of operations: Powers take precedence over multiplication and division:

$5\cdot 2^3 = 5\cdot 8 = 40$ $\qquad$ $(5\cdot 2)^3 = 10^3 = 1{,}000$

$\dfrac{2^3}{5} = \dfrac{8}{5} = 1.6$ $\qquad$ $\left(\dfrac{2}{5}\right)^3 = (0.4)^3 = 0.064$

$-4^2 = -(4\cdot 4) = -16$ $\qquad$ $(-4)^2 = (-4)(-4) = 16$

Order of Operations
(A) *If no grouping symbols are present:* **1.** Perform any multiplication and division, proceeding from left to right. **2.** Perform any addition and subtraction, proceeding from left to right. **(B)** *If symbols of grouping are present:* **1.** Simplify above and below any fraction bars following the steps in (A). **2.** Simplify within other symbols of grouping, generally starting with the innermost and working outward, following the steps in (A).

A constant that is present as a factor in a term is called the **numerical coefficient**, or simply the **coefficient** of the term.

Two terms are called **like terms** if they have exactly the same variable factors and these variables are raised to the same powers in each term. Like terms are combined by adding their numerical coefficients.

---

1. $3^2 \cdot 3^5 = 3^{2+5} = 3^7$. 7   3. $2^5 2^3 = 2^8$. 3   5. $y^2 y^7 = y^{2+7} = y^9$. 9

7. $y^8 = y^3 y^5$. 5   9. $10^5 10^6 10^7 = 10^{5+6+7} = 10^{18}$. 18   11. $10^{13} 10^5 10^1 = 10^{19}$. 5

13. $(5x^2)(2x^9) = 5 \cdot 2x^2 x^9 = 10x^{2+9} = 10x^{11}$

15. $(4y^3)(3y)(y^6) = (4 \cdot 3)y^3 y y^6 = 12y^{3+1+6} = 12y^{10}$

17. $(4 \times 10^5)(5 \times 10^4) = (4 \cdot 5) \times 10^5 \times 10^4 = 20 \times 10^{5+4} = 20 \times 10^9$

19. $(5 \times 10^8)(7 \times 10^9) = (5 \cdot 7) \times 10^8 \times 10^9 = 35 \times 10^{8+9} = 35 \times 10^{17}$

21. $9x + 8x = (9 + 8)x = 17x$ 23. $9x - 8x = (9 - 8)x = 1x$ or x

25. $5x + x + 2x = (5 + 1 + 2)x = 8x$

27. $4t - 8t - 9t = (4 - 8 - 9)t = -13t$

29. $4y + 3x + y = 3x + 4y + y = 3x + (4 + 1)y = 3x + 5y$

31. $8 + 4x - 4 = 8 - 4 + 4x = 4 + 4x$

33. $5m + 3n - m - 9n = 5m - m + 3n - 9n = (5 - 1)m + (3 - 9)n = 4m - 6n$

35. $3(u - 2v) + 2(3u + v) = 3u - 6v + 6u + 2v = 9u - 4v$

37. $4(m - 3n) - 3(2m + 4n) = 4m - 12n - 6m - 12n = -2m - 24n$

39. $(2u - v) + (3u - 5v) = 1(2u - v) + 1(3u - 5v) = 2u - v + 3u - 5v = 5u - 6v$

41. $(3x + 2) + (x - 5) = 1(3x + 2) + 1(x - 5) = 3x + 2 + x - 5 = 4x - 3$

43. $2(x + 5) - (x - 1) = 2(x + 5) + (-1)(x - 1) = 2x + 10 - x + 1 = x + 11$

45. $-ab^2 = -(3)(-2)^2 = -(3)(4) = -12$

47. $-(a - b)^2 = -[3 - (-2)]^2 = -[5]^2 = -25$

49. $ab^2 - bc^2 = (3)(-2)^2 - (-2)(6)^2 = (3)(4) - (-2)(36) = 12 - (-72) = 84$

51. $x^k x^m = x^{k+m}$. $k + m$ 53. $x^k x^1 = x^{k+1}$. 1

55. $x^{2m} x^{5m} = x^{2m+5m} = x^{7m}$. $7m$ 57. $x^{m+1} x^{2m+3} = x^{m+1+2m+3} = x^{3m+4}$. $3m + 4$

59. $x^{2m+5} x^{3m+2} = x^{5m+7}$. $2m + 5$

61. $(3x^m)(2x^{m+1}) = (3 \cdot 2)x^m x^{m+1} = 6x^{m+m+1} = 6x^{2m+1}$

63. $(5x^{2m})(2x^{5m}) = (5\cdot2)x^{2m}x^{5m} = 10x^{2m+5m} = 10x^{7m}$

65. $x^2y + 3x^2y - 5x^2y = (-1 + 3 - 5)x^2y = -3x^2y$

67. $y^3 + 4y^2 - 10 + 2y^3 - y + 7 = y^3 + 2y^3 + 4y^2 - y - 10 + 7$
$= (1 + 2)y^3 + 4y^2 - y - 3 = 3y^3 + 4y^2 - y - 3$

69. $a^2 - 3ab + b^2 + 2a^2 + 3ab - 2b^2 = a^2 + 2a^2 - 3ab + 3ab + b^2 - 2b^2$
$= (1 + 2)a^2 + (-3 + 3)ab + (1 - 2)b^2$
$= 3a^2 + 0ab + (-1)b^2$
$= 3a^2 - b^2$

71. $x - 3y - 4(2x - 3y) = x - 3y - 8x + 12y = -7x + 9y$

73. $y - 2(x - y) - 3x = y - 2x + 2y - 3x = -5x + 3y$

75. $-2(-3x + 1) - (2x + 4) = -2(-3x + 1) + (-1)(2x + 4) = 6x - 2 - 2x - 4 = 4x - 6$

77. $2(x - 1) - 3(2x - 3) - (4x - 5) = 2(x - 1) - 3(2x - 3) + (-1)(4x - 5)$
$= 2x - 2 - 6x + 9 - 4x + 5 = -8x + 12$

79. $4t - 3[4 - 2(t - 1)] = 4t - 3[4 - 2t + 2] = 4t - 3[-2t + 6]$
$= 4t + 6t - 18 = 10t - 18$

81. $3[x - 2(x + 1)] - 4(2 - x) = 3[x - 2x - 2] - 4(2 - x) = 3[-x - 2] - 4(2 - x)$
$= -3x - 6 - 8 + 4x = x - 14$

83. $x^2y - xy^2 = (3)^2(-2) - (3)(-2)^2 = (9)(-2) - (3)(4) = (-18) - (12) = -30$

85. $-x^2 - yz^2 \div x = -(3)^2 - (-2)(6)^2 \div (3) = (-9) - (-2)(36) \div (3)$
$= (-9) - (-72) \div (3) = (-9) - (-24) = 15$

87. $2a - 3\{a + 2[a - (a + 5)] + 1\} = 2a - 3\{a + 2[a - a - 5] + 1\}$
$= 2a - 3\{a + 2[-5] + 1\} = 2a - 3\{a - 10 + 1\}$
$= 2a - 3\{a - 9\} = 2a - 3a + 27 = -a + 27$ or
$27 - a$

89. $a - \{b - [c - (a - b) - c] - (c - a)\} + b = a - \{b - [c - a + b - c] - c + a\} + b$
$= a - \{b - [-a + b] - c + a\} + b = a - \{b + a - b - c + a\} + b$
$= a - \{2a - c\} + b = a - 2a + c + b = b + c - a$

91. $\{[(c - 1) - 1] - c\} - 1 = \{[c - 1 - 1] - c\} - 1$
$= \{[c - 2] - c\} - 1 = \{c - 2 - c\} - 1 = \{-2\} - 1 = -3$

93. $a - \{1 - [a - (1 - a)]\} = a - \{1 - [a - 1 + a]\} = a - \{1 - [2a - 1]\}$
$= a - \{1 - 2a + 1\} = a - \{2 - 2a\} = a - 2 + 2a = 3a - 2$

95. $x^2y - xy^2 = \left(\frac{1}{2}\right)^2\left(-\frac{2}{3}\right) - \left(\frac{1}{2}\right)\left(-\frac{2}{3}\right)^2 = \left(\frac{1}{4}\right)\left(-\frac{2}{3}\right) - \left(\frac{1}{2}\right)\left(\frac{4}{9}\right) = \frac{-1}{6} - \frac{2}{9} = \frac{-3}{18} - \frac{4}{18} = -\frac{7}{18}$

97. $-x^2 - yz^2 \div x = -\left(\frac{1}{2}\right)^2 - \left(-\frac{2}{3}\right)\left(\frac{1}{6}\right)^2 \div \left(\frac{1}{2}\right) = -\frac{1}{4} - \left(-\frac{2}{3}\right)\left(\frac{1}{36}\right) \div \left(\frac{1}{2}\right)$

$= -\frac{1}{4} - \left(-\frac{1}{54}\right) \div \left(\frac{1}{2}\right) = -\frac{1}{4} - \left(-\frac{1}{54}\right)\cdot\left(\frac{2}{1}\right) = -\frac{1}{4} - \left(-\frac{1}{27}\right)$

$= -\frac{27}{108} + \frac{4}{108} = -\frac{23}{108}$

99. $(x^2)^3 = x^2x^2x^2 = x^{2+2+2} = x^{3\cdot2} = x^6$. 6
$(x^3)^4 = x^3x^3x^3x^3 = x^{3+3+3+3} = x^{4\cdot3} = x^{12}$. 12
$(x^5)^2 = x^5x^5 = x^{5+5} = x^{2\cdot5} = x^{10}$. 10
$(x^m)^n = x^{m\cdot n}$. mn

Exercise 1-7 Solving Equations

Key Ideas and Formulas

Two terms are called **like terms** if they have exactly the same variable factors to the same powers. Like terms are **combined** into a single term by adding their numerical coefficients.

A **solution** or **root** of an equation involving a single variable is a replacement of the variable by a constant that makes the left side of the equation equal to the right side. The set of all solutions is called the **solution set.**

The following properties of equality produce equivalent equations when applied.

Equality Properties	
For a, b, and c any real numbers:	
1. If $a = b$, then $a + c = b + c$.	ADDITION PROPERTY
2. If $a = b$, then $a - c = b - c$.	SUBTRACTION PROPERTY
3. If $a = b$, then $ca = cb$, $c \neq 0$.	MULTIPLICATION PROPERTY
4. If $a = b$, then $\frac{a}{c} = \frac{b}{c}$, $c \neq 0$.	DIVISION PROPERTY

Equation-Solving Strategy
1. Use the multiplication property to remove fractions if present.
2. Simplify the left and right sides of the equation by removing grouping symbols and combining like terms.
3. Use the equality properties to get all variable terms on one side (usually the left) and all constant terms on the other side (usually the right). Combine like terms in the process.
4. Isolate the variable (with a coefficient of 1), using the division or multiplication property of equality.

---

1. $3x = 27$

$$\frac{3x}{3} = \frac{27}{3}$$

$x = 9$

CHECK: $3x = 27$

$3 \cdot 9 \stackrel{?}{=} 27$

$27 \stackrel{\checkmark}{=} 27$

3. $-2x = 14$

$$\frac{-2x}{-2} = \frac{14}{-2}$$

$x = -7$

CHECK: $-2x = 14$

$-2(-7) \stackrel{?}{=} 14$

$14 \stackrel{\checkmark}{=} 14$

5. $5x - 3 = 17$

$5x - 3 + 3 = 17 + 3$

$5x = 20$

$$\frac{5x}{5} = \frac{20}{5}$$

$x = 4$

CHECK: $5x - 3 = 17$

$5 \cdot 4 - 3 \stackrel{?}{=} 17$

$20 - 3 \stackrel{\checkmark}{=} 17$

7.
$$\begin{aligned} 2x + 7 &= 19 \\ 2x + 7 - 7 &= 19 - 7 \\ 2x &= 12 \\ \frac{2x}{2} &= \frac{12}{2} \\ x &= 6 \end{aligned}$$

CHECK:
$$\begin{aligned} 2x + 7 &= 19 \\ 2\cdot 6 + 7 &\stackrel{?}{=} 19 \\ 12 + 7 &\stackrel{\checkmark}{=} 19 \end{aligned}$$

9.
$$\begin{aligned} -3x + 7 &= -5 \\ -3x + 7 - 7 &= -5 - 7 \\ -3x &= -12 \\ \frac{-3x}{-3} &= \frac{-12}{-3} \\ x &= 4 \end{aligned}$$

CHECK:
$$\begin{aligned} -3x + 7 &= -5 \\ -3\cdot 4 + 7 &\stackrel{?}{=} -5 \\ -12 + 7 &\stackrel{\checkmark}{=} -5 \end{aligned}$$

11.
$$\begin{aligned} 3(x + 4) &= 9 \\ 3x + 12 &= 9 \\ 3x + 12 - 12 &= 9 - 12 \\ 3x &= -3 \\ \frac{3x}{3} &= \frac{-3}{3} \\ x &= -1 \end{aligned}$$

CHECK:
$$\begin{aligned} 3(x + 4) &= 9 \\ 3[(-1) + 4] &\stackrel{?}{=} 9 \\ 3[3] &\stackrel{\checkmark}{=} 9 \end{aligned}$$

13.
$$\begin{aligned} 3(x - 8) &= x + 6 \\ 3x - 24 &= x + 6 \\ 3x - 24 - x &= x + 6 - x \\ 2x - 24 &= 6 \\ 2x - 24 + 24 &= 6 + 24 \\ 2x &= 30 \\ \frac{2x}{2} &= \frac{30}{2} \\ x &= 15 \end{aligned}$$

CHECK:
$$\begin{aligned} 3(x - 8) &= x + 6 \\ 3(15 - 8) &\stackrel{?}{=} 15 + 6 \\ 3\cdot 7 &\stackrel{\checkmark}{=} 21 \end{aligned}$$

15.
$$\begin{aligned} -(x - 2) &= 9 \\ -x + 2 &= 9 \\ -x + 2 - 2 &= 9 - 2 \\ -x &= 7 \\ \frac{-x}{-1} &= \frac{7}{-1} \\ x &= -7 \end{aligned}$$

CHECK:
$$\begin{aligned} -(x - 2) &= 9 \\ -[(-7) - 2] &\stackrel{?}{=} 9 \\ -[-9] &\stackrel{\checkmark}{=} 9 \end{aligned}$$

17.
$$\begin{aligned} -3(x - 5) &= 21 \\ -3x + 15 &= 21 \\ -3x + 15 - 15 &= 21 - 15 \\ -3x &= 6 \\ \frac{-3x}{-3} &= \frac{6}{-3} \\ x &= -2 \end{aligned}$$

CHECK:
$$\begin{aligned} -3(x - 5) &= 21 \\ -3[(-2) - 5] &\stackrel{?}{=} 21 \\ -3[-7] &\stackrel{\checkmark}{=} 21 \end{aligned}$$

19.
$$\begin{aligned} -5(3 - x) &= 40 \\ -15 + 5x &= 40 \\ 15 - 15 + 5x &= 15 + 40 \\ 5x &= 55 \\ \frac{5x}{5} &= \frac{55}{5} \\ x &= 11 \end{aligned}$$

CHECK:
$$\begin{aligned} -5(3 - x) &= 40 \\ -5(3 - 11) &\stackrel{?}{=} 40 \\ -5(-8) &\stackrel{\checkmark}{=} 40 \end{aligned}$$

21.
$$\begin{aligned} 3(x + 2) &= 5(x - 6) \\ 3x + 6 &= 5x - 30 \\ 3x + 6 - 6 &= 5x - 30 - 6 \\ 3x &= 5x - 36 \\ 3x - 5x &= 5x - 36 - 5x \\ -2x &= -36 \\ \frac{-2x}{-2} &= \frac{-36}{-2} \\ x &= 18 \end{aligned}$$

CHECK:
$$\begin{aligned} 3(x + 2) &= 5(x - 6) \\ 3(18 + 2) &\stackrel{?}{=} 5(18 - 6) \\ 3\cdot 20 &\stackrel{?}{=} 5\cdot 12 \\ 60 &\stackrel{\checkmark}{=} 60 \end{aligned}$$

23.
$$\begin{aligned} 4(x - 2) &= 4x - 8 \\ 4x - 8 &= 4x - 8 \\ 0x &= 0 \end{aligned}$$

Every real number is a solution.

25.
$$\begin{aligned} 5 + 4(t - 2) &= 2(t + 7) + 1 \\ 5 + 4t - 8 &= 2t + 14 + 1 \\ 4t - 3 &= 2t + 15 \\ 4t - 3 + 3 &= 2t + 15 + 3 \\ 4t &= 2t + 18 \\ 4t - 2t &= 2t + 18 - 2t \\ 2t &= 18 \\ \frac{2t}{2} &= \frac{18}{2} \\ t &= 9 \end{aligned}$$

CHECK:
$$\begin{aligned} 5 + 4(t - 2) &= 2(t + 7) + 1 \\ 5 + 4(9 - 2) &\stackrel{?}{=} 2(9 + 7) + 1 \\ 5 + 4\cdot 7 &\stackrel{?}{=} 2\cdot 16 + 1 \\ 5 + 28 &\stackrel{?}{=} 32 + 1 \\ 33 &\stackrel{\checkmark}{=} 33 \end{aligned}$$

27.
$$\begin{aligned} 3x - (x + 2) &= 5x - 3(x - 1) \\ 3x - x - 2 &= 5x - 3x + 3 \\ 2x - 2 &= 2x + 3 \\ 2x - 2 - 2x &= 2x + 3 - 2x \\ -2 &= 3 \end{aligned}$$

No solution.

29.
$$\begin{aligned}10x + 25(x - 3) &= 275\\ 10x + 25x - 75 &= 275\\ 35x - 75 &= 275\\ 35x - 75 + 75 &= 275 + 75\\ 35x &= 350\\ \frac{35x}{35} &= \frac{350}{35}\\ x &= 10\end{aligned}$$

CHECK:
$$\begin{aligned}10x + 25(x - 3) &= 275\\ 10\cdot 10 + 25(10 - 3) &\stackrel{?}{=} 275\\ 10\cdot 10 + 25\cdot 7 &\stackrel{?}{=} 275\\ 100 + 175 &\stackrel{?}{=} 275\\ 275 &\stackrel{\surd}{=} 275\end{aligned}$$

31.
$$\begin{aligned}5x - (7x - 4) - 2 &= 5 - (3x + 2)\\ 5x - 7x + 4 - 2 &= 5 - 3x - 2\\ -2x + 2 &= -3x + 3\\ -2x + 2 - 2 &= -3x + 3 - 2\\ -2x &= -3x + 1\\ -2x + 3x &= -3x + 1 + 3x\\ x &= 1\end{aligned}$$

CHECK:
$$\begin{aligned}5x - (7x - 4) - 2 &= 5 - (3x + 2)\\ 5\cdot 1 - (7\cdot 1 - 4) - 2 &\stackrel{?}{=} 5 - (3\cdot 1 + 2)\\ 5 - (7 - 4) - 2 &\stackrel{?}{=} 5 - (3 + 2)\\ 5 - 3 - 2 &\stackrel{?}{=} 5 - 5\\ 5 - 5 &\stackrel{?}{=} 0\\ 0 &\stackrel{\surd}{=} 0\end{aligned}$$

33.
$$\begin{aligned}2(3x + 1) - 8x &= 2(1 - x)\\ 6x + 2 - 8x &= 2 - 2x\\ -2x + 2 &= 2 - 2x\\ -2x + 2 - 2 &= 2 - 2x - 2\\ -2x &= -2x\\ 0x &= 0\end{aligned}$$

Every real number is a solution.

35.
$$\begin{aligned}x + 5 &= 3x - 4\\ x + 5 - 3x &= 3x - 4 - 3x\\ -2x + 5 &= -4\\ -2x + 5 - 5 &= -4 - 5\\ -2x &= -9\\ \frac{-2x}{-2} &= \frac{-9}{-2}\\ x &= \frac{9}{2}\end{aligned}$$

CHECK:
$$\begin{aligned}x + 5 &= 3x - 4\\ \frac{9}{2} + 5 &\stackrel{?}{=} 3\left(\frac{9}{2}\right) - 4\\ \frac{9}{2} + \frac{10}{2} &\stackrel{?}{=} \frac{27}{2} - \frac{8}{2}\\ \frac{19}{2} &\stackrel{\surd}{=} \frac{19}{2}\end{aligned}$$

37.
$$\begin{aligned}3x + 7 &= 1 - (x + 4)\\ 3x + 7 &= 1 - x - 4\\ 3x + 7 &= -x - 3\\ 3x + 7 + x &= x - x - 3\\ 4x + 7 &= -3\\ 4x + 7 - 7 &= -3 - 7\\ 4x &= -10\\ \frac{4x}{4} &= \frac{-10}{4}\\ x &= -\frac{5}{2}\end{aligned}$$

CHECK:
$$\begin{aligned}3x + 7 &= 1 - (x + 4)\\ 3\left(-\frac{5}{2}\right) + 7 &\stackrel{?}{=} 1 - \left[\left(-\frac{5}{2}\right) + 4\right]\\ -\frac{15}{2} + 7 &\stackrel{?}{=} 1 - \left[-\frac{5}{2} + \frac{8}{2}\right]\\ -\frac{15}{2} + \frac{14}{2} &\stackrel{?}{=} 1 - \frac{3}{2}\\ -\frac{1}{2} &\stackrel{\surd}{=} \frac{2}{2} - \frac{3}{2}\end{aligned}$$

39. $\frac{1}{2}x = 5$ LCM = 2
$$\begin{aligned}2\cdot\frac{1}{2}x &= 2\cdot 5\\ x &= 10\end{aligned}$$

CHECK: $\frac{1}{2}x = 5$
$$\frac{1}{2}(10) \stackrel{\surd}{=} 5$$

41. $\frac{3}{5}x = \frac{1}{2}$ LCM = 10
$$\begin{aligned}10\cdot\frac{3}{5}x &= 10\cdot\frac{1}{2}\\ 6x &= 5\\ \frac{6x}{6} &= \frac{5}{6}\\ x &= \frac{5}{6}\end{aligned}$$

CHECK: $\frac{3}{5}x = \frac{1}{2}$
$$\begin{aligned}\frac{3}{5}\cdot\frac{5}{6} &\stackrel{?}{=} \frac{1}{2}\\ \frac{3}{6} &\stackrel{\surd}{=} \frac{1}{2}\end{aligned}$$

43.
$$\begin{aligned}
\frac{2}{3}x + \frac{3}{4} &= \frac{5}{6} \quad \text{LCM} = 12 \\
12\cdot\frac{2}{3}x + 12\cdot\frac{3}{4} &= 12\cdot\frac{5}{6} \\
8x + 9 &= 10 \\
8x + 9 - 9 &= 10 - 9 \\
8x &= 1 \\
\frac{8x}{8} &= \frac{1}{8} \\
x &= \frac{1}{8}
\end{aligned}$$

CHECK:
$$\begin{aligned}
\frac{2}{3}x + \frac{3}{4} &= \frac{5}{6} \\
\frac{2}{3}\cdot\frac{1}{8} + \frac{3}{4} &\stackrel{?}{=} \frac{5}{6} \\
\frac{1}{12} + \frac{3}{4} &\stackrel{?}{=} \frac{5}{6} \\
\frac{1}{12} + \frac{9}{12} &\stackrel{\surd}{=} \frac{10}{12}
\end{aligned}$$

45.
$$\begin{aligned}
\frac{1}{2}x + 3 &= \frac{3}{4}x - 5 \quad \text{LCM} = 4 \\
4\cdot\frac{1}{2}x + 4\cdot 3 &= 4\cdot\frac{3}{4}x - 4\cdot 5 \\
2x + 12 &= 3x - 20 \\
2x + 12 - 3x &= 3x - 20 - 3x \\
12 - x &= -20 \\
12 - x - 12 &= -20 - 12 \\
-x &= -32 \\
\frac{-x}{-1} &= \frac{-32}{-1} \\
x &= 32
\end{aligned}$$

CHECK:
$$\begin{aligned}
\frac{1}{2}x + 3 &= \frac{3}{4}x - 5 \\
\frac{1}{2}\cdot 32 + 3 &\stackrel{?}{=} \frac{3}{4}\cdot 32 - 5 \\
16 + 3 &\stackrel{?}{=} 24 - 5 \\
19 &\stackrel{\surd}{=} 19
\end{aligned}$$

47.
$$\begin{aligned}
\frac{2}{5}x + \frac{3}{10} &= \frac{1}{2}x - \frac{1}{10} \quad \text{LCM} = 10 \\
10\cdot\frac{2}{5}x + 10\cdot\frac{3}{10} &= 10\cdot\frac{1}{2}x - 10\cdot\frac{1}{10} \\
4x + 3 &= 5x - 1 \\
4x + 3 - 5x &= 5x - 1 - 5x \\
3 - x &= -1 \\
3 - x - 3 &= -1 - 3 \\
-x &= -4 \\
\frac{-x}{-1} &= \frac{-4}{-1} \\
x &= 4
\end{aligned}$$

CHECK:
$$\begin{aligned}
\frac{2}{5}x + \frac{3}{10} &= \frac{1}{2}x - \frac{1}{10} \\
\frac{2}{5}\cdot 4 + \frac{3}{10} &\stackrel{?}{=} \frac{1}{2}\cdot 4 - \frac{1}{10} \\
\frac{8}{5} + \frac{3}{10} &\stackrel{?}{=} 2 - \frac{1}{10} \\
\frac{16}{10} + \frac{3}{10} &\stackrel{?}{=} \frac{20}{10} - \frac{1}{10} \\
\frac{19}{10} &\stackrel{\surd}{=} \frac{19}{10}
\end{aligned}$$

49.
$$\begin{aligned}
x - \frac{3}{5} &= \frac{1}{3}x + \frac{1}{2} \quad \text{LCM} = 30 \\
30\cdot x - 30\cdot\frac{3}{5} &= 30\cdot\frac{1}{3}x + 30\cdot\frac{1}{2} \\
30x - 18 &= 10x + 15 \\
30x - 18 - 10x &= 10x + 15 - 10x \\
20x - 18 &= 15 \\
20x - 18 + 18 &= 15 + 18 \\
20x &= 33 \\
\frac{20x}{20} &= \frac{33}{20} \\
x &= \frac{33}{20}
\end{aligned}$$

CHECK:
$$\begin{aligned}
x - \frac{3}{5} &= \frac{1}{3}x + \frac{1}{2} \\
\frac{33}{20} - \frac{3}{5} &\stackrel{?}{=} \frac{1}{3}\cdot\frac{33}{20} + \frac{1}{2} \\
\frac{33}{20} - \frac{12}{20} &\stackrel{?}{=} \frac{11}{20} + \frac{10}{20} \\
\frac{21}{20} &\stackrel{\surd}{=} \frac{21}{20}
\end{aligned}$$

51.
$$\begin{aligned}
x(x - 1) + 5 &= x^2 + x - 3 \\
x^2 - x + 5 &= x^2 + x - 3 \\
x^2 - x + 5 - x^2 &= x^2 + x - 3 - x^2 \\
-x + 5 &= x - 3 \\
-x + 5 - 5 &= x - 3 - 5 \\
-x &= x - 8 \\
-x - x &= x - 8 - x \\
-2x &= -8 \\
\frac{-2x}{-2} &= \frac{-8}{-2} \\
x &= 4
\end{aligned}$$

CHECK:
$$\begin{aligned}
x(x - 1) + 5 &= x^2 + x - 3 \\
4(4 - 1) + 5 &\stackrel{?}{=} 4^2 + 4 - 3 \\
4\cdot 3 + 5 &\stackrel{?}{=} 16 + 4 - 3 \\
12 + 5 &\stackrel{\surd}{=} 17
\end{aligned}$$

53.
$$\begin{aligned} x(x-4)-2 &= x^2-4(x+3) \\ x^2-4x-2 &= x^2-4x-12 \\ x^2-4x-2-x^2 &= x^2-4x-12-x^2 \\ -4x-2 &= -4x-12 \\ -4x-2+4x &= -4x-12+4x \\ -2 &= -12 \end{aligned}$$
No solution.

55.
$$\begin{aligned} (1-x)x &= (x-1)(-x) \\ x-x^2 &= -x^2+x \\ x-x^2+x^2 &= x^2-x^2+x \\ x &= x \end{aligned}$$
Since this is always true, every real number is a solution.

57.
$$\begin{aligned} (2+x)(-3) &= (3-x)(-2) \\ -6-3x &= -6+3x \\ -6-3x-3x &= -6+3x-3x \\ -6-6x &= -6 \\ 6-6-6x &= 6-6 \\ -6x &= 0 \\ \frac{-6x}{-6} &= \frac{0}{-6} \\ x &= 0 \end{aligned}$$
CHECK:
$$\begin{aligned} (2+x)(-3) &= (3-x)(-2) \\ (2+0)(-3) &\overset{?}{=} (3-0)(-2) \\ (2)(-3) &\overset{?}{=} (3)(-2) \\ -6 &\overset{\surd}{=} -6 \end{aligned}$$

59.
$$\begin{aligned} 3(x-2)x &= 1-x(3+2x)2+7x^2 \\ 3(x^2-2x) &= 1-x(6+4x)+7x^2 \\ 3x^2-6x &= 1-6x-4x^2+7x^2 \\ 3x^2-6x &= 1-6x+3x^2 \\ 3x^2-6x-3x^2+6x &= 1-6x+3x^2-3x^2+6x \\ 0 &= 1 \end{aligned}$$
No solution.

61.
$$\begin{aligned} 10x+5y &= 150 \\ 10x+5y-5y &= 150-5y \\ 10x &= 150-5y \\ \frac{10x}{10} &= \frac{150-5y}{10} \\ x &= \frac{150}{10}-\frac{5y}{10} \\ x &= 15-\frac{1}{2}y \end{aligned}$$

63.
$$\begin{aligned} 3x-4y+5 &= x-2y+3 \\ 3x-4y+5-5 &= x-2y+3-5 \\ 3x-4y &= x-2y-2 \\ 3x-4y-x &= x-2y-2-x \\ 2x-4y &= -2y-2 \\ 2x-4y+4y &= -2y-2+4y \\ 2x &= 2y-2 \\ \frac{2x}{2} &= \frac{2y-2}{2} \\ x &= y-1 \end{aligned}$$

65.
$$\begin{aligned} 2-3(x+3y) &= x-5(y+6) \\ 2-3x-9y &= x-5y-30 \\ 2-3x-9y-2 &= x-5y-30-2 \\ -3x-9y &= x-5y-32 \\ -3x-9y+5y &= x-5y-32+5y \\ -3x-4y &= x-32 \\ -3x-4y+3x &= x-32+3x \\ -4y &= 4x-32 \\ \frac{-4y}{-4} &= \frac{4x-32}{-4} \\ y &= -x+8 \text{ (or } 8-x) \end{aligned}$$

67.
$$\begin{aligned} 3(x+y)-1 &= 17-3y \\ 3x+3y-1 &= 17-3y \\ 3x+3y-1+1 &= 17-3y+1 \\ 3x+3y &= 18-3y \\ 3x+3y-3y &= 18-3y-3y \\ 3x &= 18-6y \\ \frac{3x}{3} &= \frac{18-6y}{3} \\ x &= 6-2y \end{aligned}$$

69.
$$\begin{aligned} \frac{1}{3}x &= \frac{3}{4}y \\ \frac{4}{3}\cdot\frac{1}{3}x &= \frac{4}{3}\cdot\frac{3}{4}y \\ \frac{4x}{9} &= y \\ y &= \frac{4}{9}x \end{aligned}$$

71.
$$x - \frac{2}{5} = \frac{3}{5}y + \frac{4}{5} \quad \text{LCM} = 5$$
$$5x - 5\cdot\frac{2}{5} = 5\cdot\frac{3}{5}y + 5\cdot\frac{4}{5}$$
$$5x - 2 = 3y + 4$$
$$5x - 2 - 4 = 3y + 4 - 4$$
$$5x - 6 = 3y$$
$$\frac{5x - 6}{3} = \frac{3y}{3}$$
$$\frac{5x}{3} - \frac{6}{3} = y$$
$$y = \frac{5}{3}x - 2$$

73.
$$\frac{1}{2}(x - 4) = \frac{1}{3}(y - 5) \quad \text{LCM} = 6$$
$$6\cdot\frac{1}{2}(x - 4) = 6\cdot\frac{1}{3}(y - 5)$$
$$3(x - 4) = 2(y - 5)$$
$$3x - 12 = 2y - 10$$
$$3x - 12 + 10 = 2y - 10 + 10$$
$$3x - 2 = 2y$$
$$\frac{3x - 2}{2} = \frac{2y}{2}$$
$$\frac{3x}{2} - \frac{2}{2} = y$$
$$y = \frac{3x}{2} - 1$$

75. Subtraction property

77. Transitive property or substitution principle

79. Symmetric property

81. Division property

83. Since $\frac{11}{4}$ is the solution of the given equation and of equation (A), but not of equation (B), only (A) is equivalent to the given equation.

85. Since 2 is the solution of the given equation and of equation (A), but not of equation (B), only (A) is equivalent to the given equation.

87. Since -3 is the solution of the given equation, but not of equations (A) or (B), neither is equivalent to the given equation.

In equations 89-95, the check is omitted for lack of space. The student should perform the checking steps, however.

89.
$$x - [2x - 3(x - 4)] = 5$$
$$x - [2x - 3x + 12] = 5$$
$$x - [-x + 12] = 5$$
$$x + x - 12 = 5$$
$$2x - 12 = 5$$
$$2x - 12 + 12 = 5 + 12$$
$$2x = 17$$
$$\frac{2x}{2} = \frac{17}{2}$$
$$x = \frac{17}{2}$$

91.
$$0.25 - [1.75 - (4.25 - x)] = 4x + 3.75$$
$$0.25 - [1.75 - 4.25 + x] = 4x + 3.75$$
$$0.25 - [-2.5 + x] = 4x + 3.75$$
$$0.25 + 2.5 - x = 4x + 3.75$$
$$2.75 - x = 4x + 3.75$$
$$2.75 - x - 4x = 4x + 3.75 - 4x$$
$$2.75 - 5x = 3.75$$
$$2.75 - 5x - 2.75 = 3.75 - 2.75$$
$$-5x = 1$$
$$\frac{-5x}{-5} = \frac{1}{-5}$$
$$x = -0.2$$

93.

$$\frac{1}{2} - \left[x - \left(\frac{1}{3} - \frac{1}{4}x\right)\right] = \frac{1}{6}$$
$$\frac{1}{2} - \left[x - \frac{1}{3} + \frac{1}{4}x\right] = \frac{1}{6}$$
$$\frac{1}{2} - x + \frac{1}{3} - \frac{1}{4}x = \frac{1}{6} \quad \text{LCM} = 12$$
$$12 \cdot \frac{1}{2} - 12x + 12 \cdot \frac{1}{3} - 12 \cdot \frac{1}{4}x = 12 \cdot \frac{1}{6}$$
$$6 - 12x + 4 - 3x = 2$$
$$10 - 15x = 2$$
$$10 - 15x - 10 = 2 - 10$$
$$-15x = -8$$
$$\frac{-15x}{-15} = \frac{-8}{-15}$$
$$x = \frac{8}{15}$$

95.

$$1.2 - \left[\frac{3}{5} - \left(x - \frac{4}{5}\right)\right] = \frac{1}{5} - [0.4 - (0.7 - x)]$$
$$1.2 - \left[\frac{3}{5} - x + \frac{4}{5}\right] = \frac{1}{5} - [0.4 - 0.7 + x]$$
$$1.2 - \left[\frac{7}{5} - x\right] = \frac{1}{5} - [-0.3 + x]$$
$$1.2 - \frac{7}{5} + x = \frac{1}{5} + 0.3 - x \quad \text{LCM} = 10$$
$$10(1.2) - 10 \cdot \frac{7}{5} + 10x = 10 \cdot \frac{1}{5} + 10(0.3) - 10x$$
$$12 - 14 + 10x = 2 + 3 - 10x$$
$$-2 + 10x = 5 - 10x$$
$$-2 + 10x + 10x = 5 - 10x + 10x$$
$$-2 + 20x = 5$$
$$2 - 2 + 20x = 5 + 2$$
$$20x = 7$$
$$\frac{20x}{20} = \frac{7}{20}$$
$$x = \frac{7}{20}$$

97.

$$x - [1 - (y + x)] = y - [1 - (x + y)]$$
$$x - [1 - y - x] = y - [1 - x - y]$$
$$x - 1 + y + x = y - 1 + x + y$$
$$2x - 1 + y = 2y - 1 + x$$
$$2x - 1 + y - 2y = 2y - 1 + x - 2y$$
$$2x - 1 - y = -1 + x$$
$$2x - 1 - y + 1 - 2x = -1 + x + 1 - 2x$$
$$-y = -x$$
$$\frac{-y}{-1} = \frac{-x}{-1}$$
$$y = x$$

99.

$$\frac{1}{3} - \left[x - \left(y - \frac{1}{6}\right)\right] = x - \left[y - \left(x - \frac{1}{2}\right)\right]$$
$$\frac{1}{3} - \left[x - y + \frac{1}{6}\right] = x - \left[y - x + \frac{1}{2}\right]$$
$$\frac{1}{3} - x + y - \frac{1}{6} = x - y + x - \frac{1}{2}$$
$$\frac{1}{3} - x + y - \frac{1}{6} = 2x - y - \frac{1}{2} \quad \text{LCM} = 6$$
$$6 \cdot \frac{1}{3} - 6x + 6y - 6 \cdot \frac{1}{6} = 6 \cdot 2x - 6y - 6 \cdot \frac{1}{2}$$
$$2 - 6x + 6y - 1 = 12x - 6y - 3$$
$$-6x + 6y + 1 = 12x - 6y - 3$$
$$-12x - 6x + 6y + 1 = 12x - 6y - 3 - 12x$$
$$-18x + 6y + 1 = -6y - 3$$
$$-18x + 6y + 1 - 6y - 1 = -6y - 3 - 6y - 1$$
$$-18x = -12y - 4$$
$$x = \frac{-12y - 4}{-18}$$
$$x = \frac{-12y}{-18} + \frac{-4}{-18}$$
$$x = \frac{2y}{3} + \frac{2}{9}$$

Exercise 1-8 Word Problems

Key Ideas and Formulas

The strategy for solving word problems given in the text is reprinted here for reference.

A Strategy for Solving Word Problems
1. Read the problem carefully--several times if necessary--until you understand the problem, know what is to be found, and know what is given.
2. If appropriate, draw figures or diagrams and label known and unknown parts.
3. Look for formulas connecting the known quantities with the unknown quantities.
4. Let one of the unknown quantities be represented by a variable, say x, and try to represent all other unknown quantities in terms of x. This is an important step and must be done carefully. Be sure you clearly understand what you are letting x represent.
5. Form an equation relating the unknown quantities with the known quantities. This step may involve the translation of an English sentence into an algebraic sentence, the use of relationships in a geometric figure, the use of certain formulas, and so on.
6. Solve the equation and write answers to *all* parts of the problem requested.
7. Check all solutions in the original problem.

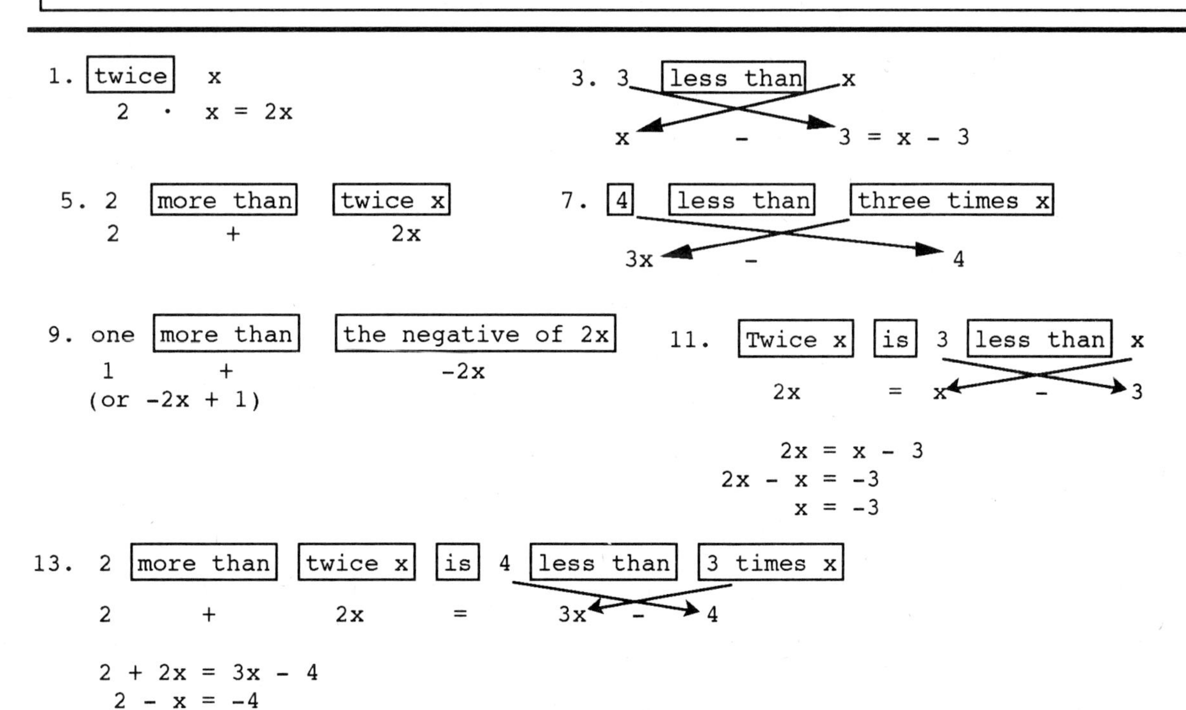

15. 1 [more than] [the negative of 2x] [is] [3 times x]

$1 \quad + \quad (-2x) \quad = \quad 3x$

$$\begin{aligned} 1 + (-2x) &= 3x \\ 1 - 2x &= 3x \\ 1 &= 5x \\ \frac{1}{5} &= x \end{aligned}$$

17. 2 [more than] x [is] 5 [less than] [twice x]

$2 \quad + \quad x \quad = \quad 2x \quad - \quad 5$

$$\begin{aligned} 2 + x &= 2x - 5 \\ 2 - x &= -5 \\ -x &= -7 \\ x &= 7 \end{aligned}$$

19. 5 [more than] [6 times x] [is] 1 [less than] [the negative of 3x]

$5 \quad + \quad 6x \quad = \quad -3x \quad - \quad 1$

$$\begin{aligned} 5 + 6x &= -3x - 1 \\ 5 + 9x &= -1 \\ 9x &= -6 \\ x &= -\frac{6}{9} \text{ or } -\frac{2}{3} \end{aligned}$$

21.

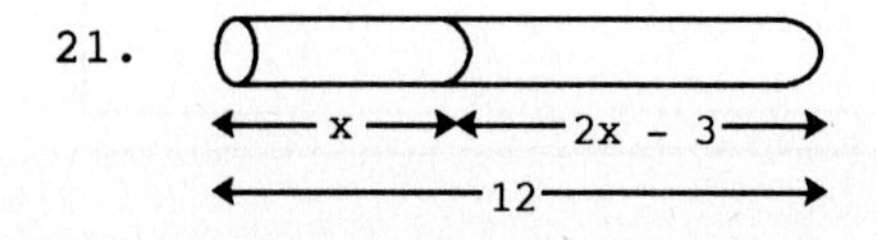

Let x = length of one piece
then 2x - 3 = length of other piece
The sum of the lengths is 12 feet.

$$\begin{aligned} (2x - 3) + x &= 12 \\ 3x - 3 &= 12 \\ 3x &= 15 \\ x &= 5 \text{ feet} \\ 2x - 3 &= 7 \text{ feet} \end{aligned}$$

23.

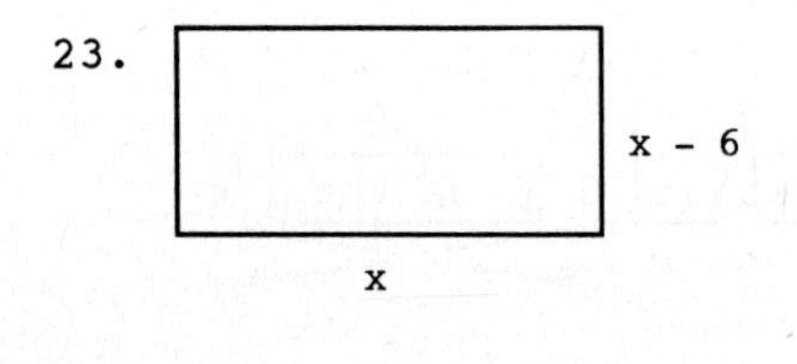

Let x = length. Then x - 6 = width. Use formula for the perimeter of a rectangle.

$$\begin{aligned} 2a + 2b &= P \\ 2x + 2(x - 6) &= 36 \\ 2x + 2x - 12 &= 36 \\ 4x - 12 &= 36 \\ 4x &= 48 \\ x &= 12 \text{ feet} = \text{length} \\ x - 6 &= 6 \text{ feet} = \text{width} \end{aligned}$$

25.

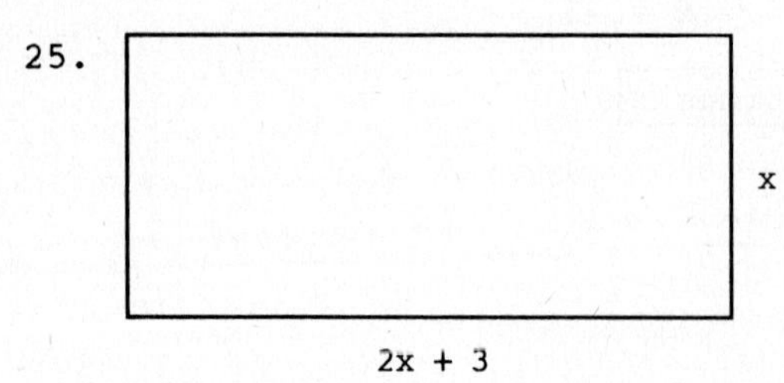

Let x = width. Then 2x + 3 = length. Use the formula for the perimeter of a rectangle:

$$\begin{aligned} 2a + 2b &= P \\ 2(2x + 3) + 2x &= 66 \\ 4x + 6 + 2x &= 66 \\ 6x + 6 &= 66 \\ 6x &= 60 \\ x &= 10 \text{ centimeters (width)} \\ 2x + 3 &= 23 \text{ centimeters (length)} \end{aligned}$$

27. Let x = the number

2 [less than] [one-sixth the number] [is] 1 [more than] [one-fourth the number]

$\frac{1}{6} \cdot x - 2 = 1 + \frac{1}{4} \cdot x$

$$\frac{x}{6} - 2 = 1 + \frac{x}{4} \quad \text{LCM} = 12$$

$$12 \cdot \frac{x}{6} - 12 \cdot 2 = 12 \cdot 1 + 12 \cdot \frac{x}{4}$$

$$2x - 24 = 12 + 3x$$

$$-x = 36$$

$$x = -36$$

29. Let x = 1st of three consecutive odd numbers
x + 2 = 2nd consecutive odd number
x + 4 = 3rd consecutive odd number
(Note: If x is odd, then x + 2 will also be odd. This number does not "end in a 2". Its last digit will merely differ by 2 from the last digit of x.)

[the sum of] [the first] and [second] [is] 5 [more than] [the third]

$x + (x + 2) = 5 + (x + 4)$

$$x + x + 2 = 5 + x + 4$$

$$2x + 2 = x + 9$$

$$\left.\begin{aligned} x &= 7 \\ x + 2 &= 9 \\ x + 4 &= 11 \end{aligned}\right\} \text{three consecutive odd numbers}$$

31. (Rectangle with length x and width $\frac{x}{6}$)

Let x = length. Then $\frac{1}{6} \cdot x = \frac{x}{6}$ = width.

$$2a + 2b = P$$

$$2x + 2 \cdot \frac{x}{6} = 84 \quad \text{LCM} = 6$$

$$2x + \frac{2x}{6} = 84$$

$$6 \cdot 2x + 6 \cdot \frac{2x}{6} = 6 \cdot 84$$

$$12x + 2x = 504$$

$$14x = 504$$

$$x = 36 \text{ (length)}$$

$$\frac{x}{6} = 6 \text{ (width)}$$

33. Let x = the number

4 [less than] [three-fifths the number] [is] 8 [more than] [one-third the number]

$\frac{3}{5} \cdot x - 4 = 8 + \frac{1}{3} \cdot x$

$$\frac{3x}{5} - 4 = 8 + \frac{x}{3} \quad \text{LCM} = 15$$

$$15 \cdot \frac{3x}{5} - 15 \cdot 4 = 15 \cdot 8 + 15 \cdot \frac{x}{3}$$

$$9x - 60 = 120 + 5x$$

$$4x = 180$$

$$x = 45$$

35. Let x = the number

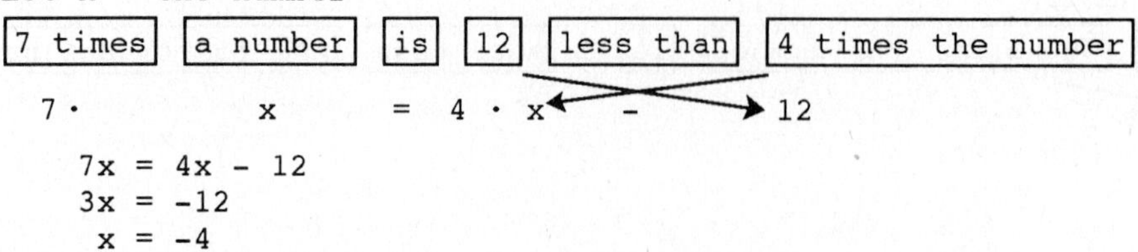

$$7x = 4x - 12$$
$$3x = -12$$
$$x = -4$$

37. Let x = the number

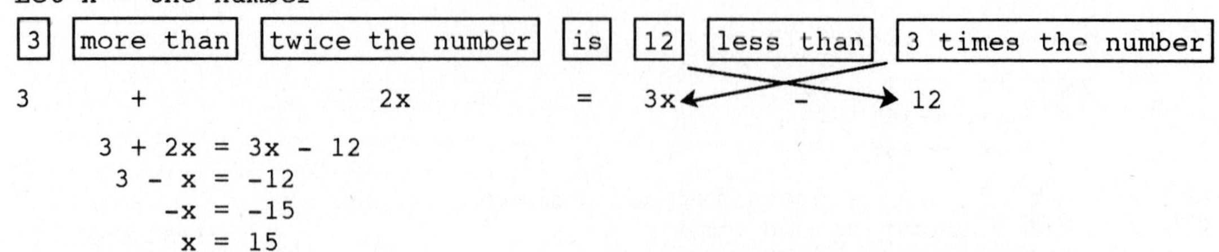

$$3 + 2x = 3x - 12$$
$$3 - x = -12$$
$$-x = -15$$
$$x = 15$$

39. Use the given formula $A = \frac{1}{2}h(B + b)$ with $A = 100$, $h = 10$, and $B = 12$.

$$100 = \frac{1}{2} \cdot 10(12 + b)$$
$$100 = 5(12 + b)$$
$$100 = 60 + 5b$$
$$40 = 5b$$
$$8 = b$$
$$b = 8 \text{ inches}$$

41. Let x = 1st of three consecutive integers
$x + 1$ = 2nd consecutive integer
$x + 2$ = 3rd consecutive integer

[whose sum] [is] 96

$$x + (x + 1) + (x + 2) = 96$$
$$x + x + 1 + x + 2 = 96$$
$$3x + 3 = 96$$
$$3x = 93$$
$$\left.\begin{aligned} x &= 31 \\ x + 1 &= 32 \\ x + 2 &= 33 \end{aligned}\right\} \text{three consecutive integers}$$

43. Let x = 1st of three consecutive even numbers
$x + 2$ = 2nd consecutive even number
$x + 4$ = 3rd consecutive even number

[whose sum] [is] 42

$$x + (x + 2) + (x + 4) = 42$$
$$x + x + 2 + x + 4 = 42$$
$$3x + 6 = 42$$
$$3x = 36$$
$$\left.\begin{aligned} x &= 12 \\ x + 2 &= 14 \\ x + 4 &= 16 \end{aligned}\right\} \text{three consecutive even numbers}$$

45. Using the hint, let b = one number and $4b$ = the other number.

$$b + 4b = 55$$
$$5b = 55$$
$$\left.\begin{aligned} b &= 11 \\ 4b &= 44 \end{aligned}\right\} \text{the two numbers}$$

47. Using the hint for Problem 45, let b = one number and $5b$ = the other number.

$$b + 5b = 48$$
$$6b = 48$$
$$\left.\begin{aligned} b &= 8 \\ 5b &= 40 \end{aligned}\right\} \text{the two numbers}$$

49.

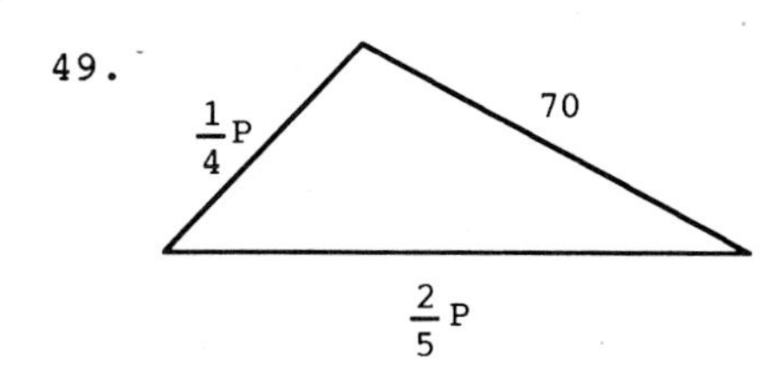

Let P = perimeter of the triangle.
Use a + b + c = P.

$$\begin{aligned}
\text{one side} &= \frac{2P}{5} \\
\text{second side} &= 70 \\
\text{third side} &= \frac{P}{4} \\
\frac{2P}{5} + 70 + \frac{P}{4} &= P \quad \text{LCM} = 20 \\
20 \cdot \frac{2P}{5} + 20 \cdot 70 + 20 \cdot \frac{P}{4} &= 20 \cdot P \\
8P + 1400 + 5P &= 20P \\
13P + 1400 &= 20P \\
1400 &= 7P \\
200 &= P \\
P &= 200 \text{ centimeters}
\end{aligned}$$

51. Let D = total length of the trip

Then $\frac{1}{3}D$ = distance traveled by mule

6 = distance traveled by boat

$\frac{1}{2}D$ = distance traveled by foot

$$\begin{aligned}
\frac{D}{3} + 6 + \frac{D}{2} &= D \quad \text{LCM} = 6 \\
6 \cdot \frac{D}{3} + 6 \cdot 6 + 6 \cdot \frac{D}{2} &= 6 \cdot D \\
2D + 36 + 3D &= 6D \\
5D + 36 &= 6D \\
36 &= D \\
D &= 36 \text{ kilometers}
\end{aligned}$$

53. Let x = the smaller number

40 − x = the larger number

3 times the smaller number	is	10	more than	twice the larger
$3 \cdot x$	=	10	+	$2(40 - x)$

$$\begin{aligned}
3x &= 10 + 2(40 - x) \\
3x &= 10 + 80 - 2x \\
3x &= 90 - 2x \\
5x &= 90 \\
x &= 18
\end{aligned}$$

55. Let x = the smaller number

70 − x = the larger number

"3 times the smaller number is as much more than the greater number as 3 times the greater exceeds 7 times the smaller" can be rewritten as:

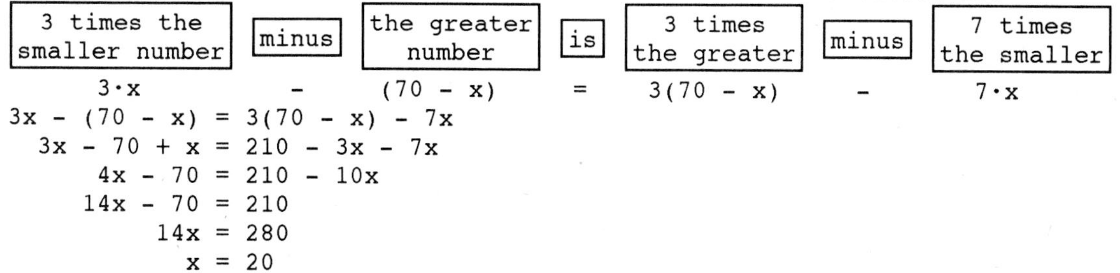

3 times the smaller number	minus	the greater number	is	3 times the greater	minus	7 times the smaller
$3 \cdot x$	−	$(70 - x)$	=	$3(70 - x)$	−	$7 \cdot x$

$$\begin{aligned}
3x - (70 - x) &= 3(70 - x) - 7x \\
3x - 70 + x &= 210 - 3x - 7x \\
4x - 70 &= 210 - 10x \\
14x - 70 &= 210 \\
14x &= 280 \\
x &= 20
\end{aligned}$$

57. Let x = the width
 3x = the length
Then if the width is increased by 20 and the length is decreased by 20, the resulting quantities will be equal (the rectangle becomes a square). Thus

$$x + 20 = 3x - 20$$
$$-2x + 20 = -20$$
$$-2x = -40$$
$$x = 20 \text{ feet}$$

59. Let x = the original angle
180 - x = the supplement

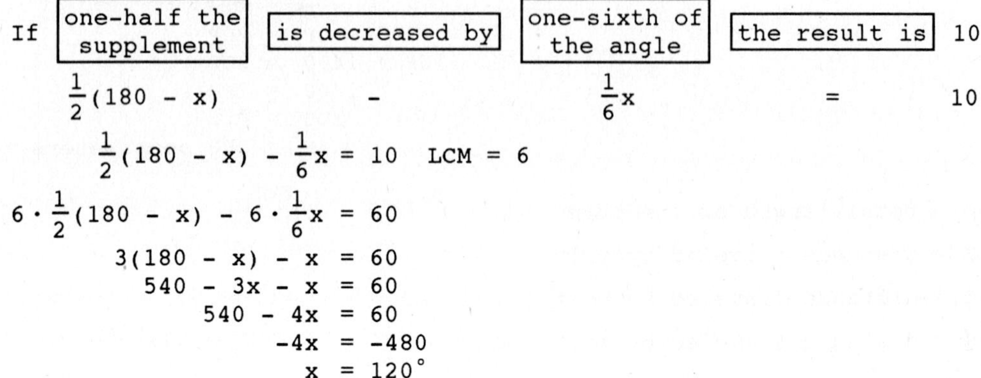

If [one-half the supplement] [is decreased by] [one-sixth of the angle] [the result is] 10

$$\frac{1}{2}(180 - x) \quad - \quad \frac{1}{6}x \quad = \quad 10$$

$$\frac{1}{2}(180 - x) - \frac{1}{6}x = 10 \quad \text{LCM} = 6$$

$$6 \cdot \frac{1}{2}(180 - x) - 6 \cdot \frac{1}{6}x = 60$$
$$3(180 - x) - x = 60$$
$$540 - 3x - x = 60$$
$$540 - 4x = 60$$
$$-4x = -480$$
$$x = 120^\circ$$

Exercise 1-9 REVIEW EXERCISE

1. (A) 143, 12 (B) 143, 0, 12 (C) 143, -1, 0, 12
 (D) $3.\overline{127}$, 143, -1.43, $\frac{2}{3}$, -1, 0, 12, $-\frac{3}{7}$ (E) $\sqrt{5}$, π

2. (a) The solution set for $x < -1$ is the set of all real numbers less than -1. This includes all points to the left of -1. Graphically,

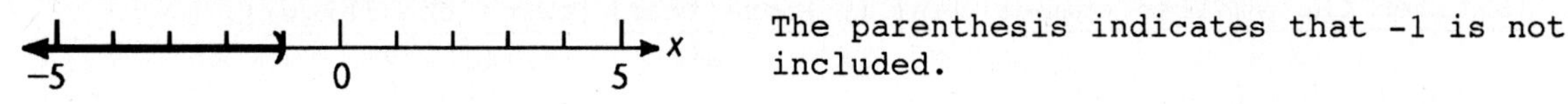

The parenthesis indicates that -1 is not included.

 (b) The solution set for $-4 \le x < 3$ is the set of all real numbers between -4 and 3, including -4 but not 3. Graphically,

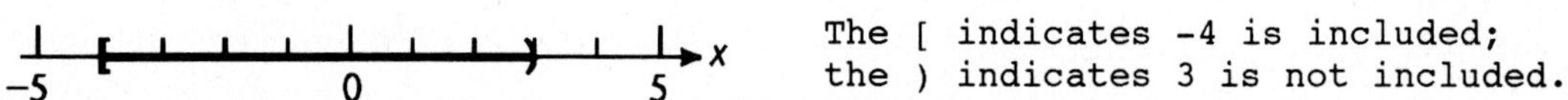

The [indicates -4 is included; the) indicates 3 is not included.

3. $5^6 5^7 = 5^{6+7} = 5^{13}$

4. $x^2 x^8 = x^{2+8} = x^{10}$

5. $3 \cdot 7 - 4 = 21 - 4 = 17$

6. $7 + 2 \cdot 3 = 7 + 6 = 13$

7. -5

8. -13

9. (-3) - (-9) = (-3) + (+9) = 6

10. 4 - 7 = 4 + (-7) = -3

11. 0 - (-3) = 0 + 3 = 3

12. -12

13. 28

14. -18

15. -4

16. 6

17. Not defined

18. 0

19. 10 - 3(6 - 4) = 10 - 3(2) = 10 - 6 = 4

20. (-8) - (-2)(-3) = (-8) - (+6) = (-8) + (-6) = -14

21. $(-9) - \left[\frac{(-12)}{3}\right] = (-9) - [-4] = (-9) + (+4) = -5$

22. $(4 - 8) + 4(-2) = (-4) + (-8) = -12$

23. 8

24. 5

25. $-[-(-3)] = -[+3] = -3$

26. $-|-(-2)| = -|+2| = -(+2) = -2$

27. $-(|(-3)| + |-2|) = -[(+3) + (+2)] = -[+5] = -5$

28. $-(|-8| - |3|) = -[(+8) - (3)] = -[5] = -5$

29. $7 + (x + 3) = 7 + x + 3 = x + 7 + 3 = x + 10$

30. $(3x)5 = 3x5 = 5(3x) = 15x$

31. $(2x)(4y) = 2x4y = 2 \cdot 4xy = 8xy$

32. $(y + 7) + (x + 2) + (z + 3) = y + 7 + x + 2 + z + 3$
$= y + x + z + 7 + 2 + 3$
$= y + x + z + 12$ or $x + y + z + 12$

33. $0 + (1x + 2) = 1x + 2 = x + 2$

34. $(x + 0) + 0y = x + 0 + 0y = x + 0 + 0 = x$

35. $3(x - 5) = 3x - 3 \cdot 5 = 3x - 15$

36. $(a + b)7 = a \cdot 7 + b \cdot 7 = 7a + 7b$

37. $2 + 3(x + 4) = 2 + 3 \cdot x + 3 \cdot 4 = 2 + 3x + 12 = 3x + 2 + 12 = 3x + 14$

38. $(x - 1)2 + 3 = x \cdot 2 - 1 \cdot 2 + 3 = 2x - 2 + 3 = 2x + 1$

39. $>$ 40. $<$ 41. $<$ 42. $<$ 43. $>$ 44. $<$

45. $3a + 8a - 5a = [3 + 8 + (-5)]a = 6a$

46. $3a - (7a - 9a) = 3a - 7a + 9a = [3 + (-7) + 9]a = 5a$

47. $2x + 3y - 4x + 5y - 6x = 2x - 4x - 6x + 3y + 5y$
$= [2 + (-4) + (-6)]x + (3 + 5)y = -8x + 8y$

48. $5x - 3y + 2z - 1$ (no like terms to be combined)

49. $2x^3y - 6xy^3 + 10x^3y + x^2y^2 = 2x^3y + 10x^3y - 6xy^3 + x^2y^2$
$= (2 + 10)x^3y - 6xy^3 + x^2y^2 = 12x^3y - 6xy^3 + x^2y^2$

50. $2xy - 3(xy - 2xy^2) - 3xy^2 = 2xy - 3xy + 6xy^2 - 3xy^2$
$= [2 + (-3)]xy + [6 + (-3)]xy^2 = -xy + 3xy^2$

51. $2[9 - 3(3 - 1)] = 2[9 - 3(2)] = 2[9 - 6] = 2[3] = 6$

52. $6 - 2 - 3 - 4 + 5 = 6 + (-2) + (-3) + (-4) + 5$
$= 6 + 5 + (-2) + (-3) + (-4) = 11 + (-9) = 2$

53. $[(-3) - (-3)] - (-4) = [(-3) + (+3)] - (-4) = [0] - (-4) = [0] + (+4) = 4$

54. $[-(-4)] + (-|-3|) = [+4] + [-(+3)] = [+4] + [-3] = 1$

55. $\left[\frac{(-16)}{2}\right] - (-3)(4) = [-8] - (-12) = [-8] + (+12) = 4$

56. $(-2)(-4)(-3) - \frac{-36}{(-2)(9)} = (-24) - \frac{-36}{-18} = (-24) - (+2) = (-24) + (-2) = -26$

57. This requires starting with the innermost grouping symbols and working outward:
$2\{9 - 2[(3 + 1) - (1 + 1)]\} = 2\{9 - 2[4 - 2]\} = 2\{9 - 2[2]\}$
$= 2\{9 - 4\} = 2\{5\} = 10$

58. $(-3) - 2\{5 - 3[2 - 2(3 - 6)]\} = (-3) - 2\{5 - 3[2 - 2(-3)]\}$
$= (-3) - 2\{5 - 3[2 - (-6)]\}$
$= (-3) - 2\{5 - 3[8]\}$
$= (-3) - 2\{5 - 24\}$
$= (-3) - 2\{-19\}$
$= (-3) - (-38)$
$= (-3) + (+38)$
$= 35$

59. The first steps here are to simplify above and below fraction bars.

$$\frac{12 - (-4)(-5)}{4 + (-2)} - \frac{-14}{7} = \frac{12 - (+20)}{4 + (-2)} - \frac{-14}{7}$$
$$= \frac{-8}{2} - \frac{-14}{7}$$
$$= (-4) - (-2)$$
$$= -2$$

60. $\dfrac{24}{(-4) + (4)} - \dfrac{24}{(-4) + 4} = \dfrac{24}{0} - \dfrac{24}{0}$ This is not defined (division by zero).

61. $-3 \cdot 5^2 + 5(-3)^2 = -3 \cdot 25 + 5 \cdot 9 = -75 + 45 = -30$

62. $-(3 \cdot 5)^2 + 5(-3)^2 = -(15)^2 + 5(-3)^2 = -225 + 5 \cdot 9 = -225 + 45 = -180$

63. $15 \div (-5) \cdot 6^2 = (-3) \cdot 6^2 = (-3) \cdot 36 = -108$

64. We replace grouping symbols as follows:
$3x \div [(-x) \cdot (x + 1)]^2 = 3(\ ) \div \{[-(\ )] \cdot [(\ ) + 1]\}^2$
Then we insert the value of x into the parentheses:
$= 3(5) \div \{[-(5)] \cdot [(5) + 1]\}^2$
$= 15 \div \{[-5] \cdot [5 + 1]\}^2$
$= 15 \div \{[-5][6]\}^2$
$= 15 \div \{-30\}^2$
$= 15 \div 900$
$= \dfrac{15}{900}$
$= \dfrac{1}{60}$

65. We replace grouping symbols as follows:
$3[14 - x(x + 1)] = 3\{14 - (\ )[(\ ) + 1]\}$
Then we insert the value of x into the parentheses:
$= 3\{14 - (3)[(3) + 1]\}$
$= 3\{14 - (3)[4]\}$
$= 3\{14 - 12\}$
$= 3\{2\}$
$= 6$

66. We replace grouping symbols, then insert the value of x:
$-(-x) = -[-(\ )] = -[-(-2)] = -[2] = -2$

67. $-(|x| - |w|) = -(|-2| - |-10|) = -[(+2) - (+10)] = -[(+2) + (-10)] = -[-8] = 8$

68. We replace grouping symbols, then insert the values of the variables:
$(x + y) - z = [(x) + (y)] - (z) = [(6) + (-8)] - (4)$
$= [-2] - (4) = [-2] + [-4] = -6$

69. We replace grouping symbols, then insert the values of the variables:

$$\left(2x - \frac{z}{x}\right) - \frac{w}{x} = \left[2(x) - \frac{z}{x}\right] - \frac{w}{x} = \left[2(-2) - \frac{0}{-2}\right] - \frac{-10}{-2}$$
$$= [(-4) - 0] - 5 = [-4] - 5 = [-4] + (-5) = -9$$

70. We replace grouping symbols, then insert the values of the variables:

$$\frac{(xyz + xz) - z}{z} = \frac{[(x)(y)(z) + (x)(z)] - (z)}{(z)}$$
$$= \frac{[(-6)(0)(-3) + (-6)(-3)] - (-3)}{(-3)}$$
$$= \frac{[0 + 18] - (-3)}{(-3)}$$
$$= \frac{[18] - (-3)}{(-3)}$$
$$= \frac{[18] + (+3)}{(-3)}$$
$$= \frac{21}{(-3)}$$
$$= -7$$

71. We insert grouping symbols, then insert the values of the variables:

$$\frac{x - 3y}{z - x} - \frac{z}{xy} = \frac{(x) - 3(y)}{(z) - (x)} - \frac{(z)}{(x)(y)}$$
$$= \frac{(-3) - 3(2)}{(-12) - (-3)} - \frac{(-12)}{(-3)(2)}$$
$$= \frac{(-3) - 6}{(-12) - (-3)} - \frac{(-12)}{-6}$$
$$= \frac{-9}{-9} - \frac{(-12)}{-6}$$
$$= 1 - 2$$
$$= -1$$

72.
$$2x - 3 = 4x + 5$$
$$2x - 3 - 4x = 4x + 5 - 4x$$
$$-2x - 3 = 5$$
$$-2x - 3 + 3 = 5 + 3$$
$$-2x = 8$$
$$\frac{-2x}{-2} = \frac{8}{-2}$$
$$x = -4$$

73.
$$x - y = 2x + 3y + 4$$
$$x - y - 2x = 2x + 3y + 4 - 2x$$
$$-x - y = 3y + 4$$
$$-x - y + y = 3y + 4 + y$$
$$-x = 4y + 4$$
$$\frac{-x}{-1} = \frac{4y + 4}{-1}$$
$$x = -4y - 4$$

74.
$$\frac{1}{3}x + \frac{1}{4}x = \frac{5}{6} \quad \text{LCM} = 12$$
$$12 \cdot \frac{1}{3}x + 12 \cdot \frac{1}{4}x = 12 \cdot \frac{5}{6}$$
$$4x + 3x = 10$$
$$7x = 10$$
$$x = \frac{10}{7}$$

75.
$$x - (5 - 3x) = 2x - 3 - (2 - 2x)$$
$$x - 5 + 3x = 2x - 3 - 2 + 2x$$
$$-5 + 4x = 4x - 5$$
$$-5 = -5$$

Since this statement is always true, the solution set is all real numbers.

? How do you get this

$4x - 4x = 5 - 5$

$0 = 0$

This is my answer

76. $3x - 4 - (5 - 6x) = 2x - 1 - (7x - 8)$
$$3x - 4 - 5 + 6x = 2x - 1 - 7x + 8$$
$$9x - 9 = -5x + 7$$
$$14x - 9 = 7$$
$$14x = 16$$
$$x = \frac{16}{14} \text{ or } \frac{8}{7}$$

77. $x + y = xy$
$$x + y - x = xy - x$$
$$y = xy - x \cdot 1$$
$$y = x(y - 1)$$
$$\frac{y}{y - 1} = \frac{x(y - 1)}{y - 1}$$
$$\frac{y}{y - 1} = x$$
$$x = \frac{y}{y - 1}$$

78. $x - 1 > 0$

79. $2x + 3 \geq 0$

80. Let x = a certain number, then the statement translates as follows:

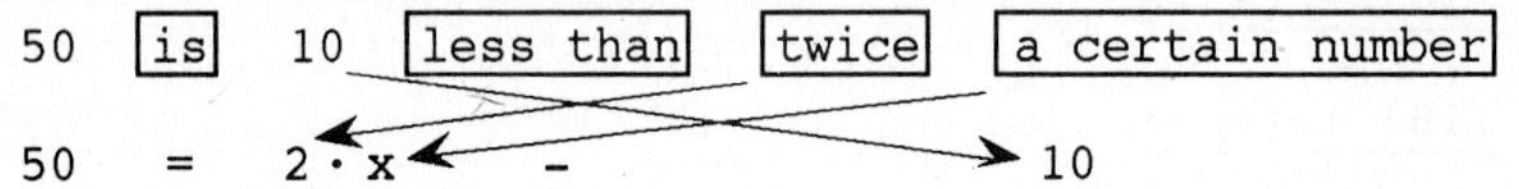

Or, more compactly:
$50 = 2x - 10$

81.
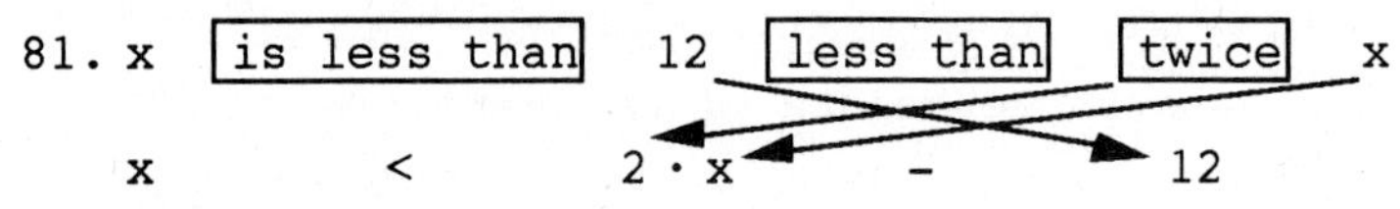

Or, more compactly:
$x < 2x - 12$

82.
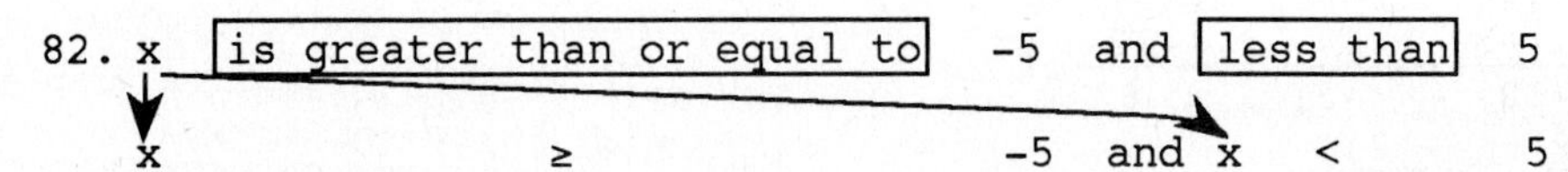

Or, more compactly, since x is between -5 (inclusive) and 5 (exclusive),
$-5 \leq x < 5$

83. Let x = a certain number. Then the number which is 6 less than the certain number must be 6 less than x, or x - 6. We recast the statement as:

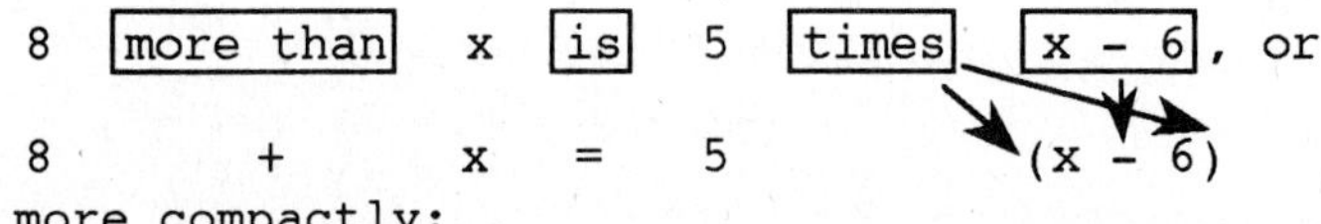

Or, more compactly:
$8 + x = 5(x - 6)$

84. In this problem there are three quantities: area, length, and width. They are related by the fact that area = length × width. We use the variable x to represent the length. Then the width is represented by translating the phrase:

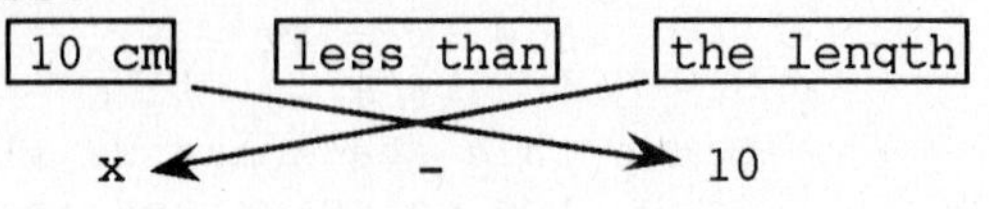

The area is represented by the constant 1200. Substituting these symbols into the area relation, we have:
$1200 = x(x - 10)$ or $x(x - 10) = 1200$

85. In this problem there are three quantities: perimeter, length, and width. They are related by the perimeter formula $2\ell + 2w = P$. We use the variable x to represent the width. Then the length is represented by translating the phrase: [5 meters] [longer than] [the width]

$$\underset{5}{\boxed{\text{5 meters}}}\ \underset{+}{\boxed{\text{longer than}}}\ \underset{x}{\boxed{\text{the width}}}$$

The perimeter is represented by the constant 43. Substituting these symbols into the perimeter formula, we have: $43 = 2(5 + x) + 2x$ or $2x + 2(x + 5) = 43$

86. Commutative property for +

87. Associative property for +

88. Commutative property for ·

89. Associative property for ·

90. Additive identity

91. Additive inverse

92. Distributive property

93. We replace grouping symbols, then insert the values of the variables:

$$\begin{aligned}
&uv - 3\{x - 2[(x + y) - (x - y)] + u\}\\
&= (u)(v) - 3[(x) - 2\{[(x) + (y)] - [(x) - (y)]\} + (u)]\\
&= (-2)(3) - 3[(2) - 2\{[(2) + (-3)] - [(2) - (-3)]\} + (-2)]\\
&= -6 - 3[(2) - 2\{[-1] - [5]\} + (-2)]\\
&= -6 - 3[(2) - 2\{-6\} + (-2)]\\
&= -6 - 3[(2) - \{-12\} + (-2)]\\
&= -6 - 3[(2) + (+12) + (-2)]\\
&= -6 - 3[12]\\
&= -6 - 36\\
&= -42
\end{aligned}$$

94. We introduce grouping symbols, then insert the values of the variables:

$$\begin{aligned}
\frac{5w}{x - 7} - \frac{wx - 4}{x - w} &= \frac{5(w)}{(x) - 7} - \frac{(w)(x) - 4}{(x) - (w)}\\
&= \frac{5(-4)}{(2) - 7} - \frac{(-4)(2) - 4}{(2) - (-4)}\\
&= \frac{-20}{-5} - \frac{(-8) - 4}{2 - (-4)}\\
&= \frac{-20}{-5} - \frac{-12}{6}\\
&= 4 - (-2)\\
&= 6
\end{aligned}$$

95. We introduce grouping symbols, then insert the values of the variables:

$$1 - \cfrac{(x)}{(x) + \cfrac{3}{(x) - \frac{1}{2}}} + \cfrac{(x)}{\cfrac{5}{(x) + \frac{1}{2}} - 1} + 1$$

$$= 1 - \cfrac{(-1)}{(-1) + \cfrac{3}{(-1) - \frac{1}{2}}} + \cfrac{(-1)}{\cfrac{5}{(-1) + \frac{1}{2}} - 1} + 1$$

$$= 1 - \cfrac{(-1)}{(-1) + \cfrac{3}{-\frac{3}{2}}} + \cfrac{(-1)}{\cfrac{5}{-\frac{1}{2}} - 1} + 1 = 1 - \frac{(-1)}{(-1) + 3 \div \left(-\frac{3}{2}\right)} + \frac{(-1)}{5 \div \left(-\frac{1}{2}\right) - 1} + 1$$

$$= 1 - \frac{(-1)}{(-1) + 3\left(-\frac{2}{3}\right)} + \frac{(-1)}{5 \cdot \left(-\frac{2}{1}\right) - 1} + 1 = 1 - \frac{(-1)}{(-1) + (-2)} + \frac{(-1)}{(-10) - 1} + 1$$

$$= 1 - \frac{(-1)}{(-3)} + \frac{(-1)}{(-11)} + 1 = 1 - \frac{1}{3} + \frac{1}{11} + 1 = \frac{33}{33} - \frac{11}{33} + \frac{3}{33} + \frac{33}{33} = \frac{58}{33}$$

96. Let x = the number

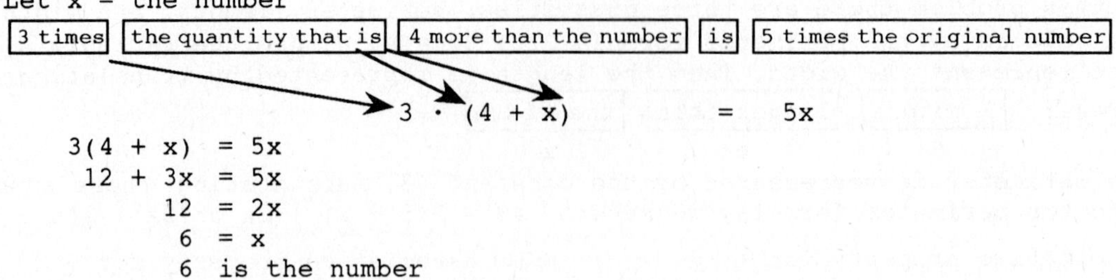

$$3(4 + x) = 5x$$
$$12 + 3x = 5x$$
$$12 = 2x$$
$$6 = x$$

6 is the number

97. Let x = the first of the three consecutive natural numbers
Then x + 1 = the second of the three consecutive natural numbers
x + 2 = the third of the three consecutive natural numbers.

twice	the sum of the first two	is	11	more than	3 times the third
2	[x + (x + 1)]	=	11	+	3(x + 2)

$$2[x + (x + 1)] = 11 + 3(x + 2)$$
$$2[2x + 1] = 11 + 3x + 6$$
$$4x + 2 = 3x + 17$$
$$x + 2 = 17$$
$$\left.\begin{aligned} x &= 15 \\ x + 1 &= 16 \\ x + 2 &= 17 \end{aligned}\right\} \text{ three consecutive natural numbers}$$

15, 16, 17 are the numbers.

98. Let x = the first of the two consecutive natural numbers
then x + 1 = the second of the two consecutive natural numbers

their product is	seven	less than	the product of the smaller times the successor of the larger
x(x + 1) =	7 (subtracted)	−	x[(x + 1) + 1]

$$x(x + 1) = x[(x + 1) + 1] - 7$$

$$x(x + 1) = x[(x + 1) + 1] - 7$$
$$x^2 + x = x[x + 2] - 7$$
$$x^2 + x = x^2 + 2x - 7$$
$$x^2 + x = x^2 + 2x - 7$$
$$x = 2x - 7$$
$$-x = -7$$
$$\left.\begin{aligned} x &= 7 \\ x + 1 &= 8 \end{aligned}\right\} \text{ two consecutive natural numbers}$$

99. Let x = the length of the shortest piece
then 3x = the length of the longest piece
then 66 - (x + 3x) = the length of the remaining pieces

Since	the longest piece	is	twice	the length of the remaining piece,
	3x	=	2	[66 - (x + 3x)]

$$3x = 2[66 - 4x]$$
$$3x = 132 - 8x$$
$$11x = 132$$
$$x = 12 \text{ in}$$
$$3x = 36 \text{ in}$$
$$66 - (x + 3x) = 66 - (12 + 36) = 18 \text{ in}$$

100.

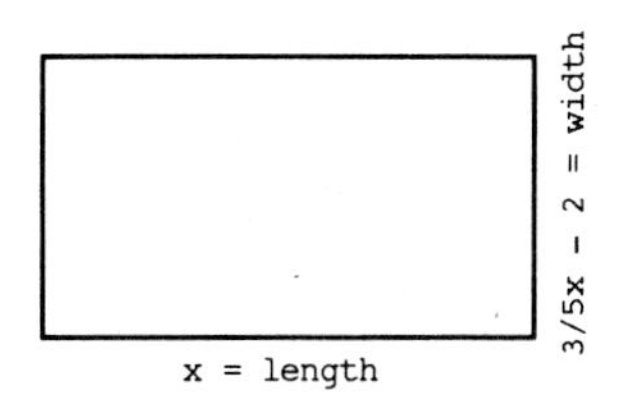

Let x = length

Then [2 cm] [less than] [three-fifths the length] is translated into

$\frac{3}{5}x \quad - \quad 2 \quad = \quad$ width.

Use 2a + 2b = P with P = 76. Then

$$2x + 2\left(\frac{3}{5}x - 2\right) = 76 \quad \text{LCM} = 5$$

$$5\cdot 2x + 5\left[2\left(\frac{3}{5}x - 2\right)\right] = 5\cdot 76$$

$$10x + 10\left(\frac{3}{5}x - 2\right) = 380$$

$$10x + 6x - 20 = 380$$

$$16x - 20 = 380$$

$$16x = 400$$

$$x = 25$$

$$\frac{3}{5}x - 2 = \frac{3}{5}(25) - 2 = 13 \quad 25 \times 13 \text{ centimeters}$$

COMMON ERROR: $5\cdot 2x + 2(3x - 2) = 5\cdot 76$. This is incorrect "cancelling"; the 5 must be multiplied times each term in parentheses, using the distributive property.

Practice Test 1

1. $(-3)(-5) + [(-3) - (-5)] - (-3)[5 - (-3)] = 15 + [(-3) + (+5)] - (-3)[5 + (+3)]$
$= 15 + [2] - (-3)[8] = 15 + [2] - [-24] = 15 + 2 + 24 = 41$

2. $|-5^2| - |-5|^2 - (-|5|^2) = |-25| - (+5)^2 - (-(5)^2) = 25 - (25) - (-25)$
$= 25 - 25 + 25 = 25$

3. $(-3)(-6)^2 + (-3)(-6)^2 - [(-3)(-6)]^2 = (-3)(36) + (-3)(36) - [18]^2$
$= -108 + (-108) - 324$
$= -216 - 324 = -540$

4. $(-6)^2 \div (-3) - [(-6) \div 3]^2 + (-6)^2 \div (-3)^2 = 36 \div (-3) - [-2]^2 + 36 \div 9$
$= -12 - (4) + 4 = -12 - 4 + 4 = -12$

5. $xy^2 - x(x - y) + x \div y^2 = (x)(y)^2 - (x)[(x) - (y)] + (x) \div (y)^2$
$= (12)(-2)^2 - (12)[(12) - (-2)] + (12) \div (-2)^2$
$= (12)(4) - (12)[14] + (12) \div 4$
$= 48 - 168 + 3 = -117$

6. $x^3x^4x^2 = x^{3+4+2} = x^9$

7. $ab + ac - ad$

8. $3x - 6xy + 9x^2 = 3x \cdot 1 - 3x \cdot 2y + 3x \cdot 3x = 3x(1 - 2y + 3x)$

9. $4x^2y + 2xy - 5x^2y - 3xy + 6x^2y = 4x^2y - 5x^2y + 6x^2y + 2xy - 3xy$
$= (4 - 5 + 6)x^2y + (2 - 3)xy = 5x^2y - xy$

10. The solution set for $x > -2$ is the set of all real numbers greater than -2. This includes all the points to the right of -2. The parenthesis through -2 indicates that -2 is not included. Graphically,

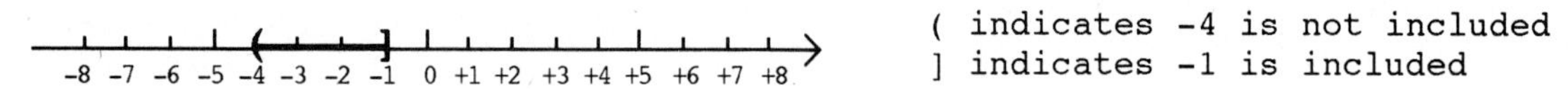

11. The solution set for $-4 < x \le -1$ is the set of all real numbers between -4 and -1, including -1 but not -4. Graphically,

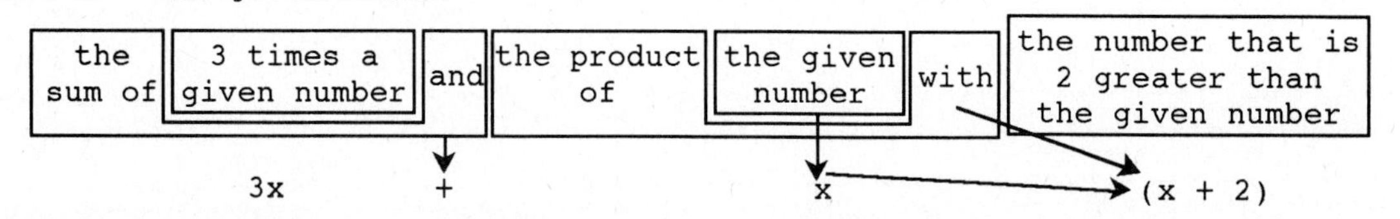

(indicates -4 is not included
] indicates -1 is included

12. Let x = the given number

or, more compactly, $3x + x(x + 2)$

13. Let x = the number

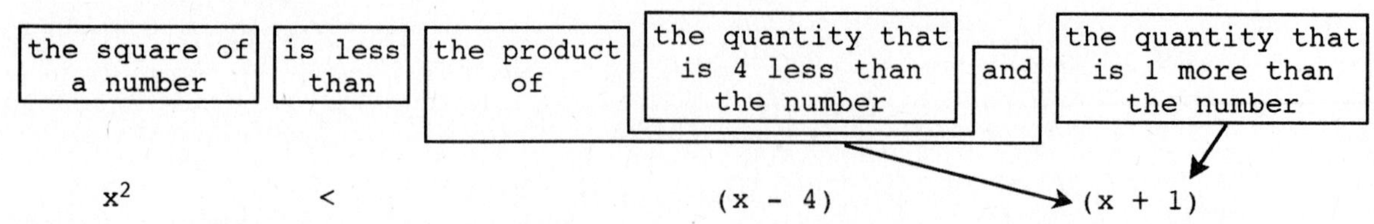

or, more compactly, $x^2 < (x - 4)(x + 1)$

14. Let x = the number

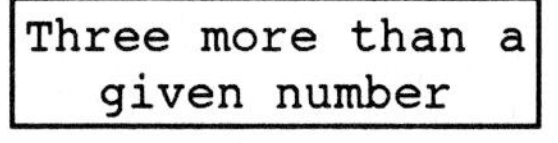

Three more than a given number	is	2 times	the number that is 3 less than the given number
$x + 3$	$=$	2	$(x - 3)$

15.
$$
\begin{aligned}
3x - 4(x + 5) &= 6 - (7 - x)\\
3x - 4x - 20 &= 6 - 7 + x\\
-x - 20 &= -1 + x\\
-2x - 20 &= -1\\
-2x &= 19\\
x &= -\frac{19}{2}
\end{aligned}
$$

16.
$$
\begin{aligned}
x(x + 2) &= x^2 - 4x + 12\\
x^2 + 2x &= x^2 - 4x + 12\\
2x &= -4x + 12\\
6x &= 12\\
x &= 2
\end{aligned}
$$

17.
$$
\begin{aligned}
5x - 4(3 - x) &= 4x + 5(x - 2) - 2\\
5x - 12 + 4x &= 4x + 5x - 10 - 2\\
9x - 12 &= 9x - 12
\end{aligned}
$$
Since this statement is always true, the solution set is all real numbers.

18.
$$
\begin{aligned}
a + b + c &= abc\\
a + b + c - a &= abc - a\\
b + c &= abc - a\\
b + c &= a(bc - 1)\\
\frac{b + c}{bc - 1} &= \frac{a(bc - 1)}{bc - 1}\\
\frac{b + c}{bc - 1} &= a\\
a &= \frac{b + c}{bc - 1}
\end{aligned}
$$

19. Let x = the number

4	times	the quantity that is 5 less than the number	is	1 more than the original number
4		$(x - 5)$	$=$	$1 + x$

$$
\begin{aligned}
4(x - 5) &= 1 + x\\
4x - 20 &= 1 + x\\
3x - 20 &= 1\\
3x &= 21\\
x &= 7
\end{aligned}
$$

20.

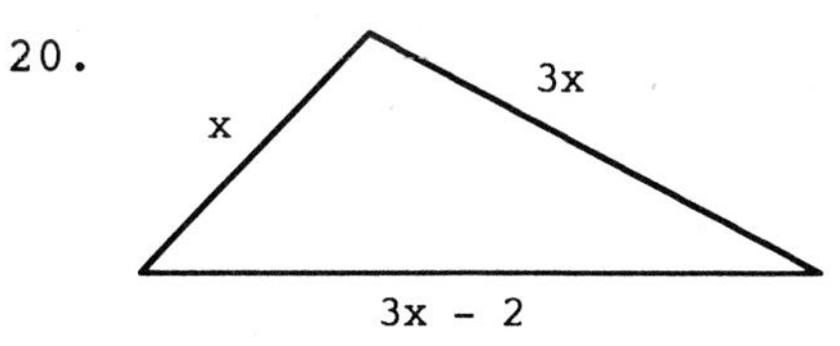

Let x = the length of one of the remaining sides. Then $3x$ = the length of the longest side. Since the longest side is two inches longer than the third side, the third side is two inches shorter than the longest side. Hence, $3x - 2$ = the length of the third side.

Use $a + b + c = P$

$$
\begin{aligned}
x + 3x + 3x - 2 &= 40\\
7x - 2 &= 40\\
7x &= 42\\
x &= 6 \text{ in}\\
3x &= 18 \text{ in}\\
3x - 2 &= 16 \text{ in}
\end{aligned}
$$

CHAPTER 2 Polynomials and Factoring

Exercise 2-1 Addition and Subtraction of Polynomials

Key Ideas and Formulas

A **polynomial** is an algebraic expression involving only the operations of addition, subtraction, and multiplication on variables and constants.

The **degree of a term** in a polynomial is the sum of the powers of the variables in the term, or the power of the only variable, if there is only one. A nonzero constant term is assigned degree 0.

The **degree of a polynomial** is the degree of its term with the highest degree. A one-term polynomial is called a **monomial**; a two-termed polynomial is called a **binomial**; a three-termed polynomial is called a **trinomial**. To add or subtract polynomials, remove parentheses using the distributive property, then combine like terms.

---

1. This is a binomial. The degrees of the terms, in order, are 2 and 0; the degree of the polynomial is 2.

3. This is a trinomial. The degrees of the terms, in order, are 5, 6, and 4. Therefore, the degree of the polynomial is 6.

5. This is a binomial. The degrees of the terms, in order, are 3 and 0; the degree of the polynomial is 3.

7. This is a monomial. The degree is $2 + 3 + 1 + 2 = 8$.

9. This is a monomial. Since it is a nonzero constant, the degree is 0.

11. -3

13. 3

15. 1

17. $(6x + 5) + (3x - 8) = 6x + 5 + 3x - 8 = 9x - 3$

19. $(7x - 5) + (-x + 3) + (-8x - 2) = 7x - 5 - x + 3 - 8x - 2 = -2x - 4$

21. $(5x^2 + 2x - 7) + (2x^2 + 3) + (-3x - 8) = 5x^2 + 2x - 7 + 2x^2 + 3 - 3x - 8$
$= 7x^2 - x - 12$

23. $(3) + (4 + 5a) + (6a + 7b) = 3 + 4 + 5a + 6a + 7b = 7 + 11a + 7b$

25. $(2x + 3y) + (4x - 5y) + (6x + 7y) = 2x + 3y + 4x - 5y + 6x + 7y = 12x + 5y$

27. $(-2x) - (3x) = -2x - 3x = -5x$

29. $(x^2) - (4x^2) = x^2 - 4x^2 = -3x^2$

31. $(6x^2) - (-3x^2) = 6x^2 + 3x^2 = 9x^2$

33. $(2x - 7) - (3x - 8) =$
$(2x - 7) + (-1)(3x - 8) =$
$2x - 7 - 3x + 8 = -x + 1$

COMMON ERRORS:

a. $(3x - 8) - (2x - 7)$.
Order is reversed in translating "subtract from"

b. $2x - 7 - 3x - 8$
Parentheses must be supplied.

35. $(y^2 - 6y - 1) - (2y^2 - 6y + 1) = (y^2 - 6y - 1) + (-1)(2y^2 - 6y + 1)$
$= y^2 - 6y - 1 - 2y^2 + 6y - 1 = -y^2 - 2$

37. $-x^2y + 3x^2y - 5x^2y = (-1 + 3 - 5)x^2y = -3x^2y$

39. $y^3 + 4y^2 - 10 + 2y^3 - y + 7 = y^3 + 2y^3 + 4y^2 - y - 10 + 7$
$= (1 + 2)y^3 + 4y^2 - y - 3 = 3y^3 + 4y^2 - y - 3$

41. $a^2 - 3ab + b^2 + 2a^2 + 3ab - 2b^2 = a^2 + 2a^2 - 3ab + 3ab + b^2 - 2b^2$
$= (1 + 2)a^2 + (-3 + 3)ab + (1 - 2)b^2$
$= 3a^2 + 0ab + (-1)b^2$
$= 3a^2 - b^2$

43. $x - 3y - 4(2x - 3y) = x - 3y - 8x + 12y = -7x + 9y$

45. $y - 2(x - y) - 3x = y - 2x + 2y - 3x = -5x + 3y$

47. $-2(-3x + 1) - (2x + 4) = -2(-3x + 1) + (-1)(2x + 4) = 6x - 2 - 2x - 4 = 4x - 6$

49. $2(x - 1) - 3(2x - 3) - (4x - 5) = 2(x - 1) - 3(2x - 3) + (-1)(4x - 5)$
$= 2x - 2 - 6x + 9 - 4x + 5 = -8x + 12$

51. $4t - 3[4 - 2(t - 1)] = 4t - 3[4 - 2t + 2] = 4t - 3[-2t + 6]$
$= 4t + 6t - 18 = 10t - 18$

53. $3[x - 2(x + 1)] - 4(2 - x) = 3[x - 2x - 2] - 4(2 - x)$
$= 3[-x - 2] - 4(2 - x) = -3x - 6 - 8 + 4x = x - 14$

55. $5 + m - 2n = 5 + (-1)(-m + 2n) = 5 - (-m + 2n)$.
$-m + 2n$ is the required expression.

57. $2 - x - y = 2 - x + (-y)$. $-y$ is the required expression.

59. $2 - x - y = 2 - x - (y)$. y is the required expression.

61. $2 - x - y = 2 + (-x - y)$. $-x - y$ is the required expression.

63. $2 - x - y = 2 + (-1)(x + y) = 2 - (x + y)$. $x + y$ is the required expression.

65. $w^2 - y - z = w^2 - 1(y + z) = w^2 - (y + z)$. $y + z$ is the required expression.

67. This is most conveniently done vertically.

$$\begin{array}{rrrrr} 2x^4 & & -\ x^2 & & -\ 7 \\ & 3x^3 & +\ 7x^2 & +\ 2x & \\ & & x^2 & -\ 3x & -\ 1 \\ \hline 2x^4 & +\ 3x^3 & +\ 7x^2 & -\ x & -\ 8 \end{array}$$

69. $(2x^3 - 3x^4 + 4x^5) + (1 - x + x^2) + (-4x^4 - 3x^3 + 2x^2 + x)$
$= 2x^3 - 3x^4 + 4x^5 + 1 - x + x^2 - 4x^4 - 3x^3 + 2x^2 + x$
$= 4x^5 - 7x^4 - x^3 + 3x^2 + 1$

71. This is most conveniently done vertically.

$$\begin{array}{rrrr} xy & -\ x^2y & +\ xy^2 & -\ x^2y^2 \\ 3xy & -\ 4x^2y & +\ 2xy^2 & -\ 5x^2y^2 \\ 6xy & +\ 2x^2y & -\ 3xy^2 & +\ 5x^2y^2 \\ \hline 10xy & -\ 3x^2y & & -\ x^2y^2 \end{array}$$

73. $(2x^3 + x^2 - 1) - (5x^3 - 3x + 1) = 2x^3 + x^2 - 1 - 5x^3 + 3x - 1$
$= -3x^3 + x^2 + 3x - 2$

75. $(3xy - 4x^2y + 2xy^2 - 5x^2y^2) - (xy - x^2y + xy^2 - x^2y^2)$
$= 3xy - 4x^2y + 2xy^2 - 5x^2y^2 - xy + x^2y - xy^2 + x^2y^2$
$= 2xy - 3x^2y + xy^2 - 4x^2y^2$

77. $(-4x^4 - 3x^3 + 2x^2 + x) - (1 - x + x^2) = -4x^4 - 3x^3 + 2x^2 + x - 1 + x - x^2$
$= -4x^4 - 3x^3 + x^2 + 2x - 1$

79. $(6xy + 2x^2y - 3xy^2 + 5x^2y^2) - (3xy - 4x^2y + 2xy^2 - 5x^2y^2)$
$= 6xy + 2x^2y - 3xy^2 + 5x^2y^2 - 3xy + 4x^2y - 2xy^2 + 5x^2y^2$
$= 3xy + 6x^2y - 5xy^2 + 10x^2y^2$

81. $(-4x^4 - 3x^3 + 2x^2 + x) - (2x^3 - 3x^4 + 4x^5) = -4x^4 - 3x^3 + 2x^2 + x - 2x^3 + 3x^4 - 4x^5$
$= -4x^5 - x^4 - 5x^3 + 2x^2 + x$

83. $(6xy + 2x^2y - 3xy^2 + 5x^2y^2) - (xy - x^2y + xy^2 - x^2y^2)$
$= 6xy + 2x^2y - 3xy^2 + 5x^2y^2 - xy + x^2y - xy^2 + x^2y^2$
$= 5xy + 3x^2y - 4xy^2 + 6x^2y^2$

85. $2t - 3\{t + 2[t - (t + 5)] + 1\} = 2t - 3\{t + 2[t - t - 5] + 1\}$
$= 2t - 3\{t + 2[-5] + 1\}$
$= 2t - 3\{t - 10 + 1\}$
$= 2t - 3\{t - 9\}$
$= 2t - 3t + 27 = -t + 27$

87. $w - \{x - [z - (w - x) - z] - (x - w)\} + x$
$= w - \{x - [z - w + x - z] - (x - w)\} + x$
$= w - \{x - [-w + x] - (x - w)\} + x$
$= w - \{x + w - x - x + w\} + x$
$= w - \{2w - x\} + x$
$= w - 2w + x + x$
$= -w + 2x$ or $2x - w$

89. $\{[(x - 1) - 1] - x\} - 1 = \{[x - 1 - 1] - x\} - 1 = \{[x - 2] - x\} - 1$
$= \{x - 2 - x\} - 1$
$= \{-2\} - 1$
$= -3$

91. $x - \{1 - [x - (1 - x)]\} = x - \{1 - [x - 1 + x]\} = x - \{1 - [2x - 1]\}$
$= x - \{1 - 2x + 1\}$
$= x - \{2 - 2x\}$
$= x - 2 + 2x$
$= 3x - 2$

93. There are three quantities in this problem: perimeter, length, and width. They are related by the perimeter formula $P = 2\ell + 2w$. Since x = length of the rectangle, and the width is 5 meters less than the length, $x - 5$ = width of the rectangle. So $P = 2x + 2(x - 5)$ represents the perimeter of the rectangle. Simplifying:
$P = 2x + 2x - 10 = 4x - 10$

95. There are several quantities involved in this problem. It is important to keep them distinct by using enough words. We write:
x = number of nickels
$x - 5$ = number of dimes
$(x - 5) + 2$ = number of quarters

This follows because there are five less dimes than nickels $(x - 5)$ and 2 more quarters than dimes (2 more than $x - 5$). Each nickel is worth 5 cents, each dime worth 10 cents, and each quarter worth 25 cents. Hence, the value of the nickels is 5 times the number of nickels, the value of the dimes is 10 times the number of dimes, and the value of the quarters is 25 times the number of quarters.
value of nickels $= 5x$
value of dimes $= 10(x - 5)$
value of quarters $= 25[(x - 5) + 2]$

The value of the pile = (value of nickels) + (value of dimes) + (value of quarters)
$= 5x + 10(x - 5) + 25[(x - 5) + 2]$

Simplifying this expression, we get:

The value of the pile $= 5x + 10x - 50 + 25[x - 5 + 2]$
$= 5x + 10x - 50 + 25[x - 3]$
$= 5x + 10x - 50 + 25x - 75$
$= 40x - 125$

97.

x 2x 2x 3x

A drawing is helpful in this problem.

Let x = length of smallest piece
Then $3x$ = length of largest piece
$2x$ = length of other two pieces

The total length of the board is the sum of the lengths of the pieces.
Hence, total length $= x + 2x + 2x + 3x = 8x$.

99. Let t = time run at slower pace.
Then $2t$ = time run at faster pace.
Using $D = rt$, we can find the three distances involved in the problem:

distance at slower pace = slower pace × time at slower pace
$= 8 \cdot t$
distance at faster pace = faster pace × time at faster pace
$= 12 \cdot 2t$

Thus, the total distance run = (distance at slower pace) + (distance at faster pace)
$= 8 \cdot t + 12 \cdot 2t$

Simplifying this expression, we get $8t + 24t = 32t$.

Exercise 2-2 Multiplication of Polynomials

Key Ideas and Formulas

Multiplication of monomials: $(ax^m)(bx^n) = abx^{m+n}$

Multiplication of polynomials: multiply each term of one by each term of the other, then add like terms.

Multiplication of binomials: use FOIL method

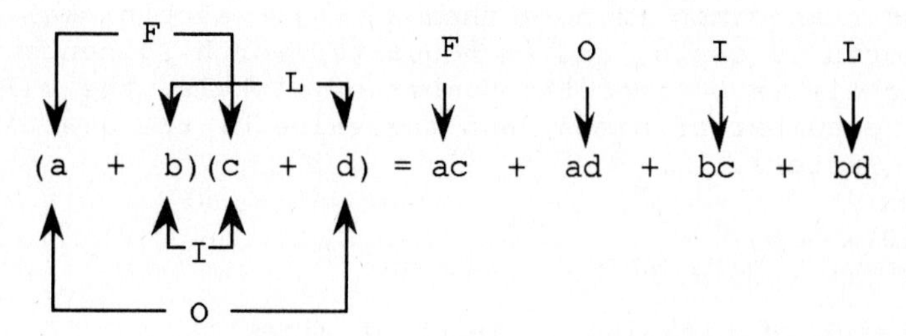

Special cases of FOIL:

1. Frequently, the Outer and Inner Terms are like terms, which are combined mentally. Examples:

$$(x + 7)(x + 2) = x^2 + 9x + 14$$

$$(2x + 3y)(5x + y) = 10x^2 + 17xy + 3y^2$$

The terms 2x and 7x are combined mentally.
The terms 2xy and 15xy are combined mentally.

2. $(a + b)(a - b) = a^2 - b^2$ The terms -ab and ab are combined mentally to yield 0.

3. Squaring binomials:
   $(a + b)^2 = (a + b)(a + b) = a^2 + 2ab + b^2$
   $(a - b)^2 = (a - b)(a - b) = a^2 - 2ab + b^2$
   The Outer and Inner terms are combined mentally to yield 2ab or -2ab.

COMMON ERRORS:

$(a + b)^2 \neq a^2 + b^2$ $(a - b)^2 \neq a^2 - b^2$ These errors must be avoided!

---

1. $y^2y^3 = y^{2+3} = y^5$

3. $(5y^4)(2y) = 5y^4 \cdot 2y = 5 \cdot 2y^4y = 10y^{4+1} = 10y^5$

5. $(8x^{11})(-3x^9) = 8(-3)x^{11+9} = -24x^{20}$

7. $(-3u^4)(2u^5)(-u^7) = (-3)(2)(-1)u^{4+5+7} = 6u^{16}$

9. $(cd^2)(c^2d^2) = c^{1+2}d^{2+2} = c^3d^4$

11. $(-3xy^2z^3)(-5xyz^2) = (-3)(-5)x^{1+1}y^{2+1}z^{3+2} = 15x^2y^3z^5$

13. $y(y + 7) = y \cdot y + y \cdot 7 = y^2 + 7y$

15. $5y(2y - 7) = 5y \cdot 2y - 5y \cdot 7 = 10y^2 - 35y$

17. $3a^2(a^3 + 2a^2) = 3a^2 \cdot a^3 + 3a^2 \cdot 2a^2 = 3a^5 + 6a^4$

19. $2y(y^2 + 2y - 3) = 2y \cdot y^2 + 2y \cdot 2y - 2y \cdot 3 = 2y^3 + 4y^2 - 6y$

21. $7m^3(m^3 - 2m^2 - m + 4) = 7m^3 \cdot m^3 - 7m^3 \cdot 2m^2 - 7m^3 \cdot m + 7m^3 \cdot 4 = 7m^6 - 14m^5 - 7m^4 + 28m^3$

23. $5uv^2(2u^3v - 3uv^2) = 5uv^2 \cdot 2u^3v - 5uv^2 \cdot 3uv^2 = 10u^4v^3 - 15u^2v^4$

25. $2cd^3(c^2d - 2cd + 4c^3d^2) = 2cd^3 \cdot c^2d - 2cd^3 \cdot 2cd + 2cd^3 \cdot 4c^3d^2 = 2c^3d^4 - 4c^2d^4 + 8c^4d^5$

27.
$$\begin{array}{r} 2y^2 + 5y - 3 \\ 3y + 2 \\ \hline 6y^3 + 15y^2 - 9y \\ 4y^2 + 10y - 6 \\ \hline 6y^3 + 19y^2 + y - 6 \end{array}$$

29.
$$\begin{array}{r} m^2 - 4mn - n^2 \\ m + 2n \\ \hline m^3 - 4m^2n - mn^2 \\ 2m^2n - 8mn^2 - 2n^3 \\ \hline m^3 - 2m^2n - 9mn^2 - 2n^3 \end{array}$$

31.
$$\begin{array}{r} 3m^2 - 2m + 1 \\ 2m^2 + 2m - 1 \\ \hline 6m^4 - 4m^3 + 2m^2 \\ 6m^3 - 4m^2 + 2m \\ - 3m^2 + 2m - 1 \\ \hline 6m^4 + 2m^3 - 5m^2 + 4m - 1 \end{array}$$

33.
$$\begin{array}{r} a^2 - ab + b^2 \\ a + b \\ \hline a^3 - a^2b + ab^2 \\ a^2b - ab^2 + b^3 \\ \hline a^3 + b^3 \end{array}$$

35.
$$
\begin{array}{r}
2x^2 - 3xy + y^2 \\
x^2 + 2xy - y^2 \\
\hline
2x^4 - 3x^3y + x^2y^2 \qquad\qquad\qquad \\
4x^3y - 6x^2y^2 + 2xy^3 \qquad\quad \\
- 2x^2y^2 + 3xy^3 - y^4 \\
\hline
2x^4 + x^3y - 7x^2y^2 + 5xy^3 - y^4
\end{array}
$$

37. $(x + 3)(x + 2) = x^2 + 5x + 6$

39. $(a + 8)(a - 4) = a^2 + 4a - 32$

41. $(t + 4)(t - 4) = t^2 - 16$

43. $(m - n)(m + n) = m^2 - n^2$

45. $(4t - 3)(t - 2) = 4t^2 - 11t + 6$

47. $(3x + 2y)(x - 3y) = 3x^2 - 7xy - 6y^2$

49. $(2m - 7)(2m + 7) = 4m^2 - 49$

51. $(6x - 4y)(5x + 3y) = 30x^2 - 2xy - 12y^2$

53. $(2s - 3t)(3s - t) = 6s^2 - 11st + 3t^2$

55. $(x^3 + y^3)(x + y) = x^4 + x^3y + xy^3 + y^4$ (F, O, I, L)

57. 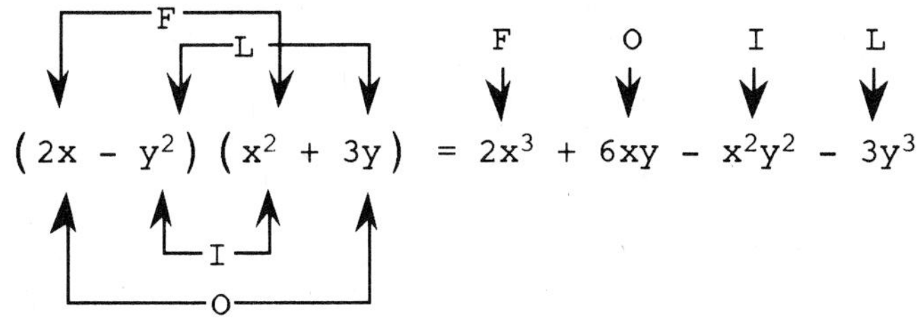 $(2x - y^2)(x^2 + 3y) = 2x^3 + 6xy - x^2y^2 - 3y^3$

F: $2x^3$, O: $6xy$, I: $-x^2y^2$, L: $-3y^3$

59. $(3x + 2)^2 = (3x)^2 + 2(3x)(2) + (2)^2$
$= 9x^2 + 12x + 4$

61. $(2x - 5y)^2 = (2x)^2 - 2(2x)(5y) + (5y)^2$
$= 4x^2 - 20xy + 25y^2$

COMMON ERRORS:

Neglecting the middle term in these results.
$(a + b)^2 \neq a^2 + b^2$
$(a - b)^2 \neq a^2 - b^2$

63. $(6u + 5v)^2 = (6u)^2 + 2(6u)(5v) + (5v)^2 = 36u^2 + 60uv + 25v^2$

65. $(2m - 5n)^2 = (2m)^2 - 2(2m)(5n) + (5n)^2 = 4m^2 - 20mn + 25n^2$

67. $(x^2 - 1)^2 = (x^2)^2 - 2(x^2)(1) + (1)^2 = x^4 - 2x^2 + 1$

69. $(x^2 + y^2)^2 = (x^2)^2 + 2(x^2)(y^2) + (y^2)^2 = x^4 + 2x^2y^2 + y^4$

71. $(x + 2y)^3 = (x + 2y)^2(x + 2y)$
$= \{(x)^2 + 2(x)(2y) + (2y)^2\}(x + 2y)$
$= \{x^2 + 4xy + 4y^2\}(x + 2y)$

$$\begin{array}{rrrr} x^2 & + 4xy & + 4y^2 & \\ x & + 2y & & \\ \hline x^3 + 4x^2y & + 4xy^2 & & \\ & 2x^2y & + 8xy^2 & + 8y^3 \\ \hline x^3 + 6x^2y & + 12xy^2 & + 8y^3 & \end{array}$$

73. We evaluate $(3x - 1)(x + 2)$ using the FOIL method and $(2x - 3)^2$ using the mechanical rule for squaring a binomial.

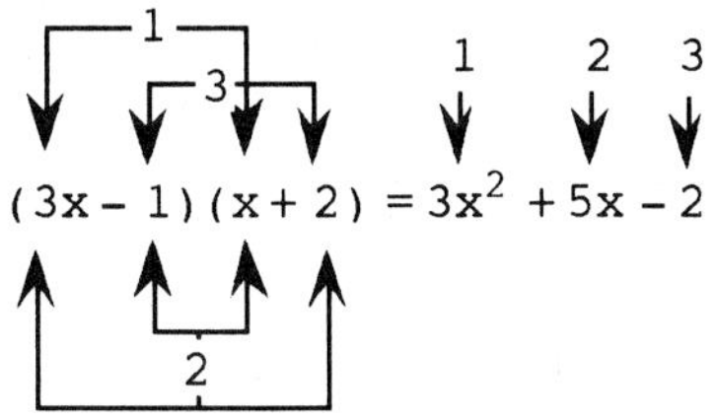

$(3x - 1)(x + 2) = 3x^2 + 5x - 2$ (1: $3x^2$, 2 and 3: $5x$, -2)

$(2x - 3)^2 = (2x)^2 - 2(2x)(3) + (3)^2$
$= 4x^2 - 12x + 9$

Therefore:
$(3x - 1)(x + 2) - (2x - 3)^2 = (3x^2 + 5x - 2) - (4x^2 - 12x + 9)$
$= 3x^2 + 5x - 2 - 4x^2 + 12x - 9$
$= -x^2 + 17x - 11$

75. Since $(x - 2)^2 = (x)^2 - 2(x)(2) + (2)^2 = x^2 - 4x + 4$,
$(x - 2)^3 = (x - 2)^2(x - 2) = (x^2 - 4x + 4)(x - 2)$

$$\begin{array}{rrrr} x^2 & - 4x & + 4 & \\ x & - 2 & & \\ \hline x^3 & - 4x^2 & + 4x & \\ & - 2x^2 & + 8x & - 8 \\ \hline x^3 & - 6x^2 & + 12x & - 8 \end{array}$$

Therefore:
$2(x - 2)^3 - (x - 2)^2 - 3(x - 2) - 4$
$= 2(x^3 - 6x^2 + 12x - 8) - (x^2 - 4x + 4) - 3(x - 2) - 4$
$= 2x^3 - 12x^2 + 24x - 16 - x^2 + 4x - 4 - 3x + 6 - 4$
$= 2x^3 - 13x^2 + 25x - 18$

77. We will multiply $(x + 2)(x^2 - 3)$ using the FOIL method.

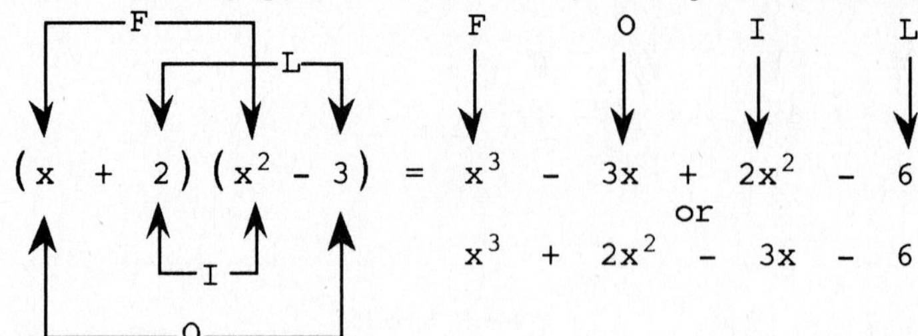

Then

$$\begin{aligned}-3x\{x[x - x(2 - x)] - (x + 2)(x^2 - 3)\} &= -3x\{x[x - 2x + x^2] - [x^3 + 2x^2 - 3x - 6]\}\\ &= -3x\{x[-x + x^2] - [x^3 + 2x^2 - 3x - 6]\}\\ &= -3x\{-x^2 + x^3 - x^3 - 2x^2 + 3x + 6\}\\ &= -3x\{-3x^2 + 3x + 6\}\\ &= 9x^3 - 9x^2 - 18x\end{aligned}$$

79. We will multiply $(2x - 1)(2x + 1)$ using the FOIL method.

1, 3, 2

$$(2x - 1)(2x + 1) = 4x^2 - 1$$

Then

$$(2x - 1)(2x + 1)(3x^3 - 4x + 3) = (4x^2 - 1)(3x^3 - 4x + 3)$$

$$\begin{array}{r} 3x^3 - 4x + 3 \\ 4x^2 - 1 \\ \hline 12x^5 - 16x^3 + 12x^2 \qquad\qquad \\ - 3x^3 \qquad\quad + 4x - 3 \\ \hline 12x^5 - 19x^3 + 12x^2 + 4x - 3 \end{array}$$

81. $$\begin{aligned}[(3x - 2) + y]^2 &= (3x - 2)^2 + 2(3x - 2)y + y^2\\ &= (3x)^2 - 2(3x)(2) + (2)^2 + 2(3x - 2)y + y^2\\ &= 9x^2 - 12x + 4 + 6xy - 4y + y^2\end{aligned}$$

83. F, L, I, O; F O I L

$$\begin{aligned}[(x + y) - 2][(x + y) + 1] &= (x + y)^2 + 1(x + y) - 2(x + y) - 2\\ &= (x + y)^2 - 1(x + y) - 2\\ &= x^2 + 2xy + y^2 - x - y - 2\end{aligned}$$

85. $(x - 2)(x + 3) = x^2 + x - 6$ $\qquad$ $(x + 4)(x - 5) = x^2 - x - 20$

Thus

$$
\begin{aligned}
(x - 2)(x + 3) &= (x + 4)(x - 5) \quad \text{becomes} \\
x^2 + x - 6 &= x^2 - x - 20 \\
x - 6 &= -x - 20 \\
2x - 6 &= -20 \\
2x &= -14 \\
x &= -7
\end{aligned}
$$

87. 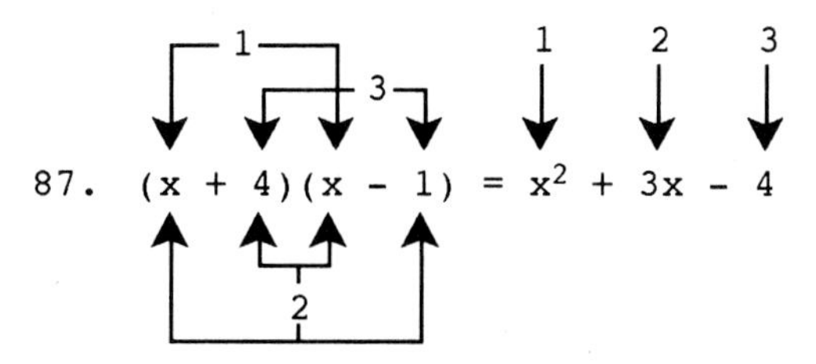 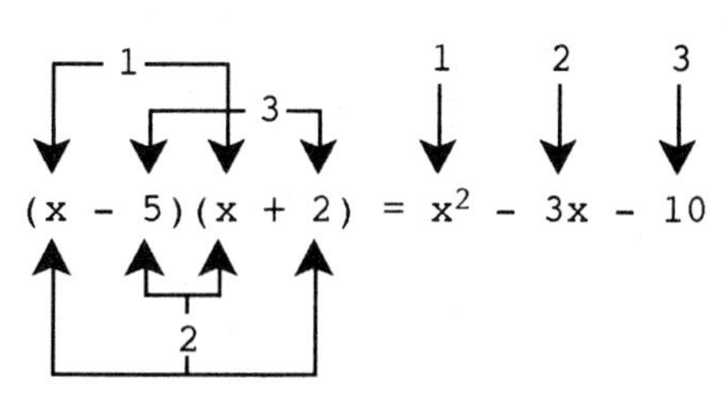

$(x + 4)(x - 1) = x^2 + 3x - 4$ $\qquad$ $(x - 5)(x + 2) = x^2 - 3x - 10$

Thus

$$
\begin{aligned}
(x + 4)(x - 1) &= (x - 5)(x + 2) \quad \text{becomes} \\
x^2 + 3x - 4 &= x^2 - 3x - 10 \\
3x - 4 &= -3x - 10 \\
6x - 4 &= -10 \\
6x &= -6 \\
x &= -1
\end{aligned}
$$

89. Since $x^m x^n = x^{m+n}$, the degree will be $m + n$.

91. There are three quantities in this problem: length, width, and area. They are related by area = (width) × (length). Since y is to represent the length of the rectangle and the length is 8 meters more than the width, the width must be 8 meters less than the length, or $y - 8$. Therefore, the area is $y(y - 8)$ or $y^2 - 8y$.

93. Let x = the number of quarters by which the price is increased. Then the price = $8 + 0.25x$. The decrease in sales will equal 100x units, hence the number of units sold = $5000 - 100x$. Since the revenue = (number of units sold) × (price per unit), the revenue = $(5000 - 100x)(8 + 0.25x)$.

95. Let x = the number of trees planted, then $x - 120$ is the number of trees in excess of 120. The decrease in yield is 0.5 times this number, or $0.5(x - 120)$. Hence the yield per tree will equal $40 - 0.5(x - 120)$. Since the yield = (number of trees planted) × (yield per tree), the yield = $x[40 - 0.5(x - 120)]$.

97. Let x = the number
then $x + 1$ = one more than the number
$x - 1$ = one less than the number
$x + 3$ = three more than the number
Then the equation can be written:
$x(x + 1) = (x - 1)(x + 3)$

99. Solving $x(x + 1) = (x - 1)(x + 3)$, we note:

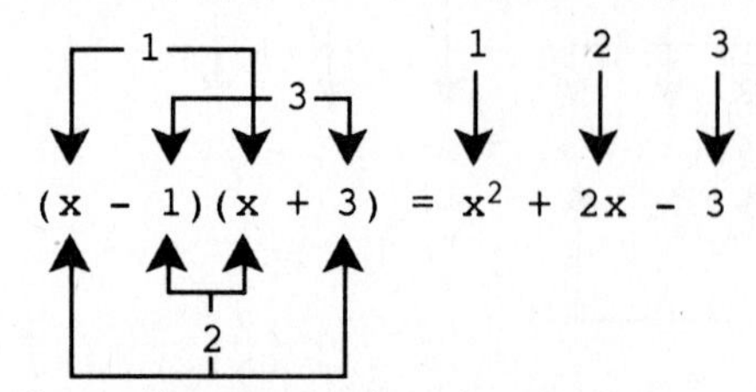

Thus the equation becomes

$$x^2 + x = x^2 + 2x - 3$$
$$x = 2x - 3$$
$$-x = -3$$
$$x = 3$$

Exercise 2-3 Basic Factoring Methods

Key Ideas and Formulas

The distributive property, viewed as $ab + ac = a(b + c)$, is used to take out a common factor in several ways. The common factor may be a monomial or polynomial.

Monomial example: $ab + ac = a(b + c)$ — common factor a

Binomial example: $a(b + c) + x(b + c) = (a + x)(b + c)$ — common factor $b + c$

Factoring by grouping: it may sometimes be possible to take a polynomial with no apparent common factor and find one when the terms are grouped.

Example: $ax + ay + bx + by = (ax + ay) + (bx + by)$
$= a(x + y) + b(x + y)$
$= (a + b)(x + y)$

Products of the form $x^2 + 2bx + b^2$ or $x^2 + 2bxy + b^2y^2$ can be recognized as perfect squares, since the third term is the square of half the second term. They are factored using $x^2 + 2bx + b^2 = (x + b)^2$ or $x^2 + 2bxy + b^2y^2 = (x + by)^2$.

Similarly, products of the form, $a^2x^2 + 2abxy + b^2y^2$ are recognized as $(ax + by)^2$.

The sum $A^2 + B^2$ of two squares cannot be factored unless there are common factors. The difference of two squares is factored

$$A^2 - B^2 = (A - B)(A + B)$$

COMMON ERRORS:

Confusing $A^2 - B^2$ with $(A - B)^2$
Confusing $A^2 + B^2$ with $(A + B)^2$

---

1. $3xz - 6x = 3x \cdot z - 3x \cdot 2 = 3x(z - 2)$

3. $8x + 2y = 2 \cdot 4x + 2 \cdot y = 2(4x + y)$

5. $6x^2 + 9x = 3x \cdot 2x + 3x \cdot 3 = 3x(2x + 3)$

7. $14x^2y - 7xy^2 = 7xy \cdot 2x - 7xy \cdot y = 7xy(2x - y)$

9. $(2x + z)(x - 3)$

11. $(a + b)(c - d)$

13. $(x - y)(x - y)$ or $(x - y)^2$

15. $ab(c + d) - (c + d) = ab(c + d) - 1(c + d) = (ab - 1)(c + d)$

17. $x^5 + x^4 + x^3 = x^3 \cdot x^2 + x^3 \cdot x + x^3 \cdot 1 = x^3(x^2 + x + 1)$

19. $a^2b - a^3b^2 - a^4b = a^2b \cdot 1 - a^2b \cdot ab - a^2b \cdot a^2 = a^2b(1 - ab - a^2)$

21. $x^2 - 10x + 25$. This is a perfect square. The constant term is the square of one-half the coefficient of x:

$$25 = \left[\frac{1}{2}(-10)\right]^2 = (-5)^2$$

$x^2 - 10x + 25 = [x + (-5)]^2 = (x - 5)^2$

23. $x^2 + 12x + 36$. This is a perfect square. The constant term is the square of one-half the coefficient of x:

$$36 = \left(\frac{1}{2} \cdot 12\right)^2 = 6^2$$

$x^2 + 12x + 36 = (x + 6)^2$

25. The last term, -36, is not the square of half the coefficient of x. The polynomial is not a perfect square.

27. $x^2 - 20x + 100$. This is a perfect square. The constant term is the square of one-half the coefficient of x:

$$100 = \left[\frac{1}{2}(-20)\right]^2 = (-10)^2$$

$x^2 - 20x + 100 = [x + (-10)]^2 = (x - 10)^2$

29. $x^2 + 24x + 144$. This is a perfect square. The constant term is the square of one-half the coefficient of x:

$$144 = \left(\frac{1}{2} \cdot 24\right)^2 = 12^2$$

$x^2 + 24x + 144 = (x + 12)^2$

31. The last term, 288, is not the square of half the coefficient of x. The polynomial is not a perfect square.

33. $abc + bcd + cde = c \cdot ab + c \cdot bd + c \cdot de = c(ab + bd + de)$

35. $xy^2z^3 - x^2y^3z + x^3yz^2 = xyz \cdot yz^2 - xyz \cdot xy^2 + xyz \cdot x^2z = xyz(yz^2 - xy^2 + x^2z)$

37. Use difference of two squares: $v^2 - 25 = v^2 - (5)^2 = (v - 5)(v + 5)$

39. Use difference of two squares: $9x^2 - 4 = (3x)^2 - (2)^2 = (3x - 2)(3x + 2)$

41. Use sum of two squares: not factorable

43. Use difference of two squares: $9x^2 - 16y^2 = (3x)^2 - (4y)^2 = (3x - 4y)(3x + 4y)$

45. Use difference of two squares: $4x^2 - 9y^2 = (2x)^2 - (3y)^2 = (2x - 3y)(2x + 3y)$

47. Use difference of two squares: $25x^2 - 64y^2 = (5x)^2 - (8y)^2 = (5x - 8y)(5x + 8y)$

49. Use sum of two squares: not factorable

51. $a(a - 2) + 3(2 - a) = a(a - 2) - 3(a - 2) = (a - 3)(a - 2)$

53. $x(x - 1) - (1 - x) = x(x - 1) + (x - 1) = x(x - 1) + 1(x - 1) = (x + 1)(x - 1)$

55. $(x - 2) - 5(2 - x) = 1(x - 2) + 5(x - 2) = 6(x - 2)$

57. $3(x - 1) + 6(1 - x) = 3(x - 1) - 6(x - 1) = -3(x - 1)$ or $3(1 - x)$

59. $(2x^3 + x^2) + (6x + 3) = x^2(2x + 1) + 3(2x + 1) = (x^2 + 3)(2x + 1)$

61. $(abc - a^2b^2) - (c^3 - abc^2) = ab(c - ab) - c^2(c - ab) = (ab - c^2)(c - ab)$

63. $(3xy + 6x) - (y^2 + 2y) = 3x(y + 2) - y(y + 2) = (3x - y)(y + 2)$

65. $2x^3 + x^2 + 6x + 3 = (2x^3 + x^2) + (6x + 3) = (x^2 + 3)(2x + 1)$
(See problem 59)

67. $abc - a^2b^2 - c^3 + abc^2 = (abc - a^2b^2) - (c^3 - abc^2) = (ab - c^2)(c - ab)$
(See problem 61)

69. $3xy + 6x - y^2 - 2y = (3xy + 6x) - (y^2 + 2y) = (3x - y)(y + 2)$
(See problem 63)

71. First, remove common factors: $3x^2 - 27 = 3(x^2 - 9)$
Now, use difference of two squares: $3(x^2 - 9) = 3(x - 3)(x + 3)$

73. First, remove common factors: $5x^2 - 125 = 5(x^2 - 25)$
Now, use difference of two squares: $5(x^2 - 25) = 5(x - 5)(x + 5)$

75. First, remove common factors: $3x^2 - 12y^2 = 3(x^2 - 4y^2)$
Now, use difference of two squares: $3(x^2 - 4y^2) = 3(x - 2y)(x + 2y)$

77. Use difference of two squares: $x^2y^2 - 1 = (xy)^2 - (1)^2 = (xy - 1)(xy + 1)$

79. Use difference of two squares: $a^2b^2 - c^2d^2 = (ab)^2 - (cd)^2 = (ab - cd)(ab + cd)$

81. $$\begin{aligned} 2x^4 - 3x^2 + 6x^2 - 9 &= (2x^4 - 3x^2) + (6x^2 - 9) \\ &= x^2(2x^2 - 3) + 3(2x^2 - 3) \\ &= (x^2 + 3)(2x^2 - 3) \end{aligned}$$

83. $$\begin{aligned} x^3y^2 - x^2y^3 + 3x - 3y &= (x^3y^2 - x^2y^3) + (3x - 3y) \\ &= x^2y^2(x - y) + 3(x - y) \\ &= (x^2y^2 + 3)(x - y) \end{aligned}$$

85. $$\begin{aligned} 4ab - b^2 - 4a^2 + ab &= (4ab - b^2) - (4a^2 - ab) \\ &= b(4a - b) - a(4a - b) \\ &= (b - a)(4a - b) \end{aligned}$$

87. $$\begin{aligned} xy^2z - xyz^2 - xy + xz &= (xy^2z - xyz^2) - (xy - xz) \\ &= xyz(y - z) - x(y - z) \\ &= (xyz - x)(y - z) \text{ or } x(yz - 1)(y - z) \end{aligned}$$

89. $$\begin{aligned} 2x - 6y - 9yz + 3xz &= (2x - 6y) - (9yz - 3xz) \\ &= 2(x - 3y) - 3z(3y - x) \\ &= 2(x - 3y) + 3z(x - 3y) \\ &= (2 + 3z)(x - 3y) \end{aligned}$$

91. $$\begin{aligned} 3a^2 + bc + ab + 3ac &= 3a^2 + ab + 3ac + bc \\ &= (3a^2 + ab) + (3ac + bc) \\ &= a(3a + b) + c(3a + b) \\ &= (a + c)(3a + b) \end{aligned}$$

93. $$\begin{aligned} 12 + xy^2 - 4x - 3y^2 &= 12 - 4x - 3y^2 + xy^2 \\ &= (12 - 4x) - (3y^2 - xy^2) \\ &= 4(3 - x) - y^2(3 - x) \\ &= (3 - x)(4 - y^2) \text{ or } (3 - x)(2 - y)(2 + y) \end{aligned}$$

95. $$\begin{aligned} ac + bd + ab + cd &= ac + ab + cd + bd \\ &= (ac + ab) + (cd + bd) \\ &= a(c + b) + d(c + b) \\ &= (a + d)(c + b) \text{ or } (a + d)(b + c) \end{aligned}$$

97. $yz + 2xy - 2xz - y^2 = yz - 2xz - y^2 + 2xy$
$= (yz - 2xz) - (y^2 - 2xy)$
$= z(y - 2x) - y(y - 2x)$
$= (z - y)(y - 2x)$

Exercise 2-4 Factoring Second-Degree Polynomials

Key Ideas and Formulas

Products of the forms $Ax^2 + Bx + C$ or $Ax^2 + Bxy + Cy^2$ are factored by trial and error by looking for combinations of factors of A and C whose products can be added to yield B. Then the factors are inserted into the format $(?x + ?)(?x + ?)$ or $(?x + ?y)(?x + ?y)$ and the product is checked.

---

1. $x^2 + 3x + 2 = (x +\ \ )(x +\ \ )$

2: [1·2] $1 + 2 = 3$; 2·1

$(x + 1)(x + 2)$

3. $x^2 + 7x + 10 = (x +\ \ )(x +\ \ )$

10: 1·10; [2·5] $2 + 5 = 7$; 5·2; 10·1

$(x + 2)(x + 5)$

5. $x^2 - 2x - 3 = (x +\ \ )(x -\ \ )$

3: [1·3] $1 + (-3) = -2$; 3·1

$(x + 1)(x - 3)$

7. $x^2 + 3x - 4 = (x +\ \ )(x -\ \ )$

4: 1·4; 2·2; [4·1] $4 + (-1) = 3$

$(x + 4)(x - 1)$

9. $x^2 - 9x + 20 = (x -\ \ )(x -\ \ )$

20: 1·20; 2·10; [4·5] $4 + 5 = 9$; 5·4; 10·2; 20·1

$(x - 4)(x - 5)$

11. $x^2 - 6x + 5 = (x -\ \ )(x -\ \ )$

5: [1·5] $1 + 5 = 6$; 5·1

$(x - 1)(x - 5)$

13. $x^2 - 6x + 9$. This is a perfect square. The constant term is the square of one half the coefficient of x:

$$9 = \left[\frac{1}{2}(-6)\right]^2 = (-3)^2$$

$x^2 - 6x + 9 = [x + (-3)]^2 = (x - 3)^2$

15. $x^2 + 14x + 49$. This is a perfect square. The constant term is the square of one half the coefficient of x:

$$49 = \left(\frac{1}{2} \cdot 14\right)^2 = 7^2$$

$x^2 + 14x + 49 = (x + 7)^2$

17. $x^2 - 6xy + 9y^2$. This is a perfect square. The coefficient of y^2 is the square of one half the coefficient of xy:

$$9 = \left[\frac{1}{2}(-6)\right]^2 = (-3)^2$$

$x^2 - 6xy + 9y^2 = (x - 3y)^2$

19. $x^2 + 4x - 5 = (x + \quad)(x - \quad)$

5: $1 \cdot 5$, $\boxed{5 \cdot 1}$ $\quad 5 + (-1) = 4$

$(x + 5)(x - 1)$

21. $x^2 - 3x - 18 = (x + \quad)(x - \quad)$

18: $1 \cdot 18$, $2 \cdot 9$, $\boxed{3 \cdot 6}$ $\quad 3 + (-6) = -3$, $6 \cdot 3$, $9 \cdot 2$, $18 \cdot 1$

$(x + 3)(x - 6)$

23. $x^2 + 4x - 12 = (x + \quad)(x - \quad)$

12: $1 \cdot 12$, $2 \cdot 6$, $3 \cdot 4$, $4 \cdot 3$, $\boxed{6 \cdot 2}$ $\quad 6 + (-2) = 4$, $12 \cdot 1$

$(x + 6)(x - 2)$

25. $x^2 - 9x + 8 = (x - \quad)(x - \quad)$

8: $\boxed{1 \cdot 8}$ $\quad 1 + 8 = 9$, $2 \cdot 4$, $4 \cdot 2$, $8 \cdot 1$

$(x - 1)(x - 8)$

27. $x^2 + 2x + 4 = (x + \quad)(x + \quad)$

4:

$1 \cdot 4 \quad 1 + 4 = 5$

$2 \cdot 2 \quad 2 + 2 = 4$

$4 \cdot 1 \quad 4 + 1 = 5$ No choice produces the middle term, so this is not factorable.

29. $2x^2 - x - 1 = (2x + \quad)(x + \quad)$

-1: $\boxed{1(-1)}$, $(-1)1$

$(2x + 1)(x - 1) = 2x^2 - x - 1$ Correct.

31. $2x^2 + 7x + 5 = (2x + \quad)(x + \quad)$

5:

$1 \cdot 5 \qquad (2x + 1)(x + 5) = 2x^2 + 11x + 5$ No.

$\boxed{5 \cdot 1} \qquad (2x + 5)(x + 1) = 2x^2 + 7x + 5$ Correct.

33. $3x^2 + 7x - 4 = (3x + \quad)(x + \quad)$

-4:

$1(-4) \qquad (3x + 1)(x - 4) = 3x^2 - 11x - 4$ No.

$2(-2) \qquad (3x + 2)(x - 2) = 3x^2 - 4x - 4$ No.

$(-2)2 \qquad (3x - 2)(x + 2) = 3x^2 + 4x - 4$ No.

$(-4)1 \qquad (3x - 4)(x + 1) = 3x^2 - x - 4$ No.

No choice produces the middle term, hence this is not factorable.

35. No combination of coefficients can produce $9x^2 + 8$; this is not factorable.

37. Remove common factors: $4x^2 - 40 = 4(x^2 - 10)$. This is not further factorable.

39. First, remove common factors: $2x^2 - 200 = 2(x^2 - 100)$.
Now, use difference of two squares: $2(x^2 - 100) = 2(x - 10)(x + 10)$.

41. $2x^2 - xy + y^2 = (2x - ?y)(x - ?y)$

1	
1·1	$(2x - y)(x - y) = 2x^2 - 3xy + y^2$

The only possible choice fails to produce the middle term, so this is not factorable.

43. $2x^2 - 9xy + 10y^2 = (2x - ?y)(x - ?y)$

10		
1·10	$(2x - y)(x - 10y) = 2x^2 - 21xy + 10y^2$	No.
2· 5	$(2x - 2y)(x - 5y) = 2x^2 - 12xy + 10y^2$	No.
[5· 2]	$(2x - 5y)(x - 2y) = 2x^2 - 9xy + 10y^2$	Correct.
10· 1		

45. $4x^2 + 4x - 3 = (?x + ?)(?x - ?)$

4	-3
1·4	1(-3)
[2·2]	[3(-1)]
4·1	

After considerable trial and error, $(2x + 3)(2x - 1)$.

47. $4x^2 + 3x - 1 = (?x + ?)(?x - ?)$

4	-1
[1·4]	[1(-1)]
2·2	
4·1	

$(x + 1)(4x - 1) = 4x^2 + 3x - 1$

49. $2x^2 + 4x + 3 = (2x + \quad)(x + \quad)$

3		
1·3	$(2x + 1)(x + 3) = 2x^2 + 7x + 3$	No.
3·1	$(2x + 3)(x + 1) = 2x^2 + 5x + 3$	No.

No choice produces the middle term, hence this is not factorable.

51. $4x^2 + 12x + 9 = (2x)^2 + 2(2x)(3) + (3)^2 = (2x + 3)^2$
Perfect square.

53. $x^2 - x - 2 = (x + \quad)(x - \quad)$

2
[1(-2)]
(-1)2

$1 + (-2) = -1$

$(x + 1)(x - 2)$

55. $9x^2 + 12x + 4 = (3x)^2 + 2(3x)(2) + (2)^2 = (3x + 2)^2$
Perfect square.

57. $6x^2 - 7x + 2 = (?x - \quad)(?x - \quad)$

6	2
1·6	[1·2]
[2·3]	2·1

After considerable trial and error,
$(2x - 1)(3x - 2)$

59. $6x^2 - xy - y^2 = (?x - ?y)(?x - ?y)$

6	-1
$1 \cdot 6$	$1(-1)$
$\boxed{2 \cdot 3}$	$\boxed{(-1)1}$

After considerable trial and error,
$(2x - y)(3x + y)$

61. No combination of coefficients can produce $x^2 + 24$; this is not factorable.

63. Use sum of two squares: $x^2 + y^2$ is not factorable.

65. $9x^2 + 6xy + y^2 = (3x)^2 + 2(3x)y + y^2 = (3x + y)^2$
Perfect square.

67. $4x^2 - 4xy - y^2 = (?x + ?y)(?x - ?y)$

4	-1
$1 \cdot 4$	$1(-1)$
$2 \cdot 2$	$(-1)1$
$4 \cdot 1$	

After considerable trial and error, we find that no choice produces the middle term, hence this is not factorable.

69. $8x^2 + 14x + 7 = (?x + \quad)(?x + \quad)$

8	7
$1 \cdot 8$	$1 \cdot 7$
$2 \cdot 4$	$7 \cdot 1$

After considerable trial and error, we find that no choice produces the middle term, hence this is not factorable.

71. $8x^2 - 14x - 15 = (?x + \quad)(?x - \quad)$

8	-15
$1 \cdot 8$	$1(-15)$
$2 \cdot 4$	$\boxed{3(-5)}$
$\boxed{4 \cdot 2}$	$5(-3)$
$8 \cdot 1$	$15(-1)$

After considerable trial and error,
$(4x + 3)(2x - 5)$

73. $6x^2 + 10xy + 5y^2 = (?x + ?y)(?x + ?y)$

6	5
$1 \cdot 6$	$1 \cdot 5$
$2 \cdot 3$	$5 \cdot 1$

After considerable trial and error, we find that no choice produces the middle term, hence this is not factorable.

75. We note first that $6x^2 + 21xy - 12y^2 = 3(2x^2 + 7xy - 4y^2) = 3(?x + ?y)(?x - ?y)$

2	-4
$1 \cdot 2$	$1 \cdot -4$
$\boxed{2 \cdot 1}$	$2 \cdot -2$
	$-2 \cdot 2$
	$\boxed{-1 \cdot 4}$

After considerable trial and error,
$3(2x - y)(x + 4y)$

77. $x^4 + 4x^2 + 3 = (x^2)^2 + 4(x^2) + 3 = (x^2 + \quad)(x^2 + \quad)$

3
[1·3] 1 + 3 = 4
3·1

$(x^2 + 1)(x^2 + 3)$

79. Use difference of two squares:
$x^4 - 1 = (x^2)^2 - (1)^2 = (x^2 + 1)(x^2 - 1) = (x^2 + 1)(x + 1)(x - 1)$

81. $x^4 + 2x^2 + 1 = (x^2)^2 + 2(x^2)(1) + (1)^2 = (x^2 + 1)^2$
Perfect square.

83. Use difference of two squares: $4x^4 - 1 = (2x^2)^2 - (1)^2 = (2x^2 - 1)(2x^2 + 1)$.

85. $9x^4 - 12x^2 + 4 = (3x^2)^2 - 2(3x^2)(2) + (2)^2 = (3x^2 - 2)^2$
Perfect square.

87. Use difference of two squares: $x^8 - 1 = (x^4)^2 - (1)^2 = (x^4 + 1)(x^4 - 1)$
$= (x^4 + 1)(x^2 + 1)(x^2 - 1) = (x^4 + 1)(x^2 + 1)(x + 1)(x - 1)$

89. $x^6 - 2x^3 + 1 = (x^3)^2 - 2(x^3)(1) + (1)^2 = (x^3 - 1)^2$.
This polynomial is a perfect square. The alert student may notice that $(x^3 - 1)^2$ is further factorable to $(x - 1)^2(x^2 + x + 1)^2$, but not by methods so far developed.

91. No combination of coefficients can produce $x^6 + 36$; this is not factorable.

93. $x^6 - 2x^3 - 8 = (x^3)^2 - 2(x^3) - 8 = (x^3 + \quad)(x^3 - \quad)$

8
1·8
[2·4] 2 + (−4) = −2
4·2
8·1

$(x^3 + 2)(x^3 - 4)$

95. $x^6 + 5x^3 + 6 = (x^3)^2 + 5(x^3) + 6 = (x^3 + \quad)(x^3 + \quad)$

6
1·6
[2·3] 2 + 3 = 5
3·2
6·1

$(x^3 + 2)(x^3 + 3)$

97. $2x^4 - x^2 - 3 = 2(x^2)^2 - (x^2) - 3 = (2x^2 + \quad)(x^2 + \quad)$

2	−3
1·2	1(−3)
[2·1]	3(−1)
	(−1)3
	[(−3)1]

After considerable trial and error,
$(2x^2 - 3)(x^2 + 1)$

99. $x^6 + x^3 - 6 = (x^3)^2 + (x^3) - 6 = (x^3 + \quad)(x^3 - \quad)$

−6
1(−6)
2(−3)
[3(−2)] 3 + −2 = 1
6(−1)

$(x^3 + 3)(x^3 - 2)$

Exercise 2-5 Factoring with the ac Test

Key Ideas and Formulas

To factor polynomials of the form

$$ax^2 + bx + c \quad \text{or} \quad ax^2 + bxy + cy^2 \qquad (1)$$

we can apply the ac test for factorability:

ac Test for Factorability
If in polynomials of type (1) the product ac has two integer factors p and q whose sum is the coefficient of the middle term b; that is, if integers p and q exist so that $$pq = ac \quad \text{and} \quad p + q = b \qquad (2)$$ then polynomials of type (1) have first-degree factors with integer coefficients. If no integers p and q exist that satisfy Equations (2), then polynomials of type (1) will not have first-degree factors with integer coefficients.

If we can find integers p and q in the ac test, we can then write the polynomials of type (1), splitting the middle term, in the forms

$$ax^2 + px + qx + c \quad \text{or} \quad ax^2 + pxy + qxy + cy^2$$

and complete the factoring using the grouping method of the previous section.

---

1. Compute ac for $4x^2 + 11x - 3$: $ac = 4(-3) = -12$
   We need two integers, p and q, whose product is −12, and whose sum is 11 = b

$pq = -12$
1(−12)
(−1)12
2(−6)
(−2)6
3(−4)
(−3)4

   −1 and 12 will work, so we split the middle term $11x = -x + 12x$ and factor by grouping:

   $$\begin{aligned} 4x^2 + 11x - 3 &= 4x^2 - x + 12x - 3 \\ &= (4x^2 - x) + (12x - 3) \\ &= x(4x - 1) + 3(4x - 1) \\ &= (x + 3)(4x - 1) \end{aligned}$$

3. Compute ac for $4x^2 + 4x - 3$: $ac = 4(-3) = -12$
   We need two integers, p and q, whose product is −12, and whose sum is 4 = b

$pq = -12$
1(−12)
(−1)12
2(−6)
(−2) 6
3(−4)
(−3) 4

   −2 and 6 will work, so we split the middle term $4x = -2x + 6x$ and factor by grouping:

   $$\begin{aligned} 4x^2 + 4x - 3 &= 4x^2 - 2x + 6x - 3 \\ &= (4x^2 - 2x) + (6x - 3) \\ &= 2x(2x - 1) + 3(2x - 1) \\ &= (2x + 3)(2x - 1) \end{aligned}$$

5. Compute ac for $4x^2 + 11x + 8$: $ac = 4\cdot 8 = 32$
We need two integers, p and q, whose product is 32, and whose sum is $11 = b$

$pq = 32$
$1\cdot 32$
$2\cdot 16$
$4\cdot 8$

None of these add up to 11; we conclude that $4x^2 + 11x + 8$ is not factorable (using integer coefficients).

7. Compute ac for $4x^2 - 25x + 6$: $ac = 4\cdot 6 = 24$
We need two integers, p and q, whose product is 24, and whose sum is $-25 = b$. In this case, both p and q will be negative.

$pq = 24$
$(-1)(-24)$
$(-2)(-12$
$(-3)(-8)$
$(-4)(-6)$

-1 and -24 will work, so we split the middle term $-25x = -x - 24x$ and factor by grouping:

$$\begin{aligned} 4x^2 - 25x + 6 &= 4x^2 - x - 24x + 6 \\ &= (4x^2 - x) - (24x - 6) \\ &= x(4x - 1) - 6(4x - 1) \\ &= (x - 6)(4x - 1) \end{aligned}$$

9. Compute ac for $6x^2 - 25x + 4$: $ac = 6\cdot 4 = 24$
We need two integers, p and q, whose product is 24, and whose sum is $-25 = b$. In this case, both p and q will be negative.

$pq = 24$
$(-1)(-24)$
$(-2)(-12)$
$(-3)(-8)$
$(-4)(-6)$

-1 and -24 will work, so we split the middle term $-25x = -x - 24x$ and factor by grouping:

$$\begin{aligned} 6x^2 - 25x + 4 &= 6x^2 - x - 24x + 4 \\ &= (6x^2 - x) - (24x - 4) \\ &= x(6x - 1) - 4(6x - 1) \\ &= (x - 4)(6x - 1) \end{aligned}$$

11. Compute ac for $4x^2 + 4x - 15$: $ac = 4(-15) = -60$
We need two integers, p and q, whose product is -60, and whose sum is $4 = b$

$pq = -60$	
$1(-60)$	$(-1)60$
$2(-30)$	$(-2)30$
$3(-20)$	$(-3)20$
$4(-15)$	$(-4)15$
$5(-12)$	$(-5)12$
$6(-10)$	$(-6)10$

-6 and 10 will work, so we split the middle term $4x = -6x + 10x$ and factor by grouping:

$$\begin{aligned} 4x^2 + 4x - 15 &= 4x^2 - 6x + 10x - 15 \\ &= (4x^2 - 6x) + (10x - 15) \\ &= 2x(2x - 3) + 5(2x - 3) \\ &= (2x + 5)(2x - 3) \end{aligned}$$

13. Compute ac for $5x^2 - 4x + 3$: $ac = 5\cdot 3 = 15$
We need two integers, p and q, whose product is 15 and whose sum is $-4 = b$.

$pq = 15$
$(-1)(-15)$
$(-3)(-5)$

None of these add up to -4, we conclude that $5x^2 - 4x + 3$ is not factorable.

15. Compute ac for $2x^2 - 7x - 15$: $ac = 2(-15) = -30$
We need two integers, p and q, whose product is -30, and whose sum is $-7 = b$.

$pq = -30$	
1(-30)	(-1)30
2(-15)	(-2)15
3(-10)	(-3)10
5(-6)	(-5)6

3 and -10 will work, so we split the middle term $-7x = -10x + 3x$ and factor by grouping:

$$\begin{aligned} 2x^2 - 7x - 15 &= 2x^2 - 10x + 3x - 15 \\ &= (2x^2 - 10x) + (3x - 15) \\ &= 2x(x - 5) + 3(x - 5) \\ &= (x - 5)(2x + 3) \end{aligned}$$

17. Compute ac for $3x^2 + 8x - 3$: $ac = 3(-3) = -9$
We need two integers, p and q, whose product is -9, and whose sum is $8 = b$.

$pq = -9$	
1(-9)	(-1)9
3(-3)	(-3)3

-1 and 9 will work, so we split the middle term $8x = -x + 9x$ and factor by grouping:

$$\begin{aligned} 3x^2 + 8x - 3 &= 3x^2 - x + 9x - 3 \\ &= (3x^2 - x) + (9x - 3) \\ &= x(3x - 1) + 3(3x - 1) \\ &= (3x - 1)(x + 3) \end{aligned}$$

19. Compute ac for $4x^2 - 3x + 2$: $ac = 4 \cdot 2 = 8$
We need two integers, p and q, whose product is 8, and whose sum is $-3 = b$.

$pq = 8$
(-1)(-8)
(-2)(-4)

None of these add up to -3; we conclude that $4x^2 - 3x + 2$ is not factorable.

21. Compute ac for $7x^2 + 26x - 8$: $ac = 7(-8) = -56$
We need two integers, p and q, whose product is -56, and whose sum is $26 = b$.

$pq = -56$	
1(-56)	(-1)56
2(-28)	(-2)28
4(-14)	(-4)14
7(-8)	(-7)8

-2 and 28 will work, so we split the middle term $26x = -2x + 28x$ and factor by grouping:

$$\begin{aligned} 7x^2 + 26x - 8 &= 7x^2 - 2x + 28x - 8 \\ &= (7x^2 - 2x) + (28x - 8) \\ &= x(7x - 2) + 4(7x - 2) \\ &= (7x - 2)(x + 4) \end{aligned}$$

23. Compute ac for $3x^2 + 11x + 10$: $ac = 3 \cdot 10 = 30$
We need two integers, p and q, whose product is 30, and whose sum is $11 = b$.

$pq = 30$
1·30
2·15
3·10
5·6

5 and 6 will work, so we split the middle term $11x = 5x + 6x$ and factor by grouping:

$$\begin{aligned} 3x^2 + 11x + 10 &= 3x^2 + 5x + 6x + 10 \\ &= (3x^2 + 5x) + (6x + 10) \\ &= x(3x + 5) + 2(3x + 5) \\ &= (3x + 5)(x + 2) \end{aligned}$$

25. Compute ac for $4x^2 + 4x + 3$: $ac = 4 \cdot 3 = 12$
We need two integers, p and q, whose product is 12, and whose sum is $4 = b$.

pq = 12
$1 \cdot 12$
$2 \cdot 6$
$3 \cdot 4$

None of these add up to 4; we conclude that $4x^2 + 4x + 3$ is not factorable.

27. Compute ac for $6x^2 - 29x + 20$: $ac = 6 \cdot 20 = 120$
We need two integers, p and q, whose product is 120, and whose sum is $-29 = b$.
In this case, both p and q will be negative.

pq = 120	
(-1)(-120)	(-5)(-24)
(-2)(-60)	(-6)(-20)
(-3)(-40)	(-8)(-15)
(-4)(-30)	(-10)(-12)

-5 and -24 will work, so we split the middle term $-29x = -5x - 24x$ and factor by grouping:

$$\begin{aligned} 6x^2 - 29x + 20 &= 6x^2 - 5x - 24x + 20 \\ &= (6x^2 - 5x) - (24x - 20) \\ &= x(6x - 5) - 4(6x - 5) \\ &= (x - 4)(6x - 5) \end{aligned}$$

29. Compute ac for $6x^2 + 7x - 3$: $ac = 6(-3) = -18$
We need two integers, p and q, whose product is -18, and whose sum is $7 = b$.

pq = -18	
1(-18)	(-1)18
2(-9)	(-2) 9
3(-6)	(-3) 6

-2 and 9 will work, so we split the middle term $7x = -2x + 9x$ and factor by grouping:

$$\begin{aligned} 6x^2 + 7x - 3 &= 6x^2 - 2x + 9x - 3 \\ &= (6x^2 - 2x) + (9x - 3) \\ &= 2x(3x - 1) + 3(3x - 1) \\ &= (2x + 3)(3x - 1) \end{aligned}$$

31. Compute ac for $6x^2 + 7x - 20$: $ac = 6(-20) = -120$
We need two integers, p and q, whose product is -120, and whose sum is $7 = b$.

pq = -120	
1(-120)	(-1)120
2(-60)	(-2) 60
3(-40)	(-3) 40
4(-30)	(-4) 30
5(-24)	(-5) 24
6(-20)	(-6) 20
8(-15)	(-8) 15
10(-12)	(-10) 12

-8 and 15 will work, so we split the middle term $7x = -8x + 15x$ and factor by grouping:

$$\begin{aligned} 6x^2 + 7x - 120 &= 6x^2 - 8x + 15x - 20 \\ &= (6x^2 - 8x) + (15x - 20) \\ &= 2x(3x - 4) + 4(3x - 4) \\ &= (2x + 5)(3x - 4) \end{aligned}$$

33. Compute ac for $6x^2 + x - 15$: $ac = 6(-15) = -90$
We need two integers, p and q, whose product is -90, and whose sum is $1 = b$.

pq = -90	
1(-90)	(-1)90
2(-45)	(-2)45
3(-30)	(-3)30
5(-18)	(-5)18
6(-15)	(-6)15
9(-10)	(-9)10

-9 and 10 will work, so we split the middle term $x = -9x + 10x$ and factor by grouping:

$$\begin{aligned}6x^2 + x - 15 &= 6x^2 - 9x + 10x - 15\\ &= (6x^2 - 9x) + (10x - 15)\\ &= 3x(2x - 3) + 5(2x - 3)\\ &= (3x + 5)(2x - 3)\end{aligned}$$

35. Compute ac for $6x^2 + 20x + 5$: $ac = 6 \cdot 5 = 30$
We need two integers, p and q, whose product is 30, and whose sum is $20 = b$.

pq = 30
$1 \cdot 30$
$2 \cdot 15$
$3 \cdot 10$
$5 \cdot 6$

None of these add up to 20; so we conclude that $6x^2 + 20x + 5$ is not factorable.

37. Compute ac for $8x^2 + 2x - 3$: $ac = 8(-3) = -24$
We need two integers, p and q, whose product is -24, and whose sum is $2 = b$.

pq = -24	
1(-24)	(-1)24
2(-12)	(-2)12
3(-8)	(-3)8
4(-6)	(-4)6

-4 and 6 will work, so we split the middle term $2x = -4x + 6x$ and factor by grouping:

$$\begin{aligned}8x^2 + 2x - 3 &= 8x^2 - 4x + 6x - 3\\ &= (8x^2 - 4x) + (6x - 3)\\ &= 4x(2x - 1) + 3(2x - 1)\\ &= (4x + 3)(2x - 1)\end{aligned}$$

39. Compute ac for $4x^2 - 7x + 10$: $ac = 4 \cdot 10 = 40$
We need two integers, p and q, whose product is 40, and whose sum is -7.

pq = 40
(-1)(-40)
(-2)(-20)
(-4)(-10)
(-5)(-8)

None of these add up to -7; so we conclude that $4x^2 - 7x + 10$ is not factorable.

41. Compute ac for $4x^2 + 7x - 10$: $ac = 4(-10) = -40$
We need two integers, p and q, whose product is -40, and whose sum is 7.

pq = -40	
1(-40)	(-1)40
2(-20)	(-2)20
4(-10)	(-4)10
5(-8)	(-5)8

None of these add up to 7; so we conclude that $4x^2 + 7x - 10$ is not factorable.

43. Compute ac for $8x^2 - 34x - 9$: $ac = 8(-9) = -72$
We need two integers, p and q, whose product is -72, and whose sum is -34.

pq = -72	
1(-72)	(-1)72
2(-36)	(-2)36
3(-24)	(-3)24
4(-18)	(-4)18
6(-12)	(-6)12
8(-9)	(-8)9

2 and -36 will work, so we split the middle term $-34x = -36x + 2x$ and factor by grouping:

$$8x^2 - 34x - 9 = 8x^2 - 36x + 2x - 9$$
$$= (8x^2 - 36x) + (2x - 9)$$
$$= 4x(2x - 9) + 1(2x - 9)$$
$$= (2x - 9)(4x + 1)$$

45. Compute ac for $4x^2 - 13xy + 10y^2$: $ac = 4 \cdot 10 = 40$
We need two integers, p and q, whose product is 40, and whose sum is $-13 = b$.
In this case, both p and q will be negative.

pq = -13
(-1)(-40)
(-2)(-20)
(-4)(-10)
(-5)(-8)

-5 and -8 will work, so we split the middle term $-13x = -5x - 8x$ and factor by grouping:

$$4x^2 - 13xy + 10y^2 = 4x^2 - 5xy - 8xy + 10y^2$$
$$= (4x^2 - 5xy) - (8xy - 10y^2)$$
$$= x(4x - 5y) - 2y(4x - 5y)$$
$$= (x - 2y)(4x - 5y)$$

47. Compute ac for $4x^2 - 9xy - 5y^2$: $ac = 4(-5) = -20$
We need two integers, p and q, whose product is -20, and whose sum is $-9 = b$.

pq = -20
(-1)(20)
(-2)(10)
(-4)(5)

None of these add up to -9; we conclude that $4x^2 - 9xy - 5y^2$ is not factorable.

49. Factor out the common factor of 2: $6x^2 + 14xy - 12y^2 = 2(3x^2 + 7xy - 6y^2)$.
Now compute ac for $3x^2 + 7xy - 6y^2$: $ac = 3(-6) = -18$.
We need two integers, p and q, whose product is -18 and whose sum is $7 = b$.

pq = -18	
(-1)18	1(-18)
(-2) 9	2(-9)
(-3) 6	3(-6)

-2 and 9 will work, so we split the middle term $7xy = -2xy + 9xy$ and factor by grouping:

$$3x^2 + 7xy - 6y^2 = 3x^2 - 2xy + 9xy - 6y^2$$
$$= x(3x - 2y) + 3y(3x - 2y)$$
$$= (x + 3y)(3x - 2y)$$
$$6x^2 + 14xy - 12y^2 = 2(x + 3y)(3x - 2y)$$

COMMON ERROR: Forgetting the 2 in the final answer.

51. Compute ac for $8x^2 + 6x - 9$: $ac = 8(-9) = -72$
We need two integers, p and q, whose product is -72, and whose sum is $6 = b$.

pq = -72	
1(-72)	(-1)72
2(-36)	(-2)36
3(-24)	(-3)24
4(-18)	(-4)18
6(-12)	(-6)12
8(-9)	(-8)9

-6 and 12 will work, so we split the middle term $6x = -6x + 12x$ and factor by grouping:

$$8x^2 + 6x - 9 = 8x^2 - 6x + 12x - 9 = (8x^2 - 6x) + (12x - 9)$$
$$= 2x(4x - 3) + 3(4x - 3) = (4x - 3)(2x + 3)$$

53. Compute ac for $6x^2 - 31x - 30$: $ac = 6(-30) = -180$
We need two integers, p and q, whose product is -180, and whose sum is $-31 = b$.

pq = -180	
1(-180)	(-1)180
2(-90)	(-2)90
3(-60)	(-3)60
4(-45)	(-4)45
5(-36)	(-5)36
6(-30)	(-6)30
9(-20)	(-9)20
10(-18)	(-10)18
12(-15)	(-12)15

5 and -36 will work, so we split the middle term $-31x = 5x - 36x$ and factor by grouping:

$$6x^2 - 31x - 30 = 6x^2 + 5x - 36x - 30$$
$$= (6x^2 + 5x) - (36x + 30)$$
$$= x(6x + 5) - 6(6x + 5) = (6x + 5)(x - 6)$$

55. Compute ac for $24x^2 + 38x + 15$: $ac = 24 \cdot 15 = 360$.
We need two integers, p and q, whose product is 360, and whose sum is $38 = b$.

pq = 360	
1·360	18·20
2·180	15·24
3·120	12·30
4·90	10·36
5·72	9·40
6·60	8·45

18 and 20 will work, so we split the middle term $38x = 18x + 20x$ and factor by grouping:

$$24x^2 + 38x + 15 = 24x^2 + 18x + 20x + 15$$
$$= (24x^2 + 18x) + (20x + 15)$$
$$= 6x(4x + 3) + 5(4x + 3) = (4x + 3)(6x + 5)$$

57. Compute ac for $24x^2 + 2x - 15$: $ac = 24(-15) = -360$.
We need two integers, p and q, whose product is -360, and whose sum is $2 = b$.

pq = -360	
1(-360)	(-1)360
2(-180)	(-2)180
3(-120)	(-3)120
4(-90)	(-4)90
5(-72)	(-5)72
6(-60)	(-6)60
8(-45)	(-8)45
9(-40)	(-9)40
10(-36)	(-10)36
12(-30)	(-12)30
15(-24)	(-15)24
18(-20)	(-18)20

-18 and 20 will work, so we split the middle term $2x = -18x + 20x$ and factor by grouping:

$$24x^2 + 2x - 15 = 24x^2 - 18x + 20x - 15$$
$$= (24x^2 - 18x) + (20x - 15)$$
$$= 6x(4x - 3) + 5(4x - 3) = (4x - 3)(6x + 5)$$

59. Compute ac for $24x^2 + 106x - 9$: $ac = 24(-9) = -216$.
We need two integers, p and q, whose product is -216, and whose sum is $106 = b$.

pq = -216	
1(-216)	(-1)216
2(-108)	(-2)108
3(-72)	(-3)72
4(-54)	(-4)54
6(-36)	(-6)36
8(-27)	(-8)27
9(-24)	(-9)24
12(-18)	(-12)18

-2 and 108 will work, so we split the middle term $106x = -2x + 108x$ and factor by grouping:

$$24x^2 + 106x - 9 = 24x^2 - 2x + 108x - 9$$
$$= (24x^2 - 2x) + (108x - 9)$$
$$= 2x(12x - 1) + 9(12x - 1)$$
$$= (12x - 1)(2x + 9)$$

61. Factor out the common factor of 3: $24x^2 + 30x - 9 = 3(8x^2 + 10x - 3)$. Now compute ac for $8x^2 + 10x - 3$: $ac = 8(-3) = -24$.
We need two integers, p and q, whose product is -24 and whose sum is 10 = b.

pq = -24	
1(-24)	(-1)24
2(-12)	(-2)12
3(-8)	(-3)8
4(-6)	(-4)6

-2 and 12 will work, so we split the middle term $10x = -2x + 12x$ and factor by grouping:

$$8x^2 + 10x - 3 = 8x^2 - 2x + 12x - 3 = (8x^2 - 2x) + (12x - 3)$$
$$= 2x(4x - 1) + 3(4x - 1) = (4x - 1)(2x + 3)$$
$$24x^2 + 30x - 9 = 3(4x - 1)(2x + 3)$$

63. Compute ac for $12x^2 + 19xy + 5y^2$: $ac = 12 \cdot 5 = 60$
We need two integers, p and q, whose product is 60, and whose sum is 19 = b.

pq = 60	
1·60	4·15
2·30	5·12
3·20	6·10

4 and 15 will work, so we split the middle term $19xy = 4xy + 15xy$ and factor by grouping:

$$12x^2 + 19xy + 5y^2 = 12x^2 + 4xy + 15xy + 5y^2$$
$$= 4x(3x + y) + 5y(3x + y)$$
$$= (4x + 5y)(3x + y)$$

65. Compute ac for $6x^2 - 20xy - 25y^2$: $ac = 6(-25) = -150$
We need two integers, p and q, whose product is -150, and whose sum is -20 = b.

pq = -150	
1(-150)	(-1)150
2(-75)	(-2) 75
3(-50)	(-3) 50
5(-30)	(-5) 30
6(-25)	(-6) 25
10(-15)	(-10) 15

None of these add up to -20; we conclude that $6x^2 - 20xy - 25y^2$ is not factorable.

67. Compute ac for $4x^2 - 17xy - 15y^2$: $ac = 4(-15) = -60$
We need two integers, p and q, whose product is -60, and whose sum is -17 = b.

pq = -60	
1(-60)	(-1)60
2(-30)	(-2)30
3(-20)	(-3)20
4(-15)	(-4)15
5(-12)	(-5)12
6(-10)	(-6)10

3 and -20 will work, so we split the middle term $-17xy = -20xy + 3xy$ and factor by grouping:

$$4x^2 - 17xy - 15y^2 = 4x^2 - 20xy + 3xy - 15y^2$$
$$= 4x(x - 5y) + 3y(x - 5y)$$
$$= (4x + 3y)(x - 5y)$$

69. Compute ac for $6x^2 + 28x - 5$: $ac = 6(-5) = -30$
We need two integers, p and q, whose product is -30, and whose sum is 28 = b.

pq = -30	
1(-30)	(-1)30
2(-15)	(-2)15
3(-10)	(-3)10
5(-6)	(-5) 6

None of these add up to 28; we conclude that $6x^2 + 28x - 5$ is not factorable.

71. Compute ac for $18x^2 + 37x - 20$: $ac = 18(-20) = -360$
We need two integers, p and q, whose product is -360, and whose sum is 37 = b.
Using the table in problem 57, we find that -8 and 45 will work, so we split the middle term $37x = -8x + 45x$ and factor by grouping:

$$18x^2 + 37x - 20 = 18x^2 - 8x + 45x - 20 = 2x(9x - 4) + 5(9x - 4)$$
$$= (2x + 5)(9x - 4)$$

73. Compute ac for $18x^2 + 9xy - 20y^2$: $ac = 18(-20) = -360$
We need two integers, p and q, whose product is -360, and whose sum is $9 = b$. Using the table in problem 57, we find that -15 and 24 will work, so we split the middle term $9xy = -15xy + 24xy$ and factor by grouping:

$$\begin{aligned} 18x^2 + 9xy - 20y^2 &= 18x^2 - 15xy + 24xy - 20y^2 \\ &= 3x(6x - 5y) + 4y(6x - 5y) \\ &= (3x + 4y)(6x - 5y) \end{aligned}$$

75. For $x^2 + bx - 18$ to be factorable, $ac = (1)(-18) = -18$ must have two factors that add up to b. If we list all possible pairs of factors of -18, the sum of each pair will give us a possible value for b, and there are no others.

$$\begin{aligned} -18 &= 1(-18) & 1 + (-18) &= -17 \\ &= 2(-9) & 2 + (-9) &= -7 \\ &= 3(-6) & 3 + (-6) &= -3 \\ &= 6(-3) & 6 + (-3) &= 3 \\ &= 9(-2) & 9 + (-2) &= 7 \\ &= 18(-1) & 18 + (-1) &= 17 \end{aligned}$$

So the possible values for b are -17, -7, -3, 3, 7, and 17.

77. For $x^2 + 5x + c$ to be factorable, $ac = 1c = c$ must have two factors that add up to $5 = b$. If we list all possible pairs of integers between 0 and 5 that add up to 5, the product of each pair will give us a possible value for c, and there are no others.

$$\begin{aligned} 0 + 5 &= 5 & 0 \cdot 5 &= 0 \\ 1 + 4 &= 5 & 1 \cdot 4 &= 4 \\ 2 + 3 &= 5 & 2 \cdot 3 &= 6 \end{aligned}$$

So the possible values for c are 0, 4, and 6.

79. Compute ac for $6x^4 + 7x^2 + 2$: $ac = 6 \cdot 2 = 12$
We need two integers, p and q, whose product is 12, and whose sum is $7 = b$.

pq = 12
1·12
2· 6
3· 4

3 and 4 will work, so we split the middle term $7x^2 = 3x^2 + 4x^2$ and factor by grouping:

$$\begin{aligned} 6x^4 + 7x^2 + 2 &= 6x^4 + 3x^2 + 4x^2 + 2 = (6x^4 + 3x^2) + (4x^2 + 2) \\ &= 3x^2(2x^2 + 1) + 2(2x^2 + 1) = (3x^2 + 2)(2x^2 + 1) \end{aligned}$$

81. Compute ac for $2x^4 + 5x^2y^2 + 3y^4$: $ac = 2 \cdot 3 = 6$
We need two integers, p and q, whose product is 6, and whose sum is $5 = b$.

pq = 6
1·6
2·3

2 and 3 will work, so we split the middle term $5x^2y^2 = 2x^2y^2 + 3x^2y^2$ and factor by grouping:

$$\begin{aligned} 2x^4 + 5x^2y^2 + 3y^4 &= 2x^4 + 2x^2y^2 + 3x^2y^2 + 3y^4 \\ &= (2x^4 + 2x^2y^2) + (3x^2y^2 + 3y^4) \\ &= 2x^2(x^2 + y^2) + 3y^2(x^2 + y^2) = (2x^2 + 3y^2)(x^2 + y^2) \end{aligned}$$

83. Compute ac for $2x^6 - x^3 - 3$: $ac = 2(-3) = -6$
We need two integers, p and q, whose product is -6, and whose sum is $-1 = b$.

pq = -6	
1(-6)	(-1)6
2(-3)	(-2)3

2 and -3 will work, so we split the middle term $-x^3 = 2x^3 - 3x^3$ and factor by grouping:

$$\begin{aligned} 2x^6 - x^3 - 3 &= 2x^6 + 2x^3 - 3x^3 - 3 = (2x^6 + 2x^3) - (3x^3 + 3) \\ &= 2x^3(x^3 + 1) - 3(x^3 + 1) \\ &= (2x^3 - 3)(x^3 + 1) \end{aligned}$$

, or using the methods of the next section,

$$(2x^3 - 3)(x + 1)(x^2 - x + 1)$$

85. Compute ac for $12x^4 + 25x^2 + 12$: ac = $12 \cdot 12 = 144$
We need two integers, p and q, whose product is 144, and whose sum is 25 = b.

pq = 144
1·144
2· 72
3· 48
4· 36
6· 24
8· 18
9· 16
12· 12

9 and 16 will work, so we split the middle term $25x^2 = 9x^2 + 16x^2$ and factor by grouping:
$12x^4 + 25x^2 + 12 = 12x^4 + 9x^2 + 16x^2 + 12 = (12x^4 + 9x^2) + (16x^2 + 12)$
$= 3x^2(4x^2 + 3) + 4(4x^2 + 3) = (4x^2 + 3)(3x^2 + 4)$

87. Factor out the common factor of 2: $12x^4 + 70x^2 - 12 = 2(6x^4 + 35x^2 - 6)$. Now compute ac for $6x^4 + 35x^2 - 6$: ac = $6(-6) = -36$.
We need two integers, p and q, whose product is -36 and whose sum is 35 = b.

pq = -36	
1(-36)	(-1)36
2(-18)	(-2)18
3(-12)	(-3)12
4(-9)	(-4)9
6(-6)	(-6)6

-1 and 36 will work, we we split the middle term $35x^2 = -1x^2 + 36x^2$ and factor by grouping:
$6x^4 + 35x^2 - 6 = 6x^4 - x^2 + 36x^2 - 6 = (6x^4 - x^2) + (36x^2 - 6)$
$= x^2(6x^2 - 1) + 6(6x^2 - 1) = (6x^2 - 1)(x^2 + 6)$
$12x^4 + 70x^2 - 12 = 2(6x^2 - 1)(x^2 + 6)$

89. Factor out the common factor of 3: $24x^6 - 69x^3 - 9 = 3(8x^6 - 23x^3 - 3)$. Now compute ac for $8x^6 - 23x^3 - 3$: ac = $8(-3) = -24$.
We need two integers, p and q, whose product is -24 and whose sum is -23 = b.

pq = -24	
1(-24)	(-1)24
2(-12)	(-2)12
3(-8)	(-3)8
4(-6)	(-4)6

1 and -24 will work, so we split the middle term $23x^3 = 1x^3 - 24x^3$ and factor by grouping:
$8x^6 - 23x^3 - 3 = 8x^6 + x^3 - 24x^3 - 3 = (8x^6 + x^3) - (24x^3 + 3)$
$= x^3(8x^3 + 1) - 3(8x^3 + 1) = (8x^3 + 1)(x^3 - 3)$
$24x^6 - 69x^3 - 9 = 3(8x^3 + 1)(x^3 - 3)$
or, using the methods of the next section, $3(2x + 1)(4x^2 - 2x + 1)(x^3 - 3)$

Exercise 2-6 More Factoring

Key Ideas and Formulas

The sum and difference of two cubes are factored

$$A^3 + B^3 = (A + B)(A^2 - AB + B^2)$$
$$A^3 - B^3 = (A - B)(A^2 + AB + B^2)$$

COMMON ERRORS:
Confusing $A^2 - AB + B^2$ with $A^2 - 2AB + B^2$
Confusing $A^2 + AB + B^2$ with $A^2 + 2AB + B^2$
These factoring forms should be memorized correctly.

A general strategy for factoring is

1. Remove common factors
2. If the polynomial has two terms, look for a difference of two squares or a sum of difference of two cubes.

3. If the polynomial has three terms,
   (A) See if it is a perfect square
   (B) Try trial and error
   (C) Or use the ac test
4. If a polynomial has more than three terms, try grouping.
5. Check that each factor has been completely factored.

---

1. First, remove common factors: $5v^2 - 125 = 5(v^2 - 25)$. Now, use difference of two squares: $5(v^2 - 25) = 5(v^2 - 5^2) = 5(v - 5)(v + 5)$.

3. Use difference of two cubes: $v^3 - 125 = v^3 - 5^3 = (v - 5)(v^2 + v\cdot 5 + 5^2)$
$= (v - 5)(v^2 + 5v + 25)$

5. First, remove common factors: $84m^2 - 21 = 21(4m^2 - 1)$. Now, use difference of two squares: $21(4m^2 - 1) = 21[(2m)^2 - 1^2] = 21(2m - 1)(2m + 1)$

7. Use sum of two cubes: $y^3 + 64 = (y + 4)(y^2 - y\cdot 4 + 4^2) = (y + 4)(y^2 - 4y + 16)$

9. Use sum of two cubes: $x^3 + 1 = x^3 + 1^3 = (x + 1)(x^2 - x + 1)$

11. Use difference of two cubes: $m^3 - n^3 = (m - n)(m^2 + mn + n^2)$

13. Use sum of two cubes: $8x^3 + 27 = (2x)^3 + (3)^3$
$= (2x + 3)[(2x)^2 - (2x)(3) + (3)^2]$
$= (2x + 3)(4x^2 - 6x + 9)$

15. First, remove common factors: $6u^2v^2 - 3uv^3 = 3uv^2(2u - v)$. In this case, there are no further steps.

17. First, remove common factors: $2x^2 - 8 = 2(x^2 - 4)$. Now, use difference of two squares: $2(x^2 - 4) = 2[x^2 - (2)^2] = 2(x - 2)(x + 2)$.

19. First, remove common factors: $2x^3 + 8x = 2x(x^2 + 4)$. In this case, there are no further steps (sum of two squares).

21. First, remove common factors: $12x^3 - 3xy^2 = 3x(4x^2 - y^2)$. Now, use difference of two squares: $3x(4x^2 - y^2) = 3x[(2x)^2 - y^2] = 3x(2x - y)(2x + y)$

23. First, remove common factors: $2x^4 + 2x = 2x(x^3 + 1)$. Now, use sum of two cubes: $2x(x^3 + 1) = 2x(x + 1)(x^2 - x + 1)$

25. First, remove common factors: $6x^2 + 36x + 48 = 6(x^2 + 6x + 8)$. Now, use trinomial factoring: $6(x^2 + 6x + 8) = 6(x + 2)(x + 4)$.

27. First, remove common factors: $3x^3 - 6x^2 + 15x = 3x(x^2 - 2x + 5)$. In this case, there are no further steps (trinomial not factorable).

29. First, use difference of two squares: $x^4 - 9 = (x^2)^2 - 3^2 = (x^2 - 3)(x^2 + 3)$. In this case, there are no further steps.

31. Use sum of two cubes:
$x^6 + 27y^6 = (x^2)^3 + (3y^2)^3 = (x^2 + 3y^2)[(x^2)^2 - x^2\cdot 3y^2 + (3y^2)^2]$
$= (x^2 + 3y^2)(x^4 - 3x^2y^2 + 9y^4)$

33. $x^4 + 4x^2 + 4$ is a perfect square: $(x^2)^2 + 2(x^2)(2) + 2^2 = (x^2 + 2)^2$

35. $x^6 + 6x^3 + 9$ is a perfect square: $(x^3)^2 + 2(x^3)(3) + 3^2 = (x^3 + 3)^2$

37. Use difference of two squares: $x^2y^2 - 16 = (xy)^2 - (4)^2 = (xy - 4)(xy + 4)$

39. Use sum of two cubes: $a^3b^3 + 8 = (ab)^3 + (2)^3$
$= (ab + 2)[(ab)^2 - (ab)(2) + (2)^2]$
$= (ab + 2)(a^2b^2 - 2ab + 4)$

41. First, remove common factors: $4x^3y + 14x^2y^2 + 6xy^3 = 2xy(2x^2 + 7xy + 3y^2)$. Now, use trinomial factoring methods: $2xy(2x^2 + 7xy + 3y^2) = 2xy(2x + y)(x + 3y)$

43. First, remove common factors: $4u^3 + 32v^3 = 4(u^3 + 8v^3)$. Now, use the sum of two cubes: $4(u^3 + 8v^3) = 4[u^3 + (2v)^3] = 4(u + 2v)[(u)^2 - (u)(2v) + (2v)^2]$
$= 4(u + 2v)(u^2 - 2uv + 4v^2)$

45. First, remove common factors: $60x^2y^2 - 200xy^3 - 35y^4 = 5y^2(12x^2 - 40xy - 7y^2)$
Now, use trinomial factoring methods:
$5y^2(12x^2 - 40xy - 7y^2) = 5y^2(2x - 7y)(6x + y)$

47. Use factoring by grouping: $xy + 2x + y^2 + 2y = (xy + 2x) + (y^2 + 2y)$
$= x(y + 2) + y(y + 2)$
$= (y + 2)(x + y)$

49. Use factoring by grouping: $x^2 - 5x + xy - 5y = (x^2 - 5x) + (xy - 5y)$
$= x(x - 5) + y(x - 5)$
$= (x - 5)(x + y)$

51. Use factoring by grouping: $ax - 2bx - ay + 2by = (ax - 2bx) - (ay - 2by)$
$= x(a - 2b) - y(a - 2b)$
$= (a - 2b)(x - y)$

53. Use factoring by grouping:
$15ac - 20ad + 3bc - 4bd = (15ac - 20ad) + (3bc - 4bd)$
$= 5a(3c - 4d) + b(3c - 4d)$
$= (3c - 4d)(5a + b)$

55. Use factoring by grouping: $x^3 - 2x^2 - x + 2 = (x^3 - 2x^2) - (x - 2)$
$= x^2(x - 2) - 1(x - 2)$
$= (x - 2)(x^2 - 1)$
Now, use difference of two squares $= (x - 2)(x - 1)(x + 1)$

57. Use factoring by grouping: $(y - x)^2 - y + x = (y - x)^2 - (y - x)$
$= (y - x)(y - x) - 1(y - x)$
$= (y - x)(y - x - 1)$

59. Use trinomial factoring methods: $x^2y^2 - xy - 6 = (xy)^2 - (xy) - 6$
$= (xy - 3)(xy + 2)$

61. Use trinomial factoring methods: $z^4 - z^2 - 6 = (z^2)^2 - (z^2) - 6$
$= (z^2 - 3)(z^2 + 2)$

63. Use factoring by grouping:
$x(x + 1)^2 - x^2(x + 1) = x(x + 1)\cdot(x + 1) - x(x + 1)\cdot x$
$= x(x + 1)\cdot[(x + 1) - x]$
$= x(x + 1)\cdot[1]$
$= x(x + 1)$

65. Use factoring by grouping:
$x^3(x - 1) - x(x - 1)^3 = x(x - 1)\cdot x^2 - x(x - 1)\cdot(x - 1)^2$
$= x(x - 1)\cdot[x^2 - (x - 1)^2]$
$= x(x - 1)\cdot[x^2 - x^2 + 2x - 1]$
$= x(x - 1)(2x - 1)$

67. Use factoring by grouping: $(x + 4)^2(x - 1)^4 - (x + 4)^3(x - 1)^3$
$= (x + 4)^2(x - 1)^3\cdot(x - 1) - (x + 4)^2(x - 1)^3\cdot(x + 4)$
$= (x + 4)^2(x - 1)^3\cdot[(x - 1) - (x + 4)] = (x + 4)^2(x - 1)^3[x - 1 - x - 4]$
$= -5(x + 4)^2(x - 1)^3$

69. Use factoring by grouping: $(x + 2)^4(x - 4)^4 - (x + 2)^3(x - 4)^5$
$= (x + 2)^3(x - 4)^4\cdot(x + 2) - (x + 2)^3(x - 4)^4\cdot(x - 4)$
$= (x + 2)^3(x - 4)^4\cdot[(x + 2) - (x - 4)] = (x + 2)^3(x - 4)^4[x + 2 - x + 4]$
$= 6(x + 2)^3(x - 4)^4$

71. Use difference of two squares: $x^8 - 4 = (x^4)^2 - (2)^2 = (x^4 - 2)(x^4 + 2)$

73. First, use difference of two squares:
$r^4 - s^4 = (r^2)^2 - (s^2)^2 = (r^2 - s^2)(r^2 + s^2)$
Now, use the same method to factor $r^2 - s^2$:
$(r^2 - s^2)(r^2 + s^2) = (r - s)(r + s)(r^2 + s^2)$

75. First, use trinomial factoring methods:
$x^4 - 3x^2 - 4 = (x^2)^2 - 3(x^2) - 4 = (x^2 - 4)(x^2 + 1)$
Now, use difference of two squares:
$(x^2 - 4)(x^2 + 1) = (x - 2)(x + 2)(x^2 + 1)$

77. Use difference of two squares: $(x - 3)^2 - 16y^2 = (x - 3)^2 - (4y)^2$
$= [(x - 3) - 4y][(x - 3) + 4y]$
$= (x - 3 - 4y)(x - 3 + 4y)$

79. Use difference of two squares:
$(a - b)^2 - 4(c - d)^2 = (a - b)^2 - [2(c - d)]^2$
$= [(a - b) - 2(c - d)][(a - b) + 2(c - d)]$

81. Use difference of two squares:
$25(4x^2 - 12xy + 9y^2) - 9a^2b^2 = [5(2x - 3y)]^2 - (3ab)^2$
$= [5(2x - 3y) - 3ab][5(2x - 3y) + 3ab]$

83. First, use difference of two squares:
$x^6 - 1 = (x^3)^2 - (1)^2 = (x^3 - 1)(x^3 + 1)$
Now, use sum of two cubes and difference of two cubes:
$(x^3 - 1)(x^3 + 1) = (x - 1)(x^2 + x + 1)(x + 1)(x^2 - x + 1)$

85. First, use factoring by grouping: $2x^3 - x^2 - 8x + 4 = (2x^3 - x^2) - (8x - 4)$
$= x^2(2x - 1) - 4(2x - 1)$
$= (2x - 1)(x^2 - 4)$
Now, use difference of two squares:
$(2x - 1)(x^2 - 4) = (2x - 1)(x - 2)(x + 2)$

87. First, group terms: $25 - a^2 - 2ab - b^2 = 25 - (a^2 + 2ab + b^2)$
Now, use difference of two squares:
$25 - (a^2 + 2ab + b^2) = (5)^2 - (a + b)^2 = [5 - (a + b)][5 + (a + b)]$

89. First, use difference of two squares: $x^4 - 1 = (x^2)^2 - 1^2 = (x^2 + 1)(x^2 - 1)$
Now, use difference of two squares again:
$(x^2 + 1)(x^2 - 1) = (x^2 + 1)(x - 1)(x + 1)$

91. First, use difference of two squares:
$x^6 - y^6 = (x^3)^2 - (y^3)^2 = (x^3 - y^3)(x^3 + y^3)$
Now, use sum of two cubes and difference of two cubes:
$(x^3 - y^3)(x^3 + y^3) = (x - y)(x^2 + xy + y^2)(x + y)(x^2 - xy + y^2)$

93. First, use difference of two squares: $x^6 - 64 = (x^3)^2 - 8^2 = (x^3 - 8)(x^3 + 8)$
Now, use sum of two cubes and difference of two cubes:
$(x^3 - 8)(x^3 + 8) = (x^3 - 2^3)(x^3 + 2^3)$
$= (x - 2)(x^2 + x \cdot 2 + 2^2)(x + 2)(x^2 - x \cdot 2 + 2^2)$
$= (x - 2)(x^2 + 2x + 4)(x + 2)(x^2 - 2x + 4)$

95. Use difference of two squares three times:
$x^8 - y^8 = (x^4)^2 - (y^4)^2 = (x^4 + y^4)(x^4 - y^4) = (x^4 + y^4)[(x^2)^2 - (y^2)^2]$
$= (x^4 + y^4)(x^2 + y^2)(x^2 - y^2)$
$= (x^4 + y^4)(x^2 + y^2)(x + y)(x - y)$

97. First, group terms: $16x^4 - x^2 + 6xy - 9y^2 = 16x^4 - (x^2 - 6xy + 9y^2)$
Now, use difference of two squares:
$16x^4 - (x^2 - 6xy + 9y^2) = (4x^2)^2 - (x - 3y)^2 = [4x^2 - (x - 3y)][4x^2 + (x - 3y)]$

99. Use factoring by grouping: $x^3 - 2x^2 + 3x - 6 = (x^3 - 2x^2) + (3x - 6)$
$= x^2(x - 2) + 3(x - 2)$
$= (x - 2)(x^2 + 3)$

101. Use factoring by grouping: $x^5 - x^4 + x - 1 = (x^5 - x^4) + (x - 1)$
$= x^4(x - 1) + 1(x - 1)$
$= (x - 1)(x^4 + 1)$

103. Use factoring by grouping: $3x^3 - x^2 + 12x - 4 = (3x^3 - x^2) + (12x - 4)$
$= x^2(3x - 1) + 4(3x - 1)$
$= (3x - 1)(x^2 + 4)$

105. First rearrange terms, then group terms and use difference of two squares:
$x^2 + 4x - y^2 + 4 = x^2 + 4x + 4 - y^2$
$= (x^2 + 4x + 4) - y^2$
$= (x + 2)^2 - y^2$
$= [(x + 2) - y][(x + 2) + y]$
$= (x + 2 - y)(x + 2 + y)$

107. To factor $a^2 - ab + b^2 = 1a^2 - 1ab + 1b^2$, we need two integers, p and q, whose product is $1 \cdot 1 = 1$, and whose sum is -1. Since the only integer factors of 1 are 1 and 1, whose sum is 2, and -1 and -1, whose sum is -2, we conclude that $a^2 - ab + b^2$ is not factorable using integer coefficients.

Exercise 2-7 Solving Equations by Factoring

Key Ideas and Formulas

Zero Factor Property: For real numbers a and b, $a \cdot b = 0$ if and only if $a = 0$ or $b = 0$ (or both).

We apply this property to solving equations that are in the form $a \cdot b = 0$ by solving the simpler equations $a = 0$ or $b = 0$.

---

1. $x^2 + 3x + 2 = 0$
$(x + 1)(x + 2) = 0$
$x + 1 = 0$ or $x + 2 = 0$
$x = -1$ $\quad$ $x = -2$

3. $x^2 + 7x + 10 = 0$
$(x + 2)(x + 5) = 0$
$x + 2 = 0$ or $x + 5 = 0$
$x = -2$ $\quad$ $x = -5$

5. $x^2 - 2x - 3 = 0$
$(x - 3)(x + 1) = 0$
$x - 3 = 0$ or $x + 1 = 0$
$x = 3$ $\quad$ $x = -1$

7. $x^2 + 3x - 4 = 0$
$(x + 4)(x - 1) = 0$
$x + 4 = 0$ or $x - 1 = 0$
$x = -4$ $\quad$ $x = 1$

9. $x^2 - 9x + 20 = 0$
$(x - 5)(x - 4) = 0$
$x - 5 = 0$ or $x - 4 = 0$
$x = 5$ $\quad$ $x = 4$

11. $x^2 - 6x + 5 = 0$
$(x - 5)(x - 1) = 0$
$x - 5 = 0$ or $x - 1 = 0$
$x = 5$ $\quad$ $x = 1$

13. $x^2 - 6x + 9 = 0$
$(x - 3)^2 = 0$
$x - 3 = 0$
$x = 3$

15. $x^2 + 14x + 49 = 0$
$(x + 7)^2 = 0$
$x + 7 = 0$
$x = -7$

17. $2x^2 - x - 1 = 0$
$(2x + 1)(x - 1) = 0$
$2x + 1 = 0$ or $x - 1 = 0$
$2x = -1$ $\quad$ $x = 1$
$x = -\frac{1}{2}$

19. $2x^2 + 7x + 5 = 0$
$(2x + 5)(x + 1) = 0$
$2x + 5 = 0$ or $x + 1 = 0$
$2x = -5$ $\quad x = -1$
$x = -\frac{5}{2}$

21. $u^2 + 5u = 0$
$u(u + 5) = 0$
$u = 0$ or $u + 5 = 0$
$u = 0$ $\quad u = -5$

23. $3A^2 = -12A$
$A^2 = -4A$
$A^2 + 4A = 0$
$A(A + 4) = 0$
$A = 0$ or $A + 4 = 0$
$A = 0$ $\quad A = -4$

25. $x^2 - 11x - 12 = 0$
$(x + 1)(x - 12) = 0$
$x + 1 = 0$ or $x - 12 = 0$
$x = -1$ $\quad x = 12$

27. $x^2 + 4x - 5 = 0$
$(x - 1)(x + 5) = 0$
$x - 1 = 0$ or $x + 5 = 0$
$x = 1$ $\quad x = -5$

29. $3Q^2 - 10Q - 8 = 0$
$(3Q + 2)(Q - 4) = 0$
$3Q + 2 = 0$ or $Q - 4 = 0$
$3Q = -2$ $\quad Q = 4$
$Q = -\frac{2}{3}$

31. $4x^2 + 4x - 3 = 0$
$(2x - 1)(2x + 3) = 0$
$2x - 1 = 0$ or $2x + 3 = 0$
$2x = 1$ $\quad 2x = -3$
$x = \frac{1}{2}$ $\quad x = -\frac{3}{2}$

33. $4x^2 + 3x - 1 = 0$
$(4x - 1)(x + 1) = 0$
$4x - 1 = 0$ or $x + 1 = 0$
$4x = 1$ $\quad x = -1$
$x = \frac{1}{4}$

35. $2x^2 + 4x + 3 = 0$
This is not factorable in the integers, so the factoring method fails.

37. $4x^2 + 12x + 9 = 0$
$(2x + 3)^2 = 0$
$2x + 3 = 0$
$2x = -3$
$x = -\frac{3}{2}$

39. $x^2 - x - 2 = 0$
$(x - 2)(x + 1) = 0$
$x - 2 = 0$ or $x + 1 = 0$
$x = 2$ $\quad x = -1$

41. $9x^2 + 12x + 4 = 0$
$(3x + 2)^2 = 0$
$3x + 2 = 0$
$3x = -2$
$x = -\frac{2}{3}$

43. $6x^2 - 7x + 2 = 0$
$(3x - 2)(2x - 1) = 0$
$3x - 2 = 0$ or $2x - 1 = 0$
$3x = 2$ $\quad 2x = 1$
$x = \frac{2}{3}$ $\quad x = \frac{1}{2}$

45. $u^2 = 2u + 3$
$u^2 - 2u - 3 = 0$
$(u - 3)(u + 1) = 0$
$u - 3 = 0$ or $u + 1 = 0$
$u = 3$ $\quad u = -1$

47. $3x^2 = x + 2$
$3x^2 - x - 2 = 0$
$(3x + 2)(x - 1) = 0$
$3x + 2 = 0$ or $x - 1 = 0$
$3x = -2$ $\quad x = 1$
$x = -\frac{2}{3}$

49. $y^2 = 5y - 2$
$y^2 - 5y + 2 = 0$
This is not factorable in the integers, so the factoring method fails.

51.
$$2x(x - 1) = 3(x + 1)$$
$$2x^2 - 2x = 3x + 3$$
$$2x^2 - 5x - 3 = 0$$
$$(2x + 1)(x - 3) = 0$$
$2x + 1 = 0$ $\quad$ $x - 3 = 0$
$2x = -1$ $\quad$ $x = 3$
$x = -\frac{1}{2}$

53.
$$t^2 = 4$$
$$t^2 - 4 = 0$$
$$(t - 2)(t + 2) = 0$$
$t - 2 = 0$ or $t + 2 = 0$
$t = 2$ $\quad$ $t = -2$

55.
$$m^2 + 4m = 12$$
$$m^2 + 4m - 12 = 0$$
$$(m + 6)(m - 2) = 0$$
$m + 6 = 0$ or $m - 2 = 0$
$m = -6$ $\quad$ $m = 2$

COMMON ERROR:

Writing $m(m + 4) = 12$
and concluding that $m = 12$ or $m + 4 = 12$
The zero factor property is only applicable if the right-hand side of the equation is 0.

57.
$$2y^2 = 2 + 3y$$
$$2y^2 - 3y - 2 = 0$$
$$(2y + 1)(y - 2) = 0$$
$2y + 1 = 0$ or $y - 2 = 0$
$2y = -1$ $\quad$ $y = 2$
$y = -\frac{1}{2}$

59.
$$2x^2 + 2 = 5x$$
$$2x^2 - 5x + 2 = 0$$
$$(2x - 1)(x - 2) = 0$$
$2x - 1 = 0$ or $x - 2 = 0$
$2x = 1$ $\quad$ $x = 2$
$x = \frac{1}{2}$

61.
$$x^2 + x = 6$$
$$x^2 + x - 6 = 0$$
$$(x + 3)(x - 2) = 0$$
$x + 3 = 0$ or $x - 2 = 0$
$x = -3$ $\quad$ $x = 2$

63. $8x^2 + 14x + 7 = 0$
This is not factorable in the integers, so the factoring method fails.

65.
$$8x^2 - 14x - 15 = 0$$
$$(4x + 3)(2x - 5) = 0$$
$4x + 3 = 0$ or $2x - 5 = 0$
$4x = -3$ $\quad$ $2x = 5$
$x = -\frac{3}{4}$ $\quad$ $x = \frac{5}{2}$

67.
$$x^4 - 81 = 0$$
$$(x^2 - 9)(2^2 + 9) = 0$$
$$(x - 3)(x + 3)(x^2 + 9) = 0$$
$x - 3 = 0$ or $x + 3 = 0$ or $x^2 + 9 = 0$
$x = 3$ $\quad$ $x = -3$ $\quad$ not factorable using integer coefficients
The only solutions we can find at this time are 3, -3.

69.
$$(x^3 - 1)(x^2 - 4) = 0$$
$$(x - 1)(x^2 + x + 1)(x - 2)(x + 2) = 0$$
$x - 1 = 0$ or $x^2 + x + 1 = 0$ or $x - 2 = 0$ or $x + 2 = 0$
$x = 1$ $\quad$ not factorable using integer coefficients $\quad$ $x = 2$ $\quad$ $x = -2$
The only solutions we can find at this time are 1, 2, -2.

71.
$$x^4 - 18x^2 + 81 = 0$$
$$(x^2 - 9)^2 = 0$$
$$x^2 - 9 = 0$$
$$(x - 3)(x + 3) = 0$$
$x - 3 = 0$ or $x + 3 = 0$
$x = 3$ $\quad$ $x = -3$

73.
$$x^6 + 7x^3 - 8 = 0$$
$$(x^3 - 1)(x^3 + 8) = 0$$
$$(x - 1)(x^2 + x + 1)(x + 2)(x^2 - 2x + 4) = 0$$
$x - 1 = 0$ or $x^2 + x + 1 = 0$ or $x + 2 = 0$ or $x^2 - 2x + 4 = 0$
$x = 1$ $\quad$ not factorable using integer coefficients $\quad$ $x = -2$ $\quad$ not factorable using integer coefficients
The only solutions we can find at this time are 1, -2.

75.

$$x^2 + 3x = 1 + 3x$$
$$x^2 = 1$$
$$x^2 - 1 = 0$$
$$(x - 1)(x + 1) = 0$$
$x - 1 = 0$ or $x + 1 = 0$

$x = 1$ $\quad$ $x = -1$

77.

$$(x + 1)(x + 2) - 6x = (2x - 1)(x - 1)$$
$$x^2 + 3x + 2 - 6x = 2x^2 - 3x + 1$$
$$x^2 - 3x + 2 = 2x^2 - 3x + 1$$
$$0 = x^2 - 1$$
$$x^2 - 1 = 0$$
$$(x - 1)(x + 1) = 0$$
$x - 1 = 0$ or $x + 1 = 0$

$x = 1$ $\quad$ $x = -1$

79. Let x = the integer

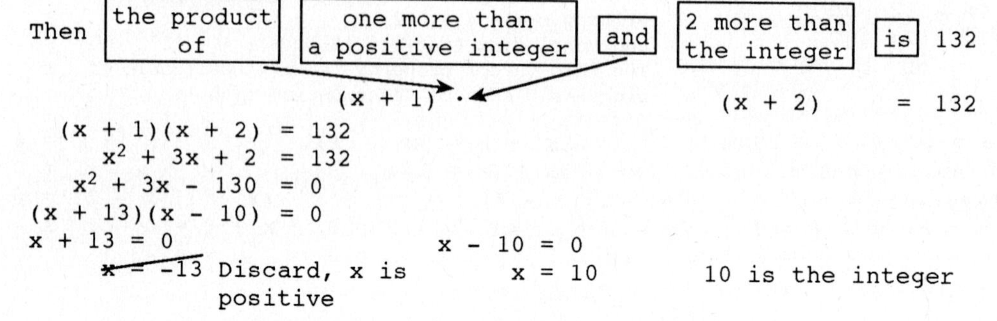

$$(x + 1)(x + 2) = 132$$
$$x^2 + 3x + 2 = 132$$
$$x^2 + 3x - 130 = 0$$
$$(x + 13)(x - 10) = 0$$
$x + 13 = 0$ $\quad$ $x - 10 = 0$

~~$x = -13$~~ Discard, x is positive $\quad$ $x = 10$ $\quad$ 10 is the integer

81. Let x = the first of these negative integers

Then $x + 1$ = the second of these negative integers

Since their product is 72, we have

$$x(x + 1) = 72$$
$$x^2 + x = 72$$
$$x^2 + x - 72 = 0$$
$$(x - 8)(x + 9) = 0$$
$x - 8 = 0$ $\quad$ $x + 9 = 0$

$x = 8$ Discard, x is negative $\quad$ $x = -9$, $x + 1 = -8$ $\quad$ The integers are -9, -8.

83. Let x = the number

Then [twice] [the square of the number] [is] [21 more than the number]

$2 \cdot x^2 = x + 21$

$$2x^2 = x + 21$$
$$2x^2 - x - 21 = 0$$
$$(2x - 7)(x + 3) = 0$$
$2x - 7 = 0$ or $x + 3 = 0$

$2x = 7$ $\quad$ $x = -3$

$x = \frac{7}{2}$ $\quad$ -3 or $\frac{7}{2}$

85. Let h = the height

Then $\frac{1}{2}h$ = the base

Since $A = \frac{1}{2}bh$, we have $36 = \frac{1}{2}\left(\frac{1}{2}h\right)h$

$$36 = \frac{1}{4}h^2$$
$$144 = h^2$$
$$0 = h^2 - 144$$
$$0 = (h + 12)(h - 12)$$
$h + 12 = 0$ or $h - 12 = 0$

~~$h = -12$~~ not possible $\quad$ $h = 12$ cm (height)

$\frac{1}{2}h = 6$ cm (base)

87.

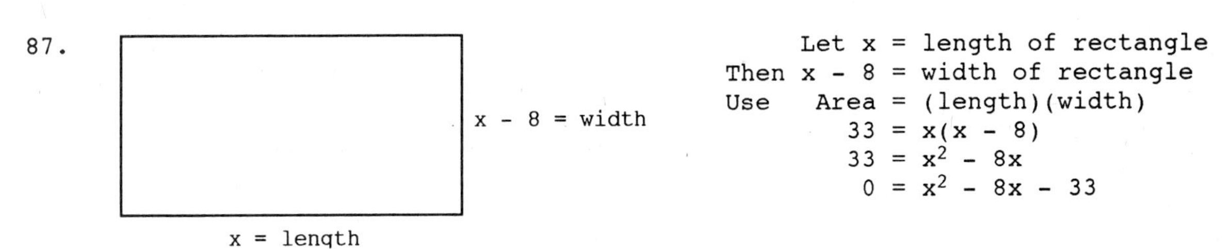

Let x = length of rectangle
Then $x - 8$ = width of rectangle
Use Area = (length)(width)

$$33 = x(x - 8)$$
$$33 = x^2 - 8x$$
$$0 = x^2 - 8x - 33$$

$$x^2 - 8x - 33 = 0$$
$$(x - 11)(x + 3) = 0$$
$$x - 11 = 0 \quad \text{or} \quad x + 3 = 0$$
$x = 11$ in (length) $\qquad$ ~~$x = -3$~~ Not possible
$x - 8 = 3$ in (width)

COMMON ERROR:

Confusing area with perimeter.
Area = (length)(width)
Perimeter = 2 × length + 2 × width
Perimeter is not relevant to this problem.

89. Let x = the length of rectangle
Then $21 - x$ = width of rectangle
Use Area = (length)(width)

$$108 = x(21 - x)$$
$$108 = 21x - x^2$$
$$x^2 - 21x + 108 = 0$$
$$(x - 9)(x - 12) = 0$$
$$x - 9 = 0 \quad \text{or} \quad x - 12 = 0$$
$$x = 9 \qquad\qquad x = 12$$
$$21 - x = 12 \qquad 21 - x = 9$$

The dimensions are 9 inches by 12 inches.

Exercise 2-8 REVIEW EXERCISE

1. 2

2. The terms have degrees $5 + 3 = 8$, $3 + 4 = 7$, $6 + 2 = 8$, $5 + 4 = 9$, so the smallest degree is 7.

3. (A) 5 (B) 3

4. (A) The degree of the first factor is $3 + 2 = 5$.
   (B) The degree of the product is $3 + 2 + 1 + 2 + 3 = 11$.

5. $(2x + 5) + (x^2 - 4) = 2x + 5 + x^2 - 4 = x^2 + 2x + 1$

6. $(2x + 5) - (x^2 - 4) = 2x + 5 - x^2 + 4 = -x^2 + 2x + 9$

7.

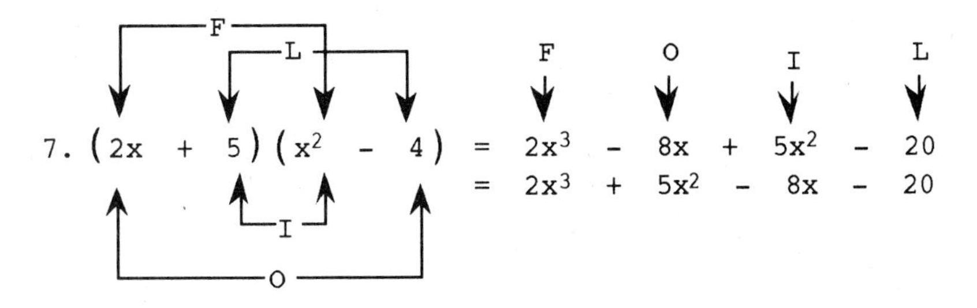

$$(2x + 5)(x^2 - 4) = 2x^3 - 8x + 5x^2 - 20$$
$$= 2x^3 + 5x^2 - 8x - 20$$

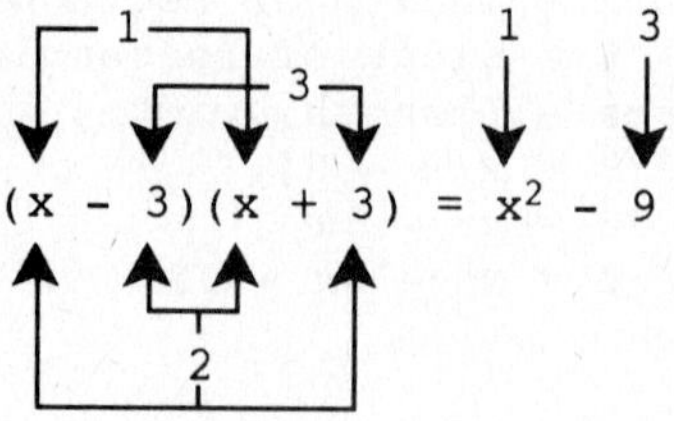

8. $(x - 3)(x + 3) = x^2 - 9$

9. $(x - 3) - (x + 3) = x - 3 - x - 3 = -6$

10. $(x - 3) + (x + 3) = x - 3 + x + 3 = 2x$

11. $(x + 4) - (x^2 - 5x + 6) = x + 4 - x^2 + 5x - 6 = -x^2 + 6x - 2$

12.
$$\begin{array}{r} x^2 - 5x + 6 \\ x + 4 \\ \hline x^3 - 5x^2 + 6x \\ 4x^2 - 20x + 24 \\ \hline x^3 - x^2 - 14x + 24 \end{array}$$

13. $(x + 4) + (x^2 - 5x + 6)$
$= x + 4 + x^2 - 5x + 6 = x^2 - 4x + 10$

14. $(3x^2 + 2x + 1) - (2x^2 - 3x + 4) = 3x^2 + 2x + 1 - 2x^2 + 3x - 4 = x^2 + 5x - 3$

15. $(3x^2 + 2x + 1) + (2x^2 - 3x + 4) = 3x^2 + 2x + 1 + 2x^2 - 3x + 4 = 5x^2 - x + 5$

16.
$$\begin{array}{r} 2x^2 - 3x + 4 \\ 3x^2 + 2x + 1 \\ \hline 6x^4 - 9x^3 + 12x^2 \\ 4x^3 - 6x^2 + 8x \\ 2x^2 - 3x + 4 \\ \hline 6x^4 - 5x^3 + 8x^2 + 5x + 4 \end{array}$$

17. $[x^2 - (1 - x - x^2)] - (x + 1)$
$= [x^2 - 1 + x + x^2] - (x + 1)$
$= x^2 - 1 + x + x^2 - x - 1 = 2x^2 - 2$

18. $(x^2 + 2)(x - 3)$ is multiplied using the FOIL method.

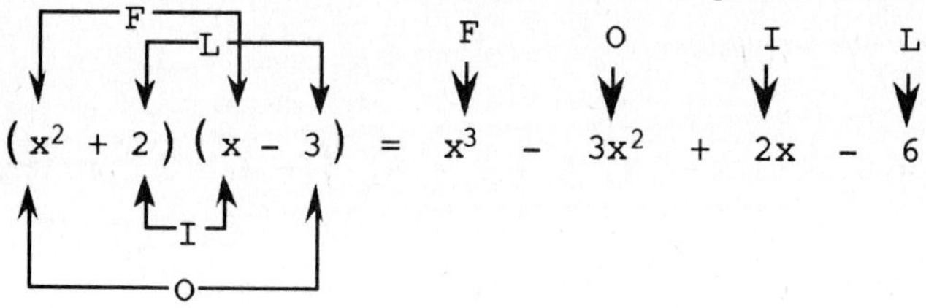

$(x^2 + 2)(x - 3) = x^3 - 3x^2 + 2x - 6$

$$\begin{aligned} -2x\{(x^2 + 2)(x - 3) - x[x - x(3 - x)]\} &= -2x\{(x^3 - 3x^2 + 2x - 6) - x[x - 3x + x^2]\} \\ &= -2x\{(x^3 - 3x^2 + 2x - 6) - x[x^2 - 2x]\} \\ &= -2x\{x^3 - 3x^2 + 2x - 6 - x^3 + 2x^2\} \\ &= -2x\{-x^2 + 2x - 6\} \\ &= 2x^3 - 4x^2 + 12x \end{aligned}$$

19. $\{x - [x - (x - 1)]\} - \{1 - [1 - (1 - x)]\}$
$= \{x - [x - x + 1]\} - \{1 - [1 - 1 + x]\} = \{x - [1]\} - \{1 - [x]\}$
$= \{x - 1\} - \{1 - x\} = x - 1 - 1 + x = 2x - 2$

20. A straight-forward way would be to multiply the various quantities, then subtract. An alternative approach that is more efficient is to factor by grouping first:
$(x - 2)^2(x - 3) - (x - 2)(x - 3)^2$
$= (x - 2)(x - 3)\cdot(x - 2) - (x - 2)(x - 3)\cdot(x - 3)$
$= (x - 2)(x - 3)[(x - 2) - (x - 3)]$
$= (x - 2)(x - 3)[x - 2 - x + 3]$
$= (x - 2)(x - 3)1$ or $(x - 2)(x - 3)$.
Now, $(x - 2)(x - 3)$ can be multiplied by FOIL to obtain $x^2 - 5x + 6$.

21. First, remove common factors: $3x^4 + 9x^3 = 3x^3(x + 3)$. In this case, there are no further steps.

22. First, remove common factors: $2x^2y^3 + x^3y = x^2y(2y^2 + x)$. In this case, there are no further steps.

23. First, remove common factors: $6x^2 + 15x = 3x(2x + 5)$. In this case, there are no further steps.

24. Use difference of two squares: $x^2 - 16 = x^2 - 4^2 = (x - 4)(x + 4)$

25. Use sum of two squares; this is not factorable.

26. First, remove common factors: $14x^2 - 56 = 14(x^2 - 4)$. Now, use difference of two squares: $14(x^2 - 4) = 14(x^2 - 2^2) = 14(x - 2)(x + 2)$.

27. First, remove common factors: $6x^2y + 9xy + 12xy^2 = 3xy(2x + 3 + 4y)$. In this case, there are no further steps.

28. Use factoring by grouping:
$$\begin{aligned} 6x^2y + 9xy + 8x + 12 &= (6x^2y + 9xy) + (8x + 12) \\ &= 3xy(2x + 3) + 4(2x + 3) = (2x + 3)(3xy + 4) \end{aligned}$$

29. Use factoring by grouping:
$$\begin{aligned} 2x^2 - 10x + x - 5 &= (2x^2 - 10x) + (x - 5) \\ &= 2x(x - 5) + 1(x - 5) \\ &= (x - 5)(2x + 1) \end{aligned}$$

30. Use trinomial factoring methods: For example, calculate $ac = (4)(-3) = -12$. Two factors of -12 whose sum is $11 = b$ are 12 and -1.
$$\begin{aligned} 4x^2 + 11x - 3 &= 4x^2 - x + 12x - 3 = (4x^2 - x) + (12x - 3) \\ &= x(4x - 1) + 3(4x - 1) = (4x - 1)(x + 3) \end{aligned}$$

31. Use difference of two squares:
$$4x^2 - 25 = (2x)^2 - (5)^2 = (2x - 5)(2x + 5)$$

32. Use trinomial factoring methods: For example, start with $(x + ?y)(x + ?y)$.

$$\begin{array}{c} 2 \\ \hline \boxed{1 \cdot 2} \\ 2 \cdot 1 \end{array} \qquad 1 + 2 = 3 \quad \text{Therefore, } x^2 + 3xy + 2y^2 = (x + y)(x + 2y)$$

33. Use sum of two squares; this is not factorable.

34. Rearrange terms, then use factoring by grouping:
$$\begin{aligned} x^2 + 2x + 4 + 2x &= x^2 + 2x + 2x + 4 \\ &= (x^2 + 2x) + (2x + 4) \\ &= x(x + 2) + 2(x + 2) \\ &= (x + 2)(x + 2) \text{ or } (x + 2)^2 \end{aligned}$$

35. Use trinomial factoring methods: For example, compute $ac = (2)(-3) = -6$. Two factors of -6 whose sum is $-5 = b$ are -6 and 1.
$$\begin{aligned} 2x^2 - 5x - 3 &= 2x^2 + x - 6x - 3 = (2x^2 + x) - (6x + 3) \\ &= x(2x + 1) - 3(2x + 1) = (2x + 1)(x - 3) \end{aligned}$$

36. Rearrange terms, then use factoring by grouping:
$$\begin{aligned} x^2 - 3x + 9 - 3x &= x^2 - 3x - 3x + 9 \\ &= (x^2 - 3x) - (3x - 9) \\ &= x(x - 3) - 3(x - 3) \\ &= (x - 3)(x - 3) \text{ or } (x - 3)^2 \end{aligned}$$

37. Use factoring by grouping: $3x^2 + 2x - 3x - 2 = (3x^2 + 2x) - (3x + 2)$
$= x(3x + 2) - 1(3x + 2)$
$= (3x + 2)(x - 1)$

38. Use difference of two cubes: $a^3 - 8 = (a)^3 - (2)^3$
$= (a - 2)[(a)^2 + (a)(2) + (2)^2]$
$= (a - 2)(a^2 + 2a + 4)$

39. Use trinomial factoring methods: For example, start with $(x + ?)(x + ?)$.

8
1·8
[2·4]

$2 + 4 = 6$ Therefore, $x^2 + 6x + 8 = (x + 2)(x + 4)$

40. Use trinomial factoring methods: For example, compute ac = (2)(2) = 4. Two factors of -5 whose sum is 4 are -1 and -4.
$2x^2 - 5xy + 2y^2 = 2x^2 - 4xy - xy + 2y^2 = (2x^2 - 4xy) - (xy - 2y^2)$
$= 2x(x - 2y) - y(x - 2y) = (x - 2y)(2x - y)$

41. Use sum of two cubes: $x^3 + 64 = x^3 + (4)^3 = (x + 4)[(x)^2 - (x)(4) + (4)^2]$
$= (x + 4)(x^2 - 4x + 16)$

42. Use difference of two squares: $a^2 - 9b^2 = a^2 - (3b)^2 = (a - 3b)(a + 3b)$

43. $x^2 + 6x + 9$ is a perfect square: $x^2 + 6x + 9 = x^2 + 2 \cdot x \cdot 3 + 3^2 = (x + 3)^2$

44. $x^2 - 8x + 16$ is a perfect square: $x^2 - 8x + 16 = x^2 - 2 \cdot x \cdot 4 + 4^2 = (x - 4)^2$

45. Use trinomial factoring methods: For example, start with $(x - ?)(x - ?)$

12
1·12
[2·6]
3·4

$2 + 6 = 8$ Therefore, $x^2 - 8x + 12 = (x - 2)(x - 6)$

46. Use trinomial factoring methods: For example, start with $(x + ?)(x + ?)$

18
1·18
2·9
[3·6]

$3 + 6 = 9$ Therefore, $x^2 + 9x + 18 = (x + 3)(x + 6)$

47. Use trinomial factoring methods: For example, start with $(x + ?)(x - ?)$

-11
[1(-11)]
11(-1)

$1 + (-11) = -10$ Therefore, $x^2 - 10x - 11 = (x + 1)(x - 11)$

48. Use trinomial factoring methods: For example, start with $(x + ?)(x - ?)$

-26
1(-26)
2(-13)
13(-2)
26(-1)

$13 + (-2) = 11$ Therefore, $x^2 + 11x - 26 = (x + 13)(x - 2)$

49. $x^2 - 12x + 36$ is a perfect square. $x^2 - 12x + 36 = x^2 - 2\cdot x\cdot 6 + 6^2 = (x - 6)^2$

50. Use trinomial factoring methods: For example, start with $(x + ?)(x + ?)$

6	
$1\cdot 6$	$(x + 1)(x + 6) = x^2 + 7x + 6$. No
$2\cdot 3$	$(x + 2)(x + 3) = x^2 + 5x + 6$. No
$3\cdot 2$	$(x + 3)(x + 2) = x^2 + 5x + 6$. No
$6\cdot 1$	$(x + 6)(x + 1) = x^2 + 7x + 6$. No

No combination of coefficients produces the middle term, hence this is not factorable.

51. Use difference of two cubes:

$$8x^3 - 125 = (2x)^3 - 5^3 = (2x - 5)[(2x)^2 + (2x)5 + 5^2]$$
$$= (2x - 5)(4x^2 + 10x + 25)$$

52. Use difference of two squares twice:

$$x^4 - 1 = (x^2)^2 - 1^2 = (x^2 + 1)(x^2 - 1) = (x^2 + 1)(x + 1)(x - 1)$$

53. Use sum of two squares; this is not factorable.

54. $4x^2 + 12x + 9$ is a perfect square.

$$4x^2 + 12x + 9 = (2x)^2 + 2(2x)3 + 3^2 = (2x + 3)^2$$

55. $9x^2 - 12x + 4$ is a perfect square.

$$9x^2 - 12x + 4 = (3x)^2 - 2(3x)2 + 2^2 = (3x - 2)^2$$

56. Use trinomial factoring methods: For example, compute $ac = (2)(1) = 2$. Two factors of 2 whose sum is 3 are 2 and 1.

$$2x^2 + 3x + 1 = 2x^2 + 2x + x + 1 = (2x^2 + 2x) + (x + 1)$$
$$= 2x(x + 1) + (x + 1) = (x + 1)(2x + 1)$$

57. Use trinomial factoring methods: For example, compute $ac = (2)(3) = 6$. Two factors of 6 whose sum is -5 are -2 and -3.

$$2x^2 - 5x + 3 = 2x^2 - 2x - 3x + 3 = (2x^2 - 2x) - (3x - 3)$$
$$= 2x(x - 1) - 3(x - 1) = (x - 1)(2x - 3)$$

58. Use trinomial factoring methods: For example, compute $ac = (6)(-1) = -6$. Two factors of -6 whose sum is 1 are 3 and -2.

$$6x^2 + x - 1 = 6x^2 - 2x + 3x - 1 = (6x^2 - 2x) + (3x - 1)$$
$$= 2x(3x - 1) + (3x - 1) = (3x - 1)(2x + 1)$$

59. Use difference of two cubes: $x^3 - y^3 = (x - y)(x^2 + xy + y^2)$

60. $x^2 + 4xy + 4y^2$ is a perfect square:

$$x^2 + 4xy + 4y^2 = x^2 + 2(x)(2y) + (2y)^2 = (x + 2y)^2$$

61. Use trinomial factoring methods: For example, start with $(x + ?)(x + ?)$

4	
$1\cdot 4$	$(x + 1)(x + 4) = x^2 + 5x + 4$. No
$2\cdot 2$	$(x + 2)(x + 2) = x^2 + 4x + 4$. No
$4\cdot 1$	$(x + 4)(x + 1) = x^2 + 5x + 4$. No

No combination of coefficients produces the middle term, hence this is not factorable.

62. Use trinomial factoring methods: For example, start with $(x + ?)(x - ?)$

-6	
1(-6)	$(x + 1)(x - 6) = x^2 - 5x - 6$. No.
2(-3)	$(x + 2)(x - 3) = x^2 - x - 6$. No.
3(-2)	$(x + 3)(x - 2) = x^2 + x - 6$. No.
6(-1)	$(x + 6)(x - 1) = x^2 + 5x - 6$. No.

No combination of coefficients produces the middle term, hence this is not factorable.

63. $x^2 + 7x + 12 = 0$
$(x + 3)(x + 4) = 0$
$x + 3 = 0$ or $x + 4 = 0$
$x = -3$ $\quad x = -4$

64. $x^4 - 16 = 0$
$(x^2 - 4)(x^2 + 4) = 0$
$(x - 2)(x + 2)(x^2 + 4) = 0$
$x - 2 = 0$ or $x + 2 = 0$ or $x^2 + 4 = 0$
$x = 2$ $\quad x = -2$ $\quad$ not factorable using integer coefficients.
The only solutions we can find at this time are 2, -2.

65. $x^2 + 3x - 10 = 0$
$(x + 5)(x - 2) = 0$
$x + 5 = 0$ or $x - 2 = 0$
$x = -5$ $\quad x = 2$

66. $x^2 - 5x = 6$
$x^2 - 5x - 6 = 0$
$(x - 6)(x + 1) = 0$
$x - 6 = 0$ or $x + 1 = 0$
$x = 6$ $\quad x = -1$

67. $x^2 + x = 20$
$x^2 + x - 20 = 0$
$(x + 5)(x - 4) = 0$
$x + 5 = 0$ or $x - 4 = 0$
$x = -5$ $\quad x = 4$

68. $x^2 - 4x - 12 = 0$
$(x - 6)(x + 2) = 0$
$x - 6 = 0$ or $x + 2 = 0$
$x = 6$ $\quad x = -2$

69. $x^3 - 1 = 0$
$(x - 1)(x^2 + x + 1) = 0$
$x - 1 = 0$ or $x^2 + x + 1 = 0$
$x = 1$ $\quad$ not factorable using integer coefficients
The only solution we can find at ths time is 1.

70. $x^2 - 1 = 0$
$(x - 1)(x + 1) = 0$
$x - 1 = 0$ or $x + 1 = 0$
$x = 1$ $\quad x = -1$

71. $x^2 + 8 = 0$
This is not factorable in the integers, so the factoring method fails.

72. $x^3 + 8 = 0$
$(x + 2)(x^2 - 2x + 4) = 0$
$x + 2 = 0$ or $x^2 - 2x + 4 = 0$
$x = -2$ $\quad$ not factorable using integer coefficients
The only solution we can find at this time is -2.

73. $x^2 + 2x = 1$
$x^2 + 2x - 1 = 0$
This is not factorable in the integers, so the factoring method fails.

74. $x^2 - 2x = -1$
$x^2 - 2x + 1 = 0$
$(x - 1)^2 = 0$
$x - 1 = 0$
$x = 1$

75. $2x - x^2 = 1$
$0 = x^2 - 2x + 1$
$0 = (x - 1)^2$
$x - 1 = 0$
$x = 1$

76. $x^2 = 2x + 1$
$x^2 - 2x - 1 = 0$
This is not factorable in the integers, so the factoring method fails.

77. Use sum of two cubes: $x^6 + 1 = (x^2)^3 + 1^3 = (x^2 + 1)[(x^2)^2 - (x^2)(1) + (1)^2]$
$= (x^2 + 1)(x^4 - x^2 + 1)$

78. First, remove common factors:
$2x^4 + 6x^3 + 3x^2 + x^3 = x^2(2x^2 + 6x + 3 + x)$
Now, rearrange terms and use factoring by grouping:

$$x^2(2x^2 + 6x + 3 + x) = x^2(2x^2 + 6x + x + 3)$$
$$= x^2[(2x^2 + 6x) + (x + 3)]$$
$$= x^2[2x(x + 3) + 1(x + 3)]$$
$$= x^2(x + 3)(2x + 1)$$

79. First, remove common factors: $x^2a^3 - 4xa^3 + 4a^3 = a^3(x^2 - 4x + 4)$
Now, use trinomial factoring methods: $a^3(x^2 - 4x + 4) = a^3(x - 2)(x - 2)$ or $a^3(x - 2)^2$

80. First, remove common factors: $3x^3y - 3xy^3 = 3xy(x^2 - y^2)$
Now, use difference of two squares: $3xy(x^2 - y^2) = 3xy(x - y)(x + y)$

81. Use factoring by grouping:
$$x^3 - 3x^2 + x - 3 = (x^3 - 3x^2) + (x - 3) = x^2(x - 3) + 1(x - 3) = (x - 3)(x^2 + 1)$$

82. First, remove common factors: $x^6 + x^5 + x^2 + x = x(x^5 + x^4 + x + 1)$
Now, use factoring by grouping:
$$x(x^5 + x^4 + x + 1) = x[(x^5 + x^4) + (x + 1)] = x[x^4(x + 1) + 1(x + 1)] = x(x + 1)(x^4 + 1)$$

83. First, remove common factors: $2x^3 - 2x^2 - 4x = 2x(x^2 - x - 2)$
Now, use trinomial factoring methods: $2x(x^2 - x - 2) = 2x(x + 1)(x - 2)$

84. First, group terms: $x^2 + 2xy + y^2 - 1 = (x^2 + 2xy + y^2) - 1$
Now, use difference of two squares:
$$(x^2 + 2xy + y^2) - 1 = (x + y)^2 - 1 = [(x + y) - 1][(x + y) + 1] = (x + y - 1)(x + y + 1)$$

85. First, use difference of two squares:
$$x^6 - 1 = (x^3)^2 - (1)^2 = (x^3 - 1)(x^3 + 1)$$
Now, use the difference of two cubes and sum of two cubes:
$$(x^3 - 1)(x^3 + 1) = (x - 1)(x^2 + x + 1)(x + 1)(x^2 - x + 1)$$

86. Use sum of two cubes: $8x^3 + 125 = (2x)^3 + (5)^3 = (2x + 5)[(2x)^2 - (2x)(5) + (5)^2] = (2x + 5)(4x^2 - 10x + 25)$

87. Use factoring by grouping: $3x^3 + 2x^2 - 15x - 10 = (3x^3 + 2x^2) - (15x + 10) = x^2(3x + 2) - 5(3x + 2) = (3x + 2)(x^2 - 5)$

88. First, remove common factors: $x^4 + 2x^3 - 3x^2 = x^2(x^2 + 2x - 3)$
Now, use trinomial factoring methods: $x^2(x^2 + 2x - 3) = x^2(x + 3)(x - 1)$

89. First, remove common factors: $-x^5y^3 - 2x^4y^2 - x^3y = -x^3y(x^2y^2 + 2xy + 1)$
Now, use trinomial factoring methods:
$$-x^3y(x^2y^2 + 2xy + 1) = -x^3y[(xy)^2 + 2(xy) + 1] = -x^3y(xy + 1)(xy + 1) \text{ or } -x^3y(xy + 1)^2$$

90. First, remove common factors: $2a^4 + 2a = 2a(a^3 + 1)$
Now, use sum of two cubes: $2a(a^3 + 1) = 2a(a + 1)(a^2 - a + 1)$

91.
$$2x^2 - x = 3$$
$$2x^2 - x - 3 = 0$$
$$(2x - 3)(x + 1) = 0$$
$2x - 3 = 0$ or $x + 1 = 0$
$2x = 3$ $\quad x = -1$
$x = \frac{3}{2}$

92.
$$x^2 - x = 20$$
$$x^2 - x - 20 = 0$$
$$(x - 5)(x + 4) = 0$$
$x - 5 = 0$ or $x + 4 = 0$
$x = 5$ or $x = -4$

93.
$$6x^2 + 5x - 6 = 0$$
$$(2x + 3)(3x - 2) = 0$$
$2x + 3 = 0$ or $3x - 2 = 0$
$2x = -3$ $\quad 3x = 2$
$x = -\frac{3}{2}$ $\quad x = \frac{2}{3}$

94.
$$12x^2 + 29x - 8 = 0$$
$$(4x - 1)(3x + 8) = 0$$
$4x - 1 = 0$ or $3x + 8 = 0$
$4x = 1$ $\quad 3x = -8$
$x = \frac{1}{4}$ $\quad x = -\frac{8}{3}$

95. Let x = the integer

Then $x + 4$ = the number that is 4 greater than the integer

We have

$$x(x + 4) = 96$$
$$x^2 + 4x = 96$$
$$x^2 + 4x - 96 = 0$$
$$(x + 12)(x - 8) = 0$$

$x + 12 = 0$ $\quad$ $x - 8 = 0$

$x = -12$ $\quad$ $x = 8$ $\quad$ The positive integer is 8.

Discard, x is to be positive $\quad$ $x + 4 = 12$

96. Let x = the number

Twice	the square of the number	is	10	more than	8 times the number
$2 \cdot$	x^2	=	10	+	$8x$

$$2x^2 = 10 + 8x$$
$$2x^2 - 8x - 10 = 0$$
$$x^2 - 4x - 5 = 0$$
$$(x - 5)(x + 1) = 0$$

$x - 5 = 0$ or $x + 1 = 0$

$x = 5$ $\quad$ $x = -1$ Discard, x is to be positive.

The number is 5.

97. Let x = the integer

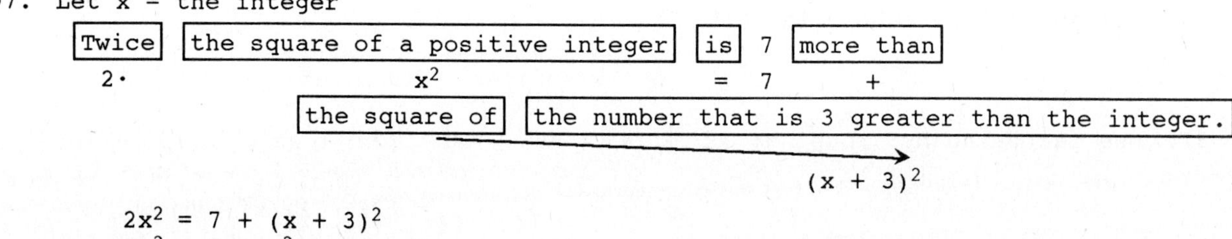

$$2x^2 = 7 + (x + 3)^2$$
$$2x^2 = 7 + x^2 + 6x + 9$$
$$x^2 - 6x - 16 = 0$$
$$(x - 8)(x + 2) = 0$$

$x - 8 = 0$ or $x + 2 = 0$

$x = 8$ $\quad$ $x = -2$ Discard, x is to be positive

The integer is 8.

98. Let x = the width of the rectangle

Then $5x$ = the length of rectangle

Use (length) × (width) = Area

$$x(5x) = 180$$
$$5x^2 = 180$$
$$x^2 = 36$$
$$x^2 - 36 = 0$$
$$(x - 6)(x + 6) = 0$$

$x - 6 = 0$ or $x + 6 = 0$

$x = 6$ in (width) $\quad$ ~~$x = -6$~~ not possible

$5x = 30$ in (length)

99. Let b = the base of the triangle

Then $\frac{1}{3}b$ = the height of the triangle

Use $\frac{1}{2}bh = \text{Area}$

$$\frac{1}{2}b\left(\frac{1}{3}b\right) = 150$$

$$\frac{1}{6}b^2 = 150$$

$$b^2 = 900$$

$$b^2 - 900 = 0$$

$$(b - 30)(b + 30) = 0$$

$b - 30 = 0$ or $b + 30 = 0$

$b = 30$ cm (base) ~~$b = -30$~~ not possible

$\frac{1}{3}b = 10$ cm (height)

100.

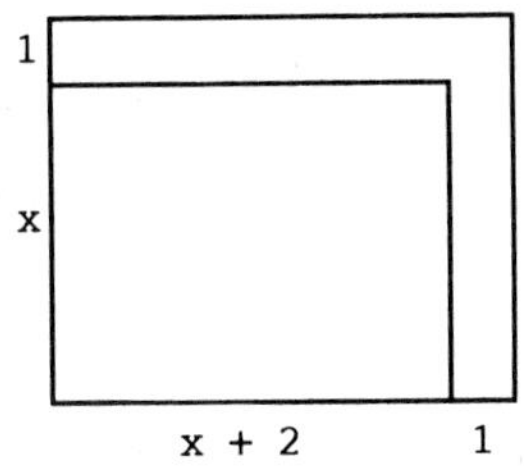

Let x = original width; then $x + 1$ = increased width

$x + 2$ = original length; then $x + 3$ = increased length

Use (length)(width) = Area

$$(x + 3)(x + 1) = 48$$

$$x^2 + 4x + 3 = 48$$

$$x^2 + 4x - 45 = 0$$

$$(x + 9)(x - 5) = 0$$

$x + 9 = 0$ or $x - 5 = 0$

~~$x = -9$~~ not possible; $x = 5$ feet

$x + 2 = 7$ feet

Practice Test 2

1. (A) 3 (B) -4 (C) 4

2. (A) The degree of the first factor is 3 + 3 = 6.
   (B) The degree of the product is 3 + 3 + 2 + 5 = 13
   (C) $(-2x^3y^3)(3x^2y^5) = (-2)3x^3x^2y^3y^5 = -6x^5y^8$

3. $(3x^2 - x + 2) - (2x^2 - 3x - 4) = 3x^2 - x + 2 - 2x^2 + 3x + 4 = x^2 + 2x + 6$

4.
$$\begin{array}{r} x^2 + 3x + 5 \\ 4x + 5 \\ \hline 4x^3 + 12x^2 + 20x \\ 5x^2 + 15x + 25 \\ \hline 4x^3 + 17x^2 + 35x + 25 \end{array}$$

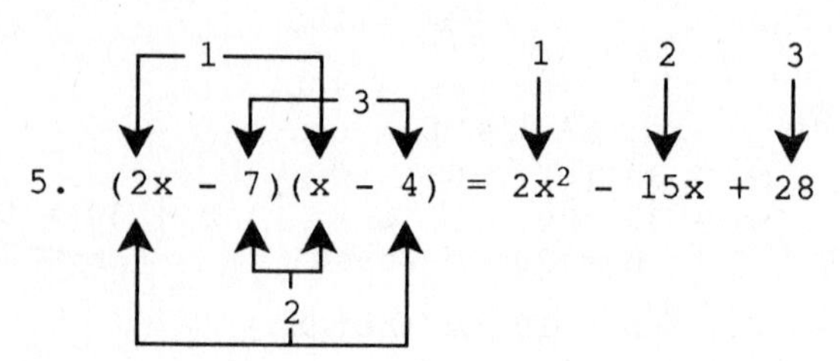

5. $(2x - 7)(x - 4) = 2x^2 - 15x + 28$

6. $(x^3 + x + 2) + (x^2 - 4x + 6) + (x^3 - x^2 - 9)$
   $= x^3 + x + 2 + x^2 - 4x + 6 + x^3 - x^2 - 9 = 2x^3 - 3x - 1$

7. $\{3 - [x - (3 - x)]\} - \{x + [3 - (x - 3)]\}$
   $= \{3 - [x - 3 + x]\} - \{x + [3 - x + 3]\}$
   $= \{3 - [2x - 3]\} - \{x + [6 - x]\}$
   $= \{3 - 2x + 3\} - \{x + 6 - x\}$
   $= \{6 - 2x\} - \{6\}$
   $= 6 - 2x - 6$
   $= -2x$

8. First, remove common factors: $2ab^2 + 4a^2b^2 + 6a^2b = 2ab(b + 2ab + 3a)$. In this case, there are no further steps.

9. First, remove common factors: $5x^2 - 80 = 5(x^2 - 16)$. Now, use difference of two squares: $5(x^2 - 16) = 5(x - 4)(x + 4)$

10. $x^2 + 12x + 36$ is a perfect square.
    $x^2 + 12x + 36 = x^2 + 2(x)(6) + 6^2 = (x + 6)^2$

11. Use trinomial factoring methods: For example, start with (x - ?)(x - ?)

32
1·32
2·16
4· 8

4 + 8 = 12 Therefore, $x^2 - 12x + 32 = (x - 4)(x - 8)$

12. Use trinomial factoring methods: For example, start with (x + ?)(x - ?)

-45
1(-45)
3(-15)
5(-9)
9(-5)
15(-3)
45(-1)

15 + (-3) = 12 Therefore, $x^2 + 12x - 45 = (x + 15)(x - 3)$

13. Use difference of two cubes:
    $x^3 - 64 = x^3 - 4^3 = (x - 4)(x^2 + x\cdot4 + 4^2) = (x - 4)(x^2 + 4x + 16)$

14. Use factoring by grouping:
    $x^2 + 4xy - 3xy - 12y^2 = (x^2 + 4xy) - (3xy + 12y^2) = x(x + 4y) - 3y(x + 4y)$
    $= (x + 4y)(x - 3y)$

15. Use trinomial factoring methods: For example, compute $ac = (6)(-20) = -120$. Two factors of -120 whose sum is -7 are -15 and 8.

$$6x^2 - 7x - 20 = 6x^2 - 15x + 8x - 20 = (6x^2 - 15x) + (8x - 20)$$
$$= 3x(2x - 5) + 4(2x - 5) = (2x - 5)(3x + 4)$$

16. Use trinomial factoring methods: For example, start with $(x^2 + ?)(x^2 - ?)$

-24
1(-24)
2(-12)
3(-8)
4(-6)
6(-4)
[8(-3)]
12(-2)
24(-1)

$8 + (-3) = 5$ Therefore, $x^4 + 5x^2 - 24 = (x^2 + 8)(x^2 - 3)$

17.
$$x^2 - 7x + 10 = 0$$
$$(x - 5)(x - 2) = 0$$
$$x - 5 = 0 \quad \text{or} \quad x - 2 = 0$$
$$x = 5 \qquad x = 2$$

18.
$$2x^2 - 3x = 2$$
$$2x^2 - 3x - 2 = 0$$
$$(2x + 1)(x - 2) = 0$$
$$2x + 1 = 0 \quad \text{or} \quad x - 2 = 0$$
$$2x = -1 \qquad x = 2$$
$$x = -\frac{1}{2}$$

19. $x^2 + 3x + 18 = 0$
This is not factorable in the integers, so the factoring method fails.

20. Let x = the number

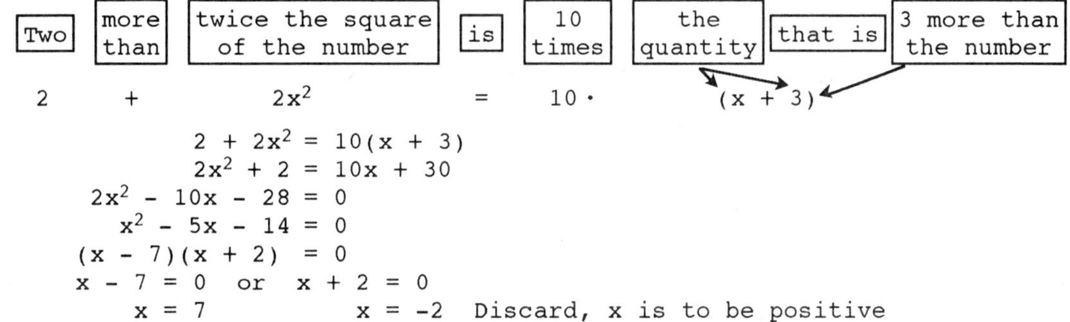

$$2 + 2x^2 = 10(x + 3)$$
$$2x^2 + 2 = 10x + 30$$
$$2x^2 - 10x - 28 = 0$$
$$x^2 - 5x - 14 = 0$$
$$(x - 7)(x + 2) = 0$$
$$x - 7 = 0 \quad \text{or} \quad x + 2 = 0$$
$$x = 7 \qquad x = -2$$ Discard, x is to be positive

7 is the number.

21. Let x = the width
Then $3x - 2$ = the length
Use (length)(width) = Area

$$(3x - 2)x = 5$$
$$3x^2 - 2x = 5$$
$$3x^2 - 2x - 5 = 0$$
$$(3x - 5)(x + 1) = 0$$
$$3x - 5 = 0 \qquad x + 1 = 0$$
$$3x = 5 \qquad \cancel{x = -1}$$ Not possible
$$x = \frac{5}{3}$$
$$x = 1\tfrac{2}{3} \text{ feet (width)}$$
$$3x - 2 = 3 \text{ feet (length)}$$

CHAPTER 3 ALGEBRAIC FRACTIONS

Exercise 3-1 Rational Expressions

Key Ideas and Formulas

Fractional forms in which the numerator and denominator are polynomials are called **rational expressions.**

Rational expressions are rewritten using:

The Fundamental Principle of Fractions
For all polynomials P, Q, and K with Q, K ≠ 0:
$\frac{PK}{QK} = \frac{P}{Q}$ We may divide out a common factor K from both the numerator and denominator. This is called **reducing to lower terms.**
$\frac{P}{Q} = \frac{PK}{QK}$ We may multiply the numerator and denominator by the same nonzero factor K. This is called **raising to higher terms.**

1. $\frac{36}{90} = \frac{\cancel{18}\cdot 2}{\cancel{18}\cdot 5} = \frac{2}{5}$

3. $\frac{75}{45} = \frac{\cancel{15}\cdot 5}{\cancel{15}\cdot 3} = \frac{5}{3}$

5. $\frac{3x^3}{6x^5} = \frac{\cancel{(3x^3)}1}{\cancel{(3x^3)}(2x^2)} = \frac{1}{2x^2}$

7. $\frac{14x^3y}{21xy^2} = \frac{\cancel{(7xy)}(2x^2)}{\cancel{(7xy)}(3y)} = \frac{2x^2}{3y}$

9. $\frac{-x^4y^2}{xy^6} = \frac{\cancel{xy^2}(-x^3)}{\cancel{xy^2}(y^4)} = -\frac{x^3}{y^4}$

11. $\frac{(-a)^5b^2}{a^3(-b)^4} = \frac{-a^5b^2}{a^3b^4} = \frac{\cancel{a^3b^2}(-a^2)}{\cancel{a^3b^2}(b^2)} = -\frac{a^2}{b^2}$

13. $\frac{15y^3(x-9)^3}{5y^4(x-9)^2} = \frac{\cancel{5y^3(x-9)^2}\cdot 3(x-9)}{\cancel{5y^3(x-9)^2}\cdot y} = \frac{3(x-9)}{y}$

15. $\frac{(2x-1)\cancel{(2x+1)}}{3x\cancel{(2x+1)}} = \frac{2x-1}{3x}$

17. $\frac{(x-1)^2(x+3)^5}{x(x-1)^4} = \frac{\cancel{(x-1)^2}(x+3)^5}{\cancel{(x-1)^2}x(x-1)^2} = \frac{(x+3)^5}{x(x-1)^2}$

19. $\dfrac{x^3(x+5)^4}{x^5(x+5)} = \dfrac{\cancel{x^3(x+5)}\cdot(x+5)^3}{\cancel{x^3(x+5)}x^2} = \dfrac{(x+5)^3}{x^2}$

21. $\dfrac{x^2-2x}{2x-4} = \dfrac{x\cancel{(x-2)}}{2\cancel{(x-2)}} = \dfrac{x}{2}$

23. $\dfrac{m^2-mn}{m^2n-mn^2} = \dfrac{\cancel{m}\cancel{(m-n)}}{\cancel{m}n\cancel{(m-n)}} = \dfrac{1}{n}$

25. $\dfrac{x^2-3x}{x^3+2x^2} = \dfrac{\cancel{x}(x-3)}{\cancel{x^2}(x+2)} = \dfrac{x-3}{x(x+2)}$

27. $\dfrac{x^3+4x^2+4x}{x^2(x+2)} = \dfrac{x(x^2+4x+4)}{x^2(x+2)} = \dfrac{\cancel{x}\cancel{(x+2)^2}}{\cancel{x^2}\cancel{(x+2)}} = \dfrac{x+2}{x}$

29. $\dfrac{3}{2x} = \dfrac{3(4xy)}{(2x)(4xy)} = \dfrac{12xy}{8x^2y}$. $12xy$

31. $\dfrac{7}{3y} = \dfrac{y(2x^3y)}{(3y)(2x^3y)} = \dfrac{14x^3y}{6x^3y^2}$. $14x^3y$

33. $\dfrac{6xy}{16x^2y^2} = \dfrac{2x(3y)}{2x(8xy^2)} = \dfrac{3y}{8xy^2}$. $3y$

35. $\dfrac{30a^4b^2}{20ab^3} = \dfrac{5ab(6a^3b)}{5ab(4b^2)} = \dfrac{6a^3b}{4b^2}$. $4b^2$

37. $\dfrac{x^2+6x+8}{3x^2+12x} = \dfrac{(x+2)\cancel{(x+4)}}{3x\cancel{(x+4)}} = \dfrac{x+2}{3x}$

39. $\dfrac{x^2-9}{x^2+6x+9} = \dfrac{(x-3)\cancel{(x+3)}}{(x+3)\cancel{(x+3)}} = \dfrac{x-3}{x+3}$

COMMON ERROR: "cancelling" the x^2. "Cancelling" is dividing out a common factor from numerator and denominator. x^2 is a term.

41. $\dfrac{4x^2-9y^2}{4x^2y+6xy^2} = \dfrac{(2x-3y)\cancel{(2x+3y)}}{2xy\cancel{(2x+3y)}} = \dfrac{2x-3y}{2xy}$

43. $\dfrac{x^2+2x+1}{x^2+3x+2} = \dfrac{\cancel{(x+1)}(x+1)}{\cancel{(x+1)}(x+2)} = \dfrac{x+1}{x+2}$

45. $\dfrac{x^2+4x+4}{x^2-4} = \dfrac{\cancel{(x+2)}(x+2)}{\cancel{(x+2)}(x-2)} = \dfrac{x+2}{x-2}$

47. $\dfrac{x^2+3x+2}{x^2+5x+6} = \dfrac{\cancel{(x+2)}(x+1)}{\cancel{(x+2)}(x+3)} = \dfrac{x+1}{x+3}$

49. $\dfrac{x^2-xy+2x-2y}{x^2-y^2} = \dfrac{x(x-y)+2(x-y)}{(x-y)(x+y)} = \dfrac{\cancel{(x-y)}(x+2)}{\cancel{(x-y)}(x+y)} = \dfrac{x+2}{x+y}$

COMMON ERROR: "cancelling" the $x - y$ in $x(x - y)$ but not the $x - y$ in $2(x - y)$. Only common factors of the **entire** numerator and denominator are divided out.

51. $\dfrac{6x^3+28x^2-10x}{12x^3-4x^2} = \dfrac{2x(3x^2+14x-5)}{4x^2(3x-1)} = \dfrac{\cancel{2x}\cancel{(3x-1)}(x+5)}{\cancel{4x^2}\cancel{(3x-1)}} = \dfrac{x+5}{2x}$

53. $\dfrac{x^3 - 8}{x^2 - 4} = \dfrac{\overset{1}{\cancel{(x - 2)}}(x^2 + 2x + 4)}{\underset{1}{\cancel{(x - 2)}}(x + 2)} = \dfrac{x^2 + 2x + 4}{x + 2}$

55. $\dfrac{2x^3 + 4x^2 - 6x}{6x^3 - 6x^2} = \dfrac{\overset{1}{\cancel{2}}x(x^2 + 2x - 3)}{\underset{3x}{\cancel{6x^2}}(x - 1)} = \dfrac{(x + 3)\overset{1}{\cancel{(x - 1)}}}{3x\underset{1}{\cancel{(x - 1)}}} = \dfrac{x + 3}{3x}$

57. $\dfrac{3x^3 - 9x^2}{6x^3 - 36x^2 + 54x} = \dfrac{3x^2(x - 3)}{6x(x^2 - 6x + 9)} = \dfrac{\overset{x}{\cancel{3x^2}}\overset{1}{\cancel{(x - 3)}}}{\underset{2}{\cancel{6x}}\underset{1}{\cancel{(x - 3)}}(x - 3)} = \dfrac{x}{2(x - 3)}$

59. $\dfrac{3x + x^2 + 15 + 5x}{x^2 + 13x + 30} = \dfrac{x(3 + x) + 5(3 + x)}{(x + 10)(x + 3)} = \dfrac{\overset{1}{\cancel{(3 + x)}}(x + 5)}{(x + 10)\underset{1}{\cancel{(x + 3)}}} = \dfrac{x + 5}{x + 10}$

61. $\dfrac{x^3 + 1}{x^2 - 1} = \dfrac{\overset{1}{\cancel{(x + 1)}}(x^2 - x + 1)}{\underset{1}{\cancel{(x + 1)}}(x - 1)} = \dfrac{x^2 - x + 1}{x - 1}$

63. $\dfrac{x^2 + 10x + 21}{x^2 + 5x - 14} = \dfrac{(x + 3)\overset{1}{\cancel{(x + 7)}}}{(x - 2)\underset{1}{\cancel{(x + 7)}}} = \dfrac{x + 3}{x - 2}$

COMMON ERROR: "cancelling" the x^2. See comment, problem 39.

65. $\dfrac{6x^2z + 4xyz}{3xyz + 2y^2z} = \dfrac{2x\overset{1}{\cancel{z}}\overset{1}{\cancel{(3x + 2y)}}}{y\underset{1}{\cancel{z}}\underset{1}{\cancel{(3x + 2y)}}} = \dfrac{2x}{y}$

67. $\dfrac{x^3 - 3x^2 + 2x - 6}{x^2 - 5x + 6} = \dfrac{x^2(x - 3) + 2(x - 3)}{(x - 3)(x - 2)} = \dfrac{\overset{1}{\cancel{(x - 3)}}(x^2 + 2)}{\underset{1}{\cancel{(x - 3)}}(x - 2)} = \dfrac{x^2 + 2}{x - 2}$

69. $\dfrac{6a^2b + 3ab^2}{8a^3b + 4a^2b^2} = \dfrac{3\overset{1}{\cancel{ab}}\overset{1}{\cancel{(2a + b)}}}{4\underset{a}{\cancel{a^2b}}\underset{1}{\cancel{(2a + b)}}} = \dfrac{3}{4a}$

71. $\dfrac{2x^2 - x - 1}{2x^2 + 3x + 1} = \dfrac{\overset{1}{\cancel{(2x + 1)}}(x - 1)}{\underset{1}{\cancel{(2x + 1)}}(x + 1)} = \dfrac{x - 1}{x + 1}$

73. $\dfrac{2x^2 - 3x + 2x - 3}{2x + 3 + 2x^2 + 3x} = \dfrac{x(2x - 3) + (2x - 3)}{(2x + 3) + x(2x + 3)} = \dfrac{(2x - 3)\overset{1}{\cancel{(x + 1)}}}{(2x + 3)\underset{1}{\cancel{(1 + x)}}} = \dfrac{2x - 3}{2x + 3}$

75. $\dfrac{3x}{4y} = \dfrac{3x(x + y)}{4y(x + y)}$. $3x(x + y)$ or $3x^2 + 3xy$

77. $\dfrac{x - 2y}{x + y} = \dfrac{(x + 2y)(x - y)}{(x + y)(x - y)}$. $(x + y)(x - y)$ or $x^2 - y^2$

79. $\dfrac{4x^2 + 8x}{12x^2 + 48x + 48} = \dfrac{4x(x + 2)}{12(x + 2)(x + 2)} = \dfrac{2x \cdot 2(x + 2)}{6(x + 2)2(x + 2)} = \dfrac{2x}{6(x + 2)}$. $2x$

81. $\dfrac{x^3 + 3x^2 + 2x}{x^3 + x^2} = \dfrac{x(x + 1)(x + 2)}{x^2(x + 1)} = \dfrac{x(x + 2)}{x^2}$. $x(x + 2)$

83. $$\frac{x^3 - y^3}{3x^3 + 3x^2y + 3xy^2} = \frac{(x - y)\overset{1}{\cancel{(x^2 + xy + y^2)}}}{3x\underset{1}{\cancel{(x^2 + xy + y^2)}}} = \frac{x - y}{3x}$$

85. $$\frac{ux + vx - uy - vy}{2ux + 2vx + uy + vy} = \frac{x(u + v) - y(u + v)}{2x(u + v) + y(u + v)} = \frac{\overset{1}{\cancel{(u + v)}}(x - y)}{\underset{1}{\cancel{(u + v)}}(2x + y)} = \frac{x - y}{2x + y}$$

87. $$\frac{x^4 - y^4}{(x^2 - y^2)(x + y)^2} = \frac{\overset{1}{\cancel{(x^2 - y^2)}}(x^2 + y^2)}{\underset{1}{\cancel{(x^2 - y^2)}}(x + y)^2} = \frac{x^2 + y^2}{(x + y)^2}$$

COMMON ERROR: "cancelling" further. $(x + y)^2 = x^2 + 2xy + y^2$. It has no **factors** in common with $x^2 + y^2$.

89. $$\frac{x^4 + x}{x^3 - x^2 + x} = \frac{x(x^3 + 1)}{x(x^2 - x + 1)} = \frac{(x + 1)\overset{1}{\cancel{(x^2 - x + 1)}}}{1\underset{1}{\cancel{(x^2 - x + 1)}}} = \frac{x + 1}{1} = x + 1$$

91. $$\frac{x^2 + y^2}{x^3 + y^3} = \frac{x^2 + y^2}{(x + y)(x^2 - xy + y^2)}.$$ There are no factors common to both numerator and denominator, hence, nothing to remove. $\frac{x^2 + y^2}{x^3 + y^3}$ is in lowest terms.

93. $$\frac{x^2u + uxy + vxy + vy^2}{xu^2 + uvy + uvx + yv^2} = \frac{ux(x+y) + vy(x+y)}{xu^2 + uvx + uvy + yv^2} = \frac{ux(x+y) + vy(x+y)}{ux(u+v) + vy(u+v)}$$

$$= \frac{(x+y)\overset{1}{\cancel{(ux+vy)}}}{(u+v)\underset{1}{\cancel{(ux+vy)}}} = \frac{x+y}{u+v}$$

95. $$\frac{(1 + h)^3 - 1}{h} = \frac{[(1 + h) - 1][(1 + h)^2 + (1 + h) + 1]}{h}$$

$$= \frac{[1 + h - 1][1 + 2h + h^2 + 1 + h + 1]}{h}$$

$$= \frac{\overset{1}{\cancel{h}}[h^2 + 3h + 3]}{\underset{1}{\cancel{h}}}$$

$$= h^2 + 3h + 3$$

97. $$\frac{2(x + 3)(x - 2)^4 - 4(x + 3)^2(x - 2)^3}{(x - 2)^8} = \frac{2(x + 3)(x - 2)^3[(x - 2) - 2(x + 3)]}{(x - 2)^8}$$

$$= \frac{2(x + 3)\overset{1}{\cancel{(x - 2)^3}}[x - 2 - 2x - 6]}{\underset{(x-2)^5}{\cancel{(x - 2)^8}}}$$

$$= \frac{2(x + 3)(-x - 8)}{(x - 2)^5} \text{ or } -\frac{2(x + 3)(x + 8)}{(x - 2)^5}$$

99. $$\frac{(x + h)^2 + 2(x + h) - (x^2 + 2x)}{h} = \frac{x^2 + 2xh + h^2 + 2x + 2h - x^2 - 2x}{h}$$
$$= \frac{2xh + h^2 + 2h}{h}$$
$$= \frac{\overset{1}{\cancel{h}}(2x + h + 2)}{\underset{1}{\cancel{h}}}$$
$$= 2x + h + 2$$

Exercise 3-2 Multiplication and Division

Key Ideas and Formulas

Multiplication
If P, Q, R, and S are polynomials ($Q, S \neq 0$), then $\frac{P}{Q} \cdot \frac{R}{S} = \frac{P \cdot R}{Q \cdot S}$

Division
If P, Q, R, and S are polynomials ($Q, R, S \neq 0$), then Divisor is R/S; Reciprocal of divisor is S/R $\frac{P}{Q} \div \frac{R}{S} = \frac{P}{Q} \cdot \frac{S}{R}$ That is, to divide one fraction by another, multiply by the reciprocal of the divisor.

Dividing Out Opposites

$$\frac{a - b}{b - a} = -1$$

---

1. $$\frac{\overset{2}{\cancel{10}}}{\underset{3}{\cancel{9}}} \cdot \frac{\overset{4}{\cancel{12}}}{\underset{3}{\cancel{15}}} = \frac{8}{9}$$

3. $$\frac{\overset{3}{\cancel{18}}}{\underset{7}{\cancel{35}}} \cdot \frac{\overset{9}{\cancel{45}}}{\underset{4}{\cancel{24}}} = \frac{27}{28}$$

5. $$\frac{2\overset{1}{\cancel{a}}}{\underset{1}{\cancel{3}}b\cancel{c}} \cdot \frac{\overset{3}{\cancel{9}}\cancel{c}}{\underset{1}{\cancel{a}}} = \frac{6}{b}$$

7. $$\frac{\overset{1 \cdot 1}{\cancel{3}\cancel{x^2}}}{\underset{1}{\cancel{4}}} \cdot \frac{\overset{\overset{1}{\cancel{4}}}{\cancel{16}}y}{\underset{\underset{1}{\cancel{4}x}}{\cancel{12}\cancel{x^3}}} = \frac{y}{x}$$

9. $$\frac{\overset{2\ 1}{\cancel{6}x\cancel{y}}}{z} \cdot \frac{x}{\underset{3 \cdot 1}{\cancel{9}\cancel{y}}z} = \frac{2x^2}{3z^2}$$

11. $$\frac{\overset{1}{\cancel{8}}x^2}{\underset{1}{\cancel{5y}}} \cdot \frac{\overset{3xy}{\cancel{15xy^2}}}{\underset{2}{\cancel{16}}} = \frac{3x^3y}{2}$$

13. $12xy \cdot \frac{x}{4yz} = \frac{\overset{3}{\cancel{12}}x\overset{1}{\cancel{y}}}{1}\cdot\frac{x}{\underset{1}{\cancel{4}}\underset{1}{\cancel{y}}z} = \frac{3x^2}{z}$

15. $\frac{18}{35}\div\frac{45}{24} = \frac{\overset{2}{\cancel{18}}}{35}\cdot\frac{24}{\underset{5}{\cancel{45}}} = \frac{48}{175}$

17. $\frac{11}{5}\div\frac{22}{25} = \frac{\overset{1}{\cancel{11}}}{\underset{1}{\cancel{5}}}\cdot\frac{\overset{5}{\cancel{25}}}{\underset{2}{\cancel{22}}} = \frac{5}{2}$

19. $\frac{9m}{8n}\div\frac{3m}{4n} = \frac{\overset{3\cdot1}{\cancel{9m}}}{\underset{2\cdot1}{\cancel{8n}}}\cdot\frac{\overset{1\cdot1}{\cancel{4n}}}{\underset{1\cdot1}{\cancel{3m}}} = \frac{3}{2}$

21. $\frac{a}{4c}\div\frac{a^2}{12c^2} = \frac{\overset{1}{\cancel{a}}}{\underset{1}{\cancel{4c}}}\cdot\frac{\overset{3c}{\cancel{12c^2}}}{\underset{a}{\cancel{a^2}}} = \frac{3c}{a}$

23. $\frac{x}{3y}\div 3y = \frac{x}{3y}\div\frac{3y}{1} = \frac{x}{3y}\cdot\frac{1}{3y} = \frac{x}{9y^2}$

25. $\frac{6xy}{z}\div\frac{x}{9yz} = \frac{6\overset{1}{\cancel{x}}y}{\underset{1}{\cancel{z}}}\cdot\frac{9y\overset{1}{\cancel{z}}}{\underset{1}{\cancel{x}}} = \frac{54y^2}{1} = 54y^2$

27. $\frac{8x^2}{5y}\div\frac{15xy^2}{16} = \frac{8\overset{x}{\cancel{x^2}}}{5y}\cdot\frac{16}{15\underset{1}{\cancel{x}}y^2} = \frac{128x}{75y^3}$

29. $12xy\div\frac{x}{4yz} = \frac{12\overset{1}{\cancel{x}}y}{1}\cdot\frac{4yz}{\underset{1}{\cancel{x}}} = \frac{48y^2z}{1} = 48y^2z$

31. $\frac{x}{4yz}\div 12xy = \frac{x}{4yz}\div\frac{12xy}{1} = \frac{\overset{1}{\cancel{x}}}{4yz}\cdot\frac{1}{12\underset{1}{\cancel{x}}y} = \frac{1}{48y^2z}$

33. $\frac{8\overset{x}{\cancel{x^2}}}{\underset{1\cdot1}{\cancel{3xy}}}\cdot\frac{\overset{2\overset{y}{\cancel{y^2}}}{\cancel{12y^3}}}{\underset{1\cdot1}{\cancel{6y}}} = \frac{16xy}{3}$

35. $\frac{21x^2y^2}{12cd}\div\frac{14xy}{9d} = \frac{\overset{3xy}{\cancel{21x^2y^2}}}{\underset{4}{\cancel{12}}c\cancel{d}}\cdot\frac{\overset{3}{\cancel{9d}}}{\underset{2}{\cancel{14xy}}} = \frac{9xy}{8c}$

37. $\frac{9u^4}{4v^3}\div\frac{-12u^2}{15v} = \frac{\overset{-3u^2}{\cancel{9u^4}}}{4\underset{v^2}{\cancel{v^3}}}\cdot\frac{15\cancel{v}}{\underset{4}{\cancel{-12u^2}}} = \frac{-45u^2}{16v^2}$

39. $\frac{3c^2d}{a^3b^3}\div\frac{3a^3b^3}{cd} = \frac{\overset{1}{\cancel{3}}c^2d}{a^3b^3}\cdot\frac{cd}{\underset{1}{\cancel{3}}a^3b^3} = \frac{c^3d^2}{a^6b^6}$

41. $\frac{3(-x)^2y}{2z}\cdot\frac{4z}{x} = \frac{3\overset{x}{\cancel{x^2}}y}{\underset{1}{\cancel{2z}}}\cdot\frac{\overset{2}{\cancel{4z}}}{\underset{1}{\cancel{x}}} = \frac{6xy}{1} = 6xy$

43. $\frac{3(-x)y^2}{2z}\div 6xy = \frac{-3xy^2}{2z}\div\frac{6xy}{1} = \frac{\overset{-y}{\cancel{3xy^2}}}{2z}\cdot\frac{1}{\underset{2}{\cancel{6xy}}} = \frac{-y}{4z}$

45. $3xy\div\frac{2x(-y)}{3z} = \frac{3xy}{1}\div\frac{-2xy}{3z} = \frac{3\overset{1}{\cancel{xy}}}{1}\cdot\frac{3z}{-2\underset{1}{\cancel{xy}}} = -\frac{9z}{2}$

47. $\frac{-3abc^2}{2d} \div \frac{3a^2b}{4cd^2} = \frac{\overset{-1}{\cancel{-3}abc^2}}{\underset{1}{\cancel{2d}}} \cdot \frac{\overset{2cd}{\cancel{4cd^2}}}{\underset{a}{\cancel{3a^2b}}} = \frac{-2c^3d}{a}$

49. $\frac{\overset{x}{\cancel{3x^2y}}}{\underset{1}{\cancel{x-y}}} \cdot \frac{\overset{1}{\cancel{x-y}}}{\underset{2}{\cancel{6xy}}} = \frac{x}{2}$

51. $\frac{x+3}{x^3+3x^2} \cdot \frac{x^3}{x-3} = \frac{\overset{1}{\cancel{x+3}}}{\underset{1\cdot 1}{\cancel{x^2}\cancel{(x+3)}}} \cdot \frac{\overset{x}{\cancel{x^3}}}{x-3} = \frac{x}{x-3}$

53. $\frac{x-2}{4y} \div \frac{x^2+x-6}{12y^2} = \frac{x-2}{4y} \cdot \frac{12y^2}{x^2+x-6} = \frac{\overset{1}{\cancel{x-2}}}{\underset{1}{\cancel{4y}}} \cdot \frac{\overset{3y}{\cancel{12y^2}}}{(x+3)\underset{1}{\cancel{(x-2)}}} = \frac{3y}{x+3}$

55. $\frac{6x^2}{4x^2y-12xy} \cdot \frac{x^2+x-12}{3x^2+12x} = \frac{\overset{\cancel{2x}}{\cancel{6x^2}}}{\underset{2}{\cancel{4x}}y\underset{1}{\cancel{(x-3)}}} \cdot \frac{\overset{1}{\cancel{(x+4)}}\overset{1}{\cancel{(x-3)}}}{\underset{1\cdot 1}{\cancel{3x}\cancel{(x+4)}}} = \frac{1}{2y}$

57. $(t^2-t-12) \div \frac{t^2-9}{t^2-3t} = \frac{t^2-t-12}{1} \cdot \frac{t^2-3t}{t^2-9}$

$= \frac{(t-4)\overset{1}{\cancel{(t+3)}}}{1} \cdot \frac{t\overset{1}{\cancel{(t-3)}}}{\underset{1\ \cdot\ 1}{\cancel{(t-3)}\cancel{(t+3)}}} = \frac{t(t-4)}{1} = t(t-4)$

59. $\frac{x^2-4}{x+3} \cdot \frac{x^2-9}{x+2} = \frac{(x-2)\overset{1}{\cancel{(x+2)}}}{\underset{1}{\cancel{(x+3)}}} \cdot \frac{\overset{1}{\cancel{(x+3)}}(x-3)}{\underset{1}{\cancel{x+2}}} = \frac{(x-2)(x-3)}{1} = x^2-5x+6$

61. $\frac{x^2+3x+2}{x+3} \cdot \frac{x^2+5x+6}{x+1} = \frac{\overset{1}{\cancel{(x+1)}}(x+2)}{\underset{1}{\cancel{(x+3)}}} \cdot \frac{(x+2)\overset{1}{\cancel{(x+3)}}}{\underset{1}{\cancel{x+1}}}$

$= \frac{(x+2)^2}{1} = x^2+4x+4$

63. $\frac{x^2+3x+2}{x+3} \div \frac{x^2+5x+6}{x+1} = \frac{(x+1)(x+2)}{x+3} \div \frac{(x+2)(x+3)}{x+1}$

$= \frac{(x+1)\overset{1}{\cancel{(x+2)}}}{x+3} \cdot \frac{x+1}{\underset{1}{\cancel{(x+2)}}(x+3)} = \frac{(x+1)^2}{(x+3)^2}$

65. $\frac{x+1}{x^2+8x+15} \cdot \frac{x^2+6x+5}{x+3} = \frac{x+1}{(x+3)\underset{1}{\cancel{(x+5)}}} \cdot \frac{(x+1)\overset{1}{\cancel{(x+5)}}}{x+3} = \frac{(x+1)^2}{(x+3)^2}$

67. $\frac{x+1}{x^2+8x+15} \div \frac{x^2+6x+5}{x+3} = \frac{\overset{1}{\cancel{x+1}}}{\underset{1}{\cancel{(x+3)}}(x+5)} \cdot \frac{\overset{1}{\cancel{x+3}}}{\underset{1}{\cancel{(x+1)}}(x+5)} = \frac{1}{(x+5)^2}$

69. $\frac{m+n}{m^2-n^2} \cdot \frac{m^2-mn}{m^2-2mn+n^2} = \frac{\overset{1}{\cancel{m+n}}}{\underset{1}{\cancel{(m+n)}}(m-n)} \cdot \frac{m\overset{1}{\cancel{(m-n)}}}{\underset{1}{\cancel{(m-n)}}(m-n)} = \frac{m}{(m-n)^2}$

71. $\frac{m + n}{m^2 - n^2} \div \frac{m^2 - mn}{m^2 - 2mn + n^2} = \frac{m + n}{m^2 - n^2} \cdot \frac{m^2 - 2mn + n^2}{m^2 - mn}$

$$= \frac{\overset{1}{\cancel{m + n}}}{\underset{1}{\cancel{(m + n)}}\underset{1}{\cancel{(m - n)}}} \cdot \frac{\overset{1}{\cancel{(m - n)}}\overset{1}{\cancel{(m - n)}}}{m\underset{1}{\cancel{(m - n)}}} = \frac{1}{m}$$

73. $-(x^2 - 3x) \cdot \frac{x - 2}{x - 3} = \frac{-x\overset{1}{\cancel{(x - 3)}}}{1} \cdot \frac{x - 2}{\underset{1}{\cancel{x - 3}}} = -x(x - 2)$ or $2x - x^2$

75. $\left(\frac{d^5}{3a} \div \frac{d^2}{6a^2}\right) \cdot \frac{a}{4d^3} = \left(\frac{\overset{d^3}{\cancel{d^5}}}{\underset{1}{\cancel{3a}}} \cdot \frac{\overset{2a}{\cancel{6a^2}}}{\underset{1}{\cancel{d^2}}}\right) \cdot \frac{a}{4d^3} = \frac{\overset{1}{\cancel{2}}a\overset{1}{\cancel{d^3}}}{1} \cdot \frac{a}{\underset{2}{\cancel{4d^3}}} = \frac{a^2}{2}$

77. $\frac{\overset{x}{\cancel{2x^2}}}{\underset{y^2}{\cancel{3y^3}}} \cdot \frac{\overset{-2\cdot 1}{\cancel{-6yz}}}{\cancel{2x}} \cdot \frac{y}{-x\cancel{z}} = \frac{-2xy}{-xy^2} = \frac{2}{y}$

79. $\frac{3xy}{2z} \cdot \frac{-4xz}{5y} \cdot \frac{-5yz}{6x} = \frac{\overset{xyz}{\cancel{60x^2y^2z^2}}}{\underset{1}{\cancel{60xyz}}} = \frac{xyz}{1} = xyz$

81. $\frac{x}{2y} \cdot \frac{3z}{4w} \cdot \frac{6y}{x} \cdot \frac{-w}{3z} = \frac{-\overset{3}{\cancel{18}}\overset{1}{\cancel{xyzw}}}{\underset{4}{\cancel{24}}\underset{1}{\cancel{xyzw}}} = -\frac{3}{4}$

83. $\frac{9 - x^2}{x^2 + 5x + 6} \cdot \frac{x + 2}{x - 3} = \frac{(3 - x)\overset{1}{\cancel{(3 + x)}}}{\underset{1}{\cancel{(x + 2)}}\underset{1}{\cancel{(x + 3)}}} \cdot \frac{\overset{1}{\cancel{x + 2}}}{x - 3} = \frac{3 - x}{x - 3} = -\frac{\overset{1}{\cancel{(x - 3)}}}{\underset{1}{\cancel{x - 3}}} = -1$

85. $\frac{x^2 + 2x + 1}{x - 2} \cdot \frac{x^2 - 4}{x^2 - 2x - 3} \cdot \frac{x^2 - 5x + 6}{x + 1}$

$$= \frac{\overset{1}{\cancel{(x + 1)}}\overset{1}{\cancel{(x + 1)}}}{\underset{1}{\cancel{x - 2}}} \cdot \frac{\overset{1}{\cancel{(x - 2)}}(x + 2)}{\underset{1}{\cancel{(x + 1)}}\underset{1}{\cancel{(x - 3)}}} \cdot \frac{\overset{1}{\cancel{(x - 3)}}(x - 2)}{\underset{1}{\cancel{x + 1}}} = \frac{(x + 2)(x - 2)}{1} = x^2 - 4$$

87. $\frac{x^2 - 4x + 3}{x - 2} \cdot \frac{x^2 - 5x + 6}{x^2 - 3x + 2} \cdot \frac{x^2 - x - 2}{x^2 - 6x + 9}$

$$= \frac{\overset{1}{\cancel{(x - 1)}}\overset{1}{\cancel{(x - 3)}}}{\underset{1}{\cancel{x - 2}}} \cdot \frac{\overset{1}{\cancel{(x - 2)}}\overset{1}{\cancel{(x - 3)}}}{\underset{1}{\cancel{(x - 2)}}\underset{1}{\cancel{(x - 1)}}} \cdot \frac{\overset{1}{\cancel{(x - 2)}}(x + 1)}{\underset{1}{\cancel{(x - 3)}}\underset{1}{\cancel{(x - 3)}}} = \frac{x + 1}{1} = x + 1$$

89. $\frac{x^2 - x - 2}{x^2 + 6x + 9} \div \left(\frac{x^2 - 4x + 3}{x^2 + x - 6} \cdot \frac{x^2 - 4x + 4}{x^2 - 9}\right)$

$$= \frac{x^2 - x - 2}{x^2 + 6x + 9} \div \left(\frac{(x - 1)\overset{1}{\cancel{(x - 3)}}}{(x + 3)\underset{1}{\cancel{(x - 2)}}} \cdot \frac{\overset{1}{\cancel{(x - 2)}}(x - 2)}{\underset{1}{\cancel{(x - 3)}}(x + 3)}\right)$$

$$= \frac{x^2 - x - 2}{x^2 + 6x + 9} \div \frac{(x - 1)(x - 2)}{(x + 3)(x + 3)} = \frac{\cancel{(x - 2)}(x + 1)}{\cancel{(x + 3)}\cancel{(x + 3)}} \cdot \frac{\cancel{(x + 3)}\cancel{(x + 3)}}{(x - 1)\cancel{(x - 2)}} = \frac{x + 1}{x - 1}$$

91. $$\frac{x^2 - xy}{xy + y^2} \div \left(\frac{x^2 - y^2}{x^2 + 2xy + y^2} \div \frac{x^2 - 2xy + y^2}{x^2y + xy^2}\right) = \frac{x^2 - xy}{xy + y^2} \div \left(\frac{x^2 - y^2}{x^2 + 2xy + y^2} \cdot \frac{x^2y + xy^2}{x^2 - 2xy + y^2}\right)$$

$$= \frac{x^2 - xy}{xy + y^2} \div \left(\frac{\cancel{(x - y)}\cancel{(x + y)}}{\cancel{(x + y)}\cancel{(x + y)}} \cdot \frac{xy\cancel{(x + y)}}{\cancel{(x - y)}(x - y)}\right) = \frac{x^2 - xy}{xy + y^2} \div \frac{xy}{x - y}$$

$$= \frac{x^2 - xy}{xy + y^2} \cdot \frac{x - y}{xy} = \frac{\cancel{x}(x - y)}{y(x + y)} \cdot \frac{x - y}{\cancel{x}y} = \frac{(x - y)^2}{y^2(x + y)}$$

93. $$\frac{x^2 - 1}{x - 1} = \frac{(x + 1)\cancel{(x - 1)}}{\cancel{(x - 1)}} = x + 1 \quad \text{if} \quad x - 1 \neq 0, \text{ that is, for } x \neq 1$$

95. $$\frac{x^2 + x - 6}{x + 3} = \frac{(x - 2)\cancel{(x + 3)}}{\cancel{(x + 3)}} = x - 2 \quad \text{if} \quad x + 3 \neq 0, \text{ that is, for } x \neq -3$$

97. $$\frac{x^3 - 2x^2 + x}{x^2 - x} = \frac{\cancel{x(x - 1)}(x - 1)}{\cancel{x(x - 1)}} = x - 1 \quad \text{if} \quad x(x - 1) \neq 0, \text{ that is, for } x \neq 0, 1$$

99. $$\frac{x^4 + x^3}{x^4 - x^3} = \frac{\cancel{x^3}(x + 1)}{\cancel{x^3}(x - 1)} = \frac{x + 1}{x - 1} \quad \text{if} \quad x^3(x - 1) \neq 0, \text{ that is, for } x \neq 0, 1.$$

Exercise 3-3 Addition and Subtraction

Key Ideas and Formulas

Addition and Subtraction of Rational Expressions	
If P, D, and Q are polynomials ($D \neq 0$), then	
$\frac{P}{D} + \frac{Q}{D} = \frac{P + Q}{D}$	(1)
$\frac{P}{D} - \frac{Q}{D} = \frac{P - Q}{D}$	(2)

If the expressions to be added or subtracted have different denominators, the fundamental principle of fractions is used to change the form of each fraction so that they have a common denominator, then property (1) or (2) is used. Usually the common denominator is chosen to be the least common multiple (LCM) of the denominators, called the least common denominator (LCD).

If the LCD is not obvious, then it is found as follows:

Finding the Least Common Denominator (LCD)	
Step 1:	Factor each denominator completely using integer coefficients.
Step 2:	The LCD must contain each *different* factor that occurs in each of the denominators to the highest power it occurs in any one denominator.

---

1. $3x$

3. x

5. v^3

7. $3x = 3x$
$6x^2 = 2 \cdot 3x^2$
$4 = 2^2$
$\text{lcm} = 2^2 \cdot 3x^2 = 12x^2$

9. $(x + 1)(x - 2)$

11. $3y(y + 3)$

13. $\dfrac{7x + 2}{5x^2}$

15. $\dfrac{4x}{2x - 1} - \dfrac{2}{2x - 1} = \dfrac{4x - 2}{2x - 1} = \dfrac{2(2x - 1)}{2x - 1} = \dfrac{2}{1} = 2$

17. $\dfrac{y}{y^2 - 9} - \dfrac{3}{y^2 - 9} = \dfrac{y - 3}{y^2 - 9} = \dfrac{y - 3}{(y - 3)(y + 3)} = \dfrac{1}{y + 3}$

19. $\dfrac{5}{3k} - \dfrac{6x - 4}{3k} = \dfrac{5}{3k} - \dfrac{(6x - 4)}{3k} = \dfrac{5 - (6x - 4)}{3k} = \dfrac{5 - 6x + 4}{3k}$

$= \dfrac{9 - 6x}{3k} = \dfrac{3(3 - 2x)}{3k} = \dfrac{3 - 2x}{k}$

21. $\dfrac{3x}{y} + \dfrac{1}{4}$ lcd $= 4y$

$\dfrac{3x}{y} + \dfrac{1}{4} = \dfrac{4 \cdot 3x}{4y} + \dfrac{1 \cdot y}{4y} = \dfrac{12x}{4y} + \dfrac{y}{4y} = \dfrac{12x + y}{4y}$

COMMON ERRORS: "cancelling" the y. y is not a common factor of numerator and denominator in problem 21 or 23. In 21 it is a factor of the denominator, but a term in the numerator. In 23 it is the entire denominator (therefore a factor) but a term in the numerator.

23. $\dfrac{2}{y} + 1$ lcd $= y$

$\dfrac{2}{y} + 1 = \dfrac{2}{y} + \dfrac{1}{1} = \dfrac{2}{y} + \dfrac{y}{y} = \dfrac{2 + y}{y}$

25. $\dfrac{u}{v^2} - \dfrac{1}{v} + \dfrac{u^3}{v^3}$ lcd $= v^3$

$\dfrac{u}{v^2} - \dfrac{1}{v} + \dfrac{u^3}{v^3} = \dfrac{u \cdot v}{v^2 \cdot v} - \dfrac{1 \cdot v^2}{v \cdot v^2} + \dfrac{u^3}{v^3} = \dfrac{uv}{v^3} - \dfrac{v^2}{v^3} + \dfrac{u^3}{v^3} = \dfrac{uv - v^2 + u^3}{v^3} = \dfrac{u^3 + uv - v^2}{v^3}$

27. $\dfrac{2}{3x} - \dfrac{1}{6x^2} + \dfrac{3}{4}$ lcd $= 12x^2$ (see problem 7)

$\dfrac{2}{3x} - \dfrac{1}{6x^2} + \dfrac{3}{4} = \dfrac{2 \cdot 4x}{3x \cdot 4x} - \dfrac{1 \cdot 2}{6x^2 \cdot 2} + \dfrac{3 \cdot 3x^2}{4 \cdot 3x^2} = \dfrac{8x}{12x^2} - \dfrac{2}{12x^2} + \dfrac{9x^2}{12x^2}$

$= \dfrac{8x - 2 + 9x^2}{12x^2} = \dfrac{9x^2 + 8x - 2}{12x^2}$

29. $\dfrac{2}{x + 1} + \dfrac{3}{x - 2}$ lcd $= (x + 1)(x - 2)$

$\dfrac{2}{x + 1} + \dfrac{3}{x - 2} = \dfrac{2(x - 2)}{(x + 1)(x - 2)} + \dfrac{3(x + 1)}{(x + 1)(x - 2)}$

$$= \frac{(2x - 4)}{(x + 1)(x - 2)} + \frac{(3x + 3)}{(x + 1)(x - 2)} = \frac{(2x - 4) + (3x + 3)}{(x + 1)(x - 2)}$$

$$= \frac{2x - 4 + 3x + 3}{(x + 1)(x - 2)} = \frac{5x - 1}{(x + 1)(x - 2)}$$

31. $\frac{3}{y + 3} - \frac{2}{3y}$ lcd $= 3y(y + 3)$

$$\frac{3}{y + 3} - \frac{2}{3y} = \frac{3 \cdot 3y}{3y(y + 3)} - \frac{2(y + 3)}{3y(y + 3)} = \frac{9y}{3y(y + 3)} - \frac{(2y + 6)}{3y(y + 3)}$$

$$= \frac{9y - (2y + 6)}{3y(y + 3)} = \frac{9y - 2y - 6}{3y(y + 3)} = \frac{7y - 6}{3y(y + 3)}$$

33. $$\frac{x}{y}\left(\frac{1}{x} + \frac{1}{y}\right) = \frac{x}{y}\left(\frac{y}{xy} + \frac{x}{xy}\right) = \frac{\overset{1}{x}}{y} \cdot \frac{y + x}{\underset{1}{x}y} = \frac{y + x}{y^2}$$

35. $$\frac{x - y}{x} \div \left(\frac{y}{x} - \frac{x}{y}\right) = \frac{x - y}{x} \div \left(\frac{y^2}{xy} - \frac{x^2}{xy}\right) = \frac{x - y}{x} \div \frac{y^2 - x^2}{xy}$$

$$= \frac{\overset{-1}{x - y}}{\underset{1}{x}} \cdot \frac{\overset{1}{x}y}{\underset{1}{(y - x)}(y + x)} = \frac{-y}{y + x}$$

37. $$\frac{x}{y} \div \frac{y}{z} - \frac{y}{x} \div \frac{y}{z} = \frac{x}{y} \cdot \frac{z}{y} - \frac{\overset{1}{y}}{x} \cdot \frac{z}{\underset{1}{y}} = \frac{xz}{y^2} - \frac{z}{x} \qquad \text{lcd} = xy^2$$

$$= \frac{xz \cdot x}{xy^2} - \frac{z \cdot y^2}{xy^2} = \frac{x^2z - y^2z}{xy^2}$$

39. $12x^3 = 2^2 \cdot 3x^3$
$8x^2y^2 = 2^3x^2y^2$
$3xy^2 = 3xy^2$ lcm $= 2^3 \cdot 3x^3y^2$
$= 24x^3y^2$

41. $15x^2y = 3 \cdot 5x^2y$
$25xy = 5^2xy$
$5y^2 = 5y^2$ lcm $= 3 \cdot 5^2x^2y^2$
$= 75x^2y^2$

43. $6(x - 1) = 2 \cdot 3(x - 1)$
$9(x - 1)^2 = 3^2(x - 1)^2$ lcm $= 2 \cdot 3^2(x - 1)^2 = 18(x - 1)^2$

45. $6(x - 7)(x + 7) = 2 \cdot 3(x - 7)(x + 7)$
$8(x + 7)^2 = 2^3(x + 7)^2$ lcm $= 2^3 \cdot 3(x - 7)(x + 7)^2 = 24(x - 7)(x + 7)^2$

47. $x^2 - 4 = (x - 2)(x + 2)$
$x^2 + 4x + 4 = (x + 2)^2$ lcm $= (x - 2)(x + 2)^2$

49. $3x^2 + 3x = 3x(x + 1)$
$4x^2 = 2^2x^2$
$3x^2 + 6x + 3 = 3(x^2 + 2x + 1) = 3(x + 1)^2$ lcm $= 2^2 \cdot 3x^2(x + 1)^2 = 12x^2(x + 1)^2$

51. $\dfrac{2}{9u^3v^2} - \dfrac{1}{6uv} + \dfrac{1}{12v^3}$

$9u^3v^2 = 3^2u^3v^2$

$6uv = 2 \cdot 3uv$

$12v^3 = 2^2 \cdot 3v^3 \qquad \text{lcm} = 2^2 \cdot 3^2u^3v^3 = 36u^3v^3$

$$\frac{2}{9u^3v^2} - \frac{1}{6uv} + \frac{1}{12v^3} = \frac{2 \cdot 4v}{9u^3v^2 \cdot 4v} - \frac{1 \cdot 6u^2v^2}{6uv \cdot 6u^2v^2} + \frac{1 \cdot 3u^3}{12v^3 \cdot 3u^3}$$

$$= \frac{8v}{36u^3v^3} - \frac{6u^2v^2}{36u^3v^3} + \frac{3u^3}{36u^3v^3} = \frac{8v - 6u^2v^2 + 3u^3}{36u^3v^3}$$

53. $\dfrac{4t - 3}{18t^3} + \dfrac{3}{4t} - \dfrac{2t - 1}{6t^2}$

$18t^3 = 2 \cdot 3^2t^3$

$4t = 2^2t$

$6t^2 = 2 \cdot 3t^2 \qquad \text{lcm} = 2^2 \cdot 3^2 \cdot t^3 = 36t^3$

$$\frac{4t - 3}{18t^3} + \frac{3}{4t} - \frac{2t - 1}{6t^2} = \frac{(4t - 3) \cdot 2}{18t^3 \cdot 2} + \frac{3 \cdot 9t^2}{4t \cdot 9t^2} - \frac{(2t - 1) \cdot 6t}{6t^2 \cdot 6t}$$

$$= \frac{(8t - 6)}{36t^3} + \frac{27t^2}{36t^3} - \frac{(12t^2 - 6t)}{36t^3}$$

$$= \frac{(8t - 6) + 27t^2 - (12t^2 - 6t)}{36t^3} = \frac{8t - 6 + 27t^2 - 12t^2 + 6t}{36t^3}$$

$$= \frac{15t^2 + 14t - 6}{36t^3}$$

55. $\dfrac{t + 1}{t - 1} - 1 = \dfrac{t + 1}{t - 1} - \dfrac{1}{1} \qquad \text{lcd} = t - 1$

$$= \frac{(t + 1)}{t - 1} - \frac{(t - 1)}{t - 1} = \frac{(t + 1) - (t - 1)}{t - 1} = \frac{t + 1 - t + 1}{t - 1} = \frac{2}{t - 1}$$

57. $5 + \dfrac{a}{a + 1} - \dfrac{a}{a - 1} = \dfrac{5}{1} + \dfrac{a}{a + 1} - \dfrac{a}{a - 1} \qquad \text{lcd} = (a + 1)(a - 1)$

$$= \frac{5(a + 1)(a - 1)}{1(a + 1)(a - 1)} + \frac{a(a - 1)}{(a + 1)(a - 1)} - \frac{a(a + 1)}{(a + 1)(a - 1)}$$

$$= \frac{5(a + 1)(a - 1) + a(a - 1) - a(a + 1)}{(a + 1)(a - 1)}$$

$$= \frac{5(a^2 - 1) + a^2 - a - a^2 - a}{(a + 1)(a - 1)} = \frac{5a^2 - 5 + a^2 - a - a^2 - a}{(a + 1)(a - 1)}$$

$$= \frac{5a^2 - 2a - 5}{(a + 1)(a - 1)}$$

59. $\dfrac{2}{3(x - 5)^2} - \dfrac{1}{4(x + 5)(x - 5)}$

$3(x - 5)^2 = 3(x - 5)^2$

$4(x + 5)(x - 5) = 2^2(x + 5)(x - 5) \qquad \text{lcm} = 2^2 \cdot 3(x + 5)(x - 5)^2$

$= 12(x + 5)(x - 5)^2$

$$\frac{2}{3(x - 5)^2} - \frac{1}{4(x + 5)(x - 5)} = \frac{2 \cdot 4(x + 5)}{3(x - 5)^2 \cdot 4(x + 5)} - \frac{1 \cdot 3(x - 5)}{4(x + 5)(x - 5) \cdot 3(x - 5)}$$

$$= \frac{8(x + 5)}{12(x + 5)(x - 5)^2} - \frac{3(x - 5)}{12(x + 5)(x - 5)^2}$$

$$= \frac{8(x + 5) - 3(x - 5)}{12(x + 5)(x - 5)^2} = \frac{8x + 40 - 3x + 15}{12(x + 5)(x - 5)^2}$$

$$= \frac{5x + 55}{12(x + 5)(x - 5)^2}$$

61. $\frac{5}{6(x - 1)} + \frac{2}{9(x - 1)^2}$ lcd $= 18(x - 1)^2$ (see problem 43)

$$\frac{5}{6(x - 1)} + \frac{2}{9(x - 1)^2} = \frac{5 \cdot 3(x - 1)}{6(x - 1) \cdot 3(x - 1)} + \frac{2 \cdot 2}{9(x - 1)^2 \cdot 2}$$

$$= \frac{15(x - 1)}{18(x - 1)^2} + \frac{4}{18(x - 1)^2} = \frac{15(x - 1) + 4}{18(x - 1)^2}$$

$$= \frac{15x - 15 + 4}{18(x - 1)^2} = \frac{15x - 11}{18(x - 1)^2}$$

63. $\frac{3}{x + 3} - \frac{3x + 1}{(x - 1)(x + 3)}$ lcd $= (x - 1)(x + 3)$

$$\frac{3}{x + 3} - \frac{3x + 1}{(x - 1)(x + 3)} = \frac{3(x - 1)}{(x - 1)(x + 3)} - \frac{(3x + 1)}{(x - 1)(x + 3)} = \frac{3(x - 1) - (3x + 1)}{(x - 1)(x + 3)}$$

$$= \frac{3x - 3 - 3x - 1}{(x - 1)(x + 3)} = \frac{-4}{(x - 1)(x + 3)}$$

COMMON ERROR: $3(x - 1) - 3x + 1$ in numerator. Numerators with more than one term should be placed in parentheses to avoid sign errors.

65. $\frac{3s}{3s^2 - 12} + \frac{1}{2s^2 + 4s}$

$3s^2 - 12 = 3(s^2 - 4) = 3(s - 2)(s + 2)$

$2s^2 + 4s = 2s(s + 2)$ lcm $= 2 \cdot 3s(s - 2)(s + 2) = 6s(s - 2)(s + 2)$

$$\frac{3s}{3s^2 - 12} + \frac{1}{2s^2 + 4s} = \frac{3s}{3(s - 2)(s + 2)} + \frac{1}{2s(s + 2)}$$

$$= \frac{3s \cdot 2s}{3(s - 2)(s + 2) \cdot 2s} + \frac{1 \cdot 3(s - 2)}{2s(s + 2) \cdot 3(s - 2)}$$

$$= \frac{6s^2}{6s(s - 2)(s + 2)} + \frac{3(s - 2)}{6s(s - 2)(s + 2)}$$

$$= \frac{6s^2 + 3(s - 2)}{6s(s - 2)(s + 2)} = \frac{6s^2 + 3s - 6}{6s(s - 2)(s + 2)}$$

$$= \frac{\overset{1}{\cancel{3}}(2s^2 + s - 2)}{\underset{1}{\cancel{3}} \cdot 2s(s - 2)(s + 2)} = \frac{2s^2 + s - 2}{2s(s - 2)(s + 2)}$$

Alternatively, it would be correct to reduce $\frac{3s}{3s^2 - 12}$ to $\frac{s}{s^2 - 4}$ at the outset.

67. $\dfrac{3}{x^2 - 4} - \dfrac{1}{x^2 + 4x + 4}$ lcd $= (x - 2)(x + 2)^2$ (see problem 47)

$$\frac{3}{x^2 - 4} - \frac{1}{x^2 + 4x + 4} = \frac{3}{(x - 2)(x + 2)} - \frac{1}{(x + 2)^2}$$
$$= \frac{3(x + 2)}{(x - 2)(x + 2)^2} - \frac{1(x - 2)}{(x - 2)(x + 2)^2}$$
$$= \frac{3(x + 2) - 1(x - 2)}{(x - 2)(x + 2)^2} = \frac{3x + 6 - x + 2}{(x - 2)(x + 2)^2}$$
$$= \frac{2x + 8}{(x - 2)(x + 2)^2} = \frac{2(x + 4)}{(x - 2)(x + 2)^2}$$

69. $\dfrac{2}{x + 3} - \dfrac{1}{x - 3} + \dfrac{2x}{x^2 - 9}$

$x + 3 = x + 3$

$x - 3 = x - 3$

$x^2 - 9 = (x - 3)(x + 3)$ lcm $= (x - 3)(x + 3)$

$$\frac{2}{x + 3} - \frac{1}{x - 3} + \frac{2x}{x^2 - 9} = \frac{2(x - 3)}{(x - 3)(x + 3)} - \frac{1(x + 3)}{(x - 3)(x + 3)} + \frac{2x}{(x - 3)(x + 3)}$$
$$= \frac{2(x - 3) - 1(x + 3) + 2x}{(x - 3)(x + 3)} = \frac{2x - 6 - x - 3 + 2x}{(x - 3)(x + 3)}$$
$$= \frac{3x - 9}{(x - 3)(x + 3)} = \frac{3\cancel{(x - 3)}^{1}}{\cancel{(x - 3)}_{1}(x + 3)} = \frac{3}{x + 3}$$

71. $$\frac{1}{x^2}\left(\frac{x}{x + 1} + \frac{x}{x - 1}\right) = \frac{1}{x^2}\left(\frac{x(x - 1)}{(x + 1)(x - 1)} + \frac{x(x + 1)}{(x + 1)(x - 1)}\right)$$
$$= \frac{1}{x^2} \cdot \frac{x(x - 1) + x(x + 1)}{(x + 1)(x - 1)} = \frac{1}{x^2} \cdot \frac{x^2 - x + x^2 + x}{(x + 1)(x - 1)} = \frac{1}{\cancel{x^2}_{1}} \cdot \frac{2\cancel{x^2}^{1}}{(x + 1)(x - 1)}$$
$$= \frac{2}{(x + 1)(x - 1)} \text{ or } \frac{2}{x^2 - 1}$$

73. $$\frac{(x + 1)^2}{x} \div \left(\frac{x + 1}{x} - \frac{x}{x - 1}\right) = \frac{(x + 1)^2}{x} \div \left(\frac{(x + 1)(x - 1)}{x(x - 1)} - \frac{x \cdot x}{x(x - 1)}\right)$$
$$= \frac{(x + 1)^2}{x} \div \frac{(x + 1)(x - 1) - x^2}{x(x - 1)} = \frac{(x + 1)^2}{x} \div \frac{x^2 - 1 - x^2}{x(x - 1)} = \frac{(x + 1)^2}{x} \div \frac{-1}{x(x - 1)}$$
$$= \frac{(x + 1)^2}{\cancel{x}_{1}} \cdot \frac{\cancel{x}^{1}(x - 1)}{-1} = \frac{(x + 1)^2(x - 1)}{-1} = -(x + 1)^2(x - 1)$$

75. $\frac{x}{(x+1)^2}\left(\frac{2}{x} + \frac{x}{x+\frac{1}{2}}\right) = \frac{x}{(x+1)^2}\left(\frac{2(x+\frac{1}{2})}{x(x+\frac{1}{2})} + \frac{x\cdot x}{x(x+\frac{1}{2})}\right)$

$= \frac{x}{(x+1)^2}\cdot\frac{2(x+\frac{1}{2}) + x^2}{x(x+\frac{1}{2})} = \frac{x}{(x+1)^2}\cdot\frac{x^2+2x+1}{x(x+\frac{1}{2})} = \frac{\cancel{x}}{\cancel{(x+1)^2}}\cdot\frac{\cancel{(x+1)^2}}{\cancel{x}(x+\frac{1}{2})}$

$= \frac{1}{x+\frac{1}{2}}$

77. $\frac{5}{y-3} - \frac{2}{3-y} = \frac{5}{y-3} - \frac{2}{-(y-3)} = \frac{5}{y-3} + \frac{2}{y-3} = \frac{5+2}{y-3} = \frac{7}{y-3}$

79. $\frac{3}{x-3} + \frac{x}{3-x} = \frac{3}{x-3} + \frac{x}{-(x-3)} = \frac{3}{x-3} - \frac{x}{x-3} = \frac{3-x}{x-3} = \frac{-\cancel{(x-3)}}{\cancel{x-3}} = -1$

81. $\frac{1}{5x-5} - \frac{1}{3x-3} + \frac{1}{1-x} = \frac{1}{5x-5} - \frac{1}{3x-3} + \frac{1}{-(x-1)} = \frac{1}{5x-5} - \frac{1}{3x-3} - \frac{1}{x-1}$

$5x - 5 = 5(x-1)$
$3x - 3 = 3(x-1)$
$x - 1 = x - 1 \qquad \text{lcm} = 3\cdot5(x-1) = 15(x-1)$

$\frac{1}{5x-5} - \frac{1}{3x-3} - \frac{1}{x-1} = \frac{1\cdot3}{5(x-1)\cdot3} - \frac{1\cdot5}{3(x-1)\cdot5} - \frac{1\cdot15}{15(x-1)}$

$= \frac{3}{15(x-1)} - \frac{5}{15(x-1)} - \frac{15}{15(x-1)} = \frac{3-5-15}{15(x-1)}$

$= \frac{-17}{15(x-1)}$

83. $\frac{x}{x^2-x-2} - \frac{1}{x^2+5x-14} - \frac{2}{x^2+8x+7}$

$x^2 - x - 2 = (x-2)(x+1)$
$x^2 + 5x - 14 = (x-2)(x+7)$
$x^2 + 8x + 7 = (x+1)(x+7) \qquad \text{lcm} = (x+1)(x-2)(x+7)$

$\frac{x}{x^2-x-2} - \frac{1}{x^2+5x-14} - \frac{2}{x^2+8x+7}$

$= \frac{x(x+7)}{(x+1)(x-2)(x+7)} - \frac{1(x+1)}{(x+1)(x-2)(x+7)} - \frac{2(x-2)}{(x+1)(x-2)(x+7)}$

$= \frac{x(x+7) - 1(x+1) - 2(x-2)}{(x+1)(x-2)(x+7)} = \frac{x^2+7x-x-1-2x+4}{(x+1)(x-2)(x+7)}$

$= \frac{x^2+4x+3}{(x+1)(x-2)(x+7)} = \frac{\cancel{(x+1)}(x+3)}{\cancel{(x+1)}(x-2)(x+7)} = \frac{x+3}{(x-2)(x+7)}$

85. $\dfrac{1}{3x^2 + 3x} + \dfrac{1}{4x^2} - \dfrac{1}{3x^2 + 6x + 3}$ $\quad$ lcd $= 12x^2(x + 1)^2$ (see problem 49)

$$\frac{1}{3x^2 + 3x} + \frac{1}{4x^2} - \frac{1}{3x^2 + 6x + 3} = \frac{1}{3x(x + 1)} + \frac{1}{4x^2} - \frac{1}{3(x + 1)^2}$$

$$= \frac{1 \cdot 4x(x + 1)}{3x(x + 1) \cdot 4x(x + 1)} + \frac{1 \cdot 3(x + 1)^2}{4x^2 \cdot 3(x + 1)^2} - \frac{1 \cdot 4x^2}{3(x + 1)^2 \cdot 4x^2}$$

$$= \frac{4x(x + 1)}{12x^2(x + 1)^2} + \frac{3(x + 1)^2}{12x^2(x + 1)^2} - \frac{4x^2}{12x^2(x + 1)^2}$$

$$= \frac{4x(x + 1) + 3(x + 1)^2 - 4x^2}{12x^2(x + 1)^2} = \frac{4x^2 + 4x + 3(x^2 + 2x + 1) - 4x^2}{12x^2(x + 1)^2}$$

$$= \frac{4x^2 + 4x + 3x^2 + 6x + 3 - 4x^2}{12x^2(x + 1)^2} = \frac{3x^2 + 10x + 3}{12x^2(x + 1)^2} = \frac{(3x + 1)(x + 3)}{12x^2(x + 1)^2}$$

87. $\dfrac{xy^2}{x^3 - y^3} - \dfrac{y}{x^2 + xy + y^2}$

$$x^3 - y^3 = (x - y)(x^2 + xy + y^2)$$

$x^2 + xy + y^2 = x^2 + xy + y^2$ $\qquad$ lcm $= (x - y)(x^2 + xy + y^2)$

$$\frac{xy^2}{x^3 - y^3} - \frac{y}{x^2 + xy + y^2} = \frac{xy^2}{(x - y)(x^2 + xy + y^2)} - \frac{y(x - y)}{(x - y)(x^2 + xy + y^2)}$$

$$= \frac{xy^2 - y(x - y)}{(x - y)(x^2 + xy + y^2)} = \frac{xy^2 - xy + y^2}{(x - y)(x^2 + xy + y^2)}$$

$$= \frac{xy^2 - xy + y^2}{x^3 - y^3}$$

89. $\dfrac{x + 1}{x + 3}\left(\dfrac{2}{x + 1} - \dfrac{1}{x + 2}\right) = \dfrac{x + 1}{x + 3}\left(\dfrac{2(x + 2)}{(x + 1)(x + 2)} - \dfrac{1(x + 1)}{(x + 1)(x + 2)}\right)$

$$= \frac{x + 1}{x + 3} \cdot \frac{2(x + 2) - 1(x + 1)}{(x + 1)(x + 2)} = \frac{x + 1}{x + 3} \cdot \frac{2x + 4 - x - 1}{(x + 1)(x + 2)}$$

$$= \frac{\overset{1}{\cancel{x + 1}}}{\underset{1}{\cancel{x + 3}}} \cdot \frac{\overset{1}{\cancel{(x + 3)}}}{\underset{1}{\cancel{(x + 1)}}(x + 2)} = \frac{1}{x + 2}$$

91. This problem can be solved exactly like the previous problem. Here is an alternative approach, however.

$$\frac{x^2 + 3x + 2}{x + 3}\left(\frac{x + 3}{x + 1} - \frac{x + 3}{x + 2}\right) = \frac{(x + 1)(x + 2)}{x + 3}\left(\frac{x + 3}{x + 1} - \frac{x + 3}{x + 2}\right)$$

$$= \frac{(x + 1)(x + 2)}{x + 3} \cdot \frac{x + 3}{x + 1} - \frac{(x + 1)(x + 2)}{x + 3} \cdot \frac{x + 3}{x + 2}$$ using the distributive property

$$= \frac{\overset{1}{\cancel{(x + 1)}}(x + 2)}{\underset{1}{\cancel{(x + 3)}}} \cdot \frac{\overset{1}{\cancel{(x + 3)}}}{\underset{1}{\cancel{(x + 1)}}} - \frac{(x + 1)\overset{1}{\cancel{(x + 2)}}}{\underset{1}{\cancel{(x + 3)}}} \cdot \frac{\overset{1}{\cancel{(x + 3)}}}{\underset{1}{\cancel{(x + 2)}}} = (x + 2) - (x + 1)$$

$$= x + 2 - x - 1 = 1$$

93. $\dfrac{x}{x^2 - 1} \div \left(\dfrac{1}{x + 1} + \dfrac{1}{x - 1}\right) = \dfrac{x}{x^2 - 1} \div \left(\dfrac{(x - 1)}{(x + 1)(x - 1)} + \dfrac{(x + 1)}{(x + 1)(x - 1)}\right)$

$= \dfrac{x}{x^2 - 1} \div \dfrac{(x - 1) + (x + 1)}{(x + 1)(x - 1)} = \dfrac{x}{x^2 - 1} \div \dfrac{x - 1 + x + 1}{(x^2 - 1)}$

$= \dfrac{x}{x^2 - 1} \div \dfrac{2x}{x^2 - 1} = \dfrac{\overset{1}{\cancel{x}}}{\underset{1}{\cancel{x^2 - 1}}} \cdot \dfrac{\overset{1}{\cancel{x^2 - 1}}}{\underset{1}{2\cancel{x}}} = \dfrac{1}{2}$

95. $\dfrac{1}{x} + \dfrac{x}{x + 1} - \dfrac{x + 1}{x + 2} = \dfrac{(x + 1)(x + 2)}{x(x + 1)(x + 2)} + \dfrac{x \cdot x(x + 2)}{x(x + 1)(x + 2)} - \dfrac{(x + 1) \cdot x(x + 1)}{x(x + 1)(x + 2)}$

$= \dfrac{(x + 1)(x + 2) + x^2(x + 2) - x(x + 1)^2}{x(x + 1)(x + 2)}$

$= \dfrac{x^2 + 3x + 2 + x^3 + 2x^2 - x(x^2 + 2x + 1)}{x(x + 1)(x + 2)}$

$= \dfrac{x^2 + 3x + 2 + x^3 + 2x^2 - x^3 - 2x^2 - x}{x(x + 1)(x + 2)}$

$= \dfrac{x^2 + 2x + 2}{x(x + 1)(x + 2)}$

97. $\left(\dfrac{1}{a} - \dfrac{1}{b}\right)\left(\dfrac{1}{a} + \dfrac{1}{b}\right) = \left(\dfrac{b}{ab} - \dfrac{a}{ab}\right)\left(\dfrac{b}{ab} + \dfrac{a}{ab}\right) = \dfrac{b - a}{ab} \cdot \dfrac{b + a}{ab} = \dfrac{(b - a)(b + a)}{a^2b^2} = \dfrac{b^2 - a^2}{a^2b^2}$

99. $\left(\dfrac{a}{b} + \dfrac{b}{a}\right) \div \left(\dfrac{a}{b} - \dfrac{b}{a}\right) = \left(\dfrac{a^2}{ab} + \dfrac{b^2}{ab}\right) \div \left(\dfrac{a^2}{ab} - \dfrac{b^2}{ab}\right) = \dfrac{a^2 + b^2}{ab} \div \dfrac{a^2 - b^2}{ab}$

$= \dfrac{a^2 + b^2}{\underset{1}{\cancel{ab}}} \cdot \dfrac{\overset{1}{\cancel{ab}}}{a^2 - b^2} = \dfrac{a^2 + b^2}{a^2 - b^2}$

Exercise 3-4 Quotients of Polynomials

Key Ideas and Formulas

Quotients of polynomials are sometimes found by a process similar to long division in arithmetic.

In the special case where a polynomial $P = Ax^n + Bx^{n-1} + \ldots + Tx + U$ is divided by a polynomial of form $x - a$, the following applies:

The remainder R is the value of the polynomial at $x = a$.

The division can be shortened by using the **synthetic division** scheme:

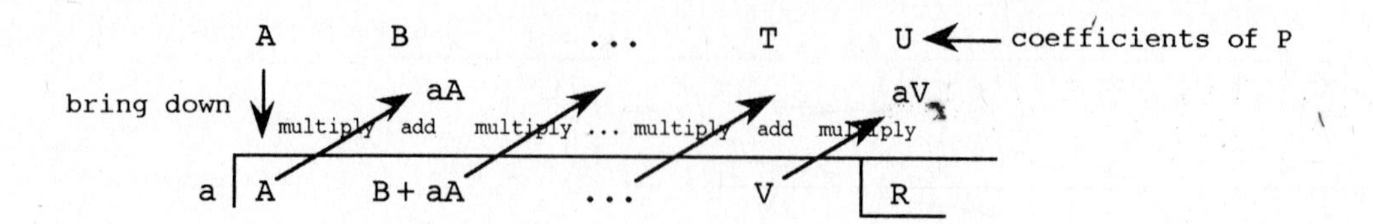

Then $P = (x - a)(Ax^{n-1} + \dots + V) + R$.

The polynomial $Q = Ax^{n-1} + \dots + V$ is the quotient. R is the remainder.

1.
$$\begin{array}{r|l} & 3x + 1 \\ \hline x - 2 & 3x^2 - 5x - 2 \\ & \underline{3x^2 - 6x} \\ & \qquad\ \ x - 2 \\ & \qquad\ \ \underline{x - 2} \\ & \qquad\qquad 0 \end{array}$$

CHECK: $(x - 2)(3x + 1) = 3x^2 - 5x - 2$

3.
$$\begin{array}{r|l} & 2y^2 + y - 3 \\ \hline y + 2 & 2y^3 + 5y^2 - y - 6 \\ & \underline{2y^3 + 4y^2} \\ & \qquad\quad y^2 - y \\ & \qquad\quad \underline{y^2 + 2y} \\ & \qquad\qquad - 3y - 6 \\ & \qquad\qquad \underline{- 3y - 6} \\ & \qquad\qquad\qquad 0 \end{array}$$

CHECK:
$$\begin{array}{r} 2y^2 + y - 3 \\ \underline{y + 2} \\ 2y^3 + y^2 - 3y \\ \underline{4y^2 + 2y - 6} \\ 2y^3 + 5y^2 - y - 6 \end{array}$$

5.
$$\begin{array}{r|l} & 3x + 1 \\ \hline x - 4 & 3x^2 - 11x - 1 \\ & \underline{3x^2 - 12x} \\ & \qquad\quad x - 1 \\ & \qquad\quad \underline{x - 4} \\ & \qquad\qquad 3 = R \end{array}$$

CHECK:
$$(x - 4)(3x + 1) + 3 \stackrel{?}{=} 3x^2 - 11x - 1$$
$$3x^2 - 11x - 4 + 3 \stackrel{?}{=} 3x^2 - 11x - 1$$
$$3x^2 - 11x - 1 \stackrel{\checkmark}{=} 3x^2 - 11x - 1$$

7.
$$\begin{array}{r|l} & 4x - 1 \\ \hline 2x - 3 & 8x^2 - 14x + 13 \\ & \underline{8x^2 - 12x} \\ & \qquad -2x + 13 \\ & \qquad \underline{-2x + 3} \\ & \qquad\qquad 10 = R \end{array}$$

CHECK:
$$(2x - 3)(4x - 1) + 10 \stackrel{?}{=} 8x^2 - 14x + 13$$
$$8x^2 - 14x + 3 + 10 \stackrel{?}{=} 8x^2 - 14x + 13$$
$$8x^2 - 14x + 13 \stackrel{\checkmark}{=} 8x^2 - 14x + 13$$

9.
$$\begin{array}{r|l} & 3x - 4 \\ \hline 2x + 3 & 6x^2 + x - 13 \\ & \underline{6x^2 + 9x} \\ & \quad -8x - 13 \\ & \quad \underline{-8x - 12} \\ & \qquad -1 = R \end{array}$$

CHECK:

$$(2x + 3)(3x - 4) + (-1) \stackrel{?}{=} 6x^2 + x - 13$$
$$6x^2 + x - 12 - 1 \stackrel{?}{=} 6x^2 + x - 13$$
$$6x^2 + x - 13 \stackrel{\surd}{=} 6x^2 + x - 13$$

11.
$$\begin{array}{r|l} & x + 2 \\ \hline x - 2 & x^2 + 0x - 4 \\ & \underline{x^2 - 2x} \\ & \quad 2x - 4 \\ & \quad \underline{2x - 4} \\ & \qquad 0 \end{array}$$

CHECK:

$$(x - 2)(x + 2) = x^2 - 4$$

13.
$$\begin{array}{r|l} & 2x + 3 \\ \hline 3x + 2 & 6x^2 + 13x + 6 \\ & \underline{6x^2 + 4x} \\ & \quad 9x + 6 \\ & \quad \underline{9x + 6} \\ & \qquad 0 \end{array}$$

CHECK:
$(3x + 2)(2x + 3) = 6x^2 + 13x + 6$

15.
$$\begin{array}{r|l} & 3x^2 + 4x \\ \hline 2x + 3 & 6x^3 + 17x^2 + 12x \\ & \underline{6x^3 + 9x^2} \\ & \quad 8x^2 + 12x \\ & \quad \underline{4x^2 + 12x} \\ & \qquad 0 \end{array}$$

CHECK: $(2x + 3)(3x^2 + 4x) =$
$6x^3 + 8x^2 + 9x^2 + 12x = 6x^3 + 17x^2 + 12x$

17.
$$\begin{array}{r|l} & 4x^3 - x^2 \\ \hline x + 1 & 4x^4 + 3x^3 - x^2 \\ & \underline{4x^4 + 4x^3} \\ & \quad -x^3 - x^2 \\ & \quad \underline{-x^3 - x^2} \\ & \qquad 0 \end{array}$$

CHECK: $(4x^3 - x^2)(x + 1) =$
$4x^4 + 4x^3 - x^3 - x^2 = 4x^4 + 3x^3 - x^2$

19.
$$\begin{array}{r|rrrr} & 1 & 2 & 3 & 4 \\ & & 2 & 8 & 22 \\ \hline 2 & 1 & 4 & 11 & 26 \end{array}$$

$x^2 + 4x + 11$, remainder 26

CHECK: $2^3 + 2 \cdot 2^2 + 3 \cdot 2 + 4 =$
$8 + 8 + 6 + 4 = 26$

21.
$$\begin{array}{r|rrrr} & 1 & 2 & 3 & 4 \\ & & -1 & -1 & -2 \\ \hline -1 & 1 & 1 & 2 & 2 \end{array}$$

$x^2 + x + 2$, remainder 2

CHECK: $(-1)^3 + 2(-1)^2 + 3(-1) + 4 =$
$-1 + 2 - 3 + 4 = 2$

23.
$$\begin{array}{r|rrrr} & 2 & -1 & 1 & -2 \\ & & 6 & 15 & 48 \\ \hline 3 & 2 & 5 & 16 & 46 \end{array}$$

$2x^2 + 5x + 16$, remainder 46

CHECK: $2 \cdot 3^3 - 3^2 + 3 - 2 =$
$54 - 9 + 3 - 2 = 46$

25.
$$\begin{array}{r|rrrr} & 2 & -1 & 1 & -2 \\ & & -2 & 3 & -4 \\ \hline -1 & 2 & -3 & 4 & -6 \end{array}$$

$2x^2 - 3x + 4$, remainder -6

CHECK: $2(-1)^3 - (-1)^2 + (-1) - 2 =$
$-2 - 1 - 1 - 2 = -6$

27.
$$\begin{array}{r|rrrrr} & 1 & 1 & 3 & 3 & 5 \\ & & 4 & 20 & 92 & 380 \\ \hline 4 & 1 & 5 & 23 & 95 & 385 \end{array}$$

$x^3 + 5x^2 + 23x + 95$, remainder 385

CHECK: $4^4 + 4^3 + 3 \cdot 4^2 + 3 \cdot 4 + 5 =$
$256 + 64 + 48 + 12 + 5 = 385$

29.
$$\begin{array}{r|rrrrr} & 1 & 1 & 3 & 3 & 5 \\ & & -2 & 2 & -10 & 14 \\ \hline -2 & 1 & -1 & 5 & -7 & 19 \end{array}$$

$x^3 - x^2 + 5x - 7$, remainder 19

CHECK: $(-2)^4 + (-2)^3 + 3(-2)^2 + 3(-2) + 5$
$16 - 8 + 12 - 6 + 5 = 19$

31. $$\begin{array}{r} 4x + 1 \\ 3x + 2\overline{)12x^2 + 11x - 2} \\ \underline{12x^2 + 8x} \\ 3x - 2 \\ \underline{3x + 2} \\ -4 = R \end{array}$$

CHECK:

$$(3x + 2)(4x + 1) + (-4) \stackrel{?}{=} 12x^2 + 11x - 2$$
$$12x^2 + 11x + 2 - 4 \stackrel{?}{=} 12x^2 + 11x - 2$$
$$12x^2 + 11x - 2 \stackrel{\surd}{=} 12x^2 + 11x - 2$$

33. $$\begin{array}{r} 4x + 6 \\ 2x - 3\overline{)8x^2 + 0x + 7} \\ \underline{8x^2 - 12x} \\ 12x + 7 \\ \underline{12x - 18} \\ 25 = R \end{array}$$

CHECK:

$$(2x - 3)(4x + 6) + 25 \stackrel{?}{=} 8x^2 + 7$$
$$8x^2 - 18 + 25 \stackrel{?}{=} 8x^2 + 7$$
$$8x^2 + 7 \stackrel{\surd}{=} 8x^2 + 7$$

35. $$\begin{array}{r} x - 4 \\ 2x + 1\overline{)2x^2 - 7x - 1} \\ \underline{2x^2 + x} \\ -8x - 1 \\ \underline{-8x - 4} \\ 3 = R \end{array}$$

CHECK: $(2x + 1)(x - 4) + 3 \stackrel{?}{=} 2x^2 - 7x - 1$

$$2x^2 - 7x - 4 + 3 \stackrel{?}{=} 2x^2 - 7x - 1$$
$$2x^2 - 7x - 1 \stackrel{\surd}{=} 2x^2 - 7x - 1$$

37. $$\begin{array}{r} x^2 + x + 1 \\ x - 1\overline{)x^3 + 0x^2 + 0x - 1} \\ \underline{x^3 - x^2} \\ x^2 + 0x \\ \underline{x^2 - x} \\ x - 1 \\ \underline{x - 1} \end{array}$$

CHECK: $(x - 1)(x^2 + x + 1) = x^3 - 1$

39. $$\begin{array}{r} x^3 + 3x^2 + 9x + 27 \\ x - 3\overline{)x^4 + 0x^3 + 0x^2 + 0x - 81} \\ \underline{x^4 - 3x^3} \\ 3x^3 + 0x^2 \\ \underline{3x^3 - 9x^2} \\ 9x^2 + 0x \\ \underline{9x^2 - 27x} \\ 27x - 81 \\ \underline{27x - 81} \\ 0 \end{array}$$

CHECK: $$\begin{array}{r} x^3 + 3x^2 + 9x + 27 \\ \underline{x - 3} \\ x^4 + 3x^3 + 9x^2 + 27x \\ \underline{- 3x^3 - 9x^2 - 27x - 81} \\ x^4 \qquad\qquad\qquad - 81 \end{array}$$

41. $$\begin{array}{r} 4a + 5 \\ a - 3\overline{)4a^2 - 7a - 22} \\ \underline{4a^2 - 12a} \\ 5a - 22 \\ \underline{5a - 15} \\ -7 = R \end{array}$$

CHECK:

$$(a - 3)(4a + 5) + (-7) \stackrel{?}{=} 4a^2 - 7a - 22$$
$$4a^2 - 7a - 15 - 7 \stackrel{?}{=} 4a^2 - 7a - 22$$
$$4a^2 - 7a - 22 \stackrel{\surd}{=} 4a^2 - 7a - 22$$

43. $$\begin{array}{r} x^2 + 3x - 5 \\ x + 2\overline{)x^3 + 5x^2 + x - 10} \\ \underline{x^3 + 2x^2} \\ 3x^2 + x \\ \underline{3x^2 + 6x} \\ - 5x - 10 \\ \underline{- 5x - 10} \\ 0 \end{array}$$

CHECK: $$\begin{array}{r} x^2 + 3x - 5 \\ \underline{x + 2} \\ x^3 + 3x^2 - 5x \\ \underline{2x^2 + 6x - 10} \\ x^3 + 5x^2 + x - 10 \end{array}$$

45.
$$\begin{array}{r}
x^2 + 3x + 8 \\
x - 3\,\overline{)\,x^3 + 0x^2 - x + 3} \\
\underline{x^3 - 3x^2} \\
3x^2 - x \\
\underline{3x^2 - 9x} \\
8x + 3 \\
\underline{8x - 24} \\
27 = R
\end{array}$$

CHECK:

$$(x - 3)(x^2 + 3x + 8) + 27 \stackrel{?}{=} x^3 - x + 3$$

$$\begin{array}{r}
x^2 + 3x + 8 \\
\underline{x - 3} \\
x^3 + 3x^2 + 8x \\
\underline{- 3x^2 - 9x - 24} \\
x^3 - x - 24
\end{array}$$

$$x^3 - x - 24 + 27 \stackrel{?}{=} x^3 - x + 3$$

$$x^3 - x + 3 \stackrel{\surd}{=} x^3 - x + 3$$

47.
$$\begin{array}{r|rrrr}
 & 1 & 0 & -2 & 4 \\
 & & 2 & 4 & 4 \\ \hline
2 & 1 & 2 & 2 & 8
\end{array}$$

$x^2 + 2x + 2$, remainder 8

CHECK: $2^3 - 2 \cdot 2 + 4 = 8 - 4 + 4 = 8$

49.
$$\begin{array}{r|rrrr}
 & 1 & 0 & -2 & 4 \\
 & & -1 & 1 & 1 \\ \hline
-1 & 1 & -1 & -1 & 5
\end{array}$$

$x^2 - x - 1$, remainder 5

CHECK: $(-1)^3 - 2(-1) + 4 = -1 + 2 + 4 = 5$

51.
$$\begin{array}{r|rrrrr}
 & 1 & 0 & 0 & -3 & 5 \\
 & & 4 & 16 & 64 & 244 \\ \hline
4 & 1 & 4 & 16 & 61 & 249
\end{array}$$

$x^3 + 4x^2 + 16x + 61$, remainder 249

CHECK: $4^4 - 3 \cdot 4 + 5 =$
$256 - 12 + 5 = 249$

53.
$$\begin{array}{r|rrrrr}
 & 1 & 0 & 0 & -3 & 5 \\
 & & -2 & 4 & -8 & 22 \\ \hline
-2 & 1 & -2 & 4 & -11 & 27
\end{array}$$

$x^3 - 2x^2 + 4x - 11$, remainder 27

CHECK: $(-2)^4 - 3(-2) + 5 =$
$16 + 6 + 5 = 27$

55.
$$\begin{array}{r|rrrrrr}
 & 1 & 0 & -3 & 1 & 0 & -1 \\
 & & 3 & 9 & 18 & 57 & 171 \\ \hline
3 & 1 & 3 & 6 & 19 & 57 & \boxed{170}
\end{array}$$

170.

57.
$$\begin{array}{r|rrrrrr}
 & 1 & 0 & -3 & 1 & 0 & -1 \\
 & & 1 & 1 & -2 & -1 & -1 \\ \hline
1 & 1 & 1 & -2 & -1 & -1 & \boxed{-2}
\end{array}$$

-2.

59.
$$\begin{array}{r|rrrrrr}
 & 1 & 0 & -3 & 1 & 0 & -1 \\
 & & -2 & 4 & -2 & 2 & -4 \\ \hline
-2 & 1 & -2 & 1 & -1 & 2 & \boxed{-5}
\end{array}$$

-5.

61.
$$\begin{array}{r|rrrrr}
 & 5 & -4 & 2 & -2 & 1 \\
 & & -5 & 9 & -11 & 13 \\ \hline
-1 & 5 & -9 & 11 & -13 & \boxed{14}
\end{array}$$

14.

63.
$$\begin{array}{r|rrrrr}
 & 5 & -4 & 2 & -2 & 1 \\
 & & -15 & 57 & -177 & 537 \\ \hline
-3 & 5 & -19 & 59 & -179 & \boxed{538}
\end{array}$$

538.

65.
$$\begin{array}{r|rrrrr}
 & 5 & -4 & 2 & -2 & 1 \\
 & & 10 & 12 & 28 & 52 \\ \hline
2 & 5 & 6 & 14 & 26 & \boxed{53}
\end{array}$$

53.

67.

$$
\begin{array}{r|l}
 & 3x^3 + x^2 - 2 \\
\hline
3x - 1 & 9x^4 + 0x^3 - x^2 - 6x - 2 \\
 & \underline{9x^4 - 3x^3} \\
 & \quad 3x^3 - x^2 \\
 & \quad \underline{3x^3 - x^2} \\
 & \qquad 0 - 6x - 2 \\
 & \qquad \underline{- 6x + 2} \\
 & \qquad\quad - 4 = R
\end{array}
$$

CHECK:

$(3x - 1)(3x^3 + x^2 - 2) + (-4) \stackrel{?}{=} 9x^4 - x^2 - 6x - 2$

$$
\begin{array}{rrrrr}
 & 3x^3 & + x^2 & & - 2 \\
 & 3x & - 1 & & \\
\hline
9x^4 & + 3x^3 & & - 6x & \\
 & - 3x^3 & - x^2 & & + 2 \\
\hline
9x^4 & & - x^2 & - 6x & + 2
\end{array}
$$

$9x^4 - x^2 - 6x + 2 + (-4) \stackrel{?}{=} 9x^4 - x^2 - 6x - 2$

$9x^4 - x^2 - 6x - 2 \stackrel{\surd}{=} 9x^4 - x^2 - 6x - 2$

69.

$$
\begin{array}{r|l}
 & 4x^2 - 2x - 1 \\
\hline
6x^2 + 3x + 5 & 24x^4 + 0x^3 + 8x^2 - 13x - 7 \\
 & \underline{24x^4 + 12x^3 + 20x^2} \\
 & \quad - 12x^3 - 12x^2 - 13x \\
 & \quad \underline{- 12x^3 - 6x^2 - 10x} \\
 & \qquad - 6x^2 - 3x - 7 \\
 & \qquad \underline{- 6x^2 - 3x - 5} \\
 & \qquad\qquad - 2 = R
\end{array}
$$

CHECK:

$(6x^2 + 3x + 5)(4x^2 - 2x - 1) + (-2) \stackrel{?}{=} 24x^4 + 8x^2 - 13x - 7$

$$
\begin{array}{rrrrr}
 & 4x^2 & - 2x & - 1 & \\
 & 6x^2 & + 3x & + 5 & \\
\hline
24x^4 & - 12x^3 & - 6x^2 & & \\
 & 12x^3 & - 6x^2 & - 3x & \\
 & & 20x^2 & - 10x & - 5 \\
\hline
24x^4 & & + 8x^2 & - 13x & - 5
\end{array}
$$

$24x^4 + 8x^2 - 13x - 5 + (-2) \stackrel{?}{=} 24x^4 + 8x^2 - 13x - 7$

$24x^4 + 8x^2 - 13x - 7 \stackrel{\surd}{=} 24x^4 + 8x^2 - 13x - 7$

71. $9x^3 - x + 2x^5 + 9x^3 - 2 - x = 2x^5 + 18x^3 - 2x - 2$

$$
\begin{array}{r|l}
 & 2x^3 + 6x^2 + 32x + 84 \\
\hline
x^2 - 3x + 2 & 2x^5 + 0x^4 + 18x^3 + 0x^2 - 2x - 2 \\
 & \underline{2x^5 - 6x^4 + 4x^3} \\
 & \quad 6x^4 + 14x^3 + 0x^2 \\
 & \quad \underline{6x^4 - 18x^3 + 12x^2} \\
 & \qquad 32x^3 - 12x^2 - 2x \\
 & \qquad \underline{32x^3 - 96x^2 + 64x} \\
 & \qquad\quad 84x^2 - 66x - 2 \\
 & \qquad\quad \underline{84x^2 - 252x + 168} \\
 & \qquad\qquad 186x - 170 = R
\end{array}
$$

CHECK: $(x^2 - 3x + 2)(2x^3 + 6x^2 + 32x + 84) + (186x - 170) \stackrel{?}{=} 2x^5 + 18x^3 - 2x - 2$

$$
\begin{array}{rrrrrr}
 & & 2x^3 & + 6x^2 & + 32x & + 84 \\
 & & & x^2 & - 3x & + 2 \\
\hline
2x^5 & + 6x^4 & + 32x^3 & + 84x^2 & & \\
 & - 6x^4 & - 18x^3 & - 96x^2 & - 252x & \\
 & & 4x^3 & + 12x^2 & + 64x & + 168 \\
\hline
2x^5 & & + 18x^3 & & - 188x & + 168
\end{array}
$$

$$(2x^5 + 18x^3 - 188x + 168) + (186x - 170) \stackrel{?}{=} 2x^5 + 18x^3 - 2x - 2$$
$$2x^5 + 18x^3 - 188x + 168 + 186x - 170 \stackrel{?}{=} 2x^5 + 18x^3 - 2x - 2$$
$$2x^5 + 18x^3 - 2x - 2 \stackrel{\surd}{=} 2x^5 + 18x^3 - 2x - 2$$

73.

$$\begin{array}{r|l} & 3x^2 - 4x + 1 \\ \hline 2x^2 + 3x + 1 & 6x^4 + x^3 - 7x^2 - x + 1 \\ & \underline{6x^4 + 9x^3 + 3x^2} \\ & \quad -8x^3 - 10x^2 - x \\ & \quad \underline{-8x^3 - 12x^2 - 4x} \\ & \qquad 2x^2 + 3x + 1 \\ & \qquad \underline{2x^2 + 3x + 1} \\ & \qquad\qquad 0 \end{array}$$

CHECK:

$$\begin{array}{r} 2x^2 + 3x + 1 \\ \underline{3x^2 - 4x + 1} \\ 6x^4 + 9x^3 + 3x^2 \qquad\qquad \\ -8x^3 - 12x^2 - 4x \qquad \\ \underline{2x^2 + 3x + 1} \\ 6x^4 + x^3 - 7x^2 - x + 1 \end{array}$$

75.

$$\begin{array}{r|l} & x^2 - x + 1 \\ \hline 2x^2 + 5x + 3 & 2x^4 + 3x^3 + 0x^2 - 2x - 3 \\ & \underline{2x^4 + 5x^3 + 3x^2} \\ & \quad -2x^3 - 3x^2 - 2x \\ & \quad \underline{-2x^3 - 5x^2 - 3x} \\ & \qquad 2x^2 + x - 3 \\ & \qquad \underline{2x^2 + 5x + 3} \\ & \qquad\quad -4x - 6 \end{array}$$

CHECK: $(2x^2 + 5x + 3)(x^2 - x + 1) + (-4x - 6) \stackrel{?}{=} 2x^4 + 3x^3 - 2x - 3$

$$\begin{array}{r} 2x^2 + 5x + 3 \qquad\qquad \\ \underline{x^2 - x + 1} \qquad\qquad \\ 2x^4 + 5x^3 + 3x^2 \qquad\qquad \\ -2x^3 - 5x^2 - 3x \qquad \\ \underline{2x^2 + 5x + 3} \\ 2x^4 + 3x^3 + 2x + 3 \end{array}$$

$$2x^4 + 3x^3 + 2x + 3 + (-4x - 6) \stackrel{?}{=} 2x^4 + 3x^3 - 2x - 3$$
$$2x^4 + 3x^3 - 2x - 3 \stackrel{\surd}{=} 2x^4 + 3x^3 - 2x - 3$$

77.

$$\begin{array}{r|l} & x^3 + 2x^2 + 4x + 8 \\ \hline 3x^2 - 5x - 2 & 3x^5 + x^4 + 0x^3 + 0x^2 - 48x - 16 \\ & \underline{3x^5 - 5x^4 - 2x^3} \\ & \quad 6x^4 + 2x^3 + 0x^2 \\ & \quad \underline{6x^4 - 10x^3 - 4x^2} \\ & \qquad 12x^3 + 4x^2 - 48x \\ & \qquad \underline{12x^3 - 20x^2 - 8x} \\ & \qquad\quad 24x^2 - 40x - 16 \\ & \qquad\quad \underline{24x^2 - 40x - 16} \\ & \qquad\qquad\qquad 0 \end{array}$$

CHECK:

$$\begin{array}{rrrrrr}
 & & x^3 & +\,2x^2 & +\,4x & +\,8 \\
 & & & 3x^2 & -\,5x & -\,2 \\ \hline
3x^5 & +\,6x^4 & +\,12x^3 & +\,24x^2 & & \\
 & -5x^4 & -\,10x^3 & -\,20x^2 & -\,40x & \\
 & & -2x^3 & -\,4x^2 & -\,8x & -\,16 \\ \hline
3x^5 & +\,x^4 & & & -48x & -\,16
\end{array}$$

79. We want to find Q and R so that $P = DQ + R$ or $\frac{P}{D} = \frac{DQ + R}{D}$ or $\frac{P}{D} = \frac{DQ}{D} + \frac{R}{D}$ or $\frac{P}{D} = Q + \frac{R}{D}$. Thus, we must divide P by D. The quotient will be Q and the remainder will be R.

$$\begin{array}{r|l}
 & x - 3 \\ \hline
x^2 - 3x + 2 & x^3 - 6x^2 + 12x - 4 \\
 & x^3 - 3x^2 + 2x \\ \hline
 & \quad -3x^2 + 10x - 4 \\
 & \quad -3x^2 + 9x - 6 \\ \hline
 & \qquad\qquad x + 2
\end{array}$$

Therefore $Q = x - 3$ and $R = x + 2$.

81. See problem 79. We divide P by D.

$$\begin{array}{r|l}
 & 1 \\ \hline
x^3 - 4 & x^3 - 6x^2 + 12x - 4 \\
 & x^3 \qquad\qquad\qquad - 4 \\ \hline
 & \quad -6x^2 + 12x
\end{array}$$

Therefore, $Q = 1$ and $R = -6x^2 + 12x$

83. See problem 79. We divide P by D.

$$\begin{array}{r|l}
 & x^2 + x - 4 \\ \hline
x^2 - x + 1 & x^4 + 0x^3 - 4x^2 + 7x + 2 \\
 & x^4 - x^3 + x^2 \\ \hline
 & \quad x^3 - 5x^2 + 7x \\
 & \quad x^3 - x^2 + x \\ \hline
 & \qquad -4x^2 + 6x + 2 \\
 & \qquad -4x^2 + 4x - 4 \\ \hline
 & \qquad\qquad 2x + 6
\end{array}$$

Therefore, $Q = x^2 + x - 4$ and $R = 2x + 6$.

85. See problem 79. We divide P by D.

$$\begin{array}{r|l}
 & x \\ \hline
x^3 + 2 & x^4 + 0x^3 - 4x^2 + 7x + 2 \\
 & x^4 \qquad\qquad\qquad + 2x \\ \hline
 & \qquad -4x^2 + 5x + 2
\end{array}$$

Therefore, $Q = x$ and $R = -4x^2 + 5x + 2$

87.

	1	-2	-3	4
		1.1	-0.99	-4.389
1.1	1	-0.9	-3.99	-0.389

-0.389000 to six decimal places

89.

	1	-2	-3	4
		1.35	-0.8775	-5.234625
1.35	1	-0.65	-3.8775	-1.234625

-1.234625.

91.

	1	-2	-3	4
		-3.3	17.49	-47.817
-3.3	1	-5.3	14.49	-43.817

-43.817000 to six decimals

93.

	1	-2	-3	4
		3.102	3.418404	1.297889
3.102	1	1.102	0.418404	5.297889

5.297889.

95.

	2	-1	1	-1
		-2.4	4.08	-6.096
-1.2	2	-3.4	5.08	-7.096

-7.096000 to six decimal places

97.

	2	-1	1	-1
		-2.002	3.005002	-4.009007
-1.001	2	-3.002	4.005002	-5.009007

-5.009007

99.

	2	-1	1	-1
		4.02	6.0702	14.211102
2.01	2	3.02	7.0702	13.211102

13.211102

Exercise 3-5 Complex Fractions

Key Ideas and Formulas

A fractional form with fractions in its numerator or denominator is called a **complex fraction.**

To reduce a complex fraction to a simple fraction, there are two approaches.

1. The complex fraction is treated as a quotient of two fractions, which are then divided.
2. The numerator and denominator of the complex fraction are multiplied by a quantity that will clear both of fractions; the LCD of the internal fractions usually serves well for this purpose.

---

1. $\dfrac{\frac{1}{2}}{\frac{2}{3}}$ lcd of $\frac{1}{2}$ and $\frac{2}{3}$ is 6; $\dfrac{\frac{1}{2}}{\frac{2}{3}} = \dfrac{6 \cdot \frac{1}{2}}{6 \cdot \frac{2}{3}} = \dfrac{3}{4}$

3. $\dfrac{\frac{3}{8}}{\frac{5}{12}}$ lcd of $\frac{3}{8}$ and $\frac{5}{12}$: $8 = 2^3$, $12 = 2^2 \cdot 3$ lcd $= 2^3 \cdot 3 = 24$; $\dfrac{\frac{3}{8}}{\frac{5}{12}} = \dfrac{\frac{3}{8} \cdot 24}{\frac{5}{12} \cdot 24} = \dfrac{9}{10}$

5. $\dfrac{1\frac{1}{3}}{2\frac{1}{6}}$ lcd of $\frac{1}{3}$ and $\frac{1}{6}$ is 6; $\dfrac{1\frac{1}{3}}{2\frac{1}{6}} = \dfrac{1 + \frac{1}{3}}{2 + \frac{1}{6}} = \dfrac{6\left(1 + \frac{1}{3}\right)}{6\left(2 + \frac{1}{6}\right)} = \dfrac{6 \cdot 1 + 6 \cdot \frac{1}{3}}{6 \cdot 2 + 6 \cdot \frac{1}{6}} = \dfrac{6 + 2}{12 + 1} = \dfrac{8}{13}$

7. $\dfrac{1\frac{2}{9}}{2\frac{5}{6}}$ lcd of $\frac{2}{9}$ and $\frac{5}{6}$ is 18; $\dfrac{1\frac{2}{9}}{2\frac{5}{6}} = \dfrac{1 + \frac{2}{9}}{2 + \frac{5}{6}} = \dfrac{18\left(1 + \frac{2}{9}\right)}{18\left(2 + \frac{5}{6}\right)} = \dfrac{18 \cdot 1 + 18 \cdot \frac{2}{9}}{18 \cdot 2 + 18 \cdot \frac{5}{6}}$

$= \dfrac{18 + 4}{36 + 15} = \dfrac{22}{51}$

9. $\dfrac{\frac{3}{4}}{\frac{x}{y}} = \dfrac{3}{4} \div \dfrac{x}{y} = \dfrac{3}{4} \cdot \dfrac{y}{x} = \dfrac{3y}{4x}$

11. $\dfrac{\frac{-x}{10}}{\frac{y}{2}} = -\dfrac{x}{10} \div \dfrac{y}{2} = -\dfrac{x}{\cancel{10}_{5}} \cdot \dfrac{\cancel{2}^{1}}{y} = \dfrac{-x}{5y}$

13. $\dfrac{\dfrac{-ab}{c}}{\dfrac{-bc}{a}} = \dfrac{-ab}{c} \div \dfrac{-bc}{a} = \dfrac{-a\overset{1}{\cancel{b}}}{c} \cdot \dfrac{a}{-\underset{1}{\cancel{b}}c} = \dfrac{-a^2}{-c^2} = \dfrac{a^2}{c^2}$

15. $\dfrac{\dfrac{2x^2}{y^2}}{\dfrac{4x}{y}} = \dfrac{\dfrac{2x^2}{y^2} \cdot y^2}{\dfrac{4x}{y} \cdot y^2} = \dfrac{2x^2}{4xy} = \dfrac{x}{2y}$ y^2 is the lcd of $\dfrac{2x^2}{y^2}$ and $\dfrac{4x}{y}$

17. $\dfrac{\dfrac{x}{y}}{\dfrac{1}{y^2}} = \dfrac{\dfrac{x}{y} \cdot y^2}{\dfrac{1}{y^2} \cdot y^2} = \dfrac{xy}{1} = xy$ y^2 is the lcd of $\dfrac{x}{y}$ and $\dfrac{1}{y^2}$

19. $\dfrac{\dfrac{y}{2x}}{\dfrac{1}{3x^2}} = \dfrac{\dfrac{y}{2x} \cdot 6x^2}{\dfrac{1}{3x^2} \cdot 6x^2} = \dfrac{y \cdot 3x}{2} = \dfrac{3xy}{2}$ $6x^2$ is the lcd of $\dfrac{y}{2x}$ and $\dfrac{1}{3x^2}$

21. $\dfrac{\dfrac{1}{1 + x} - 1}{x} = \dfrac{(1 + x)\left(\dfrac{1}{1 + x} - 1\right)}{(1 + x)x} = \dfrac{(1 + x) \cdot \dfrac{1}{1 + x} - (1 + x)}{(1 + x)x}$

$= \dfrac{1 - (1 + x)}{(1 + x)x} = \dfrac{1 - 1 - x}{(1 + x)x} = \dfrac{-x}{(1 + x)x} = \dfrac{-1}{1 + x}$

23. $\dfrac{\dfrac{1}{3} - \dfrac{1}{x}}{x - 3} = \dfrac{3x\left(\dfrac{1}{3} - \dfrac{1}{x}\right)}{3x(x - 3)} = \dfrac{3x \cdot \dfrac{1}{3} - 3x \cdot \dfrac{1}{x}}{3x(x - 3)} = \dfrac{x - 3}{3x(x - 3)} = \dfrac{1}{3x}$

25. $\dfrac{\dfrac{a}{b} - 1}{1 - \dfrac{b}{a}} = \dfrac{ab\left(\dfrac{a}{b} - 1\right)}{ab\left(1 - \dfrac{b}{a}\right)} = \dfrac{ab \cdot \dfrac{a}{b} - ab \cdot 1}{ab \cdot 1 - ab \cdot \dfrac{b}{a}} = \dfrac{a^2 - ab}{ab - b^2} = \dfrac{a(a - b)}{b(a - b)} = \dfrac{a}{b}$

27. $\dfrac{3 - \dfrac{a}{b}}{b - \dfrac{a}{3}} = \dfrac{3b\left(3 - \dfrac{a}{b}\right)}{3b\left(b - \dfrac{a}{3}\right)} = \dfrac{3b \cdot 3 - 3b \cdot \dfrac{a}{b}}{3b \cdot b - 3b \cdot \dfrac{a}{3}} = \dfrac{9b - 3a}{3b^2 - ab} = \dfrac{3(3b - a)}{b(3b - a)} = \dfrac{3}{b}$

29. $\dfrac{\dfrac{x}{y} - z}{\dfrac{x}{z} - y} = \dfrac{yz\left(\dfrac{x}{y} - z\right)}{yz\left(\dfrac{x}{z} - y\right)} = \dfrac{yz \cdot \dfrac{x}{y} - yz \cdot z}{yz \cdot \dfrac{x}{z} - yz \cdot y} = \dfrac{xz - yz^2}{xy - y^2z} = \dfrac{z(x - yz)}{y(x - yz)} = \dfrac{z}{y}$

31. $\dfrac{\dfrac{2}{x} + \dfrac{3}{y}}{\dfrac{9}{y} + \dfrac{6}{x}} = \dfrac{xy\left(\dfrac{2}{x} + \dfrac{3}{y}\right)}{xy\left(\dfrac{9}{y} + \dfrac{6}{x}\right)} = \dfrac{xy \cdot \dfrac{2}{x} + xy \cdot \dfrac{3}{y}}{xy \cdot \dfrac{9}{y} + xy \cdot \dfrac{6}{x}} = \dfrac{2y + 3x}{9x + 6y} = \dfrac{2y + 3x}{3(3x + 2y)} = \dfrac{1}{3}$

33. $$\frac{\frac{a}{3} - \frac{2}{b}}{\frac{b}{2} - \frac{3}{a}} = \frac{6ab\left(\frac{a}{3} - \frac{2}{b}\right)}{6ab\left(\frac{b}{2} - \frac{3}{a}\right)} = \frac{6ab \cdot \frac{a}{3} - 6ab \cdot \frac{2}{b}}{6ab \cdot \frac{b}{2} - 6ab \cdot \frac{3}{a}} = \frac{2a^2b - 12a}{3ab^2 - 18b} = \frac{2a(ab - 6)}{3b(ab - 6)} = \frac{2a}{3b}$$

35. $$\frac{1 + \frac{3}{x}}{x - \frac{9}{x}} = \frac{x\left(1 + \frac{3}{x}\right)}{x\left(x - \frac{9}{x}\right)} = \frac{x \cdot 1 + x \cdot \frac{3}{x}}{x \cdot x - x \cdot \frac{9}{x}} = \frac{x + 3}{x^2 - 9} = \frac{x + 3}{(x + 3)(x - 3)} = \frac{1}{x - 3}$$

37. $$\frac{1 - \frac{y^2}{x^2}}{1 - \frac{y}{x}} = \frac{x^2\left(1 - \frac{y^2}{x^2}\right)}{x^2\left(1 - \frac{y}{x}\right)} = \frac{x^2 \cdot 1 - x^2 \cdot \frac{y^2}{x^2}}{x^2 \cdot 1 - x^2 \cdot \frac{y}{x}} = \frac{x^2 - y^2}{x^2 - xy} = \frac{(x - y)(x + y)}{x(x - y)} = \frac{x + y}{x}$$

39. $$\frac{\frac{1}{x} + \frac{1}{y}}{\frac{y}{x} - \frac{x}{y}} = \frac{xy\left(\frac{1}{x} + \frac{1}{y}\right)}{xy\left(\frac{y}{x} - \frac{x}{y}\right)} = \frac{xy \cdot \frac{1}{x} + xy \cdot \frac{1}{y}}{xy \cdot \frac{y}{x} - xy \cdot \frac{x}{y}} = \frac{y + x}{y^2 - x^2} = \frac{y + x}{(y + x)(y - x)} = \frac{1}{y - x}$$

41. $$\frac{\frac{1}{x^2} + x}{\frac{1}{x} + 1} = \frac{x^2\left(\frac{1}{x^2} + x\right)}{x^2\left(\frac{1}{x} + 1\right)} = \frac{x^2 \cdot \frac{1}{x^2} + x^2 \cdot x}{x^2 \cdot \frac{1}{x} + x^2 \cdot 1} = \frac{1 + x^3}{x + x^2} = \frac{(1 + x)(1 - x + x^2)}{(1 + x)x} = \frac{1 - x + x^2}{x}$$

43. $$\frac{\frac{9}{x^2} - \frac{x}{3}}{\frac{3}{x} - 1} = \frac{3x^2\left(\frac{9}{x^2} - \frac{x}{3}\right)}{3x^2\left(\frac{3}{x} - 1\right)}$$

$$= \frac{3x^2 \cdot \frac{9}{x^2} - 3x^2 \cdot \frac{x}{3}}{3x^2 \cdot \frac{3}{x} - 3x^2 \cdot 1} = \frac{27 - x^3}{9x - 3x^2} = \frac{(3 - x)(9 + 3x + x^2)}{3x(3 - x)} = \frac{9 + 3x + x^2}{3x}$$

45. $$\frac{\frac{1}{(1 + x)^2} - 1}{x} = \frac{(1 + x)^2\left[\frac{1}{(1 + x)^2} - 1\right]}{(1 + x)^2 x} = \frac{(1 + x)^2 \cdot \frac{1}{(1 + x)^2} - (1 + x)^2 \cdot 1}{(1 + x)^2 x}$$

$$= \frac{1 - (1 + x)^2}{(1 + x)^2 x} = \frac{1 - (1 + 2x + x^2)}{(1 + x)^2 x} = \frac{1 - 1 - 2x - x^2}{(1 + x)^2 x}$$

$$= \frac{-2x - x^2}{(1 + x)^2 x} = \frac{x(-2 - x)}{(1 + x)^2 x} = \frac{-2 - x}{(1 + x)^2} \text{ or } -\frac{2 + x}{(1 + x)^2}$$

47. $$\frac{\frac{-4}{a^2} + \frac{4}{b^2}}{a - b} = \frac{a^2b^2\left(\frac{-4}{a^2} + \frac{4}{b^2}\right)}{a^2b^2(a - b)} = \frac{a^2b^2 \cdot \frac{-4}{a^2} + a^2b^2 \cdot \frac{4}{b^2}}{a^2b^2(a - b)} = \frac{-4b^2 + 4a^2}{a^2b^2(a - b)}$$

$$= \frac{4(a^2 - b^2)}{a^2b^2(a - b)} = \frac{4(a + b)(a - b)}{a^2b^2(a - b)} = \frac{4(a + b)}{a^2b^2}$$

49. $$\frac{\frac{4a}{b} - \frac{b}{a}}{\frac{4a}{b} - 4 + \frac{b}{a}} = \frac{ab\left(\frac{4a}{b} - \frac{b}{a}\right)}{ab\left(\frac{4a}{b} - 4 + \frac{b}{a}\right)} = \frac{ab \cdot \frac{4a}{b} - ab \cdot \frac{b}{a}}{ab \cdot \frac{4a}{b} - 4ab + ab \cdot \frac{b}{a}} = \frac{4a^2 - b^2}{4a^2 - 4ab + b^2}$$

$$= \frac{(2a - b)(2a + b)}{(2a - b)(2a - b)} = \frac{2a + b}{2a - b}$$

51. $$\frac{\frac{x}{y} - 2 + \frac{y}{x}}{\frac{x}{y} - \frac{y}{x}} = \frac{xy\left(\frac{x}{y} - 2 + \frac{y}{x}\right)}{xy\left(\frac{x}{y} - \frac{y}{x}\right)}$$

$$= \frac{xy \cdot \frac{x}{y} - xy \cdot 2 + xy \cdot \frac{y}{x}}{xy \cdot \frac{x}{y} - xy \cdot \frac{y}{x}} = \frac{x^2 - 2xy + y^2}{x^2 - y^2} = \frac{(x - y)(x - y)}{(x - y)(x + y)} = \frac{x - y}{x + y}$$

53. $$\frac{1 + \frac{4}{x} + \frac{4}{x^2}}{\frac{1}{x} + \frac{2}{x^2}} = \frac{x^2\left(1 + \frac{4}{x} + \frac{4}{x^2}\right)}{x^2\left(\frac{1}{x} + \frac{2}{x^2}\right)} = \frac{x^2 \cdot 1 + x^2 \cdot \frac{4}{x} + x^2 \cdot \frac{4}{x^2}}{x^2 \cdot \frac{1}{x} + x^2 \cdot \frac{2}{x^2}} = \frac{x^2 + 4x + 4}{x + 2}$$

$$= \frac{(x + 2)(x + 2)}{x + 2} = x + 2$$

55. $$\frac{x + 3 - \frac{4}{x}}{1 + \frac{1}{x} - \frac{2}{x^2}} = \frac{x^2\left(x + 3 - \frac{4}{x}\right)}{x^2\left(1 + \frac{1}{x} - \frac{2}{x^2}\right)} = \frac{x^3 + 3x^2 - x^2 \cdot \frac{4}{x}}{x^2 + x^2 \cdot \frac{1}{x} - x^2 \cdot \frac{2}{x^2}} = \frac{x^3 + 3x^2 - 4x}{x^2 + x - 2}$$

$$= \frac{x(x^2 + 3x - 4)}{(x + 2)(x - 1)} = \frac{x(x + 4)(x - 1)}{(x + 2)(x - 1)} = \frac{x(x + 4)}{x + 2} \text{ or } \frac{x^2 + 4x}{x + 2}$$

57. $$\frac{\frac{a^2}{a - b} - a}{\frac{b^2}{a - b} + b} = \frac{(a - b)\left(\frac{a^2}{a - b} - a\right)}{(a - b)\left(\frac{b^2}{a - b} + b\right)}$$

$$= \frac{(a - b) \cdot \frac{a^2}{a - b} - (a - b)a}{(a - b) \cdot \frac{b^2}{a - b} + (a - b)b} = \frac{a^2 - (a - b)a}{b^2 + (a - b)b} = \frac{a^2 - a^2 + ab}{b^2 + ab - b^2} = \frac{ab}{ab} = 1$$

59. $$\frac{\frac{1}{x} + \frac{5}{x^2} + \frac{6}{x^3}}{\frac{8}{x^2} + \frac{6}{x} + 1} = \frac{x^3\left(\frac{1}{x} + \frac{5}{x^2} + \frac{6}{x^3}\right)}{x^3\left(\frac{8}{x^2} + \frac{6}{x} + 1\right)} = \frac{x^3 \cdot \frac{1}{x} + x^3 \cdot \frac{5}{x^2} + x^3 \cdot \frac{6}{x^3}}{x^3 \cdot \frac{8}{x^2} + x^3 \cdot \frac{6}{x} + x^3 \cdot 1}$$

$$= \frac{x^2 + 5x + 6}{8x + 6x^2 + x^3} = \frac{(x + 2)(x + 3)}{x(8 + 6x + x^2)} = \frac{(x + 2)(x + 3)}{x(x + 2)(x + 4)} = \frac{x + 3}{x(x + 4)}$$

61. $$\frac{x - 3y + \frac{2y^2}{x}}{1 + \frac{y}{x} - \frac{2y^2}{x^2}} = \frac{x^2\left(x - 3y + \frac{2y^2}{x}\right)}{x^2\left(1 + \frac{y}{x} - \frac{2y^2}{x^2}\right)} = \frac{x^3 - 3x^2y + x^2 \cdot \frac{2y^2}{x}}{x^2 + x^2 \cdot \frac{y}{x} - x^2 \cdot \frac{2y^2}{x^2}}$$

$$= \frac{x^3 - 3x^2y + 2xy^2}{x^2 + xy - 2y^2} = \frac{x(x^2 - 3xy + 2y^2)}{(x + 2y)(x - y)} = \frac{x(x - 2y)(x - y)}{(x + 2y)(x - y)} = \frac{x(x - 2y)}{x + 2y}$$

63. $$\frac{\frac{1}{2} + \frac{3}{x} + \frac{4}{x^2}}{x + 1 - \frac{2}{x}} = \frac{2x^2\left(\frac{1}{2} + \frac{3}{x} + \frac{4}{x^2}\right)}{2x^2\left(x + 1 - \frac{2}{x}\right)} = \frac{2x^2 \cdot \frac{1}{2} + 2x^2 \cdot \frac{3}{x} + 2x^2 \cdot \frac{4}{x^2}}{2x^2 \cdot x + 2x^2 \cdot 1 - 2x^2 \cdot \frac{2}{x}}$$

$$= \frac{x^2 + 6x + 8}{2x^3 + 2x^2 - 4x} = \frac{(x + 2)(x + 4)}{2x(x^2 + x - 2)} = \frac{(x + 2)(x + 4)}{2x(x + 2)(x - 1)} = \frac{x + 4}{2x(x - 1)}$$

65. $$\frac{\frac{m}{m + 2} - \frac{m}{m - 2}}{\frac{m + 2}{m - 2} - \frac{m - 2}{m + 2}} = \frac{(m + 2)(m - 2)\left(\frac{m}{m + 2} - \frac{m}{m - 2}\right)}{(m + 2)(m - 2)\left(\frac{m + 2}{m - 2} - \frac{m - 2}{m + 2}\right)}$$

$$= \frac{(m + 2)(m - 2)\frac{m}{m + 2} - (m + 2)(m - 2)\frac{m}{m - 2}}{(m + 2)(m - 2)\frac{m + 2}{m - 2} - (m + 2)(m - 2)\frac{m - 2}{m + 2}}$$

$$= \frac{(m - 2)m - (m + 2)m}{(m + 2)(m + 2) - (m - 2)(m - 2)} = \frac{m^2 - 2m - m^2 - 2m}{(m^2 + 4m + 4) - (m^2 - 4m + 4)}$$

$$= \frac{-4m}{m^2 + 4m + 4 - m^2 + 4m - 4} = \frac{-4m}{8m} = -\frac{1}{2}$$

67. $$\frac{\frac{1}{x - 1} + \frac{1}{x + 1}}{\frac{1}{x^2 - 1} + \frac{1}{x^2 + 1}} = \frac{\frac{x + 1}{(x - 1)(x + 1)} + \frac{x - 1}{(x - 1)(x + 1)}}{\frac{x^2 + 1}{(x^2 + 1)(x^2 - 1)} + \frac{x^2 - 1}{(x^2 + 1)(x^2 - 1)}} = \frac{\frac{x + 1 + x - 1}{(x - 1)(x + 1)}}{\frac{x^2 + 1 + x^2 - 1}{(x^2 + 1)(x^2 - 1)}}$$

$$= \frac{\frac{2x}{x^2 - 1}}{\frac{2x^2}{(x^2 + 1)(x^2 - 1)}} = \frac{2x}{x^2 - 1} \div \frac{2x^2}{(x^2 + 1)(x^2 - 1)}$$

$$= \frac{\overset{1}{\cancel{2x}}}{\underset{1}{\cancel{x^2 - 1}}} \cdot \frac{(x^2 + 1)\overset{1}{\cancel{(x^2 - 1)}}}{\underset{x}{\cancel{2x^2}}} = \frac{x^2 + 1}{x}$$

69. $$1 - \frac{1}{1 - \frac{1}{x}} = 1 - \frac{x \cdot 1}{x\left(1 - \frac{1}{x}\right)}$$

$$= 1 - \frac{x \cdot 1}{x \cdot 1 - x \cdot \frac{1}{x}} = 1 - \frac{x}{x - 1} = \frac{1}{1} - \frac{x}{x - 1} = \frac{x - 1}{x - 1} - \frac{x}{x - 1}$$

$$= \frac{x - 1 - x}{x - 1} = \frac{-1}{x - 1} \text{ or } \frac{-1(-1)}{-1(x - 1)} = \frac{1}{1 - x}$$

71. $1 - \dfrac{2}{1 - \dfrac{2}{1 + x}} = 1 - \dfrac{2(1 + x)}{\left(1 - \dfrac{2}{1 + x}\right)(1 + x)} = 1 - \dfrac{2(1 + x)}{(1 + x) - \dfrac{2}{1 + x}(1 + x)}$

$= 1 - \dfrac{2 + 2x}{1 + x - 2} = 1 - \dfrac{2 + 2x}{x - 1} = \dfrac{x - 1}{x - 1} - \dfrac{(2 + 2x)}{x - 1} = \dfrac{x - 1 - (2 + 2x)}{x - 1}$

$= \dfrac{x - 1 - 2 - 2x}{x - 1} = \dfrac{-x - 3}{x - 1}$

73. $\dfrac{\left(1 + \dfrac{1}{x}\right)\left(1 - \dfrac{1}{x^2}\right)}{(x + 1)\left(1 + \dfrac{1}{x}\right)} = \dfrac{\overset{1}{\cancel{\left(1 + \dfrac{1}{x}\right)}}\left(\dfrac{x^2}{x^2} - \dfrac{1}{x^2}\right)}{(x + 1)\underset{1}{\cancel{\left(1 + \dfrac{1}{x}\right)}}} = \dfrac{\dfrac{x^2 - 1}{x^2}}{x + 1} = \dfrac{x^2 - 1}{x^2} \div \dfrac{(x + 1)}{1}$

$= \dfrac{x^2 - 1}{x^2} \cdot \dfrac{1}{x + 1} = \dfrac{(x - 1)\overset{1}{\cancel{(x + 1)}}}{x^2} \cdot \dfrac{1}{\underset{1}{\cancel{x + 1}}} = \dfrac{x - 1}{x^2}$

75. $1 - \dfrac{x - \dfrac{1}{x}}{1 - \dfrac{1}{x}} = 1 - \dfrac{x\left(x - \dfrac{1}{x}\right)}{x\left(1 - \dfrac{1}{x}\right)}$

$= 1 - \dfrac{x \cdot x - x \cdot \dfrac{1}{x}}{x \cdot 1 - x \cdot \dfrac{1}{x}} = 1 - \dfrac{x^2 - 1}{x - 1} = 1 - \dfrac{(x - 1)(x + 1)}{x - 1} = 1 - (x + 1)$

$= 1 - x - 1 = -x$

77. We will write each complex fraction as a simple fraction, starting with $\dfrac{1}{1 + \dfrac{1}{1 + x}}$. This has lcd $(1 + x)$.

$1 + \dfrac{1}{1 + \boxed{\dfrac{1}{1 + \dfrac{1}{1 + x}}}} = 1 + \dfrac{1}{1 + \boxed{\dfrac{(1 + x) \cdot 1}{(1 + x) \cdot 1 + (1 + x) \cdot \dfrac{1}{1 + x}}}}$

$= 1 + \dfrac{1}{1 + \boxed{\dfrac{1 + x}{1 + x + 1}}} = 1 + \dfrac{1}{1 + \boxed{\dfrac{1 + x}{2 + x}}}$ lcd $= 2 + x$

$= 1 + \dfrac{(2 + x) \cdot 1}{(2 + x)\left(1 + \dfrac{1 + x}{2 + x}\right)} = 1 + \dfrac{(2 + x) \cdot 1}{(2 + x) \cdot 1 + (2 + x) \cdot \dfrac{1 + x}{2 + x}}$

$= 1 + \dfrac{2 + x}{2 + x + 1 + x} = 1 + \dfrac{2 + x}{3 + 2x} = \dfrac{1}{1} + \dfrac{2 + x}{3 + 2x}$

$= \dfrac{(3 + 2x)}{3 + 2x} + \dfrac{(2 + x)}{3 + 2x} = \dfrac{(3 + 2x) + (2 + x)}{3 + 2x} = \dfrac{3 + 2x + 2 + x}{3 + 2x}$

$= \dfrac{5 + 3x}{3 + 2x}$ or $\dfrac{3x + 5}{2x + 3}$

79. $r = \dfrac{2}{\dfrac{1}{r_G} + \dfrac{1}{r_R}}$ $\quad$ lcd $= r_G r_R$

$$= \frac{2 \cdot r_G r_R}{r_G r_R \left(\dfrac{1}{r_G} + \dfrac{1}{r_R}\right)} = \frac{2 r_G r_R}{r_G r_R \cdot \dfrac{1}{r_G} + r_G r_R \cdot \dfrac{1}{r_R}} = \frac{2 r_G r_R}{r_R + r_G}$$

Exercise 3-6 REVIEW EXERCISE

1. $\dfrac{42}{105} = \dfrac{21 \cdot 2}{21 \cdot 5} = \dfrac{2}{5}$

2. $\dfrac{a^4b^3c^2}{a^2b^2c^2} = \dfrac{a^2b^2c^2 \cdot a^2b}{a^2b^2c^2} = \dfrac{a^2b}{1} = a^2b$

3. $\dfrac{x^2 - 4x}{x^2 + x - 20} = \dfrac{x(x - 4)}{(x + 5)(x - 4)} = \dfrac{x}{x + 5}$

4. $\dfrac{x^3 - 4x}{x^3 + 4x^2 + 4x} = \dfrac{x(x^2 - 4)}{x(x^2 + 4x + 4)} = \dfrac{x(x + 2)(x - 2)}{x(x + 2)(x + 2)} = \dfrac{x - 2}{x + 2}$

5. $\dfrac{6xy^2}{14x^2y} = \dfrac{2x \cdot 3y^2}{2x \cdot 7xy} = \dfrac{3y^2}{7xy}$. $3y^2$

6. $\dfrac{3xy^2}{8xyz} = \dfrac{2xz \cdot 3xy^2}{2xz \cdot 9xyz} = \dfrac{6x^2y^2z}{16x^2yz^2}$. $6x^2y^2z$

7. $4x = 2^2x$
$6x^2 = 2 \cdot 3x^2$
$9x^3 = 3^2x^3$ $\quad$ lcm $= 2^2 \cdot 3^2x^3 = 36x^3$

8. $x^3y = x^3 \cdot y^1$
$2x^2y^2 = 2^1x^2y^2$
$3xy^3 = 3^1x^1y^3$ $\quad$ lcm $= 2^1 \cdot 3^1x^3y^3$
$= 6x^3y^3$

9. $x^2 = x^2$
$x + 1 = x + 1$
$x^2 - 1 = (x + 1)(x - 1)$ $\quad$ lcm $= x^2(x + 1)(x - 1) = x^2(x^2 - 1) = x^4 - x^2$

10. $\dfrac{18x^3y^2(z + 3)}{12xy^2(z + 3)^2} = \dfrac{6xy^2(z + 3) \cdot 3x^2}{6xy^2(z + 3) \cdot 2(z + 3)} = \dfrac{3x^2}{2(z + 3)}$

11. $\dfrac{x^2 + 2x + 1}{x^2 - 1} = \dfrac{(x + 1)(x + 1)}{(x - 1)(x + 1)} = \dfrac{x + 1}{x - 1}$

12. $1 + \dfrac{2}{3x} = \dfrac{1}{1} + \dfrac{2}{3x}$ $\quad$ lcd $= 3x$
$= \dfrac{3x}{3x} + \dfrac{2}{3x} = \dfrac{3x + 2}{3x}$

13. $\dfrac{2}{x} - \dfrac{1}{6x} + \dfrac{1}{3}$
$x = x$
$6x = 2 \cdot 3x$
$3 = 3$
lcd $= 6x$
$\dfrac{2}{x} - \dfrac{1}{6x} + \dfrac{1}{3} = \dfrac{6 \cdot 2}{6 \cdot x} - \dfrac{1}{6x} + \dfrac{1 \cdot 2x}{3 \cdot 2x} = \dfrac{12}{6x} - \dfrac{1}{6x} + \dfrac{2x}{6x} = \dfrac{12 - 1 + 2x}{6x} = \dfrac{2x + 11}{6x}$

14. $\dfrac{1}{6x^3} - \dfrac{3}{4x} - \dfrac{2}{3}$
$6x^3 = 2 \cdot 3x^3$
$4x = 2^2x$
$3 = 3$
lcd $= 2^2 \cdot 3x^3 = 12x^3$
$\dfrac{1}{6x^3} - \dfrac{3}{4x} - \dfrac{2}{3} = \dfrac{2 \cdot 1}{2 \cdot 6x^3} - \dfrac{3 \cdot 3x^2}{4x \cdot 3x^2} - \dfrac{2 \cdot 4x^3}{3 \cdot 4x^3} = \dfrac{2}{12x^3} - \dfrac{9x^2}{12x^3} - \dfrac{8x^3}{12x^3} = \dfrac{2 - 9x^2 - 8x^3}{12x^3}$

15. $\dfrac{4x^2y^3}{3a^2b^2} \div \dfrac{2xy^2}{3ab} = \dfrac{\overset{2xy}{\cancel{4x^2y^3}}}{\underset{ab}{\cancel{3a^2b^2}}} \cdot \dfrac{\overset{1}{\cancel{3ab}}}{\underset{1}{\cancel{2xy^2}}} = \dfrac{2xy}{ab}$

16. $\dfrac{6x^2}{3(x-1)} - \dfrac{6}{3(x-1)} = \dfrac{6x^2-6}{3(x-1)} = \dfrac{6(x-1)(x+1)}{3(x-1)} = \dfrac{2(x+1)}{1} = 2(x+1)$

17. $1 - \dfrac{m-1}{m+1} = \dfrac{1}{1} - \dfrac{(m-1)}{m+1}$ lcd $= m + 1$

$= \dfrac{(m+1)}{m+1} - \dfrac{(m-1)}{m+1} = \dfrac{(m+1)-(m-1)}{m+1} = \dfrac{m+1-m+1}{m+1} = \dfrac{2}{m+1}$

18. $\dfrac{3}{x-2} - \dfrac{2}{x+1}$ lcd $= (x-2)(x+1)$

$\dfrac{3}{x-2} - \dfrac{2}{x+1} = \dfrac{3(x+1)}{(x-2)(x+1)} - \dfrac{2(x-2)}{(x-2)(x+1)} = \dfrac{3(x+1)-2(x-2)}{(x-2)(x+1)}$

$= \dfrac{3x+3-2x+4}{(x-2)(x+1)} = \dfrac{x+7}{(x-2)(x+1)}$

19. $(d-2)^2 \div \dfrac{d^2-4}{d-2} = \dfrac{(d-2)^2}{1} \cdot \dfrac{d-2}{d^2-4} = \dfrac{(d-2)^2}{1} \cdot \dfrac{\overset{1}{\cancel{d-2}}}{\underset{1}{\cancel{(d-2)}}(d+2)} = \dfrac{(d-2)^2}{d+2}$

20. $\dfrac{x+1}{x+2} - \dfrac{x+2}{x+3}$ lcd $= (x+2)(x+3)$

$\dfrac{x+1}{x+2} - \dfrac{x+2}{x+3} = \dfrac{(x+1)(x+3)}{(x+2)(x+3)} - \dfrac{(x+2)(x+2)}{(x+3)(x+2)} = \dfrac{(x+1)(x+3)-(x+2)(x+2)}{(x+2)(x+3)}$

$= \dfrac{x^2+4x+3-(x^2+4x+4)}{(x+2)(x+3)} = \dfrac{x^2+4x+3-x^2-4x-4}{(x+2)(x+3)}$

$= \dfrac{-1}{(x+2)(x+3)}$

21. $\dfrac{1}{x} - \dfrac{1}{x+1}$ lcd $= x(x+1)$

$\dfrac{1}{x} - \dfrac{1}{x+1} = \dfrac{1(x+1)}{x(x+1)} - \dfrac{1x}{x(x+1)} = \dfrac{(x+1)-x}{x(x+1)} = \dfrac{x+1-x}{x(x+1)} = \dfrac{1}{x(x+1)}$ or $\dfrac{1}{x^2+x}$

22. $\dfrac{2}{x+4} - \dfrac{1}{2}$ lcd $= 2(x+4)$

$\dfrac{2}{x+4} - \dfrac{1}{2} = \dfrac{2\cdot 2}{2(x+4)} - \dfrac{1(x+4)}{2(x+4)} = \dfrac{4-1(x+4)}{2(x+4)} = \dfrac{4-x-4}{2(x+4)} = \dfrac{-x}{2(x+4)}$ or $\dfrac{-x}{2x+8}$

23. $\dfrac{x+2}{3} \cdot \dfrac{6x}{x^2+2x} = \dfrac{\overset{1}{\cancel{x+2}}}{\underset{1}{\cancel{3}}} \cdot \dfrac{\overset{2}{\cancel{6}}\overset{1}{\cancel{x}}}{\underset{1}{\cancel{x}}\underset{1}{\cancel{(x+2)}}} = \dfrac{2}{1} = 2$

24. $x + \dfrac{x}{x-1} = \dfrac{x}{1} + \dfrac{x}{x-1}$ lcd $= x - 1$

$= \dfrac{x(x-1)}{x-1} + \dfrac{x}{x-1} = \dfrac{x(x-1)+x}{x-1} = \dfrac{x^2-x+x}{x-1} = \dfrac{x^2}{x-1}$

25. $\dfrac{2x^3}{9}\left(\dfrac{3\overset{1}{\cancel{x}}}{\underset{2y}{\cancel{4y^2}}} \cdot \dfrac{\overset{3}{\cancel{6y}}}{\underset{x}{\cancel{x^2}}}\right) = \dfrac{\overset{x^2}{\cancel{2x^3}}}{\underset{1}{\cancel{9}}} \cdot \dfrac{\overset{1}{\cancel{9}}}{\underset{1}{\cancel{2x}}y} = \dfrac{x^2}{y}$

26. $\frac{2x^3}{9}\left(\frac{3x}{4y^2} + \frac{6y}{x^2}\right) = \frac{2x^3}{9}\left(\frac{3x^3}{4x^2y^2} + \frac{24y^3}{4x^2y^2}\right) = \frac{2x^3}{9} \cdot \frac{3x^3 + 24y^3}{4x^2y^2}$

$= \frac{\overset{x}{\cancel{2x^3}}}{\underset{3}{\cancel{9}}} \cdot \frac{\overset{1}{\cancel{3}}(x^3 + 8y^3)}{\underset{2}{\cancel{4x^2}}y^2} = \frac{x(x^3 + 8y^3)}{6y^2} = \frac{x^4 + 8xy^3}{6y^2}$

27. $\dfrac{\frac{1}{4}}{\frac{2}{3}}$ lcd of $\frac{1}{4}$ and $\frac{2}{3}$ is 12.

$\dfrac{\frac{1}{4}}{\frac{2}{3}} = \dfrac{12 \cdot \frac{1}{4}}{12 \cdot \frac{2}{3}} = \frac{3}{8}$

28. $\dfrac{2\frac{3}{4}}{1\frac{1}{2}} = \dfrac{2 + \frac{3}{4}}{1 + \frac{1}{2}}$ lcd of $\frac{3}{4}$ and $\frac{1}{2}$ is 4.

$= \dfrac{4\left(2 + \frac{3}{4}\right)}{4\left(1 + \frac{1}{2}\right)} = \dfrac{4 \cdot 2 + 4 \cdot \frac{3}{4}}{4 \cdot 1 + 4 \cdot \frac{1}{2}} = \frac{8 + 3}{4 + 2} = \frac{11}{6}$

29. $\dfrac{1 - \frac{2}{y}}{1 + \frac{1}{y}}$ lcd of $\frac{2}{y}$ and $\frac{1}{y}$ is y.

$\dfrac{1 - \frac{2}{y}}{1 + \frac{1}{y}} = \dfrac{y\left(1 - \frac{2}{y}\right)}{y\left(1 + \frac{1}{y}\right)} = \dfrac{y \cdot 1 - y \cdot \frac{2}{y}}{y \cdot 1 + y \cdot \frac{1}{y}} = \frac{y - 2}{y + 1}$

30. $\dfrac{\frac{5x}{6}}{\frac{25x^2}{24}} = \dfrac{24 \cdot \frac{5x}{6}}{24 \cdot \frac{25x^2}{24}}$ lcd = 24

$= \frac{20x}{25x^2} = \frac{4}{5x}$

31. $\dfrac{\frac{2}{3 + x} - \frac{2}{3}}{x} = \dfrac{3(3 + x)\left(\frac{2}{3 + x} - \frac{2}{3}\right)}{3(3 + x)x}$ lcd = 3(3 + x)

$= \dfrac{3(3 + x) \cdot \frac{2}{3 + x} - 3(3 + x) \cdot \frac{2}{3}}{3(3 + x)x} = \dfrac{6 - (3 + x)2}{3(3 + x)x}$

$= \dfrac{6 - 6 - 2x}{3(3 + x)x} = \dfrac{-2x}{3(3 + x)x} = \dfrac{-2}{3(3 + x)}$

32. $\dfrac{x - \frac{y}{3}}{3 - \frac{y}{x}} = \dfrac{3x\left(x - \frac{y}{3}\right)}{3x\left(3 - \frac{y}{x}\right)}$ lcd of $\frac{y}{3}$ and $\frac{y}{x}$ is 3x

$= \dfrac{3x \cdot x - 3x \cdot \frac{y}{3}}{3x \cdot 3 - 3x \cdot \frac{y}{x}} = \dfrac{3x^2 - xy}{9x - 3y} = \dfrac{x(3x - y)}{3(3x - y)} = \frac{x}{3}$

33. $\dfrac{\frac{x + 1}{y}}{\frac{x^2 - 1}{y^2}} = \dfrac{y^2 \cdot \frac{(x + 1)}{y}}{y^2 \cdot \frac{(x^2 - 1)}{y^2}}$ lcd = y^2

$= \dfrac{y(x + 1)}{x^2 - 1} = \dfrac{y(x + 1)}{(x - 1)(x + 1)} = \dfrac{y}{x - 1}$

34. $\dfrac{x^2 + 3x - 4}{x^2 - 5x + 4} = \dfrac{(x - 1)(x + 4)}{(x - 1)(x - 4)} = \dfrac{x + 4}{x - 4}$

35. $\dfrac{x^2 - 1}{x^3 - 1} = \dfrac{(x - 1)(x + 1)}{(x - 1)(x^2 + x + 1)} = \dfrac{x + 1}{x^2 + x + 1}$

36. $\dfrac{x^3 - 3x^2 + 2x}{x^3 + 2x^2 - 3x} = \dfrac{x(x^2 - 3x + 2)}{x(x^2 + 2x - 3)} = \dfrac{x(x - 1)(x - 2)}{x(x - 1)(x + 3)} = \dfrac{x - 2}{x + 3}$

37. $\dfrac{x^2 + 1}{x^3 + 1} = \dfrac{x^2 + 1}{(x + 1)(x^2 - x + 1)}$. There are no common factors of numerator and denominator; this expression is already reduced.

38. $\dfrac{x + 3}{x - 3} = \dfrac{(x + 3)(x + 3)}{(x - 3)(x + 3)} = \dfrac{x^2 + 6x + 9}{x^2 - 9}$. $x^2 + 6x + 9$

39. $\dfrac{x^2 + 2x + 1}{x^4 + x} = \dfrac{(x + 1)(x + 1)}{x(x^3 + 1)} = \dfrac{(x + 1)(x + 1)}{x(x + 1)(x^2 - x + 1)} = \dfrac{x + 1}{x(x^2 - x + 1)}$. $x + 1$

40. $\dfrac{3x^2 + 6x}{4x^3 + 4x^2} = \dfrac{3x(x + 2)}{4x^2(x + 1)} = \dfrac{2\cdot 3x(x + 2)(x + 2)}{8x^2(x + 1)(x + 2)} = \dfrac{6x(x^2 + 4x + 4)}{8x^2(x^2 + 3x + 2)}$

$= \dfrac{6x^3 + 24x^2 + 24x}{8x^4 + 24x^3 + 16x^2}$. $6x^3 + 24x^2 + 24x$

41.
$$\begin{aligned} (x + 1)^2 &= (x + 1)^2 \\ (x + 1)(x - 2) &= (x + 1)^1(x - 2)^1 \\ (x - 2)^2 &= (x - 2)^2 \qquad \text{lcm} = (x + 1)^2(x - 2)^2 \end{aligned}$$

42.
$$\begin{aligned} 3(x - 2) &= 3(x - 2) \\ 4(x + 3) &= 2^2(x + 3) \\ 5(x^2 + x - 6) &= 5(x - 2)(x + 3) \qquad \text{lcm} = 2^2\cdot 3\cdot 5(x - 2)(x + 3) = 60(x^2 + x - 6) \end{aligned}$$

43.
$$\begin{aligned} x^2 - 9 &= (x + 3)(x - 3) \\ x^2 + 6x + 9 &= (x + 3)^2 \\ x^2 - 6x + 9 &= (x - 3)^2 \qquad \text{lcm} = (x + 3)^2(x - 3)^2 = [(x + 3)(x - 3)]^2 \\ &\qquad\qquad\qquad\quad\;\; = [x^2 - 9]^2 = x^4 - 18x^2 + 81 \end{aligned}$$

44.
$$\begin{array}{r l} & x^2 - 2x + 4 \\ x + 2 \,\big)\!\!\! & \overline{x^3 + 0x^2 + 0x + 8} \\ & \underline{x^3 + 2x^2} \\ & -2x^2 + 0x \\ & \underline{-2x^2 - 4x} \\ & 4x + 8 \\ & \underline{4x + 8} \\ & 0 = \text{remainder} \end{array}$$

45.
$$\begin{array}{r l} & x + 3 \\ x - 3 \,\big)\!\!\! & \overline{x^2 + 0x + 9} \\ & \underline{x^2 - 3x} \\ & 3x + 9 \\ & \underline{3x - 9} \\ & 18 = \text{remainder} \end{array}$$

46.
$$\begin{array}{r l} & x^2 - 2x - 1 \\ x - 1 \,\big)\!\!\! & \overline{x^3 - 3x^2 + x - 3} \\ & \underline{x^3 - x^2} \\ & -2x^2 + x \\ & \underline{-2x^2 + 2x} \\ & -x - 3 \\ & \underline{-x + 1} \\ & -4 = \text{remainder} \end{array}$$

Alternatively, we can use synthetic division:

$$\begin{array}{r|rrrr} & 1 & -3 & 1 & -3 \\ & & 1 & -2 & -1 \\ \hline 1 & 1 & -2 & -1 & -4 \end{array}$$

Quotient: $x^2 - 2x - 1$
Remainder: $- 4$

47. $x^2 + 1 \overline{)\, x^3 + 0x^2 + x + 0}$ quotient: x

$x^3 + \quad x$

0 = remainder

48. $x^2 + 2 \overline{)\, x^4 + 2x^3 + 3x^2 + 4x + 5}$ quotient: $x^2 + 2x + 1$

$x^4 \quad + 2x^2$

$2x^3 + x^2 + 4x$

$2x^3 \quad + 4x$

$x^2 \quad + 5$

$x^2 \quad + 2$

3 = remainder

49. $x + 2 \overline{)\, x^4 + 0x^3 + x^2 + 0x - 1}$ quotient: $x^3 - 2x^2 + 5x - 10$

$x^4 + 2x^3$

$-2x^3 + x^2$

$-2x^3 - 4x^2$

$5x^2 + 0x$

$5x^2 + 10x$

$-10x - 1$

$-10x - 20$

19 = remainder

Alternatively, we can use synthetic division:

$$\begin{array}{r|rrrrr} & 1 & 0 & 1 & 0 & -1 \\ & & -2 & 4 & -10 & 20 \\ \hline -2 & 1 & -2 & 5 & -10 & 19 \end{array}$$

Quotient: $x^3 - 2x^2 + 5x - 10$
Remainder: 19

50. $x - 1 \overline{)\, x^4 + 0x^3 + 0x^2 + 0x - 1}$ quotient: $x^3 + x^2 + x + 1$

$x^4 - x^3$

$x^3 + 0x^2$

$x^3 - x^2$

$x^2 + 0x$

$x^2 - x$

$x - 1$

$x - 1$

0 = remainder

Alternatively, we can use synthetic division:

$$\begin{array}{r|rrrrr} & 1 & 0 & 0 & 0 & -1 \\ & & 1 & 1 & 1 & 1 \\ \hline 1 & 1 & 1 & 1 & 1 & 0 \end{array}$$

Quotient: $x^3 + x^2 + x + 1$
Remainder: 0

51. $x^2 + x + 1 \overline{)\, x^4 + x^3 + 0x^2 + x + 0}$ quotient: $x^2 - 1$

$x^4 + x^3 + x^2$

$-x^2 + x + 0$

$-x^2 - x - 1$

$2x + 1$ = remainder

52. $\dfrac{2}{5b} - \dfrac{4}{3b^3} - \dfrac{1}{6a^2b^2}$

$5b = 5b$

$3b^3 = 3b^3$

$6a^2b^2 = 2 \cdot 3a^2b^2 \qquad \text{lcd} = 2 \cdot 3 \cdot 5a^2b^3 = 30a^2b^3$

$$\frac{2}{5b} - \frac{4}{3b^3} - \frac{1}{6a^2b^2} = \frac{2 \cdot 6a^2b^2}{5b \cdot 6a^2b^2} - \frac{4 \cdot 10a^2}{3b^3 \cdot 10a^2} - \frac{1 \cdot 5b}{6a^2b^2 \cdot 5b}$$

$$= \frac{12a^2b^2}{30a^2b^3} - \frac{40a^2}{30a^2b^3} - \frac{5b}{30a^2b^3} = \frac{12a^2b^2 - 40a^2 - 5b}{30a^2b^3}$$

53. $\dfrac{2}{2x - 3} - 1 = \dfrac{2}{2x - 3} - \dfrac{1}{1}$ lcd $= 2x - 3$

$$= \frac{2}{2x - 3} - \frac{(2x - 3)}{2x - 3} = \frac{2 - (2x - 3)}{2x - 3} = \frac{2 - 2x + 3}{2x - 3} = \frac{5 - 2x}{2x - 3}$$

54. $$\frac{4x^2y}{3ab^2} \div \left(\frac{\overset{1}{\cancel{2}}a^2x^2}{b^2y} \cdot \frac{6a}{\underset{1}{\cancel{2}}y^2}\right) = \frac{4x^2y}{3ab^2} \div \left(\frac{6a^3x^2}{b^2y^3}\right) = \frac{\overset{2 \cdot 1}{\cancel{4}\cancel{x^2}}y}{3a\underset{1}{\cancel{b^2}}} \cdot \frac{\overset{1}{\cancel{b^2}}y^3}{\underset{3}{\cancel{6}}a^3\underset{1}{\cancel{x^2}}} = \frac{2y^4}{9a^4}$$

55. $\dfrac{x}{x^2 + 4x} + \dfrac{2x}{3x^2 - 48}$

$x^2 + 4x = x(x + 4)$
$3x^2 - 48 = 3(x^2 - 16) = 3(x - 4)(x + 4)$ lcd $= 3x(x - 4)(x + 4)$

$$\frac{x}{x^2 + 4x} + \frac{2x}{3x^2 - 48} = \frac{x}{x(x + 4)} + \frac{2x}{3(x - 4)(x + 4)}$$

$$= \frac{x \cdot 3(x - 4)}{x(x + 4) \cdot 3(x - 4)} + \frac{2x \cdot x}{3(x - 4)(x + 4) \cdot x} = \frac{3x(x - 4) + 2x^2}{3x(x - 4)(x + 4)}$$

$$= \frac{3x^2 - 12x + 2x^2}{3x(x - 4)(x + 4)} = \frac{5x^2 - 12x}{3x(x - 4)(x + 4)} = \frac{\overset{1}{\cancel{x}}(5x - 12)}{3\underset{1}{\cancel{x}}(x - 4)(x + 4)}$$

$$= \frac{5x - 12}{3(x - 4)(x + 4)}$$

Alternatively, it would have been correct to reduce $\dfrac{x}{x^2 + 4x}$ to $\dfrac{1}{x + 4}$ at the outset.

56. $$\frac{x^3 - x}{x^2 - x} \div \frac{x^2 + 2x + 1}{x} = \frac{x^3 - x}{x^2 - x} \cdot \frac{x}{x^2 + 2x + 1}$$

$$= \frac{\overset{1}{\cancel{x}}\cancel{(x - 1)}\overset{1}{\cancel{(x + 1)}}}{\cancel{x}\underset{1}{\cancel{(x - 1)}}} \cdot \frac{x}{(x + 1)\underset{1}{\cancel{(x + 1)}}} = \frac{x}{x + 1}$$

57. $\dfrac{\dfrac{x}{y} - \dfrac{y}{x}}{\dfrac{x}{y} + 1}$ lcd of $\dfrac{x}{y}$, $\dfrac{y}{x}$, and $\dfrac{x}{y}$ is xy.

COMMON ERROR: multiplying numerator by xy and denominator by y to obtain $\dfrac{x^2 - y^2}{x + y}$. The fundamental principle of fractions requires that numerator and denominator be multiplied by the **same** quantity.

$$\frac{\frac{x}{y} - \frac{y}{x}}{\frac{x}{y} + 1} = \frac{xy\left(\frac{x}{y} - \frac{y}{x}\right)}{xy\left(\frac{x}{y} + 1\right)} = \frac{xy \cdot \frac{x}{y} - xy \cdot \frac{y}{x}}{xy \cdot \frac{x}{y} + xy \cdot 1} = \frac{x^2 - y^2}{x^2 + xy} = \frac{\overset{1}{\cancel{(x + y)}}(x - y)}{x\underset{1}{\cancel{(x + y)}}} = \frac{x - y}{x}$$

58. $\dfrac{x}{x^3 - y^3} - \dfrac{1}{x^2 + xy + y^2}$

$x^3 - y^3 = (x - y)(x^2 + xy + y^2)$
$x^2 + xy + y^2 = x^2 + xy + y^2 \qquad \text{lcd} = (x - y)(x^2 + xy + y^2)$

$$\frac{x}{x^3 - y^3} - \frac{1}{x^2 + xy + y^2} = \frac{x}{(x - y)(x^2 + xy + y^2)} - \frac{(x - y)}{(x - y)(x^2 + xy + y^2)}$$

$$= \frac{x - (x - y)}{x^3 - y^3} = \frac{x - x + y}{x^3 - y^3} = \frac{y}{x^3 - y^3}$$

59. $\dfrac{\dfrac{y^2}{x^2 - y^2} + 1}{\dfrac{x^2}{x - y} - x}$ lcd of $\dfrac{y^2}{x^2 - y^2}$ and $\dfrac{x^2}{x - y}$ is $x^2 - y^2$

$$\frac{\dfrac{y^2}{x^2 - y^2} + 1}{\dfrac{x^2}{x - y} - x} = \frac{(x^2 - y^2)\left(\dfrac{y^2}{x^2 - y^2} + 1\right)}{(x^2 - y^2)\left(\dfrac{x}{x - y} - x\right)} = \frac{(x^2 - y^2)\cdot\dfrac{y^2}{x^2 - y^2} + (x^2 - y^2)\cdot 1}{(x^2 - y^2)\cdot\dfrac{x^2}{(x - y)} - (x^2 - y^2)\cdot x}$$

$$= \frac{y^2 + (x^2 - y^2)}{\dfrac{(x + y)(x - y)}{1}\cdot\dfrac{x^2}{x - y} - x(x^2 - y^2)} = \frac{y^2 + x^2 - y^2}{(x + y)x^2 - x(x^2 - y^2)}$$

$$= \frac{x^2}{x^3 + x^2y - x^3 + xy^2} = \frac{x^2}{x^2y + xy^2} = \frac{x^2}{xy(x + y)} = \frac{x}{y(x + y)}$$

60. $\dfrac{x^3 - 1}{x^2 + x + 1} \div \dfrac{x^2 - 1}{x^2 + 2x + 1} = \dfrac{x^3 - 1}{x^2 + x + 1}\cdot\dfrac{x^2 + 2x + 1}{x^2 - 1}$

$$= \frac{\cancel{(x - 1)}^{1}\,\cancel{(x^2 + x + 1)}^{1}}{\cancel{x^2 + x + 1}_{1}}\cdot\frac{(x+1)\cancel{(x+1)}^{1}}{\cancel{(x-1)}_{1}\,\cancel{(x+1)}_{1}} = \frac{x+1}{1} = x+1$$

61. $\dfrac{1}{3x^2 - 27} - \dfrac{x - 1}{4x^3 + 24x^2 + 36x}$

$3x^2 - 27 = 3(x^2 - 9) = 3(x - 3)(x + 3)$
$4x^3 + 24x^2 + 36x = 4x(x^2 + 6x + 9) = 4x(x + 3)(x + 3)$
$\text{lcd} = 12x(x - 3)(x + 3)^2$

$$\frac{1}{3x^2 - 27} - \frac{x - 1}{4x^3 + 24x^2 + 36x} = \frac{1}{3(x - 3)(x + 3)} - \frac{(x - 1)}{4x(x + 3)^2}$$

$$= \frac{1\cdot 4x(x + 3)}{3(x - 3)(x + 3)\cdot 4x(x + 3)} - \frac{(x - 1)\cdot 3(x - 3)}{4x(x + 3)^2\cdot 3(x - 3)}$$

$$= \frac{4x(x + 3) - 3(x - 1)(x - 3)}{12x(x - 3)(x + 3)^2} = \frac{4x^2 + 12x - 3(x^2 - 4x + 3)}{12x(x - 3)(x + 3)^2}$$

$$= \frac{4x^2 + 12x - 3x^2 + 12x - 9}{12x(x - 3)(x + 3)^2} = \frac{x^2 + 24x - 9}{12x(x - 3)(x + 3)^2}$$

62. $1 + \frac{1}{x} + \frac{1}{1 + x} = \frac{1}{1} + \frac{1}{x} + \frac{1}{1 + x}$ lcd $= x(1 + x)$

$$= \frac{x(1 + x)}{x(1 + x)} + \frac{1 + x}{x(1 + x)} + \frac{x}{x(1 + x)}$$

$$= \frac{x(1 + x) + 1 + x + x}{x(1 + x)} = \frac{x + x^2 + 1 + 2x}{x(1 + x)} = \frac{x^2 + 3x + 1}{x(1 + x)}$$

63. $\left(\frac{1 + x}{x} \div \frac{x}{1 - x}\right) \div \frac{1 - x^2}{x^2} = \left(\frac{1 + x}{x} \cdot \frac{1 - x}{x}\right) \div \frac{1 - x^2}{x^2}$

$$= \frac{1 - x^2}{x^2} \div \frac{1 - x^2}{x^2} = 1$$

64. $\frac{x + 2}{x^2 - 9} \cdot \frac{x^2 - 2x - 3}{x^2 + 3x + 2} = \frac{\overset{1}{\cancel{x + 2}}}{\underset{1}{\cancel{(x - 3)}}(x + 3)} \cdot \frac{\overset{1}{\cancel{(x - 3)}}\overset{1}{\cancel{(x + 1)}}}{\underset{1}{\cancel{(x + 2)}}\underset{1}{\cancel{(x + 1)}}} = \frac{1}{x + 3}$

65. $\frac{x}{x^2 + 3x + 2} \cdot \frac{x^2 + 5x + 6}{x^2 + 7x + 12} \cdot \frac{(x^2 + 5x + 4)}{x^2}$

$$= \frac{x}{(x + 1)\underset{1}{\cancel{(x + 2)}}} \cdot \frac{\overset{1}{\cancel{(x + 2)}}\overset{1}{\cancel{(x + 3)}}}{(x + 4)\underset{1}{\cancel{(x + 3)}}} \cdot \frac{(x^2 + 5x + 4)}{x^2}$$

$$= \frac{\overset{1}{\cancel{x}}}{\underset{1}{\cancel{(x + 1)}}\underset{1}{\cancel{(x + 4)}}} \cdot \frac{\overset{1}{\cancel{(x + 1)}}\overset{1}{\cancel{(x + 4)}}}{\underset{x}{\cancel{x^2}}} = \frac{1}{x}$$

66. $x - \frac{x + 1}{\frac{1}{x} + 1} = x - \frac{x(x + 1)}{x\left(\frac{1}{x} + 1\right)} = x - \frac{x(x + 1)}{1 + x} = x - \frac{x(x + 1)}{(x + 1)} = x - x = 0$

67. $x + \frac{1}{x} - \boxed{\frac{1}{\frac{1}{x}}} - \frac{1}{\boxed{\frac{1}{\frac{1}{x}}}}$. We note that the boxed complex fraction appears more than once. We start by simplifying it: $\frac{1}{\frac{1}{x}} = \frac{x \cdot 1}{x \cdot \frac{1}{x}} = \frac{x}{1} = x.$

Thus, $x + \frac{1}{x} - \boxed{\frac{1}{\frac{1}{x}}} - \frac{1}{\boxed{\frac{1}{\frac{1}{x}}}} = x + \frac{1}{x} - x - \frac{1}{x} = 0$

68. $\frac{4}{s^2 - 4} + \frac{1}{2 - s} = \frac{4}{(s - 2)(s + 2)} + \frac{1}{2 - s}$ $(2 - s) = -(s - 2)$

$$= \frac{4}{(s - 2)(s + 2)} + \frac{1}{-(s - 2)}$$

$$= \frac{4}{(s - 2)(s + 2)} - \frac{1}{s - 2} \quad \text{lcd} = (s - 2)(s + 2)$$

$$= \frac{4}{(s - 2)(s + 2)} - \frac{(s + 2)}{(s - 2)(s + 2)} = \frac{4 - (s + 2)}{(s - 2)(s + 2)}$$

$$= \frac{4 - s - 2}{(s-2)(s+2)} = \frac{2 - s}{(s-2)(s+2)} = \frac{-(s-2)}{(s-2)(s+2)} = \frac{-1}{s+2}$$

69. $$\frac{y^2 - y - 6}{(y+2)^2} \cdot \frac{2+y}{3-y} = \frac{(y-3)(y+2)}{(y+2)(y+2)} \cdot \frac{(2+y)}{(3-y)} \qquad \begin{array}{l} 2 + y = y + 2 \\ 3 - y = -(y - 3) \end{array}$$

$$= \frac{y-3}{y+2} \cdot \frac{(y+2)}{-(y-3)} = -1$$

70. $$\frac{y}{x^2} \div \left(\frac{x^2 + 3x}{2x^2 + 5x - 3} \div \frac{x^3y - x^2y}{2x^2 - 3x + 1}\right) = \frac{y}{x^2} \div \left(\frac{x^2 + 3x}{2x^2 + 5x - 3} \cdot \frac{2x^2 - 3x + 1}{x^3y - x^2y}\right)$$

$$= \frac{y}{x^2} \div \left(\frac{x(x+3)}{(2x-1)(x+3)} \cdot \frac{(2x-1)(x-1)}{x^2y(x-1)}\right)$$

$$= \frac{y}{x^2} \div \frac{1}{xy} = \frac{y}{x^2} \cdot \frac{xy}{1} = \frac{y^2}{x}$$

71. $$\frac{1 - \boxed{\dfrac{1}{1 + \frac{x}{y}}}}{1 - \boxed{\dfrac{1}{1 - \frac{x}{y}}}}$$

We start by changing the boxed complex fractions to simple fractions. This requires us to multiply numerator and denominator of each fraction by the lcd of all internal fractions. Coincidentally, this lcd is the same for both boxed complex fractions, that is, y.

$$\frac{1 - \dfrac{1}{1 + \frac{x}{y}}}{1 - \dfrac{1}{1 - \frac{x}{y}}} = \frac{1 - \dfrac{y \cdot 1}{y\left(1 + \frac{x}{y}\right)}}{1 - \dfrac{y \cdot 1}{y\left(1 - \frac{x}{y}\right)}} = \frac{1 - \dfrac{y}{y \cdot 1 + y \cdot \frac{x}{y}}}{1 - \dfrac{y}{y \cdot 1 - y \cdot \frac{x}{y}}} = \frac{1 - \dfrac{y}{y + x}}{1 - \dfrac{y}{y - x}}$$

Now we reduce this simpler complex fraction to a simple fraction by multiplying its numerator and denominator by $(y + x)(y - x)$, the lcd of its internal fractions.

$$\frac{1 - \dfrac{1}{1 + \frac{x}{y}}}{1 - \dfrac{1}{1 - \frac{x}{y}}} = \frac{(y + x)(y - x)\left(1 - \dfrac{y}{y + x}\right)}{(y + x)(y - x)\left(1 - \dfrac{y}{y - x}\right)}$$

$$= \frac{(y + x)(y - x)\cdot 1 - (y + x)(y - x)\cdot \dfrac{y}{y + x}}{(y + x)(y - x)\cdot 1 - (y + x)(y - x)\cdot \dfrac{y}{y - x}} = \frac{(y^2 - x^2) - (y - x)y}{(y^2 - x^2) - (y + x)y}$$

$$= \frac{y^2 - x^2 - y^2 + xy}{y^2 - x^2 - y^2 - xy} = \frac{-x^2 + xy}{-x^2 - xy} = \frac{-x(x - y)}{-x(x + y)} = \frac{x - y}{x + y}$$

72. $$\left(x - \frac{1}{1 - \frac{1}{x}}\right) \div \left(\frac{x}{x + 1} - \frac{x}{1 - x}\right) = \left(x - \frac{x\cdot 1}{x\left(1 - \frac{1}{x}\right)}\right) \div \left(\frac{x}{x + 1} - \frac{x}{1 - x}\right)$$

$$= \left(x - \frac{x}{x\cdot 1 - x\cdot \frac{1}{x}}\right) \div \left(\frac{x}{x + 1} - \frac{x}{1 - x}\right) = \left(x - \frac{x}{x - 1}\right) \div \left(\frac{x}{x + 1} - \frac{x}{1 - x}\right)$$

$$= \left(\frac{x}{1} - \frac{x}{x - 1}\right) \div \left(\frac{x}{x + 1} - \frac{x}{1 - x}\right)$$

$$= \left(\frac{x(x - 1)}{x - 1} - \frac{x}{x - 1}\right) \div \left(\frac{x(1 - x)}{(x + 1)(1 - x)} - \frac{x(x + 1)}{(x + 1)(1 - x)}\right)$$

$$= \frac{x(x - 1) - x}{x - 1} \div \frac{x(1 - x) - x(x + 1)}{(x + 1)(1 - x)} = \frac{x^2 - x - x}{x - 1} \div \frac{x - x^2 - x^2 - x}{(x + 1)(1 - x)}$$

$$= \frac{x^2 - 2x}{x - 1} \div \frac{-2x^2}{(x + 1)(1 - x)} = \frac{x(x - 2)}{x - 1} \div \frac{-2x^2}{(x + 1)(1 - x)}$$

$$= \frac{x(x - 2)}{x - 1}\cdot \frac{(x + 1)(1 - x)}{-2x^2} = \frac{\overset{1}{\cancel{x}}(x - 2)}{(-1)\underset{1}{\cancel{(1 - x)}}}\cdot \frac{(x + 1)\overset{1}{\cancel{(1 - x)}}}{-2\underset{x}{\cancel{x^2}}}$$

$$= \frac{(x - 2)(x + 1)}{(-1)(-2x)} = \frac{(x - 2)(x + 1)}{2x}$$

73. $$\frac{1}{x^2 + x} - \frac{1}{x^2 + 3x + 2} + \frac{1}{x^2 + 2x}$$

$$= \frac{1}{x(x + 1)} - \frac{1}{(x + 1)(x + 2)} + \frac{1}{x(x + 2)} \qquad \text{lcd} = x(x + 1)(x + 2)$$

$$= \frac{(x + 2)}{x(x + 1)(x + 2)} - \frac{x}{x(x + 1)(x + 2)} + \frac{(x + 1)}{x(x + 1)(x + 2)}$$

$$= \frac{x + 2 - x + x + 1}{x(x + 1)(x + 2)} = \frac{x + 3}{x(x + 1)(x + 2)}$$

74. $$\frac{x}{x^2 - 1} - \frac{x}{x^2 - 2x + 1} - \frac{2}{x^2 + 2x + 1}$$

$$= \frac{x}{(x + 1)(x - 1)} - \frac{x}{(x - 1)^2} - \frac{2}{(x + 1)^2} \qquad \text{lcd} = (x + 1)^2(x - 1)^2$$

$$= \frac{x(x + 1)(x - 1)}{(x + 1)^2(x - 1)^2} - \frac{x(x + 1)^2}{(x + 1)^2(x - 1)^2} - \frac{2(x - 1)^2}{(x + 1)^2(x - 1)^2}$$

$$= \frac{x(x + 1)(x - 1) - x(x + 1)^2 - 2(x - 1)^2}{(x + 1)^2(x - 1)^2}$$

$$= \frac{x(x^2 - 1) - x(x^2 + 2x + 1) - 2(x^2 - 2x + 1)}{[(x + 1)(x - 1)]^2}$$

$$= \frac{x^3 - x - x^3 - 2x^2 - x - 2x^2 + 4x - 2}{[x^2 - 1]^2} = \frac{-4x^2 + 2x - 2}{x^4 - 2x^2 + 1}$$

75.
$$\begin{array}{r|l} & x^3 - 3x^2 + 7x - 15 \\ \hline x + 2 & x^4 - x^3 + x^2 - x + 1 \\ & \underline{x^4 + 2x^3} \\ & -3x^3 + x^2 \\ & \underline{-3x^3 - 6x^2} \\ & 7x^2 - x \\ & \underline{7x^2 + 14x} \\ & -15x + 1 \\ & \underline{-15x - 30} \\ & 31 \end{array}$$

31 = remainder, so value of the polynomial is 31.

Alternatively, we can use synthetic division:

$$\begin{array}{r|rrrrr} & 1 & -1 & 1 & -1 & 1 \\ & & -2 & 6 & -14 & 30 \\ \hline -2 & 1 & -3 & 7 & -15 & 31 \end{array}$$

Remainder: 31, so value of the polynomial is 31.

Practice Test 3

1. $\dfrac{6xy^2z^3}{15x^2yz^2} = \dfrac{3xyz^2\cdot 2yz}{3xyz^2\cdot 5x} = \dfrac{2yz}{5x}$

2. $\dfrac{x(x^2 + x - 2)}{x^3 + 5x^2 + 6x} = \dfrac{x(x - 1)(x + 2)}{x(x^2 + 5x + 6)} = \dfrac{x(x - 1)(x + 2)}{x(x + 3)(x + 2)} = \dfrac{x - 1}{x + 3}$

3. $6xy^2z = 2\cdot 3x^2y^2z$
   $3xy^2 = 3xy^2$
   $4y^2z^4 = 2^2y^2z^4 \quad \text{lcm} = 2^2\cdot 3x^2y^2z^4 = 12x^2y^2z^4$

4. $x + 2 = x + 2$
   $x^2 + 3x + 2 = (x + 2)(x + 1)$
   $x^2 + 2x + 1 = (x + 1)^2 \quad \text{lcm} = (x + 1)^2(x + 2)$

5. $\dfrac{3x^2y}{8z^3} = \dfrac{3x^2y\cdot 5yz^3}{8z^3\cdot 5yz^3} = \dfrac{15x^2y^2z^3}{40yz^6}. \quad 40yz^6$

6. $\dfrac{x + 1}{x + 2} = \dfrac{(x + 1)(x + 3)}{(x + 2)(x + 3)} = \dfrac{x^2 + 4x + 3}{x^2 + 5x + 6}. \quad x^2 + 4x + 3$

7. $3 - \dfrac{x}{x - 1} \quad \text{lcd} = x - 1$
   $= \dfrac{3}{1} - \dfrac{x}{x - 1} = \dfrac{3(x - 1)}{x - 1} - \dfrac{x}{x - 1} = \dfrac{3(x - 1) - x}{x - 1} = \dfrac{3x - 3 - x}{x - 1} = \dfrac{2x - 3}{x - 1}$

8. $\dfrac{2}{x} + \dfrac{3}{x^2} + \dfrac{4}{x^3} = \dfrac{2x^2}{x^3} + \dfrac{3x}{x^3} + \dfrac{4}{x^3} \quad \text{lcd} = x^3$
   $= \dfrac{2x^2 + 3x + 4}{x^3}$

9. $\dfrac{\overset{1}{x\cancel{y^2}}}{\underset{1}{\cancel{3z}}}\cdot\dfrac{\overset{4x}{\cancel{12xz}}}{\underset{y}{7\cancel{y^3}}} = \dfrac{4x^2}{7y}$

10. $\dfrac{4x^2y}{5z^2} \div \dfrac{8xy^2}{15z} = \dfrac{\overset{x}{\cancel{4x^2y}}}{\underset{z}{\cancel{5z^2}}}\cdot\dfrac{\overset{3}{\cancel{15z}}}{\underset{2y}{\cancel{8xy^2}}} = \dfrac{3x}{2yz}$

11. $\dfrac{x(x - 1)}{x^2 + 2x}\cdot\dfrac{x + 2}{x^2 - 2x + 1} = \dfrac{\overset{1}{\cancel{x}}\overset{1}{\cancel{(x - 1)}}}{\underset{1}{\cancel{x}}\underset{1}{\cancel{(x + 2)}}}\cdot\dfrac{\overset{1}{\cancel{x + 2}}}{\underset{1}{\cancel{(x - 1)}}(x - 1)} = \dfrac{1}{x - 1}$

12. $\dfrac{x}{x + 3} - \dfrac{x}{x - 2} = \dfrac{x(x - 2)}{(x + 3)(x - 2)} - \dfrac{x(x + 3)}{(x + 3)(x - 2)}$
    $= \dfrac{x(x - 2) - x(x + 3)}{(x + 3)(x - 2)} = \dfrac{x^2 - 2x - x^2 - 3x}{(x + 3)(x - 2)} = \dfrac{-5x}{x^2 + x - 6}$

13. $\dfrac{1}{x} - \dfrac{1}{x - 1} + \dfrac{1}{x^2} = \dfrac{x(x - 1)}{x^2(x - 1)} - \dfrac{1x^2}{x^2(x - 1)} + \dfrac{1(x - 1)}{x^2(x - 1)}$
    $= \dfrac{x(x - 1) - x^2 + x - 1}{x^2(x - 1)} = \dfrac{x^2 - x - x^2 + x - 1}{x^2(x - 1)} = \dfrac{-1}{x^3 - x}$

14. $$\frac{\frac{1}{x-2}}{\frac{x}{x^2-4x+4}} = \frac{\frac{1}{x-2}}{\frac{x}{(x-2)^2}} \qquad \text{lcd} = (x-2)^2$$

$$= \frac{(x-2)^2\,\frac{1}{x-2}}{(x-2)^2\,\frac{x}{(x-2)^2}} = \frac{x-2}{x}$$

15. $$\frac{x}{x+1}\left(\frac{x}{2}-\frac{\frac{1}{2}}{x}\right) = \frac{x}{x+1}\left(\frac{x^2}{2x}-\frac{2\cdot\frac{1}{2}}{2x}\right)$$

$$= \frac{x}{x+1}\left(\frac{x^2-1}{2x}\right) = \frac{\overset{1}{\cancel{x}}}{\underset{1}{\cancel{x+1}}}\cdot\frac{(x-1)\overset{1}{\cancel{(x+1)}}}{2\underset{1}{\cancel{x}}} = \frac{x-1}{2}$$

16. $$\frac{x-\frac{x}{\frac{1}{x}}}{x} = \frac{x-\frac{x\cdot x}{x\cdot\frac{1}{x}}}{x} = \frac{x-\frac{x^2}{1}}{x} = \frac{x-x^2}{x} = \frac{x(1-x)}{x} = 1-x$$

17. $$\frac{\frac{1}{x+1}-\frac{1}{x-1}}{\frac{2}{x^2-1}} = \frac{(x+1)(x-1)\left(\frac{1}{x+1}-\frac{1}{x-1}\right)}{(x+1)(x-1)\,\frac{2}{(x+1)(x-1)}}$$

$$= \frac{(x+1)(x-1)\cdot\frac{1}{x+1}-(x+1)(x-1)\,\frac{1}{x-1}}{2}$$

$$= \frac{(x-1)-(x+1)}{2} = \frac{x-1-x-1}{2} = \frac{-2}{2} = -1$$

18. $$\begin{array}{r|l} & x^2+1 \\ \hline x-1 & x^3-x^2+x-1 \\ & \underline{x^3-x^2} \\ & \qquad\quad x-1 \\ & \qquad\quad \underline{x-1} \\ & \qquad\qquad 0 = \text{remainder} \end{array}$$

Alternatively, we can use synthetic division:

$$\begin{array}{r|rrrr} & 1 & -1 & 1 & -1 \\ & & 1 & 0 & 1 \\ \hline 1 & 1 & 0 & 1 & 0 \end{array}$$

Quotient: $x^2 + 1$
Remainder: 0

19. $$\begin{array}{r|l} & x+1 \\ \hline x^2+x+1 & x^3+2x^2+3x+4 \\ & \underline{x^3+x^2+x} \\ & \qquad x^2+2x+4 \\ & \qquad \underline{x^2+x+1} \\ & \qquad\qquad x+3 = \text{remainder} \end{array}$$

20.
$$
\begin{array}{r|l}
 & x^2 + 11x + 82 \\
\hline
x - 7 & x^3 + 4x^2 + 5x + 6 \\
 & \underline{x^3 - 7x^2} \\
 & \quad 11x^2 + 5x \\
 & \quad \underline{11x^2 - 77x} \\
 & \qquad 82x + 6 \\
 & \qquad \underline{82x - 574} \\
 & \qquad\qquad 580
\end{array}
$$

580 = remainder, so value of polynomial is 580

Alternatively, we can use synthetic division:

$$
\begin{array}{r|rrrr}
 & 1 & 4 & 5 & 6 \\
 & & 7 & 77 & 574 \\
\hline
7 & 1 & 11 & 82 & 580
\end{array}
$$

Remainder: 580, so value of the polynomial is 580.

CHAPTER 4 Solving Equations and Applications

Exercise 4-1 Solving Equations Involving Fractions

Key Ideas and Formulas

Equation-Solving Strategy
1. Use multiplication to remove fractions if present. a. If there are no variables in denominators, simply multiply every term on both sides by the LCM of all denominators. b. If there are variables in denominators, proceed as in part a. However, any value of the variable which makes a denominator 0 must be avoided. It is simplest to list such values at the outset, then eliminate them from any proposed solution values. 2. Simplify the left and right sides of the equation by removing grouping symbols and combining like terms. 3. Use equality properties to get all variable terms on one side (usually the left) and all constant terms on the other side (usually the right). Combine like terms in the process. 4. Isolate the variable (with a coefficient of 1), using the division or multiplication property of equality.

1. $\frac{x}{5} - 2 = \frac{3}{5}$ lcm = 5

$$5 \cdot \left(\frac{x}{5} - 2\right) = 5 \cdot \frac{3}{5}$$
$$5 \cdot \frac{x}{5} - 5 \cdot 2 = 5 \cdot \frac{3}{5}$$
$$x - 10 = 3$$
$$x = 13$$

3. $\frac{x}{3} + \frac{x}{6} = 4$ lcm = 6

$$6 \cdot \left(\frac{x}{3} + \frac{x}{6}\right) = 6 \cdot 4$$
$$6 \cdot \frac{x}{3} + 6 \cdot \frac{x}{6} = 6 \cdot 4$$
$$2x + x = 24$$
$$3x = 24$$
$$x = 8$$

5. $\frac{m}{4} - \frac{m}{3} = \frac{1}{2}$ lcm = 12

$$12 \cdot \left(\frac{m}{4} - \frac{m}{3}\right) = 12 \cdot \frac{1}{2}$$
$$12 \cdot \frac{m}{4} - 12 \cdot \frac{m}{3} = 12 \cdot \frac{1}{2}$$
$$3m - 4m = 6$$
$$-m = 6$$
$$m = -6$$

7. $\frac{5}{12} - \frac{m}{3} = \frac{4}{9}$ lcm = 36

$$36 \cdot \left(\frac{5}{12} - \frac{m}{3}\right) = 36 \cdot \frac{4}{9}$$
$$36 \cdot \frac{5}{12} - 36 \cdot \frac{m}{3} = 36 \cdot \frac{4}{9}$$
$$15 - 12m = 16$$
$$-12m = 1$$
$$m = -\frac{1}{12}$$

9.
$$0.7x = 21$$
$$10(0.7x) = 10 \cdot 21$$
$$7x = 210$$
$$x = 30$$

11.
$$0.7x + 0.9x = 32$$
$$10(0.7x) + 10(0.9x) = 10 \cdot 32$$
$$7x + 9x = 320$$
$$16x = 320$$
$$x = 20$$

13. $\frac{1}{2} - \frac{2}{x} = \frac{3}{x}$ $x \neq 0$ lcm = 2x

$$2x \cdot \frac{1}{2} - 2x \cdot \frac{2}{x} = 2x \cdot \frac{3}{x}$$
$$x - 4 = 6$$
$$x = 10$$

15. $\frac{1}{m} - \frac{1}{9} = \frac{4}{9} - \frac{2}{3m}$ $m \neq 0$ lcm = 9m

$$9m \cdot \frac{1}{m} - 9m \cdot \frac{1}{9} = 9m \cdot \frac{4}{9} - 9m \cdot \frac{2}{3m}$$
$$9 - m = 4m - 6$$
$$-5m = -15$$
$$m = 3$$

17. $\frac{x - 2}{3} + 1 = \frac{x}{7}$ lcm = 21

$$21 \cdot \frac{(x - 2)}{3} + 21 \cdot 1 = 21 \cdot \frac{x}{7}$$
$$7(x - 2) + 21 = 3x$$
$$7x - 14 + 21 = 3x$$
$$7x + 7 = 3x$$
$$4x + 7 = 0$$
$$4x = -7$$
$$x = -\frac{7}{4}$$

19. $\frac{2x - 3}{9} - \frac{x + 5}{6} = \frac{3 - x}{2} - 1$ lcm = 18

$$18 \cdot \frac{(2x - 3)}{9} - 18 \cdot \frac{(x + 5)}{6} = 18 \cdot \frac{(3 - x)}{2} - 18 \cdot 1$$
$$2(2x - 3) - 3(x + 5) = 9(3 - x) - 18$$
$$4x - 6 - 3x - 15 = 27 - 9x - 18$$
$$x - 21 = -9x + 9$$
$$10x = 30$$
$$x = 3$$

21. $0.1(x - 7) + 0.05x = 0.8$ lcm = 100

$$100[0.1(x - 7)] + 100(0.05x) = 100(0.8)$$
$$10(x - 7) + 5x = 80$$
$$10x - 70 + 5x = 80$$
$$15x - 70 = 80$$
$$15x = 150$$
$$x = 10$$

23. $0.02x - 0.5(x - 2) = 5.32$ lcm = 100

$$100(0.02x) - 100[0.5(x - 2)] = 100(5.32)$$
$$2x - 50(x - 2) = 532$$
$$2x - 50x + 100 = 532$$
$$-48x + 100 = 532$$
$$-48x = 432$$
$$x = -9$$

25. $\frac{1}{2x} - \frac{3}{5} = -\frac{7}{6x}$ $x \neq 0$ lcm = 30x

$$30x \cdot \frac{1}{2x} - 30x \cdot \frac{3}{5} = 30x \cdot \frac{-7}{6x}$$
$$15 - 18x = -35$$
$$-18x = -50$$
$$x = \frac{50}{18} \text{ or } \frac{25}{9}$$

27. $\frac{3}{x} + \frac{5}{2x} = \frac{1}{2} + \frac{7}{2x}$ $x \neq 0$ lcm = 2x

$$2x \cdot \frac{3}{x} + 2x \cdot \frac{5}{2x} = 2x \cdot \frac{1}{2} + 2x \cdot \frac{7}{2x}$$
$$6 + 5 = x + 7$$
$$11 = x + 7$$
$$x = 4$$

29. $\frac{8}{3x} - \frac{1}{15} = \frac{1}{x} + \frac{4}{3x}$ $\quad x \neq 0$, $\quad \text{lcm} = 15x$

$$15x \cdot \frac{8}{3x} - 15x \cdot \frac{1}{15} = 15x \cdot \frac{1}{x} + 15x \cdot \frac{4}{3x}$$
$$40 - x = 15 + 20$$
$$40 - x = 35$$
$$-x = -5$$
$$x = 5$$

31. $\frac{1}{2x} + \frac{1}{4} = \frac{1}{x} + \frac{1}{4x}$ $\quad x \neq 0$, $\quad \text{lcm} = 4x$

$$4x \cdot \frac{1}{2x} + 4x \cdot \frac{1}{4} = 4x \cdot \frac{1}{x} + 4x \cdot \frac{1}{4x}$$
$$2 + x = 4 + 1$$
$$2 + x = 5$$
$$x = 3$$

33. $\frac{1}{4} - \frac{1}{4x} + \frac{1}{2} = \frac{2}{x}$ $\quad x \neq 0$, $\quad \text{lcm} = 4x$

$$4x \cdot \frac{1}{4} - 4x \cdot \frac{1}{4x} + 4x \cdot \frac{1}{2} = 4x \cdot \frac{2}{x}$$
$$x - 1 + 2x = 8$$
$$3x - 1 = 8$$
$$3x = 9$$
$$x = 3$$

35. $\frac{3}{2x} + \frac{2}{x} = 1 - \frac{5}{2x}$ $\quad x \neq 0$, $\quad \text{lcm} = 2x$

$$2x \cdot \frac{3}{2x} + 2x \cdot \frac{2}{x} = 2x - 2x \cdot \frac{5}{2x}$$
$$3 + 4 = 2x - 5$$
$$7 = 2x - 5$$
$$12 = 2x$$
$$x = 6$$

37. $\frac{7}{y - 2} - \frac{1}{2} = 3$ $\quad y \neq 2$, $\quad \text{lcm} = 2(y - 2)$

$$2(y - 2) \cdot \frac{7}{y - 2} - 2(y - 2) \cdot \frac{1}{2} = 2(y - 2) \cdot 3$$
$$14 - (y - 2) = 6(y - 2)$$
$$14 - y + 2 = 6y - 12$$
$$-y + 16 = 6y - 12$$
$$-7y = -28$$
$$y = 4$$

39. $\frac{3}{2x - 1} + 4 = \frac{6x}{2x - 1}$ $\quad x \neq \frac{1}{2}$, $\quad \text{lcm} = 2x - 1$

$$(2x - 1) \cdot \frac{3}{2x - 1} + 4(2x - 1) = (2x - 1) \cdot \frac{6x}{2x - 1}$$
$$3 + 8x - 4 = 6x$$
$$8x - 1 = 6x$$
$$2x - 1 = 0$$
$$2x = 1$$
$$x = \frac{1}{2}$$

$x = \frac{1}{2}$ cannot be a solution to the original equation.

No solution.

41. $\frac{2E}{E - 1} = 2 + \frac{5}{2E}$ $\quad E \neq 1, 0$, $\quad \text{lcm} = 2E(E - 1)$

$$2E(E - 1) \cdot \frac{2E}{E - 1} = 2 \cdot 2E(E - 1) + 2E(E - 1) \cdot \frac{5}{2E}$$
$$4E^2 = 4E^2 - 4E + 5E - 5$$
$$4E^2 = 4E^2 + E - 5$$
$$0 = E - 5$$
$$-E = -5$$
$$E = 5$$

43. $\frac{n - 5}{6n - 6} = \frac{1}{9} - \frac{n - 3}{4n - 4}$

$6n - 6 = 2 \cdot 3(n - 1)$
$9 = 3^2$
$4n - 4 = 2^2(n - 1)$

$n \neq 1$
$\text{lcm} = 2^2 \cdot 3^2(n - 1) = 36(n - 1)$

$$36(n - 1) \cdot \frac{(n - 5)}{6n - 6} = 36(n - 1)\frac{1}{9} - 36(n - 1) \cdot \frac{(n - 3)}{4n - 4}$$
$$6(n - 5) = 4(n - 1) - 9(n - 3)$$
$$6n - 30 = 4n - 4 - 9n + 27$$
$$6n - 30 = -5n + 23$$
$$11n - 30 = 23$$
$$11n = 53$$
$$n = \frac{53}{11}$$

45. $5 + \frac{2x}{x - 3} = \frac{6}{x - 3}$ $x \neq 3$, $\text{lcm} = x - 3$

$$5(x - 3) + (x - 3) \cdot \frac{2}{x - 3} = (x - 3) \cdot \frac{6}{x - 3}$$
$$5(x - 3) + 2x = 6$$
$$5x - 15 + 2x = 6$$
$$7x - 15 = 6$$
$$7x = 21$$
$$x = 3$$

$x = 3$ cannot be a solution to the original equation.

No solution.

47. $\frac{x^2 + 2}{x^2 - 4} = \frac{x}{x - 2}$

$\frac{x^2 + 2}{(x - 2)(x + 2)} = \frac{x}{x - 2}$ $x \neq 2, -2$, $\text{lcm} = (x + 2)(x - 2)$

$$(x - 2)(x + 2) \cdot \frac{x^2 + 2}{(x - 2)(x + 2)} = (x - 2)(x + 2) \cdot \frac{x}{x - 2}$$
$$x^2 + 2 = (x + 2)x$$
$$x^2 + 2 = x^2 + 2x$$
$$x^2 + 2 - x^2 = x^2 + 2x - x^2$$
$$2 = 2x$$
$$1 = x$$
$$x = 1$$

49. $\frac{3}{x + 1} + \frac{4}{x - 1} = \frac{5}{x^2 - 1}$

$\frac{3}{x + 1} + \frac{4}{x - 1} = \frac{5}{(x + 1)(x - 1)}$ $x \neq 1, -1$, $\text{lcm} = (x + 1)(x - 1)$

$$(x + 1)(x - 1) \cdot \frac{3}{x + 1} + (x + 1)(x - 1) \cdot \frac{4}{x - 1} = (x + 1)(x - 1) \cdot \frac{5}{(x + 1)(x - 1)}$$
$$(x - 1)3 + (x + 1)4 = 5$$
$$3x - 3 + 4x + 4 = 5$$
$$7x + 1 = 5$$
$$7x = 4$$
$$x = \frac{4}{7}$$

51. $\frac{1}{x + 2} - \frac{1}{x - 3} = \frac{1}{x^2 - x - 6}$

$\frac{1}{x + 2} - \frac{1}{x - 3} = \frac{1}{(x + 2)(x - 3)}$ $x \neq -2, 3$ lcm $= (x + 2)(x - 3)$

$$(x + 2)(x - 3) \cdot \frac{1}{x + 2} - (x + 2)(x - 3) \cdot \frac{1}{x - 3} = (x + 2)(x - 3) \cdot \frac{1}{(x + 2)(x - 3)}$$

$$(x - 3) - (x + 2) = 1$$
$$x - 3 - x - 2 = 1$$
$$-5 = 1 \quad \text{Impossible}$$

No solution.

53. $\frac{2}{x - 3} + \frac{1}{x + 4} = \frac{5}{x^2 + x - 12}$

$\frac{2}{x - 3} + \frac{1}{x + 4} = \frac{5}{(x + 4)(x - 3)}$ $x \neq -4, 3$ lcm $= (x + 4)(x - 3)$

$$(x + 4)(x - 3) \cdot \frac{2}{x - 3} + (x + 4)(x - 3) \cdot \frac{1}{x + 4} = (x + 4)(x - 3) \cdot \frac{5}{(x + 4)(x - 3)}$$

$$2(x + 4) + (x - 3) = 5$$
$$2x + 8 + x - 3 = 5$$
$$3x + 5 = 5$$
$$3x = 0$$
$$x = 0$$

55. $\frac{3}{x + 5} = \frac{2}{x + 4} - \frac{1}{x^2 + 9x + 20}$

$\frac{3}{x + 5} = \frac{2}{x + 4} - \frac{1}{(x + 5)(x + 4)}$ $x \neq -5, -4$ lcm $= (x + 5)(x + 4)$

$$(x + 5)(x + 4) \cdot \frac{3}{x + 5} = (x + 5)(x + 4) \cdot \frac{2}{x + 4} - (x + 5)(x + 4) \cdot \frac{1}{(x + 5)(x + 4)}$$

$$(x + 4)3 = (x + 5)2 - 1$$
$$3x + 12 = 2x + 10 - 1$$
$$3x + 12 = 2x + 9$$
$$x + 12 = 9$$
$$x = -3$$

57. $\frac{4}{2x^2 + 3x + 1} = \frac{2}{x^2 + x - 2}$

$\frac{4}{(2x + 1)(x + 1)} = \frac{2}{(x + 2)(x - 1)}$ $x = 1, -1, -2, -\frac{1}{2}$

lcm $= (2x + 1)(x + 1)(x + 2)(x - 1)$

$$(2x + 1)(x + 1)(x + 2)(x - 1) \cdot \frac{4}{(2x + 1)(x + 1)}$$
$$= (2x + 1)(x + 1)(x + 2)(x - 1) \cdot \frac{2}{(x + 2)(x - 1)}$$

$$(x + 2)(x - 1)4 = (2x + 1)(x + 1)2$$
$$(x^2 + x - 2)4 = (2x^2 + 3x + 1)2$$
$$4x^2 + 4x - 8 = 4x^2 + 6x + 2$$
$$4x - 8 = 6x + 2$$
$$-2x - 8 = 2$$
$$-2x = 10$$
$$x = -5$$

59. $$\frac{3}{x^2 + 5x + 6} = \frac{6}{2x^2 + 8x - 10}$$

$$\frac{3}{(x + 2)(x + 3)} = \frac{6}{2(x + 5)(x - 1)}$$

$$\frac{3}{(x + 2)(x + 3)} = \frac{3}{(x + 5)(x - 1)} \qquad x \neq -2, -3, -5, 1$$

$$\text{lcm} = (x + 2)(x + 3)(x + 5)(x - 1)$$

$$(x + 2)(x + 3)(x + 5)(x - 1)\cdot\frac{3}{(x + 2)(x + 3)} = (x + 2)(x + 3)(x + 5)(x - 1)\cdot\frac{3}{(x + 5)(x - 1)}$$

$$\begin{aligned}
(x + 5)(x - 1)3 &= (x + 2)(x + 3)3 \\
(x^2 + 4x - 5)3 &= (x^2 + 5x + 6)3 \\
3x^2 + 12x - 15 &= 3x^2 + 15x + 18 \\
12x - 15 &= 15x + 18 \\
-3x - 15 &= 18 \\
-3x &= 33 \\
x &= -11
\end{aligned}$$

61. $$\frac{3x^2 + 5x + 7}{x^2 + 3x + 5} = 3 \qquad \text{lcm} = x^2 + 3x + 5 \quad (\text{never } 0 \text{ for real } x)$$

$$\begin{aligned}
(x^2 + 3x + 5)\frac{3x^2 + 5x + 7}{x^2 + 3x + 5} &= 3(x^2 + 3x + 5) \\
3x^2 + 5x + 7 &= 3x^2 + 9x + 15 \\
5x + 7 &= 9x + 15 \\
-4x + 7 &= 15 \\
-4x &= 8 \\
x &= -2
\end{aligned}$$

63. $$\frac{14x^2 + 15x + 13}{2x^2 + 5x + 3} = 7$$

$$\frac{14x^2 + 15x + 13}{(2x + 3)(x + 1)} = 7 \qquad x \neq -\frac{3}{2}, -1$$

$$\text{lcm} = (2x + 3)(x + 1)$$

$$\begin{aligned}
(2x + 3)(x + 1)\cdot\frac{14x^2 + 15x + 13}{(2x + 3)(x + 1)} &= 7(2x + 3)(x + 1) \\
14x^2 + 15x + 13 &= 7(2x^2 + 5x + 3) \\
14x^2 + 15x + 13 &= 14x^2 + 35x + 21 \\
15x + 13 &= 35x + 21 \\
-20x + 13 &= 21 \\
-20x &= 8 \\
x &= -\frac{8}{20} \text{ or } -\frac{2}{5}
\end{aligned}$$

65. $$\frac{x + a}{x + b} = 2 \qquad x \neq -b \qquad \text{lcm} = x + b$$

$$\begin{aligned}
(x + b)\frac{x + a}{x + b} &= 2(x + b) \\
x + a &= 2x + 2b \\
x + a - x &= 2x + 2b - x \\
a &= x + 2b \\
a - 2b &= x + 2b - 2b \\
x &= a - 2b
\end{aligned}$$

67. $$\frac{x + 1}{x - 1} = b \qquad x \neq 1 \qquad \text{lcm} = x - 1$$

$$\begin{aligned}
(x - 1)\frac{x + 1}{x - 1} &= b(x - 1) \\
x + 1 &= b(x - 1) \\
x + 1 &= bx - b \\
x + 1 - x &= bx - x - b \\
1 &= bx - x - b \\
b + 1 &= bx - x - b + b \\
b + 1 &= bx - x \\
b + 1 &= (b - 1)x \\
x &= \frac{b + 1}{b - 1}
\end{aligned}$$

69. $\dfrac{x + a}{x + b} = c \qquad x \neq -b$, lcm $= x + b$

$$\begin{aligned}
(x + b)\frac{x + a}{x + b} &= c(x + b) \\
x + a &= cx + cb \\
x - cx + a &= cx - cx + cb \\
x(1 - c) + a &= cb \\
x(1 - c) + a - a &= cb - a \\
x(1 - c) &= cb - a \\
x &= \frac{cb - a}{1 - c}
\end{aligned}$$

71. Let x = the denominator; $x - 4$ = the denominator decreased by 4

Then $x - 6$ = the numerator; $x - 6 - 4$ or $x - 10$ = the numerator decreased by 4

Then **the resulting fraction** is equal to $\boxed{\frac{1}{3}}$

$\dfrac{x - 10}{x - 4} = \dfrac{1}{3} \qquad x \neq 4$, lcm $= 3(x - 4)$

$$\begin{aligned}
3(x - 4)\cdot\frac{x - 10}{x - 4} &= 3(x - 4)\cdot\frac{1}{3} \\
3(x - 10) &= x - 4 \\
3x - 30 &= x - 4 \\
2x - 30 &= -4 \\
2x &= 26 \\
x &= 13 \\
x - 6 &= 7
\end{aligned}$$

The fraction is $\dfrac{x - 6}{x} = \dfrac{7}{13}$.

73. Let x = the denominator; $x + 4$ = the denominator increased by 4

Then $x - 7$ = the numerator; $x - 7 + 4$ or $x - 3$ = the numerator increased by 4

Then **the resulting fraction** is equal to $\boxed{\frac{1}{2}}$

$\dfrac{x - 3}{x + 4} = \dfrac{1}{2} \qquad x \neq -4$, lcm $= 2(x + 4)$

$$\begin{aligned}
2(x + 4)\cdot\frac{x - 3}{x + 4} &= 2(x + 4)\cdot\frac{1}{2} \\
2(x - 3) &= x + 4 \\
2x - 6 &= x + 4 \\
x - 6 &= 4 \\
x &= 10 \\
x - 7 &= 3
\end{aligned}$$

The fraction is $\dfrac{x - 7}{x} = \dfrac{3}{10}$.

75. $\frac{3x}{24} - \frac{2 - x}{10} = \frac{5 + x}{40} - \frac{1}{15}$ $\quad 24 = 2^2 \cdot 3,\ 10 = 2 \cdot 5,\ 40 = 2^3 \cdot 5,\ 15 = 3 \cdot 5$; lcm $= 2^3 \cdot 3 \cdot 5 = 120$

$$\begin{aligned}
120 \cdot \frac{3x}{24} - 120 \cdot \frac{(2 - x)}{10} &= 120 \cdot \frac{(5 + x)}{40} - 120 \cdot \frac{1}{15} \\
15x - 12(2 - x) &= 3(5 + x) - 8 \\
15x - 24 + 12x &= 15 + 3x - 8 \\
27x - 24 &= 3x + 7 \\
24x &= 31 \\
x &= \frac{31}{24}
\end{aligned}$$

77. $\frac{5t - 22}{t^2 - 6t + 9} - \frac{11}{t^2 - 3t} - \frac{5}{t} = 0$

$$\begin{aligned}
t^2 - 6t + 9 &= (t - 3)(t - 3) \\
t^2 - 3t &= t(t - 3) \\
t &= t \qquad \text{lcm} = t(t - 3)^2 \quad t \neq 0, 3
\end{aligned}$$

$$t(t - 3)^2 \cdot \frac{(5t - 22)}{(t - 3)^2} - t(t - 3)^2 \cdot \frac{11}{t(t - 3)} - t(t - 3)^2 \cdot \frac{5}{t} = 0 \cdot t(t - 3)^2$$

$$\begin{aligned}
t(5t - 22) - (t - 3)11 - (t - 3)^2 \cdot 5 &= 0 \\
5t^2 - 22t - 11t + 33 - 5(t^2 - 6t + 9) &= 0 \\
5t^2 - 22t - 11t + 33 - 5t^2 + 30t - 45 &= 0 \\
-3t - 12 &= 0 \\
-3t &= 12 \\
t &= -4
\end{aligned}$$

79. $5 - \frac{2x}{3 - x} = \frac{6}{x - 3}$ $\quad x - 3 = -(3 - x)$; lcm $= 3 - x,\ x \neq 3$

$$\begin{aligned}
5 - \frac{2x}{3 - x} &= \frac{6}{-(3 - x)} \\
5 - \frac{2x}{3 - x} &= -\frac{6}{3 - x} \\
(3 - x) \cdot 5 - (3 - x) \cdot \frac{2x}{3 - x} &= -(3 - x) \cdot \frac{6}{3 - x} \\
(3 - x)5 - 2x &= -6 \\
15 - 5x - 2x &= -6 \\
-7x + 15 &= -6 \\
-7x &= -21 \\
x &= 3
\end{aligned}$$

$x = 3$ cannot be a solution to the original equation.

No solution.

81. $\frac{1}{c^2 - c - 2} - \frac{3}{c^2 - 2c - 3} = \frac{1}{c^2 - 5c + 6}$

$$\begin{aligned}
c^2 - c - 2 &= (c - 2)(c + 1) \\
c^2 - 2c - 3 &= (c - 3)(c + 1) \\
c^2 - 5c + 6 &= (c - 2)(c - 3)
\end{aligned}$$

$c \neq -1, 2, 3$; lcm $= (c + 1)(c - 2)(c - 3)$

$$(c + 1)(c - 2)(c - 3)\frac{1}{(c - 2)(c + 1)} - (c + 1)(c - 2)(c - 3)\frac{3}{(c - 3)(c + 1)} = (c + 1)(c - 2)(c - 3)\frac{1}{(c - 2)(c - 3)}$$

$$\begin{aligned}
(c - 3) \cdot 1 - (c - 2) \cdot 3 &= (c + 1) \cdot 1 \\
c - 3 - 3c + 6 &= c + 1 \\
-2c + 3 &= c + 1 \\
-3c &= -2 \\
c &= \frac{2}{3}
\end{aligned}$$

83. $\frac{x + 3}{(x - 1)(x - 2)} - \frac{x + 1}{(x - 2)(x - 3)} = \frac{2}{(x - 1)(x - 3)}$ lcm = $(x - 1)(x - 2)(x - 3)$, $x \neq 1, 2, 3$

$$(x - 1)(x - 2)(x - 3)\cdot\frac{(x + 3)}{(x - 1)(x - 2)} - (x - 1)(x - 2)(x - 3)\cdot\frac{(x + 1)}{(x - 2)(x - 3)} = (x - 1)(x - 2)(x - 3)\cdot\frac{2}{(x - 1)(x - 3)}$$

$$\begin{aligned}(x - 3)(x + 3) - (x - 1)(x + 1) &= (x - 2)2\\ x^2 - 9 - (x^2 - 1) &= 2x - 4\\ x^2 - 9 - x^2 + 1 &= 2x - 4\\ -8 &= 2x - 4\\ -4 &= 2x\\ -2 &= x\\ x &= -2\end{aligned}$$

COMMON ERROR: Forgetting the parentheses around $x^2 - 1$. The multiplication must be performed, followed by the subtraction of the entire quantity: $x^2 - 1$.

85. $\frac{1}{x} - \frac{1}{x^2} = \frac{12}{x^3}$ lcm = x^3, $x \neq 0$

$$\begin{aligned}x^3\cdot\frac{1}{x} - x^3\cdot\frac{1}{x^2} &= x^3\cdot\frac{12}{x^3}\\ x^2 - x &= 12\\ x^2 - x - 12 &= 0\\ (x - 4)(x + 3) &= 0\end{aligned}$$

$x - 4 = 0$ or $x + 3 = 0$

$x = 4$ $\qquad$ $x = -3$

87. $\frac{1}{(x - 1)^2} + \frac{1}{(x + 1)^2} = \frac{2}{x^2 - 1}$ $x^2 - 1 = (x - 1)(x + 1)$, lcm = $(x - 1)^2(x + 1)^2$, $x \neq 1, -1$

$$(x - 1)^2(x + 1)^2\cdot\frac{1}{(x - 1)^2} + (x - 1)^2(x + 1)^2\cdot\frac{1}{(x + 1)^2} = (x - 1)^2(x + 1)^2\cdot\frac{2}{(x - 1)(x + 1)}$$

$$\begin{aligned}(x + 1)^2 + (x - 1)^2 &= (x - 1)(x + 1)2\\ x^2 + 2x + 1 + x^2 - 2x + 1 &= (x^2 - 1)2\\ 2x^2 + 2 &= 2x^2 - 2\\ 2 &= -2\end{aligned}$$

No solution.

89. $\frac{2x + 5}{x^2 - 1} = \frac{6x + 5}{3x^2 + 3x}$

$\frac{2x + 5}{(x + 1)(x - 1)} = \frac{6x + 5}{3x(x + 1)}$ $x \neq 0, 1, -1$, lcm = $3x(x + 1)(x - 1)$

$$\begin{aligned}3x(x + 1)(x - 1)\frac{2x + 5}{(x + 1)(x - 1)} &= 3x(x + 1)(x - 1)\frac{6x + 5}{3x(x + 1)}\\ 3x(2x + 5) &= (x - 1)(6x + 5)\\ 6x^2 + 15x &= 6x^2 + 5x - 6x - 5\\ 15x &= 5x - 6x - 5\\ 15x &= -x - 5\\ 16x &= -5\\ x &= -\frac{5}{16}\end{aligned}$$

91. $\dfrac{3x}{x+2} + \dfrac{4x}{x-3} = \dfrac{7x^2+6x+5}{x^2-x-6}$ $\quad x^2 - x - 6 = (x+2)(x-3) \quad x \neq -2, 3$
$\quad \text{lcm} = (x+2)(x-3)$

$$(x+2)(x-3)\cdot\frac{3x}{x+2} + (x+2)(x-3)\cdot\frac{4x}{x-3} = (x+2)(x-3)\cdot\frac{7x^2+6x+5}{(x+2)(x-3)}$$
$$(x-3)3x + (x+2)4x = 7x^2+6x+5$$
$$3x^2 - 9x - 4x^2 + 8x = 7x^2+6x+5$$
$$7x^2 - x = 7x^2+6x+5$$
$$-x = 6x+5$$
$$-7x = 5$$
$$x = -\frac{5}{7}$$

93. $\dfrac{x-1}{x+2} + \dfrac{x-2}{x+3} = 2 + \dfrac{x-3}{x^2+5x+6}$ $\quad x^2+5x+6 = (x+2)(x+3) \quad x \neq -2, -3$
$\quad \text{lcm} = (x+2)(x+3)$

$$(x+2)(x+3)\cdot\frac{(x-1)}{x+2} + (x+2)(x+3)\cdot\frac{(x-2)}{x+3} = 2(x+2)(x+3) + (x+2)(x+3)\cdot\frac{(x-3)}{(x+2)(x+3)}$$
$$(x+3)(x-1) + (x+2)(x-2) = 2(x+2)(x+3) + (x-3)$$
$$x^2 + 2x - 3 + x^2 - 4 = 2(x^2+5x+6) + x - 3$$
$$2x^2 + 2x - 7 = 2x^2 + 10x + 12 + x - 3$$
$$2x^2 + 2x - 7 = 2x^2 + 11x + 9$$
$$2x - 7 = 11x + 9$$
$$-9x - 7 = 9$$
$$-9x = 16$$
$$x = -\frac{16}{9}$$

95. $\dfrac{x}{ab} + \dfrac{x}{bc} + \dfrac{x}{ac} = 1 \quad \text{lcm} = abc$

$$abc\cdot\frac{x}{ab} + abc\cdot\frac{x}{bc} + abc\cdot\frac{x}{ac} = abc$$
$$cx + ax + bx = abc$$
$$(c+a+b)x = abc$$
$$x = \frac{abc}{c+a+b}$$

97. $\dfrac{ax}{a-b} - \dfrac{bx}{a+b} = 1 \quad \text{lcm} = (a-b)(a+b)$

$$(a-b)(a+b)\cdot\frac{ax}{a-b} - (a-b)(a+b)\cdot\frac{bx}{a+b} = (a-b)(a+b)$$
$$(a+b)ax - (a-b)bx = (a-b)(a+b)$$
$$(a^2+ab)x + (-ab+b^2)x = (a-b)(a+b)$$
$$(a^2+b^2)x = (a-b)(a+b)$$
$$x = \frac{a^2-b^2}{a^2+b^2}$$

99. Let x = the numerator, then $x - 3$ = the denominator

$x - 5$ = the numerator decreased by 5

$x - 3 + 4$ or $x + 1$ = the denominator increased by 4

Then the sum of [the resulting fraction] [and] [the original fraction] is equal to [2]

$$\frac{x-5}{x+1} + \frac{x}{x-3} = 2$$

$x \neq -1, 3$

lcm $= (x + 1)(x - 3)$

$$(x+1)(x-3)\cdot\frac{x-5}{x+1} + (x+1)(x-3)\frac{x}{x-3} = 2(x+1)(x-3)$$

$$(x-3)(x-5) + (x+1)x = 2(x+1)(x-3)$$

$$x^2 - 8x + 15 + x^2 + x = 2(x^2 - 2x - 3)$$

$$2x^2 - 7x + 15 = 2x^2 - 4x - 6$$

$$-7x + 15 = -4x - 6$$

$$-3x + 15 = -6$$

$$-3x = -21$$

$$x = 7$$

$$x - 3 = 4$$

The fraction is $\frac{x}{x-3} = \frac{7}{4}$.

Exercise 4-2 Application: Ratio and Proportion Problems

Key Ideas and Formulas

Ratios: The **ratio** of a to b, $b \neq 0$ is $\frac{a}{b}$.

Proportion: A statement of equality between two ratios, thus $\frac{a}{b} = \frac{c}{d}$, $b, d \neq 0$, is called a **proportion**.

---

1. $\frac{33}{22} = \frac{3}{2}$

3. $\frac{25}{10} = \frac{5}{2}$

5. $\frac{300}{24} = \frac{25}{2}$

7. $\frac{m}{16} = \frac{5}{4}$

$m = 16 \cdot \frac{5}{4}$

$m = 20$

9. $\frac{x}{13} = \frac{21}{39}$

$x = 13 \cdot \frac{21}{39}$

$x = 7$

11. $\frac{7}{36} = \frac{n}{9000}$

$9000 \cdot \frac{7}{36} = n$

$n = 1750$

13. $\frac{2}{38} = \frac{m}{1900}$

$1900 \cdot \frac{2}{38} = m$

$100 = m$

15. Let q = the number of quarters, then the ratio of quarters to dimes is q/96.

$$\frac{q}{96} = \frac{5}{8}$$
$$q = 96 \cdot \frac{5}{8}$$
$$q = 60 \text{ quarters}$$

17. Let ℓ = length of the rectangle, then the ratio of length to width is $\ell/24$.

$$\frac{\ell}{24} = \frac{5}{3}$$
$$\ell = 24 \cdot \frac{5}{3}$$
$$\ell = 40 \text{ meters}$$

19. Let x = distance traveled on 18 liters. Then,

$$\frac{x}{18} = \frac{108}{12} \qquad \frac{\text{km}}{\ell} = \frac{\text{km}}{\ell}$$
$$x = 18 \cdot \frac{108}{12}$$
$$x = 162 \text{ kilometers}$$

21. Let m = the number of male models, then the ratio of male to female models is $\frac{m}{216}$.

$$\frac{m}{216} = \frac{3}{8}$$
$$m = 216 \cdot \frac{3}{8}$$
$$m = 81 \text{ male models}$$

23. Let x = the number of applicants, the the ratio of accepted to total applicants is x/705.

$$\frac{x}{705} = \frac{4}{15}$$
$$x = 705 \cdot \frac{4}{15}$$
$$x = 188 \text{ applicants accepted}$$

25. See text, Example 4. There are 38 numbers, hence

$$\text{probability of winning} = \frac{\text{Number of favorable outcomes possible}}{\text{Total number of possible outcomes}} = \frac{12}{38}$$

27. A sum of 5 can occur in the following 4 ways:
red 1, blue 4
red 2, blue 3
red 3, blue 2
red 4, blue 1

$$\text{Hence, probability of winning} = \frac{\text{Number of favorable outcomes possible}}{\text{Total number of possible outcomes}} = \frac{4}{36}$$

29. A sum of 6 can occur in the following 5 ways:
red 1, blue 5
red 2, blue 4
red 3, blue 3
red 4, blue 2
red 5, blue 1

$$\text{Hence, odds in favor of winning} = \frac{\text{Number of favorable outcomes possible}}{\text{Number of unfavorable outcomes possible}}$$
$$= \frac{5}{36 - 5} = \frac{5}{31}$$

31. A sum of 7 can occur in the following 6 ways:
red 1, blue 6
red 2, blue 5
red 3, blue 4
red 4, blue 3
red 5, blue 2
red 6, blue 1

A sum of 11 can occur in the following 2 ways:
red 5, blue 6
red 6, blue 5

Hence there are 6 + 2 = 8 favorable outcomes.

Hence, odds in favor of winning $= \dfrac{\text{Number of favorable outcomes possible}}{\text{Number of unfavorable outcomes possible}}$

$= \dfrac{8}{36 - 8} = \dfrac{8}{28}$

33. See previous problem.

Probablity of winning $= \dfrac{\text{Number of favorable outcomes possible}}{\text{Total number of possible outcomes}} = \dfrac{8}{36}$

35. Given a:b = c:d, or $\frac{a}{b} = \frac{c}{d}$, we can cross-multiply to write ad = bc. Then we know ad = 72, b = 18, hence 72 = 18c, c = 4.

37. Given a:b = c:d, or $\frac{a}{b} = \frac{c}{d}$, we can cross-multiply to write ad = bc. Then we know bc = 120, a = 24, hence 24d = 120, d = 5.

39. $\frac{a}{A} = \frac{b}{B}$

$\frac{8}{6} = \frac{b}{11}$

$b = 11 \cdot \frac{8}{6}$

$b = \frac{44}{3}$ or $14\frac{2}{3}$

41. $\frac{a}{A} = \frac{b}{B}$

$\frac{18}{8} = \frac{30}{B}$

Cross multiply to obtain

$18B = 8 \cdot 30$

$18B = 240$

$B = \frac{240}{18}$ or $13\frac{1}{3}$

43. Let x = how many milliliters acid in 52 ml of solution.

Then $\frac{x}{52} = \frac{9}{46}$ $\quad \frac{\text{acid}}{\text{solution}} = \frac{\text{acid}}{\text{solution}}$

$x = 52 \cdot \frac{9}{46}$

x = 10.17 milliliters

45. Let x = width of the enlargement

$\frac{x}{10} = \frac{23}{35}$ $\quad \frac{\text{width}}{\text{length}} = \frac{\text{width}}{\text{length}}$

$x = 10 \cdot \frac{23}{35}$

x = 6.57 inches

47. Let x = number of liters in 5 gallons.

$\frac{x}{5} = \frac{1}{0.26}$ $\quad \frac{\text{liters}}{\text{gallons}} = \frac{\text{liters}}{\text{gallons}}$

$x = 5 \cdot \frac{1}{0.26}$

x = 19.23 liters

49. Let x = number of liters in 4 quarts

$\frac{x}{4} = \frac{1}{1.0567}$ $\quad \frac{\text{liters}}{\text{quarts}} = \frac{\text{liters}}{\text{quarts}}$

$x = 4 \cdot \frac{1}{1.0567}$

x = 3.79 liters

51. Let x = number of kilometers in 1 mile

$\frac{x}{1} = \frac{1}{0.6215}$ $\quad \frac{\text{kilo}}{\text{miles}} = \frac{\text{kilo}}{\text{miles}}$

$x = \frac{1}{0.6215}$

x = 1.61 kilometers

53. Let x = number of inches in 100 centimeters

$\frac{x}{100} = \frac{1}{2.54}$ $\quad \frac{\text{inches}}{\text{centimeters}} = \frac{\text{inches}}{\text{centimeters}}$

$x = 100 \cdot \frac{1}{2.54}$

x = 39.37 inches

55. Let x = number of pounds in $100

$\frac{x}{100} = \frac{1}{1.54}$ $\quad \frac{\text{pounds}}{\text{dollars}} = \frac{\text{pounds}}{\text{dollars}}$

$x = 100 \cdot \frac{1}{1.54}$

x = 64.94 pounds

57. Let x = number of dollars in 65 krona

$\frac{x}{65} = \frac{1.5}{10}$ $\quad \frac{\text{dollars}}{\text{krona}} = \frac{\text{dollars}}{\text{krona}}$

$x = 65 \cdot \frac{1.5}{10}$

x = $9.75

59. Let x = commission for 500 shares

$$\frac{x}{500} = \frac{240}{200} \qquad \frac{\text{commission}}{\text{shares}} = \frac{\text{commission}}{\text{shares}}$$

$$x = 500 \cdot \frac{240}{200}$$

$$x = \$600$$

61. Let x = earnings per share

$$\frac{66}{x} = \frac{12}{1} \qquad \frac{\text{price}}{\text{earnings}} = \frac{\text{price}}{\text{earnings}}$$

Cross-multiply to obtain

$$66 = 12x$$

$$x = \frac{66}{12} = \$5.50$$

63. Let a = the angle corresponding to A
b = the angle corresponding to B
c = the angle corresponding to C

Then $\frac{a}{360} = \frac{25}{150}$ $\qquad \frac{b}{360} = \frac{60}{150}$ $\qquad \frac{c}{360} = \frac{65}{150}$

$a = 360 \cdot \frac{25}{150}$ $\qquad b = 360 \cdot \frac{60}{150}$ $\qquad c = 360 \cdot \frac{65}{150}$

$a = 60^\circ$ $\qquad b = 144^\circ$ $\qquad c = 156^\circ$

65. Let a = the angle corresponding to A
b = the angle corresponding to B
c = the angle corresponding to C

Then $\frac{a}{360} = \frac{35}{100}$ $\qquad \frac{b}{360} = \frac{45}{100}$ $\qquad \frac{c}{360} = \frac{100 - (35 + 45)}{100}$

$a = 360 \cdot \frac{35}{100}$ $\qquad b = 360 \cdot \frac{45}{100}$ $\qquad c = 360 \cdot \frac{20}{100}$

$a = 126^\circ$ $\qquad b = 162^\circ$ $\qquad c = 72^\circ$

67. Let a = the angle corresponding to A
b = the angle corresponding to B
c = the angle corresponding to C
d = the angle corresponding to D

Then $\frac{a}{360} = \frac{200}{1000}$ $\qquad \frac{b}{360} = \frac{250}{1000}$ $\qquad \frac{c}{360} = \frac{400}{1000}$ $\qquad \frac{d}{360} = \frac{1000 - (200 + 250 + 400)}{1000}$

$a = 360 \cdot \frac{200}{1000}$ $\qquad b = 360 \cdot \frac{250}{1000}$ $\qquad c = 360 \cdot \frac{400}{1000}$ $\qquad d = 360 \cdot \frac{150}{1000}$

$a = 72^\circ$ $\qquad b = 90^\circ$ $\qquad c = 144^\circ$ $\qquad d = 54^\circ$

69. To convert 0.65° to minutes,

solve $\frac{x}{0.65} = \frac{60}{1}$ to obtain x = 39'

40.65° = 40°39'

71. To convert 0.43° to minutes,

solve $\frac{x}{0.43} = \frac{60}{1}$ to obtain x = 25.8'

Then convert 0.8 minute to seconds:

$$\frac{x}{0.8} = \frac{60}{1}$$

$$x = 48$$

50.43° = 50°25'48"

73. To convert 0.605° to minutes,

solve $\frac{x}{0.605} = \frac{60}{1}$ to obtain x = 36.3'

Then convert 0.3 minute to seconds:

$$\frac{x}{0.3} = \frac{60}{1}$$

$$x = 18$$

70.605° = 70°36'18"

75. $30^\circ 15'45'' = \left(30 + \frac{15}{60} + \frac{45}{3600}\right)^\circ = 30.2625^\circ$

77. $60°48'6" = \left(60 + \frac{48}{60} + \frac{6}{3600}\right)° = 60.8017°$

79. $10°42'18" = \left(10 + \frac{42}{60} + \frac{18}{3600}\right)° = 10.705°$

81. $15°12'20" = \left(15 + \frac{12}{60} + \frac{20}{3600}\right)° = 15.2056°$

83. $60°24'45" = \left(60 + \frac{24}{60} + \frac{45}{3600}\right)° = 60.4125°$

85. Let x = distance, then

$$\frac{x}{3600} = \frac{0.2}{10} \qquad \frac{\text{miles}}{\text{seconds}} = \frac{\text{miles}}{\text{seconds}}$$

$$x = 3600 \cdot \frac{0.2}{10}$$

$$x = 72 \text{ miles}$$

87. Let x = change in latitude, then

$$\frac{3600}{72} = \frac{x}{25} \qquad \frac{\text{seconds}}{\text{miles}} = \frac{\text{seconds}}{\text{miles}}$$

$$x = 25 \cdot \frac{3600}{72}$$

$$x = 1250 \text{ seconds or } 20'50"$$

89. Let x = total lake population.

$$\frac{\text{total lake population}}{\text{trout marked in first sample}} = \frac{\text{total number in second sample}}{\text{trout marked in second sample}}$$

$$\frac{x}{300} = \frac{250}{25}$$

$$x = 300 \cdot \frac{250}{25}$$

$$x = 3000 \text{ trout}$$

91. Let f = how much force required to lift 1200 kg.
Then F = 1200

$$a = \frac{1}{4}\pi d^2 = \frac{1}{4}\pi(12)^2 \text{ square millimeters}$$

$$A = \frac{1}{4}\pi D^2 = \frac{1}{4}\pi(24 \cdot 10)^2 \text{ square millimeters}$$

$$\frac{f}{1200} = \frac{\frac{1}{4}\pi(12)^2}{\frac{1}{4}\pi(240)^2}$$

$$\frac{f}{1200} = \frac{\frac{1}{4}\pi \cdot 144}{\frac{1}{4}\pi \cdot 57600}$$

$$\frac{f}{1200} = \frac{36\pi}{14400\pi}$$

$$f = 1200 \cdot \frac{36\pi}{14400\pi}$$

$$f = 3 \text{ kilograms}$$

93. Let v = volume of sphere

$$\frac{v}{5^3} = \frac{4,188.79}{10^3} \qquad \frac{\text{volume}}{\text{radius}^3} = \frac{\text{volume}}{\text{radius}^3}$$

$$v = 5^3 \cdot \frac{4,188.79}{10^3}$$

$$v = 523.6 \text{ cubic inches}$$

95. Let x = interest earned in 8 years

$$\frac{x}{8} = \frac{234}{3} \qquad \frac{\text{interest}}{\text{time}} = \frac{\text{interest}}{\text{time}}$$

$$x = 8 \cdot \frac{234}{3}$$

$$x = \$624$$

97. Let x = the height of the tree

Then $\frac{x}{3.5} = \frac{600}{10}$

$$x = 3.5 \cdot \frac{600}{10}$$

$$x = 210 \text{ ft}$$

99. Let x = the width of the river

Then $\frac{x + 30}{9} = \frac{160}{12}$

$$x + 30 = 9 \cdot \frac{160}{12}$$

$$x + 30 = 120$$

$$x = 90 \text{ yd}$$

Exercise 4-3 Rate-Time Problems

Key Ideas and Formulas

Quantity-Rate-Time Formulas:

$Q = RT$ Quantity = (Rate)(Time)

$T = \frac{Q}{R}$ $R = \frac{Q}{T}$

If Q is distance D, then

$D = RT$ Distance = (Rate)(Time)

$T = \frac{D}{R}$ $R = \frac{D}{T}$

---

1. (A) $R = \frac{Q}{T} = \frac{45.20 \text{ dollars}}{8 \text{ hours}} = 5.65$ dollars per hour

(B) $Q = RT$ = (5.65 dollars per hour)(52 hours) = \$293.80

(C) $T = \frac{Q}{R} = \frac{500 \text{ dollars}}{5.65} = 88.5$ hours

3. (A) $R = \frac{Q}{T} = \frac{691,000 \text{ elephants}}{10 \text{ years}} = 69,100$ elephants per year

(B) $Q = RT$ = (69,100 elephants per year)(15 years) = 1,036,500 elephants

(C) $T = \frac{Q}{R} = \frac{609,000 \text{ elephants}}{69,100 \text{ elephants per year}} = 8.8$ years

5. Let T = number of hours cars travel

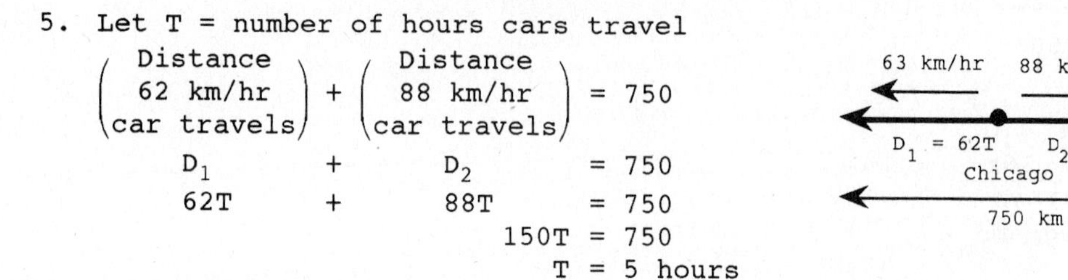

$$\begin{pmatrix}\text{Distance}\\ \text{62 km/hr}\\ \text{car travels}\end{pmatrix} + \begin{pmatrix}\text{Distance}\\ \text{88 km/hr}\\ \text{car travels}\end{pmatrix} = 750$$

$$D_1 + D_2 = 750$$

$$62T + 88T = 750$$

$$150T = 750$$

$$T = 5 \text{ hours}$$

7. Let T = number of hours until both trains meet.

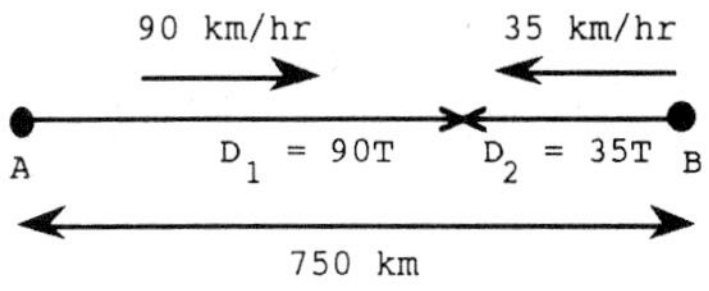

$$\begin{pmatrix}\text{Distance}\\\text{passenger}\\\text{train travels}\end{pmatrix} + \begin{pmatrix}\text{Distance}\\\text{freight train}\\\text{travels}\end{pmatrix} = 750$$

90 km/hr → ← 35 km/hr; A, B; $D_1 = 90T$, $D_2 = 35T$; 750 km

$$D_1 + D_2 = 750$$

$$90T + 35T = 750$$
$$125T = 750$$
$$T = 6 \text{ hours}$$

9. Let T = number of hours until both ships meet.

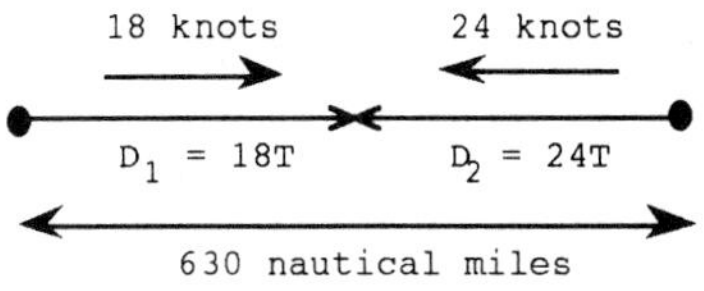

$$\begin{pmatrix}\text{Distance}\\\text{18 knot}\\\text{ship travels}\end{pmatrix} + \begin{pmatrix}\text{Distance}\\\text{24 knot}\\\text{ship travels}\end{pmatrix} = 630$$

18 knots → ← 24 knots; $D_1 = 18T$, $D_2 = 24T$; 630 nautical miles

$$18T + 24T = 630$$
$$42T = 630$$
$$T = 15 \text{ hours}$$

11. Let x = time to complete whole job

$$\begin{pmatrix}\text{Quantity produced}\\\text{by first worker}\end{pmatrix} + \begin{pmatrix}\text{Quantity produced}\\\text{by second worker}\end{pmatrix} = \begin{pmatrix}\text{Total}\\\text{quantity}\end{pmatrix}$$

$$14x + 10x = 1560$$
$$24x = 1560$$
$$x = 65 \text{ minutes}$$

13. Let x = time for all three workers together to complete task.

$\frac{10}{3}$ = time for first worker alone to complete job

$\frac{5}{2}$ = time for second worker alone to complete job

2 = time for third worker alone to complete job

Then the rate of completion for the first worker is $\frac{3}{10}$ job per hour

the rate of completion for the second worker is $\frac{2}{5}$ job per hour

the rate of completion for the third worker is $\frac{1}{2}$ job per hour

the rate of completion for all together is $\frac{1}{x}$ job per hour

$$\begin{pmatrix}\text{Rate of}\\\text{first worker}\end{pmatrix} + \begin{pmatrix}\text{Rate of}\\\text{second worker}\end{pmatrix} + \begin{pmatrix}\text{Rate of}\\\text{third worker}\end{pmatrix} = \begin{pmatrix}\text{Rate}\\\text{together}\end{pmatrix}$$

$$\frac{3}{10} + \frac{2}{5} + \frac{1}{2} = \frac{1}{x} \qquad x \neq 10, \quad \text{lcm} = 10x$$

$$10x \cdot \frac{3}{10} + 10x \cdot \frac{2}{5} + 10x \cdot \frac{1}{2} = 10x \cdot \frac{1}{x}$$
$$3x + 4x + 5x = 10$$
$$12x = 10$$
$$x = \frac{5}{6} \text{ hour or 50 minutes}$$

15. Let T = number of hours traveled by second car (how long it will take to catch up)
Then $T + 1$ = number of hours traveled by first car
$D_1 = D_2$ when the second car catches up

$$50(T + 1) = 60T$$
$$50T + 50 = 60T$$
$$50 = 10T$$
$$5 = T$$
$$T = 5 \text{ hours}$$

$D_1 = 50(T + 1)$

$D_2 = 60T$

17. Let x = time for both pipes to complete job (fill tank).
8 = time for pipe A alone to complete job
6 = time for pipe B alone to complete job

Then the rate of completion for pipe A is $\frac{1}{8}$ job per hour

the rate of completion for pipe B is $\frac{1}{6}$ job per hour

the rate of completion for both pipes together is $\frac{1}{x}$ job per hour

(Rate of pipe A) + (Rate of pipe B) = (Rate together)

$$\frac{1}{8} + \frac{1}{6} = \frac{1}{x} \qquad x \neq 0 \quad \text{lcm} = 24x$$

$$24x \cdot \frac{1}{8} + 24x \cdot \frac{1}{6} = 24x \cdot \frac{1}{x}$$
$$3x + 4x = 24$$
$$7x = 24$$
$$x = 3.43 \text{ hours}$$

19. Let x = time to complete whole job
Then $x - 15$ = time worked by second person

$$\begin{pmatrix}\text{Quantity produced}\\ \text{by first worker}\end{pmatrix} + \begin{pmatrix}\text{Quantity produced}\\ \text{by second worker}\end{pmatrix} = \begin{pmatrix}\text{Total}\\ \text{quantity}\end{pmatrix}$$

$$14x + 10(x - 15) = 1560$$
$$14x + 10x - 150 = 1560$$
$$24x - 150 = 1560$$
$$24x = 1710$$
$$x = 71.25 \text{ minutes}$$

21. Let x = time for second painter to paint house alone.
5 = time for first painter to paint house alone.
3 = time for both painters to paint house together.

Then the rate of completion for first painter is $\frac{1}{5}$ job per hour.

the rate of completion for second painter is $\frac{1}{x}$ job per hour.

the rate of completion for both painters is $\frac{1}{3}$ job per hour.

$$\begin{pmatrix}\text{Rate of}\\ \text{first painter}\end{pmatrix} + \begin{pmatrix}\text{Rate of}\\ \text{second painter}\end{pmatrix} = \begin{pmatrix}\text{Rate}\\ \text{together}\end{pmatrix}$$

$$\frac{1}{5} + \frac{1}{x} = \frac{1}{3} \qquad x \neq 0 \quad \text{lcm} = 15x$$

$$15x \cdot \frac{1}{5} + 15x \cdot \frac{1}{x} = 15x \cdot \frac{1}{3}$$
$$3x + 15 = 5x$$
$$15 = 2x$$
$$7.5 = x$$
$$x = 7.5 \text{ days}$$

23. Let x = average speed of local bus in miles per hour

Then $\frac{3}{2}x$ = average speed of express bus.

Then 60 miles $\div$ x miles per hour $= \frac{60}{x}$ hours = time of local bus

and 60 miles $\div \frac{3}{2}x$ miles per hour $= 60 \div \frac{3x}{2} = 60 \cdot \frac{2}{3x} = \frac{40}{x}$ hours = time of express bus.

Since (time of express bus) = (time of local bus) $- \frac{2}{3}$ hours (40 minutes)

$$\frac{40}{x} = \frac{60}{x} - \frac{2}{3} \qquad \text{lcm} = 3x, \quad x \neq 0$$

$$3x \cdot \frac{40}{x} = 3x \cdot \frac{60}{x} - 3x \cdot \frac{2}{3}$$
$$120 = 180 - 2x$$
$$-60 = -2x$$
$$x = 30 \text{ miles per hour (local bus)}$$
$$\frac{3}{2}x = 45 \text{ miles per hour (express bus)}$$

25. Let x = walking rate

then $6x$ = rate of bus

then $12 \div x = \frac{12}{x}$ = time walking

$12 \div 6x = \frac{12}{6x} = \frac{2}{x}$ = time on bus

Since (time walking) $= \frac{5}{2} +$ (time on bus)

$$\frac{12}{x} = \frac{5}{2} + \frac{2}{x} \qquad \text{lcm} = 2x, \quad x \neq 0$$

$$2x \cdot \frac{12}{x} = 2x \cdot \frac{5}{2} + 2x \cdot \frac{2}{x}$$
$$24 = 5x + 4$$
$$20 = 5x$$
$$x = 4 \text{ miles per hour}$$

27. Let x = rate of one worker

$x + 8$ = rate of second worker

then $\frac{260}{x}$ = time of slower worker to process 260 forms

$\frac{286}{x + 8}$ = time of faster worker to process 286 forms

Since the times are equal,

$$\frac{260}{x} = \frac{286}{x + 8} \qquad x \neq 0, -8 \quad \text{lcm} = x(x + 8)$$

$$x(x + 8)\frac{260}{x} = x(x + 8)\frac{286}{x + 8}$$
$$260(x + 8) = 286x$$
$$260x + 2080 = 286x$$
$$2080 = 26x$$
$$x = 80 \text{ forms per hour}$$
$$x + 8 = 88 \text{ forms per hour}$$

29. Let D = the distance up the trail

then $\frac{D}{1.6}$ = the time on the uphill trail

$\frac{D}{2.4}$ = the time on the downhill return

Since (time uphill) + (time downhill) = 10

$$\frac{D}{1.6} + \frac{D}{2.4} = 10$$

$$\frac{10D}{16} + \frac{10D}{24} = 10 \qquad \text{lcm} = 48$$

$$48 \cdot \frac{10D}{16} + 48 \cdot \frac{10D}{24} = 480$$

$$30D + 20D = 480$$

$$50D = 480$$

$$D = 9.6 \text{ miles}$$

31. Let x = rate of the wind

then 165 - x = rate of the plane outbound against the wind

165 + x = rate of the plane returning with the wind.

$\frac{425}{165 - x}$ = time outbound $\qquad \frac{425}{165 + x}$ = time returning

Since (time outbound) = 1.2 × (time returning)

$$\frac{425}{165 - x} = 1.2\frac{425}{165 + x}$$

$$\frac{425}{165 - x} = \frac{510}{165 + x} \qquad x \neq 165, -165 \quad \text{lcm} = (165 - x)(165 + x)$$

$$(165 - x)(165 + x) \cdot \frac{425}{165 - x} = (165 - x)(165 + x) \cdot \frac{510}{165 + x}$$

$$(165 + x)425 = (165 - x)510$$

$$70{,}125 + 425x = 84{,}150 - 510x$$

$$70{,}125 + 935x = 84{,}150$$

$$935x = 14{,}025$$

$$x = 15 \text{ miles per hour}$$

33. Let D = distance of target. Using $\frac{D}{R} = T$, we see that $\frac{D}{670}$ = number of seconds bullet travels and $\frac{D}{335}$ = number of seconds sound travels back from the target. Therefore

$$\frac{D}{670} + \frac{D}{335} = 3 \qquad \text{lcm} = 670$$

$$670 \cdot \frac{D}{670} + 670 \cdot \frac{D}{335} = 670 \cdot 3$$

$$D + 2D = 2010$$

$$3D = 2010$$

$$D = 670 \text{ meters}$$

Alternatively, we could let T = time for bullet to travel. Then 3 - T = time for sound to travel back from the target. Since the distances traveled by each are equal, using D = RT we have

(Distance of bullet) = (Distance of sound)

$$670T = 335(3 - T)$$

$$670T = 1005 - 335T$$

$$1005T = 1005$$

$$T = 1 \text{ second}$$

Then distance = (670 meters per second)(1 second)

= 670 meters

35. Let x = rate of old machine
then $x + 30$ = rate of new machine
then $\frac{1800}{x}$ = time for old machine to print 1800 copies
and $\frac{1800}{x + 30}$ = time for new machine to print 1800 copies

Since $\begin{pmatrix}\text{time for new}\\ \text{machine}\end{pmatrix} = \begin{pmatrix}\text{time for old}\\ \text{machine}\end{pmatrix} - 5$

$$\frac{1800}{x + 30} = \frac{1800}{x} - 5 \qquad x \neq 0, -30 \quad \text{lcm} = x(x + 30)$$

$$x(x + 30)\frac{1800}{x + 30} = x(x + 30)\frac{1800}{x} - 5x(x + 30)$$
$$1800x = (x + 30)1800 - 5x^2 - 150x$$
$$1800x = 1800x + 54,000 - 5x^2 - 150x$$
$$5x^2 + 150x - 54,000 = 0$$
$$x^2 + 30x - 10,800 = 0$$
$$(x - 90)(x + 120) = 0$$
$$x - 90 = 0 \quad \text{or} \quad x + 120 = 0$$
$x = 90$ copies/minute $\qquad x = -120$ discard; x cannot be negative
$x + 30 = 120$ copies/minute

37. Let x = normal driving speed
then $x + 10$ = faster driving speed
then $\frac{600}{x}$ = normal driving time
$\frac{600}{x + 10}$ = driving time at faster speed

Since $\begin{pmatrix}\text{driving time at}\\ \text{faster speed}\end{pmatrix} = \begin{pmatrix}\text{normal driving}\\ \text{time}\end{pmatrix} - 2$

$$\frac{600}{x + 10} = \frac{600}{x} - 2 \qquad x \neq 0, -10 \quad \text{lcm} = x(x + 10)$$

$$x(x + 10)\frac{600}{x + 10} = x(x + 10)\frac{600}{x} - 2x(x + 10)$$
$$600x = 600(x + 10) - 2x^2 - 20x$$
$$600x = 600x + 6000 - 2x^2 - 20x$$
$$2x^2 + 20x - 6000 = 0$$
$$x^2 + 10x - 3000 = 0$$
$$(x - 50)(x + 60) = 0$$
$$x - 50 = 0 \quad \text{or} \quad x + 60 = 0$$
$x = 50$ mph $\qquad x = -60$ discard; x cannot be negative

Exercise 4-4 Formulas and Literal Equations

Key Ideas and Formulas

When solving a formula, or an equation with more than one variable, for a variable that does not occur to a higher power than one, the process does not differ from the process of solving a linear equation in one variable:

1. Clear fractions.
2. Simplify both sides.
3. Get all terms with the variable in question to one side, all remaining terms to the other, and combine like terms.
4. Isolate the variable by dividing by its coefficient.

1. $\frac{1}{3}x + \frac{1}{4} = \frac{1}{6}$ lcm = 12

$$12 \cdot \frac{1}{3}x + 12 \cdot \frac{1}{4} = 12 \cdot \frac{1}{6}$$
$$4x + 3 = 2$$
$$4x = -1$$
$$x = -\frac{1}{4}$$

3. $1 - \frac{1}{2}x = \frac{2}{3}$ lcm = 6

$$6 - 6 \cdot \frac{1}{2}x = 6 \cdot \frac{2}{3}$$
$$6 - 3x = 4$$
$$2 = 3x$$
$$x = \frac{2}{3}$$

5. $\frac{1}{x} + 3 = \frac{5}{3}$ $x \neq 0$ lcm = 3x

$$3x \cdot \frac{1}{x} + 3x \cdot 3 = 3x \cdot \frac{5}{3}$$
$$3 + 9x = 5x$$
$$3 = -4x$$
$$x = -\frac{3}{4}$$

7. $5 - \frac{6}{x} = \frac{7}{8}$ $x \neq 0$ lcm = 8x

$$8x \cdot 5 - 8x \cdot \frac{6}{x} = 8x \cdot \frac{7}{8}$$
$$40x - 48 = 7x$$
$$-48 = -33x$$
$$x = \frac{48}{33} \text{ or } \frac{16}{11}$$

9. $\frac{1}{x} + \frac{1}{2x} = \frac{1}{3}$ $x \neq 0$ lcm = 6x

$$6x \cdot \frac{1}{x} + 6x \cdot \frac{1}{2x} = 6x \cdot \frac{1}{3}$$
$$6 + 3 = 2x$$
$$9 = 2x$$
$$x = \frac{9}{2}$$

11. $\frac{5}{x} - \frac{4}{3x} = \frac{1}{2}$ $x \neq 0$ lcm = 6x

$$6x \cdot \frac{5}{x} - 6x \cdot \frac{4}{3x} = 6x \cdot \frac{1}{2}$$
$$30 - 8 = 3x$$
$$22 = 3x$$
$$x = \frac{22}{3}$$

13. $\frac{x+1}{3} - \frac{1}{2} = \frac{x+1}{12}$ lcm = 12

$$12 \cdot \frac{(x+1)}{3} - 12 \cdot \frac{1}{2} = 12 \cdot \frac{(x+1)}{12}$$
$$4(x+1) - 6 = x + 1$$
$$4x + 4 - 6 = x + 1$$
$$4x - 2 = x + 1$$
$$3x - 2 = 1$$
$$3x = 3$$
$$x = 1$$

15. $\frac{1}{x-1} + \frac{2x}{x-1} = 3$ $x \neq 1$ lcm = x − 1

$$(x-1)\frac{1}{x-1} + (x-1)\frac{2x}{x-1} = 3(x-1)$$
$$1 + 2x = 3x - 3$$
$$1 = x - 3$$
$$4 = x$$
$$x = 4$$

17. $\frac{2}{x} + \frac{3}{x+1} = \frac{4}{x^2 + x}$

$\frac{2}{x} + \frac{3}{x+1} = \frac{4}{x(x+1)}$ $x \neq 0, -1$ lcm = x(x + 1)

$$x(x+1)\frac{2}{x} + x(x+1)\frac{3}{x+1} = x(x+1)\frac{4}{x(x+1)}$$
$$2(x+1) + 3x = 4$$
$$2x + 2 + 3x = 4$$
$$5x + 2 = 4$$
$$5x = 2$$
$$x = \frac{2}{5}$$

19. $\frac{1}{x+1} - \frac{1}{x-1} = \frac{x}{x^2 - 1}$

$\frac{1}{x+1} - \frac{1}{x-1} = \frac{x}{(x+1)(x-1)}$ $x \neq 1, -1$ lcm = (x + 1)(x − 1)

$$(x+1)(x-1)\frac{1}{x+1} - (x+1)(x-1)\frac{1}{x-1} = (x+1)(x-1)\frac{x}{(x+1)(x-1)}$$
$$x - 1 - (x+1) = x$$
$$x - 1 - x - 1 = x$$
$$-2 = x$$
$$x = -2$$

21. $A = \frac{1}{2}bh \quad \text{lcm} = 2$

$2A = bh$

$\frac{2A}{b} = h$

$h = \frac{2A}{b}$

23. $A = \frac{1}{2}(a + b)h \quad \text{lcm} = 2$

$2A = (a + b)h$

$2A = ah + bh$

$2A - bh = ah$

$\frac{2A - bh}{h} = a$

$a = \frac{2A - bh}{h}$ or $a = \frac{2A}{h} - \frac{bh}{h}$

$a = \frac{2A}{h} - b$

25. $y = 3x + 7$

$y - 7 = 3x$

$\frac{y - 7}{3} = x$

$x = \frac{y - 7}{3}$

27. $\ell = \frac{\pi}{180} r\theta \quad \text{lcm} = 180$

$180\ell = \pi r \theta$

$\frac{180\ell}{\pi r} = \theta$

$\theta = \frac{180\ell}{\pi r}$

29. $P = S(1 - dt)$

$P = S - Sdt$

$P - S = -Sdt$

$\frac{P - S}{-Sd} = t$

$\frac{S - P}{Sd} = t$

$t = \frac{S - P}{Sd}$

31. $V = \frac{4}{3}\pi ab^2 \quad \text{lcm} = 3$

$3V = 4\pi ab^2$

$\frac{3V}{4\pi b^2} = a$

$a = \frac{3V}{4\pi b^2}$

33. $P = \frac{p}{p + q} \quad \text{lcm} = p + q$

$P(p + q) = p$

$Pp + Pq = p$

$Pq = p - Pp$

$Pq = p(1 - P)$

$\frac{Pq}{1 - P} = p$

$p = \frac{Pq}{1 - P}$

35. $(ERA) = \frac{9R}{I} \quad \text{lcm} = I$

$I(ERA) = 9R$

$I = \frac{9R}{(ERA)}$

37. $F = m\frac{v - v_0}{t} \quad \text{lcm} = t$

$Ft = mt\frac{(v - v_0)}{t}$

$Ft = m(v - v_0)$

$Ft = mv - mv_0$

$Ft + mv_0 = mv$

$v = \frac{Ft + mv_0}{m}$

39. $y = Q(P - V) - F$

$y = QP - QV - F$

$y - QP + F = -QV$

$\frac{y - QP + F}{-Q} = V$

$\frac{-(y - QP + F)}{Q} = V$

$V = \frac{QP - y - F}{Q}$

41. $x(x - 2)(x - 4) = 0$

$x = 0$ or $x - 2 = 0$ or $x - 4 = 0$

$x = 2 \qquad x = 4$

43. $1 = \frac{7}{x} - \frac{12}{x^2}$ $\quad x \neq 0$ $\quad$ lcm $= x^2$

$$x^2 = x^2 \cdot \frac{7}{x} - x^2 \cdot \frac{12}{x^2}$$
$$x^2 = 7x - 12$$
$$x^2 - 7x + 12 = 0$$
$$(x - 3)(x - 4) = 0$$
$$x - 3 = 0 \text{ or } x - 4 = 0$$
$$x = 3 \qquad x = 4$$

45.
$$x(3x + 7) = x(5x + 2)$$
$$3x^2 + 7x = 5x^2 + 2x$$
$$-2x^2 + 5x = 0$$
$$x(-2x + 5) = 0$$
$$x = 0 \text{ or } -2x + 5 = 0$$
$$-2x = -5$$
$$x = \frac{5}{2}$$

47. $1 = \frac{2}{x - 1} + \frac{8}{(x - 1)^2}$ $\quad x \neq 1$ $\quad$ lcm $= (x - 1)^2$

$$(x - 1)^2 = (x - 1)^2 \cdot \frac{2}{x - 1} + (x - 1)^2 \cdot \frac{8}{(x - 1)^2}$$
$$(x - 1)^2 = (x - 1)2 + 8$$
$$x^2 - 2x + 1 = 2x - 2 + 8$$
$$x^2 - 2x + 1 = 2x + 6$$
$$x^2 - 4x - 5 = 0$$
$$(x - 5)(x + 1) = 0$$
$$x - 5 = 0 \text{ or } x + 1 = 0$$
$$x = 5 \qquad x = -1$$

49.
$$6x^2 + 5x - 6 = 0$$
$$(2x + 3)(3x - 2) = 0$$
$$2x + 3 = 0 \text{ or } 3x - 2 = 0$$
$$2x = -3 \qquad 3x = 2$$
$$x = -\frac{3}{2} \qquad x = \frac{2}{3}$$

51. $\frac{1}{f} = \frac{1}{a} + \frac{1}{b}$ $\quad$ lcm $=$ abf

$$abf \cdot \frac{1}{f} = abf \cdot \frac{1}{a} + abf \cdot \frac{1}{b}$$
$$ab = bf + af$$
$$bf + af = ab$$
$$f(b + a) = ab$$
$$\frac{f(b + a)}{b + a} = \frac{ab}{b + a}$$
$$f = \frac{ab}{b + a} \text{ or } f = \frac{ab}{a + b}$$

53.
$$C = \frac{1}{2}QC_h + \frac{D}{Q}C_0$$

Here it is more convenient to retain the fractions.

We regard $\frac{D}{Q}C_0 = \frac{DC_0}{Q} = \frac{C_0}{Q}D$ as D multiplied by a coefficient of $\frac{C_0}{Q}$. Then we multiply both sides by $\frac{Q}{C_0}$ to isolate D.

$$C - \frac{1}{2}QC_h = \frac{D}{Q}C_0$$
$$\frac{Q}{C_0}\left(C - \frac{1}{2}QC_h\right) = \frac{Q}{C_0} \cdot \frac{C_0}{Q}D$$
$$D = \frac{Q}{C_0}\left(C - \frac{1}{2}QC_h\right) \text{ or } D = \frac{Q}{C_0}\left(C - \frac{QC_h}{2}\right)$$

55. $\frac{P_1V_1}{T_1} = \frac{P_2V_2}{T_2}$ lcm $= T_1T_2$

$$T_1T_2\frac{P_1V_1}{T_1} = T_1T_2\frac{P_2V_2}{T_2}$$
$$T_2P_1V_1 = T_1P_2V_2$$
$$\frac{T_2P_1V_1}{P_1V_1} = \frac{T_1P_2V_2}{P_1V_1}$$
$$T_2 = \frac{T_1P_2V_2}{P_1V_1}$$

57. $y = \frac{2x - 3}{3x - 5}$ lcm $= 3x - 5$

$$(3x - 5)y = (3x - 5)\frac{2x - 3}{3x - 5}$$
$$(3x - 5)y = 2x - 3$$
$$3xy - 5y - 2x = 2x - 3 - 2x$$
$$3xy - 5y - 2x = -3$$
$$3xy - 5y - 2x + 5y = 5y - 3$$
$$3xy - 2x = 5y - 3$$
$$x(3y - 2) = 5y - 3$$
$$\frac{x(3y - 2)}{3y - 2} = \frac{5y - 3}{3y - 2}$$
$$x = \frac{5y - 3}{3y - 2}$$

Exercise 4-5 Miscellaneous Applications

The strategy for solving word problems, given in the text, is not repeated here for reasons of space.

1. Let x = first of these three consecutive even numbers
   $x + 2$ = second of these three consecutive even numbers
   $x + 4$ = third of these three consecutive even numbers

   Then | the sum of the first two numbers | is | 8 more than the third |

$$x + (x + 2) = 8 + (x + 4)$$
$$x + x + 2 = 8 + x + 4$$
$$2x + 2 = 12 + x$$
$$x + 2 = 12$$
$$x = 10$$
$$x + 2 = 12$$
$$x + 4 = 14$$

3. Let x = the first number, then $\frac{11}{4} - x$ = the second number. Since the product is $\frac{5}{8}$, we have

$$x\left(\frac{11}{4} - x\right) = \frac{5}{8}$$
$$\frac{11}{4}x - x^2 = \frac{5}{8} \quad \text{lcm} = 8$$
$$8 \cdot \frac{11}{4}x - 8x^2 = 8 \cdot \frac{5}{8}$$
$$22x - 8x^2 = 5$$
$$0 = 8x^2 - 22x + 5$$
$$0 = (4x - 1)(2x - 5)$$
$$4x - 1 = 0 \quad \text{or} \quad 2x - 5 = 0$$
$$4x = 1 \qquad 2x = 5$$
$$x = \frac{1}{4} \qquad x = \frac{5}{2}$$
$$\frac{11}{4} - x = \frac{5}{2} \qquad \frac{11}{4} - x = \frac{1}{4}$$

The numbers are $\frac{1}{4}$ and $\frac{5}{2}$.

5. (A) $R = \frac{Q}{T} = \frac{1{,}965 \text{ hits}}{5{,}699 \text{ at bat}} = 0.345$

(B) $Q = RT = (0.345)(9{,}000) = 3103$ (approximately)

(C) $T = \frac{Q}{R} = \frac{3{,}000}{0.345} = 8696$ at bats (approximately)

7. We are given $T = S + 2.5\left(\frac{x - 3{,}000}{100}\right)$ with $S = 20$ and $x = 4000$. Substituting, we have

$$T = 20 + 2.5\left(\frac{4000 - 3000}{100}\right)$$
$$T = 45^\circ$$

9. We are given $T = S + 2.5\left(\frac{x - 3000}{100}\right)$ with $S = 25$ and $T = 60$. Substituting, we have

$$60 = 25 + 2.5\left(\frac{x - 3000}{100}\right)$$
$$35 = 2.5\left(\frac{x - 3000}{100}\right)$$
$$35 = \frac{x - 3000}{40} \qquad \text{lcm} = 40$$
$$40(35) = 40 \cdot \frac{x - 3000}{40}$$
$$1400 = x - 3000$$
$$x = 4{,}400 \text{ meters}$$

11. We are given $T = S + 2.5\left(\frac{x - 3000}{100}\right)$ with $T = S + 60$. Substituting, we have

$$S + 60 = S + 2.5\left(\frac{x - 3000}{100}\right)$$
$$60 = 2.5\left(\frac{x - 3000}{100}\right)$$
$$60 = \frac{x - 3000}{40} \qquad \text{lcm} = 40$$
$$40(60) = 40 \cdot \frac{x - 3000}{40}$$
$$2400 = x - 3000$$
$$x = 5{,}400 \text{ meters}$$

13. We are given $w = 0.98h - 100$ with $w = 76$ kilograms. Substituting, we have

$$76 = 0.98h - 100$$
$$176 = 0.98h$$
$$h = 180 \text{ centimeters}$$

15. We are given $P = 1 + \frac{D}{33}$ with $P = 3.6$ atmospheres. Substituting, we have

$$3.6 = 1 + \frac{D}{33} \qquad \text{lcm} = 33$$
$$33(3.6) = 33(1) + 33 \cdot \frac{D}{33}$$
$$118.8 = 33 + D$$
$$85.8 = D$$
$$D = 85.8 \text{ feet}$$

17. Let N = total chipmunk population
600 = chipmunks marked in first sample
500 = chipmunks marked in second sample
60 = chipmunks found marked

Then

$$\frac{N}{600} = \frac{500}{60}$$
$$N = 600 \cdot \frac{500}{60}$$
$$N = 5000 \text{ chipmunks}$$

19. Let x = how many gallons, then

$$\frac{2}{220} = \frac{x}{(20)(27)} \qquad \frac{\text{gallons}}{\text{square feet}} = \frac{\text{gallons}}{\text{square feet}}$$
$$\frac{2}{220} = \frac{x}{540}$$
$$x = 540 \cdot \frac{2}{220}$$
$$x = \frac{54}{11} \text{ or } 4\tfrac{10}{11} \text{ or } 4.91 \text{ gallons}$$

21. Let x = the value of the house, then

$$\frac{x}{2226} = \frac{90{,}000}{840} \qquad \frac{\text{value}}{\text{tax}} = \frac{\text{value}}{\text{tax}}$$
$$x = 2226 \cdot \frac{90{,}000}{840}$$
$$x = \$238{,}500$$

23. Let x = how much insecticide, then

$$\frac{x}{140} = \frac{12}{40} \qquad \frac{\text{insecticide}}{\text{solution}} = \frac{\text{insecticide}}{\text{solution}}$$
$$x = 140 \cdot \frac{12}{40}$$
$$x = 42 \text{ ounces}$$

25. Let x = least score required for a B.
Then (total score) must equal 80% (total points).
The total score will be $72 + 85 + 78 + x$.
The total points are $100 + 100 + 100 + 250$.
Therefore

$$72 + 85 + 78 + x = 0.8(100 + 100 + 100 + 250)$$
$$235 + x = 0.8(550)$$
$$235 + x = 440$$
$$x = 205 \text{ points}$$

27. (A)

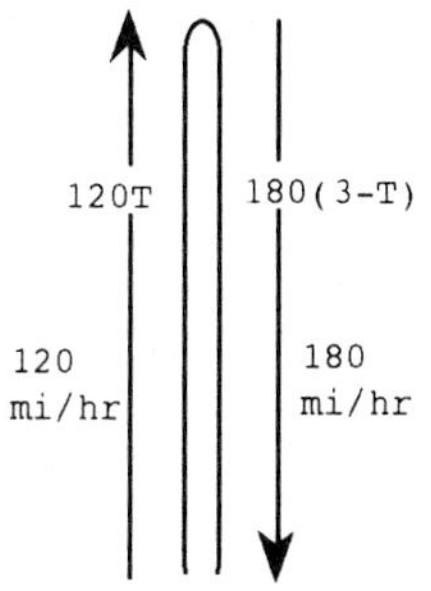

Let T = time flown north by pilot.
Then $3 - T$ = time of return trip.
Plane's rate north = 150 − 30 miles per hour = 120
Plane's rate south = 150 + 30 miles per hour = 180
Use (Distance out) = (Distance returning)

$$120T = 180(3 - T)$$
$$120T = 540 - 180T$$
$$300T = 540$$
$$T = 1.8 \text{ hours}$$
$$\text{distance north} = 120(1.8) = 216 \text{ miles}$$

(B)

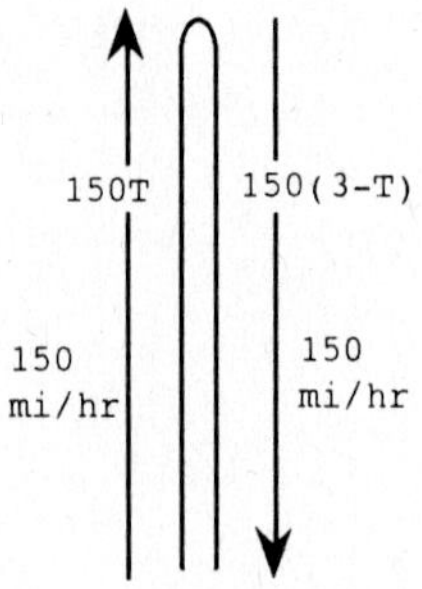

150T | 150(3-T)

150 mi/hr | 150 mi/hr

Let T = time flown north by pilot
Then 3 - T = time of return trip.
Plane's rate north = plane's rate south
= 150 miles per hour
Use (Distance out) = (Distance returning)

$$150T = 150(3 - T)$$
$$150T = 450 - 150T$$
$$300T = 450$$
$$T = 1.5 \text{ hours}$$
$$\text{distance north} = 150(1.5) = 225 \text{ miles}$$

29. Let x = the length of the desired string. Then the ratio of lengths is $\frac{x}{30}$.

(A) Solve $\frac{x}{30} = \frac{1}{2}$
$x = 30 \cdot \frac{1}{2}$
$x = 15$ inches

(B) Solve $\frac{x}{30} = \frac{2}{3}$
$x = 30 \cdot \frac{2}{3}$
$x = 20$ inches

(C) Solve $\frac{x}{30} = \frac{3}{4}$
$x = 30 \cdot \frac{3}{4}$
$x = 22.5$ inches

(D) Solve $\frac{x}{30} = \frac{4}{5}$
$x = 30 \cdot \frac{4}{5}$
$x = 24$ inches

(E) Solve $\frac{x}{30} = \frac{5}{6}$
$x = 30 \cdot \frac{5}{6}$
$x = 25$ inches

(F) Solve $\frac{x}{30} = \frac{3}{5}$
$x = 30 \cdot \frac{3}{5}$
$x = 18$ inches

(G) Solve $\frac{x}{30} = \frac{5}{8}$
$x = 30 \cdot \frac{5}{8}$
$x = 18.75$ inches

31. Let x = the price of the blouse before the discount.
Then 0.3x = the amount of the discount.

Use $\begin{pmatrix}\text{Price before}\\ \text{discount}\end{pmatrix}$ - (Discount) = (Price paid)

$$x - 0.3x = 25.06$$
$$0.7x = 25.06$$
$$x = \$35.80$$

33. Let x = the number of calendars sold at $2.50 each.
Then 2.50x = the sales revenue.
"Break even" means that the sales revenue is equal to the costs.
Solve $2.50x = 12{,}000 + 1.3x$
$1.2x = 12{,}000$
$x = 10{,}000$ calendars

35. Let x = the amount invested at 4.5%
then 0.045x = the interest on this amount at this rate for 1 year (I = Prt)
10,000 - x = the amount invested at 7%
0.07(10,000 - x) = the interest on this amount at this rate for 1 year

Use $\begin{pmatrix}\text{Interest on}\\ \text{amount at 4.5\%}\end{pmatrix} + \begin{pmatrix}\text{Interest on}\\ \text{amount at 7\%}\end{pmatrix} = \begin{pmatrix}\text{Total}\\ \text{interest}\end{pmatrix}$

$$0.045x + 0.07(10{,}000 - x) = 600$$

$$0.045x + 700 - 0.07x = 600$$
$$-0.025x + 700 = 600$$
$$-0.025x = -100$$
$$x = \$4000 \text{ at } 4.5\%$$
$$10{,}000 - x = \$6000 \text{ at } 7\%$$

37. Let x = the MSRP
then $300 + x = 11{,}200$
$x = 10{,}900$
The dealer's cost = 86% of MSRP = $0.86(10{,}900) = \$9374$

39. Let x = the original value
then (original value) - (amount depreciated) = 6400
$x - 8(3500) = 6400$
$x - 28{,}000 = 6400$
$x = \$34{,}400$

41. Let x = the price of the camera before the discount.
Then $0.2x$ = amount of the discount.
Use $\begin{pmatrix}\text{Price before}\\ \text{discount}\end{pmatrix}$ - (Discount) = (Price paid)

$$x - 0.2x = 160$$
$$0.8x = 160$$
$$x = \$200$$

43. Let x = the number of copies sold at \$19.50 per copy.
Then $19.50x$ = the sales revenue.
"Break even" means that the sales revenue is equal to the costs.
Costs = preparation costs + printing and binding costs
= 74,200 + (5.50 per copy) · (x copies)
= $74{,}200 + 5.5x$
Solve $19.5x = 74{,}200 + 5.5x$
$14x = 74{,}200$
$x = 5{,}300$ copies

45. We are given $S + Cx = Px$ with $S = 8{,}000$, $C = 18$, $P = 28$. Substituting, we obtain
$$8000 + 18x = 28x$$
$$8000 = 10x$$
$$x = 800$$

47. We are given $S + Cx = Px$ with $S = 26{,}000$, $C = 42$, $P = 55$. Substituting, we obtain
$$26{,}000 + 42x = 55x$$
$$26{,}000 = 13x$$
$$x = 2{,}000$$

49. We are to find the value of x where revenue equals cost, that is
$$x^2 + 40x = x^2 - 600x + 22{,}400$$
$$40x = -600x + 22{,}400$$
$$640x = 22{,}400$$
$$x = 35$$

51. 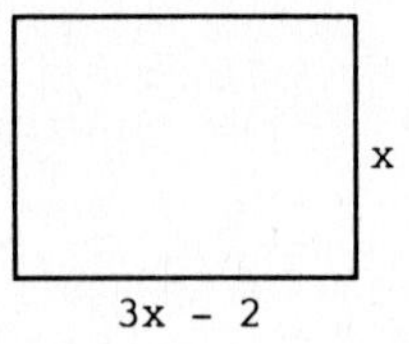

Let x = the width, then 3x - 2 = the length.
Use Perimeter = 2 × (length) + 2 × (width).

$$52 = 2(3x - 2) + 2x$$
$$52 = 6x - 4 + 2x$$
$$52 = 8x - 4$$
$$56 = 8x$$
$$x = 7 \text{ inches}$$
$$3x - 2 = 19 \text{ inches}$$

53. Let x = the number of radians

$$\frac{30}{360} = \frac{x}{2\pi}$$
$$x = 2\pi \cdot \frac{30}{360}$$
$$x = \frac{\pi}{6}$$

55. Let x = the number of degrees

$$\frac{x}{360} = \frac{\pi/4}{2\pi}$$
$$\frac{x}{360} = \frac{\pi/4 \cdot 4}{2\pi \cdot 4}$$
$$\frac{x}{360} = \frac{\pi}{8\pi}$$
$$x = 360 \cdot \frac{\pi}{8\pi}$$
$$x = 45^\circ$$

57. Let x = the number of radians

$$\frac{120}{360} = \frac{x}{2\pi}$$
$$x = 2\pi \cdot \frac{120}{360}$$
$$x = \frac{2\pi}{3}$$

59. Let x = the number of degrees

$$\frac{x}{360} = \frac{7\pi/4}{2\pi}$$
$$\frac{x}{360} = \frac{7\pi/4 \cdot 4}{2\pi \cdot 4}$$
$$\frac{x}{360} = \frac{7\pi}{8\pi}$$
$$x = 360 \cdot \frac{7\pi}{8\pi}$$
$$x = 315^\circ$$

61. Let x = arc length

$$\frac{x}{2\pi \cdot 5} = \frac{36}{360}$$
$$\frac{x}{10\pi} = \frac{1}{10}$$
$$x = 10\pi \cdot \frac{1}{10}$$
$$x = \pi$$

63. Let x = arc length

$$\frac{x}{2\pi \cdot 100} = \frac{100}{360}$$
$$\frac{x}{200\pi} = \frac{100}{360}$$
$$x = 200\pi \cdot \frac{100}{360}$$
$$x = \frac{500\pi}{9}$$

65. Let x = arc length

$$\frac{x}{2\pi \cdot 6} = \frac{40}{360}$$
$$\frac{x}{12\pi} = \frac{1}{9}$$
$$x = 12\pi \cdot \frac{1}{9}$$
$$x = \frac{4\pi}{3}$$

67. Let x = arc length

$$\frac{x}{2\pi \cdot 10} = \frac{225}{360}$$
$$\frac{x}{20\pi} = \frac{225}{360}$$
$$x = 20\pi \cdot \frac{225}{360}$$
$$x = \frac{225\pi}{18} \text{ or } \frac{25\pi}{2}$$

69. We are given v = 15 + 9.75t with v = 93 meters/sec.
Substituting, 93 = 15 + 9.75t

$$78 = 9.75t$$
$$8 = t$$
$$t = 8 \text{ seconds}$$

71. In the given relationship $\frac{F}{f} = \frac{A}{a}$, we are asked for a.

$A = 630$
$F = 2250$
$f = 25$

Substituting, $\frac{2250}{25} = \frac{630}{a}$ $\quad a \neq 0$

$90 = \frac{630}{a}$ $\quad$ lcm $= a$

$90a = 630$

$a = 7$ square centimeters

73. In this problem there is only one force on each side of the fulcrum, as shown. We are given $d_1 = 20$. Since the total length of the bar is 200, and $d_1 + d_2 = 200$, we have $d_2 = 180$, $F_2 = 50$.

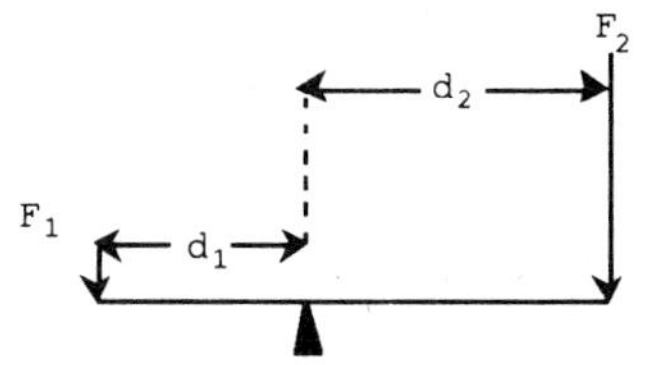

Substituting in $F_1 d_1 = F_2 d_2$ we have

$F_1 \cdot 20 = 50(180)$

$20F_1 = 9{,}000$

$F_1 = 450$ kilograms

75. Let r = the speed of light

Since $d = 5$ miles $+ 5$ miles $= 10$ miles, solve

$10 = r\left(\frac{1}{20{,}000}\right)$

$200{,}000 = r$

$r = 200{,}000$ miles per second

77. Let x = speed for 80-foot-long skid marks

$\frac{20}{30^2} = \frac{80}{x^2}$ $\qquad \frac{\text{length}}{\text{speed}^2} = \frac{\text{length}}{\text{speed}^2}$

$30^2x^2 \cdot \frac{20}{30^2} = 30^2x^2 \cdot \frac{80}{x^2}$ $\quad$ lcm $= 30^2x^2$, $x \neq 0$

$20x^2 = 30^2 \cdot 80$

$20x^2 = 72{,}000$

$x^2 = 3600$

$x^2 - 3600 = 0$

$(x - 60)(x + 60) = 0$

$x - 60 = 0$ or $x + 60 = 0$

$x = 60$ mph $\qquad x = -60$ discard; speed cannot be negative

79. Let x = the amount of tax

A rate of 0.4 mill corresponds to a tax of \$0.40 on each \$1000.

$\frac{x}{78{,}000} = \frac{0.40}{1000}$ $\qquad \frac{\text{tax}}{\text{value}} = \frac{\text{tax}}{\text{value}}$

$x = 78{,}000 \cdot \frac{0.40}{1000}$

$x = \$31.20$

81. Let x = the amount of tax

A rate of 2.8 mill corresponds to a tax of \$2.80 on each \$1000.

$\frac{x}{220{,}000} = \frac{2.80}{1000}$ $\qquad \frac{\text{tax}}{\text{value}} = \frac{\text{tax}}{\text{value}}$

$x = 220{,}000 \cdot \frac{2.80}{1000}$

$x = \$616$

83. Let C = the number of seats for California and
D = the number of seats for Delaware

Then $\dfrac{C}{29{,}760{,}000} = \dfrac{435}{248{,}700{,}000}$ and $\dfrac{D}{666{,}000} = \dfrac{435}{248{,}700{,}000}$

$\left(\text{using } \dfrac{\text{seats}}{\text{population}} = \dfrac{\text{total seats}}{\text{total population}}\right)$

$C = 29{,}760{,}000 \cdot \dfrac{435}{248{,}700{,}000}$ and $D = 666{,}000 \cdot \dfrac{435}{248{,}700{,}000}$

$C = 52.053076\ldots$ $D = 1.1648975\ldots$

Thus, California is entitled to 52 or 53 seats, Delaware to 1 or 2 seats, depending on whether the result is rounded up or down.

85. The average size of a district = $\dfrac{\text{total population}}{\text{number of seats}}$.

Thus, the average size in California would be $\dfrac{29{,}760{,}000}{52}$ or $\dfrac{29{,}760{,}000}{53}$;
that is, 572,307 with 52 seats, 561,509 with 53 seats.

The average size in Delaware would be $\dfrac{666{,}000}{1}$ or $\dfrac{666{,}000}{2}$;
that is 666,000 with 1 seat, 330,000 with 2 seats.

87.

2/3x
x
4
1/5x

Let x = total length of the pole.

Then $\dfrac{2x}{3}$ = length in air.

4 = length in water.

$\dfrac{x}{5}$ = length in sand.

$$\frac{x}{5} + 4 + \frac{2x}{3} = x \qquad \text{lcm} = 15$$

$$15 \cdot \frac{x}{5} + 15 \cdot 4 + 15 \cdot \frac{2x}{3} = 15x$$

$$3x + 60 + 10x = 15x$$

$$13x + 60 = 15x$$

$$60 = 2x$$

$$30 = x$$

$$x = 30 \text{ meters}$$

89. Let x = time of courier.
During the trip from the end to the front, the courier is traveling at a rate 25 - 5 = 20 miles per hour with respect to the column.
During the return trip, the courier is traveling at a rate 25 + 5 = 30 miles per hour with respect to the column.

Thus During the trip out $3 = 20T_1$ $\quad T_1 = \dfrac{3}{20}$ hours

During the return trip $3 = 30T_2$ $\quad T_2 = \dfrac{1}{10}$ hours

Then $x = T_1 + T_2 = \dfrac{3}{20} + \dfrac{1}{10} = \dfrac{3}{20} + \dfrac{2}{20} = \dfrac{5}{20} = \dfrac{1}{4}$ hours or 15 minutes.

Exercise 4-6 REVIEW EXERCISE

1. $3x + 8 = 1 - (5 - 9x)$

$3x + 8 = 1 - 5 + 9x$

$3x + 8 = -4 + 9x$

$-6x + 8 = -4$

$-6x = -12$

$x = 2$

2. $\frac{x}{3} + \frac{x}{4} = \frac{1}{5}$ lcm = 60

$60 \cdot \frac{x}{3} + 60 \cdot \frac{x}{4} = 60 \cdot \frac{1}{5}$

$20x + 15x = 12$

$35x = 12$

$x = \frac{12}{35}$

3. $1 - [x - (1 - x)] = x - [1 - (x - 1)]$

$1 - [x - 1 + x] = x - [1 - x + 1]$

$1 - [2x - 1] = x - [2 - x]$

$1 - 2x + 1 = x - 2 + x$

$2 - 2x = 2x - 2$

$2 - 4x = -2$

$-4x = -4$

$x = 1$

4. $x - \frac{1}{2} = \frac{x}{3}$ lcm = 6

$6x - 6 \cdot \frac{1}{2} = 6 \cdot \frac{x}{3}$

$6x - 3 = 2x$

$-3 = -4x$

$x = \frac{3}{4}$

5. $0.4x + 0.3x = 6.3$

$10(0.4x) + 10(0.3x) = 10(6.3)$

$4x + 3x = 63$

$7x = 63$

$x = 9$

6. There are two methods to solve this type of problem.

Method I: Use the lcm as usual, lcm = 15.

$-\frac{3}{5}y = \frac{2}{3}$

$15\left(-\frac{3}{5}y\right) = 15\left(\frac{2}{3}\right)$

$-9y = 10$

$y = -\frac{10}{9}$

Method II: divide both sides by $-\frac{3}{5}$, the (rational) coefficient of y.

$\left(-\frac{3}{5}y\right) \div \left(-\frac{3}{5}\right) = \frac{2}{3} \div \left(-\frac{3}{5}\right)$

$\left(-\frac{3}{5}y\right) \cdot \left(-\frac{5}{3}\right) = \frac{2}{3} \cdot \left(-\frac{5}{3}\right)$

$y = -\frac{10}{9}$

7. $\frac{x}{4} - 3 = \frac{x}{5}$ lcm = 20

$$20\left(\frac{x}{4}\right) - 20(3) = 20\left(\frac{x}{5}\right)$$
$$5x - 60 = 4x$$
$$5x = 4x + 60$$
$$x = 60$$

8. $\frac{x}{4} - \frac{x-3}{3} = 2$ lcm = 12

$$12 \cdot \frac{x}{4} - 12 \cdot \frac{(x-3)}{3} = 12 \cdot 2$$
$$3x - 4(x - 3) = 24$$
$$3x - 4x + 12 = 24$$
$$-x + 12 = 24$$
$$-x = 12$$
$$x = -12$$

9. $\frac{x}{3} + \frac{1}{2} = \frac{x}{4} + \frac{x+6}{12}$ lcm = 12

$$12 \cdot \frac{x}{3} + 12 \cdot \frac{1}{2} = 12 \cdot \frac{x}{4} + 12 \cdot \frac{x+6}{12}$$
$$4x + 6 = 3x + x + 6$$
$$4x + 6 = 4x + 6$$

This is a true statement, hence all real numbers are solutions.

10. $\frac{x}{5} + \frac{1}{2} = \frac{4x+7}{10} - \frac{x}{5}$ lcm = 10

$$10 \cdot \frac{x}{5} + 10 \cdot \frac{1}{2} = 10 \cdot \frac{4x+7}{10} - 10 \cdot \frac{x}{5}$$
$$2x + 5 = 4x + 7 - 2x$$
$$2x + 5 = 2x + 7$$
$$5 = 7$$

Impossible. No solution.

11. $\frac{1}{5}x + \frac{3}{10} = \frac{3}{4}$ lcm = 20

$$20 \cdot \frac{1}{5}x + 20 \cdot \frac{3}{10} = 20 \cdot \frac{3}{4}$$
$$4x + 6 = 15$$
$$4x = 9$$
$$x = \frac{9}{4}$$

12. $0.5x - 9.9 = 0.005x$ lcm = 1000

$$1000(0.5x) - 1000(9.9) = 1000(0.005x)$$
$$500x - 9900 = 5x$$
$$-9900 = -495x$$
$$x = 20$$

13. $0.05n + 0.1(n - 3) = 1.35$ lcm = 100

$$100(0.05n) + 100[0.1(n - 3)] = 100(1.35)$$
$$5n + 10(n - 3) = 135$$
$$5n + 10n - 30 = 135$$
$$15n - 30 = 135$$
$$15n = 165$$
$$n = 11$$

14. $A = \frac{bh}{2}$ lcm = 2

$$\frac{bh}{2} = A$$
$$2\left(\frac{bh}{2}\right) = 2A$$
$$bh = 2A$$
$$\frac{bh}{h} = \frac{2A}{h}$$
$$b = \frac{2A}{h}$$

15. $3x + 5y = 15$

$$3x + 5y - 3x = 15 - 3x$$
$$5y = 15 - 3x$$
$$\frac{5y}{5} = \frac{15 - 3x}{5}$$
$$y = \frac{15 - 3x}{5}$$

16. Let x = the number

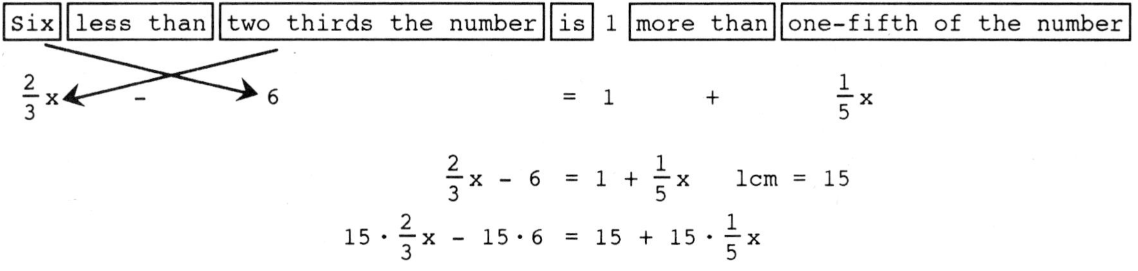

$$\frac{2}{3}x - 6 = 1 + \frac{1}{5}x \qquad \text{lcm} = 15$$
$$15 \cdot \frac{2}{3}x - 15 \cdot 6 = 15 + 15 \cdot \frac{1}{5}x$$
$$10x - 90 = 15 + 3x$$
$$7x - 90 = 15$$
$$7x = 105$$
$$x = 15$$

17. Let x = the number

Ten | more than | half the number | is | 60% of the number

$$10 \quad + \quad \frac{1}{2}x \quad = \quad 0.6x$$

$$10 + \frac{1}{2}x = 0.6x \qquad \text{lcm} = 10$$
$$10 \cdot 10 + 10 \cdot \frac{1}{2}x = 10(0.6x)$$
$$100 + 5x = 6x$$
$$100 = x$$

18. Let x = original population.

0.06x = 6% growth over x

x + 0.06x = 1.06x = population after 6% growth

0.085(1.06x) = 8.5% growth over 1.06x

1.06x + 0.085(1.06x) = population after 8.5% growth over 1.06x

Then $1.06x + 0.085(1.06x) = 253{,}022$
$$1.06x + 0.0901x = 253{,}022$$
$$1.1501x = 253{,}022$$
$$x = 220{,}000$$

19. Let x = the area
$$\frac{\text{area}}{\text{pounds}} = \frac{\text{area}}{\text{pounds}}$$
$$\frac{x}{2000} = \frac{40}{300}$$
$$x = 2000 \cdot \frac{40}{300}$$
$$x = \frac{800}{3} \text{ or } 266\tfrac{2}{3} \text{ square feet}$$

20. (A) $R = \frac{Q}{A} = \frac{2.5 \text{ gallons}}{1175 \text{ square feet}} = \frac{5}{2350} = \frac{1}{470}$ gallons/square feet

(B) $Q = RA = \left(\frac{1}{470} \text{ gallons/square foot}\right)(4000 \text{ sq ft}) = 8.5$ gallons (approx.)

(C) $A = \frac{Q}{R} = \frac{4 \text{ gallons}}{\frac{1}{470} \text{ gallons/square foot}} = 4 \div \frac{1}{470} = \frac{4}{1} \cdot \frac{470}{1} = 1880$ square feet

21. Let x = how many envelopes
then 1000 = number of envelopes at 0.11 each
and x - 1000 = number of envelopes at 0.095 each

$$\text{Use}\begin{pmatrix}\text{cost of first}\\ \text{1000 envelopes}\end{pmatrix} + \begin{pmatrix}\text{cost of}\\ \text{remaining envelopes}\end{pmatrix} = \begin{pmatrix}\text{Total}\\ \text{cost}\end{pmatrix}$$

$$\begin{aligned} 0.11(1000) + 0.095(x - 1000) &= 224 \\ 110 + 0.095x - 95 &= 224 \\ 0.095x + 15 &= 224 \\ 0.095x &= 209 \\ x &= 2200 \text{ envelopes} \end{aligned}$$

22. Let x = total sales
then x - 3000 = sales over $3000 and
$\frac{15}{1000}(x - 3000)$ = commission on sales over $3000.

$$\text{Use}\begin{pmatrix}\text{base}\\ \text{commission}\end{pmatrix} + \begin{pmatrix}\text{commission on}\\ \text{sales over \$3000}\end{pmatrix} = \begin{pmatrix}\text{total}\\ \text{commission}\end{pmatrix}$$

$$360 + \frac{15}{1000}(x - 3000) = 1065 \qquad \text{lcm} = 1000$$

$$\begin{aligned} 360{,}000 + 1000 \cdot \frac{15}{1000}(x - 3000) &= 1{,}065{,}000 \\ 360{,}000 + 15(x - 3000) &= 1{,}065{,}000 \\ 360{,}000 + 15x - 45{,}000 &= 1{,}065{,}000 \\ 15x + 315{,}000 &= 1{,}065{,}000 \\ 15x &= 750{,}000 \\ x &= \$50{,}000 \end{aligned}$$

23. Let x = the price of the stereo before the sale
Then 0.3x = amount of discount.

$$\text{Use}\begin{pmatrix}\text{Price before}\\ \text{sale}\end{pmatrix} - (\text{Discount}) = \begin{pmatrix}\text{Price}\\ \text{paid}\end{pmatrix}$$

$$\begin{aligned} x - 0.3x &= 210 \\ 0.7x &= 210 \\ x &= \$300 \end{aligned}$$

24. Let x = how many inches in 127 cm.

$$\frac{\text{inches}}{\text{centimeters}} = \frac{\text{inches}}{\text{centimeters}}$$

$$\frac{x}{127} = \frac{1}{2.54}$$

$$x = 127 \cdot \frac{1}{2.54}$$

x = 50 inches

25. Let x = how many milliliters of alcohol in 70 ml of solution.

$$\frac{\text{ml of alcohol}}{\text{ml of solution}} = \frac{\text{ml of alcohol}}{\text{ml of solution}}$$

$$\frac{x}{70} = \frac{18}{50}$$

$$x = 70 \cdot \frac{18}{50}$$

x = 25.2 milliliters

26. Let x = how many squirrels in the forest.

$$\text{Then } \frac{x}{360} = \frac{55}{6}$$

$$x = 360 \cdot \frac{55}{6}$$

x = 3,300 squirrels

27. Let x = how many milliliters in 32 fluid ounces.

$$\frac{\text{milliliters}}{\text{fluid ounces}} = \frac{\text{milliliters}}{\text{fluid ounces}}$$

$$\frac{x}{32} = \frac{100}{3.4}$$

$$x = 32 \cdot \frac{100}{3.4}$$

x = 941.18 milliliters

28. Let x = how many millimeters in $\frac{3}{4}$ inch (0.75 inch)

$$\frac{\text{millimeters}}{\text{inches}} = \frac{\text{millimeters}}{\text{inches}}$$

$$\frac{x}{0.75} = \frac{25.4}{1}$$

$$x = 0.75 \cdot \frac{25.4}{1}$$

$$x = 19.05 \text{ millimeters}$$

29. Let x = how many CI dollars in US dollars

$$\frac{\text{CI dollars}}{\text{US dollars}} = \frac{\text{CI dollars}}{\text{US dollars}}$$

$$\frac{x}{36} = \frac{1}{1.25}$$

$$x = 36 \cdot \frac{1}{1.25}$$

$$x = 28.8 \text{ CI dollars}$$

30. Let x = the central angle.

Use $\frac{\text{angle}}{360^\circ} = \frac{55}{100}$

$$\frac{x}{360} = \frac{55}{100}$$

$$x = 360 \cdot \frac{55}{100}$$

$$x = 198^\circ$$

31. Let x = the number of applicants. The ratio of applicants to accepted students is $(7 + 2):2 = \frac{9}{2}$. Thus

$$\frac{x}{148} = \frac{9}{2}$$

$$x = 148 \cdot \frac{9}{2}$$

$$x = 666 \text{ applicants}$$

32. Let x = the amount invested

$\frac{1}{3}x$ = amount in stocks

$\frac{6}{13}x$ = amount in money market funds

Then (amount in bonds) + (amount in stocks) + (amount in funds) = Total

$$16{,}000 + \frac{1}{3}x + \frac{6}{13}x = x \quad \text{lcm} = 39$$

$$39 \cdot 16{,}000 + 39 \cdot \frac{1}{3}x + 39 \cdot \frac{6}{13}x = 39x$$

$$624{,}000 + 13x + 18x = 39x$$

$$624{,}000 + 31x = 39x$$

$$624{,}000 = 8x$$

$$x = \$78{,}000$$

33. Let x = the original price

$0.2x$ = amount of first reduction

$x - 0.2x = 0.8x$ = price after first reduction

$0.19(0.8x)$ = amount of second reduction

$0.8x - 0.19(0.8x)$ = price after second reduction

Then $0.8x - 0.19(0.8x) = 158.4$

$$0.8x - 0.152x = 158.4$$

$$0.648x = 158.4$$

$$x = \$244.44$$

34. Let x = length, then $\frac{1}{4}x + 2$ = width.

Use 2(length) + 2(width) = perimeter

$$2x + 2\left(\frac{1}{4}x + 2\right) = 19$$

$$2x + 2 \cdot \frac{1}{4}x + 4 = 19$$

$$2x + \frac{1}{2}x + 4 = 19 \quad \text{lcm} = 2$$

$$4x + 2\cdot\frac{1}{2}x + 8 = 38$$
$$4x + x + 8 = 38$$
$$5x + 8 = 38$$
$$5x = 30$$
$$x = 6 \text{ m}$$
$$\frac{1}{4}x + 2 = 3.5 \text{ m}$$

35. Let x = width, then $x + 7$ = length
Use (length)(width) = Area

$$(x + 7)x = 60$$
$$x^2 + 7x = 60$$
$$x^2 + 7x - 60 = 0$$
$$(x + 12)(x - 5) = 0$$

$x + 12 = 0$ or $x - 5 = 0$

$x = -12$ discard; x cannot be negative

$x = 5$ m

$x + 7 = 12$ m

36. Let x = how many pounds in each load.
33 tons = 33(2000) pounds

Use $\frac{\text{pounds}}{\text{load}} = \frac{\text{pounds}}{\text{load}}$

$$\frac{x}{1} = \frac{33(2000)}{550}$$
$$x = 120 \text{ pounds}$$

37. Let x = how many calls to make 150 sales.

Use $\frac{\text{calls}}{\text{sales}} = \frac{\text{calls}}{\text{sales}}$

$$\frac{x}{150} = \frac{150}{4}$$
$$x = 150\cdot\frac{150}{4}$$
$$x = 5625 \text{ calls}$$

38. $\frac{2}{3m} - \frac{1}{4m} = \frac{1}{12}$ $\quad m \neq 0$, lcm = 12m

$$12m\cdot\frac{2}{3m} - 12m\cdot\frac{1}{4m} = 12m\cdot\frac{1}{12}$$
$$8 - 3 = m$$
$$5 = m$$
$$m = 5$$

39. $\frac{3x}{x - 5} - 8 = \frac{15}{x - 5}$ $\quad x \neq 5$, lcm = $x - 5$

$$(x - 5)\frac{3x}{x - 5} - 8(x - 5) = (x - 5)\frac{15}{x - 5}$$
$$3x - 8(x - 5) = 15$$
$$3x - 8x + 40 = 15$$
$$-5x + 40 = 15$$
$$-5x = -25$$
$$x = 5$$

$x = 5$ cannot be a solution to the original equation.

No solution.

40. $\frac{2}{x} + \frac{1}{x - 2} = 1$ $\quad x \neq 0, 2$, lcm = $x(x - 2)$

$$x(x - 2)\cdot\frac{2}{x} + x(x - 2)\cdot\frac{1}{x - 2} = x(x - 2)$$
$$(x - 2)2 + x = x^2 - 2x$$
$$2x - 4 + x = x^2 - 2x$$
$$3x - 4 = x^2 - 2x$$
$$0 = x^2 - 5x + 4$$
$$0 = (x - 4)(x - 1)$$

$x - 4 = 0$ or $x - 1 = 0$

$x = 4$ $\qquad x = 1$

41. $2 - \frac{3}{x} = \frac{4}{5}$ $x \neq 0$, lcm $= 5x$

$$5x \cdot 2 - 5x \cdot \frac{3}{x} = 5x \cdot \frac{4}{5}$$
$$10x - 15 = 4x$$
$$-15 = -6x$$
$$x = \frac{5}{2}$$

42. $\frac{6}{x} - \frac{5}{x - 1} = \frac{4}{x^2 - x}$

$\frac{6}{x} - \frac{5}{x - 1} = \frac{4}{x(x - 1)}$ $x \neq 0, 1$, lcm $= x(x - 1)$

$$x(x - 1) \cdot \frac{6}{x} - x(x - 1) \cdot \frac{5}{x - 1} = x(x - 1) \cdot \frac{4}{x(x - 1)}$$
$$(x - 1)6 - 5x = 4$$
$$6x - 6 - 5x = 4$$
$$x - 6 = 4$$
$$x = 10$$

43. $1 + \frac{2}{x} = \frac{3}{x^2}$ lcm $= x^2$, $x \neq 0$

$$x^2 \cdot 1 + x^2 \cdot \frac{2}{x} = x^2 \cdot \frac{3}{x^2}$$
$$x^2 + 2x = 3$$
$$x^2 + 2x - 3 = 0$$
$$(x + 3)(x - 1) = 0$$
$$x + 3 = 0 \quad \text{or} \quad x - 1 = 0$$
$$x = -3 \qquad x = 1$$

44. $x(2x - 1)(2x + 3) = 0$

$$x = 0 \text{ or } 2x - 1 = 0 \text{ or } 2x + 3 = 0$$
$$2x = 1 \qquad 2x = -3$$
$$x = \frac{1}{2} \qquad x = -\frac{3}{2}$$

45. $\frac{5}{2x + 3} - 5 = \frac{-5x}{2x + 3}$ $x \neq -\frac{3}{2}$, lcm $= 2x + 3$

$$(2x + 3)\frac{5}{2x + 3} - 5(2x + 3) = (2x + 3)\frac{-5x}{2x + 3}$$
$$5 - 5(2x + 3) = -5x$$
$$5 - 10x - 15 = -5x$$
$$-10x - 10 = -5x$$
$$-10 = 5x$$
$$-2 = x$$
$$x = -2$$

46. $\frac{3}{x} - \frac{2}{x + 1} = \frac{1}{2x}$ $x \neq 0, -1$, lcm $= 2x(x + 1)$

$$2x(x + 1)\frac{3}{x} - 2x(x + 1)\frac{2}{x + 1} = 2x(x + 1)\frac{1}{2x}$$
$$2(x + 1)3 - 2x \cdot 2 = x + 1$$
$$6x + 6 - 4x = x + 1$$
$$2x + 6 = x + 1$$
$$2x = x - 5$$
$$x = -5$$

47. $\frac{11}{9x} - \frac{1}{6x^2} = \frac{3}{2x}$ $x \neq 0$

$$9x = 3^2 x$$
$$6x^2 = 2 \cdot 3x^2$$
$$2x = 2x$$
$$\text{lcm} = 2 \cdot 3^2 x^2 = 18x^2$$
$$18x^2 \cdot \frac{11}{9x} - 18x^2 \cdot \frac{1}{6x^2} = 18x^2 \cdot \frac{3}{2x}$$
$$2x \cdot 11 - 3 \cdot 1 = 9x \cdot 3$$
$$22x - 3 = 27x$$
$$-3 = 5x$$
$$-\frac{3}{5} = x$$
$$x = -\frac{3}{5}$$

48. $\dfrac{u - 3}{2u - 2} = \dfrac{1}{6} - \dfrac{1 - u}{3u - 3}$

$$\begin{aligned}
2u - 2 &= 2(u - 1)\\
6 &= 2 \cdot 3\\
3u - 3 &= 3(u - 1)\\
\text{lcm} &= 2 \cdot 3(u - 1) = 6(u - 1) \qquad u \neq 1
\end{aligned}$$

$$\begin{aligned}
6(u - 1) \cdot \frac{(u - 3)}{2(u - 1)} &= 6(u - 1)\frac{1}{6} - 6(u - 1) \cdot \frac{(1 - u)}{3(u - 1)}\\
3(u - 3) &= u - 1 - 2(1 - u)\\
3u - 9 &= u - 1 - 2 + 2u\\
3u - 9 &= 3u - 3\\
-9 &= -3
\end{aligned}$$

No solution.

49. $\dfrac{x}{x^2 - 6x + 9} - \dfrac{1}{x^2 - 9} = \dfrac{1}{x + 3}$

$$\begin{aligned}
x^2 - 6x + 9 &= (x - 3)^2\\
x^2 - 9 &= (x - 3)(x + 3)\\
x + 3 &= x + 3\\
\text{lcm} &= (x - 3)^2(x + 3) \qquad x \neq 3, -3
\end{aligned}$$

$$(x - 3)^2(x + 3)\frac{x}{(x - 3)^2} - (x - 3)^2(x + 3)\frac{1}{(x - 3)(x + 3)}$$

$$\begin{aligned}
&= (x - 3)^2(x + 3)\frac{1}{x + 3}\\
(x + 3)x - (x - 3) &= (x - 3)^2\\
x^2 + 3x - x + 3 &= x^2 - 6x + 9\\
x^2 + 2x + 3 &= x^2 - 6x + 9\\
2x + 3 &= -6x + 9\\
2x &= -6x + 6\\
8x &= 6\\
x &= \frac{3}{4}
\end{aligned}$$

50. $\dfrac{1}{x + 1} + \dfrac{2}{x} = \dfrac{11}{4x} \qquad x \neq 0, -1$
$\qquad \text{lcm} = 4x(x + 1)$

$$\begin{aligned}
4x(x + 1) \cdot \frac{1}{x + 1} + 4x(x + 1) \cdot \frac{2}{x} &= 4x(x + 1) \cdot \frac{11}{4x}\\
4x + 4(x + 1) \cdot 2 &= (x + 1)11\\
4x + 8(x + 1) &= (x + 1)11\\
4x + 8x + 8 &= 11x + 11\\
12x + 8 &= 11x + 11\\
x + 8 &= 11\\
x &= 3
\end{aligned}$$

51. $\dfrac{x - 3}{x - 2} = \dfrac{x}{x - 2} + \dfrac{x + 1}{2 - x}$

$\dfrac{x - 3}{x - 2} = \dfrac{x}{x - 2} - \dfrac{x + 1}{x - 2} \qquad \text{lcm} = x - 2 \quad x \neq 2$

$$\begin{aligned}
(x - 2) \cdot \frac{x - 3}{x - 2} &= (x - 2) \cdot \frac{x}{x - 2} - (x - 2) \cdot \frac{(x + 1)}{x - 2}\\
x - 3 &= x - (x + 1)\\
x - 3 &= x - x - 1\\
x - 3 &= -1\\
x &= 2 \qquad x = 2 \text{ cannot be a solution to the original equation.}
\end{aligned}$$

No solution.

52.

$$S = \frac{n(a + L)}{2}$$

$$\frac{n(a + L)}{2} = S \qquad \text{lcm} = 2$$

$$2 \cdot \frac{n(a + L)}{2} = 2S$$

$$n(a + L) = 2S$$

$$na + nL = 2S$$

$$nL = 2S - na$$

$$L = \frac{2S - an}{n}$$

→ Alternatively, at this point we could divide both sides by n, to obtain

$$a + L = \frac{2S}{n}$$

Then $L = \frac{2S}{n} - a$

53.

$$P = M - Mdt$$

$$M - Mdt = P$$

$$M(1 - dt) = P$$

$$M = \frac{P}{1 - dt}$$

54. $\text{Ratio} = \dfrac{\text{number promoted}}{\text{number returned}} = \dfrac{28 - 4}{4} = \dfrac{24}{4} = \dfrac{6}{1}$

55. $\text{Ratio} = \dfrac{\text{number of red pieces}}{\text{total number of pieces}} = \dfrac{12}{40} = \dfrac{3}{10}$

56. Let x = the number of female varsity athletes.
Then 270 - x = the number of male varsity athletes.

$$\frac{x}{270 - x} = \frac{3}{2} \qquad \text{lcm} = 2(270 - x) \qquad x \neq 270$$

$$2(270 - x)\frac{x}{270 - x} = 2(270 - x)\frac{3}{2}$$

$$2x = (270 - x)3$$

$$2x = 810 - 3x$$

$$5x = 810$$

$x = 162$ female varsity athletes.

57. Recall that the lengths of sides of similar triangles are proportional. Let x = the length of the middle side and y = the length of the longest side.

Then $\dfrac{x}{10} = \dfrac{7}{5}$ $\qquad$ $\dfrac{y}{14} = \dfrac{7}{5}$

$x = 10 \cdot \dfrac{7}{5}$ $\qquad$ $y = 14 \cdot \dfrac{7}{5}$

$x = 14$ in $\qquad$ $y = 19.6$ in

58. Let T = number of hours traveled by second car (how long it will take to catch up)
Then T + 1.5 = number of hours traveled by first car

$D_1 = 56(T + 1.5)$ →

→ $D_2 = 76T$

$D_1 = D_2$ when the second car catches up.

$$56(T + 1.5) = 76T$$

$$56T + 84 = 76T$$

$$84 = 20T$$

$$4.2 = T$$

$T = 4.2$ hours

59. Let x = time to complete whole job = time worked by older press.
Then x - 10 = time worked by newer press.

$$\begin{pmatrix}\text{Quantity produced}\\ \text{by older press}\end{pmatrix} + \begin{pmatrix}\text{Quantity produced}\\ \text{by newer press}\end{pmatrix} = \begin{pmatrix}\text{Total}\\ \text{quantity}\end{pmatrix}$$

$$\begin{aligned} 45x + 55(x - 10) &= 3000 \\ 45x + 55x - 550 &= 3000 \\ 100x - 550 &= 3000 \\ 100x &= 3550 \\ x &= 35.5 \text{ minutes} \end{aligned}$$

60. Let x = grade received on third test.
Use (average) = 75

$$\begin{aligned} \frac{65 + 80 + x}{3} &= 75 \\ \frac{145 + x}{3} &= 75 \\ 145 + x &= 225 \\ x &= 80 \end{aligned}$$

61. Since we require C = 15 and $C = \frac{5}{9}(F - 32)$, we replace C with 15 and solve

$$\begin{aligned} \frac{5}{9}(F - 32) &= 15 \\ 9\left[\frac{5}{9}(F - 32)\right] &= 9 \cdot 15 \\ 5(F - 32) &= 135 \\ 5F - 160 &= 135 \\ 5F &= 295 \\ F &= 59 \end{aligned}$$

62. $\frac{x - 3}{12} - \frac{x + 2}{9} = \frac{1 - x}{6} - 1$

$12 = 2^2 \cdot 3$
$9 = 3^2$
$6 = 2 \cdot 3$
$\text{lcm} = 2^2 \cdot 3^2 = 36$

$$\begin{aligned} 36\frac{(x - 3)}{12} - 36\frac{(x + 2)}{9} &= 36\frac{(1 - x)}{6} - 36 \\ 3(x - 3) - 4(x + 2) &= 6(1 - x) - 36 \\ 3x - 9 - 4x - 8 &= 6 - 6x - 36 \\ -x - 17 &= -6x - 30 \\ -x &= -6x - 13 \\ 5x &= -13 \\ x &= \frac{-13}{5} \end{aligned}$$

63.

$$\begin{aligned} \frac{7}{2 - x} &= \frac{10 - 4x}{x^2 + 3x - 10} \\ \frac{7}{-(x - 2)} &= \frac{10 - 4x}{(x + 5)(x - 2)} \qquad \text{lcm} = (x + 5)(x - 2),\ x \neq -5,\ 2 \\ -\frac{7}{x - 2} &= \frac{10 - 4x}{(x + 5)(x - 2)} \\ -(x + 5)(x - 2)\frac{7}{x - 2} &= (x + 5)(x - 2)\frac{10 - 4x}{(x + 5)(x - 2)} \\ -7(x + 5) &= 10 - 4x \\ -7x - 35 &= 10 - 4x \\ -7x &= 45 - 4x \\ -3x &= 45 \\ x &= -15 \end{aligned}$$

64. $\frac{1}{x^2} + \frac{1}{(x + 1)^2} = \frac{2}{x^2 + x}$ $x^2 + x = x(x + 1)$
$\text{lcm} = x^2(x + 1)^2 \quad x \neq 0, -1$

$$x^2(x + 1)^2 \cdot \frac{1}{x^2} + x^2(x + 1)^2 \cdot \frac{1}{(x + 1)^2} = x^2(x + 1)^2 \cdot \frac{2}{x(x + 1)}$$
$$(x + 1)^2 + x^2 = x(x + 1)2$$
$$x^2 + 2x + 1 + x^2 = 2x^2 + 2x$$
$$2x^2 + 2x + 1 = 2x^2 + 2x$$
$$1 = 0$$

No solution.

65. $\frac{1}{x^2 + x} + \frac{1}{x^2 + 2x + 1} = \frac{2x + 1}{x(x + 1)^2}$ $x^2 + x = x(x + 1)$
$x^2 + 2x + 1 = (x + 1)^2$
$\text{lcm} = x(x + 1)^2 \quad x \neq 0, -1$

$$x(x + 1)^2 \cdot \frac{1}{x(x + 1)} + x(x + 1)^2 \cdot \frac{1}{(x + 1)^2} = x(x + 1)^2 \cdot \frac{(2x + 1)}{x(x + 1)^2}$$
$$x + 1 + x = 2x + 1$$
$$2x + 1 = 2x + 1$$

All real numbers are solutions of the original equation, with the exception of 0 and -1, which would make it meaningless.

66. $\frac{x - a}{a} - \frac{x - b}{b} = c \qquad \text{lcm} = ab$

$$ab \cdot \frac{(x - a)}{a} - ab \cdot \frac{(x - b)}{b} = abc$$
$$b(x - a) - a(x - b) = abc$$
$$bx - ab - ax + ab = abc$$
$$bx - ax = abc$$
$$(b - a)x = abc$$
$$x = \frac{abc}{b - a}$$

67. $y = \frac{4x + 3}{2x - 5} \qquad \text{lcm} = 2x - 5$

$$(2x - 5)y = (2x - 5)\frac{4x + 3}{2x - 5}$$
$$2xy - 5y = 4x + 3$$
$$2xy = 4x + 5y + 3$$
$$2xy - 4x = 5y + 3$$
$$x(2y - 4) = 5y + 3$$
$$x = \frac{5y + 3}{2y - 4}$$

68. $\frac{1}{f} = \frac{1}{f_1} + \frac{1}{f_2} \qquad \text{lcm} = ff_1f_2$

$$ff_1f_2\frac{1}{f} = ff_1f_2\frac{1}{f_1} + ff_1f_2\frac{1}{f_2}$$
$$f_1f_2 = ff_2 + ff_1$$
$$f_1f_2 - ff_1 = ff_2$$
$$f_1(f_2 - f) = ff_2$$
$$f_1 = \frac{ff_2}{f_2 - f}$$

69. Let x = how much longer for fruit to ripen at 2500 feet

$$\frac{\text{days}}{\text{feet}} = \frac{\text{days}}{\text{feet}}$$
$$\frac{x}{2500} = \frac{4}{500}$$
$$x = 2500 \cdot \frac{4}{500}$$
$$x = 20 \text{ days}$$

70. Let x = how many consecutive free throws made
x = how many more free throws attempted.
Then 72 + x = how many free throws made (in all).
96 + x = how many free throws attempted (in all).
Then the average after x free throws are made =

$$\frac{72 + x}{96 + x} = 0.8$$
$$72 + x = 0.8(96 + x)$$
$$10(72 + x) = 10(0.8)(96 + x)$$
$$720 + 10x = 8(96 + x)$$
$$720 + 10x = 768 + 8x$$
$$720 + 2x = 768$$
$$2x = 48$$
$$x = 24 \text{ consecutive free throws made}$$

71. Let x = the denominator of the original fraction
x - 1 = the numerator of the original fraction

If $\frac{1}{2}$ [is added to] [the fraction], [the result is] 1 [less than] [3 times the original]

$$\frac{1}{2} + \frac{x-1}{x} = 3 \cdot \frac{x-1}{x} - 1$$

$$\frac{1}{2} + \frac{x-1}{x} = 3 \cdot \frac{x-1}{x} - 1$$

$$\text{lcm} = 2x,\ x \neq 0$$
$$2x \cdot \frac{1}{2} + 2x \cdot \frac{(x-1)}{x} = 2x \cdot 3 \cdot \frac{(x-1)}{x} - 2x$$
$$x + 2(x - 1) = 6(x - 1) - 2x$$
$$x + 2x - 2 = 6x - 6 - 2x$$
$$3x - 2 = 4x - 6$$
$$-2 = x - 6$$
$$x = 4$$
$$x - 1 = 3$$

The fraction is $\frac{3}{4}$.

72. Let x = average speed of freight train
then 1.5x = average speed of passenger train
Then $1040 \div x = \frac{1040}{x}$ = time of freight train
$1040 \div 1.5x = \frac{1040}{1.5x} = \frac{2080}{3x}$ = time of passenger train

Since (time of passenger train) = (time of freight train) - $\left(\frac{20}{3} \text{ hours}\right)$

$$\frac{2080}{3x} = \frac{1040}{x} - \frac{20}{3} \qquad x \neq 0 \quad \text{lcm} = 3x$$

$$3x \cdot \frac{2080}{3x} = 3x \cdot \frac{1040}{x} - 3x \cdot \frac{20}{3}$$
$$2080 = 3120 - 20x$$
$$-1040 = -20x$$
$$x = 52 \text{ mph (freight train)}$$
$$1.5x = 78 \text{ mph (passenger train)}$$

73. Let T = number of hours traveled by express bus (how long it will take to catch up)

Then T + 3 = number of hours traveled by local bus

$D_1 = 48(T + 3)$ →

→ $D_2 = 66T$

$D_1 = D_2$ when the express bus catches up

$$\begin{aligned} 48(T + 3) &= 66T \\ 48T + 144 &= 66T \\ 144 &= 18T \\ T &= 8 \text{ hours} \end{aligned}$$

74. We are given r = 0.06 and asked for t when A = 2P. Substituting, we obtain

$$\begin{aligned} 2P &= P + P(0.06)t \\ 2P - P &= P(0.06)t \\ P &= P(0.06)t \\ t &= \frac{P}{P(0.06)} \\ t &= \frac{1}{0.06} \\ t &= 16\tfrac{2}{3} \text{ years} \quad \text{(independent of the value of P)} \end{aligned}$$

75. We use $F = \frac{9}{5}C + 32$ and substitute F = C.

$$\begin{aligned} C &= \frac{9}{5}C + 32 \quad \text{lcm} = 3 \\ 5C &= 5 \cdot \frac{9}{5}C + 5 \cdot 32 \\ 5C &= 9C + 160 \\ -4C &= 160 \\ C &= -40^\circ \ (= F) \end{aligned}$$

Practice Test 4

1. $1 - 2x = 3(x - 4)$
$1 - 2x = 3x - 12$
$1 - 5x = -12$
$-5x = -13$
$x = \frac{13}{5}$

2. $1 + \frac{1}{2}x = \frac{3}{4}$ lcm = 4
$4 + 4\cdot\frac{1}{2}x = 4\cdot\frac{3}{4}$
$4 + 2x = 3$
$2x = -1$
$x = -\frac{1}{2}$

3. $1 + \frac{1}{2x} = \frac{3}{4}$ $x \neq 0$ lcm = 4x
$4x + 4x\cdot\frac{1}{2x} = 4x\cdot\frac{3}{4}$
$4x + 2 = 3x$
$x + 2 = 0$
$x = -2$

4. $\frac{1}{x} + \frac{1}{2x} = \frac{3}{4}$ $x \neq 0$ lcm = 4x
$4x\cdot\frac{1}{x} + 4x\cdot\frac{1}{2x} = 4x\cdot\frac{3}{4}$
$4 + 2 = 3x$
$6 = 3x$
$x = 2$

5. $\frac{2}{x - 1} + \frac{3}{x - 2} = \frac{3}{2}$ $x \neq 1, 2$ lcm = $2(x - 1)(x - 2)$

$$2(x - 1)(x - 2)\cdot\frac{2}{x - 1} + 2(x - 1)(x - 2)\cdot\frac{3}{x - 2} = 2(x - 1)(x - 2)\cdot\frac{3}{2}$$
$$2(x - 2)2 + 2(x - 1)3 = (x - 1)(x - 2)3$$
$$4(x - 2) + 6(x - 1) = (x^2 - 3x + 2)3$$
$$4x - 8 + 6x - 6 = 3x^2 - 9x + 6$$
$$10x - 14 = 3x^2 - 9x + 6$$
$$0 = 3x^2 - 19x + 20$$
$$0 = (3x - 4)(x - 5)$$
$3x - 4 = 0$ or $x - 5 = 0$
$3x = 4$ $x = 5$
$x = \frac{4}{3}$

6. $\frac{2x}{x - 1} + 3 = 1 + \frac{2}{x - 1}$ $x \neq 1$ lcm = $x - 1$

$$(x - 1)\cdot\frac{2x}{x - 1} + 3(x - 1) = (x - 1)1 + (x - 1)\cdot\frac{2}{x - 1}$$
$$2x + 3x - 3 = x - 1 + 2$$
$$5x - 3 = x + 1$$
$$4x - 3 = 1$$
$$4x = 4$$
$x = 1$ $x = 1$ cannot be a solution to the original equation.

No solution.

7. $\frac{1}{x - 1} = \frac{x - 2}{(x - 3)(x - 4)}$ $\quad x \neq 1, 3, 4$

lcm = (x − 1)(x − 3)(x − 4)

$$(x - 1)(x - 3)(x - 4) \cdot \frac{1}{x - 1} = (x - 1)(x - 3)(x - 4) \cdot \frac{(x - 2)}{(x - 3)(x - 4)}$$

$$\begin{aligned} (x - 3)(x - 4) &= (x - 1)(x - 2) \\ x^2 - 7x + 12 &= x^2 - 3x + 2 \\ -7x + 12 &= -3x + 2 \\ -4x + 12 &= 2 \\ -4x &= -10 \\ x &= \frac{5}{2} \end{aligned}$$

8.

$$\begin{aligned} 3x + 4y &= 5 \\ 3x + 4y - 3x &= -3x + 5 \\ 4y &= -3x + 5 \\ \frac{4y}{4} &= \frac{-3x + 5}{4} \\ y &= \frac{-3x}{4} + \frac{5}{4} \end{aligned}$$

9. $\frac{3}{x} + \frac{4}{y} = 5$ $\quad$ lcm = xy

$$\begin{aligned} xy \cdot \frac{3}{x} + xy \cdot \frac{4}{y} &= 5xy \\ 3y + 4x &= 5xy \\ 3y + 4x - 4x &= 5xy - 4x \\ 3y &= 5xy - 4x \\ 3y &= x(5y - 4) \\ \frac{3y}{5y - 4} &= \frac{x(5y - 4)}{5y - 4} \\ x &= \frac{3y}{5y - 4} \end{aligned}$$

10.

$$\begin{aligned} 3x^4 &= 5y + 6 \\ 3x^4 - 6 &= 5y + 6 - 6 \\ 3x^4 - 6 &= 5y \\ \frac{3x^4 - 6}{5} &= \frac{5y}{5} \\ y &= \frac{3x^4 - 6}{5} \end{aligned}$$

11. Let x = time for newer machine to complete job alone
40 = time for older machine to complete job alone
40 − 60% (40) = 40 − 24 = 16 = time for both machines to complete job together

Then the rate of completion for the newer machine is $\frac{1}{x}$ job per hour

the rate of completion for the older machine is $\frac{1}{40}$ job per hour

the rate of completion for both machines is $\frac{1}{16}$ job per hour

$$\left(\begin{matrix}\text{Rate of newer}\\ \text{machine}\end{matrix}\right) + \left(\begin{matrix}\text{Rate of older}\\ \text{machine}\end{matrix}\right) = \left(\begin{matrix}\text{Rate}\\ \text{together}\end{matrix}\right)$$

$$\frac{1}{x} + \frac{1}{40} = \frac{1}{16}$$ $\quad x \neq 0$, lcm = 80x

$$\begin{aligned} 80x \cdot \frac{1}{x} + 80x \cdot \frac{1}{40} &= 80x \cdot \frac{1}{16} \\ 80 + 2x &= 5x \\ 80 &= 3x \\ x &= \frac{80}{3} \text{ or } 26\tfrac{2}{3} \text{ min} \end{aligned}$$

12. Let x = how many bass are in the lake

Then $\frac{x}{60} = \frac{35}{3}$

$x = 60 \cdot \frac{35}{3}$

$x = 700$ bass

13. First, since 1 pint is 2 cups, $3\frac{1}{2}$ pints is $(3\frac{1}{2})(2)$ cups = 7 cups.

Now, let x = how many tablespoons in $3\frac{1}{2}$ pints (7 cups) and use

$$\frac{\text{tablespoons}}{\text{cups}} = \frac{\text{tablespoons}}{\text{cups}}$$

$$\frac{x}{7} = \frac{1}{1/16}$$

$$\frac{x}{7} = \frac{16}{1}$$

$$x = 7 \cdot \frac{16}{1}$$

$x = 112$ tablespoons

14. Let x = how much acid in 85 milliliters of solution.

$$\frac{\text{acid}}{\text{solution}} = \frac{\text{acid}}{\text{solution}}$$

$$\frac{x}{85} = \frac{14}{50}$$

$$x = 85 \cdot \frac{14}{50}$$

$x = 23.8$ milliliters of acid

15. Let x = how many loads in $\frac{1}{2}$ ton of debris.

Use $\frac{\text{loads}}{\text{tons}} = \frac{\text{loads}}{\text{tons}}$

$$\frac{x}{1/2} = \frac{550}{33}$$

$$\frac{2x}{1} = \frac{550}{33}$$

$$x = \frac{1}{2} \cdot \frac{550}{33}$$

$x = \frac{25}{3}$ or $8\frac{1}{3}$ loads

16. Let T = number of hours traveled by second car (how long it will take to catch up)

Then $T + \frac{4}{3}$ = number of hours traveled by first car (1 hr 20 min = $1\frac{1}{3}$ or $\frac{4}{3}$ hrs)

$D_1 = 45(T + 4/3)$ →

→ $D_2 = 60T$

$D_1 = D_2$ when the second car catches up.

$$45\left(T + \frac{4}{3}\right) = 60T$$

$$45T + 45 \cdot \frac{4}{3} = 60T$$

$$45T + 60 = 60T$$

$$60 = 15T$$

$T = 4$ hours

CHAPTER 5 Linear Equations, Inequalities, and Graphs

Exercise 5-1 Graphing Linear Equations

Key Ideas and Formulas

A **Cartesian** or **rectangular coordinate system** consists of two real number lines at right angles to each other, crossing at their origins. The lines are called the **vertical axis** and the **horizontal axis**; they divide the plane into four parts called **quadrants**. Each point in the plane corresponds to a unique pair of numbers (a, b) called its **coordinates**. Each pair of numbers corresponds to a unique point in the plane called its graph.

The Graph of an Equation
The graph of an equation in two variables in a rectangular coordinate system must meet the following two conditions: 1. If an ordered pair of numbers is a solution to the equation, the corresponding point must be on the graph of the equation. 2. If a point is on the graph of an equation, its coordinates must satisfy the equation.

A **solution** of an equation in two variables is an ordered pair of real numbers, one for x, one for y, that satisfies the equation. The graph of an equation is the graph of all its solutions.

The graph of any equation of the form

$y = mx + b$ or $Ax + By = C$

Where m, b, A, B, and C are constants (A and B not both 0) and x and y are variables is a straight line.

Graphing Equations of the Form $y = mx + b$ or $Ax + By = C$	
Step 1:	Find two solutions of the equation. (A third solution is sometimes useful as a check point.)
Step 2:	Plot the solutions in a coordinate system.
Step 3:	Using a straightedge, draw a line through the points plotted in step 2.

Often, the easiest two points to find are the points where the graph crosses the x and y axes, called the **x** and **y intercepts** respectively.

Special cases: The graph of an equation of form $x = a$, thus, mentally, $0y + x = a$, is a vertical line crossing the x axis at a.

The graph of an equation of form $y = b$, thus, mentally, $0x + y = b$, is a horizontal line crossing the y axis a + b.

In applications it is often convenient to use different scales for the two axes, since the magnitudes of the variables may differ considerably.

---

1. A(-10, 10), B(10, -10), C(16, 14), D(16, 0), E(-14, -16), F(0, 4)

3. A(5, 5), B(8, 2), C(-5, 5), D(-3, 8), E(-5, -6), F(-7, -7)

5.

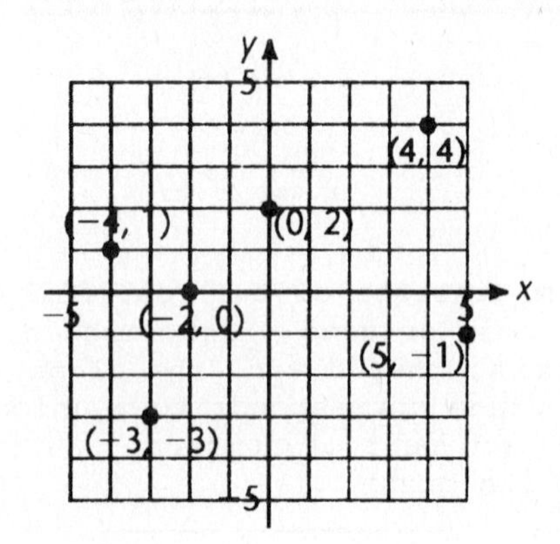

7.

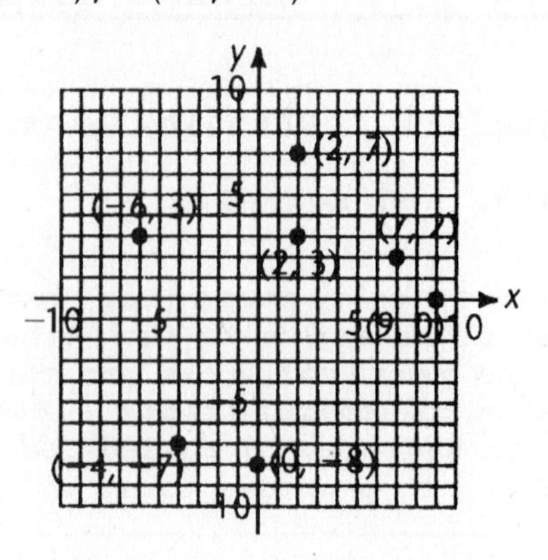

9. A($-3\frac{1}{2}$, 2), B(-2, $-4\frac{1}{2}$), C(0, $-2\frac{1}{2}$), D($2\frac{1}{2}$, 0)

11. A($2\frac{1}{2}$, 1), B($-2\frac{1}{2}$, $3\frac{1}{2}$), C(-2, $-4\frac{1}{2}$), D($3\frac{1}{3}$, -3)

13.

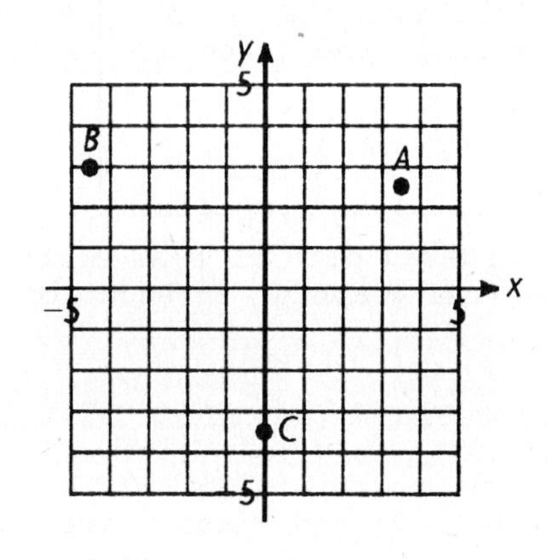

15.

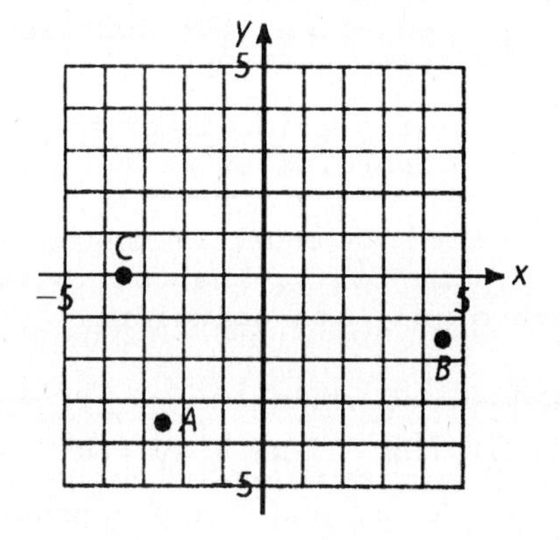

17. To find the x intercept we let y = 0 and solve for x.

$$0 = 3x - 7$$
$$3x = 7$$
$$x = \frac{7}{3}$$

To find the y intercept we let x = 0.

$$y = 0 - 7$$
$$y = -7$$

19. To find the x intercept we let y = 0 and solve for x.

$$0 = -5x + 2$$
$$5x = 2$$
$$x = \frac{2}{5}$$

To find the y intercept we let x = 0.

$$y = -5(0) + 2$$
$$y = 2$$

21. To find the x intercept we let y = 0 and solve for x.

$$2x - 3(0) = 12$$
$$2x = 12$$
$$x = 6$$

To find the y intercept we let x = 0 and solve for y.

$$2(0) - 3y = 12$$
$$-3y = 12$$
$$y = -4$$

23. To find the x intercept we let y = 0 and solve for x.

$$3x + 4(0) = -20$$
$$3x = -20$$
$$x = -\frac{20}{3}$$

To find the y intercept we let x = 0 and solve for y.

$$3(0) + 4y = -20$$
$$4y = -20$$
$$y = -5$$

Note: In graphing the following equations, three convenient solutions have been chosen and plotted. These points are not unique; any other correct solutions will lead to the same line.

25. $y = 2x$

(x,	y)
(0,	1)
(2,	4)
(-2,	-4)

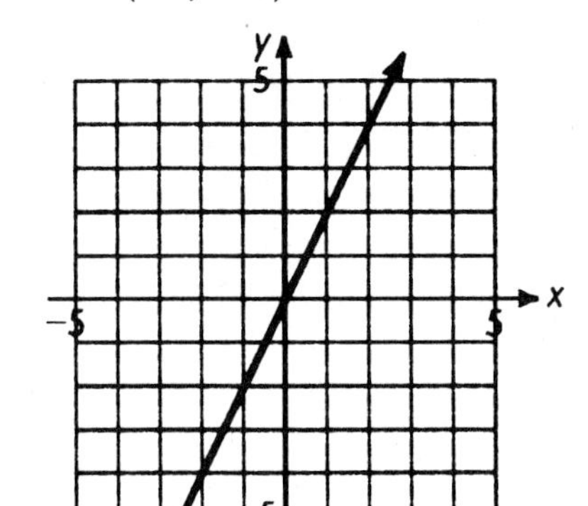

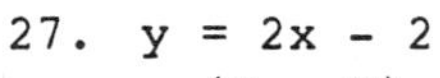
27. $y = 2x - 2$

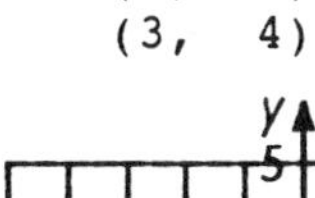

(x,	y)
(0,	-2)
(1,	0)
(3,	4)

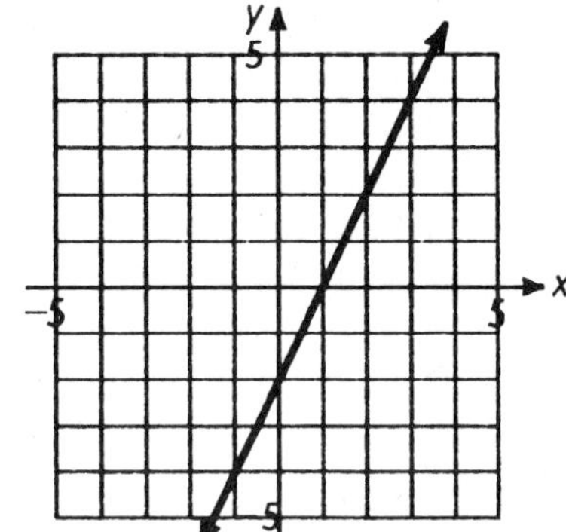

29. $x + y = -4$

(x,	y)
(0,	-4)
(-4,	0)
(-2,	-2)

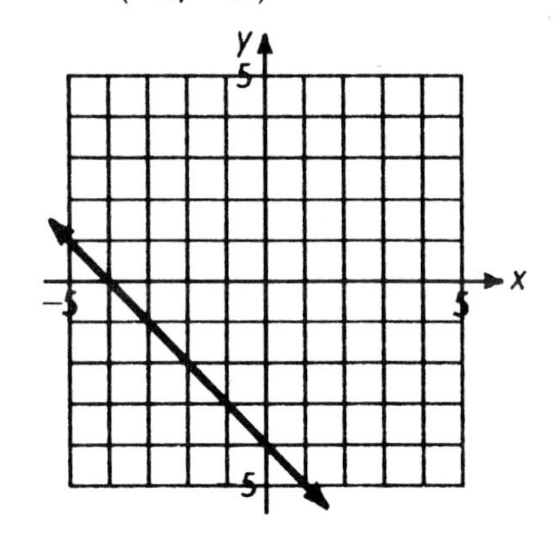

31. $x - y = 3$

(x,	y)
(0,	-3)
(3,	0)
(6,	3)

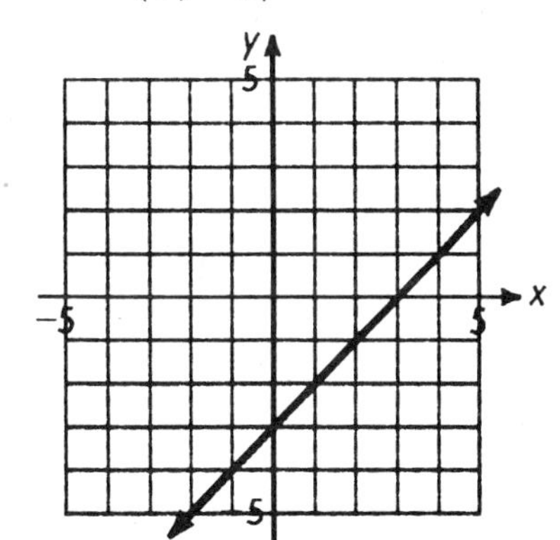

33. $3x + 4y = 12$

(x,	y)
(0,	3)
(4,	0)
(8,	-3)

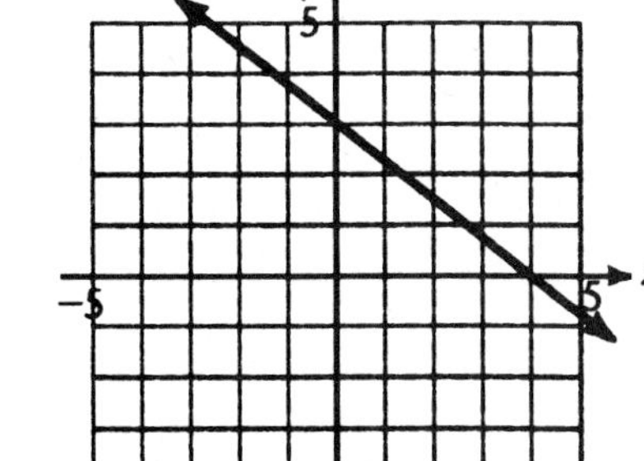

35. $8x - 3y = 24$

(x,	y)
(0,	-8)
(3,	0)
(6,	8)

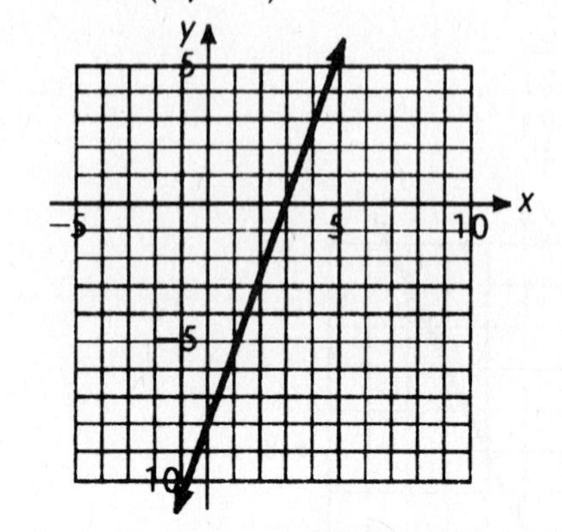

37.

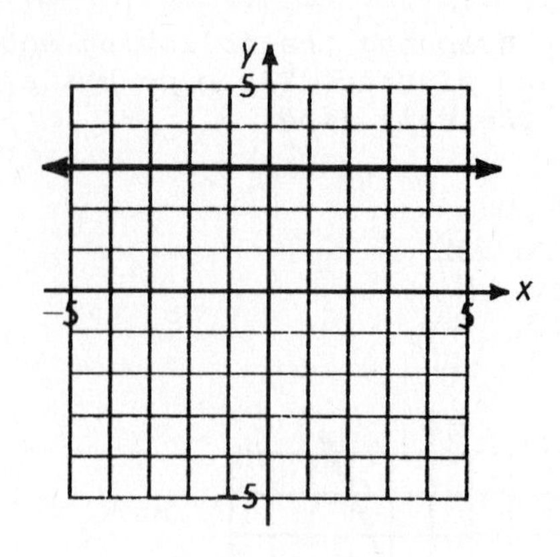

39.

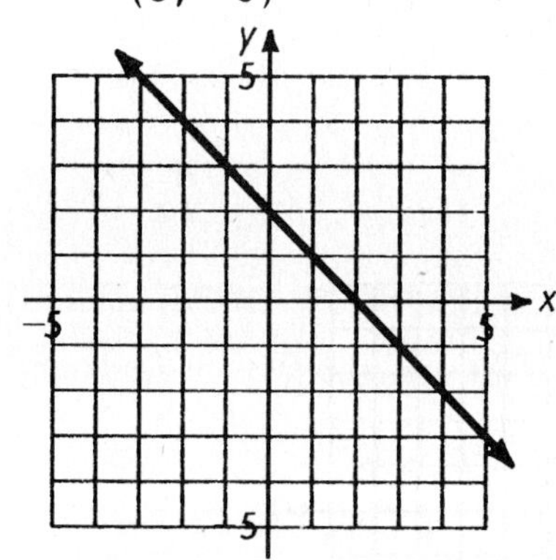

41. $y = -x + 2$

(x,	y)
(2,	0)
(0,	2)
(5,	-3)

43. $3x + 2y = 10$

(x,	y)
(0,	5)
$\left(\frac{10}{3},\right.$	$\left.0\right)$
(6,	-4)

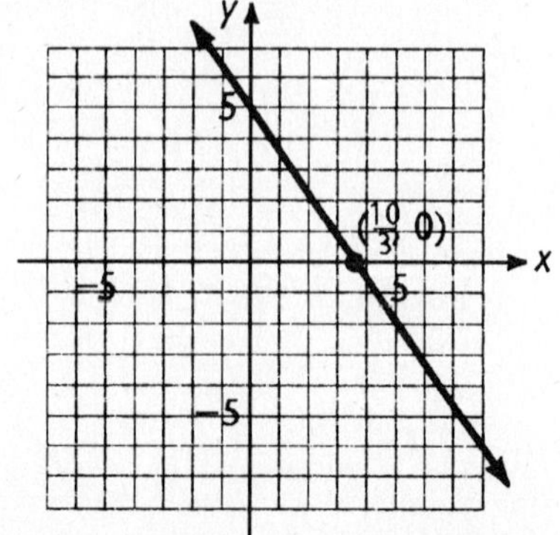

45. $5x - 6y = 15$

(x,	y)
$\left(0,\right.$	$\left.-\frac{5}{2}\right)$
(3,	0)
(-3,	-5)

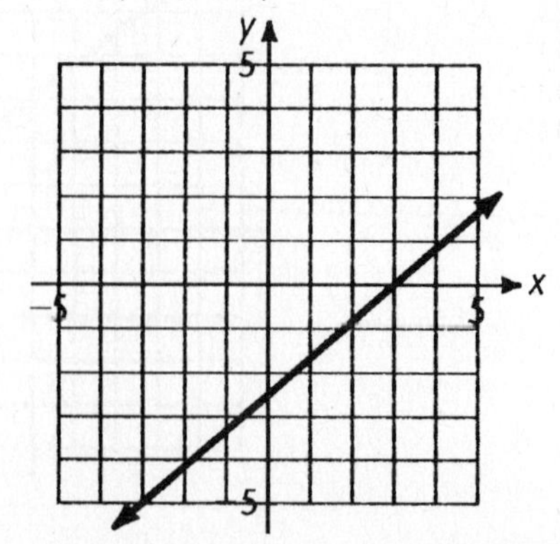

47.

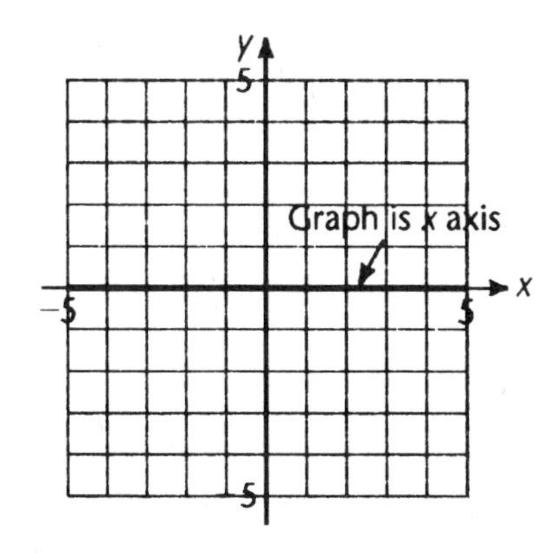

49. $y = \frac{x}{3} + 2$

(x, y)
(0, 2)
(3, 3)
(-6, 0)

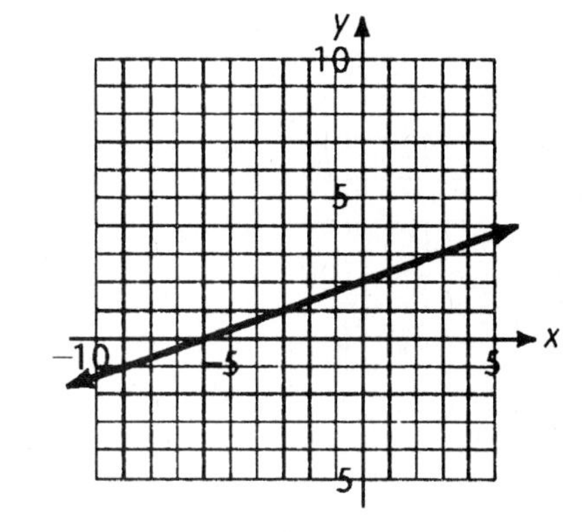

51. $y - \frac{1}{2}x = 0$

(x, y)
(0, 0)
(8, 4)
(-8, -4)

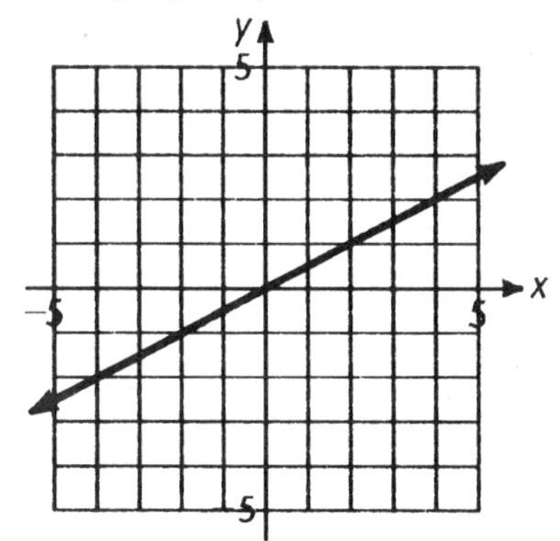

53. $y - \frac{1}{2}x = -1$

(x, y)
(0, -1)
(2, 0)
(8, 3)

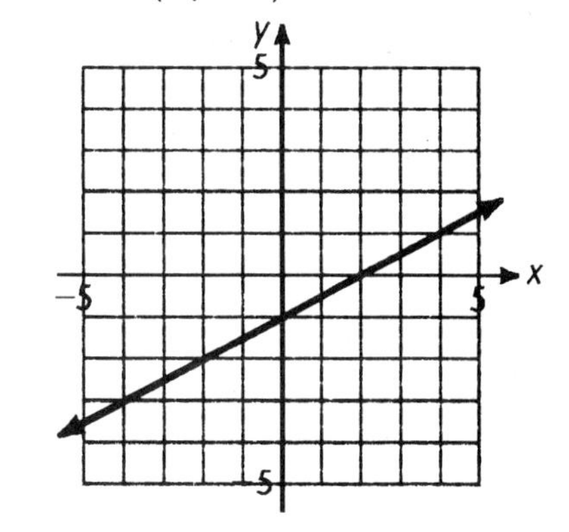

55. $\frac{1}{2}x + \frac{1}{3}y = 2$

(x, y)
(0, 6)
(4, 0)
(2, 3)

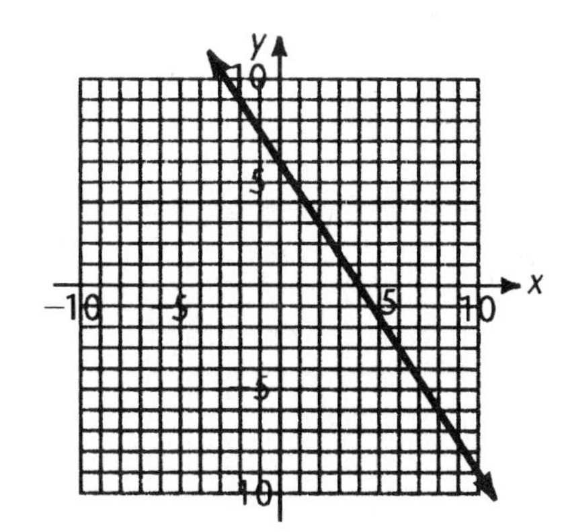

57. $\frac{1}{4}x - \frac{1}{3}y = 1$

(x,	y)
(0,	-3)
(4,	0)
(8,	3)

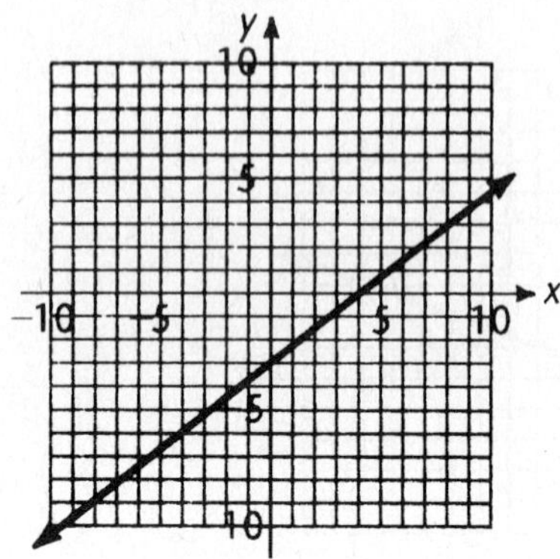

59. $\frac{1}{2}x + \frac{1}{3}y - \frac{1}{6} = 0$

(x, y)
$\left(0, \frac{1}{2}\right)$
$\left(\frac{1}{3}, 0\right)$
$\left(-\frac{1}{3}, 1\right)$

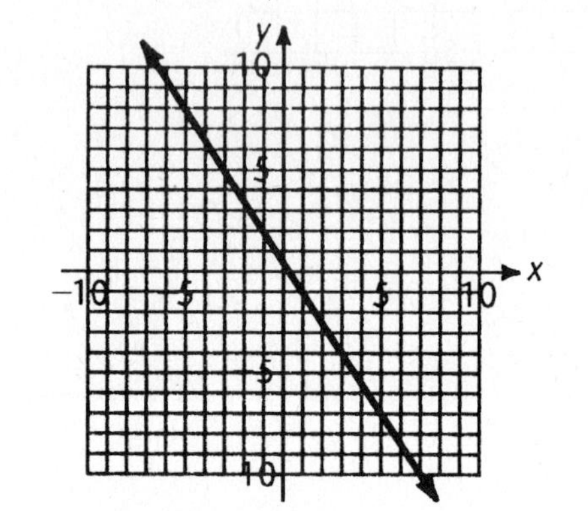

61. $3x + 4y = 6$

$3x + 4y - 3x = -3x + 6$

$4y = -3x + 6$

$\frac{4y}{4} = \frac{-3x + 6}{4}$

$y = -\frac{3}{4}x + \frac{3}{2}$

(x, y)
$\left(0, \frac{3}{2}\right)$
(2, 0)
$\left(4, -\frac{3}{2}\right)$

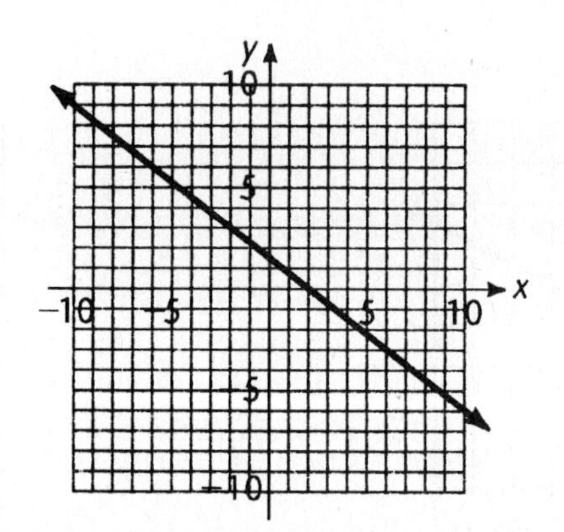

63. $x + 6 = 3x + 2 - y$

$0 = 2x - 4 - y$

$y = 2x - 4$

(x,	y)
(0,	-4)
(2,	0)
(5,	6)

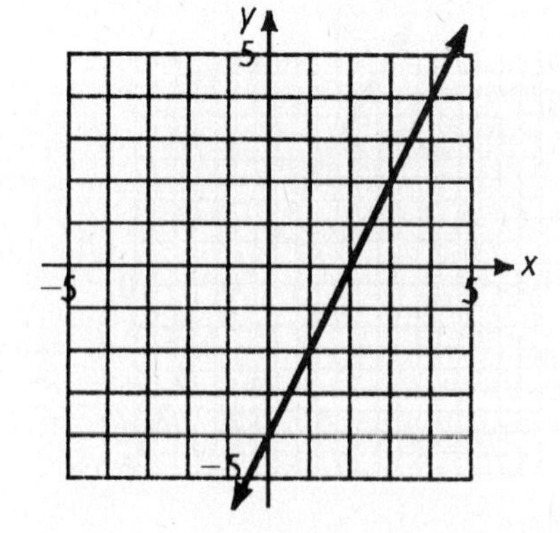

65.

$$y = 3x - 4$$
$$-3x + y = -4$$
$$(-1)(-3x + y) = (-1)(-4)$$
$$3x - y = 4$$

x	y
0	-4
$\frac{4}{3}$	0
1	-1

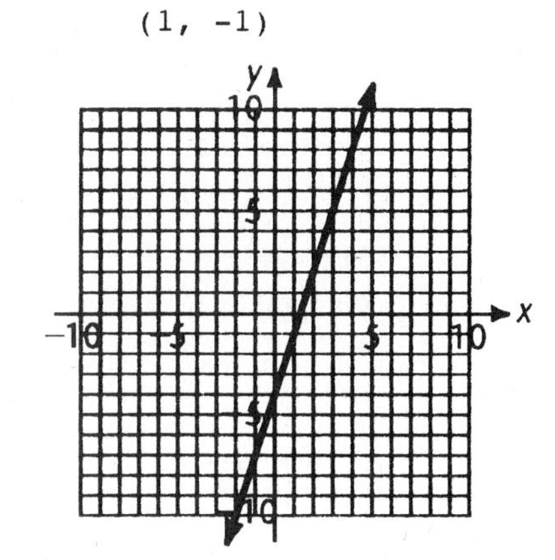

67.

$$y + 8 = 2 - x - y$$
$$x + 2y + 8 = 2$$
$$x + 2y = -6$$

x	y
0	-3
-6	0
4	-5

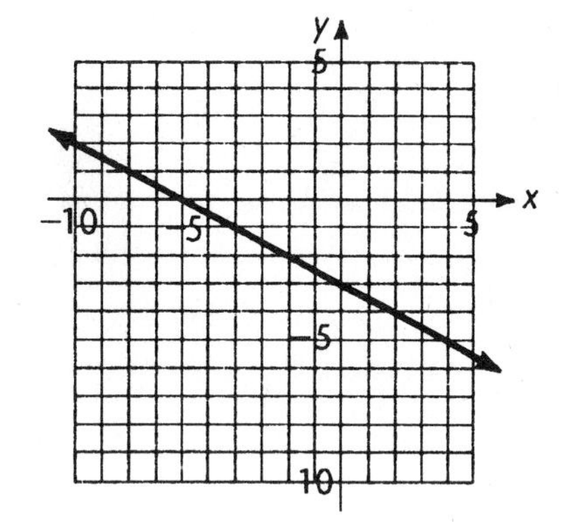

69. $I = 6t$, $0 \le t \le 10$

t	0	5	10
I	0	30	60

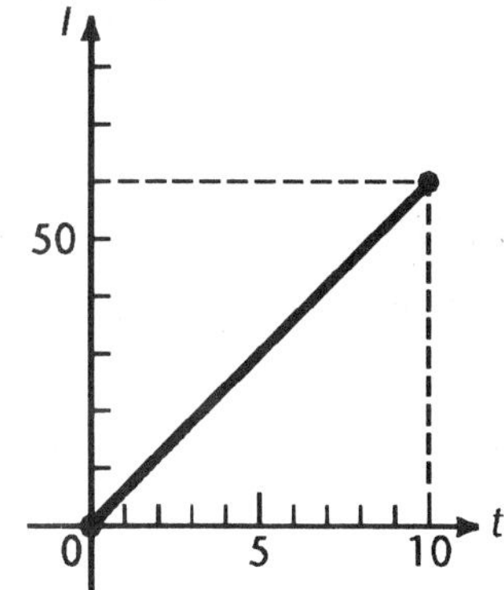

71. $v = 10 + 32t$, $0 \le t \le 5$

t	0	3	5
v	10	106	170

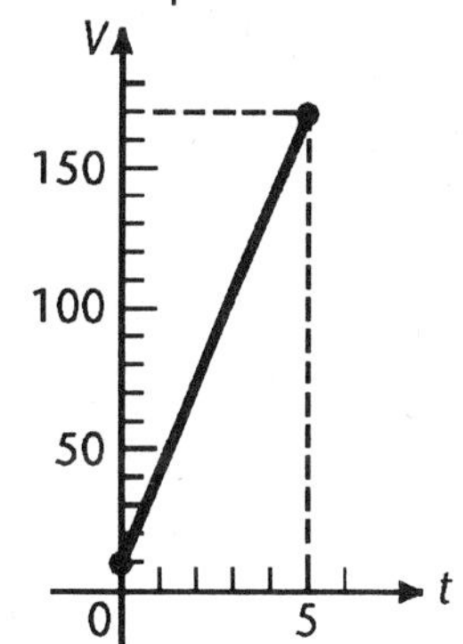

73. $y = -80x + 250$, $0 \le x \le 3$

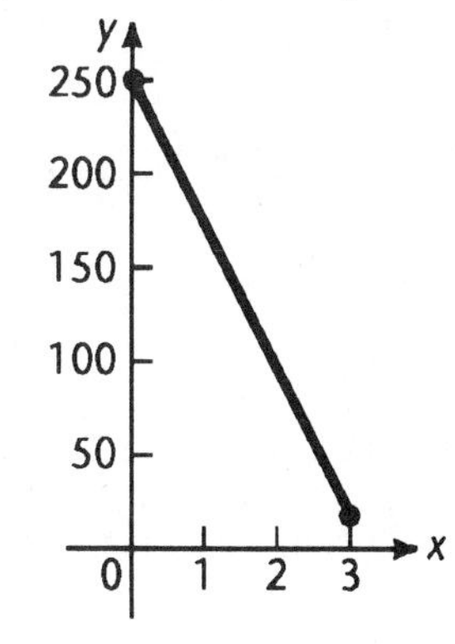

x	0	2	3
y	250	90	10

75. $y = -125x + 1000,\ 0 \le x \le 8$

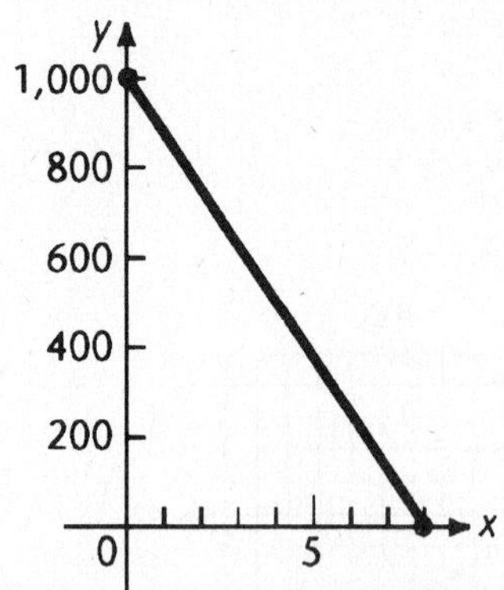

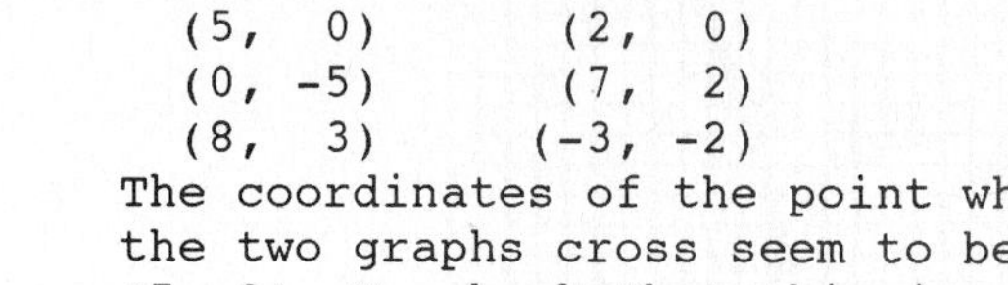

x	0	4	8
y	1000	500	0

77. $x + y = 3$ and $2x - y = 0$

(x, y)
(0, 3)
(3, 0)
(6, -3)

(x, y)
(0, 0)
(1, 2)
(3, 6)

The coordinates of the point where the two graphs cross seem to be (1, 2). To check that this is correct, we substitute (1, 2) into each equation.

$2x - y = 0$: $2 \cdot 1 - 2 \overset{\checkmark}{=} 0$ $\quad$ $x + y = 3$: $1 + 2 \overset{\checkmark}{=} 3$

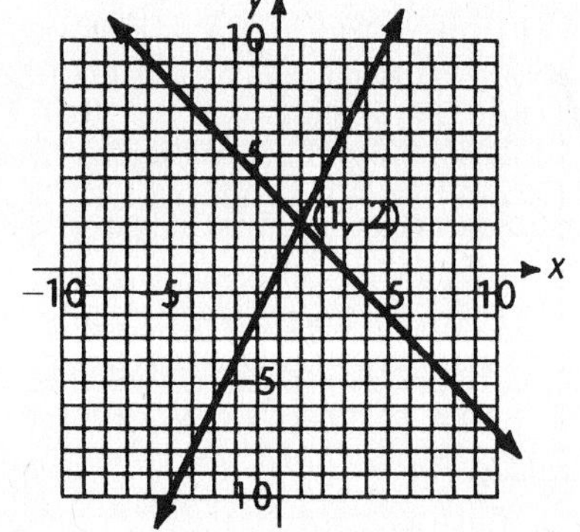

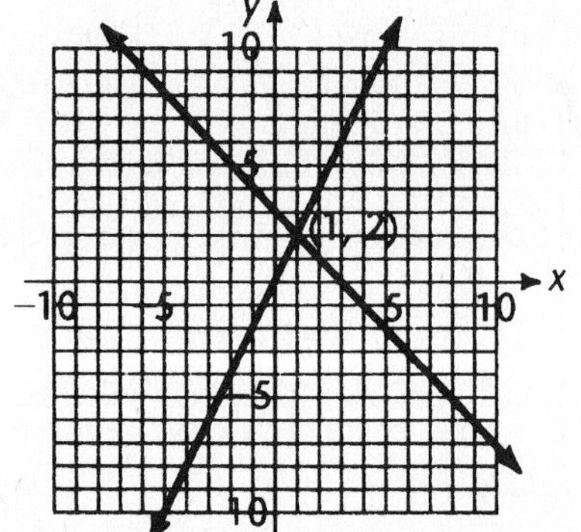

79. $x - y = 5$ and $2x - 5y = 4$

(x, y)
(5, 0)
(0, -5)
(8, 3)

(x, y)
(2, 0)
(7, 2)
(-3, -2)

The coordinates of the point where the two graphs cross seem to be (7, 2). To check that this is correct, we substitute (7, 2) into each equation.

$x - y = 5$: $7 - 2 \overset{\checkmark}{=} 5$ $\quad$ $2x - 5y = 4$: $2 \cdot 7 - 5 \cdot 2 \overset{\checkmark}{=} 4$

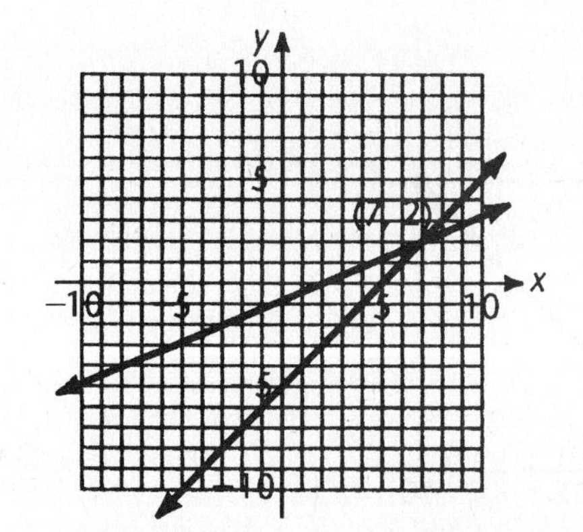

81. $y = 3x + 2$ and $y = -x + 10$

(x, y)
(0, 2)
(1, 5)
(-1, -1)

(x, y)
(0, 10)
(10, 0)
(5, 5)

The coordinates of the point where the two graphs cross seem to be (2, 8). To check that this is correct, we substitute (2, 8) into each equation.

$y = 3x + 2$: $8 \overset{\checkmark}{=} 3 \cdot 2 + 2$ $\quad$ $y = -x + 10$: $8 \overset{\checkmark}{=} -2 + 10$

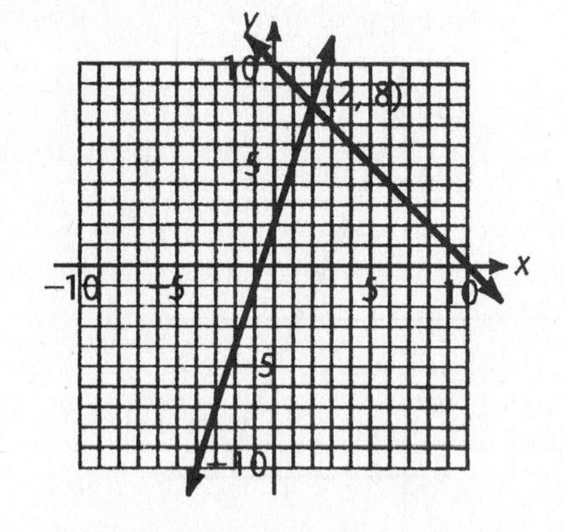

83. $y = x$ $\quad$ $x + y = 6$

(x, y)	(x, y)
(0, 0)	(0, 6)
(2, 2)	(6, 0)
(5, 5)	(2, 4)

The coordinates of the point where the two graphs cross seem to be (3, 3). To check that this is correct, we substitute (3, 3) into each equation.

$y = x$ $\quad$ $x + y = 6$

$3 \overset{\checkmark}{=} 3$ $\quad$ $3 + 3 \overset{\checkmark}{=} 6$

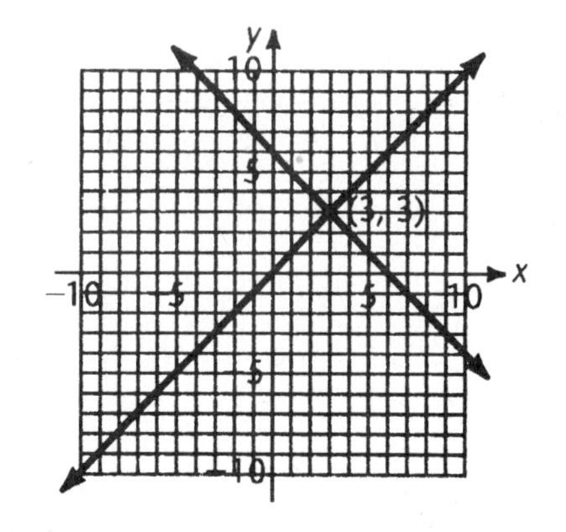

85. $y = mx - 2$

$\underline{m = 2}$ $\quad y = 2x - 2$

(x, y)
(0, -2)
(1, 0)
(2, 2)

$\underline{m = \frac{1}{2}}$ $\quad y = \frac{1}{2}x - 2$

(x, y)
(0, -2)
(2, -1)
(4, 0)

$\underline{m = 0}$ $\quad y = -2$ $\quad$ Horizontal line

$\underline{m = -\frac{1}{2}}$ $\quad y = -\frac{1}{2}x - 2$

(x, y)
(0, -2)
(2, -3)
(-4, 0)

$\underline{m = -2}$ $\quad y = -2x - 2$

(x, y)
(0, -2)
(2, -6)
(-1, 0)

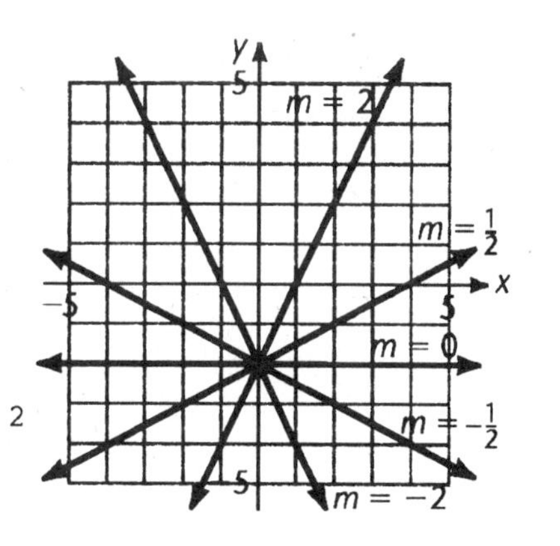

87. Following the hint, we graph

$y = x$ for $x \geq 0$

(x, y)
(0, 0)
(2, 2)
(4, 4)

$y = -x$ for $x < 0$

(x, y)
(-2, 2)
(-4, 4)
(-1, 1)

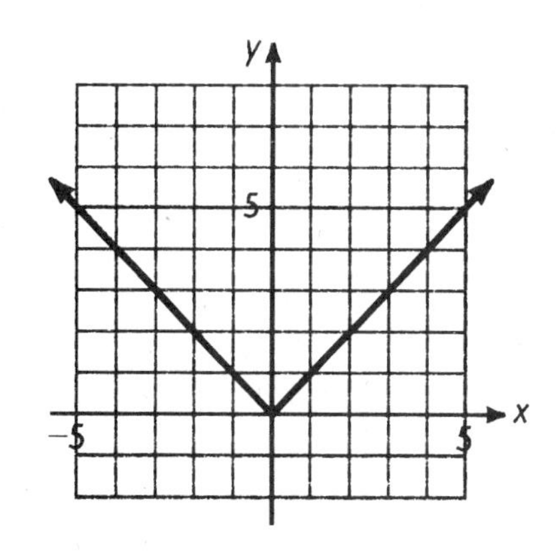

89. $y = x^2$

x	-3	-2	-1	-0.5	0	0.5	1	2	3
y	9	4	1	0.25	0	0.25	1	4	9

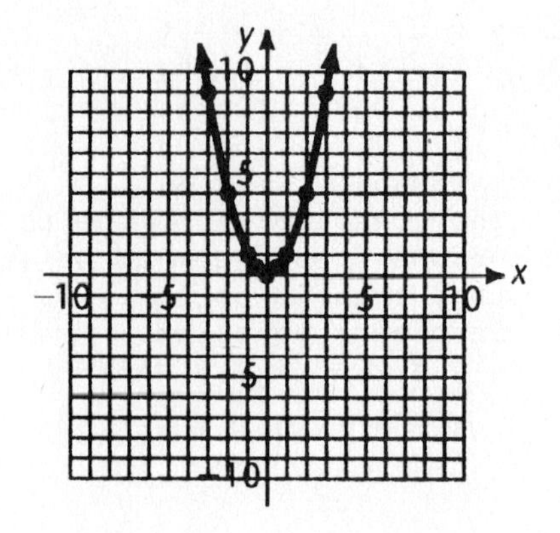

91. $y = -\sqrt{x}$

x	0	0.25	1	2	4	6	9
y	0	-0.5	-1	$-\sqrt{2} \approx -1.4$	-2	$-\sqrt{6} \approx -2.4$	-3

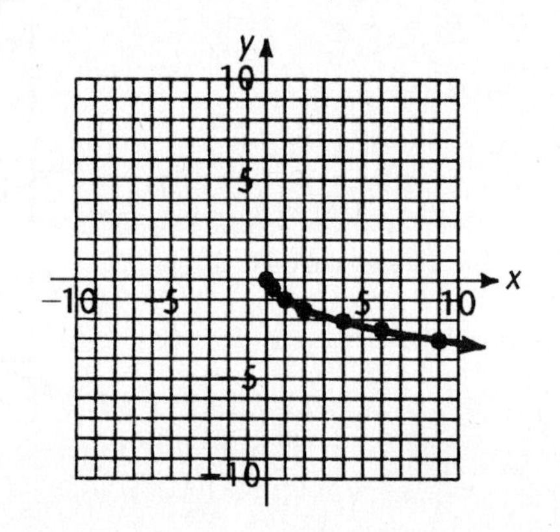

93. $A = 1000 + 60t,\ 0 \le t \le 10$

t	0	5	10
A	1000	1300	1600

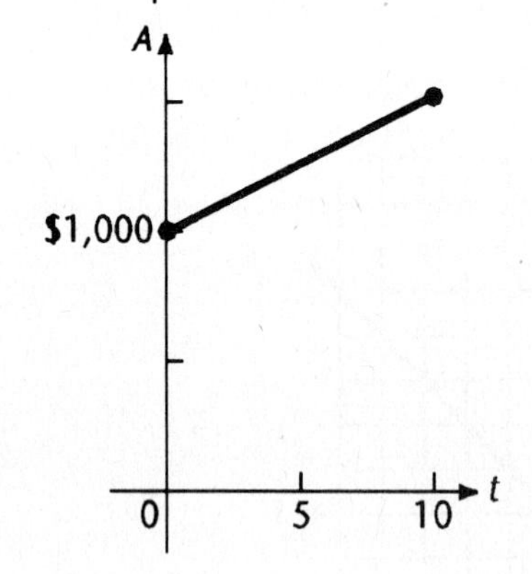

95. $d = 400 - 0.5x,\ 1 \le x \le 50$

x	2	20	50
d	399	390	375

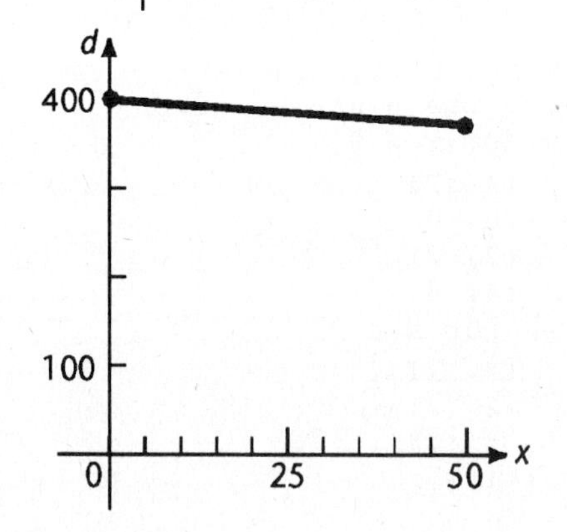

97. $p = -\frac{1}{5}d + 70,\ 30 \le d \le 175$

d	30	100	175
p	64	50	45

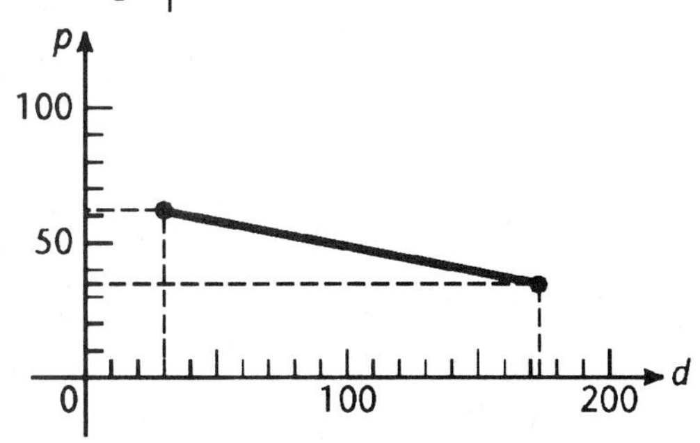

99. $d = 4\left(\frac{h}{500}\right),\ 0 \le h \le 4000$

h	0	2000	4000
d	0	16	32

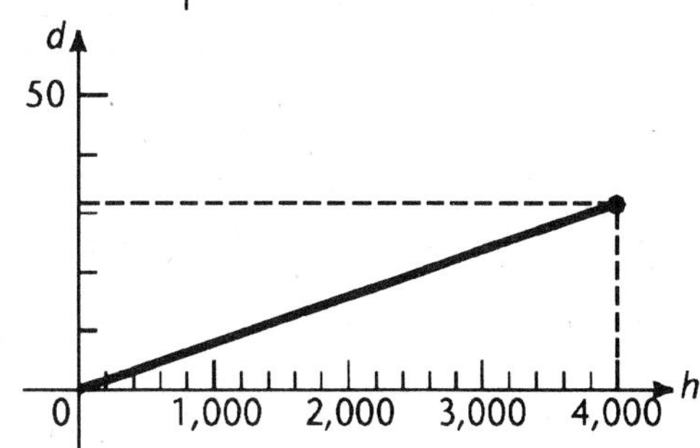

Exercise 5-2 Slope and Equations of a Line

Key Ideas and Formulas

If a line passes through $P_1(x_1, y_1)$ and $P_2(x_2, y_2)$, then its slope is given by the formula

$$m = \frac{y_2 - y_1}{x_2 - x_2} \qquad x_1 \ne x_2$$

$$= \frac{\text{Vertical change(rise)}}{\text{Horizontal change(run)}}$$

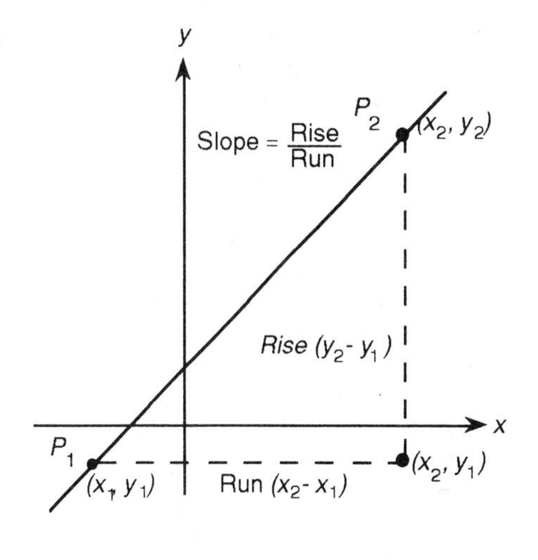

The slope is a measure of the direction and steepness of a line.

Line	Slope	Example
Rising (going from left to right)	Positive	
Falling (going from left to right)	Negative	
Horizontal	Zero	
Vertical	Not defined	

Any equation of the form $Ax + By = C$, $B \neq 0$, can always be written in the form $y = mx + b$. This form is called the **slope-intercept form** of the equation.

$y = mx + b$
m = slope
b = y intercept

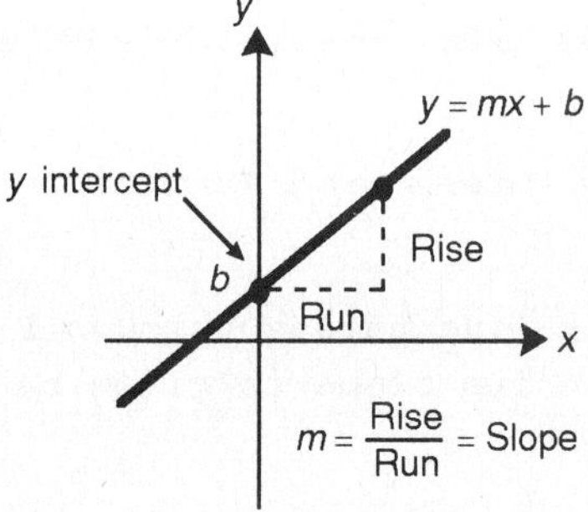

The equation of a line passing through (x_1, y_1) with slope m is given by

$$y - y_1 = m(x - x_1)$$

This form is called the **point-slope form** of the equation.

Parallel and Perpendicular Lines
Given nonvertical lines L_1 and L_2 with slopes m_1 and m_2, respectively, then $L_1 \parallel L_2$ if and only if $m_1 = m_2$ $L_1 \perp L_2$ if and only if $m_1 m_2 = -1$ or $m_2 = -\frac{1}{m_1}$ Note: $\parallel$ means "is parallel to" and $\perp$ means "is perpendicular to."

1. $y = 2x - 3$

   ↑ slope, ↑ y-intercept

(x,	y)
(0,	-3)
$\left(\frac{3}{2},\right.$	$\left.0\right)$
(4,	5)

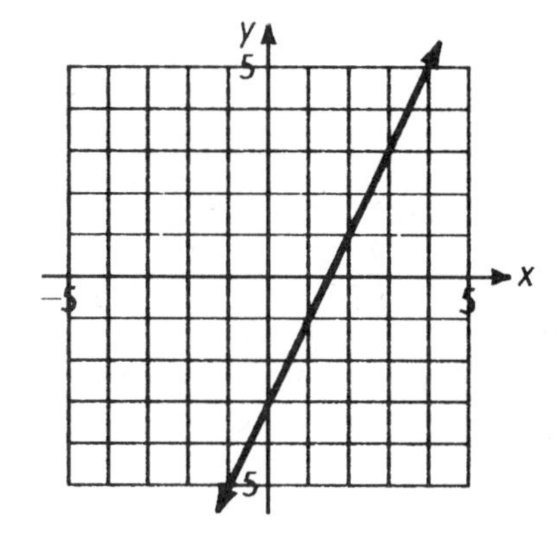

3. $y = -x + 2$

   or

   $y = -1x + 2$

   ↑ slope, ↑ y-intercept

(x,	y)
(0,	2)
(2,	0)
(4,	-2)

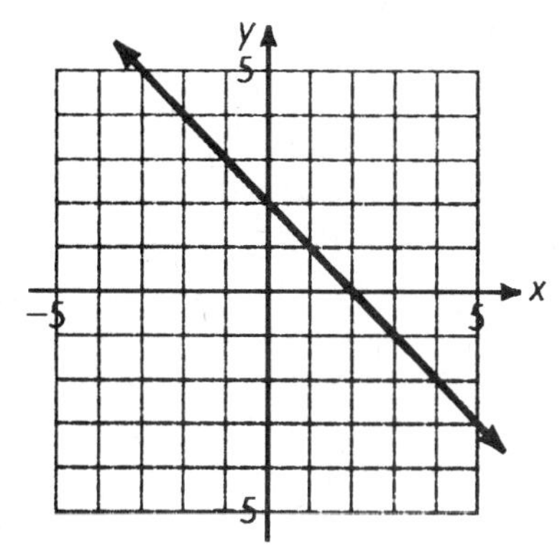

5. Since $m = 5$ and $b = -2$, $y = mx + b = 5x - 2$ is the equation.

7. Since $m = -2$ and $b = 4$, $y = mx + b = -2x + 4$ is the equation.

9. $y - y_1 = m(x - x_1)$ $\quad y_1 = 4,\ m = 2,\ x_1 = 5$

   $y - 4 = 2(x - 5)$

11. $y - y_1 = m(x - x_1)$ $\quad y_1 = 1,\ m = -2,\ x_1 = 2$

    $y - 1 = -2(x - 2)$

13. $y - y_1 = m(x - x_1)$ $\quad y_1 = -1,\ m = 1,\ x_1 = -1$

    $y - (-1) = 1[x - (-1)]$ or $y + 1 = x + 1$ or $y = x$

15. $y - y_1 = m(x - x_1)$ $\quad y_1 = -1,\ m = 2,\ x_1 = -2$

    $y - (-1) = 2[x - (-2)]$ or $y + 1 = 2[x + 2]$ or $y + 1 = 2x + 4$ or $y = 2x + 3$

17. Let $(x_1, y_1) = (3, 2)$ and $(x_2, y_2) = (5, 6)$; then

$$m = \frac{y_2 - y_1}{x_2 - x_1} = \frac{6 - 2}{5 - 3} = \frac{4}{2} = 2$$

19. Let $(x_1, y_1) = (2, 1)$ and $(x_2, y_2) = (10, 5)$; then

$$m = \frac{y_2 - y_1}{x_2 - x_1} = \frac{5 - 1}{10 - 2} = \frac{4}{8} = \frac{1}{2}$$

21. Let $(x_1, y_1) = (3, 6)$ and $(x_2, y_2) = (5, 2)$; then

$$m = \frac{y_2 - y_1}{x_2 - x_1} = \frac{2 - 6}{5 - 3} = \frac{-4}{2} = -2$$

23. Let $(x_1, y_1) = (2, 5)$ and $(x_2, y_2) = (10, 1)$; then

$$m = \frac{y_2 - y_1}{x_2 - x_1} = \frac{1 - 5}{10 - 2} = \frac{-4}{8} = -\frac{1}{2}$$

25. We have found the slope of the line to be 2 in problem 17. Now we use the point-slope form with m = 2. We can either use

$(x_1, y_1) = (3, 2)$	or	$(x_1, y_1) = (5, 6)$
$y - y_1 = m(x - x_1)$		$y - y_1 = m(x - x_1)$
$y - 2 = 2(x - 3)$		$y - 6 = 2(x - 5)$

27. We have found the slope of the line to be $\frac{1}{2}$ in problem 19. Now we use the point-slope form with $m = \frac{1}{2}$. We can either use

$(x_1, y_1) = (2, 1)$	or	$(x_1, y_1) = (10, 5)$
$y - y_1 = m(x - x_1)$		$y - y_1 = m(x - x_1)$
$y - 1 = \frac{1}{2}(x - 2)$		$y - 5 = \frac{1}{2}(x - 10)$

29. We have found the slope of the line to be -2 in problem 21. Now we use the point-slope form with m = -2. We can either use

$(x_1, y_1) = (3, 6)$	or	$(x_1, y_1) = (5, 2)$
$y - y_1 = m(x - x_1)$		$y - y_1 = m(x - x_1)$
$y - 6 = -2(x - 3)$		$y - 2 = -2(x - 5)$
$y - 6 = -2x + 6$		$y - 2 = -2x + 10$
$y = -2x + 12$		$y = -2x + 12$

31. We have found the slope of the line to be $-\frac{1}{2}$ in problem 23. Now we use the point-slope form with $m = -\frac{1}{2}$. We can either use

$(x_1, y_1) = (2, 5)$	or	$(x_1, y_1) = (10, 1)$
$y - y_1 = m(x - x_1)$		$y - y_1 = m(x - x_1)$
$y - 5 = -\frac{1}{2}(x - 2)$		$y - 1 = -\frac{1}{2}(x - 10)$
$y - 5 = -\frac{1}{2}x + 1$		$y - 1 = -\frac{1}{2}x + 5$
$y = -\frac{1}{2}x + 6$		$y = -\frac{1}{2}x + 6$

33. $y = -\frac{x}{3} + 2$

$y = -\frac{1}{3}x + 2$

↑ slope ↑ y-intercept

(x, y)
(0, 2)
(6, 0)
(-3, 3)

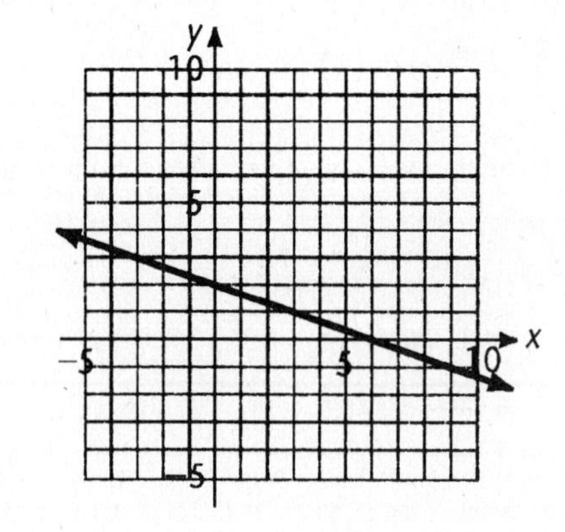

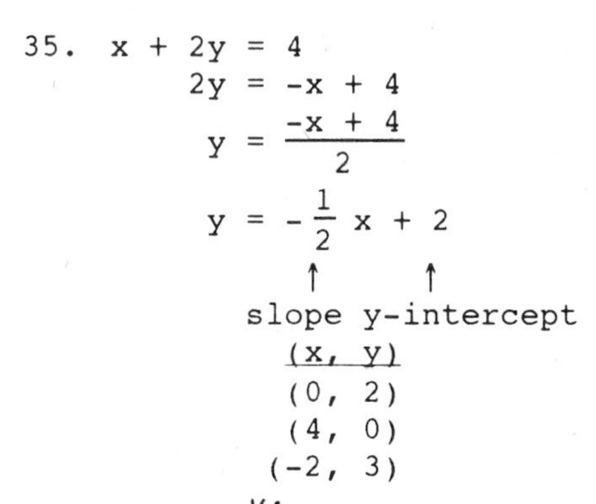

35. $x + 2y = 4$

$$2y = -x + 4$$
$$y = \frac{-x + 4}{2}$$
$$y = -\frac{1}{2}x + 2$$

↑ slope ↑ y-intercept

(x, y)
(0, 2)
(4, 0)
(-2, 3)

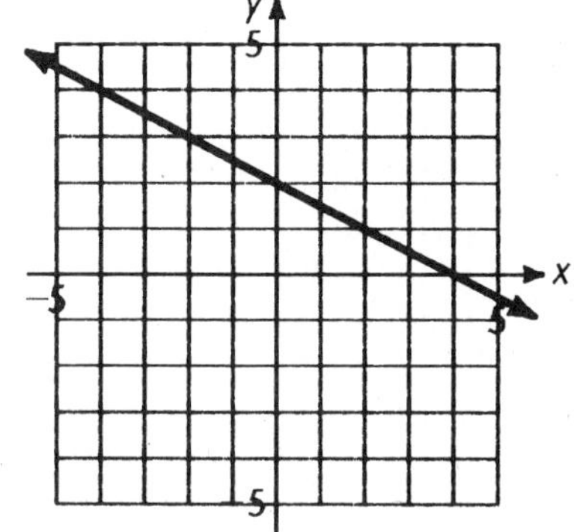

37. $2x + 3y = 4$

$$3y = -2x + 6$$
$$y = \frac{-2x + 6}{3}$$
$$y = -\frac{2}{3}x + 2$$

↑ slope ↑ y-intercept

(x, y)
(0, 2)
(3, 0)
(6, -2)

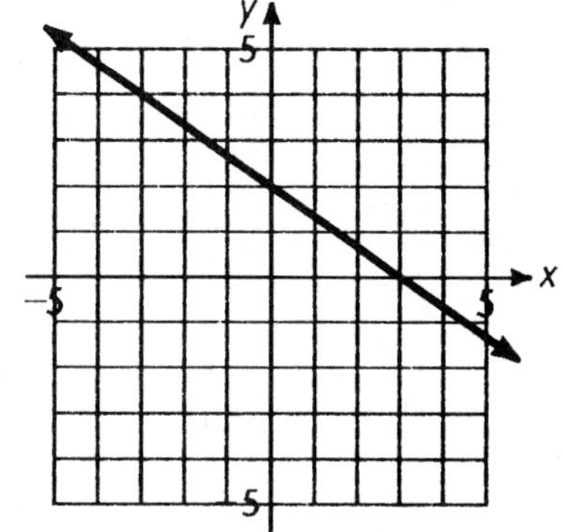

39. $y = \frac{2}{5}x - 3$

↑ slope ↑ y-intercept

(x, y)
(0, -3)
(5, -1)
(-5, -5)

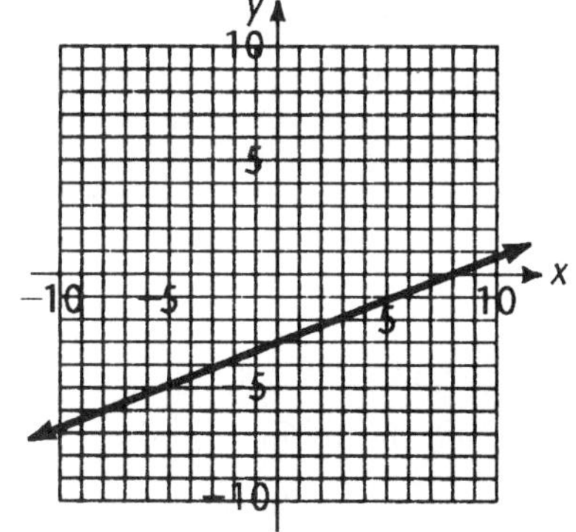

41. $y = \frac{1}{6}x + 2$

↑ slope ↑ y-intercept

(x, y)
(0, 2)
(6, 3)
(-6, 1)

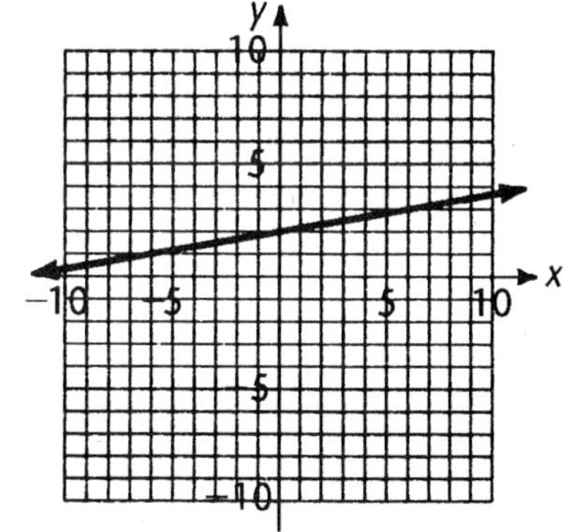

43. $2x + 3y = 5$

$$3y = -2x + 5$$

$$y = \frac{-2x + 5}{3}$$

$$y = -\frac{2}{3}x + \frac{5}{3}$$

$\uparrow$ slope $\quad \uparrow$ y-intercept

(x, y)
$\left(0, \frac{5}{3}\right)$
(1, 1)
(-2, 3)

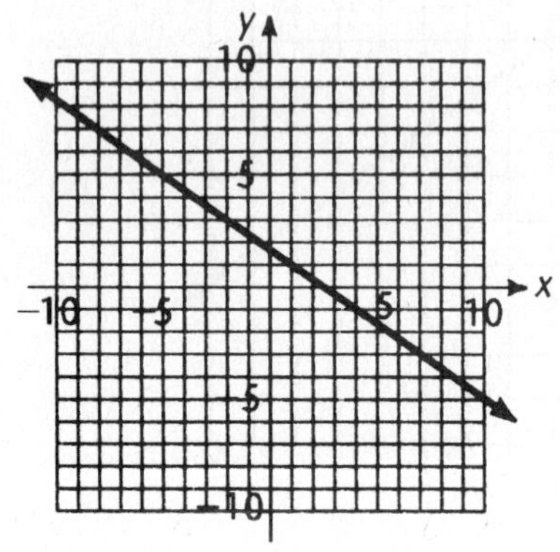

45. $x - 2y = 3$

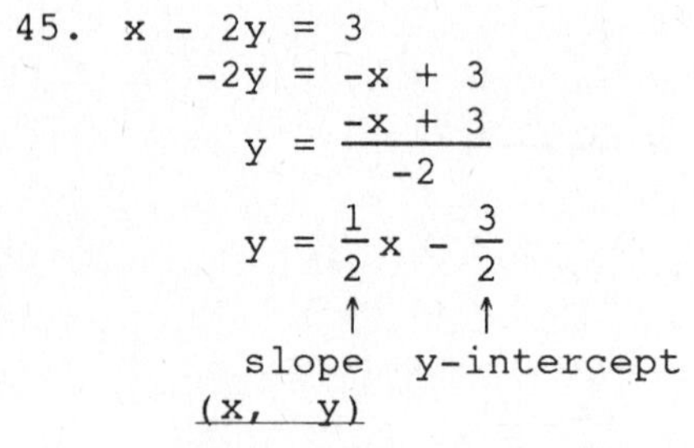

$$-2y = -x + 3$$

$$y = \frac{-x + 3}{-2}$$

$$y = \frac{1}{2}x - \frac{3}{2}$$

$\uparrow$ slope $\quad \uparrow$ y-intercept

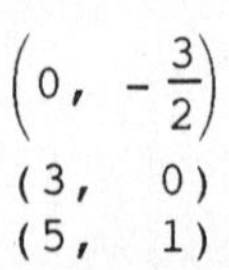

(x, y)
$\left(0, -\frac{3}{2}\right)$
(3, 0)
(5, 1)

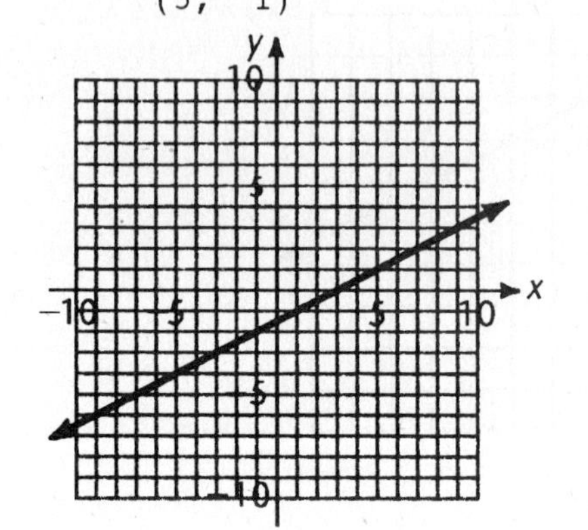

47. $\frac{1}{2}x - \frac{2}{3}y = 1$

$$6 \cdot \frac{1}{2}x - 6 \cdot \frac{2}{3}y = 6$$

$$3x - 4y = 6$$

$$-4y = -3x + 6$$

$$y = \frac{-3x + 6}{-4}$$

$$y = \frac{3}{4}x - \frac{3}{2}$$

$\uparrow$ slope $\quad \uparrow$ y-intercept

(x, y)
$\left(0, -\frac{3}{2}\right)$
(2, 0)
(6, 3)

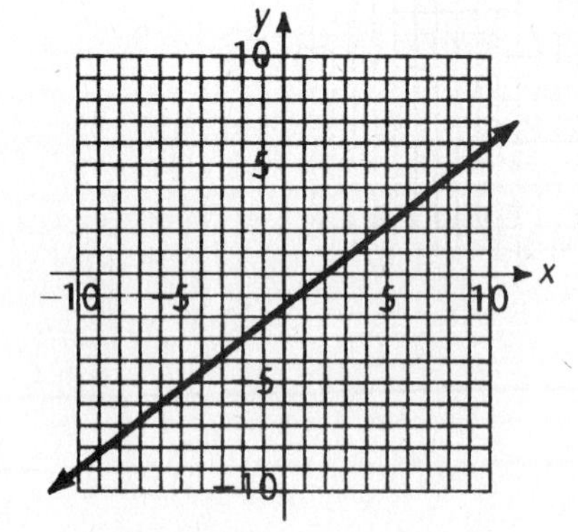

49. $\frac{1}{2}x + \frac{1}{3}y = \frac{1}{4}$

$$\frac{1}{3}y = -\frac{1}{2}x + \frac{1}{4}$$

$$3 \cdot \frac{1}{3}y = 3\left(-\frac{1}{2}x + \frac{1}{4}\right)$$

$$y = -\frac{3}{2}x + \frac{3}{4}$$

$\uparrow$ slope $\quad \uparrow$ y-intercept

(x, y)
$\left(0, \frac{3}{4}\right)$
$\left(\frac{1}{2}, 0\right)$
$\left(-1, \frac{9}{4}\right)$

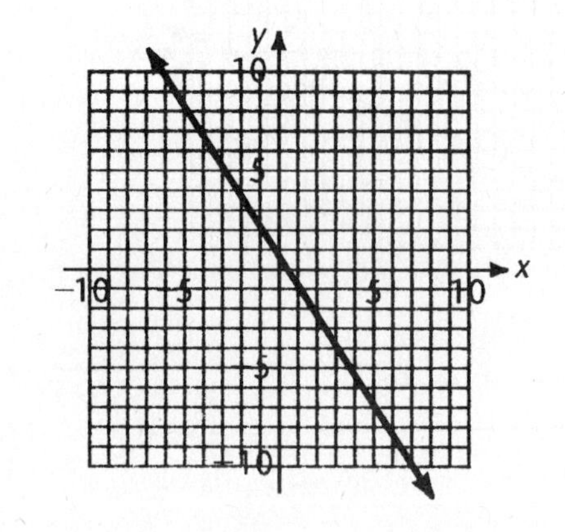

51. Since $m = -\frac{1}{2}$ and $b = -2$, $y = mx + b = -\frac{1}{2}x - 2$ or $y = -\frac{x}{2} - 2$ is the equation.

53. Since $m = \frac{2}{3}$ and $b = \frac{3}{2}$, $y = mx + b = \frac{2}{3}x + \frac{3}{2}$ is the equation.

55.
$$\begin{aligned} y - y_1 &= m(x - x_1) \\ y - 2 &= -2[x - (-3)] \\ y - 2 &= -2(x + 3) \\ y - 2 &= -2x - 6 \\ y &= -2x - 4 \end{aligned}$$

57.
$$\begin{aligned} y - y_1 &= m(x - x_1) \\ y - 3 &= \frac{1}{2}[x - (-4)] \\ y - 3 &= \frac{1}{2}(x + 4) \\ y - 3 &= \frac{x}{2} + 2 \\ y &= \frac{x}{2} + 5 \end{aligned}$$

59.
$$\begin{aligned} y - y_1 &= m(x - x_1) \\ y - 3 &= -3[x - (-1)] \\ y - 3 &= -3(x + 1) \\ y - 3 &= -3x - 3 \\ y &= -3x \end{aligned}$$

61.
$$\begin{aligned} y - y_1 &= m(x - x_1) \\ y - 2 &= -2[x - (-2)] \\ y - 2 &= -2(x + 2) \\ y - 2 &= -2x - 4 \\ y &= -2x - 2 \end{aligned}$$

63.
$$\begin{aligned} y - y_1 &= m(x - x_1) \\ y - (-1) &= \frac{1}{3}[x - (-1)] \\ y + 1 &= \frac{1}{3}(x + 1) \\ y + 1 &= \frac{1}{3}x + \frac{1}{3} \\ y &= \frac{1}{3}x - \frac{2}{3} \end{aligned}$$

65.
$$\begin{aligned} y - y_1 &= m(x - x_1) \\ y - 6 &= -\frac{1}{2}[x - (-3)] \\ y - 6 &= -\frac{1}{2}(x + 3) \\ y - 6 &= -\frac{1}{2}x - \frac{3}{2} \\ y &= -\frac{1}{2}x + \frac{9}{2} \end{aligned}$$

67.
$$\begin{aligned} y - y_1 &= m(x - x_1) \\ y - \frac{2}{3} &= \frac{1}{2}\left(x - \frac{1}{3}\right) \\ y - \frac{2}{3} &= \frac{1}{2}x - \frac{1}{6} \\ y &= \frac{1}{2}x + \frac{1}{2} \end{aligned}$$

69.
$$\begin{aligned} y - y_1 &= m(x - x_1) \\ y - \frac{3}{2} &= -\frac{3}{2}\left(x - \frac{3}{4}\right) \\ y - \frac{3}{2} &= -\frac{3}{2}x + \frac{9}{8} \\ y &= -\frac{3}{2}x + \frac{21}{8} \end{aligned}$$

71. Let $(x_1, y_1) = (3, 7)$ and $(x_2, y_2) = (-6, 4)$; then

$$m = \frac{y_2 - y_1}{x_2 - x_1} = \frac{4 - 7}{-6 - 3} = \frac{-3}{-9} = \frac{1}{3}$$

73. Let $(x_1, y_1) = (4, -2)$ and $(x_2, y_2) = (-4, 0)$; then

$$m = \frac{y_2 - y_1}{x_2 - x_1} = \frac{0 - (-2)}{-4 - 4} = \frac{2}{-8} = -\frac{1}{4}$$

75. Let $(x_1, y_1) = \left(\frac{1}{2}, \frac{2}{3}\right)$ and $(x_2, y_2) = \left(\frac{3}{4}, \frac{1}{3}\right)$; then

$$m = \frac{y_2 - y_1}{x_2 - x_1} = \frac{\frac{1}{3} - \frac{2}{3}}{\frac{3}{4} - \frac{1}{2}} = \frac{12 \cdot \frac{1}{3} - 12 \cdot \frac{2}{3}}{12 \cdot \frac{3}{4} - 12 \cdot \frac{1}{2}} = \frac{4 - 8}{9 - 6} = -\frac{4}{3}$$

77. Let $(x_1, y_1) = \left(\frac{1}{2}, -\frac{2}{3}\right)$ and $(x_2, y_2) = \left(-\frac{3}{2}, \frac{1}{6}\right)$; then

$$m = \frac{y_2 - y_1}{x_2 - x_1} = \frac{\frac{1}{6} - \left(-\frac{2}{3}\right)}{-\frac{3}{2} - \frac{1}{2}} = \frac{\frac{1}{6} + \frac{2}{3}}{-\frac{3}{2} - \frac{1}{2}} = \frac{6 \cdot \frac{1}{6} + 6 \cdot \frac{2}{3}}{6\left(-\frac{3}{2}\right) - 6\left(\frac{1}{2}\right)} = \frac{1 + 4}{-9 - 3} = -\frac{5}{12}$$

79. We have found the slope of the line to be $\frac{1}{3}$ in problem 71. Now we use the point-slope form with $m = \frac{1}{3}$. We can either use

$$\begin{aligned} (x_1, y_1) &= (3, 7) \\ y - y_1 &= m(x - x_1) \\ y - 7 &= \frac{1}{3}(x - 3) \\ y - 7 &= \frac{1}{3}x - 1 \\ y &= \frac{x}{3} + 6 \end{aligned}$$

or

$$\begin{aligned} (x_1, y_1) &= (-6, 4) \\ y - y_1 &= m(x - x_1) \\ y - 4 &= \frac{1}{3}(x + 6) \\ y - 4 &= \frac{1}{3}x + 2 \\ y &= \frac{x}{3} + 6 \end{aligned}$$

81. We have found the slope of the line to be $-\frac{1}{4}$ in problem 73. Now we use the point-slope form with $m = -\frac{1}{4}$. We can either use

$$\begin{aligned} (x_1, y_1) &= (4, -2) \\ y - y_1 &= m(x - x_1) \\ y + 2 &= -\frac{1}{4}(x - 4) \\ y + 2 &= -\frac{x}{4} + 1 \\ y &= -\frac{x}{4} - 1 \end{aligned}$$

or

$$\begin{aligned} (x_1, y_1) &= (-4, 0) \\ y - y_1 &= m(x - x_1) \\ y - 0 &= -\frac{1}{4}(x + 4) \\ y &= -\frac{x}{4} - 1 \end{aligned}$$

83. We have found the slope of the line to be $-\frac{4}{3}$ in problem 75. Now we use the point-slope form with $m = -\frac{4}{3}$. We can either use

$$\begin{aligned} (x_1, y_1) &= \left(\frac{1}{2}, \frac{2}{3}\right) \\ y - y_1 &= m(x - x_1) \\ y - \frac{2}{3} &= -\frac{4}{3}\left(x - \frac{1}{2}\right) \\ y - \frac{2}{3} &= -\frac{4}{3}x + \frac{2}{3} \\ y &= -\frac{4}{3}x + \frac{4}{3} \end{aligned}$$

or

$$\begin{aligned} (x_1, y_1) &= \left(\frac{3}{4}, \frac{1}{3}\right) \\ y - y_1 &= m(x - x_1) \\ y - \frac{1}{3} &= -\frac{4}{3}\left(x - \frac{3}{4}\right) \\ y - \frac{1}{3} &= -\frac{4}{3}x + 1 \\ y &= -\frac{4}{3}x + \frac{4}{3} \end{aligned}$$

85. We have found the slope of the line to be $-\frac{5}{12}$ in problem 77. Now we use the point-slope form with $m = -\frac{5}{12}$. We can either use

$$(x_1, y_1) = \left(\frac{1}{2}, -\frac{2}{3}\right) \quad \text{or} \quad (x_1, y_1) = \left(-\frac{3}{2}, \frac{1}{6}\right)$$

$$y - y_1 = m(x - x_1) \qquad y - y_1 = m(x - x_1)$$

$$y - \left(-\frac{2}{3}\right) = -\frac{5}{12}\left(x - \frac{1}{2}\right) \qquad y - \frac{1}{6} = -\frac{5}{12}\left[x - \left(-\frac{3}{2}\right)\right]$$

$$y + \frac{2}{3} = -\frac{5}{12}\left(x - \frac{1}{2}\right) \qquad y - \frac{1}{6} = -\frac{5}{12}\left(x + \frac{3}{2}\right)$$

$$y + \frac{2}{3} = -\frac{5}{12}x + \frac{5}{24} \qquad y - \frac{1}{6} = -\frac{5}{12}x - \frac{5}{8}$$

$$y = -\frac{5}{12}x - \frac{11}{24} \qquad y = -\frac{5}{12}x - \frac{11}{24}$$

87. vertical: $x = -3$ horizontal: $y = 5$

89. vertical: $x = -1$ horizontal: $y = 22$

91. The slope of $y = 3x + 4$ is 3. The slope of $x - 3y = 5$, that is, $y = \frac{1}{3}x - \frac{5}{3}$ is $\frac{1}{3}$. Since $3 \neq \frac{1}{3}$ and $3 \cdot \frac{1}{3} \neq -1$, the lines are neither parallel nor perpendicular.

93. The slope of $4x - 5y = 6$, that is, $y = \frac{4}{5}x - \frac{6}{5}$ is $\frac{4}{5}$. The slope of $y = -\frac{5}{4}x + \frac{3}{2}$ is $-\frac{5}{4}$. Since $\left(-\frac{5}{4}\right)\left(\frac{4}{5}\right) = -1$, the lines are perpendicular.

95. The slope of the given line is $\frac{3}{5}$.

(A) The slope of a line parallel to the given line is also $\frac{3}{5}$. We have to find the equation of a line through (1, 4) with slope $\frac{3}{5}$.

$$y - y_1 = m(x - x_1) \qquad m = \frac{3}{5} \text{ and } (x_1, y_1) = (1, 4)$$

$$y - 4 = \frac{3}{5}(x - 1)$$

$$y - 4 = \frac{3}{5}x - \frac{3}{5}$$

$$y = \frac{3}{5}x - \frac{3}{5} + 4$$

$$y = \frac{3}{5}x + \frac{17}{5}$$

(B) The slope of a line perpendicular to the given line is the negative reciprocal of $\frac{3}{5}$, that is, $-\frac{5}{3}$. We have to find the equation of a line through (1, 4) with slope $-\frac{5}{3}$.

$$y - y_1 = m(x - x_1) \qquad m = -\frac{5}{3} \text{ and } (x_1, y_1) = (1, 4)$$
$$y - 4 = -\frac{5}{3}(x - 1)$$
$$y - 4 = -\frac{5}{3}x + \frac{5}{3}$$
$$y = -\frac{5}{3}x + \frac{5}{3} + 4$$
$$y = -\frac{5}{3}x + \frac{17}{3}$$

COMMON ERROR: Using the slope of the given line. The slope of the perpendicular line must be the negative reciprocal of the slope of the given line.

97. The slope of the given line is -3
   (A) The slope of a line parallel to the given line is also -3. We have to find the equation of a line through (0, 2) with slope -3.
$$y - y_1 = m(x - x_1) \qquad m = -3 \text{ and } (x_1, y_1) = (0, 2)$$
$$y - 2 = -3(x - 0)$$
$$y - 2 = -3x$$
$$y = -3x + 2$$
   (B) The slope of a line perpendicular to the given line is the negative reciprocal of -3, that is, $\frac{1}{3}$. We have to find the equation of a line through (0, 2) with slope $\frac{1}{3}$.
$$y - y_1 = m(x - x_1) \qquad m = \frac{1}{3} \text{ and } (x_1, y_1) = (0, 2)$$
$$y - 2 = \frac{1}{3}(x - 0)$$
$$y - 2 = \frac{1}{3}x$$
$$y = \frac{1}{3}x + 2$$

99. First find the slope of the given line by writing $x + y = 3$ in the form $y = mx + b$.
$$x + y = 3$$
$$y = -x + 3$$
$$y = -1x + 3$$
   The slope of the given line is -1.
   (A) The slope of a line parallel to the given line is also -1. We have to find the equation of a line through (1, 1) with slope -1.
$$y - y_1 = m(x - x_1) \qquad m = -1 \text{ and } (x_1, y_1) = (1, 1)$$
$$y - 1 = -1(x - 1)$$
$$y - 1 = -x + 1$$
$$y = -x + 2$$
   (B) The slope of a line perpendicular to the given line is the negative reciprocal of -1, that is, 1. We have to find the equation of a line through (1, 1) with slope 1.
$$y - y_1 = m(x - x_1) \qquad m = 1 \text{ and } (x_1, y_1) = (1, 1)$$
$$y - 1 = 1(x - 1)$$
$$y - 1 = x - 1$$
$$y = x$$

101. The given line is horizontal.
   (A) A line parallel to the given line is also horizontal. We have to find the equation of a horizontal line through (-2, 5). This equation is $y = 5$.

(B) A line perpendicular to the given line is vertical. We have to find the equation of a vertical line through (-2, 5). This equation is $x = -2$.

103. (A) The assumption that the mark-up policy of the store is linear asks us to find a linear equation, that is, one of the form $R = mC + b$, that relates R and C. We are given two points on the graph of this equation, namely $(C_1, R_1) = (20, 33)$ and $(C_2, R_2) = (60, 93)$. To find the equation, we first find the slope of the line through these two points, using a slope formula.

$$m = \frac{R_2 - R_1}{C_2 - C_1} = \frac{93 - 33}{60 - 20} = \frac{60}{40} = \frac{3}{2}$$

Now we use the point-slope form with $m = \frac{3}{2}$. We can either use

$$\begin{aligned} (C_1, R_1) &= (20, 33) \\ R - R_1 &= m(C - C_1) \\ R - 33 &= \frac{3}{2}(C - 20) \\ R - 33 &= \frac{3}{2}C - 30 \\ R &= \frac{3}{2}C + 3 \end{aligned} \quad \text{or} \quad \begin{aligned} (C_2, R_2) &= (60, 93) \\ R - R_1 &= m(C - C_1) \\ R - 93 &= \frac{3}{2}(C - 60) \\ R - 93 &= \frac{3}{2}C - 90 \\ R &= \frac{3}{2}C + 3 \end{aligned}$$

(B) Let us find R for three values in the interval $10 \le C \le 300$.

C	10	60	300
R	18	93	453

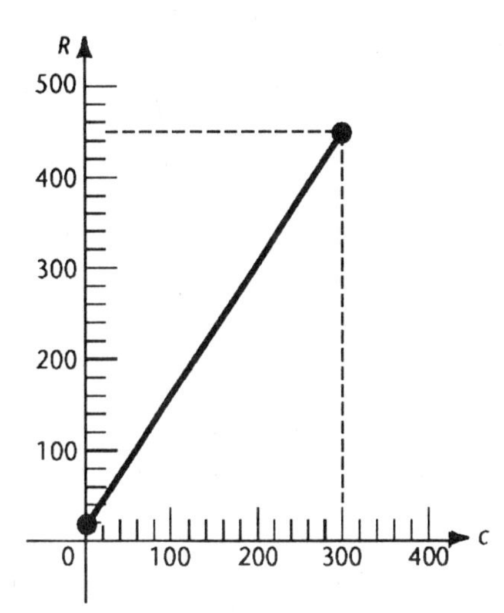

(C) We are given $R = 240$. We substitute this value into our equation

$$R = \frac{3}{2}C + 3$$

and solve for C.

$$\begin{aligned} 240 &= \frac{3}{2}C + 3 \\ 480 &= 3C + 6 \\ 474 &= 3C \\ 158 &= C \\ \text{cost} &= \$158 \end{aligned}$$

105. We are asked to find a linear equation, that is, one of the form $F = mC + b$, that relates F and C. We are given two points on the graph of this equation, namely $(C_1, F_1) = (0, 32)$ and $(C_2, F_2) = (100, 212)$. To find the equation, we first find the slope of the line through these two points, using a slope formula.

$$m = \frac{F_2 - F_1}{C_2 - C_1} = \frac{212 - 32}{100 - 0} = \frac{180}{100} = \frac{9}{5}$$

Now we use the point-slope form with $m = \frac{9}{5}$. We can either use

$(C_1, F_1) = (0, 32)$ or $(C_2, F_2) = (100, 212)$

$F - F_1 = m(C - C_1)$	$F - F_1 = m(C - C_1)$
$F - 32 = \frac{9}{5}(C - 0)$	$F - 212 = \frac{9}{5}C - 100$
$F - 32 = \frac{9}{5}C$	$F - 212 = \frac{9}{5}C - 180$
$F = \frac{9}{5}C + 32$	$F = \frac{9}{5}C + 32$

Alternatively, we can write C in terms of F.

Starting from $F = \frac{9}{5}C + 32$, we can solve for C to obtain:

$$\frac{5}{9} \cdot F = \frac{5}{9} \cdot \frac{9}{5}C + \frac{5}{9} \cdot 32$$

$$\frac{5}{9}F = C + \frac{160}{9}$$

$$C = \frac{5}{9}F - \frac{160}{9}$$

107. (A) We are asked to find a linear equation $V = mt + b$, given two points on the graph of this equation. Since $V = 20,000$ when $t = 0$ and $V = 2,000$ when $t = 10$, the two points are $(0, 20,000)$ and $(10, 2,000)$. We first find the slope of the line through these two points, using a slope formula.

$$m = \frac{V_2 - V_1}{t_2 - t_1} = \frac{20,000 - 2,000}{0 - 10} = \frac{18,000}{-10} = -1,800$$

Now we use the point-slope form with $m = -1,800$. We can either use

$(t_1, V_1) = (10, 2,000)$ or $(t_1, V_1) = (0, 20,000)$

$V - V_1 = m(t - t_1)$	$V - V_1 = m(t - t_1)$
$V - 2,000 = -1,800(t - 10)$	$V - 20,000 = -1,800(t - 0)$
$V - 2,000 = -1,800t + 18,000$	$V - 20,000 = -1,800t$
$V = -1,800t + 20,000$	$V = 1,800t + 20,000$

(B) Using our equation $V = -1,800t + 20,000$, we substitute $t = 4$, then $t = 8$, to find the corresponding V's.

value after 4 years:

$$V = -1,800(4) + 20,000 = -7,200 + 20,000 = \$12,800$$

value after 8 years:

$$V = -1,800(8) + 20,000 = -14,400 + 20,000 = \$5,600$$

(C) We have already found this to be $-1,800$ in part (A).

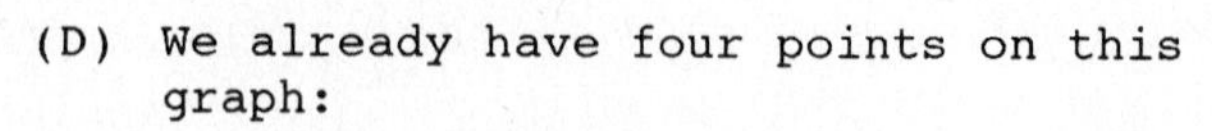

(D) We already have four points on this graph:

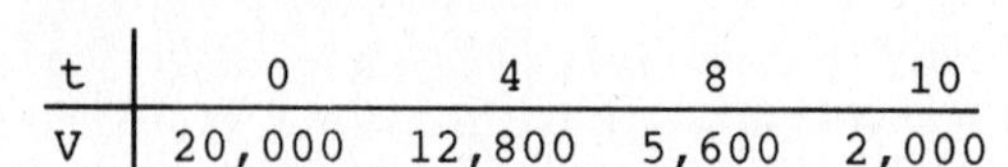

t	0	4	8	10
V	20,000	12,800	5,600	2,000

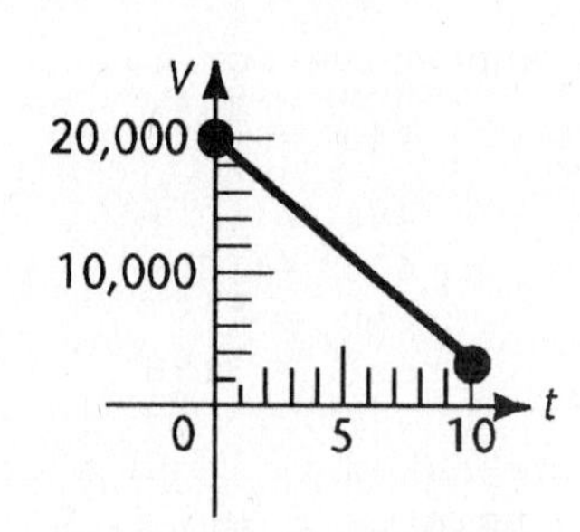

Exercise 5-3 Solving Inequalities

Key Ideas and Formulas

Interval Notation	Inequality Notation	Line Graph
$[a, b]$	$a \le x \le b$	
$[a, b)$	$a \le x < b$	
$(a, b]$	$a < x \le b$	
(a, b)	$a < x < b$	
$[b, \infty)$	$x \ge b$	
(b, ∞)	$x > b$	
$(-\infty, a]$	$x \le a$	
$(-\infty, a)$	$x < a$	

Inequality Properties

For a, b, and c any real numbers:

1. If $a < b$, then $a + c < b + c$ — Addition Property
2. If $a < b$, then $a - c < b - c$ — Subtraction Property
3. If $a < b$ and c is **positive**, then $ca < cb$ — Multiplication Property
4. If $a < b$ and c is **negative**, then $ca > cb$ — Multiplication Property
5. If $a < b$ and c is **positive**, then $\frac{a}{c} < \frac{b}{c}$ — Division Property
6. If $a < b$ and c is **negative**, then $\frac{a}{c} > \frac{b}{c}$ — Division Property

Note particularty in (4) and (6): If we multiply or divide both sides of an inequality statement by a **negative** number, the sense of the inequality **reverses.**

Similar properties hold if each inequality sign is reversed or if < is replaced with ≤ and > is replaced with ≥.

1. $-8 \le x \le 7$

3. $-6 \le x < 6$

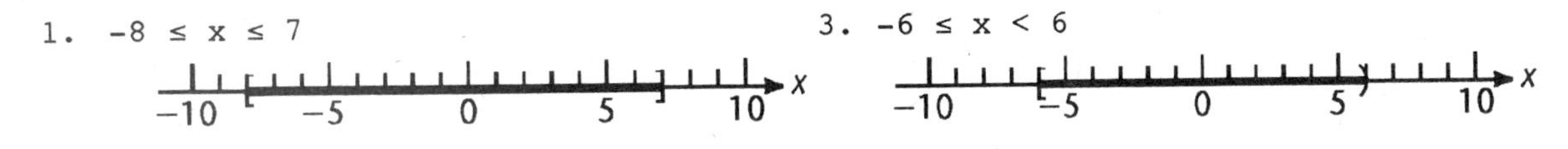

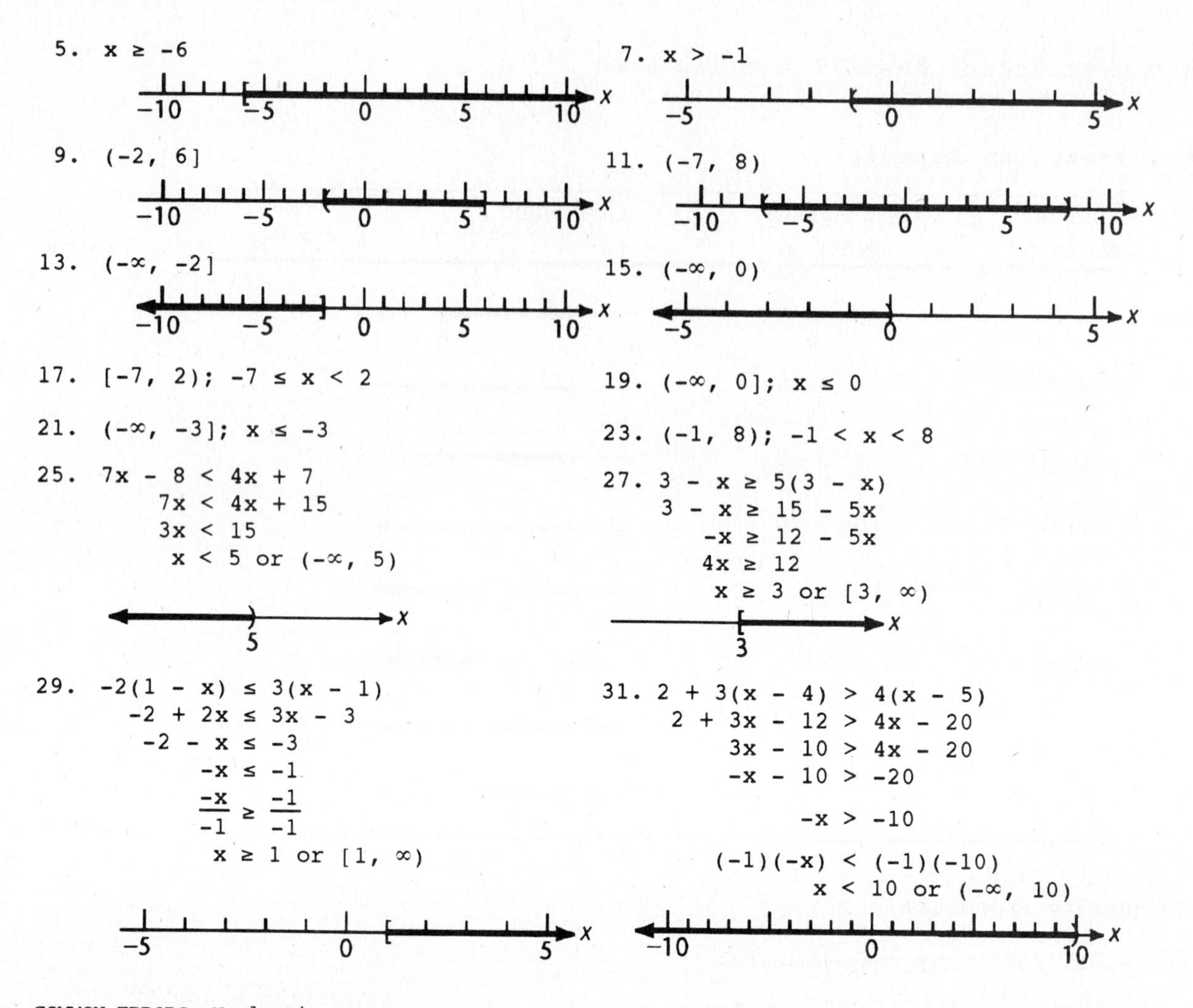

5. $x \ge -6$

7. $x > -1$

9. $(-2, 6]$

11. $(-7, 8)$

13. $(-\infty, -2]$

15. $(-\infty, 0)$

17. $[-7, 2)$; $-7 \le x < 2$

19. $(-\infty, 0]$; $x \le 0$

21. $(-\infty, -3]$; $x \le -3$

23. $(-1, 8)$; $-1 < x < 8$

25.
$$7x - 8 < 4x + 7$$
$$7x < 4x + 15$$
$$3x < 15$$
$$x < 5 \text{ or } (-\infty, 5)$$

27.
$$3 - x \ge 5(3 - x)$$
$$3 - x \ge 15 - 5x$$
$$-x \ge 12 - 5x$$
$$4x \ge 12$$
$$x \ge 3 \text{ or } [3, \infty)$$

29.
$$-2(1 - x) \le 3(x - 1)$$
$$-2 + 2x \le 3x - 3$$
$$-2 - x \le -3$$
$$-x \le -1$$
$$\frac{-x}{-1} \ge \frac{-1}{-1}$$
$$x \ge 1 \text{ or } [1, \infty)$$

31.
$$2 + 3(x - 4) > 4(x - 5)$$
$$2 + 3x - 12 > 4x - 20$$
$$3x - 10 > 4x - 20$$
$$-x - 10 > -20$$
$$-x > -10$$
$$(-1)(-x) < (-1)(-10)$$
$$x < 10 \text{ or } (-\infty, 10)$$

COMMON ERRORS: Neglecting to reverse the sense of the inequality when multiplying or dividing both sides by a negative quantity.

33.
$$2(3 - 4x) + 5 \le 6 - 5(x - 4)$$
$$6 - 8x + 5 \le 6 - 5x + 20$$
$$-8x + 11 \le -5x + 26$$
$$-3x + 11 \le 26$$
$$-3x \le 15$$
$$x \ge -5 \text{ or } [-5, \infty)$$

35.
$$3 - [2 - (1 - x)] < 1 - [2 - (3 - x)]$$
$$3 - [2 - 1 + x] < 1 - [2 - 3 + x]$$
$$3 - [1 + x] < 1 - [-1 + x]$$
$$3 - 1 - x < 1 + 1 - x$$
$$2 - x < 2 - x$$
$$2 < 2$$
This is impossible. No solution.

37.
$$\frac{N}{-2} > 4 \qquad \text{lcm} = -2$$
$$-2 \cdot \frac{N}{-2} < -2 \cdot 4$$
$$N < -8 \text{ or } (-\infty, -8)$$

39.
$$-5t < -10$$
$$\frac{-5t}{-5} > \frac{-10}{-5}$$
$$t > 2 \text{ or } (2, \infty)$$

41. $3 - m < 4(m - 3)$
$3 - m < 4m - 12$
$-m < 4m - 15$
$-5m < -15$
$\frac{-5m}{-5} > \frac{-15}{-5}$
$m > 3$ or $(3, \infty)$

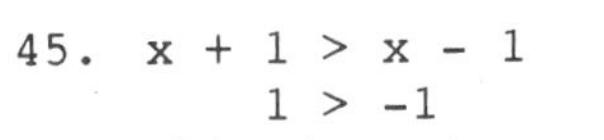

43. $-x < x - 1$
$-2x < -1$
$\frac{-2x}{-2} > \frac{-1}{-2}$
$x > \frac{1}{2}$ or $\left(\frac{1}{2}, \infty\right)$

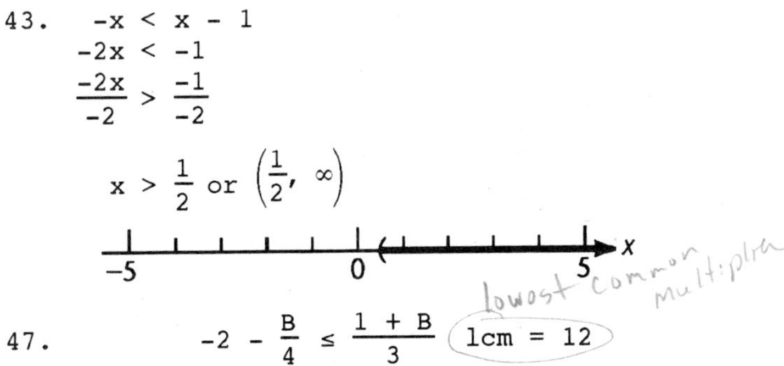

45. $x + 1 > x - 1$
$1 > -1$
This is a true statement. All real numbers are solutions. $(-\infty, \infty)$

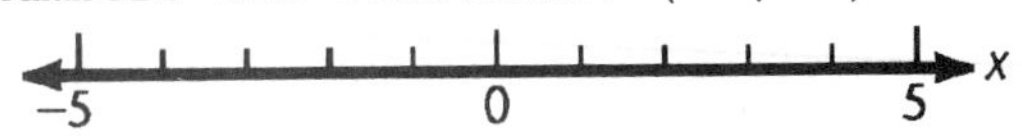

47. $-2 - \frac{B}{4} \le \frac{1 + B}{3}$ lcm = 12
$12(-2) - 12 \cdot \frac{B}{4} \le 12 \cdot \frac{(1 + B)}{3}$
$-24 - 3B \le 4(1 + B)$
$-24 - 3B \le 4 + 4B$
$-3B \le 28 + 4B$
$-7B \le 28$
$\frac{-7B}{-7} \ge \frac{28}{-7}$
$B \ge -4$ or $[-4, \infty)$

49. $-4 < 5t + 6 \le 21$
$-4 - 6 < 5t + 6 - 6 \le 21 - 6$
$-10 < 5t \le 15$
$\frac{-10}{5} < \frac{5t}{5} \le \frac{15}{5}$
$-2 < t \le 3$ or $(-2, 3]$

51. $2 < 3 - x \le 4$
$2 - 3 < 3 - x - 3 \le 4 - 3$
$-1 < -x \le 1$
$\frac{-1}{-1} > \frac{-x}{-1} \ge \frac{1}{-1}$
$1 > x \ge -1$
$-1 \le x < 1$ or $[-1, 1)$

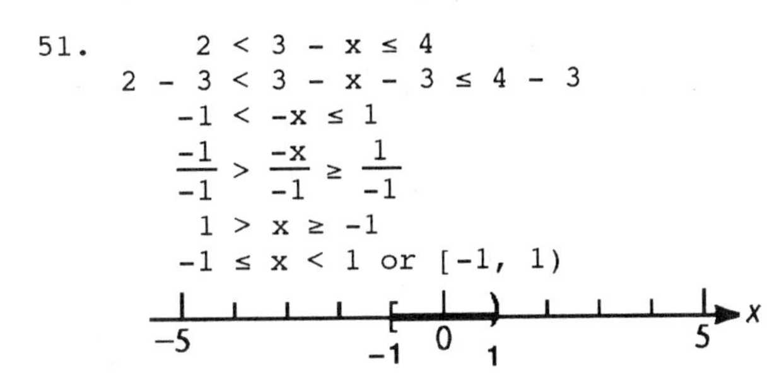

53. $0 \le 3 - 4x \le 5$
$0 - 3 \le 3 - 4x - 3 \le 5 - 3$
$-3 \le -4x \le 2$
$\frac{-3}{-4} \ge \frac{-4x}{-4} \ge \frac{2}{-4}$
$\frac{3}{4} \ge x \ge -\frac{1}{2}$
$-\frac{1}{2} \le x \le \frac{3}{4}$ or $\left[-\frac{1}{2}, \frac{3}{4}\right]$

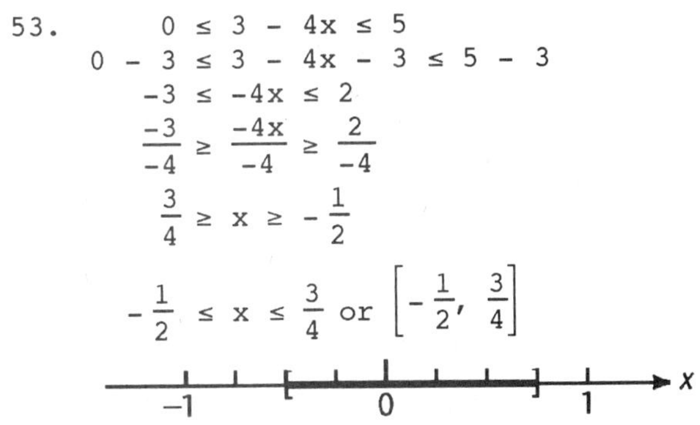

55. $16 < 7 - 3x \le 31$
$9 < -3x \le 24$
$-3 > x \ge \frac{24}{-3}$
$-3 > x \ge -8$
$-8 \le x < -3$ or $[-8, -3)$

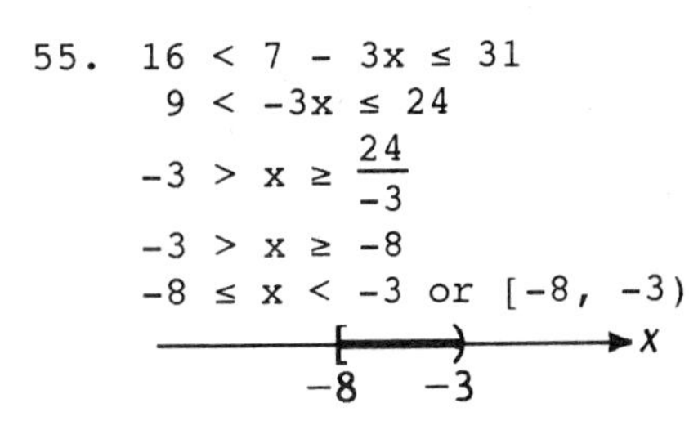

57. Let x = the number. Then translate as follows:
3 [less than] [twice the number] [is greater than or equal to] -6

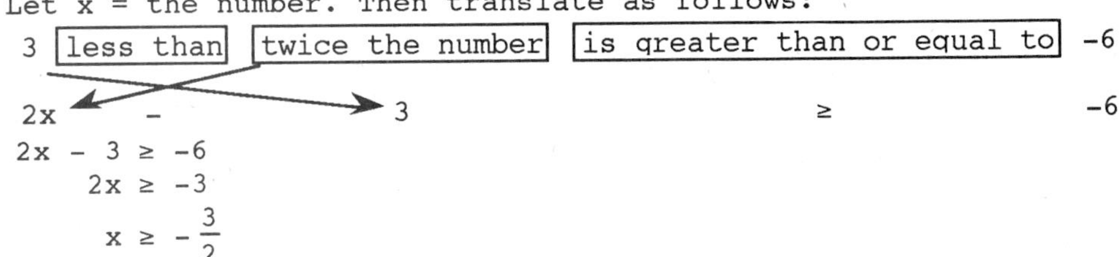

$2x - 3 \ge -6$
$2x \ge -3$
$x \ge -\frac{3}{2}$

59. Let x be the number. Then translate as follows:

15	reduced by	3	times	the number	is less than	6
15	$-$	3	$\cdot$	x	$<$	6

$$15 - 3x < 6$$
$$-3x < -9$$
$$\frac{-3x}{-3} > \frac{-9}{-3}$$
$$x > 3$$

61. Let x be the number. Then 1 more than the number is $x + 1$, and 2 more than the number is $x + 2$. Translate as follows:

The sum	is less than	75
$x + (x + 1) + (x + 2)$	$<$	75

$$x + x + 1 + x + 2 < 75$$
$$3x + 3 < 75$$
$$3x < 72$$
$$x < 24$$

63. Let x be the number. Then translate as follows:

The sum of	$\frac{1}{2}$ a number	and	$\frac{2}{5}$ the number	is at most	5	more than	the number
	$\frac{1}{2}x$	$+$	$\frac{2}{3}x$	$\le$	5	$+$	x

$$\frac{1}{2}x + \frac{2}{3}x \le 5 + x \quad \text{lcm} = 6$$
$$6 \cdot \frac{1}{2}x + 6 \cdot \frac{2}{3}x \le 6(5 + x)$$
$$3x + 4x \le 30 + 6x$$
$$7x \le 30 + 6x$$
$$x \le 30$$

65. Let x be the number. Then translate as follows:

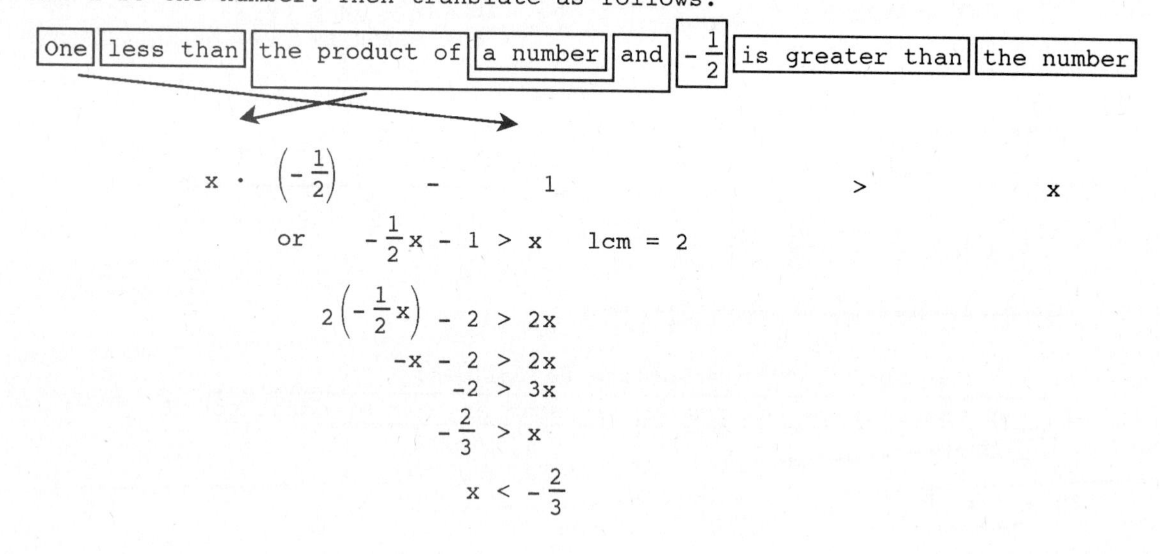

$$x \cdot \left(-\frac{1}{2}\right) - 1 > x$$
$$\text{or} \quad -\frac{1}{2}x - 1 > x \quad \text{lcm} = 2$$
$$2\left(-\frac{1}{2}x\right) - 2 > 2x$$
$$-x - 2 > 2x$$
$$-2 > 3x$$
$$-\frac{2}{3} > x$$
$$x < -\frac{2}{3}$$

67. $\frac{q}{7} - 3 > \frac{q-4}{3} + 1 \quad \text{lcm} = 21$

$$21 \cdot \frac{q}{7} - 21 \cdot 3 > 21 \cdot \frac{(q-4)}{3} + 21 \cdot 1$$
$$3q - 63 > 7(q - 4) + 21$$
$$3q - 63 > 7q - 28 + 21$$
$$3q - 63 > 7q - 7$$
$$3q > 7q + 56$$
$$-4q > 56$$
$$\frac{-4q}{-4} < \frac{56}{-4}$$
$$q < -14 \text{ or } (-\infty, -14)$$

−14 q

69. $\frac{2x}{5} - \frac{1}{2}(x - 3) \le \frac{2x}{3} - \frac{3}{10}(x + 2) \quad \text{lcm} = 30$

$$30 \cdot \frac{2x}{5} - 30 \cdot \frac{1}{2}(x - 3) \le 30 \cdot \frac{2x}{3} - 30 \cdot \frac{3}{10}(x + 2)$$
$$12x - 15(x - 3) \le 20x - 9(x + 2)$$
$$12x - 15x + 45 \le 20x - 9x - 18$$
$$-3x + 45 \le 11x - 18$$
$$-3x \le 11x - 63$$
$$-14 \le -63$$
$$\frac{-14x}{-14} \ge \frac{-63}{-14}$$
$$x \ge 4.5 \text{ or } [4.5, \infty)$$

4.5 x

71. $-4 \le \frac{9}{5}x + 32 \le 68 \quad \text{lcm} = 5$

$$5(-4) \le 5 \cdot \frac{9}{5}x + 5 \cdot 32 \le 5 \cdot 68$$
$$-20 \le 9x + 160 \le 340$$
$$-180 \le 9x \le 180$$
$$\frac{-180}{9} \le \frac{9x}{9} \le \frac{180}{9}$$
$$-20 \le x \le 20 \text{ or } [-20, 20]$$

−20 20 x

73. $-12 < \frac{3}{4}(2 - x) \le 24 \quad \text{lcm} = 4$

$$4(-12) < 4\left[\frac{3}{4}(2 - x)\right] \le 4 \cdot 24$$
$$-48 < 3(2 - x) \le 96$$
$$-48 < 6 - 3x \le 96$$
$$-54 < -3x \le 90$$
$$\frac{-54}{-3} > \frac{-3x}{-3} \ge \frac{90}{-3}$$
$$18 > x \ge -30$$
$$-30 \le x < 18 \text{ or } [-30, 18)$$

−30 18 x

75. $-6 < -\frac{2}{5}(1 - x) \le 4 \quad \text{lcm} = 5$

$$5(-6) < 5\left[-\frac{2}{5}(1 - x)\right] \le 5 \cdot 4$$
$$-30 < -2(1 - x) \le 20$$
$$-30 < -2 + 2x \le 20$$
$$-28 < 2x \le 22$$
$$\frac{-28}{2} < \frac{2x}{2} \le \frac{22}{2}$$
$$-14 < x \le 11 \text{ or } (-14, 11]$$

−14 11 x

77. $\frac{1}{3} \le \frac{1}{2}x + \frac{1}{4} < \frac{5}{6} \quad \text{lcm} = 12$

$$12 \cdot \frac{1}{3} \le 12 \cdot \frac{1}{2}x + 12 \cdot \frac{1}{4} < 12 \cdot \frac{5}{6}$$
$$4 \le 6x + 3 < 10$$
$$1 \le 6x < 7$$
$$\frac{1}{6} \le x < \frac{7}{6} \text{ or } \left[\frac{1}{6}, \frac{7}{6}\right)$$

0 1 x

79. $\frac{1}{2} < \frac{2}{3} - \frac{1}{6}x \le \frac{2}{3}$ lcm = 6

$6 \cdot \frac{1}{2} < 6 \cdot \frac{2}{3} - 6 \cdot \frac{1}{6}x \le 6 \cdot \frac{2}{3}$

$3 < 4 - x \le 4$

$-1 < -x \le 0$

$\frac{-1}{-1} > \frac{-x}{-1} \ge \frac{0}{-1}$

$1 > x \ge 0$

$0 \le x < 1$ or $[0, 1)$

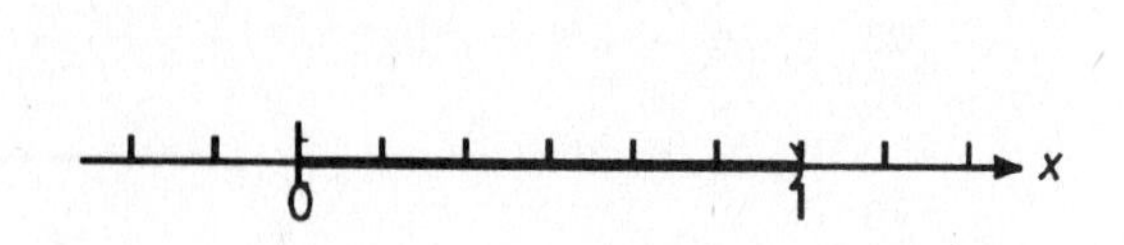

81. Since $\frac{b}{a} > 1$ and a is negative, if we multiply both sides of the inequality by a, the sense of the inequality will be reversed. Then

$a \cdot \frac{b}{a} < a \cdot 1$

$b < a$

$0 < a - b$ (subtracting b from both sides of the inequality)

Therefore $a - b$ is positive.

83. (A) F by multiplication property (part 3)
(B) T by multiplication property (part 4)
(C) T by addition property (part 1)

85. Dividing by $a - 1$ in step 4 will reverse the sense of the inequality since $a < 1$, $a - 1 < 0$.

87. 1. Def. of <; 2. product of positive numbers is positive; 3. distributive property; 4. inequality property 1

89.

1. $b - a > 0$	Def. of >
2. $c(b - a) < 0$	product of numbers of opposite sign is negative
3. $cb - ca < 0$	distributive property
4. $cb < ca$	inequality property 1

91. We are asked to find what altitude h will give T between $-40°$ and $26°$F, that is,

$-40 \le T \le 26$

Since $T = 70 - 0.0055h$, we have

$-40 \le 70 - 0.0055h \le 26$

Solving, we have

$-110 \le -0.0055h \le -44$

$\frac{-110}{-0.0055} \ge \frac{-0.0055h}{-0.0055} \ge \frac{-44}{-0.0055}$

$20{,}000 \ge h \ge 8{,}000$

$8{,}000 \le h \le 20{,}000$ or $[8{,}000, 20{,}000]$

93. A profit will result only if $R > C$. Since $R = 2x$ and $C = 300 + 1.5x$, we have

$$\begin{aligned} 2x &> 300 + 1.5x \\ 2x - 1.5x &> 300 \\ 0.5x &> 300 \\ \frac{0.5x}{0.5} &> \frac{300}{0.5} \\ x &> 600 \end{aligned}$$

More than 600 records must be sold.

95. Let x = the number of new voters registered.

Then $20{,}000 + x$ = the number of voters in the minority party

$44{,}000 + x$ = the number of voters registered.

For the number of voters in the minority party to be at least 48% of the total number of registered voters,

$$\begin{aligned} 20{,}000 + x &\ge 0.48(44{,}0000 + x) \\ 20{,}000 + x &\ge 21{,}120 + 0.48x \\ 20{,}000 + 0.52x &\ge 21{,}120 \\ 0.52x &\ge 1120 \\ x &\ge \frac{1120}{0.52} \\ x &\ge 2153.8 \end{aligned}$$

Since the solution must be a natural number, 2154 voters (or more) are needed.

97. Let x = the number of consecutive innings needed.

Then $180 + x$ = the number of innings pitched

65 = the number of earned runs allowed

$$\begin{aligned} 9 \cdot \frac{65}{180 + x} &\le 3 \\ \frac{585}{180 + x} &\le 3 \\ 585 &\le 3(180 + x) \quad \text{assuming } 180 + x \text{ is positive, as it must be since } x \text{ is positive} \\ 585 &\le 540 + 3x \\ 45 &\le 3x \\ 15 &\le x \end{aligned}$$

15 innings (or more) are needed.

99. From drawing 1, it should be clear that the worker will escape from the tunnel if he runs a distance $\frac{3}{4}L$ at his maximum rate r <u>away</u> from the train before the train covers the distance $\frac{7}{4}L$ at its rate R. Thus we require

$$\text{worker's time} < \text{train's time}$$
$$\frac{3}{4}L \div r < \frac{7}{4}L \div R$$

Drawing 1.

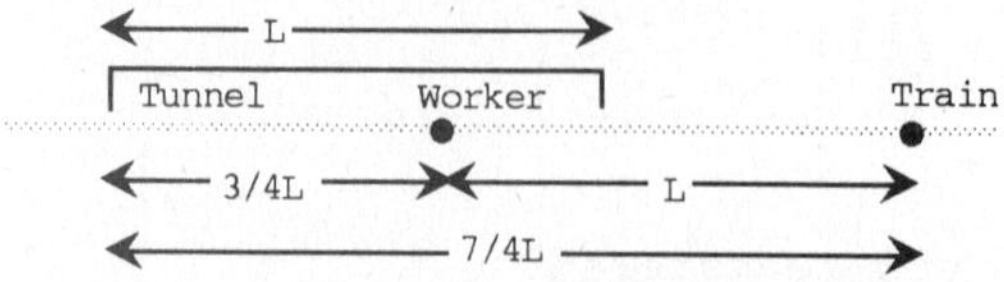

L

Tunnel Worker Train

3/4L L

7/4L

$$\frac{3L}{4} \div \frac{r}{1} < \frac{7L}{4} \div \frac{R}{1}$$
$$\frac{3L}{4} \cdot \frac{1}{r} < \frac{7L}{4} \cdot \frac{1}{R} \quad \text{lcm } 4rR > 0$$
$$4rR \cdot \frac{3L}{4} \cdot \frac{1}{r} < 4rR \cdot \frac{7L}{4} \cdot \frac{1}{R}$$
$$3LR < 7Lr$$
$$\frac{3LR}{7L} < \frac{7Lr}{7L}$$
$$\frac{3}{7}R < r$$

From drawing 2, it should be clear that the worker will also escape from the tunnel if he runs a distance $\frac{1}{4}L$ at his maximum rate r <u>toward</u> the train before the train covers the distance $\frac{3}{4}L$ at its rate R. Thus we require

$$\text{worker's time} < \text{train's time}$$
$$\frac{1}{4}L \div r < \frac{3}{4}L \div R$$

Drawing 2.

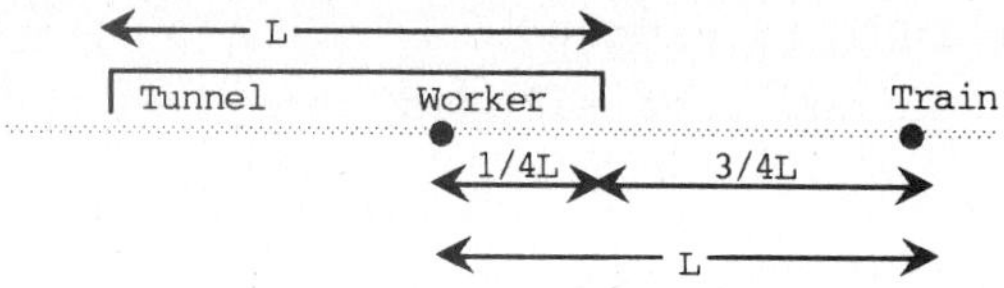

L

Tunnel Worker Train

1/4L 3/4L

L

$$\frac{L}{4} \div \frac{r}{1} < \frac{3L}{4} \div \frac{R}{1}$$
$$\frac{L}{4} \cdot \frac{1}{r} < \frac{3L}{4} \cdot \frac{1}{R} \quad \text{lcm } 4rR > 0$$
$$4rR \cdot \frac{L}{4} \cdot \frac{1}{r} < 4rR \cdot \frac{3L}{4} \cdot \frac{1}{R}$$
$$LR < 3rL$$
$$\frac{LR}{3L} < \frac{3rL}{3L}$$
$$\frac{R}{3} < r$$

Thus he will escape running toward the train if $r > \frac{R}{3} = \frac{7R}{21}$, and he will escape running away from the train if $r > \frac{3}{7}R = \frac{9R}{21}$. Since his chances are better of having speed $r > \frac{7R}{21}$ than of having speed $r > \frac{9R}{21}$, he should run toward the train.

Exercise 5-4 Absolute Value in Equations and Inequalities

Key Ideas and Formulas

$$|x| = \begin{cases} x & \text{if } x \text{ is positive} \\ 0 & \text{if } x \text{ is } 0 \\ -x & \text{(a positive quantity) if } x \text{ is negative} \end{cases}$$

The distance between two points A and B on a number line with coordinates a and b respectively is given by

$d(A, B) = |b - a|$

Since $|b - a| = |a - b|$, $d(A, B) = d(B, A)$

For $p > 0$

$|x| = p$ is equivalent to $x = p$ or $x = -p$

$|x| < p$ is equivalent to $-p < x < p$

$|x| > p$ is equivalent to $x < -p$ or $x > p$

COMMON ERROR: Writing $p < x < -p$ here. This is true for no real value of x, since $p < -p$ is false.

To solve:

$|ax + b| = p$, solve $ax + p = \pm p$

$|ax + b| < p$, solve $-p < ax + b < p$

$|ax + b| > p$, solve $ax + b < -p$ or $ax + b > p$

---

1. Since 2π is greater than 6, $6 - 2\pi$ is negative, so $|6 - 2\pi| = -(6 - 2\pi) = 2\pi - 6$

3. Since 5 is greater than $\sqrt{24}$, $5 - \sqrt{24}$ is positive, so $|5 - \sqrt{24}| = 5 - \sqrt{24}$

5. Since $\sqrt{5}$ is positive, $|\sqrt{5}| = \sqrt{5}$

7. $|(-6) - (-2)| = |-4| = 4$

9. Since $5 > \sqrt{5}$, $5 - \sqrt{5}$ is positive, so $|5 - \sqrt{5}| = 5 - \sqrt{5}$

11. Since $5 > \sqrt{5}$, $\sqrt{5} - 5$ is negative, so $|\sqrt{5} - 5| = -(\sqrt{5} - 5) = 5 - \sqrt{5}$

13. $d(A, B) = |b - a| = |5 - (-7)| = |12| = 12$

15. $d(A, B) = |b - a| = |-7 - 5| = |-12| = 12$

17. $d(A, B) = |b - a| = |-25 - (-16)| = |-9| = 9$

19. $d(A, B) = |b - a| = |14 - 11| = |3| = 3$

21. $d(A, B) = |b - a| = |-14 - (-11)| = |-3| = 3$

23. $d(B, O) = |b| = |-4| = 4$

25. $d(O, B) = |b| = |-4| = 4$

27. $d(B, C) = |c - b| = |5 - (-4)| = |9| = 9$

29. $d(A, D) = |d - a| = |8 - (-9)| = |17| = 17$

31. $d(C, A) = |a - c| = |(-9) - 5| = |-14| = 14$

33. $x = \pm 7$

35. $-7 \le x \le 7$

37. $x \le -7$ or $x \ge 7$

39. $|y - 5| = 3$
$y - 5 = \pm 3$
$y = 5 \pm 3$
$y = 2$ or 8

41. $|y - 5| < 3$
$-3 < y - 5 < 3$
$2 < y < 8$

43. $|y - 5| > 3$
$y - 5 < -3$ or $y - 5 > 3$
$y < 2$ or $y > 8$

45. $|u + 8| = 3$
$u + 8 = \pm 3$
$u = -8 \pm 3$
$u = -11$ or -5

47. $|u + 8| \le 3$
$-3 \le u + 8 \le 3$
$-11 \le u \le -5$

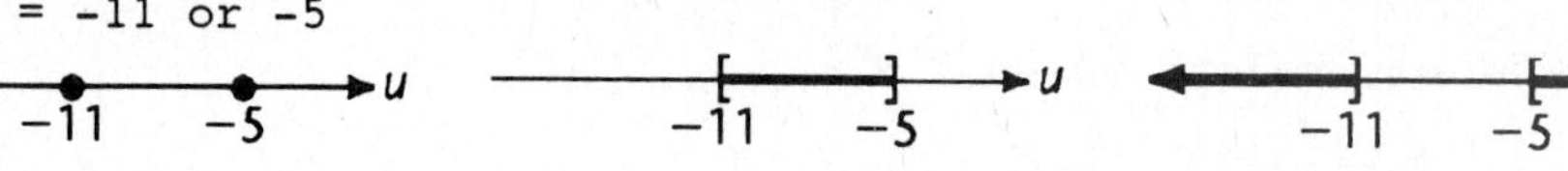

49. $|u + 8| \ge 3$
$u + 8 \le -3$ or $u + 8 \ge 3$
$u \le -11$ or $u \ge -5$

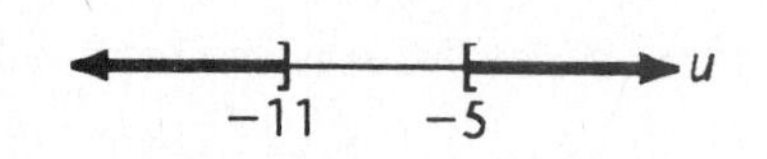

51. $|1 - x| > 1$
$1 - x < -1$ or $1 - x > 1$
$-x < -2$ or $-x > 0$
$x > 2$ or $x < 0$

53. $|4 - x| \le 3$
$-3 \le 4 - x \le 3$
$-7 \le -x \le -1$
$7 \ge x \ge 1$
$1 \le x \le 7$

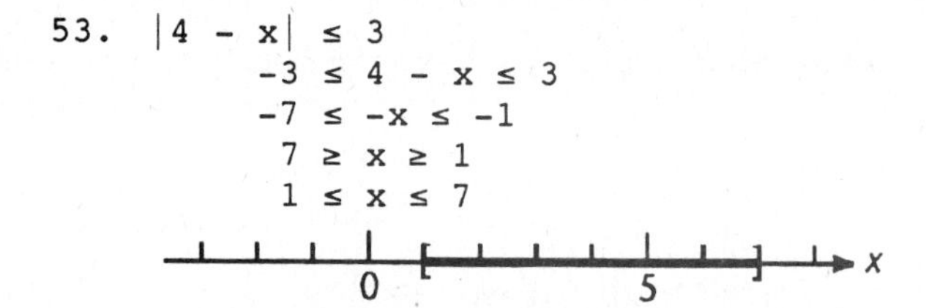

55. $|3x + 4| = 8$
$3x + 4 = \pm 8$
$3x = -4 \pm 8$
$x = \dfrac{-4 \pm 8}{3}$
$x = \dfrac{-4 - 8}{3}, \dfrac{-4 + 8}{3}$
$x = -4, \dfrac{4}{3}$

57. $|5x - 3| \le 12$
$-12 \le 5x - 3 \le 12$
$-9 \le 5x \le 15$
$-\dfrac{9}{5} \le x \le 3$

59. $|2y - 8| > 2$
$2y - 8 < -2$ or $2y - 8 > 2$
$2y < 6$ or $2y > 10$
$y < 3$ or $y > 5$

61. $|5t - 7| = 11$
$5t - 7 = \pm 11$
$5t = 7 \pm 11$
$t = \dfrac{7 \pm 11}{5}$
$t = \dfrac{7 - 11}{5}, \dfrac{7 + 11}{5}$
$t = -\dfrac{4}{5}, \dfrac{18}{5}$

63. $|9 - 7u| < 14$
$-14 < 9 - 7u < 14$
$-23 < -7u < 5$
$\dfrac{23}{7} > u > -\dfrac{5}{7}$
$-\dfrac{5}{7} < u < \dfrac{23}{7}$

65. $\left|1 - \frac{2}{3}x\right| \geq 5$

$1 - \frac{2}{3}x \leq -5$ or $1 - \frac{2}{3}x \geq 5$

$-\frac{2}{3}x \leq -6$ or $-\frac{2}{3}x \geq 4$

$3\left(-\frac{2}{3}x\right) \leq 3(-6)$ or $3\left(-\frac{2}{3}x\right) \geq 3 \cdot 4$

$-2x \leq -18$ or $-2x \geq 12$

$\frac{-2x}{-2} \geq \frac{-18}{-2}$ or $\frac{-2x}{-2} \leq \frac{12}{-2}$

$x \geq 9$ or $x \leq -6$

67. $\left|\frac{9}{5}C + 32\right| < 31$

$-31 < \frac{9}{5}C + 32 < 31$

$5(-31) < 5 \cdot \frac{9}{5}C + 5 \cdot 32 < 5 \cdot 31$

$-155 < 9C + 160 < 155$

$-315 < 9C < -5$

$-35 < C < -\frac{5}{9}$

69. x is a number whose distance from 0 is 2; that is, $|x| = 2$

71. x is a number whose distance from 0 is more than 2; that is, $|x| > 2$

73. x is a number whose distance from 0 is less than or equal to 2; that is, $|x| \leq 2$

75. x is a number whose distance from 10 is less than 0.01, that is, $|x - 10| < 0.01$

77. x is a number whose distance from -5 is less than 0.1, that is, $|x - (-5)| < 0.1$, or $|x + 5| < 0.1$

79. x is a number whose distance from -8 is less than 5, that is, $|x - (-8)| < 5$, or $|x + 8| < 5$

81. x is a number whose distance from 5 is less than 0.01, that is, $|x - 5| < 0.01$

83. $|x| = x$ if $x \geq 0$

85. $|u| = u$ if $u \geq 0$
So $|x - 5| = x - 5$ if $x - 5 \geq 0$
$x - 5 \geq 0$ if $x \geq 5$

87. $|u| = -u$ if $u \leq 0$
So $|x + 8| = -(x + 8)$ if $x + 8 \leq 0$
$x + 8 \leq 0$ if $x \leq -8$

89. $|u| = u$ if $u \geq 0$
So $|4x + 3| = 4x + 3$ if $4x + 3 \geq 0$
$4x + 3 \geq 0$ if $4x \geq -3$
$x \geq -\frac{3}{4}$

91. $|u| = -u$ if $u \leq 0$
So $|5x - 2| = -(5x - 2)$ if $5x - 2 \leq 0$
$5x - 2 \leq 0$ if $5x \leq 2$
$x \leq \frac{2}{5}$

93. Let $x = 0$ and $y = 1$. Then $|x - y| = |0 - 1| = |-1| = 1$, but $|x| - |y| = |0| - |1| = 0 - 1 = -1$. There are many other values possible for x and y for which this is not true.

95. Case 1: if $x = 0$, then both sides are 0 and the equality holds.

Case 2: suppose $x > 0$
2a: if $y > 0$, then the equality is $xy = xy$
2b: if $y < 0$, then $xy < 0$, $|xy| = -xy = x(-y) = |x| \cdot |y|$

Case 3: suppose $x < 0$
3a: if $y > 0$, then $xy < 0$, $|xy| = -xy = (-x)y = |x| \cdot |y|$
3b: if $y < 0$, then $xy > 0$, $|xy| = xy = (-x)(-y) = |x| \cdot |y|$

Exercise 5-5 Graphing Linear Inequalities

Key Ideas and Formulas

Graphing $x < a$, $x \le a$, $x > a$, or $x \ge a$

1. The graph of $x < a$ or $x \le a$ is the half-plane to the left of the boundary line $x = a$, together with the boundary line if the inequality is $\le$.
2. The graph of $x > a$ or $x \ge a$ is the half-plane to the right of the boundary line $x = a$, together with the boundary line if the inequality is $\ge$.

Graphing $y < mx + b$, $y \le mx + b$, $y > mx + b$, or $y \ge mx + b$

1. The graph of $y < mx + b$ or $y \le mx + b$ is the half-plane below the boundary line $y = mx + b$, together with the boundary line if the inequality is $\le$.
2. The graph of $y > mx + b$ or $y \ge mx + b$ is the half-plane above the boundary line $y = mx + b$, together with the boundary line if the inequality is $\ge$.

Steps in Graphing Linear Inequalities

1. Graph the corresponding equation obtained by replacing the inequality by =. Show this boundary line as a broken line if equality is not included in the original statement, or as a solid line if equality is included in the original statement.
2. Choose a test point in the plane that is not on the line—the origin is the best choice if it is not on the line—and substitute the coordinates into the inequality.
3. Determine which half-plane is the graph of the original inequality and shade it. The graph includes:
   (A) The half-plane containing the test point if the inequality is satisfied by that point.
   (B) The half-plane not containing the test point if the inequality is not satisfied by that point.

1. First graph the line $y = 2x - 3$ as a broken line, since equality is not included in $y < 2x - 3$. Choose the origin as a test point, since it is not on the line.

$$y < 2x - 3$$
$$0 \overset{?}{<} 2(0) - 3$$
$$0 \nless -3$$

The origin does not satisfy the original inequality; hence the graph consists of all points on the other side of the line. Thus, the graph is the lower half-plane.

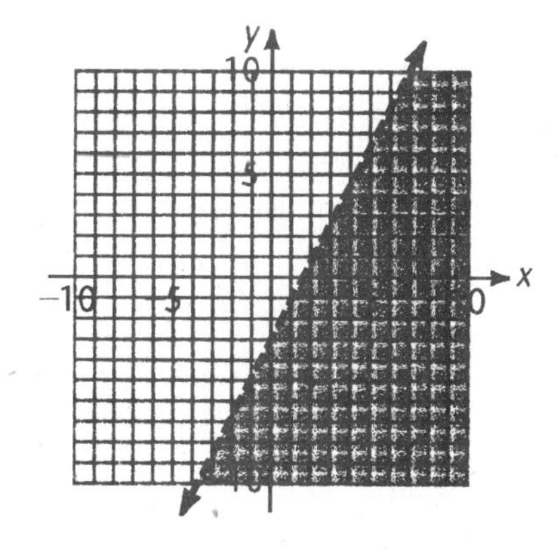

3. First graph the line $y = 4x + 2$ as a solid line, since equality is included in $y \geq 4x + 2$. Choose the origin as a test point, since it is not on the line.

$$y \geq 4x + 2$$
$$0 \overset{?}{\geq} 4(0) + 2$$
$$0 \ngeq 2$$

The origin does not satisfy the original inequality; hence the graph consists of all points on the other side of the line. Thus, the graph is the upper half-plane.

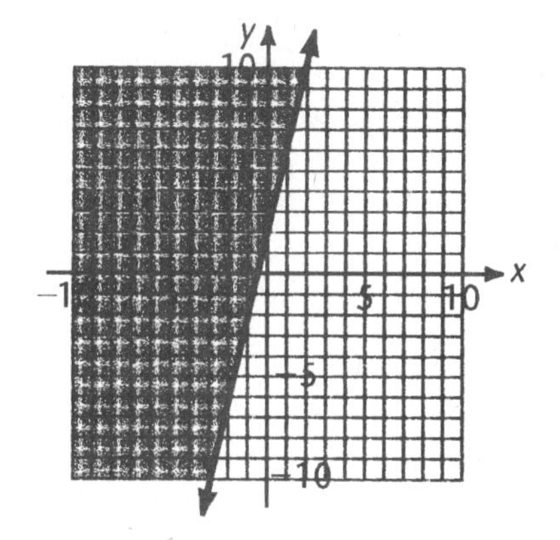

5. First graph the line $x + y = 6$ as a solid line, since equality is included in $x + y \leq 6$. Choose the origin as a test point, since it is not on the line.

$$x + y \leq 6$$
$$0 + 0 \overset{?}{\leq} 6$$
$$0 \leq 6$$

The origin satisfies the original inequality; hence all other points on the same side as the origin are also part of the graph. Thus, the graph is the lower half-plane.

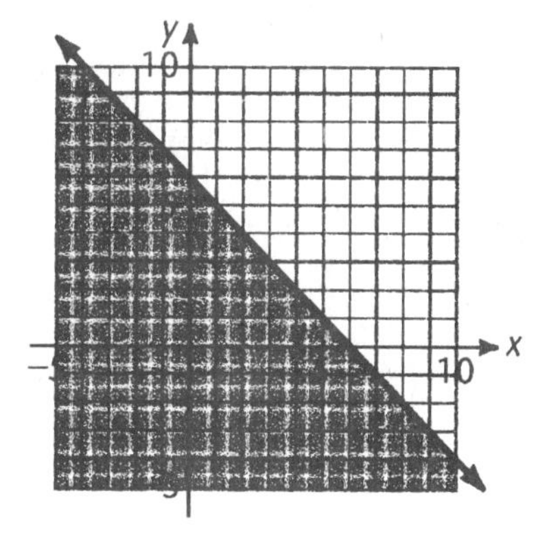

7. First graph the line $x - y = 3$ as a broken line, since equality is not included in $x - y > 3$. Choose the origin as a test point, since it is not on the line.

$$x - y > 3$$
$$0 - 0 \overset{?}{>} 3$$
$$0 \ngtr 6$$

The origin does not satisfy the original inequality; hence the graph consists of all the points on the other side of the line. Thus, the graph is the lower half-plane.

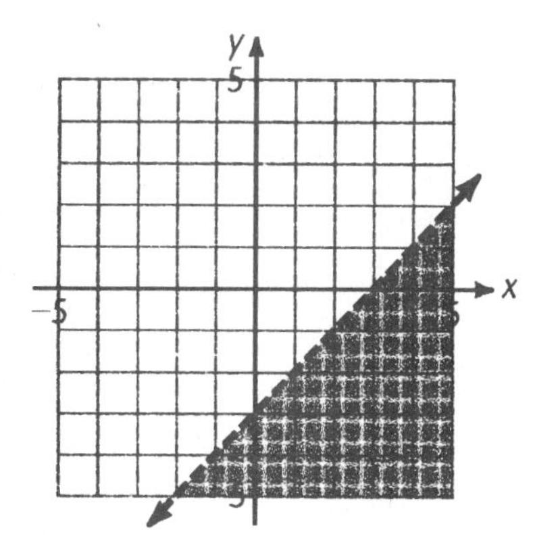

9. First graph the line $y = x - 2$ as a solid line, since equality is included in $y \geq x - 2$. Choose the origin as a test point, since it is not on the line.

$$y \geq x - 2$$
$$0 \overset{?}{\geq} 0 - 2$$
$$0 \geq 2$$

The origin satisfies the original inequality; hence all other points on the same side as the origin are also part of the graph. Thus, the graph is the upper half-plane.

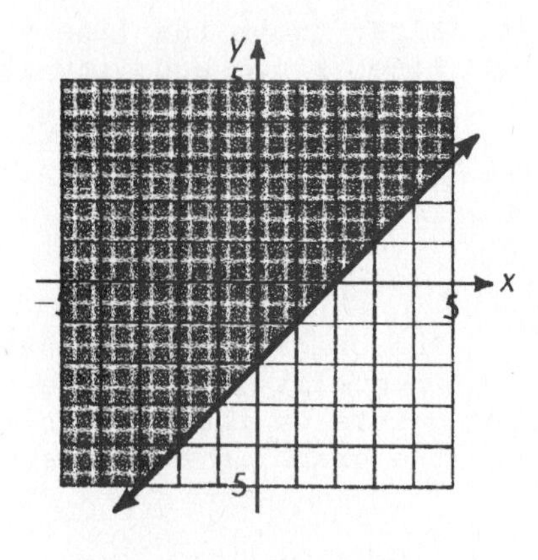

11. First graph the line $y = \frac{x}{3} - 2$ as a solid line, since equality is included in $y \geq \frac{x}{3} - 2$. Choose the origin as a test point, since it is not on the line.

$$y \geq \frac{x}{3} - 2$$
$$0 \overset{?}{\geq} \frac{0}{3} - 2$$
$$0 \geq 24$$

The origin satisfies the original inequality; hence all other points on the same side as the origin are also part of the graph. Thus, the graph is the upper half-plane.

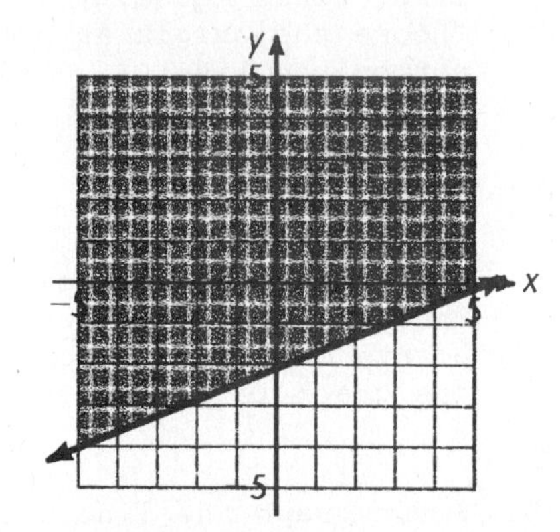

13. First graph the line $y = \frac{2}{3}x + 5$ as a solid line, since equality is included in $y \leq \frac{2}{3}x + 5$. Choose the origin as a test point, since it is not on the line.

$$y \leq \frac{2}{3}x + 5$$
$$0 \overset{?}{\leq} \frac{2}{3}(0) + 5$$
$$0 \leq 5$$

The origin satisfies the original inequality; hence all other points on the same side as the origin are also part of the graph. Thus, the graph is the lower half-plane.

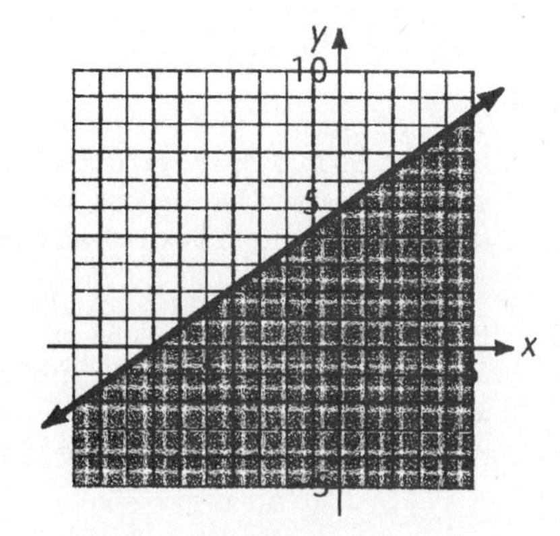

15. First graph the line $x = -5$ as a solid line, since equality is included in $x \geq -5$. Since x is required to be greater than -5, the graph is the right half-plane.

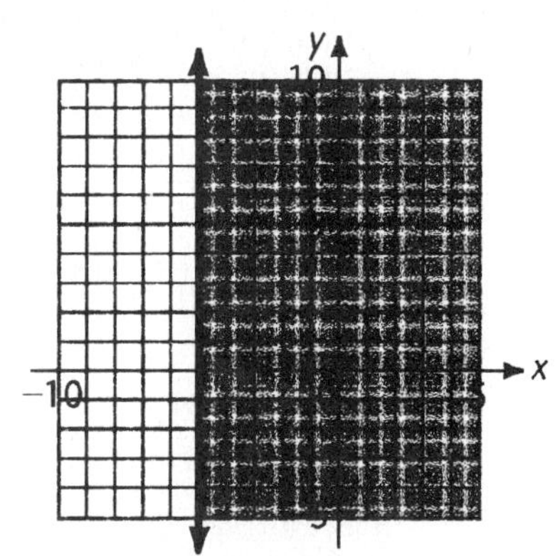

17. First graph the line $y = 0$ as a broken line (the x-axis), since equality is not included in $y < 0$. Since y is required to be less than 0, the graph is the lower half-plane.

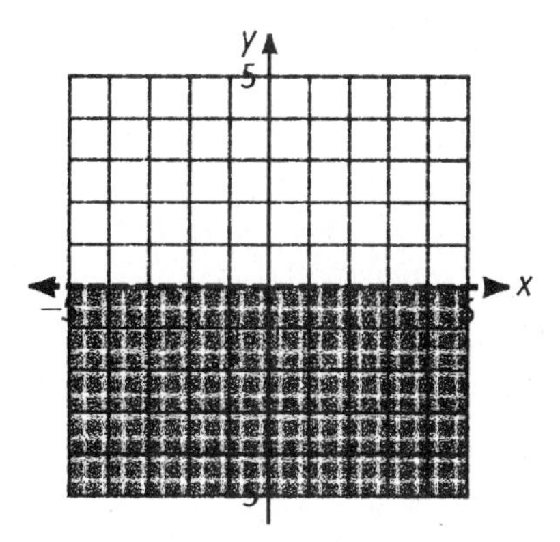

19. First graph the line $2x - 3y = 6$ as a broken line, since equality is included in $2x - 3y < 6$. Choose the origin as a test point, since it is not on the line.

$$2x - 3y < 6$$
$$2(0) - 3(0) \overset{?}{<} 6$$
$$0 < 6$$

The origin satisfies the original inequality; hence all other points on the same side as the origin are also part of the graph. Thus, the graph is the upper half-plane.

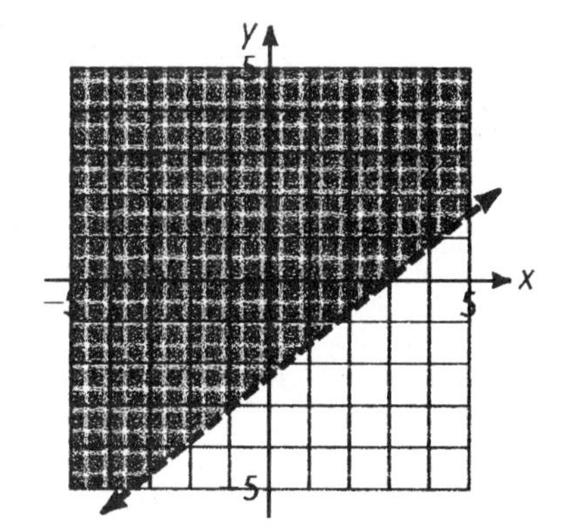

21. First graph the line $3y - 2x = 24$ as a solid line, since equality is included in $3y - 2x \geq 24$. Choose the origin as a test point, since it is not on the line.

$$3y - 2x \geq 24$$
$$3(0) - 2(0) \overset{?}{\geq} 24$$
$$0 \not\geq 24$$

The origin does not satisfy the original inequality; hence the graph consists of all points on the other side of the line. Thus, the graph is the upper half-plane.

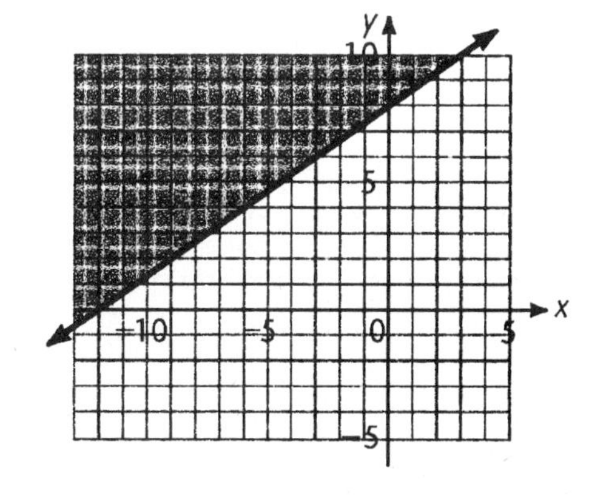

23. First graph the line $2x + 5y = 10$ as a broken line, since equality is not included in $2x + 5y > 10$. Choose the origin as a test point, since it is not on the line.

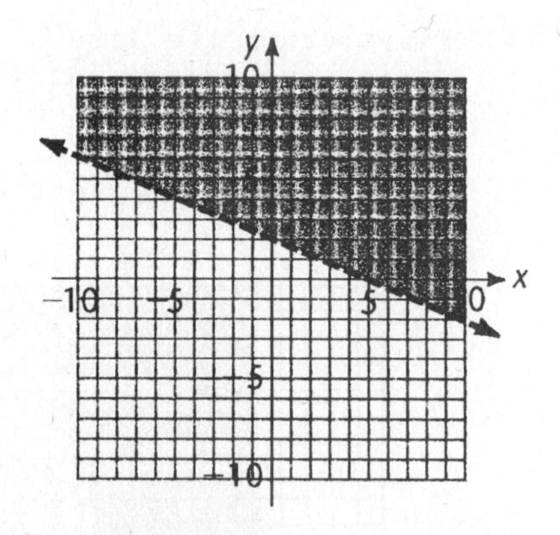

$$2x + 5y > 10$$
$$2(0) + 5(0) \overset{?}{>} 10$$
$$0 \not> 10$$

The origin does not satisfy the original inequality; hence the graph consists of all points on the other side of the line. Thus, the graph is the upper half-plane.

25. First graph the line $3x + y = 9$ as a solid line, since equality is included in $3x + y \le 9$. Choose the origin as a test point, since it is not on the line.

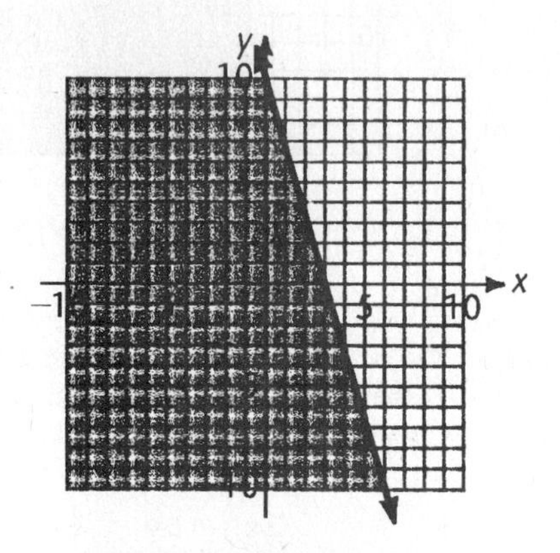

$$3x + y \le 9$$
$$3(0) + 0 \overset{?}{\le} 9$$
$$0 \le 9$$

The origin satisfies the original inequality; hence all other points on the same side as the origin are also part of the graph. Thus, the graph is the lower half-plane.

27. First graph the line $5x - 2y = 10$ as a solid line, since equality is included in $5x - 2y \le 10$. Choose the origin as a test point, since it is not on the line.

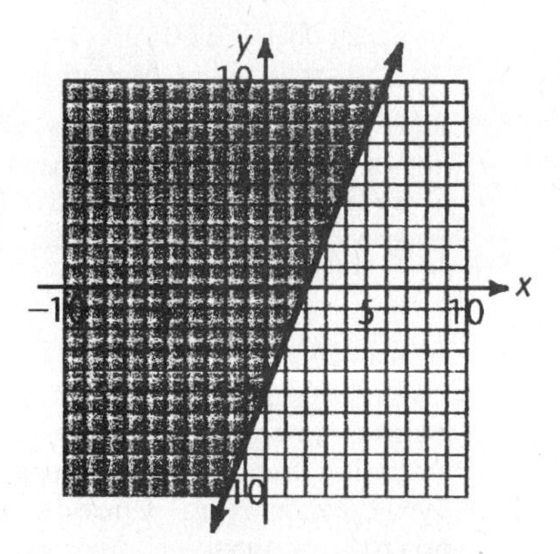

$$5x - 2y \le 10$$
$$5(0) - 2(0) \overset{?}{\le} 10$$
$$0 \le 10$$

The origin satisfies the original inequality; hence all other points on the same side as the origin are also part of the graph. Thus, the graph is the upper half-plane.

29. First graph the line $4x - y = 8$ as a solid line, since equality is included in $4x - y \ge 8$. Choose the origin as a test point, since it is not on the line.

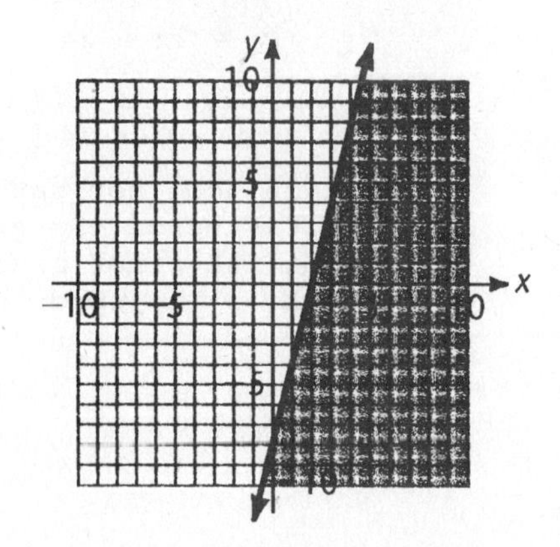

$$4x - y \ge 8$$
$$4(0) - 0 \overset{?}{\ge} 8$$
$$0 \not\ge 8$$

The origin does not satisfy the original inequality; hence the graph consists of all points on the other side of the line. Thus, the graph is the lower half-plane.

31. This is the set of points common to the graphs of the inequalities $-1 < x$ (all points to the right of the broken line $x = -1$) and $x \le 3$ (all points to the left of the solid line $x = 3$).

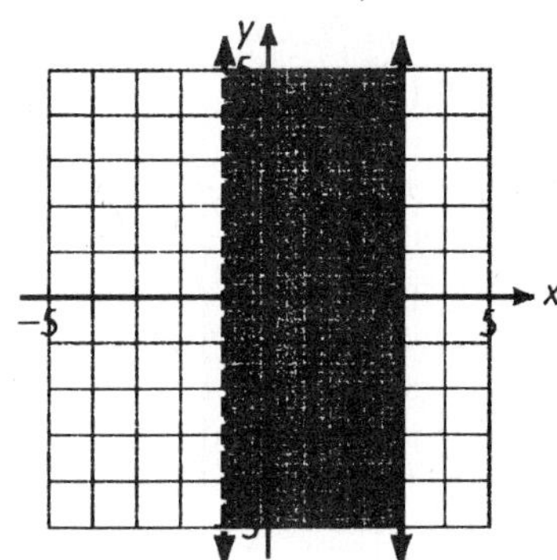

33. This is the set of points common to the graphs of the inequalities $-2 \le y$ (all points above the solid line $y = -2$) and $y \le 2$ (all points below the solid line $y = 2$).

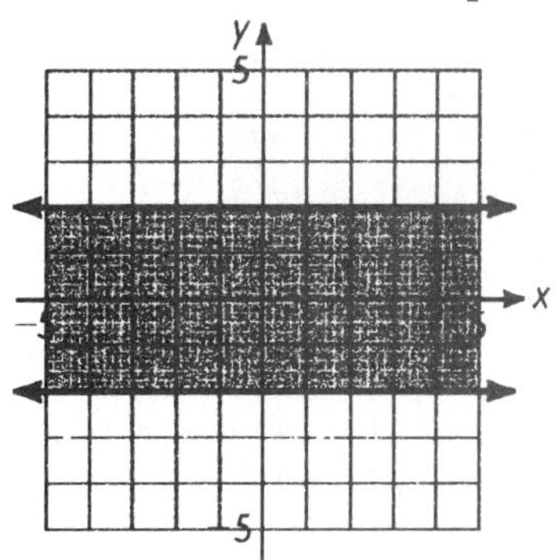

35. This is the set of points common to the graphs of the inequalities $2 \le x$ (all points to the right of the solid line $x = 2$) and $x < 3$ (all points to the left of the broken line $x = 3$).

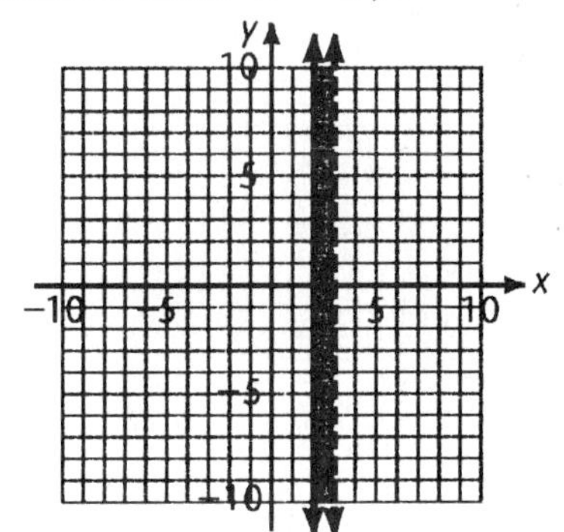

37. This is the set of points common to the graphs of the inequalities $-5 < x$ (all points to the right of the broken line $x = -5$) and $x \le -2$ (all points to the left of the solid line $x = 2$).

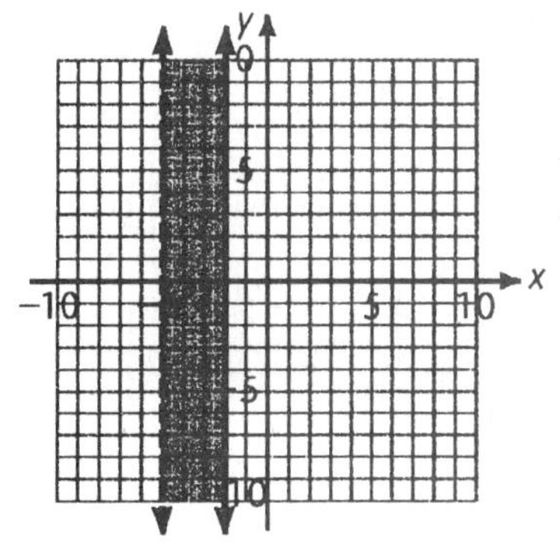

39. This is the set of points common to the graphs of the inequalities $-3 < y$ (all points above the broken line $y = -3$) and $y < 1$ (all points below the broken line $y = 1$).

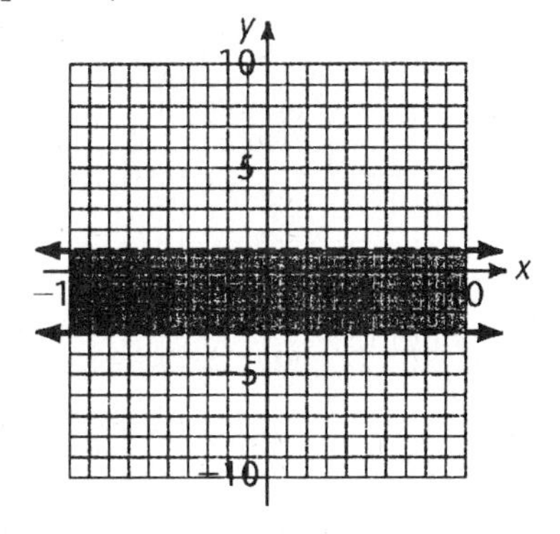

41. This is the set of points common to the graphs of the inequalities $-5 \le y$ (all points above the solid line $y = -5$) and $y < -2$ (all points below the broken line $y = -2$).

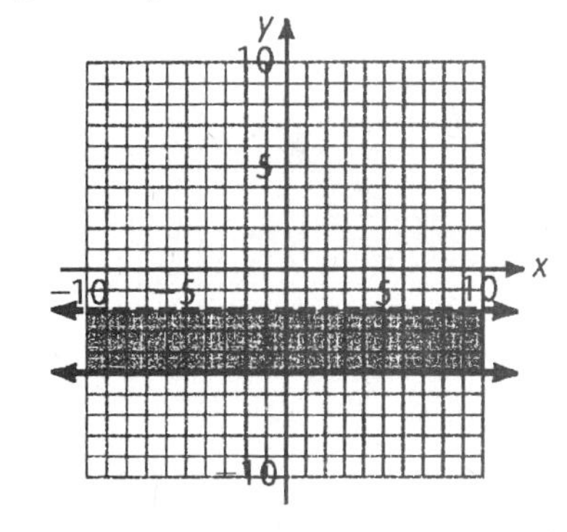

43. This is the set of points common to the solutions of the inequalities $-3 \le x$, $x \le 3$, $-1 \le y$, and $y \le 5$.

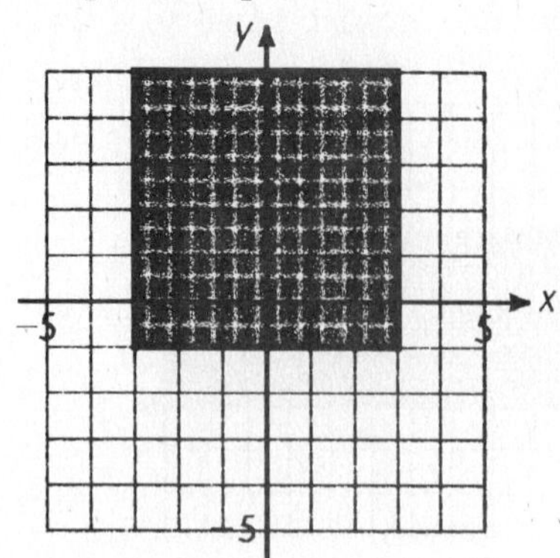

45. This is the set of points common to the solutions of the inequalities $2 \le x$, $x < 3$, $-1 < y$, and $y < 6$.

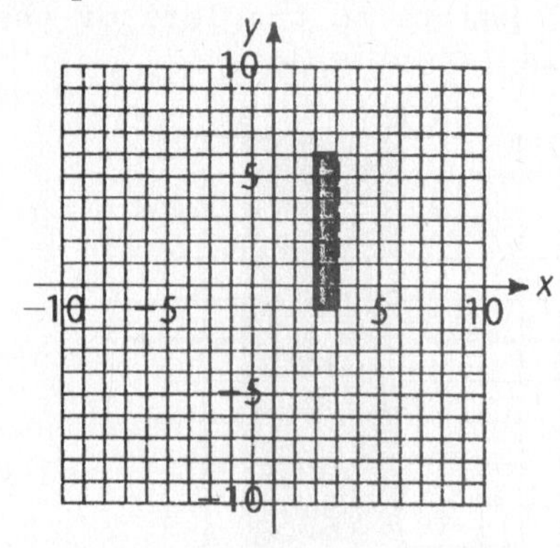

47. This is the set of points common to the solutions of the inequalities $-3 < x$, $x < 1$, $2 \le y$, and $y \le 6$.

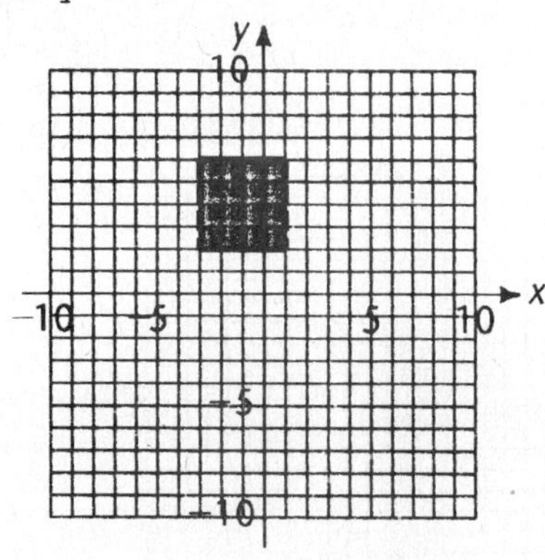

49. This is the set of points common to the solutions of the inequalities $-3 \le x$, $x \le 3$, $-3 \le y$, and $y \le 3$.

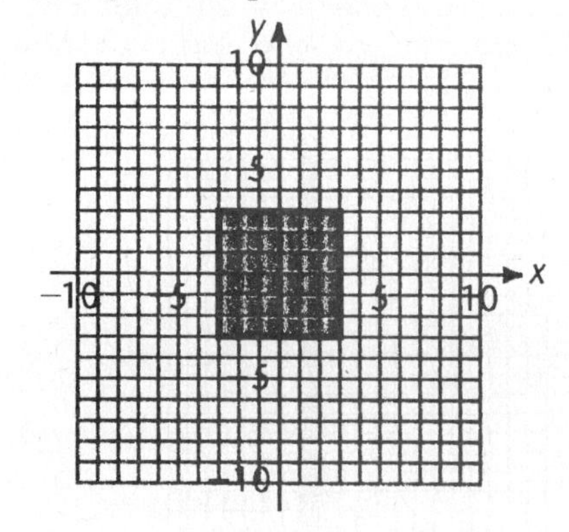

51. This is the set of points common to the solutions of the inequalities $x - 3y \ge 6$, $x \ge 0$, and $y \ge 0$.

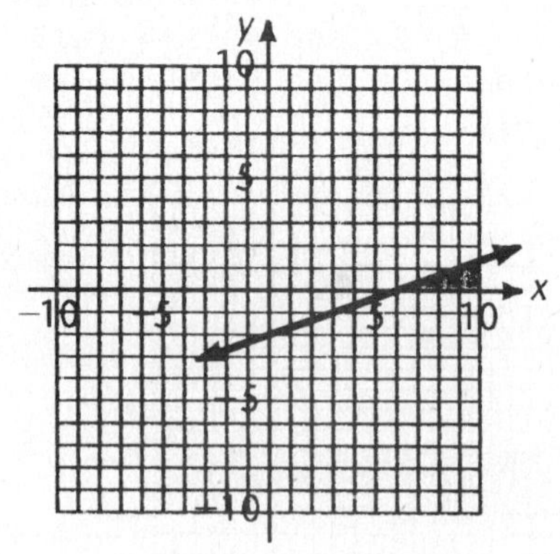

53. This is the set of points common to the solutions of the inequalities $3x - 2y < 18$, $x \ge 0$, and $y \ge 0$.

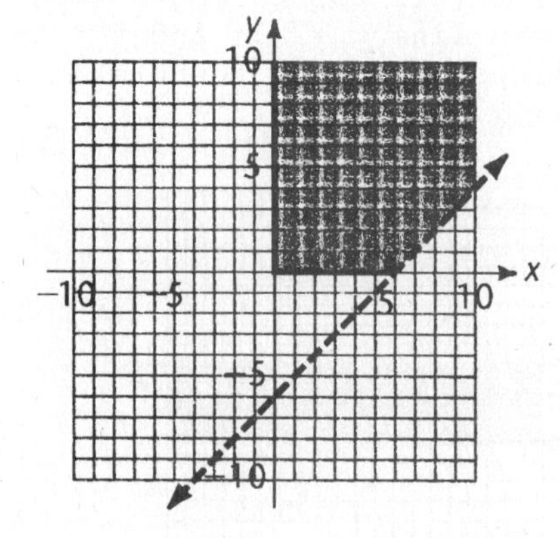

55. This is the set of points common to the solutions of the inequalities $5x + 2y < 10$, $x \geq 0$, and $y \geq 0$.

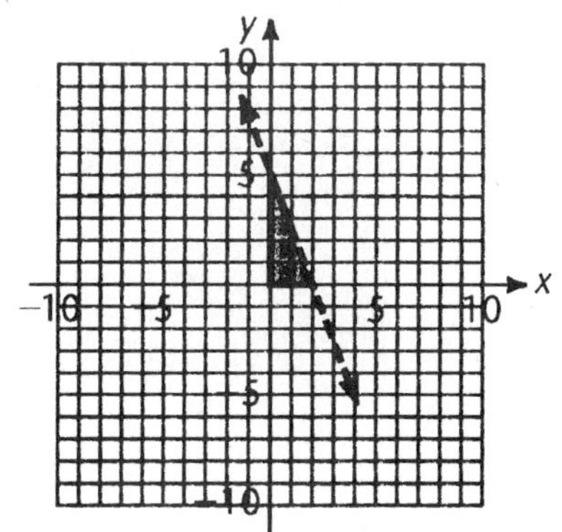

57. This is the set of all ordered pairs (x, y), such that x and y are both nonnegative and satisfy $3x + 4y \leq 12$. The result is that portion of the half-plane $3x + 4y \leq 12$ that lies in the first quadrant.

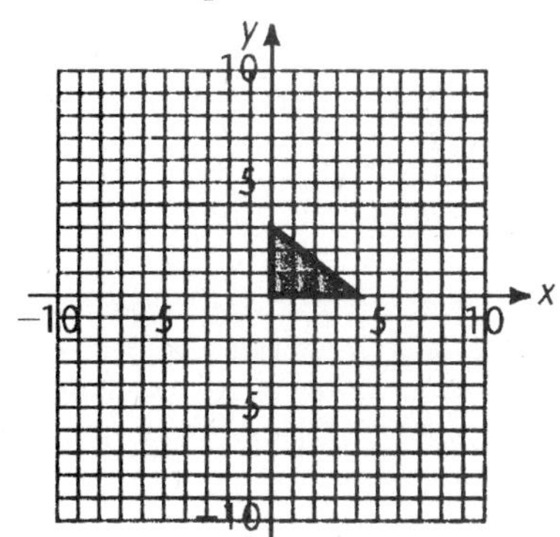

59. This is the set of all ordered pairs (x, y), such that x and y are both nonnegative and satisfy $3x + 4y \geq 12$. The result is that portion of the half-plane $3x + 4y \geq 12$ that lies in the first quadrant.

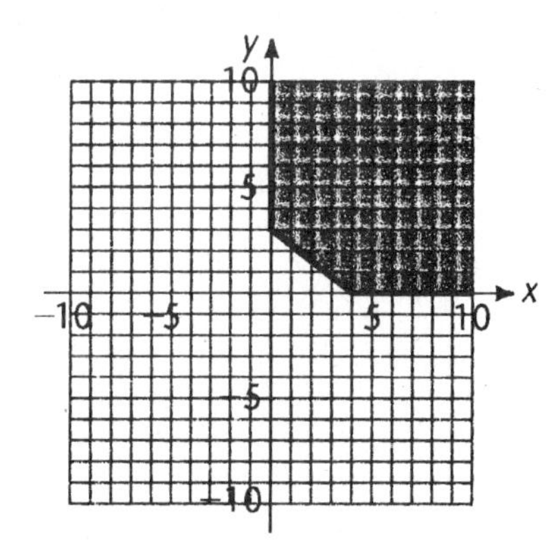

61. First graph the line $\frac{1}{5}x + \frac{2}{5}y = 1$ as a solid line, since equality is included in $\frac{1}{5}x + \frac{2}{5}y \leq 1$. Choose the origin as a test point, since it is not on the line.

$$\frac{1}{5}x + \frac{2}{5}y \leq 1$$

$$\frac{1}{5}(0) + \frac{2}{5}(0) \stackrel{?}{\leq} 1$$

$$0 \leq 1$$

The origin satisfies the original inequality; hence all other points on the same side as the origin are also part of the graph. Thus, the graph is the lower half-plane.

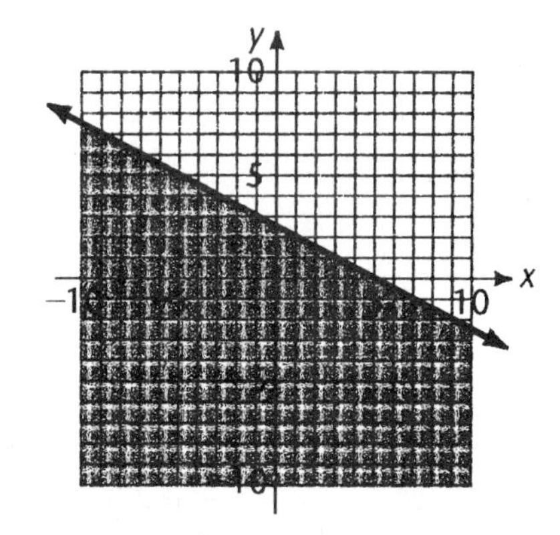

63. First graph the line $\frac{1}{8}x - \frac{1}{6}y = 1$ as a broken line, since equality is not included in $\frac{1}{8}x - \frac{1}{6}y < 1$. Choose the origin as a test point, since it is not on the line.

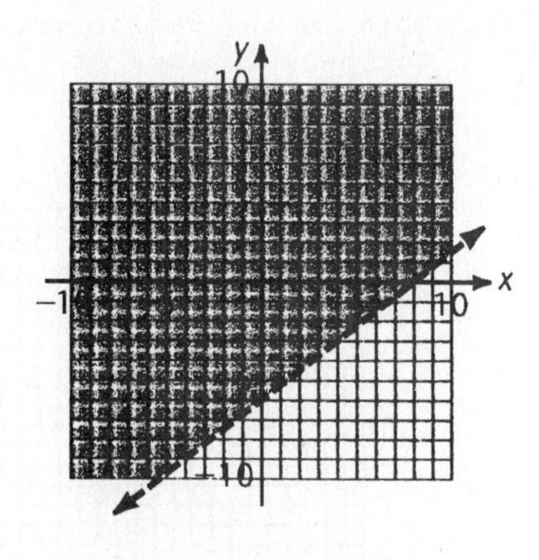

$$\frac{1}{8}x - \frac{1}{6}y < 1$$

$$\frac{1}{8}(0) - \frac{1}{6}(0) \overset{?}{<} 1$$

$$0 < 1$$

The origin satisfies the original inequality; hence all other points on the same side as the origin are also part of the graph. Thus, the graph is the upper half-plane.

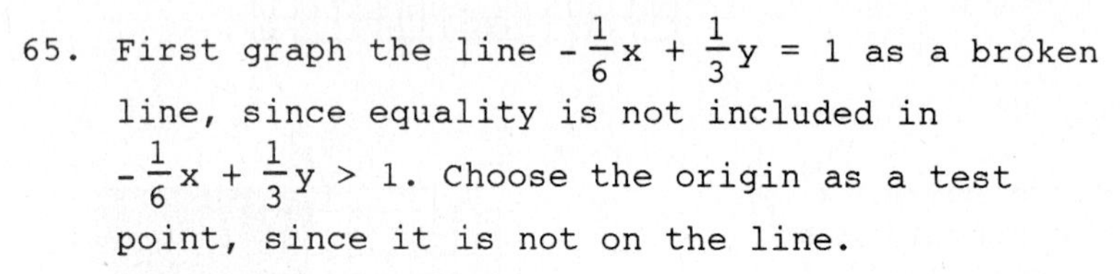

65. First graph the line $-\frac{1}{6}x + \frac{1}{3}y = 1$ as a broken line, since equality is not included in $-\frac{1}{6}x + \frac{1}{3}y > 1$. Choose the origin as a test point, since it is not on the line.

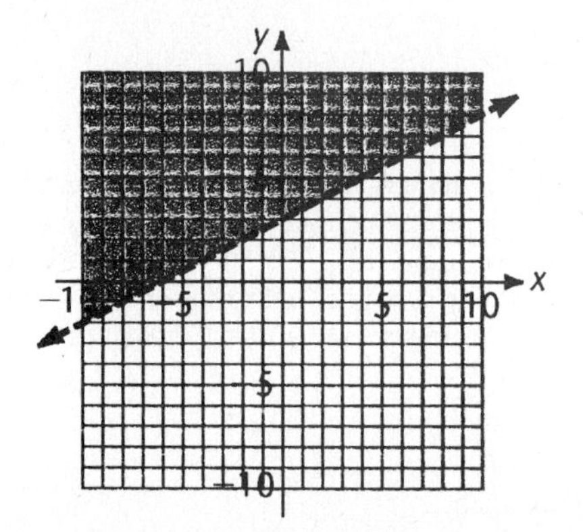

$$-\frac{1}{6}x + \frac{1}{3}y > 1$$

$$-\frac{1}{6}(0) + \frac{1}{3}(0) \overset{?}{>} 1$$

$$0 \not> 1$$

The origin does not satisfy the original inequality; hence the graph consists of all points on the other side of the line. Thus, the graph is the upper half-plane.

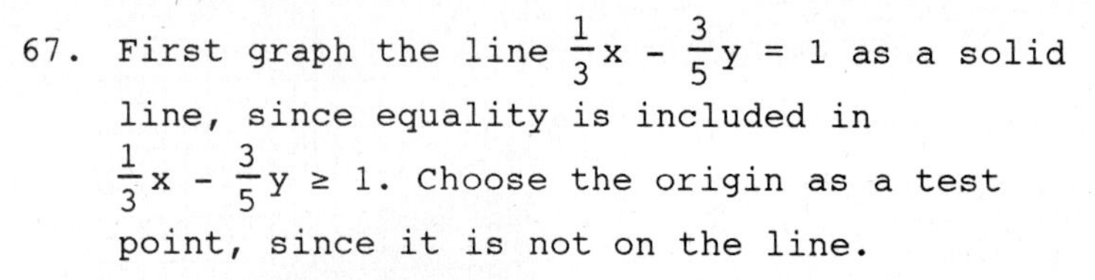

67. First graph the line $\frac{1}{3}x - \frac{3}{5}y = 1$ as a solid line, since equality is included in $\frac{1}{3}x - \frac{3}{5}y \geq 1$. Choose the origin as a test point, since it is not on the line.

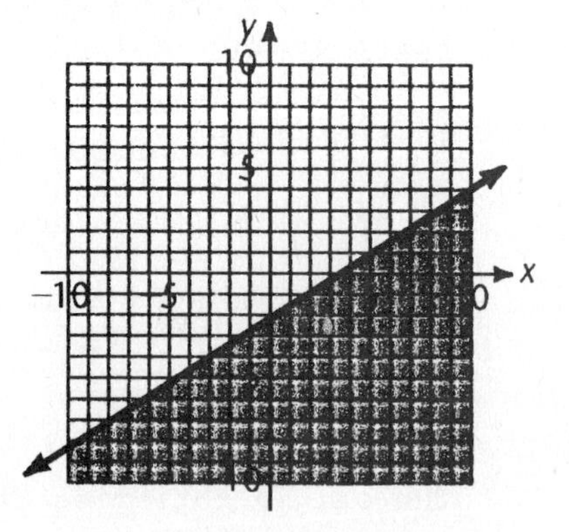

$$\frac{1}{3}x - \frac{3}{5}y \geq 1$$

$$\frac{1}{3}(0) - \frac{3}{5}(0) \overset{?}{\geq} 1$$

$$0 \not\geq 1$$

The origin does not satisfy the original inequality; hence the graph consists of all points on the other side of the line. Thus, the graph is the lower half-plane.

69. (A) Since x = how many kilograms of the tea worth \$5 a kilogram and y = how many kilograms of the tea worth \$6.50 a kilogram, the total value of the blend is $5x + 6.5y$.

(B) $5x + 6.5y \geq 133$

(C) We graph the portion of the half-plane $5x + 6.5y \geq 133$ that lies in the first quadrant.

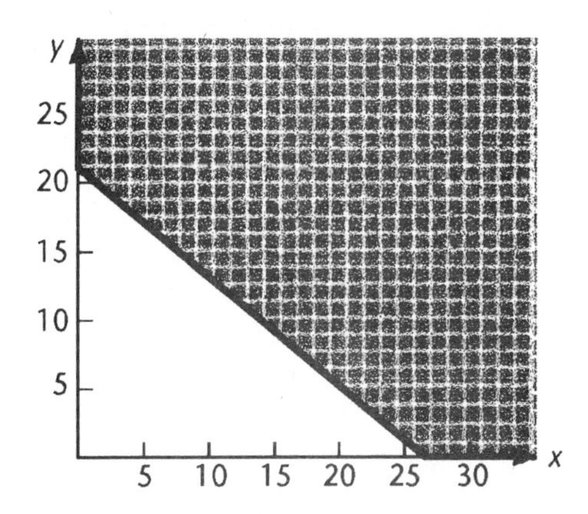

71. (A) Since x = the amount invested at 6% interest
and y = the amount invested at 9% interest
the total amount of interest is $0.06x + 0.09y$.

(B) $0.06x + 0.09y \geq 180$

(C) We graph the portion of the half-plane $0.06x + 0.09y \geq 180$ that lies in the first quadrant.

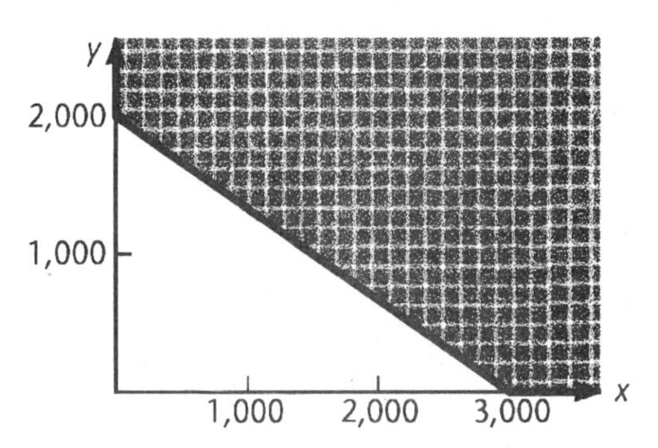

73. (A) Since x = the amount of 25% acid solution
and y = the amount of 60% acid solution,
then $0.25x$ = the amount of acid in the first solution
and $0.6y$ = the amount of acid in the second solution
Then the total amount of acid in the mixture is $0.25x + 0.6y$.

(B) $0.25x + 0.6y \geq 300$

(C) We graph the portion of the half-plane $0.25x + 0.6y \geq 300$ that lies in the first quadrant.

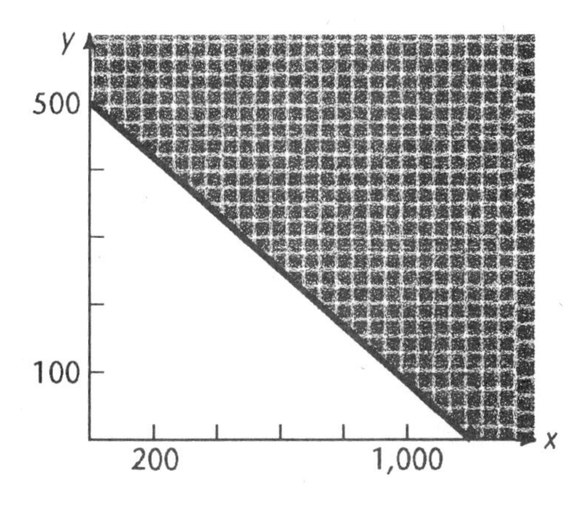

75. (A) The number of fabricating hours used in making the standard model is $6x$; the number of fabricating hours used in making the competition model is $8y$. Then the total number of fabricating hours is $6x + 8y$.

(B) $6x + 8y \leq 120$

(C) We graph the portion of the half-plane $6x + 8y \le 120$ that lies in the first quadrant.

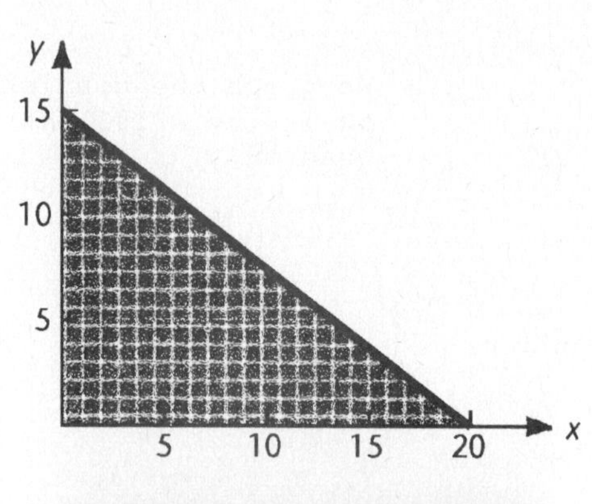

77. (A) The number of grams of protein in x grams of mix A is 0.2x; the number of grams of protein in y grams of mix B is 0.1y. Then the total number of grams of protein is $0.2x + 0.1y$.

(B) $0.2x + 0.1y \ge 23$

(C) We graph the portion of the half-plane $0.2x + 0.1y \ge 23$ that lies in the first quadrant.

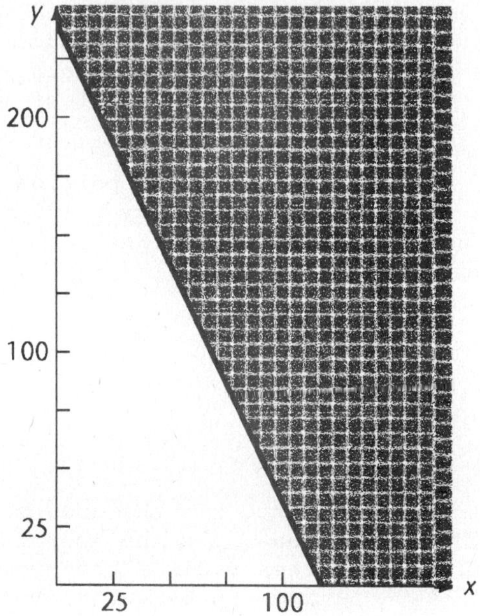

Exercise 5-6 REVIEW EXERCISE

1. $(-1, 8]$
2. $(2, \infty)$
3. $(-\infty, 10]$
4. $[-5, 10)$
5. $-4 \le x < 8$

6. $x < 3$

7. $6 \le x$ or $x \ge 6$

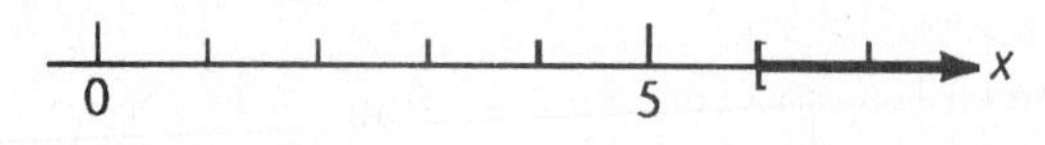

8. $2 < x \le 5$

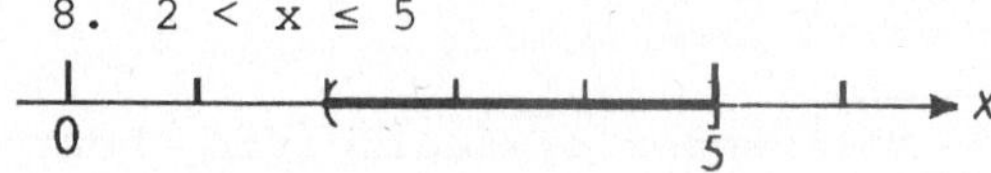

9. Let x = the number. Then translate as follows:

5 [less than] [the quantity 3 times the number] [is more than] [half the number]

$3x - 5 > \frac{1}{2}x$

or, more compactly, $3x - 5 > \frac{1}{2}x$

10. x is a number whose distance from 20 is less than 0.01, that is $|x - 20| < 0.01$. Solving, we can rewrite this as

$$-0.01 < x - 20 < 0.01$$
$$20 - 0.01 < x < 20 + 0.01$$
$$19.99 < x < 20.01$$

11. The value of x quarters is 25x. The value of y dimes is 10y. Thus, the total value is 25x + 10y.

$$25x + 10y \geq 460$$

12. $|x + 2| < 3$

$$-3 < x + 2 < 3$$
$$-5 < x < 1$$

13. The quality points for x credits of A is 4x.
The quality points for y credits of B is 3y.
The total quality points is 4x + 3y.

14. $|b - a| = |11 - (-8)| = |19| = 19$

15. A(1, 4), B(-3, 0), C(-2, -4), D(2, -3)

16.

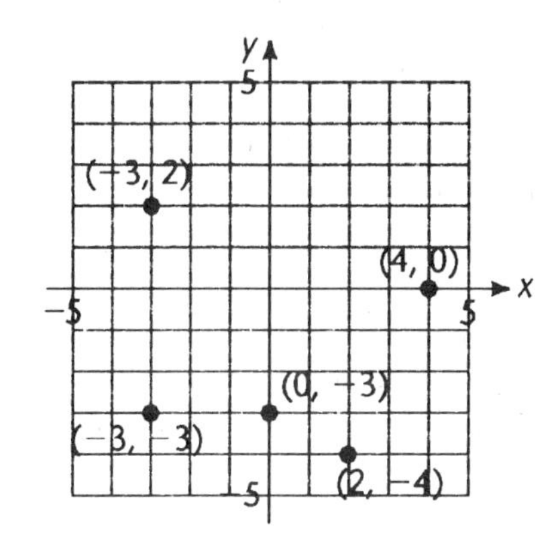

17.
$$4x - 9 = x - 15$$
$$4x - 9 + 9 = x - 15 + 9$$
$$4x = x - 6$$
$$4x - x = x - 6 - x$$
$$3x = -6$$
$$\frac{3x}{3} = \frac{-6}{3}$$
$$x = -2$$

18.
$$2x + 3(x - 1) = 5 - (x - 4)$$
$$2x + 3x - 3 = 5 - x + 4$$
$$5x - 3 = 9 - x$$
$$5x - 3 + 3 = 9 - x + 3$$
$$5x = 12 - x$$
$$5x + x = 12 - x + x$$
$$6x = 12$$
$$\frac{6x}{6} = \frac{12}{6}$$
$$x = 2$$

19.
$$4x - 9 < x - 15$$
$$4x - 9 + 9 < x - 15 + 9$$
$$4x < x - 6$$
$$4x - x < x - 6 - x$$
$$3x < -6$$
$$\frac{3x}{3} < \frac{-6}{3}$$
$$x < -2$$

20.
$$-3 < 2x - 5 < 7$$
$$-3 + 5 < 2x - 5 + 5 < 7 + 5$$
$$2 < 2x < 12$$
$$\frac{2}{2} < \frac{2x}{2} < \frac{12}{2}$$
$$1 < x < 6$$

21. $x = \pm 6$

22. $-6 < x < 6$

23. $x < -6$ or $x > 6$

24.
$$y + 9 = \pm 5$$
$$y = -9 \pm 5$$
$$y = -14, -4$$

25.
$$-5 < y + 9 < 5$$
$$-14 < y < -4$$

26. $y + 9 < -5$ or $y + 9 > 5$

$y < -14$ or $y > -4$

27.
$$\frac{x}{4} - 1 \geq \frac{x}{3} \qquad \text{lcm} = 12$$
$$12\left(\frac{x}{4}\right) - 12 \geq 12\left(\frac{x}{3}\right)$$
$$3x - 12 \geq 4x$$
$$3x \geq 4x + 12$$
$$3x - 4x \geq 4x + 12 - 4x$$
$$-1x \geq 12$$
$$\frac{-1x}{-1} \leq \frac{12}{-1}$$
$$x \leq -12$$

COMMON ERROR: Neglecting to reverse the sense of the inequality when dividing both sides by -1.

28.
$$\frac{1}{6} \leq \frac{1}{4}x + \frac{1}{5} < \frac{1}{3} \qquad \text{lcm} = 60$$
$$60 \cdot \frac{1}{6} \leq 60 \cdot \frac{1}{4}x + 60 \cdot \frac{1}{5} < 60 \cdot \frac{1}{3}$$
$$10 \leq 15x + 12 < 20$$
$$-2 \leq 15x < 8$$
$$\frac{-2}{15} \leq \frac{15x}{15} < \frac{8}{15}$$
$$-\frac{2}{15} \leq x < \frac{8}{15}$$

29. $y = -2x - 3$

↑ slope $= -2$ ↑ y-intercept $= -3$

30. We use the point-slope form with $m = -2$ and $(x_1, y_1) = (2, 4)$.
$$y - y_1 = m(x - x_1)$$
$$y - 4 = -2(x - 2)$$
$$y - 4 = -2x + 4$$
$$2x + y - 4 = 4$$
$$2x + y = 8$$

31. We use the slope formula with $(x_1, y_1) = (1, 3)$ and $(x_2, y_2) = (3, 7)$.
$$m = \frac{y_2 - y_1}{x_2 - x_1} = \frac{7 - 3}{3 - 1} = \frac{4}{2} = 2$$

32. We have found the slope of the line to be 2 in problem 31. Now we use the point-slope form with $m = 2$. We can either use

$(x_1, y_1) = (1, 3)$
$$y - y_1 = m(x - x_1)$$
$$y - 3 = 2(x - 1)$$
$$y - 3 = 2x - 2$$
$$y = 2x + 1$$

or

$(x_1, y_1) = (3, 7)$
$$y - y_1 = m(x - x_1)$$
$$y - 7 = 2(x - 3)$$
$$y - 7 = 2x - 6$$
$$y = 2x + 1$$

33. We choose three convenient solutions to plot, then draw a line through these points.

$y = 2x - 3$

(x,	y)
(0,	-3)
$\left(\frac{3}{2},\right.$	$\left.0\right)$
(3,	3)

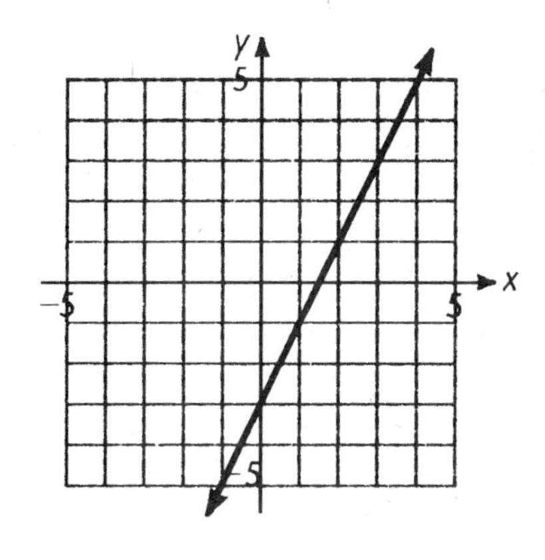

34. We choose three convenient solutions to plot, then draw a line through these points.

$2x + y = 6$

(x,	y)
(0,	6)
(3,	0)
(5,	-4)

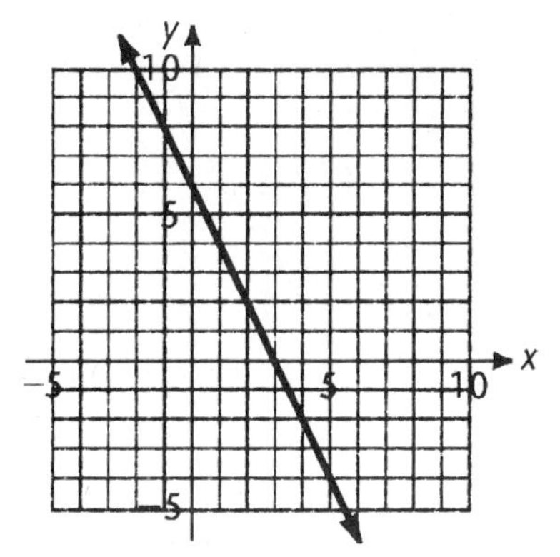

35. First graph the line $x - y = 6$ as a solid line, since equality is included in $x - y \geq 6$. Choose the origin as a test point, since it is not on the line.

$$x - y \geq 6$$
$$0 - 0 \overset{?}{\geq} 6$$
$$0 \not\geq 6$$

The origin does not satisfy the original inequality; hence the graph consists of all points on the other side of the line. Thus, the graph is the lower half-plane.

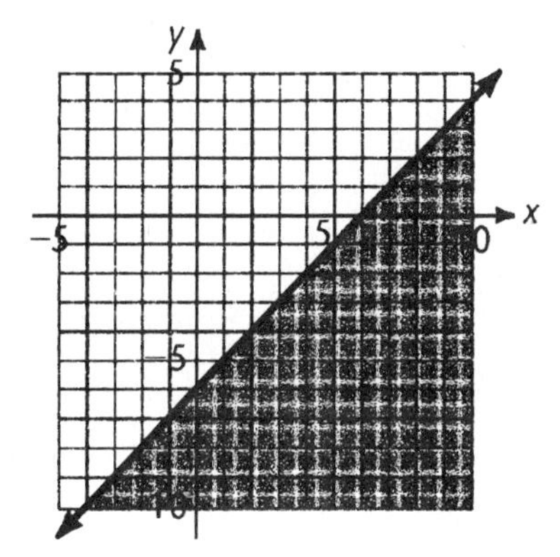

36. First graph the line $y = x - 1$ as a broken line, since equality is not included in $y > x - 1$. Choose the origin as a test point, since it is not on the line.

$$y > x - 1$$
$$0 \overset{?}{>} 0 - 1$$
$$0 > -1$$

The origin satisfies the original inequality; hence all other points on the same side as the origin are also part of the graph. Thus, the graph is the upper half-plane.

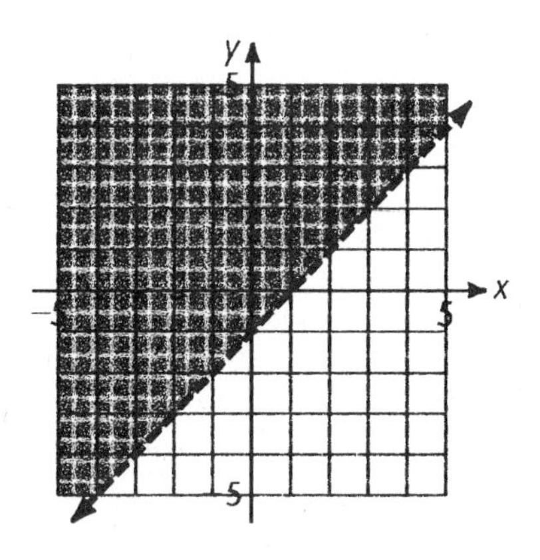

37. We choose three convenient solutions to plot, then draw a line through these points.

$$3x - 2y = 9$$

(x,	y)
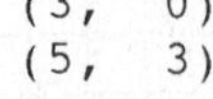 $\left(0,\right.$	$\left.-\frac{9}{2}\right)$
(3,	0)
(5,	3)

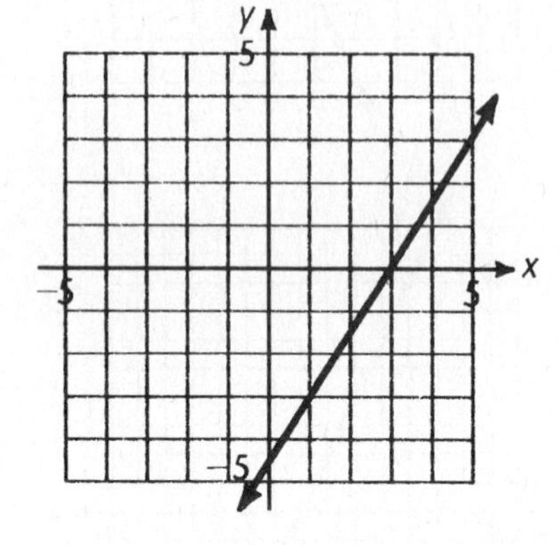

38. We choose three convenient solutions to plot, then draw a line through these points.

$$y = \frac{1}{3}x - 2$$

(x,	y)
(0,	-2)
(6,	0)
(-6,	-4)

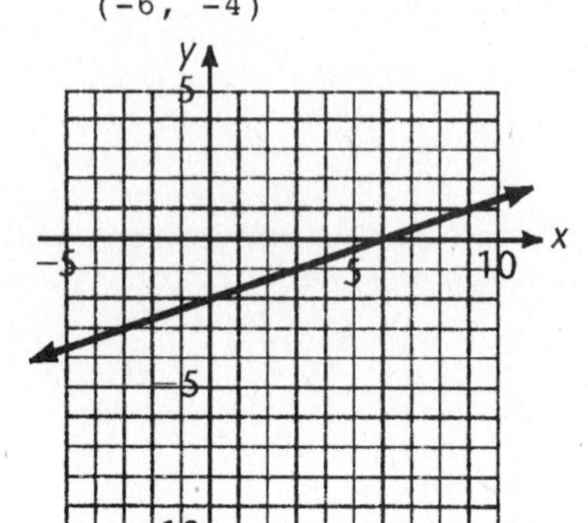

39.

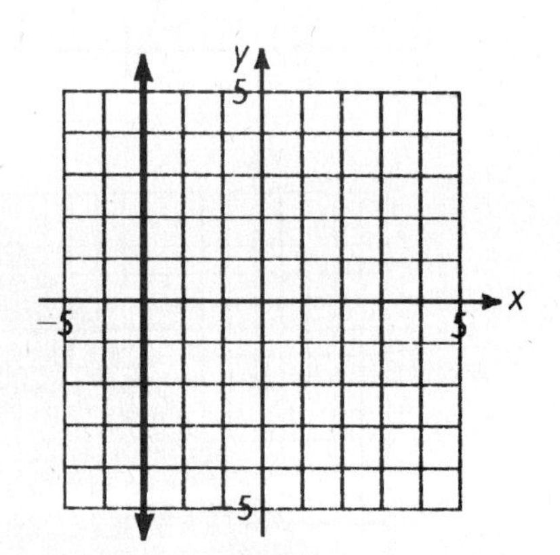

40. First we graph $4x - 5y = 20$ as a solid line, since equality is included in $4x - 5y \le 20$. Choose the origin as a test point, since it is not on the line.

$$4x - 5y \le 20$$

$$4(0) - 5(0) \stackrel{?}{\le} 20$$

$$0 \le 20$$

The origin satisfies the original inequality; hence all other points on the same side as the origin are also part of the graph. Thus, the graph is the upper half-plane.

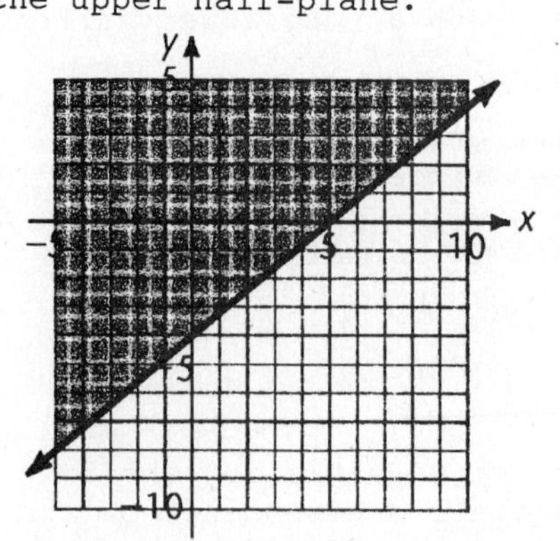

41. First we graph $y = \frac{x}{2} + 1$ as a broken line, since equality is not included in $y < \frac{x}{2} + 1$. Choose the origin as a test point, since it is not on the line.

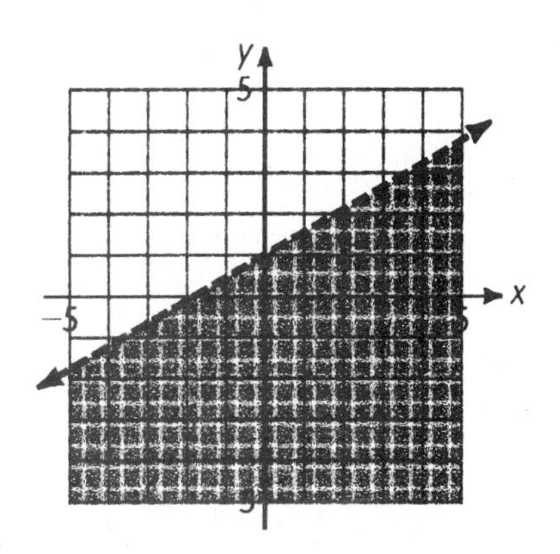

$$y < \frac{x}{2} + 1$$

$$0 \overset{?}{<} \frac{0}{2} + 1$$

$$0 < 1$$

The origin satisfies the original inequality; hence, all other points on the same side as the origin are also part of the graph. Thus, the graph is the lower half-plane.

42. First graph the line $x = -3$ as a solid line, since equality is included in $x \geq -3$. Since x is required to be <u>greater</u> than -3, the graph is the <u>right</u> half-plane.

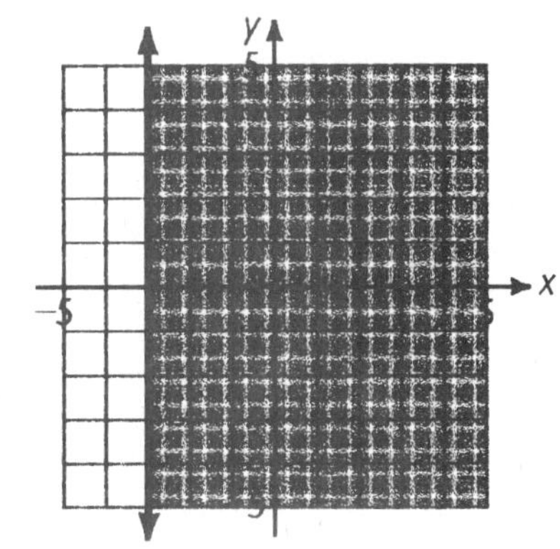

43. This is the intersection of the graphs of the inequalities $-4 \leq y$ (all points above the solid line $y = -4$) and $y < 3$ (all points below the broken line $y = 3$).

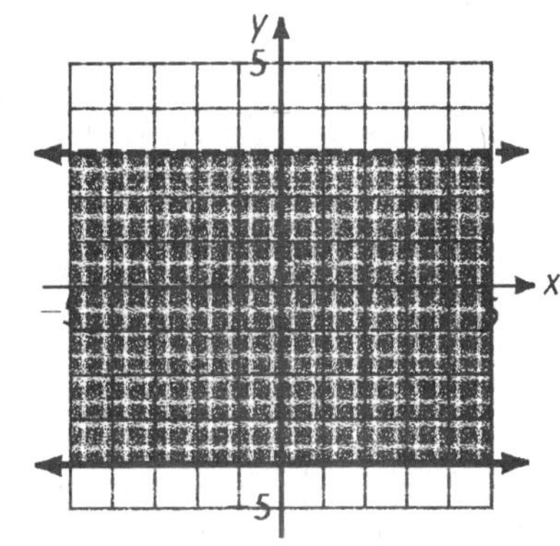

44.
$$-14 \leq 3x - 2 < 7$$
$$-14 + 2 \leq 3x - 2 + 2 < 7 + 2$$
$$-12 \leq 3x < 9$$
$$\frac{-12}{3} \leq \frac{3x}{3} < \frac{9}{3}$$
$$-4 \leq x < 3$$

−4 3 x

45.
$$-3 \leq 5 - 2x < 3$$
$$-8 \leq -2x < -2$$
$$\frac{-8}{-2} \geq \frac{-2x}{-2} > \frac{-2}{-2}$$
$$4 \geq x > 1$$
$$1 < x \leq 4$$

1 4 x

46.
$$3(2 - x) - 2 \leq 2x - 1$$
$$6 - 3x - 2 \leq 2x - 1$$
$$-3x + 4 \leq 2x - 1$$
$$-3x \leq 2x - 5$$
$$-5x \leq -5$$
$$\frac{-5x}{-5} \geq \frac{-5}{-5}$$
$$x \geq 1$$

1 x

47.
$$0.4x - 0.3(x - 3) = 5 \quad \text{lcm} = 10$$
$$10(0.4x) - 10[0.3(x - 3)] = 10(5)$$
$$4x - 3(x - 3) = 50$$
$$4x - 3x + 9 = 50$$
$$x + 9 = 50$$
$$x = 41$$

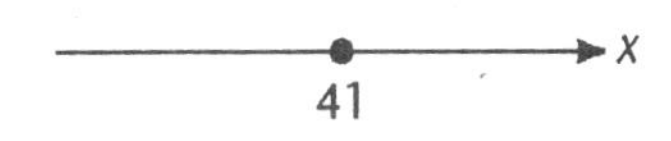

48. $|4x - 7| = 5$

$$4x - 7 = \pm 5$$
$$4x = 7 \pm 5$$
$$x = \frac{7 \pm 5}{4}$$
$$x = \frac{7 - 5}{4} \text{ or } \frac{7 + 5}{4}$$
$$x = \frac{1}{2} \text{ or } 3$$

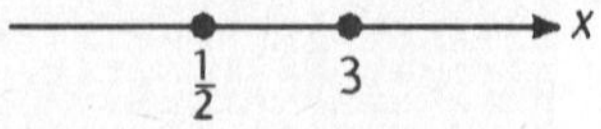

49. $|4x - 7| \le 5$

$$-5 \le 4x - 7 \le 5$$
$$2 \le 4x \le 12$$
$$\frac{1}{2} \le x \le 3$$

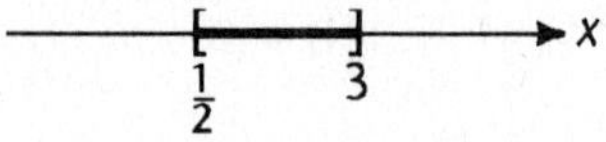

50. $|4x - 7| > 5$

$$4x - 7 < -5 \text{ or } 4x - 7 > 5$$
$$4x < 2 \text{ or } 4x > 12$$
$$x < \frac{1}{2} \text{ or } x > 3$$

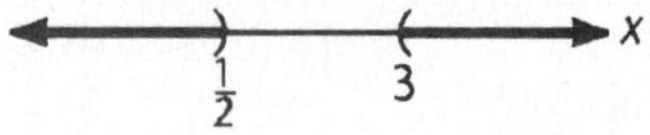

51. $\frac{x + 3}{8} \le 5 - \frac{2 - x}{3}$ lcm = 24

$$24 \cdot \frac{(x + 3)}{8} \le 24 \cdot 5 - 24 \cdot \frac{(2 - x)}{3}$$
$$3(x + 3) \le 120 - 8(2 - x)$$
$$3x + 9 \le 120 - 16 + 8x$$
$$3x + 9 \le 104 + 8x$$
$$3x \le 95 + 8x$$
$$-5x \le 95$$
$$\frac{-5x}{-5} \ge \frac{95}{-5}$$
$$x \ge -19$$

52. $-6 < \frac{3}{5}(x - 4) \le -3$

$$5(-6) \le 5\left[\frac{3}{5}(x - 4)\right] \le 5(-3)$$
$$-30 < 3(x - 4) \le -15$$
$$-30 < 3x - 12 \le -15$$
$$-18 < 3x \le -3$$
$$-6 < x \le -1$$

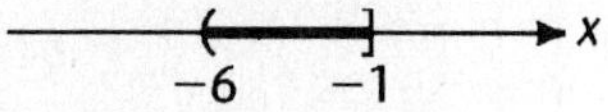

53. Starting with $x + 2y = -6$, we solve for y to obtain:

$$2y = -x - 6$$
$$\frac{1}{2}(2y) = \frac{1}{2}(-x - 6)$$
$$y = -\frac{1}{2}x - 3$$

Thus, the slope is $-\frac{1}{2}$ and the y intercept is -3.

54. $y - y_1 = m(x - x_1)$ $m = -\frac{1}{3}$, $(x_1, y_1) = (-3, 2)$

$$y - 2 = -\frac{1}{3}[x - (-3)]$$
$$y - 2 = -\frac{1}{3}(x + 3)$$
$$y - 2 = -\frac{1}{3}x - 1$$
$$y = -\frac{1}{3}x + 1$$

55. First find the slope of the line using the slope formula:

$$m = \frac{y_2 - y_1}{x_2 - x_1} = \frac{-2 - 2}{3 - (-3)} = \frac{-4}{6} = -\frac{2}{3}$$

Now use the point-slope form with $m = -\frac{2}{3}$. We can either use

$$(x_1, y_1) = (-3, 2)$$
$$y - y_1 = m(x - x_1)$$
$$y - 2 = -\frac{2}{3}[x - (-3)]$$
$$y - 2 = -\frac{2}{3}(x + 3)$$
$$3(y - 2) = -2(x + 3)$$
$$3y - 6 = -2x - 6$$
$$2x + 3y = 0$$

or

$$(x_1, y_1) = (3, -2)$$
$$y - y_1 = m(x - x_1)$$
$$y - (-2) = -\frac{2}{3}(x - 3)$$
$$y + 2 = -\frac{2}{3}(x - 3)$$
$$3(y + 2) = -2(x - 3)$$
$$3y + 6 = -2x + 6$$
$$2x + 3y = 0$$

56. First find the slope of the given line by writing $x + 2y = -6$ in the form $y = mx + b$.

$$x + 2y = -6$$
$$2y = -x - 6$$
$$\frac{1}{2}(2y) = \frac{1}{2}(-x - 6)$$
$$y = -\frac{1}{2}x - 3$$

The slope of the given line is $-\frac{1}{2}$. The slope of a line perpendicular to the given line is the negative reciprocal of $-\frac{1}{2}$, that is, 2. We have to find the equation of a line through (3, -4) with slope 2.

$$y - y_1 = m(x - x_1) \qquad m = 2,\ (x_1, y_1) = (3, -4)$$
$$y - (-4) = 2(x - 3)$$
$$y + 4 = 2x - 6$$
$$y = 2x - 10$$

57. vertical: $x = 5$, horizontal: $y = -2$

58. First find the slope of the given line by writing $3x - 2y = 5$ in the form $y = mx + b$.

$$3x - 2y = 5$$
$$-2y = -3x + 5$$
$$-\frac{1}{2}(-2y) = -\frac{1}{2}(-3x + 5)$$
$$y = \frac{3}{2}x - \frac{5}{2}$$

The slope of the given line is $\frac{3}{2}$. The slope of a line parallel to the given line is also $\frac{3}{2}$. We have to find the equation of a line through (-6, 2) with slope $\frac{3}{2}$.

$$y - y_1 = m(x - x_1) \qquad m = \frac{3}{2},\ (x_1, y_1) = (-6, 2)$$
$$y - 2 = \frac{3}{2}[x - (-6)]$$
$$y - 2 = \frac{3}{2}(x + 6)$$
$$y - 2 = \frac{3}{2}x + 9$$
$$y = \frac{3}{2}x + 11$$

59. This is the set of points common to the solutions of the inequalities $-2 < x$, $x \le 4$, $0 \le y$, and $y < 3$.

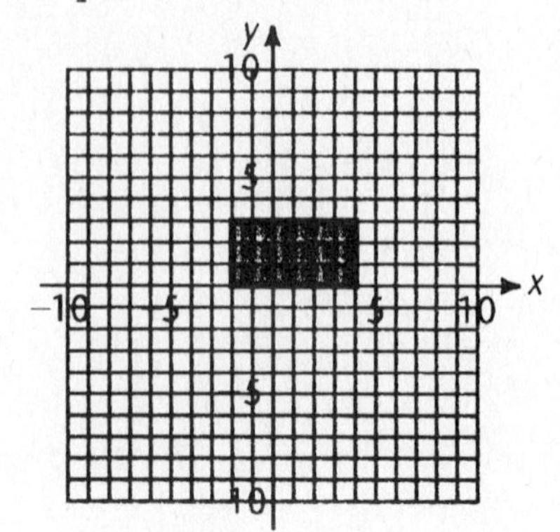

60. $$y - \frac{1}{2} = \frac{2}{3}\left(x + \frac{3}{4}\right)$$

We simplify this equation somewhat:

$$y - \frac{1}{2} = \frac{2}{3}x + \frac{2}{3} \cdot \frac{3}{4}$$

$$y - \frac{1}{2} = \frac{2}{3}x + \frac{1}{2}$$

$$y - \frac{1}{2} + \frac{1}{2} = \frac{2}{3}x + \frac{1}{2} + \frac{1}{2}$$

$$y = \frac{2}{3}x + 1$$

We now choose three convenient solutions to plot, then draw a line through these points:

$$y = \frac{2}{3}x + 1$$

(x, y)
(3, 3)
(0, 1)
(6, 5)

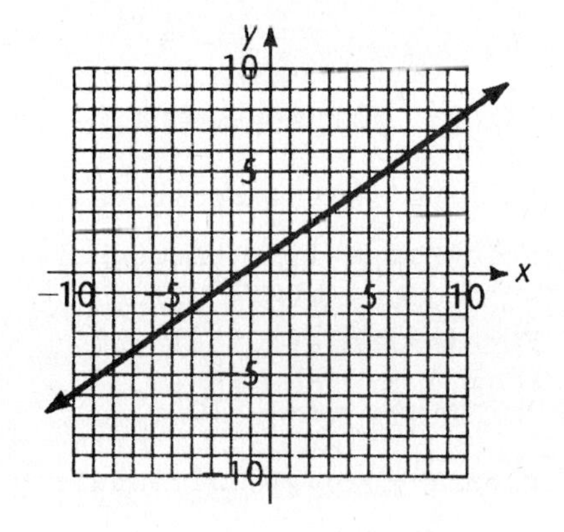

61. This is the set of points common to the solutions of the inequalities $2x + 5y \le 10$, $x \ge 0$, $y \ge 0$. The result is that portion of the half-plane $2x + 5y \le 10$ that lies in the first quadrant.

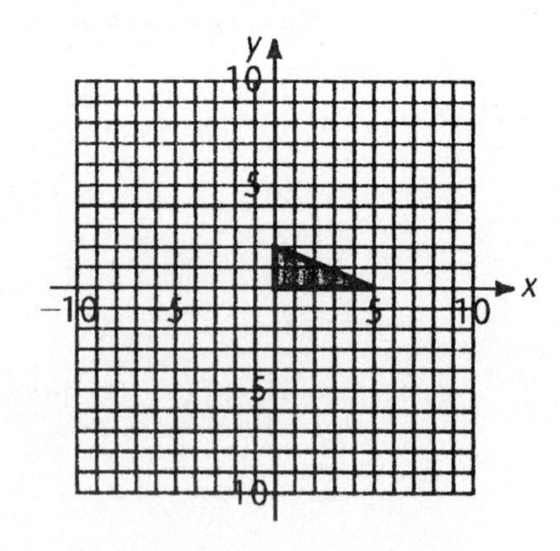

62. First graph the line $y = \frac{3}{5}x + 3$ as a solid line, since equality is included in $y \geq \frac{3}{5}x + 3$. Choose the origin as a test point, since it is not on the line.

$$y \geq \frac{3}{5}x + 3$$

$$0 \overset{?}{\geq} \frac{3}{5}(0) + 3$$

$$0 \not\geq 3$$

The origin does not satisfy the original inequality; hence the graph consists of all points on the other side of the line. Thus, the graph is the upper half-plane.

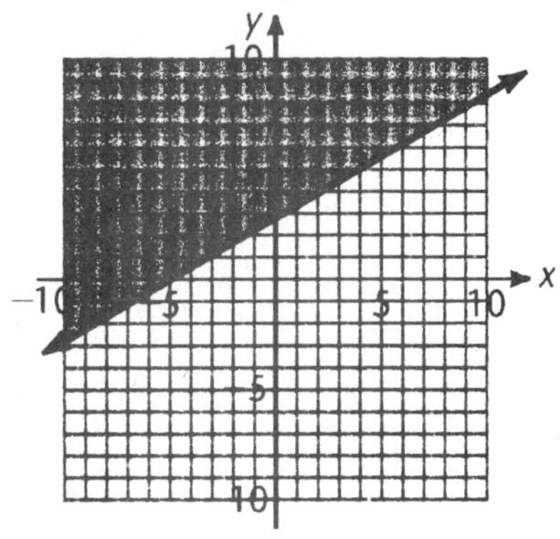

63. $|x| \leq 4$. $-4 \leq x \leq 4$. This is the set of points common to the graphs of the inequalities $-4 \leq x$ (all points to the right of the solid line $x = -4$) and $x \leq 4$ (all points to the left of the solid line $x = 4$).

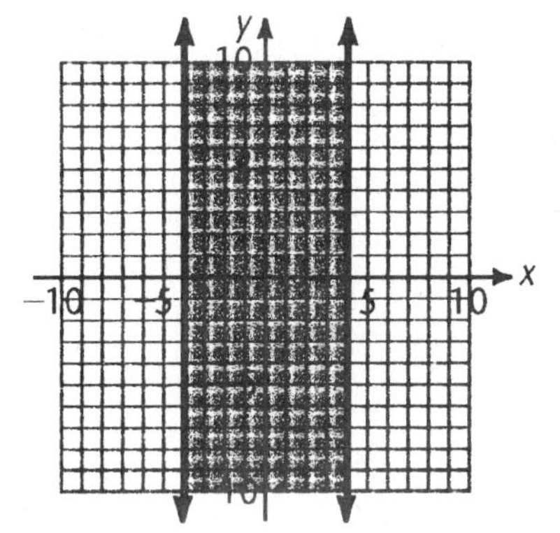

64. $\frac{1}{10}x + \frac{1}{20}y = \frac{1}{2}$

We simplify this equation somewhat:

$$20 \cdot \frac{1}{10}x + 20 \cdot \frac{1}{20}y = 20 \cdot \frac{1}{2}$$

$$2x + y = 10$$

We now choose three convenient solutions to plot, then draw a line through these points:

$2x + y = 10$

(x,	y)
(0,	10)
(5,	0)
(2,	6)

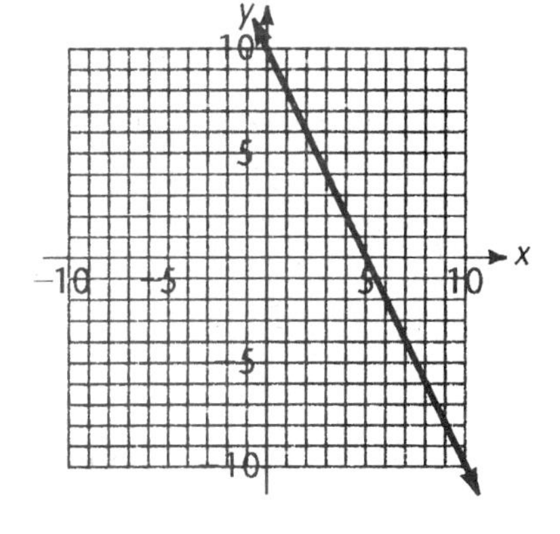

65. Let x = the number of wins needed.
Then 81 + x = the total number of wins.
81 + 61 + 20 = 162 = the total number of games played.

$$\text{The winning rate} = \frac{\text{number of wins}}{\text{number of games played}} = \frac{81 + x}{162}$$

$$\frac{81 + x}{162} \geq 0.585$$

$$81 + x \geq 0.585(162)$$

$$81 + x \geq 94.77$$

$$x \geq 13.77$$

Since the solution must be a natural number, 14 wins are needed.

66. We are asked to find a linear equation, that is, one of the form $y = mx + b$, that relates x and y. We are given two points on the graph of this equation, namely $(x_1, y_1) = (40, 1000)$ and $(x_2, y_2) = (30, 1500)$. To find the equation, we first find the slope of the line through these two points, using the slope formula

$$m = \frac{y_2 - y_1}{x_2 - x_1} = \frac{1500 - 1000}{30 - 40} = \frac{500}{-10} = -50$$

Now we use the point-slope form with $m = -50$. We can either use

$$(x_1, y_1) = (40, 1000) \quad \text{or} \quad (x_2, y_2) = (50, 1500)$$

$$\begin{aligned} y - y_1 &= m(x - x_1) \\ y - 1000 &= -50(x - 40) \\ y - 1000 &= -50x + 2000 \\ y &= -50x + 3000 \end{aligned} \qquad \begin{aligned} y - y_1 &= m(x - x_1) \\ y - 1500 &= -50(x - 30) \\ y - 1500 &= -50x + 1500 \\ y &= -50x + 3000 \end{aligned}$$

67.
$$\begin{aligned} &|3 - 2x| \le 5 \\ &-5 \le 3 - 2x \le 5 \\ &-8 \le -2x \le 2 \\ &\frac{-8}{-2} \ge \frac{-2x}{-2} \ge \frac{2}{-2} \\ &4 \ge x \ge -1 \\ &-1 \le x \le 4 \text{ or } [-1, 4] \end{aligned}$$

68.
$$\frac{3x}{5} - \frac{1}{2}(x - 3) \le \frac{1}{3}(x + 2) \qquad \text{lcm} = 30$$

$$\begin{aligned} 30 \cdot \frac{3x}{5} - 30\left[\frac{1}{2}(x - 3)\right] &\le 30\left[\frac{1}{3}(x + 2)\right] \\ 18x - 15(x - 3) &\le 10(x + 2) \\ 18x - 15x + 45 &\le 10x + 20 \\ 3x + 45 &\le 10x + 20 \\ 3x &\le 10x - 25 \\ -7x &\le -25 \\ \frac{-7x}{-7} &\ge \frac{-25}{-7} \\ x &\ge \frac{25}{7} \end{aligned}$$

69.
$$-4 \le \frac{2}{3}(6 - 2x) \le 8 \qquad \text{lcm} = 3$$

$$\begin{aligned} &3(-4) \le 3\left[\frac{2}{3}(6 - 2x)\right] \le 3 \cdot 8 \\ &-12 \le 2(6 - 2x) \le 24 \\ &-12 \le 12 - 4x \le 24 \\ &-24 \le -4x \le 12 \\ &\frac{-24}{-4} \ge \frac{-4x}{-4} \ge \frac{12}{-4} \\ &6 \ge x \ge -3 \\ &-3 \le x \le 6 \end{aligned}$$

70. $|2x - 3| < -2$. Since the absolute value of a number is never negative, there can be no numbers satisfying this condition. No solution.

71. $|u| = u$ if $u \ge 0$

So $|2x - 3| = 2x - 3$ if $2x - 3 \ge 0$

$2x - 3 \ge 0$ if $2x \ge 3$

$x \ge \frac{3}{2}$

72. $|u| = -u$ if $u \le 0$

So $|2x - 3| = -(2x - 3)$ if $2x - 3 \le 0$

$2x - 3 \le 0$ if $2x \le 3$

$x \le \frac{3}{2}$

73. (A) The slope of the first line is 2; the slope of the second line is $\frac{1}{2}$.
Since $2 \ne \frac{1}{2}$ and $2 \cdot \frac{1}{2} \ne -1$, the lines are neither parallel nor perpendicular.

(B) The slope of $y = -3x - 11$ is -3; the slope of $3x + y = -1$, that is, $y = -3x - 1$, is also -3. Therefore the lines are parallel.

74. $y = 3x + 4$

(x,	y)
(0,	4)
(-1,	1)
(-2,	-2)

$y = -x + 8$

(x,	y)
(0,	8)
(4,	4)
(8,	0)

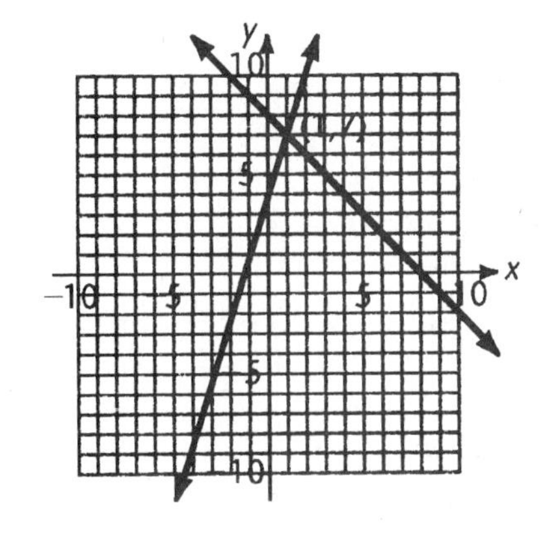

The coordinates of the point where the two graphs cross seem to be (1, 7). To check that this is correct, we substitute (1, 7) into each equation.

$y = 3x + 4$ $\qquad$ $y = -x + 8$

$7 \overset{\checkmark}{=} 3 \cdot 1 + 4$ $\qquad$ $7 \overset{\checkmark}{=} -1 + 8$

75. The investment of x at 6% gives a return of 0.06x.
The investment of y at 9% gives a return of 0.09y.
The total return is $0.06x + 0.09y$

$0.06x + 0.09y \ge 5400$

We graph the portion of the half-plane $0.06x + 0.09y \ge 5400$ that lies in the first quadrant.

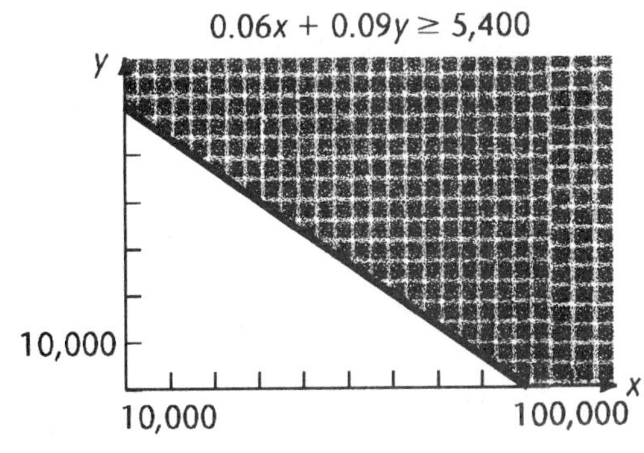

Practice Test 5

1. $\left|1 - \frac{1}{2}x\right| \le \frac{3}{4}$

$$-\frac{3}{4} \le 1 - \frac{1}{2}x \le \frac{3}{4} \quad \text{lcm} = 4$$

$$4\left(-\frac{3}{4}\right) \le 4\cdot 1 - 4\cdot\frac{1}{2}x \le 4\cdot\frac{3}{4}$$

$$-3 \le 4 - 2x \le 3$$

$$-7 \le -2x \le -1$$

$$\frac{-7}{-2} \ge \frac{-2x}{-2} \ge \frac{-1}{-2}$$

$$\frac{7}{2} \ge x \ge \frac{1}{2}$$

$$\frac{1}{2} \le x \le \frac{7}{2} \quad \text{or} \quad \left[\frac{1}{2}, \frac{7}{2}\right]$$

2. $1 - \frac{1}{2}x \ge \frac{3}{4} \quad \text{lcm} = 4$

$$4\cdot 1 - 4\cdot\frac{1}{2}x \ge 4\cdot\frac{3}{4}$$

$$4 - 2x \ge 3$$

$$-2x \ge -1$$

$$\frac{-2x}{-2} \le \frac{-1}{-2}$$

$$x \le \frac{1}{2} \quad \text{or} \quad \left(-\infty, \frac{1}{2}\right]$$

3. $-1 \le \frac{1}{2}x < \frac{3}{4}$

$$2(-1) \le 2\cdot\frac{1}{2}x < 2\cdot\frac{3}{4}$$

$$-2 \le x < \frac{3}{2} \quad \text{or} \quad \left[-2, \frac{3}{2}\right)$$

4. $-3 \le x < 6$

5. $\frac{1}{2}x + \frac{3}{4}y = 5$

$$\frac{3}{4}y = -\frac{1}{2}x + 5$$

$$\frac{4}{3}\cdot\frac{3}{4}y = \frac{4}{3}\left(-\frac{1}{2}x\right) + \frac{4}{3}\cdot 5$$

$$y = -\frac{2}{3}x + \frac{20}{3}$$

6. $1 - 2x > 3$ or $1 - 2x < -3$

7. We use the slope formula with $(x_1, y_1) = (-1, 2)$ and $(x_2, y_2) = (3, -4)$.

$$m = \frac{y_2 - y_1}{x_2 - x_1} = \frac{(-4) - 2}{3 - (-1)} = \frac{-6}{4} = -\frac{3}{2}$$

8. We use the point-slope form with $m = -4$ and $(x_1, y_1) = (5, 6)$

$$y - y_1 = m(x - x_1)$$

$$y - 6 = -4(x - 5)$$

$$y - 6 = -4x + 20$$

$$y = -4x + 26$$

9. First we use the slope formula $(x_1, y_1) = (3, -2)$ and $(x_2, y_2) = (-1, 0)$.

$$m = \frac{y_2 - y_1}{x_2 - x_1} = \frac{0 - (-2)}{(-1) - 3} = \frac{2}{-4} = -\frac{1}{2}$$

Now we use the point-slope form with $m = -\frac{1}{2}$. We can either use

$$(x_1, y_1) = (3, -2)$$
$$y - y_1 = m(x - x_1)$$
$$y - (-2) = -\frac{1}{2}(x - 3)$$
$$y + 2 = -\frac{1}{2}x + \frac{3}{2}$$
$$y = -\frac{1}{2}x - \frac{1}{2}$$

or

$$(x_1, y_1) = (-1, 0)$$
$$y - y_1 = m(x - x_1)$$
$$y - 0 = -\frac{1}{2}[x - (-1)]$$
$$y = -\frac{1}{2}(x + 1)$$
$$y = -\frac{1}{2}x - \frac{1}{2}$$

10. Since all vertical lines are parallel and also vertical lines are perpendicular to horizontal lines, we have
$x = 2$ is parallel to $x - 6 = 0$
$x = 2$ is perpendicular to $y = 3$
$x - 6 = 0$ is perpendicular to $y = 3$
Since $y = 4x + 5$ and $12x - 3y = 6$ (or $y = 4x - 2$) both have slope 4, they are parallel. There are no other pairs of parallel or perpendicular lines.

11.

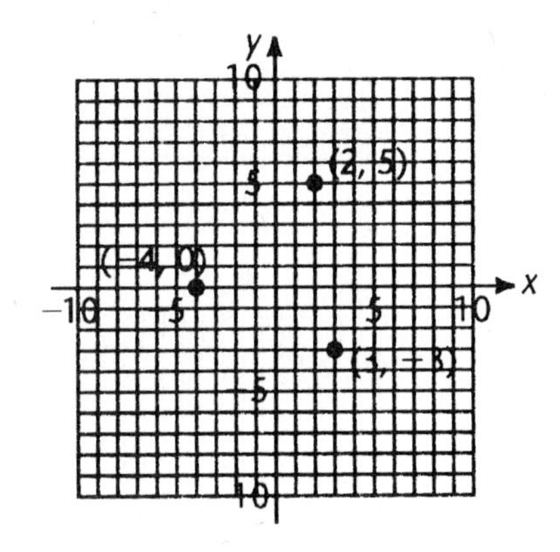

12. From the slope-intercept form, $y = mx + b$, the equation of the line is $y = 3x + 4$. We choose three convenient solutions to plot, then draw a line through these points.

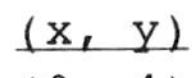

(x, y)
(0, 4)
(-1, 1)
(1, 7)

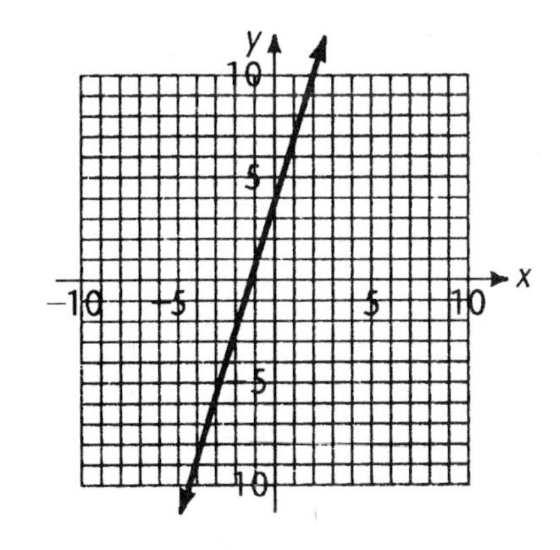

13. We choose three convenient solutions to plot, then draw a line through these points.

$$2x - 3y = 9$$

(x,	y)
(0,	-3)
(3,	-1)
(6,	1)

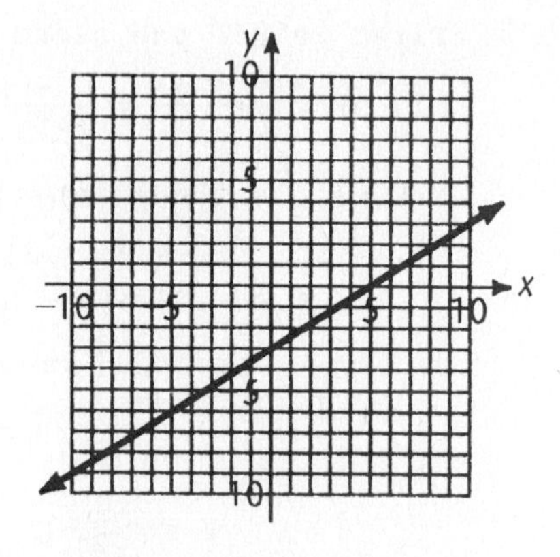

14. First we graph $4x + 5y = 20$ as a solid line, since equality is included in $4x + 5y \le 20$. Choose the origin as a test point, since it is not on the line.

$$4x + 5y \le 20$$
$$4(0) + 5(0) \overset{?}{\le} 20$$
$$0 \le 20$$

The origin satisfies the original inequality; hence, all other points on the same side as the origin are also part of the graph. Thus, the graph is the lower half-plane.

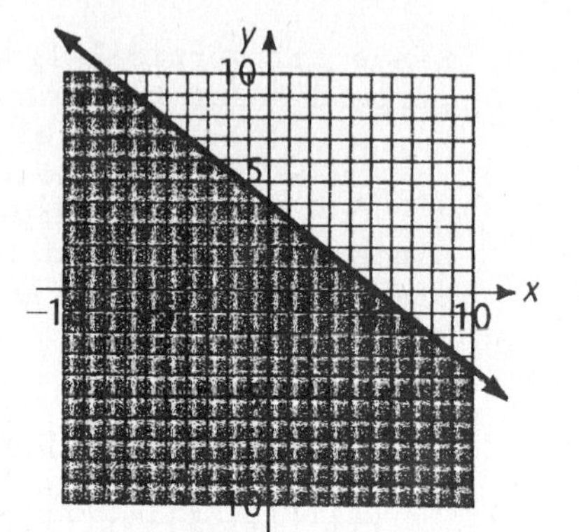

15. We choose three convenient solutions to plot, then draw a line through these points.

$$\frac{1}{2}x + \frac{1}{3}y = 2$$

(x, y)
(0, 6)
(4, 0)
(2, 3)

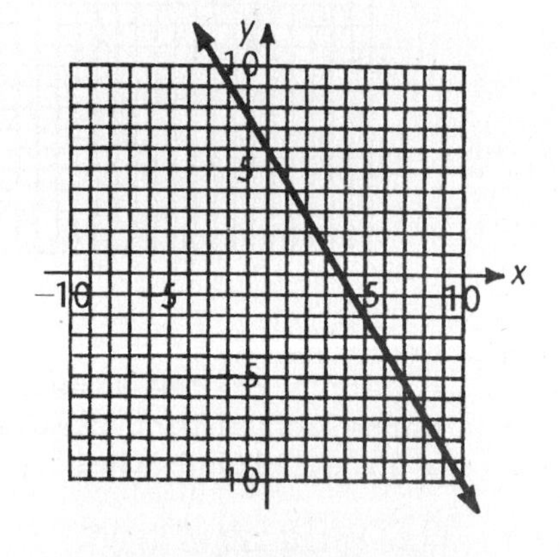

16. First graph the line $0.5x - 0.6y = 15$ or $5x - 6y = 150$ as a solid line, since equality is included in $0.5x - 0.6y \le 15$. Choose the origin as a test point, since it is not on the line.

$$0.5x - 0.6y \le 15$$
$$0.5(0) - 0.6(0) \overset{?}{\le} 15$$
$$0 \le 15$$

The origin satisfies the original inequality; hence, all other points on the same side as the origin are also part of the graph. Thus, the graph is the upper half-plane.

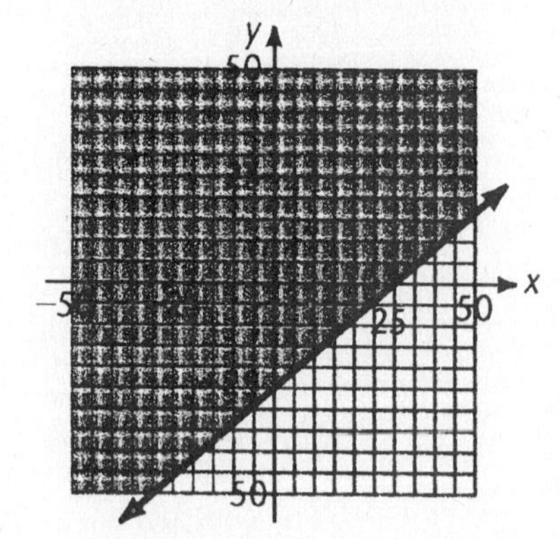

17. This is the set of points common to the solutions of the inequalities $3x + 4y \geq 12$, $x \geq 0$, and $y \geq 0$.

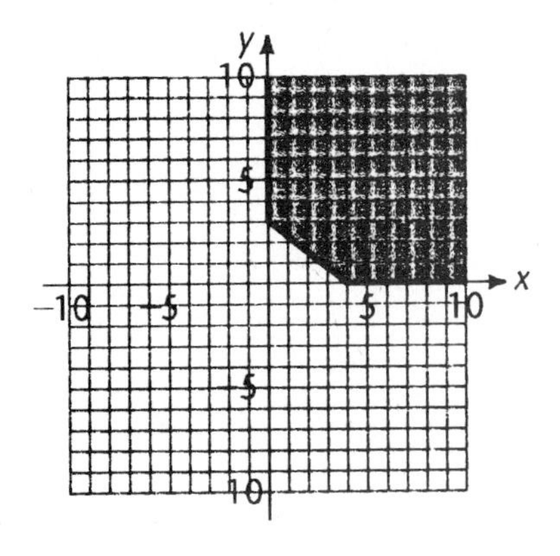

18. (A)

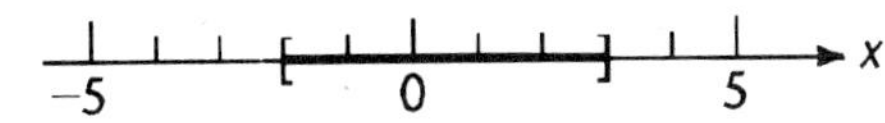

(B) This is the set of points common to the graphs of the inequalities $-2 \leq x$ (all points to the right of the solid line $x = -2$) and $x \leq 3$ (all points to the left of the solid line $x = 3$).

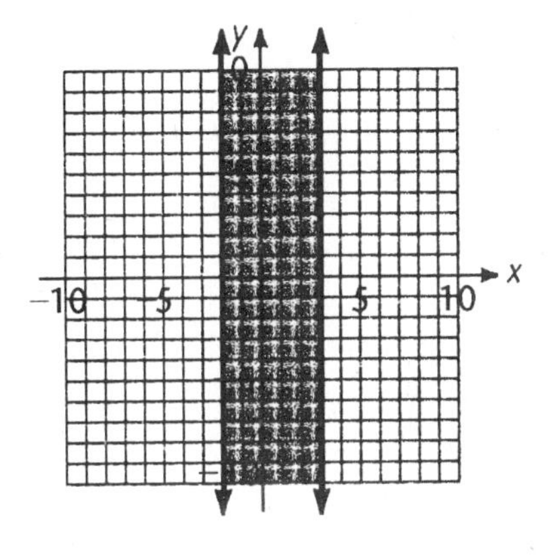

19. Let x = the number of additional minority employees.
Then x + 210 = the number of minority employees
x + 1370 = the total number of employees then in the work force.

$$\frac{x + 210}{x + 1370} \geq 0.2$$

$$x + 210 \geq 0.2(x + 1370)$$ assuming x + 1370 is positive, as it must be since x is positive

$$x + 210 \geq 0.2x + 274$$
$$0.8x + 210 \geq 274$$
$$0.8x \geq 64$$
$$x \geq 80$$

80 (or more) minority employees are needed.

20. Since x = the number of units of product A
14x = the labor cost for product A
and since y = the number of units for product B
12y = the labor cost for product B
Hence 14x + 12y = the total labor cost

$$14x + 12y \leq 1000, \quad x \geq 0, \ y \geq 0$$

We graph the portion of the half-plane $14x + 12y \leq 1000$ that lies in the first quadrant.

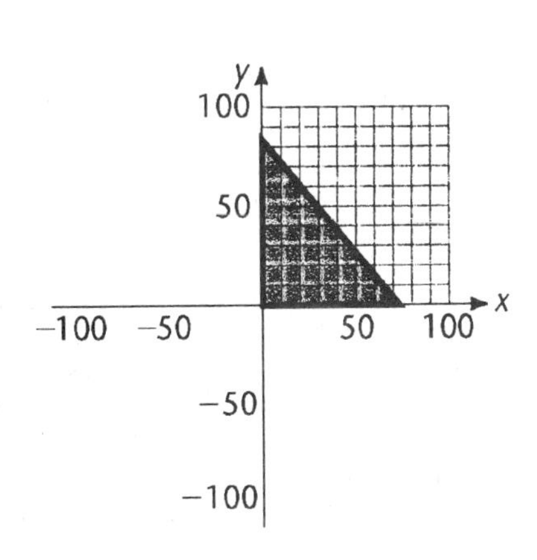

CHAPTER 6 Exponents, Radicals and Complex Numbers

Exercise 6-1 Positive-Integer Exponents

Key Ideas and Formulas

Positive Integer Exponents $a^n = \underbrace{a \cdot a \cdot \ldots \cdot a}_{n \text{ factors of } a}$

Properties of Exponents:

$$a^m a^n = a^{m+n}$$
$$(a^n)^m = a^{mn}$$
$$(ab)^m = a^m b^m$$
$$\left(\frac{a}{b}\right)^m = \frac{a^m}{b^m}$$
$$\frac{a^m}{a^n} = \begin{cases} a^{m-n} & \text{if } m \text{ is larger than } n \\ 1 & \text{if } m = n \\ \dfrac{1}{a^{n-m}} & \text{if } n \text{ is larger than } m \end{cases}$$

COMMON ERRORS: Applying properties of exponents to the wrong algebraic form.

$(a + b)^m$ is **not** the same as $a^m + b^m$

$(a - b)^m$ is **not** the same as $a^m - b^m$

---

1. $y^2y^8 = y^{2+8} = y^{10}$. 10

3. $y^9 = y^3y^6$. 6

5. $(u^4)^3 = u^{3\cdot 4} = u^{12}$. 12

7. $(x^{10}) = (x^2)^5$. 2

9. $(uv)^7 = u^7v^7$

11. $p^4q^4 = (pq)^4$. 4

13. $\left(\frac{a}{b}\right)^8 = \frac{a^8}{b^8}$

15. $\frac{m^3}{n^3} = \left(\frac{m}{n}\right)^3$. 3

17. $\frac{n^{14}}{n^8} = n^{14-8} = n^6$. 6

19. $m^6 = \frac{m^8}{m^2}$. 2

21. $\frac{x^4}{x^{11}} = \frac{1}{x^{11-4}} = \frac{1}{x^7}$. 7

23. $\frac{1}{x^8} = \frac{x^4}{x^{8+4}} = \frac{x^4}{x^{12}}$. 12

25. $(4x^2)(2x^{10}) = 4\cdot 2x^2x^{10} = 8x^{2+10} = 8x^{12}$

27. $\frac{9x^6}{3x^4} = \frac{9}{3}\cdot\frac{x^6}{x^4} = 3x^{6-4} = 3x^2$

29. $\frac{6m^5}{8m^7} = \frac{6}{8}\cdot\frac{m^5}{m^7} = \frac{3}{4}\,\frac{1}{m^{7-5}} = \frac{3}{4m^2}$

31. $(xy)^{10} = x^{10}y^{10}$

33. $\left(\frac{m}{n}\right)^5 = \frac{m^5}{n^5}$

35. $(4y^3)(5y)(y^7) = 4\cdot 5y^3yy^7 = 20y^{3+1+7} = 20y^{11}$

37. $(6 \times 10^8)(8 \times 10^9) = 6 \cdot 8 \times 10^8 \times 10^9 = 48 \times 10^{8+9} = 48 \times 10^{17}$

39. $(10^7)^2 = 10^{2 \cdot 7} = 10^{14}$

41. $(x^3)^2 = x^{2 \cdot 3} = x^6$

43. $(m^2n^5)^3 = (m^2)^3(n^5)^3 = m^{3 \cdot 2}n^{3 \cdot 5} = m^6n^{15}$

45. $\left(\frac{c^2}{d^5}\right)^3 = \frac{(c^2)^3}{(d^5)^3} = \frac{c^{3 \cdot 2}}{d^{3 \cdot 5}} = \frac{c^6}{d^{15}}$

47. $\frac{9u^8v^6}{3u^4v^8} = \frac{9}{3} \cdot \frac{u^8}{u^4} \cdot \frac{v^6}{v^8} = \frac{3}{1} \cdot \frac{u^4}{1} \cdot \frac{1}{v^2} = \frac{3u^4}{v^2}$

49. $(2s^2t^4)^4 = 2^4(s^2)^4(t^4)^4 = 2^4s^{4 \cdot 2}t^{4 \cdot 4} = 2^4s^8t^{16}$ or $16s^8t^{16}$

51. $6(xy^3)^5 = 6x^5(y^3)^5 = 6x^5y^{5 \cdot 3} = 6x^5y^{15}$

53. $\left(\frac{mn^3}{p^2q}\right)^4 = \frac{m^4(n^3)^4}{(p^2)^4q^4} = \frac{m^4n^{4 \cdot 3}}{p^{4 \cdot 2}q^4} = \frac{m^4n^{12}}{p^8q^4}$

55. $\frac{(4u^3v)^3}{(2uv^2)^6} = \frac{4^3(u^3)^3v^3}{2^6u^6(v^2)^6} = \frac{64u^{3 \cdot 3}v^3}{64u^6v^{6 \cdot 2}} = \frac{64u^9v^3}{64u^6v^{12}} = \frac{64}{64} \cdot \frac{u^9}{u^6} \cdot \frac{v^3}{v^{12}} = 1 \cdot \frac{u^3}{1} \cdot \frac{1}{v^9} = \frac{u^3}{v^9}$

57. $\frac{(9x^3)^2}{(-3x)^2} = \frac{9^2(x^3)^2}{(-3)^2x^2} = \frac{81x^{2 \cdot 3}}{9x^2} = \frac{81x^6}{9x^2} = 9x^4$

59. $\frac{(-ab^2)^3}{-(a^2b)^2} = \frac{(-a)^3(b^2)^3}{-(a^2)^2b^2} = \frac{(-a)(-a)(-a)b^{2 \cdot 3}}{-a^{2 \cdot 2}b^2} = \frac{-a^3b^6}{-a^4b^2} = \frac{a^3}{a^4} \cdot \frac{b^6}{b^2} = \frac{1}{a} \cdot \frac{b^4}{1} = \frac{b^4}{a}$

61. $\frac{-2a^2b^3}{(-2a^2b)^2} = \frac{-2a^2b^3}{(-2)^2(a^2)^2b^2} = \frac{-2a^2b^3}{(-2)(-2)a^{2 \cdot 2}b^2} = \frac{-2a^2b^3}{4a^4b^2} = \frac{-2}{4} \cdot \frac{a^2}{a^4} \cdot \frac{b^3}{b^2} = -\frac{1}{2} \cdot \frac{1}{a^2} \cdot b = \frac{-b}{2a^2}$

63. $\frac{-3^3}{(-3)^3} = \frac{-(3^3)}{(-3)(-3)(-3)} = \frac{-(27)}{-27} = \frac{-27}{-27} = 1$

65. $\frac{-x^2}{(-x)^2} = \frac{-(x)^2}{(-x)^2} = \frac{-(x^2)}{(-x)(-x)} = \frac{-(x^2)}{x^2} = \frac{-1(x^2)}{x^2} = -1$

67. $\frac{(-x^2)^2}{(-x^3)^3} = \frac{(-x^2)(-x^2)}{(-x^3)(-x^3)(-x^3)} = \frac{x^4}{-x^9} = \frac{1x^4}{-1x^9} = \frac{1}{-1x^5}$ or $-\frac{1}{x^5}$ or,

using the laws of exponents and the law $-a = (-1)\ a$,

$\frac{(-x^2)^2}{(-x^3)^3} = \frac{(-1x^2)^2}{(-1x^3)^3} = \frac{(-1)^2(x^2)^2}{(-1)^3(x^3)^3} = \frac{1x^{2 \cdot 2}}{-1x^{3 \cdot 3}} = \frac{x^4}{-1x^9} = \frac{1}{-1x^5}$ or $-\frac{1}{x^5}$

69. $\left(-\frac{x}{y}\right)^3\left(\frac{y^2}{w}\right)^2\left(\frac{w}{x^2}\right)^3 = -\frac{x^3}{y^3} \cdot \frac{(y^2)^2}{w^2} \cdot \frac{w^3}{(x^2)^3}$

$= -\frac{x^3}{y^3} \cdot \frac{y^4}{w^2} \cdot \frac{w^3}{x^6}$

$= -\frac{x^3y^4w^3}{x^6y^3w^2}$

$= -\frac{x^3}{x^6} \cdot \frac{y^4}{y^3} \cdot \frac{w^3}{w^2} = -\frac{1}{x^3} \cdot \frac{y}{1} \cdot \frac{w}{1} = -\frac{yw}{x^3}$

71. $-(a^2bc^3)\ \frac{-ab^2c}{(abc)^3} = -\frac{a^2bc^3}{1} \cdot \frac{-ab^2c}{a^3b^3c^3} = \frac{a^3b^3c^4}{a^3b^3c^3} = \frac{a^3}{a^3} \cdot \frac{b^3}{b^3} \cdot \frac{c^4}{c^3} = 1 \cdot 1 \cdot c = c$

73. $\dfrac{(a + b)^2(a - b)^2}{a^2 - b^2} = \dfrac{(a + b)^2(a - b)^2}{(a + b)(a - b)} = \dfrac{(a + b)^2}{(a + b)} \cdot \dfrac{(a - b)^2}{(a - b)} = (a + b) \cdot (a - b)$

$= a^2 - b^2$

75. $\dfrac{a^3b^4 - a^4b^3}{a^2b^2} = \dfrac{a^3b^3(b - a)}{a^2b^2} = \dfrac{a^3}{a^2} \cdot \dfrac{b^3}{b^2} \cdot (b - a) = ab(b - a)$

77. $\dfrac{5(x + 1)^4x^7 - 7(x + 1)^5x^6}{x^{14}} = \dfrac{x^6(x + 1)^4 \cdot 5x - x^6(x + 1)^4 \cdot 7(x + 1)}{x^{14}}$

$= \dfrac{x^6(x + 1)^4[5x - 7(x + 1)]}{x^{14}} = \dfrac{x^6(x + 1)^4[5x - 7x - 7]}{x^{14}}$

$= \dfrac{x^6(x + 1)^4(-2x - 7)}{x^{14}} = \dfrac{(x + 1)^4(-2x - 7)}{x^8}$

79. $\dfrac{3(x + y)^3(x - y)^4}{(x - y)^2 6(x + y)^5} = \dfrac{3(x + y)^3(x - y)^4}{6(x + y)^5(x - y)^2}$

$= \dfrac{3}{6} \cdot \dfrac{(x + y)^3}{(x + y)^5} \cdot \dfrac{(x - y)^4}{(x - y)^2}$

$= \dfrac{1}{2} \cdot \dfrac{1}{(x + y)^2} \cdot \dfrac{(x - y)^2}{1}$

$= \dfrac{(x - y)^2}{2(x + y)^2}$

81. $x^{5-n} \cdot x^{n+3} = x^{(5-n)+(n+3)} = x^{5-n+n+3} = x^8$

83. $\dfrac{x^{2n}}{x^n} = x^{2n-n} = x^n$

85. $(x^{n+1})^2 = x^{2(n+1)} = x^{2n+2}$

87. $\dfrac{u^{n+3}v^n}{u^{n+1}v^{n+4}} = \dfrac{u^{n+3}}{u^{n+1}} \cdot \dfrac{v^n}{v^{n+4}} = \dfrac{u^{(n+3)-(n+1)}}{1} \cdot \dfrac{1}{v^{(n+4)-n}} = \dfrac{u^2}{1} \cdot \dfrac{1}{v^4} = \dfrac{u^2}{v^4}$

89. $\dfrac{x^n x^{n+1}}{x^{2n}} = \dfrac{x^{n+n+1}}{x^{2n}} = \dfrac{x^{2n+1}}{x^{2n}} = x^{2n+1-2n} = x^1 = x$

91. $\dfrac{x^n y^n}{(xy)^{n-1}} = \dfrac{(xy)^n}{(xy)^{n-1}} = (xy)^{n-(n-1)} = (xy)^{n-n+1} = (xy)^1 = xy$

93. $\left(\dfrac{x^n}{y^{n+1}}\right)^2 \left(\dfrac{y^n}{x^{n+1}}\right)^2 = \dfrac{(x^n)^2}{(y^{n+1})^2} \cdot \dfrac{(y^n)^2}{(x^{n+1})^2} = \dfrac{x^{2n}}{y^{2(n+1)}} \cdot \dfrac{y^{2n}}{x^{2(n+1)}}$

$= \dfrac{x^{2n}y^{2n}}{y^{2n+2}x^{2n+2}} = \dfrac{x^{2n}}{x^{2n+2}} \cdot \dfrac{y^{2n}}{y^{2n+2}} = \dfrac{1}{x^{2n+2-2n}} \cdot \dfrac{1}{y^{2n+2-2n}} = \dfrac{1}{x^2} \cdot \dfrac{1}{y^2} = \dfrac{1}{x^2y^2}$

95. $\dfrac{(x + y)^n(x - y)^n}{(x^2 - y^2)^n} = \dfrac{[(x + y)(x - y)]^n}{(x^2 - y^2)^n} = \dfrac{[x^2 - y^2]^n}{(x^2 - y^2)^n} = 1$

97. $x^{2n} - 1 = (x^n)^2 - 1^2 = (x^n - 1)(x^n + 1)$

99. $x^{3n} + 1 = (x^n)^3 + 1^3 = (x^4 + 1)[(x^n)^2 - (x^n)(1) + 1^2] = (x^n + 1)(x^{2n} - x^n + 1)$

Exercise 6-2 Integer Exponents

Key Ideas and Formulas

For all real numbers $a \neq 0$: $a^0 = 1$. 0^0 is not defined.

If n is a positive integer and a is a nonzero real number, then

$$a^{-n} = \frac{1}{a^n}$$

$$a^n = \frac{1}{a^{-n}}$$

Laws of Exponents, m and n integers, a and b real numbers

1. $a^m a^n = a^{m+n}$
2. $(a^n)^m = a^{mn}$
3. $(ab)^m = a^m b^m$
4. $\left(\frac{a}{b}\right)^m = \frac{a^m}{b^m}$
5. $\frac{a^m}{a^n} = a^{m-n} = \frac{1}{a^{n-m}}$

---

1. 1 3. 1 5. $\frac{1}{3^3}$ 7. $\frac{1}{m^7}$ 9. 4^3 11. y^5

13. $10^7 \cdot 10^{-5} = 10^{7+(-5)} = 10^2$

15. $y^{-3}y^4 = y^{-3+4} = y^1 = y$

17. $u^5u^{-5} = u^{5+(-5)} = u^0 = 1$

19. $\frac{10^3}{10^{-7}} = 10^{3-(-7)} = 10^{10}$

21. $\frac{x^9}{x^{-2}} = x^{9-(-2)} = x^{11}$ COMMON ERROR: x^{9-2} This is an incorrect application of exponent law 5.

23. $\frac{b^{-3}}{b^5} = \frac{1}{b^{5-(-3)}} = \frac{1}{b^8}$

25. $\frac{10^{-1}}{10^6} = \frac{1}{10^{6-(-1)}} = \frac{1}{10^7}$

27. $(10^{-4})^{-3} = 10^{(-3)(-4)} = 10^{12}$

29. $(y^{-2})^{-4} = y^{(-4)(-2)} = y^8$

31. $(u^{-5}v^{-3})^{-2} = (u^{-5})^{-2}(y^{-3})^{-2} = u^{(-2)(-5)}v^{(-2)(-3)} = u^{10}v^6$

33. $(x^2y^{-3})^2 = (x^2)^2(y^{-3})^2 = x^4y^{-6} = \frac{x^4}{y^6}$

35. $(x^{-2}y^3)^{-1} = (x^{-2})^{-1}(y^3)^{-1} = x^2y^{-3} = \frac{x^2}{y^3}$

37. 1

39. $\frac{10^{-3}}{10^{-5}} = 10^{-3-(-5)} = 10^2$

41. $\frac{y^{-2}}{y^{-3}} = y^{-2-(-3)} = y^1 = y$

43. $\frac{10^{-13} \cdot 10^{-4}}{10^{-21} \cdot 10^3} = \frac{10^{-13+(-4)}}{10^{-21+3}} = \frac{10^{-17}}{10^{-18}} = 10^{-17-(-18)} = 10^1 = 10$

45. $\frac{18 \times 10^{12}}{6 \times 10^{-4}} = 3 \times 10^{12-(-4)} = 3 \times 10^{16}$

47. $\left(\frac{y}{y^{-2}}\right)^3 = \frac{y^3}{(y^{-2})^3} = \frac{y^3}{y^{-6}} = y^{3-(-6)} = y^9$

49. $\frac{1}{(3mn)^{-2}} = (3mn)^2 = 9m^2n^2$

51. $(2mn^{-3})^3 = 2^3m^3(n^{-3})^3 = 2^3m^3n^{-9}$
$= \frac{2^3m^3}{n^9}$

53. $(m^4n^{-5})^{-3} = (m^4)^{-3}(n^{-5})^{-3} = m^{-12}n^{15} = \frac{n^{15}}{m^{12}}$

55. $(2^2 3^{-3})^{-1} = (2^2)^{-1}(3^{-3})^{-1}$
$= 2^{-2}3^3 = \frac{3^3}{2^2}$

57. $(x^{-3}y^2)^{-2} = (x^{-3})^{-2}(y^2)^{-2} = x^6y^{-4} = \frac{x^6}{y^4}$

59. $\frac{8x^{-3}y^{-1}}{6x^2y^{-4}} = \frac{4y^{-1-(-4)}}{3x^{2-(-3)}} = \frac{4y^3}{3x^5}$

(Note: There are many ways to simplify this type of expression. We have chosen the most efficient way to achieve positive exponents in the answer, using $\frac{a^m}{a^n} = a^{m-n}$ if $m > n$ and $\frac{a^m}{a^n} = \frac{1}{a^{n-m}}$ if $n > m$. Since the larger exponent of x appeared in the denominator, we used the second form for x, and oppositely for y.)

61. $\frac{2a^6b^{-2}}{16a^{-3}b^2} = \frac{a^{6-(-3)}}{8b^{2-(-2)}} = \frac{a^9}{8b^4}$

63. $\left(\frac{x^{-1}}{x^{-8}}\right)^{-1} = \frac{(x^{-1})^{-1}}{(x^{-8})^{-1}} = \frac{x^1}{x^8} = \frac{1}{x^{8-1}} = \frac{1}{x^7}$

65. $\left(\frac{m^{-2}n^3}{m^4n^{-1}}\right)^2 = \frac{(m^{-2})^2(n^3)^2}{(m^4)^2(n^{-1})^2} = \frac{m^{-4}n^6}{m^8n^{-2}} = \frac{n^{6-(-2)}}{m^{8-(-4)}} = \frac{n^8}{m^{12}}$

67. $\left(\frac{6nm^{-2}}{3m^{-1}n^2}\right)^{-3} = \left(\frac{2nm^{-2}}{m^{-1}n^2}\right)^{-3} = \frac{2^{-3}n^{-3}(m^{-2})^{-3}}{(m^{-1})^{-3}(n^2)^{-3}} = \frac{2^{-3}n^{-3}m^6}{m^3n^{-6}} = \frac{m^{6-3}n^{-3-(-6)}}{2^3} = \frac{m^3n^3}{8}$

69. $\left(\frac{ab^2}{a^{-2}b^{-3}}\right)^{-1} = \frac{a^{-1}(b^2)^{-1}}{(a^{-2})^{-1}(b^{-3})^{-1}} = \frac{a^{-1}b^{-2}}{a^2b^3} = \frac{1}{a^{2-(-1)}b^{3-(-2)}} = \frac{1}{a^3b^5}$

71. $\left(\frac{3x^{-1}y^2}{4xy^{-2}}\right)^{-2} = \frac{3^{-2}(x^{-1})^{-2}(y^2)^{-2}}{4^{-2}x^{-2}(y^{-2})^{-2}} = \frac{3^{-2}x^2y^{-4}}{4^{-2}x^{-2}y^4} = \frac{4^2x^{2-(-2)}}{3^2y^{4-(-4)}} = \frac{16x^4}{9y^8}$

73. $\left(\frac{3x^{-3}y^2}{2x^2y^{-3}}\right)^{-3} = \frac{3^{-3}(x^{-3})^{-3}(y^2)^{-3}}{2^{-3}(x^2)^{-3}(y^{-3})^{-3}} = \frac{3^{-3}x^9y^{-6}}{2^{-3}x^{-6}y^9} = \frac{2^3x^{9-(-6)}}{3^3y^{9-(-6)}} = \frac{8x^{15}}{27y^{15}}$

75. $\left[\left(\frac{x^{-2}y^3t}{x^{-3}y^{-2}t^2}\right)^2\right]^{-1} = \left(\frac{x^{-2}y^3t}{x^{-3}y^{-2}t^2}\right)^{-2} = \frac{(x^{-2})^{-2}(y^3)^{-2}t^{-2}}{(x^{-3})^{-2}(y^{-2})^{-2}(t^2)^{-2}} = \frac{x^4y^{-6}t^{-2}}{x^6y^4t^{-4}} = \frac{t^{-2-(-4)}}{x^{6-4}y^{4-(-6)}} = \frac{t^2}{x^2y^{10}}$

77. $\left(\frac{2^2x^2y^0}{8x^{-1}}\right)^{-2}\left(\frac{x^{-3}}{x^{-5}}\right)^3 = \left(\frac{4x^2\cdot 1}{8x^{-1}}\right)^{-2}\left(\frac{x^{-3}}{x^{-5}}\right)^3 = \left(\frac{x^2}{2x^{-1}}\right)^{-2}\frac{(x^{-3})^3}{(x^{-5})^3}$

$= \frac{(x^2)^{-2}}{2^{-2}(x^{-1})^{-2}}\,\frac{x^{-9}}{x^{-15}} = \frac{x^{-4}}{2^{-2}x^2}\,\frac{x^{-9}}{x^{-15}} = \frac{2^2x^{-4}\cdot x^{-9}}{x^2\cdot x^{-15}} = \frac{2^2x^{-4+(-9)}}{x^{2+(-15)}} = \frac{4x^{-13}}{x^{-13}} = 4$

79. $(a^2 - b^2)^{-1} = \frac{1}{(a^2 - b^2)^1} = \frac{1}{a^2 - b^2}$

COMMON ERRORS: $a^{-2} - b^{-2}$
$(a - b)^{-2}$
Exponents do not distribute over subtraction.

Note: Errors in manipulating negative exponents abound, and are especially common in the following problems. These correct solutions should be examined carefully.

81. $\frac{x^{-1} + y^{-1}}{x + y} = \frac{\frac{1}{x} + \frac{1}{y}}{x + y} = \frac{xy\left(\frac{1}{x} + \frac{1}{y}\right)}{xy(x + y)} = \frac{y + x}{xy(x + y)} = \frac{x + y}{xy(x + y)} = \frac{1}{xy}$

83. $\frac{c - d}{c^{-1} - d^{-1}} = \frac{c - d}{\frac{1}{c} - \frac{1}{d}} = \frac{cd(c - d)}{cd\left(\frac{1}{c} - \frac{1}{d}\right)} = \frac{cd(c - d)}{d - c} = \frac{-cd(d - c)}{d - c} = -cd$

85. $(x^{-1} + y^{-1})^{-1} = \frac{1}{x^{-1} + y^{-1}} = \frac{1}{\frac{1}{x} + \frac{1}{y}}$

$= \frac{xy\cdot 1}{xy\left(\frac{1}{x} + \frac{1}{y}\right)} = \frac{xy}{y + x}$ or $\frac{xy}{x + y}$

COMMON ERROR: $x^1 + y^1$
Exponents do not distribute over addition.

87. $(x^{-1} - y^{-1})^2 = \left(\frac{1}{x} - \frac{1}{y}\right)^2 = \left(\frac{y - x}{xy}\right)^2 = \frac{(y - x)^2}{(xy)^2} = \frac{y^2 - 2xy + x^2}{x^2y^2}$ or $\frac{(y - x)^2}{x^2y^2}$

89. $\left(\frac{x^{-1}}{x^{-1} - y^{-1}}\right)^{-1} = \left(\frac{\frac{1}{x}}{\frac{1}{x} - \frac{1}{y}}\right)^{-1} = \left(\frac{xy\cdot\frac{1}{x}}{xy\left(\frac{1}{x} - \frac{1}{y}\right)}\right)^{-1}$

$= \left(\frac{y}{y - x}\right)^{-1} = \frac{1}{\frac{y}{y - x}} = \frac{y - x}{y}$

COMMON ERROR:

$\left(\frac{x^{-1}}{x^{-1} - y^{-1}}\right)^{-1} = \frac{x}{(x^{-1} - y^{-1})^{-1}} = \frac{x}{x - y}$

The first step is a correct application of exponent law 4, distributing exponents over division. However, the second step is an incorrect distribution of exponents over subtraction.

91. $a^{-4}b + a^{-3}c = a^{-4}b + a^{-4}ac = a^{-4}(b + ac) = \frac{b + ac}{a^4}$. $b + ac$

93. $a^{-1}b^{-2} + a^{-2}b^{-1} = a^{-2}ab^{-2} + a^{-2}b^{-2}b = a^{-2}b^{-2}(a + b) = \frac{a + b}{a^2b^2}$. $a + b$

95. $x^{-2}y^2 - x^2y^{-2} = x^{-2}y^4y^{-2} - x^4x^{-2}y^{-2} = x^{-2}y^{-2}(y^4 - x^4) = \frac{y^4 - x^4}{x^2y^2}$. $y^4 - x^4$

97. $a^{-3}b^{-5} + a^{-4}b^{-2} = a^{-4}ab^{-5} + a^{-4}b^3b^{-5} = a^{-4}b^{-5}(a + b^3)$

99. $x^{-1}y^{-2}z^{-3} + x^{-3}y^{-4}z^{-2} = x^{-3}x^2y^{-4}y^2z^{-3} + x^{-3}y^{-4}zz^{-3} = x^{-3}y^{-4}z^{-3}(x^2y^2 + z)$

Exercise 6-3 Scientific Notation and Applications

Key Ideas and Formulas

Scientific notation, most useful for very large and very small numbers, is a way of writing numbers as (a number between 1 and 10) × (a power of ten).

Converting Positive Numbers to Scientific Notation

1. If the number is greater than or equal to 10, the number of places the decimal point is shifted left appears as a positive exponent.
2. If the number is less than 10 but greater than or equal to 1, the decimal place is not shifted at all, and the exponent used is 0.
3. If the number is less than 1, the number of places the decimal point is shifted right appears as a negative exponent.

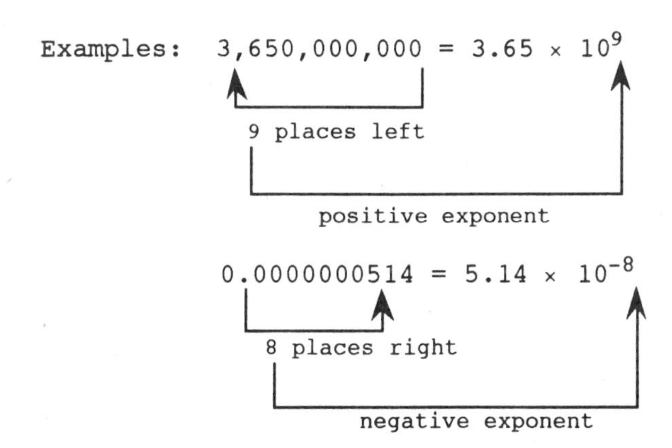

---

1. $70 = 7.0. \times 10^1 = 7 \times 10$

1 place left

positive exponent

3. $800 = 8.00. \times 10^2 = 8 \times 10^2$

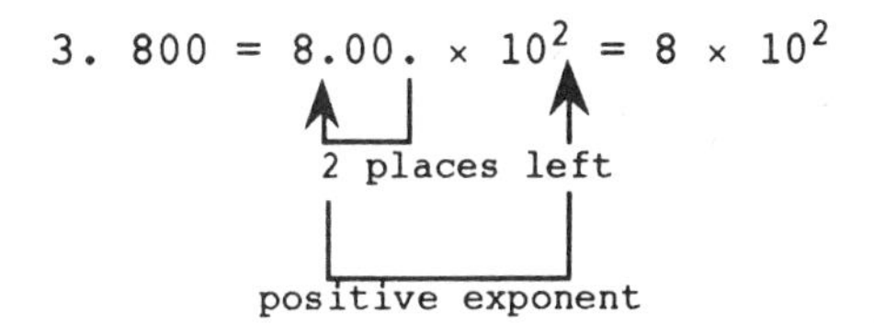

5. $80{,}000 = 8.0000. \times 10^4 = 8 \times 10^4$

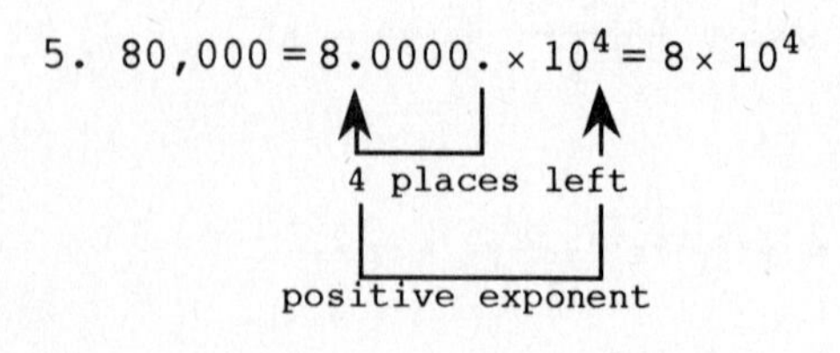

7. $0.008 = 0.008. \times 10^{-3} = 8 \times 10^{-3}$

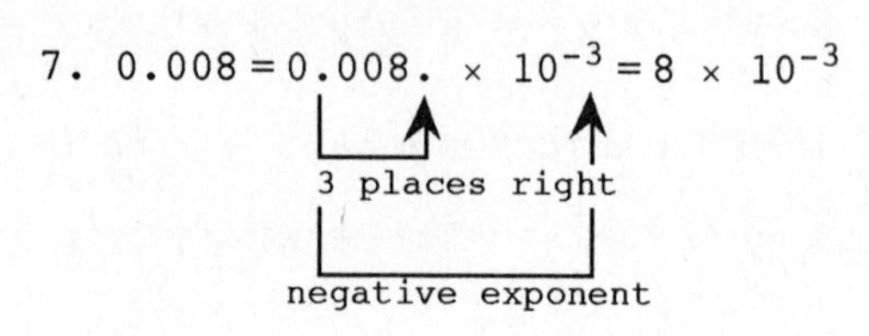

9. $0.000\ 000\ 08 = 0.00000008. \times 10^{-8} = 8 \times 10^{-8}$

8 places right

negative exponent

11. $52 = 5.2. \times 10^1 = 5.2 \times 10$

1 place left

positive exponent

13. $0.63 = 0.6.3 \times 10^{-1} = 6.3 \times 10^{-1}$

1 place right

negative exponent

15. $340 = 3.40. \times 10^2 = 3.4 \times 10^2$

2 places left

positive exponent

17. $0.085 = 0.08.5 \times 10^{-2} = 8.5 \times 10^{-2}$

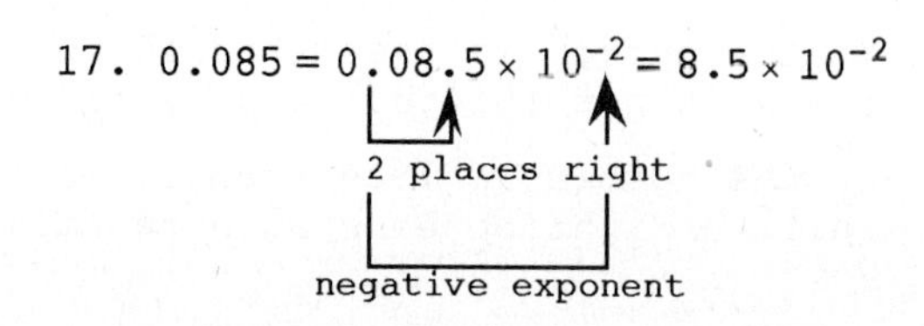

19. $6{,}300 = 6.300. \times 10^3 = 6.3 \times 10^3$

3 places left

positive exponent

21. $0.000\ 006\ 8 = 0.000006.8 \times 10^{-6} = 6.8 \times 10^{-6}$

6 places right

negative exponent

23. $8 \times 10^2 = 8 \times 100 = 800$

25. $4 \times 10^{-2} = 4 \times 0.01 = 0.04$

27. $3 \times 10^5 = 3 \times 100{,}000 = 300{,}000$

29. $9 \times 10^{-4} = 9 \times 0.0001 = 0.0009$

31. $5.6 \times 10^{-4} = 5.6 \times 10{,}000 = 56{,}000$

33. $9.7 \times 10^{-3} = 9.7 \times 0.001 = 0.0097$

35. $4.3 \times 10^5 = 4.3 \times 100{,}000 = 430{,}000$

37. $3.8 \times 10^{-7} = 3.8 \times 0.000\ 000\ 1 = 0.000\ 000\ 38$

39. $5{,}460{,}000{,}000 = 5.460\ 000\ 000. \times 10^{9} = 5.46 \times 10^{9}$

9 places left

positive exponent

41. $0.000\ 000\ 0729 = 0.000\ 000\ 07.29 \times 10^{-8} = 7.29 \times 10^{-8}$

8 places right

negative exponent

43. $0.000\ 000\ 000\ 012\ 3 = 0.000\ 000\ 000\ 01.2\ 3 \times 10^{-11} = 1.23 \times 10^{-11}$

11 places right

negative exponent

45. $6{,}789{,}000{,}000{,}000 = 6.789\ 000\ 000\ 000. \times 10^{12} = 6.789 \times 10^{12}$

12 places left

positive exponent

47. $0.102\ 003\ 004 = 0.1.02\ 003\ 004 \times 10^{-1} = 1.02003004 \times 10^{-1}$

1 place right

negative exponent

49. $1{,}234{,}000{,}567{,}000 = 1.234{,}000{,}567{,}000. \times 10^{12} = 1.234\ 000\ 567 \times 10^{12}$

12 places left

positive exponent

51. $507{,}000{,}000 = 5.07\ 000\ 000. \times 10^{8} = 5.07 \times 10^{8}$

8 places right

positive exponent

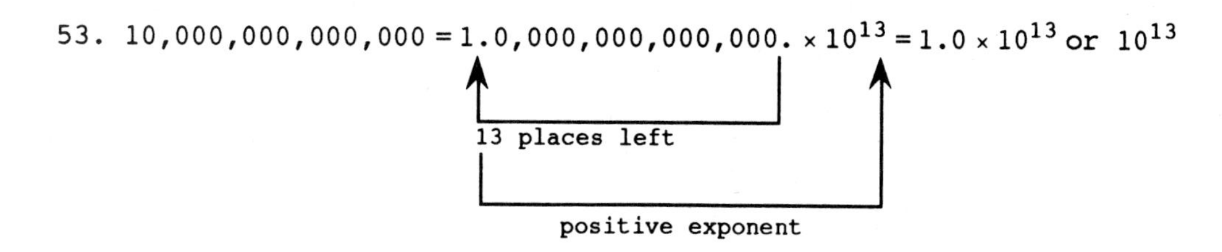

55. $\frac{1}{100,000} = 0.00001. \times 10^{-5} = 1.0 \times 10^{-5}$ or 10^{-5}

5 places right

negative exponent

57. $8.35 \times 10^{10} = 8.35 \times 10,000,000,000 = 83,500,000,000$

59. $6.14 \times 10^{-12} = 6.14 \times 0.000\,000\,000\,001 = 0.000\,000\,000\,006\,14$

61. $2.000\,01 \times 10^{9} = 2.000\,01 \times 1,000,000,000 = 2,000,010,000$

63. $1.002\,003 \times 10^{-5} = 1.002\,003 \times 0.00\,001 = 0.000\,010\,020\,03$

65. $1.539 \times 10^{-6} = 1.539 \times 0.000\,001 = 0.000\,001\,539$

67. $1.227 \times 10^{7} = 1.227 \times 10,000,000 = 12,270,000$

69. $8.65 \times 10^{5} = 8.65 \times 100,000 = 865,000$

71. $1.7 \times 10^{-24} = 1.7 \times 0.000\,000\,000\,000\,000\,000\,000\,001$
$= 0.000\,000\,000\,000\,000\,000\,000\,001\,7$

73. $(3 \times 10^{-6})(3 \times 10^{10}) = 3 \cdot 3 \times (10^{-6})(10^{10}) = 9 \times 10^{4}$

75. $(2 \times 10^{3})(3 \times 10^{-7}) = 2 \cdot 3 \times (10^{3})(10^{-7}) = 6 \times 10^{-4}$

77. $\frac{6 \times 10^{12}}{2 \times 10^{7}} = \frac{6}{2} \cdot \frac{10^{12}}{10^{7}} = 3 \times 10^{5}$

79. $\frac{15 \times 10^{-2}}{3 \times 10^{-6}} = \frac{15}{3} \cdot \frac{10^{-2}}{10^{-6}} = 5 \times 10^{4}$

81. $\frac{(90,000)(0.000\,002)}{0.006} = \frac{(9 \times 10^{4})(2 \times 10^{-6})}{6 \times 10^{-3}} = \frac{9 \cdot 2}{6} \cdot \frac{(10^{4})(10^{-6})}{10^{-3}}$
$= 3 \times 10^{1}$ or 30

83. $\frac{(60,000)(0.000\,003)}{(0.000\,4)(1,5000,000)} = \frac{(6 \times 10^{4})(3 \times 10^{-6})}{(4 \times 10^{-4})(1.5 \times 10^{6})}$

$= \frac{6 \cdot 3}{4 \cdot 1.5} \cdot \frac{(10^{4})(10^{-6})}{(10^{-4})(10^{6})} = \frac{18}{6} \cdot \frac{10^{-2}}{10^{2}} = 3 \times 10^{-4}$ or 0.0003

85. $\begin{pmatrix}\text{Mass in}\\ \text{tons}\end{pmatrix} = \begin{pmatrix}\text{Mass in}\\ \text{grams}\end{pmatrix} \times \begin{pmatrix}\text{tons in}\\ \text{1 gram}\end{pmatrix} = (6 \times 10^{27}) \times (1.1 \times 10^{-6})$
$= (6 \cdot 1.1) \cdot (10^{27})(10^{-6}) = 6.6 \times 10^{21}$ tons

87. The computer can perform 1 addition in 2.5×10^{-7} seconds, that is, $1 \div (2.5 \times 10^{-7})$ additions per second. $\frac{1}{2.5 \times 10^{-7}} = \frac{10^{7}}{2.5} = \frac{10 \times 10^{6}}{2.5} = 4 \times 10^{6}$ or 4,000,000 additions per second. 2.5×10^{-7} seconds $= 2.5 \times 10^{-7} \div 10^{-9}$ nanoseconds. This equals 2.5×10^{2} or 250 nanoseconds.

89. $\begin{pmatrix}\text{People}\\ \text{per}\\ \text{square mile}\end{pmatrix} = \begin{pmatrix}\text{Number}\\ \text{of}\\ \text{people}\end{pmatrix} \div \begin{pmatrix}\text{Number}\\ \text{of}\\ \text{square miles}\end{pmatrix} = 844,000,000 \div 1,269,000$
$= 8.44 \times 10^{8} \div 1.269 \times 10^{6} = \frac{8.44}{1.269} \times \frac{10^{8}}{10^{6}} = 6.65 \times 10^{2}$ or 665

91. $\begin{pmatrix}\text{Number of}\\ \text{violent crimes}\end{pmatrix} = \begin{pmatrix}\text{Violent crime}\\ \text{rate}\end{pmatrix} \times (\text{Population}) = \frac{731.8}{100{,}000} \times 249.6 \times 10^6$

$= 731.8 \times 10^{-5} \times 249.6 \times 10^6 = 182700 \times 10^1 = 1{,}827{,}000$

Exercise 6-4 Rational Exponents

Key Ideas and Formulas

For n a natural number, a is an nth root of b if $a^n = b$.

Number of real nth roots of a real number b:

	n even	n odd
b positive	two real nth roots	one real nth root
b negative	no real nth root	one real nth root

For n a natural number greater than 1 and b any real number:

$\sqrt[n]{b} = b^{1/n}$ If n = 2, we write $\sqrt{b}$, not $\sqrt[2]{b}$

$\sqrt{\ }$ is called a **radical**, n is the **index**, and b is the **radicand**.

	$\sqrt[n]{b}$ n even	$\sqrt[n]{b}$ n odd
b positive	$\sqrt[n]{b}$ is the positive nth root of b	$\sqrt[n]{b}$ is the unique nth root of b
b negative	$\sqrt[n]{b}$ is not a real number	$\sqrt[n]{b}$ is the unique nth root of b

$a^{1/n}$

base a	n even	n odd
Positive	$a^{1/n}$ is the positive nth root of a.	$a^{1/n}$ is the unique nth root of a.
Negative	$a^{1/n}$ is not a real number.	$a^{1/n}$ is the unique nth root of a.

For m and n natural numbers and b any real number (except b cannot be negative when n is even).

$b^{m/n} = (b^{1/n})^m \qquad b^{-m/n} = \frac{1}{b^{m/n}}$

$b^{m/n} = (b^{1/n})^m = (b^m)^{1/n}$

1. 5   3. Not a real number   5. 4   7. -4

9. $-64^{1/3} = -(64^{1/3}) = -4$

11. $16^{3/2} = (16^{1/2})^3 = 4^3 = 64$ COMMON ERROR: $16 \cdot \frac{3}{2}$ or 24. The $\frac{3}{2}$ is an exponent.

13. $8^{2/3} = (8^{1/3})^2 = 2^2 = 4$

15. $x^{1/4}x^{3/4} = x^{1/4+3/4} = x^1$ or x

17. $\frac{x^{2/5}}{x^{3/5}} = \frac{1}{x^{3/5-2/5}} = \frac{1}{x^{1/5}}$

19. $(x^4)^{1/2} = x^{1/2 \cdot 4} = x^2$

21. $(a^3b^9)^{1/3} = (a^3)^{1/3}(b^9)^{1/3} = a^{1/3 \cdot 3}b^{1/3 \cdot 9} = ab^3$

23. $\left(\frac{x^9}{y^{12}}\right)^{1/3} = \frac{(x^9)^{1/3}}{(y^{12})^{1/3}} = \frac{x^3}{y^4}$

25. $(x^{1/3}y^{1/2})^6 = (x^{1/3})^6(y^{1/2})^6 = x^2y^3$

27. $\left(\frac{4}{25}\right)^{1/2} = \frac{4^{1/2}}{25^{1/2}} = \frac{2}{5}$

29. $\left(\frac{4}{25}\right)^{3/2} = \left[\left(\frac{4}{25}\right)^{1/2}\right]^3 = \left(\frac{2}{5}\right)^3 = \frac{8}{125}$

or

$\left(\frac{4}{25}\right)^{3/2} = \frac{4^{3/2}}{25^{3/2}} = \frac{(4^{1/2})^3}{(25^{1/2})^3} = \frac{2^3}{5^3} = \frac{8}{125}$

31. $\left(\frac{1}{8}\right)^{2/3} = \frac{1^{2/3}}{8^{2/3}} = \frac{1}{(8^{1/3})^2} = \frac{1}{2^2} = \frac{1}{4}$

33. $36^{-1/2} = \frac{1}{36^{1/2}} = \frac{1}{6}$

35. $25^{-3/2} = \frac{1}{25^{3/2}} = \frac{1}{(25^{1/2})^3} = \frac{1}{5^3} = \frac{1}{125}$

37. $5^{3/2} \cdot 5^{1/2} = 5^{3/2+1/2} = 5^2 = 25$

39. $(3^6)^{-1/3} = 3^{-1/3 \cdot 6} = 3^{-2} = \frac{1}{3^2} = \frac{1}{9}$

41. $\left(-\frac{8}{27}\right)^{2/3} = \frac{(-8)^{2/3}}{27^{2/3}} = \frac{[(-8)^{1/3}]^2}{(27^{1/3})^2} = \frac{[-2]^2}{3^2} = \frac{4}{9}$

43. $\left(-\frac{64}{125}\right)^{-4/3} = 1 \div \left(-\frac{64}{125}\right)^{4/3} = 1 \div \frac{(-64)^{4/3}}{125^{4/3}} = 1 \div \frac{[(-64)^{1/3}]^4}{(125^{1/3})^4}$

$= 1 \div \frac{(-4)^4}{5^4} = 1 \div \frac{256}{625} = 1 \cdot \frac{625}{256} = \frac{625}{256}$

45. $(-16)^{-3/2} = \frac{1}{(-16)^{3/2}} = \frac{1}{((-16)^{1/2})^3}$.

$(-16)^{1/2}$ is not a real number, so this is also not a real number.

47. $x^{1/4}x^{-3/4} = x^{1/4+(-3/4)} = x^{-2/4} = x^{-1/2} = \frac{1}{x^{1/2}}$

49. $n^{3/4}n^{-2/3} = n^{3/4+(-2/3)} = n^{1/12}$

51. $(x^{-2/3})^{-6} = x^{(-6)(-2/3)} = x^4$

53. $(4u^{-2}v^4)^{1/2} = 4^{1/2}(u^{-2})^{1/2}(v^4)^{1/2} = 2u^{-1}v^2 = \frac{2v^2}{u}$ COMMON ERROR: $4u^{-1}v^2$. The exponent $\frac{1}{2}$ must be applied to each factor in parentheses, including the constant factor 4.

55. $(x^4y^6)^{-1/2} = (x^4)^{-1/2}(y^6)^{-1/2} = x^{-2}y^{-3} = \frac{1}{x^2y^3}$

57. $\left(\frac{x^{-2/3}}{y^{-1/2}}\right)^{-6} = \frac{(x^{-2/3})^{-6}}{(y^{-1/2})^{-6}} = \frac{x^4}{y^3}$

59. $(8x^6y^3)^{1/3} = 8^{1/3}(x^6)^{1/3}(y^3)^{1/3} = 2x^2y$

61. $(4x^{4/3}y^4)^{1/2} = 4^{1/2}(x^{4/3})^{1/2}(y^4)^{1/2} = 2x^{1/2\cdot 4/3}y^2 = 2x^{2/3}y^2$

63. $(8 \times 10^{-6})^{1/3} = 8^{1/3} \times (10^{-6})^{1/3} = 2 \times 10^{-2}$ or $\frac{2}{100}$ or $\frac{1}{50}$ or 0.02

65. $\left(\frac{3xy^{-4}}{12x^{-1}y^2}\right)^{1/2} = \left(\frac{x^{1-(-1)}}{4y^{2-(-4)}}\right)^{1/2} = \left(\frac{x^2}{4y^6}\right)^{1/2} = \frac{(x^2)^{1/2}}{4^{1/2}(y^6)^{1/2}} = \frac{x}{2y^3}$

67. $\left(\frac{x^{-1/5}y^{1/4}}{z^{1/2}}\right)^{10} = \frac{x^{10(-1/5)}y^{10(1/4)}}{z^{10(1/2)}} = \frac{x^{-2}y^{5/2}}{z^5} = \frac{y^{5/2}}{x^2z^5}$

69. $\left(\frac{25x^5y^{-1}}{16x^{-3}y^{-5}}\right)^{1/2} = \left(\frac{25x^8y^4}{16}\right)^{1/2} = \frac{25^{1/2}(x^8)^{1/2}(y^4)^{1/2}}{(16)^{1/2}} = \frac{5x^4y^2}{4}$ or $\frac{5}{4}x^4y^2$

71. $\left(\frac{8y^{1/3}y^{-1/4}}{y^{-1/12}}\right)^2 = \left(\frac{8y^{1/12}}{y^{-1/12}}\right)^2 = (8y^{1/12-(-1/12)})^2 = (8y^{1/6})^2 = 64y^{1/3}$

73. $x^{1/2}(x^4 - x^{1/5}) = x^{1/2}\cdot x^4 - x^{1/2}\cdot x^{1/5} = x^{9/2} - x^{7/10}$

75. $3m^{3/4}(4m^{1/4} - 2m^8) = 3m^{3/4}\cdot 4m^{1/4} - 3m^{3/4}\cdot 2m^{32/4} = 12m - 6m^{35/4}$

77. $(2x^{1/2} + y^{1/2})(x^{1/2} + y^{1/2}) = 2x^{1/2}x^{1/2} + 2x^{1/2}y^{1/2} + x^{1/2}y^{1/2} + y^{1/2}y^{1/2}$ (F, O, I, L)

$= 2x + 3x^{1/2}y^{1/2} + y$

79. $(x^{1/2} + y^{1/2})^2 = (x^{1/2})^2 + 2x^{1/2}y^{1/2} + (y^{1/2})^2 = x + 2x^{1/2}y^{1/2} + y$

81. Since $(a - b)(a^2 + ab + b^2) = a^3 - b^3$,
$(x^{1/3} - 1)(x^{2/3} + x^{1/3} + 1) = (x^{1/3})^3 - 1^3 = x - 1$

83. $(x^{3/2} - y^{3/2})(x^{3/2} + y^{3/2}) = (x^{3/2})^2 - (y^{3/2})^2 = x^3 - y^3$

85. $(a^{-1/2} + 3b^{-1/2})(2a^{-1/2} - b^{-1/2}) = 2a^{-1/2}a^{-1/2} - a^{-1/2}b^{-1/2} + 6a^{-1/2}b^{-1/2} - 3b^{-1/2}b^{-1/2}$ (F, O, I, L)

$= 2a^{-1} + 5a^{-1/2}b^{-1/2} - 3b^{-1}$

$= \frac{2}{a} + \frac{5}{a^{1/2}b^{1/2}} - \frac{3}{b}$

87. $(a^{n/2}b^{n/3})^{1/n} = a^{1/n \cdot n/2}b^{1/n \cdot n/3} = a^{1/2}b^{1/3}$

89. $\left(\frac{x^{m+2}}{x^m}\right)^{1/2} = (x^{m+2-m})^{1/2} = (x^2)^{1/2} = x$

91. $(x^{m/4}x^{m/4})^{-2} = (x^{m/4+m/4})^{-2} = (x^{m/2})^{-2} = x^{-2 \cdot m/2} = x^{-m} = \frac{1}{x^m}$

93. $\sqrt[n]{x}\,\sqrt[n]{y} = \sqrt[n]{xy}$

95. $(\sqrt[n]{x})^n = x$. $(\sqrt[n]{x^n}) = x$

97. If $x = y = -2$ and $n = 2$, $x^{1/n}$ and $y^{1/n}$ are not defined. The statement is not true in this case, and it is false in general if x and y are negative and n is even.

99. If $x = -2$ and $n = 2$, $(-2)^{1/2}$ is not defined. $((-2)^2)^{1/2} = 4^{1/2} = 2$, thus $(x^n)^{1/n} \neq x$. Neither statement is true if x is negative and n is even.

Exercise 6-5 Radical Forms and Rational Exponents

Key Ideas and Formulas

The relationships

$$a^{m/n} = (a^{1/n})^m = (a^m)^{1/n}$$

convert to the following in radical notation:

$a^{m/n} = (\sqrt[n]{a})^m$ *Remember*: When a is negative, we require that n be odd.

Square Root Properties

For positive real numbers a and b,

1. $\sqrt{a^2} = a$ and $(\sqrt{a})^2 = a$
2. $\sqrt{ab} = \sqrt{a}\sqrt{b}$
   $\sqrt{\frac{a}{b}} = \frac{\sqrt{a}}{\sqrt{b}}$

Properties of Radicals

For positive integers n, m, and k, with $n \ge 2$; and positive real numbers a and b:

1. $\sqrt[n]{a^n} = a$ and $(\sqrt[n]{a})^n = a$
2. $\sqrt[n]{ab} = \sqrt[n]{a}\,\sqrt[n]{b}$
3. $\sqrt[n]{\frac{a}{b}} = \frac{\sqrt[n]{a}}{\sqrt[n]{b}}$
4. $\sqrt[kn]{a^{km}} = \sqrt[n]{a^m}$

1. $\sqrt{11}$

3. $\sqrt[3]{5}$

5. $u^{3/5} = \sqrt[5]{u^3}$ or $(\sqrt[5]{u})^3$. The first is usually preferred, and we will use this form.

7. $4\sqrt[7]{y^3}$

9. $\sqrt[7]{(4y)^3}$

11. $\sqrt[5]{(4ab^3)^2}$

13. $\sqrt{a + b}$

15. $\sqrt{6} = \sqrt[2]{6} = 6^{1/2}$

17. $m^{1/4}$

19. $y^{3/5}$

21. $(xy)^{3/4}$

23. $(x^2 - y^2)^{1/2}$

25. $\sqrt{72} = \sqrt{36 \cdot 2} = \sqrt{36}\sqrt{2} = 6\sqrt{2}$

27. $\sqrt{76} = \sqrt{4 \cdot 19} = \sqrt{4}\sqrt{19} = 2\sqrt{19}$

29. $\dfrac{\sqrt{44}}{\sqrt{11}} = \dfrac{\sqrt{4 \cdot 11}}{\sqrt{11}} = \dfrac{\sqrt{4}\sqrt{11}}{\sqrt{11}} = \sqrt{4} = 2$

31. $\dfrac{\sqrt{48}}{\sqrt{3}} = \dfrac{\sqrt{16 \cdot 3}}{\sqrt{3}} = \dfrac{\sqrt{16}\sqrt{3}}{\sqrt{3}} = \sqrt{16} = 4$

33. $\sqrt{2^6} = \sqrt{(2^3)^2} = 2^3 = 8$

35. $\sqrt{3^4} = \sqrt{(3^2)^2} = 3^2 = 9$

37. $\sqrt{(-3)^2} = \sqrt{9} = 3$

39. $\sqrt[4]{3^2} = \sqrt[2\cdot 2]{3^{2\cdot 1}} = \sqrt[2]{3^1} = \sqrt{3}$

41. $\sqrt{x^5} = \sqrt{x^4 x} = \sqrt{x^4}\sqrt{x} = x^2\sqrt{x}$

43. $\sqrt{x^3y^5} = \sqrt{x^2y^4 \cdot xy} = \sqrt{x^2y^4}\sqrt{xy} = xy^2\sqrt{xy}$

45. $\dfrac{\sqrt{x^5}}{\sqrt{x^3}} = \sqrt{\dfrac{x^5}{x^3}} = \sqrt{x^2} = x$

47. $\dfrac{\sqrt{x^3}}{\sqrt{x^9}} = \sqrt{\dfrac{x^3}{x^9}} = \sqrt{\dfrac{1}{x^6}} = \dfrac{1}{x^3}$

49. $\sqrt{\dfrac{x^3}{24}} = \sqrt{\dfrac{x^3}{2^3 \cdot 3}} = \sqrt{\dfrac{2 \cdot 3x^3}{2^4 \cdot 3^2}} = \dfrac{\sqrt{6x^3}}{2^2 \cdot 3} = \dfrac{\sqrt{x^2 \cdot 6x}}{12} = \dfrac{\sqrt{x^2}\sqrt{6x}}{12} = \dfrac{x\sqrt{6x}}{12}$

51. $-5\sqrt[5]{y^2}$

53. $\sqrt[7]{(1 + m^2n^2)^3}$

55. $w^{-2/3} = \dfrac{1}{w^{2/3}} = \dfrac{1}{\sqrt[3]{w^2}}$

57. $(3m^2n^3)^{-3/5} = \dfrac{1}{(3m^2n^3)^{3/5}} = \dfrac{1}{\sqrt[5]{(3m^2n^3)^3}}$

59. $\sqrt{a} + \sqrt{b}$

61. $\sqrt[3]{(a^3 + b^3)^2}$

63. $(a + b)^{2/3}$

65. $-3x(a^3b)^{1/4}$

67. $(-2x^3y^7)^{1/9}$

69. $\dfrac{3}{\sqrt[3]{y}} = \dfrac{3}{y^{1/3}} = 3y^{-1/3}$

71. $\dfrac{-2x}{\sqrt{x^2 + y^2}} = \dfrac{-2x}{(x^2 + y^2)^{1/2}}$ or $-2x(x^2 + y^2)^{-1/2}$

73. $m^{2/3} - n^{1/2}$

75. $\sqrt[3]{40} = \sqrt[3]{2^3 \cdot 5} = \sqrt[3]{2^3}\sqrt[3]{5} = 2\sqrt[3]{5}$

77. $\sqrt[5]{64} = \sqrt[5]{2^6} = \sqrt[5]{2^5 \cdot 2} = \sqrt[5]{2^5}\sqrt[5]{2} = 2\sqrt[5]{2}$

79. $\sqrt[3]{375} = \sqrt[3]{5^3 \cdot 3} = \sqrt[3]{5^3}\sqrt[3]{3} = 5\sqrt[3]{3}$

81. $\sqrt[3]{256} = \sqrt[3]{2^8} = \sqrt[3]{2^6 \cdot 2^2} = \sqrt[3]{2^6}\sqrt[3]{2^2} = 2^2\sqrt[3]{2^2} = 4\sqrt[3]{4}$

83. $\dfrac{\sqrt[3]{54}}{\sqrt[3]{2}} = \dfrac{\sqrt[3]{3^2 \cdot 2}}{\sqrt[3]{2}} = \dfrac{\sqrt[3]{3^3}\sqrt[3]{2}}{\sqrt[3]{2}} = \dfrac{3\sqrt[3]{2}}{\sqrt[3]{2}} = 3$

85. $\sqrt[3]{5^6} = \sqrt[3]{(5^2)^3} = 5^2 = 25$

87. $\sqrt[4]{(-2)^4} = \sqrt[4]{16} = \sqrt[4]{2^4} = 2$

89. $\sqrt[3]{x^4} = \sqrt[3]{x^3 x} = \sqrt[3]{x^3}\sqrt[3]{x} = x\sqrt[3]{x}$

91. $\dfrac{\sqrt[4]{x^9}}{\sqrt[4]{x}} = \sqrt[4]{\dfrac{x^9}{x}} = \sqrt[4]{x^8} = \sqrt[4]{(x^2)^4} = x^2$

93. $3xy^2$

95. $\sqrt[4]{(7x^7y^3)^5} = \sqrt[4]{7^5x^{5\cdot 7}y^{5\cdot 3}} = \sqrt[4]{7^5x^{35}y^{15}} = \sqrt[4]{7^4x^{32}y^{12} \cdot 7x^3y^3}$
$= \sqrt[4]{7^4(x^8)^4(y^3)^4}\sqrt[4]{7x^3y^3} = 7x^8y^3\sqrt[4]{7x^3y^3}$

97. $\sqrt{(-2)^2} = \sqrt{4} = 2 \neq -2$. $\sqrt{x^2} = x$ is true only for $x \geq 0$.

99. $\sqrt[n]{x^{2n}} = \sqrt[n]{(x^2)^n} = x^2$ is always true, since x^2 is always $x \geq 0$.

Exercise 6-6 Rewriting Radical Expressions

Key Ideas and Formulas

Properties of Radicals

For positive integers n, m, and k, with $n \geq 2$; and positive real numbers a and b:

SQUARE ROOT PROPERTIES

1. $\sqrt{a^2} = a$, $(\sqrt{a})^2 = a$
2. $\sqrt{ab} = \sqrt{a}\sqrt{b}$
3. $\sqrt{\dfrac{a}{b}} = \dfrac{\sqrt{a}}{\sqrt{b}}$

GENERAL RADICAL PROPERTIES

1. $\sqrt[n]{a^n} = a$, $(\sqrt[n]{a})^n = a$
2. $\sqrt[n]{ab} = \sqrt[n]{a}\sqrt[n]{b}$
3. $\sqrt[n]{\dfrac{a}{b}} = \dfrac{\sqrt[n]{a}}{\sqrt[n]{b}}$
4. $\sqrt[kn]{a^{km}} = \sqrt[n]{a^m}$

Simplest Radical Form:

1. A radicand (the expression within the radical sign) contains no polynomial factor to a power greater than or equal to the index of the radical.

2. The power of the radicand and the index of the radical have no common factor other than 1.
3. No radical appears in a denominator.
4. No fraction appears within a radical.

For x <u>any</u> real number, $\sqrt{x^2} = |x|$

In general, for x *any* real number and n a positive integer greater than 1:

$$\sqrt[n]{x^n} = \begin{cases} |x| & \text{if n is even} \\ x & \text{if n is odd} \end{cases}$$

---

1. y

3. $2u$

5. $7x^2y$

7. $\sqrt{18} = \sqrt{9 \cdot 2} = \sqrt{9}\sqrt{2} = 3\sqrt{2}$

9. $\sqrt{m^3} = \sqrt{m^2 \cdot m} = \sqrt{m^2}\sqrt{m} = m\sqrt{m}$

11. $\sqrt{8x^3} = \sqrt{4x^2 \cdot 2x} = \sqrt{4x^2}\sqrt{2x} = 2x\sqrt{2x}$

13. $\frac{1}{3}$

15. $\frac{1}{y}$

17. $\frac{1}{\sqrt{5}} = \frac{1}{\sqrt{5}} \cdot \frac{\sqrt{5}}{\sqrt{5}} = \frac{\sqrt{5}}{5}$

19. $\sqrt{\frac{1}{5}} = \sqrt{\frac{1}{5} \cdot \frac{5}{5}} = \sqrt{\frac{5}{25}} = \frac{\sqrt{5}}{\sqrt{25}} = \frac{\sqrt{5}}{5}$

21. $\frac{1}{\sqrt{y}} = \frac{1}{\sqrt{y}} \cdot \frac{\sqrt{y}}{\sqrt{y}} = \frac{\sqrt{y}}{y}$

23. $\sqrt{\frac{1}{y}} = \sqrt{\frac{1}{y} \cdot \frac{y}{y}} = \sqrt{\frac{y}{y^2}} = \frac{\sqrt{y}}{\sqrt{y^2}} = \frac{\sqrt{y}}{y}$

25. $\sqrt{9x^3y^5} = \sqrt{9x^2y^4 \cdot xy} = \sqrt{9x^2y^4}\sqrt{xy} = 3xy^2\sqrt{xy}$

27. $\sqrt{18x^8y^5} = \sqrt{9x^8y^4 \cdot 2y} = \sqrt{9x^8y^4}\sqrt{2y} = 3x^4y^2\sqrt{2y}$

29. $\frac{1}{\sqrt{2x}} = \frac{1}{\sqrt{2x}} \cdot \frac{\sqrt{2x}}{\sqrt{2x}} = \frac{\sqrt{2x}}{2x}$

31. $\frac{6x^2}{\sqrt{3x}} = \frac{6x^2}{\sqrt{3x}} \cdot \frac{\sqrt{3x}}{\sqrt{3x}} = \frac{6x^2\sqrt{3x}}{3x} = 2x\sqrt{3x}$

33. $\frac{3a}{\sqrt{2ab}} = \frac{3a}{\sqrt{2ab}} \cdot \frac{\sqrt{2ab}}{\sqrt{2ab}} = \frac{3a\sqrt{2ab}}{2ab} = \frac{3\sqrt{2ab}}{2b}$

35. $\sqrt{\frac{6x}{7y}} = \sqrt{\frac{6x}{7y} \cdot \frac{7y}{7y}} = \sqrt{\frac{42xy}{49y^2}} = \frac{\sqrt{42xy}}{\sqrt{49y^2}} = \frac{\sqrt{42xy}}{7y}$

37. $\sqrt{\frac{9m^5}{2n}} = \sqrt{\frac{9m^5}{2n} \cdot \frac{2n}{2n}} = \sqrt{\frac{18m^5n}{4n^2}} = \sqrt{\frac{9m^4 \cdot 2mn}{4n^2}} = \frac{\sqrt{9m^4}\sqrt{2mn}}{\sqrt{4n^2}} = \frac{3m^2\sqrt{2mn}}{2n}$

39. $2x^2y$

41. $\sqrt[3]{2^4x^4y^7} = \sqrt[3]{2^3x^3y^6 \cdot 2xy} = \sqrt[3]{2^3x^3y^6}\sqrt[3]{2xy} = 2xy^2\sqrt[3]{2xy}$

43. $\sqrt[3]{32x^4y^6} = \sqrt[3]{2^5x^4y^6} = \sqrt[3]{2^3x^3y^6 \cdot 2^2x} = \sqrt[3]{2^3x^3y^6}\sqrt[3]{2^2x} = 2xy^2\sqrt[3]{4x}$

45. $\sqrt{32x^4y^6} = \sqrt{2^5x^4y^6} = \sqrt{2^4x^4y^6\cdot 2} = \sqrt{2^4x^4y^6}\sqrt{2} = 2^2x^2y^3\sqrt{2} = 4x^2y^3\sqrt{2}$

47. $\sqrt[5]{24x^3y^4z^5} = \sqrt[5]{z^5 2^3\cdot 3x^3y^4} = \sqrt[5]{z^5}\sqrt[5]{2^3\cdot 3x^3y^4} = z\sqrt[5]{24x^3y^4}$

49. $\sqrt[4]{24x^3y^4z^5} = \sqrt[4]{y^4z^4\cdot 2^3\cdot 3x^3z} = \sqrt[4]{y^4z^4}\sqrt[4]{2^3\cdot 3x^3z} = yz\sqrt[4]{24x^3z}$

51. $\sqrt[4]{x^2} = \sqrt[2\cdot 2]{x^{2\cdot 1}} = \sqrt[2]{x^1} = \sqrt{x}$

53. $\sqrt{2}\sqrt{8} = \sqrt{16} = 4$

55. $\sqrt{18m^3n^4}\sqrt{2m^3n^2} = \sqrt{18m^3n^4\cdot 2m^3n^2} = \sqrt{36m^6n^6} = 6m^3n^3$

57. $\dfrac{6}{\sqrt[3]{3}} = \dfrac{6}{\sqrt[3]{3}}\cdot\dfrac{\sqrt[3]{9}}{\sqrt[3]{9}} = \dfrac{6\sqrt[3]{9}}{\sqrt[3]{27}} = \dfrac{6\sqrt[3]{9}}{3} = 2\sqrt[3]{9}$

59. $\dfrac{\sqrt{4a^3}}{\sqrt{3b}} = \dfrac{\sqrt{4a^3}}{\sqrt{3b}}\cdot\dfrac{\sqrt{3b}}{\sqrt{3b}} = \dfrac{\sqrt{4a^3}\sqrt{3b}}{3b} = \dfrac{\sqrt{4a^2\cdot a}\sqrt{3b}}{3b} = \dfrac{\sqrt{4a^2}\sqrt{a}\sqrt{3b}}{3b} = \dfrac{2a\sqrt{3ab}}{3b}$

61. This expression is in simplest radical form.

COMMON ERROR: $a + b$. Exponents and Radicals do not distribute over addition.

63. $\dfrac{2x}{3y^2}$

65. $-m\sqrt[5]{3^6m^7n^{11}} = -m\sqrt[5]{3^5m^5n^{10}\cdot 3m^2n} = -m\sqrt[5]{3^5m^5n^{10}}\sqrt[5]{3m^2n} = -m\cdot 3mn^2\sqrt[5]{3m^2n} = -3m^2n^2\sqrt[5]{3m^2n}$

67. $\sqrt[6]{x^4(x-y)^2} = \sqrt[3\cdot 2]{x^{2\cdot 2}(x-y)^{2\cdot 2}} = \sqrt[3]{x^2(x-y)}$

69. $\sqrt[3]{2x^2y^3}\sqrt[3]{3x^5y} = \sqrt[3]{2x^2y^3\cdot 3x^5y} = \sqrt[3]{6x^7y^4} = \sqrt[3]{x^6y^3\cdot 6xy} = \sqrt[3]{x^6y^3}\sqrt[3]{6xy} = x^2y\sqrt[3]{6xy}$

71. $\sqrt[3]{4x^2y}\sqrt[3]{2x^4y^2} = \sqrt[3]{4x^2y\cdot 2x^4y^2} = \sqrt[3]{8x^6y^3} = 2x^2y$

73. $\sqrt[3]{9x^2y^2}\sqrt[3]{24xy^2} = \sqrt[3]{9x^2y^2\cdot 24xy^2} = \sqrt[3]{216x^3y^4} = \sqrt[3]{6^3x^3y^3y} = \sqrt[3]{6^3x^3y}\sqrt[3]{y} = 6xy\sqrt[3]{y}$

75. $\dfrac{\sqrt[3]{9x^2y^2}}{\sqrt[3]{72xy^2}} = \sqrt[3]{\dfrac{9x^2y^2}{72xy^2}} = \sqrt[3]{\dfrac{x}{8}} = \dfrac{\sqrt[3]{x}}{\sqrt[3]{8}} = \dfrac{\sqrt[3]{x}}{2}$

77. $\dfrac{4x^3y^2}{\sqrt[3]{2xy^2}} = \dfrac{4x^3y^2}{\sqrt[3]{2xy^2}}\cdot\dfrac{\sqrt[3]{2^2x^2y}}{\sqrt[3]{2^2x^2y}} = \dfrac{4x^3y^2\sqrt[3]{4x^2y}}{\sqrt[3]{2^3x^3y^3}} = \dfrac{4x^3y^2\sqrt[3]{4x^2y}}{2xy} = 2x^2y\sqrt[3]{4x^2y}$

79. $-2x\sqrt[3]{\dfrac{3y^2}{4x}} = -2x\sqrt[3]{\dfrac{3y^2}{4x}\cdot\dfrac{2x^2}{2x^2}} = -2x\sqrt[3]{\dfrac{6x^2y^2}{8x^3}} = -2x\dfrac{\sqrt[3]{6x^2y^2}}{\sqrt[3]{8x^3}} = -2x\dfrac{\sqrt[3]{6x^2y^2}}{2x} = -\sqrt[3]{6x^2y^2}$

81. $\frac{x - y}{\sqrt[3]{x - y}} = \frac{(x - y)}{\sqrt[3]{x - y}} \cdot \frac{\sqrt[3]{(x - y)^2}}{\sqrt[3]{(x - y)^2}} = \frac{(x - y)\sqrt[3]{(x - y)^2}}{\sqrt[3]{(x - y)^3}} = \frac{(x - y)\sqrt[3]{(x - y)^2}}{(x - y)} = \sqrt[3]{(x - y)^2}$

83. $\sqrt[4]{\frac{3y^3}{4x}} = \sqrt[4]{\frac{3y^3}{2^2x}} = \sqrt[4]{\frac{3y^3}{2^2x} \cdot \frac{2^2x^3}{2^2x^3}} = \sqrt[4]{\frac{12x^3y^3}{2^4x^4}} = \frac{\sqrt[4]{12x^3y^3}}{\sqrt[4]{2^4x^4}} = \frac{\sqrt[4]{12x^3y^3}}{2x}$

85. $-\sqrt{x^4 + 2x^2} = -\sqrt{x^2(x^2 + 2)} = -\sqrt{x^2}\sqrt{x^2 + 2} = -x\sqrt{x^2 + 2}$

87. $\sqrt[4]{16x^4}\sqrt[3]{16x^{24}y^4} = 2x\sqrt[3]{2^4x^{24}y} = 2x\sqrt[3]{2^3x^{24}y^3 \cdot 2y}$
$= 2x\sqrt[3]{2^3x^{24}y^3}\sqrt[3]{2y} = 2x \cdot 2x^8y\sqrt[3]{2y} = 4x^9y\sqrt[3]{2y}$

89. $\sqrt[3]{3m^2n^2}\sqrt[4]{3m^3n^2} = \sqrt[3\cdot4]{3^4m^{2\cdot4}n^{2\cdot4}}\ \sqrt[4\cdot3]{3^3m^{3\cdot3}n^{3\cdot2}} = \sqrt[12]{3^4m^8n^8}\ \sqrt[12]{3^3m^9n^6}$
$= \sqrt[12]{3^7m^{17}n^{14}} = \sqrt[12]{m^{12}n^{12} \cdot 3^7m^5n^2} = \sqrt[12]{m^{12}n^{12}}\ \sqrt[12]{3^7m^5n^2} = mn\sqrt[12]{3^7m^5n^2}$

91. $\sqrt[3]{x^{3n}(x + y)^{3n+6}} = \sqrt[3]{x^{3n}(x + y)^{3(n+2)}} = x^n(x + y)^{n+2}$

93. $\sqrt[n]{x^ny^{n^3+n}} = \sqrt[n]{x^ny^{n(n^2+1)}} = \sqrt[n]{(xy^{n^2+1})^n} = xy^{n^2+1}$

95. (A) $2\sqrt[3]{x^3} + 4\sqrt{x^2} = 2x + 4|x| = 2x + 4x = 6x$
(B) $2\sqrt[3]{x^3} + 4\sqrt{x^2} = 2x + 4|x| = 2x - 4x = -2x$

97. (A) $\sqrt[5]{x^5} + \sqrt[4]{x^4} = x + |x| = x + x = 2x$
(B) $\sqrt[5]{x^5} + \sqrt[4]{x^4} = x + |x| = x - x = 0$

99. (A) $3\sqrt[4]{x^4} - 2\sqrt[5]{x^5} = 3|x| - 2x = 3x - 2x = x$
(B) $3\sqrt[4]{x^4} - 2\sqrt[5]{x^5} = 3|x| - 2x = -3x - 2x = -5x$

Exercise 6-7 Basic Operations Involving Radicals

Key Ideas and Formulas

Algebraic expressions where two terms contain exactly the same radical—having the same index and the same radicand—can be combined into a single term.

COMMON ERROR: $\sqrt[n]{a} + \sqrt[n]{b} \neq \sqrt[n]{a + b}$. These cannot be combined into a single term.

$\sqrt[n]{a} \cdot \sqrt[n]{b} = \sqrt[n]{ab}$

When expressions like $\sqrt{a} + \sqrt{b}$ occur in a denominator, the fractional form can be converted to simplest radical form by multiplying numerator and denominator by $\sqrt{a} - \sqrt{b}$. The expressions $\sqrt{a} + \sqrt{b}$ and $\sqrt{a} - \sqrt{b}$ are called **conjugates.**

---

1. $7\sqrt{3} + 2\sqrt{3} = (7+2)\sqrt{3} = 9\sqrt{3}$

3. $2\sqrt{a} - 7\sqrt{a} = (2-7)\sqrt{a} = -5\sqrt{a}$

5. $\sqrt{n} - 4\sqrt{n} - 2\sqrt{n} = (1-4-2)\sqrt{n} = -5\sqrt{n}$

7. $\sqrt{5} - 2\sqrt{3} + 3\sqrt{5} = \sqrt{5} + 3\sqrt{5} - 2\sqrt{3} = (1+3)\sqrt{5} - 2\sqrt{3} = 4\sqrt{5} - 2\sqrt{3}$

9. $\sqrt{m} - \sqrt{n} - 2\sqrt{n} = \sqrt{m} + (-1-2)\sqrt{n} = \sqrt{m} - 3\sqrt{n}$

11. $\sqrt{18} + \sqrt{2} = \sqrt{9\cdot 2} + \sqrt{2} = \sqrt{9}\sqrt{2} + \sqrt{2} = 3\sqrt{2} + \sqrt{2} = 4\sqrt{2}$

13. $\sqrt{8} - 2\sqrt{32} = \sqrt{4\cdot 2} - 2\sqrt{16\cdot 2} = \sqrt{4}\sqrt{2} - 2\sqrt{16}\sqrt{2} = 2\sqrt{2} - 2\cdot 4\sqrt{2} = 2\sqrt{2} - 8\sqrt{2} = -6\sqrt{2}$

15. $\sqrt{7}(\sqrt{7} - 2) = \sqrt{7}\sqrt{7} - 2\sqrt{7} = 7 - 2\sqrt{7}$

17. $\sqrt{2}(3 - \sqrt{2}) = 3\sqrt{2} - \sqrt{2}\sqrt{2} = 3\sqrt{2} - 2$

19. $\sqrt{y}(\sqrt{y} - 8) = \sqrt{y}\sqrt{y} - 8\sqrt{y} = y - 8\sqrt{y}$

21. $\sqrt{n}(4 - \sqrt{n}) = 4\sqrt{n} - \sqrt{n}\sqrt{n} = 4\sqrt{n} - n$

23. $\sqrt{3}\left(\sqrt{3} + \sqrt{6}\right) = \sqrt{3}\sqrt{3} + \sqrt{3}\sqrt{6} = 3 + \sqrt{18} = 3 + \sqrt{9\cdot 2} = 3 + \sqrt{9}\sqrt{2} = 3 + 3\sqrt{2}$

25. $(2 - \sqrt{3})(3 + \sqrt{3}) = 2\cdot 3 + 2\sqrt{3} - 3\sqrt{3} - \sqrt{3}\sqrt{3} = 6 + 2\sqrt{3} - 3\sqrt{3} - 3 = 3 - \sqrt{3}$

27. $(\sqrt{5} + 2)^2 = (\sqrt{5})^2 + 2(\sqrt{5})(2) + 2^2 = 5 + 4\sqrt{5} + 4 = 9 + 4\sqrt{5}$

29. $(\sqrt{m} - 3)(\sqrt{m} - 4) = \sqrt{m}\sqrt{m} - 4\sqrt{m} - 3\sqrt{m} - 3(-4)$
$= m - 4\sqrt{m} - 3\sqrt{m} + 12 = m - 7\sqrt{m} + 12$

31. $\dfrac{1}{\sqrt{5}+2} = \dfrac{1(\sqrt{5}-2)}{(\sqrt{5}+2)(\sqrt{5}-2)} = \dfrac{\sqrt{5}-2}{5-4} = \dfrac{\sqrt{5}-2}{1} = \sqrt{5} - 2$

33. $\dfrac{2}{\sqrt{5}+1} = \dfrac{2(\sqrt{5}-1)}{(\sqrt{5}+1)(\sqrt{5}-1)} = \dfrac{2(\sqrt{5}-1)}{5-1} = \dfrac{2(\sqrt{5}-1)}{4} = \dfrac{2(\sqrt{5}-1)}{2\cdot 2} = \dfrac{\sqrt{5}-1}{2}$

35. $\dfrac{\sqrt{2}}{\sqrt{10}-2} = \dfrac{\sqrt{2}(\sqrt{10}+2)}{(\sqrt{10}-2)(\sqrt{10}+2)} = \dfrac{\sqrt{20}+2\sqrt{2}}{10-4} = \dfrac{2\sqrt{5}+2\sqrt{2}}{6}$
$= \dfrac{2(\sqrt{5}+\sqrt{2})}{2\cdot 3} = \dfrac{\sqrt{5}+\sqrt{2}}{3}$

37. $\dfrac{\sqrt{y}}{\sqrt{y} + 3} = \dfrac{\sqrt{y}(\sqrt{y} - 3)}{(\sqrt{y} + 3)(\sqrt{y} - 3)} = \dfrac{y - 3\sqrt{y}}{y - 9}$

39. $\sqrt{8mn} + 2\sqrt{18mn} = \sqrt{4 \cdot 2mn} + 2\sqrt{9 \cdot 2mn} = 2\sqrt{2mn} + 2 \cdot 3\sqrt{2mn}$

$= 2\sqrt{2mn} + 6\sqrt{2mn} = 8\sqrt{2mn}$

41. $\sqrt{8} - \sqrt{20} + 4\sqrt{2} = \sqrt{4 \cdot 2} - \sqrt{4 \cdot 5} + 4\sqrt{2} = 2\sqrt{2} - 2\sqrt{5} + 4\sqrt{2} = 6\sqrt{2} - 2\sqrt{5}$

43. $\sqrt[5]{a} - 4\sqrt[5]{a} + 2\sqrt[5]{a} = (1 - 4 + 2)\sqrt[5]{a} = -1\sqrt[5]{a}$ or $-\sqrt[5]{a}$

45. $2\sqrt[3]{x} + 3\sqrt[3]{x} - \sqrt{x} = (2 + 3)\sqrt[3]{x} - \sqrt{x} = 5\sqrt[3]{x} - \sqrt{x}$

47. $\sqrt{\dfrac{1}{8}} + \sqrt{8} = \sqrt{\dfrac{1}{8} \cdot \dfrac{2}{2}} + \sqrt{4 \cdot 2} = \sqrt{\dfrac{2}{16}} + \sqrt{4}\sqrt{2} = \dfrac{\sqrt{2}}{\sqrt{16}} + 2\sqrt{2} = \dfrac{\sqrt{2}}{4} + \dfrac{8\sqrt{2}}{4} = \dfrac{9\sqrt{2}}{4}$

49. $\sqrt{\dfrac{3uv}{2}} - \sqrt{24uv} = \sqrt{\dfrac{3uv}{2} \cdot \dfrac{2}{2}} - \sqrt{4 \cdot 6uv} = \sqrt{\dfrac{6uv}{4}} - \sqrt{4}\sqrt{6uv}$

$= \dfrac{\sqrt{6uv}}{\sqrt{4}} - 2\sqrt{6uv} = \dfrac{\sqrt{6uv}}{2} - \dfrac{2\sqrt{6uv}}{1} = \dfrac{\sqrt{6uv}}{2} - \dfrac{4\sqrt{6uv}}{2} = \dfrac{-3\sqrt{6uv}}{2}$

51. $(4\sqrt{3} - 1)(3\sqrt{3} - 2) = (4\sqrt{3})(3\sqrt{3}) - 4\sqrt{3} \cdot 2 - 1 \cdot 3\sqrt{3} + 2$

$= 4 \cdot 3\sqrt{3}\sqrt{3} - 8\sqrt{3} - 3\sqrt{3} + 2$

$= 12 \cdot 3 - 11\sqrt{3} + 2 = 36 - 11\sqrt{3} + 2 = 38 - 11\sqrt{3}$

53. $(\sqrt{x} - \sqrt{y})(\sqrt{x} + \sqrt{y}) = \sqrt{x}\sqrt{x} + \sqrt{x}\sqrt{y} - \sqrt{x}\sqrt{y} - \sqrt{y}\sqrt{y} = x - y$

55. $(5\sqrt{m} + 2)(2\sqrt{m} - 3) = (5\sqrt{m})(2\sqrt{m}) - 5\sqrt{m} \cdot 3 + 2 \cdot 2\sqrt{m} - 2 \cdot 3$

$= 5 \cdot 2\sqrt{m}\sqrt{m} - 15\sqrt{m} + 4\sqrt{m} - 6 = 10m - 11\sqrt{m} - 6$

57. $(\sqrt[3]{4} + \sqrt[3]{9})(\sqrt[3]{2} + \sqrt[3]{3}) = \sqrt[3]{4}\sqrt[3]{2} + \sqrt[3]{4}\sqrt[3]{3} + \sqrt[3]{9}\sqrt[3]{2} + \sqrt[3]{9}\sqrt[3]{3} = \sqrt[3]{8} + \sqrt[3]{12} + \sqrt[3]{18} + \sqrt[3]{27}$

$= 2 + \sqrt[3]{12} + \sqrt[3]{18} + 3 = 5 + \sqrt[3]{12} + \sqrt[3]{18}$

59.

$$\begin{array}{lll}
 & \sqrt[3]{x^2} + \sqrt[3]{x} + 1 & \\
 & \sqrt[3]{x} - 1 & \\
\hline
\sqrt[3]{x^3} + \sqrt[3]{x^2} + \sqrt[3]{x} & & \\
\quad -\sqrt[3]{x^2} - \sqrt[3]{x} - 1 & & \\
\hline
\sqrt[3]{x^3} \qquad\qquad - 1 & = & x - 1
\end{array}$$

61.

$$\begin{array}{lll}
 & \sqrt[3]{a^2} - \sqrt[3]{2a} + \sqrt[3]{4} & \\
 & \sqrt[3]{a} + \sqrt[3]{2} & \\
\hline
\sqrt[3]{a^3} - \sqrt[3]{2a^2} + \sqrt[3]{4a} & & \\
\quad + \sqrt[3]{2a^2} - \sqrt[3]{4a} + \sqrt[3]{8} & & \\
\hline
\sqrt[3]{a^3} \qquad\qquad + \sqrt[3]{8} & = & a + 2
\end{array}$$

63. $(\sqrt[6]{8} - 1)(\sqrt[6]{8} + 1) = (\sqrt[6]{8})^2 - 1^2 = \sqrt[6]{64} - 1 = \sqrt[6]{2^6} - 1 = 2 - 1 = 1$

65.
$$x^2 - 6x + 7 = 0$$
$$(3 - \sqrt{2})^2 - 6(3 - \sqrt{2}) + 7 \stackrel{?}{=} 0$$
$$3^2 - 2 \cdot 3\sqrt{2} + (\sqrt{2})^2 - 18 + 6\sqrt{2} + 7 \stackrel{?}{=} 0$$
$$9 - 6\sqrt{2} + 2 - 18 + 6\sqrt{2} + 7 \stackrel{?}{=} 0$$
$$0 \stackrel{\checkmark}{=} 0$$

67.
$$x^2 - 2x - 4 \stackrel{?}{=} 0$$
$$(1 - \sqrt{5})^2 - 2(1 - \sqrt{5}) - 4 \stackrel{?}{=} 0$$
$$1^2 - 2 \cdot 1 \cdot \sqrt{5} + (\sqrt{5})^2 - 2 + 2\sqrt{5} - 4 \stackrel{?}{=} 0$$
$$1 - 2\sqrt{5} + 5 - 2 + 2\sqrt{5} - 4 \stackrel{?}{=} 0$$
$$0 \stackrel{\checkmark}{=} 0$$

69.
$$x^2 - 8x + 1 \stackrel{?}{=} 0$$
$$(4 + \sqrt{15})^2 - 8(4 + \sqrt{15}) + 1 \stackrel{?}{=} 0$$
$$4^2 + 2 \cdot 4\sqrt{15} + (\sqrt{15})^2 - 32 - 8\sqrt{15} + 1 \stackrel{?}{=} 0$$
$$16 + 8\sqrt{15} + 15 - 32 - 8\sqrt{15} + 1 \stackrel{?}{=} 0$$
$$0 \stackrel{\checkmark}{=} 0$$

71. $$\frac{\sqrt{3} + 2}{\sqrt{3} - 2} = \frac{(\sqrt{3} + 2)(\sqrt{3} + 2)}{(\sqrt{3} - 2)(\sqrt{3} + 2)} = \frac{\sqrt{3}\sqrt{3} + 2\sqrt{3} + 2\sqrt{3} + 4}{\sqrt{3}\sqrt{3} - 4} = \frac{3 + 4\sqrt{3} + 4}{3 - 4}$$
$$= \frac{7 + 4\sqrt{3}}{-1} = -7 - 4\sqrt{3}$$

73. $$\frac{\sqrt{2} + \sqrt{3}}{\sqrt{3} - \sqrt{2}} = \frac{(\sqrt{2} + \sqrt{3})(\sqrt{3} + \sqrt{2})}{(\sqrt{3} - \sqrt{2})(\sqrt{3} + \sqrt{2})} = \frac{\sqrt{2}\sqrt{3} + \sqrt{2}\sqrt{2} + \sqrt{3}\sqrt{3} + \sqrt{3}\sqrt{2}}{\sqrt{3}\sqrt{3} - \sqrt{2}\sqrt{2}}$$
$$= \frac{\sqrt{6} + 2 + 3 + \sqrt{6}}{3 - 2} = \frac{2\sqrt{6} + 5}{1} = 2\sqrt{6} + 5$$

75. $$\frac{2 + \sqrt{x}}{\sqrt{x} - 3} = \frac{(2 + \sqrt{x})(\sqrt{x} + 3)}{(\sqrt{x} - 3)(\sqrt{x} + 3)} = \frac{2\sqrt{x} + 6 + \sqrt{x}\sqrt{x} + 3\sqrt{x}}{\sqrt{x}\sqrt{x} - 3 \cdot 3} = \frac{5\sqrt{x} + 6 + x}{x - 9} = \frac{x + 5\sqrt{x} + 6}{x - 9}$$

77. $$\frac{3\sqrt{x}}{2\sqrt{x} - 3} = \frac{3\sqrt{x}\,(2\sqrt{x} + 3)}{(2\sqrt{x} - 3)(2\sqrt{x} + 3)} = \frac{3\sqrt{x} \cdot 2\sqrt{x} + 3\sqrt{x} \cdot 3}{2\sqrt{x} \cdot 2\sqrt{x} - 3 \cdot 3} = \frac{6x + 9\sqrt{x}}{4x - 9}$$

79. $$\frac{\sqrt{3} + 2}{\sqrt{3} - 2} = \frac{(\sqrt{3} + 2)(\sqrt{3} - 2)}{(\sqrt{3} - 2)(\sqrt{3} - 2)} = \frac{\sqrt{3}\sqrt{3} - 4}{\sqrt{3}\sqrt{3} - 2\sqrt{3} - 2\sqrt{3} + 4} = \frac{3 - 4}{3 - 4\sqrt{3} + 4} = \frac{-1}{7 - 4\sqrt{3}}$$

81. $$\frac{\sqrt{2} + \sqrt{3}}{\sqrt{3} - \sqrt{2}} = \frac{(\sqrt{2} + \sqrt{3})(\sqrt{2} - \sqrt{3})}{(\sqrt{3} - \sqrt{2})(\sqrt{2} - \sqrt{3})} = \frac{\sqrt{2}\sqrt{2} - \sqrt{3}\sqrt{3}}{\sqrt{3}\sqrt{2} - \sqrt{3}\sqrt{3} - \sqrt{2}\sqrt{2} + \sqrt{2}\sqrt{3}}$$
$$= \frac{2 - 3}{\sqrt{6} - 3 - 2 + \sqrt{6}} = \frac{-1}{-5 + 2\sqrt{6}} = \frac{1}{5 - 2\sqrt{6}}$$

83. $\frac{2 + \sqrt{x}}{\sqrt{x} - 3} = \frac{(2 + \sqrt{x})(2 - \sqrt{x})}{(\sqrt{x} - 3)(2 - \sqrt{x})} = \frac{4 - \sqrt{x}\sqrt{x}}{2\sqrt{x} - \sqrt{x}\sqrt{x} - 6 + 3\sqrt{x}}$

$= \frac{4 - x}{5\sqrt{x} - x - 6} = \frac{x - 4}{x - 5\sqrt{x} + 6}$

85. $\frac{\sqrt{3}}{3} + 2\sqrt{\frac{1}{3}} + \sqrt{12} = \frac{\sqrt{3}}{3} + 2\sqrt{\frac{1}{3}\cdot\frac{3}{3}} + \sqrt{4\cdot 3} = \frac{\sqrt{3}}{3} + 2\sqrt{\frac{3}{9}} + \sqrt{4}\sqrt{3} = \frac{\sqrt{3}}{3} + 2\frac{\sqrt{3}}{\sqrt{9}} + 2\sqrt{3}$

$= \frac{\sqrt{3}}{3} + \frac{2\sqrt{3}}{3} + \frac{2\sqrt{3}}{1} = \frac{\sqrt{3}}{3} + \frac{2\sqrt{3}}{3} + \frac{6\sqrt{3}}{3} = \frac{9\sqrt{3}}{3} = 3\sqrt{3}$

87. $\sqrt[3]{\frac{1}{3}} + \sqrt[3]{3^5} = \sqrt[3]{\frac{1}{3}\cdot\frac{3^2}{3^2}} + \sqrt[3]{3^3\cdot 3^2} = \sqrt[3]{\frac{3^2}{3^3}} + \sqrt[3]{3^3}\sqrt[3]{3^2} = \frac{\sqrt[3]{9}}{3} + 3\sqrt[3]{9} = \frac{\sqrt[3]{9}}{3} + \frac{9\sqrt[3]{9}}{3} = \frac{10\sqrt[3]{9}}{3}$

89. $(\sqrt[3]{x} - \sqrt[3]{y^2})(\sqrt[3]{x^2} + 2\sqrt[3]{y}) = \sqrt[3]{x}\sqrt[3]{x^2} + \sqrt[3]{x}\cdot 2\sqrt[3]{y} - \sqrt[3]{y^2}\sqrt[3]{x^2} - \sqrt[3]{y^2}\cdot 2\sqrt[3]{y}$

$= \sqrt[3]{x^3} + 2\sqrt[3]{xy} - \sqrt[3]{x^2y^2} - 2\sqrt[3]{y^3} = x + 2\sqrt[3]{xy} - \sqrt[3]{x^2y^2} - 2y$

91.
$$\begin{array}{l}
\sqrt[3]{x^2} - \sqrt[3]{x}\sqrt[3]{y} + \sqrt[3]{y^2} \\
\quad\sqrt[3]{x} + \sqrt[3]{y} \\
\hline
\sqrt[3]{x^3} - \sqrt[3]{x^2}\sqrt[3]{y} + \sqrt[3]{x}\sqrt[3]{y^2} \\
\qquad\qquad \sqrt[3]{x^2}\sqrt[3]{y} - \sqrt[3]{x}\sqrt[3]{y^2} + \sqrt[3]{y^3} \\
\hline
\sqrt[3]{x^3} \qquad\qquad\qquad\qquad + \sqrt[3]{y^3} = x + y
\end{array}$$

93. $\frac{2\sqrt{x} + 3\sqrt{y}}{4\sqrt{x} + 5\sqrt{y}} = \frac{(2\sqrt{x} + 3\sqrt{y})(4\sqrt{x} - 5\sqrt{y})}{(4\sqrt{x} + 5\sqrt{y})(4\sqrt{x} - 5\sqrt{y})}$

$= \frac{(2\sqrt{x})(4\sqrt{x}) + (2\sqrt{x})(-5\sqrt{y}) + (3\sqrt{y})(4\sqrt{x}) + (3\sqrt{y})(-5\sqrt{y})}{(4\sqrt{x})^2 - (5\sqrt{y})^2}$

$= \frac{8x - 10\sqrt{x}\sqrt{y} + 12\sqrt{x}\sqrt{y} - 15y}{16x - 25y} = \frac{8x + 2\sqrt{xy} - 15y}{16x - 25y}$

95. Using the hint, we note from problem 59 that

$(\sqrt[3]{x} - 1)(\sqrt[3]{x^2} + \sqrt[3]{x} + 1) = x - 1$

Therefore,

$\frac{1}{\sqrt[3]{x} - 1} = \frac{1(\sqrt[3]{x^2} + \sqrt[3]{x} + 1)}{(\sqrt[3]{x} - 1)(\sqrt[3]{x^2} + \sqrt[3]{x} + 1)} = \frac{\sqrt[3]{x^2} + \sqrt[3]{x} + 1}{x - 1}$

97. Using the hint, we note from problem 91 that

$(\sqrt[3]{x} + \sqrt[3]{y})(\sqrt[3]{x^2} - \sqrt[3]{x}\sqrt[3]{y} + \sqrt[3]{y^2}) = x + y$

Therefore

$$\frac{1}{\sqrt[3]{x} + \sqrt[3]{y}} = \frac{1(\sqrt[3]{x^2} - \sqrt[3]{x}\sqrt[3]{y} + \sqrt[3]{y^2})}{(\sqrt[3]{x} + \sqrt[3]{y})(\sqrt[3]{x^2} - \sqrt[3]{x}\sqrt[3]{y} + \sqrt[3]{y^2})} = \frac{\sqrt[3]{x^2} - \sqrt[3]{x}\sqrt[3]{y} + \sqrt[3]{y^2}}{x + y}$$

99. Using the hint, we start as follows:

$$\frac{1}{\sqrt{x} + \sqrt{y} - \sqrt{z}} = \frac{1[(\sqrt{x} + \sqrt{y}) + \sqrt{z}]}{[(\sqrt{x} + \sqrt{y}) - \sqrt{z}][(\sqrt{x} + \sqrt{y}) + \sqrt{z}]} = \frac{\sqrt{x} + \sqrt{y} + \sqrt{z}}{(\sqrt{x} + \sqrt{y})^2 - (\sqrt{z})^2}$$

$$= \frac{\sqrt{x} + \sqrt{y} + \sqrt{z}}{x + 2\sqrt{xy} + y - z} = \frac{\sqrt{x} + \sqrt{y} + \sqrt{z}}{x + y - z + 2\sqrt{xy}}$$

This denominator could be rationalized if we multiplied numerator and denominator by $x + y - z - 2\sqrt{xy}$.

$$= \frac{(\sqrt{x} + \sqrt{y} + \sqrt{z})[(x + y - z) - 2\sqrt{xy}]}{[(x + y - z) + 2\sqrt{xy}][(x + y - z) - 2\sqrt{xy}]}$$

$$= \frac{(\sqrt{x} + \sqrt{y} + \sqrt{z})[(x + y - z) - 2\sqrt{xy}]}{(x + y - z)^2 - 4xy}$$

Exercise 6-8 Complex Numbers

Key Ideas and Formulas

There is a number, not a real number, whose square is -1. We write

$$i^2 = -1 \qquad i = \sqrt{-1}$$

Numbers of the form $a + bi$, where a and b are real, are called complex numbers. If $b = 0$, the number is simply the real number a. If $b \neq 0$, $a + bi$ is called an **imaginary** number; if $a = 0$, $0 + bi$ or bi is called a **pure imaginary** number.

Equality and Operations on Complex Numbers

EQUALITY	$a + bi = c + di$ if and only if $a = c$ and $b = d$
ADDITION	$(a + bi) + (c + di) = (a + c) + (b + d)i$
SUBTRACTION	$(a + bi) - (c + di) = (a - c) + (b - d)i$
MULTIPLICATION	$(a + bi)(c + di) = (ac - bd) + (ad + bc)i$
DIVISION	$\frac{a + bi}{c + di} = \frac{ac + bd}{c^2 + d^2} + \frac{-ad + bc}{c^2 + d^2} i$

If x is a negative real number, $x = -a$, where a is a positive real number,

$$\sqrt{x} = \sqrt{-a} = i\sqrt{a}$$

1. $(5 + 2i) + (3 + i) = 5 + 2i + 3 + i = 8 + 3i$

3. $(-8 + 5i) + (3 - 2i) = -8 + 5i + 3 - 2i = -5 + 3i$

5. $(8 + 5i) - (3 + 2i) = 8 + 5i - 3 - 2i = 5 + 3i$

7. $(4 + 7i) - (-2 - 6i) = 4 + 7i + 2 + 6i = 6 + 13i$

9. $(3 - 7i) + 5i = 3 - 7i + 5i = 3 - 2i$

11. $(5i)(3i) = 15i^2 = 15(-1) = -15$ or $-15 + 0i$

13. $-2i(5 - 3i) = -10i + 6i^2 = -10i + 6(-1) = -10i - 6 = -6 - 10i$

15. $(2 - 3i)(3 + 3i) = 6 + 6i - 9i - 9i^2 = 6 - 3i - 9(-1) = 6 - 3i + 9 = 15 - 3i$

17. $(7 - 6i)(2 - 3i) = 14 - 21i - 12i + 18i^2 = 14 - 33i + 18(-1) = 14 - 33i - 18$
$= -4 - 33i$

19. $(7 + 4i)(7 - 4i) = 49 - 16i^2 = 49 - 16(-1) = 49 + 16 = 65$ or $65 + 0i$

21. $(-1 + 3i)(2 - i) = -2 + i + 6i - 3i^2 = -2 + 7i - 3(-1) = -2 + 7i + 3 = 1 + 7i$

23. $(-3 + 4i)(4 - i) = -12 + 3i + 16i - 4i^2 = -12 + 19i - 4(-1)$
$= -12 + 19i + 4 = -8 + 19i$

25. $\frac{1}{2 + i} = \frac{1}{2 + i} \cdot \frac{2 - i}{2 - i} = \frac{2 - i}{4 - i^2} = \frac{2 - i}{4 - (-1)} = \frac{2 - i}{5} = \frac{2}{5} - \frac{1}{5}i$

27. $\frac{3 + i}{2 - 3i} = \frac{3 + i}{2 - 3i} \cdot \frac{2 + 3i}{2 + 3i} = \frac{6 + 9i + 2i + 3i^2}{4 - 9i^2} = \frac{6 + 11i + 3(-1)}{4 - 9(-1)}$
$= \frac{6 + 11i - 3}{4 + 9} = \frac{3 + 11i}{13} = \frac{3}{13} + \frac{11}{13}i$

29. $\frac{13 + i}{2 - i} = \frac{13 + i}{2 - i} \cdot \frac{2 + i}{2 + i} = \frac{26 + 13i + 2i + i^2}{4 - i^2} = \frac{26 + 15i + (-1)}{4 - (-1)} = \frac{25 + 15i}{5}$
$= 5 + 3i$

31. $\frac{1 + i}{1 - i} = \frac{1 + i}{1 - i} \cdot \frac{1 + i}{1 + i} = \frac{1 + i + i + i^2}{1 - i^2} = \frac{1 + 2i + (-1)}{1 - (-1)} = \frac{2i}{2} = i$

33. $\frac{3 - i}{3 + i} = \frac{3 - i}{3 + i} \cdot \frac{3 - i}{3 - i} = \frac{9 - 3i - 3i + i^2}{9 - i^2} = \frac{9 - 6i + (-1)}{9 - (-1)}$
$= \frac{8 - 6i}{10} = \frac{4 - 3i}{5}$ or $\frac{4}{5} - \frac{3}{5}i$

35. $\sqrt{-16} = i\sqrt{16} = 4i$

37. $\sqrt{-72} = i\sqrt{72} = i\sqrt{36 \cdot 2} = i\sqrt{36}\sqrt{2} = 6i\sqrt{2}$

39. $-\sqrt{-4} = -i\sqrt{4} = -2i$

41. $-\sqrt{25} = -5$

43. $(5 - \sqrt{-9}) + (2 - \sqrt{-4}) = (5 - i\sqrt{9}) + (2 - i\sqrt{4}) = 5 - 3i + 2 - 2i = 7 - 5i$

45. $(9 - \sqrt{-9}) - (12 - \sqrt{-25}) = (9 - i\sqrt{9}) - (12 - i\sqrt{25})$
$= 9 - 3i - 12 + 5i = -3 + 2i$

47. $(-2 + \sqrt{-49})(3 - \sqrt{-4}) = (-2 + i\sqrt{49})(3 - i\sqrt{4}) = (-2 + 7i)(3 - 2i)$
$= -6 + 4i + 21i - 14i^2 = -6 + 25i - 14(-1)$
$= -6 + 25i + 14 = 8 + 25i$

49. $(-1 + \sqrt{-8})(2 - \sqrt{-10}) = (-1 + i\sqrt{8})(2 - i\sqrt{10}) = -2 + i\sqrt{10} + 2i\sqrt{8} - i^2\sqrt{80}$
$= -2 + i\sqrt{10} + 2i \cdot 2\sqrt{2} - (-1)4\sqrt{5}$
$= -2 + 4\sqrt{5} + (\sqrt{10} + 4\sqrt{2})i$

51. $(\sqrt{2} + \sqrt{-2})(\sqrt{2} + \sqrt{-2}) = (\sqrt{2} + i\sqrt{2})(\sqrt{2} + i\sqrt{2})$
$= \sqrt{2}\sqrt{2} + i\sqrt{2}\sqrt{2} + i\sqrt{2}\sqrt{2} + i^2\sqrt{2}\sqrt{2}$
$= 2 + 2i + 2i + (-1)2 = 2 + 4i - 2 = 4i$

53. $\dfrac{5 - \sqrt{-4}}{3} = \dfrac{5 - i\sqrt{4}}{3} = \dfrac{5 - 2i}{3} = \dfrac{5}{3} - \dfrac{2}{3}i$

55. $\dfrac{1}{2 - \sqrt{-9}} = \dfrac{1}{2 - i\sqrt{9}} = \dfrac{1}{2 - 3i} = \dfrac{1}{2 - 3i} \cdot \dfrac{2 + 3i}{2 + 3i} = \dfrac{2 + 3i}{4 - 9i^2} = \dfrac{2 + 3i}{4 - 9(-1)}$
$= \dfrac{2 + 3i}{13} = \dfrac{2}{13} + \dfrac{3}{13}i$

57. $\dfrac{1 + \sqrt{-3}}{1 - \sqrt{-3}} = \dfrac{1 + i\sqrt{3}}{1 - i\sqrt{3}} = \dfrac{1 + i\sqrt{3}}{1 - i\sqrt{3}} \cdot \dfrac{1 + i\sqrt{3}}{1 + i\sqrt{3}} = \dfrac{1 + i\sqrt{3} + i\sqrt{3} + i^2\sqrt{3}\sqrt{3}}{1 - i^2\sqrt{3}\sqrt{3}}$
$= \dfrac{1 + 2i\sqrt{3} + (-1)3}{1 - (-1)3} = \dfrac{1 + 2i\sqrt{3} - 3}{1 + 3} = \dfrac{-2 + 2i\sqrt{3}}{4}$
$= \dfrac{-1 + i\sqrt{3}}{2} = -\dfrac{1}{2} + i\dfrac{\sqrt{3}}{2}$

59. $\dfrac{1 - \sqrt{-3}}{2 + \sqrt{-5}} = \dfrac{1 - i\sqrt{3}}{2 + i\sqrt{5}} = \dfrac{1 - i\sqrt{3}}{2 + i\sqrt{5}} \cdot \dfrac{2 - i\sqrt{5}}{2 - i\sqrt{5}} = \dfrac{2 - i\sqrt{5} - 2i\sqrt{3} + i^2\sqrt{3}\sqrt{5}}{4 - i^2\sqrt{5}\sqrt{5}}$
$= \dfrac{2 - i(\sqrt{5} + 2\sqrt{3}) + (-1)\sqrt{15}}{4 - (-1)5} = \dfrac{2 - \sqrt{15} - i(\sqrt{5} + 2\sqrt{3})}{9}$
$= \dfrac{2 - \sqrt{15}}{9} - i\dfrac{\sqrt{5} + 2\sqrt{3}}{9}$

61. $\dfrac{2}{5i} = \dfrac{2}{5i} \cdot \dfrac{i}{i} = \dfrac{2i}{5i^2} = \dfrac{2i}{5(-1)} = \dfrac{2i}{-5} = -\dfrac{2}{5}i$ or $0 - \dfrac{2}{5}i$

63. $\dfrac{1 + 3i}{2i} = \dfrac{1 + 3i}{2i} \cdot \dfrac{i}{i} = \dfrac{i + 3i^2}{2i^2} = \dfrac{i + 3(-1)}{2(-1)} = \dfrac{-3 + i}{-2} = \dfrac{3}{2} - \dfrac{1}{2}i$

65. $(2 - i)^2 + 3(2 - i) - 5 = 4 - 2 \cdot 2i + i^2 + 6 - 3i - 5$
$= 4 - 4i + (-1) + 6 - 3i - 5 = 4 - 7i$

67. $(2 + i\sqrt{3})^2 - 4(2 + i\sqrt{3}) + 7 = 4 + 2 \cdot 2i\sqrt{3} + (i\sqrt{3})^2 - 8 - 4i\sqrt{3} + 7$
$= 4 + 4i\sqrt{3} - 3 - 8 - 4i\sqrt{3} + 7 = 11 - 11 + 4i\sqrt{3} - 4i\sqrt{3} = 0$

69. $(1 - i\sqrt{2})^2 - 2(1 - i\sqrt{2}) + 3 = 1 - 2\cdot i\sqrt{2} + (i\sqrt{2})^2 - 2 + 2i\sqrt{2} + 3$
$= 1 - 2i\sqrt{2} - 2 - 2 + 2i\sqrt{2} + 3 = 0$ or $0 + 0i$

71. $x^2 - 2x + 2$ becomes $(1 - i)^2 - 2(1 - i) + 2 = 1 - 2i + i^2 - 2 + 2i + 2$
$= 1 - 2i + (-1) - 2 + 2i + 2$
$= 0$ or $0 + 0i$

73. $x^2 - 4x + 5$ becomes $(2 + i)^2 - 4(2 + i) + 5 = 4 + 2\cdot 2i + i^2 - 8 - 4i + 5$
$= 4 + 4i + (-1) - 8 - 4i + 5$
$= 0$ or $0 + 0i$

75. $x^2 - 6x + 10$ becomes $(3 + i)^2 - 6(3 + i) + 10 = 9 + 2\cdot 3i + i^2 - 18 - 6i + 10$
$= 9 + 6i + (-1) - 18 - 6i + 10$
$= 0$ or $0 + 0i$

77. $x^2 + 2x + 2$ becomes $(-1 - i)^2 + 2(-1 - i) + 2 = 1 + 2i + i^2 - 2 - 2i + 2$
$= 1 + 2i + (-1) - 2 - 2i + 2$
$= 0$ or $0 + 0i$

79. $i^2 = -1$
$i^3 = i\cdot i^2 = i(-1) = -i$
$i^4 = (i^2)^2 = (-1)^2 = 1$
$i^5 = i\cdot i^4 = i(1) = i$
$i^6 = i^2\cdot i^4 = (-1)(1) = -1$
$i^7 = i\cdot i^6 = i(-1) = -i$
$i^8 = (i^4)^2 = 1^2 = 1$

81. $(a + bi) + (c + di) = a + bi + c + di = a + c + bi + di = (a + c) + (b + d)i$

83. $(a + bi)(a - bi) = a^2 - (bi)^2 = a^2 - b^2i^2 = a^2 - b^2(-1) = a^2 + b^2$

85. $(a + bi)(c + di) = ac + adi + bci + bdi^2 = ac + adi + bci + bd(-1)$
$= ac + adi + bci - bd = ac - bd + adi + bci$
$= (ac - bd) + (ad + bc)i$

87. $(1 + i)^3 = (1 + i)(1 + i)(1 + i) = (1 + i)(1 + i)^2 = (1 + i)(1 + 2i + i^2)$
$= (1 + i)(1 + 2i - 1) = (1 + i)2i = 2i + 2i^2 = 2i + 2(-1) = -2 + 2i$

89. $\left(-\frac{1}{2} - \frac{\sqrt{3}}{2}i\right)^3 = \left(-\frac{1}{2} - \frac{\sqrt{3}}{2}i\right)\left(-\frac{1}{2} - \frac{\sqrt{3}}{2}i\right)^2$

$= \left(-\frac{1}{2} - \frac{\sqrt{3}}{2}i\right)\left[\left(-\frac{1}{2}\right)^2 + 2\left(-\frac{1}{2}\right)\left(-\frac{\sqrt{3}}{2}i\right) + \left(-\frac{\sqrt{3}}{2}i\right)^2\right]$

$= \left(-\frac{1}{2} - \frac{\sqrt{3}}{2}i\right)\left[\frac{1}{4} + \frac{2\sqrt{3}\,i}{4} + \frac{3}{4}i^2\right]$

$= \left(-\frac{1}{2} - \frac{\sqrt{3}}{2}i\right)\left[\frac{1}{4} + \frac{\sqrt{3}}{2}i + \frac{3}{4}(-1)\right]$

$= \left(-\frac{1}{2} - \frac{\sqrt{3}}{2}i\right)\left(\frac{1}{4} + \frac{\sqrt{3}}{2}i - \frac{3}{4}\right)$

$$= \left(-\frac{1}{2} - \frac{\sqrt{3}}{2}i\right)\left(-\frac{2}{4} + \frac{\sqrt{3}}{2}i\right)$$
$$= \left(-\frac{1}{2}\right)\left(-\frac{2}{4}\right) - \frac{1}{2}\left(\frac{\sqrt{3}}{2}i\right) - \frac{2}{4}\left(-\frac{\sqrt{3}}{2}i\right) - \frac{\sqrt{3}}{2}i\,\frac{\sqrt{3}}{2}i$$
$$= \frac{1}{4} - \frac{\sqrt{3}}{4}i + \frac{\sqrt{3}}{4}i - \frac{3}{4}i^2$$
$$= \frac{1}{4} - \frac{3}{4}(-1)$$
$$= \frac{1}{4} + \frac{3}{4}$$
$$= 1$$

91. $y^2 = -36$ if and only if y is a square root of -36. There are two square roots of -36, so y must be $\sqrt{-36}$ or $-\sqrt{-36}$, that is $i\sqrt{36}$ or $-i\sqrt{36}$, or 6i or -6i. This is usually written ±6i for short.

93. $(x - 9)^2 = -9$ if and only if x - 9 is a square root of -9. As in problem 91, -9 has two square roots, $i\sqrt{9}$ and $-i\sqrt{9}$, or ±3i. Therefore x - 9 = ±3i and x must be 9 ± 3i.

95. $i^{-2} = \frac{1}{i^2} = \frac{1}{-1} = -1$

$i^{-3} = \frac{1}{i^3} = \frac{1}{-i} = \frac{1}{-i}\cdot\frac{i}{i} = \frac{i}{-i^2} = \frac{i}{-(-1)} = \frac{i}{1} = i$

$i^{-4} = \frac{1}{i^4} = \frac{1}{1} = 1$

$i^{-5} = \frac{1}{i^5} = \frac{1}{i} = \frac{1}{i}\cdot\frac{-i}{-i} = \frac{-i}{-i^2} = \frac{-i}{1} = -i$

$i^{-6} = \frac{1}{i^6} = \frac{1}{-1} = -1$

$i^{-7} = \frac{1}{i^7} = \frac{1}{-i} = i$ (see above)

$i^{-8} = \frac{1}{i^8} = \frac{1}{1} = 1$

(see problem 79 for details of the second step in each case)

97. $(a + bi)\left(\frac{a}{a^2 + b^2} - \frac{b}{a^2 + b^2}i\right) = (a + bi)\left(\frac{a - bi}{a^2 + b^2}\right) = \frac{(a + bi)}{1}\,\frac{(a - bi)}{a^2 + b^2}$

$$= \frac{a^2 - b^2i^2}{a^2 + b^2} = \frac{a^2 - b^2(-1)}{a^2 + b^2} = \frac{a^2 + b^2}{a^2 + b^2} = 1$$

99. $ax^2 + bx + c$ becomes

$$a(p + qi)^2 + b(p + qi) + c = a(p^2 + 2pqi + q^2i^2) + bp + bqi + c$$
$$= ap^2 + 2apqi - aq^2 + bp + bqi + c$$
$$= (ap^2 - aq^2 + bp + c) + (2apq + bq)i$$

Exercise 6-9 REVIEW EXERCISE

1. $\frac{x^8}{x^3} = x^{8-3} = x^5$ 2. x^3y^3 3. $\frac{x^3}{y^3}$ 4. $\frac{x^3}{x^8} = \frac{1}{x^{8-3}} = \frac{1}{x^5}$

5. $(x^3)^8 = x^{8\cdot 3} = x^{24}$ 6. 1   7. $x^3x^8 = x^{3+8} = x^{11}$

8. $(-2x)^3 = (-2)^3x^3 = -8x^3$ 9. $(-2x^3)(3x^8) = (-2)(3)x^3x^8 = -6x^{3+8} = -6x^{11}$

10. 1   11. $3^{-2} = \frac{1}{3^2} = \frac{1}{9}$ 12. $\frac{1}{2^{-3}} = 2^3 = 8$ 13. $4^{-1/2} = \frac{1}{4^{1/2}} = \frac{1}{2}$

14. $(-9)^{3/2} = [(-9)^{1/2})]^3$. This is not a real number.

15. $(-8)^{2/3} = [(-8)^{1/3}]^2 = [-2]^2 = 4$

16. $4,280,000,000 = 4.280\ 000\ 000. \times 10^9 = 4.28 \times 10^9$

9 places left

positive exponent

17. $0.000\ 031\ 8 = 0.000\ 03.18 \times 10^{-5} = 3.18 \times 10^{-5}$

5 places right

negative exponent

18. $7.29 \times 10^5 = 7.29 \times 100,000 = 729,000$

19. $6.03 \times 10^{-4} = 6.03 \times 0.0001 = 0.000\ 603$ 20. $(3x^3y^2)(2xy^5) = 3\cdot 2x^{3+1}y^{2+5} = 6x^4y^7$

21. $\frac{9u^8v^6}{3u^4v^8} = \frac{3u^{8-4}}{v^{8-6}} = \frac{3u^4}{v^2}$ 22. $6(xy^3)^5 = 6x^5(y^3)^5 = 6x^5y^{5\cdot 3} = 6x^5y^{15}$

23. $\left(\frac{c^2}{d^5}\right)^3 = \frac{(c^2)^3}{(d^5)^3} = \frac{c^6}{d^{15}}$ 24. $\left(\frac{2x^2}{3y^3}\right)^2 = \frac{2^2(x^2)^2}{3^2(y^3)^2} = \frac{4x^4}{9y^6}$

25. $(x^{-3})^{-4} = x^{(-4)(-3)} = x^{12}$ 26. $\frac{y^{-3}}{y^{-5}} = y^{(-3)-(-5)} = y^2$

27. $(x^2y^{-3})^{-1} = (x^2)^{-1}(y^{-3})^{-1} = x^{-2}y^3 = \frac{1}{x^2}y^3 = \frac{y^3}{x^2}$

28. $(x^9)^{1/3} = x^{1/3(9)} = x^3$ 29. $(x^4)^{-1/2} = x^{-1/2(4)} = x^{-2} = \frac{1}{x^2}$

30. $x^{1/3}x^{-2/3} = x^{1/3-2/3} = x^{-1/3} = \frac{1}{x^{1/3}}$ 31. $\frac{u^{5/3}}{u^{2/3}} = u^{5/3-2/3} = u^1 = u$

32. $\sqrt{3m}$ 33. $3\sqrt{m}$ 34. $(2x)^{1/2}$ 35. $(a + b)^{1/2}$

36. $2 - \sqrt{-9} = 2 - i\sqrt{9} = 2 - 3i$

37. $\sqrt[3]{375} = \sqrt[3]{5^3 \cdot 3} = \sqrt[3]{5^3}\sqrt[3]{3} = 5\sqrt[3]{3}$

38. $2xy^2$

39. $\frac{5}{y}$

40. $\sqrt{36x^4y^7} = \sqrt{36x^4y^6 \cdot y} = \sqrt{36x^4y^6}\sqrt{y} = 6x^2y^3\sqrt{y}$

41. $\frac{1}{\sqrt{2y}} = \frac{1}{\sqrt{2y}} \cdot \frac{\sqrt{2y}}{\sqrt{2y}} = \frac{\sqrt{2y}}{2y}$

42. $\frac{6ab}{\sqrt{3a}} = \frac{6ab}{\sqrt{3a}} \cdot \frac{\sqrt{3a}}{\sqrt{3a}} = \frac{6ab\sqrt{3a}}{3a} = 2b\sqrt{3a}$

43. $\sqrt{2x^2y^5}\sqrt{18x^3y^2} = \sqrt{36x^5y^7} = \sqrt{36x^4y^6 \cdot xy} = \sqrt{36x^4y^6}\sqrt{xy} = 6x^2y^3\sqrt{xy}$

44. $\sqrt{\frac{y}{2x}} = \sqrt{\frac{y}{2x} \cdot \frac{2x}{2x}} = \sqrt{\frac{2xy}{4x^2}} = \frac{\sqrt{2xy}}{\sqrt{4x^2}} = \frac{\sqrt{2xy}}{2x}$

45. $4\sqrt{x} - 7\sqrt{x} = (4 - 7)\sqrt{x} = -3\sqrt{x}$

46. $\sqrt{7} + 2\sqrt{3} - 4\sqrt{3} = \sqrt{7} + (2 - 4)\sqrt{3} = \sqrt{7} - 2\sqrt{3}$

47. $\sqrt{5}(\sqrt{5} + 2) = \sqrt{5}\sqrt{5} + 2\sqrt{5} = 5 + 2\sqrt{5}$

48. $(\sqrt{3} - 1)(\sqrt{3} + 2) = \sqrt{3}\sqrt{3} + 2\sqrt{3} - 1\sqrt{3} - 2 = 3 + 2\sqrt{3} - \sqrt{3} - 2$
$= 1 + (2 - 1)\sqrt{3} = 1 + \sqrt{3}$

49. $\frac{\sqrt{5}}{3 - \sqrt{5}} = \frac{\sqrt{5}(3 + \sqrt{5})}{(3 - \sqrt{5})(3 + \sqrt{5})} = \frac{3\sqrt{5} + 5}{9 - 5} = \frac{3\sqrt{5} + 5}{4}$

50. $(-3 + 2i) + (6 - 8i) = -3 + 2i + 6 - 8i = 3 - 6i$

51. $(3 - 3i)(2 + 3i) = 6 + 9i - 6i - 9i^2 = 6 + 3i - 9(-1) = 6 + 3i + 9 = 15 + 3i$

52. $\frac{13 - i}{5 - 3i} = \frac{(13 - i)(5 + 3i)}{(5 - 3i)(5 + 3i)} = \frac{65 + 39i - 5i - 3i^2}{25 - 9i^2} = \frac{65 + 34i - 3(-1)}{25 + 9}$
$= \frac{65 + 34i + 3}{34} = \frac{68 + 34i}{34} = \frac{68}{34} + \frac{34i}{34} = 2 + i$

53. $\frac{2 - i}{2i} = \frac{(2 - i)i}{(2i)i} = \frac{2i - i^2}{2i^2} = \frac{2i - (-1)}{-2} = \frac{1 + 2i}{-2} = -\frac{1}{2} - i$

54. $\left(\frac{2x^3}{y^8}\right)^2 = \frac{(2x^3)^2}{(y^8)^2} = \frac{(2)^2(x^3)^2}{(y^8)^2} = \frac{4x^{2\cdot 3}}{y^{2\cdot 8}} = \frac{4x^6}{y^{16}}$

55. $(-x^2y)^2(-xy^2)^3 = (-1)^2(x^2)^2y^2 \cdot (-1)^3x^3(y^2)^3 = 1x^4y^2 \cdot (-1)x^3y^6$
$= -x^{4+3}y^{2+6} = -x^7y^8$

56. $\frac{-4(x^2y)^3}{(-2x)^2} = \frac{-4(x^2)^3y^3}{(-2)^2x^2} = \frac{-4x^{3\cdot 2}y^3}{4x^2} = \frac{-4x^6y^3}{4x^2} = -\frac{4}{4} \cdot \frac{x^6}{x^2} \cdot y^3 = -1x^{6-2}y^3 = -x^4y^3$

57. $(3xy^3)^2(x^2y)^3 = (3)^2x^2(y^3)^2(x^2)^3y^3 = 9x^2y^{2\cdot 3}x^{3\cdot 2}y^3 = 9x^2y^6x^6y^3 = 9x^{2+6}y^{6+3} = 9x^8y^9$

58. $\left(\frac{-2x}{y^2}\right)^3 = \frac{(-2x)^3}{(y^2)^3} = \frac{(-2)^3x^3}{(y^2)^3} = \frac{-8x^3}{y^{3\cdot 2}} = \frac{-8x^3}{y^6}$

59. $\left(\frac{3x^3y^2}{2x^2y^3}\right)^2 = \frac{(3x^3y^2)^2}{(2x^2y^3)^2} = \frac{3^2(x^3)^2(y^2)^2}{2^2(x^2)^2(y^3)^2} = \frac{9x^{2\cdot 3}y^{2\cdot 2}}{4x^{2\cdot 2}y^{2\cdot 3}} = \frac{9x^6y^4}{4x^4y^6}$

$= \frac{9}{4}\cdot\frac{x^6}{x^4}\cdot\frac{y^4}{y^6} = \frac{9}{4}\cdot x^{6-4}\cdot\frac{1}{y^{6-4}} = \frac{9}{4}x^2\,\frac{1}{y^2} = \frac{9x^2}{4y^2}$

60. $\frac{0.000\,052}{130(0.000\,2)} = \frac{5.2\times 10^{-5}}{(1.3\times 10^2)(2\times 10^4)} = \frac{5.2\times 10^{-5}}{(1.3)(2)\times 10^2\cdot 10^{-4}}$

$= \frac{5.2}{2.6}\times\frac{10^{-5}}{10^{-2}} = 2\times 10^{-3}$ or 0.002

61. $\frac{3m^4n^{-7}}{6m^3n^{-2}} = \frac{m^{4-2}}{2n^{-2-(-7)}} = \frac{m^2}{2n^5}$

62. $(x^{-3}y^2)^{-2} = (x^{-3})^{-2}(y^2)^{-2} = x^6y^{-4} = x^6\cdot\frac{1}{y^4} = \frac{x^6}{y^4}$

63. $\frac{1}{(2x^2y^{-3})^{-2}} = (2x^2y^{-3})^2 = 2^2(x^2)^2(y^{-3})^2 = 4x^4y^{-6} = 4x^4\cdot\frac{1}{y^6} = \frac{4x^4}{y^6}$

64. $\left(-\frac{a^2b}{c}\right)^2\left(\frac{c}{b^2}\right)^3\left(\frac{1}{a^3}\right)^2 = \frac{(a^2b)^2}{c^2}\cdot\frac{c^3}{(b^2)^3}\cdot\frac{1^2}{(a^3)^2} = \frac{a^4b^2}{c^2}\cdot\frac{c^3}{b^6}\cdot\frac{1}{a^6} = \frac{a^4b^2c^3}{a^6b^6c^2}$

$= \frac{c^{3-2}}{a^{6-4}b^{6-2}} = \frac{c}{a^2b^4}$

65. $\left(\frac{8u^{-1}}{2^2u^2v^0}\right)^{-2}\left(\frac{u^{-5}}{u^{-3}}\right)^3 = \left(\frac{2^3u^{-1}}{2^2u^2\cdot 1}\right)^{-2}\left(\frac{u^{-5}}{u^{-3}}\right)^3 = \frac{2^{-6}u^2}{2^{-4}u^{-4}}\cdot\frac{u^{-15}}{u^{-9}} = \frac{2^{-6}u^2u^{-15}}{2^{-4}u^{-4}u^{-9}}$

$= \frac{2^{-6}u^{-13}}{2^{-4}u^{-13}} = \frac{1}{2^{-4-(-6)}} = \frac{1}{2^2} = \frac{1}{4}$ (Note: $\frac{u^{-13}}{u^{-13}} = u^0 = 1$)

66. $\left(\frac{9m^3n^{-3}}{3m^{-2}n^2}\right)^{-2} = \left(\frac{3^2m^3n^{-3}}{3m^{-2}n^2}\right)^{-2} = \frac{3^{-4}m^{-6}n^6}{3^{-2}m^4n^{-4}} = \frac{n^{6-(-4)}}{3^{-2-(-4)}m^{4-(-6)}} = \frac{n^{10}}{3^2m^{10}} = \frac{n^{10}}{9m^{10}}$

67. $(x-y)^{-2} = \frac{1}{(x-y)^2}$ or $\frac{1}{x^2-2xy+y^2}$

68. $(9a^4b^{-2})^{1/2} = 9^{1/2}(a^4)^{1/2}(b^{-2})^{1/2} = 3a^2b^{-1} = \frac{3a^2}{b}$

69. $\left(\frac{27x^2y^{-3}}{8x^{-4}y^3}\right)^{1/3} = \left(\frac{3^3x^{2-(-4)}}{2^3y^{3-(-3)}}\right)^{1/3} = \left(\frac{3^3x^6}{2^3y^6}\right)^{1/3} = \frac{3x^2}{2y^2}$

70. $\frac{m^{-1/4}}{m^{3/4}} = \frac{1}{m^{3/4-(-1/4)}} = \frac{1}{m^{4/4}} = \frac{1}{m}$

71. $(2x^{1/2})(3x^{-1/3}) = 6x^{1/2+(-1/3)} = 6x^{3/6+(-2/6)} = 6x^{1/6}$

72. $\frac{3x^{-1/4}}{6x^{-1/3}} = \frac{x^{-1/4-(-1/3)}}{2} = \frac{x^{-3/12-(4/12)}}{2} = \frac{x^{1/12}}{2}$

73. $\frac{5^0}{3^2} + \frac{3^{-2}}{2^{-2}} = \frac{1}{3^2} + \frac{2^2}{3^2} = \frac{1}{9} + \frac{4}{9} = \frac{5}{9}$

74. $(x^{1/2} + y^{1/2})^2 = (x^{1/2})^2 + 2x^{1/2}y^{1/2} + (y^{1/2})^2$
$= x + 2x^{1/2}y^{1/2} + y$

COMMON ERROR: $x + y$
It is important to avoid distributing the square over addition. $(a + b)^2 = a^2 + 2ab + b^2$, not $a^2 + b^2$.

75. If a is a square root of b then $a^2 = b$.

76. $(2mn)^{2/3} = \sqrt[3]{(2mn)^2} = \sqrt[3]{4m^2n^2}$

77. $3\sqrt[5]{x^2}$

78. $x^{5/7}$

79. $-3(xy)^{2/3}$

80. $\sqrt[3]{(2x^2y)^3} = 2x^2y$

81. $3x\sqrt[3]{x^5y^4} = 3x\sqrt[3]{x^3y^3 \cdot x^2y} = 3x\sqrt[3]{x^3y^3}\sqrt[3]{x^2y} = 3x(xy)\sqrt[3]{x^2y} = 3x^2y\sqrt[3]{x^2y}$

82. $\frac{\sqrt{8m^3n^4}}{\sqrt{12m^2}} = \sqrt{\frac{8m^3n^4}{12m^2}} = \sqrt{\frac{2mn^4}{3}} = \sqrt{\frac{2mn^4 \cdot 3}{3 \cdot 3}} = \sqrt{\frac{6mn^4}{9}} = \frac{\sqrt{6m}\sqrt{n^4}}{\sqrt{9}} = \frac{n^2\sqrt{6m}}{3}$

83. $\sqrt[8]{y^6} = \sqrt[2 \cdot 4]{y^{2 \cdot 3}} = \sqrt[4]{y^3}$

84. $-2x\sqrt[5]{3^6x^7y^{11}} = -2x\sqrt[5]{3^5x^5y^{10} \cdot 3x^2y} = -2x\sqrt[5]{3^5x^5y^{10}}\sqrt[5]{3x^2y}$
$= -2x \cdot 3xy^2\sqrt[5]{3x^2y} = -6x^2y^2\sqrt[5]{3x^2y}$

85. $\frac{2x^2}{\sqrt[3]{4x}} = \frac{2x^2}{\sqrt[3]{2^2x}} = \frac{2x^2\sqrt[3]{2x^2}}{\sqrt[3]{2^2x}\sqrt[3]{2x^2}} = \frac{2x^2\sqrt[3]{2x^2}}{\sqrt[3]{2^3x^3}} = \frac{2x^2\sqrt[3]{2x^2}}{2x} = x\sqrt[3]{2x^2}$

86. $\sqrt[5]{\frac{3y^2}{8x^2}} = \sqrt[5]{\frac{3y^2}{2^3x^2}} = \sqrt[5]{\frac{3y^2 \cdot 2^2x^3}{2^3x^2 \cdot 2^2x^3}} = \sqrt[5]{\frac{2^2 \cdot 3x^3y^2}{2^5x^5}} = \frac{\sqrt[5]{12x^3y^2}}{\sqrt[5]{2^5x^5}} = \frac{\sqrt[5]{12x^3y^2}}{2x}$

87. $(2\sqrt{x} - 5\sqrt{y})(\sqrt{x} + \sqrt{y}) = 2\sqrt{x}\sqrt{x} + 2\sqrt{x}\sqrt{y} - 5\sqrt{y}\sqrt{x} - 5\sqrt{y}\sqrt{y}$
$= 2x + 2\sqrt{xy} - 5\sqrt{xy} - 5y = 2x - 3\sqrt{xy} - 5y$

88. $\frac{\sqrt{x} - 2}{\sqrt{x} + 2} = \frac{(\sqrt{x} - 2)(\sqrt{x} - 2)}{(\sqrt{x} + 2)(\sqrt{x} - 2)} = \frac{\sqrt{x}\sqrt{x} - 2\sqrt{x} - 2\sqrt{x} + 4}{x - 4} = \frac{x - 4\sqrt{x} + 4}{x - 4}$

89. $\frac{3\sqrt{x}}{2\sqrt{x} - \sqrt{y}} = \frac{3\sqrt{x}(2\sqrt{x} + \sqrt{y})}{(2\sqrt{x} - \sqrt{y})(2\sqrt{x} + \sqrt{y})} = \frac{3\sqrt{x} \cdot 2\sqrt{x} + 3\sqrt{x}\sqrt{y}}{(2\sqrt{x})^2 - (\sqrt{y})^2} = \frac{6x + 3\sqrt{xy}}{4x - y}$

90. $\sqrt{\frac{2}{3}} + \sqrt{\frac{3}{2}} = \sqrt{\frac{2 \cdot 3}{3 \cdot 3}} + \sqrt{\frac{3 \cdot 2}{2 \cdot 2}} = \sqrt{\frac{6}{9}} + \sqrt{\frac{6}{4}} = \frac{\sqrt{6}}{\sqrt{9}} + \frac{\sqrt{6}}{\sqrt{4}}$
$= \frac{\sqrt{6}}{3} + \frac{\sqrt{6}}{2} = \frac{2\sqrt{6}}{6} + \frac{3\sqrt{6}}{6} = \frac{5\sqrt{6}}{6}$

91. $(2 - 2\sqrt{-4}) - (3 - \sqrt{-9}) = (2 - 2i\sqrt{4}) - (3 - i\sqrt{9}) = (2 - 2i \cdot 2) - (3 - 3i)$
$= 2 - 4i - 3 + 3i = -1 - i$

92. $\dfrac{2 - \sqrt{-1}}{3 + \sqrt{-4}} = \dfrac{2 - i}{3 + i\sqrt{4}} = \dfrac{2 - i}{3 + 2i} = \dfrac{(2 - i)(3 - 2i)}{(3 + 2i)(3 - 2i)} = \dfrac{6 - 4i - 3i + 2i^2}{9 - 4i^2}$

$= \dfrac{6 - 7i - 2}{9 + 4} = \dfrac{4 - 7i}{13} = \dfrac{4}{13} - \dfrac{7}{13}i$

93. $(3 + i)^2 - 2(3 + i) + 3 = 3^2 + 2 \cdot 3i + i^2 - 6 - 2i + 3$

$= 9 + 6i - 1 - 6 - 2i + 3 = 5 + 4i$

94. $(x^{-1} + y^{-1})^{-1} = \left(\dfrac{1}{x} + \dfrac{1}{y}\right)^{-1} = \dfrac{1}{\dfrac{1}{x} + \dfrac{1}{y}} = \dfrac{xy \cdot 1}{xy \cdot \dfrac{1}{x} + xy \cdot \dfrac{1}{y}} = \dfrac{xy}{y + x}$ or $\dfrac{xy}{x + y}$

95. $\left(\dfrac{a^{-2}}{b^{-1}} + \dfrac{b^{-2}}{a^{-1}}\right)^{-1} = \left(\dfrac{b}{a^2} + \dfrac{a}{b^2}\right)^{-1} = \dfrac{1}{\dfrac{b}{a^2} + \dfrac{a}{b^2}} = \dfrac{a^2b^2 \cdot 1}{a^2b^2 \cdot \dfrac{b}{a^2} + a^2b^2 \cdot \dfrac{a}{b^2}}$

$= \dfrac{a^2b^2}{b^3 + a^3}$ or $\dfrac{a^2b^2}{a^3 + b^3}$

96. $\sqrt[9]{8x^6y^{12}} = \sqrt[9]{2^3x^6y^{12}} = \sqrt[3 \cdot 3]{2^{3 \cdot 1}x^{3 \cdot 2}y^{3 \cdot 4}} = \sqrt[3]{2x^2y^4} = \sqrt[3]{y^3 \cdot 2x^2y} = \sqrt[3]{y^3}\sqrt[3]{2x^2y} = y\sqrt[3]{2x^2y}$

97. $\sqrt[3]{3} - \dfrac{6}{\sqrt[3]{9}} + 3\sqrt[3]{\dfrac{1}{9}} = \sqrt[3]{3} - \dfrac{6\sqrt[3]{3}}{\sqrt[3]{3^2}\sqrt[3]{3}} + 3\sqrt[3]{\dfrac{1 \cdot 3}{3^2 \cdot 3}} = \sqrt[3]{3} - \dfrac{6\sqrt[3]{3}}{\sqrt[3]{3^3}} + 3\sqrt[3]{\dfrac{3}{3^3}}$

$= \sqrt[3]{3} - \dfrac{6\sqrt[3]{3}}{3} + 3\dfrac{\sqrt[3]{3}}{\sqrt[3]{3^3}} = \sqrt[3]{3} - 2\sqrt[3]{3} + \dfrac{3\sqrt[3]{3}}{3}$

$= \sqrt[3]{3} - 2\sqrt[3]{3} + \sqrt[3]{3} = (1 - 2 + 1)\sqrt[3]{3} = 0$

98. (A) $3\sqrt[3]{x^3} - 2\sqrt{x^2} = 3x - 2|x| = 3x - 2x = x$

(B) $3\sqrt[3]{x^3} - 2\sqrt{x^2} = 3x - 2|x| = 3x - 2(-x) = 3x + 2x = 5x$

99.

$$x^2 + 2x + 2 = 0$$
$$(-1 + i)^2 + 2(-1 + i) + 2 \overset{?}{=} 0$$
$$1 + 2(-1)i + i^2 - 2 + 2i + 2 \overset{?}{=} 0$$
$$1 - 2i + (-1) - 2 + 2i + 2 \overset{?}{=} 0$$
$$0 \overset{\checkmark}{=} 0$$

100.

$$x^2 - 2x - 1 = 0$$
$$(1 + \sqrt{2})^2 - 2(1 + \sqrt{2}) - 1 \overset{?}{=} 0$$
$$1 + 2\sqrt{2} + 2 - 2 - 2\sqrt{2} - 1 \overset{?}{=} 0$$
$$0 \overset{\checkmark}{=} 0$$

Practice Test 6

1. $(xy)^2 + (-x^2)^3 + \left(\frac{x}{y}\right)^4 = x^2y^2 + (-x^2)(-x^2)(-x^2) + \frac{x^4}{y^4} = x^2y^2 - x^6 + \frac{x^4}{y^4}$

2. $0.000\,034\,5 = 0.000\,03.4\,5 \times 10^{-5} = 3.45 \times 10^{-5}$

5 places right

negative exponent

3. $2.468 \times 10^{10} = 2.468 \times 10{,}000{,}000{,}000 = 24{,}680{,}000{,}000$

4. $\sqrt{9x^3y^4z^5} = \sqrt{9x^2y^4z^4 \cdot xz} = \sqrt{9x^2y^4z^4}\sqrt{xz} = 3xy^2z^2\sqrt{xz}$

5. $\frac{1}{\sqrt{3x}} = \frac{1}{\sqrt{3x}}\frac{\sqrt{3x}}{\sqrt{3x}} = \frac{\sqrt{3x}}{3x}$

6. $\frac{1}{\sqrt{x} + 3} = \frac{1}{\sqrt{x} + 3} \cdot \frac{\sqrt{x} - 3}{\sqrt{x} - 3} = \frac{\sqrt{x} - 3}{\sqrt{x}\sqrt{x} - 3 \cdot 3} = \frac{\sqrt{x} - 3}{x - 9}$

7. $\sqrt{125} + \sqrt[3]{49} = \sqrt{5^3} + \sqrt[3]{7^2} = 5^{3/2} + 7^{2/3}$

8. $3^{4/3} - 5^{2/5} = \sqrt[3]{3^4} - \sqrt[5]{5^2} = \sqrt[3]{3^3 \cdot 3} - \sqrt[5]{5^2} = \sqrt[3]{3^3}\sqrt[3]{3} - \sqrt[5]{25} = 3\sqrt[3]{3} - \sqrt[5]{25}$

9. $2^0x^{-2}yz^{-1} = 1 \cdot \frac{1}{x^2} \cdot y \cdot \frac{1}{z} = \frac{y}{x^2z}$

10. $\frac{x^{-2}}{y} + \frac{x^2}{y^{-1}} = x^{-2} \cdot \frac{1}{y} + x^2 \cdot \frac{1}{y^{-1}} = \frac{1}{x^2} \cdot \frac{1}{y} + x^2 \cdot \frac{1}{1/y} = \frac{1}{x^2y} + \frac{x^2y}{1}$

$= \frac{1}{x^2y} + \frac{x^2y \cdot x^2y}{x^2y} = \frac{1 + x^4y^2}{x^2y}$

11. $(1 + \sqrt{3})(2 - \sqrt{3}) = 2 - \sqrt{3} + 2\sqrt{3} - \sqrt{3}\sqrt{3} = 2 + \sqrt{3} - 3 = -1 + \sqrt{3}$

12. $\frac{1 + \sqrt{3}}{2 - \sqrt{3}} = \frac{1 + \sqrt{3}}{2 - \sqrt{3}} \cdot \frac{2 + \sqrt{3}}{2 + \sqrt{3}} = \frac{2 + \sqrt{3} + 2\sqrt{3} + \sqrt{3}\sqrt{3}}{4 - \sqrt{3}\sqrt{3}} = \frac{2 + 3\sqrt{3} + 3}{4 - 3}$

$= \frac{5 + 3\sqrt{3}}{1} = 5 + 3\sqrt{3}$

13. $(1 + \sqrt{3}) - (2 - \sqrt{3}) = 1 + \sqrt{3} - 2 + \sqrt{3} = -1 + 2\sqrt{3}$

14. $(1 + 3i) + (2 - 5i) = 1 + 3i + 2 - 5i = 3 - 2i$

15. $(1 + 3i)(2 - 5i) = 2 - 5i + 6i - 15i^2 = 2 + i - 15(-1) = 2 + i + 15 = 17 + i$

16. $\dfrac{1 + 3i}{2 - 5i} = \dfrac{1 + 3i}{2 - 5i}\,\dfrac{2 + 5i}{2 + 5i} = \dfrac{2 + 5i + 6i + 15i^2}{4 - 25i^2} = \dfrac{2 + 11i + 15(-1)}{4 + 25}$

$= \dfrac{-13 + 11i}{29} = -\dfrac{13}{29} + \dfrac{11}{29}i$

17. $\dfrac{1}{1 + 3i} = \dfrac{1}{1 + 3i} \cdot \dfrac{1 - 3i}{1 - 3i} = \dfrac{1 - 3i}{1 - 9i^2} = \dfrac{1 - 3i}{1 + 9} = \dfrac{1 - 3i}{10} = \dfrac{1}{10} - \dfrac{3}{10}i$

18. $(1 - 2i)^2 - 2(1 - 2i) + 6 = 1 - 2 \cdot 2i + 4i^2 - 2 + 4i + 6$

$= 1 - 4i + 4(-1) - 2 + 4i + 6$

$= 1 - 4i - 4 - 2 + 4i + 6$

$= 1$

19. $x^2 - 2x + 5$ becomes $(1 - 2i)^2 - 2(1 - 2i) + 5 = 1 - 2 \cdot 2i + 4i^2 - 2 + 4i + 5$

$= 1 - 4i + 4(-1) - 2 + 4i + 5$

$= 1 - 4i - 4 - 2 + 4i + 5$

$= 0$

20. $\sqrt[3]{(-2)^3} + \sqrt{(-2)^2} = \sqrt[3]{-8} + \sqrt{4} = -2 + 2 = 0$

CHAPTER 7 Second-Degree Equations, Inequalities and Graphs

Exercise 7-1 Solving Quadratic Equations by Square Roots and by Completing the Square

Key Ideas and Formulas

A quadratic equation in one variable is any equation that can be written in the form

$$ax^2 + bx + c = 0 \qquad a \neq 0 \text{ (standard form)}$$

To solve a quadratic equation in standard form, check if it can be solved using factoring and the zero property:

$$A \cdot B = 0 \qquad \text{if and only if } A = 0 \text{ or } B = 0 \text{ or both}$$

If $b = 0$, the square root property is used:

$$\text{if } A^2 = C, \text{ then } A = \pm\sqrt{C}$$

Otherwise, it is always possible to solve using "completing the square".

To complete the square of a quadratic of the form $x^2 + bx$, add the square of one-half of the coefficient of x, that is

$$\left(\frac{b}{2}\right)^2 \quad \text{or} \quad \frac{b^2}{4}$$

Thus $x^2 + bx + \left(\frac{b}{2}\right)^2 = \left(x + \frac{b}{2}\right)^2$

The method of completing the square:

1. Write the equation in standard form:
   $ax^2 + bx + c = 0$
2. Make the coefficient of x^2 equal to 1 by dividing both sides by the existing coefficient a.
3. Move the constant term to the right side.
4. Complete the square on the left side, adding the same amount to the right.
5. Solve by taking square roots.

---

1. $x^2 + 5x - 6 = 0$
   $(x + 6)(x - 1) = 0$
   $x + 6 = 0$ or $x - 1 = 0$
   $x = -6 \qquad x = 1$

3. $x^2 - 7x + 10 = 0$
   $(x - 5)(x - 2) = 0$
   $x - 5 = 0$ or $x - 2 = 0$
   $x = 5 \qquad x = 2$

5. $2x^2 + 7x - 4 = 0$
$(2x - 1)(x + 4) = 0$
$2x - 1 = 0$ or $x + 4 = 0$
$2x = 1 \qquad x = -4$
$x = \frac{1}{2}$

7. $x^2 - 16 = 0$
$x^2 = 16$
$x = \pm\sqrt{16}$
$x = \pm 4$

9. $x^2 + 16 = 0$
$x^2 = -16$
$x = \pm\sqrt{-16}$
$x = \pm 4i$

11. $y^2 - 45 = 0$
$y^2 = 45$
$y = \pm\sqrt{45}$
$y = \pm\sqrt{9 \cdot 5}$
$y = \pm 3\sqrt{5}$

13. $4x^2 - 9 = 0$
$4x^2 = 9$
$x^2 = \frac{9}{4}$
$x = \pm\sqrt{\frac{9}{4}}$
$x = \pm\frac{3}{2}$

15. $16y^2 = 9$
$y^2 = \frac{9}{16}$
$y = \pm\sqrt{\frac{9}{16}}$
$y = \pm\frac{3}{4}$

17. To complete the square of $x^2 + 4x$, add $\left(\frac{4}{2}\right)^2$, that is, 4; thus $x^2 + 4x + 4 = (x + 2)^2$

19. To complete the square of $x^2 - 6x$, add $\left(-\frac{6}{2}\right)^2$, that is, 9; thus $x^2 - 6x + 9 = (x - 3)^2$

21. To complete the square of $x^2 + 12x$, add $\left(\frac{12}{2}\right)^2$, that is, 36; thus $x^2 + 12x + 36 = (x + 6)^2$

23. $x^2 + 4x + 2 = 0$
$x^2 + 4x = -2 \qquad \left(\frac{4}{2}\right)^2 = 4$
$x^2 + 4x + 4 = -2 + 4$
$(x + 2)^2 = 2$
$x + 2 = \pm\sqrt{2}$
$x = -2 \pm\sqrt{2}$

25. $x^2 - 6x - 3 = 0$
$x^2 - 6x = 3 \qquad \left(-\frac{6}{2}\right)^2 = 9$
$x^2 - 6x + 9 = 3 + 9$
$(x - 3)^2 = 12$
$x - 3 = \pm\sqrt{12}$
$x = 3 \pm \sqrt{12}$
$x = 3 \pm 2\sqrt{3}$

27. $\frac{t}{2} = \frac{2}{t} \qquad t \neq 0 \quad \text{lcm} = 2t$
$2t \cdot \frac{t}{2} = 2t \cdot \frac{2}{t}$
$t^2 = 4$
$t^2 - 4 = 0$
$(t - 2)(t + 2) = 0$
$t - 2 = 0$ or $t + 2 = 0$
$t = 2 \qquad t = -2$

29. $\frac{m}{4}(m + 1) = 3 \quad \text{lcm} = 4$
$4 \cdot \frac{m}{4}(m + 1) = 4 \cdot 3$
$m(m + 1) = 12$
$m^2 + m = 12$
$m^2 + m - 12 = 0$
$(m + 4)(m - 3) = 0$
$m + 4 = 0$ or $m - 3 = 0$
$m = -4 \qquad m = 3$

31. $2y = \frac{2}{y} + 3 \quad y \neq 0, \ \text{lcm} = y$

$$y \cdot 2y = y \cdot \frac{2}{y} + y \cdot 3$$
$$2y^2 = 2 + 3y$$
$$2y^2 - 3y - 2 = 0$$
$$(2y + 1)(y - 2) = 0$$
$$2y + 1 = 0 \text{ or } y - 2 = 0$$
$$2y = -1 \qquad y = 2$$
$$y = -\frac{1}{2}$$

33. $2 + \frac{2}{x^2} = \frac{5}{x} \quad x \neq 0, \ \text{lcm} = x^2$

$$x^2 \cdot 2 + x^2 \cdot \frac{2}{x^2} = x^2 \cdot \frac{5}{x}$$
$$2x^2 + 2 = 5x$$
$$2x^2 - 5x + 2 = 0$$
$$(2x - 1)(x - 2) = 0$$
$$2x - 1 = 0 \text{ or } x - 2 = 0$$
$$2x = 1 \qquad x = 2$$
$$x = \frac{1}{2}$$

35. $\frac{x}{6} = \frac{1}{x + 1} \quad x \neq -1, \ \text{lcm} = 6(x + 1)$

$$6(x + 1) \cdot \frac{x}{6} = 6(x + 1) \cdot \frac{1}{x + 1}$$
$$(x + 1)x = 6$$
$$x^2 + x = 6$$
$$x^2 + x - 6 = 0$$
$$(x + 3)(x - 2) = 0$$
$$x + 3 = 0 \text{ or } x - 2 = 0$$
$$x = -3 \qquad x = 2$$

37. $y^2 = 2$

$$y = \pm\sqrt{2}$$

39. $16a^2 + 9 = 0$

$$16a^2 = -9$$
$$a^2 = -\frac{9}{16}$$
$$a = \pm\sqrt{-\frac{9}{16}}$$
$$a = \pm i\sqrt{\frac{9}{16}}$$
$$a = \pm\frac{3}{4}i$$

41. $9x^2 - 7 = 0$

$$9x^2 = 7$$
$$x^2 = \frac{7}{9}$$
$$x = \pm\sqrt{\frac{7}{9}}$$
$$x = \pm\frac{\sqrt{7}}{3}$$

43. $(m - 3)^2 = 25$

$$m - 3 = \pm\sqrt{25}$$
$$m - 3 = \pm 5$$
$$m = 3 \pm 5$$
$$m = 3 + 5 \text{ or } m = 3 - 5$$
$$m = 8 \qquad m = -2$$

45. $(t + 1)^2 = -9$

$$t + 1 = \pm\sqrt{-9}$$
$$t + 1 = \pm i\sqrt{9}$$
$$t + 1 = \pm 3i$$
$$t = -1 \pm 3i$$

47. $\left(x - \frac{1}{3}\right)^2 = \frac{4}{9}$

$$x - \frac{1}{3} = \pm\sqrt{\frac{4}{9}}$$
$$x = \frac{1}{3} \pm \sqrt{\frac{4}{9}}$$
$$x = \frac{1}{3} \pm \frac{2}{3}$$
$$x = \frac{1}{3} + \frac{2}{3} \text{ or } x = \frac{1}{3} - \frac{2}{3}$$
$$x = 1 \qquad x = -\frac{1}{3}$$

49. To complete the square of $x^2 + 3x$, add $\left(\frac{3}{2}\right)^2$, that is, $\frac{9}{4}$; thus

$$x^2 + 3x + \frac{9}{4} = \left(x + \frac{3}{2}\right)^2$$

51. To complete the square of $u^2 - 5u$, add $\left(-\frac{5}{2}\right)^2$, that is $\frac{25}{4}$; thus

$$u^2 - 5u + \frac{25}{4} = \left(u - \frac{5}{2}\right)^2$$

53.
$$x^2 + x - 1 = 0$$
$$x^2 + 1x = 1 \qquad \left(\frac{1}{2}\right)^2 = \frac{1}{4}$$
$$x^2 + 1x + \frac{1}{4} = 1 + \frac{1}{4}$$
$$\left(x + \frac{1}{2}\right)^2 = \frac{5}{4}$$
$$x + \frac{1}{2} = \pm\sqrt{\frac{5}{4}}$$
$$x = -\frac{1}{2} \pm \sqrt{\frac{5}{4}}$$
$$x = -\frac{1}{2} \pm \frac{\sqrt{5}}{2}$$
$$x = \frac{-1 \pm \sqrt{5}}{2}$$

55.
$$u^2 - 5u + 2 = 0$$
$$u^2 - 5u = -2 \qquad \left(-\frac{5}{2}\right)^2 = \frac{25}{4}$$
$$u^2 - 5u + \frac{25}{4} = -2 + \frac{25}{4}$$
$$\left(u - \frac{5}{2}\right)^2 = \frac{17}{4}$$
$$u - \frac{5}{2} = \pm\sqrt{\frac{17}{4}}$$
$$u = \frac{5}{2} \pm \sqrt{\frac{17}{4}}$$
$$u = \frac{5}{2} \pm \frac{\sqrt{17}}{2}$$
$$u = \frac{5 \pm \sqrt{17}}{2}$$

57.
$$m^2 - 4m + 8 = 0$$
$$m^2 - 4m = -8 \qquad \left(-\frac{4}{2}\right)^2 = 4$$
$$m^2 - 4m + 4 = -8 + 4$$
$$(m - 2)^2 = -4$$
$$m - 2 = \pm\sqrt{-4}$$
$$m - 2 = \pm i\sqrt{4}$$
$$m = 2 \pm i\sqrt{4}$$
$$m = 2 \pm 2i$$

59.
$$2y^2 - 4y + 1 = 0$$
$$y^2 - 2y + \frac{1}{2} = 0$$
$$y^2 - 2y = -\frac{1}{2} \qquad \left(-\frac{2}{2}\right)^2 = 1$$
$$y^2 - 2y + 1 = -\frac{1}{2} + 1$$
$$(y - 1)^2 = \frac{1}{2}$$
$$y - 1 = \pm\sqrt{\frac{1}{2}}$$
$$y = 1 \pm \sqrt{\frac{1}{2}}$$
$$y = 1 \pm \sqrt{\frac{2}{4}}$$
$$y = 1 \pm \frac{\sqrt{2}}{2}$$
$$y = \frac{2 \pm \sqrt{2}}{2}$$

61.
$$2u^2 + 3u - 1 = 0$$
$$u^2 + \frac{3}{2}u - \frac{1}{2} = 0$$
$$u^2 + \frac{3}{2}u = \frac{1}{2} \qquad \left(\frac{3}{2} \div 2\right)^2 = \left(\frac{3}{4}\right)^2$$
$$u^2 + \frac{3}{2}u + \frac{9}{16} = \frac{1}{2} + \frac{9}{16} \qquad = \frac{9}{16}$$
$$\left(u + \frac{3}{4}\right)^2 = \frac{17}{16}$$
$$u + \frac{3}{4} = \pm\sqrt{\frac{17}{16}}$$
$$u = -\frac{3}{4} \pm\sqrt{\frac{17}{16}}$$
$$u = -\frac{3}{4} \pm \frac{\sqrt{17}}{4}$$
$$u = \frac{-3 \pm \sqrt{17}}{4}$$

63.
$$2x^2 + 3x - 2 = 0$$
$$x^2 + \frac{3}{2}x - 1 = 0$$
$$x^2 + \frac{3}{2}x = 1 \qquad \left(\frac{3}{2} \div 2\right)^2 = \left(\frac{3}{4}\right)^2 = \frac{9}{16}$$
$$x^2 + \frac{3}{2}x + \frac{9}{16} = 1 + \frac{9}{16}$$
$$\left(x + \frac{3}{4}\right)^2 = \frac{25}{16}$$
$$x + \frac{3}{4} = \pm\sqrt{\frac{25}{16}}$$
$$x + \frac{3}{4} = \pm\frac{5}{4}$$
$$x = -\frac{3}{4} \pm \frac{5}{4}$$
$$x = -2 \text{ or } \frac{1}{2}$$

65.
$$2x^2 - 2x - 1 = 0$$
$$x^2 - x - \frac{1}{2} = 0$$
$$x^2 - x = \frac{1}{2} \qquad \left(\frac{1}{2}\right)^2 = \frac{1}{4}$$
$$x^2 - x + \frac{1}{4} = \frac{1}{2} + \frac{1}{4}$$
$$\left(x - \frac{1}{2}\right)^2 = \frac{3}{4}$$
$$x - \frac{1}{2} = \pm\sqrt{\frac{3}{4}}$$
$$x = \frac{1}{2} \pm \sqrt{\frac{3}{4}}$$
$$x = \frac{1}{2} \pm \frac{\sqrt{3}}{2}$$
$$x = \frac{1 \pm \sqrt{3}}{2}$$

67.
$$3x^2 - 2x - 1 = 0$$
$$x^2 - \frac{2}{3}x - \frac{1}{3} = 0$$
$$x^2 - \frac{2}{3}x = \frac{1}{3} \qquad \left(-\frac{2}{3} \div 2\right)^2 = \left(-\frac{1}{3}\right)^2 = \frac{1}{9}$$
$$x^2 - \frac{2}{3}x + \frac{1}{9} = \frac{1}{3} + \frac{1}{9}$$
$$\left(x - \frac{1}{3}\right)^2 = \frac{4}{9}$$
$$x - \frac{1}{3} = \pm\sqrt{\frac{4}{9}}$$
$$x - \frac{1}{3} = \pm\frac{2}{3}$$
$$x = \frac{1}{3} \pm \frac{2}{3}$$
$$x = -\frac{1}{3},\ 1$$

69. $4x^2 - 8x + 5 = 0$

$x^2 - 2x + \frac{5}{4} = 0$

$x^2 - 2x = -\frac{5}{4}$ $\quad \left(-\frac{2}{2}\right)^2 = 1$

$x^2 - 2x + 1 = -\frac{5}{4} + 1$

$(x - 1)^2 = -\frac{1}{4}$

$x - 1 = \pm\sqrt{-\frac{1}{4}}$

$x - 1 = \pm i\sqrt{\frac{1}{4}}$

$x - 1 = \pm\frac{1}{2}i$

$x = 1 \pm \frac{1}{2}i$

71. $2u^2 - 3u + 2 = 0$

$u^2 - \frac{3}{2}u + 1 = 0$

$u^2 - \frac{3}{2}u = -1$ $\quad \left(-\frac{3}{2} \div 2\right)^2 = \left(-\frac{3}{4}\right)^2 = \frac{9}{16}$

$u^2 - \frac{3}{2}u + \frac{9}{16} = -1 + \frac{9}{16}$

$\left(u - \frac{3}{4}\right)^2 = -\frac{7}{16}$

$u - \frac{3}{4} = \pm\sqrt{-\frac{7}{16}}$

$u - \frac{3}{4} = \pm i\sqrt{\frac{7}{16}}$

$u = \frac{3}{4} \pm i\sqrt{\frac{7}{16}}$

$u = \frac{3}{4} \pm i\frac{\sqrt{7}}{4}$

$u = \frac{3 \pm i\sqrt{7}}{4}$

73. $x^2 + x + 1 = 0$

$x^2 + 1x = -1$ $\quad \left(\frac{1}{2}\right)^2 = \frac{1}{4}$

$x^2 + 1x + \frac{1}{4} = -1 + \frac{1}{4}$

$\left(x + \frac{1}{2}\right)^2 = -\frac{3}{4}$

$x + \frac{1}{2} = \pm\sqrt{-\frac{3}{4}}$

$x + \frac{1}{2} = \pm i\sqrt{\frac{3}{4}}$

$x = -\frac{1}{2} \pm i\sqrt{\frac{3}{4}}$

$x = -\frac{1}{2} \pm i\frac{\sqrt{3}}{2}$

$x = \frac{-1 \pm i\sqrt{3}}{2}$

75. $\left(y + \frac{5}{2}\right)^2 = \frac{5}{2}$

$y + \frac{5}{2} = \pm\sqrt{\frac{5}{2}}$

$y = -\frac{5}{2} \pm \sqrt{\frac{5}{2}}$

$y = -\frac{5}{2} \pm\sqrt{\frac{5 \cdot 2}{2 \cdot 2}}$

$y = -\frac{5}{2} \pm\sqrt{\frac{10}{4}}$

$y = -\frac{5}{2} \pm \frac{\sqrt{10}}{\sqrt{4}}$

$y = -\frac{5}{2} \pm \frac{\sqrt{10}}{2}$

$y = \frac{-5 \pm \sqrt{10}}{2}$

77. $(x - 2)^2 = -1$

$x - 2 = \pm\sqrt{-1}$

$x - 2 = \pm i$

$x = 2 \pm i$

79. $x = \frac{1 + 3x}{x + 3}$ $\quad x \neq -3$, lcm $= x + 3$

$(x + 3)x = (x + 3) \cdot \frac{1 + 3x}{x + 3}$

$x^2 + 3x = 1 + 3x$

$x^2 - 1 = 0$

$(x - 1)(x + 1) = 0$

$x - 1 = 0$ or $x + 1 = 0$

$x = 1 \qquad x = -1$

81. $\frac{x + 2}{x + 1} - \frac{6x}{x^2 - 1} = \frac{2x - 1}{x + 1}$ $\quad x \neq -1, 1, \quad \text{lcm} = (x - 1)(x + 1)$

$$(x - 1)(x + 1)\frac{(x + 2)}{x - 1} - (x - 1)(x + 1)\frac{6x}{x^2 - 1} = (x - 1)(x + 1)\frac{(2x - 1)}{x + 1}$$
$$(x + 1)(x + 2) - 6x = (x - 1)(2x - 1)$$
$$x^2 + 3x + 2 - 6x = 2x^2 - x - 2x + 1$$
$$x^2 - 3x + 2 = 2x^2 - 3x + 1$$
$$0 = x^2 - 1$$
$$x^2 - 1 = 0$$
$$(x - 1)(x + 1) = 0$$
$$x - 1 = 0 \qquad x + 1 = 0$$
$$x = 1 \qquad x = -1$$
Neither of these can be a solution to the original equation.

No solution.

83. $x = \frac{3}{x - 2}$ $\quad x \neq 2$, lcm $= x - 2$

$$x(x - 2) = (x - 2)\frac{3}{x - 2}$$
$$x^2 - 2x = 3$$
$$x^2 - 2x - 3 = 0$$
$$(x - 3)(x + 1) = 0$$
$$x - 3 = 0 \quad \text{or} \quad x + 1 = 0$$
$$x = 3 \qquad x = -1$$

85. $x^2 + 2\sqrt{2}x - 2 = 0$

$$x^2 + 2\sqrt{2}x = 2 \qquad (\sqrt{2})^2 = 2$$
$$x^2 + 2\sqrt{2}x + 2 = 2 + 2$$
$$(x + \sqrt{2})^2 = 4$$
$$x + \sqrt{2} = \pm\sqrt{4}$$
$$x = -\sqrt{2} \pm \sqrt{4}$$
$$x = -\sqrt{2} \pm 2$$

87. $x^2 - 4\sqrt{3}x + 13 = 0$

$$x^2 - 4\sqrt{3}x = -13 \qquad (-2\sqrt{3})^2 = 12$$
$$x^2 - 4\sqrt{3}x + 12 = -13 + 12$$
$$(x - 2\sqrt{3})^2 = -1$$
$$x - 2\sqrt{3} = \pm\sqrt{-1}$$
$$x - 2\sqrt{3} = \pm i$$
$$x = 2\sqrt{3} \pm i$$

89. $x^2 - 2ix - 4 = 0$

$$x^2 - 2ix = 4 \qquad (-2i \div 2)^2 = (-i)^2 = i^2 = -1$$
$$x^2 - 2ix - 1 = 4 + (-1)$$
$$(x - i)^2 = 3$$
$$x - i = \pm\sqrt{3}$$
$$x = i \pm \sqrt{3}$$

91. $x^2 - 6ix - 9 = 0$

$$x^2 - 6ix = 9 \qquad (-6i \div 2)^2 = (-3i)^2 = 9i^2 = -9$$
$$x^2 - 6ix - 9 = 9 - 9$$
$$(x - 3i)^2 = 0$$
$$x - 3i = 0$$
$$x = 3i$$

93. $x^2 - 2ix + 2 = 0$

$$x^2 - 2ix = -2 \qquad (-2i \div 2)^2 = (-i)^2 = i^2 = -1$$
$$x^2 - 2ix + (-1) = -2 + (-1)$$
$$(x - i)^2 = -3$$
$$x - i = \pm\sqrt{-3}$$
$$x = i \pm i\sqrt{3}$$
$$x = i(1 \pm \sqrt{3})$$

95. $x^2 + mx + n = 0$

$$x^2 + mx = -n \qquad \left(\frac{m}{2}\right)^2 = \frac{m^2}{4}$$

$$x^2 + mx + \frac{m^2}{4} = -n + \frac{m^2}{4}$$

$$\left(x + \frac{m}{2}\right)^2 = -\frac{4n}{4} + \frac{m^2}{4}$$

$$\left(x + \frac{m}{2}\right)^2 = \frac{m^2 - 4n}{4}$$

$$x + \frac{m}{2} = \pm\sqrt{\frac{m^2 - 4n}{4}}$$

$$x = -\frac{m}{2} \pm \sqrt{\frac{m^2 - 4n}{4}}$$

$$x = -\frac{m}{2} \pm \frac{\sqrt{m^2 - 4n}}{2}$$

$$x = \frac{-m \pm \sqrt{m^2 - 4n}}{2}$$

97. $a^2 + b^2 = c^2$

$$a^2 = c^2 - b^2$$

$$a = \pm\sqrt{c^2 - b^2}$$

$$a = \sqrt{c^2 - b^2}$$

(positive square root only)

99. Supply [is equal to] demand.

$$p - 80 = \frac{900}{p} \qquad p \neq 0,\ \text{lcm} = p$$

$$p(p - 80) = p\frac{900}{p}$$

$$p^2 - 80p = 900$$

$$p^2 - 80p - 900 = 0$$

$$(p - 90)(p + 10) = 0$$

$p - 90 = 0$ or $p + 10 = 0$

$p = 90$ ~~$p = -10$~~ Not possible

price = 90 cents per gallon

Exercise 7-2 The Quadratic Formula

Key Ideas and Formulas

To solve $ax^2 + bx + c = 0$, $a \neq 0$, when simpler methods fail, use the Quadratic Formula

$$x = \frac{-b \pm \sqrt{b^2 - 4ac}}{2a}$$

The expression $b^2 - 4ac$ is called the **discriminant**.

Discriminant Test for how many roots of $ax^2 + bx + c = 0$, a, b, c real numbers, $a \neq 0$

$b^2 - 4ac$	ROOTS
Positive	Two real roots
0	One real root
Negative	Two nonreal roots (the roots will be complex conjugates)

Summary of Methods for solving $ax^2 + bx + c = 0$, $a \neq 0$.

1. If b = 0, solve by square roots.
2. If c = 0, solve by factoring.
3. Otherwise, try:
   a. Factoring, or
   b. Completing the square, or
   c. The quadratic formula (which always works).

---

1. $2x^2 - 5x + 3 = 0$
   $ax^2 + bx + c = 0 \quad a = 2 \quad b = -5 \quad c = 3$

3. $m = 1 - 3m^2$
   $3m^2 + m - 1 = 0$
   $ax^2 + bx + c = 0 \quad a = 3 \quad b = 1 \quad c = -1$

5. $3y^2 - 5 = 0$
   $3y^2 + 0y - 5 = 0$
   $ax^2 + bx + c = 0 \quad a = 3 \quad b = 0 \quad c = -5$

7. $x^2 + 8x + 3 = 0 \quad a = 1 \quad b = 8 \quad c = 3$

$$x = \frac{-(8) \pm \sqrt{(8)^2 - 4(1)(3)}}{2(1)} = \frac{-8 \pm \sqrt{64 - 12}}{2}$$
$$= \frac{-8 \pm \sqrt{52}}{2} = \frac{-8 \pm 2\sqrt{13}}{2} = -4 \pm \sqrt{13}$$

COMMON ERROR:

$$\frac{-8 \pm \sqrt{52}}{2} \neq \frac{\overset{-4}{\cancel{-8}} \pm \sqrt{52}}{\underset{1}{\cancel{2}}}$$

This is incorrect "cancelling". The **entire** numerator and denominator must be divided by the common factor of 2.

9. $y^2 - 10y - 3 = 0 \quad a = 1 \quad b = -10 \quad c = -3$

$$y = \frac{-(-10) \pm \sqrt{(-10)^2 - 4(1)(-3)}}{2(1)} = \frac{10 \pm \sqrt{100 + 12}}{2} = \frac{10 \pm \sqrt{112}}{2} = \frac{10 \pm 4\sqrt{7}}{2}$$
$$= 5 \pm 2\sqrt{7}$$

11. $x^2 + 3x + 5 = 0 \quad a = 1 \quad b = 3 \quad c = 5$

$$x = \frac{-(3) \pm \sqrt{(3)^2 - 4(1)(5)}}{2(1)} = \frac{-3 \pm \sqrt{9 - 20}}{2} = \frac{-3 \pm \sqrt{-11}}{2} = \frac{-3 \pm i\sqrt{11}}{2}$$
$$= -\frac{3}{2} \pm i\frac{\sqrt{11}}{2}$$

13. $x^2 + 4x + 5 = 0 \quad a = 1 \quad b = 4 \quad c = 5$

$$x = \frac{-(4) \pm \sqrt{(4)^2 - 4(1)(5)}}{2(1)} = \frac{-4 \pm \sqrt{16 - 20}}{2} = \frac{-4 \pm \sqrt{-4}}{2} = \frac{-4 \pm i\sqrt{4}}{2} = \frac{-4 \pm 2i}{2}$$
$$= -2 \pm i$$

15. $x^2 - 2x + 5 = 0 \quad a = 1 \quad b = -2 \quad c = 5$

$$x = \frac{-(-2) \pm \sqrt{(-2)^2 - 4(1)(5)}}{2(1)} = \frac{2 \pm \sqrt{4 - 20}}{2} = \frac{2 \pm \sqrt{-16}}{2} = \frac{2 \pm i\sqrt{16}}{2} = \frac{2 \pm 4i}{2}$$
$$= 1 \pm 2i$$

17. $-x^2 + 3x - 3 = 0$
$x^2 - 3x + 3 = 0 \quad a = 1 \quad b = -3 \quad c = 3$

$$x = \frac{-(-3) \pm \sqrt{(-3)^2 - 4(1)(3)}}{2(1)} = \frac{3 \pm \sqrt{9 - 12}}{2} = \frac{3 \pm \sqrt{-3}}{2} = \frac{3 \pm i\sqrt{3}}{2} = \frac{3}{2} \pm i\frac{\sqrt{3}}{2}$$

19. $-x^2 + 5x + 5 = 0$
$x^2 - 5x - 5 = 0 \quad a = 1 \quad b = -5 \quad c = -5$

$$x = \frac{-(-5) \pm \sqrt{(-5)^2 - 4(1)(-5)}}{2(1)} = \frac{5 \pm \sqrt{25 + 20}}{2} = \frac{5 \pm \sqrt{45}}{2} = \frac{5 \pm 3\sqrt{5}}{2}$$

21. $u^2 = 1 - 3u$
$u^2 + 3u - 1 = 0 \quad a = 1 \quad b = 3 \quad c = -1$

$$u = \frac{-(3) \pm \sqrt{(3)^2 - 4(1)(-1)}}{2(1)} = \frac{-3 \pm \sqrt{9 + 4}}{2} = \frac{-3 \pm \sqrt{13}}{2}$$

23. $y^2 + 3 = 2y$
$y^2 - 2y + 3 = 0 \quad a = 1 \quad b = -2 \quad c = 3$

$$y = \frac{-(-2) \pm \sqrt{(-2)^2 - 4(1)(3)}}{2(1)} = \frac{2 \pm \sqrt{4 - 12}}{2} = \frac{2 \pm \sqrt{-8}}{2} = \frac{2 \pm i\sqrt{8}}{2} = \frac{2 \pm 2i\sqrt{2}}{2}$$
$$= 1 \pm i\sqrt{2}$$

25. $2m^2 + 3 = 6m$
$2m^2 - 6m + 3 = 0 \quad a = 2 \quad b = -6 \quad c = 3$

$$m = \frac{-(-6) \pm \sqrt{(-6)^2 - 4(2)(3)}}{2(2)} = \frac{6 \pm \sqrt{36 - 24}}{4} = \frac{6 \pm \sqrt{12}}{4} = \frac{6 \pm 2\sqrt{3}}{4} = \frac{3 \pm \sqrt{3}}{2}$$

27. $p = 1 - 3p^2$
$3p^2 + p - 1 = 0 \quad a = 3 \quad b = 1 \quad c = -1$

$$p = \frac{-(1) \pm \sqrt{(1)^2 - 4(3)(-1)}}{2(3)} = \frac{-1 \pm \sqrt{1 + 12}}{6} = \frac{-1 \pm \sqrt{13}}{6}$$

29. The discriminant $b^2 - 4ac = (5)^2 - 4(4)(-6) = 121$ is positive. The equation has two real roots.

31. The discriminant $b^2 - 4ac = (-24)^2 - 4(9)(16) = 0$. The equation has one real root.

33. The discriminant $b^2 - 4ac = (-8)^2 - 4(1)(17) = -4$ is negative. The equation has two non-real complex roots.

35. The discriminant $b^2 - 4ac = (15)^2 - 4(10)(5) = 25$ is positive. The equation has two real roots.

37. The discriminant $b^2 - 4ac = (-25)^2 - 4(20)(8) = -15$ is negative. The equation has two non-real complex roots.

39. $(x - 5)^2 = 7$ This is most efficiently solved by the square root method.
$x - 5 = \pm\sqrt{7}$
$x = 5 \pm \sqrt{7}$

41. $x^2 + 2x = 2$
$x^2 + 2x - 2 = 0$ Use the quadratic formula.
$a = 1 \quad b = 2 \quad c = -2$

$$x = \frac{-(2) \pm \sqrt{(2)^2 - 4(1)(-2)}}{2(1)} = \frac{-2 \pm \sqrt{4 + 8}}{2} = \frac{-2 \pm \sqrt{12}}{2} = \frac{-2 \pm 2\sqrt{3}}{2} = -1 \pm \sqrt{3}$$

43. $2u^2 + 3u = 0$ This is most efficiently solved by factoring.
$u(2u + 3) = 0$
$u = 0$ or $2u + 3 = 0$
$2u = -3$
$u = -\frac{3}{2}$

45. $x^2 - 2x + 9 = 2x - 4$
$x^2 - 4x + 13 = 0$ Use the quadratic formula.
$a = 1 \quad b = -4 \quad c = 13$

$$x = \frac{-(-4) \pm \sqrt{(-4)^2 - 4(1)(13)}}{2(1)} = \frac{4 \pm \sqrt{16 - 52}}{2} = \frac{4 \pm \sqrt{-36}}{2} = \frac{4 \pm i\sqrt{36}}{2} = \frac{4 \pm 6i}{2}$$
$$= 2 \pm 3i$$

47. $y^2 = 10y + 3$
$y^2 - 10y - 3 = 0$ Use the quadratic formula.
$a = 1 \quad b = -10 \quad c = -3$

$$y = \frac{-(-10) \pm \sqrt{(-10)^2 - 4(1)(-3)}}{2(1)} = \frac{10 \pm \sqrt{100 + 12}}{2} = \frac{10 \pm \sqrt{112}}{2} = \frac{10 \pm 4\sqrt{7}}{2}$$
$$= 5 \pm 2\sqrt{7}$$

49. $2d^2 + 1 = 4d$
$2d^2 - 4d + 1 = 0$ Use the quadratic formula.
$a = 2 \quad b = -4 \quad c = 1$

$$d = \frac{-(-4) \pm \sqrt{(-4)^2 - 4(2)(1)}}{2(2)} = \frac{4 \pm \sqrt{16 - 8}}{4} = \frac{4 \pm \sqrt{8}}{4} = \frac{4 \pm 2\sqrt{2}}{4} = \frac{2 \pm \sqrt{2}}{2}$$

51. $\frac{2}{u} = \frac{3}{u^2} + 1 \quad u \neq 0, \text{ lcm} = u^2$
$u^2 \cdot \frac{2}{u} = u^2 \cdot \frac{3}{u^2} + u^2 \cdot 1$
$2u = 3 + u^2$
$0 = u^2 - 2u + 3$
$u^2 - 2u + 3 = 0$ Use the quadratic formula.
$a = 1 \quad b = -2 \quad c = 3$

$$u = \frac{-(-2) \pm \sqrt{(-2)^2 - 4(1)(3)}}{2(1)} = \frac{2 \pm \sqrt{4 - 12}}{2} = \frac{2 \pm \sqrt{-8}}{2} = \frac{2 \pm i\sqrt{8}}{2} = \frac{2 \pm 2i\sqrt{2}}{2}$$
$$= 1 \pm i\sqrt{2}$$

53. $\frac{1.2}{y-1} + \frac{1.2}{y} = 1 \quad y \neq 1, 0, \text{ lcm} = y(y-1)$

$$y(y-1)\frac{1.2}{y-1} + y(y-1)\frac{1.2}{y} = y(y-1) \cdot 1$$
$$1.2y + 1.2(y-1) = y(y-1)$$ Multiply by 10 to eliminate decimals.
$$10(1.2y) + 10[1.2(y-1)] = 10y(y-1)$$
$$12y + 12(y-1) = 10y(y-1)$$
$$12y + 12y - 12 = 10y^2 - 10y$$
$$24y - 12 = 10y^2 - 10y$$
$$0 = 10y^2 - 34y + 12$$
$$10y^2 - 34y + 12 = 0$$
$$5y^2 - 17y + 6 = 0$$ This is most efficiently solved by factoring.
$$(5y - 2)(y - 3) = 0$$
$$5y - 2 = 0 \quad \text{or} \quad y - 3 = 0$$
$$5y = 2 \qquad y = 3$$
$$y = \frac{2}{5}$$

55. $x^2 + 3x + 8 = 9 - 4x + x^2$
$$3x + 8 = 9 - 4x$$
$$7x + 8 = 9$$
$$7x = 1$$
$$x = \frac{1}{7}$$

57. $\frac{x^2}{x-1} - \frac{x}{x-1} = 4 \quad \text{lcm} = x - 1, x \neq 1$
$$(x-1)\frac{x^2}{x-1} - (x-1)\frac{x}{x-1} = 4(x-1)$$
$$x^2 - x = 4x - 4$$
$$x^2 - 5x + 4 = 0$$ This is most efficiently solved by factoring.
$$(x-1)(x-4) = 0$$
$$x - 1 = 0 \quad \text{or} \quad x - 4 = 0$$
$$x = 1 \qquad x = 4$$ The only solution.

1 cannot be a solution of the original equation.

59. $x - 2 = \frac{8}{x-1} - 3 \quad \text{lcm} = x - 1, x \neq 1$
$$(x-2)(x-1) = (x-1)\frac{8}{x-1} - 3(x-1)$$
$$x^2 - 3x + 2 = 8 - 3x + 3$$
$$x^2 - 3x + 2 = -3x + 11$$
$$x^2 - 9 = 0$$ This is most efficiently solved by the square root method.
$$x^2 = 9$$
$$x = \pm 3$$

61. $d = \frac{1}{2}gt^2$

$\frac{1}{2}gt^2 = d \quad \text{lcm} = 2$

$2 \cdot \frac{1}{2}gt^2 = 2d$

$gt^2 = 2d$

$\frac{gt^2}{g} = \frac{2d}{g}$

$t^2 = \frac{2d}{g}$

$t = \pm\sqrt{\frac{2d}{g}}$

$t = \sqrt{\frac{2d}{g}}$ since the positive square root is required

63. $A = P(1 + r)^2$

$P(1 + r)^2 = A$

$\frac{P(1 + r)^2}{P} = \frac{A}{P}$

$(1 + r)^2 = \frac{A}{P}$

$1 + r = \pm\sqrt{\frac{A}{P}}$

$1 + r = \sqrt{\frac{A}{P}}$ since the positive square root is required

$r = -1 + \sqrt{\frac{A}{P}}$

65. $x^2 - \sqrt{7}\,x + 2 = 0 \quad a = 1 \quad b = -\sqrt{7} \quad c = 2$

$x = \frac{-(-\sqrt{7}) \pm \sqrt{(-\sqrt{7})^2 - 4(1)(2)}}{2(1)}$

$= \frac{\sqrt{7} \pm \sqrt{7 - 8}}{2}$

$= \frac{\sqrt{7} \pm \sqrt{-1}}{2}$

$x = \frac{\sqrt{7} \pm i}{2}$

67. $\sqrt{3}\,x^2 + 4x + \sqrt{3} = 0$

$a = \sqrt{3} \quad b = 4 \quad c = \sqrt{3}$

$x = \frac{-(4) \pm \sqrt{(4)^2 - 4(\sqrt{3})(\sqrt{3})}}{2(\sqrt{3})}$

$= \frac{-4 \pm \sqrt{16 - 4(3)}}{2\sqrt{3}}$

$= \frac{-4 \pm \sqrt{4}}{2\sqrt{3}}$

$= \frac{-4 \pm 2}{2\sqrt{3}}$

$x = \frac{-4 + 2}{2\sqrt{3}}$ or $x = \frac{-4 - 2}{2\sqrt{3}}$

$= \frac{-2}{2\sqrt{3}}$ $\qquad = \frac{-6}{2\sqrt{3}}$

$= \frac{-2 \cdot \sqrt{3}}{2\sqrt{3} \cdot \sqrt{3}}$ $\qquad = \frac{-6 \cdot \sqrt{3}}{2\sqrt{3} \cdot \sqrt{3}}$

$= -\frac{\sqrt{3}}{3}$ $\qquad = \frac{-6\sqrt{3}}{6}$

$= -\sqrt{3}$

69. $2x^2 + 3ix + 2 = 0$

$a = 2 \quad b = 3i \quad c = 2$

$x = \frac{-(3i) \pm \sqrt{(3i)^2 - 4(2)(2)}}{2(2)}$

$= \frac{-3i \pm \sqrt{-9 - 16}}{4}$

$= \frac{-3i \pm \sqrt{-25}}{4}$

$= \frac{-3i \pm i\sqrt{25}}{4}$

$= \frac{-3i \pm 5i}{4}$

$x = \frac{-3i + 5i}{4}$ or $x = \frac{-3i - 5i}{4}$

$x = \frac{2i}{4}$ $\qquad x = \frac{-8i}{4}$

$x = \frac{1}{2}i$ $\qquad x = -2i$

71. $x^2 + ix - 1 = 0 \quad a = 1 \quad b = i \quad c = -1$

$$x = \frac{-(i) \pm \sqrt{(i)^2 - 4(1)(-1)}}{2(1)}$$
$$= \frac{-i \pm \sqrt{-1 + 4}}{2}$$
$$x = \frac{-i \pm \sqrt{3}}{2} \quad \text{or} \quad \frac{\pm\sqrt{3} - i}{2}$$

73. $x^2 - \frac{1}{2}x - \frac{1}{4} = 0 \quad \text{lcm} = 4$

$$4x^2 - 4 \cdot \frac{1}{2}x - 4 \cdot \frac{1}{4} = 0$$
$$4x^2 - 2x - 1 = 0 \quad a = 4 \quad b = -2 \quad c = -1$$

$$x = \frac{-(-2) \pm \sqrt{(-2)^2 - 4(4)(-1)}}{2(4)}$$
$$= \frac{2 \pm \sqrt{4 + 16}}{8} = \frac{2 \pm \sqrt{20}}{8} = \frac{2 \pm 2\sqrt{5}}{8} = \frac{2(1 \pm \sqrt{5})}{8} = \frac{1 \pm \sqrt{5}}{4}$$

75. $x^2 + x - \frac{1}{2} = 0 \quad a = 1 \quad b = 1 \quad c = -\frac{1}{2}$

$$x = \frac{-(1) \pm \sqrt{1^2 - 4(1)(-\frac{1}{2})}}{2(1)}$$
$$x = \frac{-1 \pm \sqrt{1 + 2}}{2}$$
$$x = \frac{-1 \pm \sqrt{3}}{2}$$

77. $-3.14x^2 + x + 1.07 = 0$
$3.14x^2 - x - 1.07 = 0 \quad a = 3.14 \quad b = -1 \quad c = -1.07$

$$x = \frac{-(-1) \pm \sqrt{(-1)^2 - 4(3.14)(-1.07)}}{2(3.14)}$$
$$= \frac{1 \pm 3.79989}{6.28}$$
$$x = 0.76 \text{ or } -0.45$$

79. $2.07x^2 - 3.79x + 1.34 = 0 \quad a = 2.07 \quad b = -3.79 \quad c = 1.34$

$$x = \frac{-(-3.79) \pm \sqrt{(-3.79)^2 - 4(2.07)(1.34)}}{2(2.07)}$$
$$= \frac{3.79 \pm 1.808}{4.14}$$
$$x = 1.35 \text{ or } 0.48$$

81. $4.83x^2 + 2.04x - 3.18 = 0 \quad a = 4.83 \quad b = 2.04 \quad c = -3.18$

$$x = \frac{-(2.04) \pm \sqrt{(2.04)^2 - 4(4.83)(-3.18)}}{2(4.83)}$$

$$= \frac{-2.04 \pm 8.099}{9.66}$$

$x = 0.63$ or $x = -1.05$

83. $y^2 + xy - x^2 = 0$ Write in standard form.

$$(-1)(y^2) + (-1)xy + (-1)(-x^2) = 0$$
$$-y^2 - xy + x^2 = 0$$
$$x^2 - xy - y^2 = 0 \quad \text{Use quadratic formula.}$$
$$ax^2 + bx + c = 0 \quad a = 1 \quad b = -y \quad c = -y^2$$

$$x = \frac{-(-y) \pm \sqrt{(-y)^2 - 4(1)(-y^2)}}{2(1)}$$
$$= \frac{y \pm \sqrt{y^2 + 4y^2}}{2}$$
$$x = \frac{y \pm \sqrt{5y^2}}{2}$$

85. $\frac{x + y}{x - y} = \frac{x}{y}$ $x \neq y$ $y \neq 0$ lcm $= y(x - y)$

$$y(x - y)\frac{(x + y)}{x - y} = y(x - y)\frac{x}{y} \quad \text{Write in standard form.}$$
$$y(x + y) = (x - y)x$$
$$xy + y^2 = x^2 - xy$$
$$0 = x^2 - 2xy - y^2$$
$$x^2 - 2xy - y^2 = 0 \quad \text{Use quadratic formula.}$$
$$ax^2 + bx + c = 0 \quad a = 1 \quad b = -2y \quad c = -y^2$$

$$x = \frac{-(-2y) \pm \sqrt{(-2y)^2 - 4(1)(-y^2)}}{2(1)}$$
$$= \frac{2y \pm \sqrt{4y^2 + 4y^2}}{2}$$
$$= \frac{2y \pm \sqrt{8y^2}}{2}$$
$$= \frac{2y \pm 2\sqrt{2y^2}}{2}$$
$$= y \pm \sqrt{2y^2}$$

87. $\frac{1}{x^2} + \frac{1}{y^2} = 1$ lcm $= x^2y^2$

$$x^2y^2 \cdot \frac{1}{x^2} + x^2y^2 \cdot \frac{1}{y^2} = x^2y^2$$
$$y^2 + x^2 = x^2y^2$$
$$y^2 = x^2y^2 - x^2$$
$$y^2 = x^2(y^2 - 1)$$
$$x^2 = \frac{y^2}{y^2 - 1}$$
$$x = \pm\sqrt{\frac{y^2}{y^2 - 1}}$$

89. $0.0134x^2 + 0.0414x + 0.0304 = 0$ $a = 0.0134$ $b = 0.0414$ $c = 0.0304$
The discriminant $b^2 - 4ac = (0.0414)^2 - 4(0.0134)(0.0304) = 8.45 \times 10^{-5}$ is positive. The equation has two real solutions.

91. $0.0134x^2 + 0.0214x + 0.0304 = 0 \quad a = 0.0134 \quad b = 0.0214 \quad c = 0.0304$
The discriminant $b^2 - 4ac = (0.0214)^2 - 4(0.0134)(0.0304) = -1.17 \times 10^{-3}$ is negative. The equation has no real solutions.

93. If the discriminant $b^2 - 4ac = 0$, a quadratic equation will have exactly one solution. In this case $a = 2$, $b = -3$, $c = c$.

$$\begin{aligned} b^2 - 4ac &= 0 \\ (-3)^2 - 4(2)c &= 0 \\ 9 - 8c &= 0 \\ -8c &= -9 \\ c &= \frac{9}{8} \end{aligned}$$

95. If the discriminant $b^2 - 4ac > 0$, a quadratic equation will have two real solutions. In this case, $a = a$, $b = 4$, $c = 5$.

$$\begin{aligned} b^2 - 4ac &> 0 \\ 4^2 - 4a(5) &> 0 \\ 16 - 20a &> 0 \\ -20a &> -16 \\ a &< \frac{16}{20} \\ a &< \frac{4}{5} \end{aligned}$$

97. If the discriminant $b^2 - 4ac = 0$, a quadratic equation will have exactly one solution. In this case $a = 3$, $b = 8$, $c = c$.

$$\begin{aligned} b^2 - 4ac &= 0 \\ 8^2 - 4(3)c &= 0 \\ 64 - 12c &= 0 \\ -12c &= -64 \\ c &= \frac{64}{12} \\ c &= \frac{16}{3} \end{aligned}$$

99. The two roots of $ax^2 + bx + c = 0$ are $\dfrac{-b + \sqrt{b^2 - 4ac}}{2a}$ and $\dfrac{-b - \sqrt{b^2 - 4ac}}{2a}$.

Therefore

$$\begin{aligned} r_1r_2 &= \left(\frac{-b + \sqrt{b^2 - 4ac}}{2a}\right)\left(\frac{-b - \sqrt{b^2 - 4ac}}{2a}\right) \\ &= \frac{(-b + \sqrt{b^2 - 4ac})(-b - \sqrt{b^2 - 4ac})}{4a^2} \\ &= \frac{(-b)^2 - (\sqrt{b^2 - 4ac})}{4a^2} \\ &= \frac{b^2 - (b^2 - 4ac)}{4a^2} \\ &= \frac{b^2 - b^2 + 4ac}{4a^2} \\ &= \frac{4ac}{4a^2} \\ &= \frac{c}{a} \end{aligned}$$

Exercise 7-3 Applications

Key Ideas and Formulas

The strategy for solving word problems, first given in Section 1-8, is not reprinted here for reasons of space.

Note that since quadratic equations often have two solutions, it is important to check both solutions in the original problem to see if one or the other should be rejected.

---

1. Let x = the first of the two consecutive positive even integers. Then x + 2 must be the next of the two. Then the statement is understood to mean:

The product of	x and x + 2	is	168

   $$x(x + 2) = 168$$
   $$x^2 + 2x = 168$$
   $$x^2 + 2x - 168 = 0$$

   This can be solved by factoring, but if the factors are not immediately apparent the quadratic formula is also appropriate.
   a = 1 b = 2 c = -168

   $$x = \frac{-(2) \pm \sqrt{(2)^2 - 4(1)(-168)}}{2(1)}$$
   $$= \frac{-2 \pm \sqrt{4 + 672}}{2}$$
   $$= \frac{-2 \pm \sqrt{676}}{2}$$
   $$= \frac{-2 \pm 26}{2}$$
   $$x = \frac{-2 + 26}{2} \text{ or } 12$$

   Disregard the negative answer, since we are only interested in positive integers.

   $x + 2 = 14$

   The desired integers are 12 and 14.

3. Let x = the number. Then the statement is interpreted to mean:

The number	added to	itself	is the same as	the number	multiplied by	itself
x	+	x	=	x	·	x

   $$x + x = x^2$$
   $$2x = x^2$$
   $$0 = x^2 - 2x$$
   $$x^2 - 2x = 0$$
   $$x(x - 2) = 0$$
   $$x = 0 \text{ or } x - 2 = 0$$
   $$x = 2$$

   The possible numbers are 0 and 2.

5. Let x = the original numerator
Then $x + 2$ = the original denominator
$x + 3$ = the numerator increased by 3
$(x + 2) + 3 = x + 5$ = the denominator increased by 3
Then [the resulting fraction] [is equal to] [twice the original fraction]

$$\frac{x+3}{x+5} \quad = \quad 2 \cdot \frac{x}{x+2}$$

$$\frac{x+3}{x+5} = \frac{2x}{x+2} \qquad x \neq -2, -5 \quad \text{lcm} = (x+2)(x+5)$$

$$(x+2)(x+5)\frac{(x+3)}{x+5} = (x+2)(x+5)\frac{2x}{x+2}$$

$$(x+2)(x+3) = (x+5)2x$$

$$x^2 + 5x + 6 = 2x^2 + 10x$$

$$0 = x^2 + 5x - 6$$

$$0 = (x+6)(x-1)$$

$$x + 6 = 0 \quad \text{or} \quad x - 1 = 0$$

$$x = -6 \qquad\qquad x = 1$$

$$x + 2 = -4 \qquad\qquad x + 2 = 3$$

The original fraction was $\frac{-6}{-4}$ or $\frac{1}{3}$.

7. Let x = the number

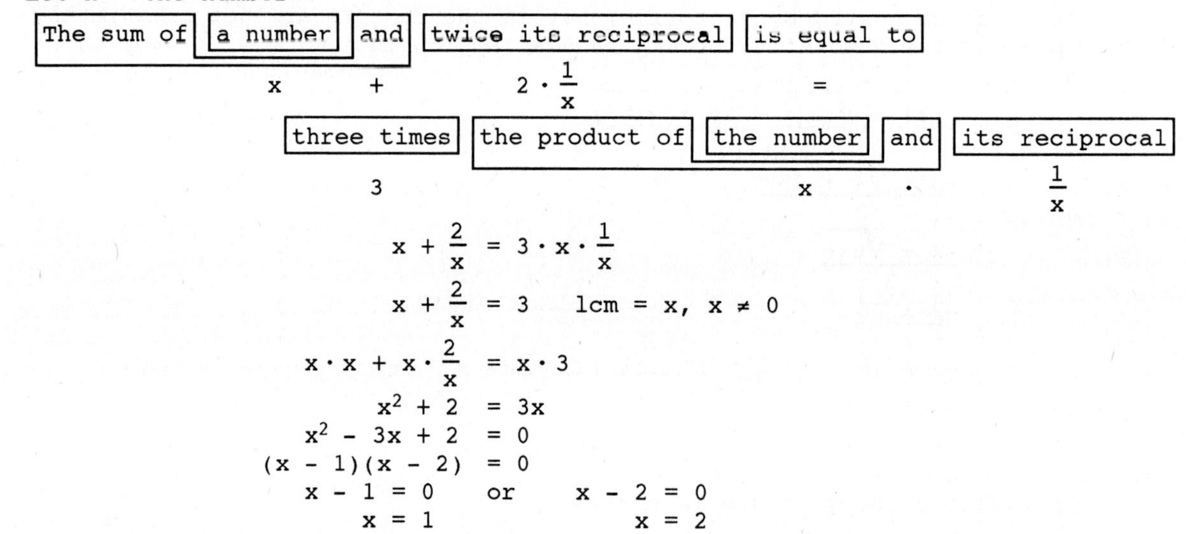

$$x \quad + \quad 2 \cdot \frac{1}{x} \quad = \quad 3 \quad x \cdot \frac{1}{x}$$

$$x + \frac{2}{x} = 3 \cdot x \cdot \frac{1}{x}$$

$$x + \frac{2}{x} = 3 \qquad \text{lcm} = x,\ x \neq 0$$

$$x \cdot x + x \cdot \frac{2}{x} = x \cdot 3$$

$$x^2 + 2 = 3x$$

$$x^2 - 3x + 2 = 0$$

$$(x-1)(x-2) = 0$$

$$x - 1 = 0 \quad \text{or} \quad x - 2 = 0$$

$$x = 1 \qquad\qquad x = 2$$

9. We are given the formula $S = \frac{n(n+1)}{2}$. We replace S by the constant value 66.

$$66 = \frac{n(n+1)}{2} \qquad \text{lcm} = 2$$

$$2 \cdot 66 = 2 \cdot \frac{n(n+1)}{2}$$

$$132 = n(n+1)$$

$$132 = n^2 + n$$

$$0 = n^2 + n - 132$$

$$0 = (n+12)(n-11)$$

$$n + 12 = 0 \quad \text{or} \quad n - 11 = 0$$

$$n = -12 \qquad\qquad n = 11$$

discard since n
is to be positive

11. Let d = the distance of the horizon from the airplane.
From the drawing it should be clear that
longest side of right triangle = 4000 + 2 = 4002
shorter sides are d, 4000.

Then

$$(4002)^2 = d^2 + (4000)^2$$
$$16{,}016{,}004 = d^2 + 16{,}000{,}000$$
$$16{,}004 = d^2$$
$$d^2 = 16{,}004$$

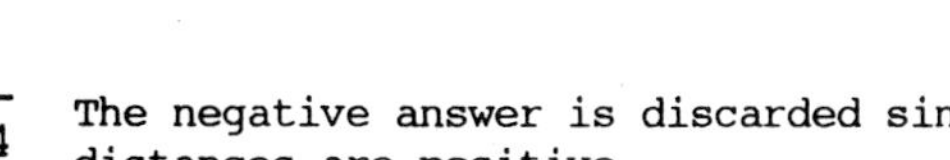

$$d = \sqrt{16{,}004}$$ The negative answer is discarded since distances are positive.

$$d \approx 127 \text{ miles}$$

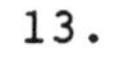

13.

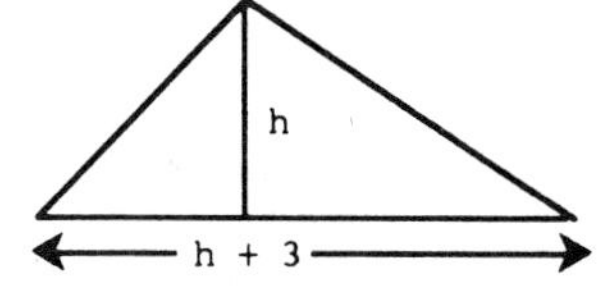

Let h = the height of the triangle
Then h + 3 = the base of the triangle

$$a = \frac{1}{2}bh$$
$$3 = \frac{1}{2}(h + 3)h \qquad \text{lcm} = 2$$
$$6 = (h + 3)h$$
$$6 = h^2 + 3h$$
$$0 = h^2 + 3h - 6 \quad \text{Use the quadratic formula}$$
$$a = 1 \quad b = 3 \quad c = 6$$
$$h = \frac{-3 \pm \sqrt{(3)^2 - 4(1)(-6)}}{2(1)}$$
$$h = \frac{-3 \pm \sqrt{9 + 24}}{2}$$
$$\text{height} = \frac{-3 + \sqrt{33}}{2}$$ The negative answer is discarded since distances are positive.

$$\text{base} = h + 3 = 3 + \frac{-3 + \sqrt{33}}{2} = \frac{6}{2} + \frac{-3 + \sqrt{33}}{2}$$
$$\text{base} = \frac{3 + \sqrt{33}}{2}$$

15.

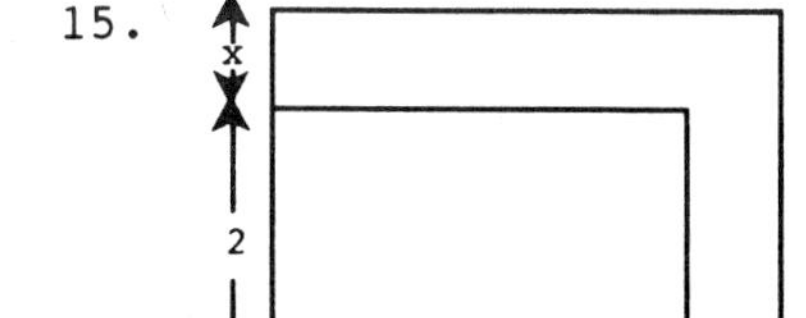

Let x = the amount of increase
Then x + 2 = new width
x + 4 = new length
(x + 2)(x + 4) = new area

Since [New area] [is] [twice] [the old area]
$(x + 2)(x + 4) = 2 \cdot 2 \cdot 4$
we have

$$(x + 2)(x + 4) = 16$$
$$x^2 + 6x + 8 = 16$$
$$x^2 + 6x - 8 = 0 \quad \text{Use the quadratic formula.}$$
$$a = 1 \quad b = 6 \quad c = -8$$
$$x = \frac{-(6) \pm \sqrt{(6)^2 - 4(1)(-8)}}{2(1)}$$

$$= \frac{-6 \pm \sqrt{36 + 32}}{2}$$

$$= \frac{-6 \pm \sqrt{68}}{2}$$

$$x = \frac{-6 + \sqrt{68}}{2} \approx 1.12 \text{ cm.}$$ The negative answer is discarded since distances are positive.

New dimensions: $x + 4$ by $x + 2$ or 5.12 by 3.12 cm.

17. Let x = length
then $x - 2$ = width

$$x(x - 2) = 24$$
$$x^2 - 2x = 24$$
$$x^2 - 2x - 24 = 0$$
$$(x - 6)(x + 4) = 0$$
$$x - 6 = 0 \quad \text{or} \quad x + 4 = 0$$
$$x = 6 \text{ m} \qquad x = -4$$ The negative answer is discarded since distances are positive.
$$x - 2 = 4 \text{ m}$$

19.

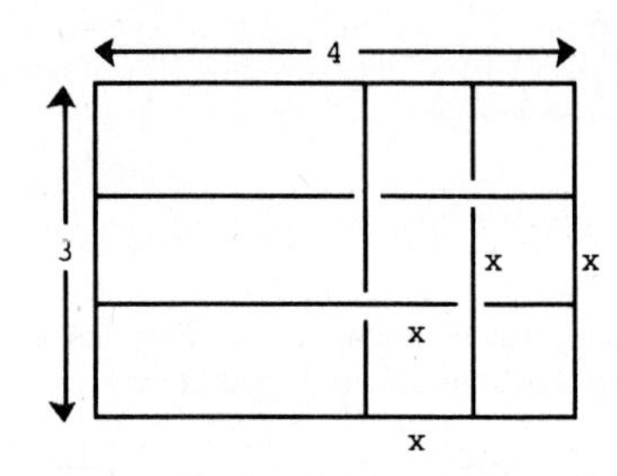

A straightforward way of solving this is to regard the cross as being made up of two overlapping rectangles as above. Then the area of the cross = (area of one rectangle) + (area of other rectangle) - (area of overlap)

$$= 3 \cdot x \quad + \quad x \cdot 4 \quad - \quad x \cdot x$$
$$= 3x + 4x - x^2$$

Since [the area of cross] [is] [one-half] [of] [the total area],

we have $3x + 4x - x^2 = \frac{1}{2} \quad 3 \cdot 4$

or

$$3x + 4x - x^2 = 6$$
$$7x - x^2 = 6$$
$$0 = x^2 - 7x + 6$$
$$x^2 - 7x + 6 = 0$$ Solve by factoring.
$$(x - 1)(x - 6) = 0$$
$$x = 1 \quad \text{or} \quad x = 6$$ Discard this answer since x must be less than both dimensions of the flag.
$$x = 1 \text{ foot}$$

21. We are given that the number of diagonals, $\frac{n(n - 3)}{2}$, is 20 in this case.

Thus,

$$\frac{n(n - 3)}{2} = 20 \qquad \text{lcm} = 2$$
$$2 \cdot \frac{n(n - 3)}{2} = 2 \cdot 20$$
$$n(n - 3) = 40$$
$$n^2 - 3n = 40$$
$$n^2 - 3n - 40 = 0$$

$(n + 5)(n - 8) = 0$

$n + 5 = 0 \qquad n - 8 = 0$

$n = -5 \qquad n = 8$

Discard since the number of sides must be positive.

23. 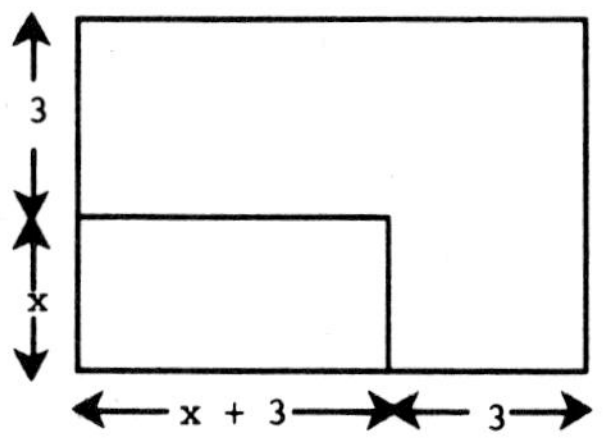 Let x = original width

Then $x + 3$ = original length

$x + 3$ = increased width

$x + 3 + 3 = x + 6$ = increased length

$x(x + 3)$ = original area

$(x + 3)(x + 6)$ = new area

Since | New area | is | 4 times | original area |

$$(x + 3)(x + 6) = 4 \cdot x(x + 3)$$

$$x^2 + 9x + 18 = 4x^2 + 12x$$

$$0 = 3x^2 + 3x - 18$$

$$0 = x^2 + x - 6$$

$$(x - 2)(x + 3) = 0$$

$x - 2 = 0$ or $x + 3 = 0$

$x = 2$ m $\qquad x = -3$ The negative answer is discarded since distances are positive.

$x + 3 = 5$ m

25. We are given $p = 0.003v^2$ with $p = 7.5$

Then $7.5 = 0.003v^2$

$$\frac{7.5}{0.003} = v^2$$

$$2500 = v^2$$

$v = \sqrt{2500}$ Discard the negative answer since speeds are positive.

$v = 50$ mi/hr

27. We are given $v^2 = 2gh$ with $g = 32$ and $h = 0.5$.

Then $v^2 = 2(32)(0.5)$

$v^2 = 32$

$v = \sqrt{32}$ Discard the negative answer since speeds are positive.

$v \approx 5.66$ ft/sec

29. We are given $h = 0.0336r^2$ with $h = 142$

Then $142 = 0.0336r^2$

$$\frac{142}{0.0336} = r^2$$

$$r^2 = \frac{142}{0.0336}$$

$r = \sqrt{\dfrac{142}{0.0336}}$ Discard the negative answer since speeds are positive.

$r \approx 65$ mi/hr

31. We are given $d = 0.044v^2 + 1.1v$ with $d = 165$.

Then $165 = 0.044v^2 + 1.1v$

$0 = 0.044v^2 + 1.1v - 165$

This can be solved by eliminating decimals and factoring, but it is more natural to use the quadratic formula and a calculator.

$a = 0.044 \quad b = 1.1 \quad c = -165$

$$v = \frac{-(1.1) \pm \sqrt{(1.1)^2 - 4(0.044)(-165)}}{2(0.044)}$$
$$= \frac{-1.1 \pm 5.5}{0.088}$$
$$v = \frac{-1.1 + 5.5}{0.088} \text{ or 50 miles per hour.}$$

The negative answer is discarded because speeds are positive.

[The factoring method is shown as an alternative:

$$0.044v^2 + 1.1v - 165 = 0$$
$$44v^2 + 1100v - 165{,}000 = 0$$
$$v^2 + 25v - 3750 = 0$$
$$(v + 75)(v - 50) = 0$$
$$v + 75 = 0 \quad \text{or} \quad v - 50 = 0$$
$$v = -75 \text{ (discarded)} \qquad v = 50]$$

33. We are given: maximum height $= \frac{v^2}{32}$.

Then
$$968 = \frac{v^2}{32}$$
$$v^2 = 32 \cdot 968$$
$$v^2 = 30{,}976$$
$$v = \sqrt{30{,}976}$$
The negative answer is discarded because speeds are positive.
$$v = 176 \text{ ft per sec}$$

35. (A) We are given $y = 128t - 16t^2$ with $y = 0$.

Then
$$0 = 128t + 16t^2$$
$$0 = 16t(8 - t)$$
$$16t = 0 \quad \text{or} \quad 8 - t = 0$$
$$t = 0 \qquad t = 8$$

The arrow is released at $t = 0$ and returns to the ground at $t = 8$.

(B) We are given $y = 128t - 16t^2$ with $y = 16$.

Then
$$16 = 128t - 16t^2$$
$$16t^2 - 128t + 16 = 0$$
$$t^2 - 8t + 1 = 0 \quad \text{Use the quadratic formula.}$$
$$a = 1 \quad b = -8 \quad c = 1$$

$$t = \frac{-(-8) \pm \sqrt{(-8)^2 - 4(1)(1)}}{2(1)}$$
$$t = \frac{8 \pm \sqrt{64 - 4}}{2}$$
$$t = \frac{8 \pm \sqrt{60}}{2}$$
$$t = 0.13 \text{ sec or } 7.87 \text{ sec}$$

37. We are given $h = \left(\sqrt{h_0} - \frac{5}{12}t\right)^2$ with $h_0 = 4$. We are asked for t = the time that corresponds to $h = \frac{1}{2}h_0 = 2$. Then

$$2 = \left(\sqrt{4} - \frac{5}{12}t\right)^2$$
$$2 = \left(2 - \frac{5}{12}t\right)^2$$

$\left(2 - \frac{5}{12}t\right)^2 = 2$ Use the square root method.

$2 - \frac{5}{12}t = \pm\sqrt{2}$

$-\frac{5}{12}t = -2 \pm \sqrt{2}$

$t = -\frac{12}{5}(-2 \pm \sqrt{2})$

$t \approx 1.41$ minutes or 8.19 minutes.

The second answer is discarded because it does not make sense. According to the formula, the tank is empty when $h = 0$, that is, when

$$0 = \left(\sqrt{4} - \frac{5}{12}t\right)^2$$

Then $2 - \frac{5}{12}t = 0$ or t would be 4.8 minutes. The formula cannot be valid for t greater than 4.8.

Thus $t = 1.41$ minutes.

39. Let 1.2 = time for both presses to do the job together
x = time for the older press to do the job alone
$x - 1$ = time for the newer press to do the job alone (one hour less)

Then $\frac{1}{1.2}$ = rate for both presses together
$\frac{1}{x}$ = rate for older press
$\frac{1}{x - 1}$ = rate for newer press

Sum of individual rates = rate together

$$\frac{1}{x} + \frac{1}{x - 1} = \frac{1}{1.2} \quad x \neq 0, 1$$

$$\frac{1}{x} + \frac{1}{x - 1} = \frac{10}{12} \quad \text{lcm} = 12x(x - 1)$$

$$12(x - 1)\frac{1}{x} + 12x(x - 1)\frac{1}{x - 1} = 12x(x - 1)\frac{10}{12}$$

$$12(x - 1) + 12x = x(x - 1)10$$

$$12x - 12 + 12x = 10x^2 - 10x$$

$$24x - 12 = 10x^2 - 10x$$

$$0 = 10x^2 - 34x + 12$$

$$10x^2 - 34x + 12 = 0$$

$$5x^2 - 17x + 6 = 0 \quad \text{Use factoring method.}$$

$$(5x - 2)(x - 3) = 0$$

$$5x - 2 = 0 \quad \text{or} \quad x - 3 = 0$$

$$5x = 2 \qquad x = 3$$

$$x = \frac{2}{5}$$

Discard the first answer because even though x is not negative, $x - 1$ would be negative.

$x = 3$ hours for the older press
$x - 1 = 2$ hours for the newer press

41.

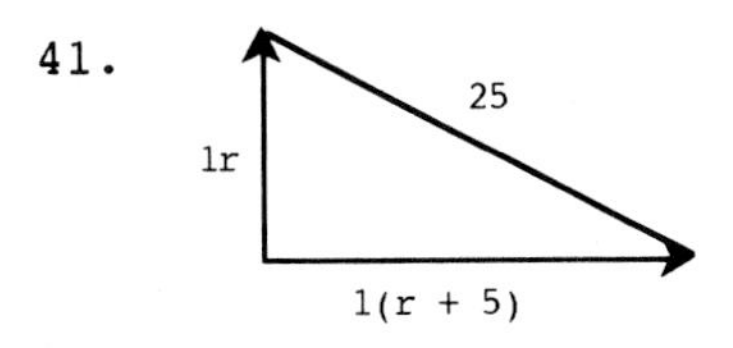

Let r = the rate of the slower boat
$r + 5$ = the rate of the faster boat
Then $1r$ = the distance the slower boat has gone
$1(r + 5)$ = the distance the faster boat has gone

From the Pythagorean theorem,

$$(1r)^2 + [1(r+5)]^2 = 25^2$$
$$r^2 + [r+5]^2 = 25^2$$
$$r^2 + r^2 + 10r + 25 = 625$$
$$2r^2 + 10r - 600 = 0$$
$$r^2 + 5r - 300 = 0$$
$$(r+20)(r-15) = 0$$

$r + 20 = 0$ $\qquad$ $r - 15 = 0$

$r = -20$ $\qquad$ $r = 15$ km/hr

discard; rates cannot be negative $\qquad$ $r + 5 = 20$ km/hr

43. Let x = the rate of the current
Then $10 - x$ = the rate of the boat upstream
$10 + x$ = the rate of the boat downstream

$\frac{D}{r} = \frac{24}{10 - x}$ = the time of the boat upstream

$\frac{D}{r} = \frac{24}{10 + x}$ = the time of the boat downstream

Then (time upstream) = 1 hour longer than (time downstream)

$$\frac{24}{10-x} = 1 + \frac{24}{10+x}$$

$x \neq 10, -10$; lcm $= (10 - x)(10 + x)$

$$(10-x)(10+x)\frac{24}{10-x} = (10-x)(10+x)1 + (10-x)(10+x)\frac{24}{10+x}$$
$$24(10+x) = (10-x)(10+x) + 24(10-x)$$
$$240 + 24x = 100 - x^2 + 240 - 24x$$
$$240 + 24x = 340 - x^2 - 24x$$
$$x^2 + 48x - 100 = 0$$
$$(x-2)(x+50) = 0$$

$x - 2 = 0$ or $x + 50 = 0$

$x = 2$ $\qquad$ $x = -50$ $\quad$ Discard this answer since rates are positive.

$x = 2$ kilometers per hour

45. We are given $A = P(1 + r)^2$ with $P = 100$ and $A = 110.25$
Then

$$110.25 = 100(1+r)^2$$
$$1.1025 = (1+r)^2$$
$$\pm\sqrt{1.1025} = 1 + r$$
$$r = -1 + \sqrt{1.1025}$$
$$r = -1 + 1.05$$
$$r = 0.05 \text{ or } 5\%$$

The negative answer is discarded because interest rates cannot be negative.

47. We are given $d = s$ with $d = \frac{3000}{p}$ and $s = 200p - 700$. Then

$$\frac{3000}{p} = 200p - 700 \qquad p \neq 0, \text{ lcm} = p$$
$$p\frac{3000}{p} = p(200p - 700)$$
$$3000 = 200p^2 - 700p$$
$$0 = 200p^2 - 700p - 3000$$
$$200p^2 - 700p - 3000 = 0 \quad \text{Use factoring method.}$$
$$2p^2 - 7p - 30 = 0$$
$$(p-6)(2p+5) = 0$$

$p - 6 = 0$ or $2p + 5 = 0$

$p = 6$ $\qquad$ $2p = -5$

$p = -\frac{5}{2}$ $\quad$ Discard this answer because prices are positive.

$p = \$6$

49.

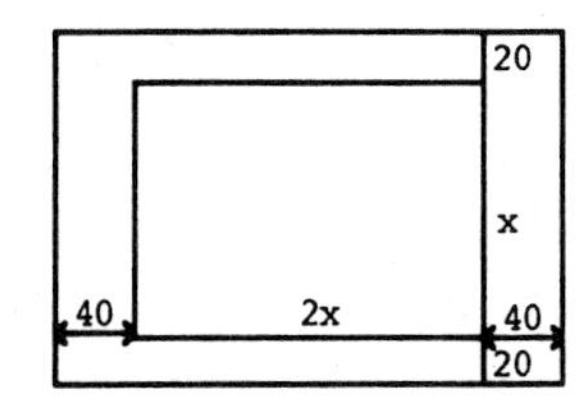

Let x = the width of the warehouse
then $2x$ = the length of the warehouse
$x + 20 + 20 = x + 40$ = the width of the property
$2x + 40 + 40 = 2x + 80$ = the length of the property

$$\text{Then } (2x + 80)(x + 40) = 72{,}200$$
$$2x^2 + 160x + 3{,}200 = 72{,}200$$
$$2x^2 + 160x - 69{,}000 = 0$$
$$x^2 + 80x - 34{,}500 = 0$$

Use the quadratic formula
$a = 1 \quad b = 80 \quad c = -34{,}500$

$$x = \frac{-(80) \pm \sqrt{(80)^2 - 4(1)(-34{,}500)}}{2(1)}$$
$$x = \frac{-80 \pm \sqrt{6{,}400 + 138{,}000}}{2}$$
$$x = \frac{-80 \pm \sqrt{144{,}400}}{2}$$
$$x = \frac{-80 \pm 380}{2}$$
$$x = \frac{-80 + 380}{2} \qquad x = \frac{-80 - 380}{2}$$
$$x = 150 \text{ ft} \qquad x = -230$$
Discard this answer because distances are positive.
$$2x = 300 \text{ ft}$$

51. Let x = the number of trees planted
Then $x - 120$ = the number of trees planted beyond 120
$\frac{1}{2}(x - 120)$ = the decrease in yield that results from these trees.
$40 - \frac{1}{2}(x - 120)$ = the yield per tree
Since the total yield = (number of trees) × (yield per tree),

$$\text{total yield} = x\left[40 - \frac{1}{2}(x - 120)\right]$$

In this situation, total yield = 4200, thus,

$$x\left[40 - \frac{1}{2}(x - 120)\right] = 4200$$
$$x\left[40 - \frac{1}{2}x + 60\right] = 4200$$
$$x\left[100 - \frac{1}{2}x\right] = 4200$$
$$100x - \frac{1}{2}x^2 = 4200 \qquad \text{lcm} = 2$$
$$200x - x^2 = 8400$$
$$0 = x^2 - 200x + 8400$$
$$0 = (x - 60)(x - 140)$$
$$x - 60 = 0 \qquad x - 140 = 0$$
$$x = 60 \qquad x = 140 \text{ trees}$$

Discard this answer
because more than
120 trees were planted.

53. Let x = the number of \$0.25 increases in price
Then $8 + 0.25x$ = the price that results per unit
$100x$ = the decrease in sales that results
$5000 - 100x$ = the number of units sold at this price.
Since revenue = (number of units sold) × (price per unit),
revenue = $(5000 - 100x)(8 + 0.25x)$

In this situation, revenue = 41,800, thus,

$$(5000 - 100x)(8 + 0.25x) = 41{,}800$$
$$40{,}000 + 1250x - 800x - 25x^2 = 41{,}800$$
$$40{,}000 + 450x - 25x^2 = 41{,}800$$
$$0 = 25x^2 - 450x + 1800$$
$$0 = x^2 - 18x + 72$$
$$0 = (x - 6)(x - 12)$$
$$x - 6 = 0 \quad \text{or} \quad x - 12 = 0$$
$$x = 6 \qquad\qquad x = 12$$

55. Let x = the number of \$10 increases in price.
Then $420 + 10x$ = the price that results per unit
$180 - x$ = the number of rentals at this price.
Since revenue = (number of rentals) × (price per unit),
revenue = $(180 - x)(420 + 10x)$
In this situation, revenue = 87,210, thus

$$87{,}210 = (180 - x)(420 + 10x)$$
$$87{,}210 = 75{,}600 + 1800x - 420x - 10x^2$$
$$87{,}210 = 75{,}600 + 1380x - 10x^2$$
$$10x^2 - 1380x + 11{,}610 = 0$$
$$x^2 - 138x + 1161 = 0 \quad \text{Use the quadratic formula}$$
$$a = 1 \quad b = -138 \quad c = 1161$$

$$x = \frac{-(-138) \pm \sqrt{(-138)^2 - 4(1)(1161)}}{2(1)}$$
$$x = \frac{138 \pm \sqrt{19{,}044 - 4644}}{2}$$
$$x = \frac{138 \pm \sqrt{14{,}400}}{2}$$
$$x = \frac{138 \pm 120}{2}$$
$$x = 129 \text{ or } 9$$

Then the price is $420 + 10x$, that is $420 + 10(129) = \$1{,}710$, or $420 + 10 \cdot 9 = \$510$.

57. Let x = the number of \$1 increases in price.
Then $7 + x$ = the price that results
$20x$ = the decrease in attendance that results
$300 - 20x$ = the attendance at this price
Since revenue = (attendance) × (price per ticket),
revenue = $(300 - 20x)(7 + x)$
In this situation, revenue = 2375, thus,

$$2375 = (300 - 20x)(7 + x)$$
$$2375 = 2100 + 300x - 140x - 20x^2$$
$$2375 = 2100 + 160x - 20x^2$$
$$20x^2 - 160x + 275 = 0$$
$$4x^2 - 32x + 55 = 0$$
$$(2x - 5)(2x - 11) = 0$$

$$2x - 5 = 0 \quad \text{or} \quad 2x - 11 = 0$$
$$2x = 5 \qquad\qquad 2x = 11$$
$$x = 2.5 \qquad\qquad x = 5.5$$

Then the price is $7 + 2.5 = \$9.50$ or $7 + 5.5 = \$12.50$.

Exercise 7-4 Radical Equations and Other Equations Reducible to Quadratic Form

Key Ideas and Formulas

An equation containing radicals, called a **radical equation**, can usually be solved by raising both sides to a natural number power.

The new equation may have solutions that are not solutions of the original equation, called **extraneous** solutions, but all solutions of the original equation will be among those of the new equation. All solutions of the new equation must be checked to eliminate the extraneous solutions.

An equation that is not quadratic, but can be transformed into the form $au^2 + bu + c = 0$, where u is an expression in some other variable, can be solved by substituting u for that expression, solving by the appropriate quadratic method, then replacing u by the original expression and solving the resulting simpler equation.

---

1. $\sqrt{x + 2} = 7$ Square both sides
   $x + 2 = 49$
   $x = 47$

   Check: $\sqrt{47 + 2} = \sqrt{49} = 7$ so 47 is a solution.

3. $\sqrt{5x + 4} = 8$ Square both sides
   $5x + 4 = 64$
   $5x = 60$
   $x = 12$

   Check: $\sqrt{5 \cdot 12 + 4} = \sqrt{64} = 8$ so 12 is a solution.

5. $x - 2 = \sqrt{x}$ Square both sides
   $x^2 - 4x + 4 = x$
   $x^2 - 5x + 4 = 0$
   $(x - 4)(x - 1) = 0$
   $x = 4, 1$

   Check: $\underline{x = 4}$ $4 - 2 = 2 = \sqrt{4}$ so 4 is a solution.
   $\underline{x = 1}$ $1 - 2 = -1 \neq \sqrt{1}$ so 1 is extraneous.

   4

7. $\sqrt{x} = x - 12$ Square both sides
   $x = x^2 - 24x + 144$
   $0 = x^2 - 25x + 144$
   $0 = (x - 16)(x - 9)$
   $x = 16, 9$

   Check: $\underline{x = 16}$ $\sqrt{16} = 4 = 16 - 12$ so 16 is a solution.
   $\underline{x = 9}$ $\sqrt{9} = 3 \neq 9 - 12$ so 9 is extraneous.

   16

9. $m - 13 = \sqrt{m + 7}$ Square both sides

$m^2 - 26m + 169 = m + 7$

$m^2 - 27m + 162 = 0$

$(m - 9)(m - 18) = 0$

$m = 9, 18$

Check: $\underline{m = 9}$ $\quad 9 - 13 = -4 \neq \sqrt{9 + 7} = 4$ so 9 is extraneous.

$\underline{m = 18}$ $\quad 18 - 13 = 5 = \sqrt{18 + 7} = 5$ so 18 is a solution.

18

11. $\sqrt{x + 3} = x - 3$ Square both sides

$x + 3 = x^2 - 6x + 9$

$0 = x^2 - 7x + 6$

$0 = (x - 6)(x - 1)$

$x = 6, 1$

Check: $\underline{x = 6}$ $\quad \sqrt{6 + 3} \stackrel{?}{=} 6 - 3$ $\qquad$ $\underline{x = 1}$ $\quad \sqrt{1 + 3} \stackrel{?}{=} 1 - 3$

$\sqrt{9} \stackrel{\checkmark}{=} 3$ $\qquad$ $\sqrt{4} \neq -2$

6 is a solution $\qquad$ 1 is extraneous

6

13. $\sqrt{x + 10} = x - 10$ Square both sides

$x + 10 = x^2 - 20x + 100$

$0 = x^2 - 21x + 90$

$0 = (x - 6)(x - 15)$

$x = 6, 15$

Check: $\underline{x = 6}$ $\quad \sqrt{6 + 10} \stackrel{?}{=} 6 - 10$ $\qquad$ $\underline{x = 15}$ $\quad \sqrt{15 + 10} \stackrel{?}{=} 15 - 10$

$\sqrt{16} \neq -4$ $\qquad$ $\sqrt{25} \stackrel{\checkmark}{=} 5$

6 is extraneous $\qquad$ 15 is a solution

15

15. $x^4 - 10x^2 + 9 = 0$

Let $u = x^2$, then $u^2 = x^4$. After substitution, the equation becomes

$u^2 - 10u + 9 = 0$

$(u - 9)(u - 1) = 0$

$u = 9, 1$

Replacing u with x^2, we have

$x^2 = 9$ or $x^2 = 1$

$x = \pm 3$ $\qquad$ $x = \pm 1$

17. $x^4 - 7x^2 - 18 = 0$

Let $u = x^2$, then $u^2 = x^4$. After substitution, the equation becomes

$u^2 - 7u - 18 = 0$

$(u - 9)(u + 2) = 0$

$u = 9, -2$

Replacing u with x^2, we have

$x^2 = 9$ or $x^2 = -2$

$x = \pm 3$ $\qquad$ $x = \pm\sqrt{-2}$

$x = \pm i\sqrt{2}$

19. $\sqrt{x^2 - 3x} = 2$ Square both sides

$x^2 - 3x = 4$

$x^2 - 3x - 4 = 0$

$(x - 4)(x + 1) = 0$

$x = 4, -1$

Check: $\underline{x = 4}$ $\quad \sqrt{(4)^2 - 3(4)} = \sqrt{16 - 12} = 2$ so 4 is a solution.

$\underline{x = -1}$ $\quad \sqrt{(-1)^2 - 3(-1)} = \sqrt{1 + 3} = 2$ so -1 is a solution.

4, -1

21.
$$\sqrt{x^2 + 6x} = 4 \quad \text{Square both sides}$$
$$x^2 + 6x = 16$$
$$x^2 + 6x - 16 = 0$$
$$(x + 8)(x - 2) = 0$$
$$x = -8, 2$$

Check: $\underline{x = -8}$ $\sqrt{(-8)^2 + 6(-8)} = \sqrt{64 - 48} = \sqrt{16} = 4$ so -8 is a solution.

$\underline{x = 2}$ $\sqrt{2^2 + 6 \cdot 2} = \sqrt{4 + 12} = \sqrt{16} = 4$ so 2 is a solution.

$-8, 2$

23.
$$x + 3 = \sqrt{x + 5} \quad \text{Square both sides}$$
$$x^2 + 6x + 9 = x + 5$$
$$x^2 + 5x + 4 = 0$$
$$(x + 1)(x + 4) = 0$$
$$x = -1, -4$$

Check: $\underline{x = -1}$ $-1 + 3 \stackrel{?}{=} \sqrt{-1 + 5}$, $2 \stackrel{\checkmark}{=} \sqrt{4}$, -1 is a solution

$\underline{x = -4}$ $-4 + 3 \stackrel{?}{=} \sqrt{-4 + 5}$, $-1 \neq \sqrt{1}$, -4 is extraneous

-1

25.
$$x - 16 = \sqrt{x + 4} \quad \text{Square both sides}$$
$$x^2 - 32x + 256 = x + 4$$
$$x^2 - 33x + 252 = 0$$
$$(x - 12)(x - 21) = 0$$
$$x = 12, 21$$

Check: $\underline{x = 12}$ $12 - 16 \stackrel{?}{=} \sqrt{12 + 4}$, $-4 \neq \sqrt{16}$, 12 is extraneous

$\underline{x = 21}$ $21 - 16 \stackrel{?}{=} \sqrt{21 + 4}$, $5 \stackrel{\checkmark}{=} \sqrt{25}$, 21 is a solution

21

27.
$$\sqrt{3x - 2} = x \quad \text{Square both sides}$$
$$3x - 2 = x^2$$
$$0 = x^2 - 3x + 2$$
$$0 = (x - 1)(x - 2)$$
$$x = 1, 2$$

Check: $\underline{x = 1}$ $\sqrt{3 \cdot 1 - 2} = \sqrt{1} = 1$ so 1 is a solution.

$\underline{x = 2}$ $\sqrt{3 \cdot 2 - 2} = \sqrt{4} = 2$ so 2 is a solution.

$1, 2$

29.
$$x + 1 = \sqrt{4x + 1} \quad \text{Square both sides}$$
$$x^2 + 2x + 1 = 4x + 1$$
$$x^2 - 2x = 0$$
$$x(x - 2) = 0$$
$$x = 0, 2$$

Check: $\underline{x = 0}$ $0 + 1 \stackrel{?}{=} \sqrt{4 \cdot 0 + 1}$, $1 \stackrel{\checkmark}{=} \sqrt{1}$, 0 is a solution

$\underline{x = 2}$ $2 + 1 \stackrel{?}{=} \sqrt{4 \cdot 2 + 1}$, $3 \stackrel{\checkmark}{=} \sqrt{9}$, 2 is a solution

$0, 2$

31. $m - 7\sqrt{m} + 12 = 0$ Isolate radical on one side

$m + 12 = 7\sqrt{m}$ Square both sides.

$m^2 + 24m + 144 = (7\sqrt{m})^2$

$m^2 + 24m + 144 = 49m$

$m^2 - 25m + 144 = 0$

$(m - 9)(m - 16) = 0$

$m = 9, 16$

COMMON ERRORS:
$m^2 - 7m + 12 = 0$
or
$m^2 - 49m + 144 = 0$
It is incorrect to square each term in a polynomial when the whole polynomial is to be squared.

Check: $\underline{m = 9}$ $9 - 7\sqrt{9} + 12 \overset{?}{=} 0$

$9 - 21 + 12 \overset{?}{=} 0$

$0 \overset{\surd}{=} 0$

9 is a solution

$\underline{m = 16}$ $16 - 7\sqrt{16} + 12 \overset{?}{=} 0$

$16 - 28 + 12 \overset{?}{=} 0$

$0 \overset{\surd}{=} 0$

16 is a solution

9, 16

33. $1 + \sqrt{x + 5} = x$ Isolate radical on one side.

$\sqrt{x + 5} = x - 1$ Square both sides.

$(\sqrt{x + 5})^2 = (x - 1)^2$

$x + 5 = x^2 - 2x + 1$

$0 = x^2 - 3x - 4$

$0 = (x + 1)(x - 4)$

$x = -1, 4$

COMMON ERROR:
$1 + x + 5 = x^2$
The square of $1 + a$ is not $1 + a^2$, but rather $1 + 2a + a^2$.

Check: $\underline{x = -1}$ $1 + \sqrt{-1 + 5} \overset{?}{=} -1$

$1 + 2 \overset{?}{=} -1$

$3 \neq -1$

-1 is extraneous

$\underline{x = 4}$ $1 + \sqrt{4 + 5} \overset{?}{=} 4$

$1 + 3 \overset{?}{=} 4$

$4 \overset{\surd}{=} 4$

4 is a solution

4

35. $\sqrt{3x + 1} = \sqrt{x} - 1$ Square both sides.

$(\sqrt{3x + 1})^2 = (\sqrt{x} - 1)^2$

$3x + 1 = x - 2\sqrt{x} + 1$ Isolate the remaining radical.

$2x = -2\sqrt{x}$ Square both sides again.

$4x^2 = 4x$

$4x^2 - 4x = 0$

$4x(x - 1) = 0$

$x = 0, 1$

Check: $\underline{x = 0}$ $\sqrt{3(0) + 1} \overset{?}{=} \sqrt{0} - 1$

$\sqrt{1} \overset{?}{=} 0 - 1$

$1 \neq -1$

0 is extraneous

$\underline{x = 1}$ $\sqrt{3(1) + 1} \overset{?}{=} \sqrt{1} - 1$

$\sqrt{4} \overset{?}{=} 1 - 1$

$2 \neq 0$

1 is also extraneous

No solution.

37. $\sqrt{3t + 4} + \sqrt{t} = -3$

This cannot have real solutions, since for any real number t both $\sqrt{3t + 4}$ and $\sqrt{t}$ must be positive, hence $\sqrt{3t + 4} + \sqrt{t}$ must be positive and cannot equal -3. Any solutions obtained by squaring both sides must be extraneous.

No solution.

39. $\sqrt{u - 2} = 2 + \sqrt{2u + 3}$ Square both sides.

$(\sqrt{u - 2})^2 = (2 + \sqrt{2u + 3})^2$

$u - 2 = 4 + 4\sqrt{2u + 3} + 2u + 3$ Isolate the remaining radical.

$u - 2 = 2u + 7 + 4\sqrt{2u + 3}$

$-u - 9 = 4\sqrt{2u + 3}$ Square both sides again.

$(-u - 9)^2 = (4\sqrt{2u + 3})^2$

$u^2 + 18u + 81 = 16(2u + 3)$

$u^2 + 18u + 81 = 32u + 48$

$u^2 - 14u + 33 = 0$

$(u - 3)(u - 11) = 0$

$u = 3, 11$

Check: <u>u = 3</u> $\sqrt{3 - 2} \stackrel{?}{=} 2 + \sqrt{2(3) + 3}$

$\sqrt{1} \stackrel{?}{=} 2 + \sqrt{9}$

$1 \stackrel{?}{=} 2 + 3$

$1 \neq 5$

3 is extraneous

<u>u = 11</u> $\sqrt{11 - 2} \stackrel{?}{=} 2 + \sqrt{2(11) + 3}$

$\sqrt{9} \stackrel{?}{=} 2 + \sqrt{25}$

$3 \stackrel{?}{=} 2 + 5$

$3 \neq 7$

11 is also extraneous

No solution.

41. $\sqrt{2x - 1} - \sqrt{x - 4} = 2$ Easier to solve with a radical on each side.

$\sqrt{2x - 1} = \sqrt{x - 4} + 2$ Square both sides

$(\sqrt{2x - 1})^2 = (\sqrt{x - 4} + 2)^2$

$2x - 1 = x - 4 + 4\sqrt{x - 4} + 4$ Isolate the remaining radical.

$2x - 1 = x + 4\sqrt{x - 4}$

$x - 1 = 4\sqrt{x - 4}$ Square both sides again.

$(x - 1)^2 = (4\sqrt{x - 4})^2$

$x^2 - 2x + 1 = 16(x - 4)$

$x^2 - 2x + 1 = 16x - 64$

$x^2 - 18x + 65 = 0$

$(x - 5)(x - 13) = 0$

$x = 5, 13$

Check: <u>x = 5</u> $\sqrt{2(5) - 1} - \sqrt{5 - 4} \stackrel{?}{=} 2$

$\sqrt{9} - \sqrt{1} \stackrel{?}{=} 2$

$3 - 1 \stackrel{?}{=} 2$

$2 \stackrel{\checkmark}{=} 2$

5 is a solution

<u>x = 13</u> $\sqrt{2(13) - 1} - \sqrt{13 - 4} \stackrel{?}{=} 2$

$\sqrt{25} - \sqrt{9} \stackrel{?}{=} 2$

$5 - 3 \stackrel{?}{=} 2$

$2 \stackrel{\checkmark}{=} 2$

13 is a solution

5, 13

43. $\sqrt{x^2 - x} = \sqrt{2x - 2}$ Square both sides

$x^2 - x = 2x - 2$

$x^2 - 3x + 2 = 0$

$(x - 1)(x - 2) = 0$

$x = 1, 2$

Check: <u>x = 1</u> $\sqrt{1^2 - 1} \stackrel{?}{=} \sqrt{2 \cdot 1 - 2}$

$\sqrt{0} \stackrel{\checkmark}{=} \sqrt{0}$

1 is a solution

<u>x = 2</u> $\sqrt{2^2 - 2} \stackrel{?}{=} \sqrt{2 \cdot 2 - 2}$

$\sqrt{2} \stackrel{\checkmark}{=} \sqrt{2}$

2 is a solution

1, 2

45.

$$\sqrt{x^2 - 3x} = \sqrt{x - 3} \quad \text{Square both sides}$$
$$x^2 - 3x = x - 3$$
$$x^2 - 4x + 3 = 0$$
$$(x - 1)(x - 3) = 0$$
$$x = 1,\ 3$$

Check: <u>x = 3</u>

$$\sqrt{3^2 - 3 \cdot 3} \stackrel{?}{=} \sqrt{3 - 3}$$
$$\sqrt{0} \stackrel{\checkmark}{=} \sqrt{0}$$

3 is a solution

<u>x = 1</u>

$$\sqrt{1^2 - 3 \cdot 1} \stackrel{?}{=} \sqrt{1 - 3}$$
$$\sqrt{-2} \stackrel{\checkmark}{=} \sqrt{-2}$$

1 is a solution

1, 3

47.

$$\sqrt{3x + 1} = \sqrt{x + 4} + 1 \quad \text{Square both sides}$$
$$(\sqrt{3x + 1})^2 = (\sqrt{x + 4} + 1)^2$$
$$3x + 1 = x + 4 + 2\sqrt{x + 4} + 1 \quad \text{Isolate the remaining radical}$$
$$3x + 1 = x + 5 + 2\sqrt{x + 4}$$
$$2x - 4 = 2\sqrt{x + 4}$$
$$x - 2 = \sqrt{x + 4} \quad \text{Square both sides again}$$
$$x^2 - 4x + 4 = x + 4$$
$$x^2 - 5x = 0$$
$$x(x - 5) = 0$$
$$x = 0,\ 5$$

Check: <u>x = 0</u>

$$\sqrt{3 \cdot 0 + 1} \stackrel{?}{=} \sqrt{0 + 4} + 1$$
$$\sqrt{1} \stackrel{?}{=} \sqrt{4} + 1$$
$$1 \neq 2 + 1$$

0 is extraneous

<u>x = 5</u>

$$\sqrt{3 \cdot 5 + 1} \stackrel{?}{=} \sqrt{5 + 4} + 1$$
$$\sqrt{16} \stackrel{?}{=} \sqrt{9} + 1$$
$$4 \stackrel{\checkmark}{=} 3 + 1$$

5 is a solution

5

49.

$$\sqrt{5x - 4} = \sqrt{x} + 2 \quad \text{Square both sides}$$
$$(\sqrt{5x - 4})^2 = (\sqrt{x} + 2)^2$$
$$5x - 4 = x + 4\sqrt{x} + 4 \quad \text{Isolate the remaining radical}$$
$$4x - 8 = 4\sqrt{x}$$
$$x - 2 = \sqrt{x}$$
$$x^2 - 4x + 4 = x$$
$$x^2 - 5x + 4 = 0$$
$$(x - 1)(x - 4) = 0$$
$$x = 1,\ 4$$

Check: <u>x = 1</u>

$$\sqrt{5 \cdot 1 - 4} \stackrel{?}{=} \sqrt{1} + 2$$
$$\sqrt{1} \stackrel{?}{=} 1 + 2$$
$$1 \neq 1 + 2$$

1 is extraneous

<u>x = 4</u>

$$\sqrt{5 \cdot 4 - 2} \stackrel{?}{=} \sqrt{4} + 2$$
$$\sqrt{16} \stackrel{?}{=} 2 + 2$$
$$4 \stackrel{\checkmark}{=} 2 + 2$$

4 is a solution

4

51.

$$\sqrt{3x - 2} = \sqrt{x} + 2 \quad \text{Square both sides}$$
$$(\sqrt{3x - 2})^2 = (\sqrt{x} + 2)^2$$
$$3x - 2 = x + 4\sqrt{x} + 4 \quad \text{Isolate the remaining radical}$$
$$2x - 6 = 4\sqrt{x}$$
$$x - 3 = 2\sqrt{x} \quad \text{Square both sides again}$$

$$x^2 - 6x + 9 = 4x$$
$$x^2 - 10x + 9 = 0$$
$$(x - 1)(x - 9) = 0$$
$$x = 1, 9$$

Check: $\underline{x = 1}$ $\sqrt{3 \cdot 1 - 2} \stackrel{?}{=} \sqrt{1} + 2$ $\qquad$ $\underline{x = 9}$ $\sqrt{3 \cdot 9 - 2} \stackrel{?}{=} \sqrt{9} + 2$

$\sqrt{1} \stackrel{?}{=} 1 + 2$ $\qquad$ $\sqrt{25} \stackrel{?}{=} 3 + 2$

$1 \neq 3$ $\qquad$ $5 \stackrel{\surd}{=} 5$

1 is extraneous $\qquad$ 9 is a solution

9

53. $x^6 - 7x^3 - 8 = 0$

Let $u = x^3$, then $u^2 = x^6$. After substituting, the original equation becomes
$$u^2 - 7u - 8 = 0$$
$$(u - 8)(u + 1) = 0$$
$$u = 8, -1$$
Replacing u with x^3, we obtain
$$x^3 = 8 \quad \text{or} \quad x^3 = -1$$
The real solutions are
$$x = \sqrt[3]{8} \text{ or } 2 \quad \text{and} \quad x = \sqrt[3]{-1} \text{ or } -1$$
$x = 2, -1$

55. $x^6 - 6x^3 + 8 = 0$

Let $u = x^3$, then $u^2 = x^6$. After substituting, the original equation becomes
$$u^2 - 6u + 8 = 0$$
$$(u - 2)(u - 4) = 0$$
$$u = 2, 4$$
Replacing u with x^3, we obtain
$$x^3 = 2 \quad \text{or} \quad x^3 = 4$$
The real solutions are
$$x = \sqrt[3]{2} \quad \text{or} \quad x = \sqrt[3]{4}$$
$\sqrt[3]{2}, \sqrt[3]{4}$

57. $y^8 - 17y^4 + 16 = 0$

Let $u = y^4$, then $u^2 = y^8$. After substituting, the original equation becomes
$$u^2 - 17u + 16 = 0$$
$$(u - 1)(u - 16) = 0$$
$$u = 1, 16$$
Replacing u with y^4, we obtain

$y^4 = 1$ $\qquad$ or $\qquad$ $y^4 = 16$

$y^4 - 1 = 0$ $\qquad$ $y^4 - 16 = 0$

$(y^2 + 1)(y^2 - 1) = 0$ $\qquad$ $(y^2 + 4)(y^2 - 4) = 0$

$y^2 + 1 = 0$ or $y^2 - 1 = 0$ $\qquad$ or $\qquad$ $y^2 + 4 = 0$ or $y^2 - 4 = 0$

$y^2 = -1$ $\quad$ $y^2 = 1$ $\qquad$ $y^2 = -4$ $\quad$ $y^2 = 4$

$y = \pm\sqrt{-1}$ $\quad$ $y = \pm 1$ $\qquad$ $y = \pm\sqrt{-4}$ $\quad$ $y = \pm 2$

$y = \pm i$ $\qquad$ $y = \pm 2i$

$y = \pm 1, \pm 2, \pm i, \pm 2i$

59. $x^4 - 13x^2 + 36 = 0$
Let $u = x^2$, then $u^2 = x^4$. After substituting, the original equation becomes,
$$u^2 - 13u + 36 = 0$$
$$(u - 4)(u - 9) = 0$$
$$u = 4, 9$$
Replacing u with x^2, we obtain
$x^2 = 4$ or $x^2 = 9$
$x = \pm 2$ $x = \pm 3$
$\pm 2, \pm 3$

61. $x^4 - 5x^2 - 36 = 0$
Let $u = x^2$, then $u^2 = x^4$. After substituting, the original equation becomes,
$$u^2 - 5u - 36 = 0$$
$$(u - 9)(u + 4) = 0$$
$$u = 9, -4$$
Replacing u with x^2, we obtain
$x^2 = 9$ or $x^2 = -4$
$x = \pm 3$ $x = \pm\sqrt{-4}$
$x = \pm 2i$
$\pm 3, \pm 2i$

63. $2x^4 - 3x^2 - 5 = 0$
Let $u = x^2$, then $u^2 = x^4$. After substituting, the original equation becomes,
$$2u^2 - 3u - 5 = 0$$
$$(2u - 5)(u + 1) = 0$$
$2u - 5 = 0$ or $u + 1 = 0$
$2u = 5$ $u = -1$
$u = \frac{5}{2}$
Replacing u with x^2, we obtain
$x^2 = \frac{5}{2}$ or $x^2 = -1$
$x = \pm\sqrt{\frac{5}{2}}$ $x = \pm\sqrt{-1}$
$x = \pm\sqrt{\frac{10}{4}}$ $x = \pm i$
$x = \pm\frac{\sqrt{10}}{2}$

65. $x^{2/3} - 3x^{1/3} - 10 = 0$
Let $u = x^{1/3}$, then $u^2 = x^{2/3}$. After substitution, the original equation becomes
$$u^2 - 3u - 10 = 0$$
$$(u - 5)(u + 2) = 0$$
$$u = 5, -2$$
Replacing u with $x^{1/3}$, we obtain
$x^{1/3} = 5$ or $x^{1/3} = -2$
$x = 125$ $x = -8$
$x = 125, -8$

COMMON ERROR: "$x = 5^{1/3}$ or $x = -2^{1/3}$." Both sides must be raised to the third power to eliminate the $\frac{1}{3}$ power from the $x^{1/3}$.

67. $y^{1/2} - 3y^{1/4} + 2 = 0$
Let $u = y^{1/4}$, then $u^2 = y^{1/2}$. After substitution, the original equation becomes
$$u^2 - 3u + 2 = 0$$
$$(u - 1)(u - 2) = 0$$
$$u = 1, 2$$
Replacing u with $y^{1/4}$, we obtain
$y^{1/4} = 1$ $y^{1/4} = 2$
$y = 1$ $y = 16$
$y = 1, 16$

69. $\frac{2}{x^2} + \frac{3}{x} + 1 = 0$

Let $u = \frac{1}{x}$, then $u^2 = \frac{1}{x^2}$. After substitution, the original equation becomes

$$2u^2 + 3u + 1 = 0$$
$$(2u + 1)(u + 1) = 0$$
$$2u + 1 = 0 \quad \text{or} \quad u + 1 = 0$$
$$2u = -1 \qquad u = -1$$
$$u = -\frac{1}{2}$$

Replacing u with $\frac{1}{x}$, we obtain

$$\frac{1}{x} = -\frac{1}{2} \quad \text{lcm} = 2x,\ x \neq 0 \qquad\qquad \frac{1}{x} = -1 \quad \text{lcm} = x,\ x \neq 0$$
$$2x \cdot \frac{1}{x} = 2x \cdot \left(-\frac{1}{2}\right) \qquad\qquad x \cdot \frac{1}{x} = x(-1)$$
$$2 = -x \qquad\qquad 1 = -x$$
$$x = -2 \qquad\qquad x = -1$$

$x = -2, -1$

71. $x^{-1} - 4x^{-2} + 3x^{-3} = 0$

Let $u = x^{-1}$, then $u^2 = x^{-2}$, $u^3 = x^{-3}$. After substitution, the original equation becomes

$$u - 4u^2 + 3u^3 = 0$$
$$u(1 - 4u + 3u^2) = 0$$
$$u(1 - u)(1 - 3u) = 0$$
$$u = 0, 1, \frac{1}{3}$$

Replacing u with x^{-1}, we obtain

$$x^{-1} = 0 \quad x \neq 0,\ \text{lcm} = x \qquad x^{-1} = 1 \quad x \neq 0,\ \text{lcm} = x \qquad x^{-1} = \frac{1}{3} \quad x \neq 0,\ \text{lcm} = 3x$$
$$\frac{1}{x} = 0 \qquad\qquad \frac{1}{x} = 1 \qquad\qquad \frac{1}{x} = \frac{1}{3}$$
$$x \cdot \frac{1}{x} = 0x \qquad\qquad x \cdot \frac{1}{x} = x \qquad\qquad 3x \cdot \frac{1}{x} = 3x\left(\frac{1}{3}\right)$$
$$1 = 0 \qquad\qquad 1 = x \qquad\qquad 3 = x$$

impossible

$x = 1, 3$

73. $6x^{-2} - 5x^{-1} - 6 = 0$

Let $u = x^{-1}$, then $u^2 = x^{-2}$. After substitution, the original equation becomes

$$6u^2 - 5u - 6 = 0$$
$$(3u + 2)(2u - 3) = 0$$
$$3u + 2 = 0 \quad \text{or} \quad 2u - 3 = 0$$
$$3u = -2 \qquad 2u = 3$$
$$u = -\frac{2}{3} \qquad u = \frac{3}{2}$$

Replacing u with x^{-1}, we obtain

$$x^{-1} = -\frac{2}{3} \qquad\qquad x^{-1} = \frac{3}{2}$$

$$\frac{1}{x} = -\frac{2}{3} \quad x \neq 0 \quad \text{lcm} = 3x \qquad\qquad \frac{1}{x} = \frac{3}{2} \quad x \neq 0 \quad \text{lcm} = 2x$$

$$3x\left(\frac{1}{x}\right) = 3x\left(-\frac{2}{3}\right) \qquad\qquad 2x\left(\frac{1}{x}\right) = 2x\left(\frac{3}{2}\right)$$

$$3 = -2x \qquad\qquad 2 = 3x$$

$$x = -\frac{3}{2} \qquad\qquad x = \frac{2}{3}$$

$$x = -\frac{3}{2}, \frac{2}{3}$$

75. $x^{-2} - 3x^{-3} - 10x^{-4} = 0$

Although we could use the method of substitution, replacing x^{-1} with u, it is far more efficient to solve using straightforward methods:

$$\frac{1}{x^2} - \frac{3}{x^3} - \frac{10}{x^4} = 0 \qquad \text{lcm} = x^4,\ x \neq 0$$

$$x^4 \cdot \frac{1}{x^2} - x^4 \cdot \frac{3}{x^3} - x^4 \cdot \frac{10}{x^4} = 0 \cdot x^4$$

$$x^2 - 3x - 10 = 0$$

$$(x - 5)(x + 2) = 0$$

$$x = 5, -2$$

77. $1 - 2x^{-2} + x^{-4} = 0$

Let $u = x^{-2}$, then $u^2 = x^{-4}$. After substitution, the original equation becomes

$$1 - 2u + u^2 = 0$$

$$(1 - u)^2 = 0$$

$$1 - u = 0$$

$$u = 1$$

Replacing u with x^{-2}, we obtain

$$x^{-2} = 1$$

$$\frac{1}{x^2} = 1 \qquad \text{lcm} = x^2,\ x \neq 0$$

$$x^2 \cdot \frac{1}{x^2} = x^2$$

$$1 = x^2$$

$$x = \pm 1$$

79. $4x^{-4} - 17x^{-2} + 4 = 0$

Let $u = x^{-2}$, then $u^2 = x^{-4}$. After substitution, the original equation becomes

$$4u^2 - 17u + 4 = 0$$

$$(4u - 1)(u - 4) = 0$$

$$4u - 1 = 0 \quad \text{or} \quad u - 4 = 0$$

$$4u = 1 \qquad u = 4$$

$$u = \frac{1}{4}$$

Replacing u by x^{-2}, we obtain

$$x^{-2} = \frac{1}{4} \qquad\qquad x^{-2} = 4$$

$$\frac{1}{x^2} = \frac{1}{4} \quad x \neq 0,\ \text{lcm} = 4x^2 \qquad\qquad \frac{1}{x^2} = 4 \quad x \neq 0,\ \text{lcm} = x^2$$

$$4x^2\left(\frac{1}{x^2}\right) = 4x^2\left(\frac{1}{4}\right) \qquad\qquad x^2\left(\frac{1}{x^2}\right) = x^2(4)$$

$$4 = x^2 \qquad\qquad 1 = 4x^2$$

$$x = \pm 2 \qquad\qquad x^2 = \frac{1}{4}$$

$$x = \pm\frac{1}{2}$$

$$x = \pm 2, \pm\frac{1}{2}$$

81. $(m^2 - m)^2 - 4(m^2 - m) = 12$
Let $u = (m^2 - m)$, then $u^2 = (m^2 - m)^2$. After substitution, the original equation becomes

$$u^2 - 4u = 12$$
$$u^2 - 4u - 12 = 0$$
$$(u - 6)(u + 2) = 0$$
$$u = 6, -2$$

Replacing u by $m^2 - m$, we obtain

$$m^2 - m = 6$$
$$m^2 - m - 6 = 0$$
$$(m - 3)(m + 2) = 0$$
$$m = 3, -2$$

$$m^2 - m = -2$$
$$m^2 - m + 2 = 0$$

This is not factorable in the integers, so we use the quadratic formula with $a = 1$, $b = -1$, $c = 2$.

$$m = \frac{-b \pm \sqrt{b^2 - 4ac}}{2a}$$
$$= \frac{-(-1) \pm \sqrt{(-1)^2 - 4(1)(2)}}{2(1)}$$
$$= \frac{1 \pm \sqrt{1 - 8}}{2}$$
$$= \frac{1 \pm \sqrt{-7}}{2}$$
$$= \frac{1 \pm i\sqrt{7}}{2}$$
$$= \frac{1}{2} \pm \frac{\sqrt{7}}{2}i$$

$m = -2, 3, \frac{1}{2} \pm \frac{\sqrt{7}}{2}i$

83. $(x - 3)^4 + 3(x - 3)^2 = 4$
Let $u = (x - 3)^2$, then $u^2 = (x - 3)^4$. After substitution, the original equation becomes

$$u^2 + 3u = 4$$
$$u^2 + 3u - 4 = 0$$
$$(u + 4)(u - 1) = 0$$
$$u = -4, 1$$

Replacing u with $(x - 3)^2$, we obtain

$$(x - 3)^2 = -4$$
$$x - 3 = \pm\sqrt{-4}$$
$$x - 3 = \pm 2i$$
$$x = 3 \pm 2i$$

$$(x - 3)^2 = 1$$
$$x - 3 = \pm\sqrt{1}$$
$$x - 3 = \pm 1$$
$$x = 3 \pm 1$$
$$x = 3 + 1 \quad \text{or} \quad x = 3 - 1$$
$$x = 4 \qquad\qquad x = 2$$

$x = 2, 4, 3 \pm 2i$

85. $\sqrt{3x + 6} - \sqrt{x + 4} = \sqrt{2}$

$$\sqrt{3x + 6} = \sqrt{x + 4} + \sqrt{2} \quad \text{Square both sides.}$$
$$(\sqrt{3x + 6})^2 = (\sqrt{x + 4} + \sqrt{2})^2$$
$$3x + 6 = x + 4 + 2\sqrt{2}\sqrt{x + 4} + 2$$
$$3x + 6 = x + 6 + 2\sqrt{2}\sqrt{x + 4}$$
$$2x = 2\sqrt{2}\sqrt{x + 4}$$
$$x = \sqrt{2}\sqrt{x + 4} \quad \text{Square both sides again.}$$

$$x^2 = (\sqrt{2}\sqrt{x + 4})^2$$
$$x^2 = 2(x + 4)$$
$$x^2 = 2x + 8$$
$$x^2 - 2x - 8 = 0$$
$$(x - 4)(x + 2) = 0$$
$$x = 4, -2$$

Check: <u>x = 4</u> $\sqrt{3\cdot 4 + 6} - \sqrt{4 + 4} = \sqrt{18} - \sqrt{8} = 3\sqrt{2} - 2\sqrt{2} = \sqrt{2}$
so 4 is a solution.

<u>x = -2</u> $\sqrt{3(-2) + 6} - \sqrt{(-2) + 4} = 0 - \sqrt{2} = -\sqrt{2} \neq \sqrt{2}$
so -2 is extraneous.

$x = 4$

87. $$\frac{1}{\sqrt{x - 2}} + \frac{2}{3} = 1$$
$$\frac{1}{\sqrt{x - 2}} = 1 - \frac{2}{3}$$
$$\frac{1}{\sqrt{x - 2}} = \frac{1}{3} \quad \text{Square both sides.}$$
$$\left(\frac{1}{\sqrt{x - 2}}\right)^2 = \left(\frac{1}{3}\right)^2$$
$$\frac{1}{x - 2} = \frac{1}{9} \qquad x \neq 2,\ \text{lcm} = 9(x - 2)$$
$$9(x - 2)\frac{1}{x - 2} = 9(x - 2)\frac{1}{9}$$
$$9 = x - 2$$
$$x = 11$$

Check: <u>x = 11</u> $\frac{1}{\sqrt{11 - 2}} + \frac{2}{3} = \frac{1}{\sqrt{9}} + \frac{2}{3} = \frac{1}{3} + \frac{2}{3} = 1$ so 11 is a solution.

$x = 11$

89. $$\frac{x}{3} + \frac{2}{x} = \frac{6x + 1}{3x} \qquad x \neq 0,\ \text{lcm} = 3x$$
$$3x\left(\frac{x}{3}\right) + 3x\left(\frac{2}{x}\right) = 3x\cdot\frac{(6x + 1)}{3x}$$
$$x^2 + 6x = 6x + 1$$
$$x^2 - 6x + 5 = 0$$
$$(x - 5)(x - 1) = 0$$
$$x = 5, 1$$

91. $$\frac{1}{x - 1} + \frac{1}{x - 2} = \frac{5}{6} \qquad x \neq 1, 2 \quad \text{lcm} = 6(x - 1)(x - 2)$$
$$6(x - 1)(x - 2)\frac{1}{x - 1} + 6(x - 1)(x - 2)\frac{1}{x - 2} = 6(x - 1)(x - 2)\frac{5}{6}$$
$$6(x - 2) + 6(x - 1) = 5(x - 1)(x - 2)$$
$$6x - 12 + 6x - 6 = 5(x^2 - 3x + 2)$$
$$12x - 18 = 5x^2 - 15x + 10$$
$$0 = 5x^2 - 27x + 28$$

This is factorable in the integers as $0 = (5x - 7)(x - 4)$, but if this is not obvious we can use the quadratic formula with
$a = 5,\ b = -27,\ c = 28$

$$x = \frac{-(-27) \pm \sqrt{(-27)^2 - 4(5)(28)}}{2(5)}$$
$$= \frac{27 \pm \sqrt{729 - 560}}{10}$$
$$= \frac{27 \pm \sqrt{169}}{10}$$
$$= \frac{27 \pm 13}{10}$$
$$x = \frac{27 + 13}{10} \quad \text{or} \quad \frac{27 - 13}{10}$$
$$x = 4 \quad \text{or} \quad 1.4$$
$$x = 4, 1.4$$

Exercise 7-5 Graphing Quadratic Polynomials

Key Ideas and Formulas

Graph of $y = x^2$

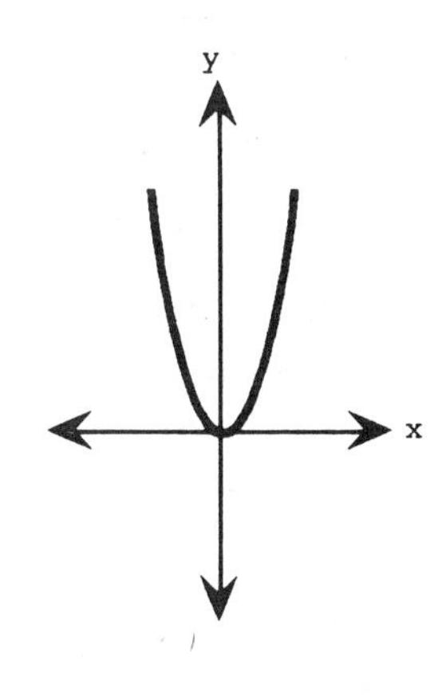

Graph of $y = -x^2$

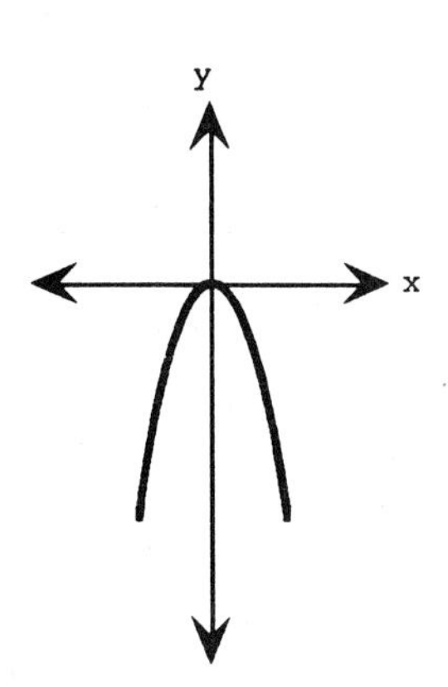

Graph of $y = ax^2 + bx + c$

The graph is a parabola.

The graph opens up if $a > 0$, down if $a < 0$.

The vertex is located at $x = -\frac{b}{2a}$, $y = -\frac{b^2 - 4ac}{4a}$.

x intercepts of the graph of $y = ax^2 + bx + c$

If $b^2 - 4ac = 0$, there is one intercept

If $b^2 - 4ac > 0$, there are two intercepts

If $b^2 - 4ac < 0$, there are no intercepts

1. $y = x^2 + 4$. The parabola opens up, since $a > 0$. The vertex is located at

$$x = -\frac{b}{2a} = \frac{-0}{2\cdot 1} = 0 \quad \text{and}$$
$$y = (0)^2 + 4 = 4$$

x	y
0	4
1	5
2	8
-1	5
-2	8

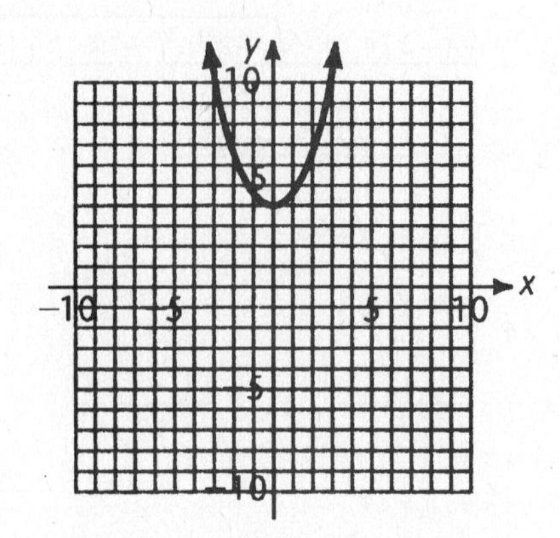

3. $y = -x^2 - 1$. The parabola opens down, since $a < 0$. The vertex is located at

$$x = -\frac{b}{2a} = -\frac{0}{2(-1)} = 0 \quad \text{and}$$
$$y = -(0)^2 - 1 = -1$$

x	y
0	-1
1	-2
2	-5
-1	-2
-2	-5

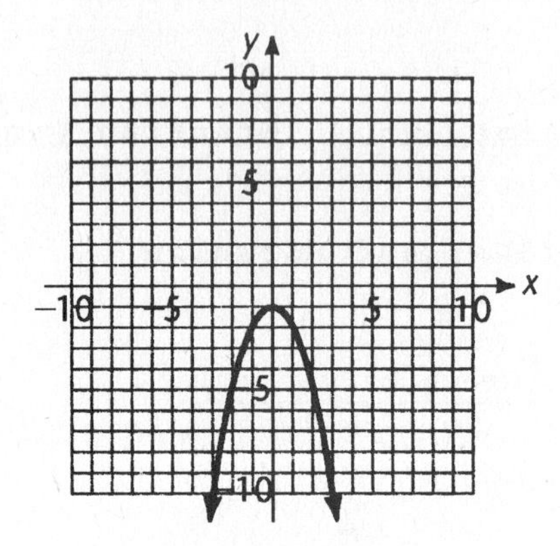

5. $y = x^2 + 2x + 1$. The parabola opens up, since $a > 0$. The vertex is located at

$$x = -\frac{b}{2a} = -\frac{2}{2\cdot 1} = -1 \quad \text{and}$$
$$y = -(1)^2 + 2(-1) + 1 = 0$$

x	y
-1	0
0	1
1	4
-2	1
-3	4

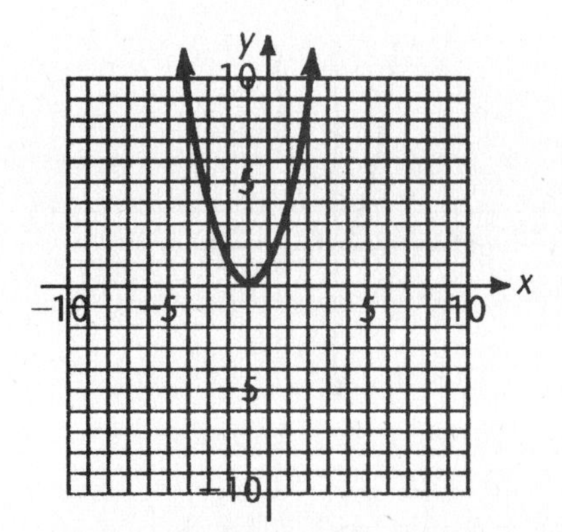

7. $y = x^2 - 5x + 4$. The parabola opens up, since $a > 0$. The vertex is located at

$$x = -\frac{b}{2a} = -\frac{(-5)}{2\cdot 1} = \frac{5}{2} \quad \text{and}$$
$$y = \left(\frac{5}{2}\right)^2 - 5\left(\frac{5}{2}\right) + 4 = -\frac{9}{4}$$

x	y
0	4
1	0
2	-2
3	-2
4	0

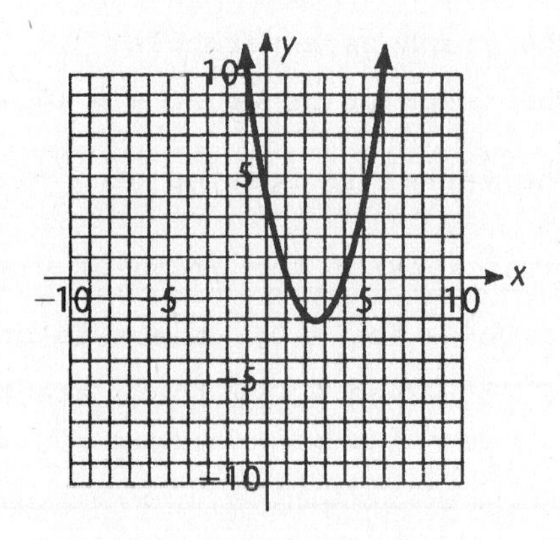

9. $y = x^2 - 5x + 6$. The parabola opens up, since $a > 0$. The vertex is located at

$$x = -\frac{b}{2a} = -\frac{(-5)}{2 \cdot 1} = \frac{5}{2} \quad \text{and}$$

$$y = \left(\frac{5}{2}\right)^2 - 5\left(\frac{5}{2}\right) + 6 = -\frac{1}{4}$$

x	y
0	6
1	2
2	0
3	0
4	2

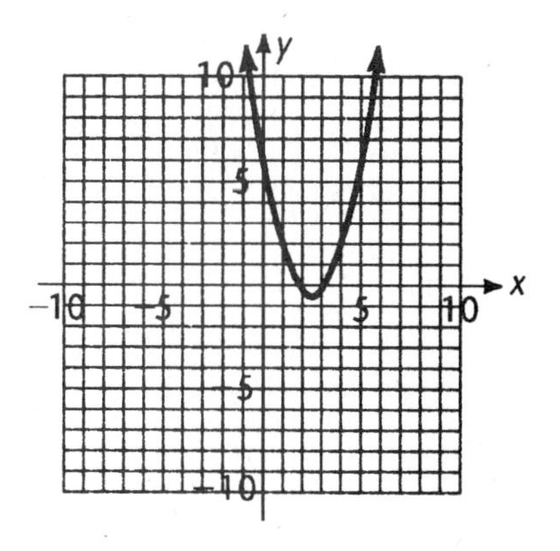

11. $y = x^2 + x - 6$. The parabola opens up, since $a > 0$. The vertex is located at

$$x = -\frac{b}{2a} = -\frac{1}{2(1)} = -\frac{1}{2} \quad \text{and}$$

$$y = \left(-\frac{1}{2}\right)^2 + \left(-\frac{1}{2}\right) - 6 = -\frac{25}{4}$$

x	y
-3	0
-2	-4
-1	-6
0	-6
1	-4
2	0

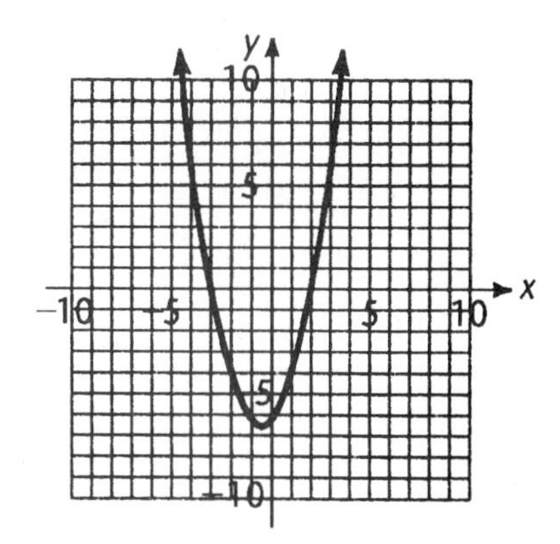

13. $y = x^2 + 4x + 3$. The parabola opens up, since $a > 0$. The vertex is located at

$$x = -\frac{b}{2a} = -\frac{4}{2 \cdot 1} = -2 \quad \text{and}$$

$$y = (-2)^2 + 4(-2) + 3 = -1$$

x	y
-4	3
-3	0
-2	-1
-1	0
0	3

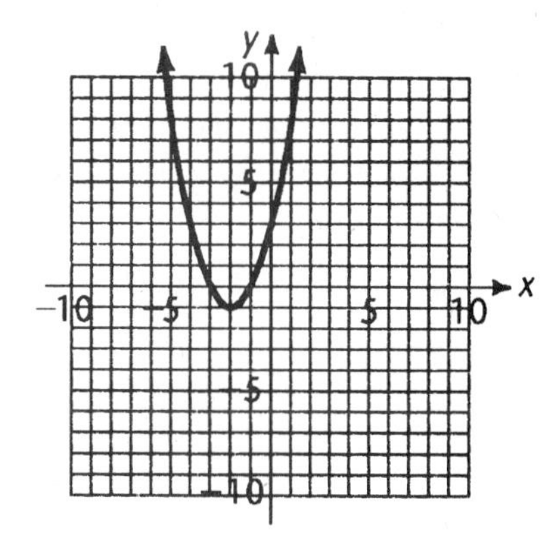

15. $y = x^2 + 6x + 5$. The parabola opens up, since $a > 0$. The vertex is located at

$$x = -\frac{b}{2a} = -\frac{6}{2 \cdot 1} = -3 \quad \text{and}$$

$$y = (-3)^2 + 6(-3) + 5 = -4$$

x	y
-5	0
-4	-3
-3	-4
-2	-3
-1	0

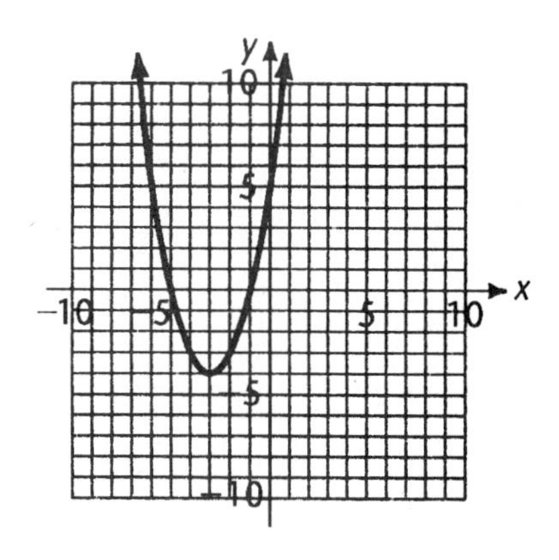

17. $y = x^2 - 4x - 5$. The parabola opens up, since $a > 0$. The vertex is located at

$$x = -\frac{b}{2a} = -\frac{(-4)}{2 \cdot 1} = 2 \text{ and}$$

$$y = 2^2 - 4(2) - 5 = -9$$

x	y
-1	0
1	-8
2	-9
3	-8
5	0

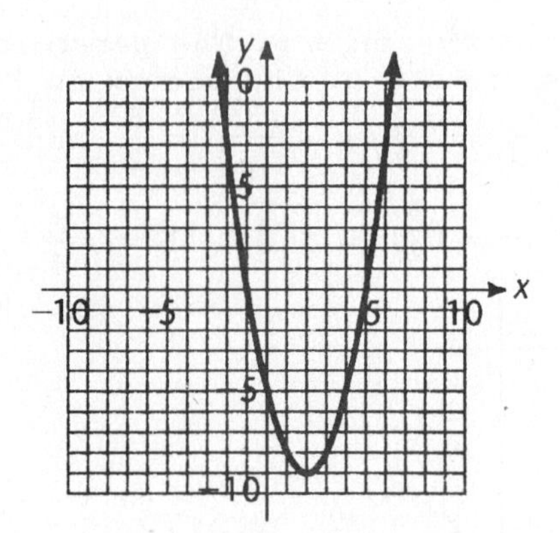

19. $y = -x^2 + 4x - 3$. The parabola opens down, since $a < 0$. The vertex is located at

$$x = -\frac{b}{2a} = -\frac{4}{2(-1)} = 2 \text{ and}$$

$$y = -(2)^2 + 4(2) - 3 = 1$$

x	y
0	-3
1	0
2	1
3	0
4	-3

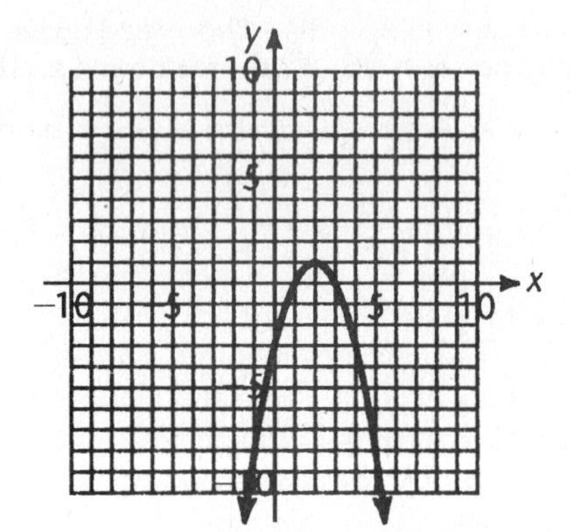

21. $y = -x^2 + x - 6$. The parabola opens down, since $a < 0$. The vertex is located at

$$x = -\frac{b}{2a} = -\frac{1}{2(-1)} = \frac{1}{2} \text{ and}$$

$$y = -\left(\frac{1}{2}\right)^2 + \frac{1}{2} - 6 = -\frac{23}{4}$$

x	y
-1	-8
0	-6
1	-6
2	-8

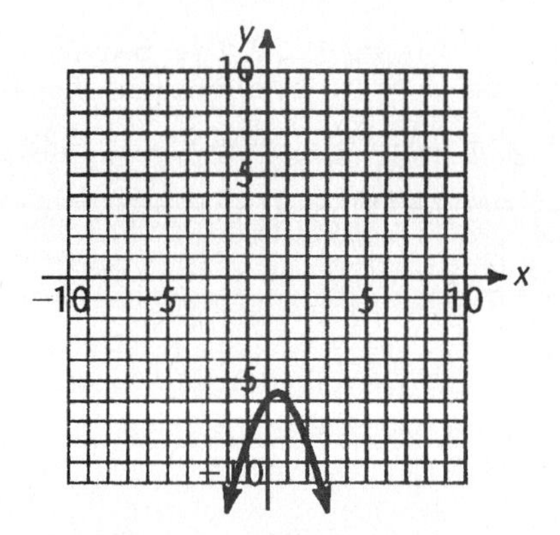

23. $y = -x^2 + 4x + 5$. The parabola opens down, since $a < 0$. The vertex is located at

$$x = -\frac{b}{2a} = -\frac{4}{2(-1)} = 2 \text{ and}$$

$$y = -2^2 + 4(2) + 5 = 9$$

x	y
-1	0
0	5
1	8
2	9
3	8
4	5
5	0

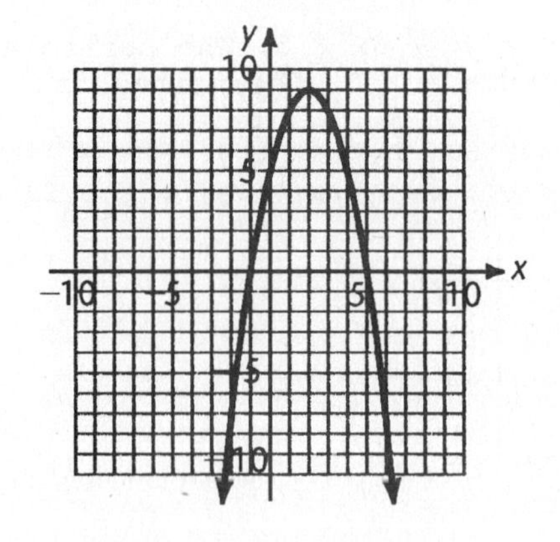

25. $y = 3x^2$. The parabola opens up, since $a > 0$. The vertex is located at

$$x = -\frac{b}{2a} = -\frac{0}{2 \cdot 3} = 0 \quad \text{and}$$

$$y = 3(0)^2 = 0$$

x	y
-2	12
-1	3
0	0
1	3
2	12

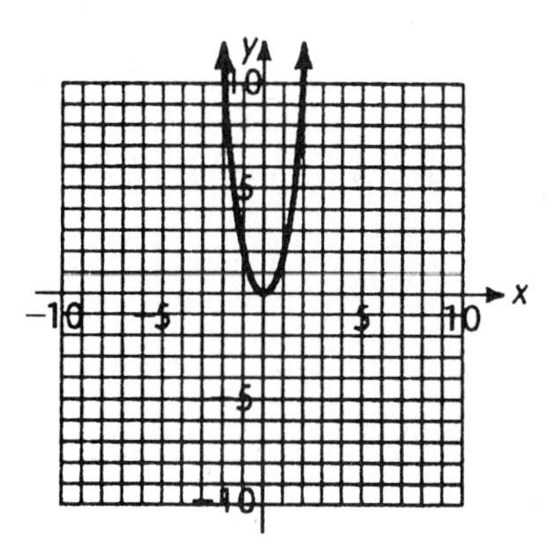

27. $y = -\frac{1}{2}x^2$. The parabola opens down, since $a < 0$. The vertex is located at

$$x = -\frac{b}{2a} = -\frac{0}{2(-\frac{1}{2})} = 0 \text{ and } y = -\frac{1}{2}(0)^2 = 0$$

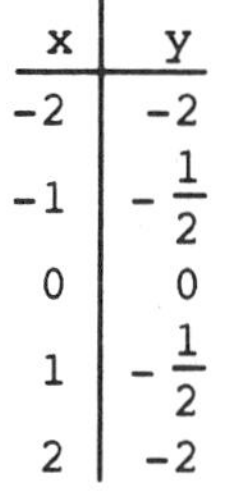

x	y
-2	-2
-1	$-\frac{1}{2}$
0	0
1	$-\frac{1}{2}$
2	-2

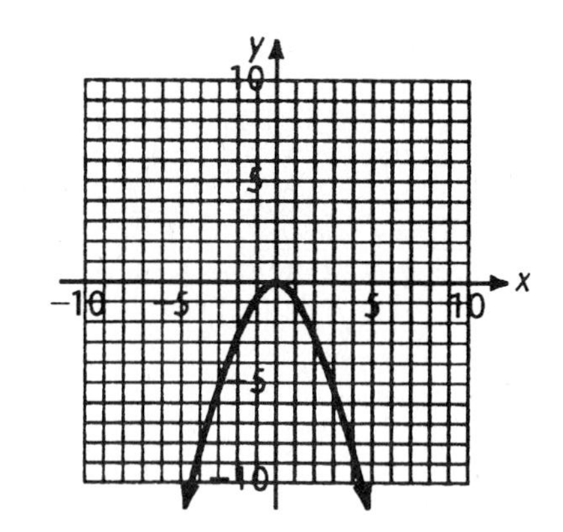

29. $y = 2x^2 + 1$. The parabola opens up, since $a > 0$. The vertex is located at

$$x = -\frac{b}{2a} = -\frac{0}{2 \cdot 2} = 0 \quad \text{and}$$

$$y = 2(0)^2 + 1 = 1$$

x	y
-2	9
-1	5
0	1
1	5
2	9

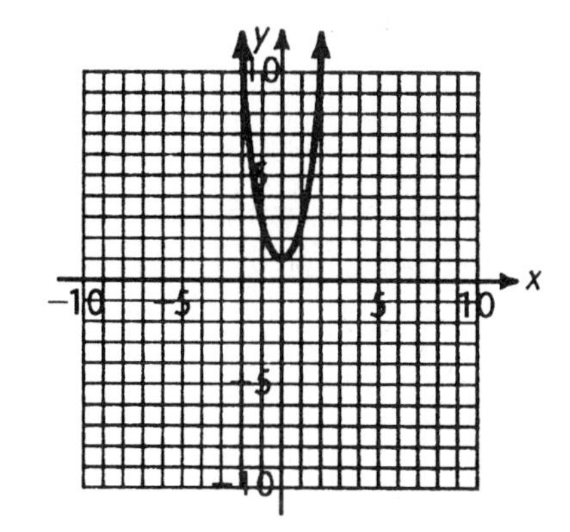

31. $y = -\frac{1}{3}x^2 - 2$. The parabola opens down, since $a < 0$. The vertex is located at

$$x = -\frac{b}{2a} = -\frac{0}{2(-\frac{1}{3})} = 0 \text{ and } y = -\frac{1}{3}(0)^2 - 2 = -2$$

x	y
-2	$-\frac{10}{3}$
-1	$-\frac{7}{3}$
0	-2
1	$-\frac{7}{3}$
2	$-\frac{10}{3}$

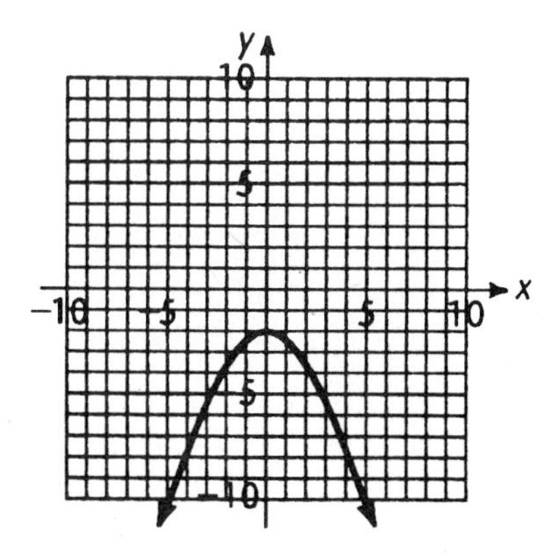

33. $y = 2x^2 + 4x - 1$. The parabola opens up, since $a > 0$. The vertex is located at

$$x = -\frac{b}{2a} = -\frac{4}{2 \cdot 2} = -1 \quad \text{and}$$

$$y = 2(-1)^2 + 4(-1) - 1 = -3$$

x	y
-3	5
-2	-1
-1	-3
0	-1
1	5

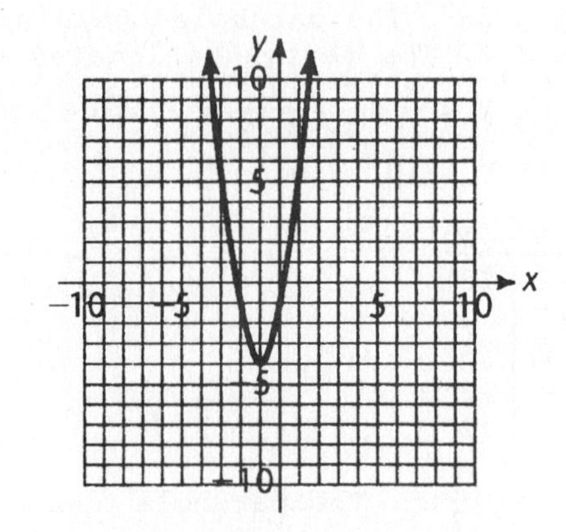

35. $y = 2x^2 - 4x + 1$. The parabola opens up, since $a > 0$. The vertex is located at

$$x = -\frac{b}{2a} = -\frac{(-4)}{2 \cdot 2} = 1 \quad \text{and}$$

$$y = 2(1)^2 - 4 \cdot 1 + 1 = -1$$

x	y
-1	7
0	1
1	-1
2	1
3	7

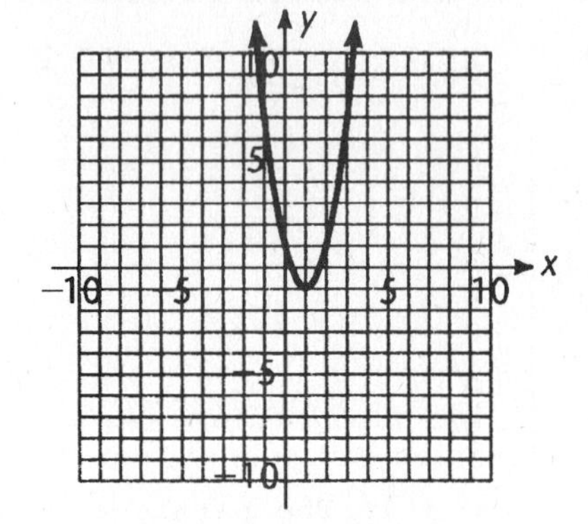

37. $y = 2x^2 - 8x + 6$. The parabola opens up, since $a > 0$. The vertex is located at

$$x = -\frac{b}{2a} = -\frac{(-8)}{2 \cdot 2} = 2 \quad \text{and}$$

$$y = 2(2)^2 - 8 \cdot 2 + 6 = -2$$

x	y
0	6
1	0
2	-2
3	0
4	6

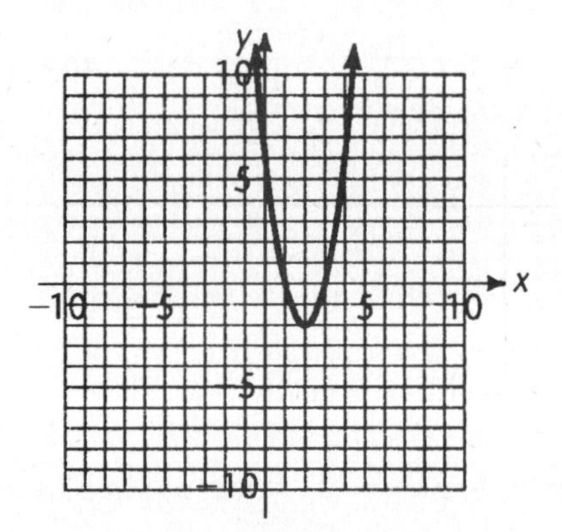

39. $y = 2x^2 - 8x + 10$. The parabola opens up, since $a > 0$. The vertex is located at

$$x = -\frac{b}{2a} = -\frac{(-8)}{2 \cdot 2} = 2 \quad \text{and}$$

$$y = 2(2)^2 - 8 \cdot 2 + 10 = 2$$

x	y
0	10
1	4
2	2
3	4
4	10

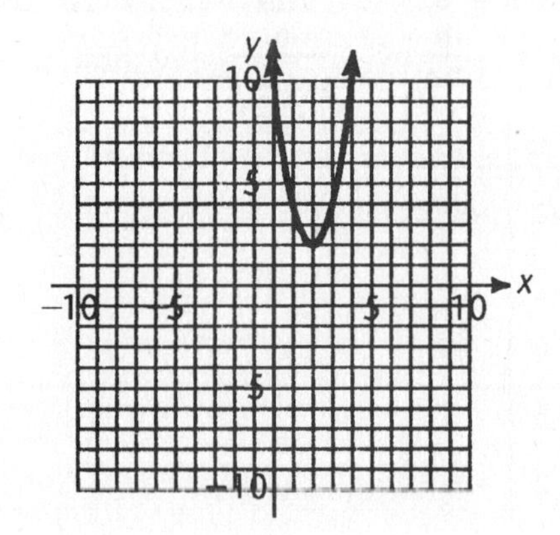

41. $y = 2x^2 - 12x + 14$. The parabola opens up, since $a > 0$. The vertex is located at

$$x = -\frac{b}{2a} = -\frac{(-12)}{2(2)} = 3 \quad \text{and}$$

$$y = 2(3)^2 - 12 \cdot 3 + 14 = -4$$

x	y
1	4
2	-2
3	-4
4	-2
5	4

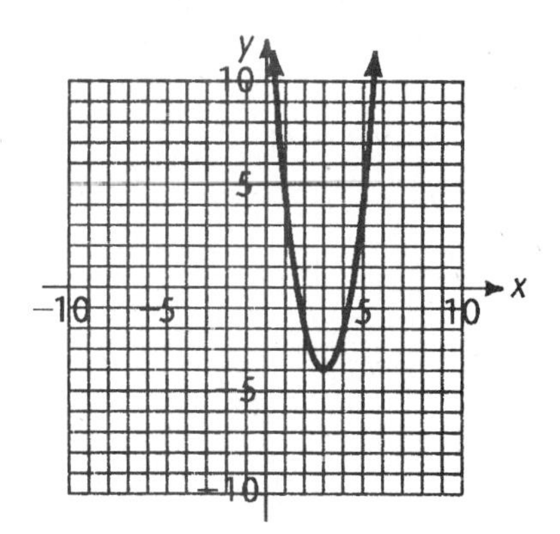

43. $y = 2x^2 + 12x + 22$. The parabola opens up, since $a > 0$. The vertex is located at

$$x = -\frac{b}{2a} = -\frac{12}{2 \cdot 2} = -3 \quad \text{and}$$

$$y = 2(-3)^2 + 12(-3) + 22 = 4$$

x	y
-5	12
-4	6
-3	4
-2	6
-1	12

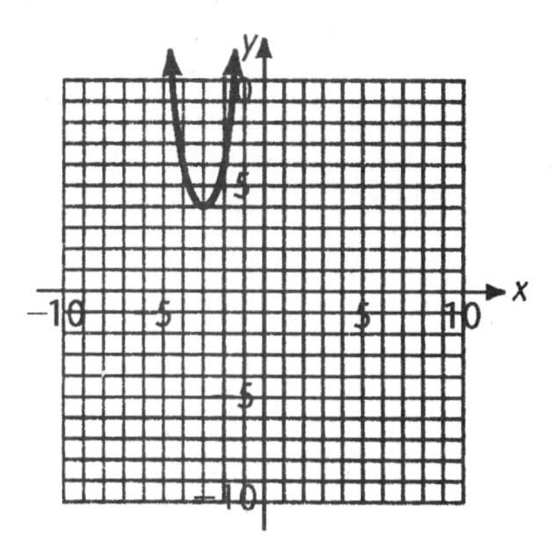

45. $y = 3x^2 - 6x + 6$. The parabola opens up, since $a > 0$. The vertex is located at

$$x = -\frac{b}{2a} = -\frac{(-6)}{2 \cdot 3} = 1 \quad \text{and}$$

$$y = 3(1)^2 - 6(1) + 6 = 3$$

x	y
-1	15
0	6
1	3
2	6
3	15

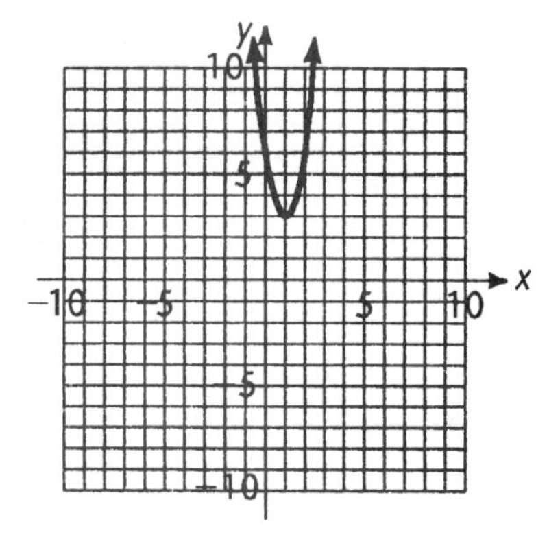

47. $y = 3x^2 - 6x$. The parabola opens up, since $a > 0$. The vertex is located at

$$x = -\frac{b}{2a} = -\frac{(-6)}{2 \cdot 3} = 1 \quad \text{and}$$

$$y = 3(1)^2 - 6 \cdot 1 = -3$$

x	y
-1	9
0	0
1	-3
2	0
3	9

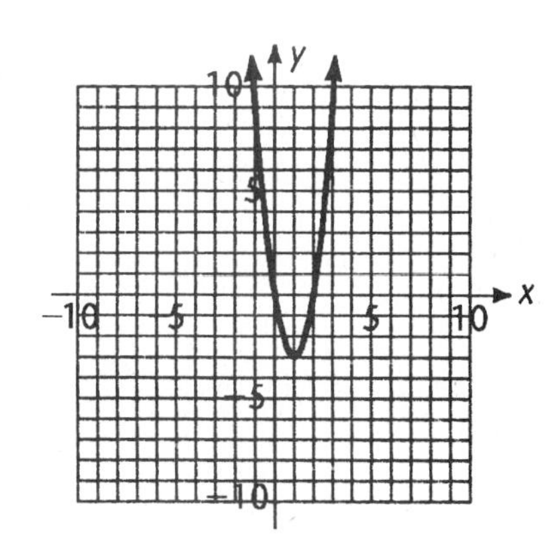

49. $y = 4x^2 - 8x + 19$. The parabola opens up, since $a > 0$. The vertex is located at

$$x = -\frac{b}{2a} = -\frac{(-8)}{2 \cdot 4} = 1 \quad \text{and}$$
$$y = 4(1)^2 - 8(1) + 19 = 15$$

x	y
0	19
1	15
2	19

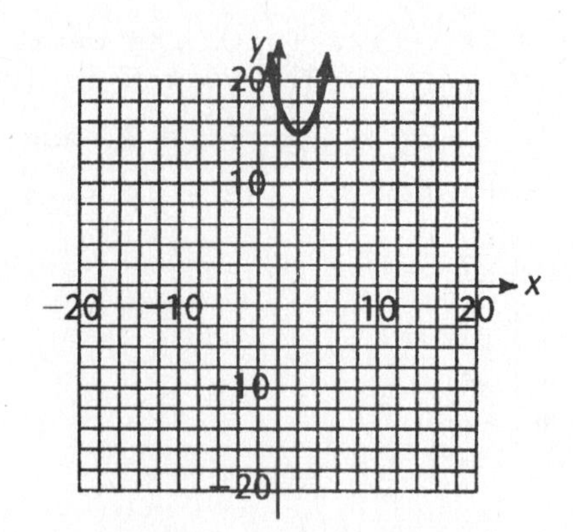

51. $y = 4x^2 - 8x + 13$. The parabola opens up, since $a > 0$. The vertex is located at

$$x = -\frac{b}{2a} = -\frac{(-8)}{2 \cdot 4} = 1 \quad \text{and}$$
$$y = 4(1)^2 - 8(1) + 13 = 9$$

x	y
0	13
1	9
2	13

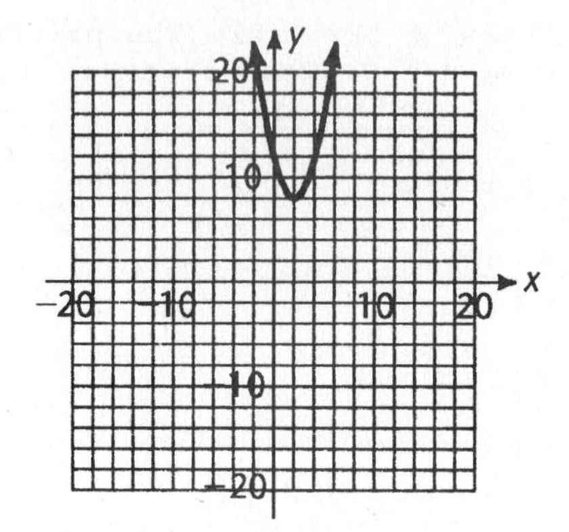

53. $y = 5x^2 - 40x + 81$. The parabola opens up, since $a > 0$. The vertex is located at

$$x = -\frac{b}{2a} = -\frac{(-40)}{2 \cdot 5} = 4 \quad \text{and}$$
$$y = 5(4)^2 - 40(4) + 81 = 1$$

x	y
2	21
3	6
4	1
5	6
6	21

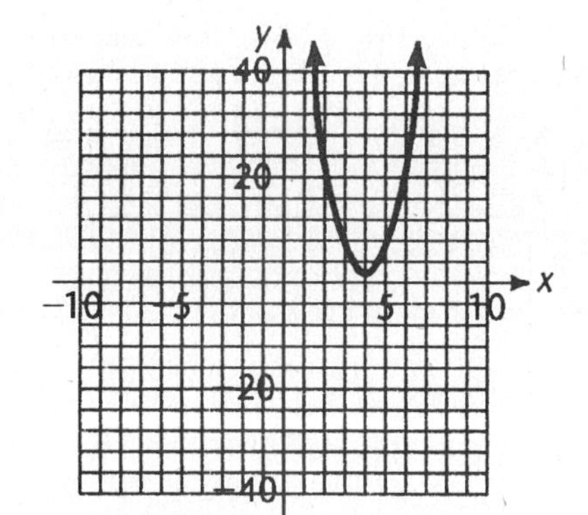

55. $y = x^2 - 2x - 3$. The parabola opens up, since $a > 0$. The vertex is located at

$$x = -\frac{b}{2a} = -\frac{(-2)}{2 \cdot 1} = 1 \quad \text{and}$$
$$y = (1)^2 - 2(1) - 3 = -4$$

x	y
-2	5
-1	0
0	-3
1	-4
2	-3
3	0

To find the x intercepts, we solve

$$x^2 - 2x - 3 = 0$$
$$(x - 3)(x + 1) = 0$$
$$x = 3 \quad \text{or} \quad x = -1$$

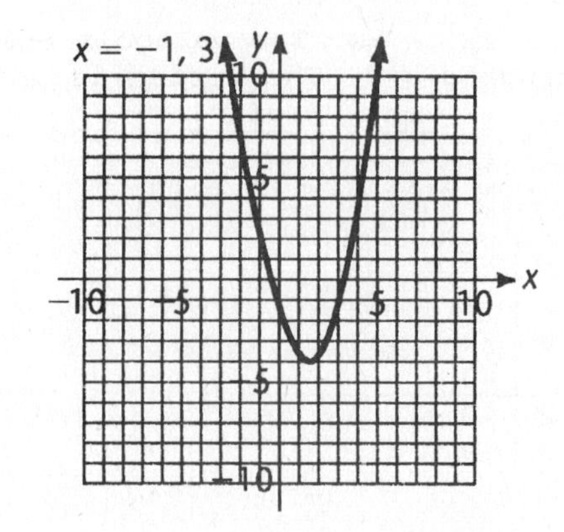

57. $y = -x^2 + x + 2$. The parabola opens down, since $a < 0$. The vertex is located at

$$x = -\frac{b}{2a} = -\frac{1}{2(-1)} = \frac{1}{2} \text{ and } y = -\left(\frac{1}{2}\right)^2 + \frac{1}{2} + 2 = \frac{9}{4}$$

x	y
-2	4
-1	0
0	-2
1	2
2	0
3	-4

To find the x intercepts, we solve

$$-x^2 + x + 2 = 0$$
$$x^2 - x - 2 = 0$$
$$(x - 2)(x + 1) = 0$$
$$x = 2 \quad \text{or} \quad x = -1$$

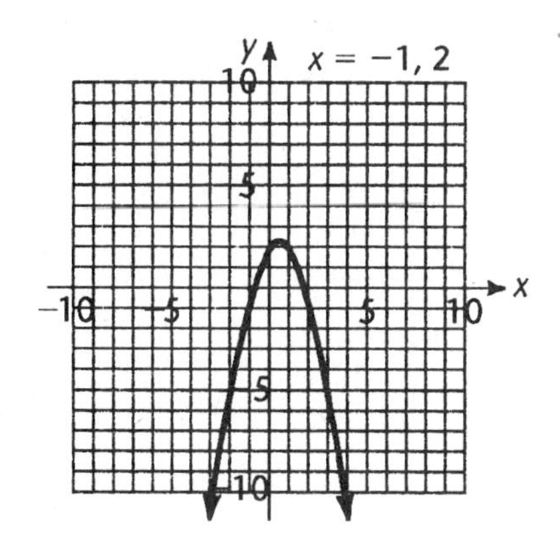

59. $y = x^2 + x - 6$. The parabola opens up, since $a > 0$. The vertex is located at

$$x = -\frac{b}{2a} = -\frac{1}{2(1)} = -\frac{1}{2} \text{ and}$$

$$y = \left(-\frac{1}{2}\right)^2 + \left(-\frac{1}{2}\right) - 6 = -\frac{25}{4}$$

x	y
-3	0
-2	-4
-1	-6
0	-6
1	-4
2	0

To find the x intercepts, we solve

$$x^2 + x - 6 = 0$$
$$(x + 3)(x - 2) = 0$$
$$x = -3 \quad \text{or} \quad x = 2$$

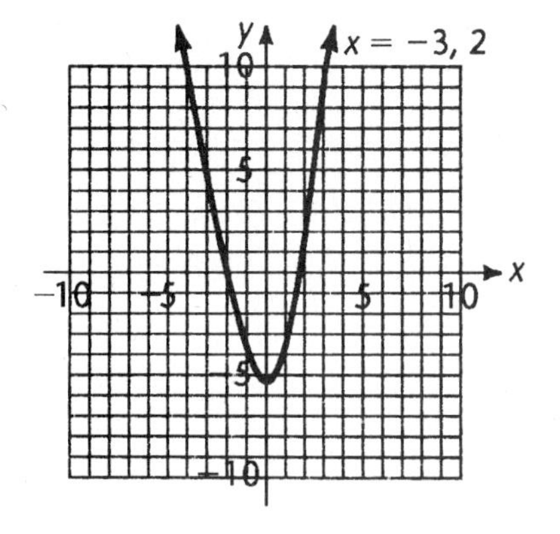

61. $y = x^2 + 4x + 4$. The parabola opens up, since $a > 0$. The vertex is located at

$$x = -\frac{b}{2a} = -\frac{4}{2(1)} = -2 \text{ and}$$

$$y = (-2)^2 + 4(-2) + 4 = 0$$

x	y
-4	4
-3	1
-2	0
-1	1
0	4

To find the x intercepts, we solve

$$x^2 + 4x + 4 = 0$$
$$(x + 2)^2 = 0$$
$$x = -2$$

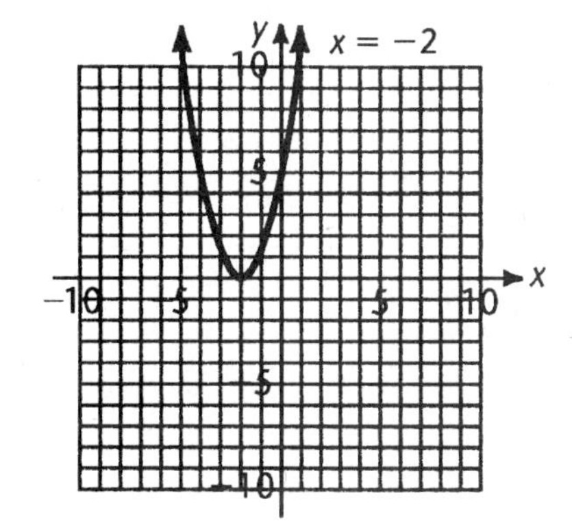

63. $y = x^2 - 3x + 3$. The parabola opens up, since $a > 0$. The vertex is located at

$$x = -\frac{b}{2a} = -\frac{(-3)}{2\cdot 1} = \frac{3}{2} \text{ and } y = \left(\frac{3}{2}\right)^2 - 3\left(\frac{3}{2}\right) + 3 = \frac{3}{4}$$

x	y
-1	7
0	3
1	1
2	1
3	3

To find the x intercepts, we would solve

$$x^2 - 3x + 3 = 0$$

However, since $b^2 - 4ac = (-3)^2 - 4(1)(3) = -3 < 0$, there are no real solutions, hence, no x intercepts.

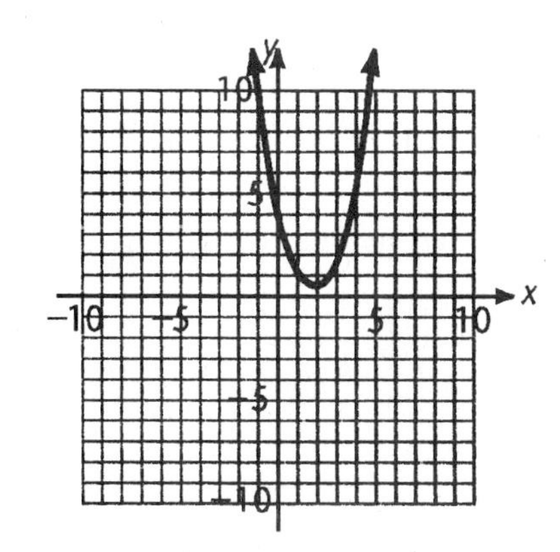

65. $y = -x^2 + 2x - 8$. The parabola opens down, since $a < 0$. The vertex is located at

$$x = -\frac{b}{2a} = -\frac{2}{2(-1)} = 1 \quad \text{and}$$
$$y = -(1)^2 + 2\cdot 1 - 8 = -7$$

x	y
0	-8
1	-7
2	-8
3	-11

To find the x intercepts, we would solve

$$-x^2 + 2x - 8 = 0$$

However, since $b^2 - 4ac = (2)^2 - 4(-1)(-8) = -28 < 0$, there are no real solutions, hence, no x intercepts.

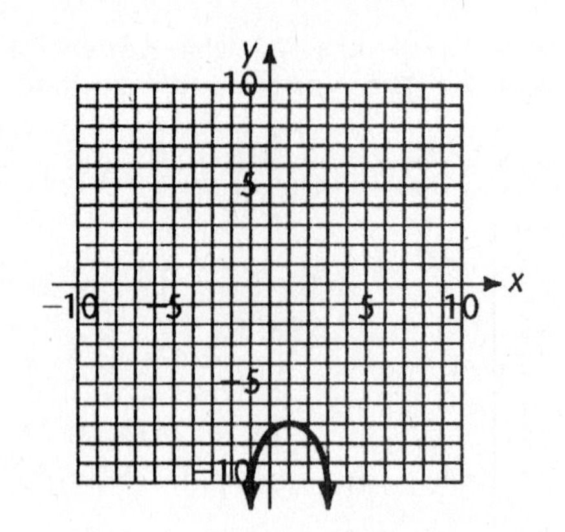

67. $y = -x^2 + 5x - 6$. The parabola opens down, since $a < 0$. The vertex is located at

$$x = -\frac{b}{2a} = -\frac{5}{2(-1)} = \frac{5}{2} \text{ and } y = -\left(\frac{5}{2}\right)^2 + 5\left(\frac{5}{2}\right) - 6 = \frac{1}{4}$$

x	y
0	-6
1	-2
2	0
3	0
4	-2
5	-6

To find the x intercepts, we solve

$$-x^2 + 5x - 6 = 0$$
$$x^2 - 5x + 6 = 0$$
$$(x - 2)(x - 3) = 0$$
$$x = 2 \quad \text{or} \quad x = 3$$

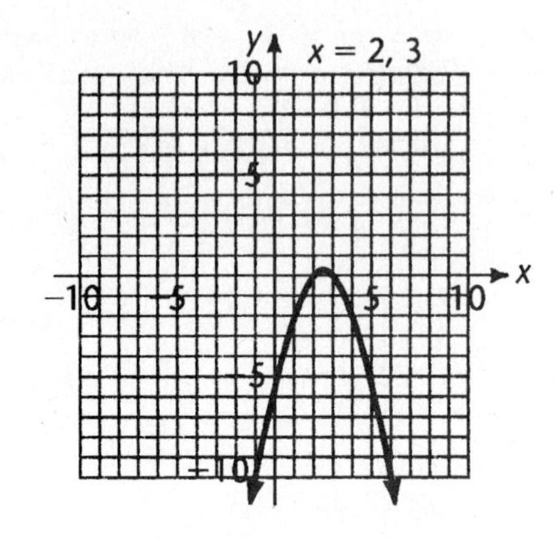

69. $y = 2x^2 - 7x + 5$. The parabola opens up, since $a > 0$. The vertex is located at

$$x = -\frac{b}{2a} = -\frac{(-7)}{2\cdot 2} = \frac{7}{4} \text{ and } y = 2\left(-\frac{7}{4}\right)^2 - 7\left(-\frac{7}{4}\right) + 5 = -\frac{9}{8}$$

x	y
0	5
1	0
2	-1
3	2
4	9

To find the x intercepts, we solve

$$2x^2 - 7x + 5 = 0$$
$$(2x - 5)(x - 1) = 0$$
$$2x - 5 = 0 \quad \text{or} \quad x - 1 = 0$$
$$x = \frac{5}{2} \qquad x = 1$$

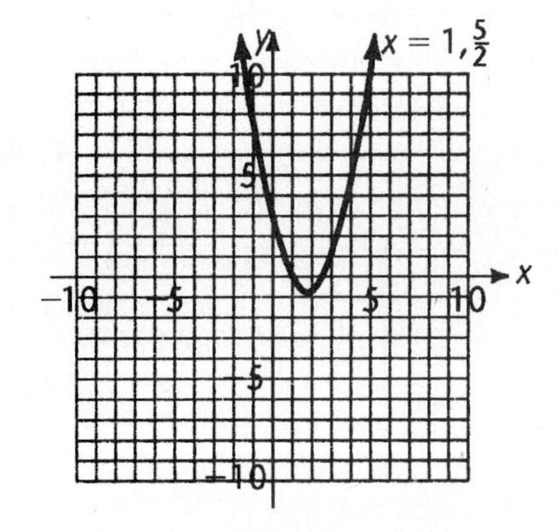

71. $y = -2x^2 - 3x + 2$. The parabola opens down, since $a < 0$. The vertex is located at

$$x = -\frac{b}{2a} = -\frac{(-3)}{2(-2)} = -\frac{3}{4} \text{ and } y = -2\left(-\frac{3}{4}\right)^2 - 3\left(-\frac{3}{4}\right) + 2 = \frac{25}{8}$$

x	y
-2	0
-1	3
0	2
1	-3

To find the x intercepts, we solve

$$-2x^2 - 3x + 2 = 0$$
$$2x^2 + 3x - 2 = 0$$
$$(2x - 1)(x + 2) = 0$$
$$2x - 1 = 0 \quad \text{or} \quad x + 2 = 0$$
$$x = \frac{1}{2} \qquad x = -2$$

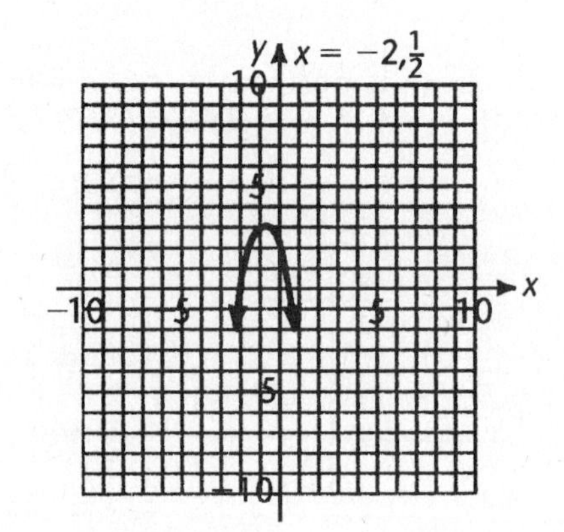

73. Since $b^2 - 4ac = (41)^2 - 4(8)(52) = 17$ is positive, there are 2 x intercepts.

75. Since $b^2 - 4ac = (-60)^2 - 4(10)(91) = -40$ is negative, there are no x intercepts.

77. Since $b^2 - 4ac = (-24)^2 - 4(23)(-25) = 2876$ is positive, there are 2 x intercepts.

79. Since $b^2 - 4ac = (52)^2 - 4(-53)(51) = 13{,}516$ is positive, there are 2 x intercepts.

Exercise 7-6 Completing the Square and Graphing

Key Ideas and Formulas

Graphs of $y = x^2$ and Related Forms

$y = x^2$

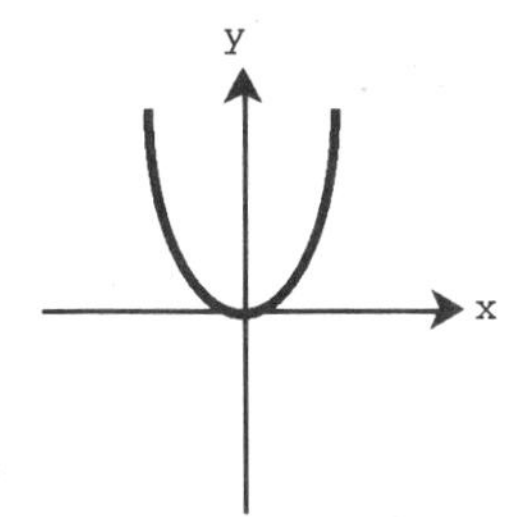

$y = -x^2$: The graph of $y = x^2$ reflected about the x-axis

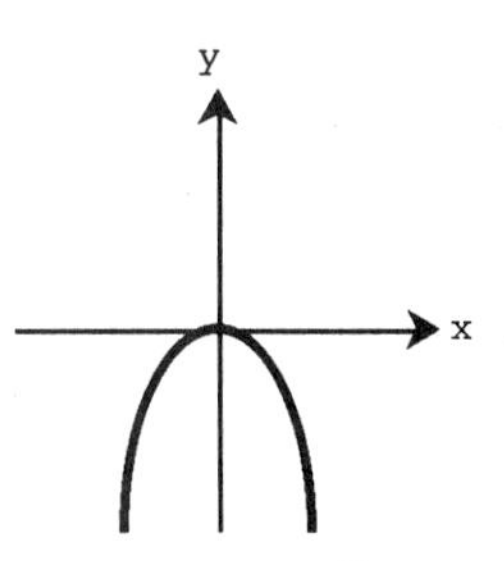

$y = ax^2$, $a \neq 0$: The graph of $y = x^2$ stretched ($a > 1$) or flattened ($a < 1$)

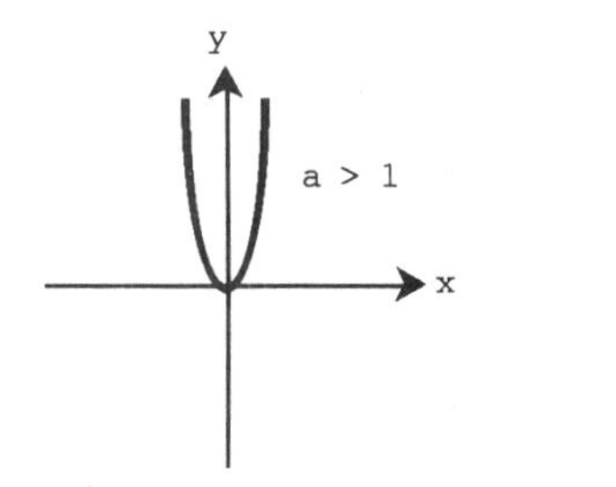

$y = x^2 + k$: The graph of $y = x^2$ raised k units

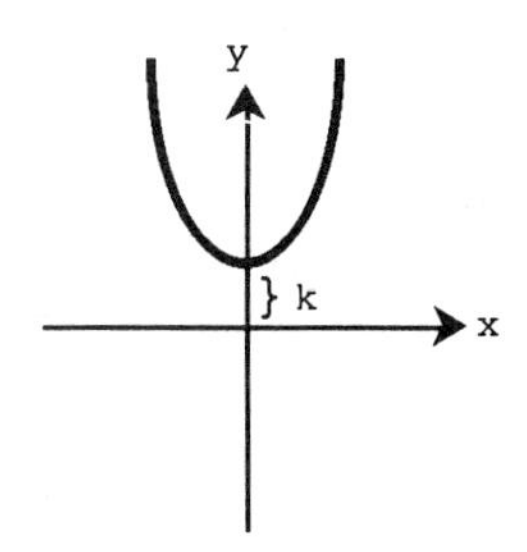

$y = (x - h)^2$: The graph of $y = x^2$ shifted h units to the right.

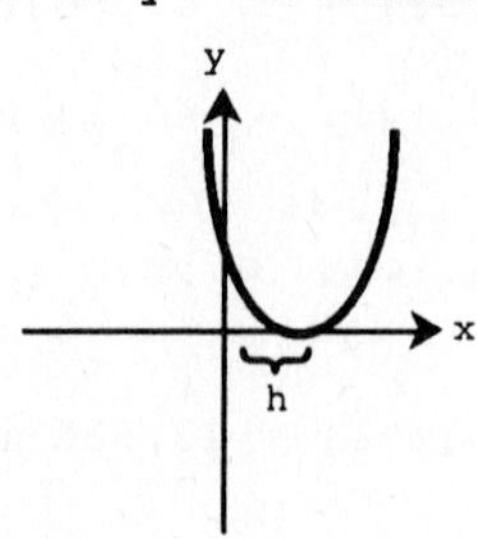

To graph $y = a(x - h)^2 + k$, in sequence:

1. Start with the graph of $y = x^2$
2. Stretch or flatten the graph using a; if a is negative also turn the graph upside down.
3. Shift the graph right h units. If h is negative, say -2, this means move left $|h|$ units.
4. Raise the graph k units. If k is negative, say -2, this means lower it $|k|$ units.

The graph of $y = ax^2 + bx + c$ is the graph of $y = x^2$ changed in the ways:

1. Stretched or flattened using a; turned over the x-axis if $a < 0$.
2. Shifted right or left using h, and up or down using k. The vertex, (0, 0) of $y = x^2$ is shifted to the point $(h, k) = \left(-\frac{b}{2a}, -\frac{b^2 - 4ac}{4a}\right)$

---

1. $y = 4x^2$

The graph is that of $y = x^2$ stretched ($a = 4$).

x	y
-1	4
0	0
1	4

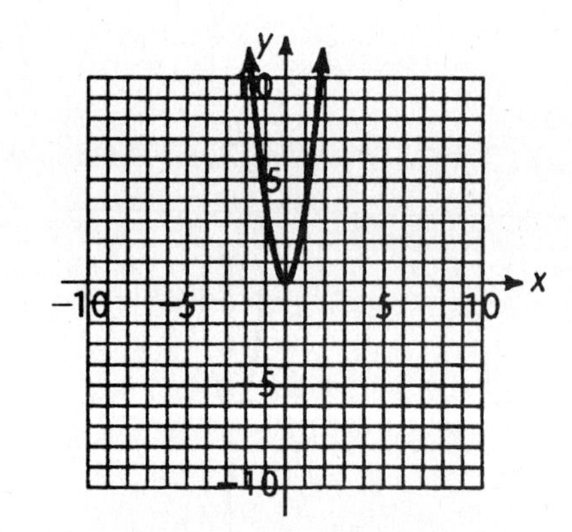

3. $y = \frac{1}{2}x^2$

The graph is that of $y = x^2$ flattened $\left(a = \frac{1}{2}\right)$.

x	y
-4	8
0	0
4	8

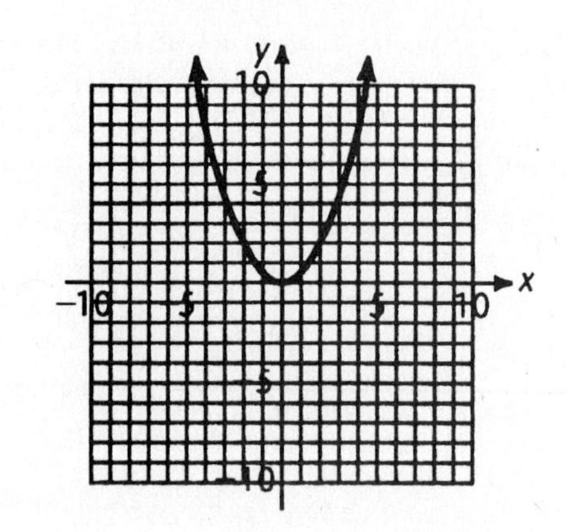

5. $y = -5x^2$

The graph is that of $y = x^2$ stretched $(a = -5)$ and turned over the x axis $(a < 0)$.

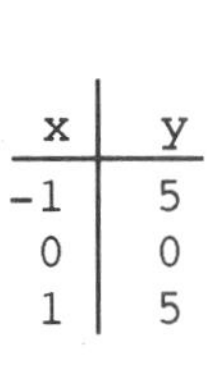

x	y
-1	5
0	0
1	5

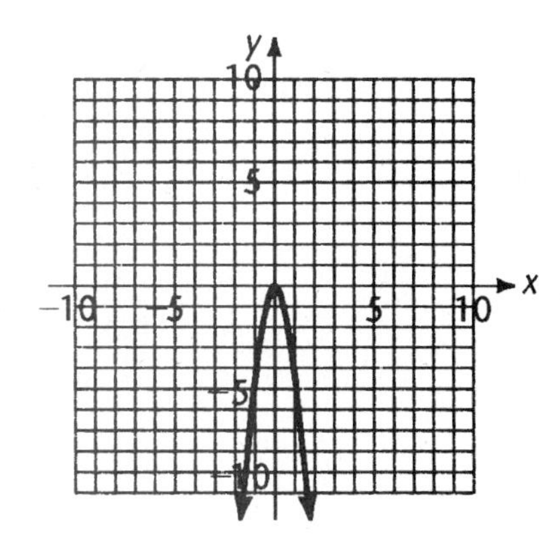

7. $y = -\frac{1}{3}x^2$

The graph is that of $y = x^2$ flattened $\left(a = -\frac{1}{3}\right)$ and turned over the x axis $(a < 0)$.

x	y
-3	-3
0	0
3	-3

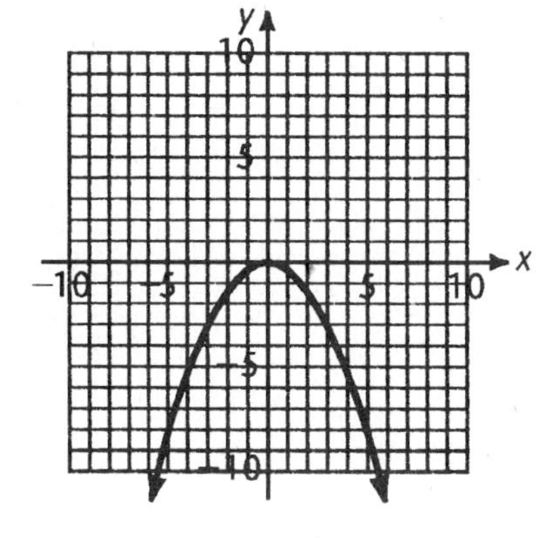

9. $y = x^2 - 4$

The graph is that of $y = x^2$ lowered 4 units $(k = -4)$.

x	y
3	5
0	-4
3	5

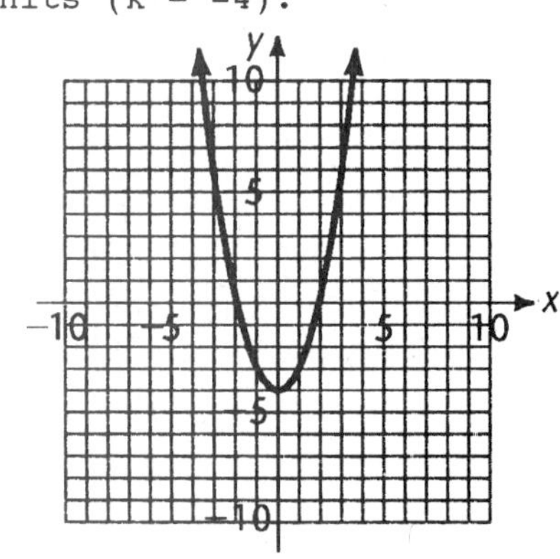

11. $y = x^2 + 6$

The graph is that of $y = x^2$ raised 6 units $(k = 6)$.

x	y
-1	7
0	6
1	7

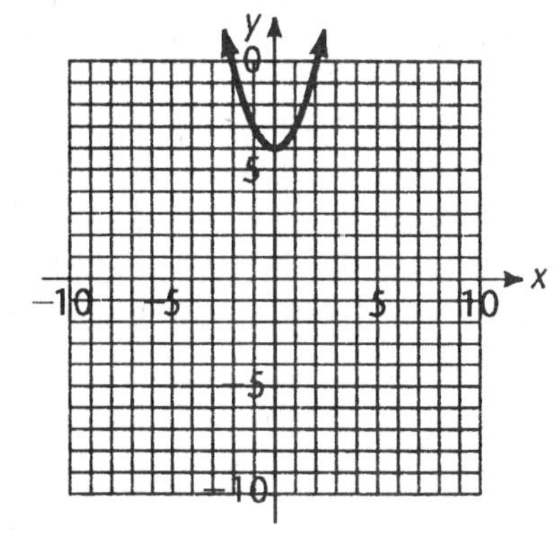

13. $y = x^2 + 5$

The graph is that of $y = x^2$ raised 5 units $(k = 5)$.

x	y
-2	9
0	5
2	9

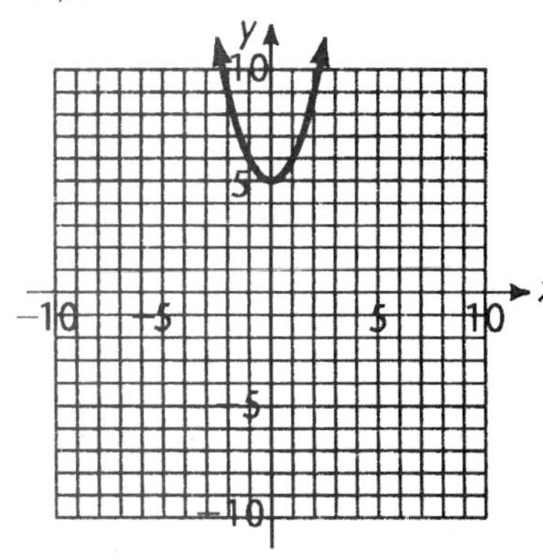

15. $y = x^2 - 7$

The graph is that of $y = x^2$ lowered 7 units $(k = -7)$.

x	y
-4	9
0	-7
4	9

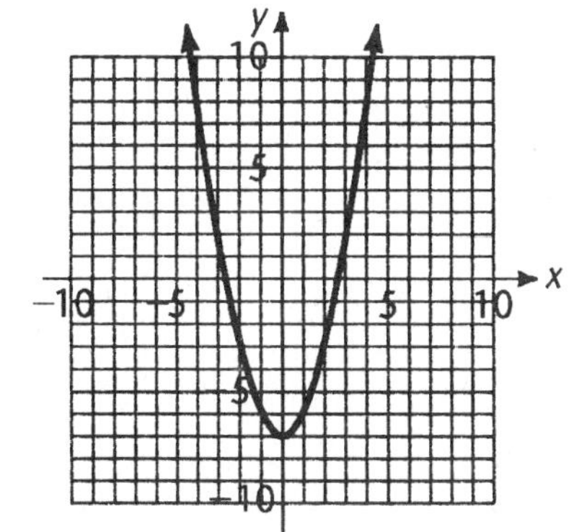

17. $y = (x - 3)^2$

The graph is that of $y = x^2$ shifted right 3 units ($h = 3$).

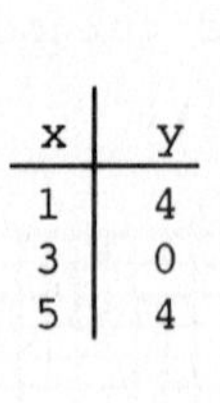

x	y
1	4
3	0
5	4

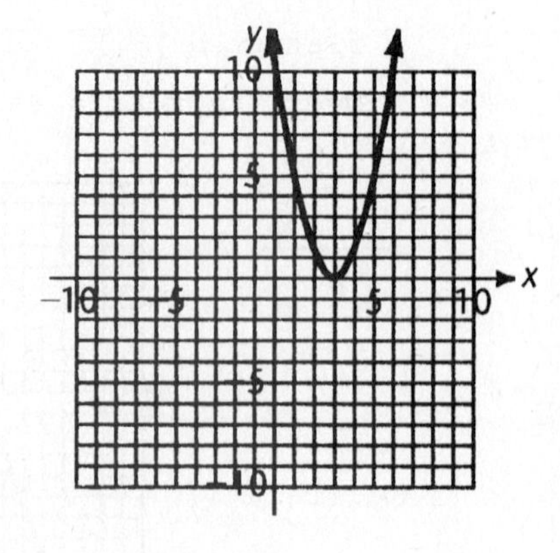

19. $y = (x + 2)^2$

The graph is that of $y = x^2$ shifted left 2 units. ($y = [x - (-2)]^2$, hence $h = -2$.)

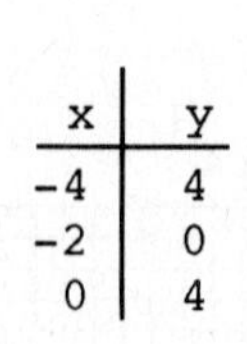

x	y
-4	4
-2	0
0	4

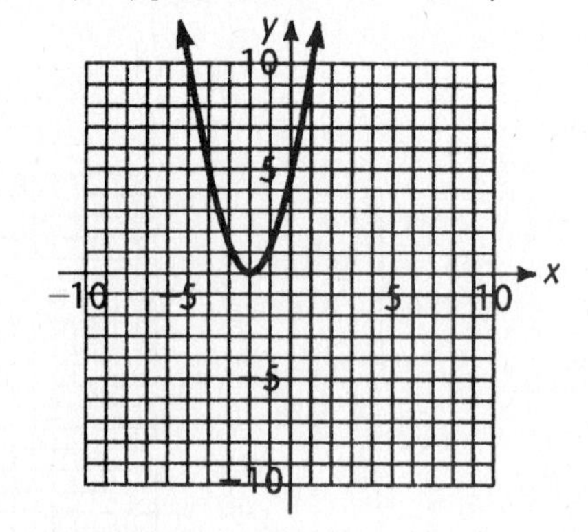

21. $y = (x - 4)^2$

The graph is that of $y = x^2$ shifted right 4 units ($h = 4$).

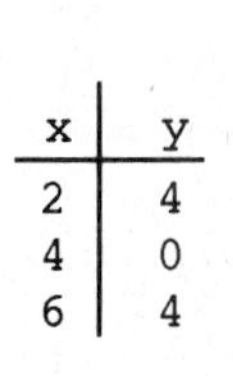

x	y
2	4
4	0
6	4

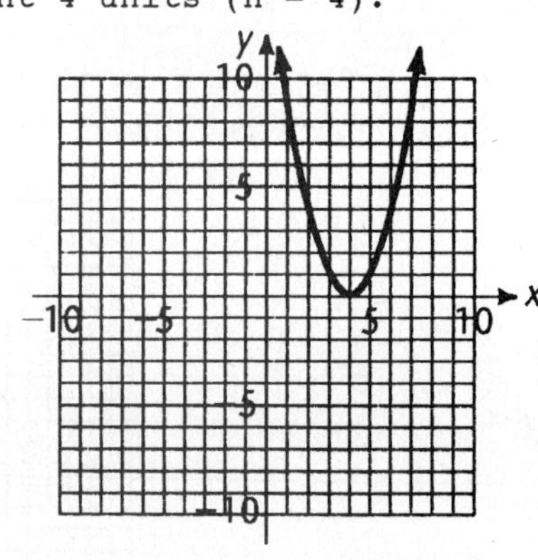

23. $y = (x + 5)^2$

The graph is that of $y = x^2$ shifted left 5 units ($h = -5$).

x	y
-7	4
-5	0
-3	4

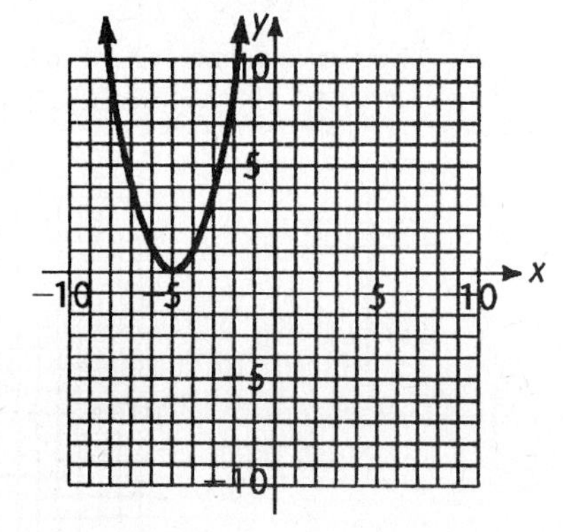

25. $y = (x - 2)^2 + 1$

The graph is that of $y = x^2$ shifted right 2 units ($h = 2$) and raised 1 unit ($k = 1$).

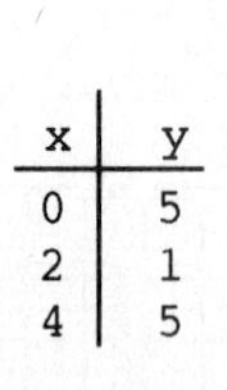

x	y
0	5
2	1
4	5

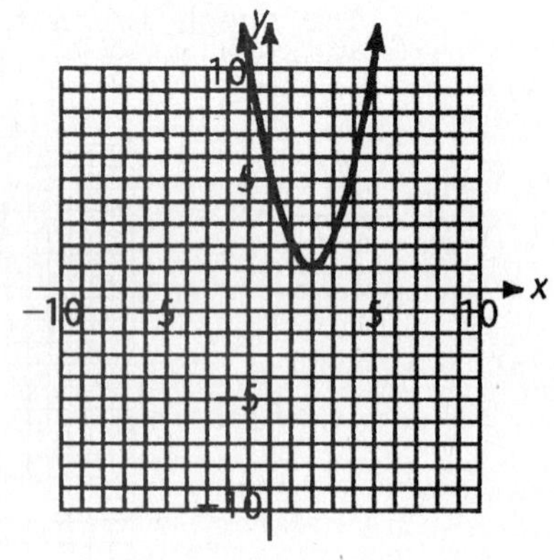

27. $y = (x + 2)^2 - 4$

The graph is that of $y = x^2$ shifted left 2 units ($h = -2$) and lowered 4 units ($k = -4$).

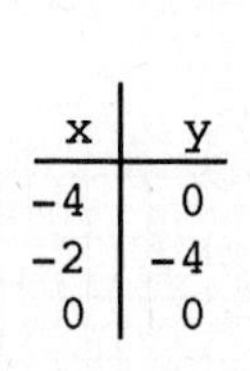

x	y
-4	0
-2	-4
0	0

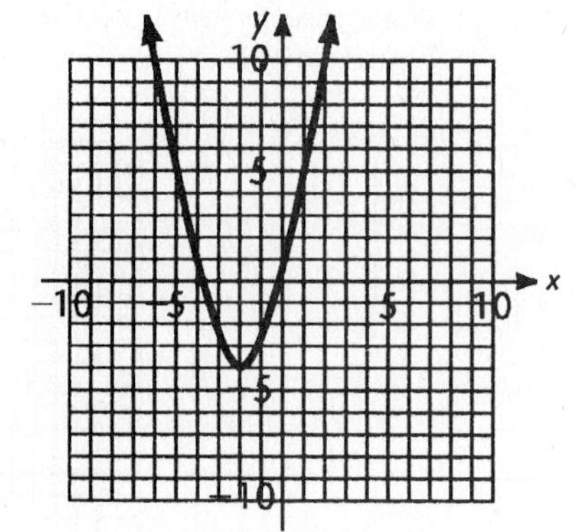

29. $y = (x + 1)^2 - 4$

The graph is that of $y = x^2$ shifted left 1 unit ($h = -1$) and lowered 4 units ($k = -4$).

x	y
-3	0
-1	-4
1	0

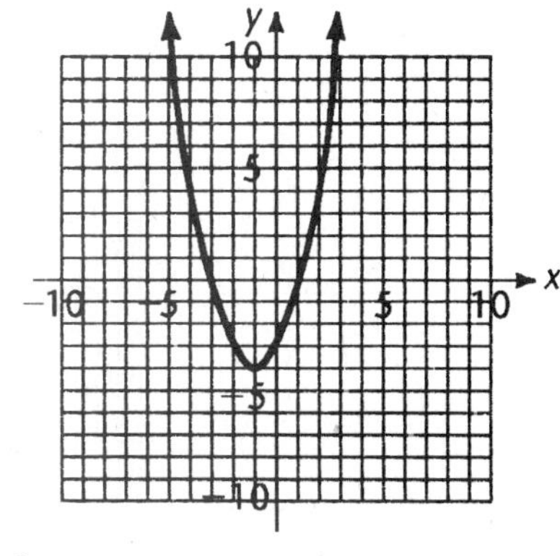

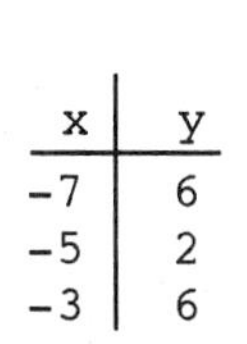

31. $y = (x + 5)^2 + 2$

The graph is that of $y = x^2$ shifted left 5 units ($h = -5$) and raised 2 units ($k = 2$).

x	y
-7	6
-5	2
-3	6

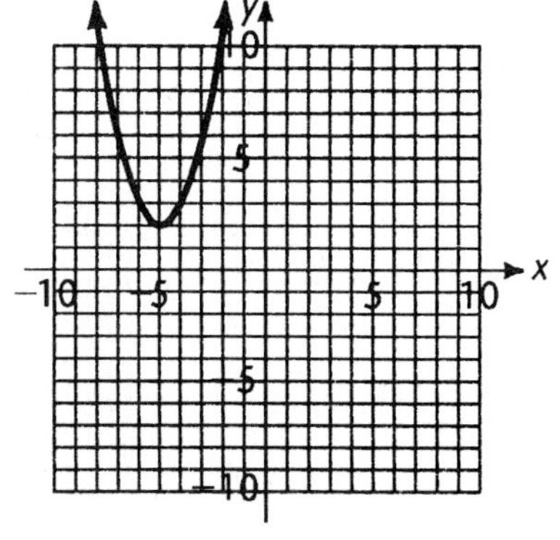

33. $y = -(x + 3)^2 - 5$

The graph is that of $y = x^2$ turned over the x axis ($a = -1$), shifted left 3 units ($h = -3$), and lowered 5 units ($k = -5$).

x	y
-5	-9
-3	-5
-1	-9

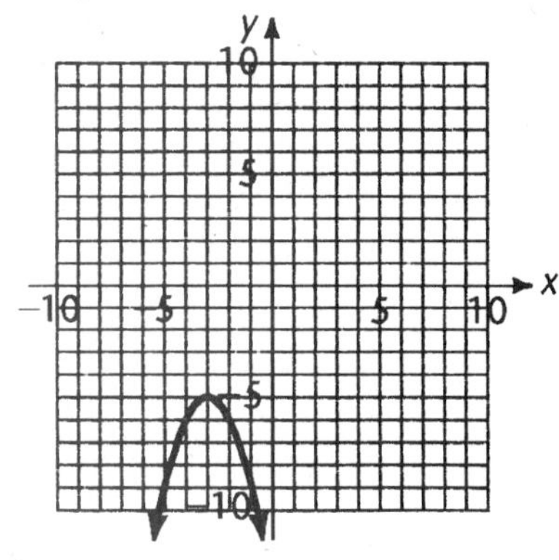

35. $y = -(x - 2)^2 - 3$

The graph is that of $y = x^2$ turned over the x axis ($a = -1$), shifted right 2 units ($h = 2$), and lowered 3 units ($k = -3$).

x	y
0	-7
2	-3
4	-7

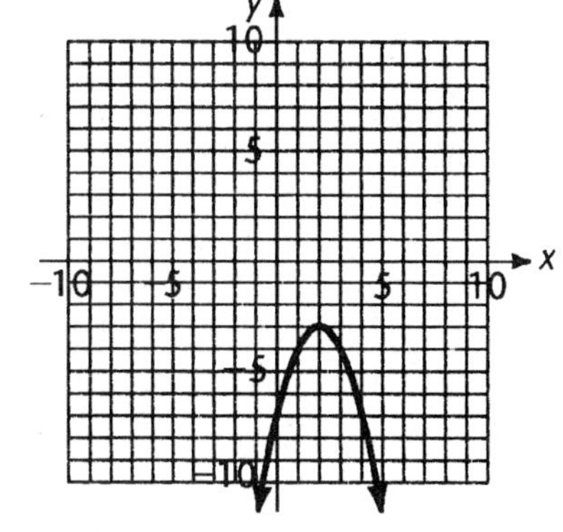

37. $y = -(x + 4)^2 + 2$

The graph is that of $y = x^2$ turned over the x axis ($a = -1$), shifted left 4 units ($h = -4$), and raised 2 units ($k = 2$).

x	y
-6	6
-4	2
-2	6

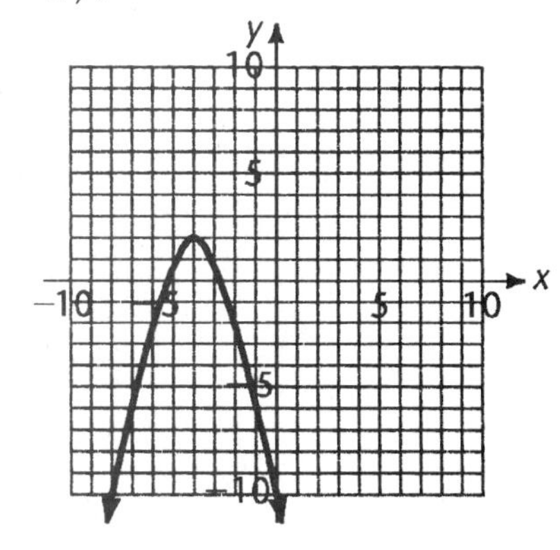

39. $y = -(x - 1)^2 - 1$

The graph is that of $y = x^2$ turned over the x axis ($a = -1$), shifted right 1 unit ($h = 1$), and lowered 1 unit ($k = -1$).

x	y
-1	-5
1	-1
3	-5

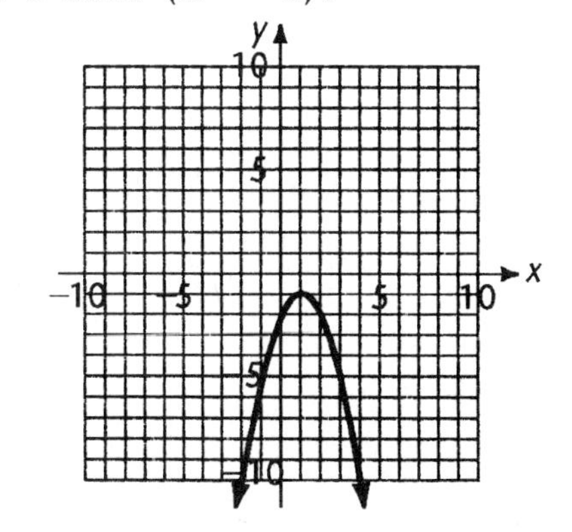

41. $y = 2(x - 1)^2 + 1$

The graph is that of $y = x^2$ stretched $(a = 2)$, shifted right 1 unit $(h = 1)$, and raised 1 unit $(k = 1)$.

x	y
-1	9
1	1
3	9

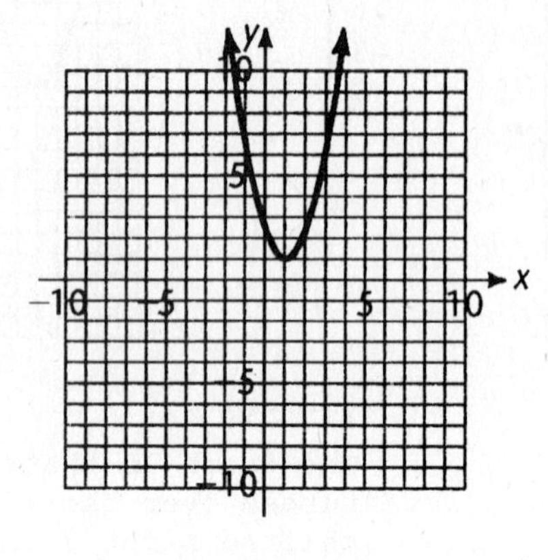

43. $y = \frac{1}{2}(x + 1)^2 + 3$

The graph is that of $y = x^2$ flattened $\left(a = \frac{1}{2}\right)$, shifted left 1 unit $(h = -1)$, and raised 3 units $(k = 3)$.

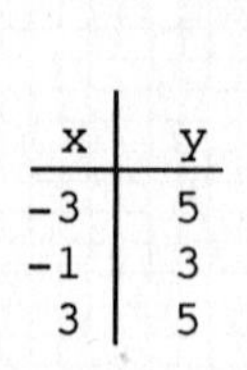

x	y
-3	5
-1	3
3	5

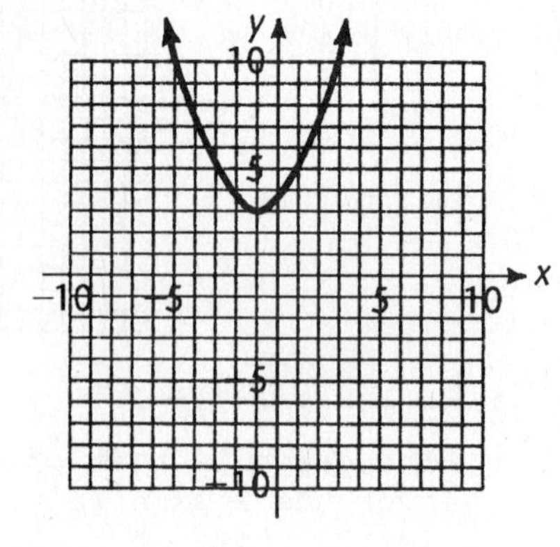

45. $y = -2(x + 2)^2 - 2$

The graph is that of $y = x^2$ stretched $(a = -2)$, turned over the x axis $(a < 0)$, shifted left 2 units $(h = -2)$, and lowered 2 units $(k = -2)$.

x	y
-4	-10
-2	-2
0	-10

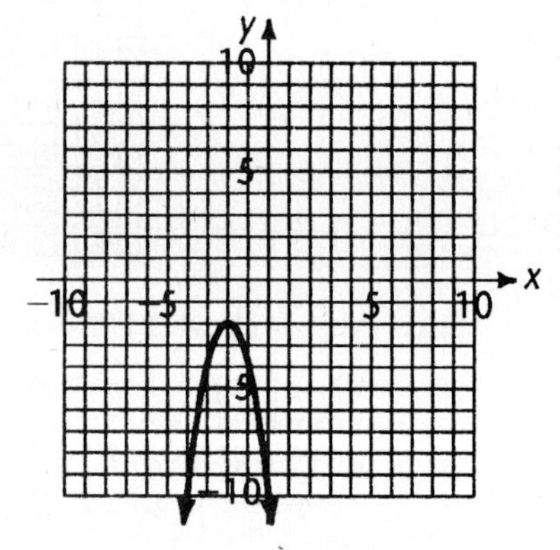

47. $y = 2(x - 5)^2 + 3$

The graph is that of $y = x^2$ stretched $(a = 2)$, shifted right 5 units $(h = 5)$, and raised 3 units $(k = 3)$.

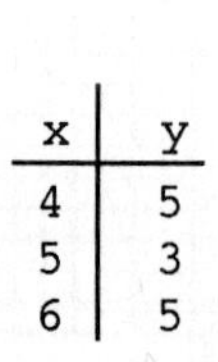

x	y
4	5
5	3
6	5

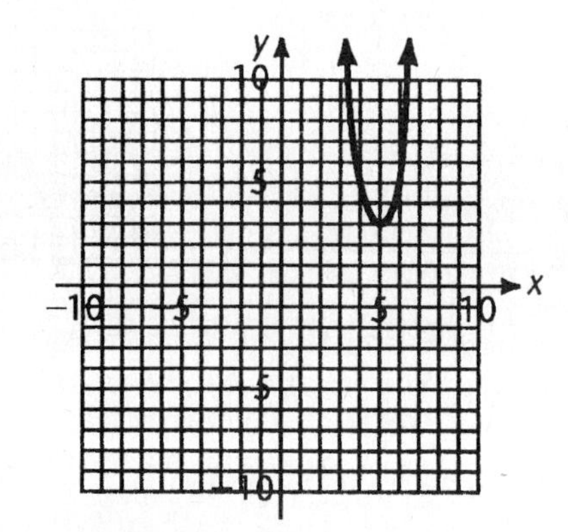

49. $y = \frac{1}{2}(x - 2)^2 + 6$

The graph is that of $y = x^2$ flattened $\left(a = \frac{1}{2}\right)$, shifted right 2 units $(h = 2)$, and raised 6 units $(k = 6)$.

x	y
0	8
2	6
4	8

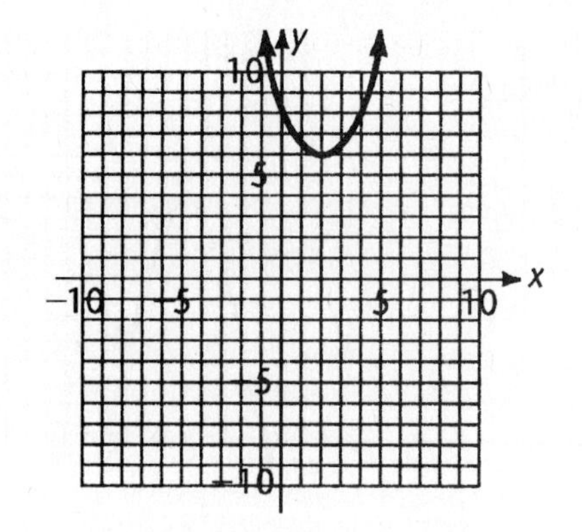

51. $y = -5(x - 1)^2 - 1$

The graph is that of $y = x^2$ stretched $(a = -5)$, turned over the x axis $(a < 0)$, shifted right 1 unit $(h = -1)$, and lowered 1 unit $(k = -1)$.

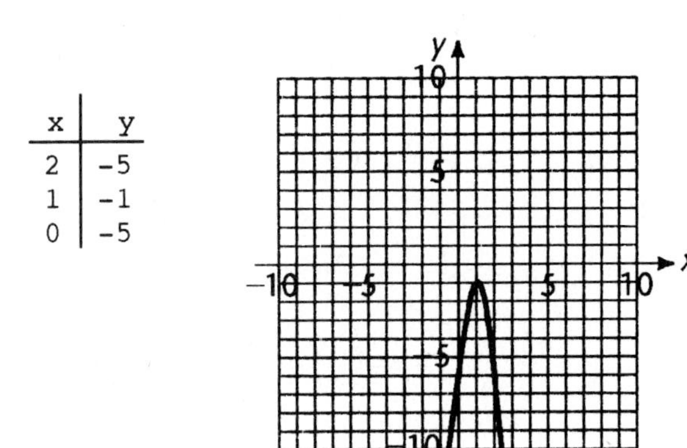

x	y
2	-5
1	-1
0	-5

53. $y = -\frac{1}{4}(x + 6)^2 + 2$

The graph is that of $y = x^2$ flattened $\left(a = -\frac{1}{4}\right)$, turned over the x axis $(a < 0)$, shifted left 6 units $(h = -6)$, and raised 2 units $(k = 2)$.

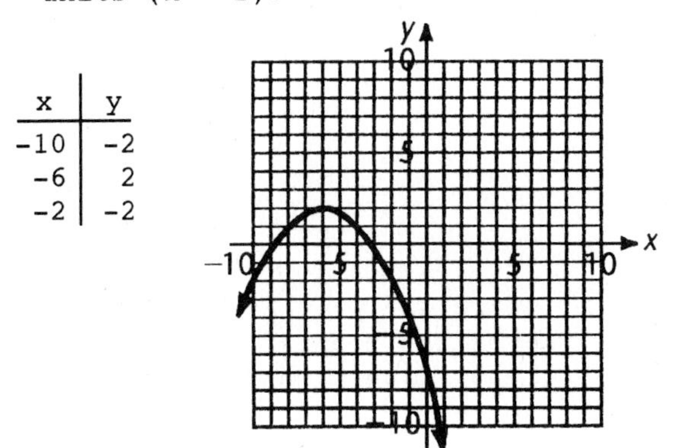

x	y
-10	-2
-6	2
-2	-2

Problems 55—71 are identical with problems 55–71 of Exercise 7-5. The graphs are not repeated for reasons of space.

55. $y = x^2 - 2x - 3$

$y = (x^2 - 2x \quad\quad) - 3$

$y = (x^2 - 2x + 1) - 1 - 3$

$y = (x - 1)^2 - 4$

The graph is that of $y = x^2$ shifted right 1 unit $(h = 1)$ and lowered 4 units $(k = -4)$.

57. $y = -x^2 + x + 2$

$y = -(x^2 - x \quad\quad) + 2$

$y = -\left(x^2 - x + \frac{1}{4}\right) + \frac{1}{4} + 2$

$y = -\left(x - \frac{1}{2}\right)^2 + \frac{9}{4}$

The graph is that of $y = x^2$ turned over the x axis $(a = -1)$, shifted right $\frac{1}{2}$ unit $\left(h = \frac{1}{2}\right)$, and raised $\frac{9}{4}$ units $\left(k = \frac{9}{4}\right)$.

59. $y = x^2 + x - 6$

$y = (x^2 + x) - 6$

$y = \left(x^2 + x + \frac{1}{4}\right) - \frac{1}{4} - 6$

$y = \left(x + \frac{1}{2}\right)^2 - \frac{25}{4}$

The graph is that of $y = x^2$ shifted left $\frac{1}{2}$ unit $\left(h = -\frac{1}{2}\right)$ and lowered $\frac{25}{4}$ unit $\left(k = -\frac{25}{4}\right)$.

61. $y = x^2 + 4x + 4$
$y = (x + 2)^2$

The graph is that of $y = x^2$ shifted left 2 units $(h = -2)$.

63. $y = x^2 - 3x + 3$
$y = (x^2 - 3x \qquad) + 3$

$$y = \left(x^2 - 3x + \frac{9}{4}\right) - \frac{9}{4} + 3$$

$$y = \left(x - \frac{3}{2}\right)^2 + \frac{3}{4}$$

The graph is that of $y = x^2$ shifted right $\frac{3}{2}$ unit $\left(h = \frac{3}{2}\right)$ and raised $\frac{3}{4}$ unit $\left(k = \frac{3}{4}\right)$.

65. $y = -x^2 + 2x - 8$
$y = -(x^2 - 2x) - 8$
$y = -(x^2 - 2x + 1) + 1 - 8$
$y = -(x - 1)^2 - 7$

The graph is that of $y = x^2$ turned over the x axis $(a = -1)$, shifted right 1 unit $(h = 1)$, and lowered 7 units $(k = -7)$.

67. $y = -x^2 + 5x - 6$
$y = -(x^2 - 5x) - 6$

$$y = -\left(x^2 - 5x + \frac{25}{4}\right) + \frac{25}{4} - 6$$

$$y = -\left(x - \frac{5}{2}\right)^2 + \frac{1}{4}$$

The graph is that of $y = x^2$ turned over the x axis $(a = -1)$, shifted right $\frac{5}{2}$ unit $\left(h = \frac{5}{2}\right)$, and raised $\frac{1}{4}$ unit $\left(k = \frac{1}{4}\right)$.

69. $y = 2x^2 - 7x + 5$

$$y = 2\left(x^2 - \frac{7}{2}x\right) + 5$$

$$y = 2\left(x^2 - \frac{7}{2}x + \frac{49}{16}\right) - \frac{49}{8} + 5$$

$$y = 2\left(x - \frac{7}{4}\right)^2 - \frac{9}{8}$$

The graph is that of $y = x^2$ stretched $(a = 2)$, shifted right $\frac{7}{4}$ unit $\left(h = \frac{7}{4}\right)$, and lowered $\frac{9}{8}$ unit $\left(k = -\frac{9}{8}\right)$.

71. $y = -2x^2 - 3x + 2$

$$y = -2\left(x^2 + \frac{3}{2}x\right) + 2$$

$$y = -2\left(x^2 + \frac{3}{2}x + \frac{9}{16}\right) + 2\left(\frac{9}{16}\right) + 2$$

$$y = -2\left(x + \frac{3}{4}\right)^2 + \frac{9}{8} + \frac{16}{8}$$

$$y = -2\left(x + \frac{3}{4}\right)^2 + \frac{25}{8}$$

The graph is that of $y = x^2$ stretched $(a = -2)$, turned over the x axis $(a < 0)$, shifted left $\frac{3}{4}$ unit $\left(h = -\frac{3}{4}\right)$, and raised $\frac{25}{8}$ unit $\left(k = \frac{25}{8}\right)$.

73. $y = 8x^2 + 41x + 52$. $a = 8$ $b = 41$ $c = 52$

$$-\frac{b^2 - 4ac}{4a} = -\frac{(41)^2 - 4(8)(52)}{4(8)} = -\frac{17}{32} = \text{the y coordinate of the vertex}$$

Since the parabola opens up $(a > 0)$ and the y coordinate of the vertex is negative, there are 2 x intercepts.

75. $y = 10x^2 - 60x + 91$. $a = 10$ $b = -60$ $c = 91$

$$-\frac{b^2 - 4ac}{4a} = -\frac{(-60)^2 - 4(10)(91)}{4(10)} = -\frac{-40}{40} = 1 = \text{the y coordinate of the vertex}$$

Since the parabola opens up ($a > 0$) and the y coordinate of the vertex is positive, there are no x intercepts.

77. $y = 23x^2 - 24x - 25$. $a = 23$ $b = -24$ $c = -25$

$$-\frac{b^2 - 4ac}{4a} = -\frac{(-24)^2 - 4(23)(-25)}{4(23)} = -\frac{2876}{92} = \text{the y coordinate of the vertex}$$

Since the parabola opens up ($a > 0$) and the y coordinate of the vertex is negative, there are 2 x intercepts.

79. $y = -53x^2 + 52x + 51$. $a = -53$ $b = 52$ $c = 51$

$$-\frac{b^2 - 4ac}{4a} = -\frac{(52)^2 - 4(-53)(51)}{4(-53)} = \frac{13,516}{212} = \text{the y coordinate of the vertex}$$

Since the parabola opens down ($a < 0$) and the y coordinate of the vertex is positive, there are 2 x intercepts.

Exercise 7-7 Conic Sections; Circles and Parabolas

Key Ideas and Formulas

The graphs of second-degree equations in two variables are plane curves, formed by the intersection of a plane and a cone, called **conic sections**. The principal conic sections are circles, parabolas, ellipses, and hyperbolas.

The Distance-Between-Two-Points formula.

Given $P_1(x_1, y_1)$ and $P_2(x_2, y_2)$, then

$$d(P_1, P_2) = d = \sqrt{(x_2 - x_1)^2 + (y_2 - y_1)^2}$$

A circle is the set of points equidistant from a fixed point. The fixed distance is called the **radius**, and the fixed point is called the **center**.

Equations of a Circle
1. Radius R and center (h, k): $(x - h)^2 + (y - k)^2 = R^2$
2. Radius R and center at the origin: $x^2 + y^2 = R^2$ since $(h, k) = (0, 0)$

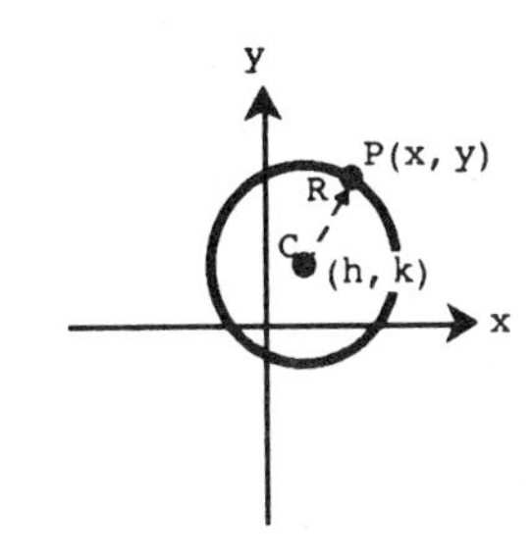

A **parabola** is the set of all points equidistant from a fixed point and a fixed line. The fixed point is called the **focus**, and the fixed line is the **directrix**.

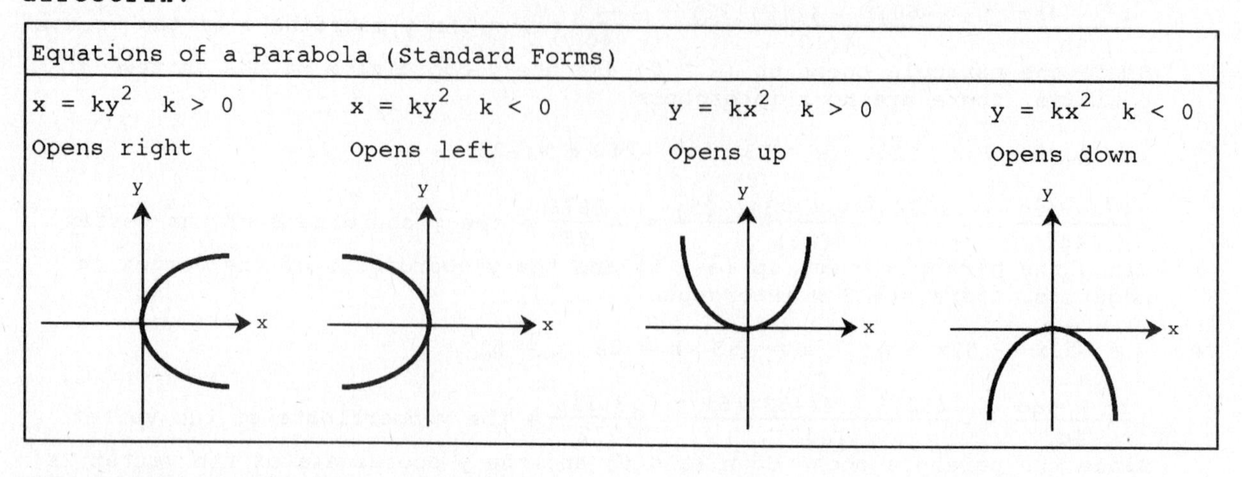

---

1. Let $P_1 = (2, 4)$ and $P_2 = (4, 5)$. Then

$$d = \sqrt{(x_2 - x_1)^2 + (y_2 - y_1)^2}$$
$$= \sqrt{(4 - 2)^2 + (5 - 4)^2}$$
$$= \sqrt{(2)^2 + (1)^2}$$
$$= \sqrt{5}$$

3. Let $P_1 = (3, -7)$ and $P_2 = (-2, 1)$. Then

$$d = \sqrt{(x_2 - x_1)^2 + (y_2 - y_1)^2}$$
$$= \sqrt{[(-2) - 3]^2 + [1 - (-7)]^2}$$
$$= \sqrt{(-5)^2 + (8)^2}$$
$$= \sqrt{89}$$

5. Let $P_1 = (-8, -2)$ and $P_2 = (-5, 1)$. Then

$$d = \sqrt{(x_2 - x_1)^2 + (y_2 - y_1)^2}$$
$$= \sqrt{[(-5) - (-8)]^2 + [1 - (-2)]^2}$$
$$= \sqrt{(3)^2 + (3)^2}$$
$$= \sqrt{18} \text{ or } 3\sqrt{2}$$

7. Let $P_1 = (-3, 0)$ and $P_2 = \left(\frac{1}{2}, \frac{9}{2}\right)$. Then

$$d = \sqrt{(x_2 - x_1)^2 + (y_2 - y_1)^2}$$
$$= \sqrt{[\tfrac{1}{2} - (-3)]^2 + [\tfrac{9}{2} - 0]^2}$$
$$= \sqrt{(\tfrac{7}{2})^2 + (\tfrac{9}{2})^2}$$
$$= \sqrt{12.25 + 20.25}$$
$$= \sqrt{32.5}$$

9. Let $P_1 = (1.3, 2.5)$ and $P_2 = (3.8, -5.4)$. Then

$$d = \sqrt{(x_2 - x_1)^2 + (y_2 - y_1)^2} = \sqrt{(3.8 - 1.3)^2 + [(-5.4) - 2.5]^2}$$
$$= \sqrt{(2.5)^2 + (-7.9)^2} = \sqrt{6.25 + 62.41} = \sqrt{68.66}$$

11. This is a circle with radius 4 and center at the origin:

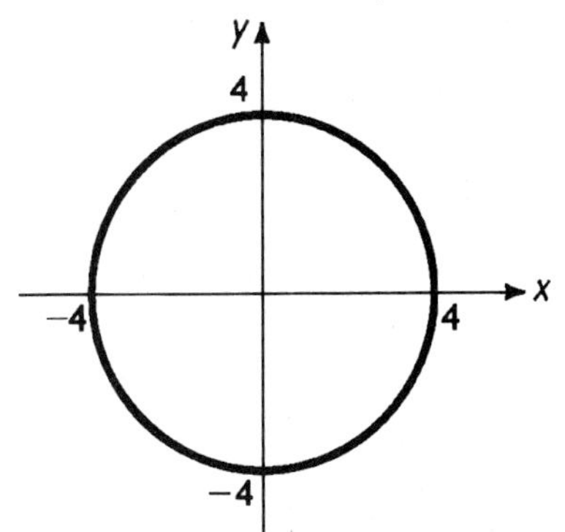

13. This is a circle with radius $\sqrt{6}$ and center at the origin:

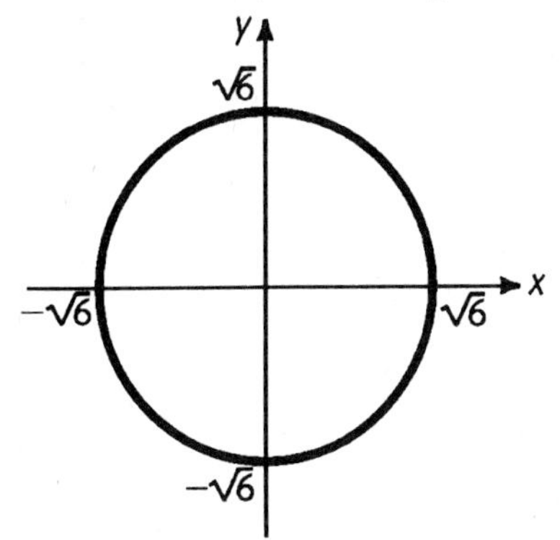

15. This is a circle with radius $\sqrt{\frac{25}{4}} = 2.5$ and center at the origin:

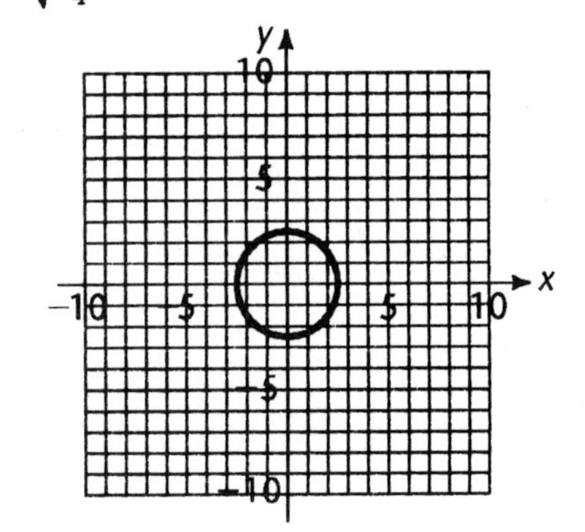

17. This is a circle with radius $\sqrt{\frac{64}{25}} = 1.6$ and center at the origin:

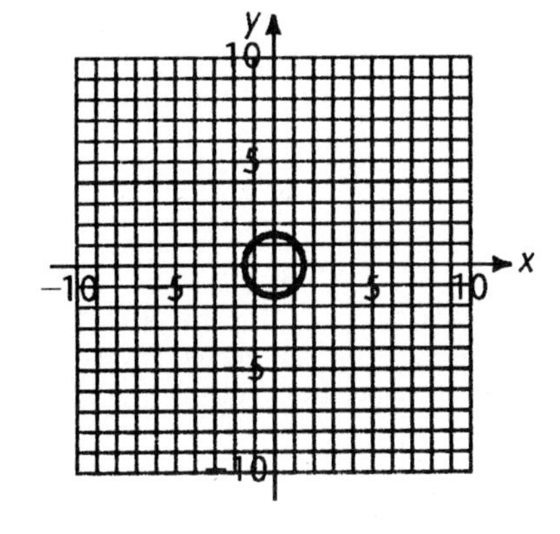

19. Since the center is at the origin, (h, k) = (0, 0); thus the equation is
$x^2 + y^2 = 7^2$ or $x^2 + y^2 = 49$

21. Since the center is at the origin, (h, k) = (0, 0); thus the equation is
$x^2 + y^2 = (\sqrt{5})^2$ or $x^2 + y^2 = 5$

23. Since the center is at the origin, (h, k) = (0, 0); thus the equation is
$x^2 + y^2 = \left(\frac{7}{2}\right)^2$ or $x^2 + y^2 = 12.25$

25. Since the center is at the origin, (h, k) = (0, 0); thus the equation is
$x^2 + y^2 = (6.2)^2$ or $x^2 + y^2 = 38.44$

27. The origin (0, 0) is part of the graph. We choose $x = 4$ to make the right side a positive perfect square; then

$$y^2 = 4(4)$$
$$y^2 = 16$$
$$y = \pm\sqrt{16} = \pm 4$$

Thus, the points (4, 4) and (4, -4) are also on the graph. We plot the three known points; then we sketch a parabola through these points.

29. The origin (0, 0) is part of the graph. We choose $x = -3$ to make the right side a positive perfect square; then

$$y^2 = -12(-3)$$
$$y^2 = 36$$
$$y = \pm\sqrt{36} = \pm 6$$

Thus, the points (-3, 6) and (-3, -6) are also on the graph. We plot the three known points; then we sketch a parabola through these points.

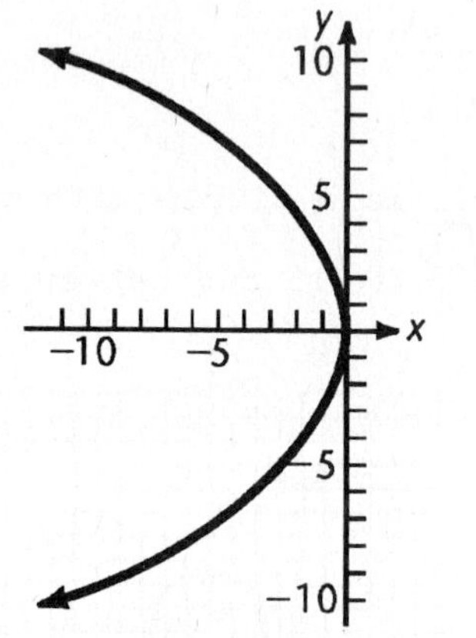

31. The origin (0, 0) is part of the graph. We choose $y = 4$ to make the right side a positive perfect square; then

$$x^2 = 4$$
$$x = \pm\sqrt{4} = \pm 2$$

Thus, the points (2, 4) and (-2, 4) are also on the graph. We plot the three known points; then we sketch a parabola through these points.

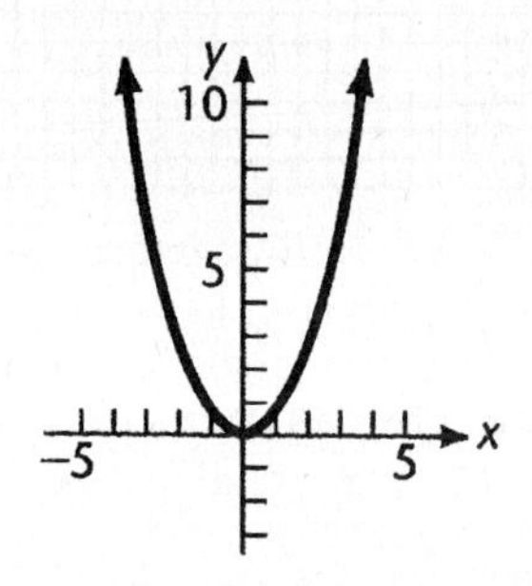

33. The origin (0, 0) is part of the graph. We choose $y = -4$ to make the right side a positive perfect square; then

$$x^2 = -16(-4)$$
$$x^2 = 64$$
$$x = \pm\sqrt{64} = \pm 8$$

Thus, the points (8, -4) and (-8, -4) are also on the graph. We plot the three known points; then we sketch a parabola through these points.

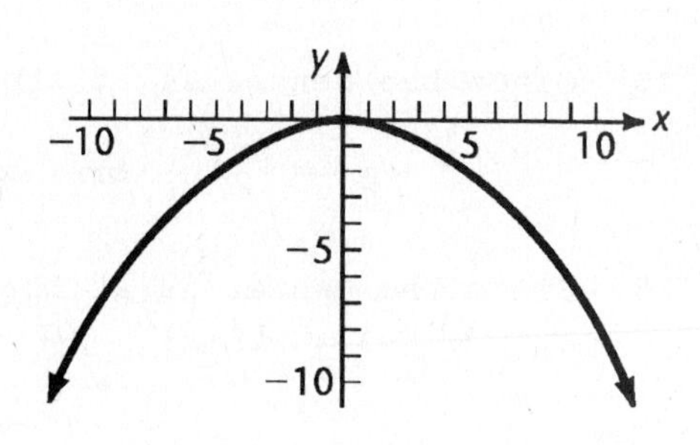

35. The origin (0, 0) is part of the graph. We choose $x = -2$ to make the right side a positive perfect square; then

$$-2 = -2y^2$$
$$1 = y^2$$
$$\pm 1 = y$$

Thus, the points (-2, 1) and (-2, -1) are also on the graph. We plot the three known points; then we sketch a parabola through these points.

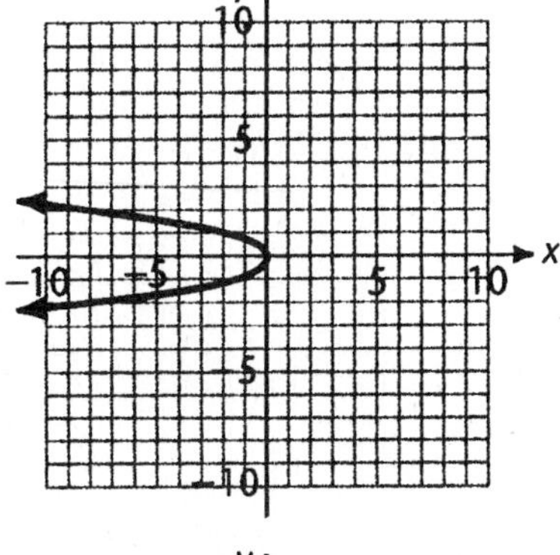

37. The origin (0, 0) is part of the graph. We choose $x = 3$ to make the left side a positive perfect square; then

$$\frac{1}{3}(3) = y^2$$
$$1 = y^2$$
$$y = \pm 1$$

Thus, the points (3, 1) and (3, -1) are also on the graph. We plot the three known points; then we sketch a parabola through these points.

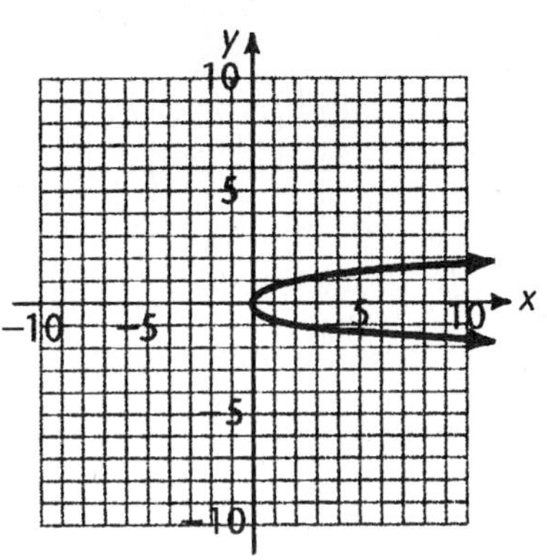

39. This is the equation of a circle with radius 4 and center at $(h, k) = (3, 4)$. Thus,

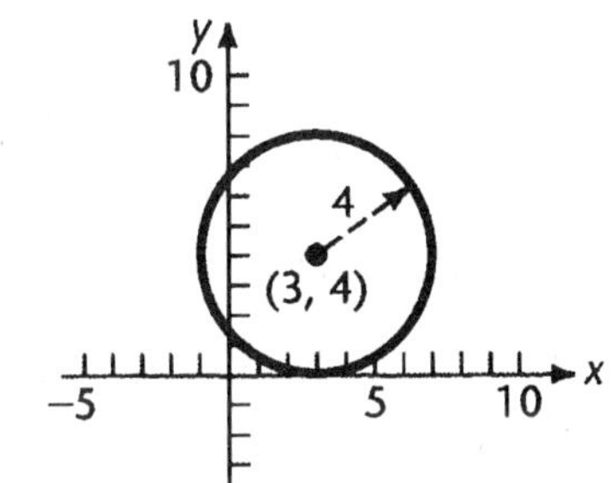

41. $(x - 4)^2 + (y + 3)^2 = 9$ is the same as $(x - 4)^2 + [y - (-3)]^2 = 3^2$, which is the equation of a circle with radius 3 and center at $(h, k) = (4, -3)$. Thus,

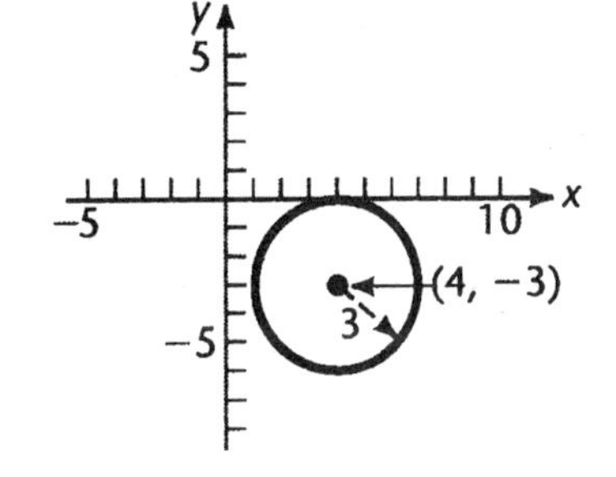

43. $(x + 3)^2 + (y + 3)^2 = 16$ is the same as $[x - (-3)]^2 + [y - (-3)]^2 = 4^2$, which is the equation of a circle with radius 4 and center at $(h, k) = (-3, -3)$. Thus,

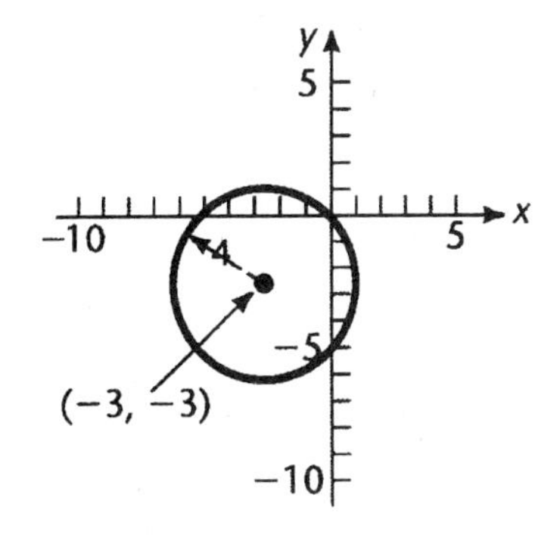

45. $(x + 2)^2 + (y + 3)^2 = 4$ is the same as $[x - (-2)]^2 + [y - (-3)]^2 = 2^2$, which is the equation of a circle with radius 2 and center at $(h, k) = (-2, -3)$. Thus,

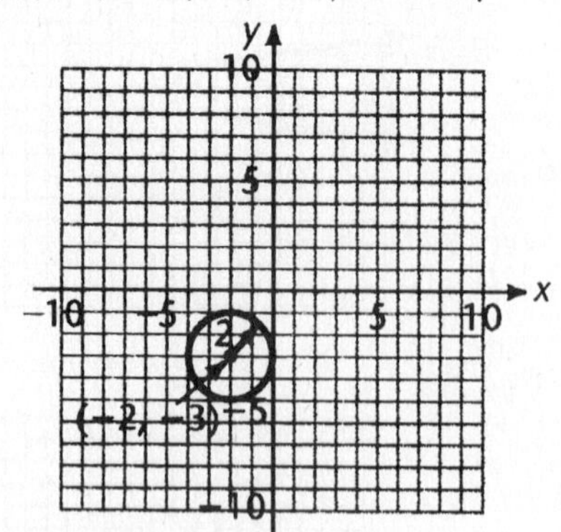

47. This is the equation of a circle with radius 3 and center at $(h, k) = (4, 3)$.

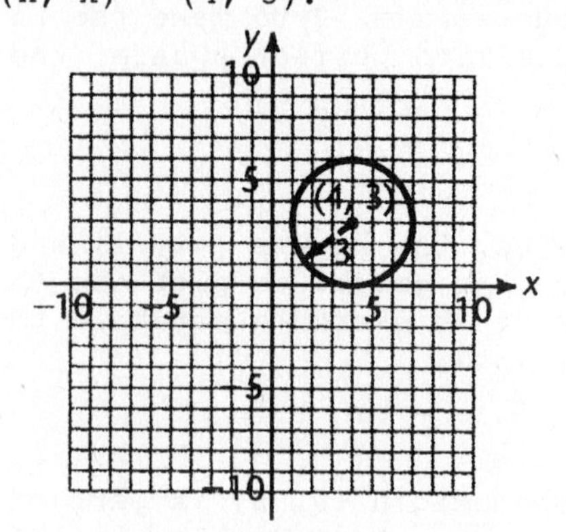

49. This is the equation of a circle with radius $\sqrt{\frac{7}{4}} = \frac{\sqrt{7}}{2}$ and center at $(h, k) = \left(\frac{1}{2}, \frac{1}{3}\right)$.

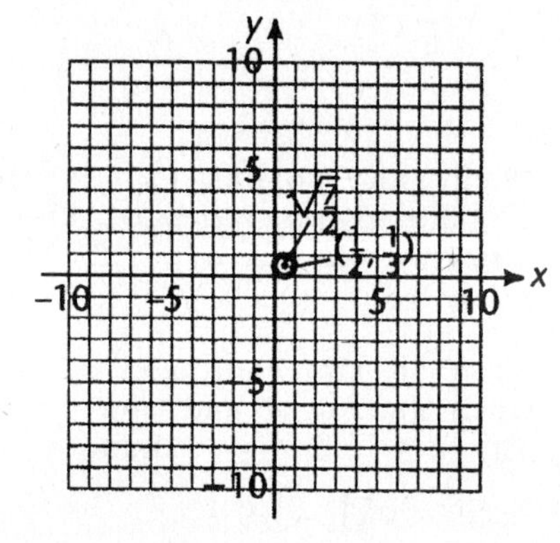

51. We have $R = 7$ and $(h, k) = (3, 5)$; hence the equation is
$$(x - 3)^2 + (y - 5)^2 = 7^2 \quad \text{or} \quad (x - 3)^2 + (y - 5)^2 = 49$$

53. We have $R = 8$ and $(h, k) = (-3, 3)$; hence the equation is
$$[x - (-3)]^2 + (y - 3)^2 = 8^2$$
$$\text{or } (x + 3)^2 + (y - 3)^2 = 64$$

COMMON ERROR:
$(x - 3)^2 + (y - 3)^2 = 8^2$
this is incorrect substitution; $h = -3$, not 3.

55. We have $R = \sqrt{3}$ and $(h, k) = (-4, -1)$; hence the equation is
$$[x - (-4)]^2 + [y - (-1)]^2 = (\sqrt{3})^2 \text{ or } (x + 4)^2 + (y + 1)^2 = 3$$

57. We have $R = \frac{9}{4}$ and $(h, k) = \left(1, \frac{1}{2}\right)$; hence the equation is
$$(x - 1)^2 + \left(y - \frac{1}{2}\right)^2 = \left(\frac{9}{4}\right)^2 \quad \text{or} \quad (x - 1)^2 + \left(y - \frac{1}{2}\right)^2 = \frac{81}{16}$$

59. We have $R = \sqrt[4]{2}$ and $(h, k) = (1.1, 2)$; hence the equation is
$$(x - 1.1)^2 + (y - 2)^2 = (\sqrt[4]{2})^2 \quad \text{or} \quad (x - 1.1)^2 + (y - 2)^2 = (2^{1/4})^2$$
or $2^{2 \cdot 1/4}$ or $2^{1/2}$,

that is, $(x - 1.1)^2 + (y - 2)^2 = \sqrt{2}$

61. $4y^2 - 8x = 0$ can be written in the standard form $x = ky^2$ as follows:

$$-8x = -4y^2$$
$$x = \frac{1}{2}y^2$$

Since $k > 0$, the x values must be positive and the parabola opens right. We look for values of y, in addition to $y = 0$, that make x easy to calculate, and then plot the graph.

x	y
0	0
8	4
8	-4

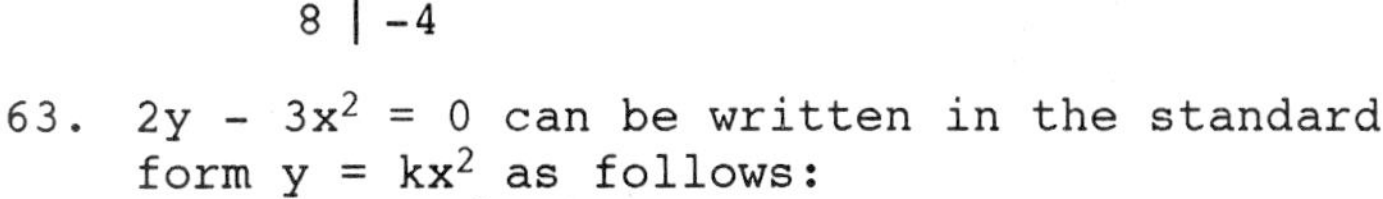

63. $2y - 3x^2 = 0$ can be written in the standard form $y = kx^2$ as follows:

$$2y = 3x^2$$
$$y = \frac{3}{2}x^2$$

Since $k > 0$, the y values must be positive and the parabola opens up. We look for values of x, in addition to $x = 0$, that make y easy to calculate, and then plot the graph.

x	y
0	0
2	6
-2	6

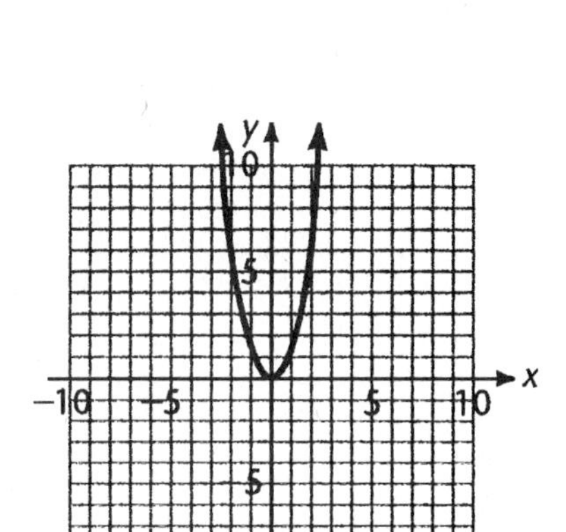

65. $7x - 2y^2 = 0$ can be written in the standard form $x = ky^2$ as follows:

$$7x = 2y^2$$
$$x = \frac{2}{7}y^2$$

Since $k > 0$, the x values must be positive and the parabola opens right. We look for values of y, in addition to $y = 0$, that make x easy to calculate, and then plot the graph.

x	y
0	0
$\frac{7}{2}$	$\frac{7}{2}$
$\frac{7}{2}$	$-\frac{7}{2}$

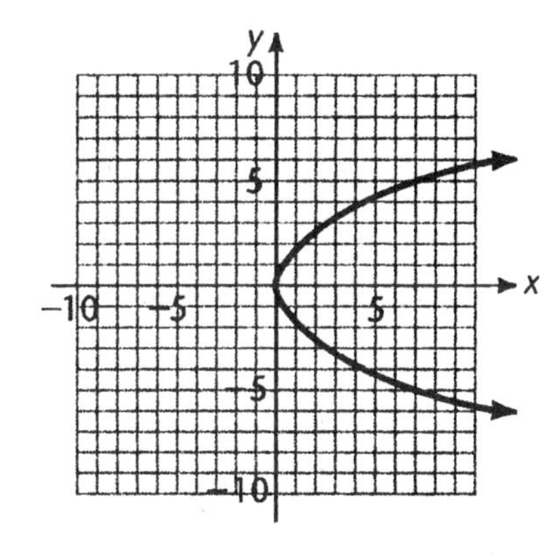

67. A triangle is isosceles if two sides have the same length. The length of each side is the distance between its endpoints, which we calculate using the distance formula:

$$d = \sqrt{(x_2 - x_1)^2 + (y_2 - y_1)^2}$$

$(x_1, y_1) = (-1, 2)$ and $(x_2, y_2) = (2, -1)$:

$$d = \sqrt{[2 - (-1)]^2 + [(-1) - 2]^2} = \sqrt{(3)^2 + (-3)^2} = \sqrt{18}$$

$(x_1, y_1) = (-1, 2)$ and $(x_2, y_2) = (3, 3)$

$$d = \sqrt{[3 - (-1)]^2 + (3 - 2)^2} = \sqrt{(4)^2 + (1)^2} = \sqrt{17}$$

$(x_1, y_1) = (2, -1)$ and $(x_2, y_2) = (3, 3)$

$$d = \sqrt{(3 - 2)^2 + [3 - (-1)]^2} = \sqrt{(1)^2 + (4)^2} = \sqrt{17}$$

Since two sides have length $\sqrt{17}$, the triangle is isosceles.

69. No, since not all three sides have the same length.

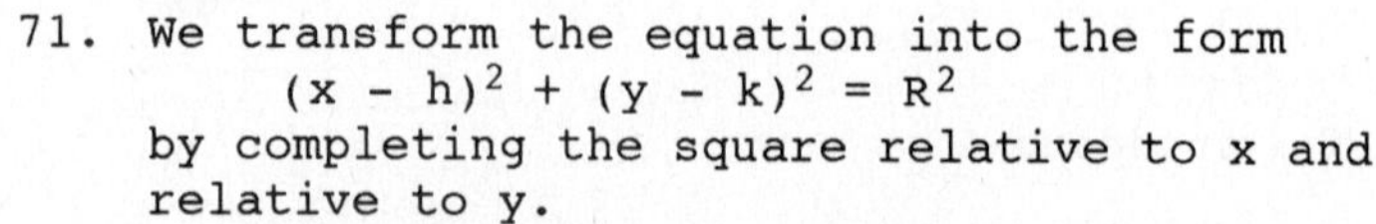

71. We transform the equation into the form

$$(x - h)^2 + (y - k)^2 = R^2$$

by completing the square relative to x and relative to y.

$$x^2 - 4x + ? + y^2 - 6y + ? = -4 + ? + ?$$
$$x^2 - 4x + 4 + y^2 - 6y + 9 = -4 + 4 + 9$$
$$(x - 2)^2 + (y - 3)^2 = 9$$

This is the equation of a circle with radius 3 and center at $(h, k) = (2, 3)$.

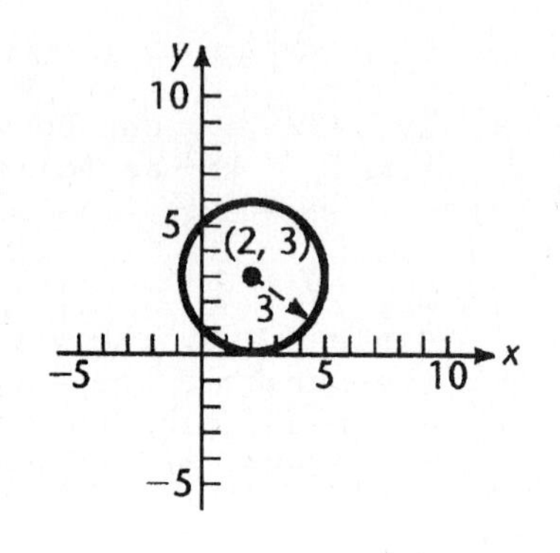

73. We transform the equation into the form

$$(x - h)^2 + (y - k)^2 = R^2$$

by completing the square relative to x and relative to y.

$$x^2 - 6x + ? + y^2 + 6y + ? = -2 + ? + ?$$
$$x^2 - 6x + 9 + y^2 + 6y + 9 = -2 + 9 + 9$$
$$(x - 3)^2 + (y + 3)^2 = 16$$

or $$(x - 3)^2 + [y - (-3)]^2 = 16$$

This is the equation of a circle with radius 4 and center at $(h, k) = (3, -3)$.

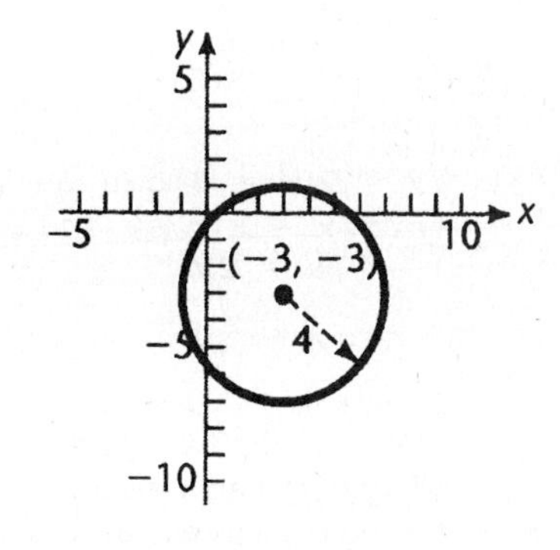

75. We transform the equation into the form

$$(x - h)^2 + (y - k)^2 = R^2$$

by completing the square relative to x and relative to y.

$$x^2 + 6x + ? + y^2 + 4y + ? = -4 + ? + ?$$
$$x^2 + 6x + 9 + y^2 + 4y + 4 = -4 + 9 + 4$$
$$(x + 3)^2 + (y + 2)^2 = 9$$

or $$[x - (-3)]^2 + [y - (-2)]^2 = 9$$

This is the equation of a circle with radius 3 and center at $(h, k) = (-3, -2)$.

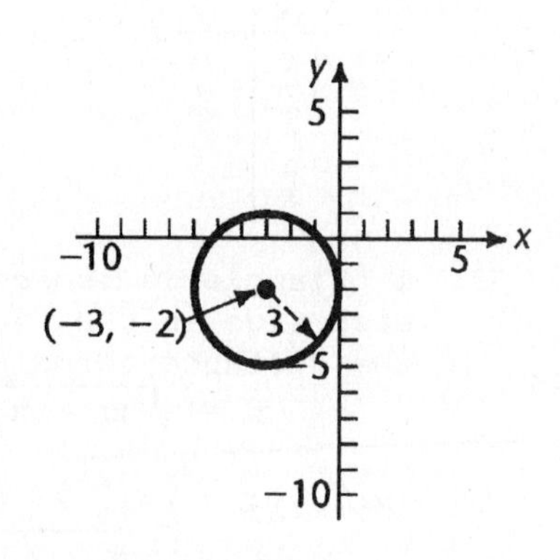

77. We transform the equation into the form
$(x - h)^2 + (y - k)^2 = R^2$
by completing the square relative to x and relative to y.

$$x^2 - 3x + ? + y^2 - 5y + ? = \frac{15}{4} + ? + ?$$

$$x^2 + 3x + \frac{9}{4} + y^2 - 5y + \frac{25}{4} = \frac{15}{4} + \frac{9}{4} + \frac{25}{4}$$

$$\left(x - \frac{3}{2}\right)^2 + \left(y - \frac{5}{2}\right)^2 = \frac{49}{4}$$

This is the equation of a circle with radius $\frac{7}{2}$ and center at $(h, k) = \left(\frac{3}{2}, \frac{5}{2}\right)$.

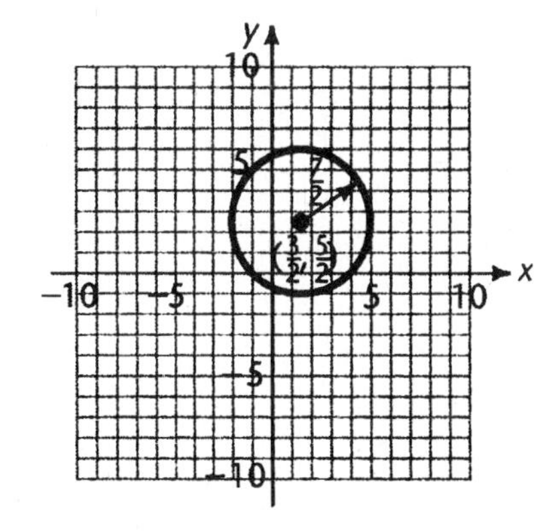

79. We transform the equation into the form
$(x - h)^2 + (y - k)^2 = R^2$
by completing the square relative to x and relative to y.

$$x^2 - x + ? + y^2 + 3y + ? = -\frac{1}{4}$$

$$x^2 - x + \frac{1}{4} + y^2 + 3y + \frac{9}{4} = -\frac{1}{4} + \frac{1}{4} + \frac{9}{4}$$

$$\left(x - \frac{1}{2}\right)^2 + \left(y + \frac{3}{2}\right)^2 = \frac{9}{4}$$

or $$\left(x - \frac{1}{2}\right)^2 + \left[y - \left(-\frac{3}{2}\right)\right]^2 = \frac{9}{4}$$

This is the equation of a circle with radius $\frac{3}{2}$ and center at $(h, k) = \left(\frac{1}{2}, -\frac{3}{2}\right)$.

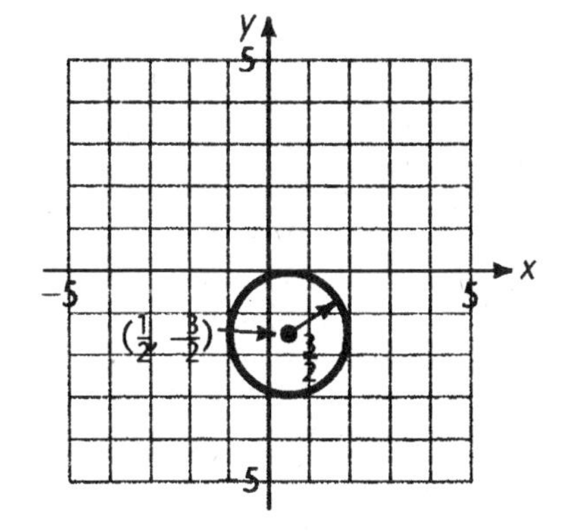

81. The focus of a parabola whose equation in standard form is $y^2 = 4ax$ is $(a, 0)$. Its directrix is the line with equation $x = -a$. The equation of this parabola is $y^2 = 4x$. To find a, we solve the equation
$4a = 4$
to find $a = 1$. Hence the focus is $(1, 0)$ and the equation of the directrix is $x = -1$.

83. The focus of a parabola whose equation in standard form is $x^2 = -4ay$ is $(0, -a)$. Its directrix is the line with equation $y = a$. The equation of this parabola is $y = -8x^2$, or $x^2 = -\frac{1}{8}y$. To find a, we solve

$$4a = \frac{1}{8}$$

to find $a = \frac{1}{32}$. Hence the focus is $\left(0, -\frac{1}{32}\right)$ and the equation of the directrix is $y = \frac{1}{32}$.

85. (x, 8) will be 13 units from (2, -4) if the distance between (x, 8) and (2, -4) is 13. Substituting into the distance formula, we obtain

$$d = \sqrt{(x_2 - x_1)^2 + (y_2 - y_1)^2}$$

$$13 = \sqrt{(2 - x)^2 + [(-4) - 8]^2} \quad \text{Let us solve this equation for x.}$$

$$13 = \sqrt{(2 - x)^2 + (-12)^2}$$

$$13 = \sqrt{4 - 4x + x^2 + 144}$$

$$13 = \sqrt{x^2 - 4x + 148}$$

$$\begin{aligned} 169 &= x^2 - 4x + 148 \\ 0 &= x^2 - 4x - 21 \\ x^2 - 4x - 21 &= 0 \\ (x - 7)(x + 3) &= 0 \\ x &= 7 \text{ or } -3 \end{aligned}$$

Check:

$\underline{x = 7}$ $\sqrt{(2 - 7)^2 + [(-4) - 8]^2} = \sqrt{(-5)^2 + (-12)^2} = 13$ so 7 is a solution

$\underline{x = -3}$ $\sqrt{[2 - (-3)]^2 + [(-4) - 8]^2} = \sqrt{(5)^2 + (-12)^2} = 13$ so -3 is a solution.

87. (x, y) will be equidistant from (-1, -1) and (0, 1) if the distance from (-1, -1) to (x, y) is equal to the distance from (0, 1) to (x, y). Substituting into the distance formula, we obtain

$$d_1 = d_2$$

$$\sqrt{[x - (-1)]^2 + [y - (-1)]^2} = \sqrt{(x - 0)^2 + (y - 1)^2}$$

Simplifying, we obtain

$$\begin{aligned} \sqrt{(x + 1)^2 + (y + 1)^2} &= \sqrt{x^2 + (y - 1)^2} \\ (x + 1)^2 + (y + 1)^2 &= x^2 + (y - 1)^2 \\ x^2 + 2x + 1 + y^2 + 2y + 1 &= x^2 + y^2 - 2y + 1 \\ 2x + 1 + 2y + 1 &= -2y + 1 \\ 2x + 4y + 1 &= 0 \end{aligned}$$

89. Since the center is at the origin, (h, k) = (0, 0), and the equation is of the form:

$$x^2 + y^2 = R^2$$

Since the circle passes through the point (4, -3), the coordinates of this point must satisfy the equation. Hence

$$\begin{aligned} (4)^2 + (-3)^2 &= R^2 \\ 25 &= R^2 \end{aligned}$$

The desired equation is $x^2 + y^2 = 25$.

91.

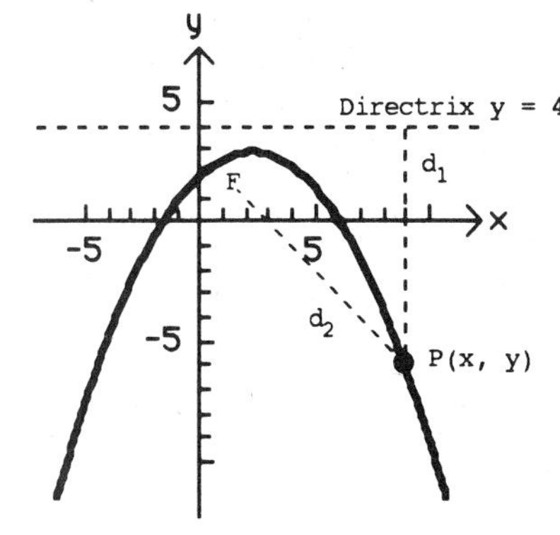

From the definition of a parabola we have

$$d_1 = d_2$$
$$|4 - y| = \sqrt{(x - 2)^2 + (y - 2)^2} \quad \text{using } F(2, 2)$$
$$(4 - y)^2 = (x - 2)^2 + (y - 2)^2$$
$$16 - 8y + y^2 = x^2 - 4x + 4 + y^2 - 4y + 4$$
$$12 - 4y = x^2 - 4x + 4$$
$$-4(y - 3) = (x - 2)^2 \quad \text{or} \quad 0 = x^2 - 4x + 4y - 8$$

93. Let $P(x, y)$ be a point on the parabola.
Then $|x|$ represents the distance from (x, y) to the directrix $x = 0$.
From the distance formula, the distance from (x, y) to the focus $(2, 2)$ is
$\sqrt{(x - 2)^2 + (y - 2)^2}$
From the definition of a parabola we have

$$|x| = \sqrt{(x - 2)^2 + (y - 2)^2}$$
$$x^2 = (x - 2)^2 + (y - 2)^2$$
$$x^2 = x^2 - 4x + 4 + y^2 - 4y + 4$$
$$4x - 4 = y^2 - 4y + 4$$
$$4(x - 1) = (y - 2)^2 \quad \text{or} \quad 4x = y^2 - 4y + 8$$

95. Following the hint, since $(40, R - 20)$ must satisfy $x^2 + y^2 = R^2$, we substitute to obtain

$$(40)^2 + (R - 20)^2 = R^2$$
$$1,600 + R^2 - 40R + 400 = R^2$$
$$R^2 - 40R + 2,000 = R^2$$
$$-40R + 2,000 = 0$$
$$-40R = -2,000$$
$$R = 50$$

Hence the desired equation is $x^2 + y^2 = 50^2$.

97. Following the hint, since $(50, -25)$ must satisfy $x^2 = -4ay$, we have

$$50^2 = -4a(-25)$$
$$2500 = 100a$$
$$25 = a$$

Therefore the equation of the parabola must be

$$x^2 = -4(25)y$$
$$x^2 = -100y$$

Exercise 7-8 Conic Sections: Ellipses and Hyperbolas

Key Ideas and Formulas

An **ellipse** is the set of all points such that the sum of the distances of each to two fixed points is constant. The fixed points are called **foci**, and each separately is a **focus**.

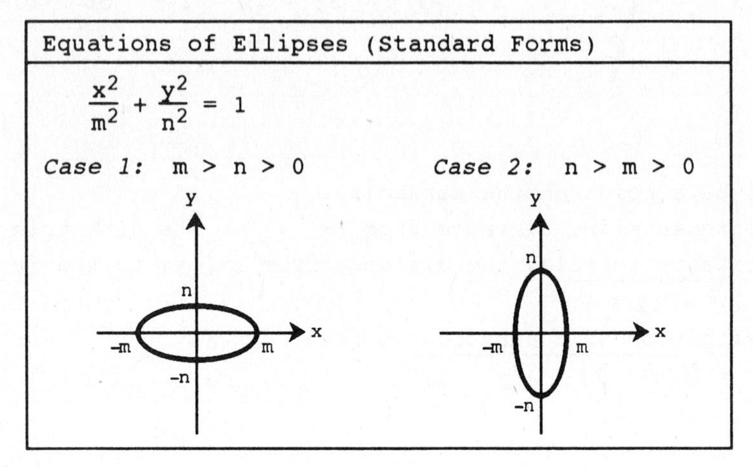

To graph $\frac{x^2}{m^2} + \frac{y^2}{n^2} = 1$:

Step 1: Find the x intercepts by letting $y = 0$ and solving for x.

Step 2: Find the y intercepts by letting $x = 0$ and solving for y.

Step 3: Sketch an ellipse passing through these intercepts.

A **hyperbola** is the set of all points such that the absolute value of the difference of the distances of each to two fixed points is constant. The two fixed points are called **foci**.

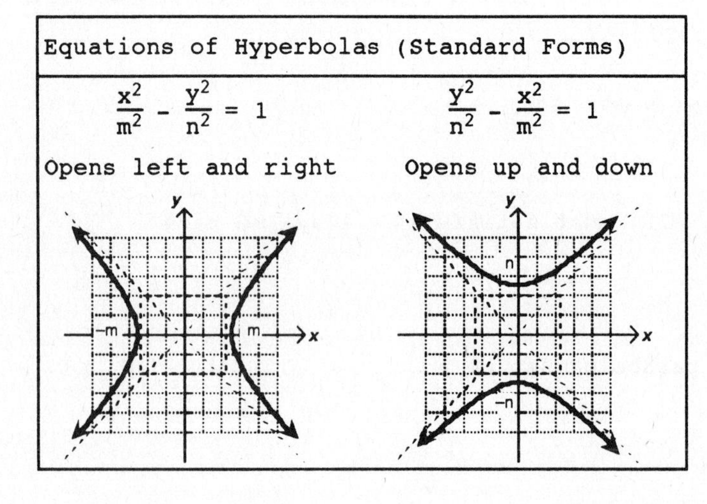

Asymptotes: (dashed lines)

$$y = \pm\frac{n}{m}x$$

To graph $\frac{x^2}{m^2} - \frac{y^2}{n^2} = 1$ and $\frac{y^2}{n^2} - \frac{x^2}{m^2} = 1$:

Step 1: Draw a dashed rectangle with intercepts $x = \pm m$ and $y = \pm n$.

Step 2: Draw dashed diagonals of the rectangle and extend to form asymptotes. (These are the graphs of $y = \pm\frac{n}{m}x$.)

Step 3: Determine the true intercepts of the hyperbola (be particularly careful in this step); then sketch in the hyperbola (both branches).

---

1. Find x intercepts:

$$\frac{x^2}{25} + \frac{0}{4} = 1$$
$$x^2 = 25$$
$$x = \pm\sqrt{25} = \pm 5$$

Find y intercepts:

$$\frac{0}{25} + \frac{y^2}{4} = 1$$
$$y^2 = 4$$
$$y = \pm\sqrt{4} = \pm 2$$

Now plot the intercepts and draw the ellipse.

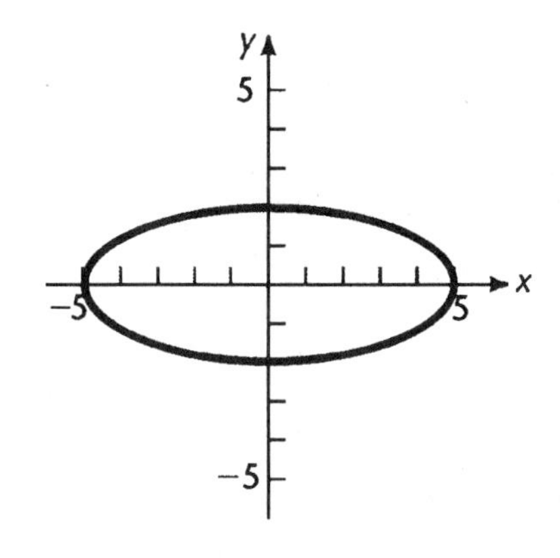

3. Find x intercepts:

$$\frac{x^2}{4} + \frac{0}{25} = 1$$
$$x^2 = 4$$
$$x = \pm\sqrt{4} = \pm 2$$

Find y intercepts:

$$\frac{0}{4} + \frac{y^2}{25} = 1$$
$$y^2 = 25$$
$$y = \pm\sqrt{25} = \pm 5$$

Now plot the intercepts and draw the ellipse.

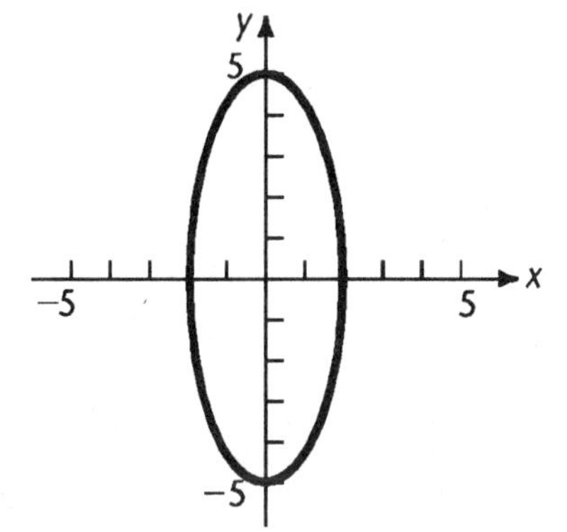

5. Find x intercepts:

$$\frac{x^2}{16} + \frac{0}{25} = 1$$
$$x^2 = 16$$
$$x = \pm\sqrt{16} = \pm 4$$

Find y intercepts:

$$\frac{0}{16} + \frac{y^2}{25} = 1$$
$$y^2 = 25$$
$$y = \pm\sqrt{25} = \pm 5$$

Now plot the intercepts and draw the ellipse.

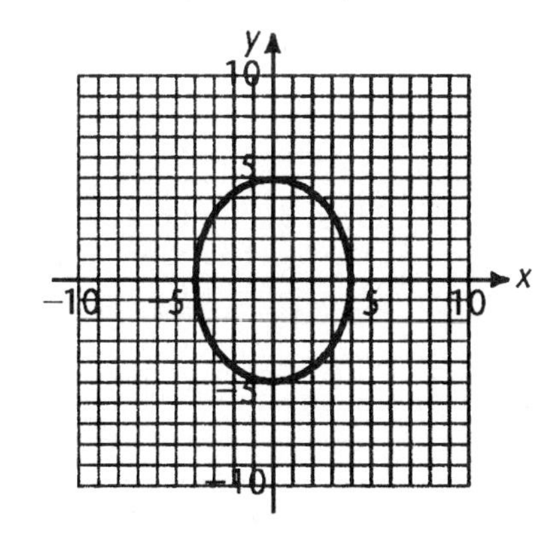

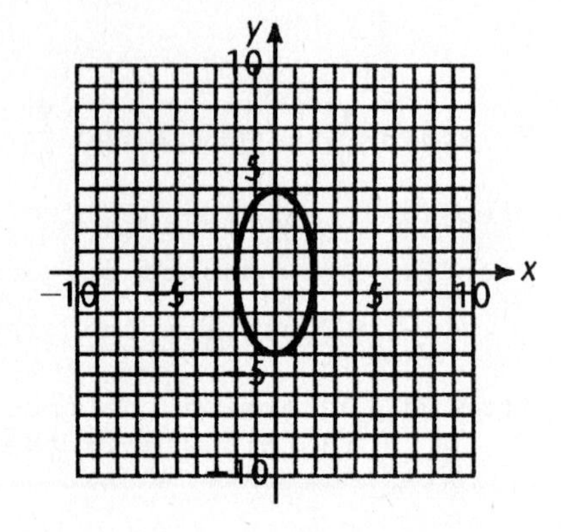

7. Find x intercepts:

$$\frac{x^2}{4} + \frac{0}{16} = 1$$
$$x^2 = 4$$
$$x = \pm\sqrt{4} = \pm 2$$

Find y intercepts:

$$\frac{0}{4} + \frac{y^2}{16} = 1$$
$$y^2 = 16$$
$$y = \pm\sqrt{16} = \pm 4$$

Now plot the intercepts and draw the ellipse.

9. First, draw a dashed rectangle with intercepts $x = \pm\sqrt{4} = \pm 2$ and $y = \pm\sqrt{25} = \pm 5$. Second, draw in asymptotes (extended diagonals of rectangle). Third, determine true intercepts for the hyperbola, as follows. Let $y = 0$; then

$$\frac{x^2}{4} - \frac{0}{25} = 1$$
$$x^2 = 4$$
$$x = \pm\sqrt{4} = \pm 2$$

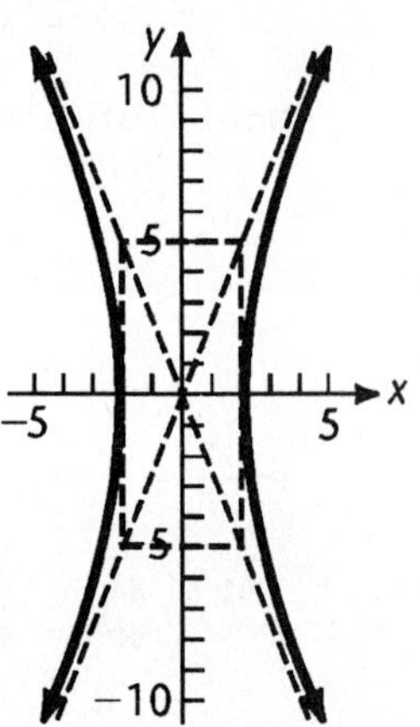

If we let $x = 0$, then

$$\frac{0}{4} - \frac{y^2}{25} = 1$$
$$y^2 = -25$$
$$y = \pm\sqrt{-25} = \pm 5i$$

These are complex numbers and do not represent real intercepts. We conclude that the only real intercepts are $x = \pm 2$, and the hyperbola opens left and right.

11. First, draw a dashed rectangle with intercepts $x = \pm\sqrt{4} = \pm 2$ and $y = \pm\sqrt{25} = \pm 5$. Second, draw in asymptotes (extended diagonals of rectangle). Third, determine true intercepts for the hyperbola, as follows. If we let $y = 0$; then

$$\frac{0}{25} - \frac{x^2}{4} = 1$$
$$x^2 = -4$$
$$x = \pm\sqrt{-4} = \pm 2i$$

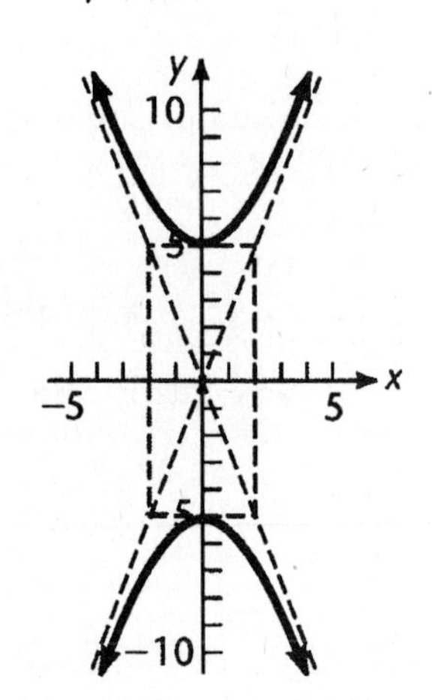

These are complex numbers and do not represent real intercepts. Let $x = 0$; then

$$\frac{y^2}{25} - \frac{0}{4} = 1$$
$$y^2 = 25$$
$$y = \pm\sqrt{25} = \pm 5$$

We conclude that the only real intercepts are $y = \pm 5$, and the hyperbola opens up and down.

13. First, draw a dashed rectangle with intercepts $x = \pm\sqrt{4} = \pm 2$ and $y = \pm\sqrt{9} = \pm 3$. Second, draw in asymptotes (extended diagonals of rectangle). Third, determine true intercepts for the hyperbola, as follows. Let $y = 0$; then

$$\frac{x^2}{4} - \frac{0}{9} = 1$$
$$x^2 = 4$$
$$x = \pm\sqrt{4} = \pm 2$$

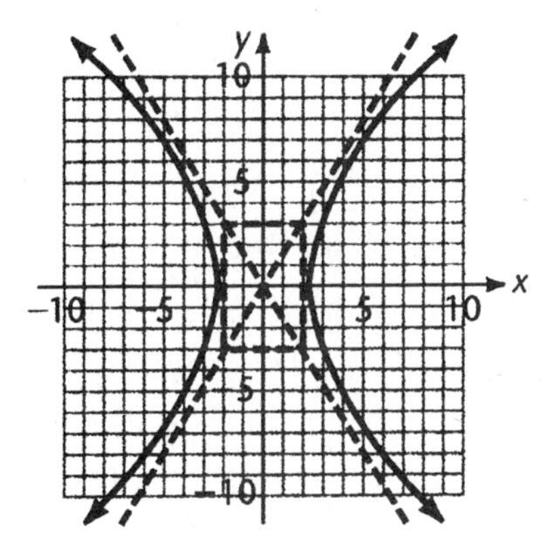

If we let $x = 0$; then

$$\frac{0}{4} - \frac{y^2}{9} = 1$$
$$y^2 = -9$$
$$y = \pm\sqrt{-9} = \pm 3i$$

These are complex numbers and do not represent real intercepts. We conclude that the only real intercepts are $x = \pm 2$, and the hyperbola opens left and right.

15. First, draw a dashed rectangle with intercepts $x = \pm\sqrt{4} = \pm 2$ and $y = \pm\sqrt{9} = \pm 3$. Second, draw in asymptotes (extended diagonals of rectangle). Third, determine true intercepts for the hyperbola, as follows. If we let $y = 0$; then

$$\frac{0}{9} - \frac{x^2}{4} = 1$$
$$x^2 = -4$$
$$x = \pm\sqrt{-4} = \pm 2i$$

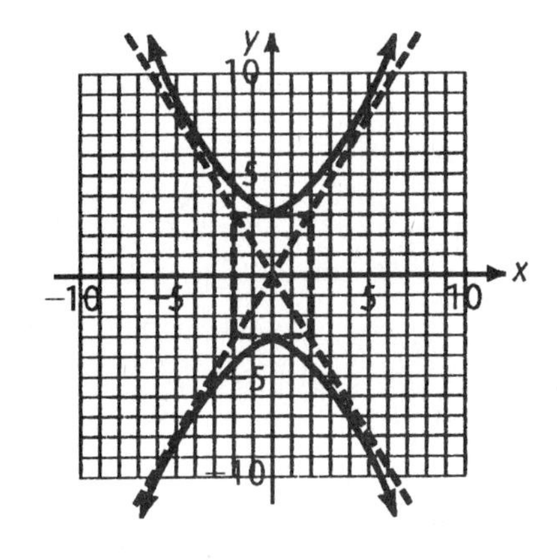

These are complex numbers and do not represent real intercepts. Let $x = 0$; then

$$\frac{y^2}{9} - \frac{0}{4} = 1$$
$$y^2 = 9$$
$$y = \pm\sqrt{9} = \pm 3$$

We conclude that the only real intercepts are $y = \pm 3$, and the hyperbola opens up and down.

17. First, draw a dashed rectangle with intercepts $x = \pm\sqrt{25} = \pm 5$ and $y = \pm\sqrt{1} = \pm 1$. Second, draw in asymptotes (extended diagonals of rectangle). Third, determine true intercepts for the hyperbola, as follows. If we let $y = 0$; then

$$0 - \frac{x^2}{25} = 1$$
$$x^2 = -25$$
$$x = \pm\sqrt{-25} = \pm 5i$$

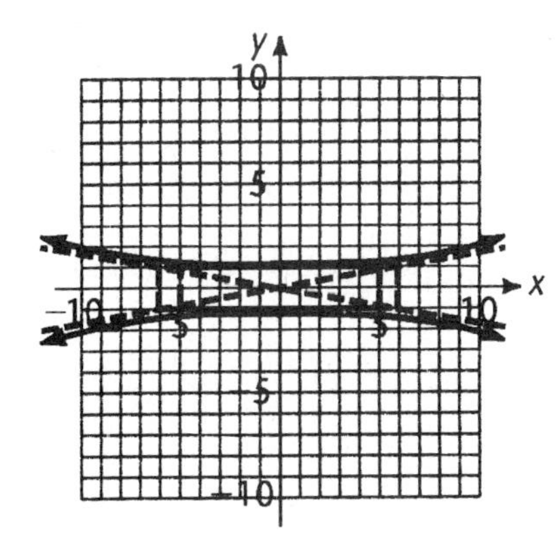

These are complex numbers and do not represent real intercepts. Let $x = 0$; then

$$y^2 - \frac{0}{25} = 1$$
$$y^2 = 1$$
$$y = \pm\sqrt{1} = \pm 1$$

We conclude that the only real intercepts are $y = \pm 1$, and the hyperbola opens up and down.

19. First, draw a dashed rectangle with intercepts $x = \pm\sqrt{1} = \pm1$ and $y = \pm\sqrt{36} = \pm6$. Second, draw in asymptotes (extended diagonals of rectangle). Third, determine true intercepts for the hyperbola, as follows. Let $y = 0$; then

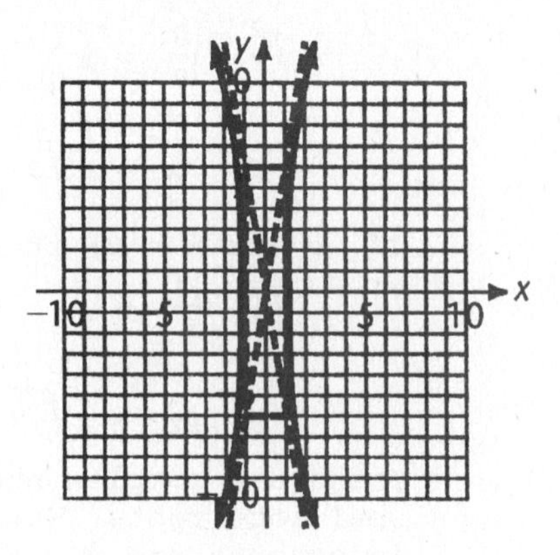

$$x^2 - \frac{0}{36} = 1$$
$$x^2 = 1$$
$$x = \pm\sqrt{1} = \pm1$$

If we let $x = 0$; then

$$0 - \frac{y^2}{36} = 1$$
$$y^2 = -36$$
$$y = \pm\sqrt{-36} = \pm6i$$

These are complex numbers and do not represent real intercepts. We conclude that the only real intercepts are $x = \pm1$, and the hyperbola opens left and right.

21. This equation is in standard form for a hyperbola. We first draw in a dashed rectangle with intercepts $x = \pm\sqrt{4} = \pm2$ and $y = \pm\sqrt{1} = \pm1$. Second, draw in asymptotes (extended diagonals of the rectangle). Third, determine true asymptotes for the hyperbola as follows. Let $y = 0$; then

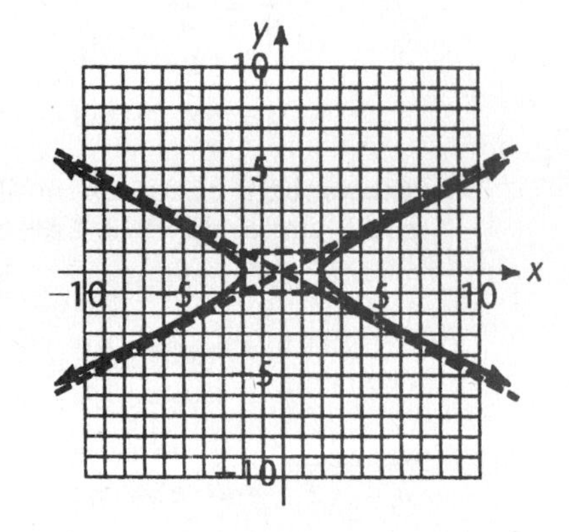

$$\frac{x^2}{4} - 0 = 1$$
$$x^2 = 4$$
$$x = \pm\sqrt{4} = \pm2$$

If we let $x = 0$; then

$$\frac{0}{4} - y^2 = 1$$
$$y^2 = -1$$
$$y = \pm\sqrt{-1} = \pm i$$

These are complex numbers and do not represent real intercepts. We conclude that the only real intercepts are $x = \pm2$, and the hyperbola opens left and right.

23. This equation is in standard form for an ellipse.

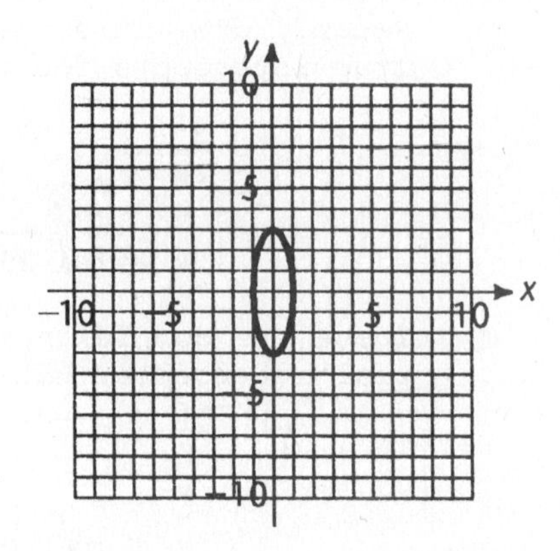

Find x intercepts:

$$x^2 + \frac{0}{9} = 1$$
$$x^2 = 1$$
$$x = \pm\sqrt{1} = \pm1$$

Find y intercepts:

$$0 + \frac{y^2}{9} = 1$$
$$y^2 = 9$$
$$y = \pm\sqrt{9} = \pm3$$

Now plot the intercepts and draw in the ellipse.

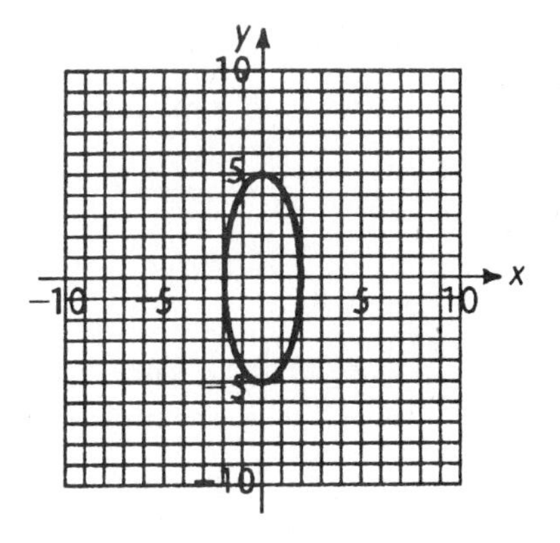

25. This equation is in standard form for an ellipse.

Find x intercepts:

$$\frac{x^2}{4} + \frac{0}{25} = 1$$
$$x^2 = 4$$
$$x = \pm\sqrt{4} = \pm 2$$

Find y intercepts:

$$\frac{0}{4} + \frac{y^2}{25} = 1$$
$$y^2 = 25$$
$$y = \pm\sqrt{25} = \pm 5$$

Now plot the intercepts and draw in the ellipse.

27. Rewriting this as $\frac{y^2}{25} - \frac{x^2}{4} = 1$ puts it in standard form for a hyperbola. We first draw a dashed rectangle with intercepts $x = \pm\sqrt{4} = \pm 2$ and $y = \pm\sqrt{25} = \pm 5$. Second, draw in asymptotes (extended diagonals of rectangle). Third, determine true intercepts for the hyperbola as follows. If we let $y = 0$, then

$$\frac{0}{25} - \frac{x^2}{4} = 1$$
$$x^2 = -4$$
$$x = \pm\sqrt{-4} = \pm 2i$$

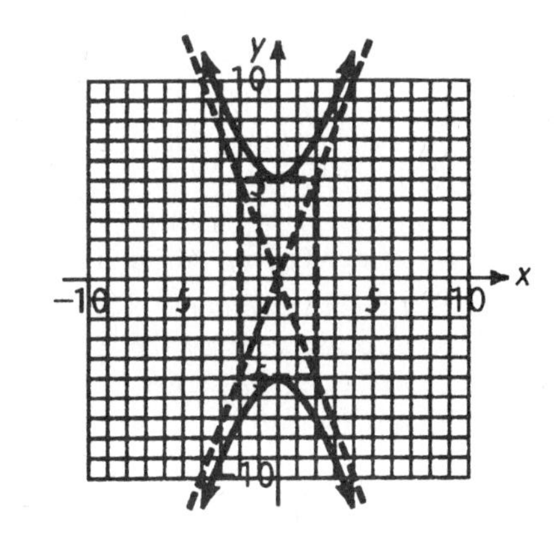

These are complex numbers and do not represent real intercepts. Let $x = 0$, then

$$\frac{y^2}{25} - \frac{0}{4} = 1$$
$$y^2 = 25$$
$$y = \pm\sqrt{25} = \pm 5$$

We conclude that the only real intercepts are $y = \pm 5$ and the hyperbola opens up and down.

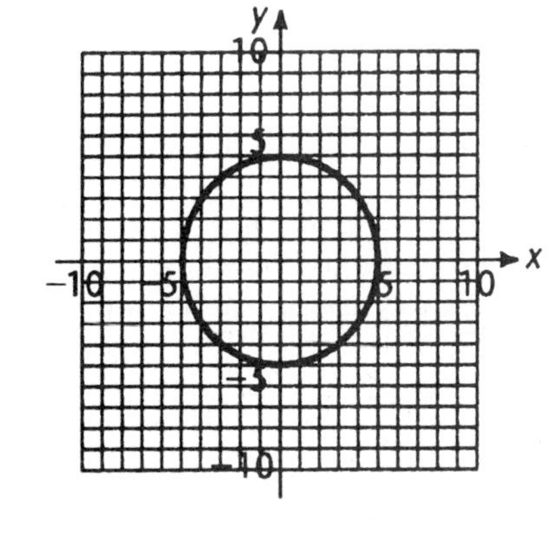

29. Rewriting $\frac{x^2}{25} + \frac{y^2}{25} = 1$ as $x^2 + y^2 = 25$, we see that this is the equation of a circle with center $(h, k) = (0, 0)$ and radius $R = 5$.

31. $4x^2 + 9y^2 = 36$ can be written in the standard form $\frac{x^2}{m^2} + \frac{y^2}{n^2} = 1$, $m > n$, by dividing through by 36.

$$\frac{4x^2}{36} + \frac{9y^2}{36} = \frac{36}{36}$$

$$\frac{x^2}{9} + \frac{y^2}{4} = 1$$

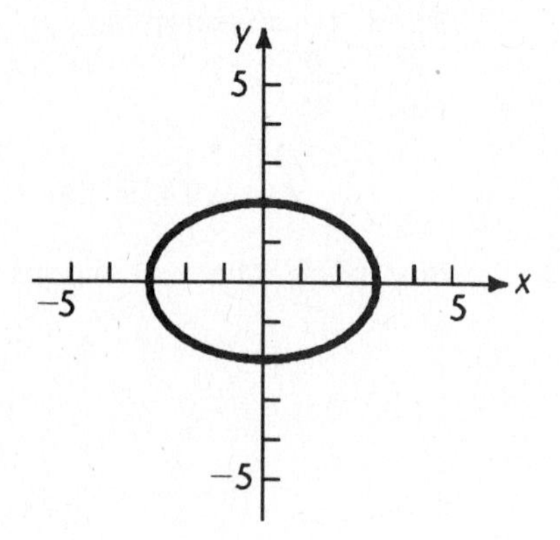

Therefore, this is the equation of an ellipse.

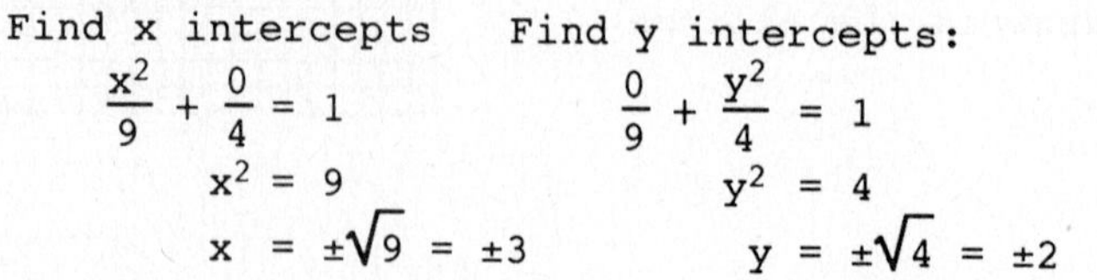

Find x intercepts

$$\frac{x^2}{9} + \frac{0}{4} = 1$$

$$x^2 = 9$$

$$x = \pm\sqrt{9} = \pm 3$$

Find y intercepts:

$$\frac{0}{9} + \frac{y^2}{4} = 1$$

$$y^2 = 4$$

$$y = \pm\sqrt{4} = \pm 2$$

Now plot the intercepts and draw the ellipse.

33. $4x^2 - 9y^2 = 36$ can be written in the standard form $\frac{x^2}{m^2} - \frac{y^2}{n^2} = 1$ by dividing through by 36.

$$\frac{4x^2}{36} - \frac{9y^2}{36} = \frac{36}{36}$$

$$\frac{x^2}{9} - \frac{y^2}{4} = 1$$

Therefore, this is the equation of a hyperbola. We graph it as follows: First, draw a dashed rectangle with intercepts $x = \pm\sqrt{9} = \pm 3$ and $y = \pm\sqrt{4} = \pm 2$. Second, draw in asymptotes (extended diagonals of the rectangle). Third, determine true intercepts for the hyperbola, as follows: Let $y = 0$; then

$$\frac{x^2}{9} - \frac{0}{4} = 1$$

$$x^2 = 9$$

$$x = \pm\sqrt{9} = \pm 3$$

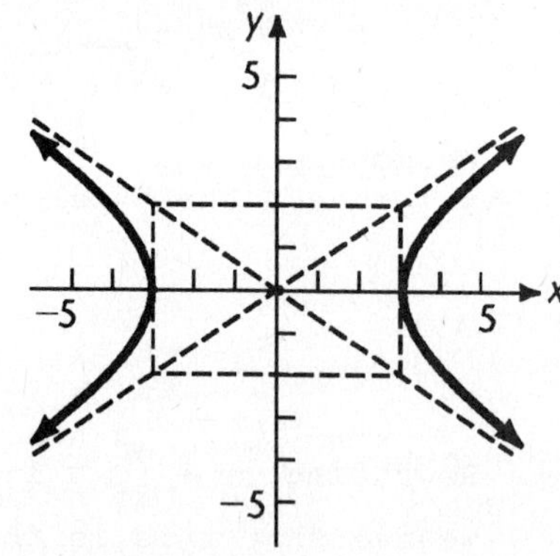

If we let $x = 0$, then

$$\frac{0}{9} - \frac{y^2}{25} = 1$$

$$y^2 = -25$$

$$y = \pm\sqrt{-25} = \pm 5i$$

These are complex numbers and do not represent real intercepts. We conclude that the only real intercepts are $x = \pm 3$, and the hyperbola opens left and right.

35. $4x^2 + 16y^2 = 16$ can be written in the standard form $\frac{x^2}{m^2} + \frac{y^2}{n^2} = 1$, $m > n$, by dividing through by 16.

$$\frac{4x^2}{16} + \frac{16y^2}{16} = \frac{16}{16}$$
$$\frac{x^2}{4} - y^2 = 1$$

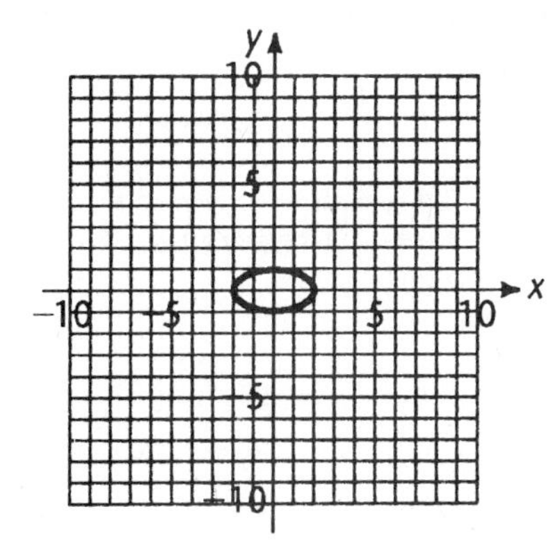

Therefore, this is the equation of an ellipse.

Find x intercepts

$$\frac{x^2}{4} + 0 = 1$$
$$x^2 = 4$$
$$x = \pm\sqrt{4} = \pm 2$$

Find y intercepts

$$\frac{0}{4} + y^2 = 1$$
$$y^2 = 1$$
$$y = \pm\sqrt{1} = \pm 1$$

Now plot the intercepts and draw the ellipse.

37. $16x^2 - 4y^2 = 16$ can be written in the standard form $\frac{x^2}{m^2} - \frac{y^2}{n^2} = 1$ by dividing through by 16.

$$\frac{16x^2}{16} - \frac{4y^2}{16} = \frac{16}{16}$$
$$x^2 - \frac{y^2}{4} = 1$$

Therefore, this is the equation of a hyperbola. We graph it as follows: First, draw a dashed rectangle with intercepts $x = \pm\sqrt{1} = \pm 1$ and $y = \pm\sqrt{4} = \pm 2$. Second, draw in asymptotes (extended diagonals of the rectangle). Third, determine true intercepts for the hyperbola, as follows: Let $y = 0$; then

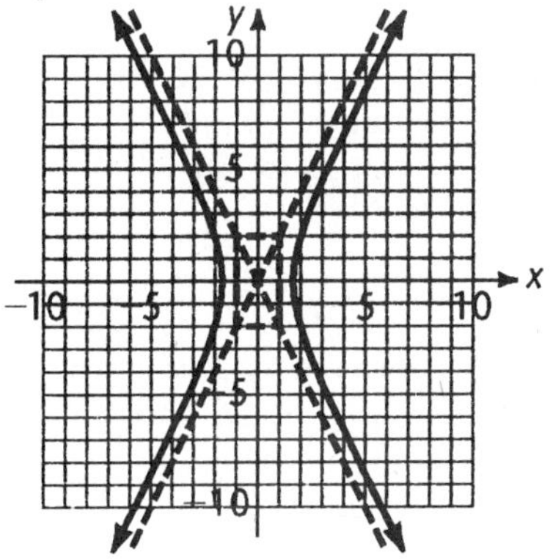

$$x^2 - \frac{0}{4} = 1$$
$$x^2 = 1$$
$$x = \pm\sqrt{1} = \pm 1$$

If we let $x = 0$, then

$$0 - \frac{y^2}{4} = 1$$
$$y^2 = -4$$
$$y = \pm\sqrt{-4} = \pm 2i$$

These are complex numbers and do not represent real intercepts. We conclude that the only real intercepts are $x = \pm 1$, and the hyperbola opens left and right.

39. $-16x^2 + 4y^2 = 16$ can be written in the standard form $\frac{y^2}{n^2} - \frac{x^2}{m^2} = 1$ by dividing through by 16.

$$\frac{-16x^2}{16} + \frac{4y^2}{16} = \frac{16}{16}$$
$$\frac{y^2}{4} - x^2 = 1$$

Therefore, this is the equation of a hyperbola. We graph it as follows: First, draw a dashed rectangle with intercepts $x = \pm\sqrt{1} = \pm 1$ and $y = \pm\sqrt{4} = \pm 2$.

Second, draw in asymptotes (extended diagonals of the rectangle). Third, determine true intercepts for the hyperbola, as follows: Let $y = 0$; then

$$\frac{0}{4} - x^2 = 1$$
$$x^2 = -1$$
$$x = \pm\sqrt{-1} = \pm i$$

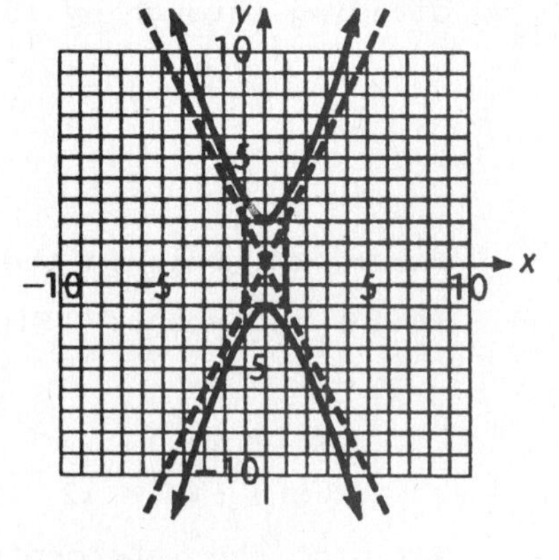

These are complex numbers and do not represent real intercepts. Let $x = 0$, then

$$\frac{y^2}{4} - 0 = 1$$
$$y^2 = 4$$
$$y = \pm\sqrt{4} = \pm 2$$

We conclude that the only real intercepts are $y = \pm 2$, and the hyperbola opens up and down.

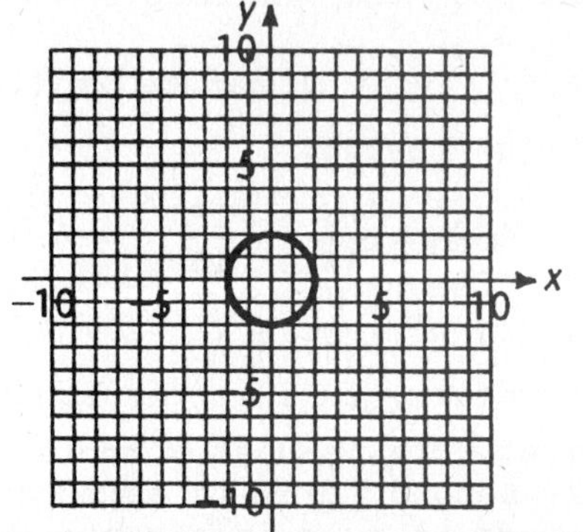

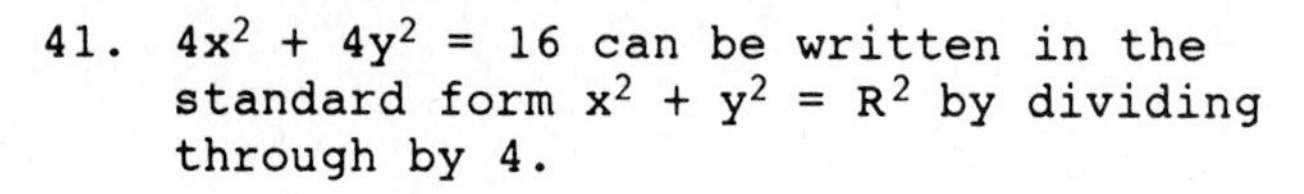

41. $4x^2 + 4y^2 = 16$ can be written in the standard form $x^2 + y^2 = R^2$ by dividing through by 4.

$$\frac{4x^2}{4} + \frac{4y^2}{4} = \frac{16}{4}$$
$$x^2 + y^2 = 4$$

This is the equation of a circle with center $(h, k) = (0, 0)$ and radius $R = 2$.

43. $4x^2 - 4y^2 = 16$ can be written in the standard form $\frac{x^2}{m^2} - \frac{y^2}{n^2} = 1$ by dividing through by 16.

$$\frac{4x^2}{16} - \frac{4y^2}{16} = \frac{16}{16}$$
$$\frac{x^2}{4} - \frac{y^2}{4} = 1$$

Therefore, this is the equation of a hyperbola. We graph it as follows: First, draw a dashed rectangle with intercepts $x = \pm\sqrt{4} = \pm 2$ and $y = \pm\sqrt{4} = \pm 2$. Second, draw in asymptotes (extended diagonals of the rectangle). Third, determine true intercepts for the hyperbola, as follows: Let $y = 0$; then

$$\frac{x^2}{4} - \frac{0}{4} = 1$$
$$x^2 = 4$$
$$x = \pm\sqrt{4} = \pm 2$$

If we let $x = 0$, then

$$\frac{0}{4} - \frac{y^2}{4} = 1$$
$$y^2 = -4$$
$$y = \pm\sqrt{-4} = \pm 2i$$

These are complex numbers and do not represent real intercepts. We conclude that the only real intercepts are $x = \pm 2$, and the hyperbola opens left and right.

45. We can rewrite $x^2 - 9y^2 = 1$ in standard form $\frac{x^2}{m^2} - \frac{y^2}{n^2} = 1$ as $x^2 - \frac{y^2}{1/9} = 1$. Therefore, this is the equation of a hyperbola. We graph it as follows: First, draw a dashed rectangle with intercepts $x = \pm\sqrt{1} = \pm 1$ and $y = \pm\sqrt{\frac{1}{9}} = \pm\frac{1}{3}$. Second, draw in asymptotes (extended diagonals of the rectangle). Third, determine true intercepts for the hyperbola, as follows: Let $y = 0$; then

$$\begin{aligned} x^2 - 9(0) &= 1 \\ x^2 &= 1 \\ x &= \pm\sqrt{1} = \pm 1 \end{aligned}$$

If we let $x = 0$, then

$$\begin{aligned} 0 - 9y^2 &= 1 \\ y^2 &= -\frac{1}{9} \\ y &= \pm\sqrt{-\frac{1}{9}} = \pm\frac{1}{3}i \end{aligned}$$

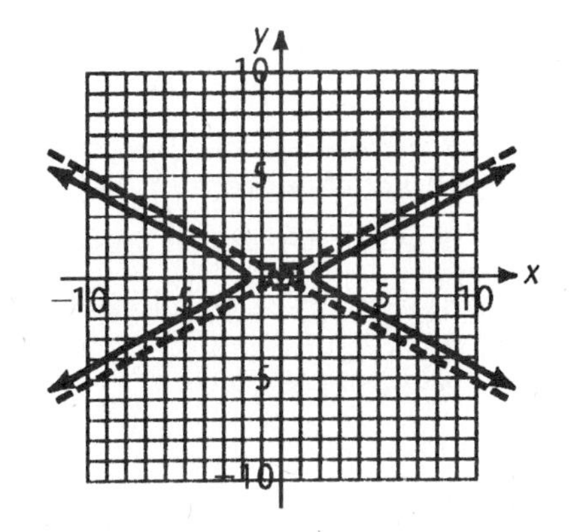

These are complex numbers and do not represent real intercepts. We conclude that the only real intercepts are $x = \pm 1$, and the hyperbola opens left and right.

47. We can rewrite $9x^2 + y^2 = 1$ in standard form $\frac{x^2}{m^2} + \frac{y^2}{n^2} = 1$ as $\frac{x^2}{1/9} + \frac{y^2}{1} = 1$.

Therefore, this is the equation of an ellipse.

Find x intercepts

$$\begin{aligned} 9x^2 + 0 &= 1 \\ x^2 &= \frac{1}{9} \\ x &= \pm\sqrt{\frac{1}{9}} = \pm\frac{1}{3} \end{aligned}$$

Find y intercepts

$$\begin{aligned} 0 + y^2 &= 1 \\ y &= \pm\sqrt{1} = \pm 1 \end{aligned}$$

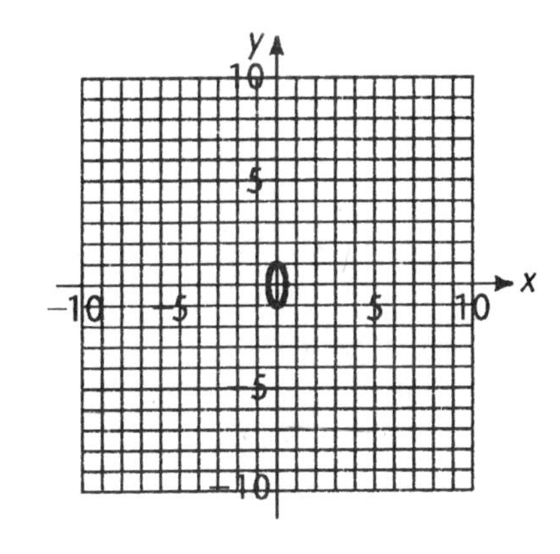

Now plot the intercepts and draw the ellipse.

49. We can rewrite $4x^2 - y^2 = 1$ in standard form $\frac{x^2}{m^2} - \frac{y^2}{n^2} = 1$ as $\frac{x^2}{1/4} - y^2 = 1$. Therefore, this is the equation of a hyperbola. We graph it as follows: First, draw a dashed rectangle with intercepts $x = \pm\sqrt{\frac{1}{4}} = \pm\frac{1}{2}$ and $y = \pm\sqrt{1} = \pm 1$. Second, draw in asymptotes (extended diagonals of the rectangle). Third, determine true intercepts for the hyperbola, as follows: Let $y = 0$; then

$$4x^2 - 0 = 1$$
$$x^2 = \frac{1}{4}$$
$$x = \pm\sqrt{\frac{1}{4}} = \pm\frac{1}{2}$$

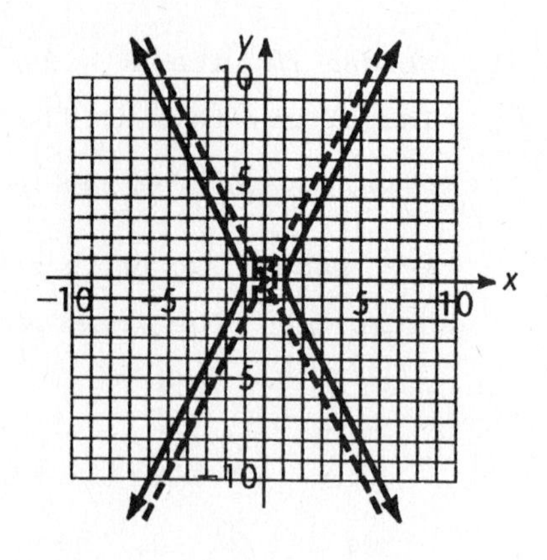

If we let $x = 0$, then

$$0 - y^2 = 1$$
$$y^2 = -1$$
$$y = \pm\sqrt{-1} = \pm i$$

These are complex numbers and do not represent real intercepts. We conclude that the only real intercepts are $x = \pm\frac{1}{2}$, and the hyperbola opens left and right.

51. We can rewrite $x^2 + 4y^2 = 1$ in standard form $\frac{x^2}{m^2} + \frac{y^2}{n^2} = 1$ as $x^2 + \frac{y^2}{1/4} = 1$.

Therefore, this is the equation of an ellipse.

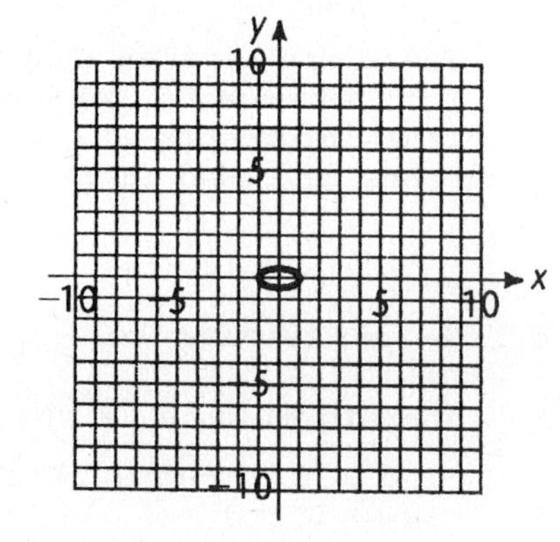

Find x intercepts
$$x^2 + 4(0) = 1$$
$$x^2 = 1$$
$$x = \pm\sqrt{1} = \pm 1$$

Find y intercepts
$$0 + 4y^2 = 1$$
$$y^2 = \frac{1}{4}$$
$$y = \pm\sqrt{\frac{1}{4}} = \pm\frac{1}{2}$$

Now plot the intercepts and draw the ellipse.

53. We can rewrite $4x^2 + 9y^2 = 1$ in standard form $\frac{x^2}{m^2} + \frac{y^2}{n^2} = 1$ as $\frac{x^2}{1/4} + \frac{y^2}{1/9} = 1$.

Therefore, this is the equation of an ellipse.

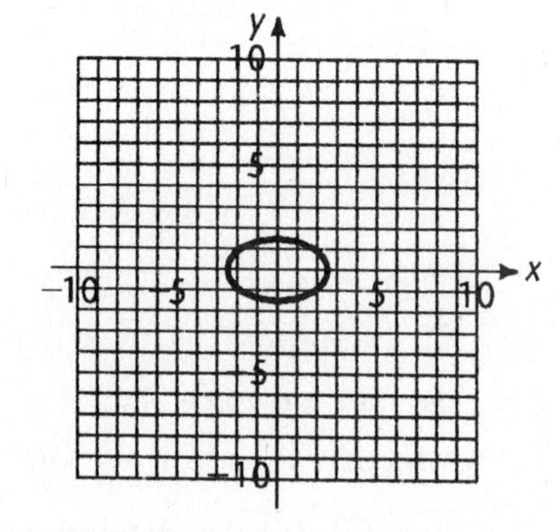

Find x intercepts
$$4x^2 + 9(0) = 1$$
$$x^2 = \frac{1}{4}$$
$$x = \pm\sqrt{\frac{1}{4}} = \pm\frac{1}{2}$$

Find y intercepts
$$4(0) + 9y^2 = 1$$
$$y^2 = \frac{1}{9}$$
$$y = \pm\sqrt{\frac{1}{9}} = \pm\frac{1}{3}$$

Now plot the intercepts and draw the ellipse.

55. This is the equation of a hyperbola in standard form $\frac{y^2}{n^2} - \frac{x^2}{m^2} = 1$.

We graph it as follows:

First, draw a dashed rectangle with intercepts $x = \pm\sqrt{1} = \pm 1$ and $y = \pm 1$. Second, draw in asymptotes (extended diagonals of the rectangle). Third, determine true intercepts for the hyperbola, as follows: If we let $y = 0$; then

$$0 - x^2 = 1$$
$$x^2 = -1$$
$$x = \pm\sqrt{-1} = \pm i$$

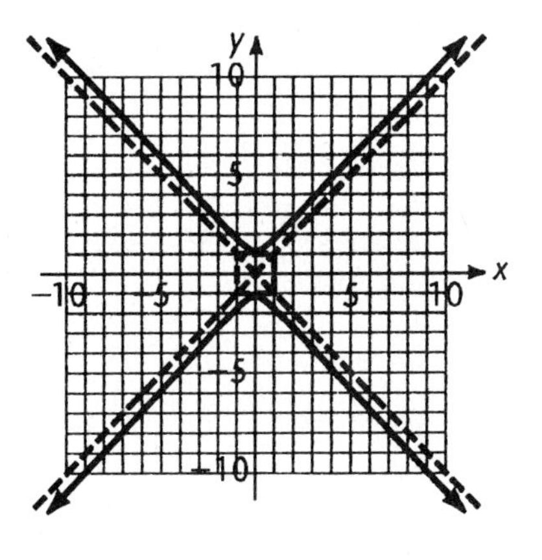

These are complex numbers and do not represent real intercepts. Let $x = 0$, then

$$y^2 - 0 = 1$$
$$y^2 = 1$$
$$y = \pm\sqrt{1} = \pm 1$$

We conclude that the only real intercepts are $y = \pm 1$, and the hyperbola opens up and down.

57. $x^2 + y^2 = 4$ is the equation in standard form of a circle with center $(h, k) = (0, 0)$ and radius $R = 2$.

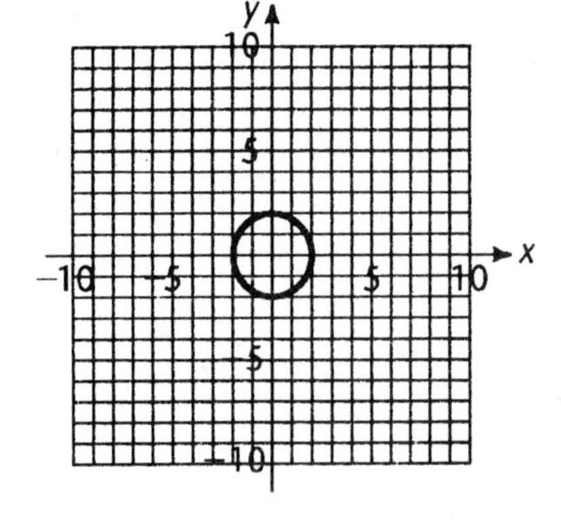

59. We can rewrite $4x^2 + 16y^2 = 1$ in standard form $\frac{x^2}{m^2} + \frac{y^2}{n^2} = 1$ as $\frac{x^2}{1/4} + \frac{y^2}{1/16} = 1$. Therefore, this is the equation of an ellipse.

Find x intercepts	Find y intercepts
$4x^2 + 16(0) = 1$	$4(0) + 16y^2 = 1$
$x^2 = \frac{1}{4}$	$y^2 = \frac{1}{16}$
$x = \pm\sqrt{\frac{1}{4}} = \pm\frac{1}{2}$	$y = \pm\sqrt{\frac{1}{16}} = \pm\frac{1}{4}$

Now plot the intercepts and draw the ellipse.

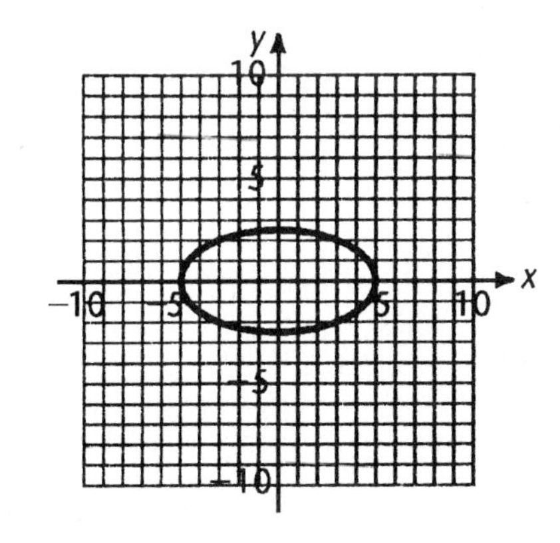

61. We can rewrite $y^2 - x = 0$ in standard form as $x = y^2$. This is the equation of a parabola with vertex (0, 0). We choose $x = 4$ to make the left side a positive perfect square, then

$$4 = y^2$$
$$y = \pm 2$$

Thus, the points (4, 2) and (4, -2) are also on the graph. We plot the three known points; then we sketch a parabola through these points.

63. We can rewrite $x^2 - y^2 = 16$ in standard form $\frac{x^2}{m^2} - \frac{y^2}{n^2} = 1$ as $\frac{x^2}{16} - \frac{y^2}{16} = 1$.

We graph it as follows:

First, draw a dashed rectangle with intercepts $x = \pm\sqrt{16} = \pm 4$ and $y = \pm 4$. Second, draw in asymptotes (extended diagonals of the rectangle). Third, determine true intercepts for the hyperbola, as follows: If we let $y = 0$; then

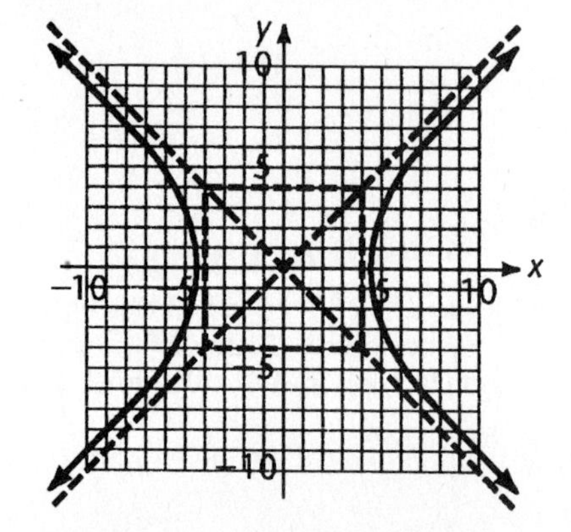

$$x^2 - 0 = 16$$
$$x^2 = 16$$
$$x = \pm\sqrt{16} = \pm 4$$

If we let $x = 0$, then

$$0 - y^2 = 16$$
$$y^2 = -16$$
$$y = \pm\sqrt{-16} = \pm 4i$$

These are complex numbers and do not represent real intercepts. We conclude that the only real intercepts are $x = \pm 4$, and the hyperbola opens left and right.

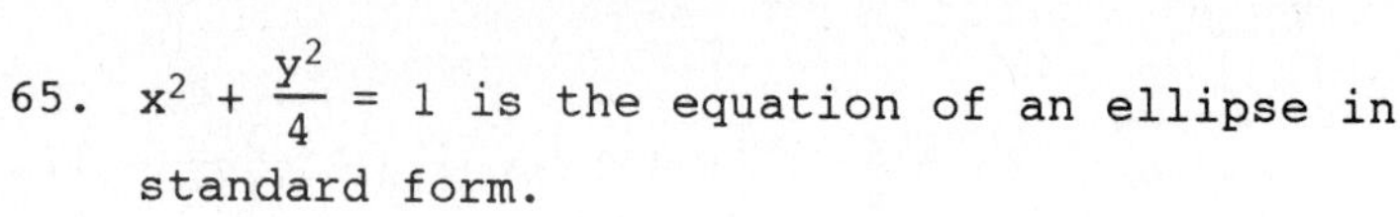

65. $x^2 + \frac{y^2}{4} = 1$ is the equation of an ellipse in standard form.

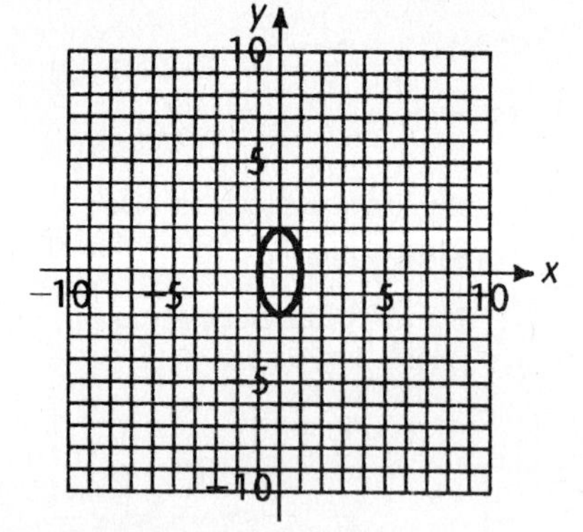

Find x intercepts

$$x^2 + \frac{0}{4} = 1$$
$$x^2 = 1$$
$$x = \pm\sqrt{1} = \pm 1$$

Find y intercepts

$$0 + \frac{y^2}{4} = 1$$
$$y^2 = 4$$
$$y = \pm\sqrt{4} = \pm 2$$

Now plot the intercepts and draw the ellipse.

67. The left side of $4x^2 + y^2 = 0$ is the sum of two squares. This can only be zero if both x and y are 0. Thus the only point on the graph of $4x^2 + y^2 = 0$ is the origin (0, 0).

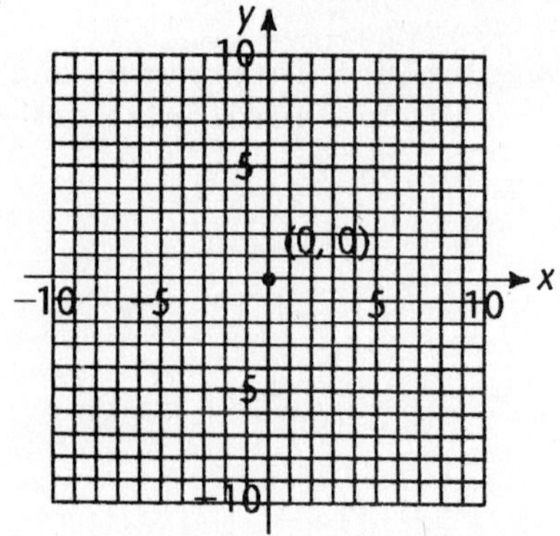

69. $x^2 + y^2 = 2$ is the equation in standard form of a circle with center $(h, k) = (0, 0)$ and radius $R = \sqrt{2}$.

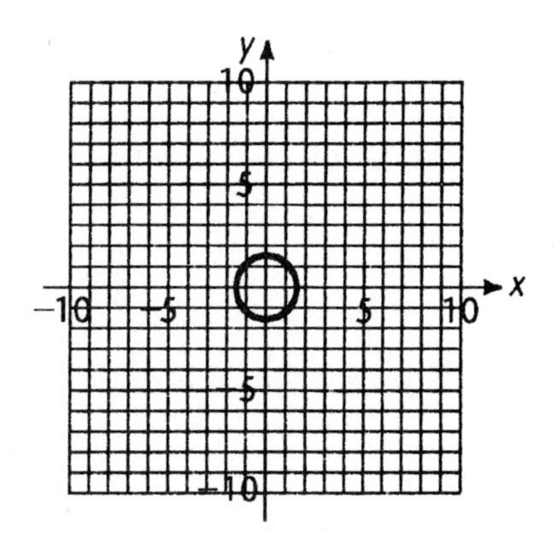

71. The foci of an ellipse whose equation in standard form is $\frac{x^2}{m^2} + \frac{y^2}{n^2} = 1$, $n > m$, are $(0, c)$ and $(0, -c)$, where $m^2 = n^2 - c^2$. The equation of this ellipse is $\frac{x^2}{4} + \frac{y^2}{25} = 1$; hence $n^2 = 25$ and $m^2 = 4$. To find c, we solve the equation

$$\begin{aligned} 4 &= 25 - c^2 \\ -21 &= -c^2 \\ c^2 &= 21 \\ c &= \pm\sqrt{21} \end{aligned}$$

We find $c = \sqrt{21}$, hence the foci are $(0, \sqrt{21})$ and $(0, -\sqrt{21})$.

73. The foci of a hyperbola whose equation in standard form is $\frac{x^2}{m^2} - \frac{y^2}{n^2} = 1$ are $(c, 0)$ and $(-c, 0)$, where $n^2 = c^2 - m^2$. The equation of this hyperbola is $\frac{x^2}{16} - \frac{y^2}{9} = 1$; hence $m^2 = 16$ and $n^2 = 9$. To find c, we solve the equation

$$\begin{aligned} 9 &= c^2 - 16 \\ 25 &= c^2 \\ c^2 &= 25 \\ c &= \pm\sqrt{25} = \pm 5 \end{aligned}$$

We have $c = 5$, hence the foci are $(5, 0)$ and $(-5, 0)$.

75. We complete the square relative to x and relative to y.

$$\begin{aligned} &x^2 - 2x + 4y^2 - 16y + 13 = 0 \\ &x^2 - 2x + ? + 4y^2 - 16y + ? = ? + ? - 13 \\ &x^2 - 2x + 1 + 4(y^2 - 4y + 4) = 1 + 4(4) - 13 \\ &(x - 1)^2 + 4(y - 2)^2 = 4 \\ &\frac{(x - 1)^2}{4} + \frac{(y - 2)^2}{1} = 1 \end{aligned}$$

This is the equation of an ellipse.

77. We complete the square relative to x and relative to y.

$$\begin{aligned} x^2 + 2x + y^2 + 4y + 1 &= 0 \\ x^2 + 2x + y^2 + 4y &= -1 \\ x^2 + 2x + ? + y^2 + 4y + ? &= -1 + ? + ? \\ x^2 + 2x + 1 + y^2 + 4y + 4 &= -1 + 1 + 4 \\ (x + 1)^2 + (y + 2)^2 &= 4 \end{aligned}$$

This is the equation of a circle.

79. We complete the square relative to x and relative to y.

$$4x^2 + 24x \quad + y^2 - 2y + 33 = 0$$
$$4x^2 + 24x + ? + y^2 - 2y + ? = -33 + ? + ?$$
$$4(x^2 + 6x + 9) + y^2 - 2y + 1 = -33 + 4(9) + 1$$
$$4(x + 3)^2 + (y - 1)^2 = 4$$
$$(x + 3)^2 + \frac{(y - 1)^2}{4} = 1$$

This is the equation of an ellipse.

81. We complete the square relative to x and relative to y.

$$4x^2 - 24x \quad - y^2 - 2y + 31 = 0$$
$$4x^2 - 24x + ? - y^2 - 2y + ? = -31 + ? + ?$$
$$4(x^2 - 6x + 9) - (y^2 + 2y + 1) = -31 + 4(9) - 1$$
$$4(x - 3)^2 - (y + 1)^2 = 4$$
$$(x - 3)^2 - \frac{(y + 1)^2}{4} = 1$$

This is the equation of a hyperbola.

83. We complete the square relative to x and relative to y.

$$-4x^2 - 24x \quad + y^2 - 2y - 27 = 0$$
$$-4x^2 - 24x + ? + y^2 - 2y + ? = 27 + ? + ?$$
$$-4(x^2 + 6x + 9) + y^2 - 2y + 1 = 27 - 4(9) + 1$$
$$-4(x + 3)^2 + (y - 1)^2 = -8$$
$$\frac{(x + 3)^2}{2} - \frac{(y - 1)^2}{8} = 1$$

This is the equation of a hyperbola.

Exercise 7-9 Nonlinear Inequalities

Key Ideas and Formulas

The value of x at which the linear expression (ax + b) is 0 is called a **critical point**. To the left of the critical point, on the real number line, (ax + b) has one sign and to the right of the critical point the opposite sign.

Sign-Analysis Method for Quadratic Inequalities

To solve a quadratic inequality of the form

$ax^2 + bx + c$ [compared to] 0 — The notation [compared to] represents the four possible inequalities $<$, $>$, $\le$, and $\ge$.

1. Factor $ax^2 + bx + c$ into linear factors.
2. Determine the critical point for each factor.
3. Analyze the sign of each factor on each interval determined by the critical points.
4. Use the signs of the factors to determine the sign of the product $ax^2 + bx + c$. If the comparison is $\le$ of $\ge$, include the critical points in the solution.

Sign analysis also can be used on other inequalities where linear terms are combined by multiplication and division and the result compared to 0.

Set Union and Intersection:

Union: $A \cup B = \{x \mid x \in A \text{ or } x \in B\}$

Intersection: $A \cap B = \{x \mid x \in A \text{ and } x \in B\}$

If $A \cap B = \varnothing$, that is, the intersection of A and B is empty, the sets A and B are said to be **disjoint**.

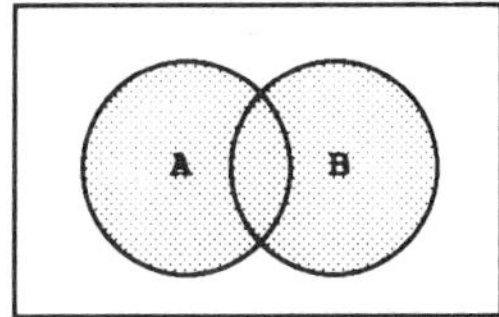

(a) Union of two sets

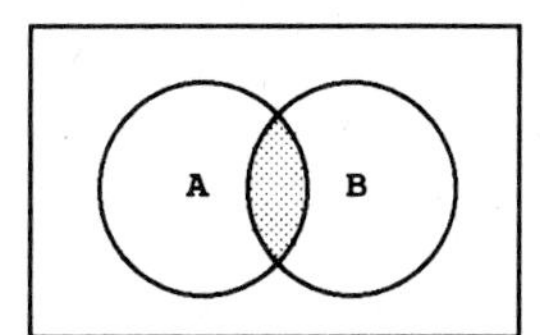

(b) Intersection of two sets

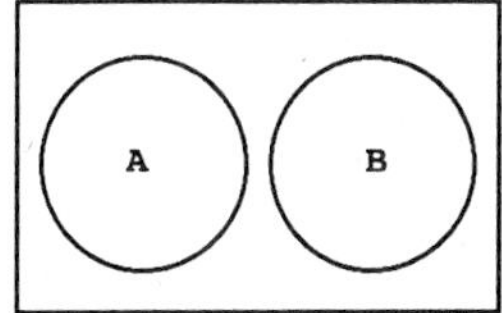

(c) Two disjoint sets

---

1. $(x-3)(x+4) < 0$
   Critical points -4, 3

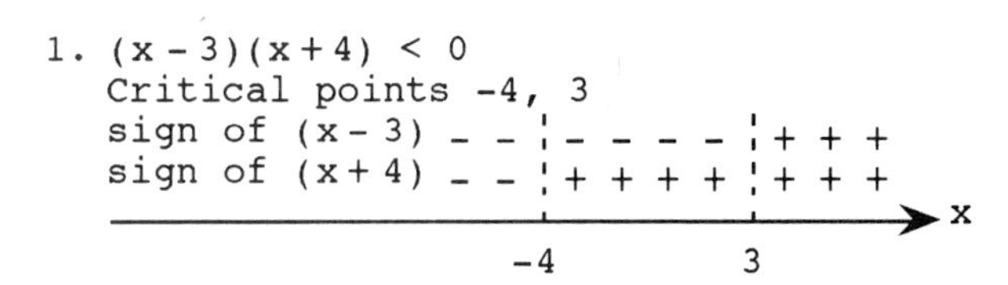

COMMON ERROR: "$x - 3 < 0$ or $x + 4 < 0$." The signs of the factors must be opposite for the product to be negative.

The product will be negative where the factors have opposite signs, between -4 and 3.

$-4 < x < 3$
$(-4, 3)$

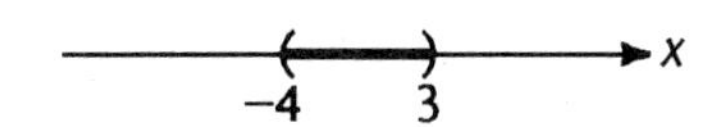

3. $(x - 3)(x + 4) \geq 0$
   Critical points -4, 3
   The signs of the factors are charted in problem 1 above. The equality part of the inequality statement is satisfied at the critical points. The inequality part is satisfied when the product is positive, that is, when the factors have the same sign, left of -4 or right of 3.

$x \leq -4$ or $x \geq 3$
$(-\infty, -4] \cup [3, \infty)$

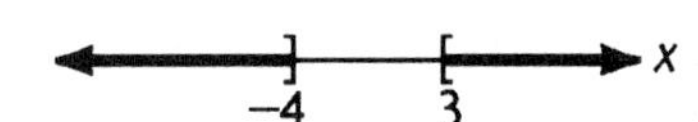

5.
$$x^2 + x < 12$$
$$x^2 + x - 12 < 0$$
$$(x - 3)(x + 4) < 0$$
The remainder of this problem is solved as problem 1 above.

7.
$$x^2 + 21 > 10x$$
$$x^2 - 10x + 21 > 0$$
$$(x - 7)(x - 3) > 0$$
Critical points 3, 7

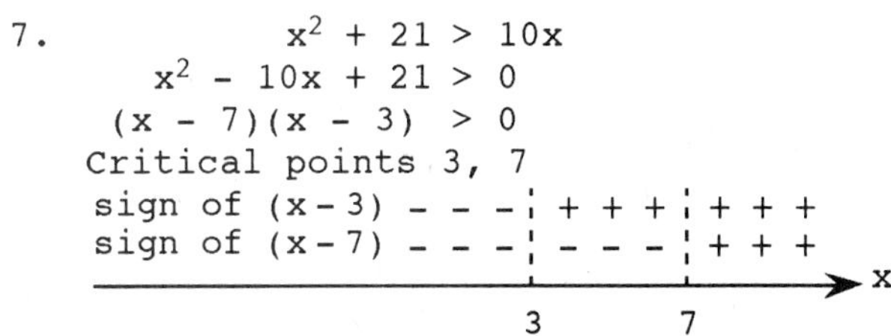

The factors must have the same sign; this occurs to the left of 3 or to the right of 7.

$x < 3$ or $x > 7$
$(-\infty, 3) \cup (7, \infty)$

9. $x^2 + 6x + 5 \le 0$

$(x + 5)(x + 1) \le 0$

Critical points -5, -1

sign of (x+5)   - - - | + + + | + + +

sign of (x+1)   - - - | - - - | + + +

x: -5, -1

The equality part of the inequality statement is satisfied at the critical points. The inequality part is satisfied when the factors have opposite signs, between -5 and -1. [-5, -1]

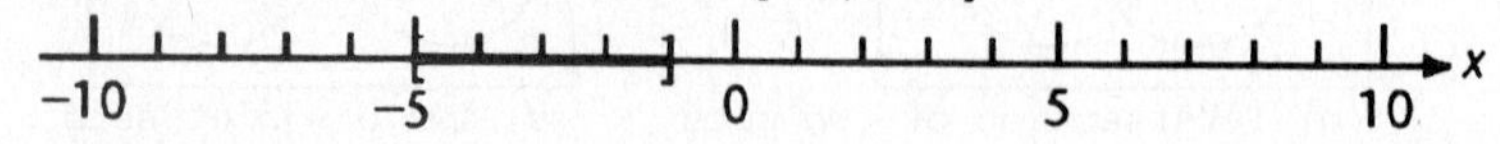

11. $x^2 - 9x + 8 < 0$

$(x - 1)(x - 8) < 0$

Critical points 1, 8

sign of (x-1)   - - - | + + + | + + +

sign of (x-8)   - - - | - - - | + + +

x: 1, 8

The factors must have opposite sign; this occurs between 1 and 8. (1, 8)

-10   -5   0   5   10   x

13. $x^2 - 3x - 10 > 0$

$(x + 2)(x - 5) > 0$

Critical points -2, 5

sign of (x+2)   - - - | + + + | + + +

sign of (x-5)   - - - | - - - | + + +

x: -2, 5

The factors must have the same sign; this occurs to the right of 5 or to the left of -2. $(-\infty, -2) \cup (5, \infty)$

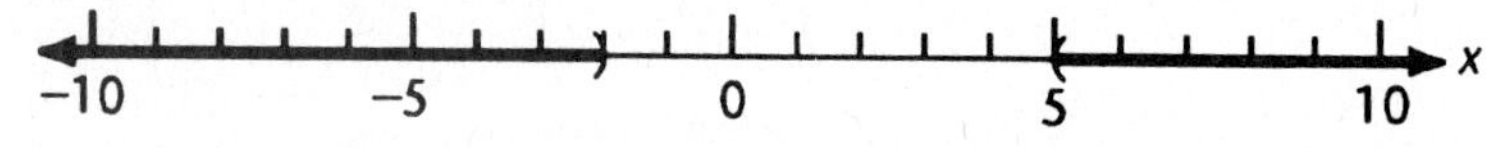

15. $x^2 + 4x - 12 \le 0$

$(x + 6)(x - 2) \le 0$

Critical points -6, 2

sign of (x+6)   - - - | + + + | + + +

sign of (x-2)   - - - | - - - | + + +

x: -6, 2

The equality part of the inequality statement is satisfied at the critical points. The inequality part is satisfied when the factors have opposite signs, between -6 and 2.

17. $x(x + 6) \geq 0$

Critical points -6, 0

sign of $(x + 6)$ − − ⋮ + + + + ⋮ + + +
sign of x − − ⋮ − − − − ⋮ + + +

x-axis: −6, 0

The equality part of the statement is satisfied at the critical points. The inequality part is satisfied when the product is positive, that is, when the factors have the same sign, left of -6 or right of 0.

$(-\infty, -6] \cup [0, \infty)$

19.

$$x^2 \geq 9$$
$$x^2 - 9 \geq 0$$
$$(x - 3)(x + 3) \geq 0$$

Critical points -3, 3

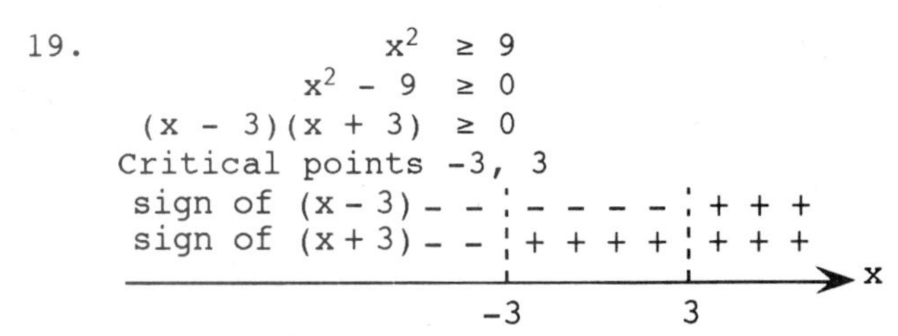

The equality part of the statement is satisfied at the critical points. The inequality part is satisfied when the product is positive, that is, when the factors have the same sign, left of -3 or right of 3.

$(-\infty, -3] \cup [3, \infty)$ — number line: −3, 3, x

21. Graphically:

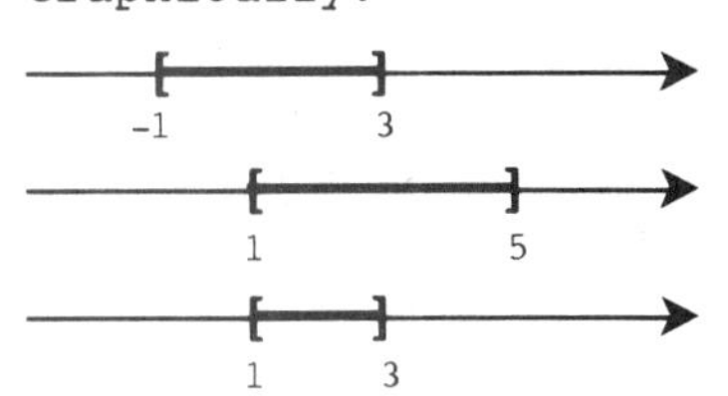

Symbolically:

$[-1, 3] \cap [1, 5] = [1, 3]$

23. Graphically:

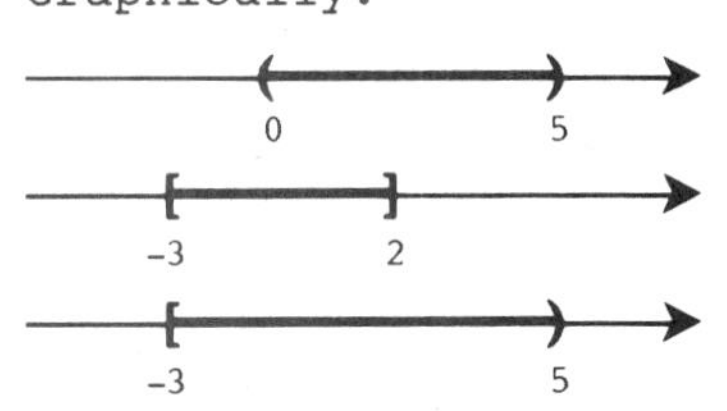

Symbolically:

$(0, 5) \cup [-3, 2] = [-3, 5)$

25. Graphically:

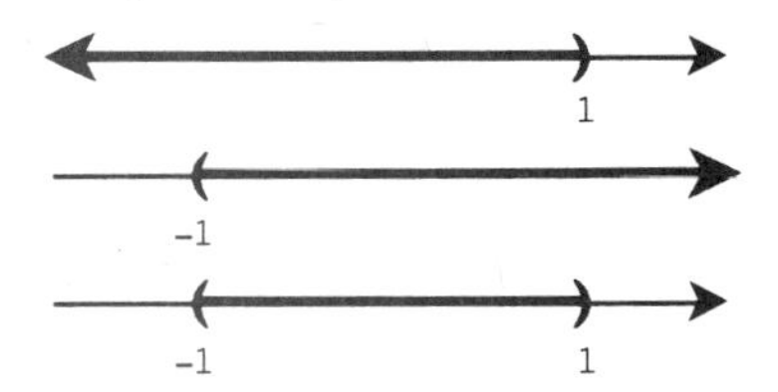

Symbolically:

$(-\infty, 1) \cap (-1, \infty) = (-1, 1)$

27. Graphically:

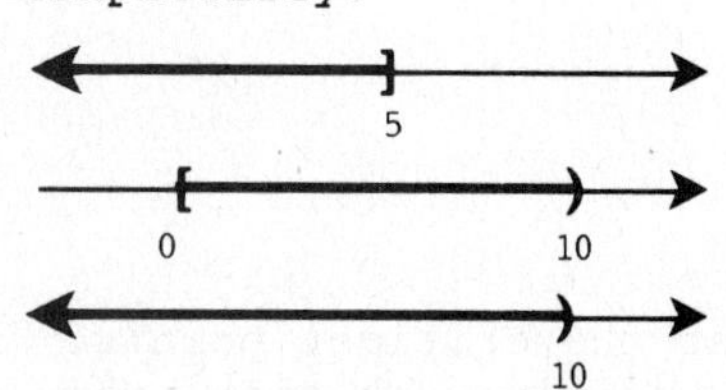

Symbolically:

$(-\infty, 5] \cup [0, 10) = (-\infty, 10)$

29. Graphically:

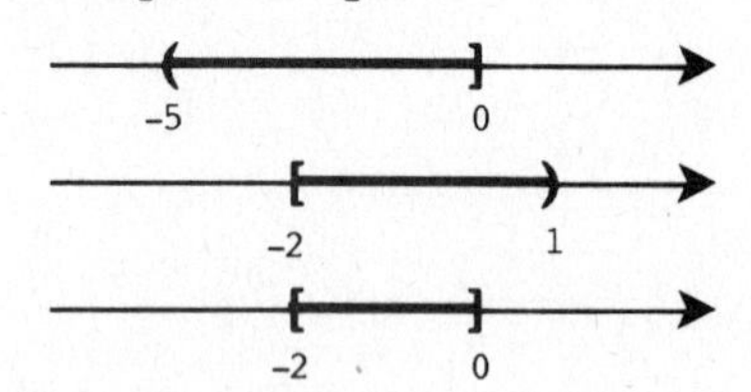

Symbolically:

$(-3, 0] \cap [-2, 1) = [-2, 0]$

31. Graphically:

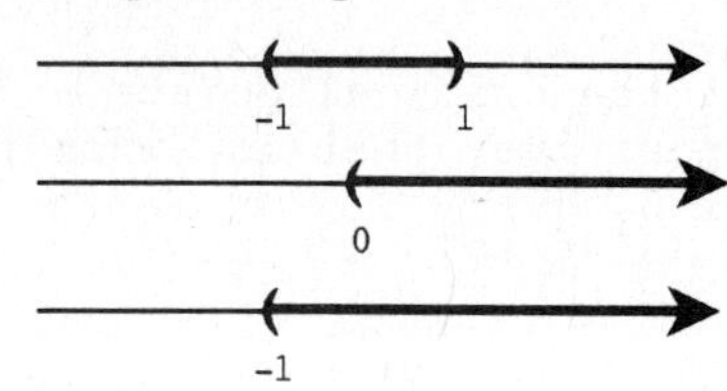

Symbolically:

$(-1, 1) \cup (0, \infty) = (-1, \infty)$

33. Graphically:

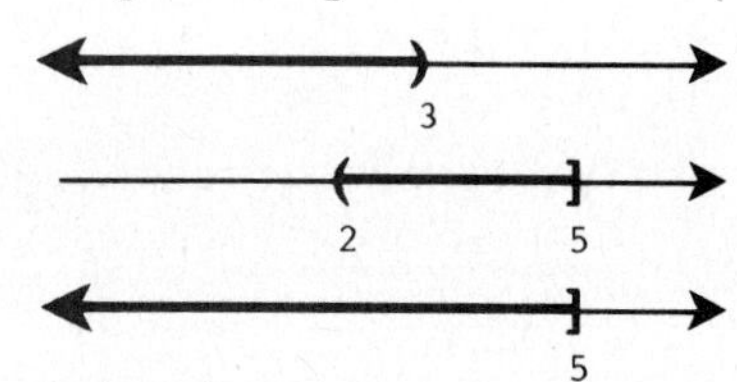

Symbolically:

$(-\infty, 3) \cup (2, 5] = (-\infty, 5]$

35. Graphically:

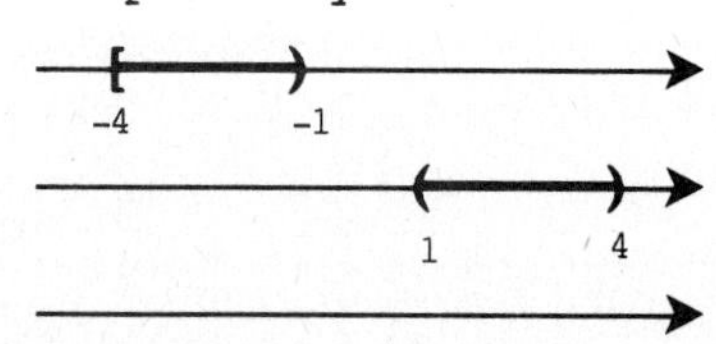

Symbolically:

$[-4, -1) \cap (1, 4) = \varnothing$

37. Graphically:

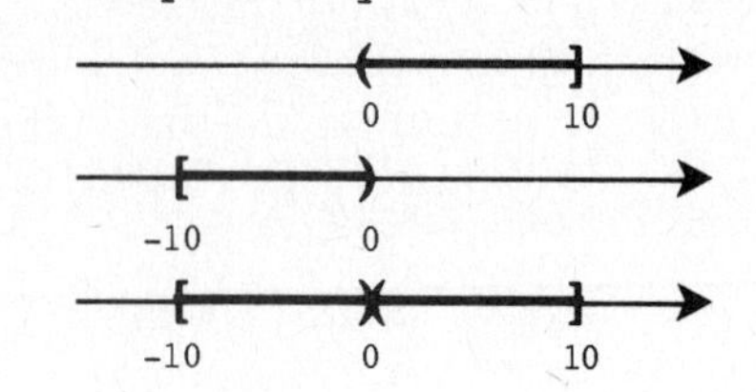

Symbolically:

$[-10, 0) \cup (0, 10]$ is not a single interval.

39. $(x - 3)(x + 4)(x - 5) < 0$
Critical points 3, −4, 5

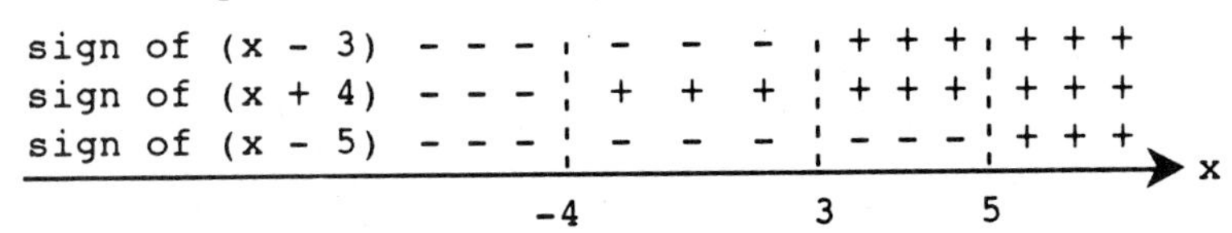

The inequality is satisfied when $(x - 3)$, $(x + 4)$, and $(x - 5)$ are all negative, or two are positive and one negative. All are negative to the left of −4 and two are positive between 3 and 5. Thus,

Solution: $(-\infty, -4) \cup (3, 5)$

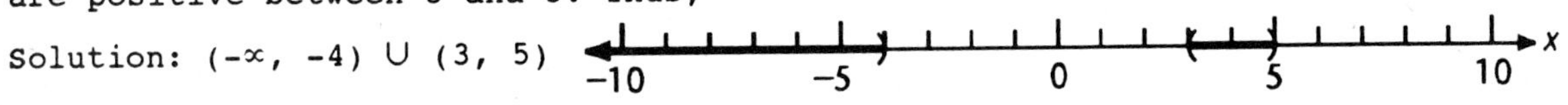

41. $x(x - 3)(x + 4) \geq 0$
Critical points 0, 3, −4

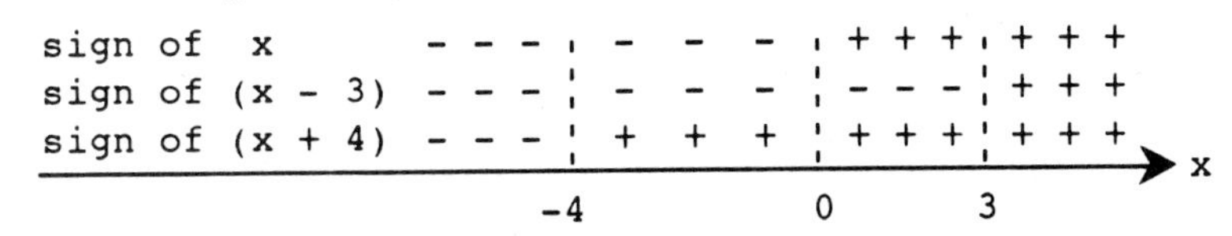

The inequality is satisfied when x, $(x - 3)$, and $(x + 4)$ are all positive, or two are negative and one positive. All are positive to the right of 3 and two are negative between −4 and 0. The equality part of the statement holds at the critical points. Thus,

Solution: $[-4, 0] \cup [3, \infty)$

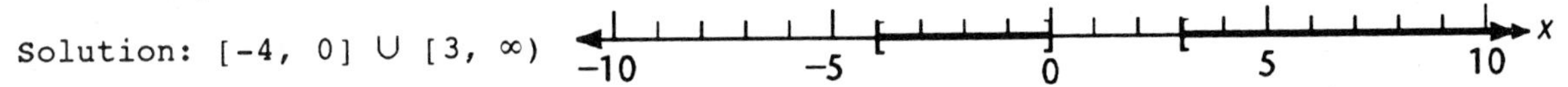

43.
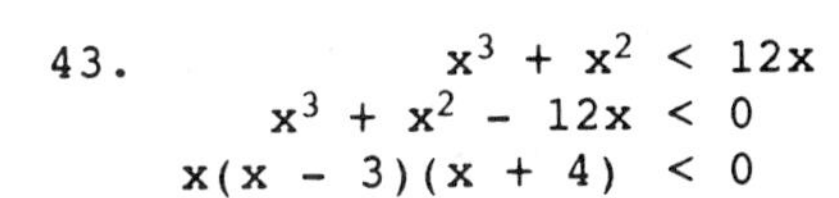

$$x^3 + x^2 < 12x$$
$$x^3 + x^2 - 12x < 0$$
$$x(x - 3)(x + 4) < 0$$

The signs of the factors are charted in problem 41 above. The inequality is satisfied when x, $(x - 3)$, and $(x + 4)$ are all negative, or two are positive and one negative. All are negative to the left of −4, and two are positive between 0 and 3. Thus,

Solution: $(-\infty, -4) \cup (0, 3)$

45.
$$x^3 + 21x > 10x^2$$
$$x^3 - 10x^2 + 21x > 0$$
$$x(x - 3)(x - 7) > 0$$
Critical points 0, 3, 7

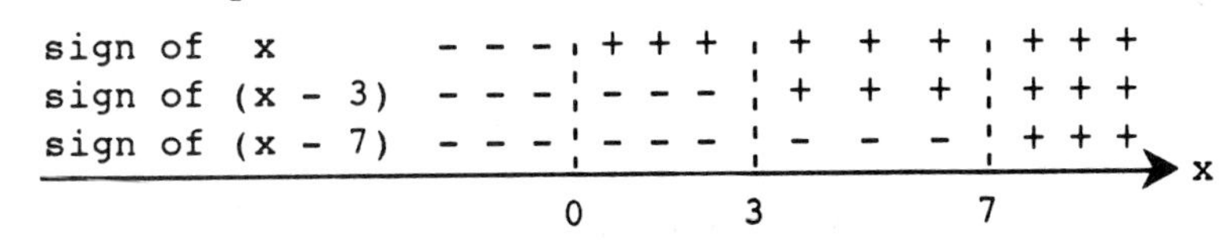

The inequality is satisfied when x, $(x - 3)$, and $(x - 7)$ are all positive, or two are negative and one positive. All are positive to the right of 7 and two are negative between 0 and 3. Thus,

Solution: $(0, 3) \cup (7, \infty)$

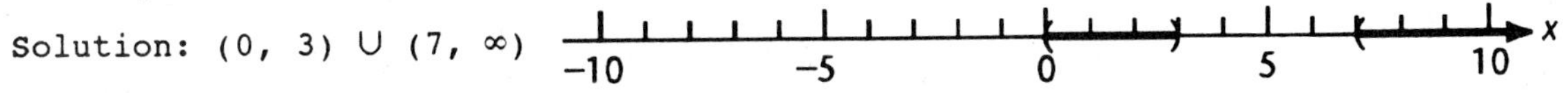

47. $\dfrac{x - 5}{x + 2} \leq 0$

Critical points -2, 5

sign of (x-5) - - ¦ - - - - - ¦ + + +
sign of (x+2) - - ¦ + + + + + ¦ + + +

-2 5 x

The equality part of the inequality statement holds when x = 5, but not when x = -2. The inequality part is satisfied when (x - 5) and (x + 2) have opposite signs, between -2 and 5.

(-2, 5]

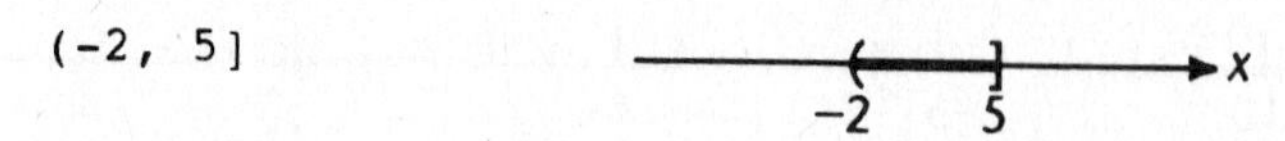

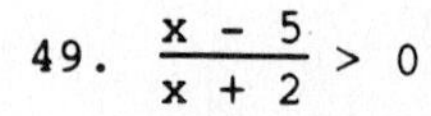

49. $\dfrac{x - 5}{x + 2} > 0$

The signs of numerator and denominator are charted in exercise 47 above. The inequality statement is satisfied when (x - 5) and (x + 2) have the same sign, to the left of -2 or to the right of 5.

$(-\infty, -2) \cup (5, \infty)$

51. $\dfrac{x - 4}{x(x + 2)} \leq 0$

Critical points -2, 0, 4

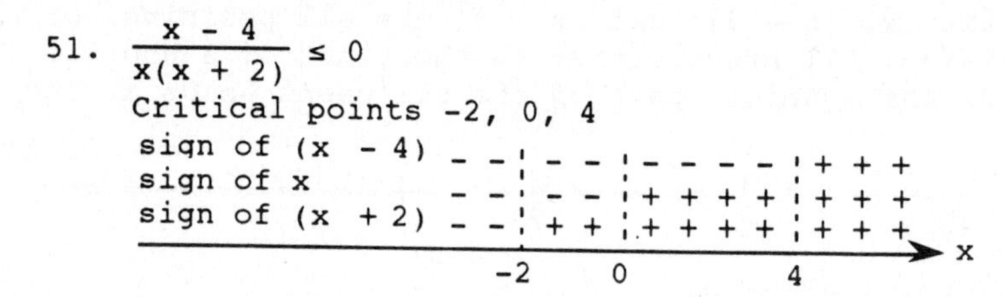

The inequality statement is satisfied when (x - 4), x and (x + 2) are all negative, two are positive and one negative, or the numerator is zero. All are negative to the left of -2 and two are positive between 0 and 4. The equality part of the inequality holds when x = 4, but not when x = 0 or -2. Thus,

Solution: $x < -2$ or $0 < x \leq 4$
$(-\infty, -2) \cup (0, 4]$

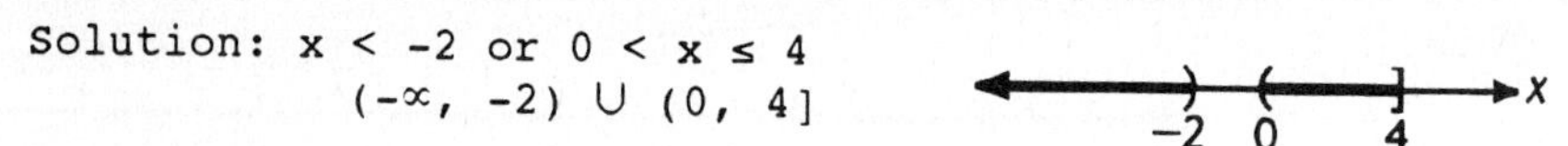

53. $\dfrac{x(x - 1)}{x + 2} \geq 0$

Critical points -2, 0, 1

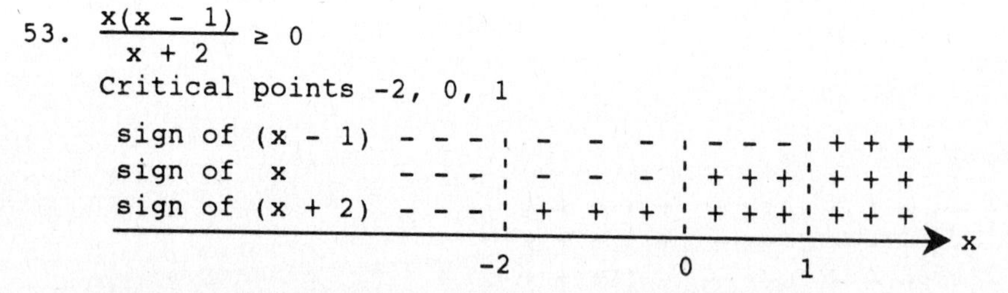

The inequality statement is satisfied when x, (x - 1), and (x + 2) are all positive, or two are negative and one positive, or the numerator is zero. All are positive to the right of 1 and two are negative between -2 and 0. The equality part of the inequality holds when x = 0 or 1, but not when x = -2. Thus,

Solution:
$-2 < x \leq 0$ or $1 \leq x$
$(-2, 0] \cup [1, \infty)$

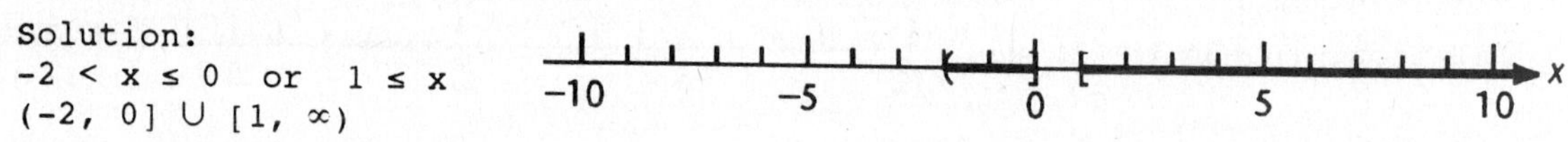

55. $\dfrac{x}{(x - 1)(x + 3)} < 0$

Critical points 0, 1, -3

	$x < -3$	$-3 < x < 0$	$0 < x < 1$	$x > 1$
sign of x	- - -	- - -	+ +	+ + +
sign of $(x - 1)$	- - -	- - -	- -	+ + +
sign of $(x + 3)$	- - -	+ + +	+ +	+ + +

The inequality statement is satisfied when x, (x - 1), and (x + 3) are all negative, or two are positive and one negative. All are negative to the left of -3 and two are positive between 0 and 1.

Solution: $(-\infty, -3) \cup (0, 1)$

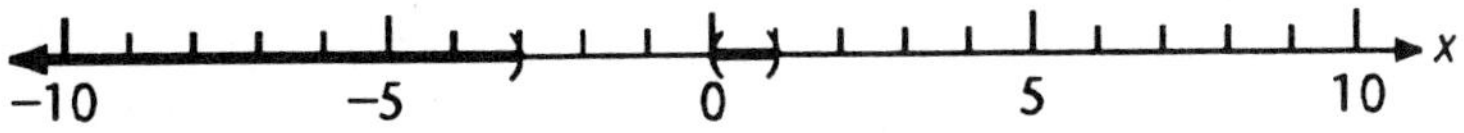

57. $\dfrac{4}{x} \geq 3$

$\dfrac{4}{x} - 3 \geq 0$

$\dfrac{4}{x} - \dfrac{3x}{x} \geq 0$

$\dfrac{4 - 3x}{x} \geq 0$

Criticial points are 0 and $\frac{4}{3}$ $(4 - 3x = 0$ if $4 = 3x$, that is, $\frac{4}{3} = x)$.

	$x < 0$	$0 < x < \frac{4}{3}$	$x > \frac{4}{3}$
sign of x	- - -	+ + +	+ + +
sign of $4 - 3x$	+ + +	+ + +	- - -

The inequality statement is satisfied when (4 - 3x) and x have the same sign or (4 - 3x) = 0. This occurs between 0 and $\frac{4}{3}$ as well as at the critical point $\frac{4}{3}$.

Solution:

$0 < x \leq \frac{4}{3}$ $\quad \left(0, \frac{4}{3}\right]$

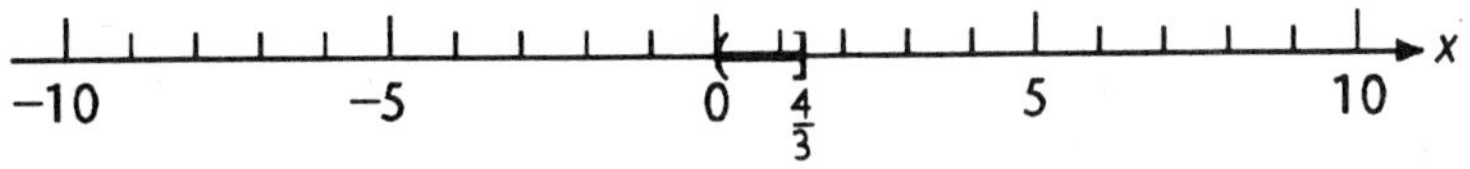

59. $\dfrac{1}{x} < 4$

$\dfrac{1}{x} - 4 < 0$

$\dfrac{1}{x} - \dfrac{4x}{x} < 0$

$\dfrac{1 - 4x}{x} < 0$

Critical points are 0 and $\frac{1}{4}$ $(1 - 4x = 0$ if $1 = 4x$, that is, $\frac{1}{4} = x)$.

	$x < 0$	$0 < x < \frac{1}{4}$	$x > \frac{1}{4}$
sign of $(1 - 4x)$	+ + +	+ +	- - -
sign of x	- - -	+ +	+ + +

The inequality statement is satisfied when $(1 - 4x)$ and x have opposite signs. This occurs to the left of 0 or to the right of $\frac{1}{4}$.

Solution: $x < 0$ or $x > \frac{1}{4}$

$(-\infty, 0) \cup \left(\frac{1}{4}, \infty\right)$

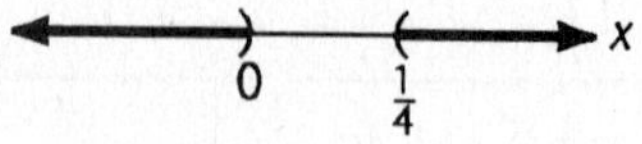

0 $\quad \frac{1}{4}$ $\quad$ x

61.
$$x^2 + 4 \geq 4x$$
$$x^2 - 4x + 4 \geq 0$$
$$(x - 2)^2 \geq 0$$
Since the square of a real number is always non-negative, thus always greater than or equal to 0, this statement is always true, regardless of the value of x. Another way of saying this is that the solution set is all real numbers; the graph is the whole real line.

63.
$$x^2 + 9 < 6x$$
$$x^2 - 6x + 9 < 0$$
$$(x - 3)^2 < 0$$
Since the square of a real number is never negative, thus never less than 0, this statement is never true. No solution.

65.
$$x^2 \geq 3$$
$$x^2 \geq 0$$
$$x^2 - (\sqrt{3})^2 \geq 0$$
$$(x - \sqrt{3})(x + \sqrt{3}) \geq 0$$
Critical points $\sqrt{3}$, $-\sqrt{3}$

sign of $(x-\sqrt{3})$ - - - | - - - | + + +
sign of $(x+\sqrt{3})$ - - - | + + + | + + +

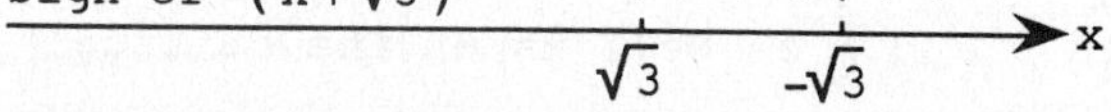

$\sqrt{3}$ $\quad -\sqrt{3}$ $\quad$ x

The equality part of the statement is satisfied at the critical points. The inequality part is satisfied when the product is positive, that is, when the factors have the same sign, left of $-\sqrt{3}$ or right of $\sqrt{3}$.

$(-\infty, -\sqrt{3}] \cup [\sqrt{3}, \infty)$

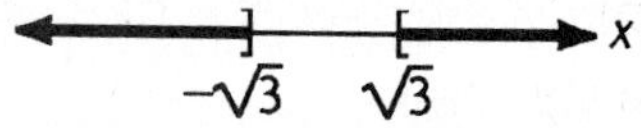

$-\sqrt{3}$ $\quad \sqrt{3}$ $\quad$ x

67.
$$\frac{2}{x - 3} \leq -2$$
$$\frac{2}{x - 3} + 2 \leq 0$$
$$\frac{2}{x - 3} + \frac{2(x - 3)}{x - 3} \leq 0$$
$$\frac{2 + 2(x - 3)}{x - 3} \leq 0$$
$$\frac{2 + 2x - 6}{x - 3} \leq 0$$
$$\frac{2x - 4}{x - 3} \leq 0$$
$$\frac{2(x - 2)}{x - 3} \leq 0$$
Critical points 2, 3

COMMON ERROR: "multiplying both sides" by $x - 3$ to get $2 \leq -2(x - 3)$.
This is incorrect; since the sign of $x - 3$ is unknown, both sides of the inequality cannot be "multiplied" by it, since the sense of the inequality might or might not reverse.

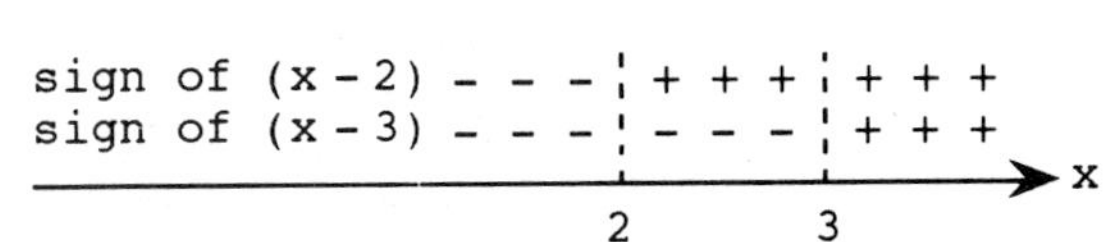

The inequality statement is satisfied when (x - 2) and (x - 3) have opposite signs, or the numerator is zero. The signs are opposite between 2 and 3. The equality part of the inequality holds when x = 2, but not when x = 3.

[2, 3)

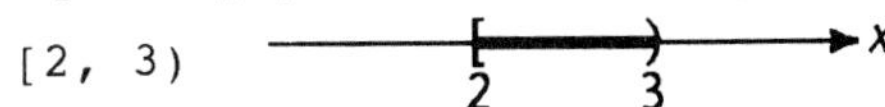

69. Graph $y = x^2 - 4$.
Find the x intercepts: $x^2 - 4 = (x + 2)(x - 2) = 0$ when $x = -2$ or 2. With this information, we can see the solution on the graph:

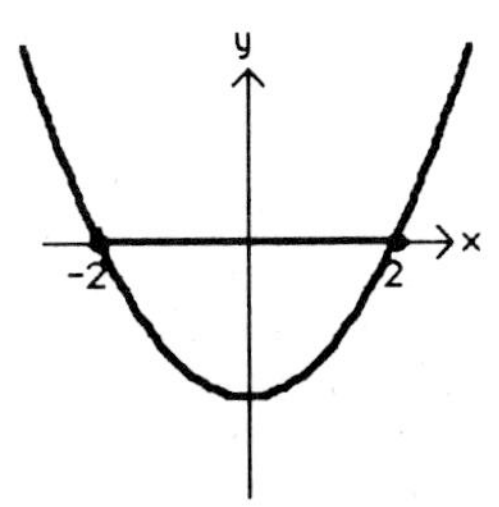

(-2, 2)

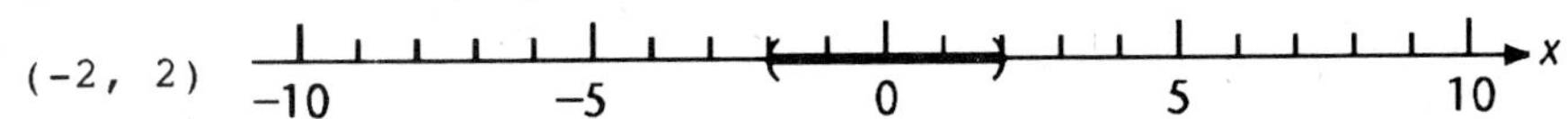

71. Graph $y = x^2 + 4x + 3$.
Find the x intercepts: $x^2 + 4x + 3 = (x + 1)(x + 3) = 0$ when $x = -1$ or -3. With this information, we can see the solution on the graph:

$x < -3$ or $x > -1$

$(-\infty, -3) \cup (-1, \infty)$

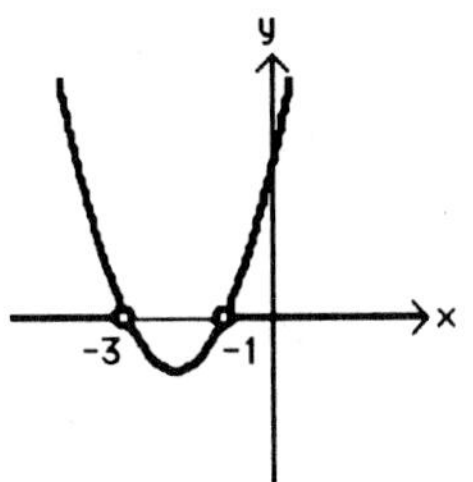

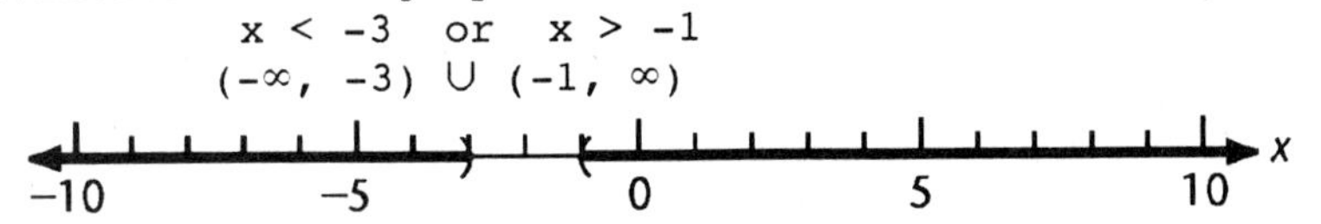

73.
$$x^2 + x \le 6$$
$$x^2 + x - 6 \le 0$$
Graph $y = x^2 + x - 6$.
Find the x intercepts: $x^2 + x - 6 = (x + 3)(x - 2) = 0$ when $x = -3$ or 2. With this information, we can see the solution on the graph:

$-3 \le x \le 2$ $[-3, 2]$

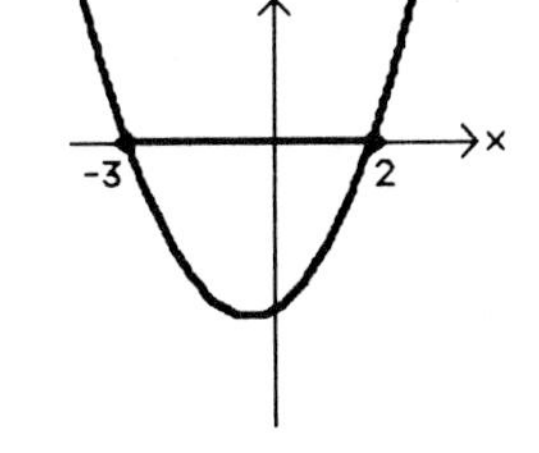

x
−10 −5 0 5 10

75.
$$\frac{2}{x - 3} \le \frac{2}{x + 2}$$
$$\frac{2}{x - 3} - \frac{2}{x + 2} \le 0$$
$$\frac{2(x + 2)}{(x - 3)(x + 2)} - \frac{2(x - 3)}{(x - 3)(x + 2)} \le 0$$
$$\frac{2(x + 2) - 2(x - 3)}{(x - 3)(x + 2)} \le 0$$
$$\frac{2x + 4 - 2x + 6}{(x - 3)(x + 2)} \le 0$$
$$\frac{10}{(x - 3)(x + 2)} \le 0$$

Critical points -2, 3

sign of (x - 3) - - ¦ - - - ¦ + + +
sign of (x + 2) - - ¦ + + + ¦ + + +
x
-2 3

The inequality part of the statement is satisfied when the quotient is negative, that is, when numerator and denominator have opposite sign, between -2 and 3. The equality part of the inequality statement is never satisfied.

Solution: $-2 < x < 3$ or $(-2, 3)$

77. $\dfrac{(x - 1)(x + 3)}{x} > 0$

Critical points 1, 0, -3

	$x<-3$	$-3<x<0$	$0<x<1$	$x>1$
sign of $(x-1)$	− −	− − − − −	− −	+ +
sign of x	− −	− − − − −	+ +	+ +
sign of $(x+3)$	− −	+ + + +	+ +	+ +

The inequality statement is satisfied when $(x - 1)$, x and $(x + 3)$ are all positive, or when two are negative and one is positive. All are positive to the right of 1 and two are negative between -3 and 0.

Solution: $-3 < x < 0$ or $1 < x$, $(-3, 0) \cup (1, \infty)$

79.

$$\frac{1}{x + 1} + \frac{1}{x - 2} \geq 0$$

$$\frac{x - 2}{(x + 1)(x - 2)} + \frac{x + 1}{(x + 1)(x - 2)} \geq 0$$

$$\frac{x - 2 + x + 1}{(x + 1)(x - 2)} \geq 0$$

$$\frac{2x - 1}{(x + 1)(x - 2)} \geq 0$$

Critical points $-1, \frac{1}{2}, 2$

	$x<-1$	$-1<x<\frac{1}{2}$	$\frac{1}{2}<x<2$	$x>2$
sign of $(2x-1)$	− − −	− −	+ +	+ + +
sign of $(x+1)$	− − −	+ +	+ +	+ + +
sign of $(x-2)$	− − −	− −	− −	+ + +

The inequality statement is satisfied when $(2x - 1)$, $(x + 1)$ and $(x - 2)$ are all positive, two are negative and one positive, or the numerator is zero. All are positive to the right of 2 and two are negative between -2 and $\frac{1}{2}$.

The equality part of the inequality holds when $x = \frac{1}{2}$, but not when $x = -1$ or 2.

Thus,

Solution: $-1 < x \leq \frac{1}{2}$ or $2 < x$, $\left(-1, \frac{1}{2}\right] \cup (2, \infty)$

81.

$$\frac{(x + 1)^2}{x^2 + 2x - 3} \leq 0$$

$$\frac{(x + 1)^2}{(x + 3)(x - 1)} \leq 0$$

Critical points -3, -1, 1

	$x<-3$	$-3<x<-1$	$-1<x<1$	$x>1$
sign of $(x+1)^2$	+ +	+ + +	+ + +	+ +
sign of $(x+3)$	− −	+ + +	+ + +	+ +
sign of $(x-1)$	− −	− − −	− − −	+ +

The inequality statement is satisfied when two of $(x + 1)^2$, $(x + 3)$ and $(x - 1)$ are positive and one negative, or the numerator is zero. (It can never happen that all three are negative—see sign chart.) Two are positive between -3 and 1, and the numerator is zero at $x = -1$, which also lies between -3 and 1. Thus,

Solution: $-3 < x < 1$ or $(-3, 1)$

83. $\sqrt{x^2 - 3x + 2}$ will be a real number when $x^2 - 3x + 2$ is non-negative, that is when $x^2 - 3x + 2 \geq 0$. To solve this inequality, we write

$$x^2 - 3x + 2 \geq 0$$
$$(x - 1)(x - 2) \geq 0$$

Critical points 1, 2

		1		2	
sign of $(x - 1)$	- -		+ + +		+ +
sign of $(x - 2)$	- -		- - -		+ +

The equality part of the statement is satisfied at the critical points. The inequality part is satisfied when the product is positive, that is, when the factors have the same sign, to the left of 1 or right of 2.

Solution: $x \geq 2$ or $x \leq 1$

85. $\sqrt{\dfrac{x + 1}{x - 1}}$ will be a real number when $\dfrac{x + 1}{x - 1}$ is non-negative, that is when

$$\frac{x + 1}{x - 1} \geq 0$$

Critical points -1, 1

		-1		1	
sign of $(x + 1)$	- -		+ + +		+ + +
sign of $(x - 1)$	- -		- - -		+ + +

The equality part of the statement is satisfied when the numerator is zero, that is, at $x = -1$. the inequality part is satisfied when the quotient is positive, that is, when the factors have the same sign, to the left of -1 or right of 1.

Solution: $x \leq -1$ or $x > 1$ $(-\infty, -1] \cup (1, \infty)$

87. $\sqrt{x^2 + x + 1}$ will be a real number when $x^2 + x + 1$ is non-negative, that is when $x^2 + x + 1 \geq 0$. Since this is equivalent to $\left(x^2 + x + \frac{1}{4}\right) + \frac{3}{4} \geq 0$, or $\left(x + \frac{1}{2}\right)^2 \geq -\frac{3}{4}$, and since the square of a real number is always positive, and thus always greater than any negative number, this is always true, regardless of the value of x. Another way of saying this is that the solution set is all real numbers.

89. $\sqrt{-1 + 2x - 3x^2}$ will be a real number when $-1 + 2x - 3x^2$ is non-negative, that is when $-1 + 2x - 3x^2 \ge 0$. This is equivalent, in turn, to

$$-\frac{1}{3}(-1 + 2x - 3x^2) \le -\frac{1}{3}(0)$$ Multiply both sides by $-\frac{1}{3}$, reversing the sense of the inequality.

$$x^2 - \frac{2}{3}x + \frac{1}{3} \le 0$$

$$\left(x^2 - \frac{2}{3}x + \frac{1}{9}\right) - \frac{1}{9} + \frac{1}{3} \le 0$$

$$\left(x - \frac{1}{3}\right)^2 \le \frac{1}{9} - \frac{1}{3}$$

$$\left(x - \frac{1}{3}\right)^2 \le -\frac{2}{9}$$

Since the square of a real number is always positive, and thus never less than any negative number, this statement is never true. No solution.

91. $d \ge 256$ will be satisfied for those values of t for which the inequality $160t - 16t^2 \ge 256$ holds. We solve this inequality as follows:

$$160t - 16t^2 \ge 256$$

$$-16t^2 + 160t - 256 \ge 0$$

$$-16(t^2 - 10t + 16) \ge 0$$ Divide both sides by -16, reversing the sense of the inequality.

$$t^2 - 10t + 16 \le 0$$

$$(t - 2)(t - 8) \le 0$$

Critical points 2, 8

		2		8	
sign of $(t-2)$	− −		+ + +		+ +
sign of $(t-8)$	− −		− − −		+ +

(number line in t, marked at 2 and 8)

The equality part of the inequality statement is satisfied at the critical points. The inequality part is satisfied when the product is negative, that is, when the factors have opposite signs, between 2 and 8.

Solution: $d \ge 256$ while $2 \le t \le 8$

Exercise 7-10 REVIEW EXERCISE

1. $4x = 2 - 3x^2$
$3x^2 + 4x - 2 = 0$ Compare with
$ax^2 + bx + c = 0$
$a = 3,\ b = 4,\ c = -2$

2. $x = \dfrac{-b \pm \sqrt{b^2 - 4ac}}{2a}$

3. $x^2 - 3x = 0$
$x(x - 3) = 0$
$x = 0$ or $x - 3 = 0$
$x = 3$
0, 3

4. $x^2 = 25$
$x = \pm\sqrt{25}$
$x = -5, 5$

5. $x^2 - 5x + 6 = 0$
$(x - 2)(x - 3) = 0$
$x - 2 = 0$ or $x - 3 = 0$
$x = 2$ $x = 3$
2, 3

6. $x^2 - 2x - 15 = 0$
$(x - 5)(x + 3) = 0$
$x - 5 = 0$ or $x + 3 = 0$
$x = 5$ $x = -3$
-3, 5

7. $x^2 - 7 = 0$
$x^2 = 7$
$x = \pm\sqrt{7}$

8. $x^2 - 6x + 8 = 0$
$(x - 2)(x - 4) = 0$
$x - 2 = 0$ or $x - 4 = 0$
$x = 2$ $\quad x = 4$
2, 4

9. $x^2 - 1 = 0$
$x^2 = 1$
$x = \pm\sqrt{1}$
$x = \pm 1$

10. $x^2 - x - 2 = 0$
$(x - 2)(x + 1) = 0$
$x - 2 = 0$ or $x + 1 = 0$
$x = 2$ $\quad x = -1$
2, -1

11. $x^2 + 3x + 1 = 0$
$a = 1,\ b = 3,\ c = 1$

$$x = \frac{-3 \pm \sqrt{(3)^2 - 4(1)(1)}}{2(1)} = \frac{-3 \pm \sqrt{9 - 4}}{2} = \frac{-3 \pm \sqrt{5}}{2}$$

12. $x^2 + 3x - 28 = 0$
$(x + 7)(x - 4) = 0$
$x + 7 = 0$ or $x - 4 = 0$
$x = -7$ $\quad x = 4$
-7, 4

13. $x^2 - 5x - 6 = 0$
$(x - 6)(x + 1) = 0$
$x - 6 = 0$ or $x + 1 = 0$
$x = 6$ $\quad x = -1$
6, -1

14. $3x^2 + 13x + 14 = 0$
$(3x + 7)(x + 2) = 0$
$3x + 7 = 0$ or $x + 2 = 0$
$3x = -7$ $\quad x = -2$
$x = -\frac{7}{3}$
$-\frac{7}{3}$, -2

15. $2x^2 - 9x + 10 = 0$
$(2x - 5)(x - 2) = 0$
$2x - 5 = 0$ or $x - 2 = 0$
$2x = 5$ $\quad x = 2$
$x = \frac{5}{2}$
$\frac{5}{2}$, 2

16. $x^2 + 9x + 20 = 0$
$(x + 4)(x + 5) = 0$
$x + 4 = 0$ or $x + 5 = 0$
$x = -4$ $\quad x = -5$
-4, -5

17. $x^2 + 3x - 28 < 0$
$(x + 7)(x - 4) < 0$
Critical points -7, 4

sign of $(x + 7)$ - - ⋮ + + + + ⋮ + +
sign of $(x - 4)$ - - ⋮ - - - - - ⋮ - -
(x axis: -7, 4)

COMMON ERROR: "$x + 7 < 0$ or $x - 4 < 0$" The signs of the factors must be opposite for the product to be negative.

The product will be negative where the factors have opposite signs, between -7 and 4. $-7 < x < 4$. $(-7, 4)$.

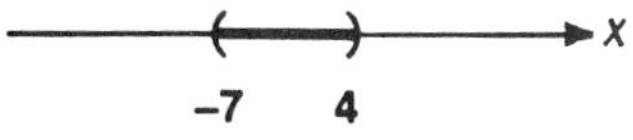

18. $x^2 - 5x + 6 > 0$
$(x - 2)(x - 3) > 0$
Critical points 2, 3

sign of $(x - 2)$ - - ⋮ + + ⋮ + + + +
sign of $(x - 3)$ - - ⋮ - - ⋮ + + + +
(x axis: 2, 3)

The product will be positive where the factors have the same sign, left of 2 or right of 3. $x < 2$ or $x > 3$ $\quad (-\infty, 2) \cup (3, \infty)$

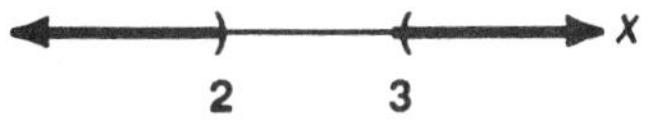

19. $x^2 + x < 20$

$x^2 + x - 20 < 0$

$(x + 5)(x - 4) < 0$

Critical points -5, 4

	$x < -5$	$-5 < x < 4$	$x > 4$
sign of $(x+5)$	- -	+ + + +	+ +
sign of $(x-4)$	- -	- - - -	+ +

(number line: critical points -5 and 4 marked on the x-axis)

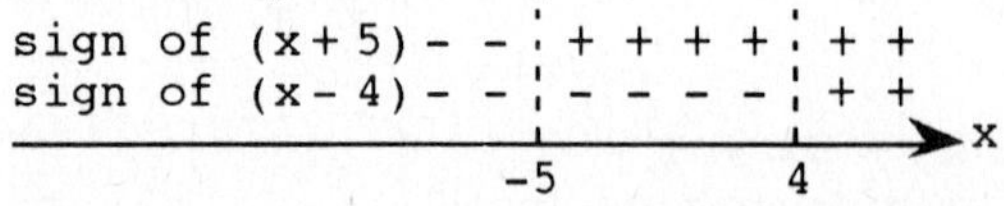

The product will be negative where the factors have opposite signs, between -5 and 4.

$-5 < x < 4$ $(-5, 4)$ (graph: open interval from -5 to 4 on the x-axis)

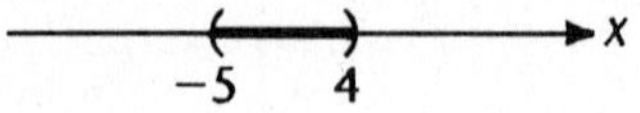

20. $x^2 + x \geq 20$

$x^2 + x - 20 \geq 0$

$(x + 5)(x - 4) \geq 0$

Critical points -5, 4

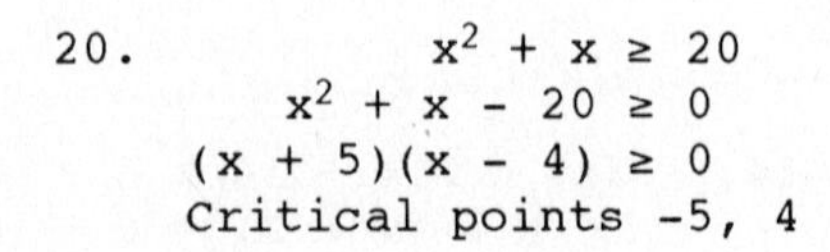

The signs of the factors are charted in exercise 19 above. The equality part of the inequality statement is satisfied at the critical points. The inequality part is satisfied when the product is positive, that is, when the factors have the same sign, left of -5 or right of 4.

$x \leq -5$ or $x \geq 4$ $(-\infty, -5] \cup [4, \infty)$ (graph: rays to the left of -5 and right of 4 on the x-axis, endpoints included)

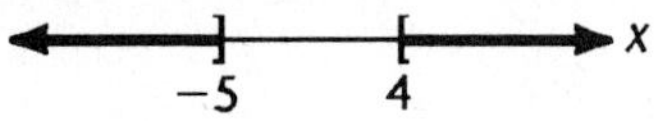

21. $\sqrt{x + 3} = 8$ Square both sides

$x + 3 = 64$

$x = 61$

Check: $\sqrt{61 + 3} = \sqrt{64} = 8$ so 61 is a solution.

61

22. $\sqrt{x + 5} = x - 1$ Square both sides

$x + 5 = x^2 - 2x + 1$

$0 = x^2 - 3x - 4$

$0 = (x - 4)(x + 1)$

$x = 4, -1$

Check: $\underline{x = 4}$ $\sqrt{4 + 5} \stackrel{?}{=} 4 - 1$

$\sqrt{9} \stackrel{\checkmark}{=} 3$

4 is a solution

$\underline{x = -1}$ $\sqrt{-1 + 5} \stackrel{?}{=} -1 - 1$

$\sqrt{4} \neq -2$

-1 is extraneous

4

23. $x = \sqrt{3x + 4}$ Square both sides

$x^2 = 3x + 4$

$x^2 - 3x - 4 = 0$

$(x - 4)(x + 1) = 0$

$x = 4, -1$

Check: $\underline{x = 4}$ $4 \stackrel{?}{=} \sqrt{3 \cdot 4 + 4}$

$4 \stackrel{\checkmark}{=} \sqrt{16}$

4 is a solution

$\underline{x = -1}$ $-1 \stackrel{?}{=} \sqrt{3(-1) + 4}$

$-1 \neq \sqrt{1}$

-1 is extraneous

4

24.
$$
\begin{aligned}
x &= \sqrt{2 - x} \quad \text{Square both sides} \\
x^2 &= 2 - x \\
x^2 + x - 2 &= 0 \\
(x + 2)(x - 1) &= 0 \\
x &= -2, 1
\end{aligned}
$$

Check: $\underline{x = -2}$ $\quad -2 \stackrel{?}{=} \sqrt{2 - (-2)}$ $\qquad$ $\underline{x = 1}$ $\quad 1 \stackrel{?}{=} \sqrt{2 - 1}$

$-2 \neq \sqrt{4}$ $\qquad$ $1 \stackrel{\checkmark}{=} \sqrt{1}$

-2 is extraneous $\qquad$ 1 is a solution

1

25. The discriminant $b^2 - 4ac = (-11)^2 - 4(7)(5) = 121 - 140 = -19$ is negative. The equation has two nonreal complex roots.

26. Let $P_1 = (1, 3)$ and $P_2 = (3, 7)$; then

$$
\begin{aligned}
d &= \sqrt{(x_2 - x_1)^2 + (y_2 - y_1)^2} = \sqrt{(3 - 1)^2 + (7 - 3)^2} = \sqrt{(2)^2 + (4)^2} \\
&= \sqrt{20} \text{ or } 2\sqrt{5}
\end{aligned}
$$

27. If the center is at the origin, then $(h, k) = (0, 0)$; thus the equation is $x^2 + y^2 = 5^2$ or $x^2 + y^2 = 25$.

28. If the center is at $(-3, 4)$, then $h = -3$ and $k = 4$, and the equation is $[x - (-3)]^2 + (y - 4)^2 = 7^2$ or $(x + 3)^2 + (y - 4)^2 = 49$

29. We transform the equation into the form

$$(x - h)^2 + (y - k)^2 = R^2$$

by completing the square relative to x.

$$
\begin{aligned}
x^2 - 2x + ? + y^2 &= ? \\
x^2 - 2x + 1 + y^2 &= 1 \\
(x - 1)^2 + y^2 &= 1
\end{aligned}
$$

This is the equation of a circle with radius 1 and center at $(h, k) = (1, 0)$.

30. (A) Graphically:

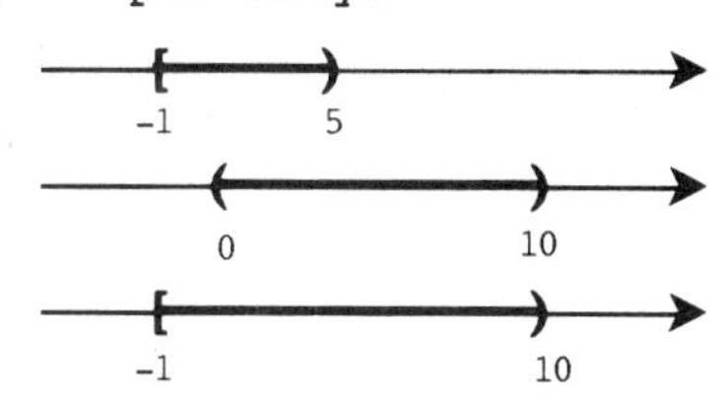

Symbolically:

$[-1, 5) \cup (0, 10) = [-1, 10)$

(B) Graphically:

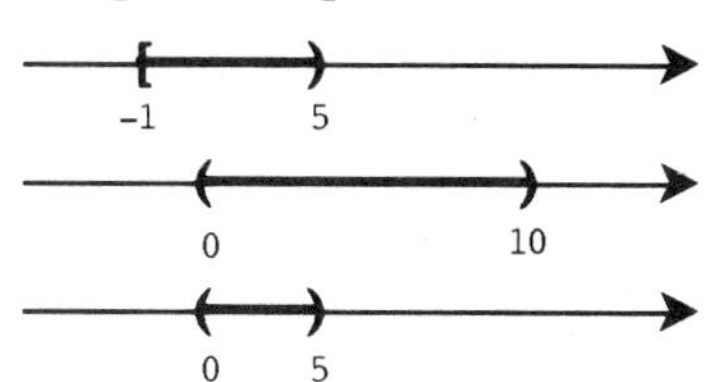

Symbolically:

$[-1, 5) \cap (0, 10) = (0, 5)$

31. $y = x^2 + 2$. This is a parabola, opening up, since the coefficient of x^2, $1 = a > 0$. The vertex is located at

$$x = -\frac{b}{2a} = -\frac{0}{2 \cdot 1} = 0 \text{ and } y = 0^2 + 2 = 2.$$

x	y
-2	6
-1	3
0	2
1	3
2	6

32. $y = x^2 + 2x + 2$. This is a parabola, opening up, since the coefficient of x^2, $1 = a > 0$. The vertex is located at

$$x = -\frac{b}{2a} = -\frac{2}{2 \cdot 1} = -1 \text{ and}$$
$$y = (-1)^2 + 2(-1) + 2 = 1.$$

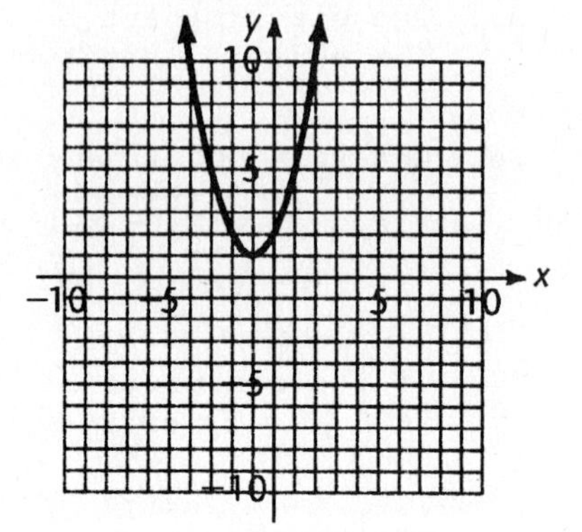

x	y
-3	5
-2	2
-1	1
0	2
1	5

33. $y = -x^2 + 3x - 2$. This is a parabola, opening down, since the coefficient of x^2, $-1 = a < 0$. The vertex is located at

$$x = -\frac{b}{2a} = -\frac{3}{2(-1)} = \frac{3}{2} \text{ and}$$
$$y = -\left(\frac{3}{2}\right)^2 + 3\left(\frac{3}{2}\right) - 2 = \frac{1}{4}$$

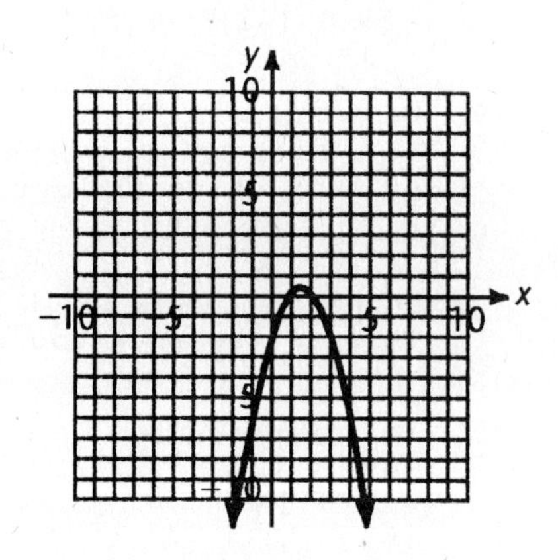

x	y
0	-2
1	0
2	0
3	-2
4	-6

34. This is a circle with radius 3 and center at the origin.

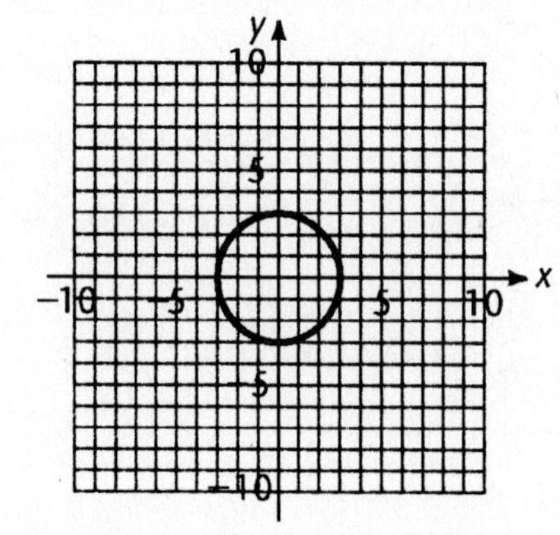

35. Rewriting this as $\frac{x^2}{9} - \frac{y^2}{9} = 1$ puts it in standard form for a hyperbola.

We first draw a dashed rectangle with intercepts $x = \pm\sqrt{9} = \pm 3$ and $y = \pm 3$. Second, draw in asymptotes (extended diagonals of the rectangle). Third, determine true intercepts for the hyperbola, as follows: Let $y = 0$; then

$$x^2 - 0 = 9$$
$$x^2 = 9$$
$$x = \pm\sqrt{9} = \pm 3$$

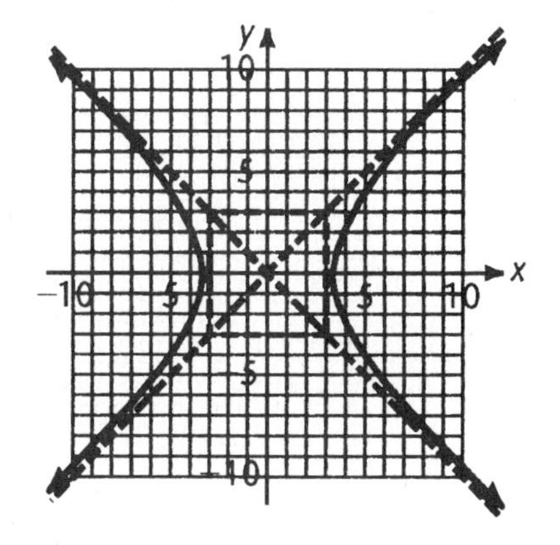

If we let $x = 0$, then

$$0 - y^2 = 9$$
$$y^2 = -9$$
$$y = \pm\sqrt{-9} = \pm 3i$$

These are complex numbers and do not represent real intercepts. We conclude that the only intercepts for the hyperbola are $x = \pm 3$, and the hyperbola opens left and right.

36. Rewriting this as $\frac{x^2}{9} - y^2 = 1$ puts it in standard form for a hyperbola.

We first draw a dashed rectangle with intercepts $x = \pm\sqrt{9} = \pm 3$ and $y = \pm\sqrt{1} = \pm 1$. Second, draw in asymptotes (extended diagonals of the rectangle). Third, determine true intercepts for the hyperbola, as follows: Let $y = 0$; then

$$x^2 - 0 = 9$$
$$x^2 = 9$$
$$x = \pm\sqrt{9} = \pm 3$$

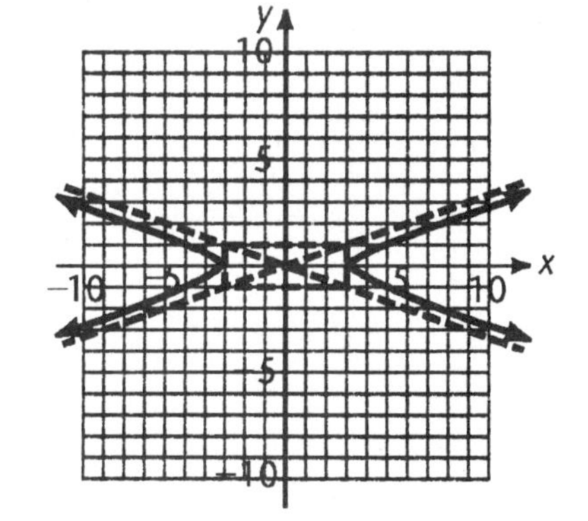

If we let $x = 0$, then

$$0 - 9y^2 = 9$$
$$y^2 = -1$$
$$y = \pm\sqrt{-1} = \pm i$$

These are complex numbers and do not represent real intercepts. We conclude that the only intercepts for the hyperbola are $x = \pm 3$, and the hyperbola opens left and right.

37. Rewriting this as $\frac{x^2}{9} + y^2 = 1$ puts it in standard form for an ellipse.

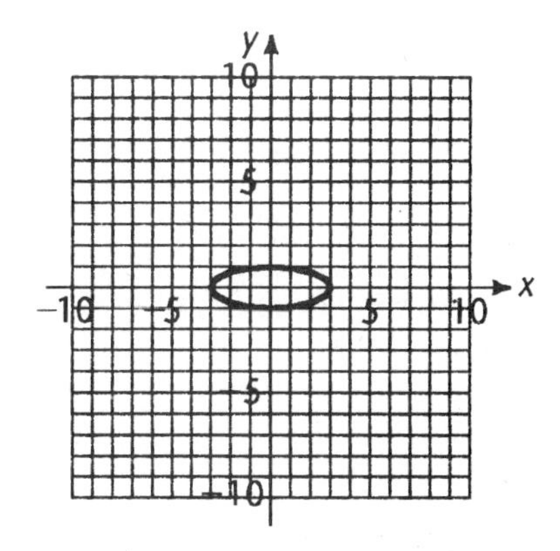

Find x intercepts	Find y intercepts
$x^2 + 0 = 9$	$0 + 9y^2 = 9$
$x^2 = 9$	$y^2 = 1$
$x = \pm\sqrt{9} = \pm 3$	$y = \pm\sqrt{1} = \pm 1$

Now plot the intercepts and draw the ellipse.

38. $y^2 = 1 - x^2$.
Rewriting this as $x^2 + y^2 = 1$, we see that this is the equation of a circle with center $(h, k) = (0, 0)$ and radius $R = 1$.

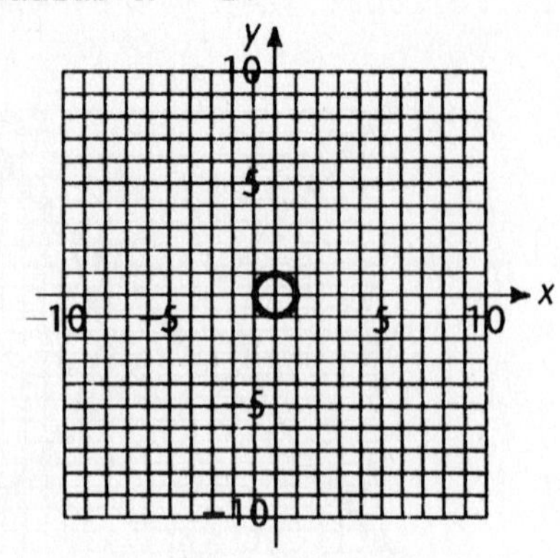

39. This is the equation, in standard form, of a circle with radius 7 and center at the origin.

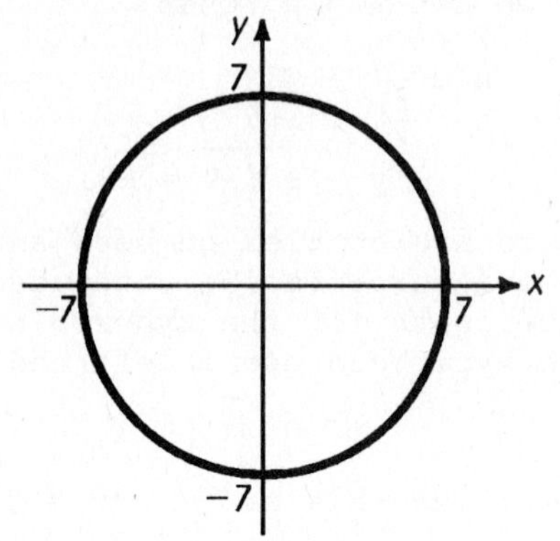

40. $(x - 2)^2 + (y + 3)^2 = 16$ is the same as $(x - 2)^2 + [y - (-3)]^2 = 4^2$, which is the equation of a circle with radius 4 and center at $(h, k) = (2, -3)$.

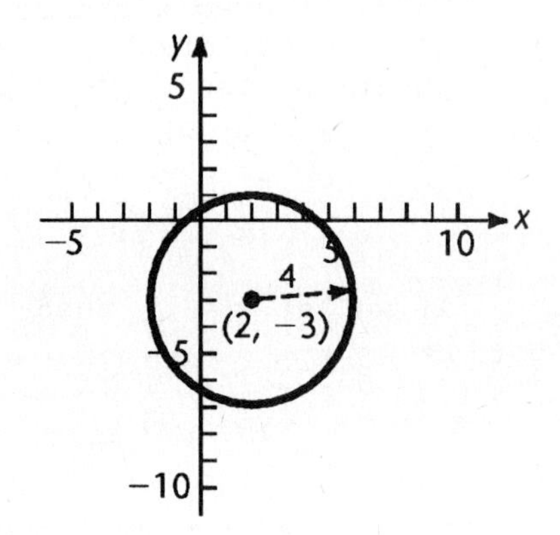

41. This is the equation of a parabola in standard form. The origin (0, 0) is part of the graph. We choose $x = -2$ to make the right side a positive perfect square; then

$$y^2 = -2(-2)$$
$$y^2 = 4$$
$$y = \pm\sqrt{4} = \pm 2$$

Thus, the points (-2, 2) and (-2, -2) are also on the graph. We plot the three known points; then we sketch a parabola through these points.

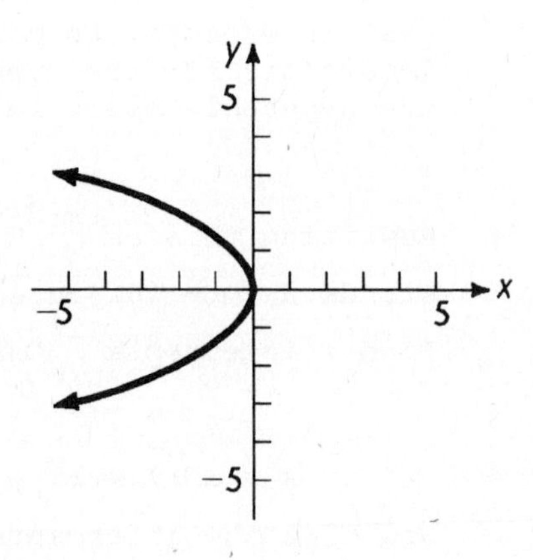

42. This is the equation of an ellipse in standard form. We graph it as follows:

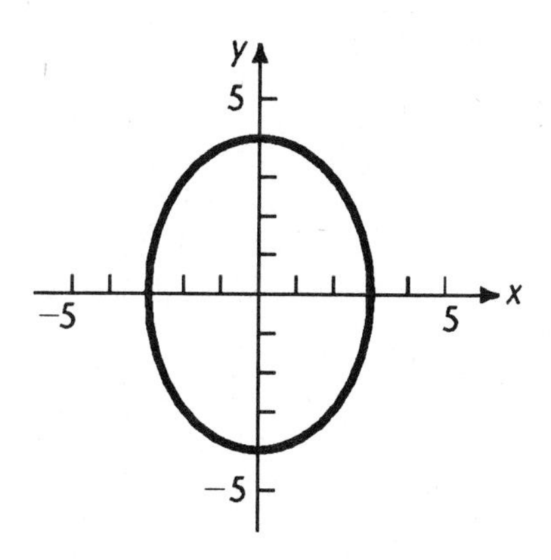

Find x intercepts:

$$\frac{x^2}{9} + \frac{0}{16} = 1$$
$$x^2 = 9$$
$$x = \pm\sqrt{9} = \pm 3$$

Find y intercepts:

$$\frac{0}{9} + \frac{y^2}{16} = 1$$
$$y^2 = 16$$
$$y = \pm\sqrt{16} = \pm 4$$

Now plot the intercepts and draw the ellipse.

43. This is the equation of a hyperbola in standard form. We graph it as follows: First, draw a dashed rectangle with intercepts $x = \pm\sqrt{9} = \pm 3$ and $y = \pm\sqrt{16} = \pm 4$. Second, draw in asymptotes (extended diagonals of the rectangle). Third, determine true intercepts for the hyperbola, as follows.

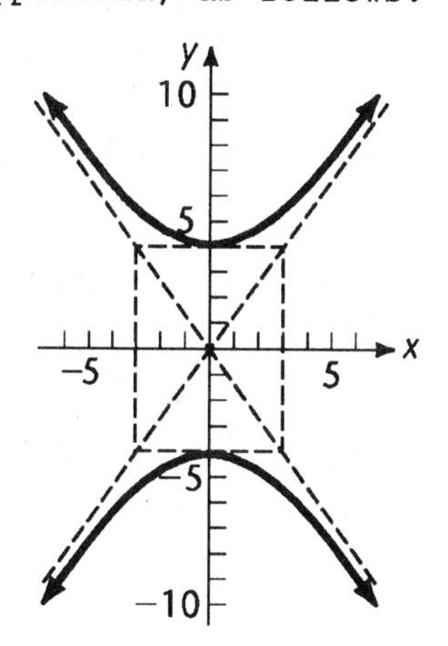

If we let $y = 0$; then

$$\frac{0}{16} - \frac{x^2}{9} = 1$$
$$x^2 = -9$$
$$x = \pm\sqrt{-9} = \pm 3i$$

These are complex numbers and do not represent real intercepts. Let $x = 0$; then

$$\frac{y^2}{16} - \frac{0}{9} = 1$$
$$y^2 = 16$$
$$y = \pm\sqrt{16} = \pm 4$$

We conclude that the only real intercepts are $y = \pm 4$, and the hyperbola opens up and down.

44. Let x = one of the numbers. Then the other number is 6 more than x.
$6 + x$ = second number
Since their product is 27, we translate this to the equation

$$x(6 + x) = 27$$
$$6x + x^2 = 27$$
$$x^2 + 6x = 27$$
$$x^2 + 6x - 27 = 0$$
$$(x + 9)(x - 3) = 0$$
$$x + 9 = 0 \quad \text{or} \quad x - 3 = 0$$
$$x = -9 \qquad\qquad x = 3$$
$$6 + x = -3 \qquad 6 + x = 9$$

We discard the negative solution since positive numbers were required. The numbers are 3 and 9.

45. Let x = the number
Then translate as follows:

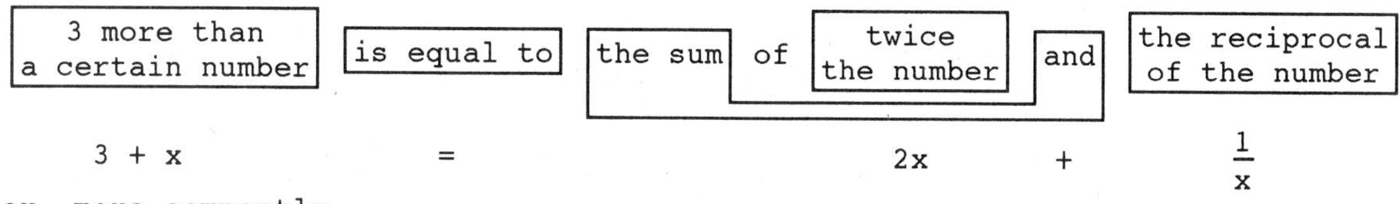

$$3 + x \quad = \quad 2x \quad + \quad \frac{1}{x}$$

or, more compactly

$$3 + x = 2x + \frac{1}{x} \qquad \text{lcm} = x,\ x \neq 0$$
$$x(3 + x) = x \cdot 2x + x \cdot \frac{1}{x}$$
$$3x + x^2 = 2x^2 + 1$$
$$0 = x^2 - 3x + 1$$
$$a = 1 \quad b = -3 \quad c = 1$$
$$x = \frac{-(-3) \pm \sqrt{(-3)^2 - 4(1)(1)}}{2(1)}$$
$$x = \frac{3 \pm \sqrt{9 - 4}}{2}$$
$$x = \frac{3 \pm \sqrt{5}}{2}$$

46.

x

x + 3

Let x = the width
then $x + 3$ = the length
Since (width) × (length) = area, we have

$$x(x + 3) = 108$$
$$x^2 + 3x = 108$$
$$x^2 + 3x - 108 = 0$$
$$(x + 12)(x - 9) = 0$$
$$x + 12 = 0 \quad \text{or} \quad x - 9 = 0$$
$$x = -12 \qquad x = 9 \text{ ft}$$

Discard; distances cannot be negative $\qquad x + 3 = 12$ ft

47. Let x = the speed of the slower ship
then $x + 5$ = the speed of the faster ship
Using distance = (rate) × (time), we have: $2x$ = distance traveled by slower ship
$2(x + 5)$ = distance traveled by faster ship

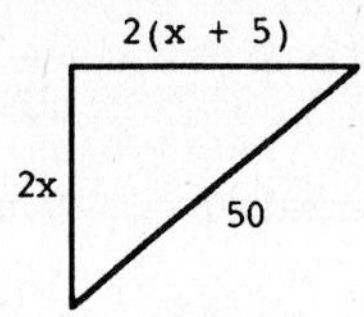

Using the Pythagorean theorem, we can write:

$$(2x)^2 + [2(x + 5)]^2 = 50^2$$
$$4x^2 + 4(x + 5)^2 = 2500$$
$$4x^2 + 4(x^2 + 10x + 25) = 2500$$
$$4x^2 + 4x^2 + 40x + 100 = 2500$$
$$8x^2 + 40x - 2400 = 0$$
$$x^2 + 5x - 300 = 0$$
$$(x - 15)(x + 20) = 0$$
$$x - 15 = 0 \quad \text{or} \quad x + 20 = 0$$
$$x = 15 \text{ mph} \qquad x = -20$$ discard; rates cannot be negative
$$x + 5 = 20 \text{ mph}$$

48.
$$10x^2 = 20x$$
$$10x^2 - 20x = 0$$
$$10x(x - 2) = 0$$
$$10x = 0 \quad \text{or} \quad x - 2 = 0$$
$$x = 0 \qquad x = 2$$
0, 2

49.
$$3x^2 = 36$$
$$x^2 = 12$$
$$x = \pm\sqrt{12}$$
$$x = \pm 2\sqrt{3}$$

50.
$$3x^2 + 27 = 0$$
$$3x^2 = -27$$
$$x^2 = -9$$
$$x = \pm\sqrt{-9}$$
$$x = \pm 3i$$

51.
$$(x - 2)^2 = 16$$
$$x - 2 = \pm\sqrt{16}$$
$$x - 2 = \pm 4$$
$$x = 2 \pm 4$$
$$x = 2 + 4 \quad \text{or} \quad x = 2 - 4$$
$$x = 6 \qquad x = -2$$
$-2, 6$

52.
$$3t^2 - 8t - 3 = 0$$
$$(3t + 1)(t - 3) = 0$$
$$3t + 1 = 0 \quad \text{or} \quad t - 3 = 0$$
$$3t = -1 \qquad t = 3$$
$$t = -\frac{1}{3}$$
$-\frac{1}{3}, 3$

53.
$$2x = \frac{3}{x} - 5 \quad x \neq 0, \text{ lcm} = x$$
$$2x \cdot x = x \cdot \frac{3}{x} - 5x$$
$$2x^2 = 3 - 5x$$
$$2x^2 + 5x - 3 = 0$$
$$(2x - 1)(x + 3) = 0$$
$$2x - 1 = 0 \quad \text{or} \quad x + 3 = 0$$
$$2x = 1 \qquad x = -3$$
$$x = \frac{1}{2}$$
$\frac{1}{2}, -3$

54. $2x^2 - 3x + 6 = 0$ This is not factorable in the integers. Use quadratic formula.

$a = 2 \quad b = -3 \quad c = 6$
$$x = \frac{-(-3) \pm \sqrt{(-3)^2 - 4(2)(6)}}{2(2)}$$
$$x = \frac{3 \pm \sqrt{9 - 48}}{4}$$
$$x = \frac{3 \pm \sqrt{-39}}{4}$$
$$x = \frac{3 \pm i\sqrt{39}}{4}$$

55.
$$x^3 - 4x = 0$$
$$x(x^2 - 4) = 0$$
$$x(x - 2)(x + 2) = 0$$
$$x = 0 \quad \text{or} \quad x - 2 = 0 \quad \text{or} \quad x + 2 = 0$$
$$x = 2 \qquad x = -2$$
$0, 2, -2$

56.
$$x^3 - 4x^2 + 4x = 0$$
$$x(x^2 - 4x + 4) = 0$$
$$x(x - 2)^2 = 0$$
$$x = 0 \quad \text{or} \quad (x - 2)^2 = 0$$
$$x - 2 = 0$$
$$x = 2$$
$0, 2$

57. $8x^2 - 9x - 10 = 0$ This is not factorable in the integers. Use quadratic formula.

$a = 8 \quad b = -9 \quad c = -10$
$$x = \frac{-(-9) \pm \sqrt{(-9)^2 - 4(8)(-10)}}{2(8)}$$
$$x = \frac{9 \pm \sqrt{81 + 320}}{16}$$
$$x = \frac{9 \pm \sqrt{401}}{16}$$

58.
$$x^4 - 3x^2 = 0$$
$$x^2(x^2 - 3) = 0$$
$$x^2 = 0 \quad \text{or} \quad x^2 - 3 = 0$$
$$x = 0 \qquad x^2 = 3$$
$$x = \pm\sqrt{3}$$
$x = 0, \pm\sqrt{3}$

59.
$$3x^2 = 2(x + 1)$$
$$3x^2 = 2x + 2$$
$$3x^2 - 2x - 2 = 0$$
$a = 3$, $b = -2$, $c = -2$
$$x = \frac{-(-2) \pm \sqrt{(-2)^2 - 4(3)(-2)}}{2(3)} = \frac{2 \pm \sqrt{4 + 24}}{6} = \frac{2 \pm \sqrt{28}}{6} = \frac{2 \pm 2\sqrt{7}}{6} = \frac{2(1 \pm \sqrt{7})}{6} = \frac{1 \pm \sqrt{7}}{3}$$

60.
$$2x(x - 1) = 3$$
$$2x^2 - 2x = 3$$
$$2x^2 - 2x - 3 = 0$$
$a = 2$ $b = -2$ $c = -3$
$$x = \frac{-(-2) \pm \sqrt{(-2)^2 - 4(2)(-3)}}{2(2)} = \frac{2 \pm \sqrt{4 + 24}}{4} = \frac{2 \pm \sqrt{28}}{4} = \frac{2 \pm 2\sqrt{7}}{4} = \frac{2(1 \pm \sqrt{7})}{4} = \frac{1 \pm \sqrt{7}}{2}$$

COMMON ERROR: "$2x = 3$ or $x - 1 = 3$." The rule: If $ab = 0$ then $a = 0$ or $b = 0$, can only be applied if 0 is on the right side.

61.
$$2x^2 - 2x = 40$$
$$2x^2 - 2x - 40 = 0$$
$$x^2 - x - 20 = 0$$
$$(x - 5)(x + 4) = 0$$
$x - 5 = 0$ or $x + 4 = 0$
$x = 5$ $\qquad x = -4$
5, -4

62.
$$\frac{8m^2 + 15}{2m} = 13 \quad m \neq 0,\ \text{lcm} = 2m$$
$$2m \cdot \frac{8m^2 + 15}{2m} = 2m \cdot 13$$
$$8m^2 + 15 = 26m$$
$$8m^2 - 26m + 15 = 0$$
$$(4m - 3)(2m - 5) = 0$$
$4m - 3 = 0$ or $2m - 5 = 0$
$4m = 3$ $\qquad 2m = 5$
$m = \frac{3}{4}$ $\qquad m = \frac{5}{2}$
$\frac{3}{4}, \frac{5}{2}$

63. $m^2 + m - 1 = 0$ This is not factorable in the integers. Use quadratic formula.
$a = 1$ $b = 1$ $c = -1$
$$m = \frac{-(1) \pm \sqrt{(1)^2 - 4(1)(-1)}}{2(1)} = \frac{-1 \pm \sqrt{1 + 4}}{2} = \frac{-1 \pm \sqrt{5}}{2}$$

64. $u + \frac{3}{u} = 2 \quad u \neq 0,\ \text{lcm} = u$

$$u \cdot u + u \cdot \frac{3}{u} = u \cdot 2$$
$$u^2 + 3 = 2u$$
$$u^2 - 2u + 3 = 0$$ This is not factorable in the integers. Use quadratic formula.

$a = 1 \quad b = -2 \quad c = 3$

$$u = \frac{-(-2) \pm \sqrt{(-2)^2 - 4(1)(3)}}{2(1)}$$
$$= \frac{2 \pm \sqrt{4 - 12}}{2}$$
$$= \frac{2 \pm \sqrt{-8}}{2}$$
$$= \frac{2 \pm i\sqrt{8}}{2}$$
$$= \frac{2 \pm 2i\sqrt{2}}{2}$$
$$= \frac{2(1 \pm i\sqrt{2})}{2}$$
$$= 1 \pm i\sqrt{2}$$

65. $\sqrt{5x - 6} - x = 0$

$$\sqrt{5x - 6} = x$$
$$5x - 6 = x^2$$
$$0 = x^2 - 5x + 6$$
$$x^2 - 5x + 6 = 0$$
$$(x - 2)(x - 3) = 0$$
$x - 2 = 0$ or $x - 3 = 0$

$x = 2 \qquad x = 3$

Check: $\sqrt{5(2) - 6} - 2 \stackrel{?}{=} 0$

$\sqrt{4} - 2 \stackrel{?}{=} 0$

$2 - 2 \stackrel{?}{=} 0$

$0 \stackrel{\checkmark}{=} 0$

2 is a solution

$\sqrt{5(3) - 6} - 3 \stackrel{?}{=} 0$

$\sqrt{9} - 3 \stackrel{?}{=} 0$

$3 - 3 \stackrel{?}{=} 0$

$0 \stackrel{\checkmark}{=} 0$

3 is a solution

2, 3

66. $8\sqrt{x} = x + 15$

$$(8\sqrt{x})^2 = (x + 15)^2$$
$$64x = x^2 + 30x + 225$$
$$0 = x^2 - 34x + 225$$
$$x^2 - 34x + 225 = 0$$
$$(x - 9)(x - 25) = 0$$
$x - 9 = 0$ or $x - 25 = 0$

$x = 9 \qquad x = 25$

Check: $8\sqrt{9} \stackrel{?}{=} 9 + 15$

$24 \stackrel{\checkmark}{=} 24$

9 is a solution

$8\sqrt{25} \stackrel{?}{=} 25 + 15$

$40 \stackrel{\checkmark}{=} 40$

25 is a solution

9, 25

67. $m^4 + 5m^2 - 36 = 0$

Let $u = m^2$, then $u^2 = m^4$. After substitution, the original equation becomes

$$u^2 + 5u - 36 = 0$$
$$(u + 9)(u - 4) = 0$$
$$u = -9, 4$$

Replacing u with m^2, we obtain

$m^2 = -9$ $\quad$ $m^2 = 4$

$m = \pm\sqrt{-9}$ $\quad$ $m = \pm\sqrt{4}$

$m = \pm 3i$ $\quad$ $m = \pm 2$

$\pm 2, \pm 3i$

68. $2x^{2/3} - 5x^{1/3} - 12 = 0$

Let $u = x^{1/3}$, then $u^2 = x^{2/3}$. After substitution, the original equation becomes

$$2u^2 - 5u - 12 = 0$$
$$(2u + 3)(u - 4) = 0$$

$2u + 3 = 0$ or $u - 4 = 0$

$2u = -3$ $\quad$ $u = 4$

$u = -\frac{3}{2}$

Replacing u with $x^{1/3}$, we obtain

$x^{1/3} = -\frac{3}{2}$ $\quad$ $x^{1/3} = 4$

$x = \left(-\frac{3}{2}\right)^3$ $\quad$ $x = (4)^3$

$x = -\frac{27}{8}$ $\quad$ $x = 64$

COMMON ERROR: "$x = \left(-\frac{3}{2}\right)^{1/3}$ or $x = 4^{1/3}$". Both sides must be raised to the third power to eliminate the $\frac{1}{3}$ power from the $x^{1/3}$.

$64, -\frac{27}{8}$

69. $\sqrt{x} + \sqrt{x + 5} = 5$

$$\sqrt{x + 5} = 5 - \sqrt{x} \quad \text{Square both sides}$$
$$(\sqrt{x + 5})^2 = (5 - \sqrt{x})^2$$
$$x + 5 = 25 - 10\sqrt{x} + x$$
$$-20 = -10\sqrt{x}$$
$$2 = \sqrt{x} \quad \text{Square both sides again}$$
$$4 = x$$

Check: $\sqrt{4} + \sqrt{4 + 5} = \sqrt{4} + \sqrt{9} = 2 + 3 = 5$, so 4 is a solution.

4

70. $\sqrt{x - 1} + \sqrt{x - 4} = 3$

$$\sqrt{x - 4} = 3 - \sqrt{x - 1} \quad \text{Square both sides}$$
$$(\sqrt{x - 4})^2 = (3 - \sqrt{x - 1})^2$$
$$x - 4 = 9 - 6\sqrt{x - 1} + x - 1$$
$$x - 4 = x + 8 - 6\sqrt{x - 1}$$
$$-12 = -6\sqrt{x - 1}$$
$$2 = \sqrt{x - 1} \quad \text{Square both sides again}$$
$$4 = x - 1$$
$$x = 5$$

Check: $\sqrt{5 - 1} + \sqrt{5 - 4} = \sqrt{4} + \sqrt{1} = 2 + 1 = 3$ so 5 is a solution.

71. $2 + \sqrt{x - 3} = 1$

$\sqrt{x - 3} = -1$

This equation can have no solution since the left side, $\sqrt{x - 3}$, is non-negative and can never equal -1.

No solution.

72.
$$\frac{3}{\sqrt{x - 4}} = \sqrt{x - 12} \quad \text{Square both sides}$$
$$\frac{9}{x - 4} = x - 12 \qquad \text{lcm: } x - 4,\ x \neq 4$$
$$(x - 4)\cdot\frac{9}{x - 4} = (x - 12)(x - 4)$$
$$9 = x^2 - 16x + 48$$
$$0 = x^2 - 16x + 39$$
$$0 = (x - 3)(x - 13)$$
$x - 3 = 0$ or $x - 13 = 0$

$x = 3$ $\qquad$ $x = 13$

Check: <u>x = 3</u>
$$\frac{3}{\sqrt{3 - 4}} \stackrel{?}{=} \sqrt{3 - 12}$$
$$\frac{3}{\sqrt{-1}} \stackrel{?}{=} \sqrt{-9}$$
$$\frac{3}{i} \stackrel{?}{=} 3i$$
$$-3i \neq 3i$$
3 is extraneous

<u>x = 13</u>
$$\frac{3}{\sqrt{13 - 4}} \stackrel{?}{=} \sqrt{13 - 12}$$
$$\frac{3}{\sqrt{9}} \stackrel{?}{=} \sqrt{1}$$
$$\frac{3}{3} \stackrel{\checkmark}{=} 1$$
13 is a solution

13

73.
$$x^2 \geq 4x + 21$$
$$x^2 - 4x - 21 \geq 0$$
$$(x + 3)(x - 7) \geq 0$$
Critical points -3, 7

sign of (x - 7) - - - ⁞ - - - ⁞ + + +
sign of (x + 3) - - - ⁞ + + + ⁞ + + +

(number line: x, with marks at -3 and 7)

The equality holds at the critical points. The inequality statement holds when both factors have the same sign, to the left of -3 or to the right of 7.

Solution: $x \leq -3$ or $x \geq 7$

$(-\infty, -3] \cup [7, \infty)$

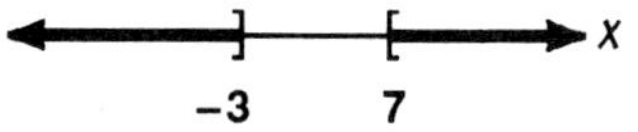

74.
$$\frac{1}{x} < 2$$
$$\frac{1}{x} - 2 < 0$$
$$\frac{1 - 2x}{x} < 0$$
Critical points $0, \frac{1}{2}$

sign of (1 - 2x) + + + ⁞ + + + ⁞ - - -
sign of x - - - ⁞ + + + ⁞ + + +

(number line: x, with marks at 0 and $\frac{1}{2}$)

The inequality holds when $(1 - 2x)$ and x have opposite signs, left of 0 or right of $\frac{1}{2}$.

Solution: $x < 0$ or $x > \frac{1}{2}$

$(-\infty, 0) \cup \left(\frac{1}{2}, \infty\right)$

75. $10x > x^2 + 25$
$0 > x^2 - 10x + 25$
$0 > (x - 5)^2$

Since the square of no real number is negative, this statement is never true. No solution.

76. $x^2 + 16 \geq 8x$
$x^2 - 8x + 16 \geq 0$
$(x - 4)^2 \geq 0$

Since the square of every real number is non-negative, this statement is true for all real numbers. The graph is the entire real line.

77. $y = (x - 2)^2 + 3$
$y = x^2 - 4x + 4 + 3$
$y = x^2 - 4x + 7$

This is the equation of a parabola, opening up, since the coefficient of x^2, $1 = a > 0$. The vertex is located at

$x = -\frac{b}{2a} = -\frac{-4}{2\cdot 1} = 2$ and $y = 2^2 - 4\cdot 2 + 7 = 3$

x	y
0	7
1	4
2	3
3	4
4	7

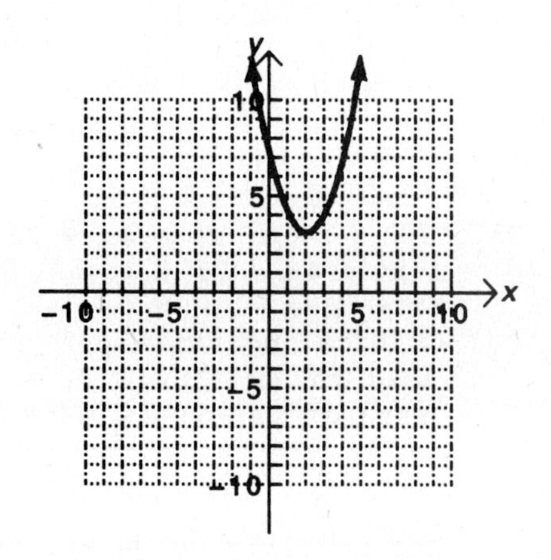

78. $y = -(x + 1)^2 - 2$
$y = -(x^2 + 2x + 1) - 2$
$y = -x^2 - 2x - 1 - 2$
$y = -x^2 - 2x - 3$

This is the equation of a parabola, opening down, since the coefficient of x^2, $-1 = a < 0$. The vertex is located at

$x = -\frac{b}{2a} = -\frac{-2}{2(-1)} = -1$ and

$y = -(-1)^2 - 2(-1) - 3 = -2$

x	y
-3	-6
-2	-3
-1	-2
0	-3
1	-6

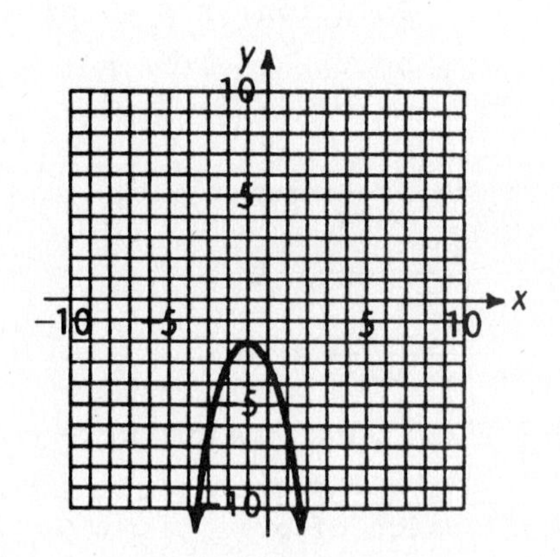

79. $4x^2 + 25y^2 = 100$

Rewriting this as $\frac{4x^2}{100} + \frac{25y^2}{100} = 1$, that is, $\frac{x^2}{25} + \frac{y^2}{4} = 1$, puts it in standard form for an ellipse.

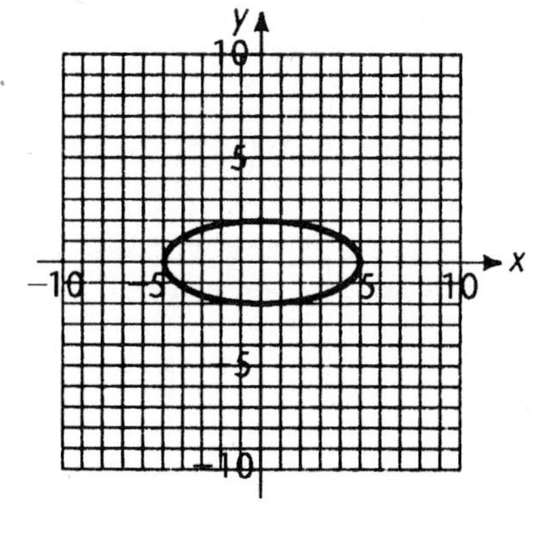

Find x intercepts	Find y intercepts
$4x^2 + 0 = 100$	$0 + 25y^2 = 100$
$4x^2 = 100$	$25y^2 = 100$
$x^2 = 25$	$y^2 = 4$
$x = \pm\sqrt{25} = \pm 5$	$y = \pm\sqrt{4} = \pm 2$

Now plot the intercepts and draw the ellipse.

80. $4x^2 - 25y^2 = 100$

Rewriting this as $\frac{4x^2}{100} - \frac{25y^2}{100} = 1$, that is, $\frac{x^2}{25} - \frac{y^2}{4} = 1$, puts it in standard form for a hyperbola. We first draw a dashed rectangle with intercepts $x = \pm\sqrt{25} = \pm 5$ and $y = \pm\sqrt{4} = \pm 2$. Second, draw in asymptotes (extended diagonals of the rectangle). Third, determine true intercepts for the hyperbola, as follows: Let $y = 0$; then

$$\frac{x^2}{25} - 0 = 1$$
$$x^2 = 25$$
$$x = \pm\sqrt{25} = \pm 5$$

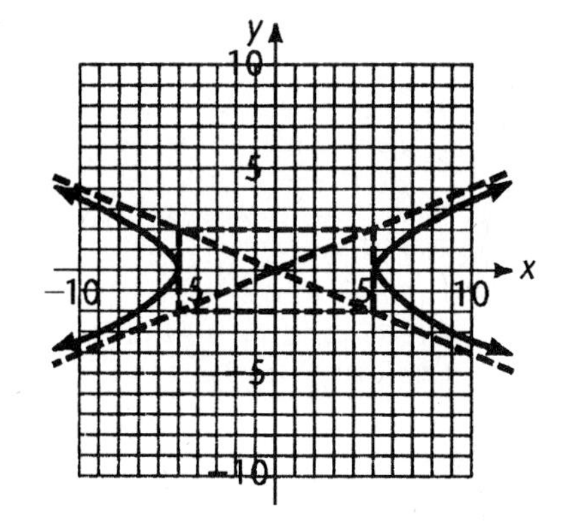

If we let $x = 0$, then

$$0 - \frac{y^2}{4} = 1$$
$$y^2 = -4$$
$$y = \pm\sqrt{-4} = \pm 2i$$

These are complex numbers and do not represent real intercepts. We conclude that the only intercepts for the hyperbola are $x = \pm 5$, and the hyperbola opens left and right.

81. We rewrite $x^2 - 6x - 3 = 0$ by completing the square.

$$x^2 - 6x = 3 \qquad \left(-\frac{6}{2}\right)^2 = (-3)^2 = 9$$
$$x^2 - 6x + 9 = 3 + 9$$
$$(x - 3)^2 = 12$$

82. Since the center is at the origin, $(h, k) = (0, 0)$, and the equation is of the form

$$x^2 + y^2 = R^2$$

Since the circle passes through the point $(12, -5)$, the coordinates of this point must satisfy the equation. Hence

$$(12)^2 + (-5)^2 = R^2$$
$$169 = R^2$$

The desired equation is $x^2 + y^2 = 169$.

83. $y = 3x^2 + 5x - 2$ is the equation of a parabola, opening up, since the coefficient of x^2, $3 = a > 0$. The vertex is located at $x = -\frac{b}{2a} = -\frac{5}{2(3)} = -\frac{5}{6}$

and $y = 3\left(-\frac{5}{6}\right)^2 + 5\left(-\frac{5}{6}\right) - 2 = -\frac{49}{12}$

To find the values where the graph crosses the x axis, we solve:

$$3x^2 + 5x - 2 = 0$$
$$(3x - 1)(x + 2) = 0$$
$$3x - 1 = 0 \quad \text{or} \quad x + 2 = 0$$
$$x = \frac{1}{3} \quad \text{and} \quad x = -2$$

are the x-values where the graph crosses the x-axis

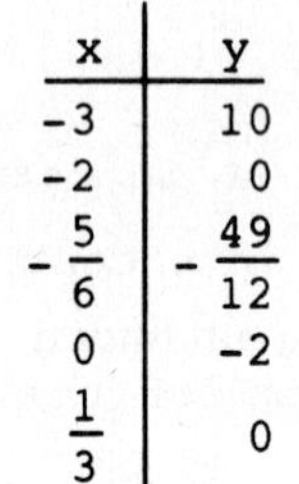

x	y
-3	10
-2	0
$-\frac{5}{6}$	$-\frac{49}{12}$
0	-2
$\frac{1}{3}$	0

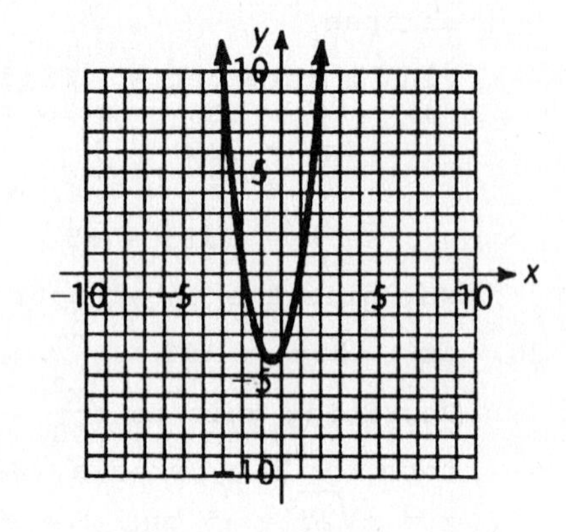

84. We transform the equation into the form

$$(x - h)^2 + (y - k)^2 = R^2$$

by completing the square relative to x and relative to y:

$$x^2 + 6x + ? + y^2 - 8y + ? = 0 + ? + ?$$
$$x^2 + 6x + 9 + y^2 - 8y + 16 = 0 + 9 + 16$$
$$(x + 3)^2 + (y - 4)^2 = 25$$

or $[x - (-3)]^2 + (y - 4)^2 = 5^2$

This is the equation of a circle with radius 5 and center at $(h, k) = (-3, 4)$.

85. We rewrite $3x^2 - 9y^2 = 36$ in standard form for a hyperbola, $\frac{x^2}{m^2} - \frac{y^2}{n^2} = 1$.

$$\frac{3x^2}{36} - \frac{9y^2}{36} = 1$$
$$\frac{x^2}{12} - \frac{y^2}{4} = 1$$

We first note that $m^2 = 12$, $n^2 = 4$, so $m = \sqrt{12} = 2\sqrt{3}$ and $n = 2$. The equations of the asymptotes of a hyperbola, whose equation in standard form is $\frac{x^2}{m^2} - \frac{y^2}{n^2}$, are given by $y = \pm\frac{n}{m}x$. In this case, $y = \pm\frac{2}{2\sqrt{3}}x$, or $y = \pm\frac{\sqrt{3}}{3}x$, are the equations of the asymptotes.

86.

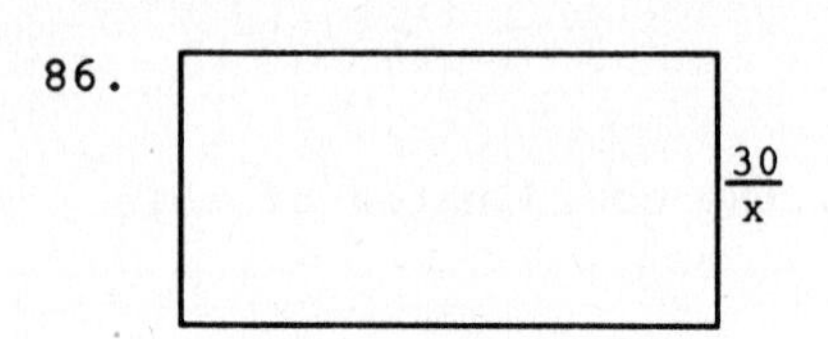

Let x = length of one side.
Then, since the area of the rectangle is 30 in^2, we have

x(length of other side) = 30

length of other side $= \frac{30}{x}$

Since $P = 2a + 2b$, we have $P = 22$,

$a = x$, $b = \frac{30}{x}$. The equation desired is

$$22 = 2x + 2 \cdot \frac{30}{x}$$

$$22 = 2x + \frac{60}{x} \quad x \neq 0 \quad \text{lcm} = x$$

$$22x = 2x \cdot x + \frac{60}{x} \cdot x$$

$$22x = 2x^2 + 60$$

$$0 = 2x^2 - 22x + 60$$

$$0 = x^2 - 11x + 30$$

$$x^2 - 11x + 30 = 0$$

$$(x - 6)(x - 5) = 0$$

$x = 5$ or $x = 6$

$\frac{30}{x} = 6$ $\qquad$ $\frac{30}{x} = 5$

Solution: 6×5 inches

87. Let x = time for boy to complete job alone
$x + 1$ = time for friend to complete job alone
$3\frac{44}{60} = \frac{224}{60} = \frac{56}{15}$ = time for both to complete job together.

Then the rate of completion for the boy is $\frac{1}{x}$ job per hour.

the rate of completion for the friend is $\frac{1}{x+1}$ job per hour.

the rate of completion for both together is $\frac{15}{56}$ job per hour.

$$\begin{pmatrix}\text{Rate of}\\ \text{boy}\end{pmatrix} + \begin{pmatrix}\text{Rate of}\\ \text{friend}\end{pmatrix} = \begin{pmatrix}\text{Rate}\\ \text{together}\end{pmatrix}$$

$$\frac{1}{x} + \frac{1}{x+1} = \frac{15}{56} \qquad \text{lcm} = 56x(x + 1);\ x \neq 0, -1$$

$$56x(x + 1)\frac{1}{x} + 56x(x + 1)\frac{1}{x+1} = 56x(x + 1)\frac{15}{56}$$

$$56(x + 1) + 56x = 15x(x + 1)$$

$$56x + 56 + 56x = 15x^2 + 15x$$

$$0 = 15x^2 - 97x - 56$$

This is factorable in the integers, but if this is not apparent use the quadratic formula.

$a = 15 \quad b = -97 \quad c = -56$

$$x = \frac{-(-97) \pm \sqrt{(-97)^2 - 4(15)(-56)}}{2(15)}$$

$$x = \frac{97 \pm \sqrt{9409 + 3360}}{30}$$

$$x = \frac{97 \pm \sqrt{12769}}{30}$$

$$x = \frac{97 \pm 113}{30}$$

$x = -\frac{16}{30}$ or $x = 7$

discard; times cannot be negative

7 hr = time for boy; 8 hr = time for friend

88. $\left(t - \frac{3}{2}\right)^2 = -\frac{3}{2}$ Use square root method

$$t - \frac{3}{2} = \pm\sqrt{-\frac{3}{2}}$$
$$t - \frac{3}{2} = \pm i\sqrt{\frac{3}{2}}$$
$$t = \frac{3}{2} \pm i\sqrt{\frac{3}{2}}$$
$$= \frac{3}{2} \pm i\sqrt{\frac{6}{4}}$$
$$= \frac{3}{2} \pm i\frac{\sqrt{6}}{2} \text{ or } \frac{3 \pm i\sqrt{6}}{2}$$

89. $3x - 1 = \frac{2(x + 1)}{x + 2}$ $x \neq -2$, lcm $= x + 2$

$$(3x - 1)(x + 2) = (x + 2)\frac{2(x + 1)}{x + 2}$$
$$3x^2 + 5x - 2 = 2(x + 1)$$
$$3x^2 + 5x - 2 = 2x + 2$$
$3x^2 + 3x - 4 = 0$ This is not factorable in the integers; use quadratic formula.

$a = 3$ $b = 3$ $c = -4$

$$x = \frac{-(3) \pm \sqrt{(3)^2 - 4(3)(-4)}}{2(3)}$$
$$= \frac{-3 \pm \sqrt{9 + 48}}{6}$$
$$= \frac{-3 \pm \sqrt{57}}{6}$$

90. $y^8 - 17y^4 + 16 = 0$

Let $u = y^4$, then $u^2 = y^8$. After substitution, the original equation becomes

$$u^2 - 17u + 16 = 0$$
$$(u - 16)(u - 1) = 0$$
$$u = 16 \text{ or } u = 1$$

Replacing u with y^4, we obtain

$$y^4 = 16 \qquad\qquad y^4 = 1$$
$$y^4 - 16 = 0 \qquad\qquad y^4 - 1 = 0$$
$$(y^2 + 4)(y^2 - 4) = 0 \qquad\qquad (y^2 + 1)(y^2 - 1) = 0$$
$$y^2 + 4 = 0 \text{ or } y^2 - 4 = 0 \text{ or } y^2 + 1 = 0 \text{ or } y^2 - 1 = 0$$
$$y^2 = -4 \qquad y^2 = 4 \qquad y^2 = -1 \qquad y^2 = 1$$
$$y = \pm\sqrt{-4} \qquad y = \pm 2 \qquad y = \pm\sqrt{-1} \qquad y = \pm 1$$
$$y = \pm 2i \qquad\qquad y = \pm i$$

Solution: $\pm 1, \pm 2, \pm i, \pm 2i$

91. $\sqrt{y - 2} - \sqrt{5y + 1} = -3$

$\sqrt{y - 2} = \sqrt{5y + 1} - 3$ Square both sides.

$$y - 2 = 5y + 1 - 2 \cdot 3\sqrt{5y + 1} + 9$$
$$y - 2 = 5y + 10 - 6\sqrt{5y + 1}$$
$$-4y - 12 = -6\sqrt{5y + 1}$$

$2y + 6 = 3\sqrt{5y + 1}$ Square both sides again.

$$4y^2 + 24y + 36 = 9(5y + 1)$$

$$4y^2 + 24y + 36 = 45y + 9$$
$$4y^2 - 21y + 27 = 0$$
$$(y - 3)(4y - 9) = 0$$
$$y - 3 = 0 \quad \text{or} \quad 4y - 9 = 0$$
$$y = 3 \qquad 4y = 9$$
$$y = \frac{9}{4}$$

Check: $\underline{y = 3}$ $\sqrt{3 - 2} - \sqrt{5(3) + 1} = \sqrt{1} - \sqrt{16} = 1 - 4 = -3$

so 3 is a solution.

$\underline{y = \frac{9}{4}}$ $\sqrt{\frac{9}{4} - 2} - \sqrt{5(\frac{9}{4}) + 1} = \sqrt{\frac{9}{4} - \frac{8}{4}} - \sqrt{\frac{45}{4} + \frac{4}{4}} = \sqrt{\frac{1}{4}} - \sqrt{\frac{49}{4}} = \frac{1}{2} - \frac{7}{2} = -3$

so $\frac{9}{4}$ is a solution.

$3, \frac{9}{4}$

92.
$$\sqrt{x - 3} + \sqrt{x + 2} = \sqrt{3x + 4} \quad \text{Square both sides}$$
$$(\sqrt{x - 3} + \sqrt{x + 2})^2 = (\sqrt{3x + 4})^2$$
$$x - 3 + 2\sqrt{x - 3}\sqrt{x + 2} + x + 2 = 3x + 4$$
$$2\sqrt{x - 3}\sqrt{x + 2} + 2x - 1 = 3x + 4$$
$$2\sqrt{x - 3}\sqrt{x + 2} = x + 5 \quad \text{Square both sides again}$$
$$4(x - 3)(x + 2) = x^2 + 10x + 25$$
$$4(x^2 - x - 6) = x^2 + 10x + 25$$
$$4x^2 - 4x - 24 = x^2 + 10x + 25$$
$$3x^2 - 14x - 49 = 0$$
$$(x - 7)(3x + 7) = 0$$
$$x - 7 = 0 \quad \text{or} \quad 3x + 7 = 0$$
$$x = 7 \qquad x = -\frac{7}{3}$$

Check: $\underline{x = 7}$ $\sqrt{7 - 3} + \sqrt{7 + 2} \stackrel{?}{=} \sqrt{3 \cdot 7 + 4}$

$$\sqrt{4} + \sqrt{9} \stackrel{?}{=} \sqrt{25}$$
$$2 + 3 \stackrel{\surd}{=} 5$$

7 is a solution

$\underline{x = -\frac{7}{3}}$ $\sqrt{-\frac{7}{3} - 3} + \sqrt{-\frac{7}{3} + 2} \stackrel{?}{=} \sqrt{3(-\frac{7}{3}) + 4}$

$$\sqrt{-\frac{16}{3}} + \sqrt{-\frac{1}{3}} \stackrel{?}{=} \sqrt{-3}$$
$$\frac{4i}{\sqrt{3}} + \frac{i}{\sqrt{3}} \neq \frac{3i}{\sqrt{3}}$$

$-\frac{7}{3}$ is extraneous

7

93.
$$x^4 - 1 = 0$$
$$(x^2 - 1)(x^2 + 1) = 0$$
$$x^2 - 1 = 0 \quad \text{or} \quad x^2 + 1 = 0$$
$$x^2 = 1 \qquad x^2 = -1$$
$$x = \pm\sqrt{1} \qquad x = \pm\sqrt{-1}$$
$$x = \pm 1 \qquad x = \pm i$$

$\pm 1, \pm i$

94. $x^{4/3} - 1 = 0$

Let $u = x^{1/3}$. After substitution, the original equation becomes

$$u^4 - 1 = 0$$

From the previous problem, the solutions of this equation are

$$u = 1, \ u = -1, \ u = i, \ u = -i$$

Replacing u with $x^{1/3}$, we obtain

$x^{1/3} = 1$	$x^{1/3} = -1$	$x^{1/3} = i$	$x^{1/3} = -i$
$x = 1^3$	$x = (-1)^3$	$x = i^3$	$x = (-i)^3$
$x = 1$	$x = -1$	$x = -i$	$x = i$

$\pm 1, \pm i$

95.

$$\frac{3}{x - 4} \le \frac{2}{x - 3}$$

$$\frac{3}{x - 4} - \frac{2}{x - 3} \le 0$$

$$\frac{3(x - 3) - 2(x - 4)}{(x - 4)(x - 3)} \le 0$$

$$\frac{3x - 9 - 2x + 8}{(x - 4)(x - 3)} \le 0$$

$$\frac{x - 1}{(x - 4)(x - 3)} \le 0$$

Critical points 1, 3, 4

	$x<1$	$1<x<3$	$3<x<4$	$x>4$
sign of $(x - 1)$	− − −	+ + +	+ +	+ + +
sign of $(x - 3)$	− − −	− − −	+ +	+ + +
sign of $(x - 4)$	− − −	− − −	− −	+ + +

The inequality statement is satisfied when $(x - 1)$, $(x - 3)$, and $(x - 4)$ are all negative, two are positive and one is negative, or the numerator is 0. Two are positive between 3 and 4 and all are negative to the left of 1. The equality part of the inequality holds when $x = 1$, but not when $x = 3$ or 4. Thus

Solution: $x \le 1$ or $3 < x < 4$

$(-\infty, 1] \cup (3, 4)$

96. $\dfrac{x^2 + 1}{x - 1} > 0$

Critical point: 1

	$x<1$	$x>1$
sign of $(x^2 + 1)$	+ + +	+ + +
sign of $x - 1$	− − −	+ + +

(sum of two squares is always positive)

The inequality statement is satisfied when $(x - 1)$ is positive, to the right of 1. Solution: $x > 1$ or $(1, \infty)$

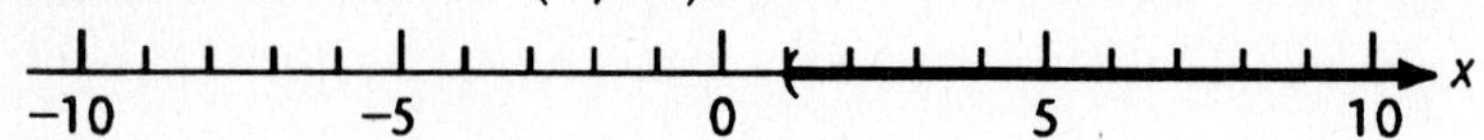

97.

$$\frac{x^2 - 1}{x^2 + 2x} < 0$$

$$\frac{(x - 1)(x + 1)}{x(x + 2)} < 0$$

Critical points 1, −1, 0, −2

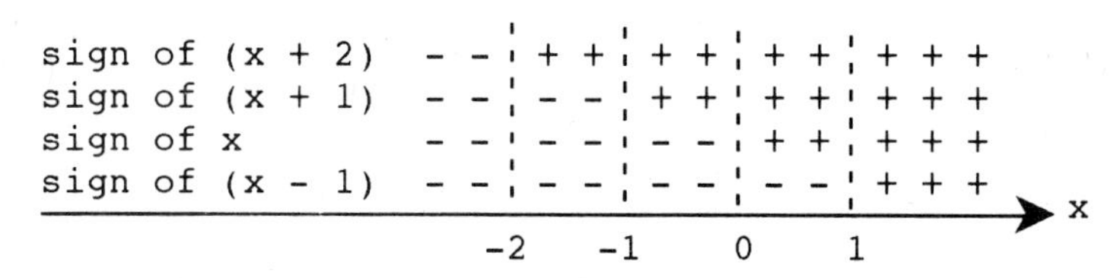

The inequality statement is satisfied when three factors are negative and one is positive, or when three are positive and one is negative. Three are negative between -2 and -1 and three are positive between 0 and 1. Thus, solution: $-2 < x < -1$ or $0 < x < 1$ $(-2, -1) \cup (0, 1)$

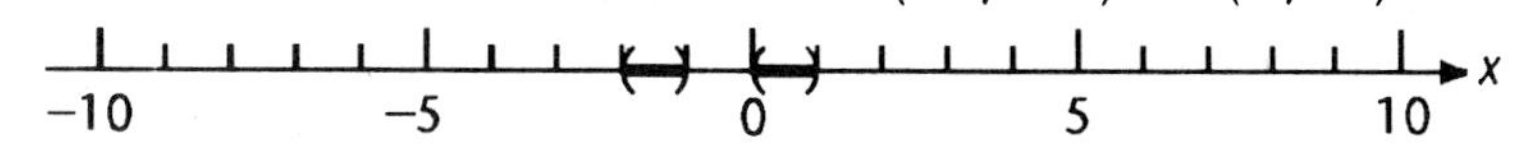

98.

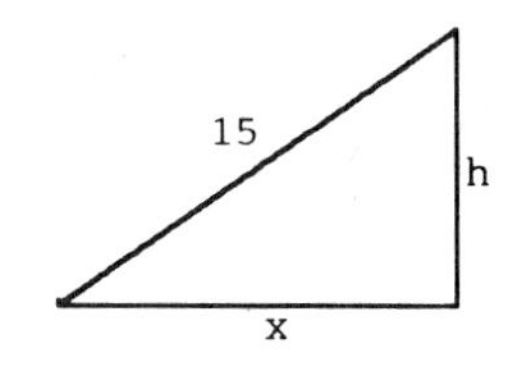

Following the hint,
let x = length of one side.
h = length of other side.
From the Pythagorean theorem, we have

$$h^2 + x^2 = (15)^2$$
$$h^2 + x^2 = 225$$
$$h^2 = 225 - x^2$$
$$h = \sqrt{225 - x^2} \quad \text{since h is positive.}$$

Now use the formula for the area of a triangle, $A = \frac{1}{2}bh$, with $A = 54$, $b = x$, and $h = \sqrt{225 - x^2}$.

$$54 = \frac{1}{2}x\sqrt{225 - x^2}$$
$$108 = x\sqrt{225 - x^2} \quad \text{Square both sides.}$$
$$11664 = x^2(225 - x^2)$$
$$11664 = 225x^2 - x^4$$
$$x^4 - 225x^2 + 11664 = 0$$

Let $u = x^2$, then $u^2 = x^4$. After substitution, the original equation becomes

$$u^2 - 225u + 11664 = 0$$

This is factorable in the integers as $(u - 81)(u - 144)$, but if this is not apparent, it can also be solved by the quadratic formula.

$a = 1$ $b = -225$ $c = 11664$

$$u = \frac{-(-225) \pm \sqrt{(-225)^2 - 4(1)(11664)}}{2(1)}$$
$$= \frac{225 \pm 63}{2}$$

$u = 144$ or $u = 81$

Replacing u with x^2, we obtain

$$x^2 = 144 \quad \text{or} \quad x^2 = 81$$
$$x = 12 \qquad\qquad x = 9$$
$$\sqrt{225 - x^2} = 9 \qquad \sqrt{225 - x^2} = 12$$

Check: <u>x = 12</u> $54 \stackrel{?}{=} \frac{1}{2} \cdot 12\sqrt{225 - (12)^2}$

$$54 \stackrel{\checkmark}{=} 54$$

Since the solutions of the equation give rise to the same triangle, there is no need to check the second solution.

Solution: 9 cm and 12 cm

99. (A) For a \$15,000 weekly cost, we substitute $C = 15$ into the cost equation to obtain

$$15 = x^2 - 10x + 31$$
$$0 = x^2 - 10x + 16$$
$$x^2 - 10x + 16 = 0$$
$$(x - 2)(x - 8) = 0$$
$$x = 2, 8$$

This corresponds to outputs of 2000 or 8000 units.

(B) For a \$6,000 weekly cost, we substitute $C = 6$ into the cost equation to obtain

$$6 = x^2 - 10x + 31$$
$$0 = x^2 - 10x + 25$$
$$x^2 - 10x + 25 = 0$$
$$(x - 5)^2 = 0$$
$$x - 5 = 0$$
$$x = 5$$

This corresponds to the output of 5000 units.

100. Let x = the number of \$0.25 increases in price.
Then $1.5 + 0.25x$ = the price that results per dozen ears
$50 - 4x$ = the number of dozen ears sold at this price.
Since revenue = (number sold) × (price per unit),
revenue = $(50 - 4x)(1.5 + 0.25x)$
In this situation, revenue = 82.5, thus

$$82.5 = (50 - 4x)(1.5 + 0.25x)$$
$$82.5 = 75 + 12.5x - 6x - x^2$$
$$82.5 = 75 + 6.5x - x^2$$
$$x^2 - 6.5x + 7.5 = 0$$
$$2x^2 - 13x + 15 = 0$$
$$(2x - 3)(x - 5) = 0$$

$2x - 3 = 0$ or $x - 5 = 0$

$2x = 3$ $\qquad x = 5$ $\qquad$ Price = $1.5 + 0.25(5)$ = \$2.75

$x = \frac{3}{2}$

discard since the price is an integer number of cents

Practice Test 7

1. $x^2 - 5x = 0$
   $x(x - 5) = 0$
   $x = 0$ or $x - 5 = 0$
   $x = 5$

   0, 5

2. $x^2 - 9 = 0$
   $x^2 = 9$
   $x = \pm\sqrt{9}$
   $x = \pm 3$

3. $x^2 - 5x - 9 = 0$ This is not factorable in the integers; use the quadratic formula.
   $a = 1 \quad b = -5 \quad c = -9$

   $$x = \frac{-(-5) \pm \sqrt{(-5)^2 - 4(1)(-9)}}{2(1)}$$
   $$x = \frac{5 \pm \sqrt{25 + 36}}{2}$$
   $$x = \frac{5 \pm \sqrt{61}}{2}$$

4. $\sqrt{x - 5} = x - 7$ Square both sides
   $x - 5 = x^2 - 14x + 49$
   $0 = x^2 - 15x + 54$
   $0 = (x - 9)(x - 6)$
   $x - 9 = 0$ or $x - 6 = 0$
   $x = 9 \qquad x = 6$

   Check: $\underline{x = 9}$ $\sqrt{9 - 5} \stackrel{?}{=} 9 - 7$ $\qquad$ $\underline{x = 6}$ $\sqrt{6 - 5} \stackrel{?}{=} 6 - 7$
   $\sqrt{4} \stackrel{\checkmark}{=} 2$ $\qquad$ $\sqrt{1} \neq -1$
   9 is a solution $\qquad$ 6 is extraneous

   9

5. $x^2 - 5x < 0$
   $x(x - 5) < 0$
   Critical points 0, 5

	$x < 0$	$0 < x < 5$	$x > 5$
sign of x	− −	+ + +	+ + +
sign of $(x - 5)$	− −	− − −	+ + +

   (number line with critical points 0 and 5 on the x-axis)

   The factors must have opposite signs; this occurs between 0 and 5.
   Solution: $0 < x < 5$

6. $x^4 - 5x^2 - 6 = 0$
   Let $u = x^2$, then $u^2 = y^4$. After substitution, the original equation becomes
   $u^2 - 5u - 6 = 0$
   $(u - 6)(u + 1) = 0$
   $u = 6$ or $u = -1$

   Replacing u with x^2, we obtain
   $x^2 = 6 \qquad x^2 = -1$
   $x = \pm\sqrt{6} \qquad x = \pm\sqrt{-1}$
   $x = \pm i$

   $\pm\sqrt{6}, \pm i$

7. $x(x + 1)(x - 4) \ge 0$
Critical points 0, -1, 4

	$x < -1$	$-1 < x < 0$	$0 < x < 4$	$x > 4$
sign of x	- -	- -	+ + +	+ +
sign of $(x + 1)$	- -	+ +	+ + +	+ +
sign of $(x - 4)$	- -	- -	- - -	+ +

(critical points on the x-axis: -1, 0, 4)

The inequality is satisfied when x, $(x + 1)$, and $(x - 4)$ are all positive, or two are negative and one positive. All are positive to the right of 4, and two are negative between -1 and 0. The equality part of the statement holds at the critical points. Thus,

Solution: $-1 \le x \le 0$ or $x \ge 4$

8. We rewrite $x^2 + 4x + 7 = 0$ by completing the square

$$x^2 + 4x = -7 \qquad \left(\frac{4}{2}\right)^2 = 2^2 = 4$$

$$x^2 + 4x + 4 = -7 + 4$$
$$(x + 2)^2 = -3$$
$$[x - (-2)]^2 + 3 = 0$$

9. We transform the equation into the form
$$(x - h)^2 + (y - k)^2 = R^2$$
by completing the square relative to x and relative to y
$$x^2 - 4x + ? + y^2 + 6y + ? = -3 + ? + ?$$
$$x^2 - 4x + 4 + y^2 + 6y + 9 = -3 + 4 + 9$$
$$(x - 2)^2 + (y + 3)^2 = 10$$
or $$(x - 2)^2 + [y - (-3)]^2 = (\sqrt{10})^2$$
This is the equation of a circle with radius $\sqrt{10}$ and center at $(h, k) = (2, -3)$.

10. Let $P_1 = (2, 3)$ and $P_2 = (-4, 5)$, then
$$d = \sqrt{(x_2 - x_1)^2 + (y_2 - y_1)^2} = \sqrt{[(-4) - 2]^2 + (5 - 3)^2} = \sqrt{(-6)^2 + (2)^2}$$
$$= \sqrt{40} \text{ or } 2\sqrt{10}$$

11. $y = x^2 + 4$

This is the equation of a parabola, opening up, since the coefficient of x^2, $1 = a > 0$. The vertex is located at
$$x = -\frac{b}{2a} = -\frac{0}{2(1)} = 0 \text{ and}$$
$$y = 0^2 + 4 = 4$$

x	y
2	8
1	5
0	4
-1	5
-2	8

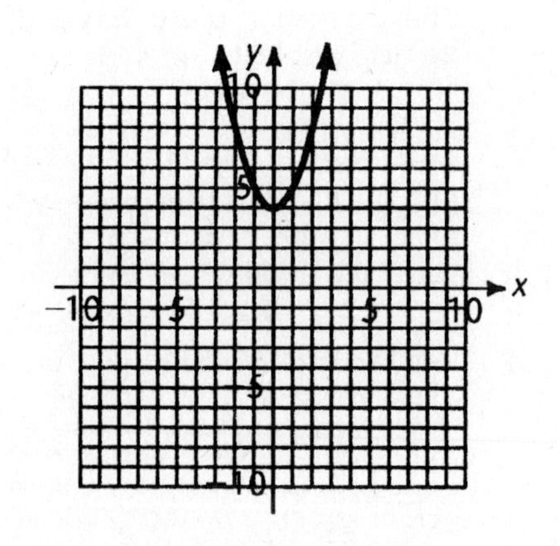

12. $y = 2x^2 + 3x + 4$

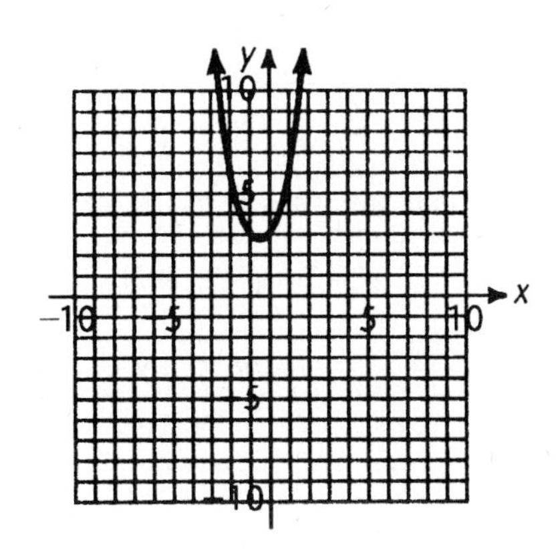

This is the equation of a parabola, opening up, since the coefficient of x^2, $2 = a > 0$. The vertex is located at

$$x = -\frac{b}{2a} = -\frac{3}{2(2)} = -\frac{3}{4} \quad \text{and}$$

$$y = 2\left(-\frac{3}{4}\right)^2 + 3\left(-\frac{3}{4}\right) + 4 = \frac{23}{8}$$

x	y
-2	6
-1	3
0	4
1	9

13. $x^2 + 4y^2 = 16$

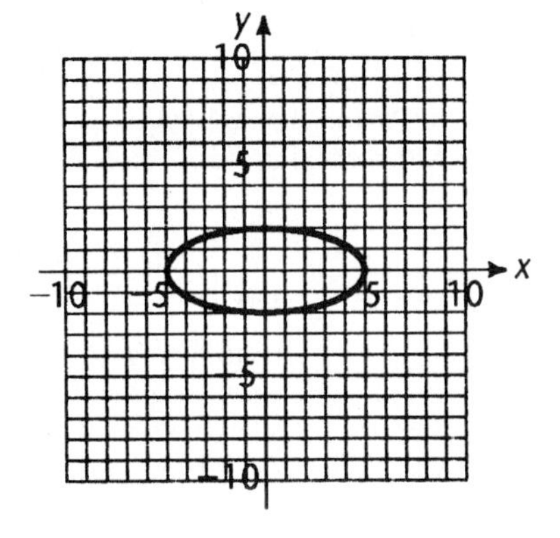

Rewriting this as $\frac{x^2}{16} + \frac{4y^2}{16} = 1$, that is,

$\frac{x^2}{16} + \frac{y^2}{4} = 1$, puts it in standard form for an ellipse.

Find x intercepts
$x^2 + 0 = 16$
$x^2 = 16$
$x = \pm\sqrt{16} = \pm 4$

Find y intercepts
$0 + 4y^2 = 16$
$y^2 = 4$
$y = \pm\sqrt{4} = \pm 2$

Now plot the intercepts and draw the ellipse.

14. This is a circle with radius 2 and center at the origin.

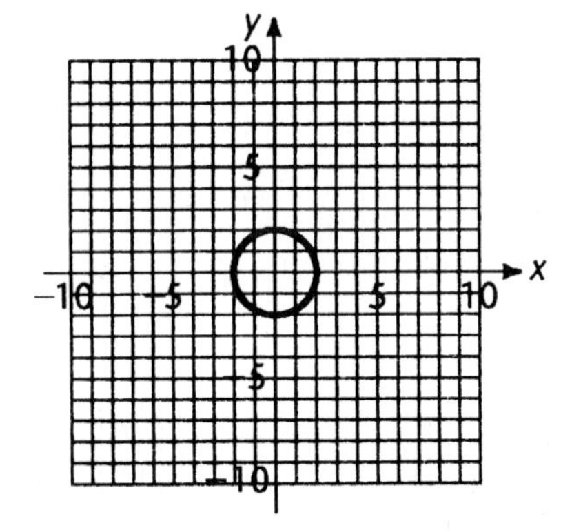

15. Rewriting this as $\frac{x^2}{4} - \frac{y^2}{4} = 1$ puts it in standard form for a hyperbola.

We first draw a dashed rectangle with intercepts $x = \pm\sqrt{4} = \pm 2$ and $y = \pm 2$. Second, draw in asymptotes (extended diagonals of the rectangle). Third, determine true intercepts for the hyperbola, as follows: Let $y = 0$; then

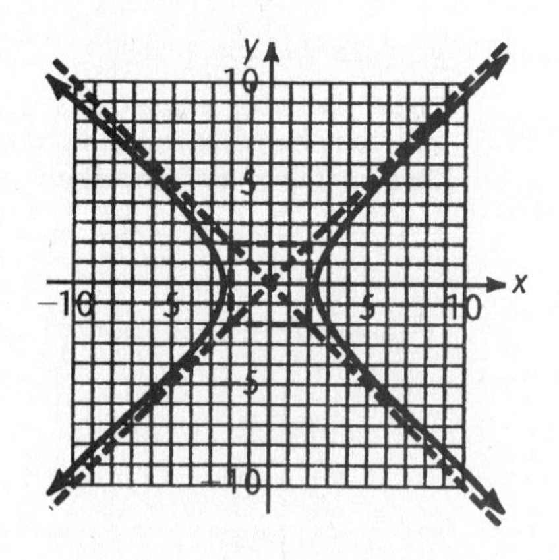

$$x^2 - 0 = 4$$
$$x^2 = 4$$
$$x = \pm\sqrt{4} = \pm 2$$

If we let $x = 0$, then

$$0 - y^2 = 4$$
$$y^2 = -4$$
$$y = \pm\sqrt{-4} = \pm 2i$$

These are complex numbers and do not represent real intercepts. We conclude that the only intercepts for the hyperbola are $x = \pm 2$, and the hyperbola opens left and right.

16. Rewriting this as $-\frac{4x^2}{16} + \frac{y^2}{16} = 1$, that is, $\frac{y^2}{16} - \frac{x^2}{4} = 1$ puts it in standard form for a hyperbola. We first draw a dashed rectangle with intercepts $x = \pm\sqrt{4} = \pm 2$ and $y = \pm\sqrt{16} = \pm 4$. Second, draw in asymptotes (extended diagonals of the rectangle). Third, determine true intercepts for the hyperbola, as follows: If we let $y = 0$; then

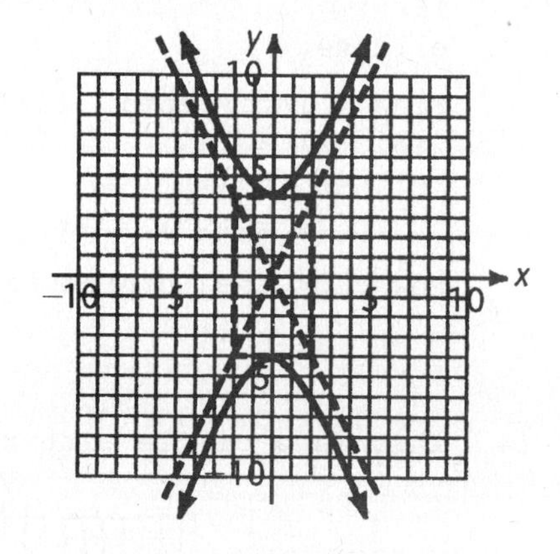

$$0 - \frac{x^2}{4} = 1$$
$$x^2 = -4$$
$$x = \pm\sqrt{-4} = \pm 2i$$

These are complex numbers and do not represent real intercepts. Let $x = 0$, then

$$\frac{y^2}{16} - 0 = 1$$
$$y^2 = 16$$
$$y = \pm\sqrt{16} = \pm 4$$

We conclude that the only intercepts for the hyperbola are $y = \pm 4$, and the hyperbola opens up and down.

17. $x + 4y^2 = 0$ can be written in the standard form $x = ky^2$ as follows:

$$x = -4y^2$$

Since $k < 0$, the x values must be negative and the parabola opens left. We look for values of y in addition to $y = 0$, that make x easy to calculate, then plot the graph.

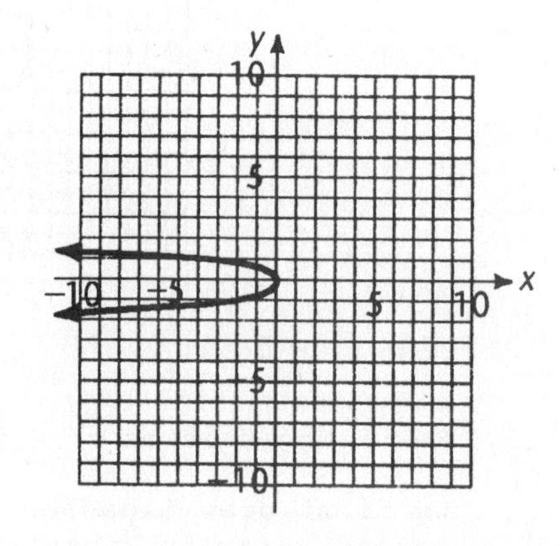

x	y
0	0
-4	1
-4	-1

18. $y + 4x^2 = 9$ can be rewritten as $y = -4x^2 + 9$.

This is the equation of a parabola, opening down, since the coefficient of x^2, $-4 = a < 0$. The vertex is located at

$$x = -\frac{b}{2a} = -\frac{0}{2(-4)} = 0 \quad \text{and}$$

$$y = -4(0)^2 + 9 = 9$$

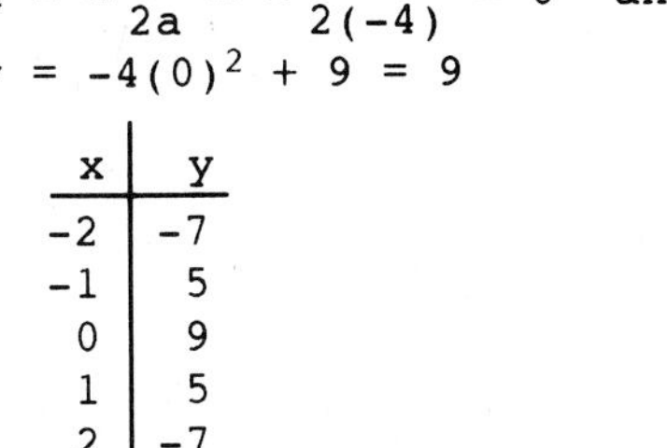

x	y
-2	-7
-1	5
0	9
1	5
2	-7

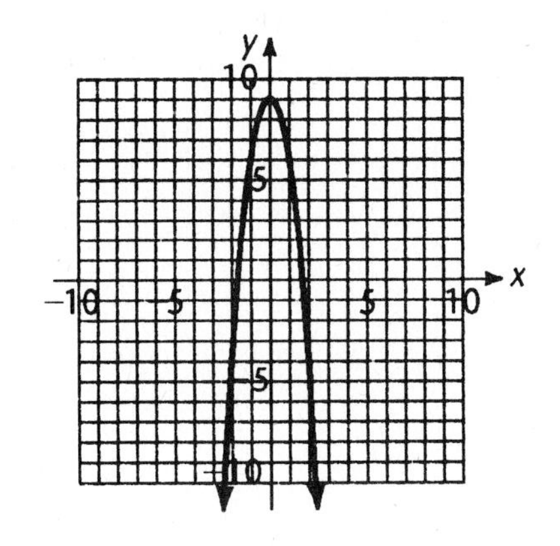

19. Let x = the number. Then we translate as follows:

the sum of	the number	and	its reciprocal	is equal to	twice the number
	x	+	$\frac{1}{x}$	=	$2x$

or, more compactly

$$x + \frac{1}{x} = 2x \qquad \text{lcm} = x,\ x \neq 0$$

$$x \cdot x + x \cdot \frac{1}{x} = x \cdot 2x$$

$$x^2 + 1 = 2x^2$$

$$1 = x^2$$

$$\pm 1 = x$$

20. (rectangle with sides x and 3x + 1)

x

3x + 1

Let x = the width
then 3x + 1 = the length
Since (length) × (width) = area, we have

$$(3x + 1)x = 80$$

$$3x^2 + x = 80$$

$$3x^2 + x - 80 = 0$$

$$(3x + 16)(x - 5) = 0$$

$$3x + 16 = 0 \quad \text{or} \quad x - 5 = 0$$

$$3x = -16 \qquad x = 5 \text{ cm (width)}$$

$$x = -\frac{16}{3} \qquad 3x + 1 = 16 \text{ cm (length)}$$

discard, lengths cannot be negative

CHAPTER 8 Systems of Equations and Inequalities

Exercise 8-1 Systems of Linear Equations in Two Variables

Key Ideas and Formulas

A **linear system** of two equations in two variables has the standard form:

$ax + by = m$	a, b, c, d, m, and n are constants
$cx + dy = n$	x and y are variables
	a, b, c, d not all 0

Solutions of the system are ordered pairs of real numbers that satisfy both equations at the same time.

Three common methods of solving systems are: graphing, substitution, and elimination using addition.

Graphing method: Graph both equations on the same coordinate system. The coordinates of any points that the graphs have in common are solutions to the system. It is necessary to check points suggested by the graph in the original equations.

Substitution method: Choose one of the two equations and solve for one variable in terms of the other. Then substitute the result into the other and solve the resulting equation in one variable. Now substitute this result back into either of the original equations to find the value of the second variable.

Elimination using addition: Use operations resulting in equivalent systems so that a variable can be eliminated when a constant multiple of one equation is added to another equation. Then solve the resulting equation in one variable, and substitute this result back into either of the original equations to find the value of the second variable.

Equivalent systems of equations result if:

(A) Two equations are interchanged.

(B) An equation is multiplied by a nonzero constant.

(C) A constant multiple of another equation is added to a given equation.

All three methods share the common result:

For systems of linear equations, there are three possibilities:

1. Consistent and Independent System: Exactly one solution, corresponding to two intersecting lines.

2. Inconsistent System: No solutions, corresponding to two parallel lines.
3. Dependent System: Infinite solutions, corresponding to two coincident lines, the same line repeated. Every solution of one equation is a solution of the other.

1.

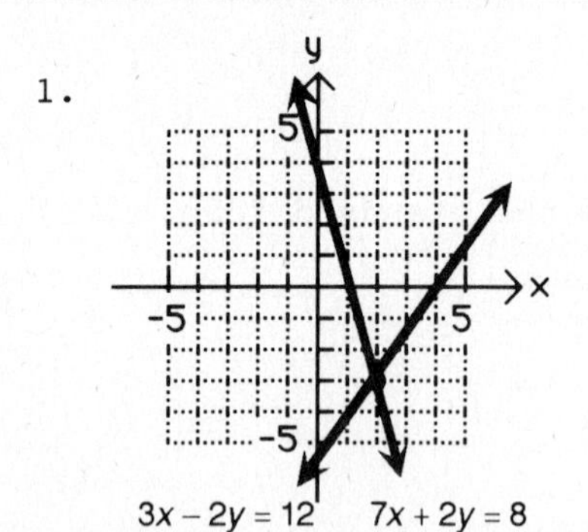

Solution: $x = 2$ or $(x, y) = (2, -3)$
$y = -3$

Check:
$$3x - 2y = 12$$
$$3(2) - 2(-3) \stackrel{?}{=} 12$$
$$6 + 6 \stackrel{\surd}{=} 12$$
$$7x + 2y = 8$$
$$7(2) + 2(-3) \stackrel{?}{=} 8$$
$$14 - 6 \stackrel{\surd}{=} 8$$

3.

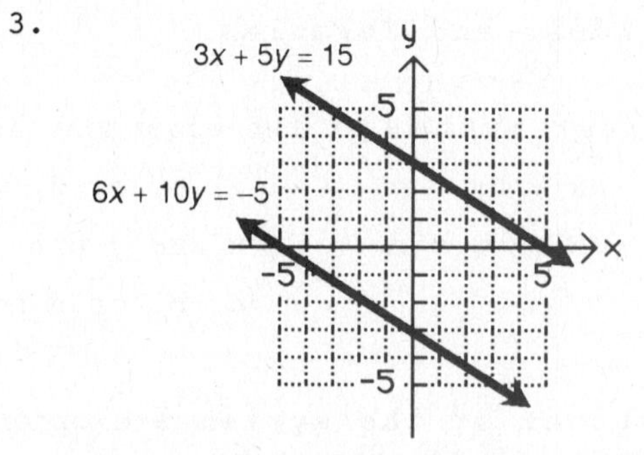

The lines are parallel.
No solution.

5.

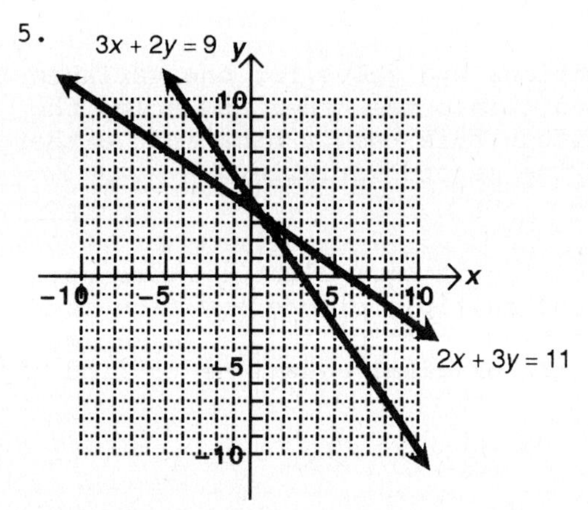

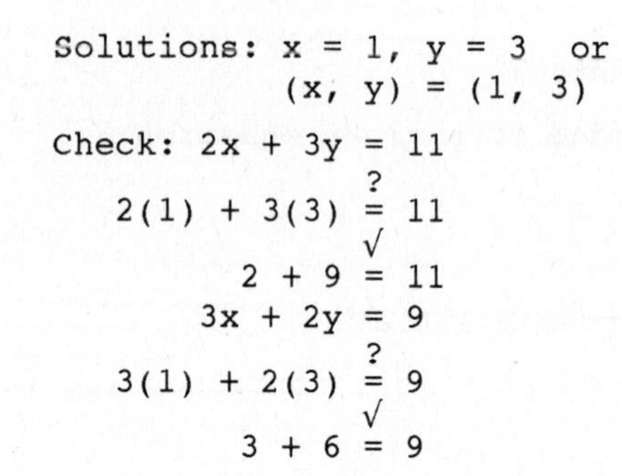

Solutions: $x = 1$, $y = 3$ or $(x, y) = (1, 3)$

Check:
$$2x + 3y = 11$$
$$2(1) + 3(3) \stackrel{?}{=} 11$$
$$2 + 9 \stackrel{\surd}{=} 11$$
$$3x + 2y = 9$$
$$3(1) + 2(3) \stackrel{?}{=} 9$$
$$3 + 6 \stackrel{\surd}{=} 9$$

7.

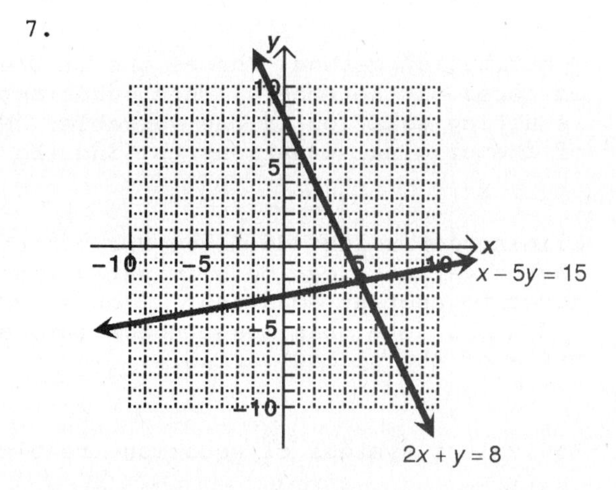

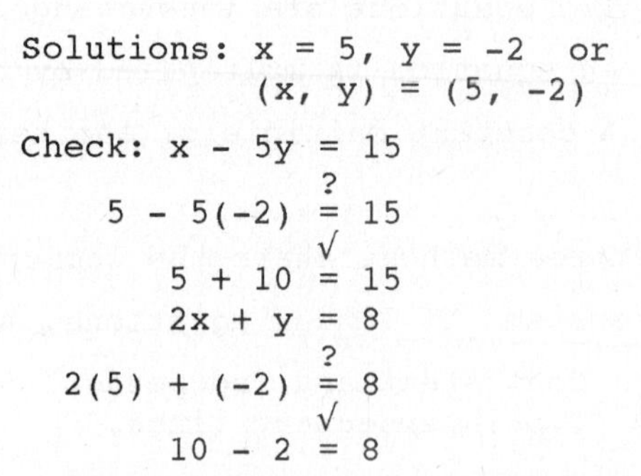

Solutions: $x = 5$, $y = -2$ or $(x, y) = (5, -2)$

Check:
$$x - 5y = 15$$
$$5 - 5(-2) \stackrel{?}{=} 15$$
$$5 + 10 \stackrel{\surd}{=} 15$$
$$2x + y = 8$$
$$2(5) + (-2) \stackrel{?}{=} 8$$
$$10 - 2 \stackrel{\surd}{=} 8$$

9.

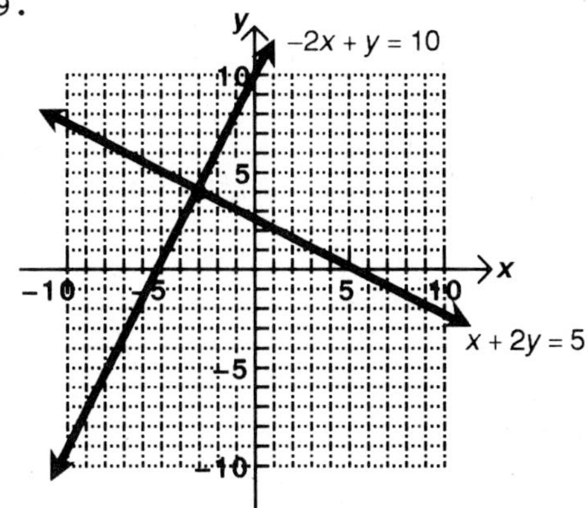

Solutions: $x = -3$, $y = 4$ or $(x, y) = (-3, 4)$

Check: $-2x + y = 10$

$$-2(-3) + 4 \overset{?}{=} 10$$
$$6 + 4 \overset{\checkmark}{=} 10$$

$$x + 2y = 5$$
$$-3 + 2(4) \overset{?}{=} 5$$
$$-3 + 8 \overset{\checkmark}{=} 5$$

11.

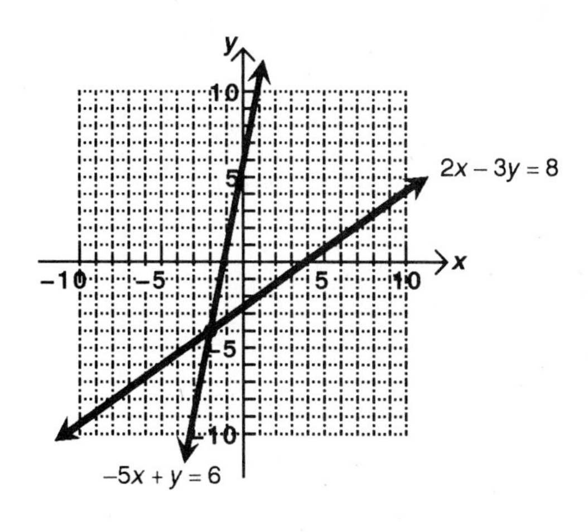

Solutions: $x = -2$, $y = -4$ or $(x, y) = (-2, -4)$

Check: $2x + 3y = 8$

$$2(-2) - 3(-4) \overset{?}{=} 8$$
$$-4 + 12 \overset{\checkmark}{=} 8$$

$$-5x + y = 6$$
$$-5(-2) + (-4) \overset{?}{=} 6$$
$$10 - 4 \overset{\checkmark}{=} 6$$

13. $2x + y = 6$ (1)

$-x + y = 3$ (2)

Solve (2) for y in terms of x.

$$y = x + 3$$

Substitute into (1) to eliminate y.

$$2x + (x + 3) = 6$$
$$2x + x + 3 = 6$$
$$3x + 3 = 6$$
$$3x = 3$$
$$x = 1$$

Now replace x with 1 in (2) to find y.

$$-1 + y = 3$$
$$y = 4$$

(1, 4)

Check: (1, 4) must satisfy both equations

$$2x + y = 6 \qquad -x + y = 3$$
$$2(1) + 4 \overset{?}{=} 6 \qquad 1 + 4 \overset{\checkmark}{=} 3$$
$$2 + 4 \overset{\checkmark}{=} 6$$

15. $3x - y = -3$ (1)

$5x + 3y = -19$ (2)

Solve (1) for y in terms of x.

$$-y = -3x - 3$$
$$y = 3x + 3$$

Substitute into (2) to eliminate y.

$$5x + 3(3x + 3) = -19$$
$$5x + 9x + 9 = -19$$
$$14x + 9 = -19$$
$$14x = -28$$
$$x = -2$$

Now replace x with −2 in (1) to find y.

$$3(-2) - y = -3$$
$$-6 - y = -3$$
$$-y = 3$$
$$y = -3$$

(−2, −3)

Check: (−2, −3) must satisfy both equations

$$3x - y = -3 \qquad 5x + 3y = -19$$
$$3(-2) - (-3) \overset{?}{=} -3 \qquad 5(-2) + 3(-3) \overset{?}{=} -19$$
$$-6 + 3 \overset{\checkmark}{=} -3 \qquad -10 - 9 \overset{\checkmark}{=} -19$$

17. $2x - 3y = 9$ (1)
$x + 4y = 10$ (2)
Solve (2) for x in terms of y.
$x = 10 - 4y$

Substitute into (1) to eliminate x.
$2(10 - 4y) - 3y = 9$
$20 - 8y - 3y = 9$
$20 - 11y = 9$
$-11y = -11$
$y = 1$

Now replace x with 1 in (2) to find x.
$x + 4(1) = 10$
$x = 6$
(6, 1)

Check: (6, 1) must satisfy both equations
$2x - 3y = 9$ $\quad$ $x + 4y = 10$
$2(6) - 3(1) \stackrel{?}{=} 9$ $\quad$ $6 + 4(1) \stackrel{?}{=} 10$
$12 - 3 \stackrel{\checkmark}{=} 9$ $\quad$ $6 + 4 \stackrel{\checkmark}{=} 10$

19. $2x + y = -1$ (1)
$x + 2y = 4$ (2)
Solve (1) for y in terms of x.
$y = -1 - 2x$

Substitute into (2) to eliminate y.
$x + 2(-1 - 2x) = 4$
$x - 2 - 4x = 4$
$-3x - 2 = 4$
$-3x = 6$
$x = -2$

Now replace x with -2 in (1) to find y.
$2(-2) + y = -1$
$-4 + y = -1$
$y = 3$
(-2, 3)

Check: (-2, 3) must satisfy both equations
$2x + y = -1$ $\quad$ $x + 2y = 4$
$2(-2) + 3 \stackrel{?}{=} -1$ $\quad$ $(-2) + 2(3) \stackrel{?}{=} 4$
$-4 + 3 \stackrel{\checkmark}{=} -1$ $\quad$ $-2 + 6 \stackrel{\checkmark}{=} 4$

21. $x - y = 9$ (1)
$x + y = 1$ (2)
Solve (1) for x in terms of y.
$x = y + 9$

Substitute into (2) to eliminate x.
$y + 9 + y = 1$
$2y + 9 = 1$
$2y = -8$
$y = -4$

Now replace y with -4 in (1) to find x.
$x - (-4) = 9$
$x + 4 = 9$
$x = 5$
(5, -4)

Check: (5, -4) must satisfy both equations
$x - y = 9$ $\quad$ $x + y = 1$
$5 - (-4) \stackrel{\checkmark}{=} 9$ $\quad$ $5 + (-4) \stackrel{\checkmark}{=} 1$

23. $8x + y = 0$ (1)
$7x - y = 15$ (2)
Solve (1) for y in terms of x.
$y = -8x$

Substitute into (2) to eliminate y.
$7x - (-8x) = 15$
$15x = 15$
$x = 1$

Now replace x with 1 in (1) to find y.
$8(1) + y = 0$
$y = -8$
(1, -8)

Check: (1, -8) must satisfy both equations.
$8x + y = 0$ $\quad$ $7x - y = 15$
$8(1) + (-8) \stackrel{?}{=} 0$ $\quad$ $7(1) - (-8) \stackrel{?}{=} 15$
$8 - 8 \stackrel{\checkmark}{=} 0$ $\quad$ $7 + 8 \stackrel{\checkmark}{=} 15$

25. $3p + 8q = 4$ (1)
$15p + 10q = -10$ (2)
If we multiply (1) by -5 and add, we can eliminate p.
$-15p - 40q = -20$
$\underline{15p + 10q = -10}$
$-30q = -30$
$q = 1$

Now substitute q = 1 back into (1) and solve for p.

$$3p + 8(1) = 4$$
$$3p + 8 = 4$$
$$3p = -4$$
$$p = -\frac{4}{3}$$

$\left(-\frac{4}{3}, 1\right)$

Check:

$$3p + 8q = 4 \qquad 15p + 10q = -10$$
$$3\left(-\frac{4}{3}\right) + 8(1) \overset{?}{=} 4 \qquad 15\left(-\frac{4}{3}\right) + 10(1) \overset{?}{=} -10$$
$$-4 + 8 \overset{\checkmark}{=} 4 \qquad -20 + 10 \overset{\checkmark}{=} -10$$

27. $6x - 2y = 18$ (1)
$-3x + y = -9$ (2)

If we multiply (2) by 2 and add, we will eliminate both x and y.

$$6x - 2y = 18$$
$$-6x + 2y = -18$$
$$0 = 0$$

There are infinitely many solutions.
Since $-3x + y = -9$, we can solve for y in terms of x to obtain $y = 3x - 9$. Then all solutions can be expressed as $(x, 3x - 9)$ for any real number x.

COMMON ERROR:
Confusing 0 = 0, indicating a dependent system, with 0 = nonzero, which would indicate an inconsistent system.

29. $3x - 2y = 5$ (1)
$-5x + 2y = -3$ (2)

If we add (1) and (2), we can eliminate y.

$$-2x = 2$$
$$x = -1$$

Now substitute x = -1 back into (2) and solve for y.

$$-5(-1) + 2y = -3$$
$$5 + 2y = -3$$
$$2y = -8$$
$$y = -4$$

(-1, -4)

Check:

$$3x - 2y = 5 \qquad -5x + 2y = -3$$
$$3(-1) - 2(-4) \overset{?}{=} 5 \qquad -5(-1) + 2(-4) \overset{?}{=} -3$$
$$-3 + 8 \overset{\checkmark}{=} 5 \qquad 5 - 8 \overset{\checkmark}{=} 3$$

31. $-3x + 2y = 16$ (1)
$2x + 3y = 11$ (2)

If we multiply (1) by 2 and (2) by 3 and add, we can eliminate x.

$$-6x + 4y = 32$$
$$6x + 9y = 33$$
$$13y = 65$$
$$y = 5$$

Now substitute y = 5 back into (2) and solve for x.

$$2x + 3(5) = 11$$
$$2x + 15 = 11$$
$$2x = -4$$
$$x = -2$$

(-2, 5)

Check:

$$-3x + 2y = 16 \qquad 2x + 3y = 11$$
$$-3(-2) + 2(5) \overset{?}{=} 16 \qquad 2(-2) + 3(5) \overset{?}{=} 11$$
$$6 + 10 \overset{\checkmark}{=} 16 \qquad -4 + 15 \overset{\checkmark}{=} 11$$

33. $2x + 5y = -1$ (1)
$-x + y = -10$ (2)

If we multiply (2) by 2 and add to (1), we can eliminate x.

$$\begin{aligned} -2x + 2y &= -20 \\ 2x + 5y &= -1 \\ \hline 7y &= -21 \\ y &= -3 \end{aligned}$$

Now substitute $y = -3$ back into (2) and solve for x.

$$\begin{aligned} -x + (-3) &= -10 \\ -x - 3 &= -10 \\ -x &= -7 \\ x &= 7 \end{aligned}$$

(7, -3)

Check:

$$2x + 5y = -1 \qquad 2(7) + 5(-3) \stackrel{?}{=} -1 \qquad 14 - 15 \stackrel{\checkmark}{=} -1$$

$$-x + y = -10 \qquad (-7) + (-3) \stackrel{\checkmark}{=} -10$$

35. $3x - 5y = -2$ (1)
$2x - 7y = 6$ (2)

If we multiply (1) by 2 and (2) by -3 and add, we can eliminate x.

$$\begin{aligned} 6x - 10y &= -4 \\ -6x + 21y &= -18 \\ \hline 11y &= -22 \\ y &= -2 \end{aligned}$$

Now substitute $y = -2$ back into (2) and solve for x.

$$\begin{aligned} 2x - 7(-2) &= 6 \\ 2x + 14 &= 6 \\ 2x &= -8 \\ x &= -4 \end{aligned}$$

(-4, -2)

Check:

$$3x - 5y = -2 \qquad 3(-4) - 5(-2) \stackrel{?}{=} -2 \qquad -12 + 10 \stackrel{\checkmark}{=} -2$$

$$2x - 7y = 6 \qquad 2(-4) - 7(-2) \stackrel{?}{=} 6 \qquad -8 + 14 \stackrel{\checkmark}{=} 6$$

37. $x - 3y = -11$ (1)
$2x + 5y = 11$ (2)

Solution by graphing:

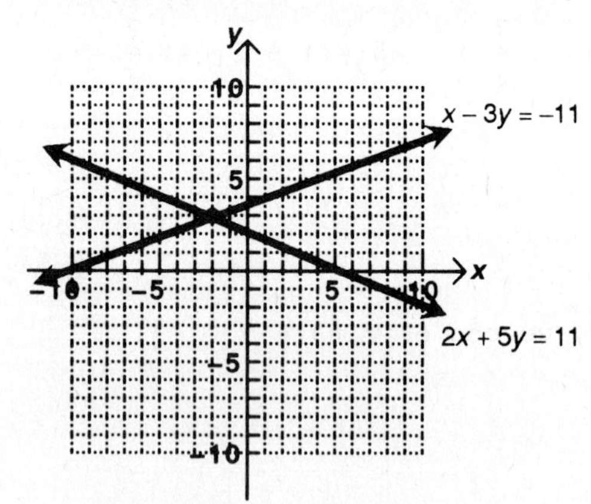

Solution: (-2, 3)

Check:

$$x - 3y = -11 \qquad -2 - 3(3) \stackrel{?}{=} -11 \qquad -2 - 9 \stackrel{\checkmark}{=} -11$$

$$2x + 5y = 11 \qquad 2(-2) + 5(3) \stackrel{?}{=} 11 \qquad -4 + 15 \stackrel{\checkmark}{=} 11$$

Solution by substitution: Solve (1) for x in terms of y.

$$\begin{aligned} x - 3y &= -11 \\ x &= 3y - 11 \end{aligned}$$

Substitute into (2) to eliminate x.

$$\begin{aligned} 2(3y - 11) + 5y &= 11 \\ 6y - 22 + 5y &= 11 \end{aligned}$$

$$11y - 22 = 11$$
$$11y = 33$$
$$y = 3$$

Replace y by 3 in (1) to find x.

$$x - 3(3) = -11$$
$$x - 9 = -11$$
$$x = -2$$

(-2, 3)

Solution by elimination using addition: We multiply (1) by -2 and add.

$$\begin{array}{r} -2x + 6y = 22 \\ \underline{2x + 5y = 11} \\ 11y = 33 \\ y = 3 \end{array}$$

Replace y by 3 in (1) to find x.

$$x - 3(3) = -11$$
$$x = -2$$

(-2, 3)

39. $11x + 2y = 1$ (1)
$9x - 3y = 24$ (2)

Solution by graphing:

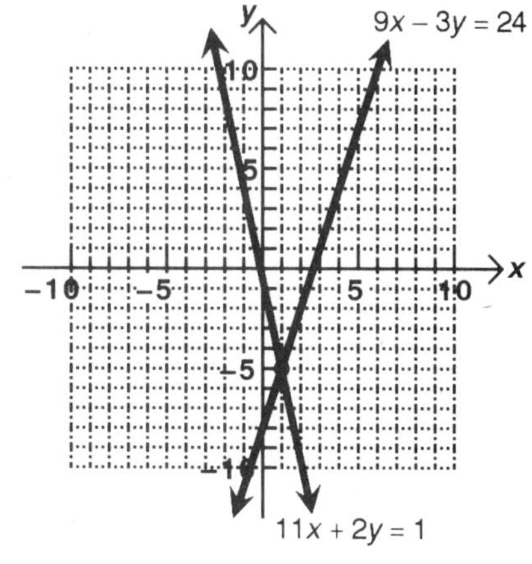

Check: $11x + 2y = 1$

$$11(1) + 2(-5) \stackrel{?}{=} 1$$
$$11 - 10 \stackrel{\checkmark}{=} 1$$

$9x - 3y = 24$

$$9(1) - 3(-5) \stackrel{?}{=} 24$$
$$9 + 15 \stackrel{\checkmark}{=} 24$$

Solution: (1, -5)

Solution by substitution: Solve (2) for y in terms of x.

$$9x - 3y = 24$$
$$-3y = -9x + 24$$
$$y = 3x - 8$$

Substitute into (1) to eliminate y.

$$11x + 2(3x - 8) = 1$$
$$11x + 6x - 16 = 1$$
$$17x - 16 = 1$$
$$17x = 17$$
$$x = 1$$

Replace x by 1 in (2) to find y.

$$9(1) - 3y = 24$$
$$9 - 3y = 24$$
$$-3y = 15$$
$$y = -5$$

(1, -5)

Solution by elimination using addition: We multiply (1) by 3 and (2) by 2, then add.

$$\begin{aligned} 33x + 6y &= 3 \\ \underline{18x - 6y} &\underline{= 48} \\ 51x &= 51 \\ x &= 1 \end{aligned}$$

Replace x by 1 in (1) to find y.

$$\begin{aligned} 11(1) + 2y &= 1 \\ 11 + 2y &= 1 \\ 2y &= -10 \\ y &= -5 \end{aligned}$$

(1, -5)

41. $y = 3x - 3$ (1)
$6x = 8 + 3y$ (2)
We choose elimination by substitution, replacing y in (2) by the expression from (1).

$$\begin{aligned} 6x &= 8 + 3(3x - 3) \\ 6x &= 8 + 9x - 9 \\ -3x &= -1 \\ x &= \frac{1}{3} \end{aligned}$$

$y = 3x - 3$ Check:

$y = 3\left(\frac{1}{3}\right) - 3$ $\quad y = 3x - 3$ $\quad 6x = 8 + 3y$

$y = 1 - 3$ $\quad -2 \stackrel{?}{=} 3\left(\frac{1}{3}\right) - 3$ $\quad 6\left(\frac{1}{3}\right) \stackrel{?}{=} 8 + 3(-2)$

$y = -2$ $\quad -2 \stackrel{\checkmark}{=} 1 - 3$ $\quad 2 \stackrel{\checkmark}{=} 8 - 6$

$\left(\frac{1}{3}, -2\right)$

43. $\frac{1}{2}x - y = -3$ (1)
$-x + 2y = 6$ (2)
We choose elimination using addition, multiplying (1) by 2 and adding.

$$\begin{aligned} x - 2y &= -6 \\ \underline{-x + 2y} &\underline{= 6} \\ 0 &= 0 \end{aligned}$$

There are infinitely many solutions. Since $\frac{1}{2}x - y = -3$, we can solve for y in terms of x as follows:

$$\begin{aligned} \frac{1}{2}x - y &= -3 \\ -y &= -\frac{1}{2}x - 3 \\ y &= \frac{1}{2}x + 3 \end{aligned}$$

Therefore all solutions can be expressed as $\left(x, \frac{1}{2}x + 3\right)$ for any real number x.

45. $2x + 3y = 2y - 2$ (1)
$3x + 2y = 2x + 2$ (2)
Since this system is not in so-called standard form, we eliminate the variables on the right side by subtraction.
$2x + y = -2$ (3)
$x + 2y = 2$ (4)

We now choose elimination by substitution, solving (3) for y in terms of x.

$y = -2x - 2$

We now substitute this expression for y into (4).

$$\begin{aligned} x + 2(-2x - 2) &= 2 \\ x - 4x - 4 &= 2 \\ -3x - 4 &= 2 \\ -3x &= 6 \\ x &= -2 \end{aligned}$$

Replace x by -2 in (3) to find y.

$$\begin{aligned} 2x + y &= -2 \\ 2(-2) + y &= -2 \\ -4 + y &= -2 \\ y &= 2 \end{aligned}$$

(-2, 2)

Check:

$$\begin{aligned} 2x + 3y &= 2y - 2 \\ 2(-2) + 3(2) &\stackrel{?}{=} 2(2) - 2 \\ -4 + 6 &\stackrel{?}{=} 4 - 2 \\ 2 &\stackrel{\checkmark}{=} 2 \end{aligned} \qquad \begin{aligned} 3x + 2y &= 2x + 2 \\ 3(-2) + 2(2) &\stackrel{?}{=} 2(-2) + 2 \\ -6 + 4 &\stackrel{?}{=} -4 + 2 \\ -2 &\stackrel{\checkmark}{=} -2 \end{aligned}$$

47. $x - 4y = -1$ (1)

$3x + 8y = 7$ (2)

We choose elimination using addition, multiplying the top equation by 2.

$$\begin{aligned} 2x - 8y &= -2 \\ 3x + 8y &= 7 \\ \hline 5x \quad &= 5 \\ x &= 1 \end{aligned}$$

Replace x in (2) by 1

$$\begin{aligned} 3(1) + 8y &= 7 \\ 3 + 8y &= 7 \\ 8y &= 4 \\ y &= \frac{1}{2} \end{aligned}$$

$\left(1, \frac{1}{2}\right)$

Check:

$$\begin{aligned} x - 4y &= -1 \\ 1 - 4\left(\frac{1}{2}\right) &\stackrel{?}{=} -1 \\ 1 - 2 &\stackrel{\checkmark}{=} -1 \end{aligned} \qquad \begin{aligned} 3x + 8y &= 7 \\ 3(1) + 8\left(\frac{1}{2}\right) &\stackrel{?}{=} 7 \\ 3 + 4 &\stackrel{\checkmark}{=} 7 \end{aligned}$$

49. $3x + 2y = 3$ (1)

$-6x + 3y = 8$ (2)

We choose elimination using addition, multiplying the top equation by 2.

$$\begin{aligned} 6x + 4y &= 6 \\ -6x + 3y &= 8 \\ \hline 7y &= 14 \\ y &= 2 \end{aligned}$$

Replace y in (1) by 2

$$\begin{aligned} 3x + 2(2) &= 3 \\ 3x + 4 &= 3 \\ 3x &= -1 \\ x &= -\frac{1}{3} \end{aligned}$$

$\left(-\frac{1}{3}, 2\right)$

Check:

$$\begin{aligned} 3x + 2y &= 3 \\ 3\left(-\frac{1}{3}\right) + 2(2) &\stackrel{?}{=} 3 \\ -1 + 4 &\stackrel{\checkmark}{=} 3 \end{aligned} \qquad \begin{aligned} -6x + 3y &= 8 \\ -6\left(-\frac{1}{3}\right) + 3(2) &\stackrel{?}{=} 8 \\ 2 + 6 &\stackrel{\checkmark}{=} 8 \end{aligned}$$

51. $2x + 3y = 2$ (1)
$4x - 9y = -1$ (2)

We choose elimination using addition, multiplying the top equation by 3.

$$\begin{array}{r} 6x + 9y = 6 \\ \underline{4x - 9y = -1} \\ 10x \quad = 5 \\ x = \frac{1}{2} \end{array}$$

Replace x in (1) by $\frac{1}{2}$

$$2\left(\frac{1}{2}\right) + 3y = 2$$
$$1 + 3y = 2$$
$$3y = 1$$
$$y = \frac{1}{3}$$

$\left(\frac{1}{2}, \frac{1}{3}\right)$

Check:

$2x + 3y = 2$: $2\left(\frac{1}{2}\right) + 3\left(\frac{1}{3}\right) \stackrel{?}{=} 2$; $1 + 1 \stackrel{\checkmark}{=} 2$

$4x - 9y = -1$: $4\left(\frac{1}{2}\right) - 9\left(\frac{1}{3}\right) \stackrel{?}{=} -1$; $2 - 3 \stackrel{\checkmark}{=} -1$

53. $4x + 3y = 1$ (1)
$-8x + 18y = 14$ (2)

We choose elimination using addition, multiplying the top equation by 2.

$$\begin{array}{r} 8x + 6y = 2 \\ \underline{-8x + 18y = 14} \\ 24y = 16 \\ y = \frac{2}{3} \end{array}$$

Replace y in (1) by $\frac{2}{3}$

$$4x + 3\left(\frac{2}{3}\right) = 1$$
$$4x + 2 = 1$$
$$4x = -1$$
$$x = -\frac{1}{4}$$

$\left(-\frac{1}{4}, \frac{2}{3}\right)$

Check:

$4x + 3y = 1$: $4\left(-\frac{1}{4}\right) + 3\left(\frac{2}{3}\right) \stackrel{?}{=} 1$; $-1 + 2 \stackrel{\checkmark}{=} 1$

$-8x + 18y = 14$: $-8\left(-\frac{1}{4}\right) + 18\left(\frac{2}{3}\right) \stackrel{?}{=} 14$; $2 + 12 \stackrel{\checkmark}{=} 14$

55. $0.2x - 0.5y = 0.07$ (1)
$0.8x - 0.3y = 0.79$ (2)

We choose elimination using addition, multiplying the top equation by -4.

$$\begin{array}{r} -0.8x + 2.0y = -0.28 \\ \underline{0.8x - 0.3y = 0.79} \\ 1.7y = 0.51 \\ y = \frac{0.51}{1.7} \\ y = 0.3 \end{array}$$

Replace y in (1) by 0.3.

$$0.2x - 0.5(0.3) = 0.07$$
$$0.2x - 0.15 = 0.07$$
$$0.2x = 0.22$$

$$x = \frac{0.22}{0.2}$$
$$x = 1.1$$

(1.1, 0.3)

Alternatively, we could eliminate decimals by multiplying both (1) and (2) through by 100 first:

$20x - 50y = 7$
$80x - 30y = 79$

Then we would proceed as before.

Check: $0.2x - 0.5y = 0.07$ $\qquad$ $0.8x - 0.3y = 0.79$

$0.2(1.1) - 0.5(0.3) \overset{?}{=} 0.07$ $\qquad$ $0.8(1.1) - 0.3(0.3) \overset{?}{=} 0.79$

$0.22 - 0.15 \overset{\surd}{=} 0.07$ $\qquad$ $0.88 - 0.09 \overset{\surd}{=} 0.79$

57. $\frac{1}{4}x - \frac{2}{3}y = -2 \quad (1)$

$\frac{1}{2}x - y = -2 \quad (2)$

We choose to eliminate fractions first, by multiplying (1) by its lcd 12 and multiplying (2) by its lcd 2.

$3x - 8y = -24 \quad (3)$
$x - 2y = -4 \quad (4)$

Now we proceed using elimination by addition, multiplying (4) by -3 and adding.

$3x - 8y = -24$
$-3x + 6y = -12$
$-2y = -12$
$y = 6$

Replace y in (4) by 6.

$x - 2(6) = -4$
$x - 12 = -4$
$x = 8$

(8, 6)

Check: $\frac{1}{4}x - \frac{2}{3}y = -2$ $\qquad$ $\frac{1}{2}x - y = -2$

$\frac{1}{4}(8) - \frac{2}{3}(6) \overset{?}{=} -2$ $\qquad$ $\frac{1}{2}(8) - 6 \overset{?}{=} -2$

$2 - 4 \overset{\surd}{=} -2$ $\qquad$ $4 - 6 \overset{\surd}{=} -2$

59.

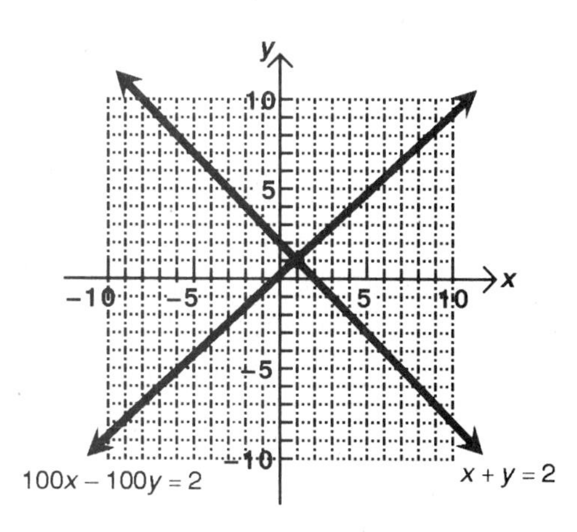

The solution appears to be approximately (1, 1).

$x + y = 2 \quad (1)$

$100x - 100y = 2 \quad (2)$

We solve by elimination using addition.
Multiply the top equation by 100.

$100x + 100y = 200$

$100x - 100y = 2$

$200x = 202$

$x = \frac{101}{100}$

Replace x in (1) by $\frac{101}{100}$

$\frac{101}{100} + y = \frac{200}{100}$

$y = \frac{99}{100}$

$\left(\frac{101}{100}, \frac{99}{100}\right)$

Check: $x + y = 2$

$\frac{101}{100} + \frac{99}{100} \overset{\checkmark}{=} \frac{200}{100}$

$100x - 100y = 2$

$100\left(\frac{101}{100}\right) - 100\left(\frac{99}{100}\right) \overset{?}{=} 2$

$101 - 99 \overset{\checkmark}{=} 2$

61.

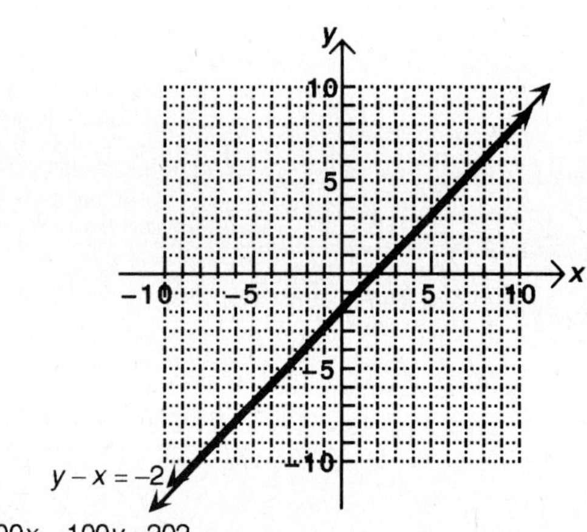

The graphs appear to be parallel lines. (No solution)

$y - x = -2 \quad (1)$

$100x - 100y = 202 \quad (2)$

We solve by substitution, solving (1) for y in terms of x.

$y = x - 2$

We now substitute this expression for y into (2).

$100x - 100(x - 2) = 200$

$100x - 100x + 200 = 202$

$200 = 202$

Impossible. No solution.

63.

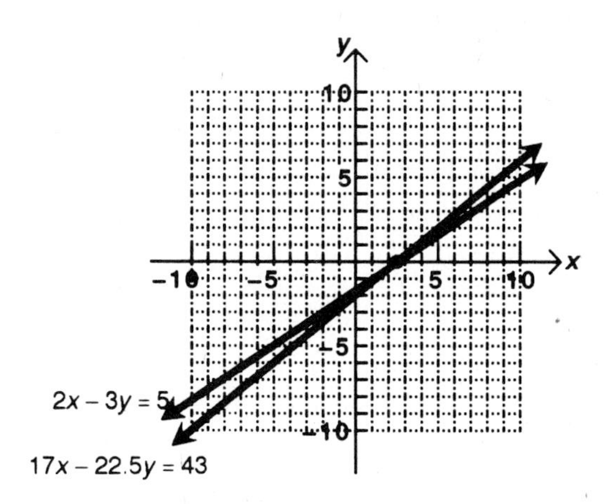

The solution appears to be approximately $(2\frac{1}{2}, 0)$.

$2x - 3y = 5$ (1)
$17x - 22.5y = 43$ (2)

We solve by elimination using addition.
Multiply the top equation by -7.5.

$$\begin{aligned} -15x + 22.5y &= -37.5 \\ 17x - 22.5y &= 43 \\ \hline 2x \quad &= 5.5 \\ x &= 2.75 \end{aligned}$$

Replace x in (1) by 2.75

$$\begin{aligned} 2(2.75) - 3y &= 5 \\ 5.5 - 3y &= 5 \\ -3y &= -0.5 \\ -3y &= -\frac{1}{2} \\ y &= \frac{1}{6} \end{aligned}$$

Check: $2x - 3y = 5$

$$2(2.75) - 3\left(\frac{1}{6}\right) \stackrel{?}{=} 5$$

$$5.5 - 0.5 \stackrel{\sqrt{}}{=} 5$$

$17x - 22.5y = 43$

$$17(2.75) - 22.5\left(\frac{1}{6}\right) \stackrel{?}{=} 43$$

$$46.75 - 3.75 \stackrel{\sqrt{}}{=} 43$$

$\left(2.75, \frac{1}{6}\right)$

65. If two lines have different slopes, they intersect. 1 solution.

67. If two lines have the same slope and the same y intercept, they coincide. Infinite number of solutions.

69. $1020 + 3210y = 1000$ can be written in slope-intercept form as follows:

$$\begin{aligned} 3210y &= -1020x + 1000 \\ y &= -\frac{1020x + 1000}{3210} \\ y &= -\frac{1020}{3210}x + \frac{1000}{3210} \end{aligned} \qquad \text{Slope: } -\frac{1020}{3210}$$

$80x + 94y = 100$ can be written in slope-intercept form as follows:

$$\begin{aligned} 94y &= -80x + 100 \\ y &= \frac{-80x + 100}{94} \\ y &= -\frac{80}{94}x + \frac{100}{94} \end{aligned} \qquad \text{Slope: } -\frac{80}{94}$$

Since the two lines have different slopes, there is 1 solution to the system.

71. $312x - 1560y = 3275$ can be written in slope-intercept form as follows:

$$-1560y = -312x + 3275$$
$$y = \frac{-312x + 3275}{-1560}$$
$$y = \frac{1}{5}x - \frac{655}{312} \qquad \text{Slope: } \frac{1}{5} \qquad \text{y-intercept: } -\frac{655}{312}$$

$-45x + 225y = -625$ can be written in slope-intercept form as follows:

$$225y = 45x - 625$$
$$y = \frac{45x - 625}{225}$$
$$y = \frac{1}{5}x - \frac{29}{9} \qquad \text{Slope: } \frac{1}{5} \qquad \text{y-intercept: } -\frac{29}{9}$$

Since the two lines have the same slope but different y-intercepts, they are parallel. No solution.

73. $97x - 61y = 429$ can be written in slope-intercept form as follows:

$$-61y = -97x + 429$$
$$y = \frac{-97x + 429}{-61}$$
$$y = \frac{97}{61}x - \frac{429}{61} \qquad \text{Slope: } \frac{97}{61} \qquad \text{y-intercept: } -\frac{429}{61}$$

$-679x + 427y = -3003$ can be written in slope-intercept form as follows:

$$427y = 679x - 3003$$
$$y = \frac{679x - 3003}{427}$$
$$y = \frac{97}{61}x - \frac{429}{61} \qquad \text{Slope: } \frac{97}{61} \qquad \text{y-intercept: } -\frac{429}{61}$$

Since the lines coincide, there are an infinite number of solutions.

Exercise 8-2 Application: Mixture Problems

Key Ideas and Formulas

Various types of applications, known as mixture problems, can be solved using either one variable or two variable (systems of equations) approaches.

Note: If one unknown quantity can easily be expressed in terms of the other, both one and two variable approaches can be used with equal ease. We have chosen between the two methods arbitrarily, in order to illustrate both.

If the two unknown quantities are not easily expressed this way, the two variable approach is preferable.

---

1. We choose the one variable approach.
   Let x = number of dimes. Then 100 - x = number of quarters.

$$\begin{pmatrix}\text{Value of}\\ \text{dimes}\end{pmatrix} + \begin{pmatrix}\text{Value of}\\ \text{quarters}\end{pmatrix} = (\text{Total value})$$
$$0.10x + 0.25(100 - x) = 14.50$$
$$100(.10x) + 100[0.25(100 - x)] = 100(14.50)$$

$$\begin{aligned}
10x + 25(100 - x) &= 1450 \\
10x + 2500 - 25x &= 1450 \\
-15x + 2500 &= 1450 \\
-15x &= -1050 \\
x &= 70 \text{ dimes} \\
100 - x &= 30 \text{ quarters}
\end{aligned}$$

Check: $0.1(70) + 0.25(30) \stackrel{?}{=} 14.50$

$$14.50 \stackrel{\checkmark}{=} 14.50$$

3. We choose the one variable approach.
   Let x = number of \$2 tickets sold.
   Then $3500 - x$ = number of \$4 tickets sold.

$$\begin{pmatrix}\text{Value of} \\ \text{\$2 tickets sold}\end{pmatrix} + \begin{pmatrix}\text{Value of} \\ \text{\$4 tickets sold}\end{pmatrix} = (\text{Total value})$$

$$\begin{aligned}
2x + 4(3{,}500 - x) &= 12{,}600 \\
2x + 14{,}000 - 4x &= 12{,}600 \\
-2x + 14{,}000 &= 12{,}600 \\
-2x &= -1{,}400 \\
x &= 700 \text{ \$2 tickets} \\
3{,}500 - x &= 2{,}800 \text{ \$4 tickets}
\end{aligned}$$

Check: $2(700) + 4(2{,}800) \stackrel{?}{=} 12{,}600$

$$12{,}600 \stackrel{\checkmark}{=} 12{,}600$$

5. We choose the one variable approach, since there is only one variable apparent.
   Let x = the number of deciliters of alcohol added.
   100 = the number of deciliters of 40% solution.
   Then $100 + x$ = the number of deciliters of 50% solution.

$$\begin{pmatrix}\text{Amount of} \\ \text{alcohol added}\end{pmatrix} + \begin{pmatrix}\text{Amount of alcohol} \\ \text{in} \\ \text{40\% solution}\end{pmatrix} = \begin{pmatrix}\text{Amount of alcohol} \\ \text{in} \\ \text{50\% solution}\end{pmatrix}$$

$$\begin{aligned}
x + 0.4(100) &= 0.5(100 + x) \\
x + 40 &= 50 + 0.5x \\
10x + 10 \cdot 40 &= 10 \cdot 50 + 10(0.5x) \\
10x + 400 &= 500 + 5x \\
5x &= 100 \\
x &= 20 \text{ deciliters}
\end{aligned}$$

Check: $20 + 0.4(100) \stackrel{?}{=} 0.5(100 + 20)$

$$60 \stackrel{\checkmark}{=} 60$$

7. We choose the two variable approach.
   Let x = number of nickels
   y = number of dimes
   $x + y = 22$ (number of coins)
   $5x + 10y = 150$ (value of coins in cents)
   We choose elimination by addition, multiplying the top equation by -5 and adding.

$$\begin{aligned}
-5x - 5y &= -110 \\
\underline{5x + 10y} &= \underline{150} \\
5y &= 40 \\
y &= 8
\end{aligned}$$

Replace y by 8 in the top equation.

$x + 8 = 22$
$x = 14$

Solution: 14 nickels, 8 dimes

Check: $14 + 8 = 22$ coins
$5(14) + 10(8) = 70 + 80 = 150$ cents or \$1.50

9. We choose the two variable approach.
   Let x = one angle
   y = other angle

Then
$$x + y = 90$$
$$\underline{x - y = 14}$$
Adding, we obtain immediately
$$2x = 104$$
$$x = 52$$
Replacing x by 52 in the top equation, we obtain
$$52 + y = 90$$
$$y = 38$$
Solution: $52°$, $38°$

Check: $52 + 38 = 90$
$52 - 38 = 14$

11. We choose the one variable approach, since there is only one variable apparent.
   Let x = the number of liters of distilled water added (0% alcohol)
   140 = the number of liters of 80% solution
   Then $140 + x$ = the number of liters of 70% solution

$$\begin{pmatrix}\text{Amount of alcohol}\\\text{in first solution}\end{pmatrix} + \begin{pmatrix}\text{Amount of alcohol}\\\text{in second solution}\end{pmatrix} = \begin{pmatrix}\text{Amount of alcohol}\\\text{in mixture}\end{pmatrix}$$
$$0.0x + 0.8(140) = 0.7(140 + x)$$
$$112 = 98 + 0.7x$$
$$14 = 0.7x$$
$$x = 20 \text{ liters}$$

COMMON ERROR: Using x to represent the amount of alcohol in the first solution. There is no alcohol (0%) in distilled water.

Check: $0.8(140) \stackrel{?}{=} 0.7(140 + 20)$
$112 \stackrel{\checkmark}{=} 112$

13. We choose the one variable approach.
   Let x = the number of deciliters of 20% solution.
   Then $90 - x$ = the number of deciliters of 50% solution.
   90 = the number of deciliters of 30% solution.

$$\begin{pmatrix}\text{Amount of alcohol}\\\text{in first solution}\end{pmatrix} + \begin{pmatrix}\text{Amount of alcohol}\\\text{in second solution}\end{pmatrix} = \begin{pmatrix}\text{Amount of alcohol}\\\text{in mixture}\end{pmatrix}$$
$$0.2x + 0.5(90 - x) = 0.3(90)$$
$$0.2x + 45 - 0.5x = 27$$
$$-0.3x + 45 = 27$$
$$-0.3x = -18$$
$$x = 60 \text{ deciliters of 20\% solution}$$
$$90 - x = 30 \text{ deciliters of 50\% solution}$$

Check: $0.2(60) + 0.5(30) \stackrel{?}{=} 0.3(90)$
$27 \stackrel{\checkmark}{=} 27$

15\. We choose the one variable approach.

Let x = how many kilograms of \$5-per-kilogram tea

Then $75 - x$ = how many kilograms of \$6.50-per-kilogram tea

75 = how many kilograms of \$6-per-kilogram tea

$$\binom{\text{Value of}}{\text{\$5-per-kg tea}} + \binom{\text{Value of}}{\text{\$6.50-per-kg tea}} = \binom{\text{Value of}}{\text{mixture}}$$

$$\begin{aligned} 5x + 6.50(750 - x) &= 6(75) \\ 5x + 487.5 - 6.5x &= 450 \\ -1.5x + 487.5 &= 450 \\ -1.5x &= -37.5 \\ x &= 25 \text{ kilograms of \$5-per-kilogram tea} \\ 75 - x &= 50 \text{ kilograms of \$6.50 per-kilogram tea} \end{aligned}$$

Check: $5(25) + 6.50(50) \stackrel{?}{=} 6(75)$

$450 \stackrel{\surd}{=} 450$

17\. We choose the one variable approach.

Let x = how much invested at 10%

$20{,}000 - x$ = how much invested at 15%

20,000 = total invested at 13%

Yield from actual investment = Yield desired

$$\binom{\text{Yield from}}{\text{10\% investment}} + \binom{\text{Yield from}}{\text{15\% investment}} = \binom{\text{Yield from}}{\text{13\% investment}}$$

$$\begin{aligned} 0.1x + 0.15(20{,}000 - x) &= 0.13(20{,}000) \\ 100(0.1x) + 100\,[0.15(20{,}000 - x)] &= 100[0.13(20{,}000)] \\ 10x + 15(20{,}000 - x) &= 13(20{,}000) \\ 10x + 300{,}000 - 15x &= 260{,}000 \\ -5x + 300{,}000 &= 260{,}000 \\ -5x &= -40{,}000 \\ x &= \$8{,}000 \text{ at } 10\% \\ 20{,}000 - x &= \$12{,}000 \text{ at } 15\% \end{aligned}$$

Check: $0.1(8{,}000) + 0.15(12{,}000) \stackrel{?}{=} 0.13(20{,}000)$

$2{,}600 \stackrel{\surd}{=} 2{,}600$

19\. We choose the two variable approach.

Let x = the number of $\frac{1}{4}$-pound packages

y = number of $\frac{1}{2}$-pound packages

Then $x + y = 144$ (number of packages)

$\frac{1}{4}x + \frac{1}{2}y = 51$ (total weight of contents)

We eliminate fractions by multiplying the bottom equation by its lcd 4.

$$\begin{aligned} x + y &= 144 \\ x + 2y &= 204 \end{aligned}$$

We now eliminate x by multiplying the top equation by -1 and adding.

$$\begin{aligned} -x - y &= -144 \\ \underline{x + 2y} &= \underline{204} \\ y &= 60 \end{aligned}$$

Replace y by 60 in the top equation.

$$\begin{aligned} x + 60 &= 144 \\ x &= 84 \end{aligned}$$

Solution: 84 $\frac{1}{4}$-pound packages; 60 $\frac{1}{2}$-pound packages

Check: $84 + 60 = 144$

$\frac{1}{4}(84) + \frac{1}{2}(60) = 21 + 30 = 51$

21. We choose the two variable approach.

Let x = amount of 50% solution
y = amount of 80% solution

100 milliliters are required, hence

$$x + y = 100 \quad (1)$$

68% of the 100 milliliters must be acid, hence

$$0.50x + 0.80y = 0.68(100) \quad (2)$$

We solve the system of equations (1), (2) by elimination using addition.

$$\begin{aligned} -0.50x - 0.50y &= -0.50(100) && -0.50[\text{equation } (1)] \\ 0.50x + 0.80y &= 0.68(100) && \text{Equation } (2) \\ \hline 0.30y &= -0.50(100) + 0.68(100) \\ 0.30y &= 18 \\ y &= 60 \\ x + y &= 100 \\ x &= 40 \end{aligned}$$

40 milliliters of 50% solution and 60 milliliters of 80% solution

Check: $40 + 60 = 100$

$0.50(40) + 0.80(60) = 20 + 48 = 68$

23. We choose the two variable approach.

Let u = time of sound underwater
a = time of sound above water

We know that (time underwater) = 6 seconds less than (time above water). Hence

$$u = a - 6 \quad (1)$$

To find a second equation, we use $D = rt$.

$5,000u$ = distance underwater
$1,100a$ = distance above water

These distances are equal, hence

$$5,000u = 1,100a \quad (2)$$

To solve the system of equations (1), (2) we substitute the expression for u from equation (1) into equation (2).

$$\begin{aligned} 5,000(a - 6) &= 1,100a \\ 5,000a - 30,000 &= 1,100a \\ -30,000 &= -3,900a \\ a &= \frac{-30,000}{-3,900} = \frac{100}{13} \text{ or } 7\tfrac{9}{13} \text{ seconds} \\ u &= a - 6 = 7\tfrac{9}{13} - 6 = 1\tfrac{9}{13} \text{ seconds} \end{aligned}$$

(A) $1\frac{9}{13}$ seconds under water, $7\frac{9}{13}$ seconds above water.

(B) We have already identified

$5,000u$ = distance under water
$1,100a$ = distance above water

$$5,000(1\tfrac{9}{13}) = 5,000\left(\frac{22}{13}\right) = \frac{110,000}{13} \approx 8462 \text{ feet}$$

As a check, $1,100(7\frac{9}{13}) = 1,100\left(\frac{100}{3}\right) = \frac{110,000}{13}$ also.

25. We choose the one variable approach, since there is only one variable apparent.

Let x = amount drained
Then $9 - x$ = amount of 50% solution remaining
x = amount of 100% (pure) antifreeze added
9 = amount of 70% solution desired

$$\begin{pmatrix}\text{Amount of antifreeze}\\ \text{after draining}\end{pmatrix} + \begin{pmatrix}\text{Amount of antifreeze}\\ \text{added}\end{pmatrix} = \begin{pmatrix}\text{Amount of antifreeze}\\ \text{desired}\end{pmatrix}$$

$$0.5(9 - x) + 1.0(x) = 0.7(9)$$

$$4.5 - 0.5x + 1.0x = 6.3$$

$$0.5x + 4.5 = 6.3$$

$$0.5x = 1.8$$

$$x = 3.6 \text{ liters}$$

Check: $0.5(9 - 3.6) + 1.0(3.6) \stackrel{?}{=} 0.7(9)$

$$6.3 \stackrel{?}{=} 6.3$$

27. Let x = sales per week
y = pay per week

Then the straight commission company pays

$y = 0.08x$ (1)

the other company pays

$y = 0.05x + 51$ (2)

We graph these for $0 \le x \le 4{,}000$

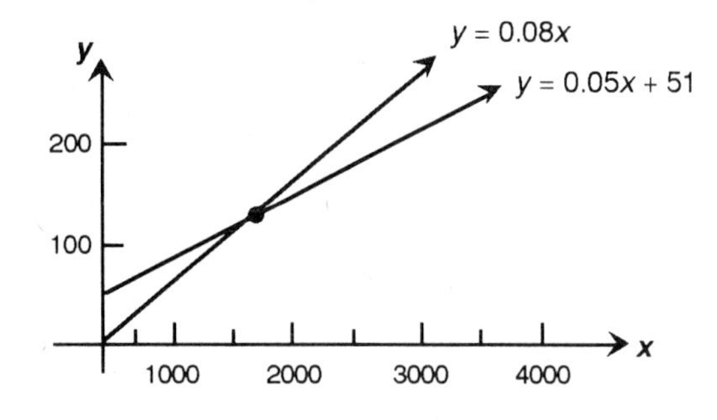

However, the disadvantage of the graphical method is evident here. It would seem that the graphs intersect near $1,700 or $1,800, but the exact point cannot be determined from the graph alone. In this case we can check by substitution, however, that

if $x = 1{,}700$, $y = 0.08(1{,}700) = 136$
if $x = 1{,}700$, $y = 0.05(1{,}700) + 51 = 85 + 51 = 136$ also.

Hence both companies pay $136 on sales of $1,700. The straight commission company pays better to the right of this point (y is greater), and the other company pays better to the left.

Algebraically: We substitute the expression for y from equation (1) into equation (2) to obtain

$$0.08x = 0.05x + 51$$

$$0.03x = 51$$

$$x = \frac{51}{0.03}$$

$x = 1{,}700$ as before.

29. (A)

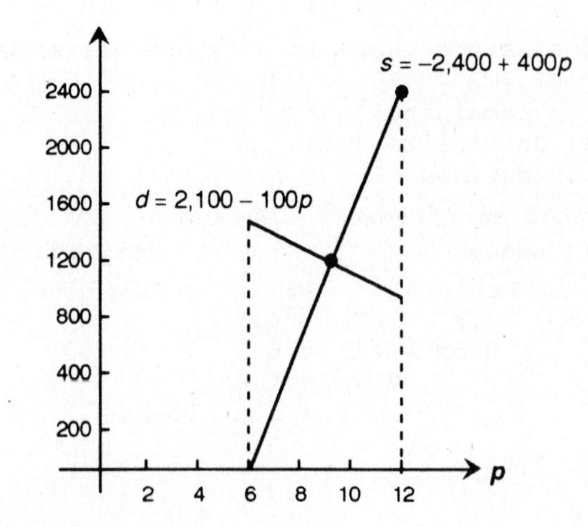

The graphs appear to intersect near p = \$9. Checking by substitution, we have

$$d = 2,100 - 100(9) = 1,200$$
$$s = -2,400 + 400(9) = 1,200$$

At price p = \$9, d = s = 1,200

(B) Algebraically, we have

$$d = 2,100 - 100p$$
$$s = -2,400 + 400p$$
$$d = s$$

Therefore,

$$2,100 - 100p = -2,400 + 400p$$
$$4,500 = 500p$$
$$p = \$9$$

31. We choose the two variable approach.

Let x = the number of $\frac{1}{2}$-pound packages

y = the number of $\frac{1}{4}$-pound packages

Then x + y = 110 (number of packages)

$\frac{1}{2}x + \frac{1}{4}y = 38$ (total weight of contents)

We eliminate fractions by multiplying the bottom equation by its lcd 4.

$$2x + y = 152$$
$$x + y = 110$$

We now eliminate y by multiplying the bottom equation by -1 and adding.

$$2x + y = 152$$
$$\underline{-x - y = -110}$$
$$x \quad = 42$$

Replace x by 42 in the top equation.

$$42 + y = 110$$
$$y = 68$$

Solution: 42 $\frac{1}{2}$-pound packages, 68 $\frac{1}{4}$-pound packages

Check: 42 + 68 = 110

$\frac{1}{2}(42) + \frac{1}{4}(68) = 21 + 17 = 38$

33. We choose the one-variable approach

Let x = number of \$15 tickets sold

Then $4300 - x$ = number of \$18 tickets sold

$$\begin{pmatrix}\text{Value of}\\ \text{\$15 tickets sold}\end{pmatrix} + \begin{pmatrix}\text{Value of}\\ \text{\$18 tickets sold}\end{pmatrix} = (\text{Total value})$$

$$\begin{aligned} 15x + 18(4300 - x) &= 69{,}000 \\ 15x + 77{,}400 - 18x &= 69{,}000 \\ -3x + 77{,}400 &= 69{,}000 \\ -3x &= -8{,}400 \\ x &= 2800 \text{ \$15 tickets} \\ 4300 - x &= 1500 \text{ \$18 tickets} \end{aligned}$$

35. Let x = how much invested at 7%

y = how much invested at 4.5%

Then $x + y = 100{,}000$ (total investment) (1)

$0.07x + 0.045y = 6{,}000$ (total yield) (2)

We solve the system of equations (1), (2) by elimination using addition

$$\begin{aligned} -0.045x - 0.045y &= -4500 \\ 0.07x + 0.045y &= 6000 \\ \hline 0.025x &= 1500 \\ x &= \$60{,}000 \text{ at } 7\% \end{aligned}$$

Replace x by \$60,000 in the top equation.

$$\begin{aligned} 60{,}000 + y &= 100{,}000 \\ y &= \$40{,}000 \text{ at } 4.5\% \end{aligned}$$

Check: $60{,}000 + 40{,}000 = 100{,}000$

$0.07(60{,}000) + 0.045(40{,}000) = 4200 + 1800 = 6000$

37. Let x = average speed of local bus in miles per hour

Then $\frac{3}{2}x$ = average speed of express bus

Let y = time of local bus

Then $y - \frac{2}{3}$ = time of express bus

Since rate × time = distance

$xy = 60$ local bus (1)

$\frac{3}{2}x\left(y - \frac{2}{3}\right) = 60$ express bus

This can be rewritten as: $\frac{3}{2}xy - x = 60$ (2)

The system (1), (2) is not a linear system, but it can be solved by substitution, replacing xy by 60 in equation (2).

$$\begin{aligned} \frac{3}{2}(60) - x &= 60 \\ 90 - x &= 60 \\ x &= 30 \text{ miles per hour (local bus)} \\ \frac{3}{2}x &= 45 \text{ miles per hour (express bus)} \end{aligned}$$

39. Let x = rate of one worker

$x + 8$ = rate of second worker

y = time of slower worker to process 260 forms

y = time of faster worker to process 286 forms

Since rate × time = quantity

$xy = 260$ (1)

$(x + 8)y = 286$

This can be rewritten as

$$xy + 8y = 286 \qquad (2)$$

The system (1), (2) is not a linear system, but it can be solved by substitution, replacing xy by 260.

$$260 + 8y = 286$$
$$8y = 26$$
$$y = \frac{13}{4} \text{ hours to process 260 (or 286) forms}$$
$$x = 260 \div y = 260 \div \frac{13}{4} = 80 \text{ forms per hour (slower worker)}$$
$$x + 8 = 88 \text{ forms per hour (faster worker)}$$

Exercise 8-3 Systems of Linear Equations in Three Variables

Key Ideas and Formulas

Solutions of systems of three equations with three variables are ordered triplets (x, y, z) of numbers which satisfy each equation. Two such systems are **equivalent** if they have the same **solution set** (set of all solutions.)

Steps in Solving Systems of Three Equations in Three Variables	
Step 1:	Choose two equations from the system and eliminate one of the three variables, using elimination by addition or subtraction. The result is generally one equation in two unknowns.
Step 2:	Now eliminate the same variable from the unused equation and one of those used in step 1. We (generally) obtain another equation in two variables.
Step 3:	The two equations from steps 1 and 2 form a system of two equations and two unknowns. Solve as in the preceding section.
Step 4:	Substitute the solution from step 3 into any of the three original equations and solve for the third variable to complete the solution of the original system.
Step 5:	Check the solution in each of the three equations.

As for systems of equations in two variables, systems of linear equations in three variables can be either

a. Consistent and Independent: one solution.
b. Inconsistent: no solution.
c. Dependent: Infinite number of solutions.

---

1. $x - y - 2z = 1 \qquad (1)$
   $y + 4z = 7 \qquad (2)$
   $z = 3 \qquad (3)$

   From equation (3), we have z = 3. Substituting this into equation (2), we have

$$y + 4(3) = 7$$
$$y + 12 = 7$$
$$y = -5$$

Substituting $y = -5$ and $z = 3$ into equation (1), we have

$$x - (-5) - 2(3) = 1$$
$$x - 1 = 1$$
$$x = 2$$

$x = 2,\ y = -5,\ z = 3$

Check: $x - y - 2z = 1$ $\qquad$ $y + 4z = 7$ $\qquad$ $z = 3$

$$2 - (-5) - 2(3) \overset{?}{=} 1 \qquad -5 + 4(3) \overset{?}{=} 7 \qquad 3 \overset{\surd}{=} 3$$
$$1 \overset{\surd}{=} 1 \qquad 7 \overset{\surd}{=} 7$$

3. $x + y + 4z = 3$ (1)
$y - 3z = -2$ (2)
$2z = 1$ (3)

From equation (3), we have $z = \frac{1}{2}$. Substituting this into equation (2), we have

$$y - 3\left(\frac{1}{2}\right) = -2$$
$$y - \frac{3}{2} = -\frac{4}{2}$$
$$y = -\frac{1}{2}$$

Substituting $y = -\frac{1}{2}$ and $z = \frac{1}{2}$ into equation (1), we have

$$x + \left(-\frac{1}{2}\right) + 4\left(\frac{1}{2}\right) = 3$$
$$x + \frac{3}{2} = \frac{6}{2}$$
$$x = \frac{3}{2}$$

$x = \frac{3}{2},\ y = -\frac{1}{2},\ z = \frac{1}{2}$

Check: $x + y + 4z = 3$ $\qquad$ $y - 3z = -2$ $\qquad$ $2z = 1$

$$\frac{3}{2} + \left(-\frac{1}{2}\right) + 4\left(\frac{1}{2}\right) \overset{?}{=} 3 \qquad \left(-\frac{1}{2}\right) - 3\left(\frac{1}{2}\right) \overset{?}{=} -2 \qquad 2\left(\frac{1}{2}\right) \overset{\surd}{=} 1$$
$$\frac{3}{2} - \frac{1}{2} + \frac{4}{2} \overset{\surd}{=} 3 \qquad -\frac{1}{2} - \frac{3}{2} \overset{\surd}{=} -\frac{4}{2}$$

5. $x + 2y + 9z = 4$ (1)
$3y - 6z = -1$ (2)
$3z = 2$ (3)

From equation (3), we have $z = \frac{2}{3}$. Substituting this into equation (2), we have

$$3y - 6\left(\frac{2}{3}\right) = -1$$
$$3y - 4 = -1$$
$$3y = 3$$
$$y = 1$$

Substituting $y = 1$ and $z = \frac{2}{3}$ into equation (1), we have

$$x + 2(1) + 9\left(\frac{2}{3}\right) = 4$$
$$x + 8 = 4$$
$$x = -4$$

$x = -4$, $y = 1$, $z = \frac{2}{3}$

Check:

$$x + 2y + 9z = 4 \qquad -4 + 2(1) + 9\left(\frac{2}{3}\right) \stackrel{?}{=} 4 \qquad -4 + 2 + 6 \stackrel{\surd}{=} 4$$

$$3y - 6z = -1 \qquad 3(1) - 6\left(\frac{2}{3}\right) \stackrel{?}{=} -1 \qquad 3 - 4 \stackrel{\surd}{=} -1$$

$$3z = 2 \qquad 3\left(\frac{2}{3}\right) \stackrel{\surd}{=} 2$$

7.
$$-2x = 2 \quad (1)$$
$$x - 3y = 2 \quad (2)$$
$$-x + 2y + 3z = -7 \quad (3)$$

From equation (1), we have $x = -1$. Substituting this into equation (2), we have

$$-1 - 3y = 2$$
$$-3y = 3$$
$$y = -1$$

Substituting $x = -1$ and $y = -1$ into equation (3), we have

$$-(-1) + 2(-1) + 3z = -7$$
$$-1 + 3z = -7$$
$$3z = -6$$
$$z = -2$$

$x = -1$, $y = -1$, $z = -2$

Check:

$$-2x = 2 \qquad -2(-1) \stackrel{?}{=} 2 \qquad 2 \stackrel{\surd}{=} 2$$

$$x - 3y = 2 \qquad -1 - 3(-1) \stackrel{?}{=} 2 \qquad -1 + 3 \stackrel{\surd}{=} 2$$

$$-x + 2y + 3z = -7 \qquad -(-1) + 2(-1) + 3(-2) \stackrel{?}{=} -7 \qquad 1 + (-2) + (-6) \stackrel{?}{=} -7 \qquad -7 \stackrel{\surd}{=} -7$$

9.
$$4y - z = -13 \quad (1)$$
$$3y + 2z = 4 \quad (2)$$
$$6x - 5y - 2z = 0 \quad (3)$$

We eliminate z from equation (2).

$$8y - 2z = -26 \qquad 2[\text{equation (1)}]$$
$$\underline{3y + 2z = 4}$$
$$11y = -22$$
$$y = -2$$

Substitute $y = -2$ into equation (1).

$$4(-2) - z = -13$$
$$-8 - z = -13$$
$$-z = -5$$
$$z = 5$$

Substitute $y = -2$ and $z = 5$ into equation (3).

$$6x - 5(-2) - 2(5) = 0$$
$$6x = 0$$

$x = 0$, $y = -2$, $z = 5$

Check:

$$4y - z = -13 \qquad 4(-2) - 5 \stackrel{?}{=} -13 \qquad -13 \stackrel{\checkmark}{=} -13$$

$$3y + 2z = 4 \qquad 3(-2) + 2(5) \stackrel{?}{=} 4 \qquad 4 \stackrel{\checkmark}{=} 4$$

$$6x - 5y - 2z = 0 \qquad 6(0) - 5(-2) - 2(5) \stackrel{?}{=} 0 \qquad 0 \stackrel{\checkmark}{=} 0$$

11.

$$\begin{aligned} 2x + y - z &= 5 && (1) \\ x - 2y - 2z &= 4 && (2) \\ 3x + 4y + 3z &= 3 && (3) \end{aligned}$$

We eliminate z from equations (2) and (3).

$$\begin{aligned} -4x - 2y + 2z &= -10 && -2[\text{equation } (1)] \\ x - 2y - 2z &= 4 && \text{Equation } (2) \\ \hline -3x - 4y &= -6 && (4) \end{aligned}$$

$$\begin{aligned} 6x + 3y - 3z &= 15 && 3[\text{equation } (1)] \\ 3x + 4y + 3z &= 3 && \text{Equation } (3) \\ \hline 9x + 7y &= 18 && (5) \end{aligned}$$

We now solve the system

$$\begin{aligned} -3x - 4y &= -6 && (4) \\ 9x + 7y &= 18 && (5) \end{aligned}$$

We eliminate x.

$$\begin{aligned} -9x - 12y &= -18 && 3[\text{equation } (4)] \\ 9x + 7y &= 18 && \text{Equation } (5) \\ \hline -5y &= 0 \\ y &= 0 \end{aligned}$$

Substituting $y = 0$ into equation (5), we have

$$\begin{aligned} 9x + 7(0) &= 18 \\ 9x &= 18 \\ x &= 2 \end{aligned}$$

Substituting $x = 2$ and $y = 0$ into equation (3), we have

$$\begin{aligned} 3(2) + 4(0) + 3z &= 3 \\ 6 + 3z &= 3 \\ 3z &= -3 \\ z &= -1 \end{aligned}$$

$x = 2$, $y = 0$, $z = -1$

Check:

$$2x + y - z = 5 \qquad 2(2) + 0 - (-1) \stackrel{?}{=} 5 \qquad 4 + 1 \stackrel{\checkmark}{=} 5$$

$$x - 2y - 2z = 4 \qquad 2 - 2(0) - 2(-1) \stackrel{?}{=} 4 \qquad 2 + 2 \stackrel{\checkmark}{=} 4$$

$$3x + 4y + 3z = 3 \qquad 3(2) + 4(0) + 3(-1) \stackrel{?}{=} 3 \qquad 6 - 3 \stackrel{\checkmark}{=} 3$$

13.

$$\begin{aligned} 2a + 4b + 3c &= 6 && (1) \\ a - 3b + 2c &= -7 && (2) \\ -a + 2b - c &= 5 && (3) \end{aligned}$$

We eliminate a from equations (1) and (2).

$$\begin{aligned} 2a + 4b + 3c &= 6 && \text{Equation } (1) \\ -2a + 4b - 2c &= 10 && 2[\text{equation } (3)] \\ \hline 8b + c &= 16 && (4) \end{aligned}$$

$$\begin{aligned} a - 3b + 2c &= -7 && \text{Equation } (2) \\ -a + 2b - c &= 5 && \text{Equation } (3) \\ \hline -b + c &= -2 && (5) \end{aligned}$$

We now solve the system

$$\begin{aligned} 8b + c &= 16 && (4) \\ -b + c &= -2 && (5) \end{aligned}$$

We eliminate b.

$$\begin{array}{rl} 8b + c = 16 & \text{Equation (4)} \\ -8b + 8c = -16 & \text{8[equation (5)]} \\ \hline 9c = 0 & \\ c = 0 & \end{array}$$

Substituting c = 0 into equation (4), we have

$$8b + 0 = 16$$
$$b = 2$$

Substituting b = 2 and c = 0 into equation (2), we have

$$a - 3(2) + 2(0) = -7$$
$$a - 6 = -7$$
$$a = -1$$

$a = -1,\ b = 2,\ c = 0$

Check: $2a + 4b + 3c = 6$ $\quad a - 3b + 4c = -7$ $\quad -a + 2b - c = 5$

$2(-1) + 4(2) + 3(0) \overset{?}{=} 6$ $\quad -1 - 3(2) + 2(0) \overset{?}{=} -7$ $\quad -(-1) + 2(2) - (0) \overset{?}{=} 5$

$-2 + 8 \overset{\surd}{=} 6$ $\quad -1 - 6 \overset{\surd}{=} -7$ $\quad 1 + 4 \overset{\surd}{=} 5$

15. $2x - 3y + 3z = -15$ (1)
$3x + 2y - 5z = 19$ (2)
$5x - 4y - 2z = -2$ (3)

We eliminate y from equations (1) and (3).

$$\begin{array}{rl} 4x - 6y + 6z = -30 & \text{2[equation (1)]} \\ 9x + 6y - 15z = 57 & \text{3[equation (2)]} \\ \hline 13x - 9z = 27 & (4) \end{array}$$

$$\begin{array}{rl} 6x + 4y - 10z = 38 & \text{2[equation (2)]} \\ 5x - 4y - 2z = -2 & \text{Equation (3)} \\ \hline 11x - 12z = 36 & (5) \end{array}$$

We now solve the system

$13x - 9z = 27$ (4)
$11x - 12z = 36$ (5)

We eliminate z.

$$\begin{array}{rl} 52x - 36z = 108 & \text{4[equation (4)]} \\ -33x + 36z = -108 & \text{-3[equation (5)]} \\ \hline 19x = 0 & \\ x = 0 & \end{array}$$

Substituting x = 0 into equation (4), we have

$$13(0) - 9z = 27$$
$$-9z = 27$$
$$z = -3$$

Substituting x = 0 and z = -3 into equation (2), we have

$$3(0) + 2y - 5(-3) = 19$$
$$2y + 15 = 19$$
$$2y = 4$$
$$y = 2$$

$x = 0,\ y = 2,\ z = -3$

Check: $2x - 3y + 3z = -15$ $\quad 3x + 2y - 5z = 19$ $\quad 5x - 4y - 2z = -2$

$2(0) - 3(2) + 3(-3) \overset{?}{=} -15$ $\quad 3(0) + 2(2) - 5(-3) \overset{?}{=} 19$ $\quad 5(0) - 4(2) - 2(-3) \overset{?}{=} -2$

$-6 - 9 \overset{\surd}{=} -15$ $\quad 4 + 15 \overset{\surd}{=} 19$ $\quad -8 + 6 \overset{\surd}{=} -2$

17. $5x - 3y + 2z = 13$ (1)
$2x + 4y - 3z = -9$ (2)
$4x - 2y + 5z = 13$ (3)

We eliminate x from equations (1) and (3).

$$\begin{array}{rl} -10x + 6y - 4z = -26 & -2[\text{equation (1)}] \\ \underline{10x + 20y - 15z = -45} & 5[\text{equation (2)}] \\ 26y - 19z = -71 & (4) \end{array}$$

$$\begin{array}{rl} -4x - 8y + 6z = 18 & -2[\text{equation (2)}] \\ \underline{4x - 2y + 5z = 13} & \text{Equation (3)} \\ -10y + 11z = 31 & (5) \end{array}$$

We now solve the system

$$\begin{array}{rl} 26y - 19z = -71 & (4) \\ -10y + 11z = 31 & (5) \end{array}$$

We eliminate y.

$$\begin{array}{rl} 130y - 95z = -355 & 5[\text{equation (4)}] \\ \underline{-130y + 143z = 403} & 13[\text{equation (5)}] \\ 48z = 48 & \\ z = 1 & \end{array}$$

Substituting z = 1 into equation (4), we have

$$\begin{aligned} 26y - 19(1) &= -71 \\ 26y - 19 &= -71 \\ 26y &= -52 \\ y &= -2 \end{aligned}$$

Substituting y = -2 and z = 1 into equation (1), we have

$$\begin{aligned} 5x - 3(-2) + 2(1) &= 13 \\ 5x + 8 &= 13 \\ 5x &= 5 \\ x &= 1 \end{aligned}$$

x = 1, y = -2, z = 1

Check:

$$\begin{array}{lll} 5x - 3y + 2z = 13 & 2x + 4y - 3z = -9 & 4x - 2y + 5z = 13 \\ 5(1) - 3(-2) + 2(1) \stackrel{?}{=} 13 & 2(1) + 4(-2) - 3(1) \stackrel{?}{=} -9 & 4(1) - 2(-2) + 5(1) \stackrel{?}{=} 13 \\ 5 + 6 + 2 \stackrel{\checkmark}{=} 13 & 2 - 8 - 3 \stackrel{\checkmark}{=} -9 & 4 + 4 + 5 \stackrel{\checkmark}{=} 13 \end{array}$$

19.

$$\begin{array}{rl} 2x - 3y + 4z = 10 & (1) \\ 3x - y - z = 8 & (2) \\ 4x + 5y - 6z = 6 & (3) \end{array}$$

We eliminate x from equations (2) and (3)

$$\begin{array}{rl} -6x + 9y - 12z = -30 & -3[\text{equation (1)}] \\ \underline{6x - 2y - 2z = 16} & 2[\text{equation (2)}] \\ 7y - 14z = -14 & (4) \end{array}$$

$$\begin{array}{rl} -4x + 6y - 8z = -20 & -2[\text{equation (1)}] \\ \underline{4x + 5y - 6z = 6} & \text{Equation (3)} \\ 11y - 14z = -14 & (5) \end{array}$$

We now solve the system

$$\begin{array}{rl} 7y - 14z = -14 & (4) \\ 11y - 14z = -14 & (5) \end{array}$$

We eliminate z

$$\begin{array}{rl} -7y + 14z = 14 & -1[\text{equation (4)}] \\ \underline{11y - 14z = -14} & \text{Equation (5)} \\ 4y = 0 & \\ y = 0 & \end{array}$$

Substituting y = 0 into equation (4), we have

$$\begin{aligned} 7(0) - 14z &= -14 \\ -14z &= -14 \\ z &= 1 \end{aligned}$$

Substituting $y = 0$ and $z = 1$ into equation (2), we have

$$3x - 0 - 1 = 8$$
$$3x = 9$$
$$x = 3$$

$x = 3,\ y = 0,\ z = 1$

Check:

$$2x - 3y + 4z = 10 \qquad 3x - y - z = 8 \qquad 4x + 5y - 6z = 6$$
$$2(3) - 3(0) + 4(1) \stackrel{?}{=} 10 \qquad 3(3) - 0 - 1 \stackrel{?}{=} 8 \qquad 4(3) + 5(0) - 6(1) \stackrel{?}{=} 6$$
$$6 + 4 \stackrel{\checkmark}{=} 10 \qquad 9 - 1 \stackrel{\checkmark}{=} 8 \qquad 12 - 6 \stackrel{\checkmark}{=} 6$$

21. $-x + 2y + 3z = 5$ (1)
$3x + 4y + 5z = 5$ (2)
$6x + 5y + z = 4$ (3)

We eliminate x from equations (2) and (3)

$-3x + 6y + 9z = 15$ 3[equation (1)]
$3x + 4y + 5z = 5$ Equation (2)
$10y + 14z = 20$ (5)

$-6x + 12y + 18z = 30$ 6[equation (1)]
$6x + 5y + z = 4$ Equation (3)
$17y + 19z = 34$ (5)

We now solve the system
$10y + 14z = 20$ (4)
$17y + 19z = 34$ (5)

We eliminate y
$-170y - 238z = -340$ -17[equation (4)]
$170y + 190z = 340$ 10[equation (5)]
$-48z = 0$
$z = 0$

Substituting $z = 0$ into equation (4), we have

$$10y + 14(0) = 20$$
$$10y = 20$$
$$y = 2$$

Substitututing $y = 2$ and $z = 0$ into equation (1), we have

$$-x + 2(2) + 3(0) = 5$$
$$-x + 4 = 5$$
$$-x = 1$$
$$x = -1$$

$x = -1,\ y = 2,\ z = 0$

Check:

$$-x + 2y + 3z = 5 \qquad 3x + 4y + 5z = 5 \qquad 6x + 5y + z = 4$$
$$-(-1) + 2(2) + 3(0) \stackrel{?}{=} 5 \qquad 3(-1) + 4(2) + 5(0) \stackrel{?}{=} 5 \qquad 6(-1) + 5(2) + 0 \stackrel{?}{=} 4$$
$$1 + 4 \stackrel{\checkmark}{=} 5 \qquad -3 + 8 \stackrel{\checkmark}{=} 5 \qquad -6 + 10 \stackrel{\checkmark}{=} 4$$

23. $x - 8y + 2z = -1$ (1)
$x - 3y + z = 1$ (2)
$2x - 11y + 3z = 2$ (3)

We eliminate x from equations (2) and (3).
$-x + 8y - 2z = 1$ -1[equation (1)]
$x - 3y + z = 1$ Equation (2)
$5y - z = 2$ (4)

$$\begin{array}{rl} -2x + 16y - 4z = 2 & -2[\text{equation (1)}] \\ \underline{2x - 11y + 3z = 2} & \text{Equation (3)} \\ 5y - z = 4 & (5) \end{array}$$

We now attempt to solve the system

$$\begin{array}{rl} 5y - z = 2 & (4) \\ 5y - z = 4 & (5) \end{array}$$

However, if we eliminate z by multiplying equation (4) by -1 and adding, we also eliminate y.

$$\begin{array}{rl} -5y + z = -2 & -1[\text{equation (4)}] \\ \underline{5y - z = 4} & \text{Equation (5)} \\ 0 = 2 & \end{array}$$

Hence the system is inconsistent. No solution.

25. $$\begin{array}{rl} x + 2y + 3z = 4 & (1) \\ 2x + 3y + 4z = -5 & (2) \\ 3x + 5y + 7z = 6 & (3) \end{array}$$

We eliminate x from equations (2) and (3).

$$\begin{array}{rl} -2x - 4y - 6z = -8 & -2[\text{equation (1)}] \\ \underline{2x + 3y + 4z = -5} & \text{Equation (2)} \\ -y - 2z = -13 & (4) \end{array}$$

$$\begin{array}{rl} -3x - 6y - 9z = -12 & -3[\text{equation (1)}] \\ \underline{3x + 5y + 7z = 6} & \text{Equation (3)} \\ -y - 2z = -6 & (5) \end{array}$$

We now attempt to solve the system

$$\begin{array}{rl} -y - 2z = -13 & (4) \\ -y - 2z = -6 & (5) \end{array}$$

However, if we eliminate z by multiplying equation (4) by -1 and adding, we also eliminate y.

$$\begin{array}{rl} y + 2z = 13 & -1[\text{equation (4)}] \\ \underline{-y - 2z = -6} & \\ 0 = 7 & \end{array}$$

Hence the system is inconsistent. No solution.

27. $$\begin{array}{rl} w - x + y + z = 0 & (1) \\ x - y - z = 1 & (2) \\ y + z = 1 & (3) \\ z = 4 & (4) \end{array}$$

From equation (4), $z = 4$. Substituting $z = 4$ into equation (3), we have

$$\begin{aligned} y + 4 &= 1 \\ y &= -3 \end{aligned}$$

Substituting $y = -3$ and $z = 4$ into equation (2), we have

$$\begin{aligned} x - (-3) - 4 &= 1 \\ x - 1 &= 1 \\ x &= 2 \end{aligned}$$

Substituting $x = 2$, $y = -3$, and $z = 4$ into equation (1), we have

$$\begin{aligned} w - 2 + (-3) + 4 &= 0 \\ w - 1 &= 0 \\ w &= 1 \end{aligned}$$

$w = 1$, $x = 2$, $y = -3$, $z = 4$

Check:

$$w - x + y + z = 0 \qquad 1 - 2 + (-3) + 4 \stackrel{?}{=} 0 \qquad 5 - 5 \stackrel{\checkmark}{=} 0$$

$$x - y - z = 1 \qquad 2 - (-3) - 4 \stackrel{?}{=} 1 \qquad 5 - 4 \stackrel{\checkmark}{=} 1$$

$$y + z = 1 \qquad -3 + 4 \stackrel{\checkmark}{=} 1$$

$$z = 4 \qquad 4 \stackrel{\checkmark}{=} 4$$

29. $$\begin{aligned} w + 2x + 3y + 4z &= 5 \quad (1) \\ x + 2y + 3z &= 4 \quad (2) \\ y + 2z &= 3 \quad (3) \\ z &= 2 \quad (4) \end{aligned}$$

From equation (4), $z = 2$. Substituting $z = 2$ into equation (3), we have

$$\begin{aligned} y + 2(2) &= 3 \\ y + 4 &= 3 \\ y &= -1 \end{aligned}$$

Substituting $y = -1$ and $z = 2$ into equation (2), we have

$$\begin{aligned} x + 2(-1) + 3(2) &= 4 \\ x + 4 &= 4 \\ x &= 0 \end{aligned}$$

Substituting $x = 0$, $y = -1$, and $z = 2$ into equation (1), we have

$$\begin{aligned} w + 2(0) + 3(-1) + 4(2) &= 5 \\ w + 5 &= 5 \\ w &= 0 \end{aligned}$$

$w = 0$, $x = 0$, $y = -1$, $z = 2$

Check: $w + 2x + 3y + 4z = 5$ $x + 2y + 3z = 4$ $y + 2z = 3$ $z = 2$

$0 + 2(0) + 3(-1) + 4(2) \stackrel{?}{=} 5$ $0 + 2(-1) + 3(2) \stackrel{?}{=} 4$ $-1 + 2(2) \stackrel{?}{=} 3$ $2 \stackrel{\surd}{=} 2$

$-3 + 8 \stackrel{\surd}{=} 5$ $-2 + 6 \stackrel{\surd}{=} 4$ $3 \stackrel{\surd}{=} 3$

31. $$\begin{aligned} w - 2x + 3y - 4z &= 5 \quad (1) \\ w - x + y - z &= 9 \quad (2) \\ w - 2x + 4y - 5z &= 10 \quad (3) \\ x - y + 2z &= 9 \quad (4) \end{aligned}$$

We eliminate w from equations (2) and (3)

$$\begin{aligned} -w + 2x - 3y + 4z &= -5 \qquad -1[\text{equation } (1)] \\ \underline{w - x + y - z} &\underline{= 9} \qquad \text{Equation } (2) \\ x - 2y + 3z &= 4 \qquad (5) \end{aligned}$$

$$\begin{aligned} -w + 2x - 3y + 4z &= -5 \qquad -1[\text{equation } (1)] \\ \underline{w - 2x + 4y - 5z} &\underline{= 10} \qquad \text{Equation } (3) \\ y - z &= 5 \end{aligned}$$

We now solve the system

$$\begin{aligned} x - y + 2z &= 9 \quad (4) \\ x - 2y + 3z &= 4 \quad (5) \\ y - z &= 5 \quad (6) \end{aligned}$$

We eliminate x from equation (5)

$$\begin{aligned} -x + y - 2z &= -9 \qquad -1[\text{equation } (4)] \\ \underline{x - 2y + 3z} &\underline{= 4} \qquad [\text{Equation } (5)] \\ -y + z &= -5 \qquad (7) \end{aligned}$$

Thus the original system is equivalent to the system (4), (6), (7)

$$\begin{aligned} x - y + 2z &= 9 \quad (4) \\ y - z &= 5 \quad (6) \\ -y + z &= -5 \quad (7) \end{aligned}$$

If we add equations (6) and (7), we obtain

$$0 = 0$$

We conclude that the original system has an infinite number of solutions. We can write them all in terms of z as follows:

Solving equation (6) for y in terms of z, we obtain

$$y = z + 5$$

Substituting $y = z + 5$ into equation (4), we obtain

$$\begin{aligned} x - (z + 5) + 2z &= 9 \\ x - z - 5 + 2z &= 9 \\ x + z - 5 &= 9 \\ x &= 14 - z \end{aligned}$$

Substituting $x = 14 - z$ and $y = z + 5$ into equation (2), we obtain

$$\begin{aligned} w - (14 - z) + (z + 5) - z &= 9 \\ w - 14 + z + z + 5 - z &= 9 \\ w - 9 + z &= 9 \\ w &= 18 - z \end{aligned}$$

Thus, all solutions can be written as $(18 - z, 14 - z, z + 5, z)$ for z any real number.

33.

$$\begin{aligned} 4w - x &= 5 && (1) \\ -3w + 2x - y &= -5 && (2) \\ 2w - 5x + 4y + 3z &= 13 && (3) \\ 2w + 2x - 2y - z &= -2 && (4) \end{aligned}$$

We eliminate z from equation (3).

$$\begin{aligned} 2w - 5x + 4y + 3z &= 13 && \text{Equation (3)} \\ \underline{6w + 6x - 6y - 3z} &\underline{= -6} && \text{3[equation (4)]} \\ 8w + x - 2y &= 7 && (5) \end{aligned}$$

We now solve the system

$$\begin{aligned} 4w - x &= 5 && (1) \\ -3w + 2x - y &= -5 && (2) \\ 8w + x - 2y &= 7 && (5) \end{aligned}$$

We eliminate y from equation (5).

$$\begin{aligned} 6w - 4x + 2y &= 10 && \text{-2[equation (2)]} \\ \underline{8w + x - 2y} &\underline{= 7} && \text{Equation (5)} \\ 14w - 3x &= 17 && (6) \end{aligned}$$

We now solve the system

$$\begin{aligned} 4w - x &= 5 && (1) \\ 14w - 3x &= 17 && (6) \end{aligned}$$

We eliminate x.

$$\begin{aligned} -12w + 3x &= -15 && \text{-3[equation (1)]} \\ \underline{14w - 3x} &\underline{= 17} && \text{Equation (6)} \\ 2w &= 2 \\ w &= 1 \end{aligned}$$

Substituting $w = 1$ into equation (1), we have

$$\begin{aligned} 4(1) - x &= 5 \\ 4 - x &= 5 \\ -x &= 1 \\ x &= -1 \end{aligned}$$

Substituting $w = 1$ and $x = -1$ into equation (2), we have

$$\begin{aligned} -3(1) = 2(-1) - y &= -5 \\ -5 - y &= -5 \\ -y &= 0 \\ y &= 0 \end{aligned}$$

Substituting $w = 1$, $x = -1$, and $y = 0$ into equation (3), we have

$$\begin{aligned} 2(1) - 5(-1) + 4(0) + 3z &= 13 \\ 7 + 3z &= 13 \\ 3z &= 6 \\ z &= 2 \end{aligned}$$

$w = 1$, $x = -1$, $y = 0$, $z = 2$

Check: $4w - x = 5$ $\qquad$ $-3w + 2x - y = -5$

$$4(1) - (-1) \stackrel{?}{=} 5 \qquad -3(1) + 2(-1) - 0 \stackrel{?}{=} -5$$

$$4 + 1 = 5 \ \checkmark \qquad -3 - 2 = -5 \ \checkmark$$

$$2w - 5x + 4y + 3z = 13 \qquad 2w + 2x - 2y - z = -2$$

$$2(1) - 5(-1) + 4(0) + 3(2) \stackrel{?}{=} 13 \qquad 2(1) + 2(-1) - 2(0) - 2 \stackrel{?}{=} -2$$

$$2 + 5 + 6 = 13 \ \checkmark \qquad 2 - 2 - 2 = -2 \ \checkmark$$

35.
$$x + 2y + 3z = 4 \quad (1)$$
$$2x - 3y + 4z = -5 \quad (2)$$
$$4x + y + 10z = 3 \quad (3)$$

Given that there are an infinite number of solutions, it suffices to eliminate x from equation (2) and solve the resulting equation for y in terms of z, then substitute the result into equation (1) to solve for x in terms of z.

$$-2x - 4y - 6z = -8 \qquad -2[\text{equation (1)}]$$
$$\underline{2x - 3y + 4z = -5} \qquad \text{Equation (2)}$$
$$-7y - 2z = -13 \quad (4)$$
$$-7y = 2z - 13$$
$$y = \frac{13 - 2z}{7}$$

Substituting this into equation (1), we obtain

$$x + 2 \cdot \frac{13 - 2z}{7} + 3z = 4$$
$$x + \frac{26 - 4z}{7} + \frac{21z}{7} = \frac{28}{7}$$
$$x + \frac{26 + 17z}{7} = \frac{28}{7}$$
$$x = \frac{2 - 17z}{7}$$

Thus, all solutions can be written as $\left(\frac{2 - 17z}{7}, \frac{13 - 2z}{7}, z\right)$ for z any real number.

37.
$$x + y + z = 4 \quad (1)$$
$$-x + y - z = 2 \quad (2)$$
$$x + 5y + z = 16 \quad (3)$$

Given that there are an infinite number of solutions, it suffices to eliminate x from equation (2) and solve the resulting equation for y in terms of z, then substitute the result into equation (1) to solve for x in terms of z.

$$x + y + z = 4 \quad (1)$$
$$\underline{-x + y - z = 2} \quad (2)$$

Adding
$$2y = 6$$
$$y = 3$$

Substituting this into equation (1), we obtain

$$x + 3 + z = 4$$
$$x = 1 - z$$

Thus, all solutions can be written as $(1 - z, 3, z)$ for z any real number.

Note: *In the following application problems, we have omitted the checking for reasons of space. The reader should perform these steps, however.*

39. Since the circle passes through (-2, -1), (-1, -2), and (6, -1), the coordinates of each of these points satisfy the equation of the circle. Substituting into

$$x^2 + y^2 + Dx + Ey + F = 0$$

we have

$$(-2)^2 + (-1)^2 + D(-2) + E(-1) + F = 0$$
$$5 - 2D - E + F = 0 \quad (1)$$
$$(-1)^2 + (-2)^2 + D(-1) + E(-2) + F = 0$$
$$5 - D - 2E + F = 0 \quad (2)$$
$$(6)^2 + (-1)^2 + D(6) + E(-1) + F = 0$$
$$37 + 6D - E + F = 0 \quad (3)$$

We solve the system of equations (1), (2), (3). For convenience we can rewrite with the constant terms on the right.

$$-2D - E + F = -5 \quad (1a)$$
$$-D - 2E + F = -5 \quad (2a)$$
$$6D - E + F = -37 \quad (3a)$$

We eliminate F from equations (2a) and (3a).

$$2D + E - F = 5 \qquad -1[\text{equation (1a)}]$$
$$\underline{-D - 2E + F = -5} \qquad \text{Equation (2a)}$$
$$D - E = 0 \quad (4)$$

$$2D + E - F = 5 \qquad -1[\text{equation (1a)}]$$
$$\underline{6D - E + F = -37} \qquad \text{Equation (3a)}$$
$$8D = -32 \quad (5)$$

We now solve the system

$$D - E = 0 \quad (4)$$
$$8D = -32 \quad (5)$$

From equation (5) we have D = -4. Substituting this into equation (4) gives

$$-4 - E = 0$$
$$E = -4$$

Substituting D = -4 and E = -4 into equation (1), we have

$$5 - 2(-4) - (-4) + F = 0$$
$$17 + F = 0$$
$$F = -17$$

D = -4, E = -4, F = -17

41. Since the parabola passes through (1, 2) (3, 6), and (-2, 11), the coordinates of each of these points satisfy the equation of the parabola. Substituting into $y = ax^2 + bx + c$, we have

$$2 = a(1)^2 + b(1) + c$$
$$a + b + c = 2 \quad (1)$$
$$6 = a(3)^2 + b(3) + c$$
$$9a + 3b + c = 6 \quad (2)$$
$$11 = a(-2)^2 + b(-2) + c$$
$$4a - 2b + c = 11 \quad (3)$$

We solve the system of equations (1), (2), (3). We eliminate c from equations (2) and (3)

$$-a - b - c = -2 \qquad -[\text{equation (1)}]$$
$$\underline{9a + 3b + c = 6} \qquad \text{Equation (2)}$$
$$8a + 2b = 4 \quad (4)$$

$$\begin{array}{rl} -a - b - c = -2 & -[\text{equation } (1)] \\ \underline{4a - 2b + c = 11} & \text{Equation } (3) \\ 3a - 3b \quad = 9 & (5) \end{array}$$

We now solve the system

$$\begin{array}{rl} 8a + 2b = 4 & (4) \\ 3a - 3b = 9 & (5) \end{array}$$

We eliminate b

$$\begin{array}{rl} 8a + 2b = 4 & \text{Equation } (4) \\ \underline{2a - 2b = 6} & \frac{2}{3}[\text{equation } (5)] \\ 10a \quad = 10 & \\ a = 1 & \end{array}$$

Substituting $a = 1$ into equation (5), we obtain

$$\begin{aligned} 3(1) - 3b &= 9 \\ -3b &= 6 \\ b &= -2 \end{aligned}$$

Substituting $a = 1$ and $b = -2$ into equation (1), we obtain

$$\begin{aligned} 1 + (-2) + c &= 2 \\ -1 + c &= 2 \\ c &= 3 \end{aligned}$$

$a = 1$, $b = -2$, $c = 3$

43. Let x = number of shirt A produced per week
y = number of shirt B produced per week
z = number of shirt C produced per week

We have

$$\begin{array}{rl} 0.2x + 0.4y + 0.3z = 1160 & \text{Cutting department} \\ 0.3x + 0.5y + 0.4z = 1560 & \text{Sewing department} \\ 0.1x + 0.2y + 0.1z = 480 & \text{Packaging department} \end{array}$$

Clearing of decimals for convenience:

$$\begin{array}{rl} 2x + 4y + 3z = 11600 & (1) \\ 3x + 5y + 4z = 15600 & (2) \\ x + 2y + z = 4800 & (3) \end{array}$$

We eliminate x from equations (1) and (2)

$$\begin{array}{rl} 2x + 4y + 3z = 11600 & \text{Equation } (1) \\ \underline{-2x - 4y - 2z = -9600} & -2[\text{equation } (3)] \\ z = 2000 & (4) \end{array}$$

$$\begin{array}{rl} 3x + 5y + 4z = 15600 & \text{Equation } (2) \\ \underline{-3x - 6y - 3z = -14400} & -3[\text{equation } (3)] \\ -y + z = 1200 & (5) \end{array}$$

Solving the system (4), (5) we have

$$\begin{array}{rl} z = 2000 & (4) \\ -y + 2000 = 1200 & \\ -y = -800 & \\ y = 800 & \end{array}$$

Substituting $y = 800$, $z = 2000$ into equation (3), we have

$$\begin{aligned} x + 2(800) + 2000 &= 4800 \\ x &= 1200 \end{aligned}$$

1200 style A; 800 style B; 2000 style C

45. Let x = amount of Mix A
y = amount of Mix B
z = amount of Mix C

The amounts of protein must add up to 23 grams.

$$\begin{pmatrix}\text{Protein in}\\ \text{Mix A}\end{pmatrix} + \begin{pmatrix}\text{Protein in}\\ \text{Mix B}\end{pmatrix} + \begin{pmatrix}\text{Protein in}\\ \text{Mix C}\end{pmatrix} = \begin{pmatrix}\text{Total}\\ \text{Protein}\end{pmatrix}$$
$$0.20x + 0.10y + 0.15z = 23$$

The amounts of fat must add up to 6.2 grams.

$$\begin{pmatrix}\text{Fat in}\\ \text{Mix A}\end{pmatrix} + \begin{pmatrix}\text{Fat in}\\ \text{Mix B}\end{pmatrix} + \begin{pmatrix}\text{Fat in}\\ \text{Mix C}\end{pmatrix} = \begin{pmatrix}\text{Total}\\ \text{Fat}\end{pmatrix}$$
$$0.02x + 0.06y + 0.05z = 6.2$$

The amounts of moisture must add up to 16 grams.

$$\begin{pmatrix}\text{Moisture in}\\ \text{Mix A}\end{pmatrix} + \begin{pmatrix}\text{Moisture in}\\ \text{Mix B}\end{pmatrix} + \begin{pmatrix}\text{Moisture in}\\ \text{Mix C}\end{pmatrix} = \begin{pmatrix}\text{Total}\\ \text{Moisture}\end{pmatrix}$$
$$0.15x + 0.10y + 0.05z = 16$$

We thus have the system of equations

$$\begin{aligned} 0.20x + 0.10y + 0.15z &= 23\\ 0.02x + 0.06y + 0.05z &= 6.2\\ 0.15x + 0.10y + 0.05z &= 16 \end{aligned}$$

Clearing decimals for convenience, we have

$$\begin{aligned} 20x + 10y + 15z &= 2300 && (1)\\ 2x + 6y + 5z &= 620 && (2)\\ 15x + 10y + 5z &= 1600 && (3) \end{aligned}$$

We eliminate z from equations (1) and (2)

$$\begin{aligned} 20x + 10y + 15z &= 2300 && \text{Equation (1)}\\ -45x - 30y - 15z &= -4800 && -3[\text{equation (3)}]\\ \hline -25x - 20y &= -2500 && (4) \end{aligned}$$

$$\begin{aligned} -2x - 6y - 5z &= -620 && -1[\text{equation (2)}]\\ 15x + 10y + 5z &= 1600 && \text{Equation (3)}\\ \hline 13x + 4y &= 980 && (5) \end{aligned}$$

We solve the system (4), (5) by elimination by addition.

$$\begin{aligned} -25x - 20y &= -2500 && \text{Equation (4)}\\ 65x + 20y &= 4900 && 5[\text{equation (5)}]\\ \hline 40x &= 2400\\ x &= 60 \end{aligned}$$

Substituting x = 60 into equation (5), we have

$$\begin{aligned} 13(60) + 4y &= 980\\ 4y &= 200\\ y &= 50 \end{aligned}$$

Substituting x = 60, y = 50 into equation (2), we have

$$\begin{aligned} 2(60) + 6(50) + 5z &= 620\\ 420 + 5z &= 620\\ 5z &= 200\\ z &= 40 \end{aligned}$$

60 grams Mix A, 50 grams Mix B, 40 grams Mix C

47. Since w refrigerators from warehouse I and y refrigerators from warehouse II are shipped to store A for a total of 80, we have

$$w + y = 80$$

Since x refrigerators from warehouse I and z refrigerators from warehouse II are shipped to store B for a total of 120, we have

$$x + z = 120$$

Since w refrigerators are shipped from warehouse I to store A and x refrigerators to store B for a total of 140, we have

$$w + x = 140$$

Since y refrigerators are shipped from warehouse II to store A and z refrigerators to store B for a total of 60, we have

$$y + z = 60$$

49\. Let x = time for oldest press to print the paper, working alone
y = time for middle press to print the paper, working alone
z = time for newest press to print the paper, working alone

Then

$\frac{1}{x}$ = portion of printing done by oldest press in 1 hour

$\frac{1}{y}$ = portion of printing done by middle press in 1 hour

$\frac{1}{z}$ = portion of printing done by newest press in 1 hour

If all three presses work 2 hours, the entire paper is printed.

$$\begin{pmatrix}\text{Portion printed}\\ \text{by oldest press}\end{pmatrix} + \begin{pmatrix}\text{Portion printed}\\ \text{by middle press}\end{pmatrix} + \begin{pmatrix}\text{Portion printed}\\ \text{by newest press}\end{pmatrix} = \begin{pmatrix}\text{Total}\\ \text{job}\end{pmatrix}$$

$$\frac{2}{x} + \frac{2}{y} + \frac{2}{z} = 1$$

If the two older presses work 4 hours, the entire paper is printed.

$$\frac{4}{x} + \frac{4}{y} = 1$$

If the oldest and newest presses work 3 hours, the entire paper is printed.

$$\frac{3}{x} + \frac{3}{z} = 1$$

We have a system of three equations in three variables.

$$\frac{2}{x} + \frac{2}{y} + \frac{2}{z} = 1 \qquad (1)$$

$$\frac{4}{x} + \frac{4}{y} = 1 \qquad (2)$$

$$\frac{3}{x} + \frac{3}{z} = 1 \qquad (3)$$

By multiplying equation (1) by -2 and adding to equation (2), we will have eliminated both x and y.

$$-\frac{4}{x} - \frac{4}{y} - \frac{4}{z} = -2 \qquad -2[\text{equation (1)}]$$

$$\frac{4}{x} + \frac{4}{y} = 1 \qquad \text{Equation (2)}$$

$$-\frac{4}{z} = -1 \qquad \text{lcm} = z,\ z \neq 0$$

$$-4 = -1z$$

$$z = 4$$

Substituting z = 4 into equation (3) gives

$$\frac{3}{x} + \frac{3}{4} = 1 \qquad \text{lcm} = 4x,\ x \neq 0$$

$$4x \cdot \frac{3}{x} + 4x \cdot \frac{3}{4} = 4x$$

$$12 + 3x = 4x$$

$$12 = x$$

$$x = 12$$

The original system is therefore equivalent to

$x + 2y = 2$
$y = 5$

Thus, $y = 5$, and $x + 2(5) = 2$ so $x = -8$.

Solution: $x = -8$, $y = 5$

Check:

$x + 2y = 2$ $\qquad$ $2x + 3y = -1$

$-8 + 2(5) \overset{\checkmark}{=} 2$ $\qquad$ $2(-8) + 3(15) \overset{\checkmark}{=} -1$

The original system is therefore equivalent to

$x + y = -3$
$y = 7$

Thus, $y = 7$, and $x + 7 = -3$ so $x = -10$.

Solution: $x = -10$, $y = 7$

Check:

$3x + 4y = -2$ $\qquad$ $2x + 3y = 1$

$3(-10) + 4(7) \overset{\checkmark}{=} -2$ $\qquad$ $2(-10) + 3(7) \overset{\checkmark}{=} 1$

33.

$$\left[\begin{array}{cc|c} 4 & 5 & 0 \\ 3 & 4 & 4 \end{array}\right] \quad R_1 + (-1)R_2 \Rightarrow R_1$$

Need a 1 here

-3 -4 -4

$$\sim \left[\begin{array}{cc|c} 1 & 1 & -4 \\ 3 & 4 & 4 \end{array}\right] \quad R_2 + (-3)R_1 \Rightarrow R_2$$

Need a 0 here

-3 -3 12

$$\sim \left[\begin{array}{cc|c} 1 & 1 & -4 \\ 0 & 1 & 16 \end{array}\right]$$

The original system is therefore equivalent to

$x + y = -4$
$y = 16$

Thus, $y = 16$, and $x + 16 = -4$, so $x = -20$.

Solution: $x = -20$, $y = 16$

Check:

$4x + 5y = 0$ $\qquad$ $3x + 4y = 4$

$4(-20) + 5(16) \overset{\checkmark}{=} 0$ $\qquad$ $3(-20) + 4(16) \overset{\checkmark}{=} 4$

35.

$$\sim \left[\begin{array}{cc|c} 1 & 2 & 4 \\ 2 & 4 & -8 \end{array}\right] R_2 + (-2)R_1 \Rightarrow R_2$$

Need a 0 here

-2 -4 -8

$$\sim \left[\begin{array}{cc|c} 1 & 2 & 4 \\ 0 & 0 & -16 \end{array}\right]$$

This matrix corresponds to the system

$x + 2y = 4$
$0x + 0y = -16$

This system has no solution.

37.

$$\left[\begin{array}{cc|c} 2 & 1 & 6 \\ 1 & -1 & -3 \end{array}\right] R_1 \Leftrightarrow R_2$$

Need a 1 here

$$\sim \left[\begin{array}{cc|c} 1 & -1 & -3 \\ 2 & 1 & 6 \end{array}\right] R_2 + (-2)R_1 \Rightarrow R_2$$

Need a 0 here

-2 2 6

$$\sim \left[\begin{array}{cc|c} 1 & -1 & -3 \\ 0 & 3 & 12 \end{array}\right] \frac{1}{3}R_2 \Rightarrow R_2$$

Need a 1 here

$$\sim \left[\begin{array}{cc|c} 1 & -1 & -3 \\ 0 & 1 & 4 \end{array}\right]$$

37. *(continued)*
The original system is therefore equivalent to

$$\begin{aligned} x - y &= -3 \\ y &= 4 \end{aligned}$$

Thus, $y = 4$, and $x - 4 = -3$ so $x = 1$.
Solution: $x = 1$, $y = 4$
Check:

$$2x + y = 6 \qquad x - y = -3$$
$$2(1) + 4 \stackrel{\checkmark}{=} 6 \qquad 1 - 4 \stackrel{\checkmark}{=} -3$$

39.

$$\left[\begin{array}{cc|c} 3 & -6 & -9 \\ -2 & 4 & 6 \end{array}\right] \tfrac{1}{3}R_1 \Rightarrow R_1$$

Need a 1 here

$$\sim \left[\begin{array}{cc|c} 1 & -2 & -3 \\ -2 & 4 & 6 \end{array}\right] R_2 + 2R_1 \Rightarrow R_2$$

Need a 0 here

2 -4 -6

$$\sim \left[\begin{array}{cc|c} 1 & -2 & -3 \\ 0 & 0 & 0 \end{array}\right]$$

This matrix corresponds to the system

$$\begin{aligned} x - 2y &= -3 \\ 0x + 0y &= 0 \end{aligned}$$

Thus $x = 2y - 3$.

Hence there are infinitely many solutions: for any real numbers, $y = s$, $x = 2s - 3$ is a solution.

41.

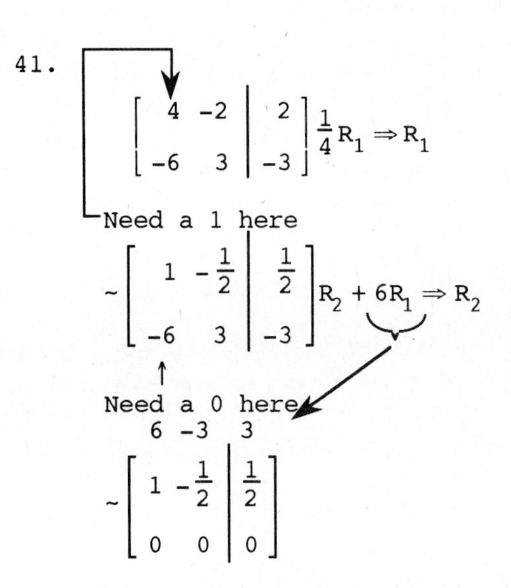

$$\left[\begin{array}{cc|c} 4 & -2 & 2 \\ -6 & 3 & -3 \end{array}\right] \tfrac{1}{4}R_1 \Rightarrow R_1$$

Need a 1 here

$$\sim \left[\begin{array}{cc|c} 1 & -\frac{1}{2} & \frac{1}{2} \\ -6 & 3 & -3 \end{array}\right] R_2 + 6R_1 \Rightarrow R_2$$

Need a 0 here

6 -3 3

$$\sim \left[\begin{array}{cc|c} 1 & -\frac{1}{2} & \frac{1}{2} \\ 0 & 0 & 0 \end{array}\right]$$

This matrix corresponds to the system

$$\begin{aligned} x - \tfrac{1}{2}y &= \tfrac{1}{2} \\ 0x + 0y &= 0 \end{aligned}$$

Thus $x = \frac{1}{2}y + \frac{1}{2}$.

Hence there are infinitely many solutions: for any real number s, $y = s$, $x = \frac{1}{2}s + \frac{1}{2}$ is a solution.

43.

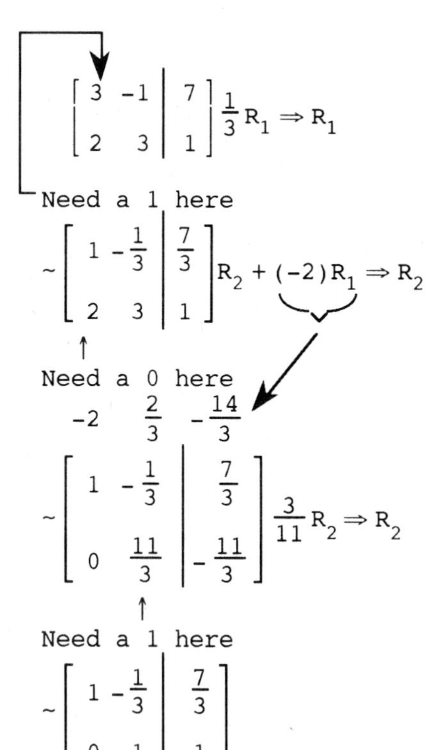

The original system is therefore equivalent to

$$x - \frac{1}{3}y = \frac{7}{3}$$
$$y = -1$$

Thus, $y = -1$, and $x - \frac{1}{3}(-1) = \frac{7}{3}$ so $x + \frac{1}{3} = \frac{7}{3}$ and $x = 2$.

Solution: $x = 2$, $y = -1$

Check:

$$3x - y \overset{\surd}{=} 7 \qquad 2x + 3y \overset{\surd}{=} 1$$
$$3(2) - (-1) = 7 \qquad 2(2) + 3(-1) = 1$$

45.

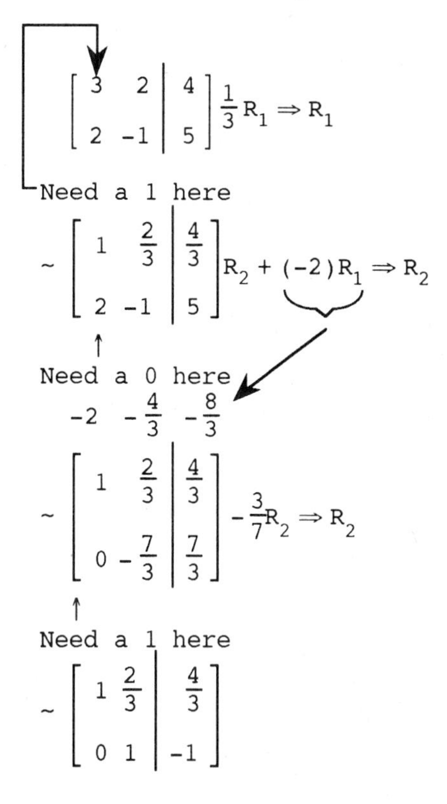

The original system is therefore equivalent to

$$x + \frac{2}{3}y = \frac{4}{3}$$
$$y = -1$$

Thus, $y = -1$, and $x + \frac{2}{3}(-1) = \frac{4}{3}$ so $x - \frac{2}{3} = \frac{4}{3}$ and $x = 2$.

Solution: $x = 2$, $y = -1$

Check:

$$3x + 2y \overset{\surd}{=} 4 \qquad 2x - y \overset{\surd}{=} 5$$
$$3(2) + 2(-1) = 4 \qquad 2(2) - (-1) = 5$$

47.

$$\left[\begin{array}{cc|c} 3 & 6 & 4 \\ 1 & 2 & 2 \end{array}\right] R_1 \Leftrightarrow R_2$$

Need a 1 here

$$\sim \left[\begin{array}{cc|c} 1 & 2 & 2 \\ 3 & 6 & 4 \end{array}\right] R_2 + (-3)R_1 \Rightarrow R_2$$

Need a 0 here

$$-3 \quad -6 \quad -6$$

$$\sim \left[\begin{array}{cc|c} 1 & 2 & 2 \\ 0 & 0 & -2 \end{array}\right]$$

This matrix corresponds to the system

$$x + 2y = 2$$
$$0x + 0y = -2$$

No solution.

49.

$$\left[\begin{array}{cc|c} 2 & -5 & -4 \\ 4 & -10 & -8 \end{array}\right] \quad R_2 + \underbrace{(-2)R_1} \Rightarrow R_2$$

↑ Need a 0 here

-4 10 8

$$\left[\begin{array}{cc|c} 2 & -5 & -4 \\ 0 & 0 & 0 \end{array}\right]$$

This matrix corresonds to the system

$2x - 5y = -4$

$0x + 0y = 0$

Thus, $x = \dfrac{5y - 4}{2}$

Hence there are infinitely many solutions: for any real number y, $\left(\dfrac{5y - 4}{2}, y\right)$ is a solution.

51.

$$\left[\begin{array}{cc|c} 0.2 & -0.5 & 0.07 \\ 0.8 & -0.3 & 0.79 \end{array}\right] 5R_1 \Rightarrow R_1$$

Need a 1 here

$$\sim \left[\begin{array}{cc|c} 1 & -2.5 & 0.35 \\ 0.8 & -0.3 & 0.79 \end{array}\right] R_2 + \underbrace{(-0.8)R_1} \Rightarrow R_2$$

↑ Need a 0 here

-0.8 2 -0.28

$$\sim \left[\begin{array}{cc|c} 1 & -2.5 & 0.35 \\ 0 & 1.7 & 0.51 \end{array}\right] \frac{1}{1.7}R_2 \Rightarrow R_2$$

↑ Need a 1 here

$$\sim \left[\begin{array}{cc|c} 1 & -2.5 & 0.35 \\ 0 & 1 & 0.3 \end{array}\right]$$

The original system is therefore equivalent to

$x - 2.5y = 0.35$

$y = 0.3$

Thus, $y = 0.3$, and $x - 2.5(0.3) = 0.35$, so $x - 0.75 = 0.35$ and $x = 1.1$.

Solution: $x = 1.1$, $y = 0.3$

Check:

$0.2x - 0.5y = 0.07$

$0.2(1.1) - 0.5(0.3) \stackrel{?}{=} 0.07$

$0.22 - 0.15 \stackrel{\checkmark}{=} 0.07$

$0.8x - 0.3y = 0.79$

$0.8(1.1) - 0.3(0.3) \stackrel{?}{=} 0.79$

$0.88 - 0.09 \stackrel{\checkmark}{=} 0.79$

53.

$$\left[\begin{array}{ccc|c} 2 & 4 & -10 & -2 \\ 3 & 9 & -21 & 0 \\ 1 & 5 & -12 & 1 \end{array}\right] R_1 \Leftrightarrow R_3$$

Need a 1 here

$$\sim \left[\begin{array}{ccc|c} 1 & 5 & -12 & 1 \\ 3 & 9 & -21 & 0 \\ 2 & 4 & -10 & -2 \end{array}\right] \begin{array}{l} R_2 + (-3)R_1 \Rightarrow R_2 \\ R_3 + (-2)R_1 \Rightarrow R_3 \end{array}$$

Need 0's here

$$\sim \left[\begin{array}{ccc|c} 1 & 5 & -12 & 1 \\ 0 & -6 & 15 & -3 \\ 0 & -6 & 14 & -4 \end{array}\right] -\frac{1}{6}R_2 \Rightarrow R_2$$

Need a 1 here

$$\sim \left[\begin{array}{ccc|c} 1 & 5 & -12 & 1 \\ 0 & 1 & -\frac{5}{2} & \frac{1}{2} \\ 0 & -6 & 14 & -4 \end{array}\right] R_3 + 6R_2 \Rightarrow R_3$$

Need a 0 here

$$\sim \left[\begin{array}{ccc|c} 1 & 5 & -12 & 1 \\ 0 & 1 & -\frac{5}{2} & \frac{1}{2} \\ 0 & 0 & -1 & -1 \end{array}\right] -R_3 \Rightarrow R_3$$

Need a 1 here

$$\sim \left[\begin{array}{ccc|c} 1 & 5 & -12 & 1 \\ 0 & 1 & -\frac{5}{2} & \frac{1}{2} \\ 0 & 0 & 1 & 1 \end{array}\right]$$

The resulting equivalent system

$$\begin{aligned} x + 5y - 12z &= 1 \\ y - \frac{5}{2}z &= \frac{1}{2} \\ z &= 1 \end{aligned}$$

is solved by substitution.

$$z = 1$$

$$y - \frac{5}{2}z = y - \frac{5}{2} = \frac{1}{2}$$

so $y = 3$

$$x + 5y - 12z = x + 15 - 12 = 1$$

so $x + 3 = 1$

$$x = -2$$

Solution: $x = -2$, $y = 3$, $z = 1$

Check:

$$2(-2) + 4(3) - 10(1) \stackrel{?}{=} -2$$

$$-4 + 12 - 10 \stackrel{\checkmark}{=} -2$$

$$3(-2) + 9(3) - 21(1) \stackrel{?}{=} 0$$

$$-6 + 27 - 21 \stackrel{\checkmark}{=} 0$$

$$1(-2) + 5(3) - 12(1) \stackrel{?}{=} 1$$

$$-2 + 15 - 12 \stackrel{\checkmark}{=} 1$$

55. $$\left[\begin{array}{ccc|c} 3 & 8 & -1 & -18 \\ 2 & 1 & 5 & 8 \\ 2 & 4 & 2 & -4 \end{array}\right] \quad \begin{array}{l} \frac{1}{2}R_3 \Rightarrow R_3 \\ R_3 \Leftrightarrow R_1 \end{array}$$

Need a 1 here

$$\sim \left[\begin{array}{ccc|c} 1 & 2 & 1 & -2 \\ 2 & 1 & 5 & 8 \\ 3 & 8 & -1 & -18 \end{array}\right] \quad \begin{array}{l} R_2 + (-2)R_1 \Rightarrow R_2 \\ R_3 + (-3)R_1 \Rightarrow R_3 \end{array}$$

Need 0's here

$$\sim \left[\begin{array}{ccc|c} 1 & 2 & 1 & -2 \\ 0 & -3 & 3 & 12 \\ 0 & 2 & -4 & -12 \end{array}\right] \quad -\frac{1}{3}R_2 \Rightarrow R_2$$

Need a 1 here

$$\sim \left[\begin{array}{ccc|c} 1 & 2 & 1 & -2 \\ 0 & 1 & -1 & -4 \\ 0 & 2 & -4 & -12 \end{array}\right] \quad R_3 + (-2)R_2 \Rightarrow R_3$$

Need a 0 here

$$\sim \left[\begin{array}{ccc|c} 1 & 2 & 1 & -2 \\ 0 & 1 & -1 & -4 \\ 0 & 0 & -2 & -4 \end{array}\right] \quad -\frac{1}{2}R_3 \Rightarrow R_3$$

Need a 1 here

$$\sim \left[\begin{array}{ccc|c} 1 & 2 & 1 & -2 \\ 0 & 1 & -1 & -4 \\ 0 & 0 & 1 & 2 \end{array}\right]$$

The resulting equivalent system

$$\begin{aligned} x + 2y + z &= -2 \\ y - z &= -4 \\ z &= 2 \end{aligned}$$

is solved by substitution.

$z = 2$

$y - z = y - 2 = -4$

so $y = -2$

$x + 2y + z = x + 2(-2) + 2 = -2$

so $x - 2 = -2$

$x = 0$

Solution: $x = 0,\ y = -2,\ z = 2$

Check:

$3(0) + 8(-2) - 2 \stackrel{?}{=} -18$

$-16 - 2 \stackrel{\surd}{=} -18$

$2(0) + (-2) + 5(2) \stackrel{?}{=} 8$

$-2 + 10 \stackrel{\surd}{=} 8$

$2(0) + 4(-2) + 2(2) \stackrel{?}{=} -4$

$-8 + 4 \stackrel{\surd}{=} -4$

57. $$\left[\begin{array}{ccc|c} 3 & 1 & -2 & 4 \\ 0 & 2 & 5 & 1 \\ 1 & 0 & 3 & -2 \end{array}\right] \quad R_1 \Leftrightarrow R_3$$

Need a 1 here

$$\sim \left[\begin{array}{ccc|c} 1 & 0 & 3 & -2 \\ 0 & 2 & 5 & 1 \\ 3 & 1 & -2 & 4 \end{array}\right] \quad R_3 + (-3)R_1 \Rightarrow R_3$$

Need a 0 here

$$\sim \left[\begin{array}{ccc|c} 1 & 0 & 3 & -2 \\ 0 & 2 & 5 & 1 \\ 0 & 1 & -11 & 10 \end{array}\right] \quad R_2 \Leftrightarrow R_3$$

Need a 1 here

$$\sim \left[\begin{array}{ccc|c} 1 & 0 & 3 & -2 \\ 0 & 1 & -11 & 10 \\ 0 & 2 & 5 & 1 \end{array}\right] \quad R_3 + (-2)R_2 \Rightarrow R_3$$

Need a 0 here

$$\sim \left[\begin{array}{ccc|c} 1 & 0 & 3 & -2 \\ 0 & 1 & -11 & 10 \\ 0 & 0 & 27 & -19 \end{array}\right] \quad \frac{1}{27}R_3 \Rightarrow R_3$$

Need a 1 here

$$\sim \left[\begin{array}{ccc|c} 1 & 0 & 3 & -2 \\ 0 & 1 & -11 & 10 \\ 0 & 0 & 1 & -\frac{19}{27} \end{array}\right]$$

This matrix corresponds to the system

$$x + 3z = -2$$
$$y - 11z = 10$$
$$z = -\frac{19}{27}$$

We solve by substitution.

$$z = -\frac{19}{27}$$

$$y - 11z = y - 11\left(-\frac{19}{27}\right) = 10$$

so $\quad y = -\frac{209}{27} + \frac{270}{27} = \frac{61}{27}$

$$x + 3z = x + 3\left(-\frac{19}{27}\right) = -2$$

so $\quad x = \frac{57}{27} - \frac{54}{27} = \frac{3}{27}$

Solution: $x = \frac{3}{27}$, $y = \frac{61}{27}$, $z = -\frac{19}{27}$

Check: $3\left(\frac{3}{27}\right) + \frac{61}{27} - 2\left(-\frac{19}{27}\right) \overset{?}{=} 4$ $\quad 0 + 2\left(\frac{61}{27}\right) + 5\left(-\frac{19}{27}\right) \overset{?}{=} 1$ $\quad \frac{3}{27} + 0 + 3\left(-\frac{19}{27}\right) \overset{?}{=} -2$

$\frac{9}{27} + \frac{61}{27} + \frac{38}{27} \overset{\checkmark}{=} \frac{108}{27}$ $\quad \frac{122}{27} - \frac{95}{27} \overset{\checkmark}{=} \frac{27}{27}$ $\quad -\frac{54}{27} \overset{\checkmark}{=} -2$

59.

$$\left[\begin{array}{ccc|c} 1 & 2 & 3 & -1 \\ 1 & 1 & 2 & 3 \\ 2 & 4 & 7 & 5 \end{array}\right] \quad \begin{array}{l} R_2 + (-1)R_1 \Rightarrow R_2 \\ R_3 + (-2)R_1 \Rightarrow R_3 \end{array}$$

Need 0's here

$$\sim \left[\begin{array}{ccc|c} 1 & 2 & 3 & -1 \\ 0 & -1 & -1 & 4 \\ 0 & 0 & 1 & 7 \end{array}\right] \quad -R_2 \Rightarrow R_2$$

Need a 1 here

$$\sim \left[\begin{array}{ccc|c} 1 & 2 & 3 & -1 \\ 0 & 1 & 1 & -4 \\ 0 & 0 & 1 & 7 \end{array}\right]$$

This matrix corresponds to the system

$$x + 2y + 3z = -1$$
$$y + z = -4$$
$$z = 7$$

We solve by substitution.

$$z = 7$$
$$y + z = y + 7 = -4$$

so $\quad y = -11$

$$x + 2y + 3z = x + 2(-11) + 3(7) = -1$$

so $\quad x - 1 = -1$

$$x = 0$$

Solution: $x = 0$, $y = -11$, $z = 7$

Check: $0 + 2(-11) + 3(7) \overset{?}{=} -1$ $\quad 0 + (-11) + 2(7) \overset{?}{=} 3$ $\quad 0 + 4(-11) + 7(7) \overset{?}{=} 5$

$-22 + 21 \overset{\checkmark}{=} -1$ $\quad -11 + 14 \overset{\checkmark}{=} 3$ $\quad -44 + 49 \overset{\checkmark}{=} 5$

61. $\left[\begin{array}{ccc|c} 1 & 1 & 0 & 2 \\ 0 & 1 & 1 & 3 \\ 1 & 0 & 1 & 4 \end{array}\right] R_3 + (-1)R_1 \Rightarrow R_3$

↑ Need a 0 here

$\sim \left[\begin{array}{ccc|c} 1 & 1 & 0 & 2 \\ 0 & 1 & 1 & 3 \\ 0 & -1 & 1 & 2 \end{array}\right] R_3 + R_2 \Rightarrow R_3$

↑ Need a 0 here

$\sim \left[\begin{array}{ccc|c} 1 & 1 & 0 & 2 \\ 0 & 1 & 1 & 3 \\ 0 & 0 & 2 & 5 \end{array}\right] \frac{1}{2}R_3 \Rightarrow R_3$

↑ Need a 1 here

$\sim \left[\begin{array}{ccc|c} 1 & 1 & 0 & 2 \\ 0 & 1 & 1 & 3 \\ 0 & 0 & 1 & \frac{5}{2} \end{array}\right]$

This matrix corresponds to the system

$$\begin{aligned} x + y \quad &= 2 \\ y + z &= 3 \\ z &= \frac{5}{2} \end{aligned}$$

We solve by substitution.

$$z = \frac{5}{2}$$

$$y + z = y + \frac{5}{2} = 3$$

$$y = \frac{1}{2}$$

$$x + y = x + \frac{1}{2} = 2$$

$$x = \frac{3}{2}$$

Solution: $x = \frac{3}{2}$, $y = \frac{1}{2}$, $z = \frac{5}{2}$

Check: $\frac{3}{2} + \frac{1}{2} + 0 \overset{\checkmark}{=} 2$ $\quad 0 + \frac{1}{2} + \frac{5}{2} \overset{\checkmark}{=} 3$

$\frac{3}{2} + 0 + \frac{5}{2} \overset{\checkmark}{=} 4$

63. $\left[\begin{array}{cccc|c} 1 & -2 & 0 & 1 & 1 \\ -2 & 5 & 3 & -1 & -5 \\ 3 & -6 & 1 & 7 & 2 \\ 1 & -1 & 4 & 7 & -3 \end{array}\right] \begin{array}{l} R_2 + 2R_1 \Rightarrow R_2 \\ R_3 + (-3)R_1 \Rightarrow R_3 \\ R_4 + (-1)R_1 \Rightarrow R_4 \end{array}$

Need 0's here

$$\sim \left[\begin{array}{cccc|c} 1 & -2 & 0 & 1 & 1 \\ 0 & 1 & 3 & 1 & -3 \\ 0 & 0 & 1 & 4 & -1 \\ 0 & 1 & 4 & 6 & -4 \end{array}\right] \quad R_4 + (-1)R_2 \Rightarrow R_4$$

↑ Need a 0 here

$$\sim \left[\begin{array}{cccc|c} 1 & -2 & 0 & 1 & 1 \\ 0 & 1 & 3 & 1 & -3 \\ 0 & 0 & 1 & 4 & -1 \\ 0 & 0 & 1 & 5 & -1 \end{array}\right] \quad R_4 + (-1)R_3 \Rightarrow R_4$$

↑ Need a 0 here

$$\sim \left[\begin{array}{cccc|c} 1 & -2 & 0 & 1 & 1 \\ 0 & 1 & 3 & 1 & -3 \\ 0 & 0 & 1 & 4 & -1 \\ 0 & 0 & 0 & 1 & 0 \end{array}\right]$$

The resulting equivalent system

$$\begin{aligned} x - 2y + \phantom{3z + {}} w &= 1 \\ y + 3z + w &= -3 \\ z + 4w &= -1 \\ w &= 0 \end{aligned}$$

is solved by substitution

$w = 0$

so $z = -1$

$y + 3z + w = y - 3 = -3$

so $y = 0$

so $x - 2y + w = x = 1$

Solution: $x = 1$, $y = 0$, $z = -1$, $w = 0$.

Check:

$$1 - 2(0) + 0 \overset{?}{=} 1$$
$$1 \overset{\surd}{=} 1$$

$$-2(1) + 5(0) + 3(-1) - 0 \overset{?}{=} -5$$
$$-5 \overset{\surd}{=} -5$$

$$3(1) - 6(0) + (-1) + 7(0) \overset{?}{=} 2$$
$$2 \overset{\surd}{=} 2$$

$$1 - 0 + 4(-1) + 7(0) \overset{?}{=} -3$$
$$-3 \overset{\surd}{=} -3$$

65.

$$\left[\begin{array}{cccc|c} 1 & 4 & 7 & 10 & 13 \\ 0 & 1 & 2 & 3 & 4 \\ 1 & -3 & -4 & -4 & -4 \\ 0 & 1 & 1 & 1 & 1 \end{array}\right] \quad R_3 + (-1)R_1 \Rightarrow R_3$$

Need a 0 here

$$\sim \left[\begin{array}{cccc|c} 1 & 4 & 7 & 10 & 13 \\ 0 & 1 & 2 & 3 & 4 \\ 0 & -7 & -11 & -14 & -17 \\ 0 & 1 & 1 & 1 & 1 \end{array}\right] \quad \begin{array}{l} R_3 + 7R_2 \Rightarrow R_3 \\ R_4 + (-1)R_2 \Rightarrow R_4 \end{array}$$

Need 0's here

$$\sim \left[\begin{array}{cccc|c} 1 & 4 & 7 & 10 & 13 \\ 0 & 1 & 2 & 3 & 4 \\ 0 & 0 & 3 & 7 & 11 \\ 0 & 0 & -1 & -2 & -3 \end{array}\right] \quad \begin{array}{l} R_3 \Leftrightarrow R_4 \\ -R_4 \Rightarrow R_4 \end{array}$$

Need a 1 here

$$\sim \left[\begin{array}{cccc|c} 1 & 4 & 7 & 10 & 13 \\ 0 & 1 & 2 & 3 & 4 \\ 0 & 0 & 1 & 2 & 3 \\ 0 & 0 & 3 & 7 & 11 \end{array}\right] \quad R_4 + (-3)R_3 \Rightarrow R_4$$

↑ Need a 0 here

$$\left[\begin{array}{cccc|c} 1 & 4 & 7 & 10 & 13 \\ 0 & 1 & 2 & 3 & 4 \\ 0 & 0 & 1 & 2 & 3 \\ 0 & 0 & 0 & 1 & 2 \end{array}\right]$$

The resulting equivalent system

$$\begin{aligned} w + 4x + 7y + 10z &= 13 \\ x + 2y + 3z &= 4 \\ y + 2z &= 3 \\ z &= 2 \end{aligned}$$

is solved by substitution.

$z = 2$

So $y + 2z = y + 2(2) = 3$

$y = -1$

so

$x + 2y + 3z = x + 2(-1) + 3(2) = 4$

$x + 4 = 4$

$x = 0$

So

$w + 4x + 7y + 10z$

$= w + 4(0) + 7(-1) + 10(2) = 13$

$w + 13 = 13$

$w = 0$

Solution: $w = 0,\ x = 0,\ y = -1,\ z = 2$

Check:

$0 + 4(0) + 7(-1) + 10(2) \stackrel{?}{=} 13$

$\sqrt{}$ $-7 + 20 = 13$

$0 + 2(-1) + 3(2) \stackrel{?}{=} 4$

$\sqrt{}$ $-2 + 6 = 4$

$-1 + 2(2) \stackrel{?}{=} 3$

$\sqrt{}$ $-1 + 4 = 3$

$\sqrt{}$ $2 = 2$

Exercise 8-5 Determinants and Cramer's Rule

Key Ideas and Formulas

A determinant is a square array of real numbers, written between parallel lines (not between square brackets--that would be a matrix).

Second-order determinant:

$$\begin{vmatrix} a_{11} & a_{12} \\ a_{21} & a_{22} \end{vmatrix} = a_{11}a_{22} - a_{21}a_{12}$$

Third-order determinant:

$$\begin{vmatrix} a_{11} & a_{12} & a_{13} \\ a_{21} & a_{22} & a_{23} \\ a_{31} & a_{32} & a_{33} \end{vmatrix} = a_{11}a_{22}a_{33} - a_{11}a_{32}a_{23} + a_{21}a_{32}a_{13} - a_{21}a_{12}a_{33} + a_{31}a_{12}a_{23} - a_{31}a_{22}a_{13}$$

The minor of an element a_{ij} is the determinant array obtained by deleting the row and column that contain a_{ij}, that is, row i and column j.

The cofactor of $a_{ij} = (-1)^{i+j}$ (Minor of a_{ij}).

The value of a determinant of order 3 is the sum of the three products obtained by multiplying each element of any one row (or each element of any one column) by its cofactor.

Cramer's Rule for Two Equations and Two Variables

Given the system

$$a_{11}x + a_{12}y = k_1$$
$$a_{21}x + a_{22}y = k_2$$

with

$$D = \begin{vmatrix} a_{11} & a_{12} \\ a_{21} & a_{22} \end{vmatrix} \neq 0$$

then

$$x = \frac{\begin{vmatrix} k_1 & a_{12} \\ k_2 & a_{22} \end{vmatrix}}{D} \quad \text{and} \quad y = \frac{\begin{vmatrix} a_{11} & k_1 \\ a_{21} & k_2 \end{vmatrix}}{D}$$

Cramer's Rule for Three Equations and Three Variables

Given the system

$$a_{11}x + a_{12}y + a_{13}z = k_1$$
$$a_{21}x + a_{22}y + a_{23}z = k_2$$
$$a_{31}x + a_{32}y + a_{33}z = k_3$$

with

$$D = \begin{vmatrix} a_{11} & a_{12} & a_{13} \\ a_{21} & a_{22} & a_{23} \\ a_{31} & a_{32} & a_{33} \end{vmatrix} \neq 0$$

then

$$x = \frac{\begin{vmatrix} k_1 & a_{12} & a_{13} \\ k_2 & a_{22} & a_{23} \\ k_3 & a_{32} & a_{33} \end{vmatrix}}{D} \quad y = \frac{\begin{vmatrix} a_{11} & k_1 & a_{13} \\ a_{21} & k_2 & a_{23} \\ a_{31} & k_3 & a_{33} \end{vmatrix}}{D} \quad z = \frac{\begin{vmatrix} a_{11} & a_{12} & k_1 \\ a_{21} & a_{22} & k_2 \\ a_{31} & a_{32} & k_3 \end{vmatrix}}{D}$$

The determinant D is called the coefficient determinant. If $D \neq 0$, the system has exactly one solution, given by Cramer's rule. If $D = 0$, the system is either inconsistent or dependent. Cramer's rule does not apply.

---

1. $\begin{vmatrix} 3 & 2 \\ 4 & 3 \end{vmatrix} = 3 \cdot 3 - 4 \cdot 2 = 9 - 8 = 1$

3. $\begin{vmatrix} 2 & 8 \\ 1 & 4 \end{vmatrix} = 2 \cdot 4 - 1 \cdot 8 = 8 - 8 = 0$

5. $\begin{vmatrix} 2 & 4 \\ 3 & -1 \end{vmatrix} = 2(-1) - 3\cdot 4 = -2 - 12 = -14$

7. $\begin{vmatrix} 5 & -4 \\ -2 & 2 \end{vmatrix} = 5\cdot 2 - (-2)(-4) = 10 - 8 = 2$

9. $\begin{vmatrix} 3 & -3.1 \\ -2 & 1.2 \end{vmatrix} = 3(1.2) - (-2)(-3.1) = 3.6 - (6.2) = -2.6$

11. $\begin{vmatrix} \frac{1}{2} & 2 \\ \frac{1}{4} & 3 \end{vmatrix} = \frac{1}{2}(3) - \frac{1}{4}(2) = \frac{3}{2} - \frac{1}{2} = 1$

13. $\begin{vmatrix} \frac{2}{3} & \frac{1}{2} \\ \frac{1}{6} & \frac{1}{4} \end{vmatrix} = \frac{2}{3}\cdot\frac{1}{4} - \frac{1}{6}\cdot\frac{1}{2} = \frac{2}{12} - \frac{1}{12} = \frac{1}{12}$

15. $\begin{vmatrix} a_{11} & a_{12} & a_{13} \\ a_{21} & a_{22} & a_{23} \\ a_{31} & a_{32} & a_{33} \end{vmatrix} = \begin{vmatrix} a_{22} & a_{23} \\ a_{32} & a_{33} \end{vmatrix}$

17. $\begin{vmatrix} a_{11} & a_{12} & a_{13} \\ a_{21} & a_{22} & a_{23} \\ a_{31} & a_{32} & a_{33} \end{vmatrix} = \begin{vmatrix} a_{11} & a_{12} \\ a_{31} & a_{32} \end{vmatrix}$

19. $(-1)^{1+1} \begin{vmatrix} a_{22} & a_{23} \\ a_{32} & a_{33} \end{vmatrix}$

21. $(-1)^{2+3} \begin{vmatrix} a_{11} & a_{12} \\ a_{31} & a_{32} \end{vmatrix}$

23. $\begin{vmatrix} -2 & 3 & 0 \\ 5 & 1 & -2 \\ 7 & -4 & 8 \end{vmatrix} = \begin{vmatrix} 1 & -2 \\ -4 & 8 \end{vmatrix}$

25. $\begin{vmatrix} -2 & 3 & 0 \\ 5 & 1 & -2 \\ 7 & -4 & 8 \end{vmatrix} = \begin{vmatrix} -2 & 0 \\ 5 & -2 \end{vmatrix}$

27. $(-1)^{1+1} \begin{vmatrix} 1 & -2 \\ -4 & 8 \end{vmatrix} = (-1)^2[1\cdot 8 - (-2)(-4)] = 0$

29. $(-1)^{3+2} \begin{vmatrix} -2 & 0 \\ 5 & -2 \end{vmatrix} = (-1)^5[(-2)(-2) - 0(-5)] = (-1)(4) = -4$

31. $D = \begin{vmatrix} 1 & 2 \\ 1 & 3 \end{vmatrix} = 1$

$x = \dfrac{\begin{vmatrix} 1 & 2 \\ -1 & 3 \end{vmatrix}}{D} = \dfrac{5}{1} = 5$

$y = \dfrac{\begin{vmatrix} 1 & 1 \\ 1 & -1 \end{vmatrix}}{D} = \dfrac{-2}{1} = -2$

$x = 5,\ y = -2$

33. $D = \begin{vmatrix} 2 & 1 \\ 5 & 3 \end{vmatrix} = 1$

$x = \dfrac{\begin{vmatrix} 1 & 1 \\ 2 & 3 \end{vmatrix}}{D} = \dfrac{1}{1} = 1$

$y = \dfrac{\begin{vmatrix} 2 & 1 \\ 5 & 2 \end{vmatrix}}{D} = \dfrac{-1}{1} = -1$

$x = 1,\ y = -1$

35. $D = \begin{vmatrix} 2 & -1 \\ -1 & 3 \end{vmatrix} = 5$

$x = \dfrac{\begin{vmatrix} -3 & -1 \\ 4 & 3 \end{vmatrix}}{5} = \dfrac{-5}{5} = -1$

$y = \dfrac{\begin{vmatrix} 2 & -3 \\ -1 & 4 \end{vmatrix}}{5} = \dfrac{5}{5} = 1$

$x = -1,\ y = 1$

37. We expand by the first row.

$$\begin{vmatrix} 1 & 0 & 0 \\ -2 & 4 & 3 \\ 5 & -2 & 1 \end{vmatrix} = a_{11}\begin{pmatrix}\text{cofactor}\\ \text{of } a_{11}\end{pmatrix} + a_{12}\begin{pmatrix}\text{cofactor}\\ \text{of } a_{12}\end{pmatrix} + a_{13}\begin{pmatrix}\text{cofactor}\\ \text{of } a_{13}\end{pmatrix}$$

$$= 1\left((-1)^{1+1}\begin{vmatrix} 4 & 3 \\ -2 & 1 \end{vmatrix}\right) + 0\left(\quad\right) + 0\left(\quad\right)$$

$$= (-1)^2[4 \cdot 1 - (-2)3]$$
$$= 10$$

It is unnecessary to evaluate these since they are multiplied by 0.

39. We expand by the first column.

$$\begin{vmatrix} 0 & 1 & 5 \\ 3 & -7 & 6 \\ 0 & -2 & -3 \end{vmatrix} = a_{11}\begin{pmatrix}\text{cofactor}\\ \text{of } a_{11}\end{pmatrix} + a_{21}\begin{pmatrix}\text{cofactor}\\ \text{of } a_{21}\end{pmatrix} + a_{31}\begin{pmatrix}\text{cofactor}\\ \text{of } a_{31}\end{pmatrix}$$

$$= 0\left(\quad\right) + 3\left((-1)^{2+1}\begin{vmatrix} 1 & 5 \\ -2 & -3 \end{vmatrix}\right) + 0\left(\quad\right)$$

It is unnecessary to evaluate these since they are multiplied by 0.

$$= 3(-1)^3[1(-3) - (-2)5]$$
$$= 3(-1)(7)$$
$$= -21$$

41. We expand by the third row.

$$\begin{vmatrix} 4 & -4 & 6 \\ 2 & 8 & -3 \\ 0 & -5 & 0 \end{vmatrix} = a_{31}\begin{pmatrix}\text{cofactor}\\ \text{of } a_{31}\end{pmatrix} + a_{32}\begin{pmatrix}\text{cofactor}\\ \text{of } a_{32}\end{pmatrix} + a_{33}\begin{pmatrix}\text{cofactor}\\ \text{of } a_{33}\end{pmatrix}$$

$$= 0\left(\quad\right) + (-5)\left((-1)^{3+2}\begin{vmatrix} 4 & 6 \\ 2 & -3 \end{vmatrix}\right) + 0\left(\quad\right)$$

It is unnecessary to evaluate these since they are multiplied by 0.

$$= (-5)(-1)^5[4(-3) - 6 \cdot 2]$$
$$= (-5)(-1)(-24)$$
$$= -120$$

43. We expand by the second column.

$$\begin{vmatrix} -1 & 2 & -3 \\ -2 & 0 & -6 \\ 4 & -3 & 2 \end{vmatrix} = a_{21}\begin{pmatrix}\text{cofactor}\\ \text{of } a_{21}\end{pmatrix} + a_{22}\begin{pmatrix}\text{cofactor}\\ \text{of } a_{22}\end{pmatrix} + a_{23}\begin{pmatrix}\text{cofactor}\\ \text{of } a_{23}\end{pmatrix}$$

$$= 2(-1)^{2+1}\begin{vmatrix} -2 & -6 \\ 4 & 2 \end{vmatrix} + 0\left(\quad\right) + (-3)(-1)^{2+3}\begin{vmatrix} -1 & -3 \\ -2 & -6 \end{vmatrix}$$

$$= 2(-1)^3[(-2)2 - 4(-6)] + (-3)(-1)^5[(-1)(-6) - (-2)(-3)]$$
$$= (-2)(20) + 3(0) = -40$$

45. We expand by the first row.

$$\begin{vmatrix} 1 & 4 & 1 \\ 1 & 1 & -2 \\ 2 & 1 & -1 \end{vmatrix} = a_{11}\begin{pmatrix}\text{cofactor}\\ \text{of } a_{11}\end{pmatrix} + a_{12}\begin{pmatrix}\text{cofactor}\\ \text{of } a_{12}\end{pmatrix} + a_{13}\begin{pmatrix}\text{cofactor}\\ \text{of } a_{13}\end{pmatrix}$$

$$= 1(-1)^{1+1}\begin{vmatrix} 1 & -2 \\ 1 & -1 \end{vmatrix} + 4(-1)^{1+2}\begin{vmatrix} 1 & -2 \\ 2 & -1 \end{vmatrix} + 1(-1)^{1+3}\begin{vmatrix} 1 & 1 \\ 2 & 1 \end{vmatrix}$$

$$= (-1)^2[1(-1) - 1(-2)] + 4(-1)^3[1(-1) - 2(-2)] + (-1)^4[1\cdot 1 - 2\cdot 1]$$
$$= 1 + (-12) + (-1)$$
$$= -12$$

47. We expand by the first row.

$$\begin{vmatrix} 1 & 4 & 3 \\ 2 & 1 & 6 \\ 3 & -2 & 9 \end{vmatrix} = a_{11}\begin{pmatrix}\text{cofactor}\\ \text{of } a_{11}\end{pmatrix} + a_{12}\begin{pmatrix}\text{cofactor}\\ \text{of } a_{12}\end{pmatrix} + a_{13}\begin{pmatrix}\text{cofactor}\\ \text{of } a_{13}\end{pmatrix}$$

$$= 1(-1)^{1+1}\begin{vmatrix} 1 & 6 \\ -2 & 9 \end{vmatrix} + 4(-1)^{1+2}\begin{vmatrix} 2 & 6 \\ 3 & 9 \end{vmatrix} + 3(-1)^{1+3}\begin{vmatrix} 2 & 1 \\ 3 & -2 \end{vmatrix}$$

$$= (-1)^2[1\cdot 9 - (-2)6] + 4(-1)^3[2\cdot 9 - 3\cdot 6] + 3(-1)^4[2(-2) - 1\cdot 3]$$
$$= 21 + 0 - 21$$
$$= 0$$

49. $D = \begin{vmatrix} 1 & 1 & 0 \\ 0 & 2 & 1 \\ -1 & 0 & 1 \end{vmatrix} = 1$

$$x = \frac{\begin{vmatrix} 0 & 1 & 0 \\ -5 & 2 & 1 \\ -3 & 0 & 1 \end{vmatrix}}{D} = \frac{2}{1} = 2$$

$$y = \frac{\begin{vmatrix} 1 & 0 & 1 \\ 0 & -5 & 1 \\ -1 & -3 & 1 \end{vmatrix}}{D} = \frac{-2}{1} = -2$$

$$z = \frac{\begin{vmatrix} 1 & 1 & 0 \\ 0 & 2 & -5 \\ -1 & 0 & -3 \end{vmatrix}}{D} = \frac{-1}{1} = -1$$

51. $D = \begin{vmatrix} 1 & 1 & 0 \\ 0 & 2 & 1 \\ -1 & 0 & 1 \end{vmatrix} = 1$

$$x = \frac{\begin{vmatrix} 1 & 1 & 0 \\ 0 & 2 & 0 \\ 0 & 0 & 1 \end{vmatrix}}{D} = \frac{2}{1} = 2$$

$$y = \frac{\begin{vmatrix} 1 & 1 & 0 \\ 0 & 0 & 1 \\ -1 & 0 & 1 \end{vmatrix}}{D} = \frac{-1}{1} = -1$$

$$z = \frac{\begin{vmatrix} 1 & 1 & 1 \\ 0 & 2 & 0 \\ -1 & 0 & 0 \end{vmatrix}}{D} = \frac{2}{1} = 2$$

53. $D = \begin{vmatrix} 0 & 1 & 1 \\ 1 & 0 & 2 \\ 1 & -1 & 0 \end{vmatrix} = 1$

$x = \dfrac{\begin{vmatrix} -4 & 1 & 1 \\ 0 & 0 & 2 \\ 5 & -1 & 0 \end{vmatrix}}{D} = \dfrac{2}{1} = 2$

$y = \dfrac{\begin{vmatrix} 0 & -4 & 1 \\ 1 & 0 & 2 \\ 1 & 5 & 0 \end{vmatrix}}{D} = \dfrac{-3}{1} = -3$

$z = \dfrac{\begin{vmatrix} 0 & 1 & -4 \\ 1 & 0 & 0 \\ 1 & -1 & 5 \end{vmatrix}}{D} = \dfrac{-1}{1} = -1$

55. $D = \begin{vmatrix} 0 & 2 & -1 \\ 1 & -1 & -1 \\ 1 & -1 & 2 \end{vmatrix} = -6$

$x = \dfrac{\begin{vmatrix} -4 & 2 & -1 \\ 0 & -1 & -1 \\ 6 & -1 & 2 \end{vmatrix}}{-6} = \dfrac{-6}{-6} = 1$

$y = \dfrac{\begin{vmatrix} 0 & -4 & -1 \\ 1 & 0 & -1 \\ 1 & 6 & 2 \end{vmatrix}}{-6} = \dfrac{6}{-6} = -1$

$z = \dfrac{\begin{vmatrix} 0 & 2 & -4 \\ 1 & -1 & 0 \\ 1 & -1 & 6 \end{vmatrix}}{-6} = \dfrac{-12}{-6} = 2$

57. $D = \begin{vmatrix} 2 & 4 & 3 \\ 1 & -3 & 2 \\ -1 & 2 & -1 \end{vmatrix} = -9$

$a = \dfrac{\begin{vmatrix} 6 & 4 & 3 \\ -7 & -3 & 2 \\ 5 & 2 & -1 \end{vmatrix}}{D} = \dfrac{9}{-9} = -1$

$b = \dfrac{\begin{vmatrix} 2 & 6 & 3 \\ 1 & -7 & 2 \\ -1 & 5 & -1 \end{vmatrix}}{D} = \dfrac{-18}{-9} = 2$

$c = \dfrac{\begin{vmatrix} 2 & 4 & 6 \\ 1 & -3 & -7 \\ -1 & 2 & 5 \end{vmatrix}}{D} = \dfrac{0}{-9} = 0$

59. We expand by the first column. Clearly the only non-zero term will be a_{21} (cofactor of a_{21}), thus

$$\begin{vmatrix} 0 & 1 & 0 & 1 \\ 2 & 4 & 7 & 6 \\ 0 & 3 & 0 & 1 \\ 0 & 6 & 2 & 5 \end{vmatrix} = 2(-1)^{2+1} \begin{vmatrix} 1 & 0 & 1 \\ 3 & 0 & 1 \\ 6 & 2 & 5 \end{vmatrix} + 0 \text{ terms}$$

The order 3 determinant is expanded by the second column. Again there is only one non-zero term, a_{32} (cofactor of a_{32}). So the original determinant is reduced to

$$2(-1)^{2+1}2(-1)^{3+2} \begin{vmatrix} 1 & 1 \\ 3 & 1 \end{vmatrix} = 4(-1)^8(1 \cdot 1 - 3 \cdot 1) = 4(-2) = -8$$

61. $$\begin{vmatrix} 2 & 0 & 0 & 0 & 0 \\ 0 & 3 & 0 & 0 & 0 \\ 0 & 0 & 2 & 0 & 0 \\ 0 & 0 & 0 & 1 & 0 \\ 0 & 0 & 0 & 0 & 4 \end{vmatrix} = 2\left((-1)^{1+1} \begin{vmatrix} 3 & 0 & 0 & 0 \\ 0 & 2 & 0 & 0 \\ 0 & 0 & 1 & 0 \\ 0 & 0 & 0 & 4 \end{vmatrix} + 0 \text{ terms}\right)$$

$$= 2 \begin{vmatrix} 3 & 0 & 0 & 0 \\ 0 & 2 & 0 & 0 \\ 0 & 0 & 1 & 0 \\ 0 & 0 & 0 & 4 \end{vmatrix}$$

$$= 2\left(3(-1)^{1+1} \begin{vmatrix} 2 & 0 & 0 \\ 0 & 1 & 0 \\ 0 & 0 & 4 \end{vmatrix} + 0 \text{ terms}\right)$$

$$= 2 \cdot 3 \begin{vmatrix} 2 & 0 & 0 \\ 0 & 1 & 0 \\ 0 & 0 & 4 \end{vmatrix}$$

$$= 2 \cdot 3\left(2(-1)^{1+1} \begin{vmatrix} 1 & 0 \\ 0 & 4 \end{vmatrix} + 0 \text{ terms}\right)$$

$$= 2 \cdot 3 \cdot 2 \begin{vmatrix} 1 & 0 \\ 0 & 4 \end{vmatrix}$$

$$= 2 \cdot 3 \cdot 2(1 \cdot 4 - 0 \cdot 0)$$

$$= 2 \cdot 3 \cdot 2 \cdot 1 \cdot 4$$

$$= 48$$

63. First we show that $\begin{vmatrix} x & y & 1 \\ 2 & 3 & 1 \\ -1 & 2 & 1 \end{vmatrix} = 0$ is the equation of a line by expanding the determinant, using the first row.

$$\begin{vmatrix} x & y & 1 \\ 2 & 3 & 1 \\ -1 & 2 & 1 \end{vmatrix} = x\left((-1)^{1+1}\begin{vmatrix} 3 & 1 \\ 2 & 1 \end{vmatrix}\right) + y\left((-1)^{1+2}\begin{vmatrix} 2 & 1 \\ -1 & 1 \end{vmatrix}\right) + 1\left((-1)^{1+3}\begin{vmatrix} 2 & 3 \\ -1 & 2 \end{vmatrix}\right)$$

$$\begin{aligned} &= x[(-1)^2(3\cdot 1 - 2\cdot 1)] + y\{(-1)^3[2\cdot 1 - (-1)1]\} \\ &\qquad + 1\{(-1)^4[2\cdot 2 - (-1)3]\} \\ &= x + y(-1)(3) + 1(1)(7) \\ &= x - 3y + 7 \end{aligned}$$

$x - 3y + 7 = 0$ is in the standard form for the equation of a line. To show that the line passes through (2, 3) and (-1, 2), we show that the coordinates of each point satisfy the equation

$$\begin{aligned} x - 3y + 7 &= 0 \\ 2 - 3(3) + 7 &\overset{?}{=} 0 \\ 2 - 9 + 7 &\overset{\checkmark}{=} 0 \end{aligned} \qquad \begin{aligned} x - 3y + 7 &= 0 \\ -1 - 3(2) + 7 &\overset{?}{=} 0 \\ -1 - 6 + 7 &\overset{\checkmark}{=} 0 \end{aligned}$$

Therefore $\begin{vmatrix} x & y & 1 \\ 2 & 3 & 1 \\ -1 & 2 & 1 \end{vmatrix}$ is the equation of a line that passes through (2, 3) and (-1, 2).

65. The required area is $A = \left|\frac{1}{2}\begin{vmatrix} -1 & 4 & 1 \\ 4 & 8 & 1 \\ 1 & 1 & 1 \end{vmatrix}\right|$ which we expand by the third column.

$$\begin{aligned} A &= \left|\frac{1}{2}\left[1\left((-1)^{1+3}\begin{vmatrix} 4 & 8 \\ 1 & 1 \end{vmatrix}\right) + 1\left((-1)^{2+3}\begin{vmatrix} -1 & 4 \\ 1 & 1 \end{vmatrix}\right) + 1\left((-1)^{3+3}\begin{vmatrix} -1 & 4 \\ 4 & 8 \end{vmatrix}\right)\right]\right| \\ &= \left|\frac{1}{2}\{(-1)^4(4\cdot 1 - 1\cdot 8) + (-1)^5[(-1)\cdot 1 - 1\cdot 4] + (-1)^6[(-1)\cdot 8 - 4\cdot 4]\}\right| \\ &= \left|\frac{1}{2}[(1)(-4) + (-1)(-5) + (1)(-24)]\right| \\ &= \left|\frac{1}{2}[-4 + 5 - 24]\right| \\ &= \left|-\frac{23}{2}\right| \\ &= \frac{23}{2} \end{aligned}$$

67. The value of the determinant $\begin{vmatrix} a_{11} & a_{12} & a_{13} \\ a_{21} & a_{22} & a_{23} \\ a_{31} & a_{32} & a_{33} \end{vmatrix}$

is defined in formula (1). We will expand the determinant by the first row and show that the expression on the right of formula (1) is obtained.

$$a_{11}\left((-1)^{1+1}\begin{vmatrix} a_{22} & a_{23} \\ a_{32} & a_{33} \end{vmatrix}\right) + a_{12}\left((-1)^{1+2}\begin{vmatrix} a_{21} & a_{23} \\ a_{31} & a_{33} \end{vmatrix}\right) + a_{13}\left((-1)^{1+3}\begin{vmatrix} a_{21} & a_{22} \\ a_{31} & a_{32} \end{vmatrix}\right)$$

$$= a_{11}\left[(-1)^2(a_{22}a_{33} - a_{32}a_{23})\right] + a_{12}\left[(-1)^3(a_{21}a_{33} - a_{31}a_{23})\right] + a_{13}\left[(-1)^4(a_{21}a_{32} - a_{31}a_{22})\right]$$

$$= a_{11}(1)(a_{22}a_{33} - a_{32}a_{23}) + a_{12}(-1)(a_{21}a_{33} - a_{31}a_{23}) + a_{13}(1)(a_{21}a_{32} - a_{31}a_{22})$$

$$= \underset{\boxed{1}}{a_{11}a_{22}a_{33}} - \underset{\boxed{2}}{a_{11}a_{32}a_{23}} - \underset{\boxed{3}}{a_{12}a_{21}a_{33}} + \underset{\boxed{4}}{a_{12}a_{31}a_{23}} + \underset{\boxed{5}}{a_{13}a_{21}a_{32}} - \underset{\boxed{6}}{a_{13}a_{31}a_{22}}$$

$$= \underset{\boxed{1}}{a_{11}a_{22}a_{33}} - \underset{\boxed{2}}{a_{11}a_{32}a_{23}} + \underset{\boxed{5}}{a_{21}a_{32}a_{13}} - \underset{\boxed{3}}{a_{21}a_{12}a_{33}} + \underset{\boxed{4}}{a_{31}a_{12}a_{23}} - \underset{\boxed{6}}{a_{31}a_{22}a_{13}}$$

which is the desired expression.

69.
$$D = \begin{vmatrix} 1 & -4 & 9 \\ 4 & -1 & 6 \\ 1 & -1 & 3 \end{vmatrix}$$

We expand the first column.

$$D = a_{11}\begin{pmatrix}\text{cofactor} \\ \text{of } a_{11}\end{pmatrix} + a_{21}\begin{pmatrix}\text{cofactor} \\ \text{of } a_{21}\end{pmatrix} + a_{31}\begin{pmatrix}\text{cofactor} \\ \text{of } a_{31}\end{pmatrix}$$

$$= 1\left((-1)^{1+1}\begin{vmatrix} -1 & 6 \\ -1 & 3 \end{vmatrix}\right) + 4\left((-1)^{2+1}\begin{vmatrix} -4 & 9 \\ -1 & 3 \end{vmatrix}\right) + 1\left((-1)^{3+1}\begin{vmatrix} -4 & 9 \\ -1 & 6 \end{vmatrix}\right)$$

$$= (-1)^2[(-1)\cdot 3 - (-1)6] + 4(-1)^3[(-4)3 - (-1)9] + (-1)^4[(-4)6 - (-1)9]$$

$$= 3 + 4(-1)(-3) + (-15) = 0$$

Since $D = 0$, the system either has no solution or infinitely many. Since $x = 0$, $y = 0$, $z = 0$ is a solution, the second case must hold.

Exercise 8-6 Systems of Linear Inequalities

Key Ideas and Formulas

The graph of a linear equation divides the plane into two half-planes. The solution of a linear inequality is one of the half-planes and a test point will indicate which.

The graph of a system of inequalities is the intersection of the graphs of each inequality in the system, that is, the set of all points that belong to the graphs.

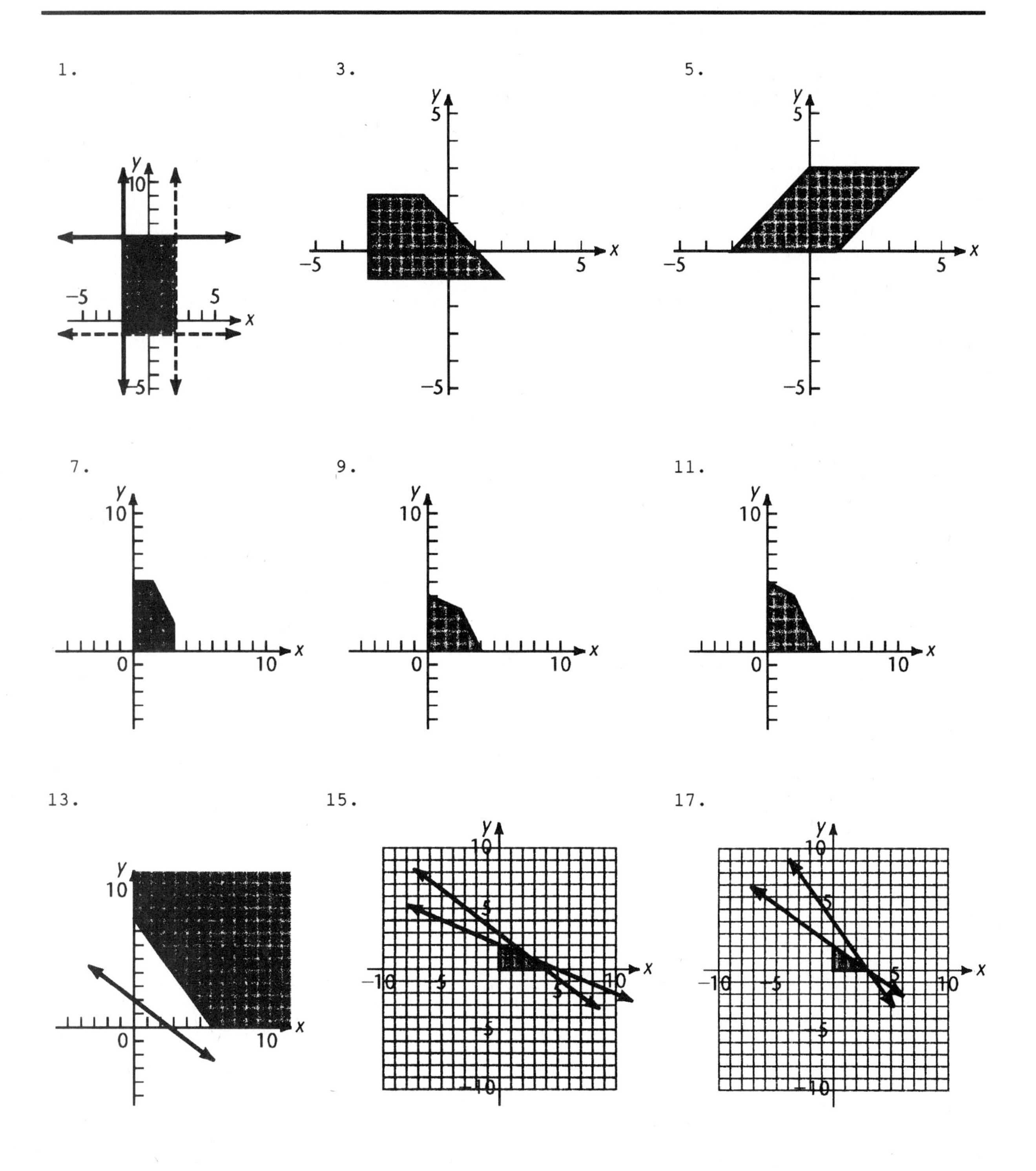

19.

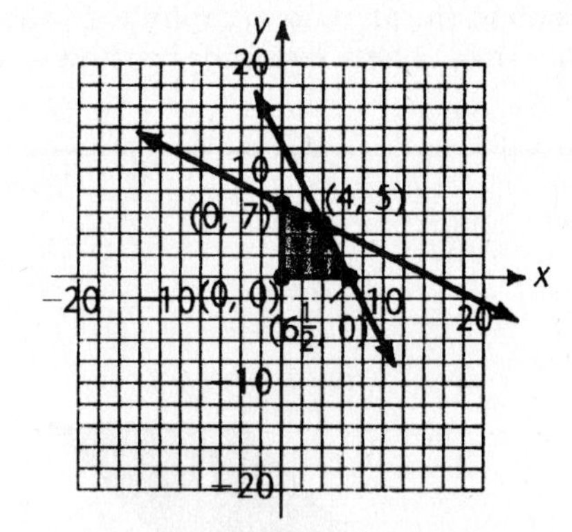

The corners are found by solving

$2x + y = 13$ $x + 2y = 14$	$x + 2y = 14$ $x = 0$	$x = 0$ $y = 0$	$y = 0$ $2x + y = 13$
to get (4, 5)	to get (0, 7)	to get (0, 0)	to get $(6\frac{1}{2}, 0)$

21.

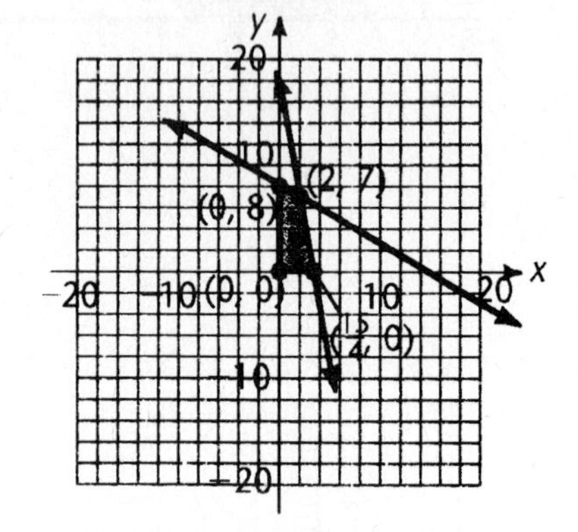

The corners are found by solving

$4x + y = 15$ $x + 2y = 16$	$x + 2y = 16$ $x = 0$	$x = 0$ $y = 0$	$y = 0$ $4x + y = 15$
to get (2, 7)	to get (0, 8)	to get (0, 0)	to get $(3\frac{3}{4}, 0)$

23.

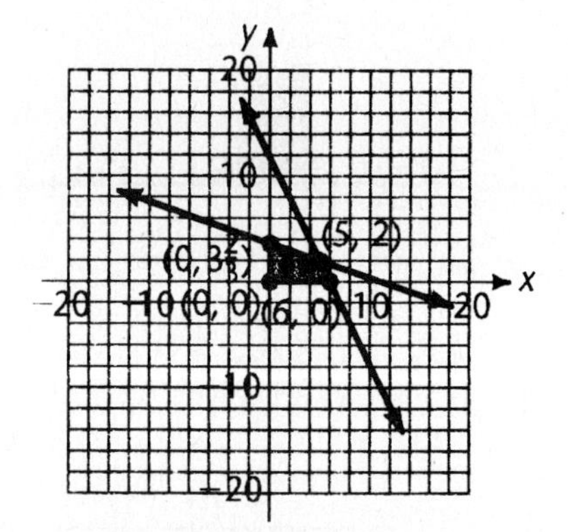

The corners are found by solving

$x + 3y = 11$ $2x + y = 12$	$x + 3y = 11$ $x = 0$	$x = 0$ $y = 0$	$y = 0$ $2x + y = 12$
to get (5, 2)	to get $(0, 3\frac{2}{3})$	to get (0, 0)	to get (6, 0)

25.

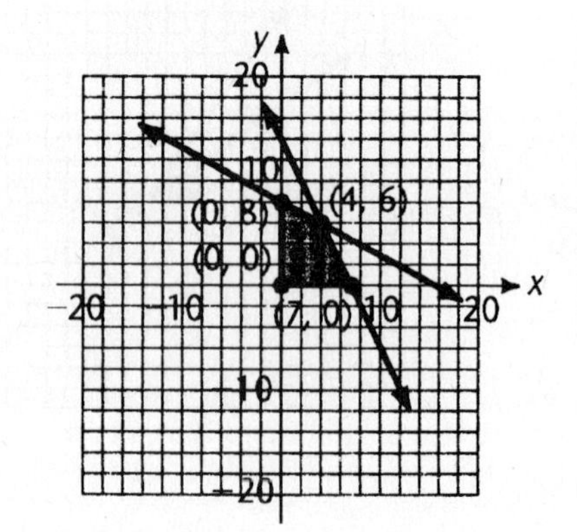

The corners are found by solving

$2x + y = 14$ $x + 2y = 16$	$x + 2y = 16$ $x = 0$	$x = 0$ $y = 0$	$y = 0$ $2x + y = 14$
to get (4, 6)	to get (0, 8)	to get (0, 0)	to get (7, 0)

27.

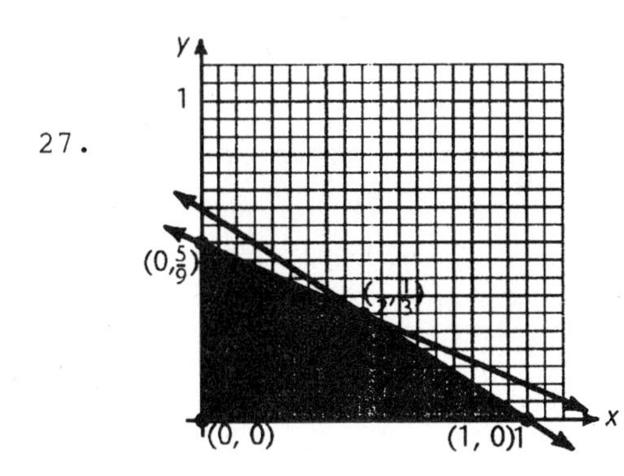

The corners are found by solving

$2x + 3y = 2$	$4x + 9y = 5$	$x = 0$	$y = 0$
$4x + 9y = 5$	$x = 0$	$y = 0$	$2x + 3y = 2$
to get	to get	to get	to get
$\left(\frac{1}{2}, \frac{1}{3}\right)$	$\left(0, \frac{5}{9}\right)$	$(0, 0)$	$(1, 0)$

29.

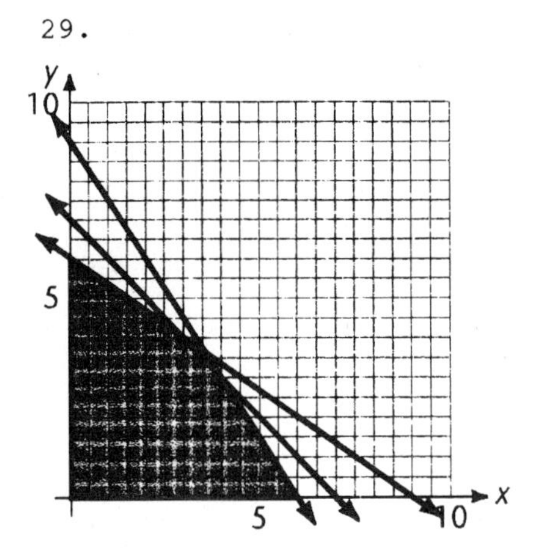

31.

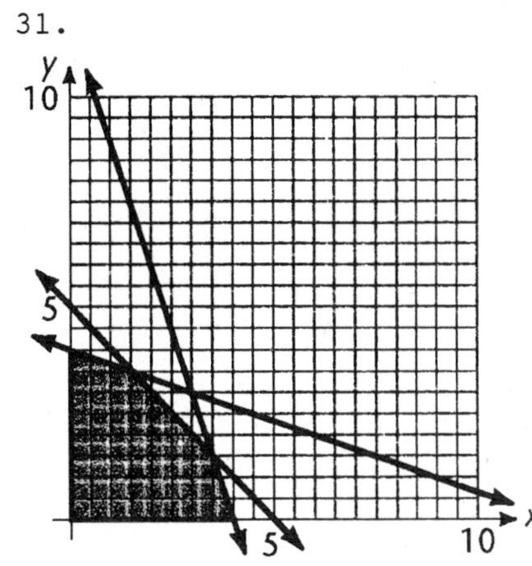

33.

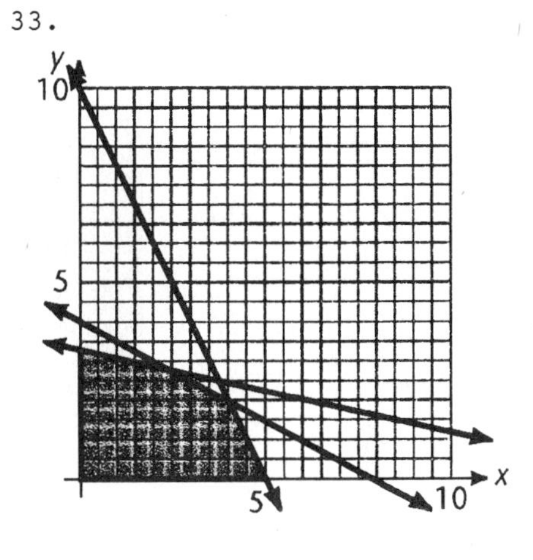

35.

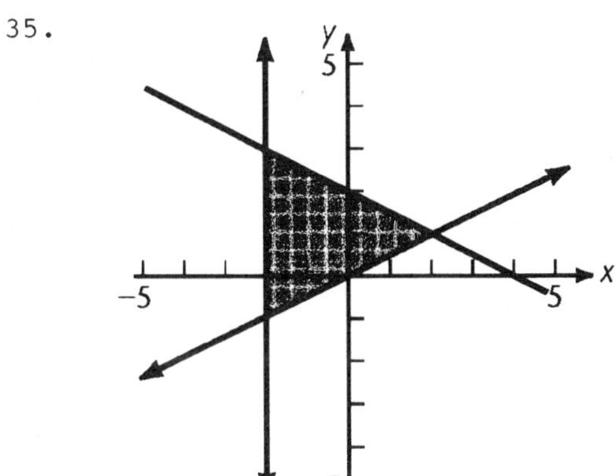

37.

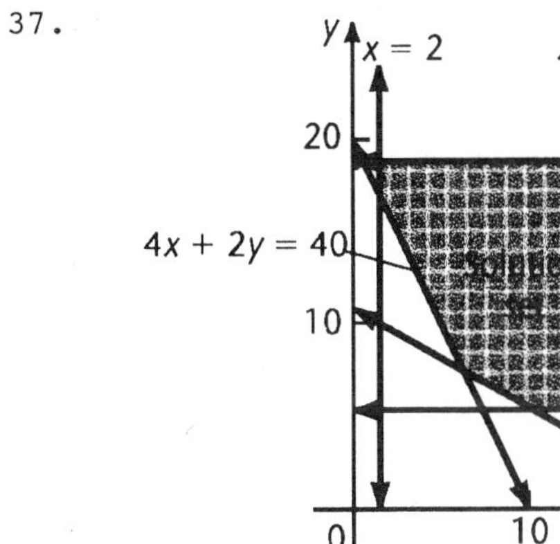

39.

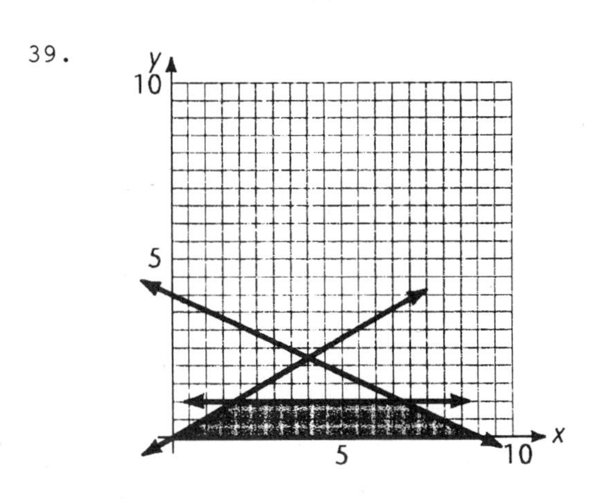

41.

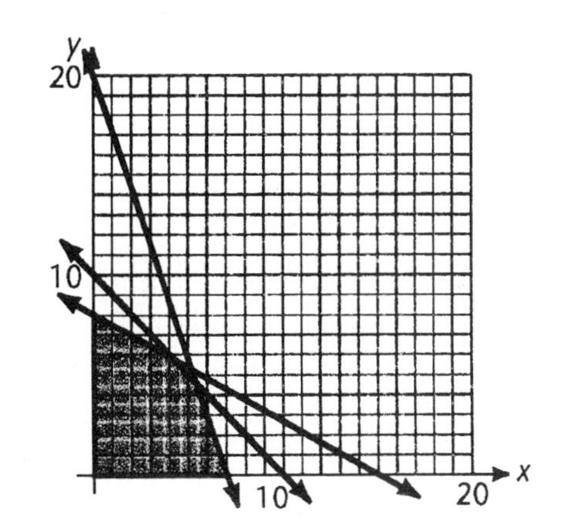

43. Let x = number of trick skis produced per day
y = number of slalom skis produced per day
Clearly x and y must be positive. Hence

$x \geq 0$ (1)

$y \geq 0$ (2)

To fabricate x trick skis requires 6x hours. To fabricate y slalom skis requires 4y hours. 108 hours are available for fabricating; hence

$6x + 4y \leq 108$ (3)

To finish x trick skis requires 1x hours. To finish y slalom skis requires 1y hours. 24 hours are available for finishing, hence

$x + y \leq 24$ (4)

Graphing the inequality system (1), (2), (3), (4), we have

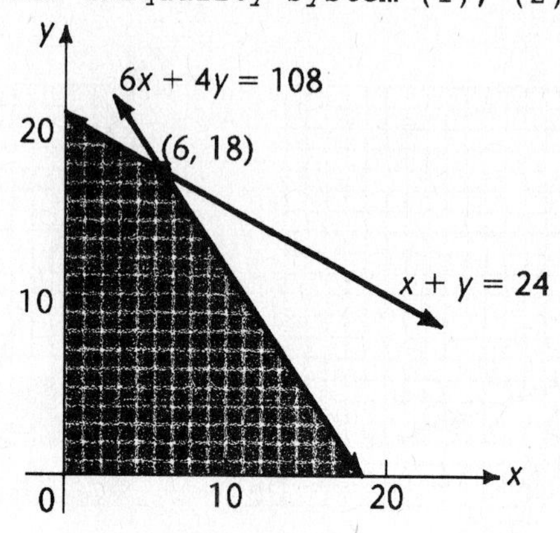

45. We can find the most profitable solution by comparing the profits at each of the corner points.

Corner	Profit = 15x + 12y	
(6, 18)	$15 \cdot 6 + 12 \cdot 18 = 306$	Most profitable combination
(0, 24)	$15 \cdot 0 + 12 \cdot 24 = 288$	
(0, 0)	$15 \cdot 0 + 12 \cdot 0 = 0$	
(18, 0)	$15 \cdot 18 + 12 \cdot 0 = 270$	

The most profitable combination is 6 trick skis, 18 slalom skis.

47. Let x = number of ounces of food M
y = number of ounces of food N
Clearly x and y must be positive. Hence

$x \geq 0$ (1)

$y \geq 0$ (2)

The amount of calcium in the diet must be at least 360 units, hence

$30x + 10y \geq 360$ (3)

The amount of iron in the diet must be at least 160 units, hence

$10x + 10y \geq 160$ (4)

The amount of vitamin A in the diet must be at least 240 units, hence

$10x + 30y \geq 240$ (5)

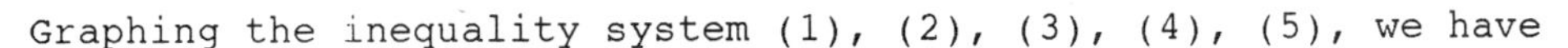

Graphing the inequality system (1), (2), (3), (4), (5), we have

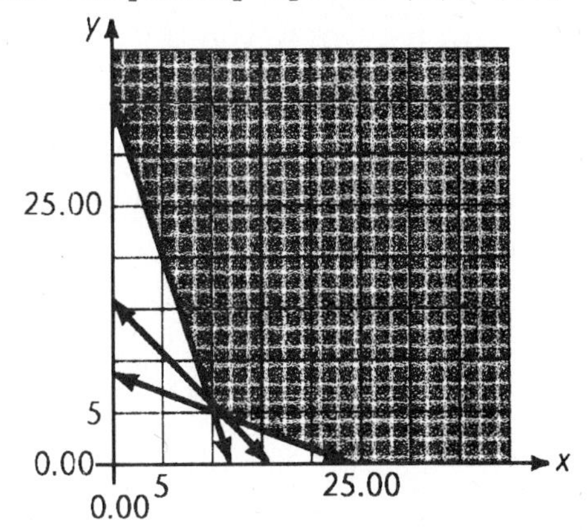

49. We can find the least costly solution by comparing the costs at each of the corner points.

Corner	Cost = 8x + 10y	
(36, 0)	$8 \cdot 36 + 10 \cdot 0 = 288$	
(10, 6)	$8 \cdot 10 + 10 \cdot 6 = 140$	
(12, 4)	$8 \cdot 12 + 10 \cdot 4 = 136$	Least costly combination
(0, 24)	$8 \cdot 0 + 10 \cdot 24 = 240$	

The least costly combination is 12 ounces of M, 4 ounces of N.

Exercise 8-7 Systems Involving Second Degree Equations

Key Ideas and Formulas

The substitution, and, occasionally, the elimination using addition methods of solving systems of two equations can be applied when the systems contain second degree equations. In this case, there may be no solution, one solution, two, three, or four solutions, or possibly an infinite number of solutions. Solutions may also involve non-real complex numbers.

---

1. $x^2 + y^2 = 25$
   $y = -4$
   Substitute $y = -4$ into $x^2 + y^2 = 25$.
   $x^2 + (-4)^2 = 25$
   $x^2 + 16 = 25$
   $x^2 = 9$
   $x = \pm 3$
   $(-3, -4)$, $(3, -4)$

3. $x^2 + y^2 = 36$
   $x = 4$
   Substitute $x = 4$ into $x^2 + y^2 = 36$.
   $4^2 + y^2 = 36$
   $16 + y^2 = 36$
   $y^2 = 20$
   $y = \pm 2\sqrt{5}$
   $(4, 2\sqrt{5})$, $(4, -2\sqrt{5})$

5. $y^2 = 2x$

$x = y - \frac{1}{2}$

Substitute $x = y - \frac{1}{2}$ for x in $y^2 = 2x$.

$$y^2 = 2\left(y - \frac{1}{2}\right)$$
$$y^2 = 2y - 1$$
$$y^2 - 2y + 1 = 0$$
$$(y - 1)^2 = 0$$
$$y = 1$$

For $y = 1$, $x = y - \frac{1}{2} = 1 - \frac{1}{2} = \frac{1}{2}$.

$\left(\frac{1}{2}, 1\right)$

7. $x^2 - y^2 = 9$

$x - y = 3$

Solve $x - y = 3$ for x in terms of y.

$x = y + 3$

Substitute this expression for x in $x^2 - y^2 = 9$.

$$(y + 3)^2 - y^2 = 9$$
$$y^2 + 6y + 9 - y^2 = 9$$
$$6y + 9 = 9$$
$$6y = 0$$
$$y = 0$$

For $y = 0$, $x = y + 3 = 0 + 3 = 3$

(3, 0)

9. $x^2 + 4y^2 = 32$

$x + 2y = 0$

Solve $x + 2y = 0$ for x in terms of y.

$x = -2y$

Substitute this expression for x in $x^2 + 4y^2 = 32$.

$$(-2y)^2 + 4y^2 = 32$$
$$4y^2 + 4y^2 = 32$$
$$8y^2 = 32$$
$$y^2 = 4$$
$$y = \pm 2$$

For $y = 2$, $x = -2(2) = -4$

For $y = -2$, $x = -2(-2) = 4$.

(-4, 2), (4, -2)

11. $\frac{x^2}{9} + y^2 = 1$

$x + y = 1$

Solve $x + y = 1$ for x in terms of y.

$x = 1 - y$

Substitute this expression for x in $\frac{x^2}{9} + y^2 = 1$.

$$\frac{(1 - y)^2}{9} + y^2 = 1 \quad \text{lcm} = 9$$
$$9 \cdot \frac{(1 - y)^2}{9} + 9y^2 = 9$$
$$(1 - y)^2 + 9y^2 = 9$$
$$1 - 2y + y^2 + 9y^2 = 9$$
$$1 - 2y + 10y^2 = 9$$
$$10y^2 - 2y - 8 = 0$$
$$5y^2 - y - 4 = 0$$

Solve by factoring.

$$(5y + 4)(y - 1) = 0$$

$5y + 4 = 0$ or $y - 1 = 0$

$y = -\frac{4}{5}$ $\qquad$ $y = 1$

$x = 1 - \left(-\frac{4}{5}\right)$ $\qquad$ $x = 1 - 1$

$x = \frac{9}{5}$ $\qquad$ $x = 0$

$\left(\frac{9}{5}, -\frac{4}{5}\right)$ $\qquad$ (0, 1)

13. $x^2 = 2y$

$3x = y + 5$

Solve $3x = y + 5$ for y in terms of x.

$3x - 5 = y$

$y = 3x - 5$

Substitute this expression for y in $x^2 = 2y$.

$$x^2 = 2(3x - 5)$$
$$x^2 = 6x - 10$$
$$x^2 - 6x + 10 = 0$$

This equation is not factorable, so we use the quadratic formula with $a = 1$, $b = -6$, $c = 10$.

$$\begin{aligned} x &= \frac{-(-6) \pm \sqrt{(-6)^2 - 4(1)(10)}}{2(1)} \\ &= \frac{6 \pm \sqrt{36 - 40}}{2} \\ &= \frac{6 \pm \sqrt{-4}}{2} \\ &= \frac{6 \pm 2i}{2} \end{aligned}$$

$x = 3 \pm 2i$

For $x = 3 + i$, $y = 3(3 + i) - 5 = 9 + 3i - 5 = 4 + 3i$

For $x = 3 - i$, $y = 3(3 - i) - 5 = 9 - 3i - 5 = 4 - 3i$

$(3 + i, 4 + 3i)$, $(3 - i, 4 - 3i)$

15.

$$\begin{aligned} x^2 - y^2 &= 3 \\ x^2 + y^2 &= 5 \\ \hline 2x^2 &= 8 \\ x^2 &= 4 \\ x &= \pm 2 \end{aligned}$$

We have added to eliminate y.

For $x = 2$:

$$\begin{aligned} x^2 + y^2 &= 5 \\ (2)^2 + y^2 &= 5 \\ 4 + y^2 &= 5 \\ y^2 &= 1 \\ y &= \pm 1 \end{aligned}$$

For $x = -2$:

$$\begin{aligned} x^2 + y^2 &= 5 \\ (-2)^2 + y^2 &= 5 \\ 4 + y^2 &= 5 \\ y^2 &= 1 \\ y &= \pm 1 \end{aligned}$$

$(2, 1)$, $(2, -1)$, $(-2, 1)$, $(-2, -1)$

17.

$$\begin{aligned} x^2 - 2y^2 &= 1 \\ x^2 + 4y^2 &= 25 \end{aligned}$$

We multiply the top equation by 2 and add to eliminate y.

$$\begin{aligned} 2x^2 - 4y^2 &= 2 \\ x^2 + 4y^2 &= 25 \\ \hline 3x^2 &= 27 \\ x^2 &= 9 \\ x &= \pm 3 \end{aligned}$$

For $x = 3$:

$$\begin{aligned} x^2 + 4y^2 &= 25 \\ (3)^2 + 4y^2 &= 25 \\ 9 + 4y^2 &= 25 \\ 4y^2 &= 16 \\ y^2 &= 4 \\ y &= \pm 2 \end{aligned}$$

For $x = -3$:

$$\begin{aligned} x^2 + 4y^2 &= 25 \\ (-3)^2 + 4y^2 &= 25 \\ 9 + 4y^2 &= 25 \\ 4y^2 &= 16 \\ y^2 &= 4 \\ y &= \pm 2 \end{aligned}$$

$(3, 2)$, $(3, -2)$, $(-3, 2)$, $(-3, -2)$

19. The graph of $x^2 + y^2 = 25$ is a circle with center $(0, 0)$ and radius 5.

The graph of $y = -4$ is a horizontal straight line.

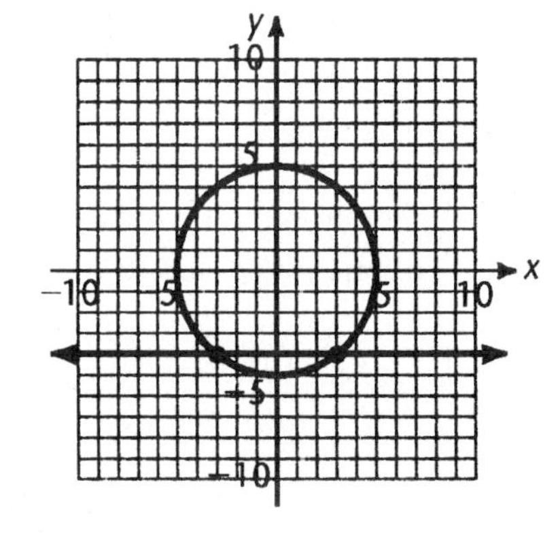

21. The graph of $x^2 + y^2 = 36$ is a circle with center $(0, 0)$ and radius 6.

    The graph of $x = 4$ is a vertical straight line.

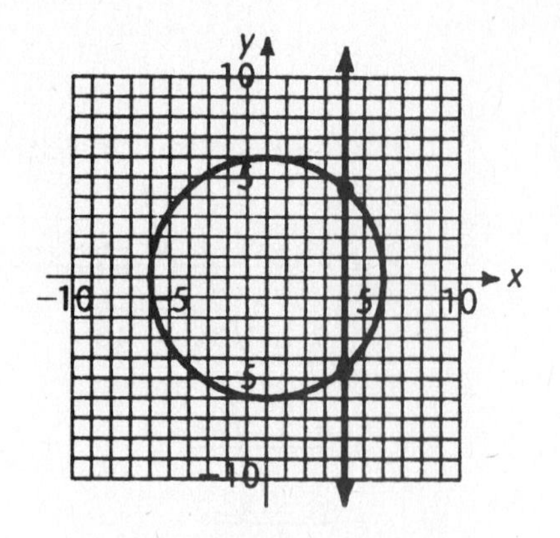

23. The graph of $y^2 = 2x$ is a parabola with vertex $(0, 0)$, opening right.

    The graph of $x = y - \frac{1}{2}$ is a straight line with slope 1.

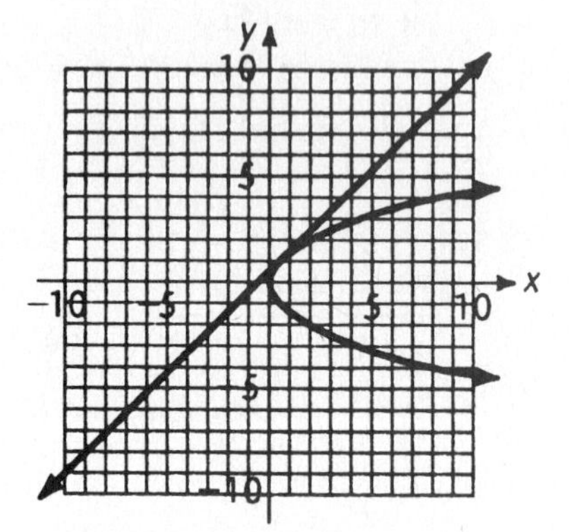

25. The graph of $x^2 - y^2 = 9$ is a hyperbola in standard position with intercepts $x = \pm 3$, opening right and left.

    The graph of $x - y = 3$ is a straight line with slope 1.

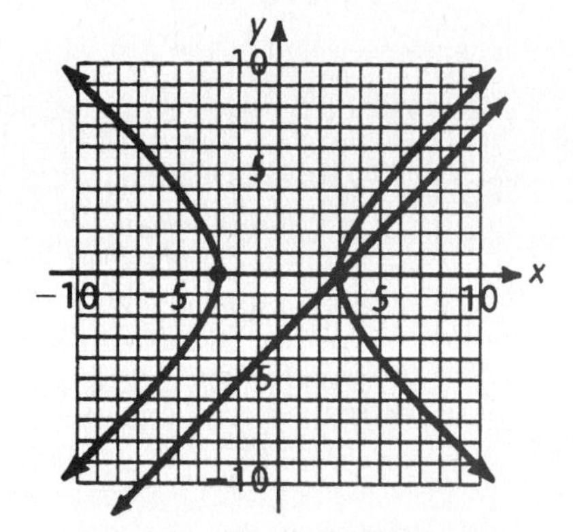

27. The graph of $x^2 + 4y^2 = 32$ is an ellipse in standard position with intercepts $x = \pm 4\sqrt{2}$ and $y = \pm 2\sqrt{2}$.

    The graph of $x + 2y = 0$ is a straight line with slope $-\frac{1}{2}$.

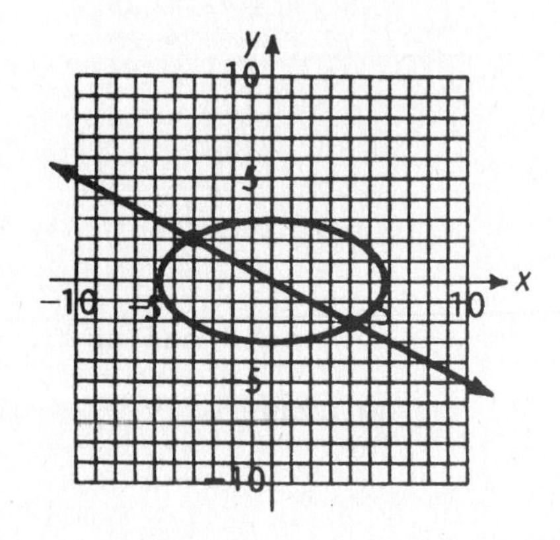

29. $y = x^2$
$x - y = 3$
Substitute $y = x^2$ into $x - y = 3$.
$x - x^2 = 3$
$0 = x^2 - x + 3$
This equation is not factorable, so we use the quadratic formula with $a = 1$, $b = -1$, and $c = 3$.

$$x = \frac{-(-1) \pm \sqrt{(-1)^2 - 4(1)(3)}}{2(1)}$$
$$= \frac{1 \pm \sqrt{1 - 12}}{2}$$
$$= \frac{1 \pm \sqrt{-11}}{2}$$
$$= \frac{1 \pm i\sqrt{11}}{2}$$

Since there are only complex solutions, we conclude that there are no points of intersection.

The graph of $y = x^2$ is a parabola with vertex (0, 0), opening up.

The graph of $x - y = 3$ is a straight line with slope 1.

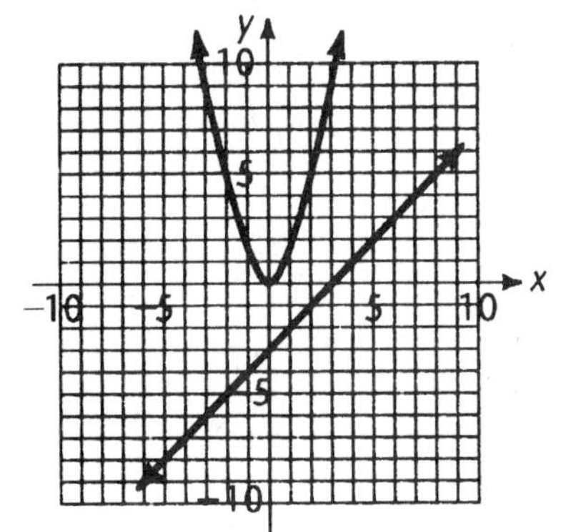

There are no real points of intersection.

31. $x^2 + \frac{y^2}{25} = 1$
$x - y = 8$
Solve $x - y = 8$ for x in terms of y
$x = y + 8$
Substitute this expression for x in $x^2 + \frac{y^2}{25} = 1$.

$$(y + 8)^2 + \frac{y^2}{25} = 1 \quad \text{lcm} = 25$$
$$25(y + 8)^2 + 25\left(\frac{y^2}{25}\right) = 25$$
$$25(y^2 + 16y + 64) + y^2 = 25$$
$$25y^2 + 400y + 1600 + y^2 = 25$$
$$26y^2 + 400y + 1575 = 0$$

This equation is not factorable, so we use the quadratic formula with $a = 26$, $b = 400$, $c = 1575$.

$$y = \frac{-400 \pm \sqrt{(400)^2 - 4(26)(1575)}}{2(24)}$$
$$y = \frac{-400 \pm \sqrt{160{,}000 - 163{,}800}}{48}$$
$$y = \frac{-400 \pm \sqrt{-3{,}800}}{48}$$
$$y = \frac{-400 \pm 2i\sqrt{950}}{48}$$
$$y = \frac{-200 \pm i\sqrt{950}}{24}$$

Since there are only complex solutions, we conclude that there are no points of intersection.

The graph of $x^2 + \frac{y^2}{25} = 1$ is an ellipse in standard position with intercepts $x = \pm 1$, $y = \pm 5$.

The graph of $x - y = 8$ is a straight line with slope 1.

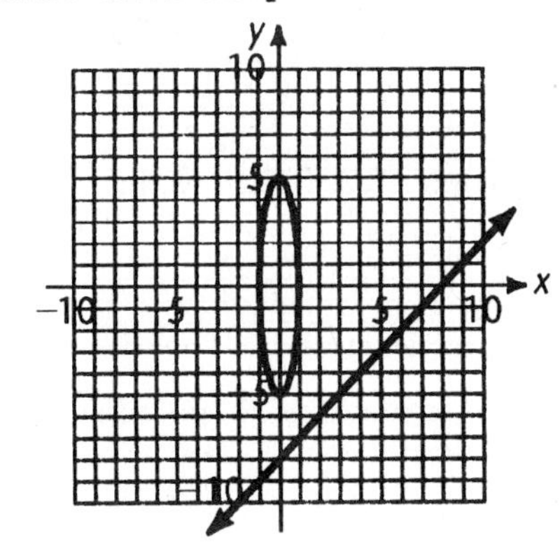

There are no real points of intersection.

33. $x^2 + \frac{y^2}{4} = 1$

$\frac{x^2}{4} - y^2 = 1$

We multiply the top equation by 16 and the bottom by 4 and add to eliminate y.

$$\begin{array}{rl} 16x^2 + 4y^2 &= 16 \\ x^2 - 4y^2 &= 4 \\ \hline 17x^2 &= 20 \end{array}$$

$$x^2 = \frac{20}{17}$$

$$x = \pm\sqrt{\frac{20}{17}} = \pm\frac{2\sqrt{85}}{17}$$

For either value of x:

$$\frac{20}{17} + \frac{y^2}{4} = 1$$

$$\frac{20}{17} + \frac{y^2}{4} = \frac{17}{17}$$

$$\frac{y^2}{4} = -\frac{3}{17}$$

$$y^2 = -\frac{12}{17}$$

$$y = \pm\sqrt{-\frac{12}{17}} = \pm\frac{2i\sqrt{51}}{17}$$

Since there are only complex solutions, we conclude that there are no ponts of intersection.

The graph of $x^2 + \frac{y^2}{4} = 1$ is an ellipse with intercepts $x = \pm1$, $y = \pm2$.

The graph of $\frac{x^2}{4} - y^2 = 1$ is a hyperbola with intercepts $x = \pm2$, opening left and right.

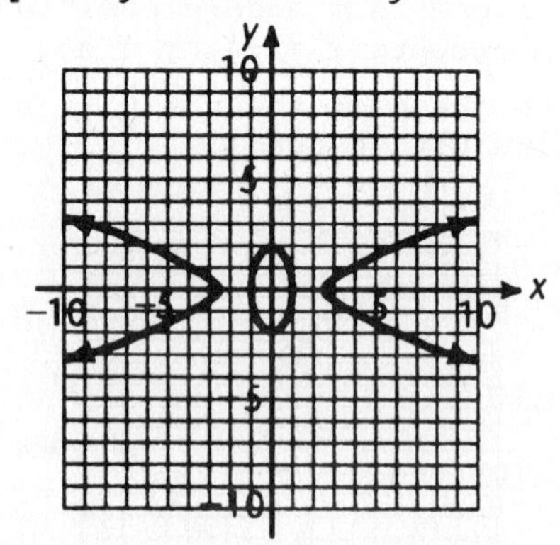

There are no real points of intersection.

35. $x^2 + y^2 = 36$

$\frac{x^2}{25} + \frac{y^2}{16} = 1$

We multiply the top equation by -25 and the bottom by 400 and add to eliminate y.

$$\begin{array}{rl} -25x^2 - 25y^2 &= -900 \\ 16x^2 + 25y^2 &= 400 \\ \hline -9x^2 &= -500 \end{array}$$

$$x^2 = \frac{500}{9}$$

$$x = \pm\sqrt{\frac{500}{9}} = \pm\frac{10\sqrt{5}}{3}$$

For either value of x

$$\frac{500}{9} + y^2 = 36$$

$$\frac{500}{9} + y^2 = \frac{324}{9}$$

$$y^2 = -\frac{176}{9}$$

$$y = \pm\sqrt{-\frac{176}{9}} = \pm\frac{4i\sqrt{11}}{3}$$

Since there are only complex solutions, we conclude ther there are no points of intersection.

The graph of $x^2 + y^2 = 36$ is a circle with center (0, 0) and radius 6.

The graph of $\frac{x^2}{25} + \frac{y^2}{16} = 1$ is an ellipse with intercepts $x = \pm5$ and $y = \pm4$.

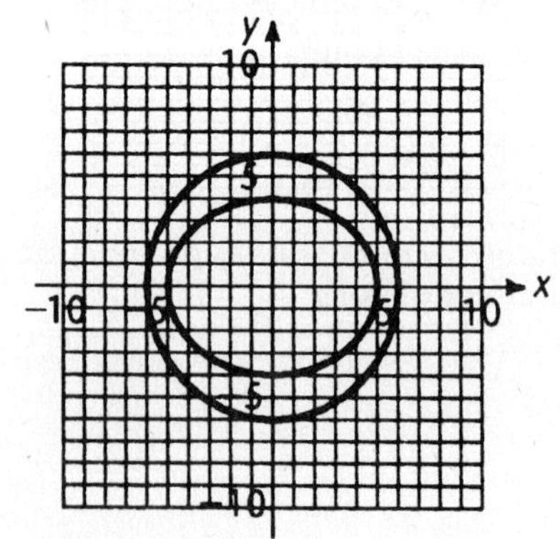

There are no real points of intersection.

37. $xy - 6 = 0$
$x - y = 4$
Solve the bottom equation for x in terms of y.
$x = y + 4$
Substitute this expression for x in the top equation.
$(y + 4)y - 6 = 0$
$y^2 + 4y - 6 = 0$
This equation is not factorable, so we use the quadratic formula with a = 1, b = 4, c = -6.

$$y = \frac{-(4) \pm \sqrt{(4)^2 - 4(1)(-6)}}{2(1)}$$
$$= \frac{-4 \pm \sqrt{16 + 24}}{2}$$
$$= \frac{-4 \pm \sqrt{40}}{2}$$
$$= \frac{-4 \pm 2\sqrt{10}}{2}$$
$$= -2 \pm \sqrt{10}$$

For $y = -2 + \sqrt{10}$, $x = y + 4 = -2 + \sqrt{10} + 4 = 2 + \sqrt{10}$
For $y = -2 - \sqrt{10}$, $x = y + 4 = -2 - \sqrt{10} + 4 = 2 - \sqrt{10}$
$(2 + \sqrt{10}, -2 + \sqrt{10})$, $(2 - \sqrt{10}, -2 - \sqrt{10})$

39. $x^2 + xy - y^2 = -5$
$y - x = 3$
Solve the bottom equation for y in terms of x.
$y = x + 3$
Substitute this expression for y in the top equation.
$$x^2 + x(x + 3) - (x + 3)^2 = -5$$
$$x^2 + x^2 + 3x - (x^2 + 6x + 9) = -5$$
$$x^2 + x^2 + 3x - x^2 - 6x - 9 = -5$$
$$x^2 - 3x - 9 = -5$$
$$x^2 - 3x - 4 = 0$$

Solve by factoring.
$(x - 4)(x + 1) = 0$
$x - 4 = 0 \qquad x + 1 = 0$
$x = 4 \qquad x = -1$
For $x = 4$, $y = x + 3 = 4 + 3 = 7$
For $x = -1$, $y = x + 3 = -1 + 3 = 2$
(4, 7), (-1, 2)

41. $2x^2 - 3y^2 = 10$ (1)
$x^2 + 4y^2 = -17$ (2)
We multiply the bottom equation by -2 and add to eliminate x.
$2x^2 - 3y^2 = 10$ (1)
$-2x^2 - 8y^2 = 34$ -2[equation (2)]
$-11y^2 = 44$
$y^2 = -4$
$y = \pm 2i$

For $y = 2i$

$$x^2 + 4(2i)^2 = -17$$
$$x^2 + 4(-4) = -17$$
$$x^2 - 16 = -17$$
$$x^2 = 1$$
$$x = \pm i$$

For $y = -2i$:

$$x^2 + 4(-2i)^2 = -17$$
$$x^2 + 4(-4) = -17$$
$$x^2 - 16 = -17$$
$$x^2 = -1$$
$$x = \pm i$$

$(i, 4i)$, $(i, -4i)$, $(-i, 4i)$, $(-i, -4i)$

43. $x^2 + y^2 = 20$

$x^2 = y$

Substitute $y = x^2$ from the second equation into the first equation.

$$y + y^2 = 20$$
$$y^2 + y - 20 = 0$$
$$(y + 5)(y - 4) = 0$$
$$y + 5 = 0 \quad \text{or} \quad y - 4 = 0$$
$$y = -5 \qquad\qquad y = 4$$

For $y = -5$:

$$x^2 = -5$$
$$x = \pm\sqrt{-5}$$
$$x = \pm i\sqrt{5}$$

For $y = 4$:

$$x^2 = 4$$
$$x = \pm 2$$

$(i\sqrt{5}, -5)$, $(-i\sqrt{5}, 5)$, $(2, 4)$, $(-2, 4)$

45. $x^2 + y^2 = 16$

$y^2 = 4 - x$

Substitute $y^2 = 4 - x$ from the second equation into the first equation.

$$x^2 + 4 - x = 16$$
$$x^2 - x + 4 = 16$$
$$x^2 - x - 12 = 0$$
$$(x - 4)(x + 3) = 0$$
$$x - 4 = 0 \quad \text{or} \quad x + 3 = 0$$
$$x = 4 \qquad\qquad x = -3$$

For $x = 4$:

$$y^2 = 4 - x$$
$$y^2 = 4 - 4$$
$$y^2 = 0$$
$$y = 0$$

For $x = -3$:

$$y^2 = 4 - x$$
$$y^2 = 4 - (-3)$$
$$y^2 = 7$$
$$y = \pm\sqrt{7}$$

$(4, 0)$, $(-3, \sqrt{7})$, $(-3, -\sqrt{7})$

47. a b

Let a = width of rectangle

b = length of rectangle

From the area formula we have $ab = 32$. From the perimeter formula we have $2a + 2b = 36$. To solve the system

$$ab = 32$$
$$2a + 2b = 36$$

we solve the second equation for a in terms of b.

$$2a = 36 - 2b$$
$$a = 18 - b$$

Now we substitute this expression for a into the first equation.

$$(18 - b)b = 32$$
$$18b - b^2 = 32$$
$$0 = b^2 - 18b + 32$$
$$b^2 - 18b + 32 = 0$$
$$(b - 16)(b - 2) = 0$$
$$b - 16 = 0 \quad \text{or} \quad b - 2 = 0$$
$$b = 16 \qquad b = 2$$

For $b = 16$:
$$a = 18 - b$$
$$a = 18 - 16$$
$$a = 2$$

For $b = 2$:
$$a = 18 - b$$
$$a = 18 - 2$$
$$a = 16$$

The dimensions are 2 feet by 16 feet.

49.
$$2x^2 + y^2 = 18$$
$$xy = 4$$

Solve the bottom equation for y in terms of x.

$$y = \frac{4}{x}$$

Substitute this expression for y in the top equation.

$$2x^2 + \left(\frac{4}{x}\right)^2 = 18$$
$$2x^2 + \frac{16}{x^2} = 18 \qquad \text{lcd} = x^2,\ x \neq 0$$

Multiply both sides of this equation by x^2.

$$2x^4 + 16 = 18x^2$$
$$2x^4 - 18x^2 + 16 = 0$$
$$x^4 - 9x^2 + 8 = 0$$

The equation is quadratic in x^2. Let $u = x^2$, then $u^2 = x^4$.

$$u^2 - 9u + 8 = 0$$
$$(u - 8)(u - 1) = 0$$
$$u - 8 = 0 \quad \text{or} \quad u - 1 = 0$$
$$u = 8 \qquad u = 1$$

Replacing u by x^2, we obtain

$$x^2 = 8 \quad \text{or} \quad x^2 = 1$$
$$x = \pm\sqrt{8} \qquad x = \pm 1$$
$$x = \pm 2\sqrt{2}$$

For $x = 2\sqrt{2}$: $y = \dfrac{4}{2\sqrt{2}} = \dfrac{4\sqrt{2}}{2\sqrt{2}\sqrt{2}} = \dfrac{4\sqrt{2}}{2 \cdot 2} = \sqrt{2}$

For $x = -2\sqrt{2}$: $y = \dfrac{4}{-2\sqrt{2}} = \dfrac{4\sqrt{2}}{-2\sqrt{2}\sqrt{2}} = \dfrac{4\sqrt{2}}{-2 \cdot 2} = \dfrac{4\sqrt{2}}{-4} = -\sqrt{2}$

For $x = 1$: $y = \dfrac{4}{1} = 4$

For $x = -1$: $y = \dfrac{4}{-1} = -4$

$(2\sqrt{2}, \sqrt{2})$, $(-2\sqrt{2}, -\sqrt{2})$, $(1, 4)$, $(-1, -4)$

51. $x^2 + 2xy + y^2 = 36$ (1)

$x^2 - xy = 0$ (2)

Factor equation (2).

$x(x - y) = 0$

$x = 0$ or $x - y = 0$

$x = y$

For $x = 0$ [replace x with 0 in equation (1) and solve for y]:

$(0)^2 + (0)y + y^2 = 36$

$y^2 = 36$

$y = \pm 6$

For $x = y$ [replace x with y in equation (1) and solve for y]:

$y^2 + 2yy + y^2 = 36$

$4y^2 = 36$

$y^2 = 9$

$y = \pm 3$

For $y = 3$, $x = 3$; for $y = -3$, $x = -3$.

(0, 6), (0, -6), (3, 3), (-3, -3)

53. $x^2 - 2xy + 2y^2 = 16$ (1)

$x^2 - y^2 = 0$ (2)

Factor equation (2).

$(x - y)(x + y) = 0$

$x - y = 0$ or $x + y = 0$

$x = y$ $\quad$ $x = -y$

For $x = y$ [replace x with y in equation (1) and solve for y]:

$y^2 - 2yy + 2y^2 = 16$

$y^2 = 16$

$y = \pm 4$

For $y = 4$, $x = 4$; for $y = -4$, $x = -4$.

For $x = -y$ [replace x with -y in equation (1) and solve for y]:

$(-y)^2 - 2(-y)y + 2y^2 = 16$

$y^2 + 2y^2 + 2y^2 = 16$

$5y^2 = 16$

$y^2 = \frac{16}{5}$

$y = \pm\sqrt{\frac{16}{5}} = \pm\frac{\sqrt{16}}{\sqrt{5}} = \pm\frac{4}{\sqrt{5}} = \pm\frac{4\sqrt{5}}{\sqrt{5}\sqrt{5}} = \pm\frac{4\sqrt{5}}{5}$ or $\pm\frac{4}{5}\sqrt{5}$

For $y = \frac{4}{5}\sqrt{5}$, $x = -\frac{4}{5}\sqrt{5}$; for $y = -\frac{4}{5}\sqrt{5}$, $x = \frac{4}{5}\sqrt{5}$

$\left(\frac{4}{5}\sqrt{5}, -\frac{4}{5}\sqrt{5}\right)$, $\left(-\frac{4}{5}\sqrt{5}, \frac{4}{5}\sqrt{5}\right)$, (4, 4), (-4, -4)

Exercise 8-8 REVIEW EXERCISE

1.

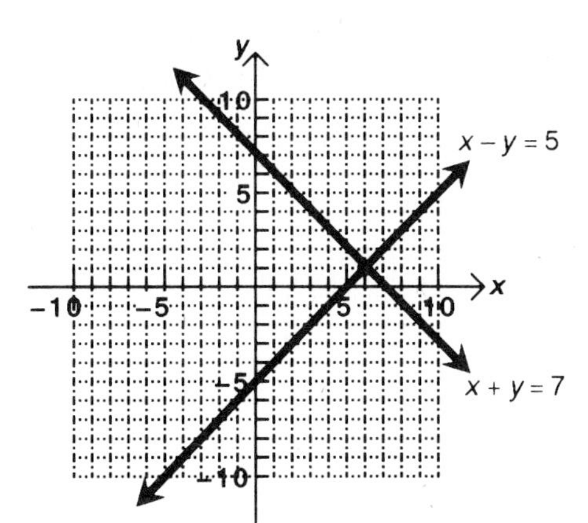

The graphs appear to intersect at (6, 1).

Check: $x - y = 5$ $\quad$ $x + y = 7$

$6 - 1 \stackrel{?}{=} 5$ $\quad$ $6 + 1 \stackrel{?}{=} 7$

$5 \stackrel{\checkmark}{=} 5$ $\quad$ $7 \stackrel{\checkmark}{=} 7$

Solution: (6, 1)

2.

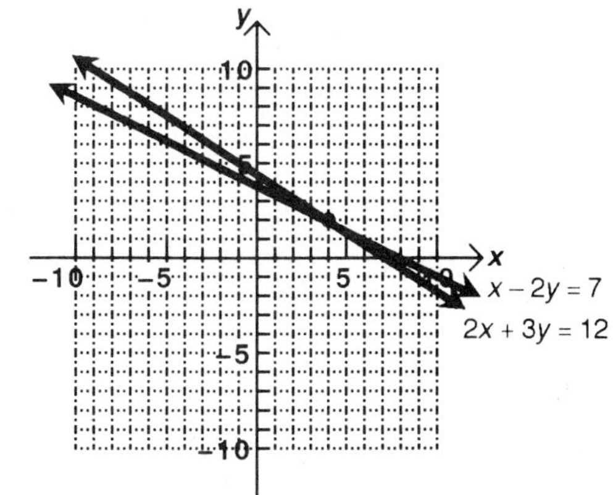

The graphs appear to intersect at (3, 2).

Check: $x + 2y = 7$ $\quad$ $2x + 3y = 12$

$3 + 2(2) \stackrel{?}{=} 7$ $\quad$ $2(3) + 3(2) \stackrel{?}{=} 12$

$7 \stackrel{\checkmark}{=} 7$ $\quad$ $6 + 6 \stackrel{\checkmark}{=} 12$

Solution: (3, 2)

3.

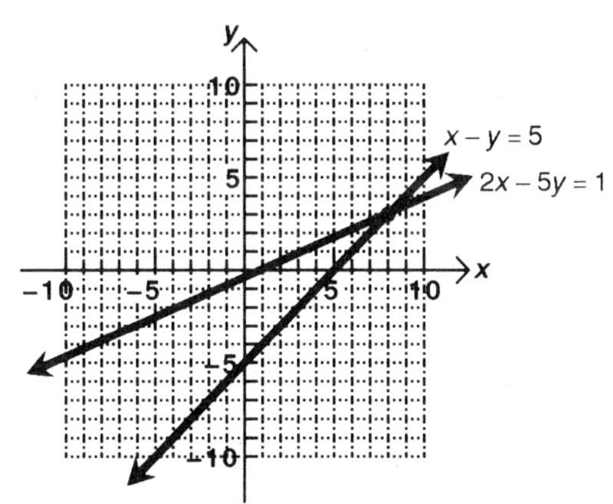

The graphs appear to intersect at (8, 3).

Check: $x - y = 5$ $\quad$ $2x - 5y = 1$

$8 - 3 \stackrel{\checkmark}{=} 5$ $\quad$ $2(8) - 5(3) \stackrel{?}{=} 1$

$16 - 15 \stackrel{\checkmark}{=} 1$

4.

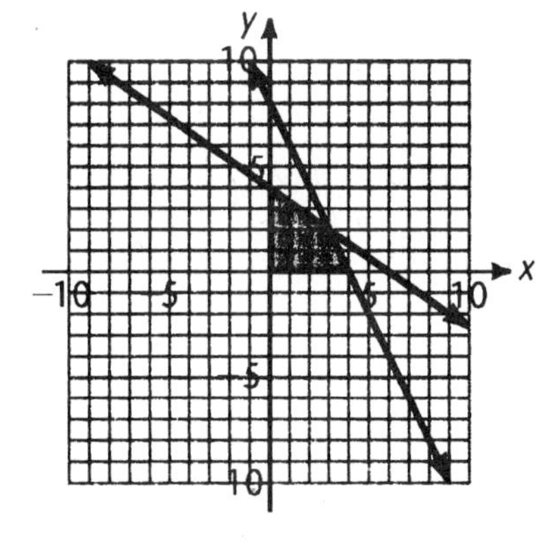

5.

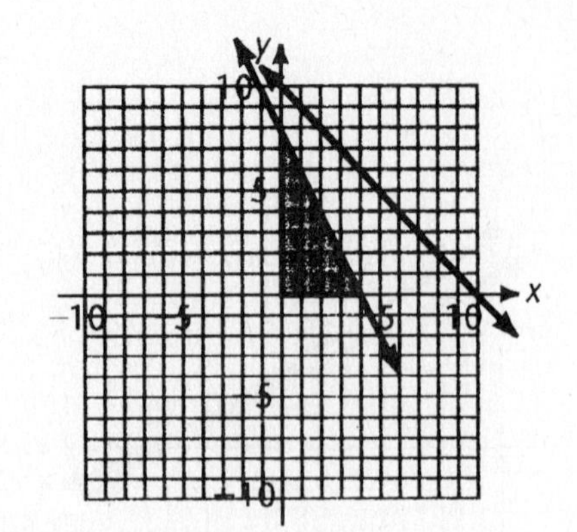

6.

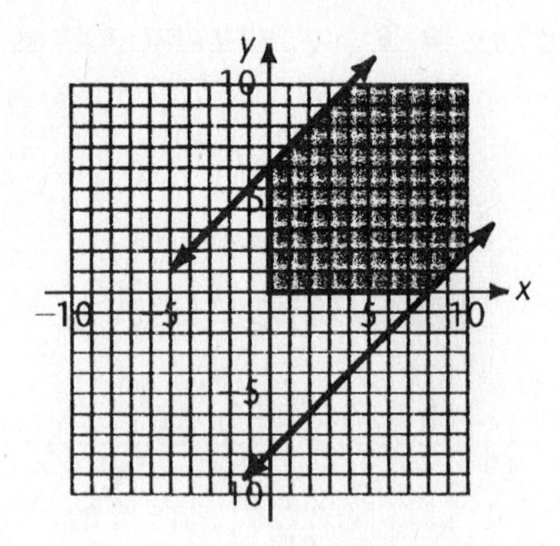

7. $2x + 3y = 7$ (1)
$3x - y = 5$ (2)
Solve (2) for y in terms of x.
$-y = -3x + 5$
$y = 3x - 5$
Substitute into (1) to eliminate y.
$2x + 3(3x - 5) = 7$
$2x + 9x - 15 = 7$
$11x - 15 = 7$
$11x = 22$
$x = 2$
Now replace x with 2 in (2) to find y.
$3(2) - y = 5$
$6 - y = 5$
$-y = -1$
$y = 1$
Solution: (2, 1)

Check: $2x + 3y = 7$ $\quad$ $3x - y = 5$
$2(2) + 3(1) \stackrel{?}{=} 7$ $\quad$ $3(2) - 1 \stackrel{?}{=} 5$
$7 \stackrel{\surd}{=} 7$ $\quad$ $5 \stackrel{\surd}{=} 5$

8. $x - 3y = 14$ (1)
$2x + y = 7$ (2)
Solve (2) for y in terms of x.
$y = 7 - 2x$
Substitute into (1) to eliminate y
$x - 3(7 - 2x) = 14$
$x - 21 + 6x = 14$
$7x - 21 = 14$
$7x = 35$
$x = 5$
Now replace x with 5 in (2) to find y.
$2(5) + y = 7$
$10 + y = 7$
$y = -3$
Solution: (5, -3)

Check: $x - 3y = 14$ $\quad$ $2x + y = 7$
$5 - 3(-3) \stackrel{?}{=} 14$ $\quad$ $2(5) + (-3) \stackrel{?}{=} 7$
$5 + 9 \stackrel{\surd}{=} 14$ $\quad$ $10 - 3 \stackrel{\surd}{=} 7$

9. $2x + 3y = 7$ (1)
$3x - y = 5$ (2)

We multiply (2) by 3 and add.
$2x + 3y = 7$ (1)
$\underline{9x - 3y = 15}$ 3[equation (2)]
$11x = 22$
$x = 2$
Replace x by 2 in (2) to find y.
$3(2) - y = 5$
$6 - y = 5$
$-y = -1$
$y = 1$
Solution: (2, 1)
This has already been checked in problem 7.

10. $5x + 2y = -1$ (1)
$2x - 3y = -27$ (2)

We multiply (2) by 2 and (1) by 3 and add.

$$\begin{aligned} 15x + 6y &= -3 \quad 3[\text{equation (1)}] \\ 4x - 6y &= -54 \\ \hline 19x &= -57 \\ x &= -3 \end{aligned}$$

Replace x by -3 in (1) to find y.

$$\begin{aligned} 5(-3) + 2y &= -1 \\ -15 + 2y &= -1 \\ 2y &= 14 \\ y &= 7 \end{aligned}$$

Solution: (-3, 7)

Check:

$$5x + 2y = -1 \qquad 2x - 3y = -27$$
$$5(-3) + 2(7) \stackrel{?}{=} -1 \qquad 2(-3) - 3(7) \stackrel{?}{=} -27$$
$$-15 + 14 \stackrel{\surd}{=} -1 \qquad -6 - 21 \stackrel{\surd}{=} -27$$

11. $5x + 3y = 26$ (1)
$8x - 2y = -6$ (2)

We multiply (1) by 2 and (2) by 3 and add.

$$\begin{aligned} 10x + 6y &= 52 \quad 2[\text{equation (1)}] \\ 24x - 6y &= -18 \quad 3[\text{equation (2)}] \\ \hline 34x &= 34 \\ x &= 1 \end{aligned}$$

Replace x by 1 in (1) to find y.

$$\begin{aligned} 5(1) + 3y &= 26 \\ 5 + 3y &= 26 \\ 3y &= 21 \\ y &= 7 \end{aligned}$$

Solution: (1, 7)

Check:

$$5x + 3y = 26 \qquad 8x - 2y = -6$$
$$5(1) + 3(7) \stackrel{?}{=} 26 \qquad 8(1) - 2(7) \stackrel{?}{=} -6$$
$$5 + 21 \stackrel{\surd}{=} 26 \qquad 8 - 14 \stackrel{\surd}{=} -6$$

12. $3x - 5y = 9$ (1)
$-2x + 3y = -5$ (2)

We multiply (1) by 2 and (2) by 3 and add.

$$\begin{aligned} 6x - 10y &= 18 \quad 2[\text{equation (1)}] \\ -6x + 9y &= -15 \quad 3[\text{equation (2)}] \\ \hline -y &= 3 \\ y &= -3 \end{aligned}$$

Replace y by -3 in (1) to find x.

$$\begin{aligned} 3x - 5(-3) &= 9 \\ 3x + 15 &= 9 \\ 3x &= -6 \\ x &= -2 \end{aligned}$$

Solution: (-2, -3)

Check:

$$3x - 5y = 9 \qquad -2x + 3y = -5$$
$$3(-2) - 5(-3) \stackrel{?}{=} 9 \qquad -2(-2) + 3(-3) \stackrel{?}{=} -5$$
$$-6 + 15 \stackrel{\surd}{=} 9 \qquad 4 - 9 \stackrel{\surd}{=} -5$$

13. $y + 2z = 4$ (1)
$x - z = -2$ (2)
$x + y = 1$ (3)

We eliminate z from equation (1).

$$\begin{aligned} y + 2z &= 4 \quad (1) \\ 2x - 2z &= -4 \quad 2[\text{equation (2)}] \\ \hline 2x + y &= 0 \end{aligned}$$

We now solve the system (3), (4).

$$\begin{aligned} x + y &= 1 \quad (3) \\ 2x + y &= 0 \quad (4) \\ -x - y &= -1 \quad -1[\text{equation (3)}] \\ 2x + y &= 0 \\ \hline x &= -1 \end{aligned}$$

Substituting x = -1 into equation (3), we have

$$\begin{aligned} -1 + y &= 1 \\ y &= 2 \end{aligned}$$

Substituting x = -1 into equation (2), we have

$$\begin{aligned} -1 - z &= -2 \\ -z &= -1 \\ z &= 1 \end{aligned}$$

Solution: x = -1, y = 2, z = 1

14. $x^2 + y^2 = 2$
$2x - y = 3$
We solve the first-degree equation for y in terms of x, then substitute into the second-degree equation.

$$\begin{aligned} 2x - y &= 3 \\ -y &= 3 - 2x \\ y &= 2x - 3 \\ x^2 + (2x - 3)^2 &= 2 \\ x^2 + 4x^2 - 12x + 9 &= 2 \\ 5x^2 - 12x + 7 &= 0 \\ (5x - 7)(x - 1) &= 0 \\ x &= \frac{7}{5}, 1 \end{aligned}$$

For $x = \frac{7}{5}$:

$$y = 2\left(\frac{7}{5}\right) - 3 = -\frac{1}{5}$$

For $x = 1$:

$$y = 2(1) - 3 = -1$$

Solutions: $(1, -1)$, $\left(\frac{7}{5}, -\frac{1}{5}\right)$

15. $x^2 - y^2 = 7$
$x^2 + y^2 = 25$

We use elimination by addition.

$$\begin{aligned} x^2 - y^2 &= 7 \\ x^2 + y^2 &= 25 \\ \hline 2x^2 &= 32 \\ x^2 &= 16 \\ x &= \pm 4 \end{aligned}$$

For $x = 4$:

$$\begin{aligned} x^2 + y^2 &= 25 \\ 4^2 + y^2 &= 25 \\ 16 + y^2 &= 25 \\ y^2 &= 9 \\ y &= \pm 3 \end{aligned}$$

For $x = -4$:

$$\begin{aligned} x^2 + y^2 &= 25 \\ (-4)^2 + y^2 &= 25 \\ 16 + y^2 &= 25 \\ y^2 &= 9 \\ y &= \pm 3 \end{aligned}$$

Solutions: $(4, 3)$, $(-4, 3)$, $(4, -3)$, $(-4, -3)$

16.
$$\begin{aligned} x + 2y + z &= 3 && (1) \\ 2x + 3y + 4z &= 3 && (2) \\ x + 2y + 3z &= 1 && (3) \end{aligned}$$

We eliminate x from equations (2) and (3)

$$\begin{aligned} 2x + 3y + 4z &= 3 && \text{Equation (2)} \\ -2x - 4y - 2z &= -6 && -2[\text{equation (1)}] \\ -y + 2z &= -3 && (4)\ (\text{adding}) \\ x + 2y + 3z &= 1 && \text{Equation (3)} \\ -x - 2y - z &= -3 && -1[\text{equation (1)}] \\ 2z &= -2 && (5)\ (\text{adding}) \end{aligned}$$

We solve the system (4), (5) by substituting 2z from equation (5) into equation (4).

$$\begin{aligned} -y + 2z &= -3 \\ 2z &= -2 \\ -y - 2 &= -3 \\ -y &= -1 \\ y &= 1 \\ z &= -1 \end{aligned}$$

Substituting $y = 1$, $z = -1$ into equation (1) we have

$$\begin{aligned} x + 2(1) - 1 &= 3 \\ x &= 2 \end{aligned}$$

Solution: $(2, 1, -1)$

Check:

$$x + 2y + z = 3$$
$$2 + 2(1) + (-1) \overset{\checkmark}{=} 3$$

$$2x + 3y + 4z = 3$$
$$2(2) + 3(1) + 4(-1) \overset{?}{=} 3$$
$$4 + 3 - 4 \overset{\checkmark}{=} 3$$

$$x + 2y + 3z = 1$$
$$2 + 2(1) + 3(-1) \overset{?}{=} 1$$
$$2 + 2 - 3 \overset{\checkmark}{=} 1$$

17. $x + y + z = 8$ (1)
$2x - y + z = 3$ (2)
$y - 2z = -5$ (3)

We eliminate x from equation (2).

$$\begin{array}{rl} -2x - 2y - 2z = -16 & -2[\text{equation (1)}] \\ \underline{2x - y + z = 3} & \text{Equation (2)} \\ -3y - z = -13 & \text{(4) (adding)} \end{array}$$

We now solve the system (3), (4).

$$\begin{array}{rl} 3y - 6z = -15 & 3[\text{equation (3)}] \\ \underline{-3y - z = -13} & \text{Equation (4)} \\ -7z = -28 & \\ z = 4 & \end{array}$$

Substituting z = 4 into equation (3), we have

$$y - 2(4) = -5$$
$$y - 8 = -5$$
$$y = 3$$

Substituting y = 3 and z = 4 into equation (1), we have

$$x + 3 + 4 = 8$$
$$x = 1$$

Solution: (1, 3, 4)

Check: $x + y + z = 8$, $1 + 3 + 4 \overset{\checkmark}{=} 8$; $2x - y + z = 3$, $2(1) - 3 + 4 \overset{?}{=} 3$, $2 - 3 + 4 \overset{\checkmark}{=} 3$; $y - 2z = -5$, $3 - 2(4) \overset{?}{=} -5$, $3 - 8 \overset{\checkmark}{=} -5$

18. $x + y + z = 8$ (1)
$x - 2y + 3z = 4$ (2)
$2x + 5z = 12$ (3)

We eliminate x from equations (2) and (3).

$$\begin{array}{rl} -x - y - z = -8 & (-1)[\text{equation (1)}] \\ \underline{x - 2y + 3z = 4} & \text{Equation (2)} \\ -3y + 2z = -4 & (4) \end{array}$$

$$\begin{array}{rl} -2x - 2y - 2z = -16 & (-2)[\text{equation (1)}] \\ \underline{2x + 5z = 12} & \text{Equation (3)} \\ -2y + 3z = -4 & (5) \end{array}$$

We now solve the system (4), (5).

$$\begin{array}{rl} -6y + 4z = -8 & 2[\text{equation (4)}] \\ \underline{6y - 9z = 12} & -3[\text{equation (5)}] \\ -5z = 4 & \\ z = -\frac{4}{5} & \end{array}$$

Substituting $z = -\frac{4}{5}$ into equation (5), we have

$$-2y + 3\left(-\frac{4}{5}\right) = -4$$
$$-2y - \frac{12}{5} = -\frac{20}{5}$$

$$-2y = -\frac{8}{5}$$
$$y = \frac{4}{5}$$

Substituting $y = \frac{4}{5}$ and $z = -\frac{4}{5}$ into equation (1), we have

$$x + \frac{4}{5} + \left(-\frac{4}{5}\right) = 8$$
$$x = 8$$

Solution: $\left(8, \frac{4}{5}, -\frac{4}{5}\right)$

Check: $x + y + z = 8$ $\quad$ $x - 2y + 3z = 4$ $\quad$ $2x + 5z = 12$

$$8 + \frac{4}{5} + \left(-\frac{4}{5}\right) \stackrel{\surd}{=} 8 \qquad 8 - 2\left(\frac{4}{5}\right) + 3\left(-\frac{4}{5}\right) \stackrel{?}{=} 4 \qquad 2(8) + 5\left(-\frac{4}{5}\right) \stackrel{?}{=} 12$$
$$8 - \frac{8}{5} - \frac{12}{5} \stackrel{?}{=} 4 \qquad 16 - 4 \stackrel{\surd}{=} 12$$
$$8 - 4 \stackrel{\surd}{=} 4$$

19. $x - 3y = 8$ (1)
$-4x + 12y = 32$ (2)

If we multiply (1) by 4 and add, we will eliminate both x and y.

$$\begin{aligned} 4x - 12y &= 32 \\ \underline{-4x + 12y} &\underline{= 32} \\ 0 &= 64 \end{aligned}$$

No solution.

20. $x^2 + y^2 = 25$
$4x - 3y = 0$

Solve $4x - 3y = 0$ for x in terms of y.

$$4x = 3y$$
$$x = \frac{3}{4}y$$

Substitute this expression for x in $x^2 + y^2 = 25$

$$\left(\frac{3}{4}y\right)^2 + y^2 = 25$$
$$\frac{9y^2}{16} + y^2 = 25 \quad \text{lcm} = 16$$
$$16 \cdot \frac{9y^2}{16} + 16y^2 = 16(25)$$
$$9y^2 + 26y^2 = 400$$
$$25y^2 = 400$$
$$y^2 = 16$$
$$y = \pm 4$$

For $y = 4$:
$$x = \frac{3}{4}(4)$$
$$x = 3$$

For $y = -4$:
$$x = \frac{3}{4}(-4)$$
$$x = -3$$

(3, 4), (-3, -4)

21. $x^2 + y = 2$
$x^2 + y^2 = 4$

Solve $x^2 + y = 2$ for x^2 in terms of y.

$$x^2 = 2 - y$$

Substitute this expression for x^2 in $x^2 + y^2 = 4$

$$2 - y + y^2 = 4$$
$$y^2 - y - 2 = 0$$

Solve by factoring

$$(y - 2)(y + 1) = 0$$
$$y - 2 = 0 \quad \text{or} \quad y + 1 = 0$$
$$y = 2 \qquad\qquad y = -1$$

For $y = 2$:
$$x^2 = 2 - 2$$
$$x^2 = 0$$
$$x = 0$$

For $y = -1$:
$$x^2 = 2 - (-1)$$
$$x^2 = 3$$
$$x = \pm\sqrt{3}$$

$(0, 2), (\sqrt{3}, -1), (-\sqrt{3}, -1)$

22. $x + 2y + z = 4$ (1)
$x - z = 0$ (2)
$x + y = 2$ (3)

We solve equation (2) for z in terms of equation (3) for y in terms of x, then substitute the expressions obtained into (1).

$$x - z = 0$$
$$x = z$$
$$x + y = 2$$
$$y = 2 - x$$

Subtituting into equation (1), we obtain

$$x + 2(2 - x) + x = 4$$
$$x + 4 - 2x + x = 4$$
$$4 = 4$$

We conclude that there are an infinite number of solutions. Since $x = z$ and $y = 2 - x = 2 - z$ for every solution, we can write the solutions as $(z, 2 - z, z)$ for z any real number.

23. $x + 2y + z = 4$ (1)
$x + z = 0$ (2)
$x + y = 2$ (3)

We solve equation (2) for z in terms of x and equation (3) for y in terms of x, then substitute the expressions obtained into (1).

$$x + z = 0$$
$$z = -x$$
$$x + y = 2$$
$$y = 2 - x$$

Subtituting into equation (1), we obtain

$$x + 2(2 - x) + (-x) = 4$$
$$x + 4 - 2x - x = 4$$
$$-2x = 0$$
$$x = 0$$

Since $z = -x$, $z = 0$. Since $y = 2 - x$, $y = 2$.
Solution: $(0, 2, 0)$

Check: $x + 2y + z = 4$ $\quad$ $x + z = 0$ $\quad$ $x + y = 2$

$0 + 2(2) + 0 \stackrel{?}{=} 4$ $\quad$ $0 + 0 \stackrel{\surd}{=} 0$ $\quad$ $0 + 2 \stackrel{\surd}{=} 2$

$4 \stackrel{\surd}{=} 4$

24. $x + 2y = 6$ (1)
$3x + 4z = 0$ (2)
$-x + z = -4$ (3)

We multiply equation (3) by 3 and add to equation (2) to eliminate z.

$3x + 4z = 0$ $\quad$ Equation (2)
$-3x + 3z = -12$ $\quad$ 3[equation (3)]

$$7z = -12$$
$$z = -\frac{12}{7}$$

Subtituting $z = -\frac{12}{7}$ into equation (3), we obtain

$$
\begin{aligned}
-x + \left(-\frac{12}{7}\right) &= -4 \\
-x &= -\frac{28}{7} + \frac{12}{7} \\
-x &= -\frac{16}{7} \\
x &= \frac{16}{7}
\end{aligned}
$$

Substituting $x = \frac{16}{7}$ into equation (1), we obtain

$$
\begin{aligned}
\frac{16}{7} + 2y &= 6 \\
12y &= \frac{42}{7} - \frac{16}{7} \\
2y &= \frac{26}{7} \\
y &= \frac{13}{7}
\end{aligned}
$$

Solution: $\left(\frac{16}{7}, \frac{13}{7}, -\frac{12}{7}\right)$

Check: $x + 2y = 6$ $\qquad$ $3x + 4z = 0$ $\qquad$ $-x + z = -4$

$$\frac{16}{7} + 2\left(\frac{13}{7}\right) \stackrel{?}{=} 6 \qquad 3\left(\frac{16}{7}\right) + 4\left(-\frac{12}{7}\right) \stackrel{?}{=} 0 \qquad -\frac{16}{7} + \left(-\frac{12}{7}\right) \stackrel{?}{=} -4$$

$$\frac{42}{7} \stackrel{\surd}{=} 6 \qquad \frac{48}{7} - \frac{48}{7} \stackrel{\surd}{=} 0 \qquad -\frac{28}{7} \stackrel{\surd}{=} -4$$

25. $x^2 - 9y^2 = 36$ (1)

$\underline{x + 9y^2 = 6}$ (2)

$x^2 + x = 42$ We have added to eliminate y.

Solve by factoring:

$$
\begin{aligned}
x^2 + x - 42 &= 0 \\
(x - 6)(x + 7) &= 0
\end{aligned}
$$

$x - 6 = 0$ or $x + 7 = 0$

$x = 6$ $\qquad$ $x = -7$

For $x = 6$

$$
\begin{aligned}
6 + 9y^2 &= 6 \\
9y^2 &= 0 \\
y &= 0
\end{aligned}
$$

For $x = -7$

$$
\begin{aligned}
-7 + 9y^2 &= 6 \\
9y^2 &= 13 \\
y^2 &= \frac{13}{9} \\
y &= \pm\frac{\sqrt{13}}{3}
\end{aligned}
$$

Solution: $(6, 0)$, $\left(-7, -\frac{\sqrt{13}}{3}\right)$, $\left(-7, \frac{\sqrt{13}}{3}\right)$

26. $x^2 - 9y^2 = 36$

$x - 9y^2 = 16$

We multiply the bottom equation by -1 and add to the top equation to eliminate y.

$$\begin{aligned} x^2 - 9y^2 &= 36 \\ -x + 9y^2 &= -16 \\ \hline x^2 - x &= 20 \end{aligned}$$

Solve by factoring:

$$x^2 - x - 20 = 0$$
$$(x - 5)(x + 4) = 0$$
$$x - 5 = 0 \quad \text{or} \quad x + 4 = 0$$
$$x = 5 \qquad\qquad x = -4$$

For $x = 5$

$$5 - 9y^2 = 16$$
$$-9y^2 = 11$$
$$y^2 = -\frac{11}{9}$$
$$y = \pm\frac{i\sqrt{11}}{3}$$

For $x = -4$

$$-4 - 9y^2 = 16$$
$$-9y^2 = 20$$
$$y^2 = -\frac{20}{9}$$
$$y = \pm\frac{2i\sqrt{5}}{3}$$

Solution: $\left(5, \frac{i\sqrt{11}}{3}\right), \left(5, -\frac{i\sqrt{11}}{3}\right), \left(-4, \frac{2i\sqrt{5}}{3}\right), \left(-4, \frac{-2i\sqrt{5}}{3}\right)$

27. $x^2 - 9y^2 = 36$
$3x - y = 0$

Solve $3x - y = 0$ for y in terms of x.

$$-y = -3x$$
$$y = 3x$$

Substitute this expression for y in $x^2 - 9y^2 = 36$.

$$x^2 - 9(3x)^2 = 36$$
$$x^2 - 9(9x^2) = 36$$
$$x^2 - 81x^2 = 36$$
$$-80x^2 = 36$$
$$x^2 = -\frac{9}{20}$$
$$x = \pm i\sqrt{\frac{9}{20}}$$
$$x = \pm\frac{3i\sqrt{5}}{10}$$

For $x = \frac{3i\sqrt{5}}{10}$:

$$y = 3\left(\frac{3i\sqrt{5}}{10}\right)$$
$$y = \frac{9i\sqrt{5}}{10}$$

For $x = -\frac{3i\sqrt{5}}{10}$:

$$y = 3\left(-\frac{3i\sqrt{5}}{10}\right)$$
$$y = \frac{-9i\sqrt{5}}{10}$$

Solution: $\left(\frac{3i\sqrt{5}}{10}, \frac{9i\sqrt{5}}{10}\right), \left(-\frac{3i\sqrt{5}}{10}, -\frac{9i\sqrt{5}}{10}\right)$

28.
$$\begin{aligned} x + 3y + 5z &= 10 && (1) \\ 2x - 2y + z &= 9 && (2) \\ 5x + z &= 8 && (3) \end{aligned}$$

We eliminate y from equations (1) and (2).

$$\begin{aligned} 2x + 6y + 10z &= 20 && 2[\text{equation (1)}] \\ 6x - 6y + 3z &= 27 && 3[\text{equation (2)}] \\ \hline 8x + 13z &= 47 && (4) \end{aligned}$$

We now solve the system.

$5x + z = 8$ (3)

$8x + 13z = 47$ (4)

We eliminate z.

$-65x - 13z = -104$ -13[equation (3)]

$8x + 13z = 47$ Equation (4)

$-57x = -57$

$x = 1$

Substituting x = 1 into equation (3) gives

$5(1) + z = 8$

$5 + z = 8$

$z = 3$

Substituting x = 1 and z = 3 into equation (1) gives

$1 + 3y + 5(3) = 10$

$3y + 16 = 10$

$3y = -6$

$y = -2$

Solution: x = 1, y = -2, z = 3

Check:

$x + 3y + 5z = 10$

$1 + 3(-2) + 5(3) \stackrel{?}{=} 10$

$1 - 6 + 15 = 10$ √

$2x - 2y + z = 9$

$2(1) - 2(-2) + 3 \stackrel{?}{=} 9$

$2 + 4 + 3 = 9$ √

$5x + z = 8$

$5(1) + 3 = 8$ √

29. $2x + 5y = 10$ (1)

$3x + 8y = 5$ (2)

We multiply (1) by 3 and (2) by -2 and add.

$6x + 15y = 30$ 3[equation (1)]

$-6x - 16y = -10$ -2[equation (2)]

$-y = 20$

$y = -20$

Replace y by -20 in (1) to find x.

$2x + 5(-20) = 10$

$2x - 100 = 10$

$2x = 110$

$x = 55$

Solution: (55, -20)

Check:

$2x + 5y = 10$

$2(55) + 5(-20) \stackrel{?}{=} 10$

$110 - 100 = 10$ √

$3x + 8y = 5$

$3(55) + 8(-20) \stackrel{?}{=} 5$

$165 - 160 = 5$ √

30. $2x + 3y = 6$ (1)

$5x + 7y = -2$ (2)

We multiply (1) by 7 and (2) by -3 and add.

$14x + 21y = 42$ 7[equation (1)]

$-15x - 21y = 6$ -3[equation (2)]

$-x = 48$

$x = -48$

Replace x by -48 in (1) to find y.

$2(-48) + 3y = 6$

$-96 + 3y = 6$

$3y = 102$

$y = 34$

Solution: (-48, 34)

Check: $2x + 3y = 6$

$2(-48) + 3(34) \stackrel{?}{=} 6$

$-96 + 102 = 6$ √

$5x + 7y = -2$

$5(-48) + 7(34) \stackrel{?}{=} -2$

$-240 + 238 = -2$ √

31. $x^2 + y^2 = 9$
$x - y = 0$

Solve $x - y = 0$ for x in terms of y.
$x = y$

Substitute this expression for x in $x^2 + y^2 = 9$

$$y^2 + y^2 = 9$$
$$2y^2 = 9$$
$$y^2 = \frac{9}{2}$$
$$y = \pm\sqrt{\frac{9}{2}}$$
$$y = \pm\frac{3\sqrt{2}}{2}$$

Since $x = y$, $x = \pm\frac{3\sqrt{2}}{2}$.

$\left(\frac{3\sqrt{2}}{2}, \frac{3\sqrt{2}}{2}\right), \left(-\frac{3\sqrt{2}}{2}, -\frac{3\sqrt{2}}{2}\right)$

32. $y = 4x^2$
$y = 11x + 3$

Relace y in the bottom expression with $4x^2$.

$$4x^2 = 11x + 3$$
$$4x^2 - 11x - 3 = 0$$

Solve by factoring.

$(4x + 1)(x - 3) = 0$

$4x + 1 = 0$ or $x - 3 = 0$
$4x = -1$ $\quad x = 3$
$x = -\frac{1}{4}$

For $x = -\frac{1}{4}$

$$y = 4\left(-\frac{1}{4}\right)^2$$
$$y = 4\left(\frac{1}{16}\right)$$
$$y = \frac{1}{4}$$

For $x = 3$

$$y = 4(3)^2$$
$$y = 36$$

$\left(-\frac{1}{4}, \frac{1}{4}\right)$, $(3, 36)$

33. $\begin{vmatrix} 1 & 2 \\ 3 & 4 \end{vmatrix} = 1\cdot 4 - 3\cdot 2 = 4 - 6 = -2$

34. $\begin{vmatrix} -3 & 5 \\ 9 & -15 \end{vmatrix} = (-3)(-15) - 9\cdot 5 = 45 - 45 = 0$

35. $\begin{vmatrix} 9 & 7 \\ 6 & 8 \end{vmatrix} = 9\cdot 8 - 6\cdot 7 = 72 - 42 = 30$

36. $\begin{vmatrix} \frac{1}{2} & 2 \\ \frac{1}{12} & \frac{2}{3} \end{vmatrix} = \frac{1}{2}\cdot\frac{2}{3} - \frac{1}{12}\cdot 2 = \frac{2}{6} - \frac{1}{6} = \frac{1}{6}$

37. $R_1 \Leftrightarrow R_2$ means interchange Rows 1 and 2.

$$\left[\begin{array}{cc|c} 4 & 5 & 6 \\ 1 & 2 & 3 \end{array}\right]$$

38. $R_2 + (-1)R_1 \Rightarrow R_2$ means replace Row 2 by itself plus -1 times Row 1.

$$\left[\begin{array}{cc|c} 1 & 2 & 3 \\ 4 & 5 & 6 \end{array}\right] \Rightarrow \left[\begin{array}{cc|c} 1 & 2 & 3 \\ 3 & 3 & 3 \end{array}\right]$$

-1 -2 -3

39. $\frac{1}{4}R_2 \Rightarrow R_2$ means multiply Row 2 by $\frac{1}{4}$.

$$\left[\begin{array}{cc|c} 1 & 2 & 3 \\ 1 & \frac{5}{4} & \frac{3}{2} \end{array}\right]$$

40. $R_1 + \frac{1}{2}R_2 \Rightarrow R_1$ means replace Row 1 by itself plus $\frac{1}{2}$ times Row 2.

$2 \quad \frac{5}{2} \quad 3$

$$\left[\begin{array}{cc|c} 1 & 2 & 3 \\ 4 & 5 & 6 \end{array}\right] \Rightarrow \left[\begin{array}{cc|c} 3 & \frac{9}{2} & 6 \\ 4 & 5 & 6 \end{array}\right]$$

41. $D = \begin{vmatrix} 3 & 8 \\ 2 & 5 \end{vmatrix} = -1$

$$x = \frac{\begin{vmatrix} 9 & 8 \\ -3 & 5 \end{vmatrix}}{D} = \frac{69}{-1} = -69$$

$$y = \frac{\begin{vmatrix} 3 & 9 \\ 2 & -3 \end{vmatrix}}{D} = \frac{-27}{-1} = 27$$

$x = -69,\ y = 27$

42. $D = \begin{vmatrix} 4 & 3 \\ 3 & 4 \end{vmatrix} = 7$

$$x = \frac{\begin{vmatrix} 5 & 3 \\ 4 & 4 \end{vmatrix}}{D} = \frac{8}{7}$$

$$y = \frac{\begin{vmatrix} 4 & 5 \\ 3 & 4 \end{vmatrix}}{D} = \frac{1}{7}$$

$x = \frac{8}{7},\ y = \frac{1}{7}$

43. $D = \begin{vmatrix} 1 & 1 \\ 1 & -1 \end{vmatrix} = -2$

$$x = \frac{\begin{vmatrix} 3 & 1 \\ 1 & -1 \end{vmatrix}}{D} = \frac{-4}{-2} = 2$$

$$y = \frac{\begin{vmatrix} 1 & 3 \\ 1 & 1 \end{vmatrix}}{D} = \frac{-2}{-2} = 1$$

$x = 2,\ y = 1$

44. $D = \begin{vmatrix} 7 & 18 \\ 2 & 5 \end{vmatrix} = -1$

$$x = \frac{\begin{vmatrix} -2 & 18 \\ 2 & 5 \end{vmatrix}}{D} = \frac{-46}{-1} = 46$$

$$y = \frac{\begin{vmatrix} 7 & -2 \\ 2 & 2 \end{vmatrix}}{D} = \frac{18}{-1} = -18$$

$x = 46,\ y = -18$

Note: *In the applications we choose to solve using two variable methods, although each could be solved using one variable. For example, in problem 45, we could let 30 - x = number of dimes and proceed using the single equation 5x + 10(30 - x) = 230.*

45. Let x = number of nickels
y = number of dimes
$x + y = 30$ (number of coins)
$5x + 10y = 230$ (value of coins in cents)

We choose elimination by addition, multiplying the top equation by -5 and adding.

$$\begin{aligned} -5x - 5y &= -150 \\ 5x + 10y &= 230 \\ \hline 5y &= 80 \\ y &= 16 \end{aligned}$$

Replace y by 16 in the top equation.

$$\begin{aligned} x + 16 &= 30 \\ x &= 14 \end{aligned}$$

Solution: 14 nickels, 16 dimes

Check: $14 + 16 = 30$ coins
$5(14) + 10(16) = 70 + 160 = 230$ cents or \$2.30

46. Let x = amount invested at 10%. Then $0.1x$ = return on this investment.
y = amount invested at 6%. Then $0.06y$ = return on this investment.

Then $x + y = 6{,}000$ (amounts invested)
$0.1x + 0.06y = 440$ (annual return)

We choose elimination by addition, multiplying the bottom equation by 100 to clear of decimals.

$$10x + 6y = 44{,}000$$

We multiply the top equation by -6 and add

$$\begin{aligned} -6x - 6y &= -36{,}000 \\ 10x + 6y &= 44{,}000 \\ \hline 4x &= 8{,}000 \\ x &= 2{,}000 \end{aligned}$$

Replacing x by 2,000 in the top equation, we obtain

$$\begin{aligned} 2{,}000 + y &= 6{,}000 \\ y &= \$4{,}000 \end{aligned}$$

Solution: \$2,000 at 10%, \$4,000 at 6%

Check: $2{,}000 + 4{,}000 = 6{,}000$
$0.1(2{,}000) + 0.06(4{,}000) = 200 + 240 = 440$

47. Let x = number of pounds of \$2.40-per-pound candy
y = number of pounds of \$3-per-pound candy
$x + y = 10$ (amount of candy)
$2.4x + 3y = 28$ (value of candy)

We choose elimination by addition, multiplying the top equation by -2.4 and adding.

$$\begin{aligned} -2.4x - 2.4y &= -24 \\ 2.4x + 3.0y &= 28 \\ \hline 0.6y &= 4 \\ y &= \frac{4}{0.6} \\ y &= \frac{20}{3} \text{ or } 6\tfrac{2}{3} \text{ pounds of \$3-per-pound candy} \end{aligned}$$

Replace y by $\frac{20}{3}$ in the top equation.

$$x + \frac{20}{3} = 10$$

$$x = \frac{30}{3} - \frac{20}{3}$$

$x = \frac{10}{3}$ or $3\frac{1}{3}$ pounds of \$2.40-per-pound candy.

Check: $3\frac{1}{3} + 6\frac{2}{3} = 10$ pounds of candy.

$2.4\left(\frac{10}{3}\right) + 3\left(\frac{20}{3}\right) = 8 + 20 = \$28$ worth of candy.

48. Let x = number of \$8 tickets
y = number of \$10 tickets
$x + y = 3700$ (total number of tickets)
$8x + 10y = 35{,}000$ (total value of tickets)

We choose elimination by addition, multiplying the top equation by -8 and adding.

$$\begin{aligned} -8x - 8y &= -29{,}600 \\ \underline{8x + 10y} &= \underline{35{,}000} \\ 2y &= 5{,}400 \\ y &= 2700 \end{aligned}$$

Replace y by 2700 in the top equation.

$$x + 2700 = 3700$$
$$x = 1000$$

Solution: 1000 advance tickets, 2700 tickets at the door.

Check: $1000 + 2700 = 3700$
$8(1000) + 10(2700) = 8{,}000 + 27{,}000 = 35{,}000$

49.

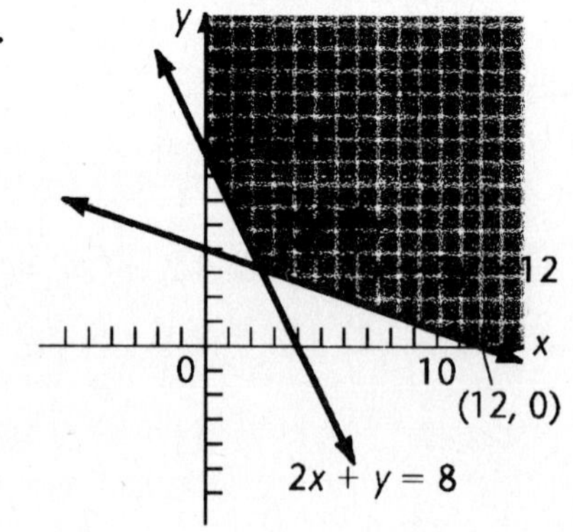

The corners are found by solving

$2x + y = 8$	$x + 3y = 12$	$x = 0$
$x + 3y = 12$	$y = 0$	$2x + y = 8$
to get	to get	to get
$\left(\frac{12}{5}, \frac{16}{5}\right)$	(12, 0)	(0, 8)

50.

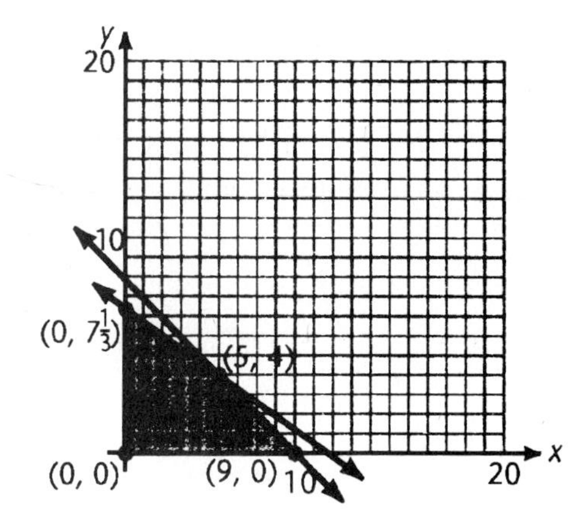

The corners are found by solving

$x + y = 9$ $2x + 3y = 22$	$x + y = 9$ $y = 0$	$x = 0$ $2x + 3y = 22$
to get	to get	to get
$(5, 4)$	$(9, 0)$	$(0, 7\frac{1}{3})$

51.

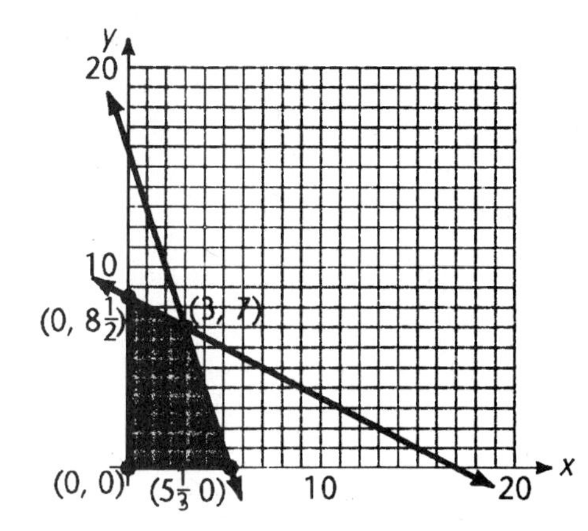

The corners are found by solving

$x + 2y = 17$ $3x + y = 16$	$x + 2y = 17$ $x = 0$	$x = 0$ $y = 0$	$y = 0$ $3x + y = 16$
to get	to get	to get	to get
$(3, 7)$	$(0, 8\frac{1}{2})$	$(0, 0)$	$(5\frac{1}{3}, 0)$

52.

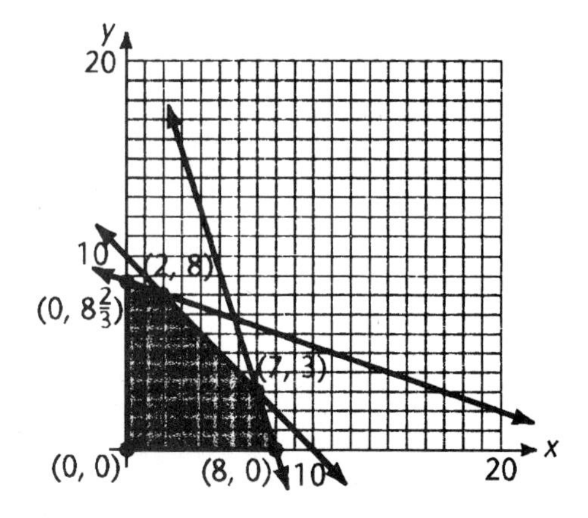

The corners are found by solving

$x + y = 10$ $x + 3y = 26$	$x + 3y = 26$ $x = 0$	$x = 0$ $y = 0$	$y = 0$ $3x + y = 24$	$3x + y = 24$ $x + y = 10$
to get	to get	to get	to get	to get
$(2, 8)$	$(0, 8\frac{2}{3})$	$(0, 0)$	$(8, 0)$	$(7, 3)$

53.

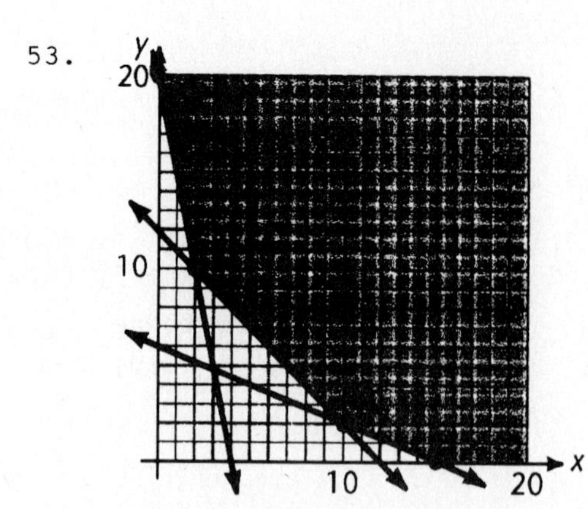

The corners are found by solving

$5x + y = 20$, $x = 0$	$5x + y = 20$, $x + y = 12$	$x + y = 12$, $2x + 5y = 30$	$2x + 5y = 30$, $y = 0$
to get (0, 20)	to get (2, 10)	to get (10, 2)	to get (15, 0)

54.
$$3x - 2y = -1 \qquad (1)$$
$$-6x + 4y = 3 \qquad (2)$$

If we multiply (1) by 2 and add, we will eliminate both x and y.

$$\begin{aligned} 6x - 4y &= -2 \\ \underline{-6x + 4y} &= \underline{3} \\ 0 &= 1 \end{aligned}$$

No solution.

55.
$$\begin{aligned} 3x - 2y - 7z &= -6 && (1) \\ -x + 3y + 2z &= -1 && (2) \\ x + 5y + 3z &= 3 && (3) \end{aligned}$$

We eliminate x from equations (1) and (3).

$$\begin{aligned} 3x - 2y - 7z &= -6 && (1) \\ \underline{-3x + 9y + 6z} &= \underline{-3} && 3[\text{equation } (2)] \\ 7y - z &= -9 && (4) \end{aligned}$$

$$\begin{aligned} -x + 3y + 2z &= -1 && (2) \\ \underline{x + 5y + 3z} &= \underline{3} && (3) \\ 8y + 5z &= 2 && (5) \end{aligned}$$

We now solve the system

$$\begin{aligned} 7y - z &= -9 && (4) \\ 8y + 5z &= 2 && (5) \end{aligned}$$

We eliminate z.

$$\begin{aligned} 35y - 5z &= -45 && 5[\text{equation } (4)] \\ \underline{8y + 5z} &= \underline{2} && (5) \\ 43y &= -43 \\ y &= -1 \end{aligned}$$

Substituting y = -1 into equation (5), we have

$$\begin{aligned} 8(-1) + 5z &= 2 \\ -8 + 5z &= 2 \\ 5z &= 10 \\ z &= 2 \end{aligned}$$

Substituting y = -1 and z = 2 into equation (3), we have

$$\begin{aligned} x + 5(-1) + 3(2) &= 3 \\ x - 5 + 6 &= 3 \\ x + 1 &= 3 \\ x &= 2 \end{aligned}$$

Solution: x = 2, y = -1, z = 2

Check:

$3x - 2y - 7z = -6$	$-x + 3y + 2z = -1$	$x + 5y + 3z = 3$
$3(2) - 2(-1) - 7(2) \stackrel{?}{=} -6$	$-(2) + 3(-1) + 2(2) \stackrel{?}{=} -1$	$2 + 5(-1) + 3(2) \stackrel{?}{=} 3$
$6 + 2 - 14 \stackrel{\checkmark}{=} -6$	$-2 - 3 + 4 \stackrel{\checkmark}{=} -1$	$2 - 5 + 6 \stackrel{\checkmark}{=} 3$

56. $3x^2 - y^2 = -6$
$2x^2 + 3y^2 = 29$

We use elimination by addition. We multiply the top equation by 3 and add.

$$\begin{aligned} 9x^2 - 3y^2 &= -18 \\ 2x^2 + 3y^2 &= 29 \\ \hline 11x^2 &= 11 \\ x^2 &= 1 \\ x &= \pm 1 \end{aligned}$$

For $x = 1$:

$$\begin{aligned} 3(1)^2 - y^2 &= -6 \\ -y^2 &= -9 \\ y^2 &= 9 \\ y &= \pm 3 \end{aligned}$$

For $x = -1$:

$$\begin{aligned} 3(-1)^2 - y^2 &= -6 \\ -y^2 &= -9 \\ y &= \pm 3 \end{aligned}$$

Solutions: (1, 3), (1, -3), (-1, 3), (-1, -3)

57. $x^2 = y$ (1)
$y = 2x - 2$ (2)

We substitute the expression for y in (2) into (1) to eliminate y.

$$x^2 = 2x - 2$$
$$x^2 - 2x + 2 = 0$$

This is not factorable in the integers, so we use the quadratic formula with $a = 1$, $b = -2$, $c = 2$.

$$\begin{aligned} x &= \frac{-(-2) \pm \sqrt{(-2)^2 - 4(1)(2)}}{2(1)} \\ &= \frac{2 \pm \sqrt{4 - 8}}{2} \\ &= \frac{2 \pm \sqrt{-4}}{2} \\ &= \frac{2 \pm 2i}{2} \\ &= 1 \pm i \end{aligned}$$

For $x = 1 + i$:

$$\begin{aligned} y &= 2(1 + i) - 2 \\ y &= 2 + 2i - 2 \\ y &= 2i \end{aligned}$$

For $x = 1 - i$

$$\begin{aligned} y &= 2(1 - i) - 2 \\ y &= 2 - 2i - 2 \\ y &= -2i \end{aligned}$$

Solutions: $(1 + i, 2i)$, $(1 - i, -2i)$

58. $2x + 3y - 4z = -25$ (1)
$5x - 4y - 3z = 16$ (2)
$-x + y + 2z = 0$ (3)

We eliminate x from equations (1) and (2).

$$\begin{aligned} 2x + 3y - 4z &= -25 && (1) \\ -2x + 2y + 4z &= 0 && 2[\text{equation } (3)] \\ \hline 5y &= -25 && (4) \end{aligned}$$

$$\begin{aligned} 5x - 4y - 3z &= 16 && (2) \\ -5x + 5y + 10z &= 0 && 5[\text{equation } (3)] \\ \hline y + 7z &= 16 && (5) \end{aligned}$$

We now solve the system

$$5y = -25 \quad (4)$$
$$y + 7z = 16 \quad (5)$$

From equation (4), we have $y = -5$. Substituting $y = -5$ into equation (5), we have

$$-5 + 7z = 16$$
$$7z = 21$$
$$z = 3$$

Substituting $y = -5$ and $z = 3$ into equation (3), we have

$$-x + (-5) + 2(3) = 0$$
$$-x + 1 = 0$$
$$-x = -1$$
$$x = 1$$

Solution: $x = 1$, $y = -5$, $z = 3$

Check: $2x + 3y - 4z = -25$ $\qquad$ $5x - 4y - 3z = 16$ $\qquad$ $-x + y + 2z = 0$

$2(1) + 3(-5) - 4(3) \stackrel{?}{=} -25$ $\qquad$ $5(1) - 4(-5) - 3(3) \stackrel{?}{=} 16$ $\qquad$ $-1 + (-5) + 2(3) \stackrel{?}{=} 0$

$2 - 15 - 12 \stackrel{\surd}{=} -25$ $\qquad$ $5 + 20 - 9 \stackrel{\surd}{=} 16$ $\qquad$ $-6 + 6 \stackrel{\surd}{=} 0$

59. $x + 3y + 2z = 2 \quad (1)$
$x - 3y - 2z = -1 \quad (2)$
$2x + 9y - 8z = 2 \quad (3)$

We eliminate y and z from equation (2) by adding equation (1).

$$x + 3y + 2z = 2 \quad (1)$$
$$\underline{x - 3y - 2z = -1} \quad (2)$$
$$2x = 1$$
$$x = \frac{1}{2}$$

Substituting $x = \frac{1}{2}$ into equations (1) and (3), we obtain

$$\frac{1}{2} + 3y + 2z = 2$$
$$3y + 2z = \frac{3}{2} \quad (4)$$

$$2\left(\frac{1}{2}\right) + 9y - 8z = 2$$
$$9y - 8z = 1 \quad (5)$$

We now solve the system (4), (5).

$$12y + 8z = 6 \quad 4[\text{equation (4)}]$$
$$\underline{9y - 8z = 1} \quad (5)$$
$$21y = 7$$
$$y = \frac{1}{3}$$

Substituting $y = \frac{1}{3}$ into equation (5), we have

$$9\left(\frac{1}{3}\right) - 8z = 1$$
$$3 - 8z = 1$$
$$-8z = -2$$
$$z = \frac{1}{4}$$

Solution: $x = \frac{1}{2}$, $y = \frac{1}{3}$, $z = \frac{1}{4}$

Check: $x + 3y + 2z = 2$ $\qquad$ $x - 3y - 2z = -1$ $\qquad$ $2x + 9y - 8z = 2$

$$\frac{1}{2} + 3\left(\frac{1}{3}\right) + 2\left(\frac{1}{4}\right) \stackrel{?}{=} 2 \qquad \frac{1}{2} - 3\left(\frac{1}{3}\right) - 2\left(\frac{1}{4}\right) \stackrel{?}{=} -1 \qquad 2\left(\frac{1}{2}\right) + 9\left(\frac{1}{3}\right) - 8\left(\frac{1}{4}\right) \stackrel{?}{=} 2$$

$$\frac{1}{2} + 1 + \frac{1}{2} \stackrel{\checkmark}{=} 2 \qquad \frac{1}{2} - 1 - \frac{1}{2} \stackrel{\checkmark}{=} -1 \qquad 1 + 3 - 2 \stackrel{\checkmark}{=} 2$$

60. $x^2 + y^2 = 25$
$y + 2x^2 = 22$

Solve $x^2 + y^2 = 25$ for x^2 in terms of y.
$x^2 = 25 - y^2$

Substitute this expression for x^2 in $y + 2x^2 = 22$
$y + 2(25 - y^2) = 22$
$y + 50 - 2y^2 = 22$
$0 = 2y^2 - y - 28$

Solve by factoring.
$0 = (2y + 7)(y - 4)$
$2y + 7 = 0$ or $y - 4 = 0$
$2y = -7$ $\qquad$ $y = 4$
$y = -\frac{7}{2}$

For $y = -\frac{7}{2}$

$x^2 + \left(-\frac{7}{2}\right)^2 = 25$
$x^2 + \frac{49}{4} = \frac{100}{4}$
$x^2 = \frac{51}{4}$
$x = \pm\frac{\sqrt{51}}{2}$

For $y = 4$

$x^2 + 4^2 = 25$
$x^2 + 16 = 25$
$x^2 = 9$
$x = \pm 3$

Solution: $\left(\frac{\sqrt{51}}{2}, -\frac{7}{2}\right)$, $\left(-\frac{\sqrt{51}}{2}, -\frac{7}{2}\right)$, $(3, 4)$, $(-3, 4)$

61. $x^2 + y^2 = 25$
$x^2 + 4y^2 = 16$

We multiply the top equation by -1 and add to eliminate x.

$-x^2 - y^2 = -25$
$x^2 + 4y^2 = 16$
$3y^2 = -9$
$y^2 = -3$
$y = \pm\sqrt{-3}$
$y = \pm\sqrt{3}\,i$

For $y^2 = -3$:
$x^2 + (-3) = 25$
$x^2 - 3 = 25$
$x^2 = 28$
$x = \pm\sqrt{28}$ or $\pm 2\sqrt{7}$

Solution: $(2\sqrt{7}, \sqrt{3}\,i)$, $(2\sqrt{7}, -\sqrt{3}\,i)$, $(-2\sqrt{7}, \sqrt{3}\,i)$, $(-2\sqrt{7}, -\sqrt{3}\,i)$

62. $x^2 - 4y = 0$
$-4x + y^2 = 0$

Solve $x^2 = 4y$ for y in terms of x.

$$\frac{x^2}{4} = y$$

Substitute this expression for y in $-4x + y^2 = 0$

$$-4x + \left(\frac{x^2}{4}\right)^2 = 0$$

$$-4x + \frac{x^4}{16} = 0 \quad \text{lcm} = 16$$

$$-64x + x^4 = 0$$

$$x^4 - 64x = 0$$

$$x(x^3 - 64) = 0$$

$$x(x - 4)(x^2 + 4x + 16) = 0$$

$x = 0$ or $x - 4 = 0$ or $x^2 + 4x + 16 = 0$

$x = 4$

Solve by the quadratic formula with $a = 1$, $b = 4$, $c = 16$

$$x = \frac{-4 \pm \sqrt{(4)^2 - 4(1)(16)}}{2(1)}$$

$$x = \frac{-4 \pm \sqrt{16 - 64}}{2}$$

$$x = \frac{-4 \pm \sqrt{-48}}{2}$$

$$x = \frac{-4 \pm 4i\sqrt{3}}{2}$$

$$x = -2 \pm 2i\sqrt{3}$$

For $x = 0$

$$y = \frac{0^2}{4}$$

$$y = 0$$

For $x = 4$

$$y = \frac{x^2}{4}$$

$$y = \frac{4^2}{4}$$

$$y = 4$$

For $x = -2 \pm 2i\sqrt{3}$

$$y = \frac{(-2 \pm 2i\sqrt{3})^2}{4}$$

$$y = \frac{4 \pm 2(-2)(2i\sqrt{3}) + (2i\sqrt{3})^2}{4}$$

$$y = \frac{4 \mp 8i\sqrt{3} - 12}{4}$$

$$y = \frac{-8 \mp 8i\sqrt{3}}{4}$$

$$y = -2 \mp 2i\sqrt{3}$$

Solution: $(0, 0)$, $(4, 4)$, $(-2 + 2i\sqrt{3}, -2 - 2i\sqrt{3})$, $(-2 - 2i\sqrt{3}, -2 + 2i\sqrt{3})$

63. $x^2 + \frac{y^2}{9} = 1$
$x - y^2 = 1$

Solve $x - y^2 = 1$ for y^2 in terms of x.

$$-y^2 = 1 - x$$

$$y^2 = x - 1$$

Substitute this expression for y^2 in $x^2 + \frac{y^2}{9} = 1$.

$$x^2 + \frac{x - 1}{9} = 1 \quad \text{lcm} = 9$$

$$9x^2 + 9 \cdot \frac{x - 1}{9} = 9$$
$$9x^2 + x - 1 = 9$$
$$9x^2 + x - 10 = 0$$

Solve by factoring

$$(9x + 10)(x - 1) = 0$$
$$9x + 10 = 0 \quad \text{or} \quad x - 1 = 0$$
$$9x = -10 \qquad x = 1$$
$$x = -\frac{10}{9}$$

For $x = \frac{-10}{9}$

$$y^2 = x - 1$$
$$y^2 = -\frac{10}{9} - 1$$
$$y^2 = -\frac{19}{9}$$
$$y = \pm\sqrt{\frac{-19}{9}}$$
$$y = \pm\frac{i\sqrt{19}}{3}$$

For $x = 1$

$$y^2 = x - 1$$
$$y^2 = 1 - 1$$
$$y^2 = 0$$
$$y = 0$$

Solution: $\left(-\frac{10}{9}, \frac{i\sqrt{19}}{3}\right)$, $\left(-\frac{10}{9}, -\frac{i\sqrt{19}}{3}\right)$, $(1, 0)$

64. $x^2 + y^2 = 9$

$x^2 + \frac{y^2}{9} = 1$

Multiply the bottom equation by -9 and add to eliminate y.

$$\begin{array}{rl} x^2 + y^2 &= 9 \\ -9x^2 - y^2 &= -9 \\ \hline -8x^2 \quad &= 0 \\ x^2 &= 0 \\ x &= 0 \end{array}$$

For $x = 0$: $x^2 + y^2 = 9$

$$0 + y^2 = 9$$
$$y = \pm 3$$

Solution: $(0, 3)$, $(0, -3)$

65. $\frac{x^2}{9} - \frac{y^2}{16} = 1$

$4x + 3y = 0$

Solve $4x + 3y = 0$ for x in terms of y.

$$4x = -3y$$
$$x = -\frac{3y}{4}$$

Substitute this expression for x in $\frac{x^2}{9} - \frac{y^2}{16} = 1$.

$$\frac{\left(-\frac{3y}{4}\right)^2}{9} - \frac{y^2}{16} = 1$$
$$\frac{9y^2}{16} \div \frac{9}{1} - \frac{y^2}{16} = 1$$
$$\frac{9y^2}{16} \cdot \frac{1}{9} - \frac{y^2}{16} = 1$$
$$\frac{y^2}{16} - \frac{y^2}{16} = 1$$
$$0 = 1$$

Impossible. No solution.

66. $9x^2 - y = 0$

$y = 4x + 13$

Substitute $4x + 13$ for y in $9x^2 - y = 0$.

$$9x^2 - (4x + 13) = 0$$
$$9x^2 - 4x - 13 = 0$$

Solve by factoring.

$(9x - 13)(x + 1) = 0$

$9x - 13 = 0$ or $x + 1 = 0$

$9x = 13$ $x = -1$

$x = \frac{13}{9}$

For $x = \frac{13}{9}$

$y = 4x + 13$

$y = 4\left(\frac{13}{9}\right) + 13$

$y = \frac{52}{9} + \frac{117}{9}$

$y = \frac{169}{9}$

For $x = -1$

$y = 4x + 13$

$y = 4(-1) + 13$

$y = 9$

Solution: $\left(\frac{13}{9}, \frac{169}{9}\right)$, $(-1, 9)$

67.

$$\left[\begin{array}{cc|c} 3 & 2 & 3 \\ 1 & 3 & 8 \end{array}\right] R_1 \Leftrightarrow R_2$$

Need a 1 here

$$\sim \left[\begin{array}{cc|c} 1 & 3 & 8 \\ 3 & 2 & 3 \end{array}\right] R_2 + (-3)R_1 \Rightarrow R_2$$

Need a 0 here

-3 -9 -24

$$\sim \left[\begin{array}{cc|c} 1 & 3 & 8 \\ 0 & -7 & -21 \end{array}\right] -\frac{1}{7}R_2 \Rightarrow R_2$$

Need a 1 here

$$\sim \left[\begin{array}{cc|c} 1 & 3 & 8 \\ 0 & 1 & 3 \end{array}\right]$$

The original system is therefore equivalent to

$x + 3y = 8$

$y = 3$

Thus, $y = 3$, and $x + 3(3) = 8$ so $x = -1$.

Solution: $x = -1$, $y = 3$

Check:

$3x + 2y = 3$ $x + 3y = 8$

$3(-1) + 2(3) \overset{\surd}{=} 3$ $-1 + 3(3) \overset{\surd}{=} 8$

68.

$$\left[\begin{array}{cc|c} 3 & -1 & 0 \\ -2 & 1 & -1 \end{array}\right] R_1 + R_2 \Rightarrow R_1$$

Need a 1 here

$$\sim \left[\begin{array}{cc|c} 1 & 0 & -1 \\ -2 & 1 & -1 \end{array}\right] R_2 + 2R_1 \Rightarrow R_2$$

Need a 0 here

2 0 -2

$$\sim \left[\begin{array}{cc|c} 1 & 0 & -1 \\ 0 & 1 & -3 \end{array}\right]$$

The original system is therefore equivalent to

$x = -1$

$y = -3$

Check:

$3x - y = 0$ $-2x + y = -1$

$3(-1) - (-3) \overset{?}{=} 0$ $-2(-1) + (-3) \overset{?}{=} -1$

$-3 + 3 \overset{\surd}{=} 0$ $2 - 3 \overset{\surd}{=} -1$

69. $\left[\begin{array}{cc|c} 1 & 2 & 5 \\ 0 & 3 & 6 \end{array}\right] \frac{1}{3}R_2 \Rightarrow R_2$

↑ Need a 1 here

$\sim \left[\begin{array}{cc|c} 1 & 2 & 5 \\ 0 & 1 & 2 \end{array}\right]$

The original system is therefore equivalent to

$x + 2y = 5$
$y = 2$

Thus, $y = 2$ and $x + 2(2) = 5$ so $x = 1$

Solution: $x = 1$, $y = 2$.

Check: $x + 2y = 5$ $3y = 6$

$1 + 2(2) \stackrel{\checkmark}{=} 5$ $3(2) \stackrel{\checkmark}{=} 6$

70. $\left[\begin{array}{cc|c} 4 & 3 & 10 \\ 1 & 2 & 5 \end{array}\right] R_1 \Leftrightarrow R_2$

Need a 1 here

$\sim \left[\begin{array}{cc|c} 1 & 2 & 5 \\ 4 & 3 & 10 \end{array}\right] R_2 + (-4)R_1 \Rightarrow R_2$

↑ Need a 0 here

$-4 \quad -8 \quad -20$

$\sim \left[\begin{array}{cc|c} 1 & 2 & 5 \\ 0 & -5 & -10 \end{array}\right] -\frac{1}{5}R_2 \Rightarrow R_2$

↑ Need a 1 here

$\sim \left[\begin{array}{cc|c} 1 & 2 & 5 \\ 0 & 1 & 2 \end{array}\right]$

The original system is therefore equivalent to

$x + 2y = 5$
$y = 2$

Thus, $y = 2$ and $x + 2(2) = 5$ so $x = 1$.

Solution: $x = 1$, $y = 2$.

Check: $4x - 3y = 10$ $x + 2y = 5$

$4(1) + 3(2) \stackrel{\checkmark}{=} 10$ $1 + 2(2) \stackrel{\checkmark}{=} 5$

71. $\left[\begin{array}{ccc|c} 1 & 2 & 3 & 4 \\ 1 & 0 & 1 & 6 \\ 0 & -1 & 1 & 5 \end{array}\right] R_2 + (-1)R_1 \Rightarrow R_2$

Need a 0 here

$\sim \left[\begin{array}{ccc|c} 1 & 2 & 3 & 4 \\ 0 & -2 & -2 & 2 \\ 0 & -1 & 1 & 5 \end{array}\right] -\frac{1}{2}R_2 \Rightarrow R_2$

Need a 1 here

$\sim \left[\begin{array}{ccc|c} 1 & 2 & 3 & 4 \\ 0 & 1 & 1 & -1 \\ 0 & -1 & 1 & 5 \end{array}\right] R_3 + R_2 \Rightarrow R_3$

↑ Need a 0 here

$\sim \left[\begin{array}{ccc|c} 1 & 2 & 3 & 4 \\ 0 & 1 & 1 & -1 \\ 0 & 0 & 2 & 4 \end{array}\right] \frac{1}{2}R_3 \Rightarrow R_3$

↑ Need a 1 here

$\sim \left[\begin{array}{ccc|c} 1 & 2 & 3 & 4 \\ 0 & 1 & 1 & -1 \\ 0 & 0 & 1 & 2 \end{array}\right]$

The resulting equivalent system

$$\begin{aligned} x + 2y + 3z &= 4 \\ y + z &= -1 \\ z &= 2 \end{aligned}$$

is solved by substitution.

$z = 2$

$y + z = y + 2 = -1$ so

$y = -3$

$x + 2y + 3z = x + 2(-3) + 3(2) = 4$

so $x = 4$

Solution: $x = 4$, $y = -3$, $z = 2$

Check:

$x + 2y + 3z = 4$

$4 + 2(-3) + 3(2) \overset{?}{=} 4$

$4 - 6 + 6 \overset{\surd}{=} 4$

$x + z = 6$

$4 + 2 \overset{\surd}{=} 6$

$-y + z = 5$

$-(-3) + 2 \overset{\surd}{=} 5$

72. $\left[\begin{array}{ccc|c} 1 & -1 & 1 & 0 \\ 0 & 1 & -1 & -1 \\ 2 & -1 & -1 & -3 \end{array}\right]$ $R_3 + (-2)R_1 \Rightarrow R_1$

↑ Need a 0 here

$\sim \left[\begin{array}{ccc|c} 1 & -1 & 1 & 0 \\ 0 & 1 & -1 & -1 \\ 0 & 1 & -3 & -3 \end{array}\right]$ $R_3 + (-1)R_2 \Rightarrow R_3$

↑ Need a 0 here

$\sim \left[\begin{array}{ccc|c} 1 & -1 & 1 & 0 \\ 0 & 1 & -1 & -1 \\ 0 & 0 & -2 & -2 \end{array}\right]$ $-\frac{1}{2}R_3 \Rightarrow R_3$

↑ Need a 1 here

$\sim \left[\begin{array}{ccc|c} 1 & -1 & 1 & 0 \\ 0 & 1 & -1 & -1 \\ 0 & 0 & 1 & 1 \end{array}\right]$

The resulting equivalent system

$$\begin{aligned} x - y + z &= 0 \\ y - z &= -1 \\ z &= 1 \end{aligned}$$

is solved by substitution.

$z = 1$

$y - z = y - 1 = -1$

so $y = 0$

$x - y + z = x - 0 + 1 = 0$

so $x = -1$

Solution: $x = -1$, $y = 0$, $z = 1$

Check: $x - y + z = 0$

$-1 + 0 + 1 \overset{\surd}{=} 0$

$y - z = -1$

$0 - 1 \overset{\surd}{=} -1$

$2x - y - z = -3$

$2(-1) - 0 - 1 \overset{\surd}{=} -3$

73. $\left[\begin{array}{ccc|c} 1 & -1 & 1 & 3 \\ 1 & 1 & -1 & 4 \\ 0 & -2 & -2 & -1 \end{array}\right]$ $R_2 + (-1)R_1 \Rightarrow R_2$

Need a 0 here

$\sim \left[\begin{array}{ccc|c} 1 & -1 & 1 & 3 \\ 0 & 2 & -2 & 1 \\ 0 & -2 & -2 & -1 \end{array}\right]$ $R_3 + R_2 \Rightarrow R_3$

Need a 0 here

$\sim \left[\begin{array}{ccc|c} 1 & -1 & 1 & 3 \\ 0 & 2 & -2 & 1 \\ 0 & 0 & -4 & 0 \end{array}\right]$ $\frac{1}{2}R_2 \Rightarrow R_2$, $-\frac{1}{4}R_3 \Rightarrow R_3$

Need 1's here

$\sim \left[\begin{array}{ccc|c} 1 & -1 & 1 & 3 \\ 0 & 1 & -1 & \frac{1}{2} \\ 0 & 0 & 1 & 0 \end{array}\right]$

The resulting equivalent system

$$x - y + z = 3$$
$$y - z = \frac{1}{2}$$
$$z = 0$$

is solved by substitution.

$$z = 0$$
$$y - z = y - 0 = \frac{1}{2}$$

so $y = \frac{1}{2}$

$$x - y + z = x - \frac{1}{2} + 0 = 3$$

so $x = \frac{7}{2}$

Solution: $x = \frac{7}{2}$, $y = \frac{1}{2}$, $z = 0$

Check: $x - y + z = 3$ $\qquad x + y - z = 4$ $\qquad -2y - 2z = -1$

$$\frac{7}{2} - \frac{1}{2} + 0 \overset{\checkmark}{=} 3 \qquad \frac{7}{2} + \frac{1}{2} - 0 \overset{\checkmark}{=} 4 \qquad -2\left(\frac{1}{2}\right) - 2(0) \overset{?}{=} -1$$

$$-1 - 0 \overset{\checkmark}{=} -1$$

74. $\left[\begin{array}{ccc|c} 1 & 2 & 3 & 4 \\ 1 & 0 & 1 & 6 \\ 2 & 6 & 8 & 10 \end{array}\right]$ $R_2 + (-1)R_1 \Rightarrow R_2$, $R_3 + (-2)R_1 \Rightarrow R_3$

Need 0's here

$\left[\begin{array}{ccc|c} 1 & 2 & 3 & 4 \\ 0 & -2 & -2 & 2 \\ 0 & 2 & 2 & 2 \end{array}\right]$ $R_3 + R_2 \Rightarrow R_3$

Need a 0 here

$\left[\begin{array}{ccc|c} 1 & 2 & 3 & 4 \\ 0 & -2 & -2 & 2 \\ 0 & 0 & 0 & 4 \end{array}\right]$

Since the last row is equivalent to

$$0x + 0y + 0z = 4$$

there is no solution.

75. We expand by the first column.

$$\begin{vmatrix} 1 & 1 & 1 \\ 0 & 1 & 2 \\ 3 & 2 & 1 \end{vmatrix} = a_{11}\begin{pmatrix}\text{cofactor}\\ \text{of } a_{11}\end{pmatrix} + a_{21}\begin{pmatrix}\text{cofactor}\\ \text{of } a_{21}\end{pmatrix} + a_{31}\begin{pmatrix}\text{cofactor}\\ \text{of } a_{31}\end{pmatrix}$$

$$= 1(-1)^{1+1}\begin{vmatrix} 1 & 2 \\ 2 & 1 \end{vmatrix} + 0\left(\;\right) + 3(-1)^{3+1}\begin{vmatrix} 1 & 1 \\ 1 & 2 \end{vmatrix}$$

$$= (-1)^2[1\cdot 1 - 2\cdot 2] + 3(-1)^4[1\cdot 2 - 1\cdot 1]$$

$$= -3 + 3 = 0$$

76. We expand by the second column.

$$\begin{vmatrix} 1 & 2 & 1 \\ 2 & 1 & 1 \\ 3 & 0 & 1 \end{vmatrix} = a_{12}\begin{pmatrix}\text{cofactor}\\ \text{of } a_{12}\end{pmatrix} + a_{22}\begin{pmatrix}\text{cofactor}\\ \text{of } a_{22}\end{pmatrix} + a_{32}\begin{pmatrix}\text{cofactor}\\ \text{of } a_{32}\end{pmatrix}$$

$$= 2(-1)\begin{vmatrix} 2 & 1 \\ 3 & 1 \end{vmatrix} + 1(-1)^{2+2}\begin{vmatrix} 1 & 1 \\ 3 & 1 \end{vmatrix} + 0\left(\;\right)$$

$$= 2(-1)^3[2\cdot 1 - 3\cdot 1] + 1(-1)^4[1\cdot 1 - 3\cdot 1]$$

$$= -2(-1) + (-2) = 0$$

77. We expand by the first column.

$$\begin{vmatrix} 13 & 14 & 15 \\ 0 & 16 & 17 \\ 0 & 0 & 18 \end{vmatrix} = a_{11}\begin{pmatrix}\text{cofactor}\\ \text{of } a_{11}\end{pmatrix} + a_{21}\begin{pmatrix}\text{cofactor}\\ \text{of } a_{21}\end{pmatrix} + a_{31}\begin{pmatrix}\text{cofactor}\\ \text{of } a_{31}\end{pmatrix}$$

$$= 13(-1)^{1+1}\begin{vmatrix} 16 & 17 \\ 0 & 18 \end{vmatrix} + 0\left(\;\right) + 0\left(\;\right)$$

$$= 13(-1)^2[16\cdot 18 - 0\cdot 17]$$

$$= 13\cdot 16\cdot 18$$

$$= 3744$$

78. We expand by the first column.

$$\begin{vmatrix} 0 & 2 & 4 \\ 1 & 3 & 5 \\ 2 & 4 & 6 \end{vmatrix} = a_{11}\begin{pmatrix}\text{cofactor}\\ \text{of } a_{11}\end{pmatrix} + a_{21}\begin{pmatrix}\text{cofactor}\\ \text{of } a_{21}\end{pmatrix} + a_{31}\begin{pmatrix}\text{cofactor}\\ \text{of } a_{31}\end{pmatrix}$$

$$= 0\left(\;\right) + 1(-1)^{2+1}\begin{vmatrix} 2 & 4 \\ 4 & 6 \end{vmatrix} + 2(-1)^{3+1}\begin{vmatrix} 2 & 4 \\ 3 & 5 \end{vmatrix}$$

$$= (-1)^3[2\cdot 6 - 4\cdot 4] + 2(-1)^4[2\cdot 5 - 3\cdot 4]$$

$$= (-1)(-4) + 2(-2)$$

$$= 0$$

79. $D = \begin{vmatrix} 1 & 1 & 1 \\ 1 & 0 & 1 \\ 0 & 6 & -5 \end{vmatrix} = 5$

$$x = \frac{\begin{vmatrix} 15 & 1 & 1 \\ 10 & 0 & 1 \\ 0 & 6 & -5 \end{vmatrix}}{D} = \frac{20}{5} = 4$$

$$y = \frac{\begin{vmatrix} 1 & 1 & 0 \\ 1 & 10 & 1 \\ 0 & 0 & -5 \end{vmatrix}}{D} = \frac{25}{5} = 5$$

$$z = \frac{\begin{vmatrix} 1 & 1 & 15 \\ 1 & 0 & 10 \\ 0 & 6 & 0 \end{vmatrix}}{D} = \frac{30}{5} = 6$$

80. $D = \begin{vmatrix} 1 & 1 & 1 \\ 2 & 0 & -1 \\ 0 & 2 & -1 \end{vmatrix} = 8$

$$x = \frac{\begin{vmatrix} 12 & 1 & 1 \\ 1 & 0 & -1 \\ 3 & 2 & -1 \end{vmatrix}}{D} = \frac{24}{8} = 3$$

$$y = \frac{\begin{vmatrix} 1 & 12 & 1 \\ 2 & 1 & -1 \\ 0 & 3 & -1 \end{vmatrix}}{D} = \frac{32}{8} = 4$$

$$z = \frac{\begin{vmatrix} 1 & 1 & 12 \\ 2 & 0 & 1 \\ 0 & 2 & 3 \end{vmatrix}}{D} = \frac{40}{8} = 5$$

81.

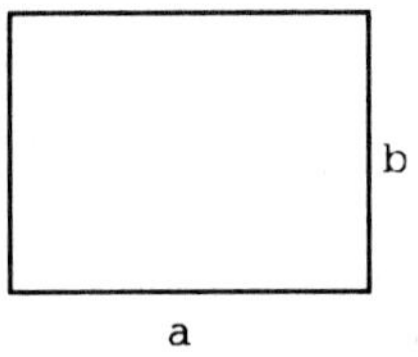

a = length of one side of rectangle
b = length of other side of rectangle

From the area formula we have ab = 30.
From the perimeter formula we have 2a + 2b = 22.

To solve the system

$$ab = 30$$
$$2a + 2b = 22$$

we solve the second equation for a in terms of b.

$$2a = 22 - 2b$$
$$a = 11 - b$$

Now we substitute this expression for a into the first equation.

$$(11 - b)b = 30$$
$$11b - b^2 = 30$$
$$0 = b^2 - 11b + 30$$
$$b^2 - 11b + 30 = 0$$
$$(b - 6)(b - 5) = 0$$

$b - 6 = 0$ or $b - 5 = 0$

$b = 6$ $\quad$ $b = 5$

For b = 6:

$a = 11 - b$
$a = 11 - 6$
$a = 5$

For b = 5:

$a = 11 - b$
$a = 11 - 6$
$a = 6$

The dimensions are 6 cm by 5 cm

82. Let x = amount of 40% solution
y = amount of 70% solution

100 grams are required, hence

$$x + y = 100 \quad (1)$$

49% of the 100 grams must be acid, hence

$$0.40x + 0.70y = 0.49(100) \quad (2)$$

We solve the system of equations (1), (2) by elimination using addition. First multiply (2) by 10 to eliminate decimals.

$$4x + 7y = 490$$

Now multiply equation (1) by -4 and add.

$$\begin{aligned} -4x - 4y &= -400 \\ \underline{4x + 7y} &= \underline{490} \\ 3y &= 90 \\ y &= 30 \\ x + y &= 100 \\ x + 30 &= 100 \\ x &= 70 \end{aligned}$$

70 grams of 40% solution and 30 grams of 70% solution.

Check: $70 + 30 = 100$

$0.40(70) + 0.70(30) = 28 + 21 = 49$

83. Since the circle passes through (5, 7), (-1, -1), and (6, 0), the coordinates of each of these points satisfy the equation of the circle. Substituting into

$$x^2 + y^2 + Dx + Ey + F = 0$$

we have

$$\begin{aligned} 5^2 + 7^2 + D(5) + E(7) + F &= 0 \\ 74 + D(5) + E(7) + F &= 0 \\ 5D + 7E + F &= -74 \quad (1) \\ (-1)^2 + (-1)^2 + D(-1) + E(-1) + F &= 0 \\ 2 - D - E + F &= 0 \\ -D - E + F &= -2 \quad (2) \\ 6^2 + 0^2 + D(6) + E(0) + F &= 0 \\ 36 + 6D + F &= 0 \\ 6D + F &= -36 \quad (3) \end{aligned}$$

We solve the system of equations (1), (2), (3).

We eliminate E from equation (1).

$$\begin{aligned} 5D + 7E + F &= -74 && \text{Equation (1)} \\ \underline{-7D - 7E + 7F} &= \underline{-14} && \text{7[equation (2)]} \\ -2D + 8F &= -88 && \text{(4)} \end{aligned}$$

We now solve the system

$$\begin{aligned} 6D + F &= -36 && (3) \\ -2D + 8F &= -88 && (4) \end{aligned}$$

We eliminate D.

$$\begin{aligned} 6D + F &= -36 && \text{Equation (3)} \\ \underline{-6D + 24F} &= \underline{-264} && \text{3[equation (4)]} \\ 25F &= -300 \\ F &= -12 \end{aligned}$$

Substituting F = -12 into equation (3) gives

$$\begin{aligned} 6D - 12 &= -36 \\ 6D &= -24 \\ D &= -4 \end{aligned}$$

Substituting D = -4 and F = -12 into equation (2) gives

$$\begin{aligned} -(-4) - E + (-12) &= -2 \\ 4 - E - 12 &= -2 \\ -E - 8 &= -2 \\ -E &= 6 \\ E &= -6 \end{aligned}$$

The equation of the circle is $x^2 + y^2 - 4x - 6y - 12 = 0$.

84. Let x = number of 1-cu ft boxes
y = number of 1.5-cu ft boxes
There are 30 boxes, hence

$x + y = 30$ (1)

Sicne x 1-cu ft boxes hold x cubic feet and y 1.5 cu-ft boxes hold 1.5y cubic feet, we have

$x + 1.5y = 39$ (2)

We solve the system (1), (2) by eliminaton using addition. We multiply the top equation by -1 and add.

$$\begin{aligned} -x - y &= -30 \\ \underline{x + 1.5y} &\underline{= 39} \\ 0.5y &= 9 \\ y &= 18 \end{aligned}$$

Substituting into equation (1), we have

$$\begin{aligned} x + 18 &= 30 \\ x &= 12 \end{aligned}$$

12 1-cu ft boxes, 18 1.5-cu ft boxes.

85. $\frac{1}{4}x - \frac{3}{4}y = -\frac{3}{8}$ (1)

$-\frac{2}{3}x + 2y = -1$ (2)

We first clear of fractions for convenience, by multiplying (1) by its lcd 8 and multiplying (2) by its lcd 3.

$2x - 6y = -3$ (3)
$-2x + 6y = -3$ (4)

If we add equations (3) and (4), we eliminate both x and y.

$0x + 0y = -6$

Hence the system is inconsistent.

No solution.

86. $x^2 + 2xy - y^2 = -4$ (1)
$x^2 - xy = 0$ (2)

Factor equation (2).

$$x(x - y) = 0$$
$$x = 0 \qquad x - y = 0$$
$$x = y$$

For x = 0 [replace x with 0 in equation (1) and solve for y]:

$$\begin{aligned} (0)^2 + 2(0)y - y^2 &= -4 \\ -y^2 &= -4 \\ y^2 &= 4 \\ y &= \pm 2 \end{aligned}$$

For $x = y$ [replace x with y in equation (1) and solve for y]:

$$y^2 + 2yy - y^2 = -4$$
$$2y^2 = -4$$
$$y^2 = -2$$
$$y = \pm\sqrt{-2}$$
$$y = \pm i\sqrt{2}$$

For $y = i\sqrt{2}$, $x = i\sqrt{2}$; for $y = -i\sqrt{2}$, $x = -i\sqrt{2}$

Solutions: $(0, 2)$, $(0, -2)$, $(i\sqrt{2}, i\sqrt{2})$, $(-i\sqrt{2}, -i\sqrt{2})$

87. $\dfrac{x^2}{9} - \dfrac{y^2}{4} = 1$

$x^2 + y^2 = 25$

We use elimination by addition, multiplying the top equation by 36 to eliminate fractions, multiplying the bottom equation by 9, and adding to eliiminate y.

$$\begin{aligned} 4x^2 - 9y^2 &= 36 \\ 9x^2 + 9y^2 &= 225 \\ \hline 13x^2 \quad &= 261 \end{aligned}$$

$$x^2 = \frac{261}{13}$$

$$x = \pm\sqrt{\frac{261}{13}} = \pm\frac{3\sqrt{377}}{13}$$

For either value of x, we replace x^2 with $\dfrac{261}{13}$ in $x^2 + y^2 = 25$ and solve for y.

$$\frac{261}{13} + y^2 = 25$$

$$y^2 = \frac{325}{13} - \frac{261}{13}$$

$$y^2 = \frac{64}{13}$$

$$y = \pm\sqrt{\frac{64}{13}} = \pm\frac{8\sqrt{13}}{13}$$

$$\left(\frac{3\sqrt{377}}{13}, \frac{8\sqrt{13}}{13}\right), \left(\frac{3\sqrt{377}}{13}, \frac{-8\sqrt{13}}{13}\right), \left(\frac{-3\sqrt{377}}{13}, \frac{8\sqrt{13}}{13}\right), \left(\frac{-3\sqrt{377}}{13}, \frac{-8\sqrt{13}}{13}\right)$$

88. $\dfrac{x^2}{9} - \dfrac{y^2}{4} = 1$

$\dfrac{x^2}{16} - \dfrac{y^2}{9} = 1$

We first clear of fractions for convenience, multiplying the top equation by its lcd 36 and the bottom equation by its lcd 144.

$$4x^2 - 9y^2 = 36$$
$$9x^2 - 16y^2 = 144$$

We now multiply the top equation by 9 and the bottom equation by -4 and add to eliminate x.

$$\begin{aligned} 36x^2 - 81y^2 &= 324 \\ -36x^2 + 64y^2 &= -576 \\ \hline -17y^2 &= -252 \end{aligned}$$

$$y^2 = \frac{252}{17}$$

$$y = \pm\sqrt{\frac{252}{17}} = \pm\sqrt{\frac{7 \cdot 36}{17}} = \pm\frac{6\sqrt{119}}{17}$$

For either value of y, we replace y^2 with $\frac{252}{17}$ in $4x^2 - 9y^2 = 36$ and solve for x.

$$4x^2 - 9\left(\frac{252}{17}\right) = 36$$

$$4x^2 - \frac{2268}{17} = \frac{612}{17}$$

$$4x^2 = \frac{2880}{17}$$

$$x^2 = \frac{720}{17}$$

$$x = \pm\sqrt{\frac{720}{17}} = \pm\sqrt{\frac{5 \cdot 144}{17}} = \pm\frac{12\sqrt{85}}{17}$$

$$\left(\frac{12\sqrt{85}}{17}, \frac{6\sqrt{119}}{17}\right), \left(\frac{-12\sqrt{85}}{17}, \frac{6\sqrt{119}}{17}\right), \left(\frac{12\sqrt{85}}{17}, \frac{-6\sqrt{119}}{17}\right), \left(\frac{-12\sqrt{85}}{17}, \frac{-6\sqrt{119}}{17}\right)$$

89. $x + y = 10$
$xy = 1$

Solve the top equation for x in terms of y.
$x = 10 - y$

Replace x by this expression in the bottom equation.

$$(10 - y)y = 1$$
$$10y - y^2 = 1$$
$$0 = y^2 - 10y + 1$$

This is not factorable in the integers, so we use the quadratic formula with $a = 1$, $b = -10$, $c = 1$

$$y = \frac{-(-10) \pm \sqrt{(10)^2 - 4(1)(1)}}{2(1)}$$

$$y = \frac{10 \pm \sqrt{100 - 4}}{2}$$

$$y = \frac{10 \pm \sqrt{96}}{2}$$

$$y = 5 \pm 2\sqrt{6}$$

For $y = 5 + 2\sqrt{6}$
$x = 10 - y$
$x = 10 - (5 + 2\sqrt{6})$
$x = -5 - 2\sqrt{6}$

For $y = 5 - 2\sqrt{6}$
$x = 10 - y$
$x = 10 - (5 - 2\sqrt{6})$
$x = 5 + 2\sqrt{6}$

$(-5 - 2\sqrt{6}, 5 + 2\sqrt{6})$, $(5 + 2\sqrt{6}, 5 - 2\sqrt{6})$

90. $x^2 + y^2 = 4$
$xy = 1$

We solve the bottom equation for y in terms of x.

$$y = \frac{1}{x}$$

We replace y with $\frac{1}{x}$ in the top equation.

$$x^2 + \left(\frac{1}{x}\right)^2 = 4$$
$$x^2 + \frac{1}{x^2} = 4 \qquad \text{lcm} = x^2,\ x \neq 0$$
$$x^2 \cdot x^2 + x^2 \cdot \frac{1}{x^2} = 4x^2$$
$$x^4 + 1 = 4x^2$$
$$x^4 - 4x^2 + 1 = 0$$

Let $u = x^2$, then $u^2 = x^4$. After substitution, the equation becomes
$$u^2 - 4u + 1 = 0$$

This is not factorable in the integers, so we use the quadratic formula with $a = 1$, $b = -4$, and $c = 1$.

$$u = \frac{-(-4) \pm \sqrt{(-4)^2 - 4(1)(1)}}{2(1)}$$
$$= \frac{4 \pm \sqrt{16 - 4}}{2}$$
$$u = \frac{4 \pm \sqrt{12}}{2}$$
$$u = 2 \pm \sqrt{3}$$

Replacing u with x^2, we have
$$x^2 = 2 \pm \sqrt{3}$$
$$x = \pm\sqrt{2 \pm \sqrt{3}}$$

Since $y = \frac{1}{x}$, we have four solutions:

$$\left(\sqrt{2 + \sqrt{3}}, \frac{1}{\sqrt{2 + \sqrt{3}}}\right), \left(-\sqrt{2 + \sqrt{3}}, -\frac{1}{\sqrt{2 + \sqrt{3}}}\right),$$
$$\left(\sqrt{2 - \sqrt{3}}, \frac{1}{\sqrt{2 - \sqrt{3}}}\right), \left(-\sqrt{2 - \sqrt{3}}, -\frac{1}{\sqrt{2 - \sqrt{3}}}\right)$$

91. $x^2 + y^2 = 9$
$x^2 - y^2 = 0$

$2x^2 = 9$ We have added to eliminate y.

$$x^2 = \frac{9}{2}$$
$$x = \pm\sqrt{\frac{9}{2}} = \pm\frac{3\sqrt{2}}{2}$$

For either value of x, we replace x^2 with $\frac{9}{2}$ in $x^2 - y^2 = 0$ and solve for y.

$$\frac{9}{2} - y^2 = 0$$
$$\frac{9}{2} = y^2$$
$$y = \pm\sqrt{\frac{9}{2}} = \pm\frac{3\sqrt{2}}{2}$$

$$\left(\frac{3\sqrt{2}}{2}, \frac{3\sqrt{2}}{2}\right), \left(\frac{3\sqrt{2}}{2}, -\frac{3\sqrt{2}}{2}\right), \left(-\frac{3\sqrt{2}}{2}, \frac{3\sqrt{2}}{2}\right), \left(-\frac{3\sqrt{2}}{2}, -\frac{3\sqrt{2}}{2}\right)$$

92. $x^2 - y = 0$
$y - x^2 = 1$

If we rewrite the top equation as $-y + x^2 = 0$ and add, we eliminate both x and y.

$$-y + x^2 = 0$$
$$\underline{y - x^2 = 1}$$
$$0 = 1$$

Hence the system is inconsistent. No solution.

93. $\frac{1}{3}x + \frac{1}{3}y + \frac{1}{3}z = 1$
$\frac{1}{4}y + \frac{1}{4}z = 1$
$\frac{1}{12}x + \frac{1}{6}y + \frac{1}{4}z = 1$

We start by eliminating fractions for convenience.

$x + y + z = 3$ (1)
$y + z = 4$ (2)
$x + 2y + 3z = 12$ (3)

We eliminate x from equation (3).

$-x - y - z = -3$ (-1)[equation (1)]
$\underline{x + 2y + 3z = 12}$ (3)
$y + 2z = 9$ (4)

We now solve the system (2), (4).

$y + z = 4$ (2)
$y + 2z = 9$ (4)

We eliminate y.

$-y - z = -4$ (-1)[equation (4)]
$\underline{y + 2z = 9}$ (4)
$z = 5$

Substituting $z = 5$ into equation (2) gives

$y + 5 = 4$
$y = -1$

Substituting $y = -1$ and $z = 5$ into equation (1) gives

$x + (-1) + 5 = 3$
$x + 4 = 3$
$x = -1$

Solution: $x = -1$, $y = -1$, $z = 5$

Check: $\frac{1}{3}x + \frac{1}{3}y + \frac{1}{3}z = 1$

$\frac{1}{3}(-1) + \frac{1}{3}(-1) + \frac{1}{3}(5) \overset{?}{=} 1$

$-\frac{1}{3} - \frac{1}{3} + \frac{5}{3} \overset{\surd}{=} \frac{3}{3}$

$\frac{1}{4}y + \frac{1}{4}z = 1$

$\frac{1}{4}(-1) + \frac{1}{4}(5) \overset{?}{=} 1$

$-\frac{1}{4} + \frac{5}{4} \overset{\surd}{=} \frac{4}{4}$

$\frac{1}{12}x + \frac{1}{6}y + \frac{1}{4}z = 1$

$\frac{1}{12}(-1) + \frac{1}{6}(-1) + \frac{1}{4}(5) \overset{?}{=} 1$

$-\frac{1}{12} - \frac{1}{6} + \frac{5}{4} \overset{?}{=} 1$

$-\frac{1}{12} - \frac{2}{12} + \frac{15}{12} \overset{\surd}{=} \frac{12}{12}$

94. $0.2x + 0.4y + 0.4z = 1.4$
$0.5x + 1.5y + 2.0z = 8.0$
$0.4x + 0.4y + 0.8z = 3.2$

We start by clearing of decimals for convenience.

$2x + 4y + 4z = 14$ (1)
$5x + 15y + 20z = 80$ (2)
$4x + 4y + 8z = 32$ (3)

We eliminate x from equations (2) and (3)

$-10x - 20y - 20z = -70$ -5[equation (1)]
$10x + 30y + 40z = 160$ 2[equation (2)]
$10y + 20z = 90$ (4)

$-4x - 8y - 8z = -28$ -2[equation (1)]
$4x + 4y + 8z = 32$ (3)
$-4y = 4$
$y = -1$

Substituting y = -1 into equation (4), we obtain

$10(-1) + 20z = 90$
$-10 + 20z = 90$
$20z = 100$
$z = 5$

Substituting y = -1 and z = 5 into equation (1) we obtain

$2x + 4(-1) + 4(5) = 14$
$2x + 16 = 14$
$2x = -2$
$x = -1$

Solution: $x = -1$, $y = -1$, $z = 5$

Check: $0.2x + 0.4y + 0.4z = 1.4$

$0.2(-1) + 0.4(-1) + 0.4(5) \overset{?}{=} 1.4$

$-0.2 - 0.4 + 2.0 \overset{\surd}{=} 1.4$

$0.4x + 0.4y + 0.8z = 3.2$

$0.4(-1) + 0.4(-1) + 0.8(5) \overset{?}{=} 3.2$

$-0.4 - 0.4 + 4.0 \overset{\surd}{=} 3.2$

$0.5x + 1.5y + 2.0z = 8.0$

$0.5(-1) + 1.5(-1) + 2.0(5) \overset{?}{=} 8.0$

$-0.5 - 1.5 + 10 \overset{\surd}{=} 8$

95. Let x and y equal the two numbers.
Then $xy = 5$ (product)
$x + y = 5$ (sum)

We solve the bottom equation for y in terms of x.
$y = 5 - x$

Now replace y with this expression in xy = 5.

$$x(5 - x) = 5$$
$$5x - x^2 = 5$$
$$0 = x^2 - 5x + 5$$

This is not factorable in the integers, so we use the quadratic formula with a = 1, b = -5, c = 5.

$$x = \frac{-(-5) \pm \sqrt{(-5)^2 - 4(1)(5)}}{2(1)}$$
$$x = \frac{5 \pm \sqrt{25 - 20}}{2}$$
$$x = \frac{5 \pm \sqrt{5}}{2}$$

For $x = \frac{5 + \sqrt{5}}{2}$

$$y = 5 - \frac{5 + \sqrt{5}}{2}$$
$$y = \frac{10 - 5 - \sqrt{5}}{2}$$
$$y = \frac{5 - \sqrt{5}}{2}$$

For $x = \frac{5 - \sqrt{5}}{2}$

$$y = 5 - \frac{5 - \sqrt{5}}{2}$$
$$y = \frac{10 - 5 + \sqrt{5}}{2}$$
$$y = \frac{5 + \sqrt{5}}{2}$$

The numbers are $\frac{5 + \sqrt{5}}{2}$ and $\frac{5 - \sqrt{5}}{2}$

96. Since the parabola passes through $\left(0, \frac{1}{4}\right)$, $\left(1, \frac{7}{4}\right)$, and $\left(2, \frac{29}{4}\right)$, the coordinates of each of these points must satisfy the equation of the parabola. Substituting into $y = ax^2 + bx + c$, we have

$$\frac{1}{4} = a(0)^2 + b(0) + c$$
$$c = \frac{1}{4} \qquad (1)$$
$$\frac{7}{4} = a(1)^2 + b(1) + c$$
$$a + b + c = \frac{7}{4} \qquad (2)$$
$$\frac{29}{4} = a(2)^2 + b(2) + c$$
$$4a + 2b + c = \frac{29}{4} \qquad (3)$$

We substitute $c = \frac{1}{4}$ into equations (2) and (3) to obtain

$$a + b + \frac{1}{4} = \frac{7}{4}$$
$$a + b = \frac{3}{2} \qquad (4)$$
$$4a + 2b + \frac{1}{4} = \frac{29}{4}$$
$$4a + 2b = 7 \qquad (5)$$

We solve the system (4), (5) by multiplying equation (4) by -2 and adding to eliminate b.

$$\begin{aligned} -2a - 2b &= -3 && -2[\text{equation } (4)] \\ 4a + 2b &= 7 && (5) \\ \hline 2a &= 4 \\ a &= 2 \end{aligned}$$

Substituting a = 2 into equation (4), we obtain

$$2 + b = \frac{3}{2}$$

$$b = -\frac{1}{2}$$

Thus, the equation of the parabola is $y = 2x^2 - \frac{1}{2}x + \frac{1}{4}$.

97. Let x = number of $\frac{1}{2}$-pound packages

y = number of $\frac{1}{3}$-pound packages

There are 120 packages. Hence

$$x + y = 120 \qquad (1)$$

Since x $\frac{1}{2}$-pound packages weigh $\frac{1}{2}x$ pounds and y $\frac{1}{3}$-pound packages weigh $\frac{1}{3}y$ pounds, we have

$$\frac{1}{2}x + \frac{1}{3}y = 48 \qquad (2)$$

We solve the system (1), (2) by elimination using addition. We multiply the bottom equation by -3 and add.

$$\begin{aligned} x + y &= 120 \\ -\frac{3}{2}x - y &= -144 \\ \hline -\frac{1}{2}x &= -24 \\ x &= 48 \end{aligned}$$

Substituting into equation (1), we have

$$48 + y = 120$$

$$y = 72$$

48 $\frac{1}{2}$-pound packages and 72 $\frac{1}{3}$-pound packages

98. Let x = number of grams of mix A

y = number of grams of mix B

z = number of grams of mix C

We have

$$\begin{aligned} 0.30x + 0.20y + 0.10z &= 27 && \text{(protein)} \\ 0.03x + 0.05y + 0.04z &= 5.4 && \text{(fat)} \\ 0.10x + 0.20y + 0.10z &= 19 && \text{(moisture)} \end{aligned}$$

Clearing of decimals for convenience, we have

$$\begin{aligned} 3x + 2y + z &= 270 \\ 3x + 5y + 4z &= 540 \\ x + 2y + z &= 190 \end{aligned}$$

Form the augmented matrix and solve.

$$\left[\begin{array}{ccc|c} 3 & 2 & 1 & 270 \\ 3 & 5 & 4 & 540 \\ 1 & 2 & 1 & 190 \end{array}\right] R_1 \Leftrightarrow R_3$$

Need a 1 here

$$\sim \left[\begin{array}{ccc|c} 1 & 2 & 1 & 190 \\ 3 & 5 & 4 & 540 \\ 3 & 2 & 1 & 270 \end{array}\right] \begin{array}{l} R_2 + (-3)R_1 \Rightarrow R_2 \\ R_3 + (-3)R_1 \Rightarrow R_3 \end{array}$$

Need 0's here

$$\sim \left[\begin{array}{ccc|c} 1 & 2 & 1 & 190 \\ 0 & -1 & 1 & -30 \\ 0 & -4 & -2 & -300 \end{array}\right] (-1)R_2 \Rightarrow R_2$$

Need a 1 here

$$\sim \left[\begin{array}{ccc|c} 1 & 2 & 1 & 190 \\ 0 & 1 & -1 & 30 \\ 0 & -4 & -2 & -300 \end{array}\right] R_3 + 4R_2 \Rightarrow R_3$$

Need a 0 here

$$\left[\begin{array}{ccc|c} 1 & 2 & 1 & 190 \\ 0 & 1 & -1 & 30 \\ 0 & 0 & -6 & -180 \end{array}\right] -\frac{1}{6}R_3 \Rightarrow R_3$$

Need a 1 here

$$\left[\begin{array}{ccc|c} 1 & 2 & 1 & 190 \\ 0 & 1 & -1 & 30 \\ 0 & 0 & 1 & 30 \end{array}\right]$$

The resulting equivalent system

$$\begin{aligned} x + 2y + z &= 190 \\ y - z &= 30 \\ z &= 30 \end{aligned}$$

is solved by substitution

$z = 30$

$y - z = y - 30 = 30$

so $y = 60$

$x + 2y + z = x + 2(60) + 30 = x + 150 = 190$

so $x = 40$

40 grams mix A, 60 grams mix B, 30 grams mix C

99. Let x = number of grams of mix A
y = number of grams of mix B

We have

$0.30x + 0.20y \ge 27$ (protein)
$0.03x + 0.05y \ge 5.4$ (fat)
$0.10x + 0.20y \ge 19$ (moisture)

Clearing of decimals, we have

$3x + 2y \ge 270$
$3x + 5y \ge 540$
$x + 2y \ge 190$

Moreover, x and y must be positive, thus

$x \ge 0$
$y \ge 0$

Graphing these inequalities, we obtain

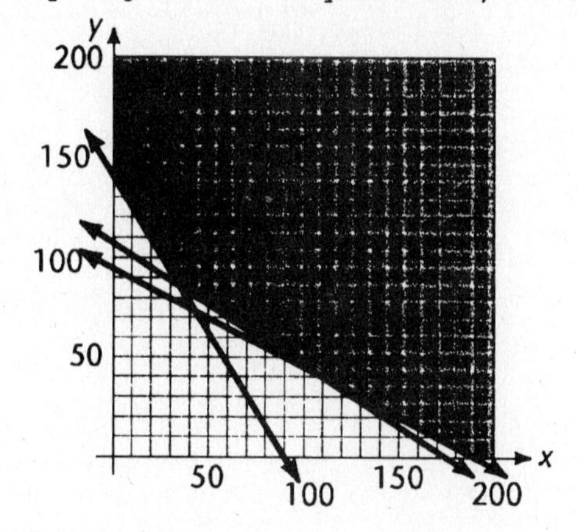

100. The corner points in problem 99 are obtained by solving

$x = 0$	$3x + 2y = 270$	$3x + 5y = 540$	$x + 2y = 190$
$3x + 2y = 270$	$3x + 5y = 540$	$x + 2y = 190$	$y = 0$
to obtain	to obtain	to obtain	to obtain
(0, 135)	(30, 90)	(130, 30)	(190, 0)

We can find the least costly solution by comparing the costs at each of the corner points.

Corner	Cost = 0.03x + 0.03y	
(0, 135)	0.03(0) + 0.03(135) = 4.05	
(30, 90)	0.03(30) + 0.03(90) = 3.60	least costly combination
(130, 30)	0.03(130) + 0.03(30) = 4.80	
(190, 0)	0.03(190) + 0.03(0) = 5.70	

The least costly combination is 30 grams of Mix A, 90 grams of mix B.

Practice Test 8

1. $x + 3y = 5$ (1)
$3x + 2y = 1$ (2)

We multiply (1) by -3 and add.

$$\begin{aligned} -3x - 9y &= -15 \quad 3[\text{equation (1)}] \\ \underline{3x + 2y} &= \underline{\ \ 1} \quad (2) \\ -7y &= -14 \\ y &= 2 \end{aligned}$$

Replace y by 2 in (1) to find x.

$$\begin{aligned} x + 3(2) &= 5 \\ x + 6 &= 5 \\ x &= -1 \end{aligned}$$

Solution (-1, 2)

Check: $x + 3y = 5$ $\quad$ $3x + 2y = 1$

$(-1) + 3(2) \overset{?}{=} 5$ $\quad$ $3(-1) + 2(2) \overset{?}{=} 1$

$-1 + 6 \overset{\surd}{=} 5$ $\quad$ $-3 + 4 \overset{\surd}{=} 1$

2. $3x + 4y = 1$ (1)
$2x - 3y = 12$ (2)

We multiply (1) by 3 and (2) by 4 and add.

$$\begin{aligned} 9x + 12y &= 3 \\ \underline{8x - 12y} &= \underline{48} \\ 17x \qquad &= 51 \\ x &= 3 \end{aligned}$$

Replace x by 3 in (1) to find y.

$$\begin{aligned} 3(3) + 4y &= 1 \\ 9 + 4y &= 1 \\ 4y &= -8 \\ y &= -2 \end{aligned}$$

Solution (3, -2)

Check: $3x + 4y = 1$ $\quad$ $2x - 3y = 12$

$3(3) + 4(-2) \overset{?}{=} 1$ $\quad$ $2(3) - 3(-2) \overset{?}{=} 12$

$9 - 8 \overset{\surd}{=} 1$ $\quad$ $6 + 6 \overset{\surd}{=} 12$

3. $x + y + z = 6$ (1)
$2x + y = 8$ (2)
$3y + 4z = 10$ (3)

We eliminate x from (2).

$$\begin{aligned} -2x - 2y - 2z &= -12 \quad -2[\text{equation (1)}] \\ \underline{2x + y \qquad} &= \underline{\ \ 8} \quad (2) \\ -y - 2z &= -4 \quad (4) \end{aligned}$$

We now solve the system (3), (4).

$$\begin{aligned} 3y + 4z &= 10 \quad (3) \\ -y - 2z &= -4 \quad (4) \end{aligned}$$

We eliminate y.

$$\begin{aligned} 3y + 4z &= 10 \quad (3) \\ \underline{-3y - 6z} &= \underline{-12} \quad 3[\text{equation (4)}] \\ -2z &= -2 \\ z &= 1 \end{aligned}$$

Substituting z = 1 into equation (3), we have

$$\begin{aligned} 3y + 4(1) &= 10 \\ 3y + 4 &= 10 \\ 3y &= 6 \\ y &= 2 \end{aligned}$$

Substituting y = 2 and z = 1 into equation (1), we have

$$\begin{aligned} x + 2 + 1 &= 6 \\ x &= 3 \end{aligned}$$

Solution: (3, 2, 1)

Check: $x + y + z = 6$ $\quad$ $2x + y = 8$ $\quad$ $3y + 4z = 10$

$3 + 2 + 1 \overset{\surd}{=} 6$ $\quad$ $2(3) + 2 \overset{\surd}{=} 8$ $\quad$ $3(2) + 4(1) \overset{\surd}{=} 10$

4. $x + 2y + 3z = 4$ (1)
$2x - y + z = 5$ (2)
$3x + y + 4z = 6$ (3)

We eliminate x from equations (2) and (3).

$$\begin{array}{rl} -2x - 4y - 6z = -8 & -2[\text{equation } (1)] \\ \underline{2x - y + z = 5} & (2) \\ -5y - 5z = -3 & (4) \end{array}$$

$$\begin{array}{rl} -3x - 6y - 9z = -12 & -3[\text{equation } (1)] \\ \underline{3x + y + 4z = 6} & (3) \\ -5y - 5z = -6 & (5) \end{array}$$

We attempt to solve the system (4), (5).
If we multiply equation (5) by -1 and add to equation (4), we eliminate both y and z.

$$\begin{array}{rl} 5y + 5z = 6 & -1[\text{equation } (5)] \\ \underline{-5y - 5z = -3} & (4) \\ 0 = 3 & \end{array}$$

Hence the system is inconsistent. No solution.

5. $x + 2y + 3z = 4$ (1)
$-x + y + 3z = 5$ (2)
$2x + y = -1$ (3)

We eliminate x from equations (2) and (3).

$$\begin{array}{rl} x + 2y + 3z = 4 & (2) \\ \underline{-x + y + 3z = 5} & (3) \\ 3y + 6z = 9 & (4) \end{array}$$

$$\begin{array}{rl} -2x - 4y - 6z = -8 & -2[\text{equation } (1)] \\ \underline{2x + y \qquad = -1} & (3) \\ -3y - 6z = -9 & (5) \end{array}$$

We now solve the system (4), (5). If we add the two equations, we eliminate x, y, and the constant term.

$$\begin{array}{rl} 3y + 6z = 9 & (4) \\ \underline{-3y - 6z = -9} & (5) \\ 0 = 0 & \end{array}$$

Therefore, the system has an infinite number of solutions. We solve for y in terms of z.

$$3y + 6z = 9$$
$$y + 2z = 3$$
$$y = 3 - 2z$$

Substituting this expression for y in equation (3), we have

$$2x + 3 - 2z = -1$$
$$2x = 2z - 4$$
$$x = \frac{2z - 4}{2}$$
$$x = z - 2$$

Thus, the solutions can be written as $(z - 2, 3 - 2z, z)$ for z any real number.

6.
$$\left[\begin{array}{cc|c} 1 & 2 & 3 \\ 3 & 2 & 1 \end{array}\right] \quad R_2 + \underbrace{(-3)R_1}_{} \Rightarrow R_2$$

↑ Need a 0 here

$-3 \quad -6 \quad -9$

$$\sim \left[\begin{array}{cc|c} 1 & 2 & 3 \\ 0 & -4 & -8 \end{array}\right] \quad -\frac{1}{4}R_2 \Rightarrow R_2$$

↑ Need a 1 here

$$\sim \left[\begin{array}{cc|c} 1 & 2 & 3 \\ 0 & 1 & 2 \end{array}\right]$$

The original system is therefore equivalent to

$$x + 2y = 3$$
$$y = 2$$

Thus, $y = 2$, and $x + 2(2) = 3$ so $x = -1$.

Solution: $(-1, 2)$

Check: $x + 2y = 3$ $\qquad$ $3x + 2y = 1$

$-1 + 2(2) \overset{\checkmark}{=} 3$ $\qquad$ $3(-1) + 2(2) \overset{?}{=} 1$

$-3 + 4 \overset{\checkmark}{=} 1$

7.
$$\left[\begin{array}{ccc|c} 1 & 2 & 2 & 4 \\ 1 & 2 & 1 & 5 \\ 2 & 1 & 1 & 6 \end{array}\right] \quad \begin{array}{l} R_2 + (-1)R_1 \Rightarrow R_2 \\ R_3 + (-2)R_1 \Rightarrow R_3 \end{array}$$

Need 0's here

$$\sim \left[\begin{array}{ccc|c} 1 & 2 & 2 & 4 \\ 0 & 0 & -1 & 1 \\ 0 & -3 & -3 & -2 \end{array}\right] \quad \begin{array}{l} R_2 \Leftrightarrow R_3 \\ -\frac{1}{3}R_3 \Rightarrow R_3 \end{array}$$

Need a 1 here

$$\sim \left[\begin{array}{ccc|c} 1 & 2 & 2 & 4 \\ 0 & 1 & 1 & \frac{2}{3} \\ 0 & 0 & -1 & 1 \end{array}\right] \quad -R_3 \Rightarrow R_3$$

↑ Need a 1 here

$$\sim \left[\begin{array}{ccc|c} 1 & 2 & 2 & 4 \\ 0 & 1 & 1 & \frac{2}{3} \\ 0 & 0 & 1 & -1 \end{array}\right]$$

The resulting equivalent system

$$x + 2y + 2z = 4$$
$$y + z = \frac{2}{3}$$
$$z = -1$$

is solved by substitution.

$$z = -1$$
$$y + z = y - 1 = \frac{2}{3}$$

so $\quad y = \frac{5}{3}$

$$x + 2y + 2z = x + 2\left(\frac{5}{3}\right) + 2(-1) = x + \frac{4}{3} = 4$$

so $\quad x = \frac{8}{3}$

Solution: $\left(\frac{8}{3}, \frac{5}{3}, -1\right)$

Check: $x + 2y + 2z = 4$ $\qquad$ $x + 2y + z = 5$ $\qquad$ $2x + y + z = 6$

$\frac{8}{3} + 2\left(\frac{5}{3}\right) + 2(-1) \overset{?}{=} 4$ $\qquad$ $\frac{8}{3} + 2\left(\frac{5}{3}\right) - 1 \overset{?}{=} 5$ $\qquad$ $2\left(\frac{8}{3}\right) + \frac{5}{3} + (-1) \overset{?}{=} 6$

$\frac{8}{3} + \frac{10}{3} - \frac{6}{3} \overset{\checkmark}{=} \frac{12}{3}$ $\qquad$ $\frac{8}{3} + \frac{10}{3} - \frac{3}{3} \overset{\checkmark}{=} \frac{15}{3}$ $\qquad$ $\frac{16}{3} + \frac{5}{3} - \frac{3}{3} \overset{\checkmark}{=} \frac{18}{3}$

8.

$$\left[\begin{array}{ccc|c} 1 & 2 & 3 & 6 \\ 2 & 3 & 1 & 6 \\ 3 & 1 & 2 & 6 \end{array}\right] \begin{array}{l} \\ R_2 + (-2)R_1 \Rightarrow R_2 \\ R_3 + (-3)R_1 \Rightarrow R_3 \end{array}$$

Need 0's here

$$\sim \left[\begin{array}{ccc|c} 1 & 2 & 3 & 6 \\ 0 & -1 & -5 & -6 \\ 0 & -5 & -7 & -12 \end{array}\right] \begin{array}{l} \\ -R_2 \Rightarrow R_2 \\ \\ \end{array}$$

Need a 1 here

$$\sim \left[\begin{array}{ccc|c} 1 & 2 & 3 & 6 \\ 0 & 1 & 5 & 6 \\ 0 & -5 & -7 & -12 \end{array}\right] \begin{array}{l} \\ \\ R_3 + 5R_2 \Rightarrow R_3 \end{array}$$

Need a 0 here

$$\sim \left[\begin{array}{ccc|c} 1 & 2 & 3 & 6 \\ 0 & 1 & 5 & 6 \\ 0 & 0 & 18 & 18 \end{array}\right] \begin{array}{l} \\ \\ \frac{1}{18}R_3 \Rightarrow R_3 \end{array}$$

Need a 1 here

$$\sim \left[\begin{array}{ccc|c} 1 & 2 & 3 & 6 \\ 0 & 1 & 5 & 6 \\ 0 & 0 & 1 & 1 \end{array}\right]$$

The resulting equivalent system

$$\begin{aligned} x + 2y + 3z &= 6 \\ y + 5z &= 6 \\ z &= 1 \end{aligned}$$

is solved by substitution.

$$z = 1$$

$$y + 5z = y + 5(1) = y + 5 = 6$$

so $y = 1$

$$x + 2(1) + 3(1) = x + 5 = 6$$

so $x = 1$

Solution: (1, 1, 1)

Check: $x + 2y + 3z = 6$; $1 + 2 + 3 \overset{\checkmark}{=} 6$ $\quad$ $2x + 3y + z = 6$; $2 + 3 + 1 \overset{\checkmark}{=} 6$ $\quad$ $3x + y + 2z = 6$; $3 + 1 + 2 \overset{\checkmark}{=} 6$

9. We solve by Cramer's rule, given that D = 5.

$$x = \frac{\begin{vmatrix} 1 & 1 & 3 \\ 0 & 2 & 7 \\ 0 & 1 & 6 \end{vmatrix}}{D} = \frac{5}{5} = 1$$

$$y = \frac{\begin{vmatrix} 2 & 1 & 3 \\ 3 & 0 & 7 \\ 4 & 0 & 6 \end{vmatrix}}{D} = \frac{10}{5} = 2$$

$$z = \frac{\begin{vmatrix} 2 & 1 & 1 \\ 3 & 2 & 0 \\ 4 & 1 & 0 \end{vmatrix}}{D} = \frac{-5}{5} = -1$$

10. $x^2 + y^2 = 16$
$x^2 - y = 4$

Solve the bottom equation for x^2 in terms of y.

$$x^2 = y + 4$$

Substitute this expression into the top equation to eliminate x.

$$y + 4 + y^2 = 16$$
$$y^2 + y - 12 = 0$$

Solve by factoring.

$$(y + 4)(y - 3) = 0$$
$$y + 4 = 0 \quad \text{or} \quad y - 3 = 0$$
$$y = -4 \qquad\qquad y = 3$$

For $y = -4$

$$x^2 = -4 + 4$$
$$x^2 = 0$$
$$x = 0$$

For $y = 3$

$$x^2 = 3 + 4$$
$$x^2 = 7$$
$$x = \pm\sqrt{7}$$

$(0, -4)$, $(\sqrt{7}, 1)(-\sqrt{7}, 1)$

11. $x^2 + y^2 = 16$
$x - y = 0$

Solve the bottom equation for x in terms of y.

$$x = y$$

Substitute this expression for x in the top equation.

$$y^2 + y^2 = 16$$
$$2y^2 = 16$$
$$y^2 = 8$$
$$y = \pm\sqrt{8} \quad \text{or} \quad \pm 2\sqrt{2}$$

Since $x = y$, the solution is $(2\sqrt{2}, 2\sqrt{2})$, $(-2\sqrt{2}, -2\sqrt{2})$

12. $x^2 + y^2 = 16$
$\dfrac{x^2}{25} + \dfrac{y^2}{36} = 1$

We choose elimination using addition, multiplying the top equation by -25 and the bottom equation by 900 to eliminate y.

$$\begin{array}{rl} -25x^2 - 25y^2 &= -400 \\ 36x^2 + 25y^2 &= 900 \\ \hline 11x^2 \qquad\quad &= 500 \end{array}$$

$$x^2 = \frac{500}{11}$$

$$x = \pm\sqrt{\frac{500}{11}} = \pm\frac{10\sqrt{55}}{11}$$

$$x^2 = \frac{500}{11}$$

$$\frac{500}{11} + y^2 = 16$$

$$y^2 = \frac{176}{11} - \frac{500}{11}$$

$$y^2 = -\frac{324}{11}$$

$$y = \pm\sqrt{-\frac{324}{11}} = \pm\frac{18i\sqrt{11}}{11}$$

$$\left(\frac{10\sqrt{55}}{11}, \frac{18i\sqrt{11}}{11}\right), \left(\frac{10\sqrt{55}}{11}, -\frac{18i\sqrt{11}}{11}\right), \left(-\frac{10\sqrt{55}}{11}, \frac{18i\sqrt{11}}{11}\right), \left(-\frac{10\sqrt{55}}{11}, -\frac{18i\sqrt{11}}{11}\right)$$

13.

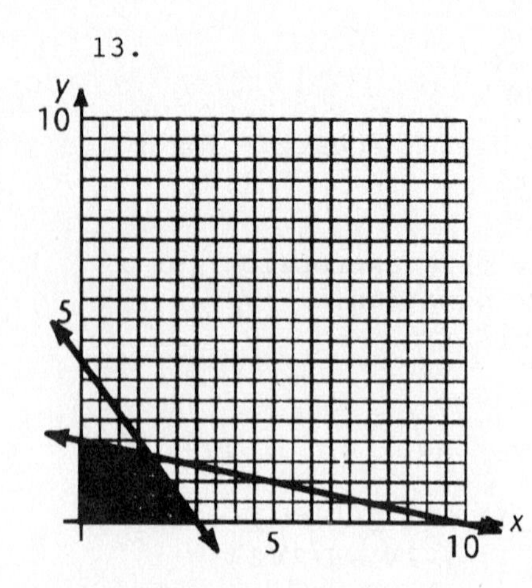

14.

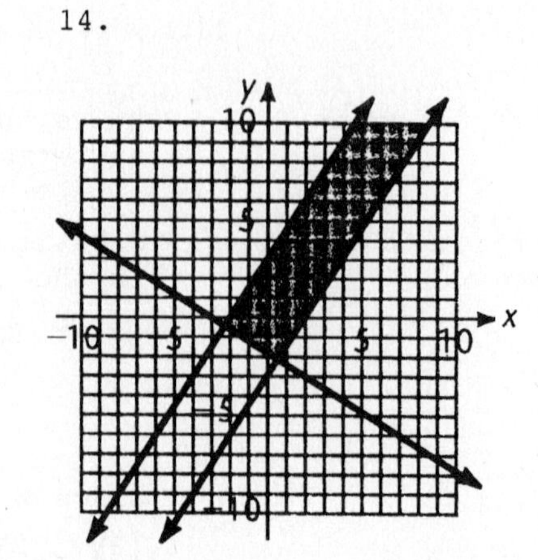

15.

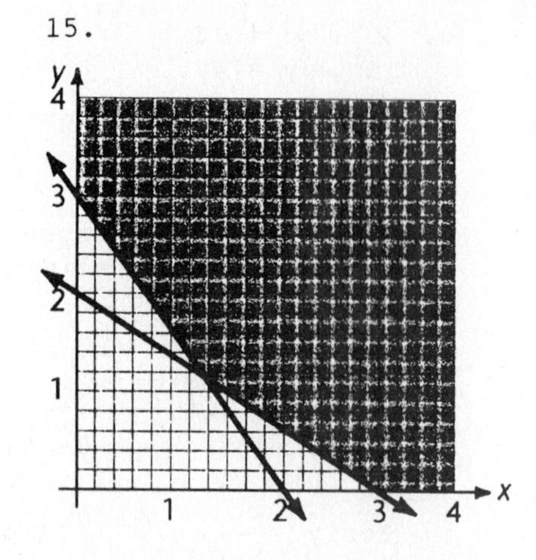

16. (A) $R_1 \Leftrightarrow R_2$ means interchange Rows 1 and 2.

$$\left[\begin{array}{cc|c} 4 & 5 & 6 \\ 1 & 2 & 3 \end{array}\right]$$

(B) $3R_2 \Rightarrow R_2$ means multiply Row 2 of the matrix in (A) by 3.

$$\left[\begin{array}{cc|c} 4 & 5 & 6 \\ 3 & 6 & 9 \end{array}\right]$$

(C) $R_1 + (-1)R_2 \Rightarrow R_1$ means replace Row 1 of the matrix in (B) by itself plus (−1) times Row 2.

$$\left[\begin{array}{cc|c} 4 & 5 & 6 \\ 3 & 6 & 9 \end{array}\right] \Rightarrow \left[\begin{array}{cc|c} 1 & -1 & -3 \\ 3 & 6 & 9 \end{array}\right]$$

$-3 \quad -6 \quad -9$

17. $$\begin{vmatrix} 2 & 3 \\ 4 & 5 \end{vmatrix} = 2\cdot 5 - 4\cdot 3 = 10 - 12 = -2$$

We expand $\begin{vmatrix} 1 & 2 & 3 \\ 0 & 4 & 5 \\ 6 & 7 & 8 \end{vmatrix}$ by the first column.

$$\begin{vmatrix} 1 & 2 & 3 \\ 0 & 4 & 5 \\ 6 & 7 & 8 \end{vmatrix} = a_{11}\begin{pmatrix}\text{cofactor} \\ \text{of } a_{11}\end{pmatrix} + a_{21}\begin{pmatrix}\text{cofactor} \\ \text{of } a_{21}\end{pmatrix} + a_{31}\begin{pmatrix}\text{cofactor} \\ \text{of } a_{31}\end{pmatrix}$$

$$= 1(-1)^{1+1}\begin{vmatrix} 4 & 5 \\ 7 & 8 \end{vmatrix} + 0\left(\;\right) + 6(-1)^{3+1}\begin{vmatrix} 2 & 3 \\ 4 & 5 \end{vmatrix}$$

$$= (-1)^2[4\cdot 8 - 7\cdot 5] + 6(-1)^4[2\cdot 5 - 4\cdot 3]$$

$$= -3 + 6(-2) = -15$$

Thus,

$$\begin{vmatrix} 2 & 3 \\ 4 & 5 \end{vmatrix} - \begin{vmatrix} 1 & 2 & 3 \\ 0 & 4 & 5 \\ 6 & 7 & 8 \end{vmatrix} = -2 - (-15) = 13$$

18. Let x and y equal the two numbers.

Then $xy = -\frac{1}{2}$ (product)

$x + y = -\frac{1}{2}$ (sum)

We solve the bottom equation for y in terms of x.

$$y = -x - \frac{1}{2}$$

Now we replace y with this expression in $xy = -\frac{1}{2}$

$$x\left(-x - \frac{1}{2}\right) = -\frac{1}{2}$$

$$-x^2 - \frac{1}{2}x = -\frac{1}{2} \quad \text{lcm} = 2$$

$$-2x^2 - x = -1$$

$$0 = 2x^2 + x - 1$$

Solve by factoring.

$$0 = (2x - 1)(x + 1)$$

$2x - 1 = 0$ or $x + 1 = 0$

$2x = 1$ $\qquad x = -1$

$x = \frac{1}{2}$

For $x = \frac{1}{2}$ $\qquad$ For $x = -1$

$\frac{1}{2} + y = -\frac{1}{2}$ $\qquad$ $-1 + y = -\frac{1}{2}$

$y = -1$ $\qquad$ $y = \frac{1}{2}$

The numbers are -1 and $\frac{1}{2}$.

19. Let x = number of \$1.50 tickets

y = number of \$2.50 tickets

$x + y = 760$ (total number of tickets)

$1.5x + 2.5y = 1280$ (total value of tickets)

We choose elimination by addition, multiplying the top equation by -1.5 and adding.

$$\begin{aligned} -1.5x - 1.5y &= -1140 \\ 1.5x + 2.5y &= 1280 \\ \hline y &= 140 \end{aligned}$$

Replace y by 140 in the top equation

$$x + 140 = 760$$

$$x = 620$$

Solution: 620 general admission, 140 reserved.

Check: $620 + 140 = 760$

$1.5(620) + 2.5(140) = 930 + 350 = 1280$

20. Let x = number of pounds of mix A
    y = number of pounds of mix B

We have

$$0.18x + 0.12y = 26.4 \text{ (nitrogen)}$$
$$0.12x + 0.15y = 24.6 \text{ (potash)}$$

Clearing of decimals for convenience, we have

$$18x + 12y = 2640$$
$$12x + 15y = 2460$$

We choose elimination using addition, multiplying the top equation by 2 and the bottom equation by -3 and adding.

$$\begin{aligned} 36x + 24y &= 5280 \\ \underline{-36x - 45y} &= \underline{-7380} \\ -21y &= -2100 \\ y &= 100 \end{aligned}$$

Substituting into the top equation, we have

$$\begin{aligned} 18x + 12(100) &= 2640 \\ 18x + 1200 &= 2640 \\ 18x &= 1440 \\ x &= 80 \end{aligned}$$

Solution: 80 pounds of mix A, 100 pounds of mix B.

CHAPTER 9 FUNCTIONS

Exercise 9-1 Functions

Key Ideas and Formulas

Informal Definition of Function

A **function** is a rule that produces a correspondence between a first variable x and a second variable y such that to each value of x there corresponds *one and only one* value for y. In this case, we say that y is a function of x.

If a variable y is a function of a variable x, we call the variable denoted by x the independent variable and the variable denoted by y the dependent variable. The value of the dependent variable is determined by, or dependent on, the value of the independent variable. The set of allowable values for the independent variable is called the domain of the function. The set of values for the dependent variable is called the range of the function.

Formal Definition of Function

A function is a rule that produces a correspondence between a first set of elements, called the **domain**, and a second set, called the **range**, such that to each element in the domain there corresponds *one and only one* element in the range.

A function can be specified in many ways, including an equation, a table, a graph, or directly as a set of ordered pairs.

The graph of all ordered pairs (x, y) in a function is called the graph of the function.

Vertical-Line Test

If y is a function of x, then the graph of all ordered pairs (x, y) belonging to the function has the property that no vertical line crosses the graph more than once. Conversely, if no vertical line crosses a graph more than once, the graph is the graph of a function.

If a domain is not specified, and unless stated to the contrary, the domain will be assumed to consist of all values of x for which the equation provides a real value of y.

---

1. Function (exactly one range value corresponds to each domain value)

3. Not a function (two range values correspond to the domain value 3)

5. Function (exactly one range value corresponds to each domain value)

7. Not a function (three range values correspond to the domain value 1)

9. Function (exactly one range value corresponds to each domain value)

11. Function (exactly one range value corresponds to each domain value)

13. Function (each vertical line passes through at most one point on the graph)

15. Not a function ($x = 0$, for example, passes through more than one point on the graph)

17. Function (each vertical line passes through at most one point on the graph)

19. Function (each vertical line passes through at most one point on the graph)

21. Not a function ($x = 0$, for example, passes through more than one point on the graph)

23. Not a function ($x = 0$, for example, passes through more than one point on the graph)

25.

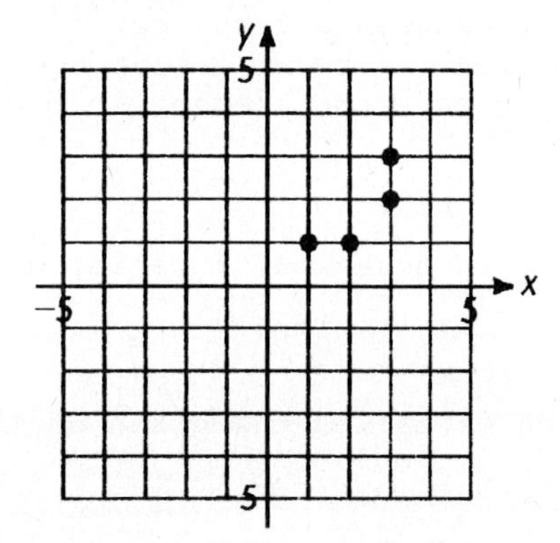

Not a function (two values of y correspond to $x = 3$)

27.

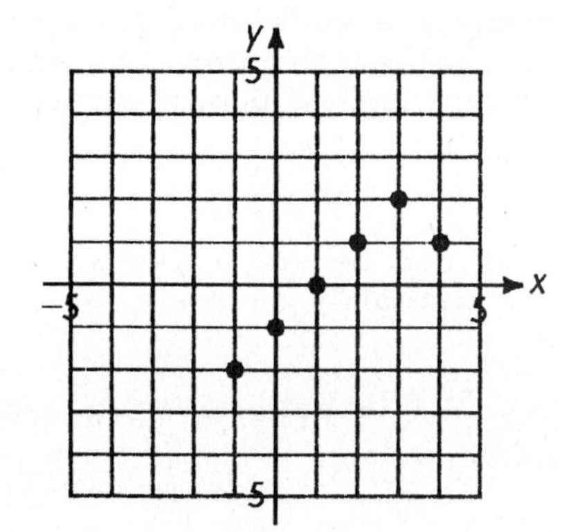

A function (each value of x corresponds to one value of y)

29.

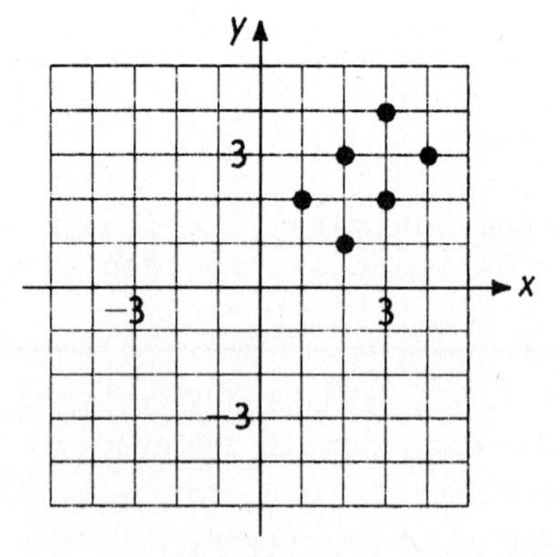

Not a function (two values of y correspond to $x = 2$, for example)

31.

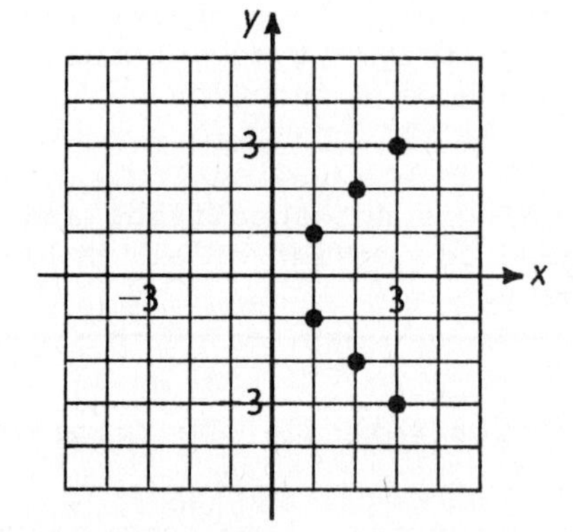

Not a function (two values of y correspond to $x = 1$, for example)

33. $\frac{y + 1}{3} = x$

$y + 1 = 3x$

$y = 3x - 1$

A function. (Each value of x corresponds to one value of y.)

35. $y - 1 = x^2 - 3x$

$y = x^2 - 3x + 1$

A function. (Each value of x corresponds to one value of y.)

37. $y^2 = x$

$y = \pm\sqrt{x}$

Not a function (two values of y correspond to x = 1, for example)

39. $x = y^2 - y$

$y^2 - y + \frac{1}{4} = x + \frac{1}{4}$

$\left(y - \frac{1}{2}\right)^2 = x + \frac{1}{4}$

$y - \frac{1}{2} = \pm\sqrt{x + \frac{1}{4}}$

$y = \frac{1}{2} \pm \sqrt{x + \frac{1}{4}}$

Not a function (two values of y correspond to x = 0, for example)

41. $x - 3 = \frac{y}{x}$

$y = x(x - 3)$

A function. (Each value of x corresponds to one value of y.)

43. $xy - x - y - 1 = 0$

$xy - y = x + 1$

$y(x - 1) = x + 1$

$y = \frac{x + 1}{x - 1}$

A function. (Each value of x corresponds to one value of y.)

45. $\frac{y - 1}{x} = x + 2$

$y - 1 = x(x + 2)$

$y = x^2 + 2x + 1$

A function. (Each value of x corresponds to one value of y.)

47. $x^2 = y^2 + 1$

$y^2 = x^2 - 1$

$y = \pm\sqrt{x^2 - 1}$

Not a function (two values of y correspond to x = 2, for example)

49. $9x^2 = 36 - 4y^2$

$9x^2 + 4y^2 = 36$

$4y^2 = 36 - 9x^2$

$y^2 = \frac{36 - 9x^2}{4}$

$y = \pm\sqrt{\frac{36 - 9x^2}{4}}$

Not a function (two values of y correspond to x = 0, for example)

51.

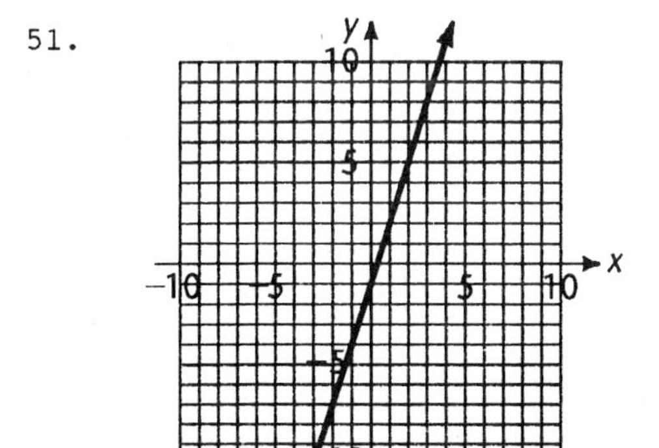

A function. Each vertical line crosses the graph at most once.

53.

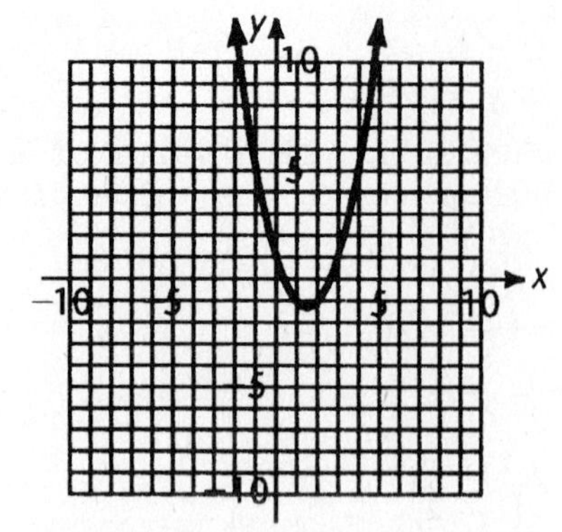

A function. Each vertical line crosses the graph at most once.

55.

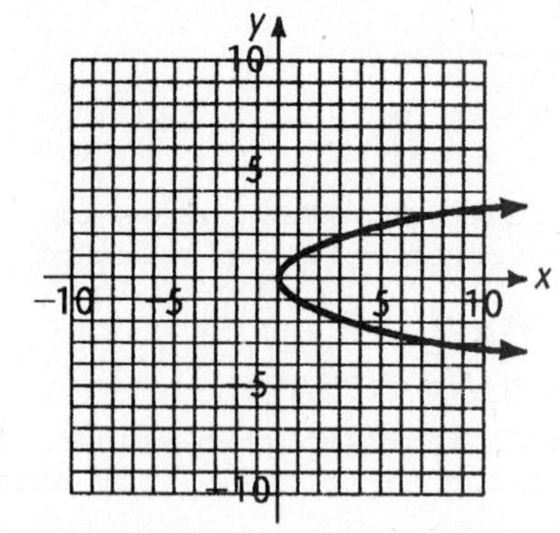

Not a function. The line x = 1, for example, crosses the graph twice.

57.

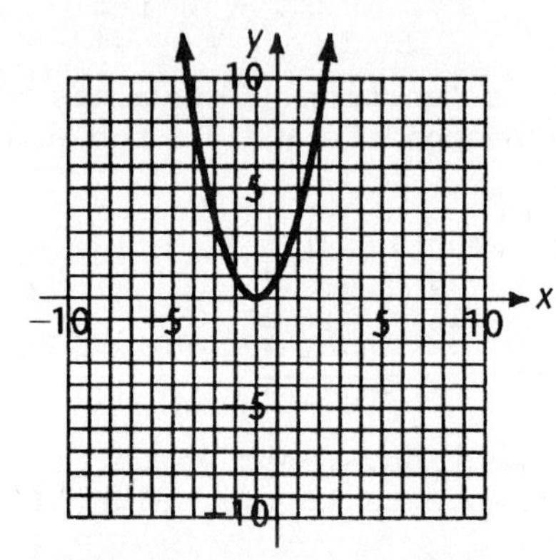

A function. Each vertical line crosses the graph at most once.

59.

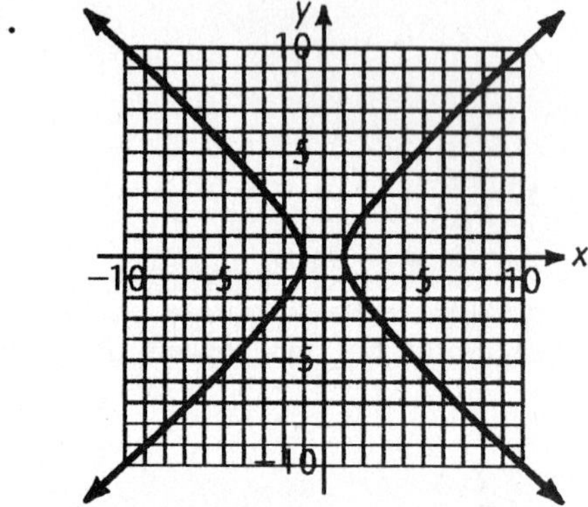

Not a function. The line x = 5, for example, crosses the graph twice.

61.

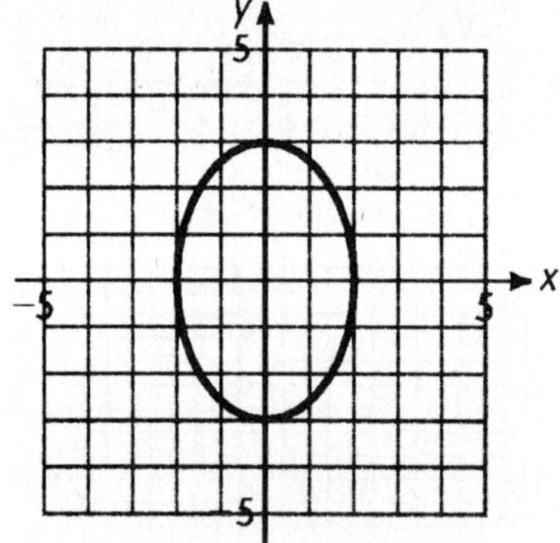

Not a function. The line x = 0, for example, crosses the graph twice.

63.

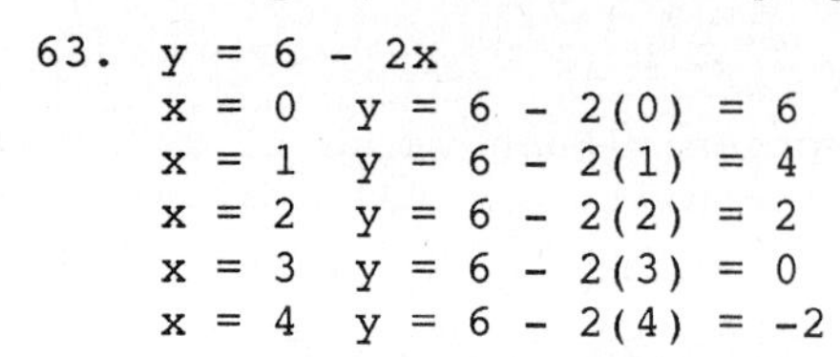

$y = 6 - 2x$

$x = 0 \quad y = 6 - 2(0) = 6$

$x = 1 \quad y = 6 - 2(1) = 4$

$x = 2 \quad y = 6 - 2(2) = 2$

$x = 3 \quad y = 6 - 2(3) = 0$

$x = 4 \quad y = 6 - 2(4) = -2$

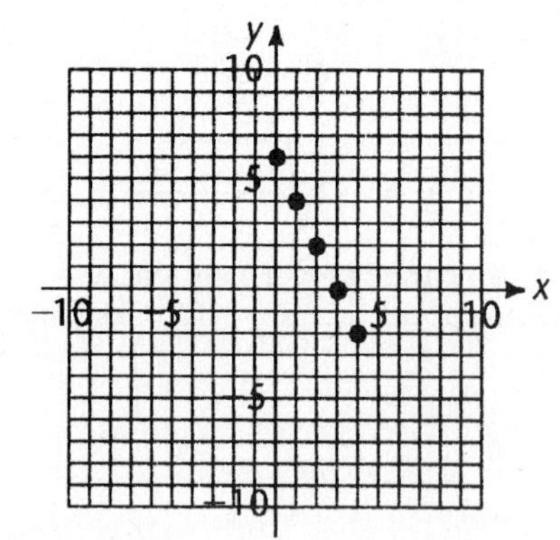

A function. (Each value of x corresponds to one value of y.)

65. $y^2 = x$

$x = 0 \quad y^2 = 0 \quad y = 0$
$x = 1 \quad y^2 = 1 \quad y = \pm 1$
$x = 4 \quad y^2 = 4 \quad y = \pm 2$

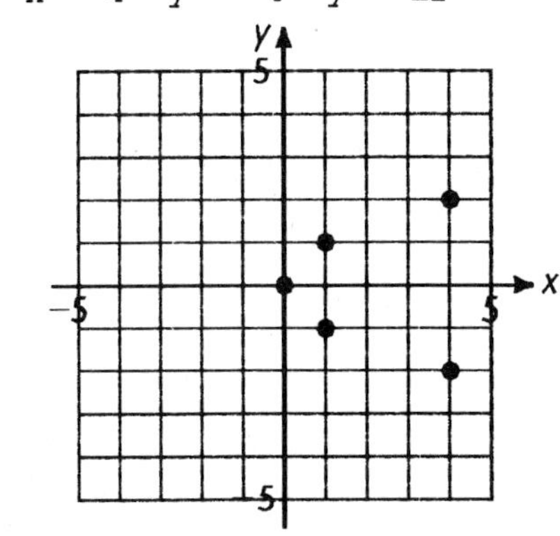

Not a function. (Two values of y correspond to x = 1, for example.)

67. $x^2 + y^2 = 4$

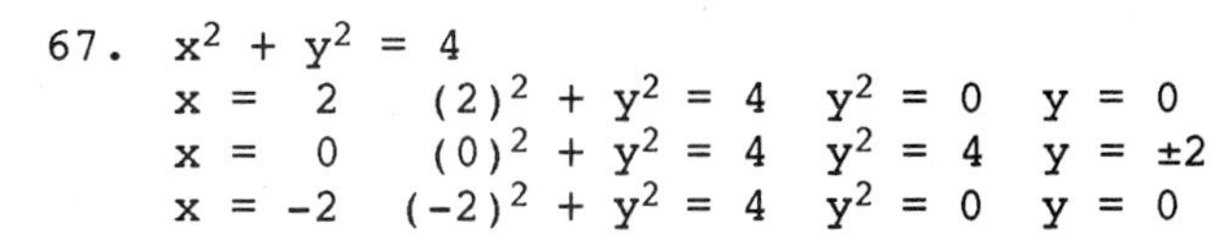

$x = 2 \quad (2)^2 + y^2 = 4 \quad y^2 = 0 \quad y = 0$
$x = 0 \quad (0)^2 + y^2 = 4 \quad y^2 = 4 \quad y = \pm 2$
$x = -2 \quad (-2)^2 + y^2 = 4 \quad y^2 = 0 \quad y = 0$

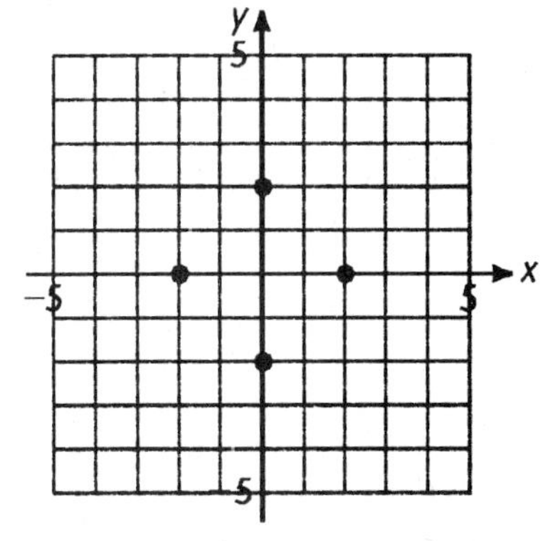

Not a function. (Two values of y correspond to x = 0, for example.)

69. $y = |x|$

$x = -2 \quad y = |-2| = 2$
$x = 0 \quad y = |0| = 0$
$x = 2 \quad y = |2| = 2$

A function. (Each value of x corresponds to one value of y.)

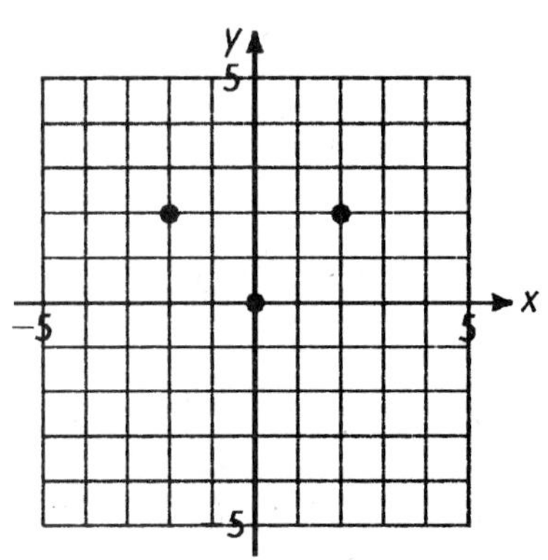

71. For each real x, y is defined and real. Domain: R.

73. For each real x, y is defined and real. Domain: R.

75. For y to be defined, x cannot be zero. Domain: All real numbers except x = 0.

77. $y = \dfrac{x - 1}{(x + 2)(x - 3)}$

For y to be defined, $(x + 2)(x - 3)$ cannot be zero.
$(x + 2)(x - 3) = 0$
$x + 2 = 0$ or $x - 3 = 0$
$x = -2 \qquad x = 3$
Domain: All real numbers except x = -2 and x = 3.

COMMON ERROR:
Deciding that x cannot = 1. The expression for y is defined and equal to 0 if x = 1.

79. $y = \dfrac{3x}{x^2 + x - 12}$

For y to be defined, $x^2 + x - 12$ cannot be zero.
$x^2 + x - 12 = 0$
$(x + 4)(x - 3) = 0$
$x + 4 = 0$ or $x - 3 = 0$
$x = -4 \qquad x = 3$
Domain: All real numbers except x = -4 and x = 3.

81. $y = \sqrt{4 - x}$

For y to be real, $4 - x$ cannot be negative; that is

$4 - x \ge 0$
$-x \ge -4$
$x \le 4$

Domain: $x \le 4$ or $(-\infty, 4]$

83. For y to be real, $\frac{x - 1}{x + 3}$ cannot be negative; that is,

$\frac{x - 1}{x + 3} \ge 0$

Sign of $(x - 1)$ − − ¦ − − − ¦ + +
Sign of $(x + 3)$ − − ¦ + + + ¦ + +
(number line: −3, 1)

Domain: $x < -3$ or $x \ge 1$, $(-\infty, -3) \cup [1, \infty)$

85. For y to be real, $x^2 - x - 6$ cannot be negative; that is.

$x^2 - x - 6 \ge 0$ or $(x - 3)(x + 2) \ge 0$

sign of $(x - 3)$ − − − ¦ − − − ¦ + + +
sign of $(x + 2)$ − − − ¦ + + + ¦ + + +
(number line x: −2, 3)

Domain: $x \le -2$ or $x \ge 3$, $(-\infty, -2] \cup [3, \infty)$

87. For y to be real, x cannot be negative; that is, $x \ge 0$.

89. For each real x, y is defined and real. Domain: R.

91. $y = \frac{x}{2}$

$x = -10 \quad y = \frac{-10}{2} = -5$

$x = -9 \quad y = \frac{-9}{2} = -4.5$

$\vdots$

$x = 9 \quad y = \frac{9}{2} = 4.5$

$x = 10 \quad y = \frac{10}{2} = 5$

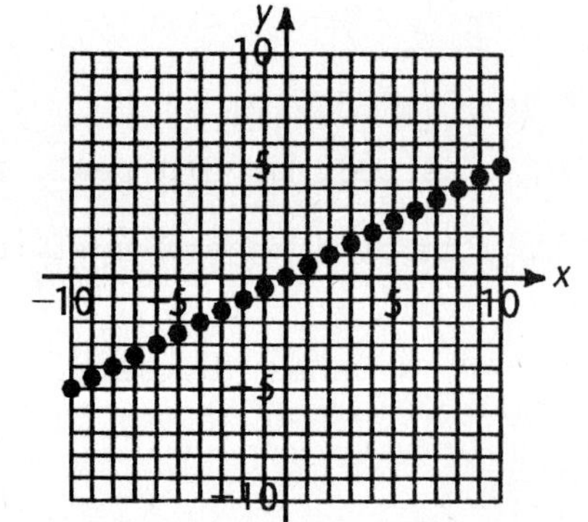

Since exactly one value of y corresponds to each value of x, this rule is a function.

93. Since $0 \le x \le 3$ and x is an integer, there are four possible values for x: 0, 1, 2, 3. Since y is an integer and $0 \le y \le x$, we have:

$x = 0$	$0 \le y \le 0$	$y = 0$
$x = 1$	$0 \le y \le 1$	$y = 0$ or 1
$x = 2$	$0 \le y \le 2$	$y = 0$ or 1 or 2
$x = 3$	$0 \le y \le 3$	$y = 0$ or 1 or 2 or 3

Since more than one value of y corresponds to some x values, this rule is not a function.

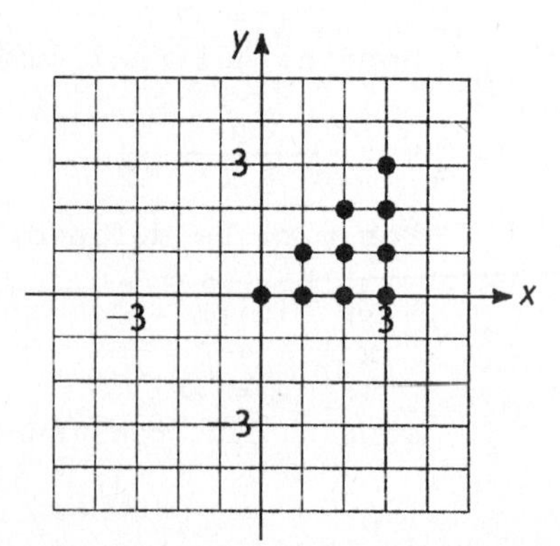

95. $y = \frac{|x|}{x}$

$x = -3 \quad y = \frac{|-3|}{-3} = \frac{3}{-3} = -1$

$x = -2 \quad y = \frac{|-2|}{-2} = \frac{2}{-2} = -1$

$x = -1 \quad y = \frac{|-1|}{-1} = \frac{1}{-1} = -1$

$x = 1 \quad y = \frac{|1|}{1} = \frac{1}{1} = 1$

$x = 2 \quad y = \frac{|2|}{2} = \frac{2}{2} = 1$

$x = 3 \quad y = \frac{|3|}{3} = \frac{3}{3} = 1$

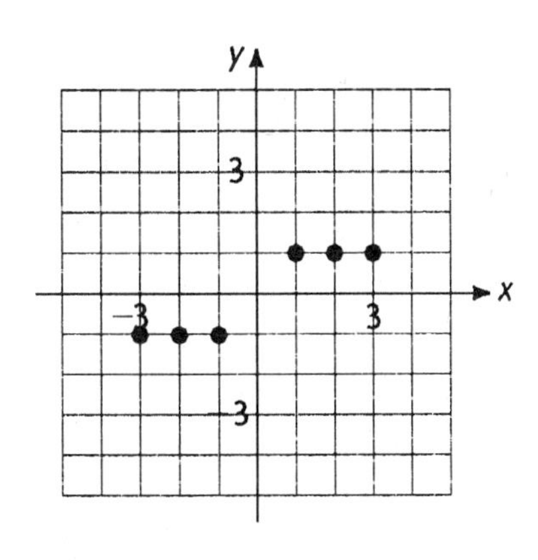

Since exactly one output value corresponds to each input value, this rule is a function.

97. Since exactly one number is associated with each letter, this rule is a function. 9 is the value that corresponds to x.

Domain: letters A through Y with Q omitted.
Range: integers 2 through 9.

99. (A) **Partial table**

t	d
0	0
1	144
2	256
3	336
4	384
5	400
6	384
7	336
8	256
9	144
10	0

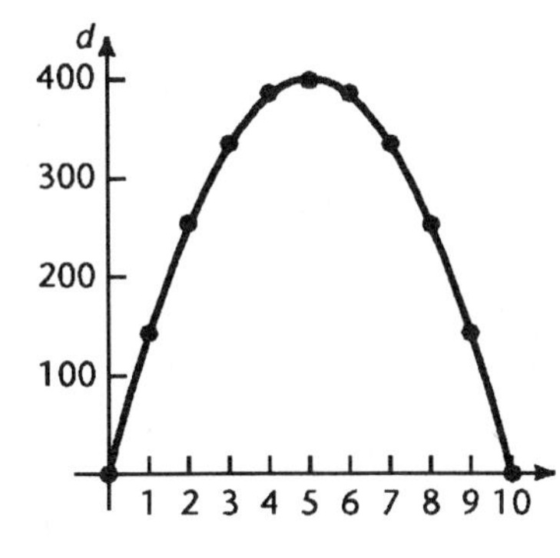

(B)

Domain Set of all real t	**Range Set of all real d**
$0 \le t \le 10$	$0 \le d \le 400$

(C) The rule is a function, since each vertical line passes through at most one point on the graph.

101. Given that the graph is a straight line, we are asked to find a linear equation, that is, one of the form $p = md + b$, that relates d and p. We choose two points on the graph arbitrarily from the long list given, say, $(d_1, p_1) = (30, 64)$ and $(d_2, p_2) = (130, 44)$. To find the equation, we first find the slope of the line through these two points, using a slope formula.

$$m = \frac{p_2 - p_1}{d_2 - d_1} = \frac{44 - 64}{130 - 30} = \frac{-20}{100} = -\frac{1}{5}$$

Now we can use the point-slope form with $m = -\frac{1}{5}$ and $(d_1, p_1) = (30, 64)$

$$p - p_1 = m(d - d_1)$$

$$p - 64 = -\frac{1}{5}(d - 30)$$

$$p - 64 = -\frac{1}{5}d + 6$$

$$p = -\frac{1}{5}d + 70$$

Therefore, the value of p corresponding to $d = 75$ is $p = -\frac{1}{5}(75) + 70 = 55$.

Exercise 9-2 Function Notation

Key Ideas and Formulas

The Function Symbol $f(x)$

For any element x in the domain of the function f, the function symbol $f(x)$ represents the element in the range of f corresponding to that x.

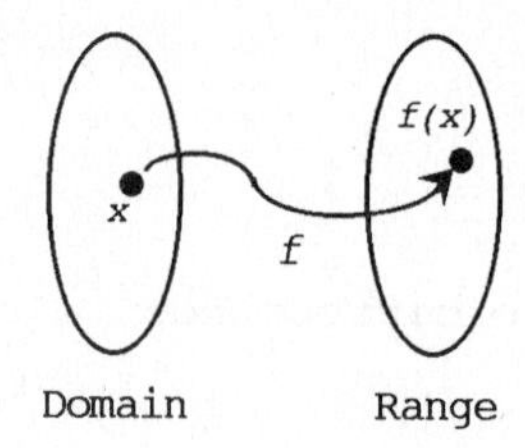

f(x) is **not** the product of f and x. The symbol f(x) is read "f of x" or "the value of f at x." The variable x is the independent variable; y and f(x) both name the dependent variable.

If x is an input value, then f(x) is an output value; symbolically $f: x \Rightarrow f(x)$.

Given a function specified by f(x) = an expression involving x, we find f(a), the range value associated with the domain value a, by replacing x by a wherever it occurs in the given expression, and evaluating the result.

For functions f(x) and g(x),

$$(f + g)(x) = f(x) + g(x)$$
$$(f - g)(x) = f(x) - g(x)$$
$$(f \cdot g)(x) = f(x) \cdot g(x)$$
$$\left(\frac{f}{g}\right)(x) = \frac{f(x)}{g(x)}$$

The domain of f + g, f - g, and f · g is the set of numbers for which both f and g are defined. The domain of f/g is the set of numbers for which both f and g are defined and g(x) is not 0.

We call g(f(x)) the **composition of f by g**. The domain of g(f(x)) is the set of all numbers in the domain of f for which f(x) is in the domain of g.

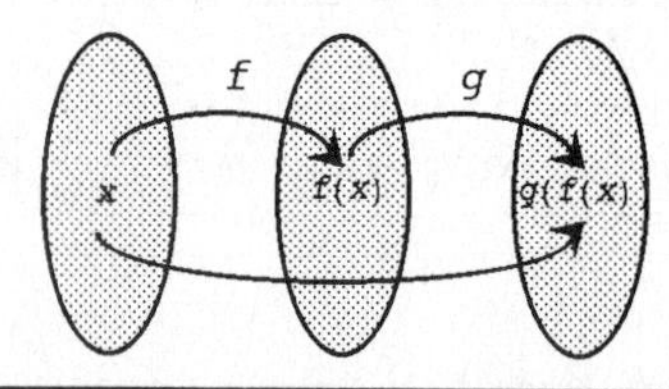

1. $f(x) = 3x - 2$
$f(2) = 3(2) - 2 = 6 - 2 = 4$

COMMON ERROR:
$f(2) \neq f \cdot 2$
$\neq (3x - 2) \cdot 2$
This represents incorrect use of the function symbol. f(2) means: replace x by 2 in the expression for f(x).

3. $f(x) = 3x - 2$
$f(-2) = 3(-2) - 2 = -6 - 2 = -8$

5. $f(x) = 3x - 2$
$f(0) = 3(0) - 2 = -2$

7. $g(x) = x - x^2$
$g(2) = (2) - (2)^2 = 2 - 4 = -2$

9. $g(x) = x - x^2$
$g(4) = (4) - (4)^2 = 4 - 16 = -12$

11. $g(x) = x - x^2$
$g(-2) = (-2) - (-2)^2 = -2 - 4 = -6$

13. $h(x) = x(x - 1)$
$h(2) = 2(2 - 1) = 2(1) = 2$

15. $h(x) = x(x - 1)$
$h(0) = 0(0 - 1) = 0$

17. $h(x) = x(x - 1)$
$h(-6) = (-6)[(-6) - 1] = (-6)(-7) = 42$

19. $f(x) = 3x - 2$
$f(x + 1) = 3(x + 1) - 2$
$= 3x + 3 - 2$
$= 3x + 1$

COMMON ERROR:
$f(x + 1) \neq f(x) + 1$
$\neq 3x - 2 + 1$
This represents incorrect use of the function symbol. f(x + 1) means: replace x by x + 1 in the expression for f(x).

21. $h(x) = x(x - 1)$
$h(x + 1) = (x + 1)[(x + 1) - 1]$
$= (x + 1)[x + 1 - 1]$
$= (x + 1)x$
$= x^2 + x$

23. $g(x) = x - x^2$
$g(x - 1) = (x - 1) - (x - 1)^2$
$= (x - 1) - (x^2 - 2x + 1)$
$= x - 1 - x^2 + 2x - 1$
$= -x^2 + 3x - 2$

25. $f(x) = 10x - 7$
$f(-2) = 10(-2) - 7 = -20 - 7 = -27$

27. $g(x) = 6 - 2x$
$g(2) = 6 - 2(2) = 6 - 4 = 2$

29. $g(x) = 6 - 2x$
$g(0) = 6 - 2(0) = 6$

31. $f(x) = 10x - 7$
$f(3) = 10(3) - 7 = 23$
$g(x) = 6 - 2x$
$g(2) = 6 - 2(2) = 2$
$f(3) + g(2) = 23 + 2 = 25$

33. $g(x) = 6 - 2x,\ G(x) = x - x^2$
$2g(-1) - 3G(-1) = 2[(6 - 2(-1)] - 3[(-1) - (-1)^2]$
$= 2[6 + 2] - 3[-1 - 1]$
$= 2[8] - 3[-2]$
$= 16 + 6$
$= 22$

35. $f(x) = 10x - 7,\ g(x) = 6 - 2x,\ G(x) = x - x^2$
$$\frac{f(2) \cdot g(-4)}{G(-1)} = \frac{[10(2) - 7] \cdot [6 - 2(-4)]}{(-1) - (-1)^2} = \frac{[20 - 7][6 + 8]}{-1 - 1} = \frac{(13)(14)}{-2} = -91$$

37. $g(x) = 6 - 2x$
$g(u - 2) = 6 - 2(u - 2)$
$= 6 - 2u + 4$
$= 10 - 2u$

39. $G(x) = x - x^2$
$G(3a) = (3a) - (3a)^2$
$= 3a - 9a^2$

41. $g(x) = 6 - 2x$
$g(2 + h) = 6 - 2(2 + h)$
$= 6 - 4 - 2h$
$= 2 - 2h$

COMMON ERRORS:
$g(2+h) \neq g(2) + g(h)$
$\neq g(2) + h$
These represent incorrect use of the function symbol and incorrect substitution.

43. $g(x) = 6 - 2x$

$$\frac{g(2 + h) - g(2)}{h} = \frac{[6 - 2(2 + h)] - [6 - 2(2)]}{h} = \frac{[6 - 4 - 2h] - [6 - 4]}{h} = \frac{2 - 2h - 2}{h} = \frac{-2h}{h} = -2$$

45. $f(x) = 10x - 7$

$$\frac{f(3 + h) - f(3)}{h} = \frac{[10(3 + h) - 7] - [10(3) - 7]}{h} = \frac{[30 + 10h - 7] - [30 - 7]}{h} = \frac{23 + 10h - 23}{h} = \frac{10h}{h} = 10$$

47. $F(x) = 3x^2$, $g(x) = 6 - 2x$
$F[g(1)] = F[6 - 2(1)] = F[4] = 3(4)^2 = 3(16) = 48$

49. $g(x) = 6 - 2x$, $f(x) = 10x - 7$
$g[f(1)] = g[10(1) - 7] = g[3] = 6 - 2(3) = 6 - 6 = 0$

51. $f(x) = 10x - 7$, $G(x) = x - x^2$
$f[G(1)] = f[(1) - (1)^2] = f(0) = 10(0) - 7 = -7$

53. $(f + g)(x) = f(x) + g(x) = (3x - 2) + (x - x^2) = -x^2 + 4x - 2$. Domain: all real numbers.

55. $(g + h)(x) = g(x) + h(x) = (x - x^2) + x(x - 1) = x - x^2 + x^2 - x = 0$. Domain: all real numbers.

57. $(g - h)(x) = g(x) - h(x) = (x - x^2) - x(x - 1) = x - x^2 - x^2 + x = 2x - 2x^2$. Domain: all real numbers.

59. $(g - f)(x) = g(x) - f(x) = (x - x^2) - (3x - 2) = -x^2 - 2x + 2$. Domain: all real numbers.

61. $(f - h)(x) = f(x) - h(x) = (3x - 2) - x(x - 1) = 3x - 2 - x^2 + x = -x^2 + 4x - 2$. Domain: all real numbers.

63. $(f \cdot h)(x) = f(x) \cdot h(x) = (3x - 2)x(x - 1) = (3x - 2)(x^2 - x) = 3x^3 - 5x^2 + 2x$. Domain: all real numbers.

65. $\left(\frac{f}{g}\right)(x) = \frac{f(x)}{g(x)} = \frac{3x - 2}{x - x^2}$. This is defined for all x except $x - x^2 = 0$ or $x(1 - x) = 0$, that is, for $x \neq 0, 1$.

67. $\left(\frac{h}{f}\right)(x) = \frac{h(x)}{f(x)} = \frac{x(x - 1)}{3x - 2}$. This is defined for all x except $3x - 2 = 0$, that is, for $x \neq \frac{2}{3}$.

69. $\left(\frac{h}{g}\right)(x) = \frac{h(x)}{g(x)} = \frac{x(x - 1)}{x - x^2} = \frac{x(x - 1)}{x(1 - x)} = -1$. This is defined for all x except $x - x^2 = 0$ or $x(1 - x) = 0$, that is, for $x \neq 0, 1$.

71. $f(g(x)) = f(6 - 2x) = 10(6 - 2x) - 7 = 60 - 20x - 7 = 53 - 20x$

73. $f(G(x)) = f(x - x^2) = 10(x - x^2) - 7 = 10x - 10x^2 - 7 = -10x^2 + 10x - 7$

75. $g(F(x)) = g(3x^2) = 6 - 2(3x^2) = 6 - 6x^2$

77. $F(f(x)) = F(10x - 7) = 3(10x - 7)^2 = 3(100x^2 - 140x + 49) = 300x^2 - 420x + 147$

79. $F(G(x)) = F(x - x^2) = 3(x - x^2)^2 = 3(x^2 - 2x^3 + x^4) = 3x^2 - 6x^3 + 3x^4$

81. $G(g(x)) = G(6 - 2x) = (6 - 2x) - (6 - 2x)^2 = 6 - 2x - (36 - 24x + 4x^2)$
$= 6 - 2x - 36 + 24x - 4x^2 = -4x^2 + 22x - 30$

83. $f(x) = 10x - 7$
$f(x + h) = 10(x + h) - 7 = 10x + 10h - 7$

85. $F(x) = 3x^2$
$F(x + h) = 3(x + h)^2 = 3(x^2 + 2xh + h^2) = 3x^2 + 6xh + 3h^2$

87. $f(x) = 10x - 7$

$$\frac{f(x + h) - f(x)}{h} = \frac{[10(x + h) - 7] - [10x - 7]}{h} = \frac{[10x + 10h - 7] - [10x - 7]}{h}$$
$$= \frac{10x + 10h - 7 - 10x + 7}{h} = \frac{10h}{h} = 10$$

89. $F(x) = 3x^2$

$$\frac{F(x + h) - F(x)}{h} = \frac{3(x + h)^2 - 3x^2}{h} = \frac{3(x^2 + 2xh + h^2) - 3x^2}{h}$$
$$= \frac{3x^2 + 6xh + 3h^2 - 3x^2}{h} = \frac{6xh + 3h^2}{h} = \frac{h(6x + 3h)}{h} = 6x + 3h$$

91. $A(w) = \frac{w - 3}{w + 5}$

$A(5) = \frac{5 - 3}{5 + 5} = \frac{2}{10} = \frac{1}{5}$

$A(0) = \frac{0 - 3}{0 + 5} = \frac{-3}{5}$

$A(-5) = \frac{-5 - 3}{-5 + 5} = \frac{-8}{0}$ is not defined

$A(x - 5) = \frac{x - 5 - 3}{x - 5 + 5} = \frac{x - 8}{x}$

93. Total sales of the two models combined.

95. $(f \cdot g)(x) = xg(x)$ = (price per unit) × (number of units) = total revenue from sales

97. The cost is the cost per record times the number of records, hence $C(x) = 8.60x$ or $8.6x$.

99. We translate:

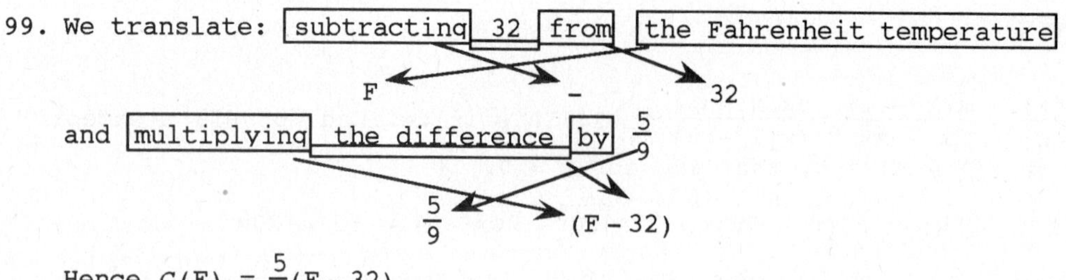

Hence $C(F) = \frac{5}{9}(F - 32)$

101. $d(t) = 30t$

(A) $d(1) = 30(1) = 30$ miles
$d(10) = 30(10) = 300$ miles

(B) $$\frac{d(2 + h) - d(2)}{h} = \frac{[30(2 + h)] - [30(2)]}{h} = \frac{[60 + 30h] - [60]}{h}$$
$$= \frac{60 + 30h - 60}{h} = \frac{30h}{h} = 30$$

Exercise 9-3 Graphing Polynomial Functions

Key Ideas and Formulas

A function defined by an equation of the form

$$f(x) = ax + b$$

where a and b are constants and x is a variable, is called a **linear function.** The graph of a linear function is a nonvertical straight line with slope a and y intercept b.

A function defined by an equation of the form

$$f(x) = ax^2 + bx + c$$

where a, b, and c are constants ($a \neq 0$) and x is a variable, is called a **quadratic function.** The graph of a quadratic function is a parabola with vertex at $x = -\frac{b}{2a}$ and y intercept c. The x intercepts, if any, are $\frac{-b \pm \sqrt{b^2 - 4ac}}{2a}$. The parabola opens up if $a > 0$ and down if $a < 0$.

A function defined by an equation of the form

$$f(x) = a_n x^n + a_{n-1} x^{n-1} + \dots + a_1 x + a_0$$

where the coefficients a_i are constants, $a_n \neq 0$, and n is a nonnegative integer, is called an nth degree polynomial function. For $n > 1$, the graph of an nth degree polynomial function is a smooth continuous curve.

1. $f(x) = 2x - 4$

x	f(x)
0	-4
2	0
4	4

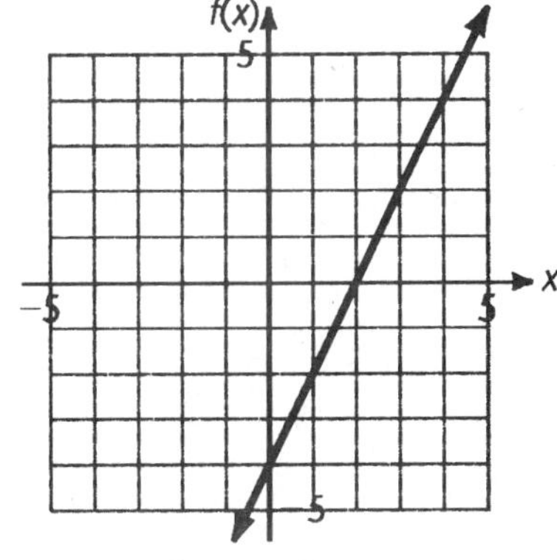

Slope: 2
y intercept: -4

3. $h(x) = 4 - 2x$

x	h(x)
0	4
2	0
4	-4

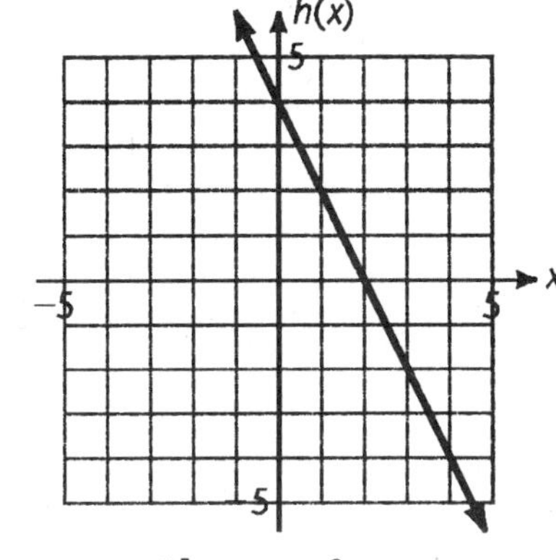

Slope: -2
y intercept: 4

5. $g(x) = -\frac{2}{3}x + 4$

x	g(x)
0	4
3	2
6	0

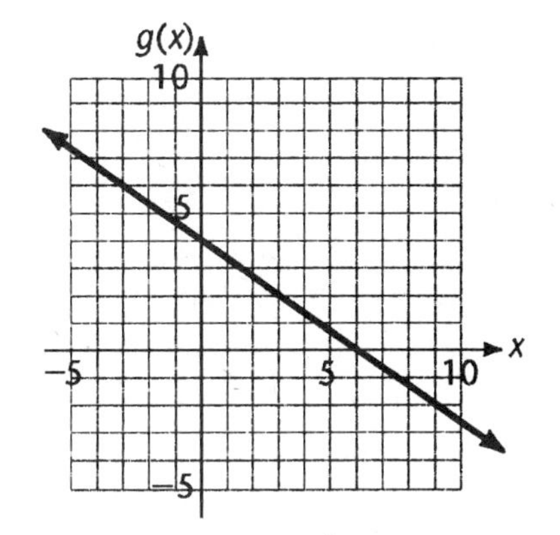

Slope: $-\frac{2}{3}$
y intercept: 4

7. $f(x) = 120 + 10x$

x	f(x)
-5	70
0	120
5	170

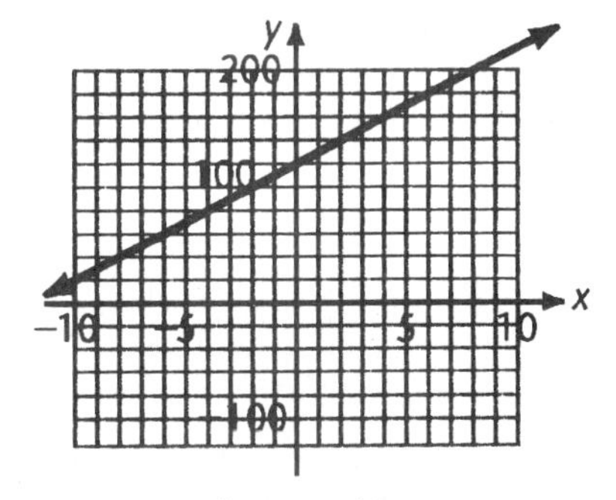

Slope: 10
y intercept: 120

9. $h(x) = 300x - 1500$

x	h(x)
0	-1500
5	0
10	1500

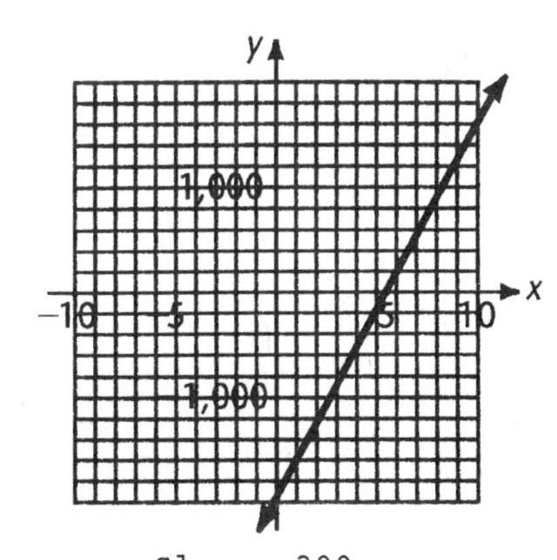

Slope: 300
y intercept: -1500

11. $g(x) = 80{,}000 - 10{,}000x$

x	g(x)
0	80,000
4	40,000
8	0

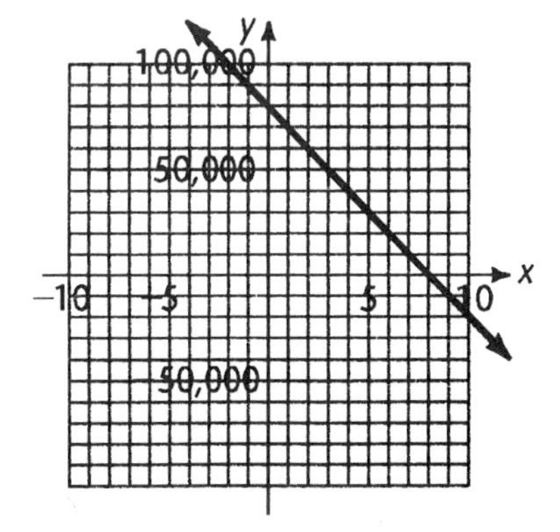

Slope: -10,000
y intercept: 80,000

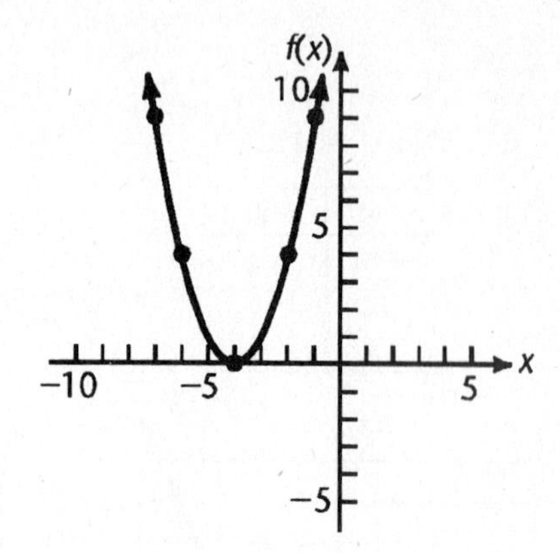

13. $f(x) = x^2 + 8x + 16$
Since a = 1, the parabola opens up. The vertex is at $x = -\frac{b}{2a} = -\frac{8}{2(1)} = -4$. At this value of x, $f(x) = (-4)^2 + 8(-4) + 16 = 0$. Thus, the vertex is (-4, 0). The y intercept is 16. The only x intercept is -4.

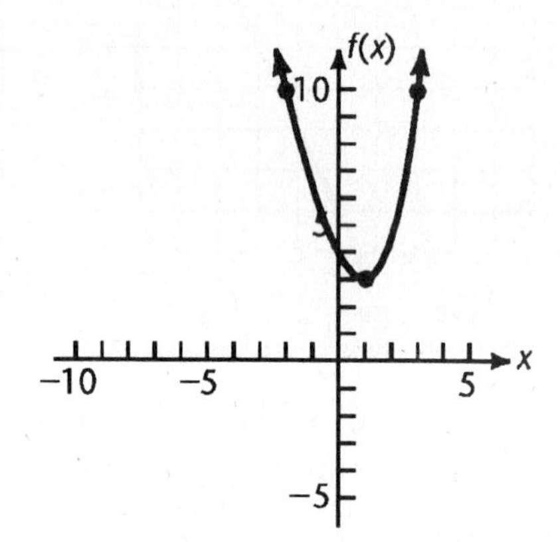

15. $f(x) = x^2 - 2x + 4$
Since a = 1, the parabola opens up. The vertex is at $x = -\frac{b}{2a} = -\frac{-2}{2(1)} = 1$. At this value of x, $f(x) = 1^2 - 2 \cdot 1 + 4 = 3$. Thus, the vertex is (1, 3). The y intercept is 4.
The x intercepts are given by
$$x = \frac{-(-2) \pm \sqrt{(-2)^2 - 4(1)(4)}}{2(1)} = \frac{2 \pm \sqrt{-12}}{2}.$$
Since these are not real numbers, there are no x intercepts.

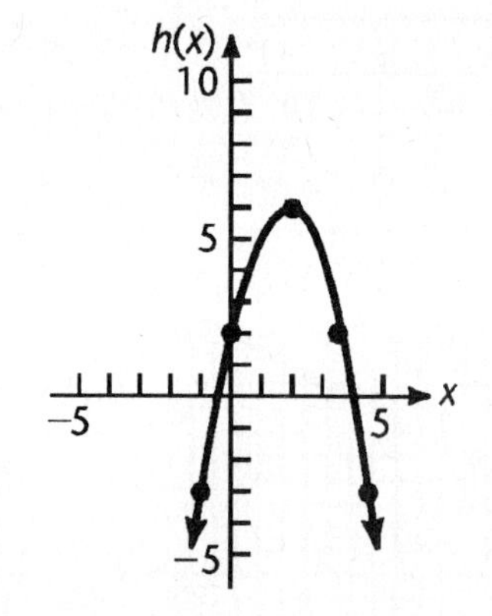

17. $h(x) = 2 + 4x - x^2 = -x^2 + 4x + 2$
Since a = -1, the parabola opens down. The vertex is at $x = -\frac{b}{2a} = -\frac{4}{2(-1)} = 2$. At this value of x, $h(x) = 2 + 4 \cdot 2 - 2^2 = 6$. Thus, the vertex is (2, 6). The y intercept is 2.
The x intercepts are given by
$$x = \frac{-(4) \pm \sqrt{(4)^2 - 4(-1)(2)}}{2(-1)} = \frac{-4 \pm \sqrt{24}}{-2} = 2 \pm \sqrt{6}$$

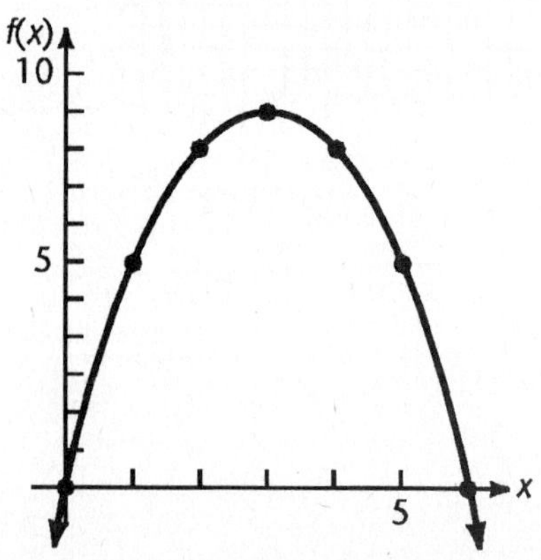

19. $f(x) = 6x - x^2 = -x^2 + 6x + 0$
Since a = -1, the parabola opens down. The vertex is at $x = -\frac{b}{2a} = -\frac{6}{2(-1)} = 3$. At this value of x, $f(x) = 6 \cdot 3 - 3^2 = 9$. Thus, the vertex is (3, 9).
The y intercept is 0.
The x intercepts are given by
$$x = \frac{-6 \pm \sqrt{(6)^2 - 4(-1)(0)}}{2(-1)} = 0, 6$$

21. $F(x) = x^2 - 4 = x^2 + 0x - 4$
Since $a = 1$, the parabola opens up. The vertex is at $x = -\frac{b}{2a} = -\frac{0}{2 \cdot 1} = 0$. At this value of x,
$F(x) = 0^2 - 4 = -4$. Thus, the vertex is $(0, -4)$.
The y intercept is -4.
The x intercepts are given by

$$x = \frac{-0 \pm \sqrt{(0)^2 - 4(1)(-4)}}{2(1)} = \pm 2$$

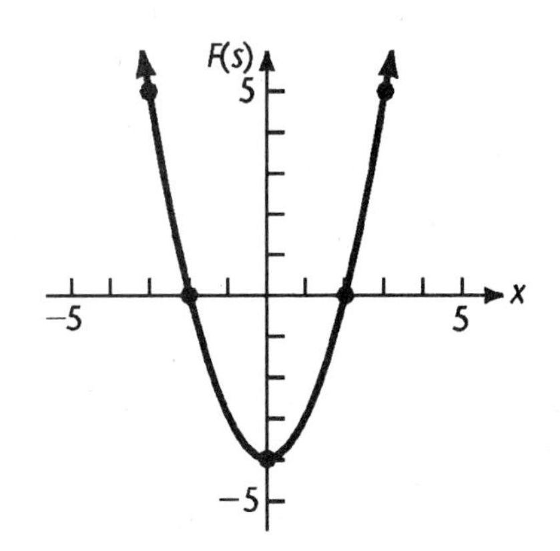

23. $F(x) = 4 - x^2 = -x^2 + 0x + 4$
Since $a = -1$, the parabola opens down.
The vertex is at $x = -\frac{b}{2a} = -\frac{0}{2(-1)} = 0$.
At this value of x, $F(x) = 4 - 0^2 = 4$.
Thus, the vertex is $(0, 4)$.
The y intercept is 4.
The x intercepts are given by

$$x = \frac{-0 \pm \sqrt{0^2 - 4(-1)4}}{2(-1)} = \pm 2$$

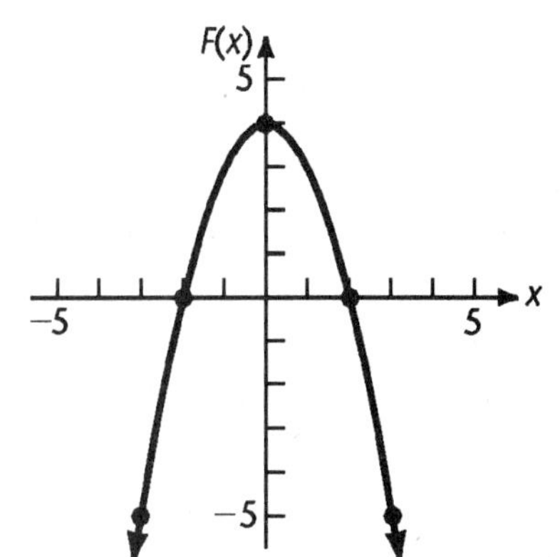

25. $f(x) = x^2 - 7x + 10$
Since $a = 1$, the parabola opens up. The vertex is at $x = -\frac{b}{2a} = \frac{-7}{2(1)} = 3.5$. At this value of x, $f(x) = (3.5)^2 - 7(3.5) + 10 = -2.25$.
Thus, the vertex is $(3.5, -2.25)$.
The y intercept is 10.
The x intercepts are given by

$$x = \frac{-(-7) \pm \sqrt{(-7)^2 - 4(1)(10)}}{2(1)} = \frac{7 \pm 3}{2} = 5,\ 2$$

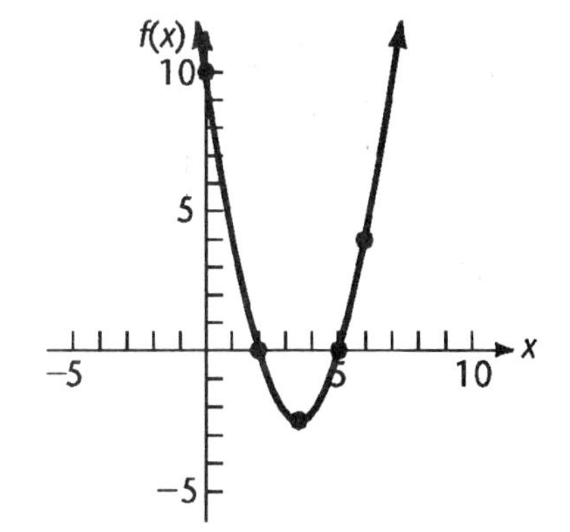

27. $g(x) = 4 + 3x - x^2 = -x^2 + 3x + 4$
Since $a = -1$, the parabola opens down.
The vertex is at $x = -\frac{b}{2a} = -\frac{3}{2(-1)} = 1.5$. At this value of x, $g(x) = 4 + 3(1.5) - (1.5)^2 = 6.25$.
Thus, the vertex is $(1.5, 6.25)$.
The y intercept is 4.
The x intercepts are given by

$$x = \frac{-3 \pm \sqrt{(3)^2 - 4(-1)(4)}}{2(-1)} = \frac{-3 \pm 5}{-2} = -1,\ 4$$

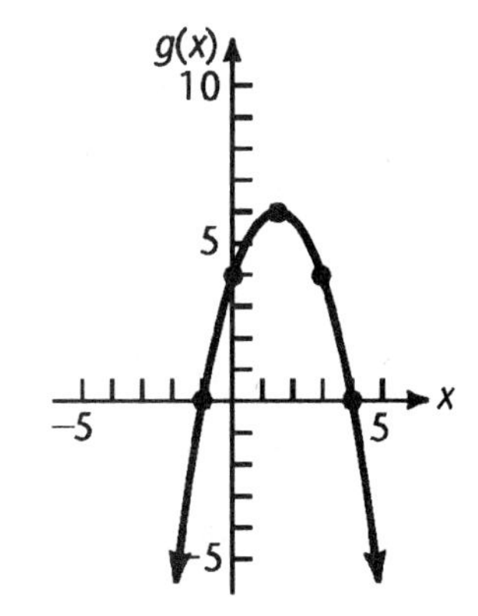

29. $f(x) = \frac{1}{2}x^2 + 2x + 0$

Since $a = \frac{1}{2}$, the parabola opens up. The vertex is at $x = -\frac{b}{2a} = -\frac{2}{2(1/2)} = -2$. At this value of x, $f(x) = \frac{1}{2}(-2)^2 + 2(-2) = -2$

Thus, the vertex is (-2, -2). The y intercept is 0. The x intercepts are given by

$$x = \frac{-2 \pm \sqrt{(2)^2 - 4(1/2)(0)}}{2(1/2)} = -4,\ 0.$$

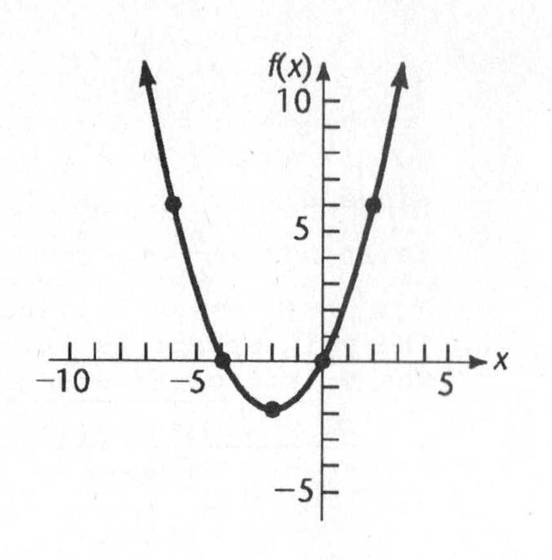

31. $f(x) = -2x^2 - 8x - 2$

Since a = -2, the parabola opens down. The vertex is at $x = -\frac{b}{2a} = -\frac{-8}{2(-2)} = -2$. At this value of x, $f(x) = -2(-2)^2 - 8(-2) - 2 = 6$. Thus, the vertex is (-2, 6). The y intercept is -2. The x intercepts are given by

$$x = \frac{-(-8) \pm \sqrt{(-8)^2 - 4(-2)(-2)}}{2(-2)} = \frac{8 \pm \sqrt{48}}{-4} = -2 \pm \sqrt{3}$$

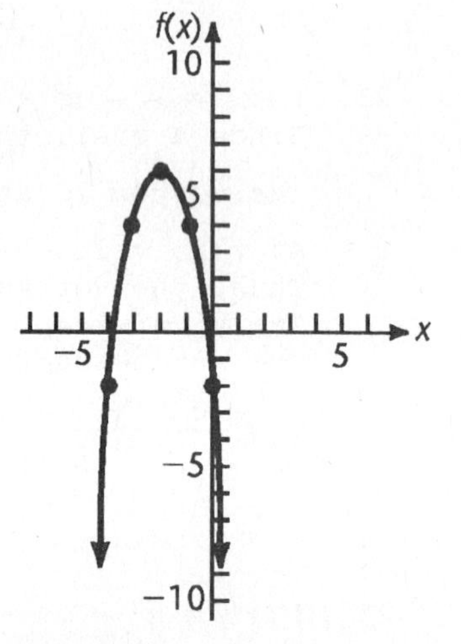

33. $P(x) = x^3 - 5x^2 + 2x + 8$, $-2 \le x \le 5$

Partial Table

x	P(x)
5	18
4	0
3	-4
2	0
1	6
0	8
-1	0
-2	-24

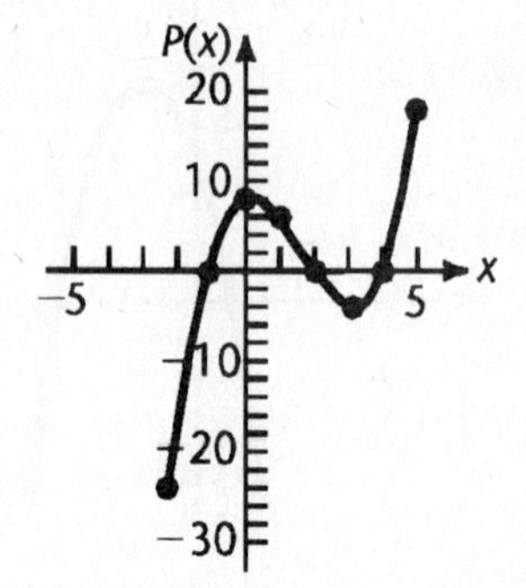

35. $P(x) = x^3 + 4x^2 - x - 4$, $-5 \le x \le 2$

Partial Table

x	P(x)
2	18
1	0
0	-4
-1	0
-2	6
-3	8
-4	0
-5	-24

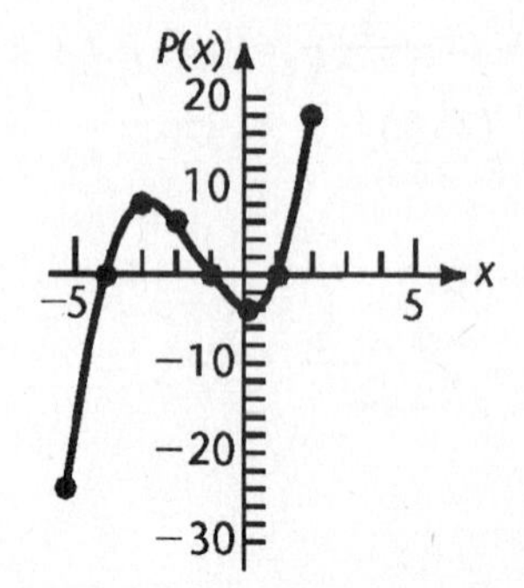

37. $P(x) = x^3 + 2$

Partial Table

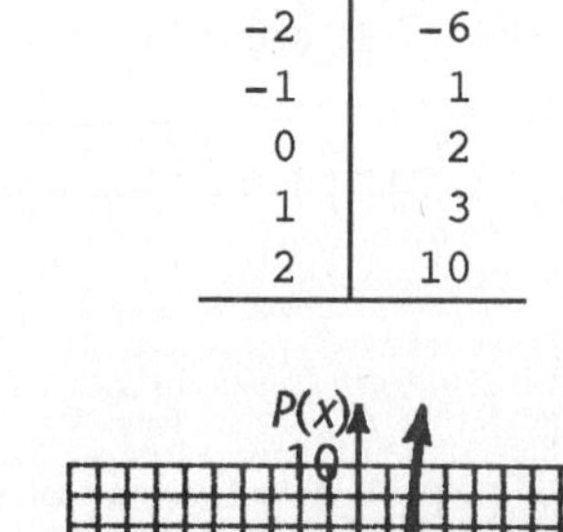

x	P(x)
-2	-6
-1	1
0	2
1	3
2	10

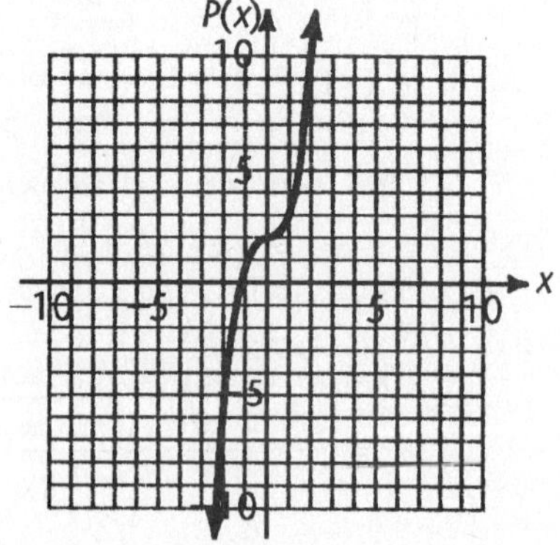

39. $P(x) = 3 - x^3$

Partial Table

x	P(x)
-2	11
-1	4
0	3
1	2
2	-5

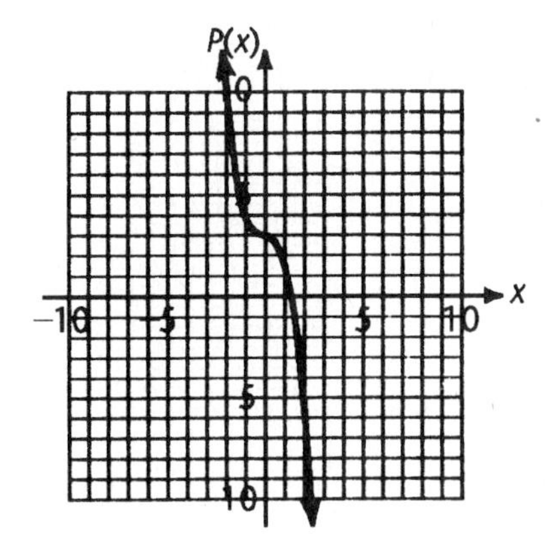

41. $P(x) = (x - 1)^3$

Partial Table

x	P(x)
-1	-8
0	-1
1	0
2	1
3	8

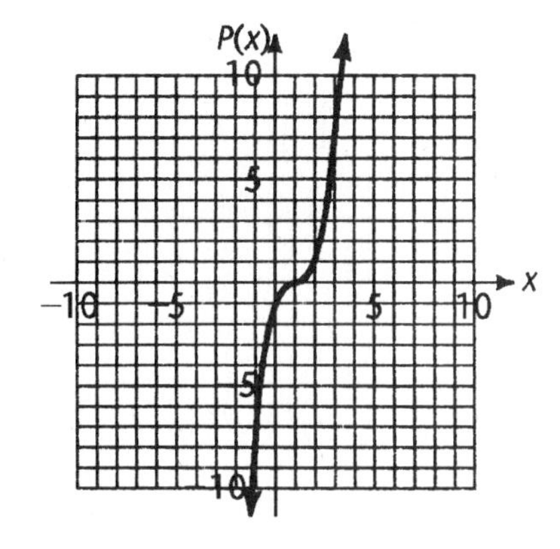

43. $P(x) = (x + 2)^3$

Partial Table

x	P(x)
-4	-8
-3	-1
-2	0
-1	1
0	8

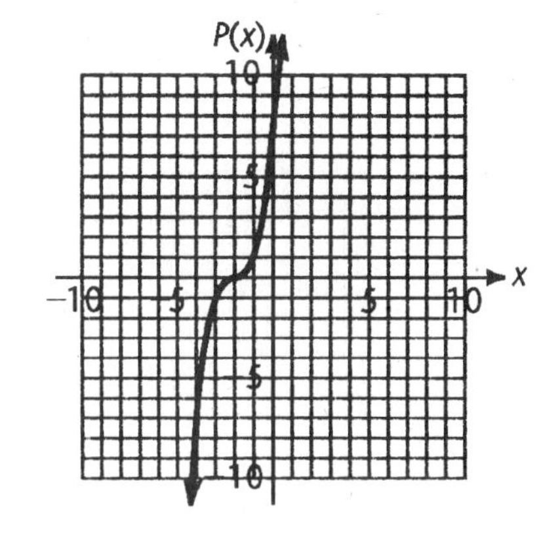

45. $P(x) = (x + 1)^3 - 2$

Partial Table

x	P(x)
-3	-10
-2	-3
-1	-2
0	-1
1	6

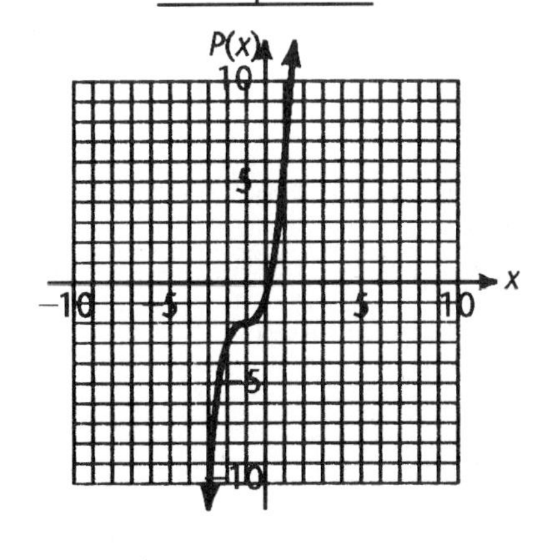

47. $P(x) = x^4 + 4x^3 - 7x^2 - 22x + 24, \ -4 \le x \le 2$

Partial Table

x	P(x)
-4	0
-3.5	-6.1875
-3	0
-2	24
-1	36
0	24
1	0
1.5	-6.1875
2	0

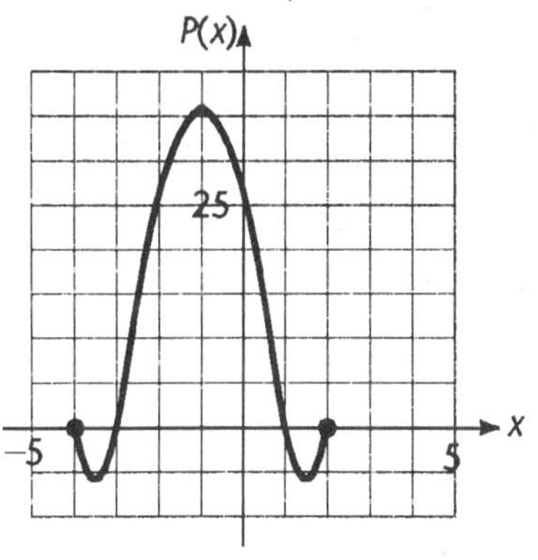

49. $P(x) = x^4 - 2x^2 + 1$, $-2 \le x \le 2$

Partial Table

x	P(x)
-2	9
-1.5	1.5625
-1	0
-0.5	0.5625
0	1
0.5	0.5625
1	0
1.5	1.5625
2	9

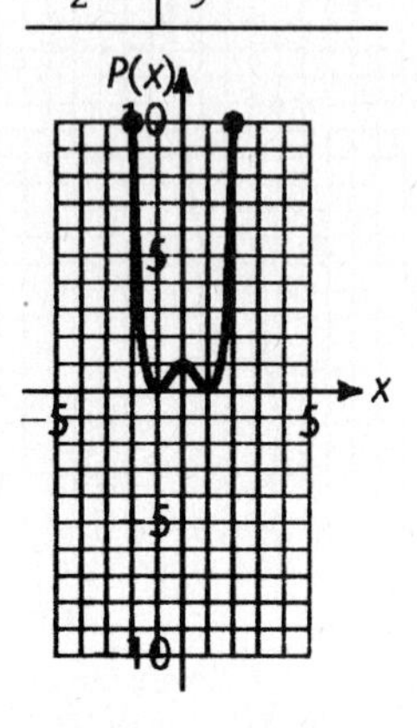

51. $P(x) = x^4 - 4x^2$

Partial Table

x	P(x)
-2.5	14.0625
-2	0
-1	-3
-0.5	-0.9375
0	0
0.5	-0.9375
1	-3
2	0
2.5	14.0625

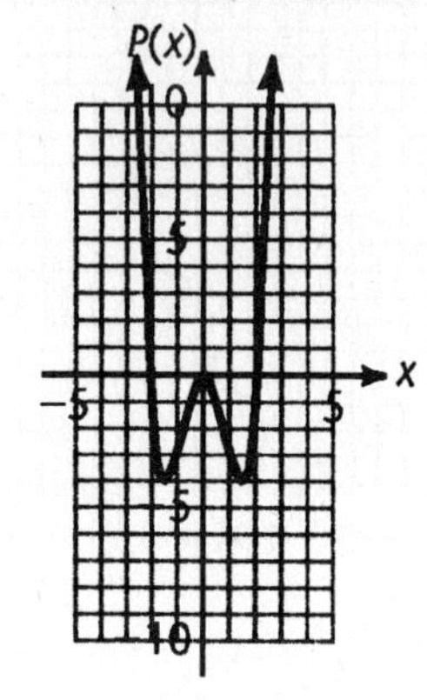

53. $P(x) = (x + 1)^4$

Partial Table

x	P(x)
-3	16
-2.5	5.0625
-2	1
-1	0
0	1
0.5	5.0625
1	16

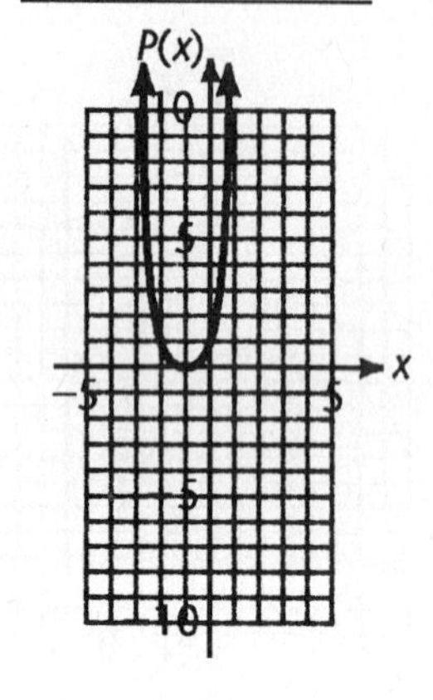

55. $P(x) = x^4 - 2x^3 - 2x^2 + 8x - 8$

Partial Table

x	P(x)
3	25
2	0
1.5	-2.1875
1	-3
0.5	-4.6875
0	-8
-1	-15
-2	0
-3	85

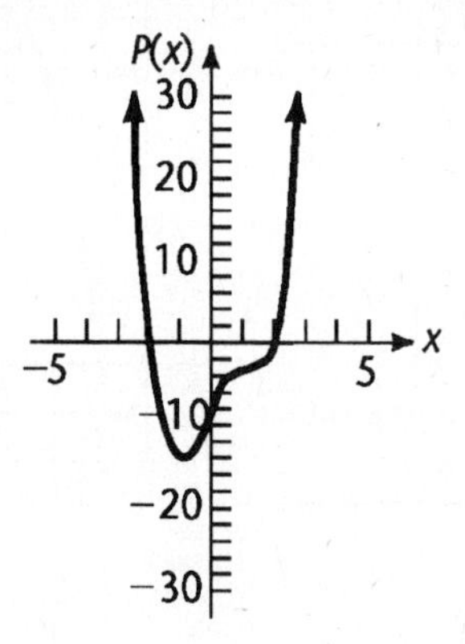

57. $P(x) = x^4 + 4x^3 - x^2 - 20x - 20$

Partial Table

x	P(x)
3	100
2	-16
1	-36
0	-20
-1	-4
-1.5	-0.6875
-2	0
-2.5	0.3125
-3	4
-4	44

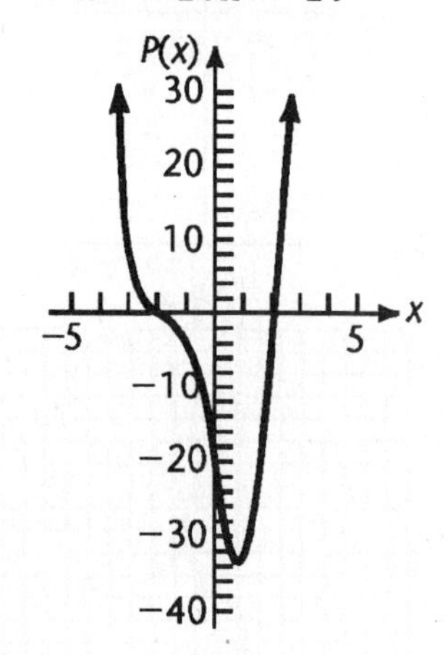

59. $f(x) = \frac{1}{x}$. Note first that the function has domain $x \neq 0$. Therefore the graph will be in two pieces, one to the left of the y axis, the other to the right.

Partial Table

x	f(x)
-10	-0.1
-5	-0.2
-4	-0.25
-3	-0.33
-2	-0.5
-1	-1
-0.5	-2
-0.1	-10

x	f(x)
10	0.1
5	0.2
4	0.25
3	0.33
2	0.5
1	1
0.5	2
0.1	10

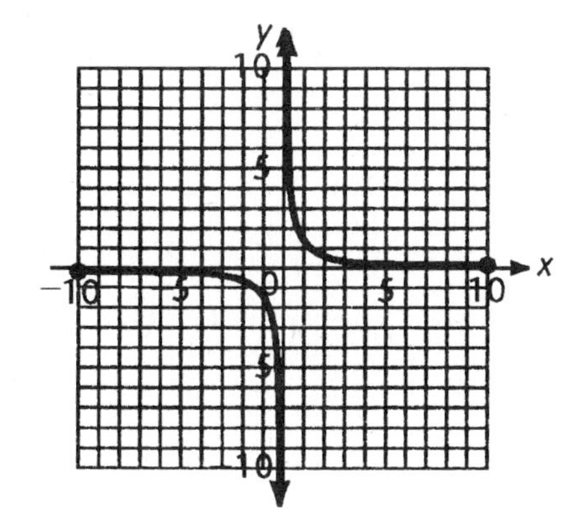

61. $f(x) = 1 - \sqrt{x}$. Note first that the function has domain $x \geq 0$, so we only examine values of x to the right of the y axis.

Partial Table

x	f(x)
0	1
1	0
2	-0.4
3	-0.7
4	-1
9	-2
10	-2.2

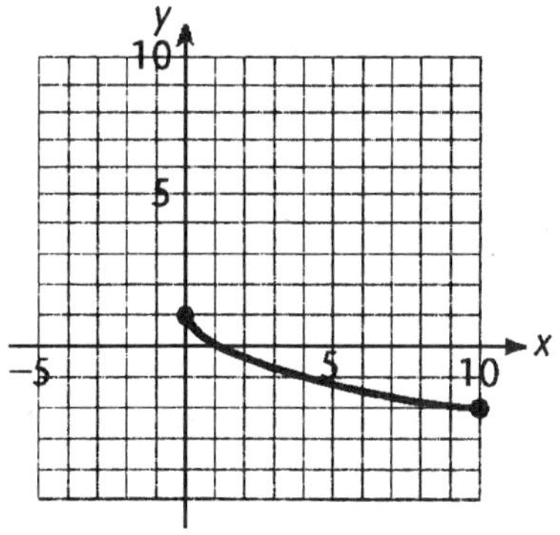

63. $f(x) = |x|$

Partial Table

x	f(x)
-10	10
-9	9
-5	5
-2	2
-1	1
0	0
1	1
2	2
5	5
9	9
10	10

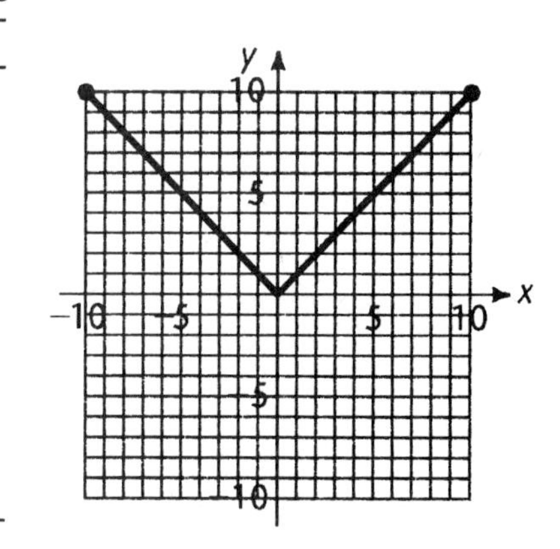

65. $g(n) = 96{,}000 + 80n$, $0 \leq n \leq 1{,}000$ is a linear function. We choose each end value and the middle value in this interval.

n	g(n)
0	96,000
500	136,000
1,000	176,000

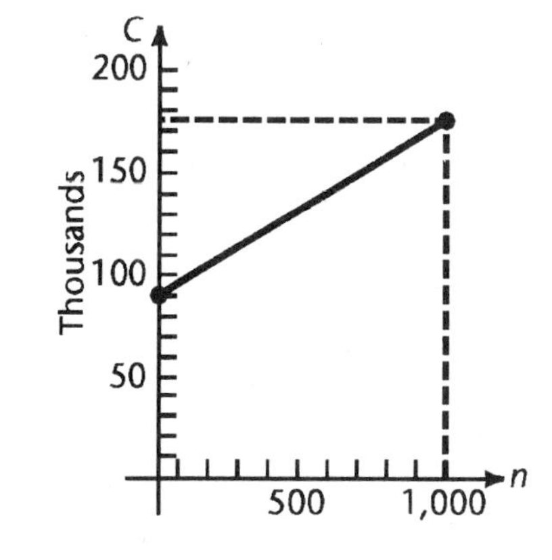

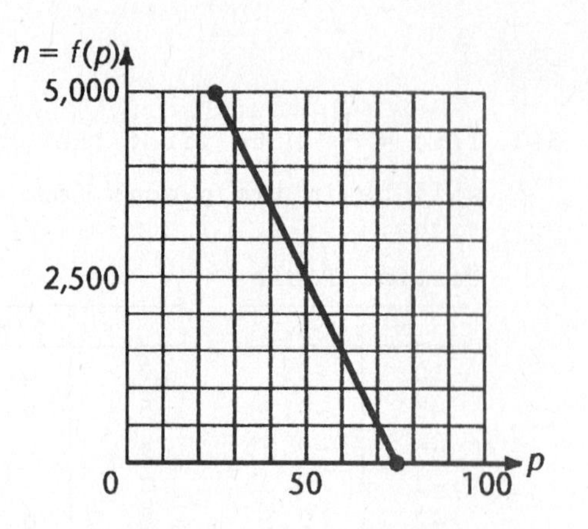

67. (A) $f(p) = 7500 - 100p$, $25 \le p \le 75$ is a linear function. We choose each end value and the middle value in this interval.

p	f(p)
25	5000
50	2500
75	0

(B) $R(p) = np$, $n = 7{,}500 - 100p$
$R(p) = p(7{,}500 - 100p)$
$R(p) = 7500p - 100p^2$

(C) Since $R(p) = 7500p - 100p^2 = -100p^2 + 7500p$ is a quadratic function, its graph is a parabola.

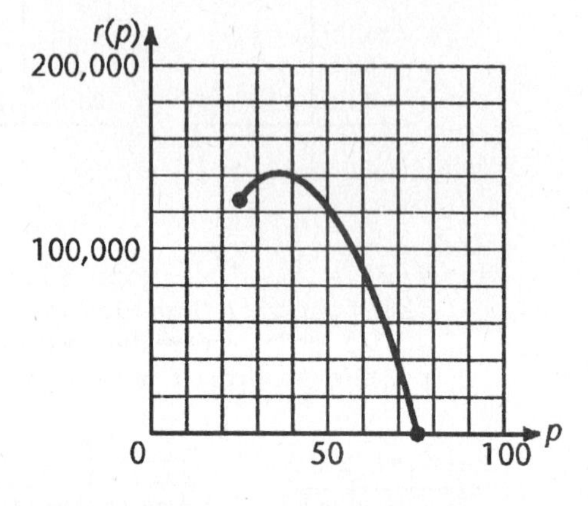

Partial Table

p	R(p)
25	125,000
40	140,000
50	125,000
75	0

The largest value of R(p) occurs at the vertex $x = -\frac{b}{2a} = -\frac{7500}{2(-100)} = 37.5$.
The largest value will then be $R(37.5) = 7500(37.5) - 100(37.5)^2 = \$140{,}625$.

69.

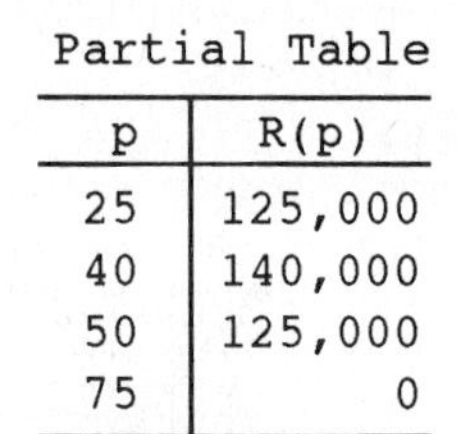

(A) Let x = width of pen
Then $2x + 2\ell = 100$
so $2\ell = 100 - 2x$
Therefore $\ell = 50 - x$
= length of pen
$A(x) = x(50 - x) = 50x - x^2$

(B) Since all distances must be positive, $x > 0$ and $50 - x > 0$. Therefore $0 < x < 50$ or $(0, 50)$ is the domain of $A(x)$.

(C) Partial Table

x	A(x)
0	0
10	400
20	600
30	600
40	400
50	0

Note: 0, 50 were excluded from the domain for geometrical reasons; but can be used to help draw the graph since the <u>polynomial</u> $50x - x^2$ has domain including these values.

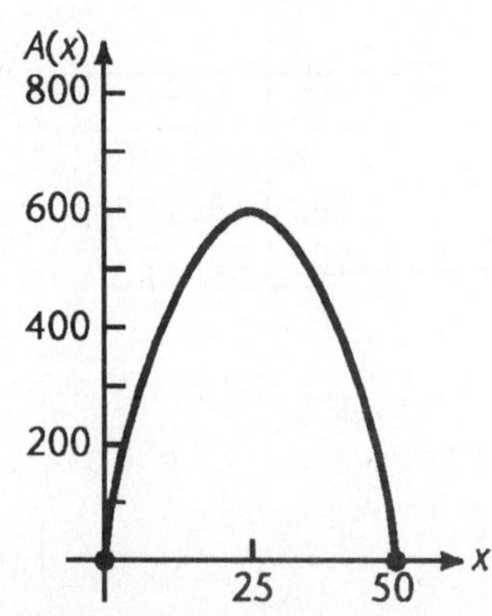

(D) Since $A(x) = 50x - x^2 = -x^2 + 50x$ is a quadratic function, its graph is a parabola, and its largest value occurs at the vertex $x = -\frac{b}{2a} = 25$. Then the dimensions are 25 meters by 25 meters, $50 - x = 25$ meters, and the rectangle will be a square. The largest area will then be $A(25) = 625$ square meters.

71. (A) $f(p) = 80 - p \quad 10 \le p \le 80$ is a linear function. We choose each end value and another value in the interval.

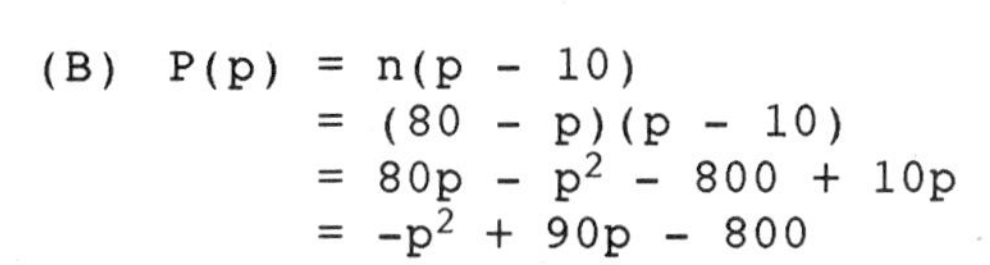

p	f(p)
10	70
40	40
80	0

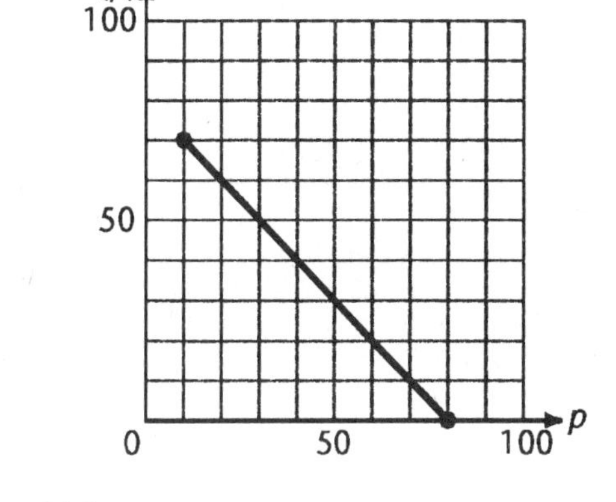

(B) $P(p) = n(p - 10)$
$= (80 - p)(p - 10)$
$= 80p - p^2 - 800 + 10p$
$= -p^2 + 90p - 800$

(C) Since P(p) is a quadratic function, its graph is a parabola.

Partial Table

p	P(p)
10	0
30	1000
40	1200
50	1200
70	1000
80	0

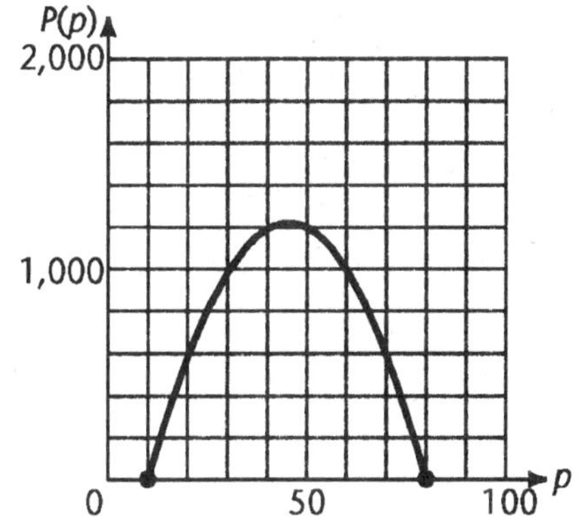

The largest value of P(p) occurs at the vertex $x = -\frac{b}{2a} = \frac{-90}{2(-1)} = 45$.
The largest value will then be $P(45) = -(45)^2 + 90(45) - 800 = \$1225$.

73.

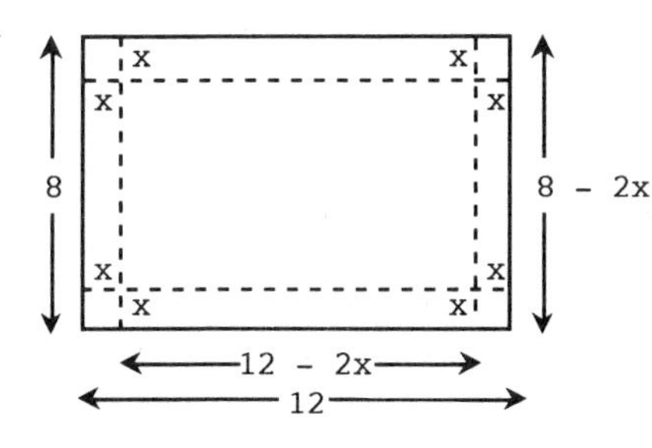

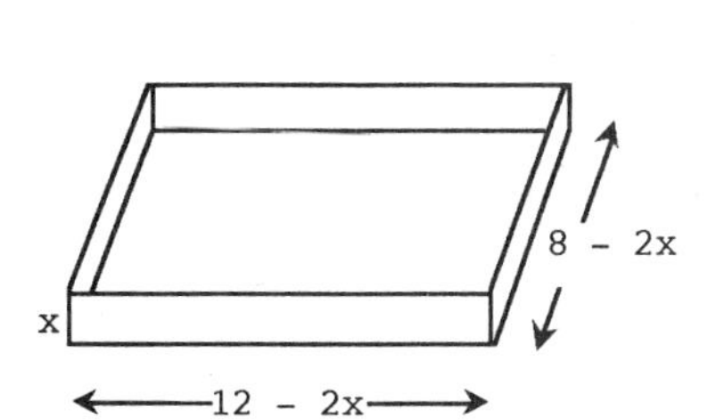

(A) From the figures it should be clear that V = length × width × height = $(12 - 2x)(8 - 2x)x$, or $V(x) = 4x^3 - 40x^2 + 96x$.

(B) Since all distances must be positive, we have $x > 0$, $8 - 2x > 0$, and $12 - 2x > 0$. Therefore $0 < x < 4$. (The condition $12 - 2x > 0$ will be met automatically if these conditions are met.)

(C) Partial Table

x	V(x)
0	0
0.5	38.5
1	60
1.5	67.5
2	64
2.5	52.5
3	36
3.5	17.5
4	0

Note: 0, 4 were excluded from the domain for geometrical reasons; but can be used to help draw the graph since the polynomial $4x^3 - 40x^2 + 96x$ has domain including these values.

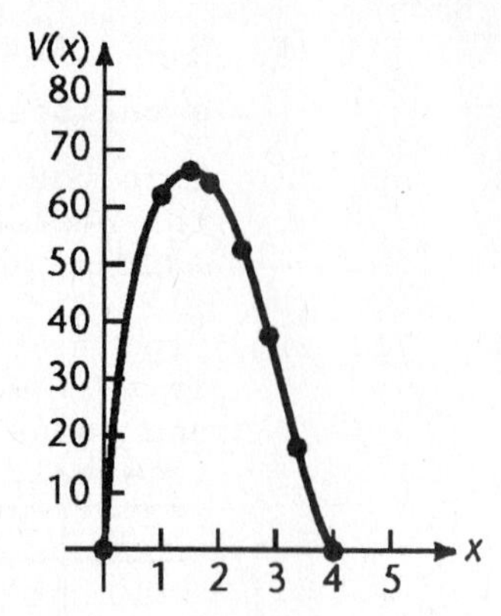

(D) Largest value of $V(x) \approx V(1.5) \approx 67.5$ cubic inches. A 1.5 inch square should be cut from each corner. As a result, if $x = 1.5$, then

$$8 - 2x = 8 - 2(1.5) = 5$$
$$12 - 2x = 12 - 2(1.5) = 9$$

The dimensions will be 5 inches × 9 inches × 1.5 inches.

75. (A) The girth = distance around = $2(2x) + 2(x) = 6x$
Therefore the length must be $108 - 6x$.
Since volume = length × width × height, we have

$$V(x) = (2x)(x)(108 - 6x)$$
$$V(x) = 2x^2(108 - 6x)$$
$$V(x) = 216x^2 - 12x^3$$

(B) Since all distances must be positive, we have $x > 0$ and $108 - 6x > 0$. Therefore $0 < x < 18$.

(C) Partial Table

x	V(x)
0	0
2	768
4	2304
6	5184
8	7680
10	9600
12	10368
14	9408
16	6144
18	0

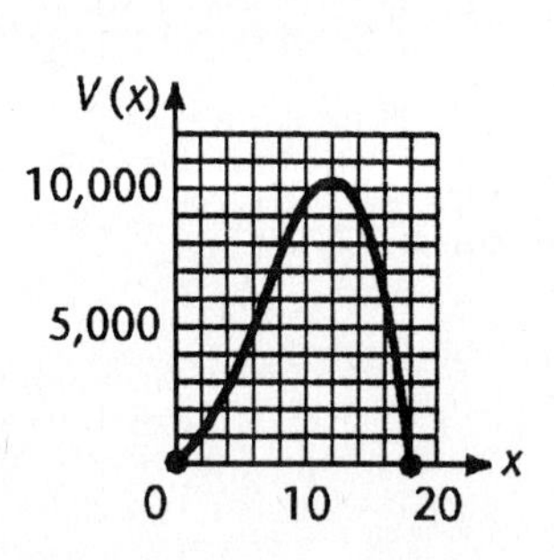

(D) Largest value of $V(x) \approx V(12) = 10{,}368$ cubic inches.
Dimensions: 12 in by 24 in by 36 in.

77. (A) Let x = number of additional trees planted
then $120 + x$ = total number of trees planted

$$40 - \frac{1}{4}x = \text{yield per tree}$$

Since the total yield $Y(x)$ = (total number of trees) × (yield per tree),

we have $Y(x) = (120 + x)\left(40 - \frac{1}{4}x\right)$

(B) $Y(x) = 4800 + 10x - \frac{1}{4}x^2$. Since this is a quadratic function, its graph is a parabola.

Partial Table

x	P(x)
0	4800
20	4900
40	4800
50	4675
100	3300
150	675
160	0

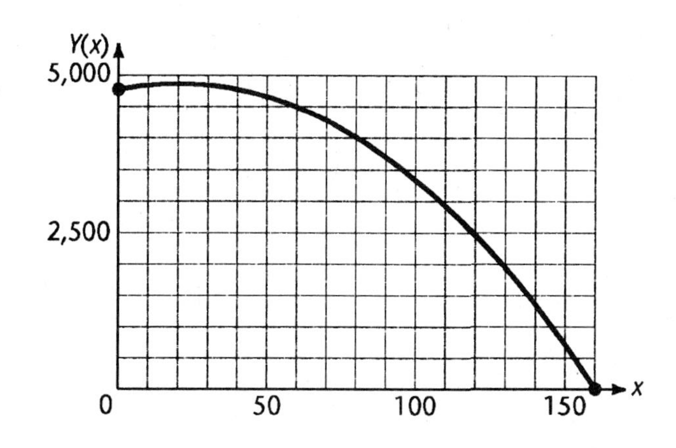

(C) The largest value of Y(x) occurs at the vertex $x = -\frac{b}{2a} = -\frac{10}{2(-1/4)} = 20$.
The largest possible total yield will then be
$Y(20) = 4800 + 10(20) - \frac{1}{4}(20)^2 = 4900$ bushels.

79. (A) Let x = number of quarters by which price is increased
then $8 + 0.25x$ = price that results per unit.
$5000 - 100x$ = number of units sold.
Since the revenue R(x) = (number of units sold) × (price per unit),
we have $R(x) = (5000 - 100x)(8 + 0.25x)$

(B) $R(x) = 40{,}000 + 450x - 25x^2$.
Since this is a quadratic function, its graph is a parabola.

Partial Table

x	R(x)
0	40,000
5	41,625
10	42,000
20	39,000
25	34,375
40	18,000
50	0

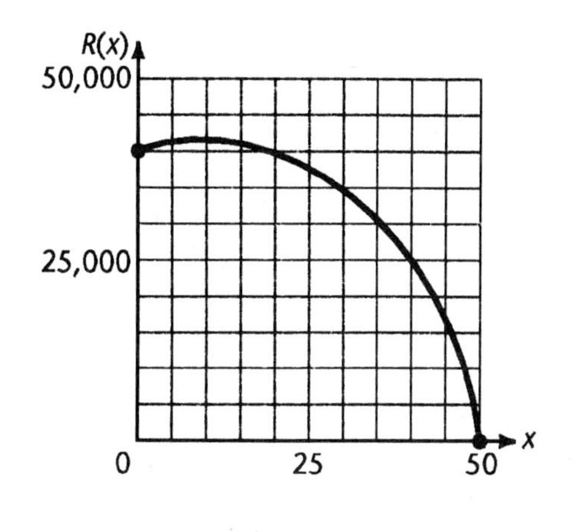

(C) The largest value of R(x) occurs at the vertex $x = -\frac{b}{2a} = -\frac{450}{2(-25)} = 9$.
The largest possible revenue will be
$R(9) = 40{,}000 + 450(9) - 25(9)^2 = \$42{,}025$.

81. (A) Let x = the number of \$10 increases in price
then $420 + 10x$ = the price that results per week
$180 - x$ = the number of units rented
Since the total revenue = (number of units rented) × (price per unit),
we have $R(x) = (420 + 10x)(180 - x)$

(B) $R(x) = 75{,}600 + 1380x - 10x^2$. Since this is a quadratic function, its graph is a parabola.

Partial Table

x	R(x)
0	75,600
20	99,200
60	122,400
100	113,600
120	97,200
180	0

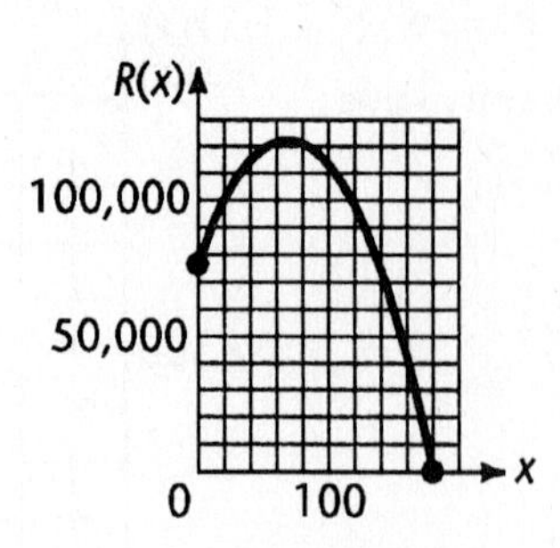

(C) The largest value of R(x) occurs at the vertex $x = -\frac{b}{2a} = -\frac{1380}{2(-10)} = 69$.
The resulting price is 420 + 10(69) = \$1110.

83. (A) Let x = the number of \$1 increases in price
then 7 + x = the price that results
300 - 20x = the number of tickets sold at this price.
Since the total revenue = (number of tickets sold) × (price per ticket),
we have $R(x) = (7 + x)(300 - 20x)$

(B) $R(x) = 2100 + 160x - 20x^2$.
Since this is a quadratic function, its graph is a parabola.

Partial Table

x	R(x)
0	2100
2	2340
4	2420
6	2340
8	2100
10	1700
15	0

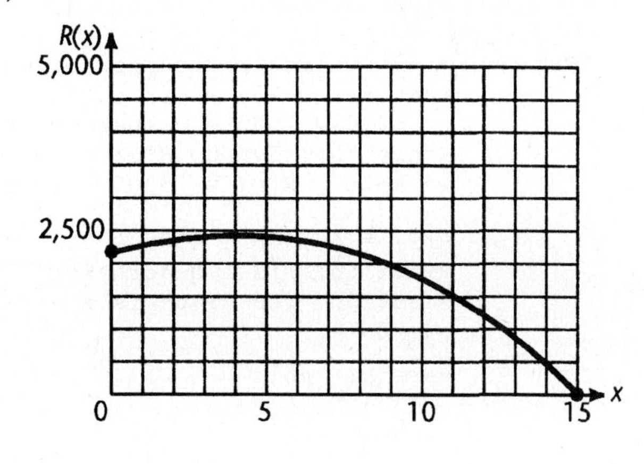

(C) The largest value of R(x) occurs at the vertex $x = -\frac{b}{2a} = -\frac{160}{2(-20)} = 4$.
The resulting price is \$11.

Exercise 9-4 Inverse Functions

Key Ideas and Formulas

A function f has an inverse function, written f^{-1}, if and only if there exists a one-to-one correspondence between domain and range values of f. In this case, we say that f is a **one-to-one function** and note that

$$f[(f^{-1}(y)] = y \quad \text{and} \quad f^{-1}[f(x)] = x$$

Then the range for f is the domain for f^{-1}. The domain for f is the range for f^{-1}.

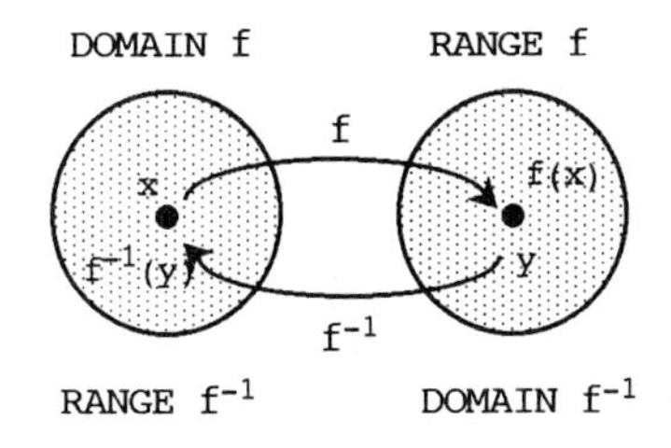

If (a, b) belongs to f, then (b, a) belongs to f^{-1}.

The graph of f^{-1} is the graph of f reflected about the 45 degree line (y = x) as a mirror image.

Horizontal-Line Test for an Inverse Function:

A function has an inverse if each **horizontal** line in the coordinate system crosses the graph of the function at most once.

To find an inverse rule for y = f(x), where f(x) is specified by an algebraic expression in x:

1. Interchange the variables, writing x = f(y).
2. Solve for y in terms of x, if possible, to obtain $y = f^{-1}(x)$.

A function f is one-to-one if and only if f(p) = f(q) implies p = q.

---

1. Since each ordered pair has a different second component, the function has an inverse.

3. Since two different ordered pairs, for example (-1, 1) and (1, 1), have the same second component, the function does not have an inverse.

5. $f^{-1} = \{(1, -2), (2, -1), (3, 0), (4, 1), (5, 2)\}$
   The domain of f^{-1} is the set of first components of f^{-1}: {1, 2, 3, 4, 5}

7. $f^{-1} = \left\{\left(\frac{1}{5}, -2\right), \left(\frac{1}{4}, -1\right), \left(\frac{1}{3}, 0\right), \left(\frac{1}{2}, 1\right), (1, 2)\right\}$

   The domain of f^{-1} is the set of first components of f^{-1}: $\left\{\frac{1}{5}, \frac{1}{4}, \frac{1}{3}, \frac{1}{2}, 1\right\}$

9.

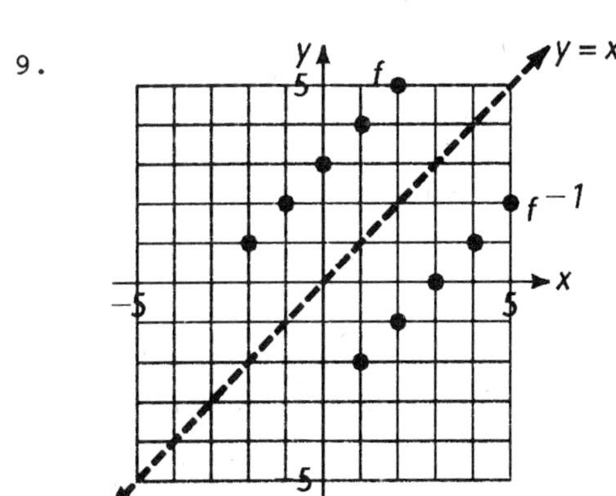

11.

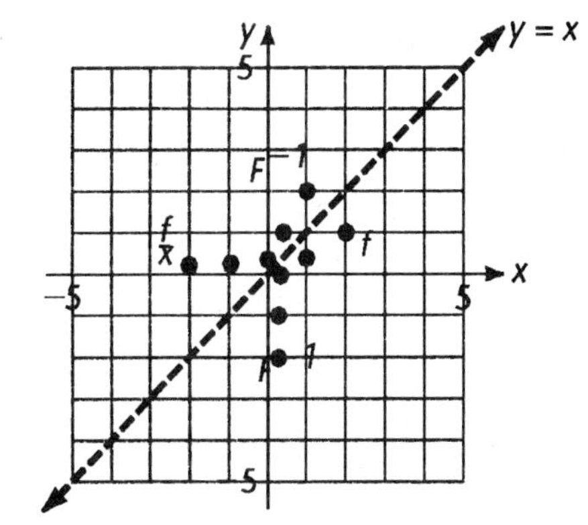

13. $f: y = 3x - 2$

$f^{-1}: x = 3y - 2$

$x + 2 = 3y$

$y = \dfrac{x + 2}{3}$

$f^{-1}(x) = \dfrac{x + 2}{3}$

$f^{-1}(-1) = \dfrac{-1 + 2}{3} = \dfrac{1}{3}$

$f^{-1}[f(-1)] = f^{-1}[3(-1) - 2]$

$= f^{-1}(-5)$

$= \dfrac{-5 + 2}{3} = \dfrac{-3}{3} = -1$

$f[f^{-1}(-1)] = f\left(\dfrac{1}{3}\right) = 3 \cdot \dfrac{1}{3} - 2 = -1$

15. $f: y = \dfrac{x}{3} - 2$

$f^{-1}: x = \dfrac{y}{3} - 2$

$x + 2 = \dfrac{y}{3}$

$y = 3(x + 2)$

$y = 3x + 6$

$f^{-1}(x) = 3x + 6$

$f^{-1}(-1) = 3(-1) + 6 = 3$

$f^{-1}[f(-1)] = f^{-1}\left[\dfrac{-1}{3} - 2\right]$

$= f^{-1}\left(-\dfrac{7}{3}\right)$

$= 3\left(-\dfrac{7}{3}\right) + 6 = -1$

$f[f^{-1}(-1)] = f(3) = \dfrac{3}{5} - 2 = -1$

17. The graph passes the horizontal line test. The function has an inverse.

19. The graph fails the horizontal line test. (The x axis, for example, crosses the line more than once.) The function does not have an inverse.

21. The graph passes the horizontal line test. The function has an inverse.

23. (A) $f: y = 3x - 2$

$f^{-1}: x = 3y - 2$

$y = \dfrac{x + 2}{3}$

$f^{-1}(x) = \dfrac{x + 2}{3}$

(B) $f^{-1}(2) = \dfrac{2 + 2}{3} = \dfrac{4}{3}$

(C) $f[(f^{-1}(3)] = 3[f^{-1}(3)] - 2 = 3\left[\dfrac{3 + 2}{3}\right] - 2$

$= 3\left[\dfrac{5}{3}\right] - 2$

$= 5 - 2$

$= 3$

25. (A) $F: y = \dfrac{x}{3} - 2$

$F^{-1}: x = \dfrac{y}{3} - 2$

$x + 2 = \dfrac{y}{3}$

$y = 3(x + 2)$

$F^{-1}(x) = 3(x + 2)$

(B) $F^{-1}(-1) = 3[(-1) + 2] = 3$

(C) $F^{-1}[F(4)] = 3[F(4) + 2]$

$= 3\left[\left(\dfrac{4}{3} - 2\right) + 2\right]$

$= 3\left[\dfrac{4}{3} - 2 + 2\right]$

$= 3\left[\dfrac{4}{3}\right]$

$= 4$

27. $f: y = \sqrt{x} + 1$
$f^{-1}: x = \sqrt{y} + 1$
$\sqrt{y} = x - 1$
$y = (x - 1)^2$
$f^{-1}(x) = (x - 1)^2$

29. $f: y = \sqrt{x - 3}$
$f^{-1}: x = \sqrt{y - 3}$
$x^2 = y - 3$
$y = x^2 + 3$
$f^{-1}(x) = x^2 + 3$

31. $f: y = \sqrt{x + 1} - 3$
$f^{-1}: x = \sqrt{y + 1} - 3$
$\sqrt{y + 1} = x + 3$
$y + 1 = (x + 3)^2$
$y = (x + 3)^2 - 1$
$f^{-1}(x) = (x + 3)^2 - 1$

33. $f: y = \frac{1}{x + 1}$
$f^{-1}: x = \frac{1}{y + 1}$
$(y + 1)x = 1$
$y + 1 = \frac{1}{x}$
$y = \frac{1}{x} - 1$
$y = \frac{1 - x}{x}$
$f^{-1}(x) = \frac{1 - x}{x}$

35. $f: y = \sqrt[3]{x}$
$f^{-1}: x = \sqrt[3]{y}$
$x^3 = y$
$y = x^3$
$f^{-1}(x) = x^3$

37. $f: y = \sqrt[3]{x - 1}$
$f^{-1}: x = \sqrt[3]{y - 1}$
$x^3 = y - 1$
$y = x^3 + 1$
$f^{-1}(x) = x^3 + 1$

39. $f: y = x^5 - 2$
$f^{-1}: x = y^5 - 2$
$y^5 = x + 2$
$y = \sqrt[5]{x + 2}$
$f^{-1}(x) = \sqrt[5]{x + 2}$

41. $f: y = 1 - \frac{1}{x}$
$f^{-1}: x = 1 - \frac{1}{y}$
$xy = y - 1$
$xy - y = -1$
$y(x - 1) = -1$
$y = \frac{-1}{x - 1}$ or $\frac{1}{1 - x}$
$f^{-1}(x) = \frac{1}{1 - x}$

43. $f: y = \frac{x + 3}{x - 1}$
$f^{-1}: x = \frac{y + 3}{y - 1}$
$x(y - 1) = y + 3$
$xy - x = y + 3$
$xy - y - x = 3$
$xy - y = x + 3$
$y(x - 1) = x + 3$
$y = \frac{x + 3}{x - 1}$
$f^{-1}(x) = \frac{x + 3}{x - 1}$

45. $f: y = -x^2 \quad x \geq 0$
$f^{-1}: x = -y^2 \quad y \geq 0$
$-x = y^2 \quad y \geq 0$
$y = \sqrt{-x}$ (Since y cannot be negative, we can discard the negative square root.)
$f^{-1}(x) = \sqrt{-x}$

47. (Problem 27)
$f(x) = \sqrt{x} + 1$
$f^{-1}(x) = (x - 1)^2$
$f^{-1}[f(x)] = f^{-1}(\sqrt{x} + 1)$
$= (\sqrt{x} + 1 - 1)^2$
$= (\sqrt{x})^2$
$= x$

49. (Problem 29)
$f(x) = \sqrt{x - 3}$
$f^{-1}(x) = x^2 + 3$
$f^{-1}[f(x)] = f^{-1}(\sqrt{x - 3})$
$= (\sqrt{x - 3})^2 + 3$
$= x - 3 + 3$
$= x$

51. (Problem 31)
$f(x) = \sqrt{x + 1} - 3$
$f^{-1}(x) = (x + 3)^2 - 1$
$f^{-1}[f(x)] = f^{-1}(\sqrt{x + 1} - 3)$
$= (\sqrt{x + 1} - 3 + 3)^2 - 1$
$= (\sqrt{x + 1})^2 - 1$
$= x + 1 - 1$
$= x$

53. (Problem 33)

$$f(x) = \frac{1}{x+1}$$
$$f^{-1}(x) = \frac{1-x}{x}$$

$$f^{-1}[f(x)] = f^{-1}\left[\frac{1}{x+1}\right]$$
$$= \frac{1 - \frac{1}{x+1}}{\frac{1}{x+1}}$$ multiply numerator and denominator by x + 1, the lcd of internal fractions
$$= \frac{x+1-1}{1}$$
$$= x$$

55. (Problem 35)

$$f(x) = \sqrt[3]{x}$$
$$f^{-1}(x) = x^3$$
$$f^{-1}[f(x)] = f^{-1}(\sqrt[3]{x})$$
$$= [\sqrt[3]{x}]^3$$
$$= x$$

57. (Problem 37)

$$f(x) = \sqrt[3]{x-1}$$
$$f^{-1}(x) = x^3 + 1$$
$$f^{-1}[f(x)] = f^{-1}(\sqrt[3]{x-1})$$
$$= (\sqrt[3]{x-1})^3 + 1$$
$$= x - 1 + 1$$
$$= x$$

59. (Problem 39)

$$f(x) = x^5 - 2$$
$$f^{-1}(x) = \sqrt[5]{x+2}$$
$$f^{-1}[f(x)] = f^{-1}(x^5 - 2)$$
$$= \sqrt[5]{x^5 - 2 + 2}$$
$$= \sqrt[5]{x^5}$$
$$= x$$

61. (Problem 41)

$$f(x) = 1 - \frac{1}{x}$$
$$f^{-1}(x) = \frac{1}{1-x}$$
$$f^{-1}[f(x)] = f^{-1}\left(1 - \frac{1}{x}\right)$$
$$= \frac{1}{1 - \left(1 - \frac{1}{x}\right)}$$
$$= \frac{1}{1 - 1 + \frac{1}{x}}$$
$$= \frac{1}{\frac{1}{x}}$$
$$= x$$

63. (Problem 43)

$$f(x) = \frac{x+3}{x-1}$$
$$f^{-1}(x) = \frac{x+3}{x-1}$$
$$f^{-1}[f(x)] = f^{-1}\left(\frac{x+3}{x-1}\right)$$
$$= \frac{\frac{x+3}{x-1} + 3}{\frac{x+3}{x-1} - 1}$$ multiply numerator and denominator by x - 1, the lcd of internal fractions
$$= \frac{x + 3 + 3(x-1)}{x + 3 - (x-1)}$$
$$= \frac{x + 3 + 3x - 3}{x + 3 - x + 1}$$
$$= \frac{4x}{4}$$
$$= x$$

65. (Problem 45)

$$f(x) = -x^2 \qquad x \ge 0$$
$$f^{-1}(x) = \sqrt{-x}$$
$$f^{-1}[f(x)] = f^{-1}(-x^2)$$
$$= \sqrt{-(-x^2)}$$
$$= \sqrt{x^2}$$
$$= |x|$$
$$= x \quad \text{if} \quad x \ge 0$$

67. $f(x) = 3x - 2$ $\quad f^{-1}(x) = \dfrac{x + 2}{3}$

x	f(x)
-2	-8
0	-2
2	4

x	$f^{-1}(x)$
-8	-2
-2	0
4	2

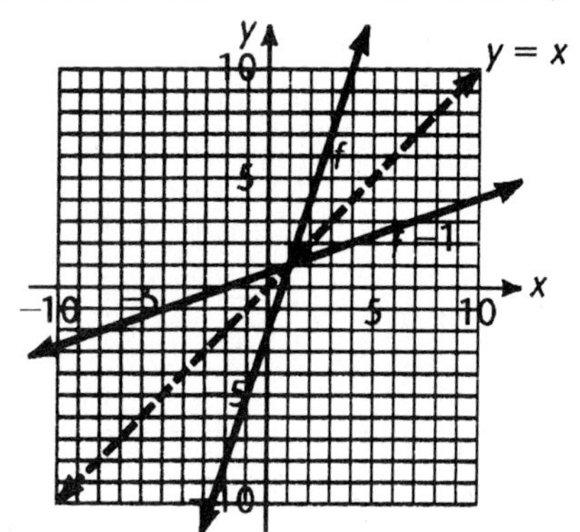

69. $f(x) = \dfrac{x}{3} - 2$ $\quad f^{-1}(x) = 3(x + 2)$

x	f(x)
-3	-3
0	-2
3	-1

x	$f^{-1}(x)$
-3	-3
-2	0
-1	3

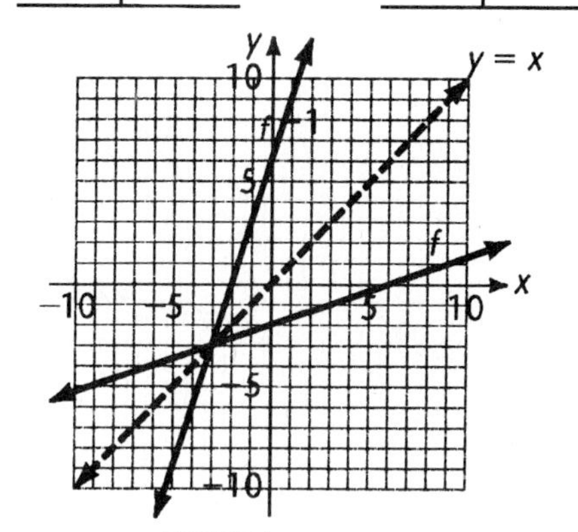

71. $f(x) = \sqrt{x} + 1$ $\quad f^{-1}(x) = (x - 1)^2$

x	f(x)
0	1
1	2
4	3
9	4

x	$f^{-1}(x)$
1	0
2	1
3	4
4	9

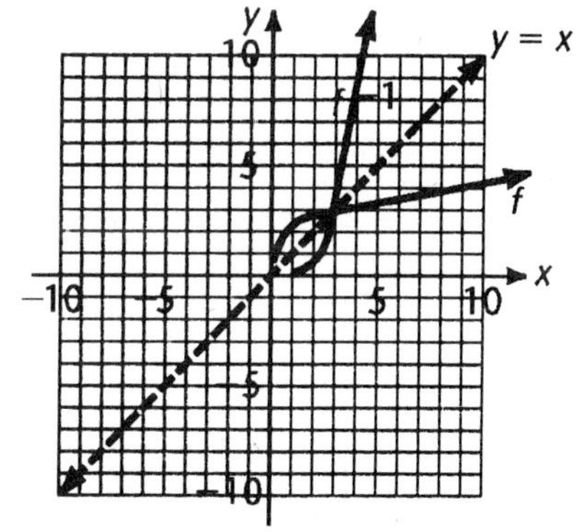

73. $f(x) = \sqrt{x - 3}$ $\quad f^{-1}(x) = x^2 + 3$

x	f(x)
3	0
4	1
7	2

x	$f^{-1}(x)$
0	3
1	4
2	7

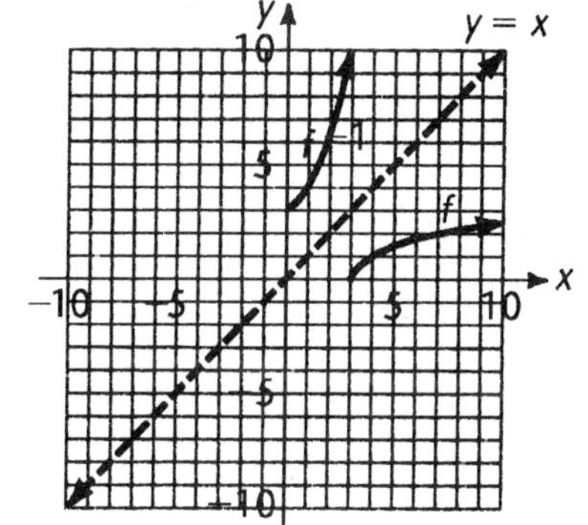

75. $f(x) = 3x - 5$

$f(p) = 3p - 5 \quad f(q) = 3q - 5$

$3p - 5 = 3q - 5 \Rightarrow 3p = 3q \Rightarrow p = q$

77. $f(x) = \sqrt{x + 1}$

$f(p) = \sqrt{p + 1} \quad f(q) = \sqrt{q + 1}$

$\sqrt{p + 1} = \sqrt{q + 1} \Rightarrow p + 1 = q + 1 \Rightarrow p = q$

79. $f(x) = \dfrac{1}{x + 1}$

$f(p) = \dfrac{1}{p + 1} \quad f(q) = \dfrac{1}{q + 1}$

$\dfrac{1}{p + 1} = \dfrac{1}{q + 1} \Rightarrow p + 1 = q + 1 \Rightarrow p = q$

81. $f(x) = \sqrt{x+1} - 3$ $f^{-1}(x) = (x+3)^2 - 1$

x	f(x)
-1	-3
0	-2
3	-1
8	0

x	$f^{-1}(x)$
-3	-1
-2	0
-1	3
0	8

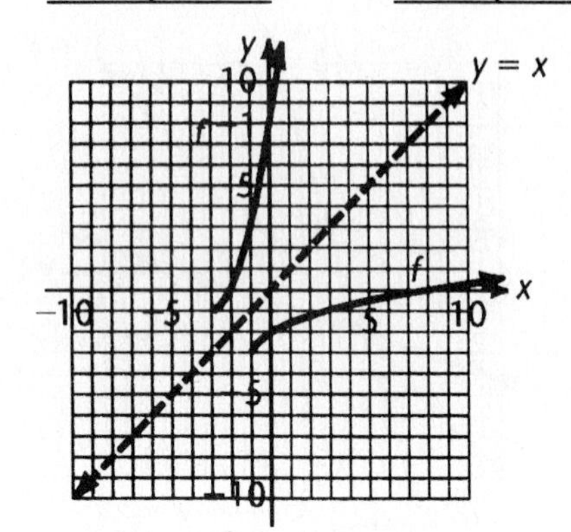

83. $f(x) = \dfrac{1}{x+1}$ $f^{-1}(x) = \dfrac{1-x}{x}$

x	f(x)
-5	-0.25
-4	-0.33
-3	-0.5
-2	-1
-1.5	-2
-1.1	-10
-0.9	10
-0.5	2
0	1
1	0.5
2	0.33
3	0.25
4	0.2

x	$f^{-1}(x)$
-0.25	-5
-0.37	-4
-0.5	-3
-1	-2
-2	-1.5
-10	-1.1
10	-0.9
2	-0.5
1	0
0.5	1
0.33	2
0.25	3
0.2	4

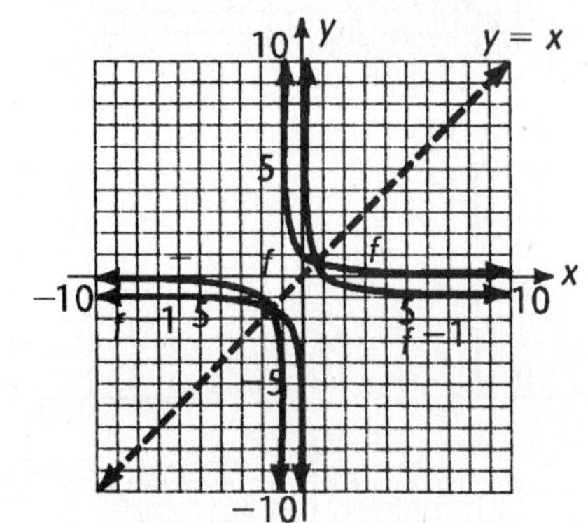

85. $f(x) = \sqrt[3]{x}$ $f^{-1}(x) = x^3$

x	f(x)
-8	-2
-1	-1
0	0
1	1
8	2

x	$f^{-1}(x)$
-2	-8
-1	-1
0	0
1	1
2	2

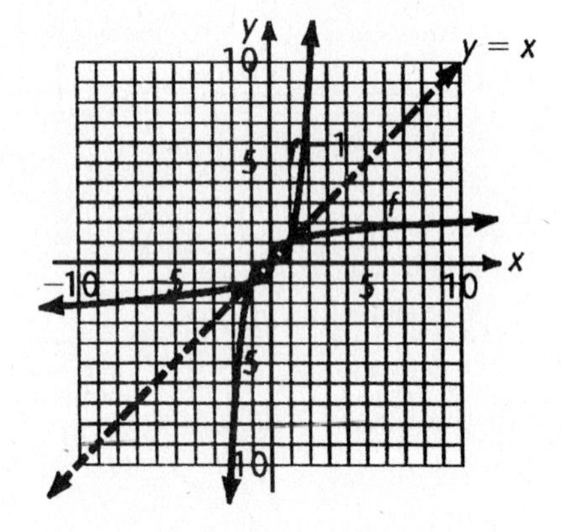

87. $f(x) = \sqrt[3]{x-1}$ $f^{-1}(x) = x^3 + 1$

x	f(x)
-7	-2
0	-1
1	0
2	1
9	2

x	$f^{-1}(x)$
-2	-7
-1	0
0	1
1	2
2	9

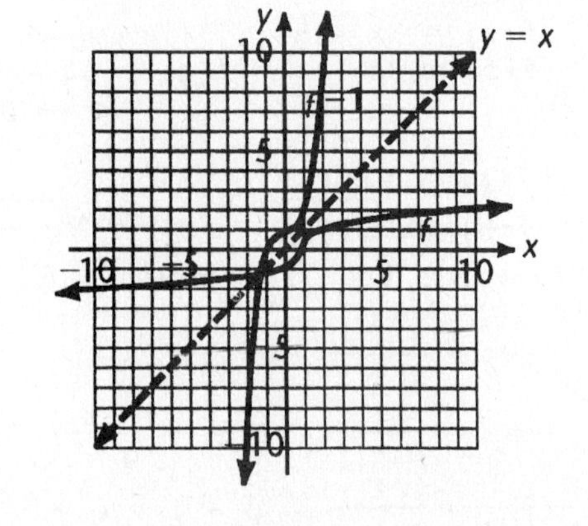

89. $f(x) = x^5 - 2$ $\qquad f^{-1}(x) = \sqrt[5]{x + 2}$

x	f(x)
-1.5	-9.6
-1	-3
0	-2
1	-1
1.5	6.6

x	$f^{-1}(x)$
-9.6	-1.5
-3	-1
-2	0
-1	1
6.6	1.5

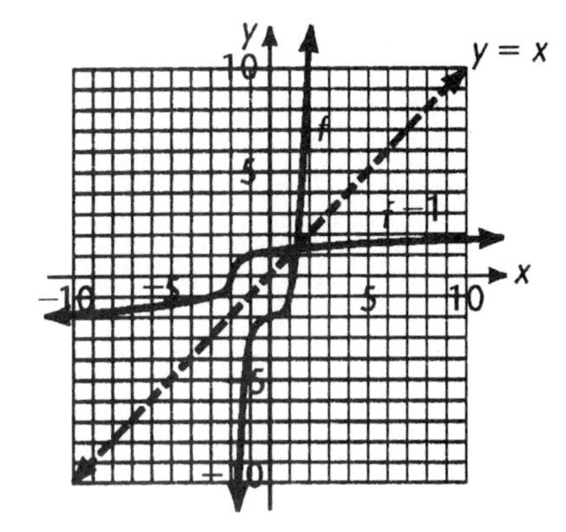

91. $f(x) = 1 - \frac{1}{x}$ $\qquad f^{-1}(x) = \frac{1}{1 - x}$

x	f(x)
-10	1.1
-5	1.2
-3	1.33
-2	1.5
-1	2
-0.5	3
-0.1	11
0.1	-9
0.5	-1
1	0
2	0.5
3	0.67
5	0.8
10	0.9

x	$f^{-1}(x)$
1.1	-10
1.2	-5
1.33	-3
1.5	-2
2	-1
3	-0.5
11	-0.1
-9	0.1
-1	0.5
0	1
0.5	2
0.67	3
0.8	5
0.9	10

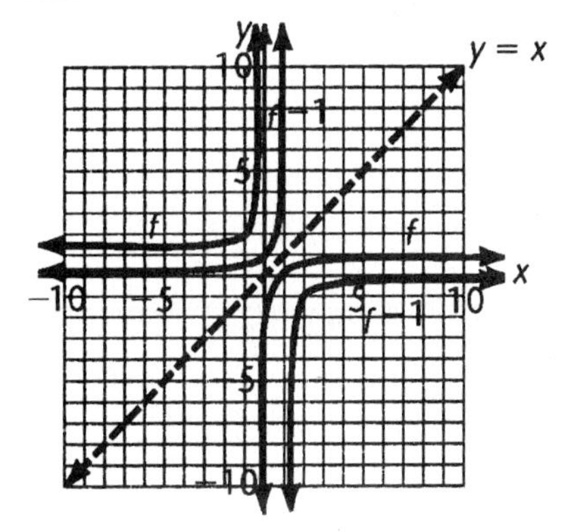

93. $f(x) = \frac{x + 3}{x - 1} = f^{-1}(x)$

x	f(x)
-10	0.63
-5	0.33
-2	-0.33
-1	-1
0	-3
0.5	-7
0.9	-39
1.1	41
1.5	9
2	5
3	3
4	2.33
5	2
10	1.44

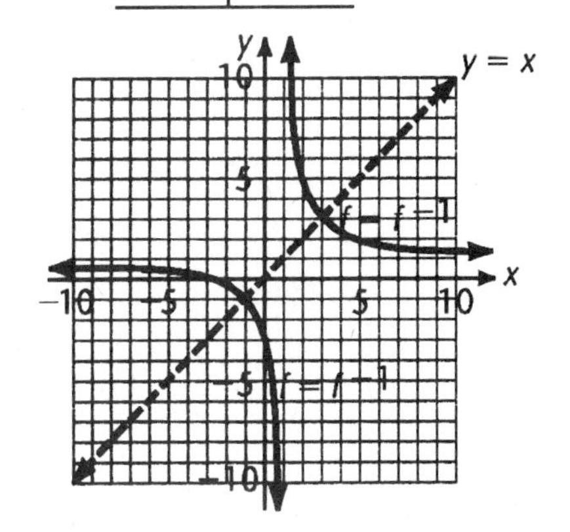

95. $f(x) = -x^2 \quad x \geq 0$ $\qquad f^{-1}(x) = \sqrt{-x}$

x	f(x)
0	0
1	-1
2	-4
3	-9

x	$f^{-1}(x)$
0	0
-1	1
-4	2
-9	3

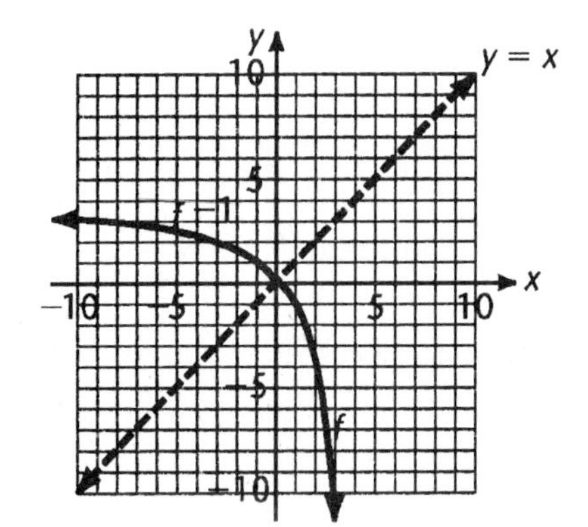

Exercise 9-5 Variation

Key Ideas and Formulas

y **varies directly as** x means $y = kx \quad k \neq 0 \quad k$ is a constant, known as the **constant of variation.**

y **varies inversely as** x means $y = \frac{k}{x} \quad k \neq 0$

w **varies jointly as** x and y means $w = kxy \quad k \neq 0$

w varies directly as x and inversely as y (combined variation) means $w = k\frac{x}{y} \quad k \neq 0$

To solve a typical variation problem, there are two methods.

Method 1:

Write the equation of variation. Find k using the first set of values. Answer the question using k and the second set of values.

Method 2:

Write the equation of variation and solve for k. Then apply subscripts to the variables and equate the results. Substitute all known values and solve for the unknown.

(Thus, if $y = kx$, $k = \frac{y}{x}$, and $k = \frac{y_1}{x_1} = \frac{y_2}{x_2}$.)

---

1. $F = kv^2$ 3. $f = k\sqrt{T}$ 5. $y = \frac{k}{\sqrt{x}}$ 7. $t = \frac{k}{T}$ 9. $R = kSTV$ 11. $V = khr^2$

We use Method 2 for all problems in this exercise, except problem 37.

13. $y = kx \quad \frac{y}{x} = k \quad \frac{y_1}{x_1} = k \quad \frac{y_2}{x_2} = k$

Hence $\frac{y_1}{x_1} = \frac{y_2}{x_2}$. Substitute and solve for the one unknown y_2.

$$\frac{7.5}{3} = \frac{y_2}{8}$$
$$8 \cdot \frac{7.5}{3} = y_2$$
$$\frac{60}{3} = y_2$$
$$y_2 = 20$$

15. $y = \frac{k}{x} \quad xy = k \quad x_1y_1 = k \quad x_2y_2 = k$

Hence $x_1y_1 = x_2y_2$. Substitute and solve for the one unknown y_2.

$$8 \cdot 9 = 6y_2$$
$$72 = 6y_2$$
$$y_2 = 12$$

17. $u = k\sqrt{v} \quad \frac{u}{\sqrt{v}} = k \quad \frac{u_1}{\sqrt{v_1}} = k \quad \frac{u_2}{\sqrt{v_2}} = k$

Hence $\frac{u_1}{\sqrt{v_1}} = \frac{u_2}{\sqrt{v_2}}$. Substitute and solve for the one unknown u_2.

$$\frac{2}{\sqrt{2}} = \frac{u_2}{\sqrt{8}}$$
$$\frac{2\sqrt{8}}{\sqrt{2}} = u_2$$
$$u_2 = \frac{2\sqrt{8}\sqrt{2}}{\sqrt{2}\sqrt{2}}$$
$$u_2 = \frac{2\sqrt{16}}{2}$$
$$u_2 = 4$$

19. $L = \frac{k}{\sqrt{M}} \quad L\sqrt{M} = k \quad L_1\sqrt{M_1} = k \quad L_2\sqrt{M_2} = k$

Hence $L_1\sqrt{M_1} = L_2\sqrt{M_2}$. Substitute and solve for the one unknown L_2.

$$9\sqrt{9} = L_2\sqrt{3}$$
$$27 = L_2\sqrt{3}$$
$$L_2 = \frac{27}{\sqrt{3}}$$
$$L_2 = \frac{27\sqrt{3}}{3}$$
$$L_2 = 9\sqrt{3}$$

21. $U = k\frac{ab}{c^3}$

23. $L = k\frac{wh^2}{\ell}$

25. $Q = k\frac{mn^2}{P} \quad k = \frac{QP}{mn^2} \quad k = \frac{Q_1P_1}{m_1n_1^2} \quad k = \frac{Q_2P_2}{m_2n_2^2}$

Hence $\frac{Q_1P_1}{m_1n_1^2} = \frac{Q_2P_2}{m_2n_2^2}$. Substitute and solve for the one unknown Q_2.

$$\frac{(-4)(12)}{(6)(2)^2} = \frac{Q_2(6)}{(4)(3)^2}$$
$$\frac{-48}{24} = \frac{6Q_2}{36}$$
$$-2 = \frac{Q_2}{6}$$
$$Q_2 = -12$$

27. $w = \frac{k}{d^2} \quad k = wd^2 \quad k = w_1d_1^2 \quad k = w_2d_2^2$

Hence $w_1d_1^2 = w_2d_2^2$. Substitute and solve for the one unknown w_2.

Note: d_1 = distance from surface of earth to center = 4000

d_2 = distance from 400 miles above surface to center = 4400

$$100(4000)^2 = w_2(4400)^2$$
$$\frac{100(4000)^2}{(4400)^2} = w_2$$
$$w_2 \approx 83 \text{ pounds}$$

29. $L = kv^2 \quad k = \frac{L}{v^2} \quad k = \frac{L_1}{v_1^2} \quad k = \frac{L_2}{v_2^2}$

Hence $\frac{L_1}{v_1^2} = \frac{L_1}{v_2^2}$. Substitute and solve for the one unknown v_2.

$$\frac{40}{(30)^2} = \frac{250}{v_2^2}$$
$$\frac{40}{900} = \frac{250}{v_2^2}$$
$$900v_2^2 \cdot \frac{40}{900} = 900v_2^2 \cdot \frac{250}{v_2^2}$$
$$40v_2^2 = 225{,}000$$
$$v_2^2 = 5625$$
$$v_2 = 75 \text{ mph}$$

31. $I = k\frac{E}{R} \quad k = \frac{RI}{E} \quad k = \frac{R_1 I_1}{E_1} \quad k = \frac{R_2 I_2}{E_2}$

Hence $\frac{R_1 I_1}{E_1} = \frac{R_2 I_2}{E_2}$. Substitute and solve for the one unknown I_2.

$$\frac{(5)(22)}{110} = \frac{11I_2}{220}$$
$$1 = \frac{11I_2}{220}$$
$$11I_2 = 220$$
$$I_2 = 20 \text{ amperes}$$

33. $L = k\frac{wh^2}{\ell} \quad k = \frac{L\ell}{wh^2} \quad k = \frac{L_1\ell_1}{w_1 h_1^2} \quad k = \frac{L_2\ell_2}{w_2 h_2^2}$

Hence $\frac{L_1\ell_1}{w_1 h_1^2} = \frac{L_2\ell_2}{w_2 h_2^2}$. Substitute and solve for the one unknown L_2.

$$\frac{(320)(8)}{2(4)^2} = \frac{L_2(12)}{2(6)^2}$$
$$\frac{2560}{32} = \frac{12L_2}{72}$$
$$80 = \frac{L_2}{6}$$
$$L_2 = 480 \text{ pounds}$$

35. Let h = the number of hours
w = the number of workers

$h = \frac{k}{w} \quad k = hw \quad k = h_1 w_1 \quad k = h_2 w_2$

Hence $h_1 w_1 = h_2 w_2$. Substitute and solve for the one unknown h_2.

$$30(6) = 9w_2$$
$$180 = 9w_2$$
$$w_2 = 20 \text{ hours}$$

37. $n = k\dfrac{P_1P_2}{d^2}$ In this problem, to avoid confusion of subscripts, we use Method 1.

We find k using n = 100,000, $P_1 = 3.2 \times 10^6$, $P_2 = 5.4 \times 10^6$, and d = 900.

$$k = \frac{nd^2}{P_1P_2}$$

$$k = \frac{(100{,}000)(900)^2}{(3.2 \times 10^6)(5.4 \times 10^6)} = \frac{9 \times 10^9}{17.28 \times 10^{12}} = \frac{1}{1.92 \times 10^3}$$

Now find n using this value for k and $P_1 = 4 \times 10^5$, $P_2 = 1.5 \times 10^6$, d = 100.

$$n = \frac{1}{1.92 \times 10^3} \frac{P_1P_2}{d^2}$$

$$= \frac{1}{1.92 \times 10^3} \frac{(4 \times 10^5)(1.5 \times 10^6)}{(100)^2}$$

$$= \frac{6 \times 10^{11}}{1.92 \times 10^7}$$

$= 3.125 \times 10^4$ or 31,250 phone calls

39. $P = kv^3$ $k = \dfrac{P}{v^3}$ $k = \dfrac{P_1}{v_1^3}$ $k = \dfrac{P_2}{v_2^3}$ Hence $\dfrac{P_1}{v_1^3} = \dfrac{P_2}{v_2^3}$

Given that the speed in the second situation (v_2) is double the speed in the first situation (v_1), we substitute to find the effect on P.

$$v_2 = 2v_1$$

Hence $$\frac{P_1}{v_1^3} = \frac{P_2}{(2v_1)^3}$$

$$\frac{P_1}{v_1^3} = \frac{P_2}{8v_1^3}$$

$$8v_1^3 \cdot \frac{P_1}{v_1^3} = 8v_1^3 \cdot \frac{P_2}{8v_1^3}$$

$$8P_1 = P_2$$

Thus the horsepower must be multiplied by 8—new horsepower must be 8 times the old.

41. $f = k\dfrac{\sqrt{T}}{L}$ $k = \dfrac{fL}{\sqrt{T}}$ $k = \dfrac{f_1L_1}{\sqrt{T_1}}$ $k = \dfrac{f_2L_2}{\sqrt{T_2}}$ Hence $\dfrac{f_1L_1}{\sqrt{T_1}} = \dfrac{f_2L_2}{\sqrt{T_2}}$

Given: new tension is 4 times old tension, that is, $T_2 = 4T_1$
new length is 2 times old length, that is, $L_2 = 2L_1$

We substitute to find the effect on f.

$$\frac{f_1L_1}{\sqrt{T_1}} = \frac{f_2(2L_1)}{\sqrt{4T_1}}$$

$$\frac{f_1L_1}{\sqrt{T_1}} = \frac{2f_2L_1}{2\sqrt{T_1}}$$

$$\frac{f_1L_1}{\sqrt{T_1}} = \frac{f_2L_1}{\sqrt{T_1}}$$

$$f_1 = f_2$$

The frequency remains the same.

43. $t^2 = kd^3$

45. $t = k\frac{r}{v}$ $k = \frac{tv}{r}$ $k = \frac{t_1 v_1}{r_1}$ $k = \frac{t_2 v_2}{r_2}$ Hence $\frac{t_1 v_1}{r_1} = \frac{t_2 v_2}{r_2}$.

Substitute and solve for the one unknown t_2.

$$\frac{(1.42)(18,000)}{(4,050)} = \frac{t_2(18,500)}{(4,300)}$$

$$t_2 = \frac{(4,300)}{(18,500)} \cdot \frac{(1.42)(18,000)}{(4,050)}$$

$$t_2 \approx 1.47 \text{ hours}$$

47. $d = kh$ $k = \frac{d}{h}$ $k = \frac{d_1}{h_1}$ $k = \frac{d_2}{h_2}$ Hence $\frac{d_1}{h_1} = \frac{d_2}{h_2}$

Substitute and solve for the one unknown d_2.

$$\frac{4}{500} = \frac{d_2}{2500}$$

$$d_2 = 2500 \cdot \frac{4}{500}$$

$$d_2 = 20 \text{ days}$$

49. $L = kv^2$ $k = \frac{L}{v^2}$ $k = \frac{L_1}{v_1^2}$ $k = \frac{L_2}{v_2^2}$ Hence $\frac{L_1}{v_1^2} = \frac{L_2}{v_2^2}$

Given: new speed is double the old speed, thus $v_2 = 2v_1$. We substitute to find the effect on L.

$$\frac{L_1}{v_1^2} = \frac{L_2}{(2v_1)^2}$$

$$\frac{L_1}{v_1^2} = \frac{L_2}{4v_1^2}$$

$$4v_1^2 \cdot \frac{L_1}{v_1^2} = 4v_1^2 \cdot \frac{L_2}{4v_1^2}$$

$$4L_1 = L_2$$

Thus, the length of the skid marks will be multiplied by 4—quadrupled.

51. $P = kAv^2$ $k = \frac{P}{Av^2}$ $k = \frac{P_1}{A_1 v_1^2}$ $k = \frac{P_2}{A_2 v_2^2}$ Hence $\frac{P_1}{A_1 v_1^2} = \frac{P_2}{A_2 v_2^2}$

We substitute and solve for the one unknown P_2.

$$\frac{120}{(100)(20)^2} = \frac{P_2}{(200)(30)^2}$$

$$\frac{120}{40,000} = \frac{P_2}{180,000}$$

$$P_2 = 180,000 \cdot \frac{120}{40,000}$$

$$P_2 = 540 \text{ pounds}$$

53. (A) $\Delta S = kS$

(B) Since $k = \frac{\Delta S}{S}$, $k = \frac{\Delta S_1}{S_1}$ and $k = \frac{\Delta S_2}{S_2}$.

Hence $\frac{\Delta S_1}{S_1} = \frac{\Delta S_2}{S_2}$. We substitute and solve for the one unknown ΔS_2.

$$\frac{1}{50} = \frac{\Delta S_2}{500}$$

$$\Delta S_2 = \frac{500}{50}$$

$$\Delta S_2 = 10 \text{ ounces}$$

(C) We use $\frac{\Delta S_1}{S_1} = \frac{\Delta S_2}{S_2}$ again.

$$\frac{1}{60} = \frac{\Delta S_2}{480}$$

$$\Delta S_2 = \frac{480}{60}$$

$$\Delta S_2 = 8 \text{ candlepower}$$

55. Let f = frequency of vibration
L = length of pipe

Then $f = \frac{k}{L}$, $k = fL$, $k = f_1L_1$, $k = f_2L_2$. Hence $f_1L_1 = f_2L_2$.

We substitute and solve for the one unknown f_2.

$$(16)(32) = f_2(16)$$
$$32 = f_2$$
$$f_2 = 32 \text{ times per second}$$

57. $N = k\frac{F}{d}$

59. $v = \frac{k}{\sqrt{w}}$ $\quad k = v\sqrt{w}$ $\quad k = v_1\sqrt{w_1}$ $\quad k = v_2\sqrt{w_2}$

Hence $v_1\sqrt{w_1} = v_2\sqrt{w_2}$.

Given: weight of hydrogen molecule is $\frac{1}{16}$ times weight of oxygen molecule: $w_2 = \frac{1}{16}w_1$

$$v_1\sqrt{w_1} = v_2\sqrt{\frac{1}{16}w_1}$$
$$v_1\sqrt{w_1} = v_2\sqrt{\frac{w_1}{16}}$$
$$v_1\sqrt{w_1} = v_2\frac{\sqrt{w_1}}{4}$$
$$4v_1\sqrt{w_1} = v_2\sqrt{w_1}$$
$$v_2 = 4v_1$$
$$v_2 = 4(0.3)$$
$$v_2 = 1.2 \text{ miles/second}$$

61. $A = kwt$ $\quad k = \frac{A}{wt}$ $\quad k = \frac{A_1}{w_1t_1}$ $\quad k = \frac{A_2}{w_2t_2}$ $\quad$ Hence $\frac{A_1}{w_1t_1} = \frac{A_2}{w_2t_2}$

Substitute and solve for the one unknown t_2 ($A_1 = A_2 = 1$ entire job).

$$\frac{1}{(10)(8)} = \frac{1}{4t_2}$$
$$80t_2 \cdot \frac{1}{80} = 80t_2 \cdot \frac{1}{4t_2}$$
$$t_2 = 20 \text{ days}$$

63. $V = kr^3$ $\quad k = \frac{V}{r^3}$ $\quad k = \frac{V_1}{r_1^3}$ $\quad k = \frac{V_2}{r_2^3}$ $\quad$ Hence $\frac{V_1}{r_1^3} = \frac{V_2}{r_2^3}$

Given: new radius is double the old radius: $r_2 = 2r_1$

Substitute to find the effect on V.

$$\frac{V_1}{r_1^3} = \frac{V_2}{(2r_1)^3}$$
$$\frac{V_1}{r_1^3} = \frac{V_2}{8r_1^3}$$
$$8r_1^3\frac{V_1}{r_1^3} = 8r_1^3\frac{V_2}{8r_1^3}$$
$$8V_1 = V_2$$

Hence the new volume is 8 times the old volume.

Exercise 9-6 REVIEW EXERCISE

1. Not a function (two range values correspond to the domain value 3)

2. Function (exactly one range value coresponds to each domain value)

3. Function (exactly one range value corresponds to each domain value)

4. Function (each vertical line passes through at most one point on the graph)

5. Not a function ($x = 0$, for example, passes through more than one point on the graph)

6. Function (each vertical line passes through at most one point on the graph)

7. Function (each value of the independent variable corresponds to one value of the dependent variable)

8. This rule is not a function since, for example, if $x = 4$, then $y^2 = 4$, so $y = \pm 2$.

9. $x^2 + y^2 = 25$ is the same as $y^2 = 25 - x^2$. This rule is not a function since, for example, if $x = 0$, then $y^2 = 25$, so $y = \pm 5$.

10. Not a function (two range values correspond to the domain value 1)

11. Function (exactly one range value corresponds to each domain value)

12. Function (exactly one range value corresponds to each domain value)

13. Domain = Set of all first components = {1, 3, 5}. Range = Set of all second components = {2, 4, 6}

14. Domain = Set of all first components = {-1, 1, 3, 5}. Range = Set of all second components = {0}

15. $f(x) = \frac{1}{x}$ is defined and real for all $x \neq 0$. Domain: $x \neq 0$

    To find the range, we set $y = \frac{1}{x}$ and solve for x to obtain $x = \frac{1}{y}$. This is defined and real for all $y \neq 0$. Range: $y \neq 0$

16. $f(x) = \frac{1}{\sqrt{x}}$ is defined and real for all $x > 0$. Domain: $x > 0$

    To find the range, we note first that $\sqrt{x}$ is always non-negative, thus the range cannot include any negative numbers. We now set $y = \frac{1}{\sqrt{x}}$ and solve for x.

    $$y = \frac{1}{\sqrt{x}}$$
    $$y^2 = \frac{1}{x}$$
    $$x = \frac{1}{y^2}$$

    This is defined and real for all $y \neq 0$. Since the range cannot include any negative numbers, we have
    Range: $y > 0$

17. $f(x) = \frac{1}{\sqrt{x - 1}}$ is defined and real if $x - 1$ is positive, that is,

$$x - 1 > 0$$
$$x > 1$$

Domain: $x > 1$ or $(1, \infty)$

To find the range, we note first that $\sqrt{x - 1}$ is always non-negative, thus the range cannot include any negative numbers. We now set $y = \frac{1}{\sqrt{x - 1}}$ and solve for x.

$$y = \frac{1}{\sqrt{x - 1}}$$
$$y^2 = \frac{1}{x - 1}$$
$$y^2(x - 1) = 1$$
$$x - 1 = \frac{1}{y^2}$$
$$x = 1 + \frac{1}{y^2}$$

This is defined and real for all $y \neq 0$. Since the range cannot include any negative numbers, we have

Range: $y > 0$

18. $f(x) = \sqrt{x - 1}$ is defined and real if $x - 1$ is non-negative, that is

$$x - 1 \geq 0$$
$$x \geq 1$$

Domain: $x \geq 1$ or $[1, \infty)$

To find the range, we note first that $\sqrt{x - 1}$ is always non-negative, that is, the range cannot include any negative numbers. We now set $y = \sqrt{x - 1}$ and solve for x.

$$y = \sqrt{x - 1}$$
$$y^2 = x - 1$$
$$x = 1 + y^2$$

This is defined and real for all y. Thus, we have

Range: $y \geq 0$ or $[0, \infty)$

19. For each real x, f(x) is defined and real. Domain: all real numbers. To find the range, we set $y = x - 1$ and solve for x to obtain $x = y + 1$. This is defined and real for all y.

Range: all real numbers

20. $f(x) = \frac{1}{x - 1}$ is defined and real for all x except $x = 1$. Domain: $x \neq 1$

To find the range, we set $y = \frac{1}{x - 1}$ and solve for x.

$$y = \frac{1}{x - 1}$$
$$(x - 1)y = 1$$
$$x - 1 = \frac{1}{y}$$
$$x = 1 + \frac{1}{y}$$

This is defined and real for all $y \neq 0$.

Range: $y \neq 0$

21. $f(x) = 6 - x$
$f(6) = 6 - 6 = 0$
$f(0) = 6 - 0 = 6$
$f(-3) = 6 - (-3) = 9$

22. $G(x) = x - 2x^2$
$G(2) = 2 - 2(2)^2 = 2 - 8 = -6$
$G(0) = 0 - 2(0)^2 = 0$
$G(-1) = (-1) - 2(-1)^2$
$= (-1) - 2(1) = -3$

23. $f(x) = 6 - x$
$f(m) = 6 - m$
$f(x + h) = 6 - (x + h) = 6 - x - h$

24. $G(x) = x - 2x^2$
$G(c) = c - 2c^2$
$G(x + h) = (x + h) - 2(x + h)^2$
$= x + h - 2(x^2 + 2xh + h^2)$
$= x + h - 2x^2 - 4xh - 2h^2$

25. $(f + G)(x) = f(x) + G(x) = 6 - x + x - 2x^2 = 6 - 2x^2$

26. $(f - G)(x) = f(x) - G(x) = (6 - x) - (x - 2x^2) = 6 - x - x + 2x^2 = 6 - 2x + 2x^2$

27. $(G - f)(x) = G(x) - f(x) = (x - 2x^2) - (6 - x) = x - 2x^2 - 6 + x = -2x^2 + 2x - 6$

28. $(f\cdot G)(x) = f(x)\cdot G(x) = (6 - x)(x - 2x^2) = 6x - 12x^2 - x^2 + 2x^3 = 6x - 13x^2 + 2x^3$

29. $\left(\frac{f}{G}\right)(x) = \frac{f(x)}{G(x)} = \frac{6 - x}{x - 2x^2}$

30. $\left(\frac{G}{f}\right)(x) = \frac{G(x)}{f(x)} = \frac{x - 2x^2}{6 - x}$

31. $f(G(x)) = f(x - 2x^2) = 6 - (x - 2x^2) = 6 - x + 2x^2$

32. $G(f(x)) = G(6 - x) = (6 - x) - 2(6 - x)^2 = 6 - x - 2(36 - 12x + x^2)$
$= 6 - x - 72 + 24x - 2x^2 = -66 + 23x - 2x^2$

33. $f(x) = 2x - 4$

x	f(x)
0	-4
2	0
4	4

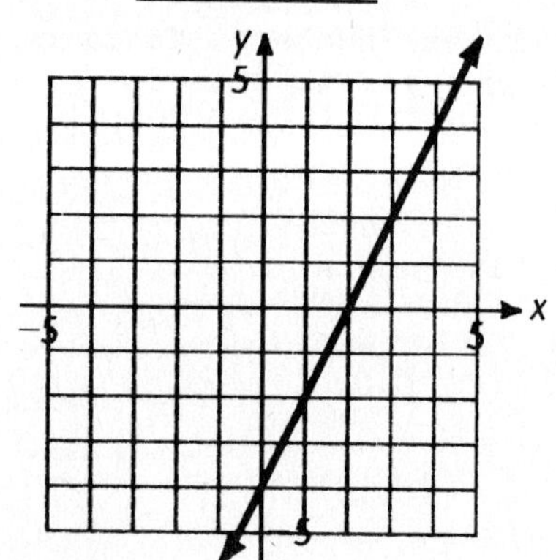

Slope: 2 y intercept: -4

34. $g(x) = \frac{x^2}{2}$

This is a quadratic function. Its graph is a parabola. Since $a = \frac{1}{2}$, the parabola opens up. The vertex is at (0, 0) and the x and y intercepts are 0.

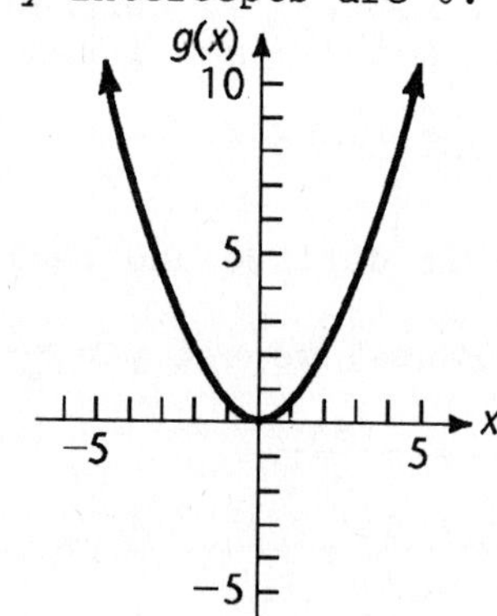

35.

x	M(x)
0	5
2	7
4	9

x	$M^{-1}(x)$
5	0
7	2
9	4

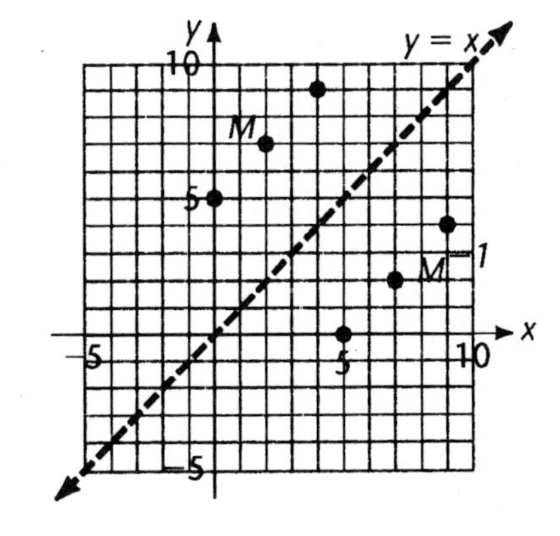

36. $M^{-1} = \{(5, 0), (7, 2), (9, 4)\}$
Domain of M^{-1} = Set of all first components = $\{5, 7, 9\}$
Range of M^{-1} = Set of all second components = $\{0, 2, 4\}$

37.
$$f: y = 3x + 5$$
$$f^{-1}: x = 3y + 5$$
$$x - 5 = 3y$$
$$\frac{x - 5}{3} = y$$
$$f^{-1}(x) = \frac{x - 5}{3}$$

38.
$$f: y = \frac{3}{x + 5}$$
$$f^{-1}: x = \frac{3}{y + 5}$$
$$x(y + 5) = 3$$
$$xy + 5x = 3$$
$$xy = 3 - 5x$$
$$y = \frac{3 - 5x}{x}$$
$$f^{-1}(x) = \frac{3 - 5x}{x}$$

39.
$$f: y = \sqrt{3x + 5}$$
$$f^{-1}: x = \sqrt{3y + 5}$$
$$x^2 = 3y + 5$$
$$x^2 - 5 = 3y$$
$$\frac{x^2 - 5}{3} = y$$
$$f^{-1}(x) = \frac{x^2 - 5}{3}$$

40.
$$f: y = 3\sqrt{x} + 5$$
$$f^{-1}: x = 3\sqrt{y} + 5$$
$$x - 5 = 3\sqrt{y}$$
$$\frac{x - 5}{3} = \sqrt{y}$$
$$\left(\frac{x - 5}{3}\right)^2 = y$$
$$f^{-1}(x) = \left(\frac{x - 5}{3}\right)^2$$

41.
$$f: y = \frac{3}{x} + 5$$
$$f^{-1}: x = \frac{3}{y} + 5$$
$$x - 5 = \frac{3}{y}$$
$$y(x - 5) = 3$$
$$y = \frac{3}{x - 5}$$
$$f^{-1}(x) = \frac{3}{x - 5}$$

42. $m = kn^2$

43. $P = \dfrac{k}{Q^3}$

44. $A = kab$

45. $m = kn \quad k = \dfrac{m}{n} \quad k = \dfrac{m_1}{n_1} \quad k = \dfrac{m_2}{n_2}$

Hence $\dfrac{m_1}{n_1} = \dfrac{m_2}{n_2}$.
Substitute and solve for the one unknown m_2.
$$\frac{27}{6} = \frac{m_2}{10}$$
$$m_2 = 10 \cdot \frac{27}{6}$$
$$m_2 = 45$$

46. $P = \dfrac{k}{Q} \quad k = PQ \quad k = P_1Q_1 \quad k = P_2Q_2$

Hence $P_1Q_1 = P_2Q_2$.
Substitute and solve for the one unknown P_2.
$$(27)(6) = P_2(10)$$
$$162 = 10P_2$$
$$P_2 = 16.2$$

47. $A = kab \qquad k = \frac{A}{ab} \qquad k = \frac{A_1}{a_1 b_1} \qquad k = \frac{A_2}{a_2 b_2}$

Hence $\frac{A_1}{a_1 b_1} = \frac{A_2}{a_2 b_2}$. Substitute and solve for the one unknown A_2.

$$\frac{162}{(6)(9)} = \frac{A_2}{(8)(7)}$$

$$\frac{162}{54} = \frac{A_2}{56}$$

$$A_2 = 56 \cdot \frac{162}{54}$$

$$A_2 = 168$$

48. Cost = (Fixed Cost) + (Cost of Copies)
$C(x) = 200 + 0.05x, \quad 0 \le x \le 3{,}000$
This is a linear function. We choose each end value and the middle value in this interval.

x	C(x)
0	200
1,500	275
3,000	350

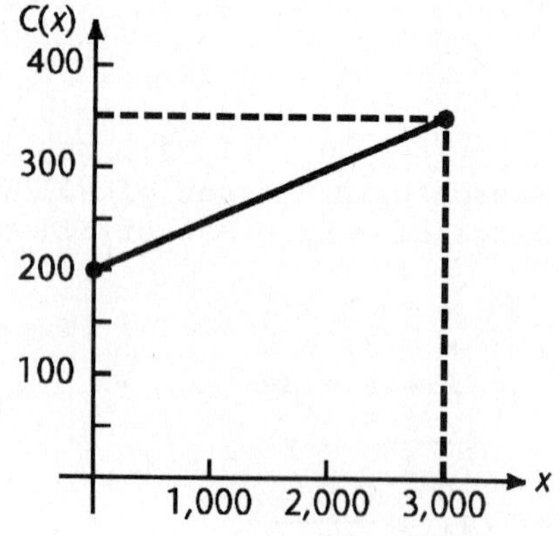

49. We are asked to find a linear function, that is, one of the form $V(t) = mt + b$. Since $V(0) = 80{,}000 = m(0) + b$, we have $b = 80{,}000$. Thus, $V(t) = mt + 80{,}000$. Since $V(10) = 5000$, we have $5000 = m(10) + 80{,}000$, that is, $10m = 5000 - 80{,}000$, $10m = -75{,}000$, or $m = -7500$. Thus, $V(t) = -7500t + 80{,}000$.

To graph this linear function, we choose the given end values and the middle value in this interval.

t	V(t)
0	80,000
5	42,500
10	5,000

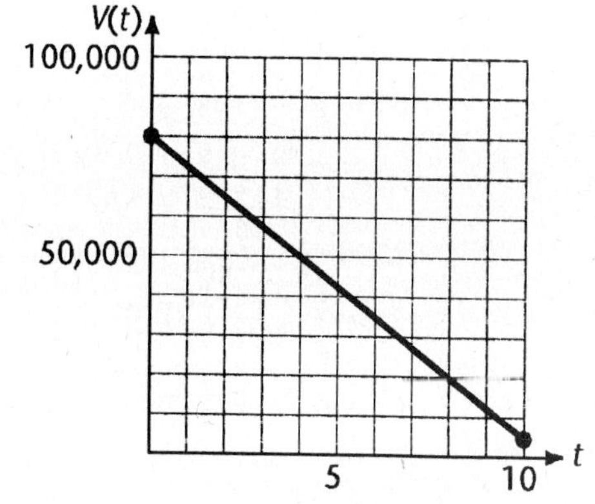

50. No. A function has an inverse only when it is one-to-one, that is, only when every element in the range corresponds to exactly one element in the domain.

51. $(g - f)(x) = g(x) - f(x) = (x - 2) - (4 - x^2) = x - 2 - 4 + x^2 = x^2 + x - 6$

52. $(f - g)(x) = f(x) - g(x) = (4 - x^2) - (x - 2) = 4 - x^2 - x + 2 = 6 - x - x^2$

53. $(f \cdot g + f)(x) = (f \cdot g)(x) + f(x) = f(x) \cdot g(x) + f(x) = (4 - x^2)(x - 2) + (4 - x^2)$
$= 4x - 8 - x^3 + 2x^2 + 4 - x^2 = -x^3 + x^2 + 4x - 4$

54. $(f \cdot g - g)(x) = (f \cdot g)(x) - g(x) - f(x) \cdot g(x) - g(x) = (4 - x^2)(x - 2) - (x - 2)$
$= 4x - 8 - x^3 + 2x^2 - x + 2 = -x^3 + 2x^2 + 3x - 6$

55. $\left(\frac{f}{g}\right)(x) = \frac{f(x)}{g(x)} = \frac{4 - x^2}{x - 2} = \frac{(2 + x)(2 - x)}{x - 2} = -(2 + x) = -x - 2$

56. $\left(\frac{g}{f}\right)(x) = \frac{g(x)}{f(x)} = \frac{x - 2}{4 - x^2} = \frac{-(2 - x)}{(2 + x)(2 - x)} = \frac{-1}{2 + x}$

57. $f(g(-2)) = f(-2 - 2) = f(-4) = 4 - (-4)^2 = 4 - 16 = -12$

58. $g(f(3)) = g(4 - 3^2) = g(-5) = (-5) - 2 = -7$

59. $f(g(x)) = f(x - 2) = 4 - (x - 2)^2 = 4 - (x^2 - 4x + 4) = 4 - x^2 + 4x - 4 = 4x - x^2$

60. $g(f(x)) = g(4 - x^2) = (4 - x^2) - 2 = 2 - x^2$

61. $f(f(x)) = f(4 - x^2) = 4 - (4 - x^2)^2 = 4 - (16 - 8x^2 + x^4) = 4 - 16 + 8x^2 - x^4$
$= -12 + 8x^2 - x^4$

62. $g(g(x)) = g(x - 2) = (x - 2) - 2 = x - 4$

63. $f(x) = 2x - 3$

(A) $f(3 + h) = 2(3 + h) - 3 = 6 + 2h - 3 = 2h + 3$

(B) $\frac{f(3 + h) - f(3)}{h} = \frac{[2(3 + h) - 3] - [2(3) - 3]}{h} = \frac{[6 + 2h - 3] - [3]}{h}$
$= \frac{6 + 2h - 3 - 3}{h} = \frac{2h}{h} = 2$

64. $g(t) = -\frac{3}{2}t + 6$

t	g(t)
0	6
2	3
4	0

Slope: $-\frac{3}{2}$

y intercept: 6

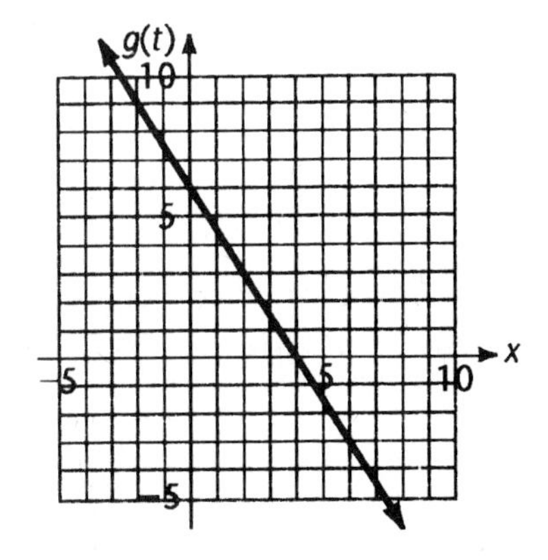

65. $f(x) = x^2 - 4x + 5$

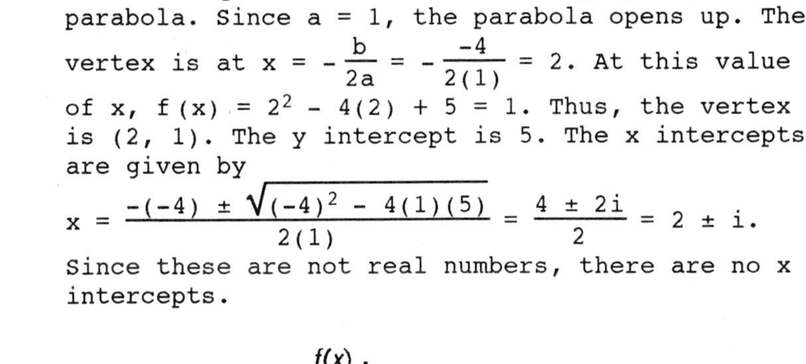
This is a quadratic function. Its graph is a parabola. Since $a = 1$, the parabola opens up. The vertex is at $x = -\frac{b}{2a} = -\frac{-4}{2(1)} = 2$. At this value of x, $f(x) = 2^2 - 4(2) + 5 = 1$. Thus, the vertex is (2, 1). The y intercept is 5. The x intercepts are given by

$$x = \frac{-(-4) \pm \sqrt{(-4)^2 - 4(1)(5)}}{2(1)} = \frac{4 \pm 2i}{2} = 2 \pm i.$$

Since these are not real numbers, there are no x intercepts.

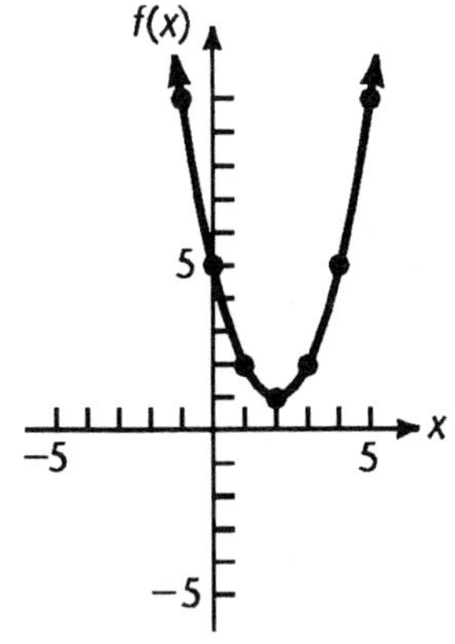

66. $P(x) = x^3 - 2x^2 - 5x + 6$

Partial Table

x	P(x)
4	18
3	0
2	-4
1	0
0	6
-1	8
-2	0
-3	-24

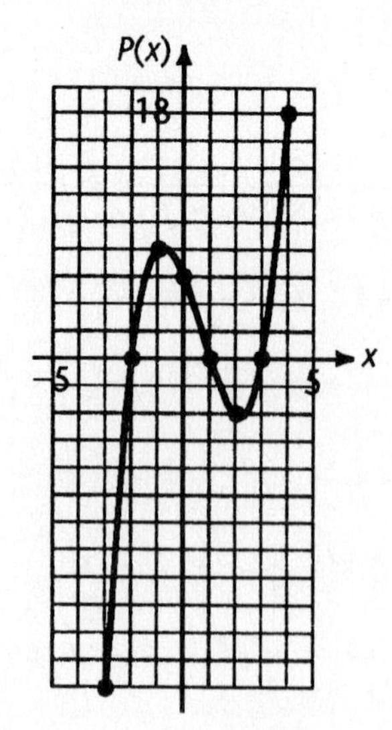

67. Problem 3 is the only one where each domain value corresponds to exactly one range value and each range value corresponds to exactly one domain value.

68. Problem 6 is the only one where each vertical line and each horizontal line crosses the graph at most once.

69. In problem 7 both $x = 0$ and $x = \sqrt{2}$ correspond to $y = 0$.
In problem 8 both $y = -1$ and $y = 1$ correspond to $x = 1$.
In problem 9 both $y = 5$ and $y = -5$ correspond to $x = 0$.
None of these are one-to-one correspondences.

70. $g(t) = 96t - 16t^2 = -16t^2 + 96t + 0$
This is a quadratic function. Its graph is a parabola. Since $a = -16$, the parabola opens down. The vertex is at $t = -\frac{b}{2a} = -\frac{96}{2(-16)} = 3$. At this value of t, $g(t) = 96(3) - 16(3)^2 = 144$. Thus, the vertex is (3, 144). The y intercept is 0. The t intercepts are given by $t = \frac{-96 \pm \sqrt{(96)^2 - 4(-16)(0)}}{2(-16)} = 0, 6$

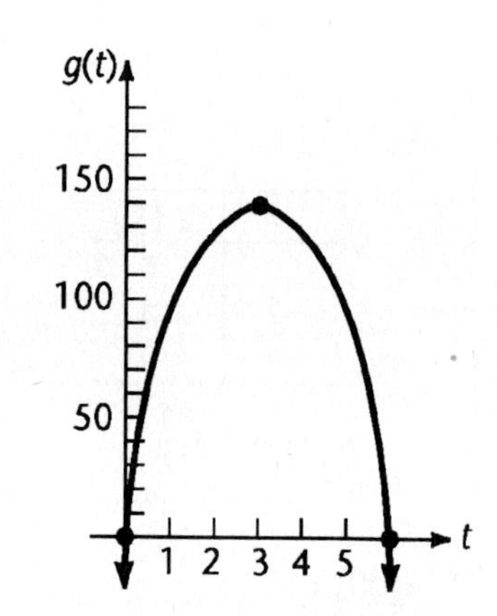

71. Each range value corresponds to exactly one domain value, so this function has an inverse, as follows:
$4 \rightarrow 1$
$5 \rightarrow 2$
$6 \rightarrow 3$

72. The range value 2 corresponds to more than one domain value, so this function has no inverse.

73. Each range value corresponds to exactly one domain value, so this function has an inverse, as follows:
$1 \rightarrow 1$
$-1 \rightarrow 2$
$-3 \rightarrow 3$

74. This function passes the horizontal line test; it has an inverse.

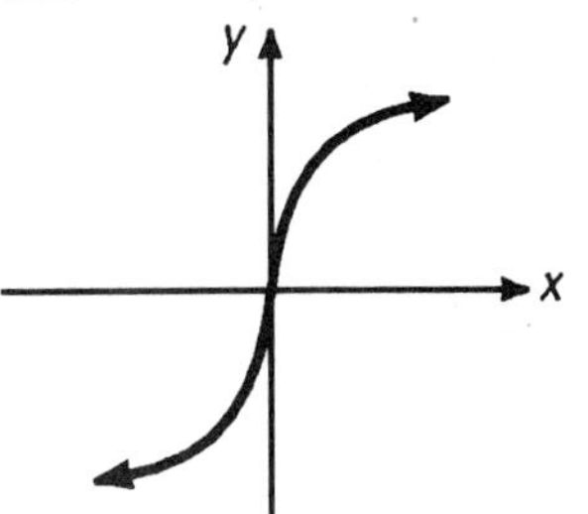

75. This function fails the horizontal line test; it has no inverse.

76. This function fails the horizontal line test; it has no inverse.

77. Each range value corresponds to exactly one domain value, so this function has an inverse, as follows: (3, 1), (7, 5), (11, 9).

78. Each range value corresponds to exactly one domain value, so this function has an inverse, as follows: (3, 1), (1, 3), (7, 5), (5, 7).

79. The range value 3 corresponds to more than one domain value, so this function has no inverse.

80. This function has an inverse:

$$f: y = x^3 + 5$$
$$f^{-1}: x = y^3 + 5$$
$$y^3 = x - 5$$
$$y = \sqrt[3]{x - 5}$$
$$f^{-1}(x) = \sqrt[3]{x - 5}$$

81. The range value -3 corresponds to the domain values -1 and 1, so this function has no inverse.

82. This function has an inverse:

$$f: y = 3x + 2$$
$$f^{-1}: x = 3y + 2$$
$$x - 2 = 3y$$
$$\frac{x - 2}{3} = y$$
$$f^{-1}(x) = \frac{x - 2}{3}$$

83. For y to be real, $\frac{x + 2}{x - 5}$ cannot be negative; that is, $\frac{x + 2}{x - 5} \geq 0$.

Sign of (x+2) − − ¦ + + + ¦ + +
Sign of (x−5) − − ¦ − − − ¦ + +
-2 5

Domain: $x \leq -2$ or $x > 5$

$(-\infty, -2] \cup (5, \infty)$

84. For each real x, f(x) is defined and real. Domain: all real numbers.

85. For f(x) to be defined, x - 5 cannot be 0.

$$x - 5 = 0$$
$$x = 5$$

Domain: all real numbers except x = 5.

86. $M: y = \frac{x + 3}{2}$

$M^{-1}: x = \frac{y + 3}{2}$

$2x = y + 3$

$y = 2x - 3$

$M^{-1}(x) = 2x - 3$

$M[M^{-1}(x)] = M(2x - 3) = \frac{2x - 3 + 3}{2} = \frac{2x}{2} = x$

87. $y = k\frac{x}{z}$

88. $y = k\frac{x^3}{\sqrt{z}}$

89. $A = k\frac{ab^3}{c^2}$

90. $y = k\frac{x}{z}$

$k = \frac{z}{x}y \quad k = \frac{z_1}{x_1}y_1 \quad k = \frac{z_2}{x_2}y_2$. Hence, $\frac{z_1}{x_1}y_1 = \frac{z_2}{x_2}y_2$.

Substitute and solve for the one unknown y_2.

$\frac{2}{6} \cdot 4 = \frac{4}{4}y_2$

$\frac{8}{6} = y_2$

$y_2 = \frac{4}{3}$

91. $y = k\frac{x^3}{\sqrt{z}} \quad k = y\frac{\sqrt{z}}{x^3} \quad k = y_1\frac{\sqrt{z_1}}{x_1^3} \quad k = y_2\frac{\sqrt{z_2}}{x_2^3}$

Hence, $y_1\frac{\sqrt{z_1}}{x_1^3} = y_2\frac{\sqrt{z_2}}{x_2^3}$. Substitute and solve for the one unknown y_2.

$12\frac{\sqrt{4}}{2^3} = y_2\frac{\sqrt{9}}{1^3}$

$12 \cdot \frac{2}{8} = y_2 \cdot 3$

$3 = 3y_2$

$y_2 = 1$

92. $R = f(p) = 6{,}000p - 30p^2, \; 0 \le p \le 200$

(A) Partial table

p	R
200	0
180	108,000
160	192,000
140	252,000
120	288,000
100	300,000
80	288,000
60	252,000
40	192,000
20	108,000
0	0

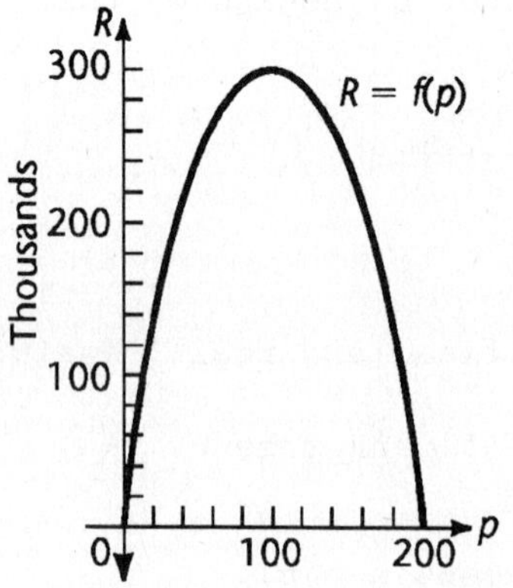

R
300
200
100
Thousands
R = f(p)
0
100
200
p

(B) Since $f(p) = 6{,}000p - 30p^2 = -30p^2 + 6{,}000p$ is a quadratic function, its graph is a parabola, and its largest value occurs at the vertex $p = -\frac{b}{2a} = 100$. Then $p = \$100$ gives $R = f(100) = \$300{,}000$.

93. $R(x) = (5 + x)(4000 - 400x)$
$= 20{,}000 + 2000x - 400x^2$

(A) Partial table

x	R(x)
0	20,000
2	22,400
4	21,600
6	17,600
8	10,400
10	0

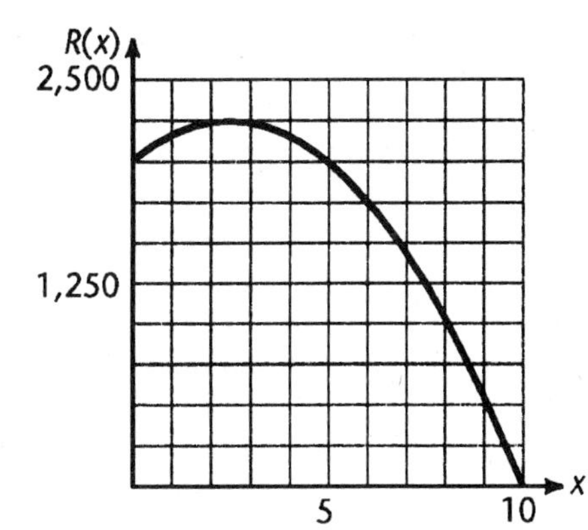

(B) Since $R(x) = -400x^2 + 2000x + 20{,}000$ is a quadratic function, its graph is a parabola, and its largest value occurs at the vertex $x = -\frac{b}{2a} = -\frac{2000}{2(-400)} = 2.5$. Then the price will be $5 + 2.5 = \$7.50$ and the largest monthly revenue will be $R(2.5) = \$22{,}500$.

94. $g(t) = 1 - t^2$

(A) $g(2 + h) = 1 - (2 + h)^2 = 1 - (4 + 4h + h^2) = 1 - 4 - 4h - h^2 = -3 - 4h - h^2$

(B) $$\frac{g(2 + h) - g(2)}{h} = \frac{[1 - (2 + h)^2] - [1 - (2)^2]}{h}$$
$$= \frac{[1 - (4 + 4h + h^2)] - [1 - 4]}{h} = \frac{[1 - 4 - 4h - h^2] - [-3]}{h}$$
$$= \frac{-3 - 4h - h^2 + 3}{h} = \frac{-4h - h^2}{h} = \frac{h(-4 - h)}{h} = -4 - h$$

95. $f(x) = \frac{1}{x}$

(A) $f(2 + h) = \frac{1}{2 + h}$

(B) $$\frac{f(2 + h) - f(2)}{h} = \frac{\frac{1}{2 + h} - \frac{1}{2}}{h}$$ Multiply numerator and denominator by $(2 + h)2$, the lcd of all internal fractions.

$$= \frac{(2 + h)2 \cdot \frac{1}{2 + h} - (2 + h)2 \cdot \frac{1}{2}}{(2 + h)2h}$$
$$= \frac{2 - (2 + h)}{(2 + h)2h}$$
$$= \frac{2 - 2 - h}{(2 + h)2h}$$
$$= \frac{-h}{(2 + h)2h}$$
$$= \frac{-1}{4 + 2h}$$

96. $f(x) = x^2,\ x \ge 0$

(A) $f: y = x^2,\ x \ge 0$

$f^{-1}: x = y^2,\ y \ge 0$

Solving for y, we find:

$y = \sqrt{x}$ since y must be nonnegative

$f^{-1}(x) = \sqrt{x}$

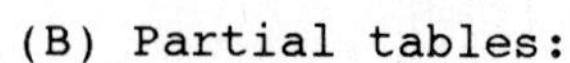

(B) Partial tables:

x	f(x)
0	0
1	1
2	4
3	9

x	$f^{-1}(x)$
0	0
1	1
4	2
9	3

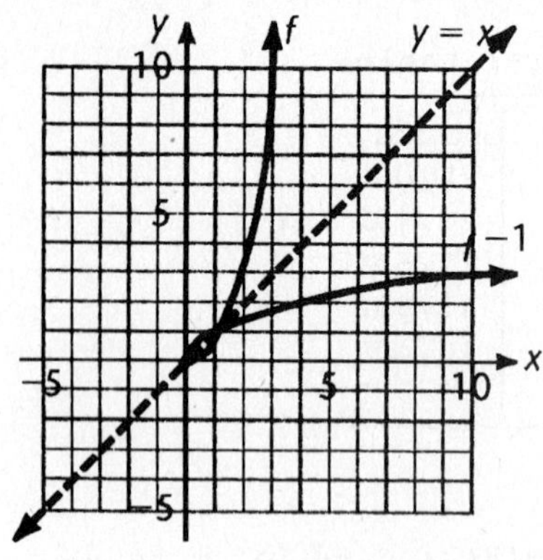

(C) $f^{-1}(x) = \sqrt{x}$

$f^{-1}(9) = \sqrt{9} = 3$

$f^{-1}[f(x)] = \sqrt{f(x)} = \sqrt{x^2} \quad (x \ge 0)$

$= x \quad (x \ge 0)$

97. $f(x) = \dfrac{1}{x}$

(A) $f: y = \dfrac{1}{x}$

$f^{-1}: x = \dfrac{1}{y}$

$xy = 1$

$y = \dfrac{1}{x}$

$f^{-1}(x) = \dfrac{1}{x}$

(B) Partial Tables

x	$f(x) = f^{-1}(x)$
-10	-0.1
-5	-0.2
-4	-0.25
-3	-0.33
-2	-0.5
-1	-1
-0.5	-2
-0.1	-10
0.1	10
0.5	5
1	1
2	0.5
3	0.33
4	0.25
5	0.2
10	0.1

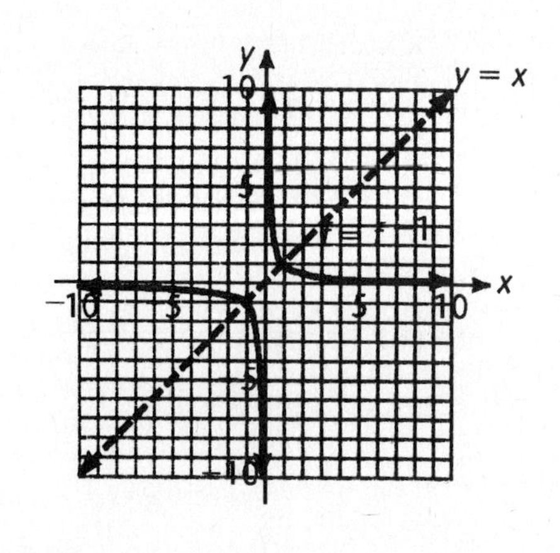

(C) $f^{-1}(9) = \dfrac{1}{9}$

$f^{-1}[f(x)] = f^{-1}\left(\dfrac{1}{x}\right) = \dfrac{1}{\frac{1}{x}} = x$

98.

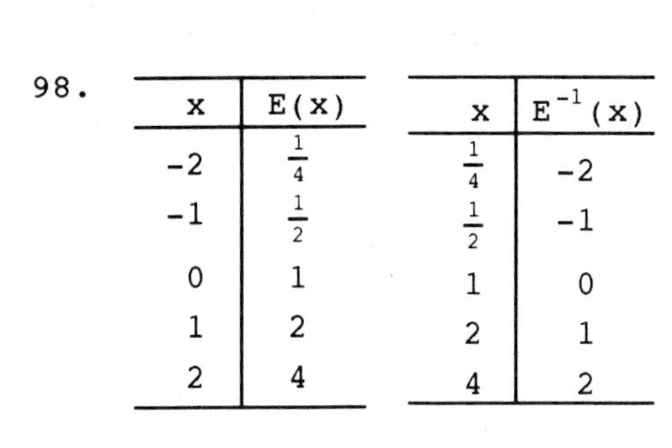

x	E(x)
-2	$\frac{1}{4}$
-1	$\frac{1}{2}$
0	1
1	2
2	4

x	$E^{-1}(x)$
$\frac{1}{4}$	-2
$\frac{1}{2}$	-1
1	0
2	1
4	2

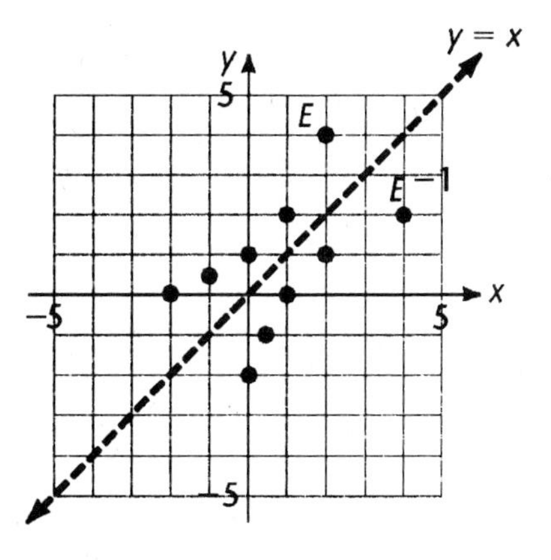

99. $t = k\dfrac{wd}{P}$

$Pt = kwd$

$k = \dfrac{Pt}{wd}$ $\quad k = \dfrac{P_1t_1}{w_1d_1}$ $\quad k = \dfrac{P_2t_2}{w_2d_2}$ $\quad$ Hence $\dfrac{P_1t_1}{w_1d_1} = \dfrac{P_2t_2}{w_2d_2}$.

Substitute and solve for the one unknown t_2.

$$\frac{(400)(4)}{(800)(8)} = \frac{400t_2}{(1600)(24)}$$
$$\frac{1,600}{6,400} = \frac{400t_2}{38,400}$$
$$\frac{1}{4} = \frac{t_2}{96}$$
$$t_2 = 24 \text{ seconds}$$

100. $F = kAv^2$

$k = \dfrac{F}{Av^2}$

$k = \dfrac{F_1}{A_1v_1^2}$ $\quad k = \dfrac{F_2}{A_2v_2^2}$ $\quad$ Hence $\dfrac{F_1}{A_1v_1^2} = \dfrac{F_2}{A_2v_2^2}$

Given: new area is half of old area—$A_2 = \dfrac{1}{2}A_1$

new velocity is double old velocity—$v_2 = 2v_1$

We substitute to find the effect on F.

$$\frac{F_1}{A_1v_1^2} = \frac{F_2}{\frac{1}{2}A_1(2v_1)^2}$$
$$\frac{F_1}{A_1v_1^2} = \frac{F_2}{\frac{1}{2}A_1 \cdot 4v_1^2}$$
$$\frac{F_1}{A_1v_1^2} = \frac{F_2}{2A_1v_1^2}$$
$$2A_1v_1^2 \cdot \frac{F_1}{A_1v_1^2} = 2A_1v_1^2 \cdot \frac{F_2}{2A_1v_1^2}$$
$$2F_1 = F_2$$

Hence, F_2 is twice F_1; the force on the wall is doubled.

Practice Test 9

1. Not a function (two range values correspond to the domain value 1).

2. Not a function (fails the vertical line test).

3. Function (exactly one value of y corresponds to each value of x).

4. $y^2 - x^2 = 4$ is the same as $y^2 = x^2 + 4$. This rule is not a function since, for example, if $x = 0$, then $y^2 = 4$, so $y = \pm 2$.

5. $f(x) = x^2 - 2$
$f(3) = 3^2 - 2 = 9 - 2 = 7$

6. $f(g(x)) = f(x + 3) = (x + 3)^2 - 2$
$= x^2 + 6x + 9 - 2 = x^2 + 6x + 7$

7. $g(x + h) = (x + h) + 3 = x + h + 3$

8. $f(x) = 3x + 4$

x	f(x)
-1	1
0	4
1	7

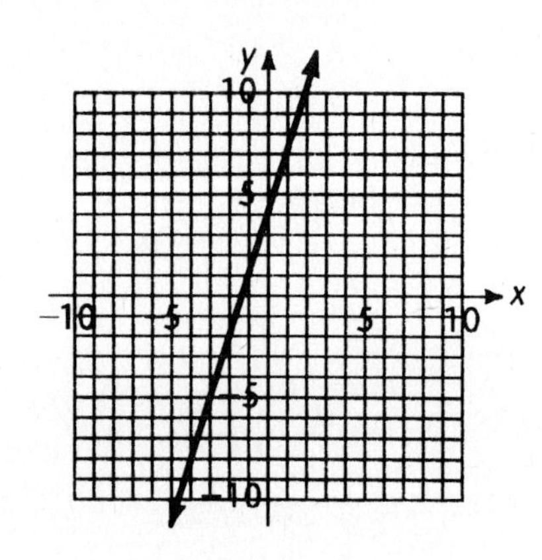

9. $f(x) = x^2 + 2$
This is a quadratic function. Its graph is a parabola. Since a = 1, the parabola opens up. The vertex is at (0, 0), and there are no x intercepts.

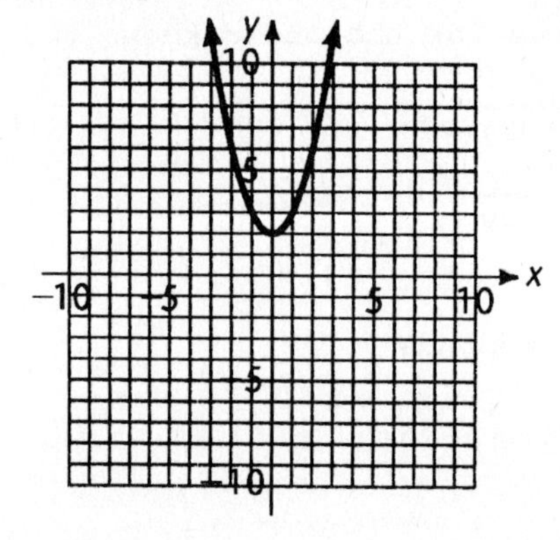

10. For y to be real, x - 4 cannot be negative; that is, $(x - 2)(x + 2) \geq 0$.

sign of (x+2) − − − ⋮ + + + ⋮ + + +
sign of (x−2) − − − ⋮ − − − ⋮ + + +
x: −2, 2

Domain: $x \leq -2$ or $x \geq 2$ $(-\infty, -2] \cup [2, \infty)$

11. For g(x) to be defined, $x^2 - 4$ cannot be 0.
$(x^2 - 4) = 0$
$(x + 2)(x - 2) = 0$
$x + 2 = 0$ or $x - 2 = 0$
$x = -2$ $x = 2$
Domain: $x \neq -2, 2$

12. Each range value corresponds to exactly one domain value, so this function has an inverse.

13. This function fails the horizontal line test; it has no inverse.

14. $h(1) = h(-1) = 0$. Since two domain values correspond to the same range value, this function has no inverse.

15. The inverse is given by (3, 1), (4, 2), (5, 3). The domain of the inverse is the set of first components of the inverse, that is, {3, 4, 5}.

16. $f: y = \sqrt{x} + 2$ Note that $y \geq 2$.

$f^{-1}: x = \sqrt{y} + 2$

$$x - 2 = \sqrt{y}$$

$$(x - 2)^2 = y$$

$f^{-1}(x) = (x - 2)^2$. Since the range of the original function was $[2, \infty)$, the domain of the inverse function is also $[2, \infty)$, that is, $x \geq 2$.

17. $f: \quad y = \dfrac{1}{x - 1}$

$f^{-1}: \quad x = \dfrac{1}{y - 1}$

$$x(y - 1) = 1$$

$$xy - x = 1$$

$$xy = x + 1$$

$$y = \frac{x + 1}{x}$$

$f^{-1}(x) = \dfrac{x + 1}{x}$ Domain: $x \neq 0$

18. $y = kx \qquad k = \dfrac{y}{x} \qquad k = \dfrac{y_1}{x_1} \qquad k = \dfrac{y_2}{x_2}$

Hence $\dfrac{y_1}{x_1} = \dfrac{y_2}{x_2}$.

Substitute and solve for the one unknown y_2.

$$\frac{18}{8} = \frac{y_2}{20}$$

$$y_2 = 20 \cdot \frac{18}{8}$$

$$y_2 = 45$$

19. $z = k\dfrac{x^2}{y} \qquad k = z\dfrac{y}{x^2} \qquad k = z_1\dfrac{y_1}{x_1^2} \qquad k = z_2\dfrac{y_2}{x_2^2}$

Hence $z_1\dfrac{y_1}{x_1^2} = z_2\dfrac{y_2}{x_2^2}$.

Substitute and solve for the one unknown z_2.

$$4 \cdot \frac{6}{2^2} = z_2 \cdot \frac{15}{3^2}$$

$$\frac{24}{4} = z_2 \cdot \frac{15}{9}$$

$$6 = \frac{15}{9} z_2$$

$$z_2 = \frac{9}{15} \cdot 6$$

$$z_2 = 3.6$$

20. $P(x) = (x - 4)(1000 - 50x)$
$= -4000 + 1200x - 50x^2$

Partial Table

x	P(x)
4	0
5	750
10	3000
12	3200
15	2750
20	0

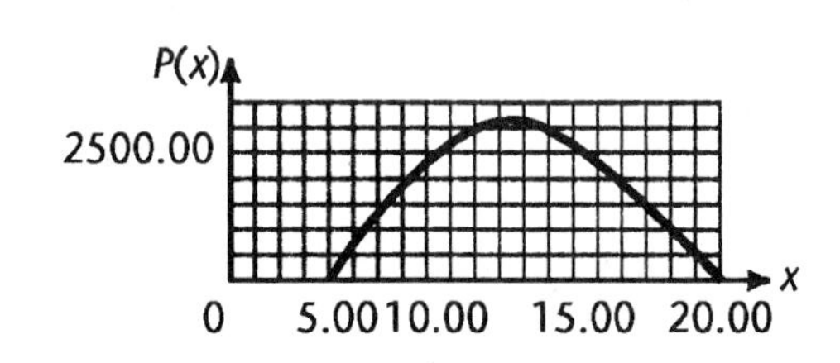

Since $P(x)$ is a quadratic function, its graph is a parabola, and its largest value occurs at the vertex $x = -\dfrac{b}{2a} = -\dfrac{1200}{2(-50)} = 12$. The largest possible profit will then be $P(12) = (12 - 4)(1000 - 50 \cdot 12) = \$3200$.

CHAPTER 10 Exponential and Logarithmic Functions

Exercise 10-1 Exponential Functions

Key Ideas and Formulas

The equation $f(x) = b^x$ $b > 0$, $b \neq 1$, defines an **exponential function** for each constant value of b. Domain of f: R. Range of f: Positive real numbers.

Typical graphs:

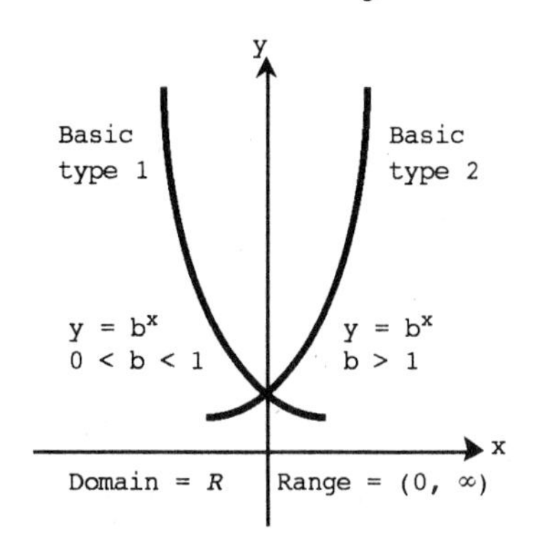

Compound Interest

$$A = P(1 + r)^n$$

where A = amount, P = principal, r = interest rate per period, and n = number of periods.

Continuously Compounded Interest

$$A = P(e^a)^t = Pe^{at}$$

where A = amount, P = principal, a = annual interest rate, t = number of years, and $e \approx 2.718\,281\,83$.

Exponent Properties

For real numbers s and t, and positive real numbers a and b:

1. $b^s b^t = b^{s+t}$
2. $(b^s)^t = b^{st}$
3. $(ab)^s = a^s b^s$
4. $\left(\frac{a}{b}\right)^s = \frac{a^s}{b^s}$ $\quad b \neq 0$
5. $\frac{b^s}{b^t} = b^{s-t}$ $\quad b \neq 0$

Exponential functions are either increasing ($b > 1$) or decreasing ($b < 1$), hence, one-to-one.

For $a, b > 0$, $a, b \neq 1$

$a^m = a^n$ if and only if $m = n$.

$a^m = b^m$, $m \neq 0$ if and only if $a = b$.

1. 4.7288…

3. 0.1026…

5. 7.3890…

7. 0.0497…

9. 1453.0403…

11. 4.1132…

13. Use $A = P\left(1 + \frac{a}{n}\right)^{nt}$ with P = 1000, a = 0.06, n = 6, t = 5

$$A = 1000\left(1 + \frac{0.06}{6}\right)^{(6)(5)}$$
$$A = \$1347.85$$

15. Use $A = P\left(1 + \frac{a}{n}\right)^{nt}$ with P = 1000, a = 0.075, n = 4, t = 8

$$A = 1000\left(1 + \frac{0.075}{4}\right)^{(4)(8)}$$
$$A = \$1812.02$$

17. Use $A = P\left(1 + \frac{a}{n}\right)^{nt}$ with P = 1000, a = 0.015, n = 2, t = 20

$$A = 1000\left(1 + \frac{0.015}{2}\right)^{(2)(20)}$$
$$A = \$1348.35$$

19. Use $A = P\left(1 + \frac{a}{n}\right)^{nt}$ with P = 1000, a = 0.18, n = 12, t = 1

$$A = 1000\left(1 + \frac{0.18}{12}\right)^{(12)(1)}$$
$$A = \$1195.62$$

21. Use $A = Pe^{at}$ with P = 1000, a = 0.05, t = 30
$$A = 1000e^{(0.05)(30)}$$
$$A = \$4481.69$$

23. Use $A = Pe^{at}$ with P = 1000, a = 0.08, t = 10
$$A = 1000e^{(0.08)(10)}$$
$$A = \$2225.54$$

25.

x	y
-3	0.03
-2	0.11
-1	0.33
0	1.00
1	3.00
2	9.00
3	27.00

27.

x	y
-3	27.00
-2	9.00
-1	3.00
0	1.00
1	0.33
2	0.11
3	0.03

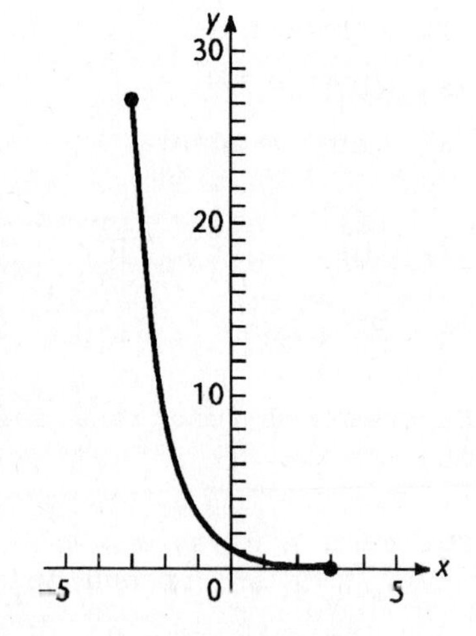

29.

x	y
-3	0.15
-2	0.44
-1	1.33
0	4.00
1	12.00
2	36.00
3	108.00

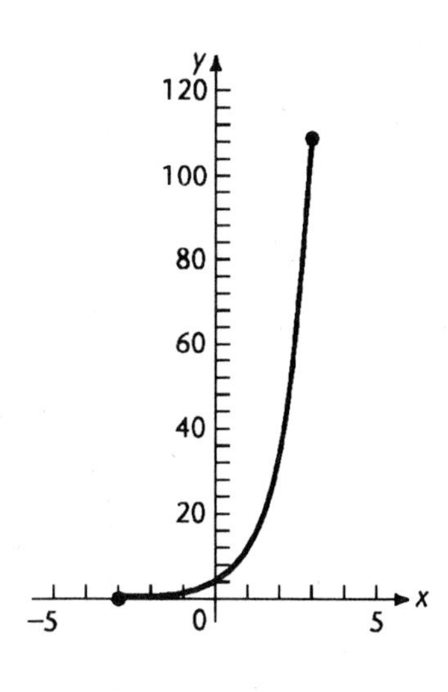

31.

x	y
-3	1
-2	2
-1	4
0	8
1	16
2	32
3	64

33.

x	y
-3	448
-2	112
-1	28
0	7
1	1.75
2	0.44
3	0.11

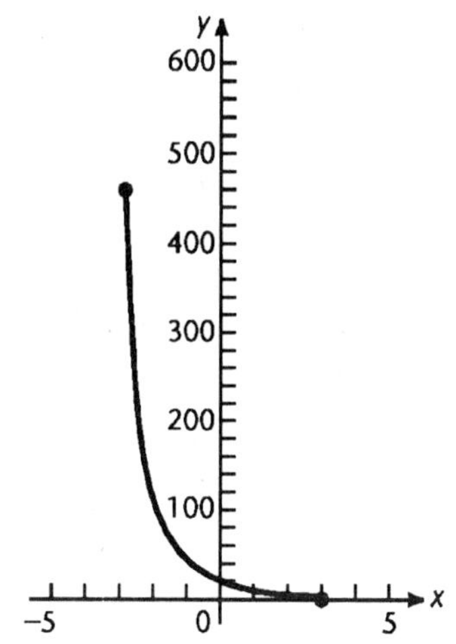

35.

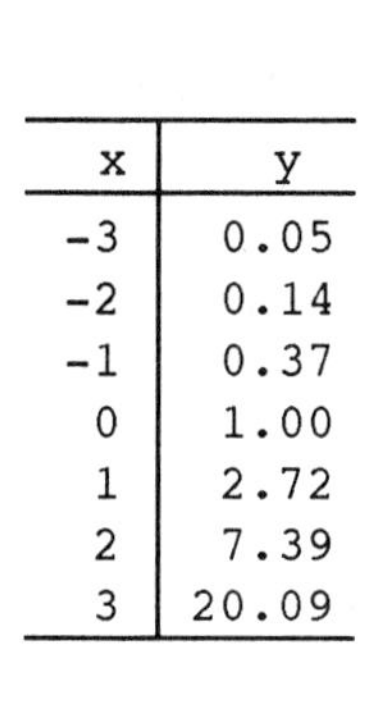

x	y
-3	0.05
-2	0.14
-1	0.37
0	1.00
1	2.72
2	7.39
3	20.09

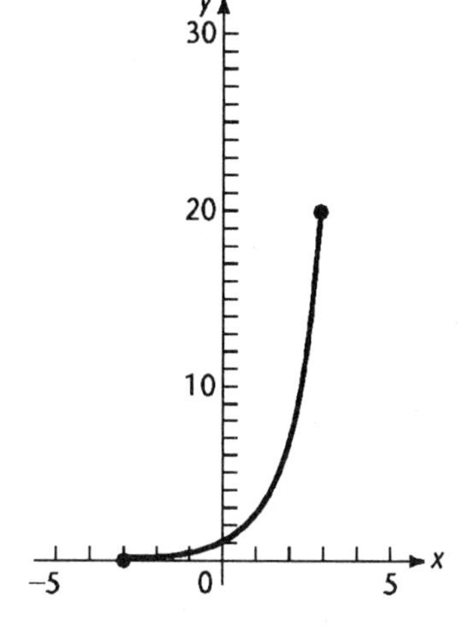

37.

x	y
-3	-2.88
-2	-2.75
-1	-2.5
0	-2
1	-1
2	1
3	5

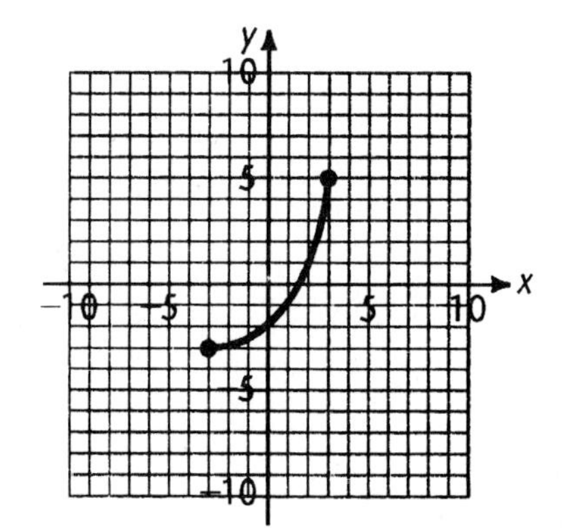

39.

x	y
-3	9
-2	5
-1	3
0	2
1	1.5
2	1.25
3	1.13

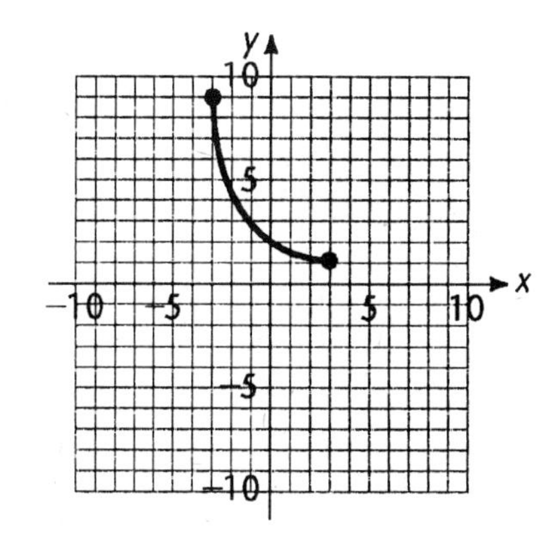

41.

x	y
-3	4.05
-2	4.13
-1	4.36
0	5
1	6.72
2	11.39
3	24.09

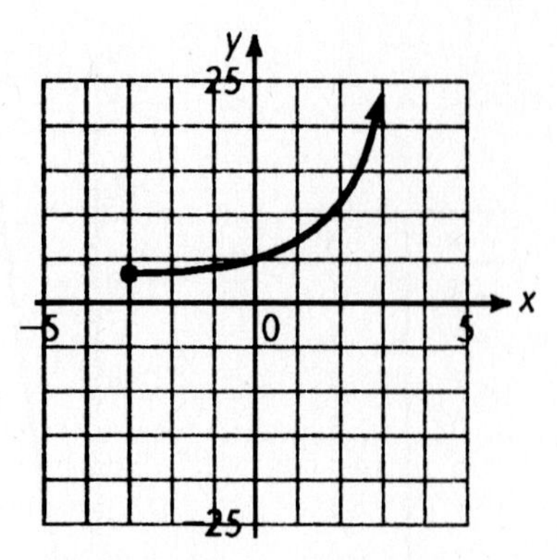

43.

x	y
-3	18.09
-2	5.39
-1	1.72
0	-1
1	-1.64
2	-1.87
3	-1.95

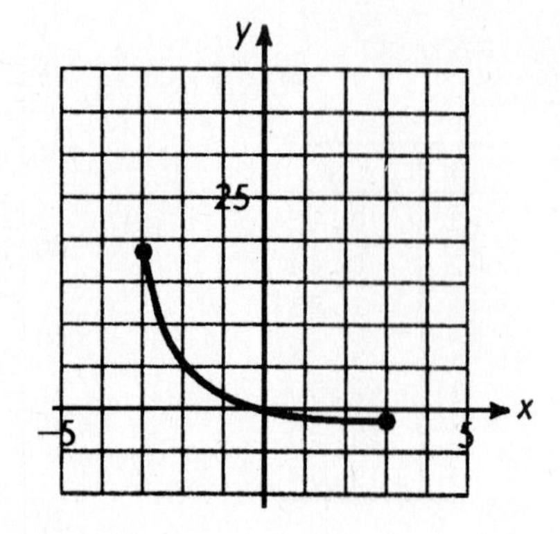

45.

x	y
-3	0.11
-2	0.33
-1	1
0	3
1	9
2	27
3	81

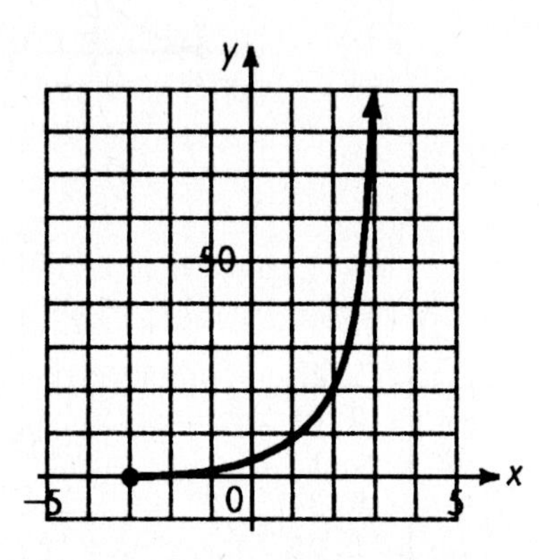

47.

x	y
-3	0.01
-2	0.02
-1	0.05
0	0.13
1	0.37
2	1
3	2.72

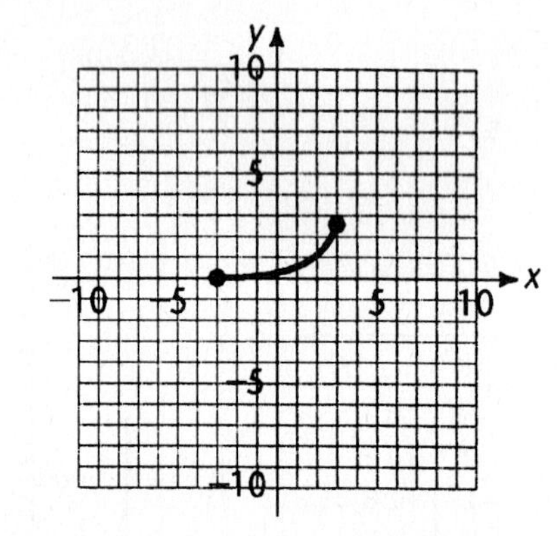

49.

x	y
-3	243
-2	81
-1	27
0	9
1	3
2	1
3	0.33

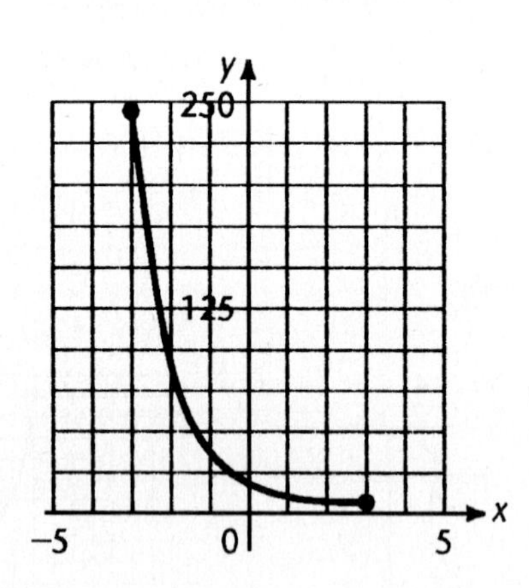

51.

x	y
-3	54.6
-2	20.1
-1	7.39
0	2.72
1	1
2	0.37
3	0.14

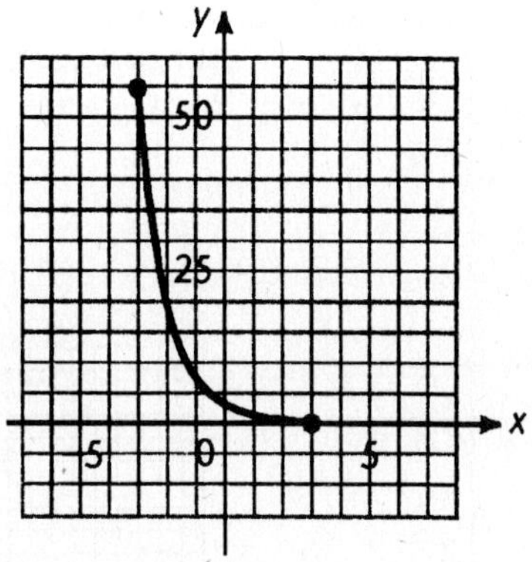

53.

x	y
-3	4.07
-2	4.22
-1	4.67
0	6
1	10
2	22
3	58

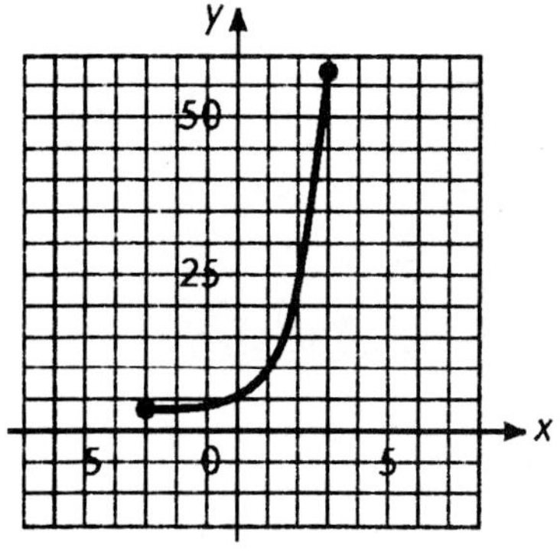

55.

x	y
-3	25
-2	13
-1	7
0	4
1	2.5
2	1.75
3	1.38

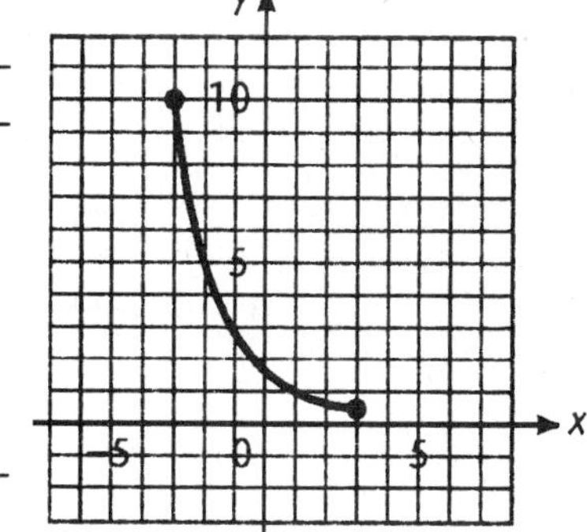

57.

x	y
-3	1.10
-2	1.27
-1	1.73
0	3
1	6.44
2	15.8
3	41.2

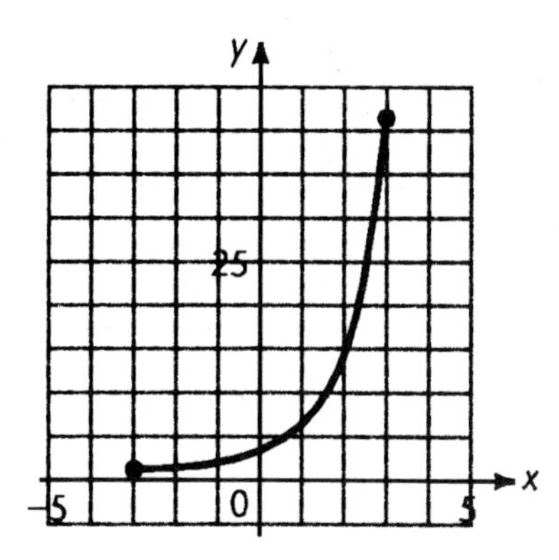

59.

x	y
-3	77.3
-2	26.6
-1	7.87
0	1
1	-1.53
2	-2.46
3	-2.80

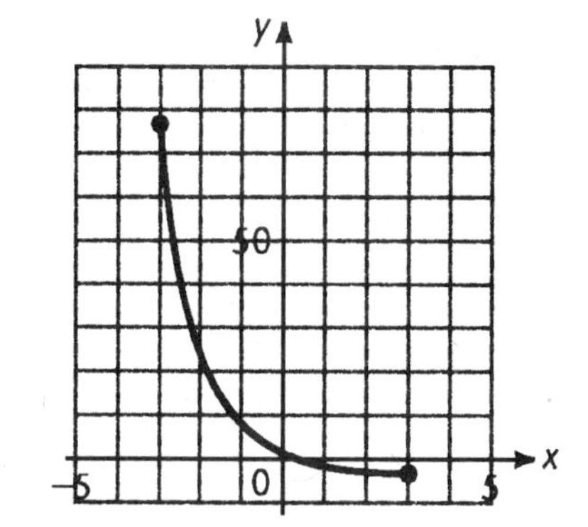

61.

x	y
-3	14.33
-2	12.71
-1	11.28
0	10.00
1	8.87
2	7.87
3	6.98

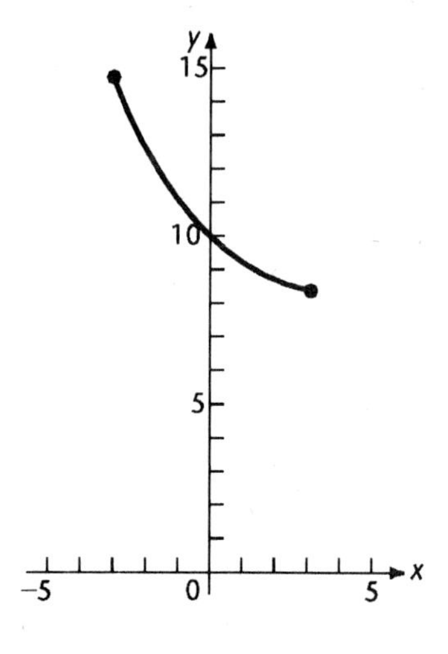

63.

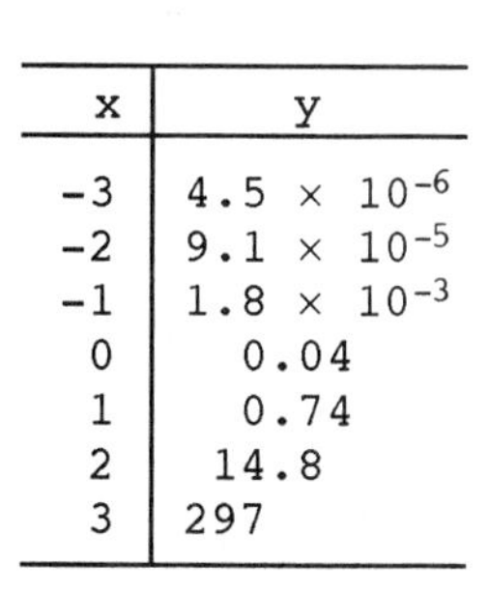

x	y
-3	4.5×10^{-6}
-2	9.1×10^{-5}
-1	1.8×10^{-3}
0	0.04
1	0.74
2	14.8
3	297

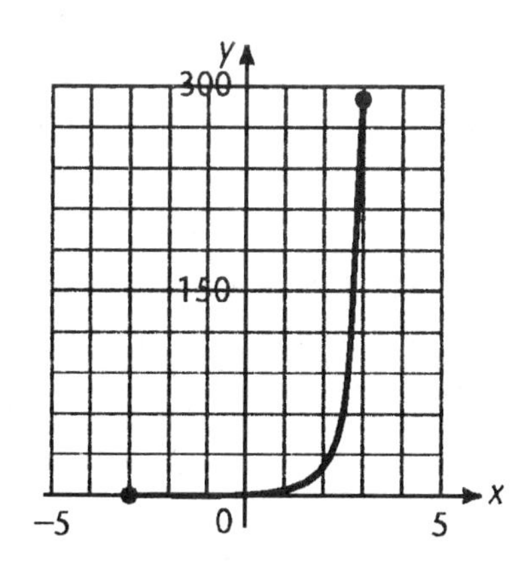

65.

x	y
-2	0.63
-1	5
0	10
1	5
2	0.63
0.5	8.41
1.5	2.10
-1.5	2.10
-0.5	8.41

These points are included because more information is needed. (refers to the rows 0.5, 1.5, -1.5, -0.5)

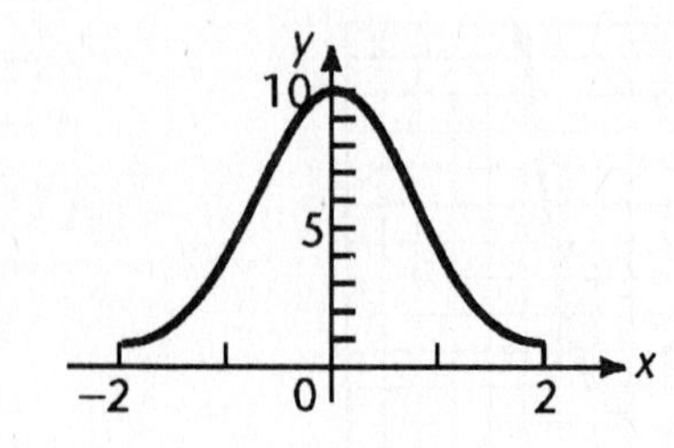

67.

x	y
-3	$\frac{a}{8} = 0.13a$
-2	0.25a
-1	0.5a
0	a
1	2a
2	4a
3	8a

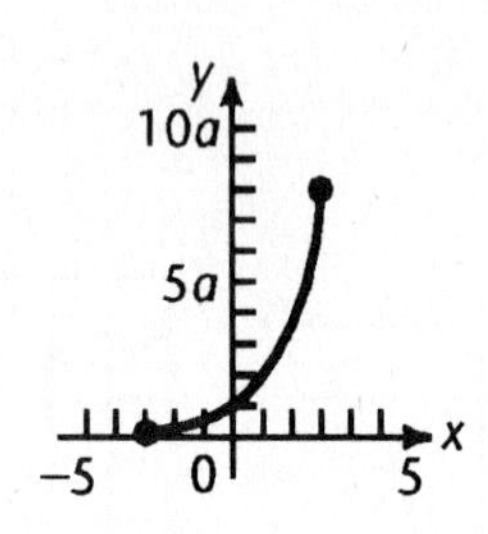

69.

x	y
-3	8a
-2	4a
-1	2a
0	a
1	$\frac{a}{2} = 0.5a$
2	0.25a
3	0.13a

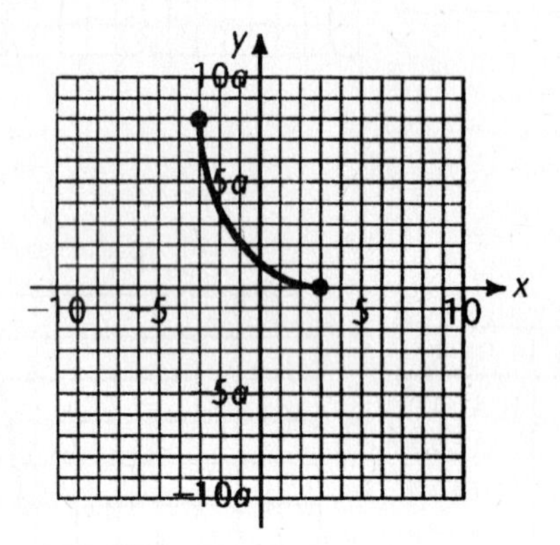

71.

x	$y = 2^x$
-3	0.13
-2	0.25
-1	0.50
0	1
1	2
2	4
3	8

$x = 2^y$	y
0.13	-3
0.25	-2
0.50	-1
1	0
2	1
4	2
8	3

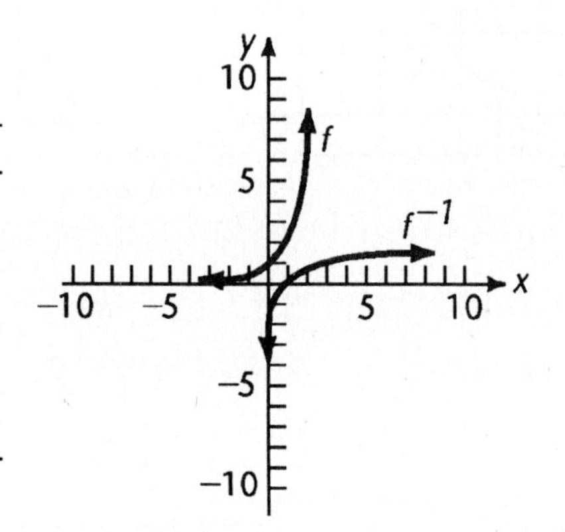

73.

x	$y = 10^{-x}$
-2	100
-1	10
0	1
1	0.1
2	0.01
3	0.001

$x = 10^{-y}$	y
100	-2
10	-1
1	0
0.1	1
0.01	2
0.001	3

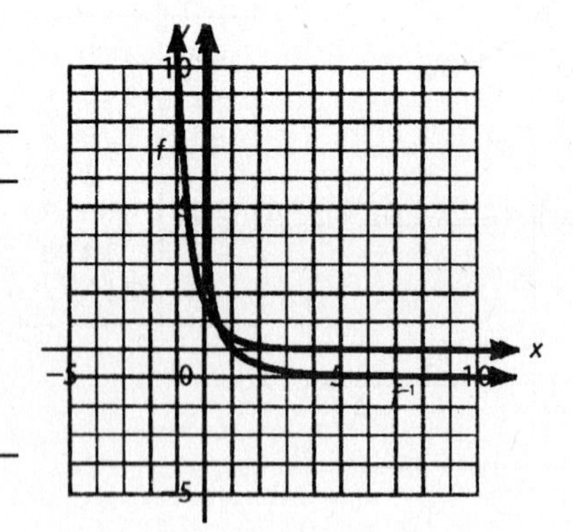

75.

x	$y = e^x$
-3	0.05
-2	0.14
-1	0.37
0	1.00
1	2.72
2	7.39
3	20.09

$x = e^y$	y
0.05	-3
0.14	-2
0.37	-1
1.00	0
2.72	1
7.39	2
20.09	3

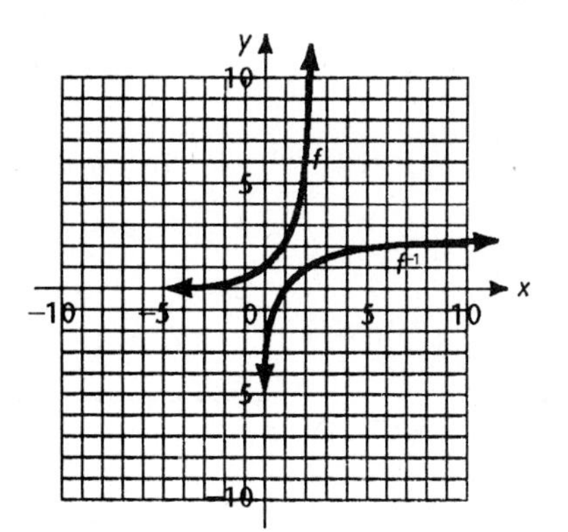

77.

x	P
-1	18.1
0	14.7
1	11.9
2	9.7
3	7.8
4	6.3
5	5.1

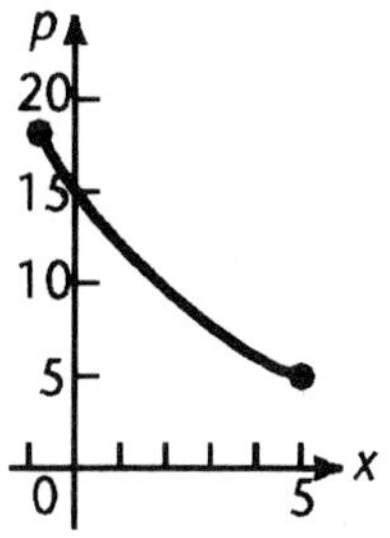

79.

t	A
0	100
28	50
2(28)	25
3(28)	12.5
4(28)	6.3
5(28)	3.1
6(28)	1.6

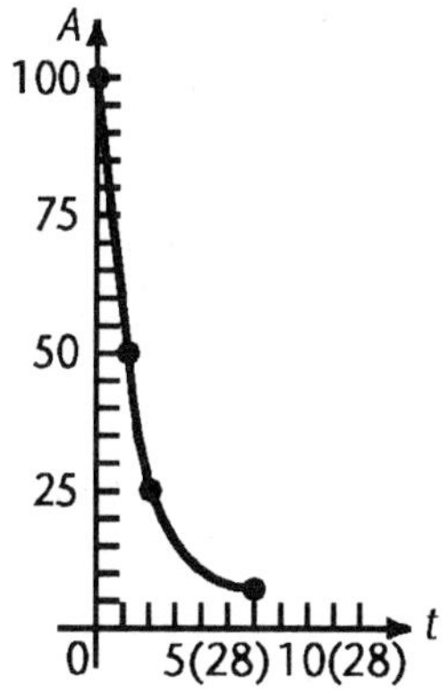

81.

i	N → rounded to the nearest integer
1	100
2	90
3	80
4	72
5	64
6	58
7	52
8	46
9	41
10	37

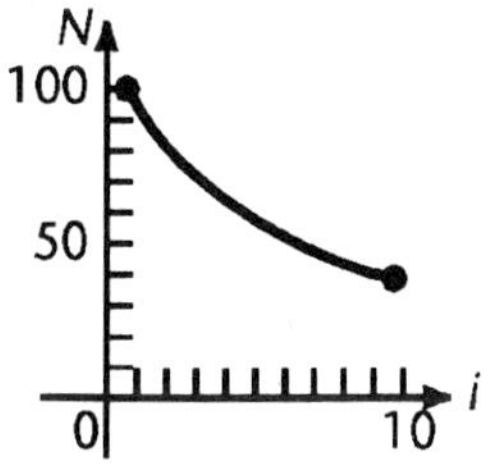

Exercise 10-2 Logarithmic Functions

Key Ideas and Formulas

Logarithmic Function with base b:

For $b > 0$ and $b \neq 1$:

$$y = \log_b x \quad \text{is equivalent to} \quad x = b^y$$

(The log to the base b of x is the power to which b must be raised to obtain x.)

Typical Logarithmic Graphs:

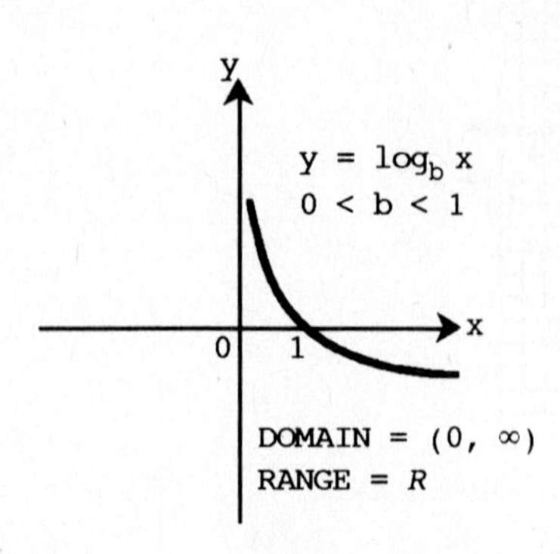

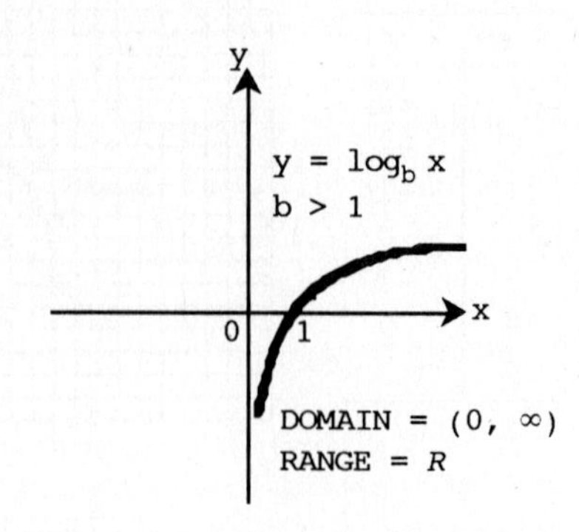

Logarithmic-Exponential Identities

For $b > 0$, $b \neq 1$ $\log_b b^x = x$

That is, the power to which b must be raised to obtain b^x is: x.

$b^{\log_b x} = x \quad x > 0$

That is, the power to which b must be raised to obtain x is: $\log_b x$.

---

1. What power of 3 yields 9? Since $3^2 = 9$. $\log_3 9 = 2$.

3. What power of 3 yields $\frac{1}{27}$? Since $3^{-3} = \frac{1}{3^3} = \frac{1}{27}$, $\log_3 \frac{1}{27} = -3$.

5. What power of 3 yields 3? Since $3^1 = 3$, $\log_3 3 = 1$.

7. What power of 3 yields $\frac{1}{\sqrt{3}}$? Since $3^{-1/2} = \frac{1}{3^{1/2}} = \frac{1}{\sqrt{3}}$, $\log_3 \frac{1}{\sqrt{3}} = -\frac{1}{2}$.

9. What power of 9 yields 81? Since $9^2 = 81$, $\log_9 81 = 2$.

11. What power of 9 yields 27? Since $9^{3/2} = (\sqrt{9})^3 = 3^3 = 27$, $\log_9 27 = \frac{3}{2}$.

13. What power of 9 yields $\frac{1}{9}$? Since $9^{-1} = \frac{1}{9}$, $\log_9 \frac{1}{9} = -1$.

15. What power of 9 yields $\frac{1}{3}$? Since $9^{-1/2} = \frac{1}{\sqrt{9}} = \frac{1}{3}$, $\log_9 \frac{1}{3} = -\frac{1}{2}$.

17. What power of 9 yields 1? Since $9^0 = 1$, $\log_9 1 = 0$.

19. $\log_{10} 10^5 = 5$

21. $\log_2 2^{-4} = -4$

23. $\log_6 36 = \log_6 6^2 = 2$

25. $\log_{10} 1000 = \log_{10} 10^3 = 3$

27. $9 = 3^2$

29. $81 = 3^4$

31. $1{,}000 = 10^3$

33. $1 = e^0$

35. $\log_8 64 = 2$

37. $\log_{10} 10{,}000 = 4$ 39. $\log_v u = x$ 41. $\log_{27} 9 = \frac{2}{3}$

43. Write $\log_2 x = 2$ in equivalent exponential form: $x = 2^2 = 4$

45. Write $\log_b 16 = 2$ in equivalent exponential form.

$16 = b^2$
$b^2 = 16$
$b = 4$ since bases are required to be positive

47. Write $\log_b 3 = \frac{1}{2}$ in equivalent exponential form.

$b^{1/2} = 3$
$b = 3^2$
$b = 9$

49. Write $\log_5 x = 3$ in equivalent exponential form.

$5^3 = x$
$x = 125$

51. Write $\log_5 x = \frac{2}{3}$ in equivalent exponential form.

$5^{2/3} = x$
$x = \sqrt[3]{5^2}$
$x = \sqrt[3]{25}$

53. $0.001 = 10^{-3}$ 55. $3 = 81^{1/4}$ 57. $16 = \left(\frac{1}{2}\right)^{-4}$ 59. $N = a^e$

61. $\log_{10} 0.01 = -2$ 63. $\log_e 1 = 0$ 65. $\log_2 \left(\frac{1}{8}\right) = -3$ 67. $\log_{81} \frac{1}{3} = -\frac{1}{4}$

69. $7 = \sqrt{49}$ is rewritten $7 = 49^{1/2}$. In equivalent logarithmic form this becomes $\log_{49} 7 = \frac{1}{2}$.

71. $\log_b b^u = u$ 73. $\log_e e^{1/2} = \frac{1}{2}$ 75. $\log_{23} 1 = \log_{23} 23^0 = 0$

77. $\ln e^3 = \log_e e^3 = 3$ 79. $\ln e^{\sqrt{2}} = \log_e e^{\sqrt{2}} = \sqrt{2}$

81. Write $\log_4 x = \frac{1}{2}$ in equivalent exponential form: $x = 4^{1/2} = \sqrt{4} = 2$.

83. Write $\log_b 1000 = \frac{3}{2}$ in equivalent exponential form.

$1000 = b^{3/2}$
$10^3 = b^{3/2}$
$(10^3)^{2/3} = (b^{3/2})^{2/3}$ [If two numbers are equal the results are equal if they are raised to the same power.]
$10^{2/3(3)} = b^{2/3(3/2)}$
$10^2 = b$
$b = 100$

85. Write $\log_b 1 = 0$ in equivalent exponential form: $1 = b^0$
This statement is true if b is any real number except 0. However, bases are required to be positive and 1 is not allowed, so the original statement is true if b is any positive real number except 1.

87. (A)

x	$y = 10^x$
-2	0.01
-1	0.1
0	1
1	10

$x = 10^y$	y
0.01	-2
0.1	-1
1	0
10	1

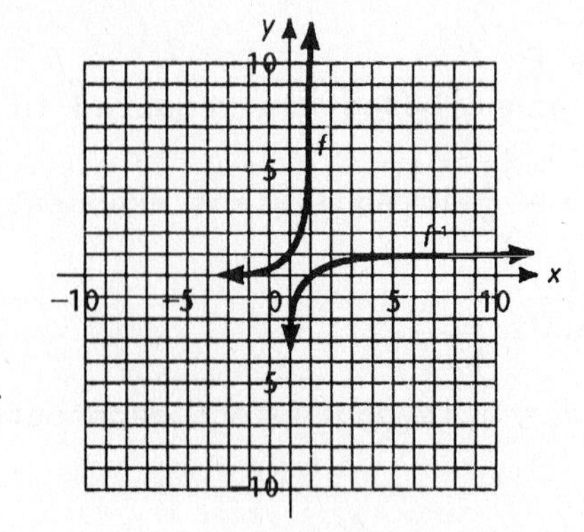

(B) Domain of f is the set of real numbers; range of f is the set of positive real numbers. The domain of f is the range of f^{-1} and the range of f is the domain of f^{-1}.

(C) f^{-1} is called the logarithmic function with base 10. $f^{-1}(x) = \log_{10} x$

89. (A)

x	$y = (\frac{1}{2})^x$
-3	8
-2	4
-1	2
0	1
1	$\frac{1}{2}$
2	$\frac{1}{4}$
3	$\frac{1}{8}$

$x = (\frac{1}{2})^y$	y
8	-3
4	-2
2	-1
1	0
$\frac{1}{2}$	1
$\frac{1}{4}$	2
$\frac{1}{8}$	3

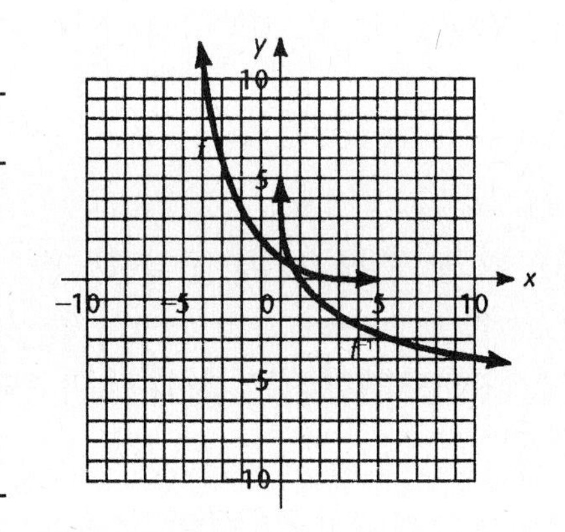

(B) Domain of f is the set of real numbers; range of f is the set of positive real numbers. The domain of f is the range of f^{-1} and the range of f is the domain of f^{-1}.

(C) f^{-1} is called the logarithmic function with base $\frac{1}{2}$. $f^{-1}(x) = \log_{1/2} x$

91. The domain of f is the set of all real numbers; the range of f is {1}. The domain of f^{-1} is {1}; the range of f^{-1} is the set of all real numbers. No, since f is not a one-to-one function. (For example, f(2) = f(3) = 1.)

93. Write $\log_b x = 3$ in equivalent exponential form: $b^3 = x$
Then $\frac{1}{x} = \frac{1}{b^3} = b^{-3}$.

Write $\frac{1}{x} = b^{-3}$ in equivalent logarithmic form: $\log_b\left(\frac{1}{x}\right) = -3$
-3

95. Start by letting $b = \ln\left(\frac{1}{x}\right)$. We want to show that $-\ln x = b$. We rewrite $b = \ln \frac{1}{x}$ in equivalent exponential form:

$$b = \log_e \frac{1}{x}$$

$$e^b = \frac{1}{x}$$

Then $xe^b = 1$

$$x = \frac{1}{e^b}$$

$$x = e^{-b}$$

Rewriting this in equivalent logarithmic form: $-b = \log_e x$. Thus, $-b = \ln x$, or $b = -\ln x$. Thus we have shown that $\ln\left(\frac{1}{x}\right) = -\ln x$.

Exercise 10-3 Properties of Logarithmic Functions

Key Ideas and Formulas

Properties of Logarithmic Functions:

If b, M and N are positive real numbers, $b \neq 1$, and p and x are real numbers, then:

1. $\log_b 1 = 0$
2. $\log_b b = 1$
3. $\log_b b^x = x$
4. $b^{\log_b x} = x$
5. $\log_b MN = \log_b M + \log_b N$
6. $\log_b \frac{M}{N} = \log_b M - \log_b N$
7. $\log_b M^p = p \log_b M$
8. $\log_b M = \log_b N$ if and only if $M = N$

COMMON ERRORS:
There is no connection between $\log_a (M + N)$ and $\log_a M + \log_a N$ or between $\log_a (M - N)$ and $\log_a M - \log_a N$. The respective expressions are not equal and should not be confused.

---

1. $\log_b u + \log_b v$

3. $\log_b A - \log_b B$

5. $5 \log_b u$

7. $\frac{3}{5} \log_b N$

9. $\log_b \sqrt{Q} = \log_b Q^{1/2} = \frac{1}{2} \log_b Q$

11. $\log_b uvw = \log_b [(uv)w] = \log_b (uv) + \log_b w = \log_b u + \log_b v + \log_b w$

13. $\log_b AB$

15. $\log_b \left(\frac{X}{Y}\right)$

17. $\log_b w + \log_b x - \log_b y = \log_b wx - \log_b y = \log_b \frac{wx}{y}$

19. $\log_b(30) = \log_b(2\cdot3\cdot5) = \log_b 2 + \log_b 3 + \log_b 5 = 0.69 + 1.10 + 1.61 = 3.40$

21. $\log_b \frac{2}{5} = \log_b 2 - \log_b 5 = 0.69 - 1.61 = -0.92$

23. $\log_b 27 = \log_b 3^3 = 3 \log_b 3 = 3(1.10) = 3.30$

25. $\log_b \frac{1}{3} = \log_b 1 - \log_b 3 = 0 - 0.683 = -0.683$

27. $\log_b \frac{2}{3} = \log_b 2 - \log_b 3 = 0.431 - 0.683 = -0.252$

29. $\log_b 6 = \log_b(2\cdot3) = \log_b 2 + \log_b 3 = 0.431 + 0.683 = 1.114$

31. $\log_b 9 = \log_b 3^2 = 2 \log_b 3 = 2(0.683) = 1.366$

33. $\log_b \sqrt{2} = \log_b 2^{1/2} = \frac{1}{2} \log_b 2 = \frac{1}{2}(0.431) = 0.216$

35. $\log_b u^2v^7 = \log_b u^2 + \log_b v^7 = 2 \log_b u + 7 \log_b v$

37. $\log_b \frac{1}{a} = \log_b 1 - \log_b a = 0 - \log_b a = -\log_b a$

Alternatively, $\log_b \frac{1}{a} = \log_b a^{-1} = -1 \log_b a = -\log_b a$

39. $\log_b \frac{\sqrt[3]{N}}{p^2q^3} = \log_b \sqrt[3]{N} - \log_b p^2q^3$
$= \log_b N^{1/3} - (\log_b p^2 + \log_b q^3)$
$= \frac{1}{3} \log_b N - 2 \log_b p - 3 \log_b q$

COMMON ERROR:
Forgetting the parentheses.
$-\log_b p^2 + \log_b q^3$ is incorrect.

41. $\log_b \sqrt[4]{\frac{x^2y^3}{\sqrt{z}}} = \log_b \left(\frac{x^2y^3}{z^{1/2}}\right)^{1/4} = \frac{1}{4} \log_b \frac{x^2y^3}{z^{1/2}} = \frac{1}{4}(\log_b x^2 + \log_b y^3 - \log_b z^{1/2})$
$= \frac{1}{4}(2 \log_b x + 3 \log_b y - \frac{1}{2} \log_b z)$

43. $2 \log_b x - \log_b y = \log_b x^2 - \log_b y = \log_b \frac{x^2}{y}$

COMMON ERROR:
$\log_b(x^2 - y)$
"Factoring out the log" is incorrect.

45. $3 \log_b x + 2 \log_b y - 4 \log_b z = \log_b x^3 + \log_b y^2 - \log_b z^4$
$= \log_b x^3y^2 - \log_b z^4 = \log_b \frac{x^3y^2}{z^4}$

47. $\frac{1}{5}(2 \log_b x + 3 \log_b y) = \frac{1}{5}(\log_b x^2 + \log_b y^3) = \frac{1}{5} \log_b x^2y^3$
$= \log_b (x^2y^3)^{1/5} = \log_b \sqrt[5]{x^2y^3}$

49. $\log_b 3 + 2 \log_b x + \log_b y = \log_b 3 + \log_b x^2 + \log_b y = \log_b 3x^2y$

51. $\frac{1}{2}(\log_b x - \log_b y) = \frac{1}{2}\log_b \frac{x}{y} = \log_b \left(\frac{x}{y}\right)^{1/2}$

53. $\log_b 2 + 3\log_b x + 4\log_b y = \log_b 2 + \log_b x^3 + \log_b y^4 = \log_b 2x^3y^4$

55. $\log_b 7.5 = \log_b \frac{15}{2} = \log_b \frac{3 \cdot 5}{2} = \log_b 3 + \log_b 5 - \log_b 2 = 1.10 + 1.61 - 0.69 = 2.02$

57. $\log_b \sqrt[3]{2} = \log_b 2^{1/3} = \frac{1}{3}\log_b 2 = \frac{1}{3}(0.69) = 0.23$

59.
$$\begin{aligned}\log_b \sqrt{0.9} = \log_b \left(\frac{9}{10}\right)^{1/2} &= \frac{1}{2}\log_b \frac{3^2}{2 \cdot 5} = \frac{1}{2}[\log_b 3^2 - \log_b 2 - \log_b 5] \\ &= \frac{1}{2}[2\log_b 3 - \log_b 2 - \log_b 5] = \frac{1}{2}[2.20 - 0.69 - 1.61] \\ &= \frac{1}{2}(-0.10) = -0.05\end{aligned}$$

61. $\log_b \frac{6}{7} = \log_b \frac{2 \cdot 3}{7} = \log_b 2 + \log_b 3 - \log_b 7 = 0.431 + 0.683 - 1.209 = -0.095$

63. $\log_b \frac{14}{3} = \log_b \frac{2 \cdot 7}{3} = \log_b 2 + \log_b 7 - \log_b 3 = 0.431 + 1.209 - 0.683 = 0.957$

65.
$$\begin{aligned}\log_b \sqrt{42} = \log_b (2 \cdot 3 \cdot 7)^{1/2} &= \frac{1}{2}(\log_b 2 + \log_b 3 + \log_b 7) \\ &= \frac{1}{2}(0.431 + 0.683 + 1.209) = 1.1615\end{aligned}$$

67.
$$\begin{aligned}\frac{3}{2}\log_b 4 - \frac{2}{3}\log_b 8 + 2\log_b 2 &= \log_b x \\ \log_b 4^{3/2} - \log_b 8^{2/3} + \log_b 2^2 &= \log_b x \\ \log_b 8 - \log_b 4 + \log_b 4 &= \log_b x \\ \log_b 8 &= \log_b x \\ 8 &= x\end{aligned}$$

69.
$$\begin{aligned}\log_b 6 + \frac{1}{2}\log_b 16 - \log_b 216 &= \log_b x \\ \log_b 6 + \log_b 16^{1/2} - \log_b 216 &= \log_b x \\ \log_b 6 + \log_b 4 - \log_b 216 &= \log_b x \\ \log_b \frac{6 \cdot 4}{216} &= \log_b x \\ \frac{6 \cdot 4}{216} &= x \\ x &= \frac{1}{9}\end{aligned}$$

71.
$$\begin{aligned}2\log_b 2 + 3\log_b 3 - 4\log_b 4 &= \log_b x \\ \log_b 2^2 + \log_b 3^3 - \log_b 4^4 &= \log_b x \\ \log_b 4 + \log_b 27 - \log_b 256 &= \log_b x \\ \log_b \frac{4 \cdot 27}{256} &= \log_b x \\ \frac{4 \cdot 27}{256} &= x \\ x &= \frac{27}{64}\end{aligned}$$

73.
$$\begin{aligned}\log_b y - \log_b c + kt &= 0 \\ \log_b \frac{y}{c} + kt &= 0 \\ \log_b \frac{y}{c} &= -kt\end{aligned}$$

Write this in equivalent exponential form.

$$\begin{aligned}\frac{y}{c} &= b^{-kt} \\ y &= cb^{-kt}\end{aligned}$$

75. Let $u = \log_b M$ and $v = \log_b N$; then $M = b^u$ and $N = b^v$. Thus, $\log_b M/N = \log_b b^u/b^v = \log_b b^{u-v} = u - v = \log_b M - \log_b N$.

77. $MN = b^{\log_b M} b^{\log_b N} = b^{\log_b M + \log_b N}$; hence, by definition of logarithm
$\log_b MN = \log_b M + \log_b N$

Exercise 10-4 Logarithms to Various Bases

Key Ideas and Formulas

Logarithmic Notation: Two important logarithms written without a base indicated:

$\log x = \log_{10} x \qquad \ln x = \log_e x$

$\log x = y$ is equivalent to $x = 10^y$

$\ln x = y$ is equivalent to $x = e^y$

Change-of-Base Formula

$$\log_b N = \frac{\log_a N}{\log_a b} = \frac{\log N}{\log b} = \frac{\ln N}{\ln b}$$

---

1. 4.9177 3. -2.8419 5. 3.7623 7. -2.5128 9. Not defined

11. -2.3010 13. -5.1549

15. 4.3429×10^{-7}, that is, 0.0000 to four decimal places

17. 200,800 19. 0.000 664 8 21. 47.73 23. 0.6760 25. 293.8

27. 3.6974×10^{-6} 29. 5.8114×10^{-5} 31. 3.4367 33. 4.959

35. 7.861 37. 3.301

39. $\log_5 372 = \frac{\ln 372}{\ln 5} = 3.6776$ or $\log_5 372 = \frac{\log_{10} 372}{\log_{10} 5} = 3.6776$

41. $\log_8 0.0352 = \frac{\ln 0.0352}{\ln 8} = -1.6094$

43. $\log_3 0.1483 = \frac{\ln 0.1483}{\ln 3} = -1.7372$

45. $\log_2 4.32 = \frac{\ln 4.32}{\ln 2} = 2.1110$

47. $\log_3 0.5 = \frac{\ln 0.5}{\ln 3} = -0.6309$

49. $\log_{12} 150 = \frac{\ln 150}{\ln 12} = 2.0164$

51. $\log_{1/2} 4 = \frac{\ln 4}{\ln 1/2} = -2.000$

53.

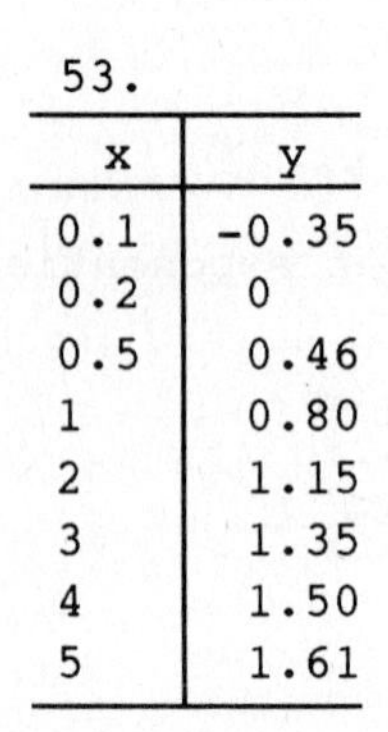

x	y
0.1	-0.35
0.2	0
0.5	0.46
1	0.80
2	1.15
3	1.35
4	1.50
5	1.61

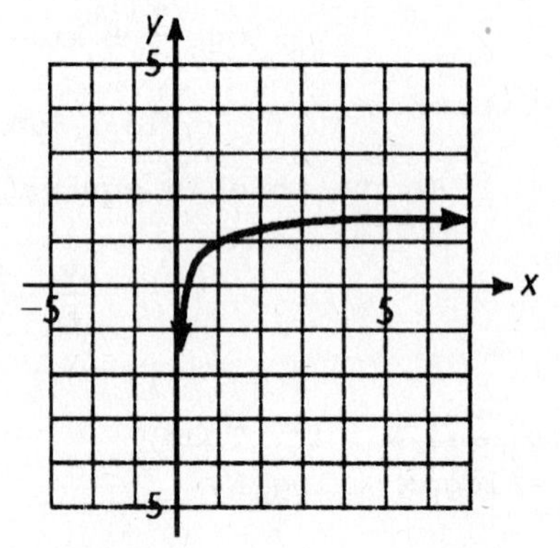

55.

x	y
0.5	-1.20
1	0
2	1.20
3	1.91
4	2.41
5	2.80
6	3.11

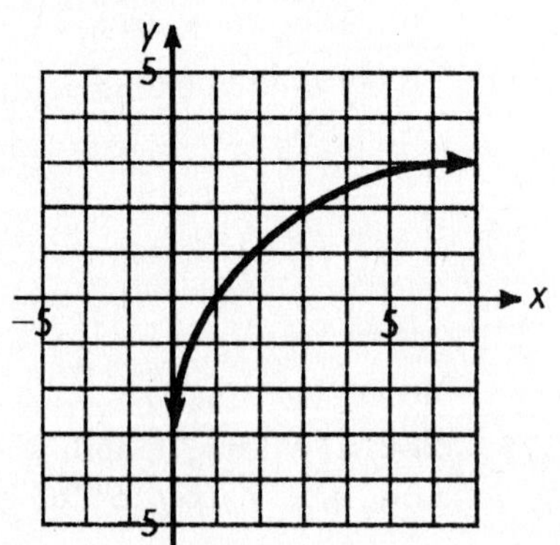

57. $y = 4 \log_5 x = \dfrac{4 \ln x}{\ln 5}$

x	y
0.2	-4
0.5	-1.72
1	0
2	1.72
3	2.73
4	3.45
5	4
6	4.45

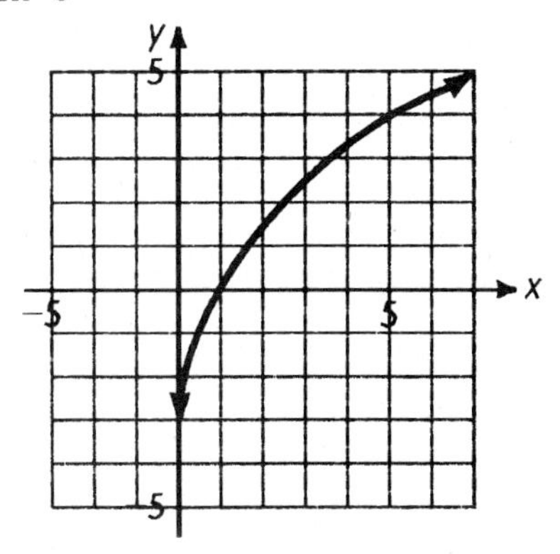

59. $x = \log(5.3147 \times 10^{12}) = \log 5.3147 + \log 10^{12} = 0.725 + 12 = 12.725$
(Round-off error is significantly reduced if we enter 5.3147 rather than 5.3147×10^{12})

61. -25.715

63. $\log x = 32.068523 = 0.068523 + 32 = \log 1.1709 + \log 10^{32} = \log(1.1709 \times 10^{32})$
Hence $x = 1.1709 \times 10^{32}$

65. 4.2672×10^{-7}

67.

x	y
-1.99	-2
-1.9	-1
-1.8	-0.7
-1.5	-0.3
-1	0
0	0.3
1	0.5
2	0.6
5	0.8
10	1

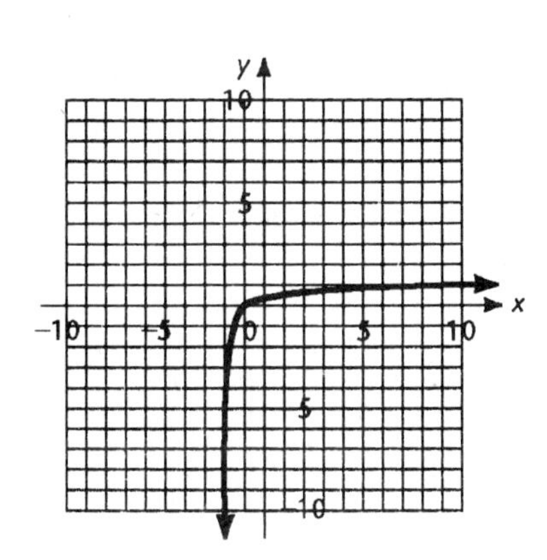

69.

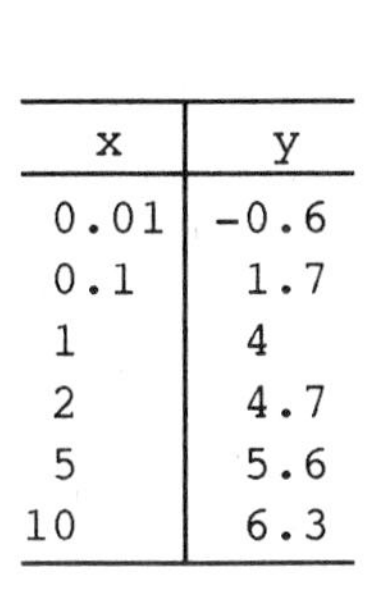

x	y
0.01	-0.6
0.1	1.7
1	4
2	4.7
5	5.6
10	6.3

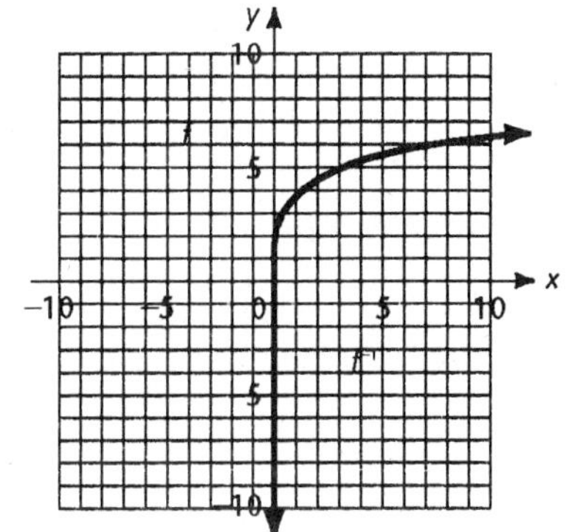

71.

x	y
0.01	-9.2
0.1	-7.6
1	-3
2	-1.6
5	0.2
10	1.6

73.

x	y
-2.99	-8
-2.9	-6
-2	-4
-1	-3.4
0	-3.0
2	-2.6
5	-2.2
10	-1.8

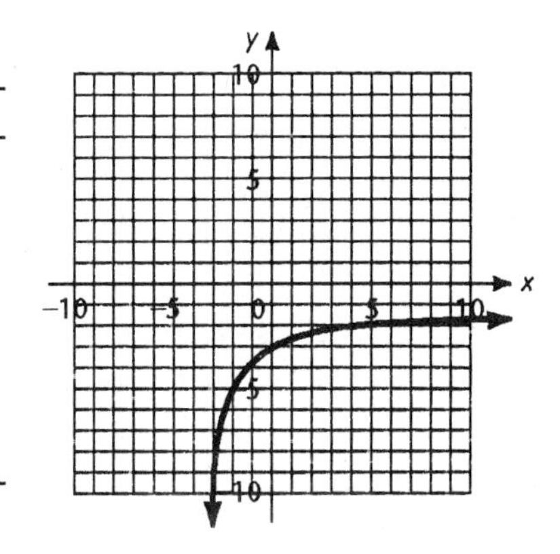

Exercise 10-5 Exponential and Logarithmic Equations

Key Ideas and Formulas

To solve equations in which the variable appears only in the exponent, use logarithms. If $a^x = b$, then $x \log a = \log b$ (or $x \ln a = \ln b$).

To solve equations in which the variable appears only in an expression whose logarithm is given, change to equivalent exponential form. If $\log_a x = b$, then $x = a^b$.

---

1. $10^{-x} = 0.0347$
$-x = \log_{10} 0.0347$
$x = -\log_{10} 0.0347$
$x = 1.46$

3. $10^{3x+1} = 92$
$3x + 1 = \log_{10} 92$
$3x = \log_{10} 92 - 1$
$x = \dfrac{\log_{10} 92 - 1}{3}$
$x = 0.321$

5. $e^x = 3.65$
$x = \ln 3.65$
$x = 1.29$

7. $e^{2x-1} = 405$
$2x - 1 = \ln 405$
$2x = 1 + \ln 405$
$x = \dfrac{1 + \ln 405}{2}$
$x = 3.50$

9. $5^x = 18$
$x \ln 5 = \ln 18$
$x = \dfrac{\ln 18}{\ln 5}$
$x = 1.80$

11. $2^{-x} = 0.238$
$-x \ln 2 = \ln 0.238$
$-x = \dfrac{\ln 0.238}{\ln 2}$
$x = -\dfrac{\ln 0.238}{\ln 2}$
$x = 2.07$

13. $5^x = 10$
$x \ln 5 = \ln 10$
$x = \dfrac{\ln 10}{\ln 5}$
$x = 1.43$

15. $10^{-3x} = 144$
$-3x = \log_{10} 144$
$x = \dfrac{\log_{10} 144}{-3}$
$x = -0.719$

17. $e^{2x} = 8$
$2x = \ln 8$
$x = \dfrac{\ln 8}{2}$
$x = 1.04$

19. $\log 5 + \log x = 2$
$\log (5x) = 2$
$5x = 10^2$
$5x = 100$
$x = 20$

21. $\log x + \log (x - 3) = 1$
$\log [x(x - 3)] = 1$
$x(x - 3) = 10^1$
$x^2 - 3x = 10$
$x^2 - 3x - 10 = 0$
$(x - 5)(x + 2) = 0$
$x = 5$ or -2

COMMON ERROR:
$\log x + \log x - \log 3 = 1$
$\log(a - b)$ cannot be rewritten as $\log a - \log b$.

Check: $\log 5 + \log (5 - 3) \overset{\surd}{=} 1$
$\log (-2) + \log (-2 - 3)$ is not defined.
$x = 5$

23. $\log 4 + \frac{1}{2} \log x = \log 3$
$\log 4 + \log x^{1/2} = \log 3$
$\log 4x^{1/2} = \log 3$
$4x^{1/2} = 3$
$x^{1/2} = \dfrac{3}{4}$
$x = \dfrac{9}{16}$

Check: $\log 4 + \frac{1}{2} \log \frac{9}{16} = \log 4 + \log \sqrt{\frac{9}{16}} = \log 4 + \log \frac{3}{4} = \log 4 \cdot \frac{3}{4} = \log 3$

25.
$$\log x = 1 - \log x$$
$$2\log x = 1$$
$$\log x = \frac{1}{2}$$
$$x = 10^{1/2} \text{ or } \sqrt{10}$$

Check: $\log\sqrt{10} = \log 10^{1/2} = \frac{1}{2} = 1 - \frac{1}{2} = 1 - \log\sqrt{10}$

27.
$$2 = 1.05^x$$
$$\ln 2 = x \ln 1.05$$
$$\frac{\ln 2}{\ln 1.05} = x$$
$$x = 14.2$$

29.
$$e^{-1.4x} = 13$$
$$-1.4x = \ln 13$$
$$x = \frac{\ln 13}{-1.4}$$
$$x = -1.83$$

31.
$$123 = 500e^{-0.12x}$$
$$\frac{123}{500} = e^{-0.12x}$$
$$\ln\left(\frac{123}{500}\right) = -0.12x$$
$$\frac{\ln\left(\frac{123}{500}\right)}{-0.12} = x$$
$$x = 11.7$$

33.
$$\log x - \log 5 = \log 2 - \log(x - 3)$$
$$\log\frac{x}{5} = \log\frac{2}{x - 3}$$
$$\frac{x}{5} = \frac{2}{x - 3}$$
excluded value: $x \neq 3$
$$5(x - 3)\frac{x}{5} = 5(x - 3)\frac{2}{x - 3}$$
$$(x - 3)x = 10$$
$$x^2 - 3x = 10$$
$$(x - 5)(x + 2) = 0$$

Check: $\log 5 - \log 5 \overset{?}{=} \log 2 - \log 2$
$\log(-2)$ is not defined
Solution: 5

35.
$$(\ln x)^3 = \ln x^4$$
$$(\ln x)^3 = 4\ln x$$
$$(\ln x)^3 - 4\ln x = 0$$
$$\ln x[(\ln x)^2 - 4] = 0$$
$$\ln x(\ln x - 2)(\ln x + 2) = 0$$
$\ln x = 0$ $\qquad$ $\ln x - 2 = 0$ $\qquad$ $\ln x + 2 = 0$
$x = 1$ $\qquad$ $\ln x = 2$ $\qquad$ $\ln x = -2$
$\qquad$ $x = e^2$ $\qquad$ $x = e^{-2}$

COMMON ERROR: Replacing $(\ln x)^3$ by $3\ln x$. Although $\ln(x^3) = 3\ln x$, $(\ln x)^3$ is not the same as either $\ln(x^3)$ or $3\ln x$.

Check:

$(\ln 1)^3 \overset{?}{=} \ln 1^4$ $\qquad$ $(\ln e^2)^3 \overset{?}{=} \ln(e^2)^4$ $\qquad$ $(\ln e^{-2})^3 \overset{?}{=} \ln(e^{-2})^4$
$0 \overset{\checkmark}{=} 0$ $\qquad$ $8 \overset{\checkmark}{=} 8$ $\qquad$ $-8 \overset{\checkmark}{=} -8$

Solution: $1, e^2, e^{-2}$

37.
$$\ln(\ln x) = 1$$
$$\ln x = e^1$$
$$\ln x = e$$
$$x = e^e$$

39. $x^{\log x} = 100x$

We start by taking logarithms of both sides.
$$\log(x^{\log x}) = \log 100x$$
$$\log x \log x = \log 100 + \log x$$
$$(\log x)^2 = 2 + \log x$$
$$(\log x)^2 - \log x - 2 = 0$$

$(\log x - 2)(\log x + 1) = 0$

$\log x - 2 = 0 \qquad \log x + 1 = 0$

$\log x = 2 \qquad \log x = -1$

$x = 10^2 \qquad x = 10^{-1}$

$x = 100 \qquad x = 0.1$

Check: $100^{\log 100} \overset{?}{=} 100 \cdot 100 \qquad (0.1)^{\log(0.1)} \overset{?}{=} 100(0.1)$

$10^4 \overset{\surd}{=} 10^4 \qquad 0.1^{-1} \overset{?}{=} 10$

$10 \overset{\surd}{=} 10$

Solution: 100, 0.1

41.

$$I = I_0 e^{-kx}$$
$$\frac{I}{I_0} = e^{-kx}$$
$$\ln \frac{I}{I_0} = -kx$$
$$-\frac{1}{k} \ln \frac{I}{I_0} = x$$
$$x = -\frac{1}{k} \ln \frac{I}{I_0}$$

43.

$$N = 10 \log \frac{I}{I_0}$$
$$\frac{N}{10} = \log \frac{I}{I_0}$$
$$\frac{I}{I_0} = 10^{N/10}$$
$$I = I_0 10^{N/10}$$

45.

$$I = \frac{E}{R}(1 - e^{-Rt/L})$$
$$RI = E(1 - e^{-Rt/L})$$
$$\frac{RI}{E} = 1 - e^{-Rt/L}$$
$$\frac{RI}{E} - 1 = -e^{-Rt/L}$$
$$-\left(\frac{RI}{E} - 1\right) = e^{-Rt/L}$$
$$-\frac{RI}{E} + 1 = e^{-Rt/L}$$
$$1 - \frac{RI}{E} = e^{-Rt/L}$$
$$\ln\left(1 - \frac{RI}{E}\right) = -\frac{Rt}{L}$$
$$-\frac{L}{R} \ln\left(1 - \frac{RI}{E}\right) = t$$
$$t = -\frac{L}{R} \ln\left(1 - \frac{RI}{E}\right)$$

47. The inequality sign should have been reversed when both sides were multiplied by $\log \frac{1}{2}$, a negative quantity.

49. To find the doubling time we replace A in $A = P(1 + 0.12)^n$ with 2P and solve for n.

$$2P = P(1.12)^n$$
$$2 = (1.12)^n$$
$$\ln 2 = n \ln 1.12$$
$$n = \frac{\ln 2}{\ln 1.12}$$
$$n = 6.116 \text{ years}$$

51. To find the tripling time we replace A in $A = P(1 + 0.04)^n$ with 3P and solve for n.

$$3P = P(1.04)^n$$
$$3 = (1.04)^n$$
$$\ln 3 = n \ln 1.04$$
$$n = \frac{\ln 3}{\ln 1.04}$$
$$n = 28.011 \text{ years}$$

53. We first solve

$$2 = \left(1 + \frac{P}{100}\right)^n$$

for n in terms of P

$$\ln 2 = n \ln\left(1 + \frac{P}{100}\right)$$

$$n = \frac{\ln 2}{\ln\left(1 + \frac{P}{100}\right)}$$

Let us test

$$\frac{\ln 2}{\ln\left(1 + \frac{P}{100}\right)} \approx \frac{72}{100P}$$

for various values of P. We try P = 1%, 6%, 9%, 12%, 24%.

P	$\frac{\ln 2}{\ln\left(1 + \frac{P}{100}\right)}$	$\frac{72}{100P}$
1%	69.7	72
6%	11.9	12
9%	8.04	8
12%	6.12	6
24%	3.22	3

Students are invited to try other values of P, and to draw their own conclusions.

55. $A = 5{,}000 \cdot 2^{2t}$

We are to find t for A = 1,000,000. Hence, we solve

$$1{,}000{,}000 = 5{,}000 \cdot 2^{2t}$$

$$\frac{1{,}000{,}000}{5{,}000} = 2^{2t}$$

$$200 = 2^{2t}$$

$$\ln 200 = 2t \ln 2$$

$$\frac{\ln 200}{2 \ln 2} = t$$

t = 3.8 hours, approximately

57. We substitute D = 6 into $L = 8.8 + 5.1 \log D$ to obtain

$$L = 8.8 + 5.1 \log 6$$

$$L = 12.8 \text{ (13th magnitude)}$$

59. Using the hint, we solve

$$2P_0 = P_0 e^{0.02t}$$

$$2 = e^{0.02t}$$

$$\ln 2 = 0.02t$$

$$\frac{\ln 2}{0.02} = t$$

t = 35 years to the nearest year

61.

$$A = Pe^{-0.0248t}$$

$$0.5P = Pe^{-0.0248t}$$

$$0.5 = e^{-0.0248t}$$

$$\ln(0.5) = -0.0248t$$

$$\frac{\ln 0.5}{-0.0248} = t$$

t = 28 years, approximately

63. Using the hint, we solve

$$\begin{aligned} 0.06A_0 &= A_0 e^{-0.000124t} \\ 0.06 &= e^{-0.000124t} \\ \ln 0.06 &= -0.000124t \\ \frac{\ln 0.06}{-0.000124} &= t \\ t &= 22{,}689 \text{ years} \end{aligned}$$

65. Given $I = I_0 10^{N/10}$, we solve for N in terms of I and I_0.

$$\begin{aligned} \frac{I}{I_0} &= 10^{N/10} \\ \log \frac{I}{I_0} &= \frac{N}{10} \\ 10 \log \frac{I}{I_0} &= N \\ N &= 10 \log \frac{I}{I_0} \end{aligned}$$

67. Using the formula $I = I_0 e^{-kd}$, we wish to find d for which $I = 0.01I_0$. We solve for d in terms of k.

$$\begin{aligned} 0.01I_0 &= I_0 e^{-kd} \\ 0.01 &= e^{-kd} \\ \ln(0.01) &= -kd \\ d &= \frac{\ln(0.01)}{-k} \end{aligned}$$

Hence if $k = 0.0485$, $d = \frac{\ln(0.01)}{-0.0485} = 95$ feet

and if $k = 0.00942$, $d = \frac{\ln(0.01)}{-0.00942} = 489$ feet

69. We substitute 60 for N in $N = 80(1 - e^{-0.08n})$ and solve for n.

$$\begin{aligned} 60 &= 80(1 - e^{-0.08n}) \\ \frac{60}{80} &= 1 - e^{-0.08n} \\ 0.75 &= 1 - e^{-0.08n} \\ -0.25 &= -e^{-0.08n} \\ 0.25 &= e^{-0.08n} \\ \ln 0.25 &= -0.08n \\ n &= \frac{\ln 0.25}{-0.08} \\ n &= 17.3 \text{ weeks} \end{aligned}$$

Exercise 10-6 REVIEW EXERCISE

1. $n = \log_{10} m$

2. $x = 10^y$

3. $\log_3 27 = \log_3 3^3 = 3$

4. $\log_4 2 = \log_4 4^{1/2} = \frac{1}{2}$

5. $\log_4 32 = \log_4 2^5 = \log_4 4^{5/2} = \frac{5}{2}$

6. $\log_4 \left(\frac{1}{4}\right) = \log_4 4^{-1} = -1$

7. $\log_3 \sqrt{3} = \log_3 3^{1/2} = \frac{1}{2}$

8. $\log_3 \left(\frac{1}{9}\right) = \log_3 \frac{1}{3^2} = \log_3 3^{-2} = -2$

9. $\log_2 x = 3$
   $x = 2^3$
   $x = 8$

10. $\log_x 25 = 2$
    $25 = x^2$
    $x = 5$ since bases are restricted positive.

11. $\log_{10} x = 2$
$x = 10^2$
$x = 100$

12. $\log_8 x = -1$
$x = 8^{-1}$
$x = \frac{1}{8}$

13. $\log_x 9 = 1$
$x^1 = 9$
$x = 9$

14. $\log_x \left(\frac{1}{8}\right) = 3$
$x^3 = \frac{1}{8}$
$x = \sqrt[3]{\frac{1}{8}}$
$x = \frac{1}{2}$

15. $10^x = 17.5$
$x = \log_{10} 17.5$
$x = 1.24$

16. $e^x = 143{,}000$
$x = \ln 143{,}000$
$x = 11.9$

17. $2^x = 10$
$x \log_{10} 2 = 1$
$x = \frac{1}{\log_{10} 2}$
$x = 3.32$

18. $3^x = \frac{1}{10}$
$x \log_{10} 3 = \log_{10} 10^{-1}$
$x \log_{10} 3 = -1$
$x = \frac{-1}{\log_{10} 3}$
$x = -2.10$

19. $\log x - 2 \log 3 = 2$
$\log x - \log 3^2 = 2$
$\log x - \log 9 = 2$
$\log \frac{x}{9} = 2$
$\frac{x}{9} = 10^2$
$x = 9 \times 10^2$
$x = 900$

20. $\log x + \log(x - 3) = 1$
$\log[(x(x - 3)] = 1$
$x(x - 3) = 10^1$
$x^2 - 3x = 10$
$x^2 - 3x - 10 = 0$
$(x - 5)(x + 2) = 0$
$x - 5 = 0 \qquad x + 2 = 0$
$x = 5 \qquad x = -2$
Solution: 5

Check:
<u>x = 5</u> $\log 5 + \log 2 \stackrel{?}{=} 1$
$1 = 1$ √
<u>x = -2</u> $\log(-2)$ is undefined

21. $\log 3 + \log x = \log 300$
$\log 3x = \log 300$
$3x = 300$
$x = 100$

22. $\frac{1}{2} \log x = \log 7$
$\log x^{1/2} = \log 7$
$x^{1/2} = 7$
$x = 7^2$
$x = 49$

23. $y = e^x$

24. $y = \ln x$

25. $\log_{1/4} 16 = \log_{1/4} 4^2 = \log_{1/4} \frac{1}{4^{-2}} = \log_{1/4} \left(\frac{1}{4}\right)^{-2} = -2$

26. $\log_{1/2} \left(\frac{1}{8}\right) = \log_{1/2} \left(\frac{1}{2}\right)^3 = 3$

27. $\log_{1/2} 32 = \log_{1/2} 2^5 = \log_{1/2} \frac{1}{2^{-5}} = \log_{1/2} \left(\frac{1}{2}\right)^{-5} = -5$

28. $\log_{1/4} 2 = \log_{1/4} \sqrt{4} = \log_{1/4} 4^{1/2} = \log_{1/4} \left(\frac{1}{4}\right)^{-1/2} = -\frac{1}{2}$

29. $\log_x 9 = -2$

$x^{-2} = 9$

$\frac{1}{x^2} = 9$

$\frac{1}{9} = x^2$

$x = \pm\sqrt{\frac{1}{9}}$

$x = \frac{1}{3}$ since bases are restricted positive

30. $\log_{16} x = \frac{3}{2}$

$16^{3/2} = x$

$64 = x$

$x = 64$

31. $\log_x e^5 = 5$

$e^4 = x^5$

$x = e$

32. $10^{\log_{10} x} = 33$

$\log_{10} x = \log_{10} 33$

$x = 33$

33. $\ln x = 0$

$e^0 = x$

$x = 1$

34. $\log_x \frac{1}{2} = 2$

$x^2 = \frac{1}{2}$

$x = \pm\sqrt{\frac{1}{2}}$

$x = \frac{1}{\sqrt{2}}$

since bases are restricted positive

35. $\ln e^x = x$

$x = x$

All real numbers are solutions.

36. $\log x = 0$

$x = 10^0$

$x = 1$

37. $25 = 5(2)^x$

$\frac{25}{5} = 2^x$

$5 = 2^x$

$\ln 5 = x \ln 2$

$x = \frac{\ln 5}{\ln 2}$

$x = 2.32$

COMMON ERROR:
$25 = 10^x$. This violates the order of operations rules; the exponent x applies only to the base 2.

38. $4{,}000 = 2{,}500e^{0.12x}$

$\frac{4{,}000}{2{,}500} = e^{0.12x}$

$0.12x = \ln \frac{4000}{2500}$

$x = \frac{1}{0.12} \ln \frac{4000}{2500}$

$x = 3.92$

39. $0.01 = e^{-0.05x}$

$-0.05x = \ln 0.01$

$x = \frac{\ln 0.01}{-0.05}$

$x = 92.1$

40. $500 = 5^x$

$\ln 500 = x \ln 5$

$\frac{\ln 500}{\ln 5} = x$

$x = 3.861$

41. $e^{-3x} = e^6$
$-3x = 6$
$x = -2$

42. $10^{-x} = 500$
$-x = \log_{10} 500$
$x = -\log_{10} 500$
$x = -2.70$

43. $\log 3x^2 - \log 9x = 2$
$\log \frac{3x^2}{9x} = 2$
$\frac{3x^2}{9x} = 10^2$
$\frac{x}{3} = 100$
$x = 300$

44. $\log x - \log 3 = \log 4 - \log(x + 4)$
$\log \frac{x}{3} = \log \frac{4}{x + 4}$
$\frac{x}{3} = \frac{4}{x + 4}$ excluded value: $x \neq -4$
$3(x + 4) \frac{x}{3} = 3(x + 4) \frac{4}{x + 4}$
$(x + 4)x = 12$
$x^2 + 4x = 12$
$x^2 + 4x - 12 = 0$
$(x + 6)(x - 2) = 0$
$x + 6 = 0$ $x - 2 = 0$
$x = -6$ $x = 2$
Solution: 2

Check:
$\underline{x = -6}$ $\log(-6)$ is not defined
$\underline{x = 2}$ $\log 2 - \log 3 \stackrel{?}{=} \log 4 - \log(2 + 4)$
$\log \frac{2}{3} \stackrel{?}{=} \log \frac{4}{6}$
$\log \frac{2}{3} \stackrel{\checkmark}{=} \log \frac{2}{3}$

45. $(\log x)^3 = \log x^9$
$(\log x)^3 = 9 \log x$
$(\log x)^3 - 9 \log x = 0$
$\log x[(\log x)^2 - 9] = 0$
$\log x(\log x - 3)(\log x + 3) = 0$

$\log x = 0$ $\log x - 3 = 0$ $\log x + 3 = 0$
$x = 1$ $\log x = 3$ $\log x = -3$
$x = 10^3$ $x = 10^{-3}$

Check: $\underline{x = 1}$ $(\log 1)^3 \stackrel{?}{=} \log 1^9$
$0 \stackrel{\checkmark}{=} 0$

$\underline{x = 10^3}$ $(\log 10^3)^3 = \log(10^3)^9$
$27 \stackrel{\checkmark}{=} 27$

$\underline{x = 10^{-3}}$ $(\log 10^{-3})^3 \stackrel{?}{=} \log(10^{-3})^9$
$-27 \stackrel{\checkmark}{=} -27$

Solution: $1, 10^3, 10^{-3}$

46. $\ln(\log x) = 1$
$\log x = e$
$x = 10^e$

47. $\log_5 23 = \frac{\ln 23}{\ln 5} = 1.95$

48. $\log_2 6 = \frac{\ln 6}{\ln 2} = 2.585$

49. $\log_{12} 10 = \frac{\ln 10}{\ln 12} = 0.927$

50. $\log_{121} 1 = 0$

51. $y = 3e^{-2x}$

x	y
-0.5	8.15
0	3
0.5	1.10
1	0.41
2	0.05
3	0.01

52. $y = 10^{x/2}$

x	y
-3	0.03
-2	0.10
-1	0.31
-0.5	0.56
0	1
0.5	1.77
1	3.16
2	10

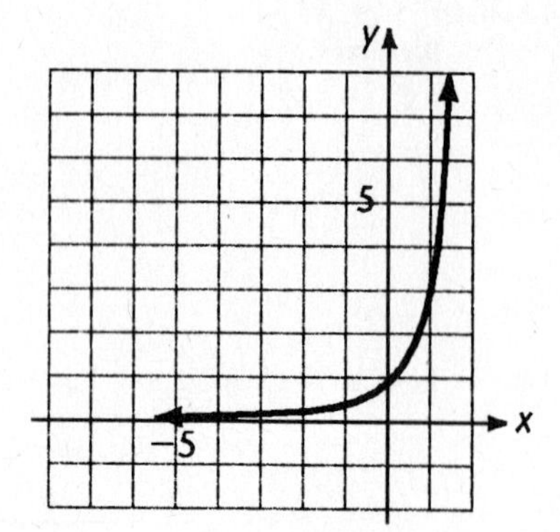

53. $y = 2 \log 3x$

x	y
0.1	-1.05
0.33	0
0.5	0.35
1	0.95
2	1.56
3	1.91
4	2.15

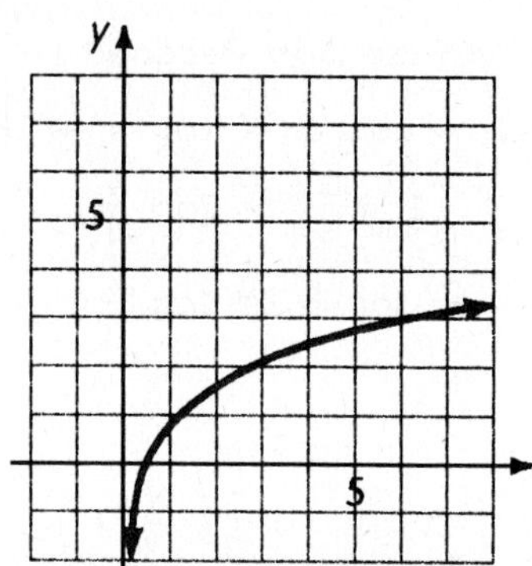

54. $y = 2 \ln \frac{x}{3}$

x	y
0.5	-3.58
1.0	-2.20
1.5	-1.38
2.0	-0.81
2.5	-0.36
3.0	0
3.5	0.31
4.0	0.58

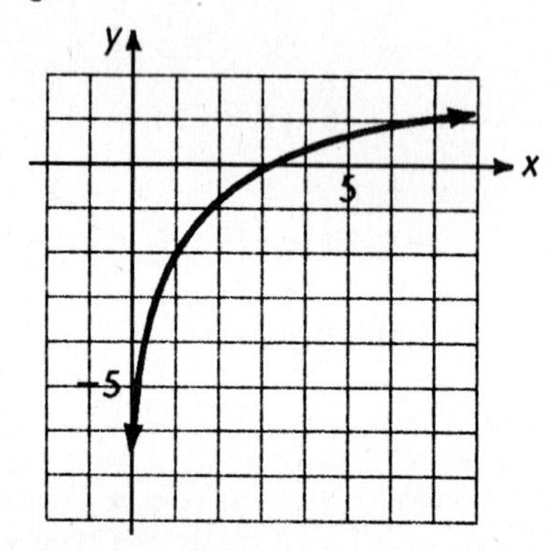

55. $y = 1.08^x$

x	y
-30	0.10
-20	0.21
-10	0.46
-5	0.68
-1	0.93
0	1
1	1.08
2	1.16
5	1.46
10	2.16
20	4.66
30	10.06

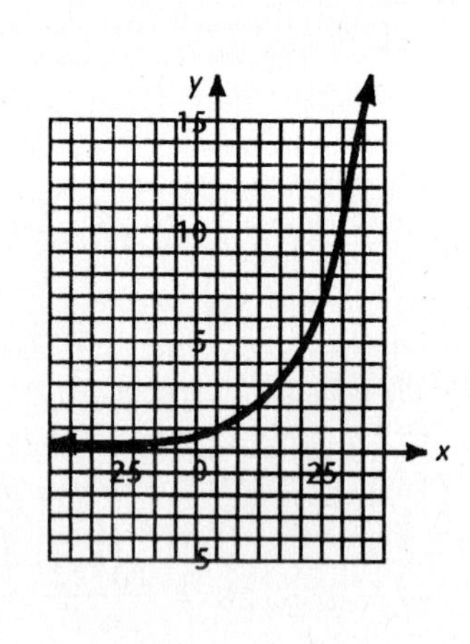

56.
$$\ln y = -5t + \ln C$$
$$\ln y - \ln C = -5t$$
$$\ln \frac{y}{C} = -5t$$
$$\frac{y}{C} = e^{-5t}$$
$$y = Ce^{-5t}$$

57. $\ln y = 3x + 4$
$$y = e^{3x+4}$$

58. $e^y = 3x + 4$
$$y = \ln(3x + 4)$$

59.

x	y = ln x	x = ln y	y
$\frac{1}{e}$	-1	-1	$\frac{1}{e}$
1	0	0	1
e	1	1	e
e^2	2	2	e^2

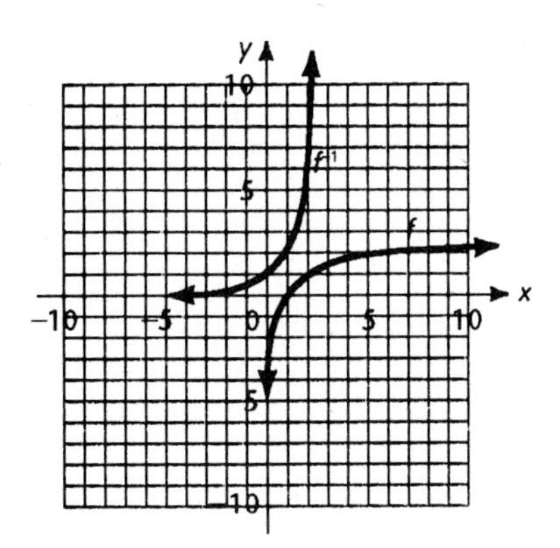

60.

x	y = log x	x = log y	y
0.1	-1	-1	0.1
1	0	0	1
5	0.69	0.69	5
10	1	1	10

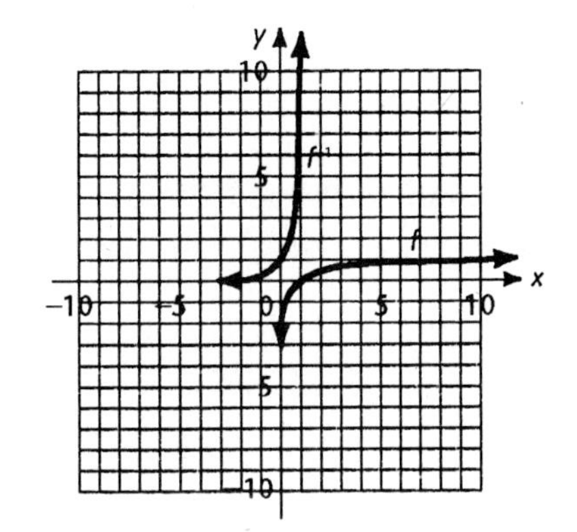

61.

x	$y = \log_{1/2} x$	$x = \log_{1/2} y$	y
$\frac{1}{4}$	2	2	$\frac{1}{4}$
$\frac{1}{2}$	1	1	$\frac{1}{2}$
1	0	0	1
2	-1	-1	2
4	-2	-2	4
8	-3	-3	8

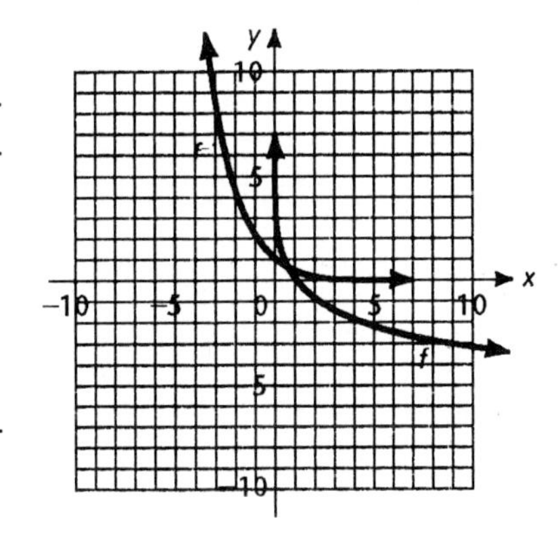

62.

x	$y = \log_2 x$	$x = \log_2 y$	y
1	0	0	1
2	1	1	2
4	2	2	4
8	3	3	8

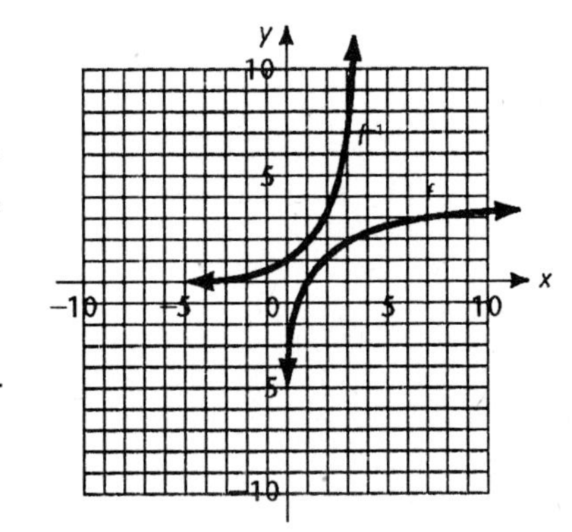

63. $\log(\log x) = 0$
$$\log x = 10^0$$
$$\log x = 1$$
$$x = 10^1$$
$$x = 10$$

64. $\ln x = e$
$$x = e^e$$

65. $-3 \log x = \log \frac{1}{x^3}$
$$\log x^{-3} = \log \frac{1}{x^3}$$
$$x^{-3} = \frac{1}{x^3}$$
This is true for all x. However, x must be restricted to the domain of log x for the original equation to hold.
Solution: $x > 0$.

66. $\log(-x) = \log x$
$$-x = x$$
$$-2x = 0$$
$$x = 0$$
However, 0 is not in the domain of log x. Thus, there is no solution.

67. If $\log_1 x = y$, then we would have to have $1^y = x$; that is, $1 = x$ for arbitrary positive x, which is impossible.

68. Let $u = \log_b M$ and $v = \log_b N$; then $M = b^u$ and $N = b^v$.
Thus, $\log_b (M/N) = \log_b (b^u/b^v) = \log_b b^{u-v} = u - v = \log_b M - \log_b N$.

69. 6% compounded quarterly = 0.06/4 = 0.015 per year.
Use $A = P(1 + a)^{nt}$ with $P = 1000$, $a = 0.015$, $n = 4$, $t = 3$
$$A = 1000(1 + 0.015)^{(4)(3)}$$
$$A = \$1195.62$$

70. Use $A = Pe^{at}$ with $P = 1000$, $a = 0.04$, $t = 10$
$$A = 1000e^{0.04(10)}$$
$$A = \$1491.82$$

71. We solve $P = P_0(1.03)^t$ for t, using $P = 2P_0$.
$$2P_0 = P_0(1.03)^t$$
$$2 = (1.03)^t$$
$$\ln 2 = t \ln 1.03$$
$$\frac{\ln 2}{\ln 1.03} = t$$
$$t = 23.4 \text{ years}$$

72. We solve $P = P_0 e^{0.03t}$ for t, using $P = 2P_0$.
$$2P_0 = P_0 3^{0.03t}$$
$$2 = e^{0.03t}$$
$$\ln 2 = 0.03t$$
$$\frac{\ln 2}{0.03} = t$$
$$t = 23.1 \text{ years}$$

73. A_0 = original amount
$0.01A_0$ = 1 percent of original amount

We solve $A = A_0 e^{-0.000124t}$ for t, using $A = 0.01A_0$.
$$0.01A_0 = A_0 e^{-0.000124t}$$
$$0.01 = e^{-0.000124t}$$
$$\ln 0.01 = -0.000124t$$
$$\frac{\ln 0.01}{-0.000124} = t$$
$$t = 37{,}100 \text{ years}$$

74.
$$x = -\frac{1}{k}\ln\frac{I}{I_0}$$
$$-kx = \ln\frac{I}{I_0}$$
$$\frac{I}{I_0} = e^{-kx}$$
$$I = I_0e^{-kx}$$

75.
$$r = P\,\frac{i}{1 - (1 + i)^{-n}}$$
$$\frac{r}{P} = \frac{i}{1 - (1 + i)^{-n}}$$
$$\frac{P}{r} = \frac{1 - (1 + i)^{-n}}{i}$$
$$\frac{Pi}{r} = 1 - (1 + i)^{-n}$$
$$\frac{Pi}{r} - 1 = -(1 + i)^{-n}$$
$$1 - \frac{Pi}{r} = (1 + i)^{-n}$$
$$\log\left(1 - \frac{Pi}{r}\right) = -n\log(1 + i)$$
$$\frac{\log\left(1 - \frac{Pi}{r}\right)}{-\log(1 + i)} = n$$
$$n = -\frac{\log\left(1 - \frac{Pi}{r}\right)}{\log(1 + i)}$$

Practice Test 10

1. $\log_3 9 - \log_4 2 = \log_3 3^2 - \log_4 \sqrt{4} = 2 - \log_4 4^{1/2} = 2 - \frac{1}{2} = \frac{3}{2}$

2. $\log_{1/2} \frac{1}{4} - \log_3 \sqrt{3} = \log_{1/2} \left(\frac{1}{2}\right)^2 - \log_3 3^{1/2} = 2 - \frac{1}{2} = \frac{3}{2}$

3. $\log_b 12 = \log_b (2^2 \cdot 3) = \log_b 2^2 + \log_b 3 = 2 \log_b 2 + \log_b 3$
$= 2(0.356) + 0.565 = 1.277$

4. $\log_2 x - 3 \log_2 (x + 1) = \log_2 x - \log_2 (x + 1)^3 = \log_2 \frac{x}{(x + 1)^3}$

5. $\log x - 2 \log y + 3 \log z = \log x - \log y^2 + \log z^3 = \log \frac{x}{y^2} + \log z^3 = \log \frac{xz^3}{y^2}$

6. $\log_2 x = 5$
$2^5 = x$
$x = 32$

7. $\log_x 5 = 2$
$x^2 = 5$
$x = \sqrt{5}$ (since bases are positive)

8. $\ln e^2 = \log x$
$2 = \log x$
$10^2 = x$
$x = 100$

9. $\log x + \log(x - 3) = 1$
$\log[(x - 3)] = 1$
$x(x - 3) = 10^1$
$x^2 - 3x = 10$
$x^2 - 3x - 10 = 0$
$(x - 5)(x + 2) = 0$
$x - 5 = 0 \qquad x + 2 = 0$
$x = 5 \qquad x = -2$
Solution: 5

Check:
$\underline{x = 5}$ $\log 5 + \log 2 \stackrel{?}{=} 1$
$1 \stackrel{\checkmark}{=} 1$
$\underline{x = 2}$ $\log(-2)$ is undefined

10. $\log x^3 = 3$
$10^3 = x^3$
$x = 10$

11. $\ln \frac{1}{x} = -1$
$\frac{1}{x} = e^{-1}$
$\frac{1}{x} = \frac{1}{e}$
$x = e$

12. $2^x = 10$
$x \log_{10} 2 = \log_{10} 10$
$x \log_{10} 2 = 1$
$x = \frac{1}{\log_{10} 2}$
$x = 3.3219$

13. $e^x = 10$
$x = \ln 10$
$x = 2.3026$

14. $1000 = 50e^{0.008x}$
$\frac{1000}{50} = e^{0.008x}$
$20 = e^{0.008x}$
$\ln 20 = 0.008x$
$\frac{\ln 20}{0.008} = x$
$x = 374.4665$

15. $10^x = 0.004$
$x = \log 0.004$
$x = -2.3979$

16. $y = \log_2 x$

x	y
$\frac{1}{8}$	-3
$\frac{1}{4}$	-2
$\frac{1}{2}$	-1
1	0
2	1
4	2
8	3

17. $y = \left(\frac{1}{2}\right)^x$

x	y
-3	8
-2	4
-1	2
0	1
1	$\frac{1}{2}$
2	$\frac{1}{4}$
3	$\frac{1}{8}$

18.

x	$y = e^x$
-2	0.13
-1	0.37
0	1
1	2.72
2	7.38

$x = e^y$	y
0.13	-2
0.37	-1
1	0
2.72	1
7.38	2

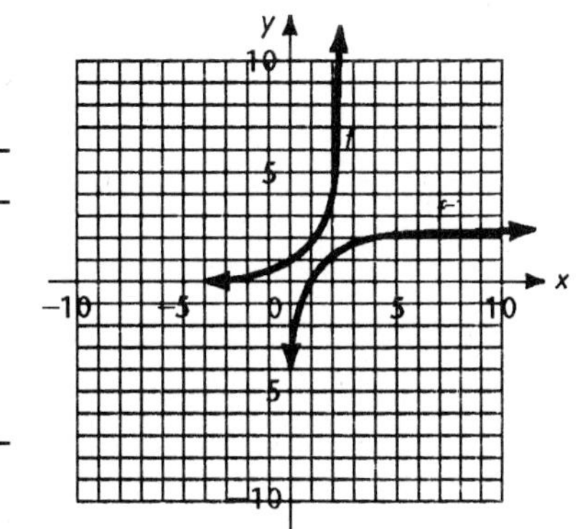

Domain of f is the set of real numbers; range of f is the set of positive real numbers. The domain of f is the range of f^{-1} and the range of f is the domain of f^{-1}.

f^{-1} is called the logarithmic function with base e. $f^{-1}(x) = \ln x$.

19. 9% compounded semiannually = 0.09/2 = 0.045 per year. Use $A = P(1 + a)^{nt}$ with $P = 100$, $a = 0.045$, $n = 2$, $t = 10$

$$A = 100(1 + 0.045)^{(2)(10)}$$
$$A = \$241.17$$

20.

$$T - M = C_0 e^{kt}$$
$$\frac{T - M}{C_0} = e^{kt}$$
$$\ln \frac{T - M}{C_0} = kt$$
$$t = \frac{1}{k} \ln \frac{T - M}{C_0}$$

CHAPTER 11 Sequences and Series

Exercise 11-1 Sequences and Series

Key Ideas and Formulas

A sequence is a function with domain the natural numbers.

Notation: for a(n) we write a_n. The elements of the range, a(n), are called terms of the sequence, a_n. a_1 is the first term, a_2 is the second term, a_n is the nth term.

Finite sequence: domain is finite set of successive natural numbers.
$a_1, a_2, a_3, \dots, a_n$

Infinite sequence: domain is infinite set of successive natural numbers.
$a_1, a_2, a_3, \dots, a_n, \dots$

Some sequences are specified by a recursion formula, a formula that defines each term in terms of one or more preceding terms.

The sum of the terms of a sequence is called a series. If $a_1, a_2, \dots, a_n$ are the terms of the sequence, the series is written

$\sum_{k=1}^{n} a_k = a_1 + a_2 + \dots + a_n$ k is called the summing index.

---

1. $a_1 = 1 - 2 = -1$ $a_2 = 2 - 2 = 0$ $a_3 = 3 - 2 = 1$ $a_4 = 4 - 2 = 2$

3. $a_1 = 2 \cdot 1 + 3 = 5$ $a_2 = 2 \cdot 2 + 3 = 7$ $a_3 = 2 \cdot 3 + 3 = 9$ $a_4 = 2 \cdot 4 + 3 = 11$

5. $a_1 = 2^1 = 2$ $a_2 = 2^2 = 4$ $a_3 = 2^3 = 8$ $a_4 = 2^4 = 16$

7. $a_1 = \frac{1 - 1}{1 + 1} = 0$ $a_2 = \frac{2 - 1}{2 + 1} = \frac{1}{3}$ $a_3 = \frac{3 - 1}{3 + 1} = \frac{2}{4} = \frac{1}{2}$ $a_4 = \frac{4 - 1}{4 + 1} = \frac{3}{5}$

9. $a_1 = \left(1 - \frac{1}{1}\right)^1 = 0$ $a_2 = \left(1 - \frac{1}{2}\right)^2 = \left(\frac{1}{2}\right)^2 = \frac{1}{4}$ $a_3 = \left(1 - \frac{1}{3}\right)^3 = \left(\frac{2}{3}\right)^3 = \frac{8}{27}$
$a_4 = \left(1 - \frac{1}{4}\right)^4 = \left(\frac{3}{4}\right)^4 = \frac{81}{256}$

11. $a_1 = (-2)^{1+1} = (-2)^2 = 4$ $a_2 = (-2)^{2+1} = (-2)^3 = -8$
$a_3 = (-2)^{3+1} = (-2)^4 = 16$ $a_4 = (-2)^{4+1} = (-2)^5 = -32$

13. $a_1 = \frac{(-1)^1}{1} = -1$ $a_2 = \frac{(-1)^2}{2} = \frac{1}{2}$ $a_3 = \frac{(-1)^3}{3} = -\frac{1}{3}$ $a_4 = \frac{(-1)^4}{4} = \frac{1}{4}$

15. $a_8 = 8 - 2 = 6$ 17. $a_{18} = 2(18) + 3 = 39$ 19. $a_6 = 2^6 = 64$

21. $a_{100} = \frac{100 - 1}{100 + 1} = \frac{99}{101}$

23. $a_8 = \left(1 - \frac{1}{8}\right)^8 = \left(\frac{7}{8}\right)^8 = \frac{5,764,801}{16,777,216} \approx 0.3436$

25. $a_7 = (-2)^{7+1} = (-2)^8 = 256$

27. $a_{20} = \frac{(-1)^{20}}{20} = \frac{1}{20}$

29. $S_6 = \frac{1}{1} + \frac{1}{2} + \frac{1}{3} + \frac{1}{4} + \frac{1}{5} + \frac{1}{6}$

31. $S_5 = 1 + 2 + 3 + 4 + 5$

33. $S_4 = (1 - 1)^2 + (2 - 1)^2 + (3 - 1)^2 + (4 - 1)^2 = 0 + 1 + 4 + 9$

35. $S_3 = \frac{1}{10^1} + \frac{1}{10^2} + \frac{1}{10^3} = \frac{1}{10} + \frac{1}{100} + \frac{1}{1000}$

37. $S_6 = \left(\frac{1}{5}\right)^1 + \left(\frac{1}{5}\right)^2 + \left(\frac{1}{5}\right)^3 + \left(\frac{1}{5}\right)^4 + \left(\frac{1}{5}\right)^5 + \left(\frac{1}{5}\right)^6$
$= \frac{1}{5} + \frac{1}{25} + \frac{1}{125} + \frac{1}{625} + \frac{1}{3125} + \frac{1}{15,625}$

39. $S_4 = (-1)^1 + (-1)^2 + (-1)^3 + (-1)^4 = (-1) + 1 + (-1) + 1 = -1 + 1 - 1 + 1$

41. $S_5 = (-1)^1 1 + (-1)^2 2 + (-1)^3 3 + (-1)^4 4 + (-1)^5 5 = (-1) + 2 + (-3) + 4 + (-5)$
$= -1 + 2 - 3 + 4 - 5$

43. $a_1 = (-1)^{1+1} 1^2 = 1$
$a_2 = (-1)^{2+1} 2^2 = -4$
$a_3 = (-1)^{3+1} 3^2 = 9$
$a_4 = (-1)^{4+1} 4^2 = -16$
$a_5 = (-1)^{5+1} 5^2 = 25$

45. $a_1 = \frac{1}{3}\left[1 - \frac{1}{10^1}\right] = \frac{1}{3} \cdot \frac{9}{10} = \frac{3}{10} = 0.3$
$a_2 = \frac{1}{3}\left[1 - \frac{1}{10^2}\right] = \frac{1}{3} \cdot \frac{99}{100} = \frac{33}{100} = 0.33$
$a_3 = \frac{1}{3}\left[1 - \frac{1}{10^3}\right] = \frac{1}{3} \cdot \frac{999}{1,000} = \frac{333}{1,000} = 0.333$
$a_4 = \frac{1}{3}\left[1 - \frac{1}{10^4}\right] = \frac{1}{3} \cdot \frac{9,999}{10,000} = \frac{3,333}{10,000} = 0.3333$
$a_5 = \frac{1}{3}\left[1 - \frac{1}{10^5}\right] = \frac{1}{3} \cdot \frac{99,999}{100,000} = \frac{33,333}{100,000} = 0.33333$

47. $a_1 = (1 + 1)[1 - (-1)^{1+1}] = 2[1 - (-1)^2] = 2[1 - 1] = 0$
$a_2 = (2 + 1)[1 - (-1)^{2+1}] = 3[1 - (-1)^3] = 3[1 + 1] = 6$
$a_3 = (3 + 1)[1 - (-1)^{3+1}] = 4[1 - (-1)^4] = 4[1 - 1] = 0$
$a_4 = (4 + 1)[1 - (-1)^{4+1}] = 5[1 - (-1)^5] = 5[1 + 1] = 10$
$a_5 = (5 + 1)[1 - (-1)^{5+1}] = 6[1 - (-1)^6] = 6[1 - 1] = 0$

49. $a_1 = 1$
$a_2 = 2a_{2-1} = 2a_1 = 2 \cdot 1 = 2$
$a_3 = 2a_{3-1} = 2a_2 = 2 \cdot 2 = 4$
$a_4 = 2a_{4-1} = 2a_3 = 2 \cdot 4 = 8$
$a_5 = 2a_{5-1} = 2a_4 = 2 \cdot 8 = 16$

COMMON ERROR:
$a_2 \neq 2a_2 - 1$
The -1 is in the subscript.

51. $a_1 = 7$

$a_2 = a_{2-1} - 4 = a_1 - 4 = 7 - 4 = 3$

$a_3 = a_{3-1} - 4 = a_2 - 4 = 3 - 4 = -1$

$a_4 = a_{4-1} - 4 = a_3 - 4 = -1 - 4 = -5$

$a_5 = a_{5-1} - 4 = a_4 - 4 = -5 - 4 = -9$

53. $a_1 = 4$

$a_2 = \frac{1}{4}a_{2-1} = \frac{1}{4}a_1 = \frac{1}{4}\cdot 4 = 1$

$a_3 = \frac{1}{4}a_{3-1} = \frac{1}{4}a_2 = \frac{1}{4}\cdot 1 = \frac{1}{4}$

$a_4 = \frac{1}{4}a_{4-1} = \frac{1}{4}a_3 = \frac{1}{4}\cdot\frac{1}{4} = \frac{1}{16}$

$a_5 = \frac{1}{4}a_{5-1} = \frac{1}{4}a_4 = \frac{1}{4}\cdot\frac{1}{16} = \frac{1}{64}$

55. $a_1 = 1$

$a_2 = 2a_{2-1} = 2a_1 = 2\cdot 1 = 2$

$a_3 = 3a_{3-1} = 3a_2 = 3\cdot 2 = 6$

$a_4 = 4a_{4-1} = 4a_3 = 4\cdot 6 = 24$

$a_5 = 5a_{5-1} = 5a_4 = 5\cdot 24 = 120$

57. $a_1 = 1$

$a_2 = a_{2-1} - 2 = a_1 - 2 = 1 - 2 = -1$

$a_3 = a_{3-1} - 3 = a_2 - 3 = (-1) - 3 = -4$

$a_4 = a_{4-1} - 4 = a_3 - 4 = (-4) - 4 = -8$

$a_5 = a_{5-1} - 5 = a_4 - 5 = (-8) - 5 = -13$

59. $a_1 = x$

$a_2 = a_{2-1}\cdot\frac{x}{2} = a_1\cdot\frac{x}{2} = x\cdot\frac{x}{2} = \frac{x^2}{2}$

$a_3 = a_{3-1}\cdot\frac{x}{3} = a_2\cdot\frac{x}{3} = \frac{x^2}{2}\cdot\frac{x}{3} = \frac{x^3}{6}$

$a_4 = a_{4-1}\cdot\frac{x}{4} = a_3\cdot\frac{x}{4} = \frac{x^3}{6}\cdot\frac{x}{4} = \frac{x^4}{24}$

$a_5 = a_{5-1}\cdot\frac{x}{5} = a_4\cdot\frac{x}{5} = \frac{x^4}{24}\cdot\frac{x}{5} = \frac{x^5}{120}$

61. $a_1 = 1$

$a_2 = 1$

$a_3 = a_{3-1} + a_{3-2} = a_2 + a_1 = 1 + 1 = 2$

$a_4 = a_{4-1} + a_{4-2} = a_3 + a_2 = 2 + 1 = 3$

$a_5 = a_{5-1} + a_{5-2} = a_4 + a_3 = 3 + 2 = 5$

63. a_n: 4, 5, 6, 7, ...

n = 1, 2, 3, 4, ...

Comparing a_n with n, we see that

$a_n = n + 3$

65. a_n: 3, 6, 9, 12, ...

n = 1, 2, 3, 4, ...

Comparing a_n with n, we see that

$a_n = 3n$

67. a_n: $\frac{1}{2}, \frac{2}{3}, \frac{3}{4}, \frac{4}{5}, \ldots$

n = 1, 2, 3, 4, ...

Comparing a_n with n, we see that

$a_n = \frac{n}{n + 1}$

69. a_n: 1, -1, 1, -1, ...

n = 1, 2, 3, 4, ...

Comparing a_n with n, we see that a_n involves (-1) to successively even and odd powers, hence to a power that depends on n. We could write

$a_n = (-1)^{n-1}$ or $a_n = (-1)^{n+1}$

or other choices. $a_n = (-1)^{n+1}$ is one of many correct answers.

71. a_n: -2, 4, -8, 16, ...

n = 1, 2, 3, 4, ...

Comparing a_n with n, we see that a_n involves -1 and 2 to successively higher powers, hence to powers that depends on n. We write

$a_n = (-1)^n(2)^n$ or $a_n = (-2)^n$

73. a_n: $x, \frac{x^2}{2}, \frac{x^3}{3}, \frac{x^4}{4}, \ldots$

or $\frac{x^1}{1}, \frac{x^2}{2}, \frac{x^3}{3}, \frac{x^4}{4}, \ldots$

$n = 1, 2, 3, 4 \ldots$

Comparing a_n with n, we see that $a_n = \frac{x^n}{n}$

75. $S_5 = \frac{(-1)^1}{1} + \frac{(-1)^2}{2} + \frac{(-1)^3}{3} + \frac{(-1)^4}{4} + \frac{(-1)^5}{5} = \frac{-1}{1} + \frac{1}{2} + \frac{-1}{3} + \frac{1}{4} + \frac{-1}{5}$

$= -1 + \frac{1}{2} - \frac{1}{3} + \frac{1}{4} - \frac{1}{5}$

77. $S_4 = \frac{(-2)^{1+1}}{1} + \frac{(-2)^{2+1}}{2} + \frac{(-2)^{3+1}}{3} + \frac{(-2)^{4+1}}{4} = \frac{4}{1} - \frac{8}{2} + \frac{16}{3} - \frac{32}{4}$

79. $S_3 = \frac{1}{1}x^{1+1} + \frac{1}{2}x^{2+1} + \frac{1}{3}x^{3+1} = x^2 + \frac{x^3}{2} + \frac{x^4}{3}$

81. $S_5 = \frac{(-1)^{1+1}}{1}x^1 + \frac{(-1)^{2+1}}{2}x^2 + \frac{(-1)^{3+1}}{3}x^3 + \frac{(-1)^{4+1}}{4}x^4 + \frac{(-1)^{5+1}}{5}x^5$

$= x - \frac{x^2}{2} + \frac{x^3}{3} - \frac{x^4}{4} + \frac{x^5}{5}$

83. $S_4 = 1^3 + 2^3 + 3^3 + 4^3$

$k = 1, 2, 3, 4, \ldots$

Clearly, $a_k = k^3$, $k = 1, 2, 3, 4$

$S_4 = \sum_{k=1}^{4} k^3$

85. $S_4 = 1^2 + 2^2 + 3^2 + 4^2$

$k = 1, 2, 3, 4, \ldots$

Clearly, $a_k = k^2$, $k = 1, 2, 3, 4$

$S_4 = \sum_{k=1}^{4} k^2$

87. $S_5 = \frac{1}{2^1} + \frac{1}{2^2} + \frac{1}{2^3} + \frac{1}{2^4} + \frac{1}{2^5}$

$k = 1, 2, 3, 4, 5$

Clearly, $a_k = \frac{1}{2^k}$, $k = 1, 2, 3, 4, 5$

$S_5 = \sum_{k=1}^{5} \frac{1}{2^k}$

89. $S_n = 1 + \frac{1}{2^2} + \frac{1}{3^2} + \ldots + \frac{1}{n^2}$

Clearly, $a_k = \frac{1}{k^2}$, $k = 1, 2, 3, \ldots, n$

$S_n = \sum_{k=1}^{n} \frac{1}{k^2}$

91. $S_n = 1 - 4 + 9 + \ldots + (-1)^{n+1}n^2$

$k = 1, 2, 3, \ldots, n$

Clearly, $a_k = (-1)^{k+1}k^2$,

$k = 1, 2, 3, \ldots, n$

$S_n = \sum_{k=1}^{n} (-1)^{k+1}k^2$

93. $a_1 = -9 \cdot 1^3 + 63 \cdot 1^2 - 132 \cdot 1 + 80 = 2$

$a_2 = -9 \cdot 2^3 + 63 \cdot 2^2 - 132 \cdot 2 + 80 = -4$

$a_3 = -9 \cdot 3^3 + 63 \cdot 3^2 - 132 \cdot 3 + 80 = 8$

$a_4 = -9 \cdot 4^3 + 63 \cdot 4^2 - 132 \cdot 4 + 80 = -16$

95. $a_1 = \frac{2^1}{48}(-1^3 + 15 \cdot 1^2 - 86 \cdot 1 + 192) = 5$

$a_2 = \frac{2^2}{48}(-2^3 + 15 \cdot 2^2 - 86 \cdot 2 + 192) = 6$

$a_3 = \frac{2^3}{48}(-3^3 + 15 \cdot 3^2 - 86 \cdot 3 + 192) = 7$

$a_4 = \frac{2^4}{48}(-4^3 + 15 \cdot 4^2 - 86 \cdot 4 + 192) = 8$

97. (A) $a_1 = 3$

$$a_2 = \frac{a_{2-1}^2 + 2}{2a_{2-1}} = \frac{a_1^2 + 2}{2a_1} = \frac{3^2 + 2}{2 \cdot 3} \approx 1.83$$

$$a_3 = \frac{a_{3-1}^2 + 2}{2a_{3-1}} = \frac{a_2^2 + 2}{2a_2} = \frac{(1.83)^2 + 2}{2(1.83)} \approx 1.46$$

$$a_4 = \frac{a_{4-1}^2 + 2}{2a_{4-1}} = \frac{a_3^2 + 2}{2a_3} = \frac{(1.46)^2 + 2}{2(1.46)} \approx 1.415$$

(B) Table $\sqrt{2} = 1.414$
Calculator $\sqrt{2} = 1.4142135...$

(C) $a_1 = 1$

$$a_2 = \frac{a_1^2 + 2}{2a_1} = \frac{1^2 + 2}{2 \cdot 1} = 1.5$$

$$a_3 = \frac{a_2^2 + 2}{2a_2} = \frac{(1.5)^2 + 2}{2(1.5)} \approx 1.417$$

$$a_4 = \frac{a_3^2 + 2}{2a_3} = \frac{(1.417)^2 + 2}{2(1.417)} \approx 1.414$$

99. $$a_1 = \frac{1}{\sqrt{5}}\left(\frac{1 + \sqrt{5}}{2}\right) - \frac{1}{\sqrt{5}}\left(\frac{1 - \sqrt{5}}{2}\right) = \frac{1}{\sqrt{5}}\left(\frac{1 + \sqrt{5}}{2} - \frac{1 - \sqrt{5}}{2}\right) = \frac{1}{\sqrt{5}} \cdot \frac{2\sqrt{5}}{2} = 1$$

$$a_2 = \frac{1}{\sqrt{5}}\left(\frac{1 + \sqrt{5}}{2}\right)^2 - \frac{1}{\sqrt{5}}\left(\frac{1 - \sqrt{5}}{2}\right)^2 = \frac{1}{\sqrt{5}}\left(\frac{1 + 2\sqrt{5} + 5}{4}\right) - \frac{1}{\sqrt{5}}\left(\frac{1 - 2\sqrt{5} + 5}{4}\right)$$

$$= \frac{1}{\sqrt{5}}\left(\frac{6 + 2\sqrt{5}}{4} - \frac{6 - 2\sqrt{5}}{4}\right) = \frac{1}{\sqrt{5}} \cdot \frac{4\sqrt{5}}{4} = 1$$

$$a_3 = \frac{1}{\sqrt{5}}\left(\frac{1 + \sqrt{5}}{2}\right)^3 - \frac{1}{\sqrt{5}}\left(\frac{1 - \sqrt{5}}{2}\right)^3$$

$$= \frac{1}{\sqrt{5}} \frac{(1 + \sqrt{5})(1 + \sqrt{5})(1 + \sqrt{5})}{8} - \frac{1}{\sqrt{5}} \frac{(1 - \sqrt{5})(1 - \sqrt{5})(1 - \sqrt{5})}{8}$$

$$= \frac{1}{\sqrt{5}} \frac{(6 + 2\sqrt{5})(1 + \sqrt{5})}{8} - \frac{1}{\sqrt{5}} \frac{(6 - 2\sqrt{5})(1 - \sqrt{5})}{8}$$

$$= \frac{1}{\sqrt{5}} \frac{6 + 8\sqrt{5} + 10}{8} - \frac{1}{\sqrt{5}} \frac{6 - 8\sqrt{5} + 10}{8}$$

$$= \frac{1}{\sqrt{5}}\left(\frac{16 + 8\sqrt{5}}{8} - \frac{16 - 8\sqrt{5}}{8}\right) = \frac{1}{\sqrt{5}} \frac{16\sqrt{5}}{8} = 2$$

Exercise 11-2 Arithmetic Sequences and Series

Key Ideas and Formulas

A sequence $a_1, a_2, a_3, \ldots, a_n, \ldots$ is called an **arithmetic sequence** if there is a constant d, called the **common difference**, such that each term is d more than preceding term.

$$a_n = a_{n-1} + d \quad \text{for every } n > 1.$$

nth Term formula: $a_n = a_1 + (n - 1)d$

Sum Formulas: $S_n = \frac{n}{2}[2a_1 + (n - 1)d]$ — Use when given the number of terms n, the first term a_1, and the common difference d.

$S_n = \frac{n}{2}(a_1 + a_n)$ — Use when given the number of terms n, the first term a_1, and the last term a_n.

---

1. $4 - 2 = 2$, $8 - 4 = 4$. Since these are different, this is not an arithmetic sequence.

3. $-16 - (-11) = -5$; $(-21) - (-16) = -5$. This is an arithmetic sequence with $d = -5$.

5. $(-1) - 5 = -6$; $(-7) - (-1) = -6$. This is an arithmetic sequence with $d = -6$.

7. $\frac{2}{3} - \frac{1}{2} = \frac{1}{6}$, $\frac{3}{4} - \frac{2}{3} = \frac{1}{12}$. Since these are different, this is not an arithmetic sequence.

9. $a_2 = a_1 + d = -5 + 4 = -1$
$a_3 = a_2 + d = -1 + 4 = 3$
$a_4 = a_3 + d = 3 + 4 = 7$

11. $a_{15} = a_1 + 14d = -3 + 14 \cdot 5 = 67$
$S_n = \frac{n}{2}[2a_1 + (n - 1)d]$
$S_{11} = \frac{11}{2}[2(-3) + (11 - 1)5]$
$= \frac{11}{2}(44)$
$= 242$

13. $a_2 - a_1 = 5 - 1 = d = 4$
$S_n = \frac{n}{2}[2a_1 + (n - 1)d]$
$S_{21} = \frac{21}{2}[2 \cdot 1 + (21 - 1)4]$
$= \frac{21}{2}(82)$
$= 861$

15. $a_2 - a_1 = 5 - 7 = -2 = d$
$a_n = a_1 + (n - 1)d$
$a_{15} = 7 + (15 - 1)(-2) = -21$

17. $a_n = a_1 + (n - 1)d$
$a_{20} = a_1 + 19d$
$117 = 3 + 19d$
So, d = 6. Therefore,
$a_{101} = a_1 + (100)d$
$= 3 + 100(6)$
$= 603$

19. $S_n = \frac{n}{2}(a_1 + a_n)$
$S_{40} = \frac{40}{2}(-12 + 22) = 200$

21. $a_2 - a_1 = d$
$\frac{1}{2} - \frac{1}{3} = \frac{1}{6} = d$
$a_n = a_1 + (n - 1)d$
$a_{11} = \frac{1}{3} + (11 - 1)\frac{1}{6} = 2$
$S_n = \frac{n}{2}(a_1 + a_n)$
$S_{11} = \frac{11}{2}\left(\frac{1}{3} + 2\right) = \frac{77}{6}$

23. $a_n = a_1 + (n - 1)d$
$a_{10} = a_1 + 9d$
$a_3 = a_1 + 2d$
Eliminating d between these two statements by addition, we have
$2a_{10} = 2a_1 + 18d$
$-9a_3 = -9a_1 - 18d$
$2a_{10} - 9a_3 = -7a_1$
$a_1 = \frac{2a_{10} - 9a_2}{-7}$
$= \frac{2(55) - 9(13)}{-7}$
$= 1$

25. $a_n = a_1 + (n - 1)d$
$d = a_2 - a_1$
$= (3\cdot2 + 3) - (3\cdot1 + 3) = 3$
$a_{51} = a_1 + (51 - 1)d$
$= (3\cdot1 + 3) + 50\cdot3 = 156$
$S_n = \frac{n}{2}(a_1 + a_n)$
$S_{51} = \frac{51}{2}(3\cdot1 + 3 + 156)$
$= \frac{51}{2}\cdot162 = 4{,}131$

27. We can see intuitively that the series has 100 terms. To check, apply the formula
$a_n = a_1 + (n - 1)d$
and solve for n. (d = 2 - 1 = 1)
$100 = 1 + (n - 1)1$
$100 = 1 + n - 1$
$n = 100$
Then $S_n = \frac{n}{2}(a_1 + a_n)$
$S_{100} = \frac{100}{2}(1 + 100)$
$= 50(101)$
$= 5050$

29. We apply the formula
$a_n = a_1 + (a - 1)d$
and solve for n. (d = 4 - 2 = 2)
$80 = 2 + (n - 1)2$
$80 = 2 + n - 2$
$n = 80$
Then $S_n = \frac{n}{2}(a_1 + a_n)$
$S_{80} = \frac{80}{2}(2 + 160)$
$= 40(162)$
$= 6480$

31. We apply the formula
$a_n = a_1 + (n - 1)d$
and solve for n. (d = 3 - 1 = 2)
$151 = 1 + (n - 1)2$
$151 = 1 + 2n - 2$
$152 = 2n$
$n = 76$
Then $S_n = \frac{n}{2}(a_1 + a_n)$
$S_{76} = \frac{76}{2}(1 + 151)$
$= 38(152)$
$= 5776$

33. We apply the formula

$$a_n = a_1 + (n - 1)d$$

and solve for n. ($d = 6 - 3 = 3$)

$$\begin{aligned} 180 &= 3 + (n - 1)3 \\ 180 &= 3 + 3n - 3 \\ 180 &= 3n \\ n &= 60 \end{aligned}$$

Then $S_n = \frac{n}{2}(a_1 + a_n)$

$$\begin{aligned} S_{60} &= \frac{60}{2}(3 + 180) \\ &= 30(183) \\ &= 5490 \end{aligned}$$

35. We apply the formula

$$a_n = a_1 + (n - 1)d$$

and solve for n. ($d = 24 - 22 = 2$)

$$\begin{aligned} 134 &= 22 + (n - 1)2 \\ 134 &= 22 + 2n - 2 \\ 114 &= 2n \\ n &= 57 \end{aligned}$$

Then $S_n = \frac{n}{2}(a_1 + a_n)$

$$\begin{aligned} S_{57} &= \frac{57}{2}(22 + 134) \\ &= \frac{57}{2} \cdot 156 \\ &= 4446 \end{aligned}$$

37. We apply the formula

$$a_n = a_1 + (n - 1)d$$

and solve for n. ($d = 84 - 81 = 3$)

$$\begin{aligned} 360 &= 81 + (n - 1)3 \\ 360 &= 81 + 3n - 3 \\ 282 &= 3n \\ n &= 94 \end{aligned}$$

Then $S_n = \frac{n}{2}(a_1 + a_n)$

$$\begin{aligned} S_{94} &= \frac{94}{2}(81 + 360) \\ &= 47(441) \\ &= 20{,}727 \end{aligned}$$

39.

$$\begin{aligned} g(t) &= 5 - t \\ g(1) &= 5 - 1 = 4 \\ g(51) &= 5 - 51 = -46 \end{aligned}$$

$$g(1) + g(2) + g(3) + \dots + g(51) = S_{51}$$

$$\begin{aligned} S_n &= \frac{n}{2}(a_1 + a_n) \\ S_{51} &= \frac{51}{2}(g(1) + g(51)) \\ &= \frac{51}{4}[4 + (-46)] = -1{,}071 \end{aligned}$$

41.

$$\begin{aligned} g(x) &= 4 - 3x \\ g(1) &= 4 - 3 \cdot 1 = 1 \\ g(100) &= 4 - 3 \cdot 100 = -296 \end{aligned}$$

$$g(1) + g(2) + g(3) + \dots + g(100) = S_{100}$$

$$\begin{aligned} S_n &= \frac{n}{2}(a_1 + a_n) \\ S_{100} &= \frac{100}{2}(g(1) + g(100)) \\ &= \frac{100}{2}[1 + (-296)] = -14{,}750 \end{aligned}$$

43. a, x, b will be an arithmetic sequence if $b - x = x - a = d$. In this case,

$$\begin{aligned} 60 - x &= x - 10 \\ 60 - 2x &= -10 \\ -2x &= -70 \\ x &= 35 \end{aligned}$$

45. a, x, b will be an arithmetic sequence if $b - x = x - a = d$. In this case,

$$\begin{aligned} 58 - x &= x - 11 \\ 58 - 2x &= -11 \\ -2x &= -69 \\ x &= 34.5 \end{aligned}$$

47. First, find n:

$$\begin{aligned} a_n &= a_1 + (n - 1)d \\ 134 &= 22 + (n - 1)2 \\ n &= 57 \end{aligned}$$

Now, find S_{57}:

$$\begin{aligned} S_n &= \frac{n}{2}(a_1 + a_n) \\ S_{57} &= \frac{57}{2}(22 + 134) = 4{,}446 \end{aligned}$$

49\. First, find n:

$a_n = a_1 + (n - 1)d$
$360 = 81 + (n - 1)3$
$279 = 3n - 3$
$n = 94$

Now, find S_{94}:

$S_n = \frac{n}{2}(a_1 + a_n)$

$S_{94} = \frac{94}{2}(81 + 360) = 20,727$

51\. The sequence of odd natural numbers is an arithmetic sequence with $d = 2$.

$S_n = \frac{n}{2}[2a_1 + (n - 1)d]$

$S_n = \frac{n}{2}[2 \cdot 1 + (n - 1)2]$

$S_n = \frac{n}{2}[2 + 2n - 2]$

$S_n = \frac{n}{2} \cdot 2n$

$S_n = n^2$

53\. The sequence of multiples of 3 is an arithmetic sequence with $d = 3$, $a_1 = 3$.

$S_n = \frac{n}{2}[2a_1 + (n - 1)d]$

$S_n = \frac{n}{2}[2 \cdot 3 + (n - 1)3]$

$S_n = \frac{n}{2} \cdot 3[2 + (n - 1)]$

$S_n = \frac{3n}{2}(n + 1)$

55\. We have an arithmetic sequence, 16, 48, 80, ..., with $a_1 = 16$, $d = 32$.

(A) This requires a_{11}: $a_n = a_1 + (n - 1)d$

$a_{11} = 16 + (11 - 1)32 = 336$ feet

(B) This requires S_{11}: $S_n = \frac{n}{2}(a_1 + a_n)$

$S_{11} = \frac{11}{2}(16 + 336) = 1,936$ feet

(C) This requires S_t: $S_n = \frac{n}{2}[2a_1 + (n - 1)d]$

$S_t = \frac{t}{2}[2a_1 + (t - 1)32]$

$= \frac{t}{2}[32 + (t - 1)32]$

$= \frac{t}{2} \cdot 32t = 16t^2$ feet

57\. $a_1 = 1^2 - (1 - 1)^2 = 1^2 - 0^2 = 1 - 0 = 1$
$a_2 = 2^2 - (2 - 1)^2 = 2^2 - 1^2 = 4 - 1 = 3$
$a_3 = 3^2 - (3 - 1)^2 = 3^2 - 2^2 = 9 - 4 = 5$
$a_4 = 4^2 - (4 - 1)^2 = 4^2 - 3^2 = 16 - 9 = 7$
$a_5 = 5^2 - (5 - 1)^2 = 5^2 - 4^2 = 25 - 16 = 9$

These are the differences between successive values of $y = x^2$, x an integer. In going from $x = 5$ to $x = 6$, we move up $6^2 - 5^2 = 36 - 25 = 11$ units.

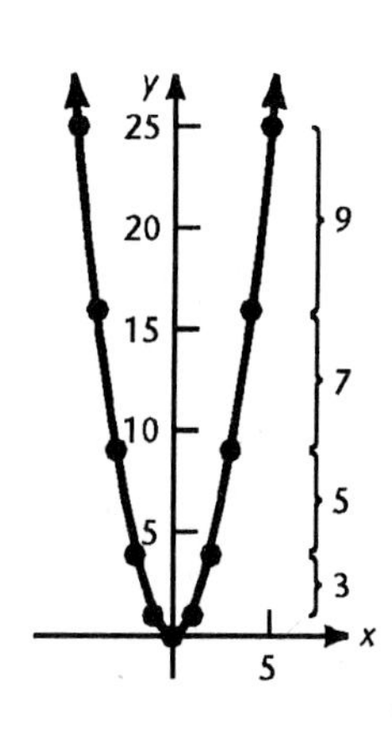

Exercise 11-3 Geometric Sequences and Series

Key Ideas and Formulas

A sequence $a_1, a_2, a_3, \dots, a_n, \dots$ is called a **geometric sequence** if there is a nonzero constant r, called the **common ratio**, such that each term is r times the preceding term.

$a_r = ra_{n-1}$ for every $n > 1$.

nth Term formula: $a_n = a_1r^{n-1}$

Sum Formulas: $S_n = \dfrac{a_1 - a_1r^n}{1 - r} = \dfrac{a_1(1 - r^n)}{1 - r}$ Use when given the number of terms n, the first term a_1, and the common ratio r.

$r \neq 1$

$S_n = \dfrac{a_1 - ra_n}{1 - r}$ $r \neq 1$ Use when given the number of terms n, the first term a_1, and the last term a_n.

Sum of an infinite geometric series: If $|r| < 1$

$$S_\infty = \frac{a_1}{1 - r}$$

If $|r| \geq 1$, an infinite geometric series has no sum.

---

1. $\frac{-4}{2} = -2$, $\frac{8}{-4} = 2$. Since these are equal, this is a geometric sequence with $r = -2$.

3. $\frac{-16}{-11} = \frac{16}{11}$, $\frac{-21}{-16} = \frac{21}{16}$. Since these are different, this is not a geometric sequence.

5. $\frac{-1}{5} = -\frac{1}{5}$, $\frac{-7}{-1} = 7$. Since these are different, this is not a geometric sequence.

7. $\frac{2}{3} \div \frac{1}{2} = \frac{4}{3}$, $\frac{3}{4} \div \frac{2}{3} = \frac{9}{8}$. Since these are different, this is not a geometric sequence.

9. $a_2 = a_1r = (-6)\left(-\frac{1}{2}\right) = 3$

$a_3 = a_2r = 3\left(-\frac{1}{2}\right) = -\frac{3}{2}$

$a_4 = a_3r = \left(-\frac{3}{2}\right)\left(-\frac{1}{2}\right) = \frac{3}{4}$

11. $a_n = a_1r^{n-1}$

$a_{10} = 81\left(\frac{1}{3}\right)^{10-1} = \frac{1}{243}$

13. $S_n = \dfrac{a_1 - ra_n}{1 - r}$

$S_7 = \dfrac{3 - 3(2{,}187)}{1 - 3} = 3{,}279$

17. $S_n = \dfrac{a_1 - a_1r^n}{1 - r}$

$S_{10} = \dfrac{5 - 5(-2)^{10}}{1 - (-2)}$

$= \dfrac{-5{,}115}{3} = -1{,}705$

21. $S_n = \dfrac{a_1 - a_1r^n}{1 - r}$

First, note $a_1 = (-3)^{1-1} = 1$

$r = \dfrac{a_2}{a_1} = \dfrac{(-3)^{2-1}}{(-3)^{1-1}} = -3$

$S_7 = \dfrac{1 - 1(-3)^7}{1 - (-3)} = \dfrac{2188}{4} = 547$

25. $r = \dfrac{a_2}{a_1} = \dfrac{-5}{1} = -5$

$S_n = \dfrac{a_1 - ra_n}{1 - r} = \dfrac{1 - (-5)(15{,}625)}{1 - (-5)}$

$= 13{,}021$

29. $r = \dfrac{1}{3} \div 1 = \dfrac{1}{3}$

$S_n = \dfrac{a_1 - ra_n}{1 - r} = \dfrac{1 - \frac{1}{3}\cdot\frac{1}{243}}{1 - \frac{1}{3}}$

$S_n = \dfrac{729 - 1}{729 - 243} = \dfrac{728}{486}$

$S_n = \dfrac{364}{243} \approx 1.4979$

15. $a_n = a_1r^{n-1}$

$1 = 100r^{6-1}$

$\dfrac{1}{100} = r^5$

$r = \sqrt[5]{0.01} = 10^{-2/5} = 0.398$

19. First, find r:

$a_n = a_1r^{n-1}$

$\dfrac{8}{3} = 9r^{4-1}$

$\dfrac{8}{27} = r^3$

$r = \dfrac{2}{3}$

$a_2 = ra_1 = \dfrac{2}{3}(9) = 6$

$a_3 = ra_2 = \dfrac{2}{3}(6) = 4$

23. $r = \dfrac{a_2}{a_1} = \dfrac{6}{2} = 3$

$S_n = \dfrac{a_1 - ra_n}{1 - r} = \dfrac{2 - 3\cdot162}{1 - 3} = 242$

27. $r = 1 \div \dfrac{1}{4} = 4$

$S_n = \dfrac{a_1 - ra_n}{1 - r} = \dfrac{\frac{1}{4} - 4(1024)}{1 - 4}$

$= 1365.25$

31. $r = 1.1 \div 1 = 1.1$

$S_n = \dfrac{a_1 - ra_n}{1 - r} = \dfrac{1 - 1.1(2.143\,58881)}{1 - 1.1}$

$S_n = 13.579\,476\,9$

33. $\frac{a_2}{a_1} = \frac{1}{3} = r, \ |r| < 1$

Therefore, this infinite geometric series has a sum.

$S_\infty = \frac{a_1}{1 - r} = \frac{3}{1 - \frac{1}{3}} = \frac{9}{2}$

35. $\frac{a_2}{a_1} = \frac{4}{2} = 2 = r \geq 1$

Therefore, this infinite geometric series has no sum.

37. $\frac{a_2}{a_1} = \left(-\frac{1}{2}\right) \div 2 = -\frac{1}{4} = r, \ |r| < 1$

Therefore, this infinite geometric series has a sum.

$S_\infty = \frac{a_1}{1 - r} = \frac{2}{1 - \left(-\frac{1}{4}\right)} = \frac{8}{5}$

39. $\frac{a_2}{a_1} = \frac{1}{4} \div 1 = \frac{1}{4} = r, \ |r| < 1$

Therefore, this infinite geometric series has a sum.

$S_\infty = \frac{a_1}{1 - r} = \frac{1}{1 - \frac{1}{4}} = \frac{4}{3}$

41. $\frac{a_2}{a_1} = \frac{5}{2} \div 2 = \frac{5}{4} \geq 1$

Therefore, this infinite geometric series has no sum.

43. $\frac{a_2}{a_1} = \frac{a_3}{a_2} = r$ for a_1, a_2, a_3 to be a geometric sequence. Hence

$\frac{x}{10} = \frac{12.1}{x}$

$x^2 = 121$

$x = \pm 11$

Since x is to be between a and b, $x = 11$.

45. $\frac{a_2}{a_1} = \frac{a_3}{a_2} = r$ for a_1, a_2, a_3 to be a geometric sequence. Hence

$\frac{x}{8} = \frac{12}{x}$

$x^2 = 96$

$x = \pm\sqrt{96} \approx \pm 9.798$

Since x is to be between a and b, $x = \sqrt{96} \approx 9.798$

47. $\frac{a_2}{a_1} = \frac{a_3}{a_2} = r$ for a_1, a_2, a_3 to be a geometric sequence. Hence

$\frac{x}{11} = \frac{12}{x}$

$x^2 = 132$

$x = \pm\sqrt{132} \approx \pm 11.489$

Since x is to be between a and b, $x = \sqrt{132} \approx 11.489$.

49. $0.777\overline{7} = 0.7 + 0.07 + 0.007 + 0.0007 + \ldots$

This is an infinite geometric series with $a_1 = 0.7$ and $r = 0.1$. Thus,

$S_\infty = \frac{a_1}{1 - r} = \frac{0.7}{1 - 0.1} = \frac{0.7}{0.9} = \frac{7}{9}$

51. $0.5454\overline{54} = 0.54 + 0.0054 + 0.000054 + \ldots$

This is an infinite geometric series with $a_1 = 0.54$ and $r = 0.01$. Thus,

$S_\infty = \frac{a_1}{1 - r} = \frac{0.54}{1 - 0.01} = \frac{0.54}{0.99} = \frac{6}{11}$

53. $3.216216\overline{216} = 3 + 0.216 + 0.000216 + 0.000000216 + \ldots$. Therefore, we note: $0.216 + 0.000216 + 0.000000216 + \ldots$ is an infinite geometric series with $a_1 = 0.216$ and $r = 0.001$. Thus,

$$3.216216\overline{216} = 3 + S_\infty = 3 + \frac{a_1}{1 - r} = 3 + \frac{0.216}{1 - 0.001} = 3 + \frac{0.216}{0.999}$$
$$= 3\tfrac{8}{37} \text{ or } \frac{119}{37}$$

55. $0.384615\overline{384615} = 0.384615 + 0.000000384615, \ldots$. Therefore, this is an infinite geometric series with $a_1 = 0.384615$ and $r = 0.000001$. Thus,

$$S_\infty = \frac{a_1}{1 - r} = \frac{0.384615}{1 - 0.000001} = \frac{0.384615}{0.999999} = \frac{384,615}{999,999} = \frac{5}{13}$$

57. $0.05\overline{5} = 0.05 + 0.005 + 0.0005 + \ldots$. Therefore, this is an infinite geometric series with $a_1 = 0.05$ and $r = 0.1$. Thus

$$S_\infty = \frac{a_1}{1 - r} = \frac{0.05}{1 - 0.1} = \frac{0.05}{0.9} = \frac{1}{18}$$

59. $0.999\overline{9} = 0.9 + 0.09 + 0.009 + \ldots$. This is an infinite geometric series with $a_1 = 0.9$ and $r = 0.1$. Thus

$$S_\infty = \frac{0.9}{1 - 0.1} = \frac{0.9}{0.9} = 1$$

61. After one year, $A(1 + r)$ is present. Hence, the geometric sequence has $a_1 = A(1 + r)$. The ratio is given as $(1 + r)$, hence

$$\begin{aligned} a_n &= a_1(1 + r)^{n-1} \\ &= A(1 + r)(1 + r)^{n-1} \\ &= A(1 + r)^n \text{ is the amount present after n years} \end{aligned}$$

The time taken for P to double is represented by n years. We set $A = 2P$, then solve

$$\begin{aligned} 2P &= P(1 + 0.06)^n \\ 2 &= (1.06)^n \\ \log 2 &= n \log 1.06 \\ n &= \frac{\log 2}{\log 1.06} \\ &\approx 12 \text{ years} \end{aligned}$$

63. We are asked for the sum of a finite geometric series.

$$a_1 = P(1 + r)^n$$

$$R = \frac{a_2}{a_1} = \frac{P(1 + r)^{n-1}}{P(1 + r)^n} = \frac{1}{1 + r} \quad \text{using R for the ratio to avoid confusion of letters}$$

There are $(n + 1)$ terms, hence

$$S_{n+1} = \frac{a_1 - Ra_{n+1}}{1 - R} = \frac{P(1 + r)^n - \frac{1}{1 + r}P}{1 - \frac{1}{1 + r}}$$

Multiply numerator and denominator by $1 + r$, the lcd of internal denominators.

$$= \frac{P(1 + r)^n(1 + r) - P}{1 + r - 1}$$

$$= \frac{P(1 + r)^{n+1} - P}{r}$$

65. We are asked for the sum of an infinite geometric series.
a_1 = number of revolutions in the first minute = 300
$r = \frac{2}{3}$, $|r| < 1$, so the series has a sum

$$S_\infty = \frac{a_1}{1 - r} = \frac{300}{1 - \frac{2}{3}} = 900 \text{ revolutions}$$

67. The vertical distance consists of

first drop = 5 ft
second drop = 0.7(5 ft)
third drop = 0.7(0.7)5 ft
first bounce = 0.7(5 ft)
second bounce = 0.7(0.7)(5 ft)
third bounce, etc.

Thus, the ball s travels

$$5 + 0.7(5) + 0.7(5) + (0.7)^2 5 + (0.7)^2 5 + (0.7)^3 5 + (0.7)^3 5 + \dots$$
$$= 5 + 0.7(10) + (0.7)^2 10 + (0.7)^3 10 + \dots$$

$0.7(10) + (0.7)^2(10) + (0.7)^3 10 + \dots$ is an infinite geometric series with $a_1 = 0.7(10)$ and $r = 0.7$. Therefore, the ball travels

$$5 + S_\infty = 5 + \frac{a_1}{1 - r} = 5 + \frac{0.7(10)}{1 - 0.7} = 5 + \frac{7}{0.3} = \frac{8.5}{0.3} = \frac{85}{3} = 28\tfrac{1}{3} \text{ feet.}$$

69. We are required to find the sum of an infinite geometric series.
a_1 = \$800,000
$r = 0.8$, $|r| < 1$, so the series has a sum

$$S_\infty = \frac{a_1}{1 - r} = \frac{\$800{,}000}{1 - 0.8} = \frac{\$800{,}000}{0.2} = \$4{,}000{,}000$$

Exercise 11-4 Binomial Formula

Key Ideas and Formulas

$$\binom{n}{r} = \frac{n!}{r!(n - r)!} = \frac{n(n - 1)(n - 2)\dots(n - r + 1)}{r(r - 1)\dots 2\cdot 1}$$

For $n \in N$, $n! = n(n - 1)\dots 2\cdot 1$, n factorial

$1! = 0! = 1$

$n! = n\cdot(n - 1)!$

Binomial Formula

$$(a + b)^n = \sum_{k=0}^{n} \binom{n}{k} a^{n-k} b^k \quad n \in N,\ n \ge 1$$

$$(a + b)^0 = 1$$

$$(a + b)^1 = a + b$$

$$(a + b)^2 = a^2 + 2ab + b^2$$

$$(a + b)^3 = a^3 + 3a^2b + 3ab^2 + b^3$$

$$(a + b)^4 = a^4 + 4a^3b + 6a^2b^2 + 4ab^3 + b^4$$

$$(a + b)^5 = a^5 + 5a^4b + 10a^3b^2 + 10a^2b^3 + 5ab^4 + b^5$$

1. $6! = 6 \cdot 5 \cdot 4 \cdot 3 \cdot 2 \cdot 1 = 720$

3. $\frac{20!}{19!} = \frac{20 \cdot 19!}{19!} = 20$ COMMON ERROR: $\frac{20}{19}$. It is incorrect to "cancel" the factorial symbol.

5. $\frac{10!}{7!} = \frac{10 \cdot 9!}{7!} = \frac{10 \cdot 9 \cdot 8!}{7!} = \frac{10 \cdot 9 \cdot 8 \cdot 7!}{7!} = 10 \cdot 9 \cdot 8 = 720$

7. $\frac{6!}{4!2!} = \frac{6 \cdot 5 \cdot 4 \cdot 3 \cdot 2 \cdot 1}{4 \cdot 3 \cdot 2 \cdot 1 \cdot 2 \cdot 1} = \frac{6 \cdot 5}{2 \cdot 1} = 15$

9. $\frac{9!}{0!(9 - 0)!} = \frac{9!}{1(9!)} = 1$

11. $\frac{8!}{2!(8 - 2)!} = \frac{8!}{2!6!} = \frac{8 \cdot 7 \cdot 6 \cdot 5 \cdot 4 \cdot 3 \cdot 2 \cdot 1}{2 \cdot 1 \cdot 6 \cdot 5 \cdot 4 \cdot 3 \cdot 2 \cdot 1} = \frac{8 \cdot 7}{2 \cdot 1} = 28$

13. Since $n! = n(n - 1)!$, $\frac{n!}{(n - 1)!} = n$. Hence $9 = \frac{9!}{(9 - 1)!} = \frac{9!}{8!}$.

15. $6 \cdot 7 \cdot 8 = \frac{1 \cdot 2 \cdot 3 \cdot 4 \cdot 5 \cdot 6 \cdot 7 \cdot 8}{1 \cdot 2 \cdot 3 \cdot 4 \cdot 5} = \frac{8!}{5!}$

17. $20 \cdot 19 = \frac{20 \cdot 19 \cdot 18 \cdot 17 \cdot 16 \cdot \ldots \cdot 1}{18 \cdot 17 \cdot 16 \cdot \ldots \cdot 1} = \frac{20!}{18!}$

19. $31 \cdot 30 \cdot 29 \cdot 28 = \frac{31 \cdot 30 \cdot 29 \cdot 28 \cdot 27 \cdot 26 \cdot \ldots \cdot 1}{27 \cdot 26 \cdot \ldots \cdot 1} = \frac{31!}{27!}$

21. $\binom{9}{5} = \frac{9!}{5!(9 - 5)!} = \frac{9!}{5!4!} = \frac{9 \cdot 8 \cdot 7 \cdot 6 \cdot 5!}{5!4 \cdot 3 \cdot 2 \cdot 1} = \frac{9 \cdot 8 \cdot 7 \cdot 6}{4 \cdot 3 \cdot 2 \cdot 1} = 126$

23. $\binom{6}{5} = \frac{6!}{5!(6 - 5)!} = \frac{6!}{5!1!} = \frac{6 \cdot 5!}{5! \cdot 1} = 6$

25. $\binom{9}{9} = \frac{9!}{9!(9 - 9)!} = \frac{9!}{9!0!} = \frac{1}{0!} = \frac{1}{1} = 1$

27. $\binom{17}{13} = \frac{17!}{13!(17 - 13)!} = \frac{17!}{13!4!} = \frac{17 \cdot 16 \cdot 15 \cdot 14 \cdot 13!}{13!4 \cdot 3 \cdot 2 \cdot 1} = \frac{17 \cdot 16 \cdot 15 \cdot 14}{4 \cdot 3 \cdot 2 \cdot 1} = 2,380$

29. $\binom{36}{2} = \frac{36!}{2!(36 - 2)!} = \frac{36!}{2!34!} = \frac{36 \cdot 35 \cdot 34!}{2 \cdot 1 \cdot 34!} = \frac{36 \cdot 35}{2 \cdot 1} = 630$

31. $\binom{32}{3} = \frac{32!}{3!(32 - 3)!} = \frac{32!}{3!29!} = \frac{32 \cdot 31 \cdot 30 \cdot 29!}{3 \cdot 2 \cdot 1 \cdot 29!} = \frac{32 \cdot 31 \cdot 30}{3 \cdot 2 \cdot 1} = 4960$

33. $(a + b)^7 = \sum_{k=0}^{7} \binom{7}{k} a^{7-k} b^k$

$= \binom{7}{0} a^7 + \binom{7}{1} a^6 b + \binom{7}{2} a^5 b^2 + \binom{7}{3} a^4 b^3 + \binom{7}{4} a^3 b^4 + \binom{7}{5} a^2 b^5 + \binom{7}{6} ab^6 + \binom{7}{7} b^7$

$= a^7 + 7a^6 b + 21a^5 b^2 + 35a^4 b^3 + 35a^3 b^4 + 21a^2 b^5 + 7ab^6 + b^7$

35. $(a + b)^{10} = \sum_{k=0}^{10} \binom{10}{k} a^{10-k} b^k$

$= \binom{10}{0} a^{10} + \binom{10}{1} a^9 b + \binom{10}{2} a^8 b^2 + \binom{10}{3} a^7 b^3 + \binom{10}{4} a^6 b^4 + \binom{10}{5} a^5 b^5$

$+ \binom{10}{6} a^4 b^6 + \binom{10}{7} a^3 b^7 + \binom{10}{8} a^2 b^8 + \binom{10}{9} ab^9 + \binom{10}{10} b^{10}$

$= a^{10} + 10a^9 b + 45a^8 b^2 + 120a^7 b^3 + 210a^6 b^4 + 252a^5 b^5 + 210a^4 b^6 + 120a^3 b^7$
$+ 45a^2 b^8 + 10ab^9 + b^{10}$

37. $(u + v)^5 = \sum_{k=0}^{5} \binom{5}{k} u^{5-k} v^k$

$= \binom{5}{0} u^5 + \binom{5}{1} u^4 v + \binom{5}{2} u^3 v^2 + \binom{5}{3} u^2 v^3 + \binom{5}{4} uv^4 + \binom{5}{5} v^5$

$= u^5 + 5u^4 v + 10u^3 v^2 + 10u^2 v^3 + 5uv^4 + v^5$

39. $(y - 1)^4 = [y + (-1)]^4 = \sum_{k=0}^{4} \binom{4}{k} y^{4-k} (-1)^k$

$= \binom{4}{0} y^4 + \binom{4}{1} y^3 (-1) + \binom{4}{2} y^2 (-1)^2 + \binom{4}{3} y(-1)^3 + \binom{4}{4} (-1)^4$

$= y^4 - 4y^3 + 6y^2 - 4y + 1$

41. $(2x - y)^5 = [2x + (-y)]^5 = \sum_{k=0}^{5} \binom{5}{k} (2x)^{5-k} (-y)^k$

$= \binom{5}{0} (2x)^5 + \binom{5}{1} (2x)^4 (-y) + \binom{5}{2} (2x)^3 (-y)^2 + \binom{5}{3} (2x)^2 (-y)^3$

$+ \binom{5}{4} (2x)(-y)^4 + \binom{5}{5} (-y)^5$

$= (2x)^5 + 5(2x)^4 (-y) + 10(2x)^3 (-y)^2 + 10(2x)^2 (-y)^3 + 5(2x)(-y)^4 + (-y)^5$

$= 32x^5 + 5(16x^4)(-y) + 10(8x^3)(y^2) + 10(4x^2)(-y^3) + 10xy^4 - y^5$

$= 32x^5 - 80x^4 y + 80x^3 y^2 - 40x^2 y^3 + 10xy^4 - y^5$

43. In the expansion of $(a + b)^n$, the exponent of b in the r-th term is r - 1 and the exponent of a is n - (r - 1). Here r = 3, n = 16. Third term =

$$\binom{16}{2}a^{14}b^2 = \frac{16!}{2!14!}a^{14}b^2 = \frac{16 \cdot 15}{2 \cdot 1}a^{14}b^2 = 120a^{14}b^2$$

45. In the expansion of $(a + b)^n$, the exponent of b in the r-th term is r - 1 and the exponent of a is n - (r - 1). Here r = 4, n = 35. Fourth term =

$$\binom{35}{3}a^{32}b^3 = \frac{35!}{3!32!}a^{32}b^3 = \frac{35 \cdot 34 \cdot 33}{3 \cdot 2 \cdot 1}a^{32}b^3 = 6545a^{32}b^3$$

47. In the expansion of $(a + b)^n$, the exponent of b in the r-th term is r - 1 and the exponent of a is n - (r - 1). Here, r = 7, n = 15.

Seventh term $= \binom{15}{6}u^9v^6 = \frac{15!}{9!6!}u^9v^6 = \frac{15 \cdot 14 \cdot 13 \cdot 12 \cdot 11 \cdot 10}{6 \cdot 5 \cdot 4 \cdot 3 \cdot 2 \cdot 1}u^9v^6 = 5{,}005u^9v^6$

49. In the expansion of $(a + b)^n$, the exponent of b in the r-th term is r - 1 and the exponent of a is n - (r - 1). Here, r = 11, n = 12.

Eleventh term $= \binom{12}{0}(2m)^2n^{10} = \frac{12!}{10!2!}4m^2n^{10} = \frac{12 \cdot 11}{2 \cdot 1}4m^2n^{10} = 264m^2n^{10}$

51. In the expansion of $(a + b)^n$, the exponent of b in the r-th term is r - 1 and the exponent of a is n - (r - 1). Here, r = 7, n = 12.

Seventh term $= \binom{12}{6}\left(\frac{w}{2}\right)(-2)^6 = \frac{12!}{6!6!}\frac{w^6}{2^6}2^6 = \frac{12 \cdot 11 \cdot 10 \cdot 9 \cdot 8 \cdot 7 \cdot 6!}{6 \cdot 5 \cdot 4 \cdot 3 \cdot 2 \cdot 1 \cdot 6!}w^6 = 924w^6$

53. $(1.1)^5 = (1 + 0.1)^5 = \sum_{k=0}^{5}\binom{5}{k}1^{5-k}(0.1)^k$ $\quad (1^{5-k} = 1$ for any k)

$$= \sum_{k=0}^{5}\binom{5}{k}(0.1)^k$$

$$= \binom{5}{0} + \binom{5}{1}(0.1) + \binom{5}{2}(0.1)^2 + \binom{5}{3}(0.1)^3 + \binom{5}{4}(0.1)^4 + \binom{5}{5}(0.1)^5$$

$$= 1 + 0.5 + 0.1 + 0.01 + 0.0005 + 0.00001$$

$$= 1.61051$$

55. $(1.01)^{10} = (1 + 0.01)^{10} = \sum_{k=0}^{10}\binom{10}{k}1^{10-k}(0.01)^k$ $\quad (1^{10-k} = 1$ for any k)

$$= \sum_{k=0}^{10}\binom{10}{k}(0.01)^k$$

$$= \binom{10}{0} + \binom{10}{1}(0.01) + \binom{10}{2}(0.01)^2 + \binom{10}{3}(0.01)^3 + \binom{10}{4}(0.01)^4 + \binom{10}{5}(0.01)^5$$

$$+ \binom{10}{6}(0.01)^6 + \binom{10}{7}(0.01)^7 + \binom{10}{8}(0.01)^8 + \binom{10}{9}(0.01)^9 + \binom{10}{10}(0.01)^{10}$$

$= 1 + 0.1 + 0.0045 + 0.00012 + 0.0000021 + 0.000000\,0252$
$+ 0.000\,000\,000\,21 + 0.000\,000\,000\,0012 + 0.000\,000\,000\,000\,000\,45$
$+ 0.000\,000\,000\,000\,000\,001 + 0.000\,000\,000\,000\,000\,000\,01$

$= 1.104\,622\,125\,411\,204\,510\,01$

57. $\binom{n}{r} = \frac{n!}{r!(n - r)!} = \frac{n!}{(n - r)!r!} = \frac{n!}{(n - r)![n - (n - r)]!} = \binom{n}{n - r}$

59. $(1 + i)^6 = \sum_{k=0}^{6} \binom{6}{k} 1^{6-k} i^6 \qquad (1^{6-k} = 1 \text{ for any } k)$

$= \sum_{k=0}^{6} \binom{6}{k} i^6$

$= \binom{6}{0} + \binom{6}{1}i + \binom{6}{2}i^2 + \binom{6}{3}i^3 + \binom{6}{4}i^4 + \binom{6}{5}i^5 + \binom{6}{6}i^6$

$= 1 + 6i + 15i^2 + 20i^3 + 15i^4 + 6i^5 + i^6$

$= 1 + 6i - 15 - 20i + 15 + 6i - 1$

$= -8i$

Exercise 11-5 REVIEW EXERCISE

1. Since $\frac{-8}{16} = \frac{4}{-8} = -\frac{1}{2}$, this is a geometric sequence. $r = -\frac{1}{2}$

2. Since $7 - 5 = 9 - 7 = 2$, this is an arithmetic sequence. $d = 2$

3. Since $-5 - (-8) = -2 - (-5) = 3$, this is an arithmetic sequence. $d = 3$

4. Since $\frac{3}{2} \neq \frac{5}{3}$ and $3 - 2 \neq 5 - 3$, this is neither an arithmetic nor a geometric sequence.

5. Since $\frac{2}{-1} = \frac{-4}{2} = -2$, this is a geometric sequence. $r = -2$

6. Since $2 - 1 \neq 4 - 2$ and $\frac{2}{1} \neq \frac{7}{4}$, this is neither an arithmetic nor a geometric sequence.

7. Since $7 - 10 = 4 - 7 = 1 - 4 = -3$, this is an arithmetic sequence. $d = -3$

8. Since $\frac{4}{3} = \frac{16}{3} \div 4 = \frac{64}{9} \div \frac{16}{3} = \frac{4}{3}$, this is a geometric sequence. $r = \frac{4}{3}$

9. Since $4 - 1 \neq 9 - 4$ and $\frac{4}{1} \neq \frac{9}{4}$, this is neither an arithmetic nor a geometric sequence.

10. Since $\frac{5}{10} = \frac{2.5}{5} = \frac{1.25}{2.5} = \frac{1}{2}$, this is a geometric sequence. $r = \frac{1}{2}$

11. $a_n = 2n + 3$
$a_1 = 2 \cdot 1 + 3 = 5$
$a_2 = 2 \cdot 2 + 3 = 7$
$a_3 = 2 \cdot 3 + 3 = 9$
$a_4 = 2 \cdot 4 + 3 = 11$

12. $a_n = 32\left(\frac{1}{2}\right)^n$
$a_1 = 32\left(\frac{1}{2}\right)^1 = 16$
$a_2 = 32\left(\frac{1}{2}\right)^2 = 8$
$a_3 = 32\left(\frac{1}{2}\right)^3 = 4$
$a_4 = 32\left(\frac{1}{2}\right)^4 = 2$

13. $a_n = n^2 - 4$
$a_1 = 1^2 - 4 = -3$
$a_2 = 2^2 - 4 = 0$
$a_3 = 3^2 - 4 = 5$
$a_4 = 4^2 - 4 = 12$

14. $a_n = n^2 + n$
$a_1 = 1^2 + 1 = 2$
$a_2 = 2^2 + 2 = 6$
$a_3 = 3^2 + 3 = 12$
$a_4 = 4^2 + 4 = 20$

15. $a_n = (-1)^n(n + 1)$
$a_1 = (-1)^1(1 + 1) = -2$
$a_2 = (-1)^2(2 + 1) = 3$
$a_3 = (-1)^3(3 + 1) = -4$
$a_4 = (-1)^4(4 + 1) = 5$

16. $a_n = \frac{(-1)^{n+1}}{n}$
$a_1 = \frac{(-1)^{1+1}}{1} = 1$
$a_2 = \frac{(-1)^{2+1}}{2} = -\frac{1}{2}$
$a_3 = \frac{(-1)^{3+1}}{3} = \frac{1}{3}$
$a_4 = \frac{(-1)^{4+1}}{4} = -\frac{1}{4}$

17. $a_1 = -8,\ a_n = a_{n-1} + 3,\ n \geq 2$
$a_1 = -8$
$a_2 = a_1 + 3 = -8 + 3 = -5$
$a_3 = a_2 + 3 = -5 + 3 = -2$
$a_4 = a_3 + 3 = -2 + 3 = 1$

18. $a_1 = -1,\ a_n = (-2)a_{n-1},\ n \geq 2$
$a_1 = -1$
$a_2 = (-2)a_1 = (-2)(-1) = 2$
$a_3 = (-2)a_2 = (-2)2 = -4$
$a_4 = (-2)a_3 = (-2)(-4) = 8$

19. This is an arithmetic sequence with $d = 2$. Hence,
$a_n = a_1 + (n - 1)d$
$a_{10} = 5 + (10 - 1)d = 23$

20. This is a geometric sequence with $r = \frac{1}{2}$. Hence,
$a_n = a_1 r^{n-1}$
$a_{10} = 16\left(\frac{1}{2}\right)^{10-1} = \frac{1}{32}$

21. $a_{10} = 10^2 - 4 = 96$

22. $a_{10} = 10^2 + 10 = 110$

23. $a_{10} = (-1)^{10}(10 + 1) = 11$

24. $a_{10} = \frac{(-1)^{10+1}}{10} = -\frac{1}{10}$

25. This is an arithmetic sequence with $d = 3$. Hence,
$a_n = a_1 + (n - 1)d$
$a_{10} = -8 + (10 - 1)3 = 19$

26. This is a geometric sequence with $r = -2$. Hence,
$a_n = a_1 r^{n-1}$
$a_{10} = (-1)(-2)^{10-1} = 512$

27. This is an arithmetic series with $n = 10$, $a_1 = 5$, $a_{10} = 23$

$$S_n = \frac{n}{2}(a_1 + a_n)$$

$$S_{10} = \frac{10}{2}(5 + 23) = 140$$

28. This is a geometric series with $r = \frac{1}{2}$, $a_1 = 16$, $a_{10} = \frac{1}{32}$

$$S_n = \frac{a_1 - ra_n}{1 - r}$$

$$S_{10} = \frac{16 - \frac{1}{2}\left(\frac{1}{32}\right)}{1 - \frac{1}{2}} = \frac{16 - \frac{1}{64}}{\frac{1}{2}} = 31\frac{31}{32}$$

29. This is an arithmetic series with $n = 10$, $a_1 = -8$, $a_{10} = 19$

$$S_n = \frac{n}{2}(a_1 + a_n)$$

$$S_{10} = \frac{10}{2}(-8 + 19) = 55$$

30. This is a geometric series with $r = -2$, $a_1 = -1$, $a_{10} = 512$

$$S_n = \frac{a_1 - ra_n}{1 - r}$$

$$S_{10} = \frac{-1 - (-2)(512)}{1 - (-2)} = 341$$

31. $S_5 = (1^2 - 4) + (2^2 - 4) + (3^2 - 4) + (4^2 - 4) + (5^2 - 4)$
$= (-3) + 0 + 5 + 12 + 21 = 35$

32. $S_5 = (1^2 + 1) + (2^2 + 2) + (3^2 + 3) + (4^2 + 4) + (5^2 + 5)$
$= 2 + 6 + 12 + 20 + 30 = 70$

33. $S_5 = (-2) + 3 + (-4) + 5 + (-6) = -4$

34. $S_5 = 1 + \left(-\frac{1}{2}\right) + \frac{1}{3} + \left(-\frac{1}{4}\right) + \frac{1}{5} = \frac{47}{60}$

35. $6! = 6 \cdot 5 \cdot 4 \cdot 3 \cdot 2 \cdot 1 = 720$

36. $\frac{22!}{19!} = \frac{22 \cdot 21 \cdot 20 \cdot 19!}{19!} = 9{,}240$

37. $\frac{7!}{2!(7 - 2)!} = \frac{7!}{2!5!} = \frac{7 \cdot 6 \cdot 5!}{2 \cdot 1 \cdot 5!} = 21$

38. $\binom{7}{2} = \frac{7!}{2!5!} = 21$ from problem 37.

39. $\binom{8}{6} = \frac{8!}{6!2!} = \frac{8 \cdot 7 \cdot 6!}{6!2 \cdot 1} = 28$

40. $\binom{6}{5} = \frac{6!}{5!1!} = \frac{6!}{5!} = 6$

41. First, find d:

$$\begin{aligned} a_n &= a_1 + (n - 1)d \\ 28 &= 3 + (11 - 1)d \\ 28 &= 3 + 10d \\ d &= 2.5 \\ a_5 &= 3 + (5 - 1)2.5 = 13 \end{aligned}$$

42.

$$\begin{aligned} a_n &= a_1 + (n - 1)d \\ a_3 &= a_1 + 2d \\ a_{103} &= a_1 + 102d \end{aligned}$$

Eliminating d between these two statements by addition, we have

$$\begin{aligned} -51a_3 &= -51a_1 - 102d \\ a_{103} &= a_1 + 102d \\ \hline -51a_3 + a_{103} &= -50a_1 \end{aligned}$$

$$a_1 = \frac{-51a_3 + a_{103}}{-50} = \frac{-51(3) + 153}{-50} = 0$$

43. First, find d:

$a_n = a_1 + (n - 1)d$
$70 = -30 + (51 - 1)d$
$70 = -30 + 50d$
$d = 2$
$a_{25} = -30 + (25 - 1)2 = 18$

44. First, find r:

$a_n = a_1 r^{n-1}$
$40.5 = \frac{1}{2} r^{5-1}$
$81 = r^4$
$r = \pm 3$
$a_7 = a_1 r^{7-1} = \frac{1}{2}(\pm 3)^6 = \frac{729}{2} = 364.5$

45.

$a_n = a_1 r^{n-1}$
$a_2 = a_1 r^1$
$a_6 = a_1 r^5$
$\sqrt{5} = a_1 r$
$25\sqrt{5} = a_1 r^5$

From $\sqrt{5} = a_1 r_1$ we see that $a_1 = \frac{\sqrt{5}}{r}$. Substituting into $25\sqrt{5} = a_1 r^5$, we obtain

$25\sqrt{5} = \frac{\sqrt{5}}{r} r^5$
$25\sqrt{5} = r^4\sqrt{5}$
$25 = r^4$
$r = \pm\sqrt[4]{25}$
$r = \pm\sqrt{5}.$

Since r is to be positive, $r = \sqrt{5}$. Thus, $a_1 = \frac{\sqrt{5}}{r} = \frac{\sqrt{5}}{\sqrt{5}} = 1.$

46. $a_n = a_1 r^{n-1}$
$6 = 2r^{3-1}$
$6 = 2r^2$
$3 = r^2$
$r = \pm\sqrt{3}$

Since r is to be positive,
$r = \sqrt{3}$

47. $a_n = a_1 + (n - 1)d$
$\frac{3}{2} = a_1 + 4d$
$4 = a_1 + 14d$

Eliminating a_1 between these two statements by addition, we have

$-\frac{3}{2} = -a_1 - 4d$
$4 = a_1 + 14d$

$\frac{5}{2} = 10d$
$d = \frac{1}{4}$

48. $a_n = a_1 + (n - 1)d$
$a_{80} = a_1 + 79d$
$a_{60} = a_1 + 59d$

Therefore,

$a_{80} - a_{60} = (a_1 + 79d) - (a_1 + 59d)$
$= 20d$

In this case, $a_{80} - a_{60} = 400$. Thus,

$400 = 20d$
$d = 20$

49. $S_4 = \frac{3}{2\cdot 1} + \frac{3}{2\cdot 2} + \frac{3}{2\cdot 3} + \frac{3}{2\cdot 4} = \frac{3}{2} + \frac{3}{4} + \frac{1}{2} + \frac{3}{8} = \frac{25}{8}$

50. $S_4 = \frac{(-1)^{1+1}}{2^1} + \frac{(-1)^{2+1}}{2^2} + \frac{(-1)^{3+1}}{2^3} + \frac{(-1)^{4+1}}{2^4} = \frac{1}{2} - \frac{1}{4} + \frac{1}{8} - \frac{1}{16} = \frac{5}{16}$

51. $S_{10} = (2\cdot 1 - 8) + (2\cdot 2 - 8) + (2\cdot 3 - 8) + (2\cdot 4 - 8) + (2\cdot 5 - 8) + (2\cdot 6 - 8) + (2\cdot 7 - 8) + (2\cdot 8 - 8) + (2\cdot 9 - 8) + (2\cdot 10 - 8)$

$= (-6) + (-4) + (-2) + 0 + 2 + 4 + 6 + 8 + 10 + 12 = 30$

52. $S_7 = \frac{16}{2^1} + \frac{16}{2^2} + \frac{16}{2^3} + \frac{16}{2^4} + \frac{16}{2^5} + \frac{16}{2^6} + \frac{16}{2^7} = 8 + 4 + 2 + 1 + \frac{1}{2} + \frac{1}{4} + \frac{1}{8} = 15\frac{7}{8}$

53. $S_n = \frac{a_1 - ra_n}{1 - r}$ $\quad a_1 = \frac{1}{2}$ $\quad r = \frac{7}{6} \div \frac{1}{2} = \frac{7}{3}$ $\quad a_n = \frac{2401}{162}$

$$S_5 = \frac{\frac{1}{2} - \frac{7}{3} \cdot \frac{2401}{162}}{1 - \frac{7}{3}} = \frac{\frac{1}{2} - \frac{16807}{486}}{1 - \frac{7}{3}}$$

Multiply numerator and denominator by 486, the lcd of internal fractions

$$= \frac{243 - 16807}{486 - 1134} = \frac{16564}{648} = \frac{4141}{162} \approx 25.56$$

54. $S_n = \frac{a_1 - ra_n}{1 - r}$ $\quad a_1 = 40$ $\quad r = \frac{1}{2}$ $\quad a_n = 40\left(\frac{1}{2}\right)^{11}$

$$S_{12} = \frac{40 - \frac{1}{2} \cdot 40\left(\frac{1}{2}\right)^{11}}{1 - \frac{1}{2}} = \frac{40 - 40\left(\frac{1}{2}\right)^{12}}{\frac{1}{2}} = \left[40 - 40\left(\frac{1}{2}\right)^{12}\right] \cdot 2$$

$$= 80 - 80\left(\frac{1}{2}\right)^{12} = 80 \cdot \frac{4095}{4096} = \frac{20,475}{256} \approx 79.98$$

55. This is an infinite geometric series with $a_1 = 40$, $r = \frac{1}{2}$ $\quad |r| < 1$.

$$S_\infty = \frac{a_1}{1 - r} = \frac{40}{1 - \frac{1}{2}} = 80$$

56. This is an infinite geometric series with $a_1 = 27$, $r = -\frac{18}{27} = -\frac{2}{3}$ $\quad |r| < 1$.

$$S_\infty = \frac{a_1}{1 - r} = \frac{27}{1 - \left(-\frac{2}{3}\right)} = \frac{81}{5}$$

57. This is an infinite geometric series with $a_1 = \frac{1}{3}$, $r = -\frac{1}{9} \div \frac{1}{3} = -\frac{1}{3}$ $\quad |r| < 1$.

$$S_\infty = \frac{a_1}{1 - r} = \frac{\frac{1}{3}}{1 - \left(-\frac{1}{3}\right)} = \frac{1}{4}$$

58. $\sum_{k=1}^{n} \frac{(-1)^{k+1}}{3^k}$

59. a_k: 10, 7, 4, 1, ...
This is an arithmetic sequence with $a_1 = 10$, $d = -3$. Thus,
$$a_k = 10 - (k - 1)3$$
To determine n, we note $a_n = -38$.
Thus,
$$-38 = 10 - (n - 1)3$$
$$n = 17$$
The sum is therefore:
$$\sum_{k=1}^{17} [10 - 3(k - 1)]$$

60. a_k: 3, 4, $\frac{16}{3}$, $\frac{64}{9}$
This is a geometric sequence with $a_1 = 3$, $r = \frac{4}{3}$, $n = 4$. The sum is therefore
$$\sum_{k=1}^{4} 3\left(\frac{4}{3}\right)^{k-1}$$

61. a_k: 1, 4, 9, 16, ...
$k = 1, 2, 3, 4, \ldots$
Comparing a_k with k, we see that $a_k = k^2$. The sum is therefore
$$\sum_{k=1}^{11} k^2$$

62. a_k: 10, 5, 2.5, ...
This is a geometric sequence with $a_1 = 10$, $r = \frac{5}{10} = \frac{1}{2}$. To determine n, we note $a_n = 0.3125$. Thus
$$a_n = a_1 r^{n-1}$$
$$0.3125 = 10\left(\frac{1}{2}\right)^{n-1}$$
$$0.03125 = \left(\frac{1}{2}\right)^{n-1}$$
$$\frac{1}{32} = \left(\frac{1}{2}\right)^{n-1}$$
$$\left(\frac{1}{2}\right)^5 = \left(\frac{1}{2}\right)^{n-1}$$
$$n - 1 = 5$$
$$n = 6$$
The sum is therefore
$$\sum_{k=1}^{6} 10\left(\frac{1}{2}\right)^{k-1}$$

63. $$\frac{20!}{18!(20 - 18)!} = \frac{20!}{18!2!} = \frac{20 \cdot 19 \cdot 18}{18!2 \cdot 1} = 190$$

64. $$\binom{16}{12} = \frac{16!}{12!(16 - 12)!} = \frac{16!}{12!4!} = \frac{16 \cdot 15 \cdot 14 \cdot 13 \cdot 12!}{12! \cdot 4 \cdot 3 \cdot 2 \cdot 1} = 1,820$$

65. $$\binom{11}{11} = \frac{11!}{11!(11 - 11)!} = \frac{11!}{11!0!} = 1$$

66. $$\binom{20}{2} = \frac{20!}{2!(20 - 2)!} = \frac{20!}{2!18!} = \frac{20 \cdot 19 \cdot 18!}{2 \cdot 1 \cdot 18!} = 190$$

67. $$\binom{15}{3} = \frac{15!}{3!(15 - 3)!} = \frac{15!}{3!12!} = \frac{15 \cdot 14 \cdot 13 \cdot 12!}{3 \cdot 2 \cdot 1 \cdot 12!} = \frac{15 \cdot 14 \cdot 13}{3 \cdot 2 \cdot 1} = 455$$

68. $(x - y)^5 = [x + (-y)]^5 = \sum_{k=0}^{5} \binom{5}{k} x^{5-k}(-y)^k$

$$= \binom{5}{0}x^5 + \binom{5}{1}x^4(-y) + \binom{5}{2}x^3(-y)^2 + \binom{5}{3}x^2(-y)^3 + \binom{5}{4}x(-y)^4 + \binom{5}{5}(-y)^5$$

$$= x^5 - 5x^4y + 10x^3y^2 - 10x^2y^3 + 5xy^4 - y^5$$

69. $(x + 1)^7 = \sum_{k=0}^{7} \binom{7}{k} x^{7-k}1^k$ $\quad (1^k = 1 \text{ for all } k)$

$$= \binom{7}{0}x^7 + \binom{7}{1}x^6 + \binom{7}{2}x^5 + \binom{7}{3}x^4 + \binom{7}{4}x^3 + \binom{7}{5}x^2 + \binom{7}{6}x + \binom{7}{7}$$

$$= x^7 + 7x^6 + 21x^5 + 35x^4 + 35x^3 + 21x^2 + 7x + 1$$

70. In the expansion of $(a + b)^n$, the exponent of b in the r-th term is $r - 1$ and the exponent of a is $n - (r - 1)$. Here, $r = 10$, $n = 12$.

$$\text{Tenth term} = \binom{12}{9}(2x)^3(-y)^9 = \frac{12!}{9!3!}(8x^3)(-y^9) = \frac{12\cdot 11\cdot 10\cdot 9!}{9!\cdot 3\cdot 2\cdot 1}(-8x^3y^9) = -1760x^3y^9$$

71. The term involving $a^{12}b^8$ has coefficient $\binom{20}{8}$. The term is

$$\binom{20}{8}a^{12}b^8 = \frac{20!}{8!(20 - 8)!}a^{12}b^8 = \frac{20!}{8!12!}a^{12}b^8 = 125{,}970a^{12}b^8$$

72. $0.7272\overline{72} = 0.72 + 0.0072 + 0.000072 + \ldots$
This is an infinite geometric sequence with $a_1 = 0.72$ and $r = 0.01$.

$$0.7272\overline{72} = S_\infty = \frac{a}{1 - r} = \frac{0.72}{1 - 0.01} = \frac{0.72}{0.99} = \frac{8}{11}$$

73. $1.234\overline{234} = 1 + 0.234 + 0.000\,234 + \ldots$. Therefore, we note:
$0.234 + 0.000234 + \ldots$ is an infinite geometric series with $a_1 = 0.234$ and $r = 0.001$. Thus,

$$1.234\overline{234} = 1 + S_\infty = 1 + \frac{a_1}{1 - r} = 1 + \frac{0.234}{1 - 0.001} = 1 + \frac{0.234}{0.999} = 1\tfrac{26}{111} \text{ or } \frac{137}{111}.$$

74. Since the second term is 8, we recognize this as the row of coefficients for $(a + b)^8$, thus:

$\binom{8}{0}, \binom{8}{1}, \binom{8}{2}, \binom{8}{3}, \binom{8}{4}, \binom{8}{5}, \binom{8}{6}, \binom{8}{7}, \binom{8}{8}$, that is,

1, 8, 28, 56, 70, 56, 28, 8, 1.

75. An arithmetic sequence is involved, with $a_1 = \frac{9}{2}$, $d = \frac{3g}{2} - \frac{g}{2} = g$.
Distance fallen during the twenty-fifth second = a_{25}.

$$a_n = a_1 + (n - 1)d$$

$$a_{25} = \frac{g}{2} + (25 - 1)g = \frac{49g}{2} \text{ feet}$$

Total distance fallen after twenty-five seconds

$$= a_1 + a_2 + a_3 + \dots + a_{25} = S_{25}$$

$$S_n = \frac{n}{2}(a_1 + a_n) = \frac{25}{2}\left(\frac{g}{2} + \frac{49g}{2}\right) = \frac{625g}{2} \text{ feet}$$

Practice Test 11

1. $a_6 = (6 - 1)6 = 30$

2. $a_1 = 6$
$a_2 = 2a_{2-1} + 1 = 2a_1 + 1 = 2\cdot 6 + 1 = 13$
$a_3 = 2a_{3-1} + 1 = 2a_2 + 1 = 2\cdot 13 + 1 = 27$
$a_4 = 2a_{4-1} + 1 = 2a_3 + 1 = 2\cdot 27 + 1 = 55$
$a_5 = 2a_{5-1} + 1 = 2a_4 + 1 = 2\cdot 55 + 1 = 111$
$a_6 = 2a_{6-1} + 1 = 2a_5 + 1 = 2\cdot 111 + 1 = 223$

3. $a_1 = 0$
$a_2 = 2$
$a_3 = a_{3-1} + a_{3-2} = a_2 + a_1 = 2$
$a_4 = a_{4-1} + a_{4-2} = a_3 + a_2 = 2 + 2 = 4$
$a_5 = a_{5-1} + a_{5-2} = a_4 + a_3 = 4 + 2 = 6$
$a_6 = a_{6-1} + a_{6-2} = a_5 + a_4 = 6 + 4 = 10$

4. $a_n = a_1 + (n - 1)d$
$a_6 = 5 + (6 - 1)3 = 20$

5. $a_n = a_1 r^{n-1}$

$$a_6 = 8\left(-\frac{1}{2}\right)^{6-1} = 8\left(-\frac{1}{2}\right)^5 = 8\left(-\frac{1}{32}\right) = -\frac{1}{4}$$

6. Since $\frac{-5}{3} = \frac{25}{3} \div 5 = -\frac{125}{9} \div \frac{25}{3} = -\frac{5}{3}$, this is a geometric sequence. $r = -\frac{5}{3}$

7. Since $5 - 3 \neq 3 - 5$ and $\frac{5}{3} \neq \frac{3}{5}$ this is neither arithmetic nor geometric.

8. $a_n = a_1 + (n - 1)d$
$a_4 = a_1 + 3d$
$a_{15} = a_1 + 14d$

Eliminating a_1 between these statements by addition, we have

$$\begin{aligned} -a_4 &= -a_1 - 3d \\ a_{15} &= a_1 + 14d \\ \hline a_{15} - a_4 &= 11d \end{aligned}$$

$$d = \frac{a_{15} - a_4}{11} = \frac{23 - 1}{11} = 2$$

9. Since the nth term, $a_n = \frac{(-1)^{n+1}}{10^n}$, we have

$$a_1 = \frac{(-1)^{1+1}}{10^1} = \frac{1}{10}$$

$$a_2 = \frac{(-1)^{2+1}}{10^2} = \frac{-1}{100}$$

$$r = a_2 \div a_1 = -\frac{1}{100} \div \frac{1}{10} = -\frac{1}{10}$$

10. $a_n = a_1 + (n - 1)d$ $\qquad$ $S_n = \frac{n}{2}(a_1 + a_n)$

$a_{50} = 3 + 49d$ $\qquad$ $250 = \frac{50}{2}(3 + a_{50})$

From the second equation we obtain $a_{50} = 7$. Substituting into the first equation, we obtain $7 = 3 + 49d$, hence $d = \frac{4}{49}$.

11. a_k: 5, 8, 11, ...

This is an arithmetic sequence with $a_1 = 5$, $d = 3$. We apply the formula

$$a_n = a_1 + (n - 1)d$$

and solve for n.

$$38 = 5 + (n - 1)3$$
$$38 = 5 + 3n - 3$$
$$36 = 3n$$
$$n = 12$$

Then $S_n = \frac{n}{2}(a_1 + a_n)$

$$S_{12} = \frac{12}{2}(5 + 38)$$
$$S_{12} = 258$$

12. This is a geometric series with $r = \frac{1}{2}$, $a_1 = 1$

$$S_n = \frac{a_1 - ra_n}{1 - r} = \frac{1 - \frac{1}{2}\cdot\frac{1}{1024}}{1 - \frac{1}{2}} = \frac{1 - \frac{1}{2048}}{1 - \frac{1}{2}}$$

Multiply numerator and denominator by 2048, the lcd of the internal denominators.

$$S_n = \frac{2048 - 1}{2048 - 1024} = \frac{2047}{1024}$$

13. This is an infinite geometric series with $r = \frac{1}{2}$, $a_1 = 1$ $\quad |r| < 1$.

$$S_\infty = \frac{a_1}{1 - r} = \frac{1}{1 - \frac{1}{2}} = 2$$

14. This is an arithmetic series.

$$\sum_{k=1}^{8} (7 - 2k) = (7 - 2\cdot1) + (7 - 2\cdot2) + \ldots + (7 - 2\cdot8)$$

We can therefore write:

$$a_1 = 7 - 2\cdot1 = 5, \quad a_n = a_8 = 7 - 2.8 = -9$$
$$d = (7 - 2\cdot2) - (7 - 2\cdot1) = 2$$
$$n = 8$$
$$S_n = \frac{n}{2}(a_1 + a_n)$$
$$= \frac{8}{2}[5 + (-9)] = -16$$

15. This is an arithmetic series with $a_1 = 1$, $d = 1$, $n = 200$, $a_n = 200$

$$S_n = \frac{n}{2}(a_1 + a_n)$$
$$= \frac{200}{2}(1 + 200)$$
$$= 100(201) = 20{,}100$$

16. $6! - \binom{9}{3} = 6! - \dfrac{9!}{3!(9-3)!} = 6! - \dfrac{9!}{3!6!} = 6! - \dfrac{9\cdot 8\cdot 7\cdot 6!}{3\cdot 2\cdot 1\cdot 6!}$

$$= 6\cdot 5\cdot 4\cdot 3\cdot 2\cdot 1 - \frac{9\cdot 8\cdot 7}{3\cdot 2\cdot 1} = 720 - 84 = 636$$

17. $a_k = 2, \frac{2}{3}, \frac{2}{9}, \ldots$

This is a geometric sequence with $a_1 = 2$, $r = \frac{1}{3}$. To determine n, we note $a_n = \frac{2}{243}$. Thus

$$a_n = a_1 r^{n-1}$$
$$\frac{2}{243} = 2\left(\frac{1}{3}\right)^{n-1}$$
$$\frac{1}{243} = \left(\frac{1}{3}\right)^{n-1}$$
$$\left(\frac{1}{3}\right)^5 = \left(\frac{1}{3}\right)^{n-1}$$
$$5 = n - 1$$
$$n = 6$$

The sum is therefore

$$\sum_{k=1}^{6} 2\left(\frac{1}{3}\right)^{k-1}$$

18. $(x - y)^6 = [x + (-y)]^6 = \sum_{k=0}^{6}\binom{6}{k}x^{6-k}(-y)^k$

$$= \binom{6}{0}x^6 + \binom{6}{1}x^5(-y) + \binom{6}{2}x^4(-y)^2 + \binom{6}{3}x^3(-y)^3 + \binom{6}{4}x^2(-y)^4 + \binom{6}{5}x(-y)^5 + (-y)^6$$

$$= x^6 - 6x^5y + 15x^4y^2 - 20x^3y^3 + 15x^2y^4 - 6xy^5 + y^6$$

19. $(a + 3)^5 = \sum_{k=0}^{5}\binom{5}{k}x^{5-k}3^k$

$$= \binom{5}{0}a^5 + \binom{5}{1}a^4(3) + \binom{5}{2}a^3(3)^2 + \binom{5}{3}a^2(3)^3 + \binom{5}{4}a(3)^4 + \binom{5}{5}(3)^5$$

$$= a^5 + 5a^4(3) + 10a^3(9) + 10a^2(27) + 5a(81) + 243$$

$$= a^5 + 15a^4 + 90a^3 + 270a^2 + 405a + 243$$

20. In the expansion of $(a + b)^n$, the exponent of b in the r-th term is $r - 1$ and the exponent of a is $n - (r - 1)$. Here, $r = 8$, $n = 18$.

8th term $= \binom{18}{7}a^{11}b^7 = \dfrac{18!}{7!(18-7)!}a^{11}b^7 = \dfrac{18!}{7!11!}a^{11}b^7 = 31{,}824a^{11}b^7$

APPENDIX A Setting Up Word Problems

A strategy for solving word problems is given in Section 1-8 of the text.

Strategy for Solving Word Problems

1. Read the problem very carefully—several times if necessary until you understand the problem, know what is to be found, and know what is given.

2. If appropriate, draw figures or diagrams and label known and unknown parts. Look for formulas connecting the known quantities with the unknown quantities.

3. Let one of the unknown quantities be represented by a variable, say x, and try to represent all other unknown quantities in terms of x. This is an important step and must be done carefully. Be sure you clearly understand what you are letting x represent.

4. Form an equation relating the unknown quantities with the known quantities. This step may involve the translation of an English sentence into an algebraic sentence, the use of relationships in a geometric figure, the use of certain formulas, and so on.

5. Solve the equation and write answers to *all* parts of the problem requested.

6. Check all solutions in the original problem.

As noted there, the key step, and often the most difficult one, is step 4, forming the equation. There is no general technique for doing this, but there is a procedure that works well for many students and can be used in a wide variety of word problems.

If you have difficulty finding the relationships that lead to an appropriate equation to solve, guess a numerical answer and try it to see if it works. This then becomes an arithmetic problem, exactly the same as checking your answer after solving the problem. Most likely your guess will be incorrect, but in checking you will better see the structure of the problem. You can then replace your guess by a variable and follow your same logic to get the appropriate equation. Some examples will illustrate the process.

Example 1

Find three consecutive even numbers such that twice the first plus the third is 10 more than the second.

Solution

Let us guess that the first number is 6 so that the numbers are 6, 8, and 10. Then

$$\text{Twice the first plus the third} = 2\cdot 6 + 10 = 22$$
$$\text{10 more than the second} = 8 + 10 = 18$$

Our guess is incorrect. However, if we replace our initial guess of 6 by x, then 8 is replaced by $x + 2$ and 10 by $x + 4$:

Twice the first plus the third $= 2x + (x + 4)$
10 more than the second $= (x + 2) + 10$

The equation we seek is, therefore,

$$2x + (x + 4) = (x + 2) + 10$$

The solution is completed in Example 3, Section 1-8.

Example 2

The length of a rectangle is 1 inch more than twice its width. If the area is 21 square inches, find its dimensions.

Solution

Suppose we guess that the width is 4 inches. Then

Width = 4
Length = $2 \cdot 4 + 1 = 9$
Area = Width · Length = $4 \cdot 9 = 36$

Our guess is incorrect, so we replace our incorrect guess by x:

Width = x
Length = $2x + 1$
Area = $x(2x + 1) = 21$

This last equation is what we sought:

$$x(2x + 1) = 21$$

The solution is completed in Example 3, Section 2-7.

Example 3

A jet plane leaves San Francisco and travels at 650 kilometers per hour toward Los Angeles. At the same time another plane leaves Los Angeles and travels at 800 kilometers per hour toward San Francisco. If the cities are 570 kilometers apart, how long will it take the jets to meet, and how far from San Francisco will they be at that time?

Solution

We guess the planes will meet in 1/2 hour. Then

Plane from SF to LA travels $650 \cdot 1/2 = 325$ km
Plane from LA to SF travels $800 \cdot 1/2 = 400$ km

The planes have thus traveled 725 km, so they have passed each other. (This also tells us that the guess of 1/2 is too large.)

Now replace the guess of 1/2 by x.

Plane from SF to LA travels $650x$ km
Plane from LA to SF travels $800x$ km

The total kilometers must be 650, so we get the equation

$$650x + 800x = 650$$

The solution is completed in Example 2, Section 4-3.

Example 4

A speedboat takes 1.5 times longer to go 120 miles up a river than to return. If the boat cruises at 25 miles per hour in still water, what is the rate of the current?

Solution

Let us guess that the current is 3 miles per hour. Then the speed of the boat upstream is 22 (that is, 25 - 3) and the speed downstream is 28 (that is, 25 + 3). Since $D = RT$, $T = D/R$, so

Time upstream is 120/22 = 5.4545...
Time downstream is 120/28 = 4.2857...

Is the time up 1.5 times the time down? No, 1.5 times 4.2857 is more than 6, so it is not equal to the time downstream. Thus, our guess is incorrect and we replace it by x. Now the speed upstream is $25 - x$ and the speed downstream is $25 + x$. Thus,

$$\text{Time upstream is } \frac{120}{25 - x}$$

$$\text{Time downstream is } \frac{120}{25 + x}$$

and our equation is

$$\frac{120}{25 - x} = 1.5 \frac{120}{(25 + x)}$$

The solution is completed in Example 11, Section 4-2.

Example 5

Five people form a glider club and decide to share the cost of a glider equally. They find, however, that if they let three more join the club, the share for each of the original five will be reduced by $480. What is the total cost of the glider?

Solution

Suppose we guess the cost of the glider to be $5,000. Then five shares would be $1,000 each, that is $5,000/5. With three additional shares, each would be $5,000/8 = $625. Since the cost of each share has not been reduced by $480 our guess is incorrect. Replace the guess by x:

$$\text{For five shares: } \frac{x}{5} \text{ each}$$

$$\text{For eight shares: } \frac{x}{8} \text{ each}$$

The cost per share for five members should be $480 more than the cost per share for eight members, that is,

$$\frac{x}{5} = \frac{x}{8} + 480 \qquad \text{or} \qquad \frac{x}{8} = \frac{x}{5} - 480$$

The solution is completed in Example 1, Section 4-5.

Example 6

A change machine changes dollar bills into quarters and nickels. If you receive 12 coins after inserting a $1 bill, how many of each type of coin did you receive?

Solution

If we guess 3 quarters and 9 nickels (the total number of coins is 12), then the value of the coins would be

Quarters: $3 \cdot 25 = 75$ cents
Nickels: $9 \cdot 5 = 45$ cents

which totals more than 100 cents, more than $1. If we replace our guesses of 3 and 9 by x and y, then

$$x + y = 12$$

and

$$x \cdot 25 + y \cdot 5 = 100$$

or

$$\begin{aligned} x + y &= 12 \\ 25x + 5y &= 100 \end{aligned}$$

Using this last system of equations, the solution is completed in Example 1, Section 8-2.

Example 7

A chemical storeroom has a 20% alcohol solution and a 50% alcohol solution. How many centiliters must be taken from each to obtain 24 centiliters of a 30% solution?

Solution

Let us guess that we need 10 centiliters of the weak (20%) solution and, therefore, 14 centiliters of the strong. Then our mixture will be 24 centiliters. How much alcohol does it contain?

The 10 centiliters is 20% alcohol, so the amount is 20% of 10, that is $0.2(10) = 2$. The 14 centiliters is 50% alcohol, so the amount there is $0.5(14) = 7$. The total amount of alcohol in the mixture is, therefore, $2 + 7 = 9$ centiliters. Is this 30% of 24?

$$30\% \text{ of } 24 = 0.3(24) = 7.2$$

Thus our mixture has too much alcohol (we have used too much of the stronger solution). Replace our guesses of 10 and 14 by x and y, respectively. Then

$$\begin{aligned} x + y &= 24 \\ 0.2x + 0.5y &= 7.2 \end{aligned}$$

The solution is completed in Example 6, Section 8-2.

There are no matched problems and no exercise set for this appendix. There are an ample number of word problems in the text—especially in Sections 1-8, 2-7, 4-1, 4-2, 7-3, 8-1, 8-2, 8-3, and 9-5—upon which to practice this technique.